Introduction to Organic Electronic and Optoelectronic Materials and Devices

OPTICAL SCIENCE AND ENGINEERING

Founding Editor
Brian J. Thompson
University of Rochester
Rochester, New York

1. Electron and Ion Microscopy and Microanalysis: Principles and Applications, *Lawrence E. Murr*
2. Acousto-Optic Signal Processing: Theory and Implementation, *edited by Norman J. Berg and John N. Lee*
3. Electro-Optic and Acousto-Optic Scanning and Deflection, *Milton Gottlieb, Clive L. M. Ireland, and John Martin Ley*
4. Single-Mode Fiber Optics: Principles and Applications, *Luc B. Jeunhomme*
5. Pulse Code Formats for Fiber Optical Data Communication: Basic Principles and Applications, *David J. Morris*
6. Optical Materials: An Introduction to Selection and Application, *Solomon Musikant*
7. Infrared Methods for Gaseous Measurements: Theory and Practice, *edited by Joda Wormhoudt*
8. Laser Beam Scanning: Opto-Mechanical Devices, Systems, and Data Storage Optics, *edited by Gerald F. Marshall*
9. Opto-Mechanical Systems Design, *Paul R. Yoder, Jr.*
10. Optical Fiber Splices and Connectors: Theory and Methods, *Calvin M. Miller with Stephen C. Mettler and Ian A. White*
11. Laser Spectroscopy and Its Applications, *edited by Leon J. Radziemski, Richard W. Solarz, and Jeffrey A. Paisner*
12. Infrared Optoelectronics: Devices and Applications, *William Nunley and J. Scott Bechtel*
13. Integrated Optical Circuits and Components: Design and Applications, *edited by Lynn D. Hutcheson*
14. Handbook of Molecular Lasers, *edited by Peter K. Cheo*
15. Handbook of Optical Fibers and Cables, *Hiroshi Murata*
16. Acousto-Optics, *Adrian Korpel*
17. Procedures in Applied Optics, *John Strong*
18. Handbook of Solid-State Lasers, *edited by Peter K. Cheo*
19. Optical Computing: Digital and Symbolic, *edited by Raymond Arrathoon*
20. Laser Applications in Physical Chemistry, *edited by D. K. Evans*
21. Laser-Induced Plasmas and Applications, *edited by Leon J. Radziemski and David A. Cremers*
22. Infrared Technology Fundamentals, *Irving J. Spiro and Monroe Schlessinger*
23. Single-Mode Fiber Optics: Principles and Applications, Second Edition, Revised and Expanded, *Luc B. Jeunhomme*
24. Image Analysis Applications, *edited by Rangachar Kasturi and Mohan M. Trivedi*
25. Photoconductivity: Art, Science, and Technology, *N. V. Joshi*
26. Principles of Optical Circuit Engineering, *Mark A. Mentzer*
27. Lens Design, *Milton Laikin*

28. Optical Components, Systems, and Measurement Techniques, *Rajpal S. Sirohi and M. P. Kothiyal*
29. Electron and Ion Microscopy and Microanalysis: Principles and Applications, Second Edition, Revised and Expanded, *Lawrence E. Murr*
30. Handbook of Infrared Optical Materials, *edited by Paul Klocek*
31. Optical Scanning, *edited by Gerald F. Marshall*
32. Polymers for Lightwave and Integrated Optics: Technology and Applications, *edited by Lawrence A. Hornak*
33. Electro-Optical Displays, *edited by Mohammad A. Karim*
34. Mathematical Morphology in Image Processing, *edited by Edward R. Dougherty*
35. Opto-Mechanical Systems Design: Second Edition, Revised and Expanded, *Paul R. Yoder, Jr.*
36. Polarized Light: Fundamentals and Applications, *Edward Collett*
37. Rare Earth Doped Fiber Lasers and Amplifiers, *edited by Michel J. F. Digonnet*
38. Speckle Metrology, *edited by Rajpal S. Sirohi*
39. Organic Photoreceptors for Imaging Systems, *Paul M. Borsenberger and David S. Weiss*
40. Photonic Switching and Interconnects, *edited by Abdellatif Marrakchi*
41. Design and Fabrication of Acousto-Optic Devices, *edited by Akis P. Goutzoulis and Dennis R. Pape*
42. Digital Image Processing Methods, *edited by Edward R. Dougherty*
43. Visual Science and Engineering: Models and Applications, *edited by D. H. Kelly*
44. Handbook of Lens Design, *Daniel Malacara and Zacarias Malacara*
45. Photonic Devices and Systems, *edited by Robert G. Hunsberger*
46. Infrared Technology Fundamentals: Second Edition, Revised and Expanded, *edited by Monroe Schlessinger*
47. Spatial Light Modulator Technology: Materials, Devices, and Applications, *edited by Uzi Efron*
48. Lens Design: Second Edition, Revised and Expanded, *Milton Laikin*
49. Thin Films for Optical Systems, *edited by Francoise R. Flory*
50. Tunable Laser Applications, *edited by F. J. Duarte*
51. Acousto-Optic Signal Processing: Theory and Implementation, Second Edition, *edited by Norman J. Berg and John M. Pellegrino*
52. Handbook of Nonlinear Optics, *Richard L. Sutherland*
53. Handbook of Optical Fibers and Cables: Second Edition, *Hiroshi Murata*
54. Optical Storage and Retrieval: Memory, Neural Networks, and Fractals, *edited by Francis T. S. Yu and Suganda Jutamulia*
55. Devices for Optoelectronics, *Wallace B. Leigh*
56. Practical Design and Production of Optical Thin Films, *Ronald R. Willey*
57. Acousto-Optics: Second Edition, *Adrian Korpel*
58. Diffraction Gratings and Applications, *Erwin G. Loewen and Evgeny Popov*
59. Organic Photoreceptors for Xerography, *Paul M. Borsenberger and David S. Weiss*
60. Characterization Techniques and Tabulations for Organic Nonlinear Optical Materials, *edited by Mark G. Kuzyk and Carl W. Dirk*
61. Interferogram Analysis for Optical Testing, *Daniel Malacara, Manuel Servin, and Zacarias Malacara*
62. Computational Modeling of Vision: The Role of Combination, *William R. Uttal, Ramakrishna Kakarala, Spiram Dayanand, Thomas Shepherd, Jagadeesh Kalki, Charles F. Lunskis, Jr., and Ning Liu*
63. Microoptics Technology: Fabrication and Applications of Lens Arrays and Devices, *Nicholas Borrelli*
64. Visual Information Representation, Communication, and Image Processing, *edited by Chang Wen Chen and Ya-Qin Zhang*
65. Optical Methods of Measurement, *Rajpal S. Sirohi and F. S. Chau*

66. Integrated Optical Circuits and Components: Design and Applications, *edited by Edmond J. Murphy*
67. Adaptive Optics Engineering Handbook, *edited by Robert K. Tyson*
68. Entropy and Information Optics, *Francis T. S. Yu*
69. Computational Methods for Electromagnetic and Optical Systems, *John M. Jarem and Partha P. Banerjee*
70. Laser Beam Shaping, *Fred M. Dickey and Scott C. Holswade*
71. Rare-Earth-Doped Fiber Lasers and Amplifiers: Second Edition, Revised and Expanded, *edited by Michel J. F. Digonnet*
72. Lens Design: Third Edition, Revised and Expanded, *Milton Laikin*
73. Handbook of Optical Engineering, *edited by Daniel Malacara and Brian J. Thompson*
74. Handbook of Imaging Materials: Second Edition, Revised and Expanded, *edited by Arthur S. Diamond and David S. Weiss*
75. Handbook of Image Quality: Characterization and Prediction, *Brian W. Keelan*
76. Fiber Optic Sensors, *edited by Francis T. S. Yu and Shizhuo Yin*
77. Optical Switching/Networking and Computing for Multimedia Systems, *edited by Mohsen Guizani and Abdella Battou*
78. Image Recognition and Classification: Algorithms, Systems, and Applications, *edited by Bahram Javidi*
79. Practical Design and Production of Optical Thin Films: Second Edition, Revised and Expanded, *Ronald R. Willey*
80. Ultrafast Lasers: Technology and Applications, *edited by Martin E. Fermann, Almantas Galvanauskas, and Gregg Sucha*
81. Light Propagation in Periodic Media: Differential Theory and Design, *Michel Nevière and Evgeny Popov*
82. Handbook of Nonlinear Optics, Second Edition, Revised and Expanded, *Richard L. Sutherland*
83. Polarized Light: Second Edition, Revised and Expanded, *Dennis Goldstein*
84. Optical Remote Sensing: Science and Technology, *Walter Egan*
85. Handbook of Optical Design: Second Edition, *Daniel Malacara and Zacarias Malacara*
86. Nonlinear Optics: Theory, Numerical Modeling, and Applications, *Partha P. Banerjee*
87. Semiconductor and Metal Nanocrystals: Synthesis and Electronic and Optical Properties, edited by *Victor I. Klimov*
88. High-Performance Backbone Network Technology, *edited by Naoaki Yamanaka*
89. Semiconductor Laser Fundamentals, *Toshiaki Suhara*
90. Handbook of Optical and Laser Scanning, *edited by Gerald F. Marshall*
91. Organic Light-Emitting Diodes: Principles, Characteristics, and Processes, *Jan Kalinowski*
92. Micro-Optomechatronics, *Hiroshi Hosaka, Yoshitada Katagiri, Terunao Hirota, and Kiyoshi Itao*
93. Microoptics Technology: Second Edition, *Nicholas F. Borrelli*
94. Organic Electroluminescence, *edited by Zakya Kafafi*
95. Engineering Thin Films and Nanostructures with Ion Beams, *Emile Knystautas*
96. Interferogram Analysis for Optical Testing, Second Edition, *Daniel Malacara, Manuel Sercin, and Zacarias Malacara*
97. Laser Remote Sensing, *edited by Takashi Fujii and Tetsuo Fukuchi*
98. Passive Micro-Optical Alignment Methods, *edited by Robert A. Boudreau and Sharon M. Boudreau*
99. Organic Photovoltaics: Mechanism, Materials, and Devices, *edited by Sam-Shajing Sun and Niyazi Serdar Saracftci*
100. Handbook of Optical Interconnects, *edited by Shigeru Kawai*
101. GMPLS Technologies: Broadband Backbone Networks and Systems, *Naoaki Yamanaka, Kohei Shiomoto, and Eiji Oki*

102. Laser Beam Shaping Applications, *edited by Fred M. Dickey, Scott C. Holswade and David L. Shealy*
103. Electromagnetic Theory and Applications for Photonic Crystals, *Kiyotoshi Yasumoto*
104. Physics of Optoelectronics, *Michael A. Parker*
105. Opto-Mechanical Systems Design: Third Edition, *Paul R. Yoder, Jr.*
106. Color Desktop Printer Technology, *edited by Mitchell Rosen and Noboru Ohta*
107. Laser Safety Management, *Ken Barat*
108. Optics in Magnetic Multilayers and Nanostructures, *Štefan Višňovský*
109. Optical Inspection of Microsystems, *edited by Wolfgang Osten*
110. Applied Microphotonics, *edited by Wes R. Jamroz, Roman Kruzelecky, and Emile I. Haddad*
111. Organic Light-Emitting Materials and Devices, *edited by Zhigang Li and Hong Meng*
112. Silicon Nanoelectronics, *edited by Shunri Oda and David Ferry*
113. Image Sensors and Signal Processor for Digital Still Cameras, *Junichi Nakamura*
114. Encyclopedic Handbook of Integrated Circuits, *edited by Kenichi Iga and Yasuo Kokubun*
115. Quantum Communications and Cryptography, *edited by Alexander V. Sergienko*
116. Optical Code Division Multiple Access: Fundamentals and Applications, *edited by Paul R. Prucnal*
117. Polymer Fiber Optics: Materials, Physics, and Applications, *Mark G. Kuzyk*
118. Smart Biosensor Technology, *edited by George K. Knopf and Amarjeet S. Bassi*
119. Solid-State Lasers and Applications, *edited by Alphan Sennaroglu*
120. Optical Waveguides: From Theory to Applied Technologies, *edited by Maria L. Calvo and Vasudevan Lakshiminarayanan*
121. Gas Lasers, *edited by Masamori Endo and Robert F. Walker*
122. Lens Design, Fourth Edition, *Milton Laikin*
123. Photonics: Principles and Practices, *Abdul Al-Azzawi*
124. Microwave Photonics, *edited by Chi H. Lee*
125. Physical Properties and Data of Optical Materials, *Moriaki Wakaki, Keiei Kudo, and Takehisa Shibuya*
126. Microlithography: Science and Technology, Second Edition, *edited by Kazuaki Suzuki and Bruce W. Smith*
127. Coarse Wavelength Division Multiplexing: Technologies and Applications, *edited by Hans Joerg Thiele and Marcus Nebeling*
128. Organic Field-Effect Transistors, *Zhenan Bao and Jason Locklin*
129. Smart CMOS Image Sensors and Applications, *Jun Ohta*
130. Photonic Signal Processing: Techniques and Applications, *Le Nguyen Binh*
131. Terahertz Spectroscopy: Principles and Applications, *edited by Susan L. Dexheimer*
132. Fiber Optic Sensors, Second Edition, *edited by Shizhuo Yin, Paul B. Ruffin, and Francis T. S. Yu*
133. Introduction to Organic Electronic and Optoelectronic Materials and Devices, *edited by Sam-Shajing Sun and Larry R. Dalton*
134. Introduction to Nonimaging Optics, *Julio Chaves*

Introduction to Organic Electronic and Optoelectronic Materials and Devices

Edited by

Sam-Shajing Sun
Larry R. Dalton

CRC Press
Taylor & Francis Group
Boca Raton London New York

CRC Press is an imprint of the
Taylor & Francis Group, an **informa** business

CRC Press
Taylor & Francis Group
6000 Broken Sound Parkway NW, Suite 300
Boca Raton, FL 33487-2742

Printed in the United States of America on acid-free paper
10 9 8 7 6 5 4 3 2 1

International Standard Book Number-13: 978-0-8493-9284-9 (Hardcover)

Library of Congress Cataloging-in-Publication Data

Introduction to organic electronic and optoelectronic materials and devices / edited by Sam-Shajing Sun and Larry R. Dalton.
p. cm. -- (Optical science and engineering ; 133)
Includes bibliographical references and index.
ISBN-13: 978-0-8493-9284-9
ISBN-10: 0-8493-9284-5
1. Semiconductors--Materials. 2. Optoelectronics--Materials. 3. Organic semiconductors. 4. Organic compounds--Electric properties. I. Sun, Sam-Shajing. II. Dalton, Larry R.

TK7871.I5847 2005
621.36--dc22 2007041305

Visit the Taylor & Francis Web site at
http://www.taylorandfrancis.com

and the CRC Press Web site at
http://www.crcpress.com

Contents

Book Abstract xiii
Preface xv
Acknowledgments xvii
Editors xix
Contributors xxi

Chapter 1 Introduction to Optoelectronic Materials 1
N. Peyghambarian and M. Fallahi

Chapter 2 Introduction to Optoelectronic Device Principles 25
Joachim Piprek

Chapter 3 Basic Electronic Structures and Charge Carrier Generation in Organic Optoelectronic Materials 47
Sam-Shajing Sun

Chapter 4 Charge Transport in Conducting Polymers 87
Vladimir N. Prigodin and Arthur J. Epstein

Chapter 5 Major Classes of Organic Small Molecules for Electronics and Optoelectronics 129
Xianle Meng, Weihong Zhu, and He Tian

Chapter 6 Major Classes of Conjugated Polymers and Synthetic Strategies 173
Yongfang Li and Jianhui Hou

Chapter 7 Low Energy Gap, Conducting, and Transparent Polymers 211
Arvind Kumar, Yogesh Ner, and Gregory A. Sotzing

Chapter 8 Conjugated Polymers, Fullerene C_{60}, and Carbon Nanotubes for Optoelectronic Devices 237
Liangti Qu, Liming Dai, and Sam-Shajing Sun

Chapter 9 Introduction of Organic Superconducting Materials 263
Hatsumi Mori

Chapter 10 Molecular Semiconductors for Organic Field-Effect Transistors 287
Antonio Facchetti

Chapter 11 Polymer Field-Effect Transistors 319
Henrik G.O. Sandberg

Chapter 12 Organic Molecular Light-Emitting Materials and Devices 351
Franky So and Jianmin Shi

Chapter 13 Polymer Light-Emitting Diodes: Devices and Materials 373
Xiong Gong and Shu Wang

Chapter 14 Organic and Polymeric Photovoltaic Materials and Devices 401
Sam-Shajing Sun and Cheng Zhang

Chapter 15 Organic Molecular Nonlinear Optical Materials and Devices 421
Mojca Jazbinsek and Peter Günter

Chapter 16 Polymeric Second-Order Nonlinear Optical Materials and Devices 467
Sei-Hum Jang and Alex K.-Y. Jen

Chapter 17 Organic and Polymeric Third-Order Nonlinear Optical Materials and Device Applications 513
Joel M. Hales and Joseph W. Perry

Chapter 18 Organic Multiphoton Absorbing Materials and Devices 573
Kevin D. Belfield, Sheng Yao, and Mykhailo V. Bondar

Chapter 19 Organic and Polymeric Photorefractive Materials and Devices 607
Oksana Ostroverkhova

Chapter 20 Organic/Metal Interface Properties 637
Yongli Gao

Chapter 21 Single-Molecule Organic Electronics and Optoelectronics 659
Ling Zang, Xiaomei Yang, and Tammene Naddo

Chapter 22 Introduction to Nonvolatile Organic Thin-Film Memory Devices 701
Yang Yang

Chapter 23 Introduction to Organic Electrochromic Materials and Devices 713
Prasanna Chandrasekhar

Chapter 24 An Introduction to Conducting Polymer Actuators 733

Geoffrey M. Spinks, Philip G. Whitten, Gordon G. Wallace, and Van-Tan Truong

Chapter 25 Organic Liquid Crystal Optoelectronic Materials and Devices 765

Sebastian Gauza

Chapter 26 Organic and Polymeric Photonic Band Gap Materials and Devices 793

Scott Meng and Thein Kyu

Chapter 27 Introduction to Polymer Photonics for Information Technology 819

Antao Chen

Chapter 28 Organic Low-Dielectric Constant Materials for Microelectronics 845

Jinghong Chen

Chapter 29 Self-Assembly of Organic Optoelectronic Materials and Devices 867

J.R. Heflin

Index 887

Book Abstract

Introduction to Organic Electronic and Optoelectronic Materials and Devices introduces students and nonexpert readers to fundamental principles and concepts of recent progress in the rapidly evolving field of organic electronic and optoelectronic materials and devices. This book is suitable for students and nonexperts with diverse backgrounds and at senior level in colleges and universities, and in different disciplines of science, mathematics, and engineering. While there have already been a number of specialized monographs or focused books on a variety of subjects related to organic/polymeric electronic and optoelectronic materials and devices, there are no books comprehensively covering the aforementioned subject matter at the basic level suitable for student/nonexpert learning and classroom teaching purposes. This textbook fills such a gap. All illustrations used in the book are available via download from www.crcpress.com/e_products/downloads/download.asp?cat_no=9284. These are intended as supplements for study and instruction.

Preface

Electronic, photonic, and optoelectronic (OE) materials and devices, including, but not limited to, conducting and semiconducting materials used in transistors and integrated circuits (ICs), light-emitting diodes and display/lighting devices, solar cells, photo detectors, electro-optical devices, optoelectronic sensors, etc., have dramatically impacted the way humans live in the twentieth and twenty-first centuries. In OE devices, electrons and photons are used to generate, process, transmit, and store information at unprecedented rates with ever-decreasing power requirements. Most of today's commercially available electronic and optoelectronic devices are fabricated from inorganic semiconductors and metal conductors. In the past several decades, however, research and development on organic/polymeric electronic and optoelectronic materials and devices has grown rapidly. Compared to their inorganic counterparts, emerging organic and polymeric optoelectronic materials have exhibited advantages such as improved speed, reduced power consumption, increased brightness (for displays), and improved processability leading to conformal and flexible devices and the potential for low-cost mass production. Plastic optoelectronic materials and devices are rapidly becoming a reality.

Though there are a number of specialized research review books relevant to selected topics of organic optoelectronic materials and devices, there are no books available covering the combined subjects of organic electronic and optoelectronic materials/devices, suitable for classroom instruction at the senior college level or suitable for providing nonexperts a convenient introduction to this research discipline. It is the objective of this book to serve as a textbook suitable for senior undergraduate or graduate level courses for students majoring in materials science, physics, chemistry, chemical engineering, electrical engineering, optical engineering, or other information/energy-related science and engineering disciplines. This book is also suitable as a desk reference for scientists and engineers involved in research and development in the fields of telecommunications, computing, defense technologies, etc.

As with all books, the publisher, the editors, and the contributing authors have tried their best to make this textbook as informative, accurate, reliable, and nonbiased as possible. However, by no means is this book error-free or inclusive of every critical item. While the editors of this book are mainly responsible for the selection of topics/chapters, contributing authors, and the components/styles of the book, it is the contributing authors who are mainly responsible for the contents, opinions, and accuracy of each topic/chapter. Any comments, suggestions, or questions about this book (particularly those from course instructors/students) are welcomed and may be directed directly to the book editors or the contributing authors. It is hoped that the subsequent editions of this textbook could be further improved after instructional activities and feedbacks.

Sam-Shajing Sun, PhD (editor)
Center for Materials Research and Chemistry Department
Norfolk State University
Norfolk, Virginia

Larry R. Dalton, PhD (coeditor)
Department of Chemistry
University of Washington
Seattle, Washington

Acknowledgments

Prof. Sam-Shajing Sun (editor) wishes to express his sincere appreciation and thanks to the following people and organizations whose contributions and/or assistances are critical and essential to the success of this textbook project.

- CRC Press, Taylor & Francis Group, in particular, Mr. Taisuke Soda (now at McGraw-Hill) for his belief and enthusiasm to this project, Ms. Stephanie J. Morkert and Mr. Richard Tressider for their highly professional assistance, and Dr. S. Vinithan (at SPi) for typesetting the book.
- All contributing authors of this textbook for their expertise, hard work, patience, suggestions, and refereeing service to the project. The editor is fully aware of the challenges and sometimes formidable tasks of translating and presenting a relatively newly developed, complex, and sometimes debatable subject matter into an easily understandable and acceptable format for students and non-experts learning purpose.
- Coeditor Professor Larry R. Dalton. It is Prof. Dalton's vision, leadership, and years of assistance/advice, including his assistance and advice to the editor that was very instrumental for the success of this project. Prof. Sun also wishes to acknowledge and thank the support and/or contributions from the multi-university involved Center on Materials and Devices for Information Technology Research (CMDITR, directed by Prof. Dalton and sponsored by the National Science Foundation. Award number DMR-0120967).

  Professor Yang Yang at University of California at Los Angeles, Professor Mikael Wasielewski at Northwestern University, Professor Jean-Luc Brédas at Georgia Institute of Technology, and particularly, Professor Rudolph A. Marcus (Nobel Laureate, Chemistry, 1992) at California Institute of Technology, and Professor Alan J. Heeger (Nobel Laureate, Chemistry, 2000) at the University of California at Santa Barbara, for their many helpful discussions with the editor on certain subjects/elements relevant to this book.
- Dr. Aloysius Hepp and Dr. Sheila Bailey (both at NASA Glenn Research Center), and Dr. Charles Lee (at DoD/AFOSR) for their years of support and assistance to the editor via research/educational grants (in particular, a NASA sponsored Center for Research and Education in Advanced Materials [CREAM], award number NCC3-1035) focusing on subjects of this textbook.
- Last but not least, the editor's family (including his daughter Marcia M. Sun) for their understanding and support to editor's numerous absences during "non-working" hours, including evenings, weekends, and holidays.

Editors

Sam-Shajing Sun, PhD (editor, seated in the photo), obtained his BS in physical chemistry from Peking University (PKU) in China, his MS in inorganic/analytical/nuclear chemistry from California State University at Northridge (CSUN), and his PhD in organic/polymer/materials chemistry from the University of Southern California (USC). Dr. Sun's PhD dissertation (under the direction of professor Larry R. Dalton) was entitled, "Design, Synthesis, and Characterization of Novel Organic Photonic Materials." After postdoctoral research experience at Loker Hydrocarbon Institute (Director, George A. Olah), Dr. Sun joined the chemistry faculty of Norfolk State University (NSU) in 1998, was promoted as associate professor (with tenure) in 2002 and full professor in 2006. Since joining NSU, Dr. Sun has won a number of U.S. government research and educational grant awards in the field of optoelectronic polymers and is currently leading several research and education projects focused on advanced optoelectronic and nanomaterials. Dr. Sun's main research interests and expertise are in the design, synthesis, processing, characterization, and modeling of novel polymers and thin-film devices for optoelectronic applications particularly photovoltaic energy conversion. More information on Professor Sun can be accessed at http://vigyan.nsu.edu/~cmr/ssun.htm

Larry R. Dalton, PhD (coeditor, standing in the photo), BS (1965) and MS (1966) from Honors College of Michigan State University, and AM, PhD (1971) from Harvard University, is the George B. Kauffman professor of chemistry and electrical engineering at the University of Washington where he also directs the National Science Foundation, Science and Technology Center on Materials and Devices for Information Technology Research. Since 2002 he has received such awards as the 2006 IEEE/LEOS William Streifer Scientific Achievement Award, the 2003 Chemistry of Materials Award of the American Chemical Society, and the Quality Education for Minorities/Mathematics, Science, and Engineering Network 2005 Giants in Science Award. During this period, Larry was elected fellow of the American Association for the Advancement of Science and became a senior member of the IEEE (2006). His research interests focus on high-performance organic electro-optic materials and new sensor materials including metamaterials and silicon photonics. More information on Professor Dalton can be accessed at http://depts.washington.edu/eooptic/.

Contributors

Kevin D. Belfield
Department of Chemistry and CREOL
College of Optics and Photonics
University of Central Florida
Orlando, Florida

Mykhailo V. Bondar
Institute of Physics
National Academy of Science of Ukraine
Kiev, Ukraine

Prasanna Chandrasekhar
Ashwin-Ushas Corporation
Lakewood, New Jersey

Antao Chen
Applied Physics Laboratory (APL)
University of Washington
Seattle, Washington

Jinghong Chen
Honeywell Electronic Materials
Sunnyvale, California

Liming Dai
Department of Chemical and Materials Engineering
School of Engineering
University of Dayton
Dayton, Ohio

Arthur J. Epstein
Department of Physics and Chemistry
Ohio State University
Columbus, Ohio

Antonio Facchetti
Department of Chemistry and the Materials Research Center
Northwestern University
Evanston, Illinois

M. Fallahi
College of Optical Science
University of Arizona
Tucson, Arizona

Yongli Gao
Department of Physics and Astronomy
University of Rochester
Rochester, New York

Sebastian Gauza
College of Optics and Photonics CREOL/FPCE
University of Central Florida
Orlando, Florida

Xiong Gong
Institute for Polymers and Organic Solids
University of California
Santa Barbara, California

Peter Günter
Nonlinear Optics Laboratory
ETH Zurich (Swiss Federal Institute of Technology)
Zurich, Switzerland

Joel M. Hales
School of Chemistry and Biochemistry
Georgia Institute of Technology
Atlanta, Georgia

J.R. Heflin
Department of Physics
Virginia Polytechnic Institute and State University
Blacksburg, Virginia

Jianhui Hou
CAS Key Laboratory of Organic Solids
Institute of Chemistry
Chinese Academy of Sciences
Beijing, China

Sei-Hum Jang
Department of Materials Science and Engineering
University of Washington
Seattle, Washington

Mojca Jazbinsek
Nonlinear Optics Laboratory
ETH Zurich (Swiss Federal Institute of Technology)
Zurich, Switzerland

Alex K.-Y. Jen
Department of Materials Science and Engineering
University of Washington
Seattle, Washington

Arvind Kumar
Department of Chemistry and the Polymer Program
Institute of Materials Science
University of Connecticut
Storrs, Connecticut

Thein Kyu
Department of Polymer Engineering
University of Akron
Akron, Ohio

Yongfang Li
CAS Key Laboratory of Organic Solids
Institute of Chemistry
Chinese Academy of Sciences
Beijing, China

Scott Meng
Department of Polymer Engineering
University of Akron
Akron, Ohio

Xianle Meng
Laboratory for Advanced Materials and Institute of Fine Chemicals
East China University of Science and Technology
Shanghai, China

Hatsumi Mori
Institute for Solid State Physics
University of Tokyo
Chiba, Japan

and

Care Research for Evolutional Science and Technology (CREST)
Japan Science and Technology Agency (JST)
Saitama, Japan

Tammene Naddo
Laboratory for Nanoscale Exploration and Single-Molecule Imaging (NESMI)
Department of Chemistry and Biochemistry
Southern Illinois University
Carbondale, Ilinois

Yogesh Ner
Department of Chemistry and the Polymer Program
Institute of Materials Science
University of Connecticut
Storrs, Connecticut

Oksana Ostroverkhova
Department of Physics
Oregon State University
Corvallis, Oregon

Joseph W. Perry
School of Chemistry and Biochemistry
Georgia Institute of Technology
Atlanta, Georgia

N. Peyghambarian
College of Optical Science
University of Arizona
Tucson, Arizona

Joachim Piprek
NUSOD Institute LLC
Newark, Delaware

Vladimir N. Prigodin
Department of Physics
Ohio State University
Columbus, Ohio

and

A.F. Ioffe Physico-Technical Institute
St. Petersburg, Russia

Liangti Qu
Department of Chemical and Materials Engineering
School of Engineering
University of Dayton
Dayton, Ohio

Henrik G.O. Sandberg
Printer Electronics and Optics
VTT Technical Research Center of Finland
Espoo, Finland

Jianmin Shi
Optical Electronics and Sensor Divisions
U.S. Army Research Laboratory
Adelphi, Maryland

Franky So
Department of Materials Science and Engineering
University of Florida
Gainesville, Florida

Gregory A. Sotzing
Department of Chemistry and the Polymer Program
Institute of Materials Science
University of Connecticut
Storrs, Connecticut

Geoffrey M. Spinks
ARC Centre of Excellence for Electromaterials Science
Intelligent Polymer Research Institute
University of Wollongong
Wollongong, New South Wales, Australia

Sam-Shajing Sun
Center for Materials Research and Chemistry Department
Norfolk State University
Norfolk, Virginia

He Tian
Laboratory for Advanced Materials and Institute of Fine Chemicals
East China University of Science and Technology
Shanghai, China

Van-Tan Truong
Maritime Platforms Division
Defence Science and Technology Organisation (DSTO)
Melbourne, Victoria, Australia

Gordon G. Wallace
ARC Centre of Excellence for Electromaterials Science
Intelligent Polymer Research Institute
University of Wollongong
Wollongong, New South Wales, Australia

Shu Wang
Key Laboratory of Organic Solids
Institute of Chemistry
Chinese Academy of Science
Beijing, China

Philip G. Whitten
ARC Centre of Excellence for Electromaterials Science
Intelligent Polymer Research Institute
University of Wollongong
Wollongong, New South Wales, Australia

Xiaomei Yang
Laboratory for Nanoscale Exploration and Single-Molecule Imaging (NESMI)
Department of Chemistry and Biochemistry
Southern Illinois University
Carbondale, Ilinois

Yang Yang
Department of Materials Science and Engineering
University of California
Los Angeles, California

Sheng Yao
Department of Chemistry and CREOL
College of Optics and Photonics
University of Central Florida
Orlando, Florida

Ling Zang
Laboratory for Nanoscale Exploration and Single-Molecule Imaging (NESMI)
Department of Chemistry and Biochemistry
Southern Illinois University
Carbondale, Ilinois

Cheng Zhang
Center for Materials Research and Chemistry Department
Norfolk State University
Norfolk, Virginia

Weihong Zhu
Laboratory for Advanced Materials and Institute of Fine Chemicals
East China University of Science and Technology
Shanghai, China

1 Introduction to Optoelectronic Materials

N. Peyghambarian and M. Fallahi

CONTENTS

1.1 Introduction 2
1.2 Types of Optoelectronic Materials 2
1.2.1 Semiconductors 2
1.2.1.1 Basic Concepts in Semiconductors 2
1.2.1.2 p–n Homojunctions and Heterojunctions 4
1.2.1.3 Alloy Semiconductors, Quantum Wells, and Strained Quantum Wells 6
1.2.1.4 Light Absorption and Emission in Semiconductors 8
1.2.1.5 Semiconductor LEDs and Lasers 10
1.2.2 Optical Glass 10
1.2.3 Electro-Optic Materials 12
1.2.4 Organic and Polymeric Materials 13
1.3 Waveguiding Principles for Optoelectronic Materials 14
1.3.1 Waveguiding in Semiconductors 16
1.3.2 Waveguiding in Glass–Ion-Exchange Glass and Optical Fibers 17
1.3.3 Waveguiding in EO Materials like $LiNbO_3$ 18
1.3.4 Silica-on-Silicon Waveguides 18
1.3.5 Waveguiding in Organic Materials 19
1.4 Challenges and Recent Developments 20
1.4.1 Material Fabrication and Compatibility 20
1.4.2 Heterogeneous Integration 21
Exercise Questions 22
List of Abbreviations 23
References 23

Abstract This chapter summarizes the principles of optoelectronic materials. Four classes of materials, including inorganic semiconductors; glassy materials; electro-optic crystals, such as lithium niobate ($LiNbO_3$); and organic and polymeric materials, are reviewed. Waveguiding approaches in these four classes of materials are also reviewed and some current examples provided. Finally some of the challenges in optoelectronic materials including compatibility issues, hybrid materials, and integration between different types of materials are discussed.

1.1 INTRODUCTION

Since the early 1980s the field of integrated optics and optoelectronics has experienced very rapid growth. Today optical technologies are extensively used in a wide range of applications such as telecommunications, medical, and security systems. Lasers and other photonic components are the key elements and are the focus of major research. The quest for higher performance, lower cost, and complex functionality has been the main motivation. Inorganic materials in general and semiconductors in particular play key roles in these developments. Among various inorganic materials, semiconductors have experienced tremendous developments because of their unique band structures, which give them superior electrical and optical properties suitable for a range of electronics, and passive and active optical elements. While the integration of photonics and electronics on the same chip has been the focus of major efforts, success has been somewhat limited owing to material and processing incompatibility. Heterogeneous integration combining organic and inorganic materials seems to be a promising approach to bypass these limitations. This chapter is dedicated to the fundamentals of inorganic materials and their properties used in integrated optoelectronics.

1.2 TYPES OF OPTOELECTRONIC MATERIALS

Optoelectronic components are fabricated from a broad range of materials. The material selection is based on a number of factors including optical properties (refractive index, absorption, and emission properties), electrical properties (mobility and conductivity), stability, and process compatibilities. As waveguides, the materials should exhibit very low absorption or scattering loss. Optical modulators and switches rely on materials with high electro-optic coefficients. Light-emitting diodes (LED) and lasers require materials with large radiative emission efficiencies and high gain. Finally, detectors require absorption at the desired wavelengths. On the basis of these requirements, optoelectronic materials are divided into three major categories: organics, inorganics, and hybrids. Crystalline solid materials can be divided into three categories: conductors, insulators, and semiconductors. A simple representation of the energy band structure of crystalline solid materials can illustrate their differences. In conductors the upper band is partially filled, allowing for high conductivity. In insulators the valence band is completely full and the conduction band is completely empty, and the separation between the valence band and the conduction band (energy gap) is very large (several electron volts), making them electrically insulating. In semiconductors, while the conduction band is typically empty, the energy gap separation between the conduction band and valence band is relatively small (~1 eV), allowing relatively easy transition of electrons from the valence band to the conduction band under thermal or optical excitation. Their electrical properties can be easily modified by doping, or with temperature or optical excitation. The band gap energy in semiconductors plays a key role in their electrical and optical properties.

Several base technologies have taken advantage of solid-state materials including semiconductors, insulators, and glassy materials. Examples of such technologies are silica-on-silicon, silicon-on-insulator (SOI), III–V semiconductors, ion in-diffused lithium niobate ($LiNbO_3$), and ion-exchanged glass. Lightwave circuits and photonic components have been fabricated using these techniques. A brief review of different types of optoelectronic materials and technologies that take advantage of them follows.

1.2.1 SEMICONDUCTORS

1.2.1.1 Basic Concepts in Semiconductors

Semiconductors are examples of crystalline solids. The potential of the crystalline lattice, $W(r)$, has the full periodicity of the lattice, $W(r) = W(r + \mathbf{R})$, with **R** being a vector in the direct lattice (see Figure 1.1). The periodic potential gives rise to the band structure of the solid with delocalized electronic wavefunctions $\psi_{nk}(r)$ given by the Bloch's theorem. The most general solution of the Schrödinger equation for an electron in the lattice periodic potential, $W(r)$, is given by the Bloch wavefunction,

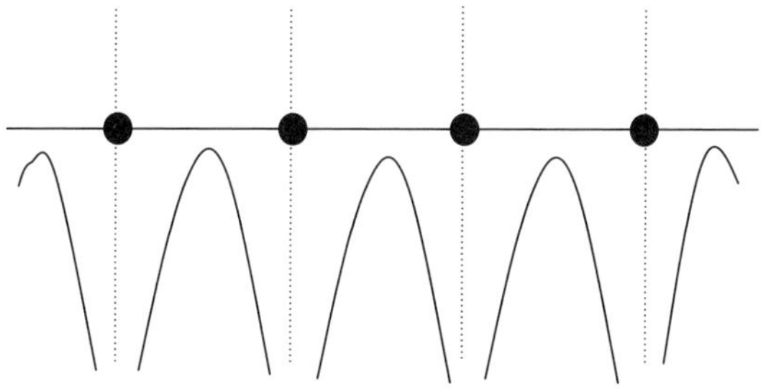

FIGURE 1.1 Schematic of the periodic lattice potential.

$$\psi_{nk}(r) = e^{ik.r} u_{nk}(r)/V^{0.5} \tag{1.1}$$

where

- $u_{nk}(r)$ is a function that has the translational symmetry of the lattice, $u_{nk}(r) = u_{nk}(r + \mathbf{R})$
- V is the volume of the crystal
- n and k are indices that label the electron with wavevector k in a given band n

Thus, the Bloch electron wavefunction has the character of a free-running wave with the amplitude modulated by the periodic lattice. The memory of the original atomic function is kept in the modulation of the Bloch function $u_{nk}(r)$.

The majority of the important semiconductors used in optoelectronics have diamond or zincblende lattice structures. In a diamond lattice the atoms within the lattice are identical, whereas in a zincblende lattice they consist of sublattices made from different atoms. The most popular diamond lattices are the group IV semiconductors such as Si and Ge, whereas most of the group III–V compounds have zincblende structures [1].

The band structure of semiconductors gives them their unique optical and electrical properties. For semiconductors to conduct current electrons from the valence band, they should be excited to the conduction band by various excitation means. We often take the top of the valence band as the reference level. The energy separation between the highest valence band state and the lowest conduction band state is called the band gap energy, E_g. The band gap wavelength can be obtained from the band gap energy

$$\lambda_g(\mu m) = \frac{1.24}{E_g(eV)} \tag{1.2}$$

Figure 1.2 shows the energy gap versus lattice constant for a range of semiconductor materials. As can be observed, the majority of semiconductors have band gap energies between 0.5 and 3 eV.

Semiconductors can be divided into two different categories: direct and indirect band gap semiconductors. In indirect band gap semiconductors the minimum of the conduction band and the maximum of the valence band occur at different k values. The elemental semiconductors Si and Ge are among the indirect band gap materials and play a very important role in electronic circuits. On the other hand, in direct band gap materials the minimum of the conduction band and the maximum of the valence band lie at the same k value. Compound semiconductors such as group III–V binary compounds GaAs- and InP-based compounds are among the direct band gap materials. Figure 1.3 shows the simplified energy versus momentum (E versus k) in these semiconductors. Direct band gap materials have high radiative recombination efficiency and are very attractive as light sources. A wide range of direct band gap materials with different absorption/emission wavelengths can be obtained from III–V ternary and quaternary compounds such as InGaAs, InGaAsP, and InAlGaAs.

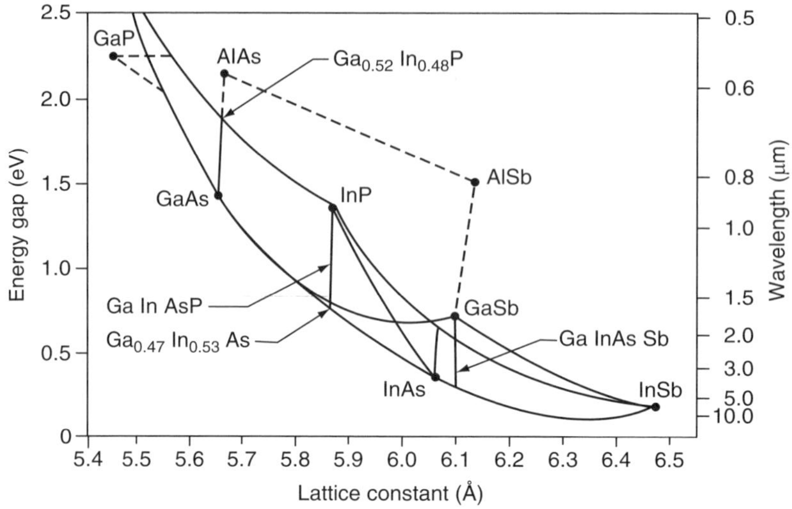

FIGURE 1.2 Energy gap and lattice constant of several semiconductors.

In addition, the valence band of most common semiconductors has its maximum at $k = 0$ and consists of three sub-bands called the heavy hole, the light hole, and the split-off. In bulk semiconductors the heavy hole and the light hole are degenerate at $k = 0$. However, the degeneracy can be removed using strained quantum.

By doping semiconductors with atoms of difference valence, we can significantly alter the electric and optical properties of the material. If the impurity atom has one or more electrons than the atom it replaced, it acts as a donor of electrons and the semiconductor is called n-type. One example is a Se atom replacing some As atoms in GaAs binary compounds. The electrons from the impurity atom can be easily excited to the conduction band of the semiconductor and provide high conductivity. Similarly if the impurity atom has fewer electrons then it becomes an acceptor (e.g., if Zn atoms replace Ga atoms in a GaAs compound) and the semiconductor is called p-type.

1.2.1.2 p–n Homojunctions and Heterojunctions

p–n Junctions are formed when p-type and n-type semiconductors are brought into contact to form a junction. p–n junctions are of major importance in a wide range of electronic and photonic components [1]. They allow electrical biasing and pumping of semiconductors. They can be divided into homojunctions and heterojunctions. In the case of homojunctions, the two materials are the same with the same band gap. In heterojunctions the materials have different band gaps.

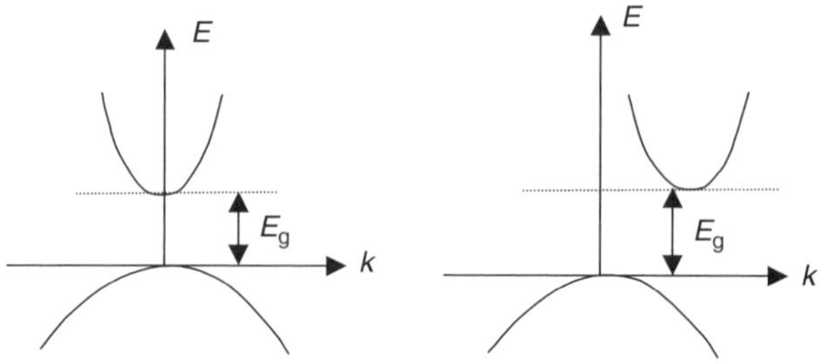

FIGURE 1.3 E versus k for direct and indirect band gap semiconductor.

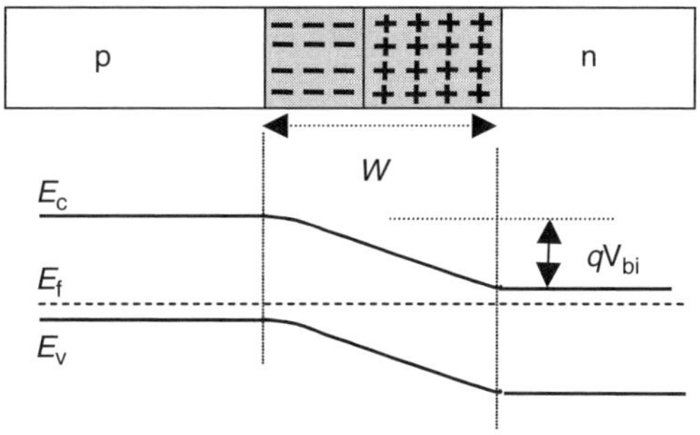

FIGURE 1.4 Charge distribution and band structure of a p–n homojunction, where V_{bi} is the built-in potential barrier and q is the electron charge.

The junction holes from the p-type semiconductor diffuse into the n-type material, leaving a negative space charge near the junction interface of the p-type material. At the same time the electrons from the n-type material diffuse into the p-type semiconductor, leaving a positive space charge near the junction. The formation of the space charge creates a field that is directed from the n-type to p-type, inhibiting further diffusion of the electrons and holes. The diffusion of the n and p majority carriers at the interface creates a space charge depleted of its majority carriers. As a result, the space charge region is usually called the depletion region. Figure 1.4 shows the depletion region and the band structure in a p–n homojunction.

Contrary to silicon electronics technology, the majority of compound semiconductor photonics components are fabricated from heterostructures with different optical and electronic characteristics. They are formed by growing a compound semiconductor over another semiconductor with a different band gap. Some very important heterojunctions are AlGaAs/GaAs and InGaAsP/InP. Figure 1.5 shows the band diagram of a p–n heterojunction under equilibrium. The parameters of great importance in heterostructures are the band gap difference, ΔE_g, and the conduction and valence band offsets, ΔE_c and ΔE_v. In the majority of heterojunctions (type I) they are related by $\Delta E_g = \Delta E_c + \Delta E_v$. A larger conduction band offset versus valence band offset is desirable for electron confinement and improved temperature behavior of semiconductor lasers and amplifiers. The conduction band offsets in the GaAs/AlGaAs and InGaAsP/InP families are around 0.67 ΔE_g and 0.4 ΔE_g, respectively. As a

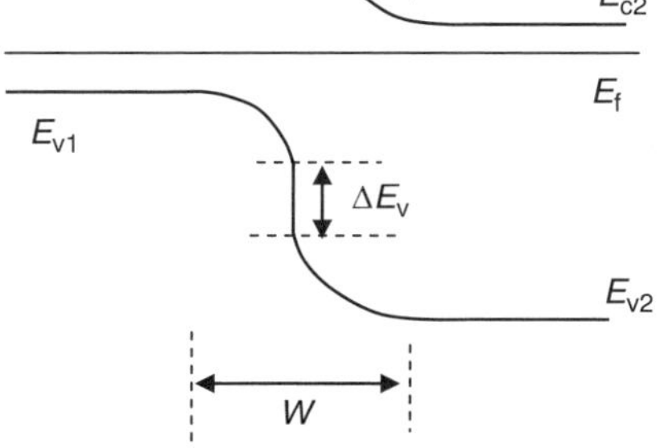

FIGURE 1.5 Band structure of a p–n heterojunction at thermal equilibrium, where V_{bi} is the built-in potential barrier and q is the electron charge.

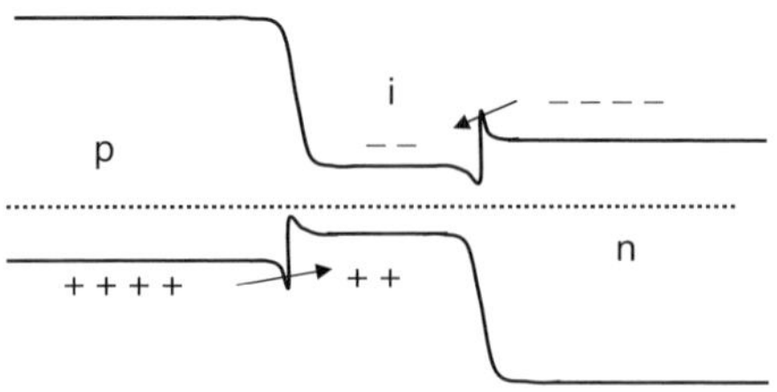

FIGURE 1.6 Band structure of a p–i–n heterostructure.

result, InGaAsP/InP laser performance is more sensitive to temperature than that of AlGaAs/GaAs lasers. A new family of long wavelength lasers, based on InGaAlAs/InP, has been developed to overcome this limitation. InAlGaAs/InP has a conduction band offset of about 0.72 ΔE_g.

A range of semiconductor photonic components, including semiconductor lasers and photodiodes, are made of p–i–n double heterostructures. In this case the p and n portions are made of large band gap compound semiconductors that surround an undoped low-band gap compound semiconductor (see Figure 1.6). These structures are very efficient for achieving carrier confinement under forward biasing as well as photon confinement, necessary for lasing operations.

1.2.1.3 Alloy Semiconductors, Quantum Wells, and Strained Quantum Wells

Alloy semiconductors can be made from I–VII, II–VI, and III–V elements. III–V Compound semiconductors exhibit a wide range of band gaps and are of great interest in optical communications. Various binaries (such as GaAs, InP), ternaries (AlGaAs, InGaAs), and quaternaries (InGaAsP, InAlGaAs) are widely used for the fabrication of lasers and photonic integrated components for optical communications in the 800–1600 nm wavelength range.

A good knowledge of the various physical and optical properties of III–V ternary and quaternary compounds is necessary in the design of photonic components [2,3]. These parameters are obtained using empirical formulae from their constituent binary compounds. Whereas for GaAs all the compositions of $Al_xGa_{1-x}As$ are lattice matched, the situation is much more complicated for InP-related compounds. For $Ga_xIn_{1-x}As_yP_{1-y}$ quaternaries the lattice match condition to InP is obtained when $x \cong 0.468y$. Interestingly, all the $Ga_xIn_{1-x}As_yP_{1-y}$ compositions lattice matched to InP have a direct band gap, with their energy gap at room temperature obtained from

$$E_g(y) = 1.35 - 0.775y + 0.149y^2 \tag{1.3}$$

For all other compositions the band gap (in electron volts) can be calculated from

$$E_g(x, y) = 1.35 + 0.642x - 1.101y + 0.758x^2 + 0.101y^2 - 0.159xy - 0.28x^2y + 0.109xy^2 \tag{1.4}$$

The band gap of semiconductors changes as a function of temperature. The temperature dependence of the band gap is generally obtained from

$$E_g(T) = E_g(0) - \frac{\alpha T^2}{T + \beta} \tag{1.5}$$

where

$E_g(0)$ is the energy gap at 0 K

α and β are constants

TABLE 1.1
Physical Parameters of Several III–V Binary Compounds

Compound	Lattice Constant (Å)	Band Gap Energy (eV)	α (10^{-4} eV/K)	β (K)	Refractive Index near E_g
GaAs	5.6532	1.424	5.405	204	3.65
InP	5.8687	1.35	3.63	162	3.41
AlAs	5.660	2.95	6.0	408	3.18
InAs	6.0583	0.354	2.5	75	3.52
GaP	5.4512	2.272	5.771	372	3.45

Table 1.1 gives some of the key parameters for important III–V binary compounds at 300 K.

Growing a thin layer of narrow band gap compounds between two wide band gap films usually forms quantum-well structures. The thickness of the layer is in the order of 10 nm. As a result, the movement of the electrons and holes will be restricted in the growth direction and the kinetic energy of the carriers will be quantized into discrete levels. For an infinite potential well the energy levels of a particle obtained from the Schrödinger equation is given by

$$E(n) = \frac{\hbar^2}{2m^*}\left(\frac{n\pi}{L}\right)^2 \tag{1.6}$$

where

n is an integer
L is the thickness of the quantum-well layer
m^* is the effective mass of the particle

It is easy to notice that the separation between different energy levels can be increased by reducing the thickness of the well. As a result, the optical (absorption, emission) and electrical properties of the layer can be modified. Another important change in the quantum well is the modification of the density of states to step-like states. All these effects result in significant performance improvement for quantum-well lasers and modulators. Today, the majority of the semiconductor lasers and electroabsorption modulators have quantum-well active regions.

One major limitation in the growth of high-quality semiconductors is the need for lattice-matching to the substrate. While this is true for the growth of thick semiconductors (above 100 nm), the growth of thin slightly latticed-mismatched layers is achieved with high quality and superior performance. This is achieved by the growth of a thin elastically strained layer, forcing the atoms of the epitaxial layer to become equal to the ones of the substrate in the direction parallel to the interface. These layers are called strained layers. Figure 1.6 illustrates this lattice deformation.

The amount of strain is defined as

$$\text{strain} = \frac{a_{\text{epi}} - a_{\text{sub}}}{a_{\text{sub}}} \tag{1.7}$$

The effect of strain can cause major changes to the band structure of compound semiconductors compared to unstrained layers. The most significant effects are change in the band gap, the removal of heavy hole and light hole degeneracy at the maximum of the valence band (see Figure 1.7), and

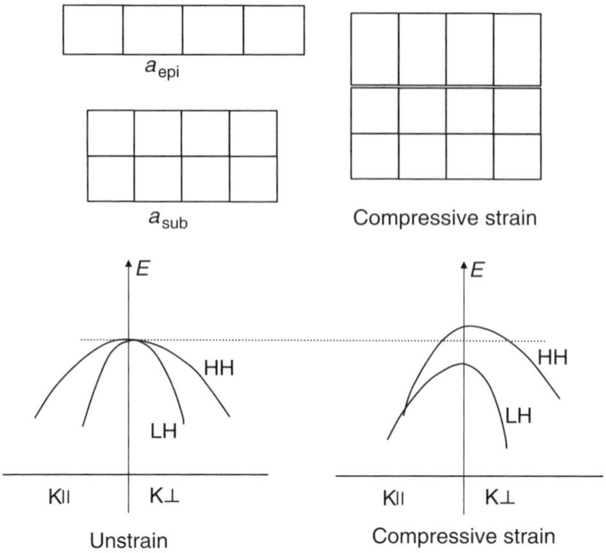

FIGURE 1.7 Lattice deformation and modification of the bulk valence band due to compressive strain, where K⊥ is the strain axis (growth direction) and K∥ is the growth plane.

the reduced effective mass of the holes. Due to these effects strained quantum-well lasers have shown improved performance such as lower threshold, higher efficiency, and faster modulation.

So far all the quantum wells we have considered were such that the sum of the conduction and valence band offsets between the two materials was equal to their band gap difference. This is not always true. In fact some material systems, such as antimony-based III–V, exhibit different band formation. In some heterostructures such as GaSbAs/InGaAs the valence band of the large band gap material is above the valence band of the narrow band gap material; alternatively, in the GaSb/InAs junctions the valence band of the large band gap material is above the conduction band of the narrow band gap material. These heterojunctions are called quasi-type II and type II, respectively. In type II materials, since the valence band of material 1 is above the conduction band of material 2, the electrons from the valence band of material 1 can easily be injected into the conduction band of material 2. Figure 1.8 shows the band diagram of quasi-type II and type II quantum-well structures. These materials are very attractive for mid-IR lasers and detectors.

1.2.1.4 Light Absorption and Emission in Semiconductors

When an electromagnetic wave propagates inside a semiconductor, its intensity typically decreases exponentially with distance. The intensity (I) of the wave after a distance L in the semiconductor is given from

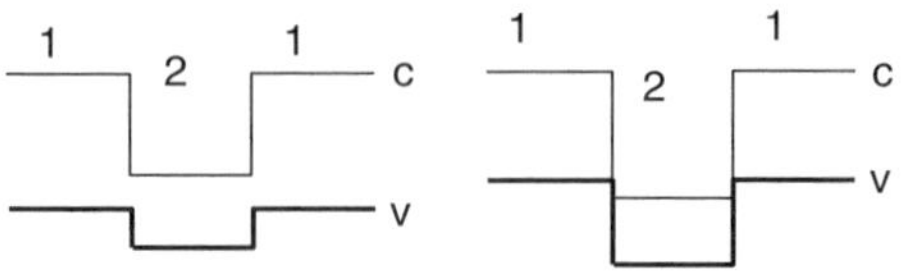

FIGURE 1.8 Quasi-type II (left) and type II (right) heterostructures.

$$I = I_0 \exp^{(-\alpha L)} \tag{1.8}$$

where $\alpha(\omega)$ is the frequency-dependent loss in the semiconductor, which is caused by two different phenomena, e.g., absorption and scattering, $\alpha(\omega) = \alpha_{\text{abs}}(\omega) + \alpha_{\text{scat}}(\omega)$.

The main mechanism of absorption in semiconductors is band-to-band absorption. In this case, a photon with energy in the vicinity of the band gap of the semiconductor gets absorbed, generating an electron–hole pair with the electron being in the conduction band and the hole remaining in the valence band. Similarly, an electron from the conduction band can recombine radiatively with a hole in the valence band, emitting a photon. In both cases the absorption or emission of photons requires the conservation of energy and momentum. They are simply obtained by

$$E_{\text{f}} - E_{\text{i}} = \hbar\omega_{\text{g}} \quad \text{and} \quad \hbar k_{\text{f}} - \hbar k_{\text{i}} = \hbar q \tag{1.9}$$

where

k_{f} (E_{f}) and k_{i} (E_{i}) are the final and initial wave vectors (energy) of the electrons
q and $\hbar\omega$ are the wave vector and the energy of the photons, respectively

The photon wave number is very small compared to the electron wave number and can be neglected. In direct band gap semiconductors near $\Gamma = 0$, k_{f} and k_{i} can be nearly identical. As a result, conservation of momentum can be easily achieved with the small photon momentum. However, for indirect band transitions the large momentum difference between the valence band maximum and conduction band minimum requires the absorption or emission of a photon to conserve momentum. As a result, the probability of light emission in indirect band gap materials, such as Si and Ge, is negligible and these materials are not suitable for LEDs and lasers.

In fact the absorption of a photon with energy near the band gap can generate electron–hole pairs, which are bound together by the Coulomb interaction [4]. Such a bound electron–hole pair is much like a hydrogen atom and is called an exciton. The binding energy (BE) of the exciton can be obtained from

$$\text{BE} = \frac{e^4 m_r^*}{2\hbar^2 \varepsilon_0^2} = \frac{e^2}{2a_B \varepsilon_0} = \frac{\hbar^2}{2m_r^* a_B^2} \tag{1.10}$$

where m_r^* is the reduced effective mass of the exciton, given by

$$\frac{1}{m_r^*} = \frac{1}{m_e^*} + \frac{1}{m_h^*}$$

and

$$a_B = \frac{\hbar^2 \varepsilon_0}{m_r^* e^2} \tag{1.11}$$

is the exciton Bohr radius. Excitons in semiconductors such as GaAs are referred to as Wannier–Mott excitons. They have large Bohr radii and small binding energy. For example in GaAs, the exciton Bohr radius is 140 Å and its binding energy is 4.2 meV. This is in contrast to excitons in

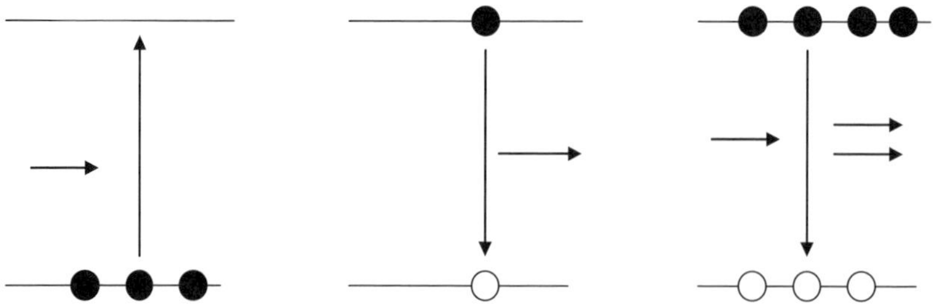

FIGURE 1.9 Mechanism of stimulated absorption (left), spontaneous emission (center), and stimulated emission (right).

organic materials, which are referred to as Frenkel excitons with small Bohr radii (5–10 Å) and large binding energy (0.5–1 eV). Excitonic absorption in inorganic semiconductors usually manifests itself as a series of strong absorption peaks just below the energy band gap. A photon may also be emitted as a result of bound electron–hole recombination, resulting in an excitonic emission.

1.2.1.5 Semiconductor LEDs and Lasers

The majority of semiconductor LEDs and lasers are made of p–i–n heterostructures. Under forward biasing, carriers (electrons and holes) are injected into the depletion region, where they can recombine radiatively to generate photons [5]. Radiative recombination occurs through two processes: spontaneous emission and stimulated emission. Spontaneous emission is the natural radiative recombination process of an electron from the conduction band to the valence band. This is the main process in LEDs. In stimulated emission a photon with a specific wavelength, direction, and polarization stimulates another photon of similar characteristics (same frequency, direction, and polarization) by forcing the recombination of an electron from the conduction band with a hole in the valence band. This is the photon amplification process involved in lasers. Figure 1.9 shows the mechanism of absorption and emission.

In a laser, in addition to the active gain material, a feedback cavity is also required. The feedback mirrors provide recirculation of photons and amplification. When the modal gain is increased to the point where it overcomes the cavity and mirror losses, then lasing emission starts. The mirrors can be fabricated by various techniques including cleaving (as in Fabry–Perot lasers) or feedback gratings as in distributed feedback (DFB) [6] or distributed Bragg reflectors (DBR) [7] lasers. Lateral mode confinement inside the cavity can be achieved by gain guiding or index guiding.

Today semiconductor lasers are widely used in various commercial and defense applications. Compact lasers with very low threshold current, high spectral purity, high power, and high beam quality are demonstrated over a wide wavelength range. Figure 1.10 shows the laser characteristics of a high-power optically pumped vertical external cavity surface-emitting laser (VECSEL). High beam quality and large wavelength tuning are demonstrated [8].

1.2.2 Optical Glass

Among all optical materials, glass has been the material most commonly and widely used for centuries as the ideal transparent material for the fabrication of lenses, telescopes, and mirrors, etc. However, it was during the second half of the twentieth century that the use of glass for the fabrication of waveguides and fibers intensified. Glass-based fibers and integrated optics have several unique features including excellent transparency; high thermal, mechanical, and chemical stability; and very high damage threshold. Glass is also an ideal host material for a range of optically active elements

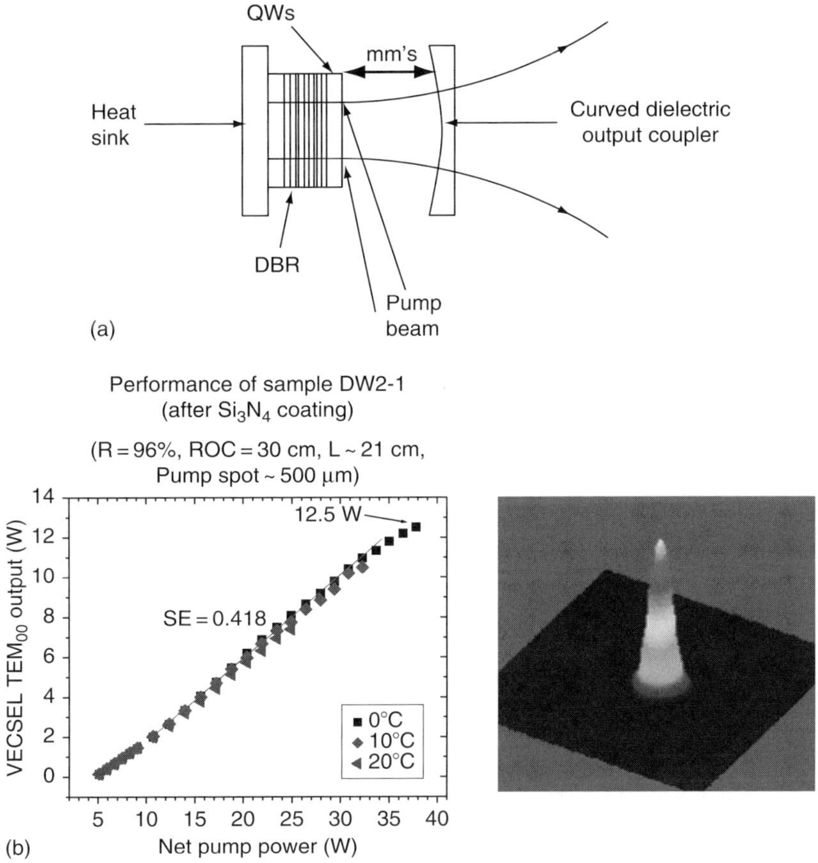

FIGURE 1.10 (a) Schematic of an optically pumped external cavity semiconductor laser, (b) output power versus pumping power for the VECSEL.

such as rare-earth and nonlinear materials. In fact during the past few years, glasses doped with rare-earth ions Er^{3+}, Yb^{3+}, Nd^{3+} elements have been widely developed for light amplification and emission. Among all these elements erbium (Er^{3+}), a three-level system, is of particular importance as it can provide light amplification and emission in the 1500–1600 nm wavelength range, particularly important for optical communication. In addition pumping is also relatively easy, since 980 nm semiconductor lasers are very suitable pumps and are commercially available. Figure 1.11 shows the energy diagram and the mechanism of absorption and emission in erbium. A 980 nm pump excites the Er ions from the ground state to the excited state. The excited ions have a very short lifetime in the $^4I_{11/2}$ and they rapidly decay to the $^4I_{13/2}$, where they have a longer lifetime and provide inversion of population and amplification of the 1550 nm signals.

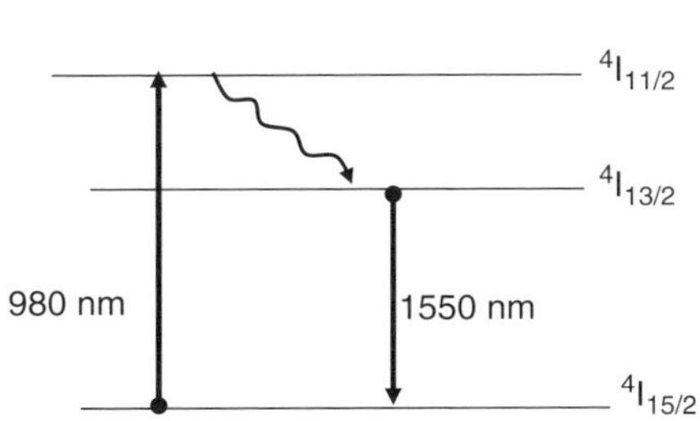

FIGURE 1.11 Schematic energy diagram of Er^{3+} ions.

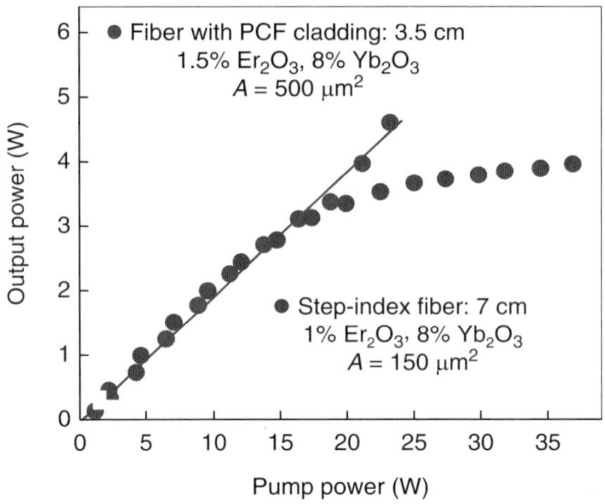

FIGURE 1.12 Examples of the output power versus pumping power for Er/Yb co-doped short fiber lasers. The red curve is for a 3.5 cm long photonic crystal fiber laser while the blue curve is for a 7 cm long step-index fiber.

To obtain more efficient pumping of the erbium ions, co-doping with ytterbium (Yb) ions is preferred. The absorption spectrum of Yb^{3+} in the 975 nm region is significantly broader than Er^{3+}. The excited carriers are then transferred to the $^4I_{11/2}$ level of Er^{3+} and consequently participate in population inversion and gain processes. Er^{3+}–Yb^{3+} co-doped fibers are widely used in the fabrication of ultrashort amplifiers and lasers [9]. Figure 1.12 shows the output of a short fiber laser exhibiting more than 5 W of power at 1.55 μm.

While the majority of electrically pumped lasers and photonic components for optical communication are made of semiconductors, a wide range of passive and active integrated optical components are developed using other inorganic materials such as glass. Having a refractive index close to that of fiber optics, these materials can provide several benefits such as low cost, high stability, low propagation loss, and low coupling loss to single-mode fibers. Glass waveguides can be fabricated using different techniques. The most popular methods are silica-on-silicon and ion-exchange glass. These techniques will be discussed later in this chapter.

1.2.3 Electro-Optic Materials

A number of inorganic materials change their optical properties in the presence of an applied electric field, caused by changing the position or orientation of its constituent atoms [10]. One very important effect is the change of refractive index in the presence of an applied field. When an electric field is applied across a film its refractive index is changed according to

$$\Delta\left(\frac{1}{n^2}\right) = rE + sE^2 \tag{1.12}$$

The first term is called the linear electro-optic or Pockels effect while the second term is the quadratic electro-optic or Kerr effect. While the quadratic electro-optic effect exists in all materials, the linear electro-optic is present only in crystals with noncentrosymmetry. A wide range of components such as optical modulators, switches, and tunable filters are fabricated using these properties.

In general the linear change in index of refraction due to an arbitrary applied field $\boldsymbol{E}(E_x, E_y, E_z)$ is defined by

$$\Delta\left(\frac{1}{n^2}\right)_i = \sum_{j=1}^{3} r_{ij}E_j \tag{1.13}$$

where r_{ij} is the electro-optic tensor, with $i = 1, \ldots, 6$ and $j = 1, \ldots, 3$. The magnitude of the index change strongly depends on the direction of the applied field and the crystal electro-optic coefficient. In centrosymmetric crystals, such as Si, all the r_{ij} coefficients are zero. As a result, they do not show a linear electro-optic effect. In a noncentrosymmetric material, several of the r_{ij} coefficients also vanish by symmetry. As a result, only a few of the coefficients are nonzero. Compound semiconductors such as GaAs exhibit a linear electro-optic effect and have been used as modulators. However, their electro-optic coefficient is relatively small, $r_{41} \cong 1.4$ pm/V. Among various inorganic materials, $LiNbO_3$ crystals are very attractive since they exhibit much larger electro-optic coefficients that make them suitable for external modulator applications. They are uniaxial crystals with $n_x = n_y = n_o$ and $n_z = n_e$. The most important electro-optic coefficients in $LiNbO_3$ are r_{33} and r_{13} with $r_{33} = 30.9$ pm/V and $r_{13} = r_{23} = 9.6$ pm/V.

One can apply an electric field parallel to E_z axis; the change in refractive index in different directions is simply obtained from

$$\Delta n_z = -\frac{n_e^3}{2} r_{33} E_z \tag{1.14}$$

$$\Delta n_x = \Delta n_y = -\frac{n_o^3}{2} r_{13} E_z \tag{1.15}$$

To achieve the largest index change for a given field, it is clearly advantageous to use the Δn_z. Now if an optical field travels at a length L of this material, it will experience a phase shift of

$$\Delta\Phi = \frac{\pi r_{33} n_e^3 V L}{d\lambda_o} \tag{1.16}$$

where

V is the applied voltage

d is the film thickness along the direction of the applied voltage

A parameter of great importance in modulators is the voltage for which a phase shift of π occurs. It is obtained from

$$V_\pi = \frac{\lambda_o d}{n^3 r L} \tag{1.17}$$

For practical modulators a V_π value below 1 V is desirable. While a longer L can help in reducing V_π, it can cause additional optical insertion loss. As a result, electro-optic materials with large electro-optic coefficients and low loss are highly desirable to meet these goals.

1.2.4 Organic and Polymeric Materials

Polymer is a term used to describe a large molecule or a macromolecule that is made by linking identical smaller units (called monomers) that are covalently bonded together. Polymers used for optics are typically amorphous with little or no order, as opposed to semiconductors that are typically crystalline. Amorphous polymers are similar to inorganic glasses. The coupling between neighboring chain units (intermolecular interactions) in polymers is weak, as opposed to inorganic solids with atoms that are strongly coupled. Randomly oriented polymers look like cooked spaghetti. They are sometimes referred to as plastics, which refers to their mechanical behavior above the glass transition temperature, where they soften dramatically and can be easily pulled and shaped.

Some polymers consist of a backbone and side chains. There are several types of polymers: homopolymers, copolymers, oligomers, and dendrimers. Only one type of monomer, A, units are used in homopolymers, AAAAA…. Two building blocks, A and B, are used to make copolymers. A structure such as AAAABBBBBAAAA… is referred to as a block copolymer. Some polymers adopt an alternating structure, ABABABAB… referred to as an alternating copolymer. Short polymer chains of 5–10 repeat units are often referred to as oligomers. Dendrimers are tree-structure macromolecules with many branches.

Organic materials are carbon-based. Usually carbon atoms are bound together by either σ-bonds or π-bonds. Single bonds like C–H bonds are σ-bonds while double bonds consist of a hybrid orbital of a σ-bond with a π-bond. σ-bonds are strong while π-bonds are weak. Electrons involved in π-bonds are delocalized over the molecule and are responsible for most of the interesting optical properties of organics. On the other hand, the σ-bonds are more localized and hold the molecule together, but do not usually play a major role in optical properties. Conjugated polymers are formed from a chain or ring of carbons with alternating single and double (or multiple) bonds, e.g., benzene molecule.

In contrast to inorganic solids where the allowed energy levels form valence and conduction bands, organic molecular materials are characterized by their HOMO, highest occupied molecular orbital, and LUMO, lowest unoccupied molecular orbital. Electronic transport occurs primarily by the hopping mechanism in organics. Electrons and holes on polymer chains hop from one chain to another for transport. This is in contrast with inorganic crystalline solids where band-type transport is dominant owing to the presence of a periodic crystalline structure and a well-defined density of states.

Electronic transitions in organics need to be considered together with vibrational states, as opposed to inorganics where the electronic transitions and the vibrational states can be easily separated. This is because the atoms in organics are often small, e.g., hydrogen has a small mass, whereas an inorganic solid like NaCl has atoms with much larger mass. The smaller mass makes the vibrational frequencies, $(f/m)^{0.5}$, much larger in organics. The vibrational states are then treated like simple harmonic oscillators with quantized energy levels, forming configurational coordinates. Optical transitions become combined electronic and vibrational transitions. Electronic transitions occur between the ground state and excited state vibrational quantized energy levels.

Optical absorption and generation of electron–hole pairs are also different in organics compared with inorganic semiconductors. The binding energies of the bound electron–hole pair excitons are much larger, while the binding energies of Bohr radii are much smaller in organics. The p–n junction concept is also different with donor–acceptor transitions being the dominant mechanism at the junctions in organics.

1.3 WAVEGUIDING PRINCIPLES FOR OPTOELECTRONIC MATERIALS

Optical waveguides are the most basic and essential elements in integrated optics and have been studied in depth [11,12]. The fundamental requirement in guided waves is the process of total internal reflection (TIR). Total internal reflection is a phenomenon through which a ray of light is totally reflected at the interface between two materials with different refractive indices. An incident ray traveling from a medium of index n_2 to a medium of index n_1 satisfies Snell's law described by

$$n_1 \sin\theta_1 = n_2 \sin\theta_2 \tag{1.18}$$

where θ_2 and θ_1 are the angles of the incident and refracted rays. The key requirement for TIR is that the incident ray should arrive from the material with the higher refractive index material ($n_1 < n_2$). In this case the incident ray will be totally reflected for a range of incident angles satisfying the condition:

$$\theta_2 > \theta_c = \sin^{-1}\left(\frac{n_1}{n_2}\right) \tag{1.19}$$

where θ_c is called the critical angle. A three layer waveguide can be simply constructed by sandwiching a layer of high refractive index material (n_2) known as the core between two lower-index layers (n_1 and n_3) called claddings. If $n_1 = n_3$, then the waveguide is called a symmetric waveguide. If the incident angle at each core-cladding interface satisfies the TIR condition then the wave can remain confined inside the core. The condition for waveguiding is that the reflected waves from each interface interfere constructively, resulting in a guided mode. In this case the propagation constant of the guided mode, β, satisfies the condition:

$$n_2 k_0 > \beta > n_1 k_0,\ n_3 k_0 \text{ with } k_0 = \frac{2\pi}{\lambda_0} \tag{1.20}$$

The effective refractive index of the guided mode is obtained from

$$n_{\text{eff}} = \frac{\beta}{k_0} \tag{1.21}$$

A limited number of guided modes are supported in a waveguide. Single-mode waveguides where only the fundamental mode is guided are of great interest in optical communication. In general, a mode becomes confined above a certain (cut-off) value of t/λ_0, where t is the thickness of the core. According to the mode conditions, cut-off values of t/λ for transverse electric (TE) and transverse magnetic (TM) modes are given by

$$\text{For TE: } \left(\frac{t}{\lambda_0}\right)_{\text{TE}} = \frac{1}{2\pi\sqrt{n_2^2 - n_3^2}}\left[m\pi + \tan^{-1}\left(\frac{n_3^2 - n_1^2}{n_2^2 - n_3^2}\right)^{\frac{1}{2}}\right] \tag{1.22}$$

$$\text{For TM: } \left(\frac{t}{\lambda_0}\right)_{\text{TM}} = \frac{1}{2\pi\sqrt{n_2^2 - n_3^2}}\left[m\pi + \tan^{-1}\frac{n_2^2}{n_1^2}\left(\frac{n_3^2 - n_1^2}{n_2^2 - n_3^2}\right)^{\frac{1}{2}}\right] \tag{1.23}$$

where m is an integer ($m = 0, 1, 2, \ldots$) that refers to the mth confined TE or TM mode.

Multimode waveguides have several shortcomings in optical communication. In these waveguides different modes propagate at different speeds. As a result, modal dispersion would be a major limiting factor for high-data-rate communication. Single-mode waveguides are required to eliminate modal dispersion. The condition for single-mode waveguiding can simply be obtained from the normalized equations as follows:

$$\tan^{-1}\sqrt{\alpha} < V_l < \pi + \tan^{-1}\sqrt{\alpha} \tag{1.24}$$

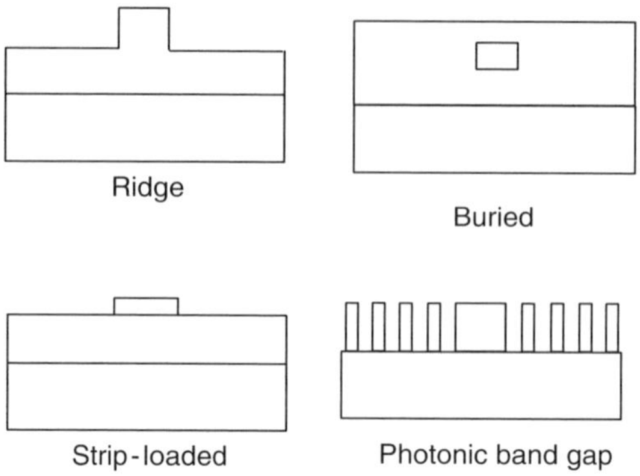

FIGURE 1.13 Different types of waveguides.

where

$V_l = \frac{2\pi t}{\lambda_0}\sqrt{n_2^2 - n_3^2}$ is the normalized frequency

$\alpha = \frac{n_3^2 - n_1^2}{n_2^2 - n_3^2}$ is the asymmetry measure of the waveguide

In a symmetric waveguide, $n_3 = n_1$, we have $\alpha = 0$ and single-mode operation is satisfied for $0 < V_l < \pi$.

Low-loss waveguides are typically desirable in integrated optics. Several types of waveguides are widely fabricated. The most common ones are shown in Figure 1.13. Among these various waveguides, buried waveguides are very attractive for their beam shaping possibilities. These waveguides can be fabricated to have a circular cross-section beam suitable for efficient coupling into single-mode fibers. However, as a drawback they are harder to fabricate compared to ridge and strip-loaded waveguides. Photonic band gap waveguides are particularly attractive for their compactness. These waveguides are expected to play a significant role in future generations of complex optoelectronics components.

1.3.1 Waveguiding in Semiconductors

Low-loss semiconductor waveguides can be fabricated by using materials with band gap energy significantly larger than the energy of the propagating wave. Silicon is presently not only the most exploited and well-known medium for integrated electronics, but is also highly transparent to light with low absorption in the near-IR spectral region making it a strong candidate for low-loss optical waveguides. Silicon waveguides generally consist of three layers: two cladding layers and the core layer. A ridge is formed to provide lateral confinement of the guided mode. A number of material combinations such as SiO–SiON–SiO, SiO–SiC–SiO, Si–SiGe–Si, and SOI amenable to silicon processing can satisfy waveguiding criteria. Among them, SOI has been developed in the last few years as an alternative technology platform to silica-on-silicon technology and its usefulness in making optoelectronic circuits has been established. High-performance components with low propagation loss in SOI waveguides have been reported [13]. The SOI structure possesses unique optical properties owing to the refractive index difference between silicon (n = 3.2) and SiO_2 (n = 1.444). Because of the strong confinement in the SOI system, the waveguides can be placed close together and have a small radius of curvature. These qualities make very dense component integration possible. Waveguides can be fabricated through well-established microelectronic processes, including photolithography and reactive-ion etching (RIE). However, the coupling between a silica fiber and a silicon waveguide typically involves mismatched modes and interfaces due to the large difference in the material indices. The guided mode is highly asymmetric, whereas the fiber mode is usually symmetric. Mode transformers are typically needed in the device design to get the optimum mode matching. The large index contrast between silicon and the silica fiber core implies a rather high Fresnel reflection of order 0.9 dB/facet, which can be eliminated by applying antireflection coatings on both facets. Moreover, crystalline silicon's birefringent response may be unacceptable in polarization sensitive applications.

1.3.2 Waveguiding in Glass–Ion-Exchange Glass and Optical Fibers

Waveguides in glass can be made by the ion-exchanged process that involves exchanging alkali ions (e.g., Na^+, K^+) originally in the glass with other ions (e.g., Ag^+, Tl^+, Cs^+), resulting in a local increase of the refractive index of the glass. A typical fabrication process is depicted in Figure 1.14. A metallic mask is usually used to control the location of ion exchange. The diffusion process is achieved at an elevated temperature of 300°C–400°C in molten salts. This thermal ion-exchanged process forms waveguides directly below the surface [14]. The waveguides may also be buried in order to change the guided mode profile; this burial process occurs through electric field assisted ion exchange. The introduced ions are driven by an applied electric field inside the glass. Ultimately, the shape of the elevated index region changes from a semicircular to a more circular form. This method results in single-mode waveguides with losses of 0.1 dB/cm and less. In addition to potassium and silver dopants, rare-earth ions like Er^{3+} can also be used.

Glass optical fibers have become critical elements in optical communication systems. They provide waveguiding of high-data-rate signals with very low loss over thousands of kilometers. They are usually cylindrical waveguides made of silica glass. In order to achieve waveguiding the core has a larger index than the cladding and it is formed by doping the silica with various oxides such as GeO_2, Al_2O_3, or P_2O_5. In addition, by doping the core with rare-earth ions such as Er^{3+} and Yb^{3+}, light amplification and lasing emission are obtained under optical pumping. Today optical fibers are manufactured with high purity and uniformity providing very low loss (<0.2 dB/km) and large bandwidth.

Single-mode fibers are fabricated using step-index or graded-index structures. For a step-index fiber the single-mode condition is given by

$$V = \frac{2\pi}{\lambda_o} a \sqrt{n_2^2 - n_1^2} < 2.405 \tag{1.25}$$

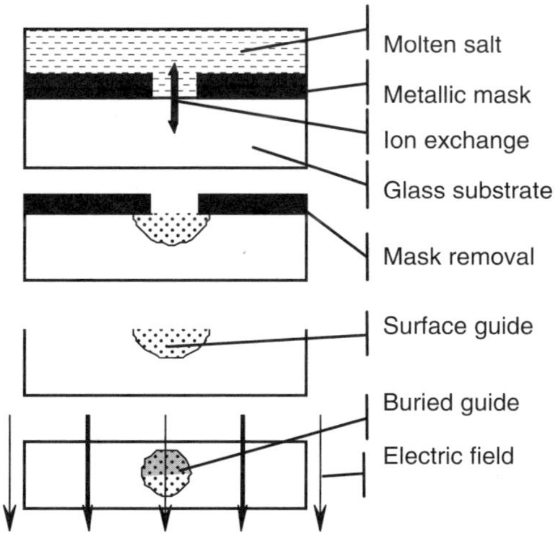

FIGURE 1.14 Waveguide fabrication by ion-exchange process.

where

a is the radius of the core

n_2 and n_1 are the refractive indices of the core and cladding, respectively

The numerical aperture of the fiber is defined as $NA = \sin\theta = \sqrt{n_2^2 - n_1^2}$.

In graded-index fiber, the refractive index in the core is a function of the radial position. As a result, the single-mode condition is different than in step-index fibers and depends on the detailed form of the core refractive index profile.

1.3.3 Waveguiding in EO Materials like $LiNbO_3$

Several methods can be used to produce a high-index waveguide core region in $LiNbO_3$ including diffusion, ion implantation, or ion exchange. Among these, titanium (Ti) diffusion is the method of choice. Titanium diffusion is a well-established technique to fabricate low-loss optical waveguides in $LiNbO_3$ crystals [15]. These waveguides are fabricated by depositing a 50–100 nm of Ti film over the $LiNbO_3$ crystals using the lift-off method. In order to achieve single-mode waveguides, the width of the Ti film is around 5 μm. Ti in-diffusion then takes place at a temperature of about 800°C–1100°C for 6–8 h under a controlled atmosphere. The best experimental waveguide samples exhibit propagation losses as low as 0.1 dB/cm. A wide range of integrated optic devices have been developed in $LiNbO_3$ crystals, including modulators and switches, by taking advantage of the high electro-optic, acousto-optic, and piezoelectric coefficients of this material. The index contrast between the core and cladding is relatively small (<0.02), which can affect the design flexibility.

Despite the great success of Ti-diffused waveguides, this technology still has major drawbacks. The high temperatures (900°C–1100°C) required for titanium in-diffusion cause lithium out-diffusion, creating an unwanted planar waveguide that competes with the waveguide structure defined to fabricate the devices. Ti:$LiNbO_3$-derived waveguides exist only as the embedded strip type, thus they are simply incompatible for integration.

$LiNbO_3$ substrates are typically either z-cut or x-cut crystals. In order to take advantage of the r_{33} coefficient, the electrodes for the applied field are fabricated differently. Figure 1.15 shows the schematic cross section of z-cut and x-cut waveguide phase modulators.

1.3.4 Silica-on-Silicon Waveguides

Silica-on-silicon technology offers many attractive features for integrated optics. Silicon wafers are of high optical quality, large size, and low cost. The large size allows the fabrication of complex optical circuitry using mature silicon microelectronic processing. In addition to their low optical loss and precise waveguide mode control, silica-on-silicon integrated optics can be easily packaged with fiber optics using silicon V-groove technology. Waveguide layers are deposited by two popular methods: flame hydrolysis deposition (FHD) and plasma-enhanced chemical vapor deposition (PECVD). The main drawback of these techniques is the requirement for postdeposition high-temperature treatment (around 1200°C) to reduce scattering loss and to densify the films. This is a

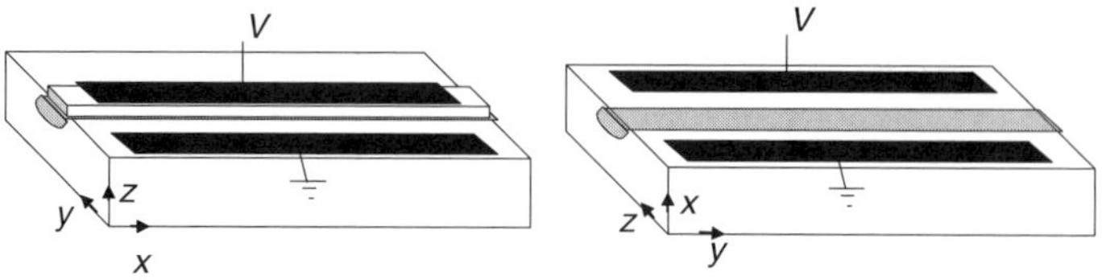

FIGURE 1.15 Schematic cross section of a z-cut and x-cut waveguide phase modulators.

limiting factor on their integration with other photonic or electronic components. Meanwhile, very low waveguide loss, below 0.1 dB/cm, has been readily obtained in silica-on-silicon waveguides.

Fabrication of waveguide devices requires multiple steps including deposition of the lower cladding to optically isolate the core from the high-index silicon substrate, core layer deposition, lithography, RIE, and deposition of the upper cladding. Both deposition technologies (FHD and PECVD) have shown commercial level success in the integrated optics industry. In both of these two approaches, high purity chemicals such as SiC_4, $GeCl_4$, and $POCl_3$ have been used as precursor reactants. These precursor chemicals react on a heated substrate to form dense glass films (PECVD) or react in a fuel and oxygen mixture flame to form glass microparticles and immediately deposit them onto a substrate (FHD). In both the cases, a second step of thermal treatment is required to produce optical quality glasses. PECVD films in general require an annealing step around 800°C–1100°C without major molecular level reorganization, whereas the films produced by FHD need a consolidation process around 1200°C–1350°C with major molecular level reorganization through melting.

In the FHD deposition scheme a mixture of gases is burnt in an oxygen/hydrogen torch to produce fine particles which stick onto substrates on a rotating turntable. The combination of the turntable rotation and traversing of the torch is designed to achieve layer uniformity. Dopants used include TiO_2, GeO_2, P_2O_5, Si_3N_4, and As_2O_3 to increase the refractive index, and fluorine and B_2O_3 to decrease the index. P_2O_5, B_2O_3, and As_2O_3 are network modifiers, and are used to lower the melting point of the deposited layers. This is essential in FHD-deposited layers, and also important in other processes to improve cladding conformality and planarization [16]. The gases $SiCl_4$, $TiCl_4$, and $GeCl_4$ are used to produce doped SiO_2. Small amounts of phosphorous and boron are added using PCl_3 and BCl_3. After deposition of a porous layer, the glass is consolidated by heating at 1200°C–1350°C. The deposition and consolidation process is intrinsically planarizing, providing excellent cladding conformality over closely spaced cores, which are used in, for example, directional couplers and Y-junctions. The waveguides produced from FHD layer have been found to have propagation losses as low as 0.05 dB/cm.

Plasma-enhanced chemical vapor deposition is a well-known technique in the microelectronic industry for the deposition of doped and undoped SiO_2 and Si_3N_4. It is very attractive for the deposition of waveguide layers since PECVD can achieve high deposition rates (0.1 μm/min) with thickness nonuniformity better than 1% and index nonuniformity within 0.0001 [17]. Waveguides produced by PECVD have demonstrated propagation losses of 0.1 dB/cm [18]. In a PECVD system the plasma is generated over a silicon wafer at frequencies ranging around 100 kHz to 13.5 MHz and powers from a few watts to a few hundred watts. The input gases are SiH_4 and N_2O (diluted by He, Ar, or N_2). A range of dopants have been used for waveguide fabrication, including phosphorus, boron, and fluorine. While PECVD has comparatively low-deposition temperatures (around 350°C), higher temperatures (800°C–1100°C) are usually required to eliminate hydrogen-related absorption bands, which arise due to hydrogen-based gaseous precursors. In addition, reflow of core and cladding at elevated temperatures is also advantageous to improve planarization of the cladding layer over the waveguide core.

1.3.5 Waveguiding in Organic Materials

Organic waveguides can be fabricated using some of the techniques described in Section 1.3.1. Here we describe a new hybrid organic–solgel waveguide fabrication method (see Figure 1.16). Organosilicate solgel solutions are prepared for the cladding and core solgel waveguide layers, which consist of methacryloxy propyltrimethoxysilane (MAPTMS) and an index modifier (zirconium(IV)-*n*-propoxide) with molar ratios of 95(MAPTMS)/5 mol% and 85(MAPTMS)/15 mol%, respectively. 0.1 N HCl is used as a catalyst to accelerate hydrolysis of the silanes with IRGACURE 184 (CIBA) used as the photoinitiator. When mercury *I*-line (365 nm) radiation is delivered to the solgel layer through a photomask, exposed regions become insoluble in isopropanol, which is used as the etchant for the wet-etching process. The under-cladding is coated on a silica (6 μm)-on-silicon substrate with a 100 nm thick Ti bottom electrode and baked at 150°C for 1 h to ensure formation of the silica network. The core is wet-etched after being coated onto the

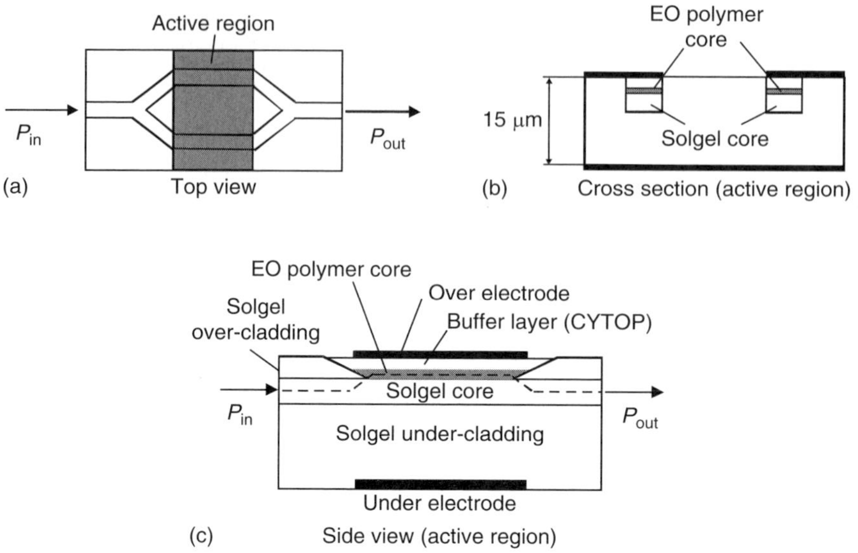

FIGURE 1.16 Schematic of an integrated polymer/solgel Mach–Zehnder waveguide modulator (see Ref. [20]). The top view (a), the cross section (b) and the side view (c) of the device are shown.

lower cladding; waveguide definition was accomplished by photolithography with a mask aligner. The over-cladding window is aligned with the core to establish an adiabatic transition region between the solgel core and the EO polymer core that is deposited in the window of the over-cladding. A vertical taper angle of 0.9°–3.6° in the over-cladding reduces the adiabatic transition loss to less than 0.5 dB. The EO polymer core is laterally confined by the solgel cladding, in order to obtain good mode confinement [19].

Such a waveguiding scheme has been employed to demonstrate optical modulation in a hybrid solgel-polymer Mach–Zehnder modulator with a V_π of 1 V (see Figure 1.17) [20].

1.4 CHALLENGES AND RECENT DEVELOPMENTS

1.4.1 Material Fabrication and Compatibility

Semiconductor heterostructures are formed by epitaxial growth techniques such as molecular beam epitaxy (MBE) or metal–organic vapor phase epitaxy (MOVPE). Over the last decades these growth

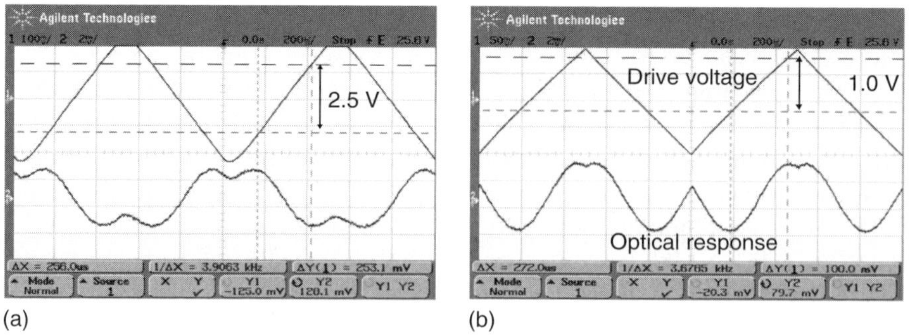

FIGURE 1.17 The oscilloscope screen shows the traces of the modulation voltage and the optical signal. Upper: Applied voltage. Lower: Optical output signal. (a) Phase modulator with 2.5 V for V_π, (b) Mach–Zehnder modulator with 1.0 V for V_π.

techniques have been improved almost to the point of perfection. Today, multiwafer growth systems are available where complex multilayer structures can be grown with monolayer accuracy.

The next generation of integrated photonics and optoelectronics requires the fabrication of different components such as laser, modulators, detectors, and waveguides on the same chip. A major challenge in such integration is the ability to grow high-quality elements with optimized composition, band gap, and doping. While multiple epitaxial regrowth provides such needs to a certain extent, it is less effective when the number of regrowth and processing steps increase. As a result, it is desirable to be able to locally grow various structures using band gap engineering. New techniques such as selective area epitaxy (SAE) and growth on patterned substrates (GPS) are used for localized band gap engineering, allowing the integration of diverse photonic components (such as lasers and modulators) on the same substrate. Epitaxial techniques such as MOVPE and chemical beam epitaxy (CBE) are suitable for SAE and GPS. Selective area epitaxy consists of localized growth through a patterned dielectric such as SiO_2, under an optimized growth condition, which inhibits deposition on the dielectric mask. By controlling the dielectric width, the composition and thickness of the quantum-well layers can be controlled and engineered. As a result, various compositions and band gaps can be grown on a single epitaxial step by varying the dielectric openings across the wafer. Using this technique high-performance integrated lasers–modulators have been fabricated.

1.4.2 Heterogeneous Integration

Monolithic integration of optoelectronic devices for high-speed optical communication systems offers the advantages of compactness, reliability, and reduced packaging costs. One of the problems in monolithic integration is the difficulty of fabricating dissimilar optical devices such as lasers, amplifiers, modulators, detectors, and waveguides on the same chip, because different devices often have incompatible processing requirements. Heterogeneous integration of different materials is the enabling approach for the development of complex and low-cost integrated optoelectronics components and subsystems. Progress in heterogeneous integration has been, however, very limited due to the epitaxial growth or processing incompatibility between the materials. In the last few years the progress in the development of organic polymers and hybrid organic–inorganic materials has made them a candidate for low-cost heterogeneous integration of optoelectronic components. The process compatibility and low-temperature processing of these materials are particularly suitable for such integration. In addition, the development of active functionality such as electro-optics and nonlinear optics with high performance combined with low optical loss makes them attractive. Hybrid organic–inorganic materials are particularly attractive as their properties are closely related in functionality to organic polymers for low temperature and ease of processing, and to conventional inorganic glasses for hardness, chemical and thermal stability, and transparency. However, several difficulties exist for their integration with semiconductors and other inorganic optoelectronic components. One major limitation in integration of organic and hybrid integrated optics with compound semiconductors and silicon is their low refractive index of around 1.5–2. One approach is to deposit a thick layer of lower-index cladding material over the semiconductor before the fabrication of organic waveguide and integrated optical elements. Over 10 μm of low cladding is needed to prevent mode radiation into the high-index semiconductor. Figure 1.18 shows the schematic cross section as well as the fabricated single-mode solgel waveguide on high-index semiconductor substrate. Single-mode waveguiding is also shown. This process has been successfully deployed for the integration of an organic–inorganic buried waveguide over high-index semiconductors [21].

Other techniques such as the slot waveguide approach have been investigated in order to facilitate the integration of low-index materials including polymers with high-index silicon waveguides [22]. In this technique, a narrow trench only 100 nm wide is created inside a submicron silicon ridge by electron beam lithography and RIE. The low refractive index material can then be deposited inside the trench. The guided field appears to peak in the trench providing a high mode confinement factor

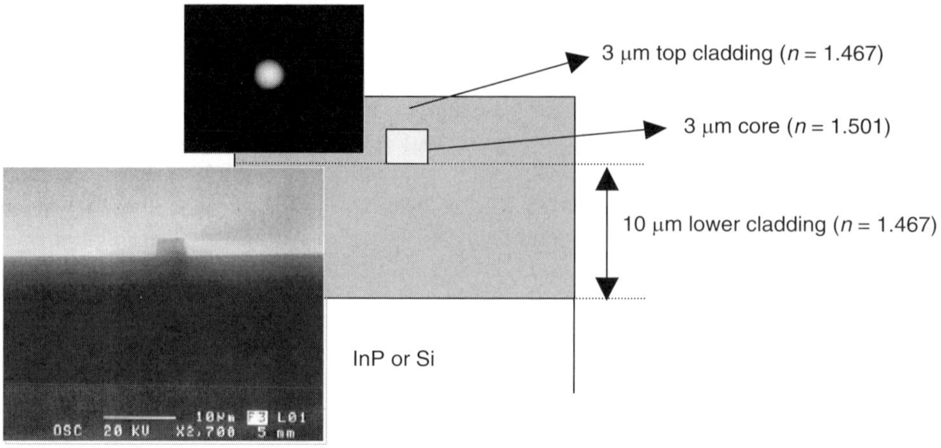

FIGURE 1.18 Integration of an organic–inorganic buried waveguide over a high-index semiconductor.

inside the low-index film. Although an attractive approach, more experimental results are needed to prove the practicality and reliability of the method.

Another promising method for the integration of organics with semiconductors is the use of photonic band gap structures [23]. As an example, a 2-D photonic band gap structure can be fabricated in semiconductors. The desired defect structure can then be filled with active organic materials. As a summary, Table 1.2 highlights the advantages and disadvantages of the various technologies for the fabrication of planar photonic circuits with the emphasis on their integration and compatibility with semiconductor technology.

EXERCISE QUESTIONS

1. A slab waveguide consists of a Ge-doped silica core with a thickness of 3 μm and an index of 1.5. The lower cladding layer is an infinitely thick silica layer with a refractive index of 1.48. The top layer is air. The waveguide is excited with a 1.3 μm wavelength laser.
   (a) What is the range of allowed propagation constant for the waveguide?
   (b) How many modes will the waveguide support?

TABLE 1.2
Major Material Contenders for Waveguide Fabrication in Integrated Optics

Material System	Deposition/Fabrication	Advantages	Disadvantages
Inorganic glass	FHD, PECVD/lithography/RIE, ion exchange	Stable, low loss, fiber matching, amorphous	High T°C (800°C–1350°C)
Polymers	Spin-coating/lithography/wet/dry etch, molding, contact print	Index control, low loss, low cost, low T°C	Weak stability, emerging technology, birefringence
III–V	Epitaxial growth/lithography/RIE	Stable, mature technology, monolithic integration, compactness	High cost, high loss, high coupling loss, anisotropic
$LiNbO_3$	Thermal diffusion of Ti, Zn, and Si	High electro- and acousto-optic coefficients	Inherent, incompatibility for integration
Solgel	Lithography/wet etch	Index control, low cost, low T°C, amorphous	Emerging technology

2. (a) Calculate the reflectivity of a bare GaAs end facet (take index of refraction of GaAs as 3.5).
   (b) Calculate the resonator mode spacing and the width of each mode for a GaAs semiconductor laser with a length of 250 mm and with uncoated facets.
3. Calculate the V_π of a $LiNbO_3$ waveguide modulator with a thickness of 1.5 μm and a length of 2.5 mm operating at a wavelength of 1.55 μm. How can one optimize the modulator in order to achieve a V_π of 1 V?

LIST OF ABBREVIATIONS

a_B	exciton Bohr radius
BE	binding energy
CBE	chemical beam epitaxy
DBR	distributed Bragg reflector
DFB	distributed feedback laser
E_g	semiconductor band gap energy
FHD	flame hydrolysis deposition
GPS	growth on patterned substrates
LED	light-emitting diode
$LiNbO_3$	lithium niobate
MAPTMS	methacryloxy propyltrimethoxysilane
MBE	molecular beam epitaxy
MOVPE	metal–organic vapor phase epitaxy
n_{eff}	effective index of a guided mode
NA	numerical aperture of a fiber
PECVD	plasma-enhanced chemical vapor deposition
SAE	selective area epitaxy
SOI	silicon-on-insulator
V	voltage number for a step-index fiber
V_π	voltage for a π-phase shift

REFERENCES

1. S.M. Sze, *Physics of Semiconductor Devices*, Wiley & Sons, New York, 1981.
2. S. Adachi, *Physical Properties of III–V Semiconductor Compounds*, Wiley & Sons, New York, 1992.
3. B. Broberg and S. Lindgren, Refractive index of InGaAsP layers and InP in the transparent wavelength region, *J. Appl. Phys.*, 55, 3376, 1984.
4. N. Peyghambarian, S.W. Koch, and A. Mysyrowics, *Introduction to Semiconductor Optics*, Prentice Hall, Englewood Cliffs, 1993.
5. L. Coldren and S. Crorzine, *Diode Lasers and Photonic Integrated Circuits*, Wiley series, New York, 1995.
6. H. Kogelnik and C.V. Shank, Stimulated emission in a periodic structure, *Appl. Phys. Lett.*, 18, 152, 1971.
7. F.K. Reinhart, R.A. Logan, and C.V. Shank, GaAs-$Al_xGa_{1-x}As$ injection lasers with distributed Bragg reflectors, *Appl. Phys. Lett.*, 27, 45, 1975.
8. L. Fan, M. Fallahi, et al., Tunable high-power high brightness linearly polarized vertical-external-cavity surface emitting lasers, *Appl. Phys. Lett.*, 88, 21105, 2006.
9. L. Li et al., Short cladding-pumped Er/Yb phosphate fiber laser with 1.5 W output power, *Appl. Phys. Lett.*, 85, 2721, 2004.
10. A. Yariv, *Optical Electronics*, 4th edition, Saunders College Publishing, Philadelphia, 1991.
11. K. Okamoto, *Fundamentals of Optical Waveguides*, Academic Press, San Diego, 2000.
12. C.R. Pollock, *Fundamentals of Optoelectronics*, Irwin, Chicago, 1995.
13. D. Marris et al., Ultralow loss successive divisions using silicon-on-insulator microwaveguides, *Appl. Phys. Lett.*, 87, 211102, 2005.
14. S. Najafi, S. Honkanen, and A. Tervonen, Recent progress in glass integrated optical circuits, *Proc. SPIE*, 2291, 6, 1994.

15. L. Thylen, Integrated optics in $LiNbO_3$: Recent developments in devices for telecommunication, *J. Lightwave Technol.*, 6, 847, 1988.
16. M.F. Grant, Glass integrated optical devices on silicon for optical communications, *Crit. Rev. Opt. Sci. Technol. SPIE*, CR53, 55, 1994.
17. G. Grant et al., Low loss PECVD silica channel waveguides for optical communications, *Electron Lett.*, 26, 2135, 1990.
18. M.D. McCourt, Commercial glass waveguide devices, *Crit. Rev. Opt. Sci. Technol. SPIE*, CR53, 200, 1994.
19. Y. Enami, G. Meredith, N. Peyghambarian, and A.K.-Y. Jen, Hybrid electro-optics polymer/sol–gel waveguide modulator fabricated by all-wet etching process, *Appl. Phys. Lett.*, 83, 4692, 2003.
20. Y. Enami, C.T. DeRose, D. Mathine, C. Loychik, C. Greenlee, R.A. Norwood, T.D. Kim, J. Luo, Y. Tian, A.K-Y. Jen, and N. Peyghambarian, Hybrid polymer/sol–gel waveguide modulators with exceptionally large electro-optic coefficients, *Nature Photonics*, 1, 180, 2007.
21. D. Lu, O. Mishechkin, and M. Fallahi, Low cost integrated optics on high index semiconductor for heterogeneous integration by hybrid sol–gel technology, Proceedings of European Conference on Optical Communication (ECOC), Italy, September 2003.
22. M. Lipson, High-confinement nanophotonics structures on chip, *Proc. SPIE*, 5729, 104, 2005.
23. A. Scherer, O. Painter, J. Vuckovic, et al., Photonic crystals for confining, guiding, and emitting light, *IEEE Trans. Nanotech.*, 1, 4, 2002.

2 Introduction to Optoelectronic Device Principles

Joachim Piprek

CONTENTS

2.1 Basic Processes 25
2.2 Photodetectors 27
  2.2.1 Pin Photodiode 27
  2.2.2 Solar Cell 31
  2.2.3 Photoconductor 33
2.3 Light-Emitting Diodes 34
2.4 Laser Diodes 36
  2.4.1 Gain, Loss, and Lasing Threshold 36
  2.4.2 Laser Characteristics 39
  2.4.3 Optical Resonator Designs 40
2.5 Optical Waveguides 41
2.6 Electro-Optic Modulators 43
2.7 Summary 45
Exercise Questions 46
List of Abbreviations 46
References 46

Abstract This chapter introduces basic principles of semiconductor optoelectronic devices. Such devices transform light into electricity and vice versa. Their operation is based on the microscopic interaction of photons and electrons. Traditional optoelectronic devices like photodetectors, light-emitting diodes (LEDs), and laser diodes are explained. Key device parameters and performance characteristics are described. The chapter concludes with a brief introduction of optical waveguide principles and electro-optic modulators.

2.1 BASIC PROCESSES

The central microscopic process in all optoelectronic devices is the absorption or the generation of a photon by an electron. The photon generation can happen spontaneously or can be triggered by another photon. We distinguish the following three key processes and corresponding device types:

- Photon absorption in photodetectors
- Spontaneous photon emission in light-emitting diodes
- Stimulated photon emission in laser diodes

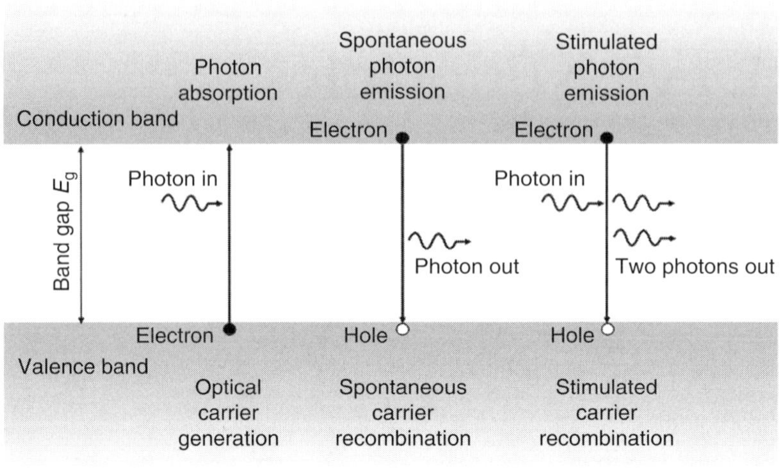

FIGURE 2.1 Illustration of key processes in optoelectronic devices.

Typically, photon absorption and generation involve the generation and recombination of an electron–hole pair, respectively. Electrons and holes carry electric charges and are often called carriers. Figure 2.1 shows all three processes schematically. The photon energy $h\nu$ must be equal or larger than the energy gap E_g between valence and conduction band (h, Planck's constant; and ν, light wave frequency). Correspondingly, the maximum light wavelength $\lambda = c/\nu$ is equal to the so-called gap wavelength

$$\lambda_g(\text{nm}) = \frac{hc}{E_g} = \frac{1241}{E_g(\text{eV})} \tag{2.1}$$

where c denotes the free-space light velocity.

Transition rates are used to quantify the number of generation or recombination events per unit volume and per second. The carrier generation rate G_{opt} due to photon absorption is proportional to the photon flux Φ_{ph} per area and per second

$$G_{opt} = \alpha \times \Phi_{ph} = \alpha \frac{P_{opt}}{A \times h\nu} \tag{2.2}$$

where the material parameter α is called absorption coefficient (unit: m^{-1}). The photon flux is equal to the optical power P_{opt} divided by the area A and the photon energy $h\nu$.

The net rate of spontaneous carrier recombination (spontaneous photon emission) is given by

$$R_{spon} = B(np - n_o p_o) \tag{2.3}$$

where B is the material parameter that is referred to as spontaneous or bimolecular recombination coefficient (unit: $m^3\ s^{-1}$). The recombination rate depends both on the availability of electrons in the conduction band (density n) and holes in valence band (density p). It vanishes at thermal equilibrium ($np = n_o p_o$), that is, the carrier densities need to be increased artificially to achieve spontaneous photon emission.

Photon emission by electron–hole recombination can also be triggered by an incident photon of equal wavelength. This process is called stimulated carrier recombination (stimulated photon emission) and, similar to photon absorption, its rate is given by

$$R_{\text{stim}} = g \times \Phi_{\text{ph}} = g \frac{P_{\text{opt}}}{A \times h\nu} \tag{2.4}$$

where the material parameter g is called the optical gain (unit: m^{-1}). More details are given in several textbooks on optoelectronic devices, e.g., in Ref. [1].

There are also nonradiative carrier recombination processes in semiconductors that do not emit photons. Instead, the electron energy is released in the form of atomic vibrations. The nonradiative recombination rate R_{nr} is given in terms of the nonradiative carrier lifetime τ_{nr} as

$$R_{\text{nr}} = \frac{n_{\text{min}}}{\tau_{\text{nr}}} \tag{2.5}$$

where n_{min} stands for the density of minority carriers (holes in n-doped material or electrons in p-doped material).

2.2 PHOTODETECTORS

Photodetectors convert light into electricity, or in other words, photons into conducting electrons. The electron absorbs the photon energy so that it can separate itself from the host atom and move freely in the conduction band following the direction of the internal electric field, thereby generating an electrical current. The electron typically leaves behind a hole in the valence band that moves in the opposite direction. There are many variations of semiconductor-based photodetectors. We introduce three common types of semiconductor-based photodetectors: pin photodiodes, solar cells, and photoconductors. For a more extensive review, the interested reader is referred to Refs. [2,3].

2.2.1 PIN PHOTODIODE

Photodetectors often utilize semiconductor p–n junctions (see Chapter 1). For enhanced photocurrent, it is desirable for the device to have a shallow junction followed by a wide depletion region where most of the photon absorption and electron–hole generation take place. Consequently, the basic p–n junction is typically modified to incorporate an undoped intrinsic (i) region that is sandwiched between the p- and the n-material resulting in a so-called pin diode. An example of a pin photodiode is shown in Figure 2.2, together with vertical profiles of ionized dopants, electric field, and light power. Owing to the presence of the electric field within the i-region, the photogenerated electron–hole pairs are immediately separated and are swept toward the electrodes. The movement of charge carriers inside the device is mirrored by an electrical current in the external circuit. The magnitude of this photocurrent I_{ph} is proportional to the intensity of the incident light. Photodiodes are typically operated under reverse bias thereby increasing the internal electric field. Light with wavelengths $\lambda > \lambda_g$ has a photon energy that is too small to elevate electrons from the valence into the conduction bands and is therefore not detected. Incident photons with wavelength shorter than λ_g become absorbed according to the absorption constant α of the semiconductor material. The optical power P_{opt} decays exponentially with the light travel distance, and the light power at the distance x from the semiconductor surface is given by

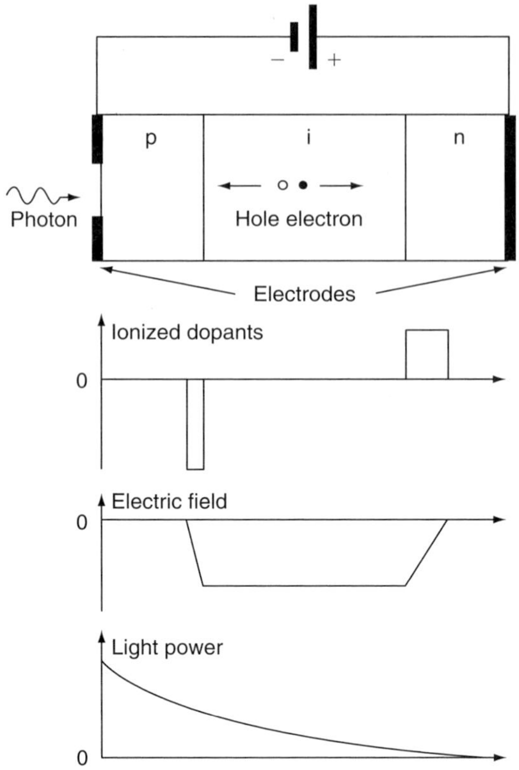

FIGURE 2.2 Schematic structure and internal physics of pin photodiodes.

$$P_{\mathrm{opt}}(x) = P_{\mathrm{in}}(1 - R)\exp(-\alpha x) \tag{2.6}$$

where

P_{in} is the incident power

R the optical reflectance of the semiconductor surface

Most of the photon absorption (63%) occurs within the so-called penetration depth $\delta = \alpha^{-1}$. Thus, the absorbing i-layer of the photodetector should be at least as thick as the light penetration depth. Figure 2.3 plots the absorption constant of traditional semiconductors as function of wavelength indicating the band gap wavelength λ_g for each material. The band gap absorption $\alpha(\lambda_g)$ of silicon at the light wavelength of $\lambda = \lambda_g = 1.1$ μm translates into a penetration depth of several 100 μm, which would require a very thick intrinsic layer. Thus, germanium with a gap wavelength of $\lambda_g = 1.87$ μm is a much better detector material for $\lambda = 1.1$ μm, giving an absorption constant of about 2×10^6 m^{-1} and a penetration depth of 0.5 μm. On the other hand, with photon energies larger than the energy band gap, the excess energy is wasted and eventually transferred into heat. The excess energy reduces the responsivity R_{ph} (A W^{-1}) of a photodiode that gives the ratio of generated photocurrent I_{ph} to incident optical power P_{in}

$$R_{\mathrm{ph}} = \frac{I_{\mathrm{ph}}}{P_{\mathrm{in}}} = \eta_{\mathrm{ph}} \frac{e}{h\nu} = \eta_{\mathrm{ph}} \frac{e\lambda}{hc} \tag{2.7}$$

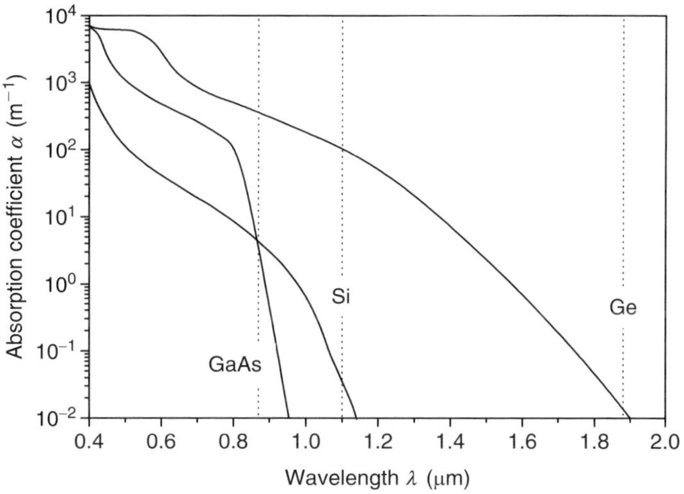

FIGURE 2.3 Optical absorption constants of GaAs, Si, and Ge as function of light wavelength. The dotted lines indicate the band gap wavelength.

The responsivity depends on the light wavelength and the spectral responsivity curve $R_{ph}(\lambda)$ is one of the key performance characteristics of photodiodes (Figure 2.4). It is related to the quantum efficiency η_{ph} of the photodetector, which gives the number of electrons collected (I_{ph}/e) relative to the number of incident photons ($P_{in}/h\nu$). The maximum possible quantum efficiency is 100% when each incident photon generates one electron that arrives in the doped regions. Practically, the quantum efficiency is limited by the surface reflectance R and by the absorption layer thickness d

$$\eta_{ph} = (1 - R)[1 - \exp(-\alpha d)] \tag{2.8}$$

It is further reduced by carrier recombination within the intrinsic layer.

Figure 2.5 shows the current–voltage (IV) characteristic of a photodiode with and without light. The dark current is the current through the diode in the absence of light, like the small current observed on reverse-biased diodes. This current is due to the thermal excitation of carriers and due to leakage. The dark current limits the minimum power detected by the photodiode since a photocurrent much smaller than the dark current would be hard to measure.

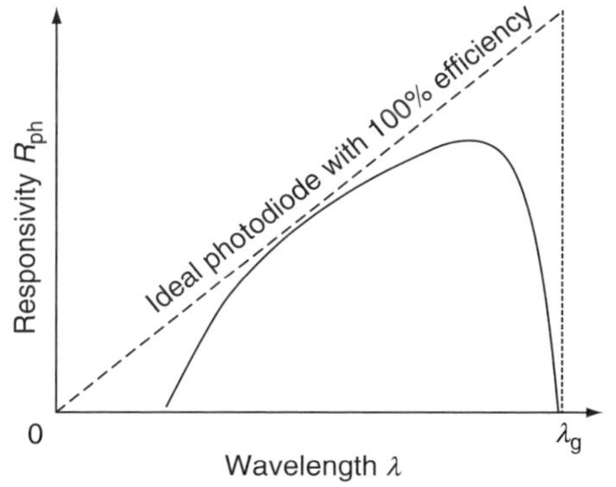

FIGURE 2.4 Schematic responsivity spectrum of photodiodes.

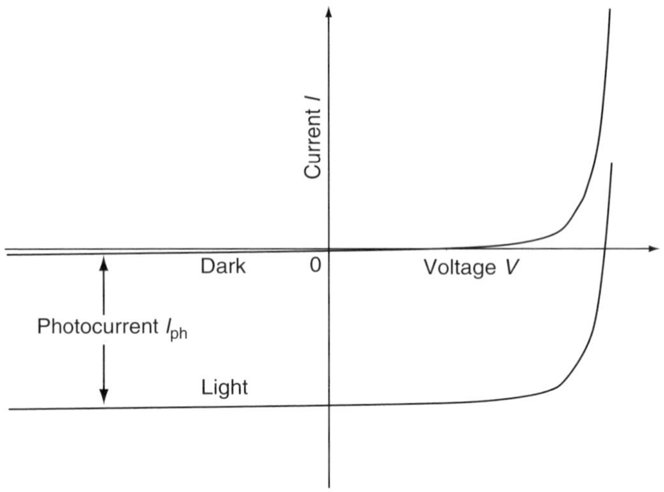

FIGURE 2.5 Current–voltage characteristic for a photodiode with and without illumination.

The separation of two thin layers of negative and positive charges by the intrinsic layer (Figure 2.2) is similar to a parallel plate capacitor. The intrinsic layer capacitance of the pin diode is given by

$$C_{\mathrm{i}} = \varepsilon_{\mathrm{o}}\varepsilon_{\mathrm{r}}\frac{A}{d_{\mathrm{i}}} \tag{2.9}$$

where

$\varepsilon_{\mathrm{o}}\varepsilon_{\mathrm{r}}$ is the permittivity
A is the cross-sectional area
d_{i} is the intrinsic layer thickness

The time required to move charges from one side to the other through the external circuit depends on the external resistance R_{ex} and it is given by the so-called resistance–capacitance (*RC*) time constant

$$\tau_{RC} = R_{\mathrm{ex}}C_{\mathrm{i}} \tag{2.10}$$

which limits the response time of the photodiode. However, the response time is often increased by the slow internal transit time of photogenerated carriers traveling across the intrinsic region. The carrier drift velocity v_{drift} rises with the internal electric field up to the saturation velocity v_{sat}, which is in the order of 10^5 m s^{-1} in traditional semiconductor materials. The carrier transit time is given by

$$\tau_{\mathrm{tr}} = \frac{d_{\mathrm{i}}}{v_{\mathrm{drift}}} \geq \frac{d_{\mathrm{i}}}{v_{\mathrm{sat}}} \tag{2.11}$$

and it is often longer than the *RC* time constant. Thus, the speed of a pin photodiode is normally limited by the transit time of photogenerated carriers across the intrinsic layer.

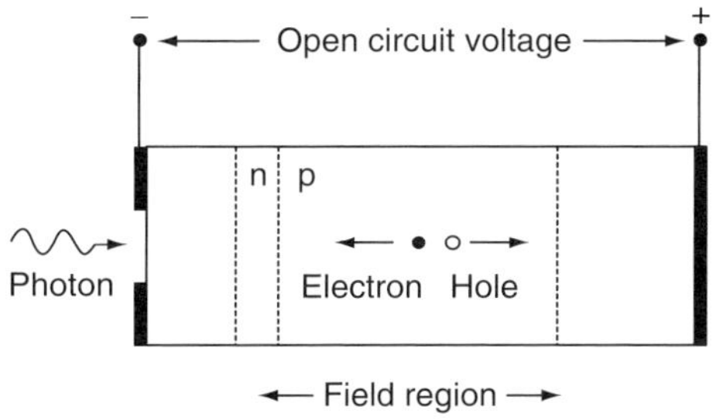

FIGURE 2.6 Solar cell.

2.2.2 Solar Cell

Solar cells generate electrical power by absorbing sunlight. Their operation is different from regular photodiodes as a forward bias is generated by the sunlight, which causes a forward current in the external circuit. Solar cells are also referred to as photovoltaic devices. Figure 2.6 illustrates the internal device physics of a p–n junction solar cell. Typically, a thin highly n-doped region is placed on top of a wide and lower doped p-region. Intrinsic regions are avoided to keep the series resistance small. Carrier diffusion creates a depletion region with a built-in electric field. This field separates electron–hole pairs generated by photon absorption. Electrons travel to the n-side and holes travel to the p-side, as in regular pn photodiodes. However, with open external circuit, the photogenerated carriers create a forward bias, called open circuit voltage V_{oc}. Figure 2.7 shows the *IV* characteristic of a solar cell with V_{oc} marking the point of zero external current. The total current inside the diode is also zero in that case, since the photogenerated forward bias V_{oc} causes an internal forward current that is exactly opposite to the photocurrent. When an external resistance R_{ex} (load) is connected to the solar cell, the photogenerated bias leads to an external current that is given by the dark current of the diode minus the photocurrent

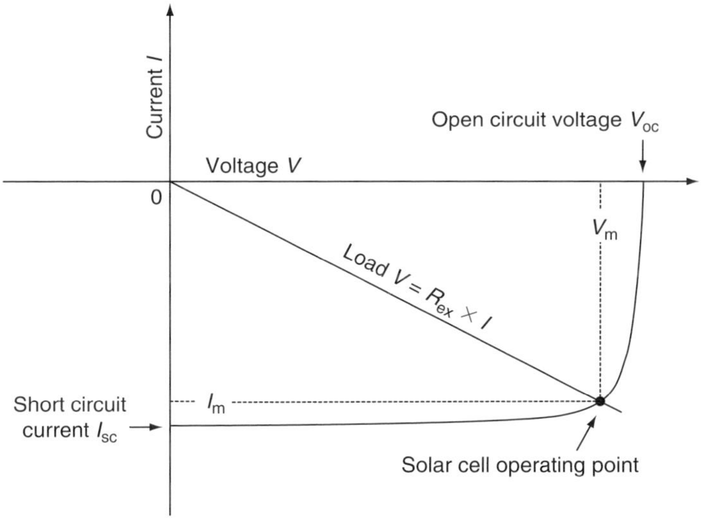

FIGURE 2.7 Solar cell current–voltage characteristic and parameters.

$$I = I_o\left[\exp\left(\frac{eV}{mkT}\right) - 1\right] - I_{ph} \tag{2.12}$$

where
- m is the diode ideality factor
- k is the Boltzmann constant
- T is the temperature

In the extreme case of zero resistance, the external short circuit current I_{sc} is equal to the photocurrent I_{ph}. The power $I \times V$ delivered to the load depends on the load resistance and can be determined graphically as shown in Figure 2.7. The optimum operating point gives the maximum area $I_m \times V_m$. Ideally, with a rectangular IV curve, the maximum power would be $I_{sc} \times V_{oc}$. Thus, the figure of merit of actual solar cells is given by the fill factor (FF)

$$\text{FF} = \frac{I_m V_m}{I_{sc} V_{oc}} \tag{2.13}$$

A key design issue of solar cells arises from the need to capture the solar spectrum over a large wavelength range. Figure 2.8 shows the free-space solar spectrum, which extends from ultraviolet (UV) to infrared (IR) radiation. The short-wavelength UV photons experience the strongest absorption (Figure 2.3) and typically disappear within a small region near the semiconductor surface. Many of these photogenerated carriers are out of reach for the electrical field within the depletion region (Figure 2.6). However, holes may diffuse toward the depletion region where they are swept away by the built-in field. The hole diffusion length L_p is limited by nonradiative carrier recombination, as holes are the minority carriers in the n-doped top region. Strong surface recombination is also detrimental to the solar cell performance. On the other side, long-wavelength IR photons penetrate deep into the p-region of the solar cell, some of them leaving the depletion region. Here, photogenerated electrons are the minority carriers and they typically exhibit a larger diffusion length L_n

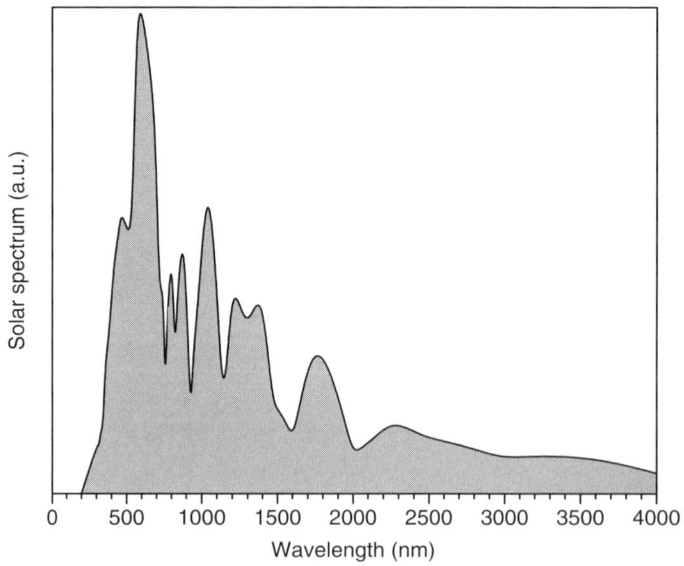

FIGURE 2.8 Solar spectrum.

than holes. Therefore, the p-region is usually located below the n-region. The cutoff wavelength λ_g is given by the band gap E_g of the semiconductor and sunlight with $\lambda > \lambda_g$ is not collected by the solar cell. Heterostructure solar cells with varying band gap are employed to harvest a larger part of the solar spectrum. Further details on solar cells are given in various texts, for instance in Ref. [4].

2.2.3 PHOTOCONDUCTOR

The key effect in photoconductors is the electrical conductivity enhancement by absorption of photons. A typical photoconductor structure is shown in Figure 2.9. The lightly doped semiconductor is sandwiched between two ohmic contacts, that is, there is no preferred direction of the current flow and the *IV* characteristic is symmetric (Figure 2.10). It is now essential that the light penetration is perpendicular to the current flow so that the conductivity enhancement is uniform between the contacts. In general, the semiconductor conductivity

$$\sigma = en\mu_n + ep\mu_p \tag{2.14}$$

is proportional to the density of electrons (n) and holes (p) as well as to their mobilities μ_n and μ_p, respectively. Without light, the equilibrium carrier concentrations are n_o and p_o and they depend on the doping of the semiconductor. Light absorption creates an excess electron concentration $\Delta n = n - n_o$, which is equal to the excess hole concentration $\Delta p = p - p_o$. The optical carrier generation rate G_{opt} is given by Equation 2.2. For simplicity, we here assume that it is uniform throughout the device. Under constant illumination, the carrier density would keep rising if there were no carrier recombination. The total recombination rate $\Delta n/\tau_{rec}$ also rises until both generation rate and recombination rate are equal so that the carrier density remains constant (τ_{rec}, recombination lifetime). The steady state excess carrier density is then given by

$$\Delta n = \Delta p = \tau_{rec} G_{opt} \tag{2.15}$$

The applied electric field F results in the photocurrent density

$$j_{ph} = \frac{I_{ph}}{A} = \Delta\sigma F = e\Delta n(\mu_n + \mu_p)F \tag{2.16}$$

The drift velocity of electrons ($\mu_n F$) is often much larger than the drift velocity of holes ($\mu_p F$). Thus, the electron arrives at the contact while the hole is still traveling in the semiconductor. Since the

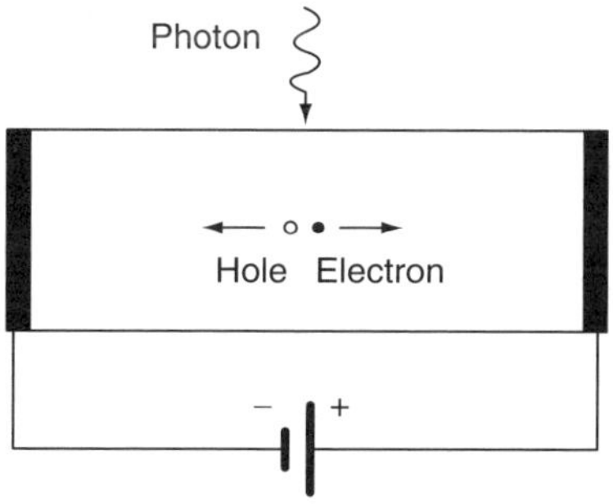

FIGURE 2.9 Photoconductor.

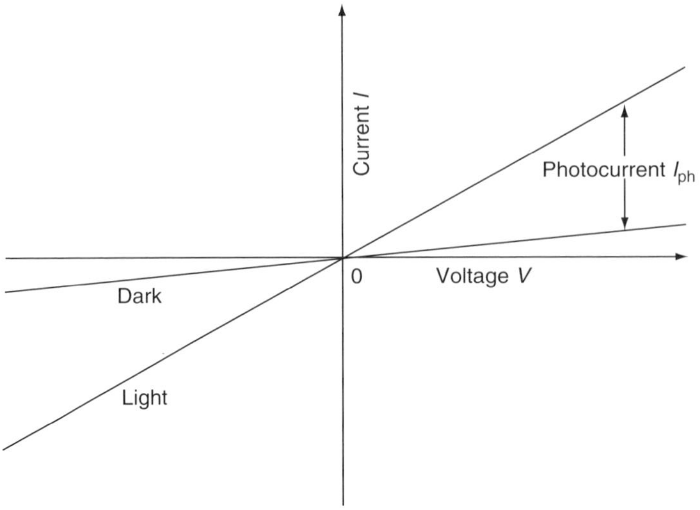

FIGURE 2.10 Current–voltage characteristic for a photoconductor with and without illumination.

photoconductor must remain neutral, another electron is injected from the opposite contact and travels across the semiconductor. Depending on the hole velocity, this process may be repeated many times until the hole recombines or reaches the contact. Many electrons may flow in the external circuit for each photogenerated hole. The photoconductive gain g_{ph} is defined as the rate I_{ph}/e of electron flow in the external circuit divided by the total rate AdG_{opt} of optical electron (hole) generation inside the device

$$g_{ph} = \frac{I_{ph}}{eAdG_{opt}} = \frac{\tau_{rec}(\mu_n + \mu_p)F}{d} = \frac{\tau_{rec}}{\tau_{tr,n}} + \frac{\tau_{rec}}{\tau_{tr,p}} = \frac{\tau_{rec}}{\tau_{tr,n}}\left(1 + \frac{\mu_p}{\mu_n}\right) \tag{2.17}$$

where

d is the electrode distance

the carrier transit time $\tau_{tr} = d/\mu F$

The gain rises with shorter electron transit time $\tau_{tr,n}$ and with longer recombination time τ_{rec}. However, slow recombination also increases the response time of the photoconductor, which is detrimental to high-speed applications. The transit time is reduced with higher fields but higher bias increases the dark current (Figure 2.10). Thus, there are several trade-offs that need to be balanced according to the demands of the specific application.

Besides the intrinsic photoconductors described above, there are also extrinsic photoconductors in which the absorption of low-energy photons (infrared radiation) triggers carrier transitions between local defect states and continuous energy bands so that typically only one type of carrier is generated (electrons from donor-type defects or holes from acceptor-type defects) [3].

2.3 LIGHT-EMITTING DIODES

Without external bias, electrons and holes at a semiconductor p–n junction are separated by a carrier depletion region (see Chapter 1). Forward bias reduces the depletion region and supports the diffusion of carriers to the other side of the junction where they can recombine spontaneously generating a photon (see Figure 2.1). This process is also called electroluminescence and it is the key process in light-emitting diodes (LEDs). A rising injection of carriers at the electrodes results in an increasing

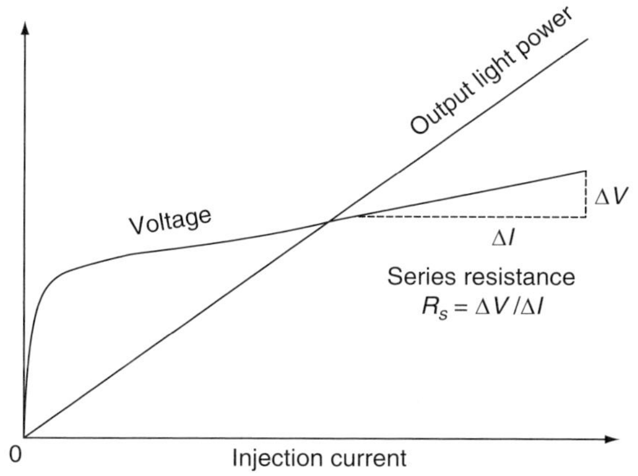

FIGURE 2.11 Typical LED characteristics: light versus current (*LI*) and voltage versus current (*IV*).

emission of light. Figure 2.11 shows the LED light–current (*LI*) and the *IV* characteristics. The optical output power P_{out} is related to the electrical injection current I by the equation

$$\frac{P_{out}}{h\nu} = \eta_{ext}\frac{I}{e} = \eta_{int}\eta_{opt}\frac{I}{e} \tag{2.18}$$

which is discussed next. Dividing the optical power by the photon energy $h\nu$ on the left-hand side of the equation gives the number of photons leaving the LED per second. On the other side, I/e is the number of electrons (and holes) injected into the LED per second. Thus, the external quantum efficiency η_{ext} is simply the ratio of emitted photons versus injected electrons. Ideally, this ratio should be 100% but it is often much smaller in real devices. To better reveal the internal processes involved, the external quantum efficiency is split into the internal efficiency η_{int} and the optical efficiency η_{opt}.

The internal quantum efficiency is the ratio of photons generated inside the LED to injected electrons. As carriers can recombine by radiative or nonradiative processes, η_{int} is mainly limited by nonradiative recombination. The internal efficiency can be written as fraction I_{spon}/I of the total injection current that feeds spontaneous photon emission

$$\eta_{int} = \frac{I_{spon}}{I} = \frac{I_{spon}}{I_{spon} + I_{nr}} \tag{2.19}$$

where I_{nr} is the nonradiative recombination current.

The optical efficiency η_{opt} gives the fraction of photons that leave the LED. This photon extraction efficiency is mainly limited by total internal reflection. Figure 2.12 schematically shows the light rays for a simple LED structure. Only light rays with close to normal incidence at the semiconductor–air interface can escape from the device. The light escape cone is confined by the total reflection angle φ_c with $\sin\varphi_c = 1/n_r$ (n_r, refractive index of the semiconductor). The optical extraction efficiency from a flat semiconductor–air surface is given by

$$\eta_{opt} = \frac{1}{2}\left(1 - \sqrt{1 - \frac{1}{n_r}}\right) \tag{2.20}$$

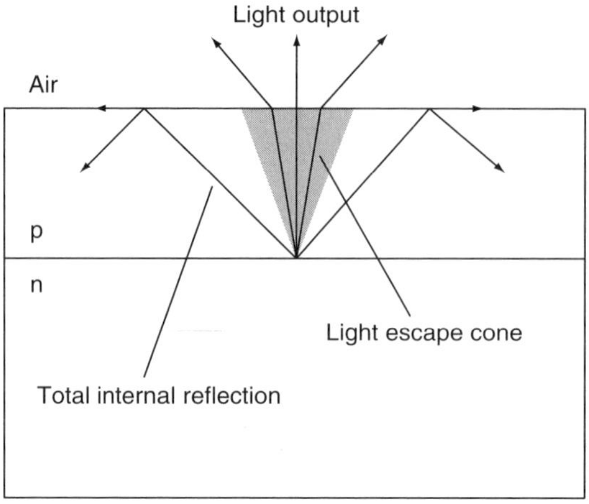

FIGURE 2.12 Illustration of light propagation in a simple LED structure.

The LED efficiency can be extracted from the slope (W A^{-1}) of the *LI* characteristic in Figure 2.11. Another popular LED parameter is the power conversion efficiency, i.e., the optical output power divided by the electrical input power

$$\eta_{con} = \frac{P_{out}}{IV} \tag{2.21}$$

Even with $\eta_{ext} = 100\%$, the conversion efficiency η_{con} is below 100% due to the series resistance of the diode, which transforms part of the electrical power into heat. The series resistance is given by the linear slope of the *IV* characteristic at higher current (Figure 2.11).

As electrons and holes may travel at various energies inside their bands, the energy $h\nu$ released by spontaneous recombination also varies. The LED emission spectrum is shown schematically in Figure 2.13. On the low-energy side, luminescence is limited by the band gap energy. More extensive details on semiconductor LEDs are given elsewhere, for example, in Ref. [5].

2.4 LASER DIODES

2.4.1 Gain, Loss, and Lasing Threshold

Traveling through a semiconductor, a single photon is able to generate an identical second photon by stimulating the recombination of an electron–hole pair. This is the basic physical mechanism of lasing. The second photon exhibits the same wavelength and the same phase as the first photon, doubling the amplitude of their monochromatic wave. Subsequent repetition of this process leads to strong light amplification. However, the competing process is the absorption of photons by generation of new electron–hole pairs (see Figure 2.1). Stimulated emission prevails when more electrons are present at the higher energy level (conduction band) than at the lower energy level (valence band). This situation is called inversion and it is one of the key requirements of lasing. Semiconductor lasers typically employ pin junctions with a small central layer of lower band gap (Figure 2.14). Very thin active layers (quantum wells) are often used in modern lasers, for details see Ref. [6]. At forward bias, electrons and holes are collected in the active layer to achieve inversion. Continuous current injection into the device leads to continuous stimulated emission of photons, but only if enough photons are constantly present in the device to trigger this process. Thus, only part of all

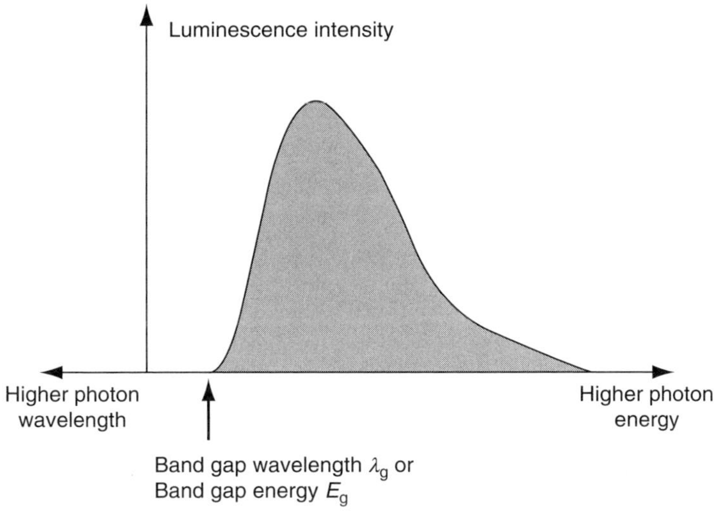

FIGURE 2.13 Schematic LED emission spectrum.

photons can be allowed to leave the laser diode as lasing beam, the rest must be reflected to remain inside the diode and to generate new photons (Figure 2.15). This optical feedback and confinement of photons in an optical resonator is the second basic requirement of lasing.

The light amplification in the active layer is measured as optical gain and it depends on wavelength, carrier density, photon density, and temperature. We here only evaluate the gain $g(N)$ as function of the carrier density $N = n = p$ within the active layer. A typical curve $g(N)$ is shown in Figure 2.16. Inversion is achieved and the gain is positive for carrier densities above the transparency density N_{tr}. The gain can be described by the logarithmic function

$$g(N) = g_o \ln\left(\frac{N}{N_{tr}}\right) \tag{2.22}$$

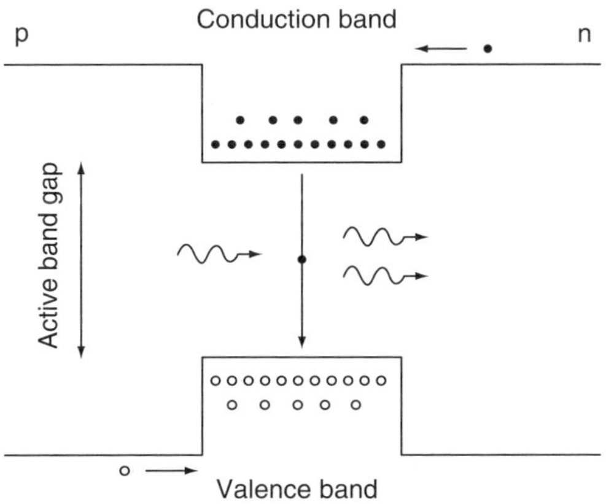

FIGURE 2.14 Energy band diagram of a pin heterojunction illustrating stimulated photon emission at forward bias.

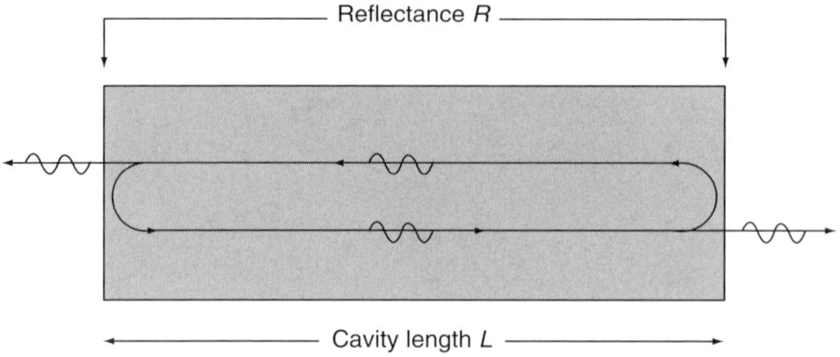

FIGURE 2.15 Optical feedback in a Fabry–Perot laser.

To reach the lasing threshold, the optical gain must compensate for the internal optical loss (α_i) and for photon emission from the device (α_m). The internal optical loss α_i is related to photon scattering or absorption. For the simple Fabry–Perot laser structure shown in Figure 2.15, the mirror loss parameter is given by

$$\alpha_m = \frac{1}{L}\ln\left(\frac{1}{R}\right) \tag{2.23}$$

At threshold, no photons return to the active layer, the photon flux in Equation 2.4 is still zero and so is the stimulated recombination rate. The threshold current I_{th} maintains the threshold carrier density N_{th} by feeding all carrier loss mechanisms, namely nonradiative and spontaneous recombination (see Section 2.1). Spontaneous recombination is needed to provide initial photons for stimulated recombination. The threshold current I_{th} can be calculated from

$$\eta_a I_{th} = eV_a\left(\frac{N_{th}}{\tau_{nr}} + BN_{th}^2\right) \tag{2.24}$$

with the injection efficiency η_a giving the fraction of electrons that recombines within the active layer of volume V_a. With stronger current injection $I > I_{th}$, the carrier density remains constant at N_{th} as additional electron–hole pairs are consumed by stimulated recombination.

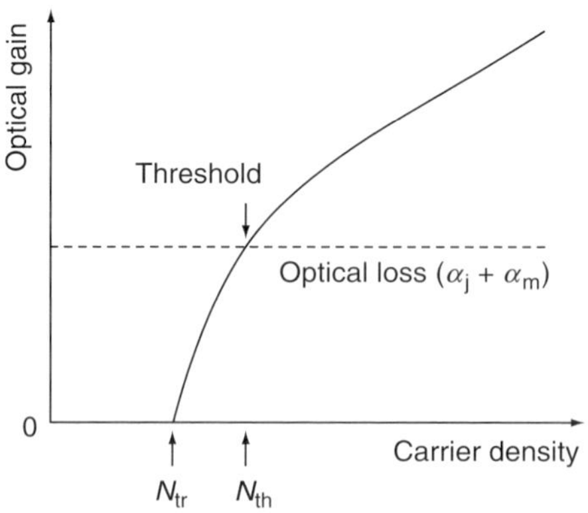

FIGURE 2.16 Gain versus carrier density $g(N)$ with transparency density N_{tr} and threshold density N_{th}.

2.4.2 Laser Characteristics

The most basic laser characteristic is the *LI* curve, which is shown schematically in Figure 2.17. At small currents, $I < I_{th}$, low intensity light is generated by spontaneous emission (not shown). Lasing starts at the threshold current I_{th} and the lasing power as function of injection current is given by

$$P_{out} = \eta_s (I - I_{th}) = \eta_d \frac{h\nu}{e}(I - I_{th}) = \eta_i \frac{\alpha_m}{\alpha_m + \alpha_i} \frac{h\nu}{e}(I - I_{th}) \tag{2.25}$$

with the slope efficiency η_s (W A^{-1}). The slope efficiency is a measure for the differential quantum efficiency η_d of the laser. This efficiency gives the fraction of carriers injected above threshold that contributes photons to the laser beam. It can be separated into the internal differential efficiency η_i and the optical efficiency η_o. The latter is equal to $\alpha_m (\alpha_m + \alpha_i)^{-1}$ and it gives the fraction of stimulated photons that leaves the laser. The internal differential efficiency η_i is the fraction of the total current increment above threshold that results in stimulated emission of photons

$$\eta_i = \frac{I_{stim}}{I - I_{th}} \tag{2.26}$$

It may be close to unity above threshold as there are no further recombination losses with constant carrier density N_{th} in the active layer. However, some carriers escape from the active region, especially at high injection currents or at high temperature.

Figure 2.17 also illustrates the *IV* curve of the laser. The series resistance is defined the same way as for LEDs (Figure 2.11). In contrast to the rather broad emission spectrum of LEDs shown in Figure 2.13, the lasing spectrum is very narrow and it peaks at a wavelength close to the band gap wavelength of the active layer.

Harmonic variation of the injection current above threshold results in the modulation of the output power. An amplitude modulation characteristic is shown in Figure 2.18. There is an upper limit of the modulation frequency beyond which the photon flux variation cannot follow the current variation any more. This limit is called modulation bandwidth. It is slightly larger than the resonance frequency f_r of the laser in

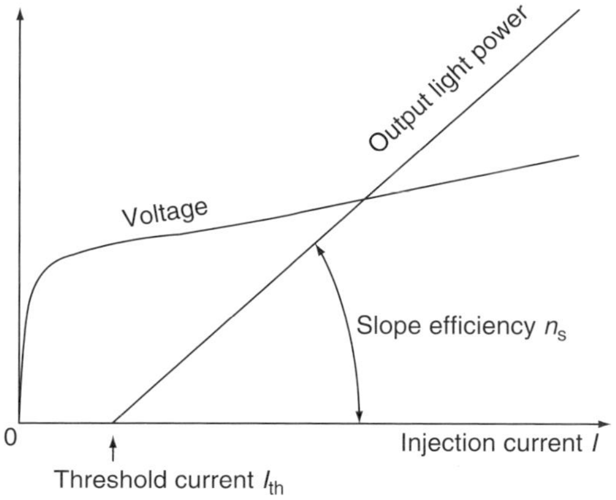

FIGURE 2.17 Laser characteristics: light power versus current (*LI*) and voltage versus current (*IV*).

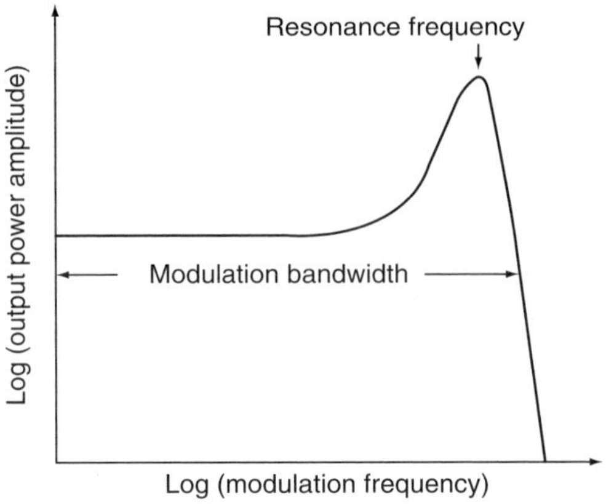

FIGURE 2.18 Analog modulation characteristic of a laser.

$$f_r^2 = \frac{a v_g \eta_i}{e V_{opt}} (I - I_{th}) \tag{2.27}$$

where

a represents the slope of the gain curve $g(N)$ in Figure 2.16
V_{opt} denotes the volume occupied by the internal optical field
v_g is the photon group velocity

Thus, the modulation bandwidth can be increased with stronger current injection (higher DC optical power), steeper gain curves, and smaller optical cavities. More details on laser modulation are given in Ref. [7].

2.4.3 Optical Resonator Designs

Modern laser diodes use heterostructures of different semiconductor materials to form the pin junction (Figure 2.14). The center part of this structure has a lower band gap and a higher refractive index than the cladding material, thus confining electron–hole pairs by energy barriers and photons by total reflection. Stimulated photons establish an optical field inside the laser diode, which is governed by the design of the optical cavity. The simplest resonator uses the reflection at the two laser facets (Figure 2.15). Constructive interference of forward and backward traveling optical waves is restricted to specific wavelengths. These wavelengths constitute the longitudinal mode spectrum of the laser. The cavity length L is typically of the order of several hundred microns, much larger than the lasing wavelength, so that many longitudinal modes may exist. The actual lasing modes are those receiving strong optical gain. Single-mode lasing is hard to achieve in simple Fabry–Perot structures, especially under modulation.

Dynamic single-mode operation is required in many applications and it is achieved using optical cavities with selective reflection (Figure 2.19). The distributed feedback (DFB) laser is widely used in single-mode fiber optic applications. Typical DFB lasers exhibit a periodic longitudinal variation of the refractive index within one layer of the edge-emitting waveguide structure. This index variation provides continuous (distributed) reflection at a wavelength given by the variation period. Facet reflection is not needed in DFB lasers; however, facet coating may be used to increase the light emission from one end of the cavity. Other laser resonators terminate the optical cavity by two distributed Bragg reflectors (DBRs) with stepwise alternating index. A special type of such DBR lasers is the vertical-cavity surface-emitting laser (VCSEL), which emits through the bottom or top surface of

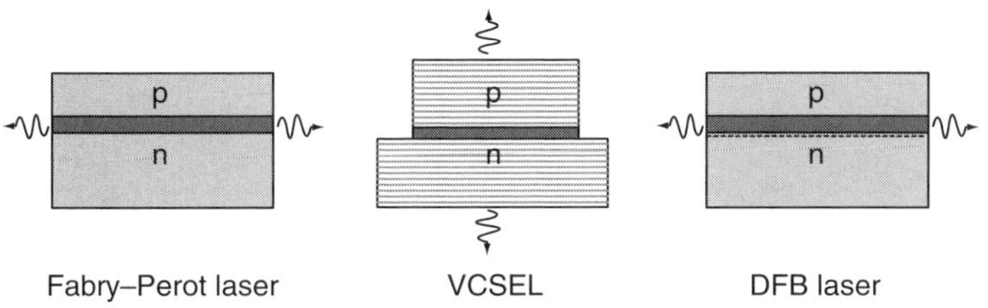

FIGURE 2.19 Optical resonator designs for semiconductor lasers.

the layered structure. The light travels perpendicular to the active layer and receives optical gain only over a very short travel distance. Thus, many more photon roundtrips are needed for lasing and highly reflective VCSEL mirrors are required with R >99%.

In transversal directions, the optical wave is typically confined by the refractive index profile, using a ridge (DFB laser) or a pillar (VCSEL) to form the waveguide. Even with restriction to one longitudinal mode, multiple transversal optical modes may occur in all three types of lasers. The first or fundamental mode has only one intensity maximum at the laser axis. Higher-order modes have multiple maximums in transverse direction.

Optical waveguide principles are introduced in Section 2.5. The various laser structures are covered in more detail in specialized texts like Ref. [7].

2.5 OPTICAL WAVEGUIDES

Devices based on electron–photon interaction are considered active devices and they are the main focus of this chapter. However, the light propagation inside optoelectronic devices is also passively affected by the optical material properties, in particular by the refractive index. A simple example is light reflection and diffraction at interfaces, governed by the Fresnel equations, which lead to Snell's law [8]. On the basis of such simple formulas, geometrical optics (ray tracing) can be used to study the internal propagation of light beams as long as the structural dimensions are much larger than the wavelength.

We here focus on smaller structures with dimensions close to the optical wavelength. Such multilayered structures are often used in active devices to confine and to guide the optical wave. In fact, the use of heterostructures for carrier confinement (see Figure 2.14) also creates a step profile of the refractive index, which can be employed for optical field confinement (Figure 2.20). The simple case of a symmetric planar waveguide is further illustrated in Figure 2.21. The light is guided by a thin core layer (refractive index n_{r1}), which is embedded in a cladding material with a slightly lower refractive index n_{r2}. The lightwave experiences total reflection if the angle θ is less than the critical angle θ_c defined by

$$\cos\theta_c = \sin\varphi_c = \frac{n_{r2}}{n_{r1}} \tag{2.28}$$

The bounce angle θ is complementary to the angle of incidence φ. The wave is guided if $\theta < \theta_c$. We here employ the simplified ray picture in Figure 2.21 to introduce some general features of guided waves. Only discrete bounce angles θ_m (discrete modes) are allowed since reflected waves need to add up constructively (mode number $m = 0, 1, 2, \ldots$). The number of guided modes is limited by $\theta < \theta_c$, which results in

$$m < \frac{2d}{\lambda}\sqrt{n_{r1}^2 - n_{r2}^2} \tag{2.29}$$

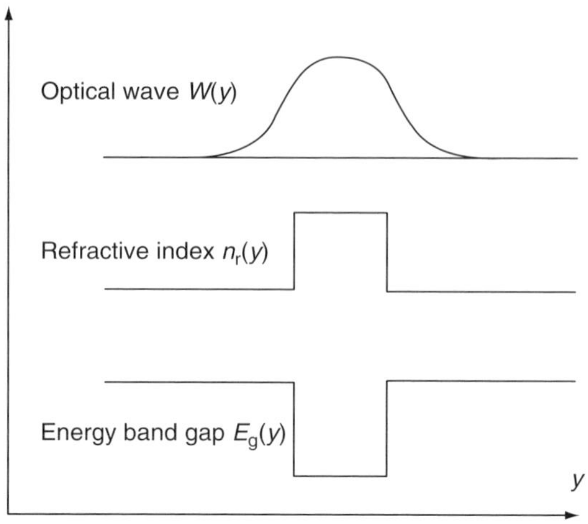

FIGURE 2.20 Vertical profile of band gap, refractive index, and optical wave within a planar symmetric heterostructure.

For any wavelength λ, the fundamental mode ($m = 0$) is always supported. However, θ_m rises with longer wavelength and comes closer to the cutoff angle θ_c resulting in weaker guiding. The optical mode is given as function of time t by

$$W(x,y,t) = W_m(y)\cos(2\pi\nu t - \beta_m x) \tag{2.30}$$

where W stands for any field component of the electromagnetic wave. The mode propagates in x-direction with the propagation constant

$$\beta_m = 2\pi \frac{n_{r1}}{\lambda} \sin\theta_m \tag{2.31}$$

The fundamental mode exhibits the smallest angle θ_m and the largest propagation constant.

The transverse field profile $W_m(y)$ also depends on the mode number. Figure 2.22 shows transverse mode profiles for our symmetric planar waveguide. Even-numbered modes exhibit maximum

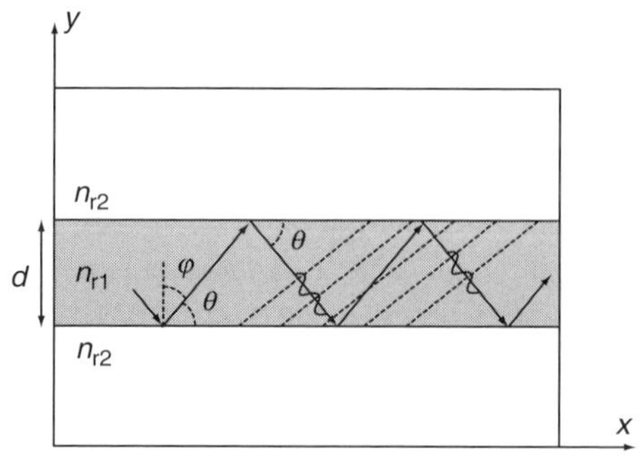

FIGURE 2.21 Light propagation in a planar waveguide. The dashed lines indicate constructive interference of reflected waves.

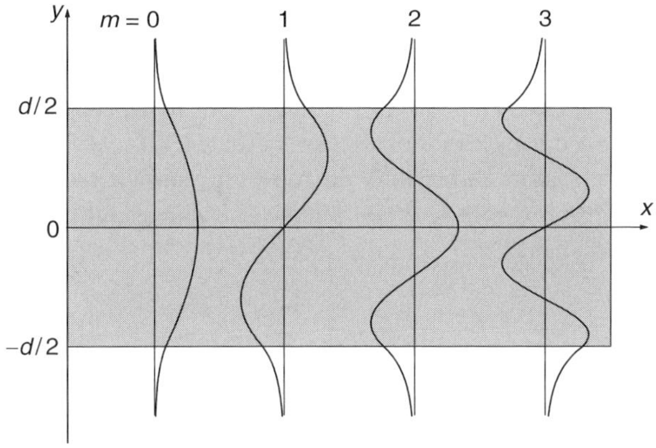

FIGURE 2.22 Transverse optical field profiles $W_m(y)$ for waveguide mode number m.

intensity in the waveguide center where odd-numbered modes are zero. The smaller the index contrast the deeper the modes penetrate into the cladding layer. Reducing the index contrast $n_{r1} - n_{r2}$ or the waveguide thickness d also lowers the number of guided modes and can be used to restrict waveguiding to a single mode with $m = 0$.

Two-dimensional optical waveguides confine the light in both transverse directions. The principles and the modal properties in each direction are similar to planar waveguides, with independent modal indexes in vertical and lateral direction [7,8].

Thus far, this section focused on traditional optical waveguides with minimum dimensions of several micrometers, depending on the light wavelength. Modern electronic circuits feature much smaller dimensions of less than 100 nm and optical waveguides of similar size are highly desirable to combine electronics and photonics. Nanoscale optical waveguides based on surface plasmons [9] are therefore receiving increasing interest. Surface plasmons are groups of electrons that collectively move back and forth at a metal surface. Such electron plasma oscillations are able to interact with light waves so that both travel together along the metal/dielectric interface (Figure 2.23). This light–matter interaction at nanostructured metallic surfaces has led to a new branch of photonics called plasmonics [10].

2.6 ELECTRO-OPTIC MODULATORS

The refractive index n_r of an optical waveguide can be changed by applying an electric field F. In linear approximation,

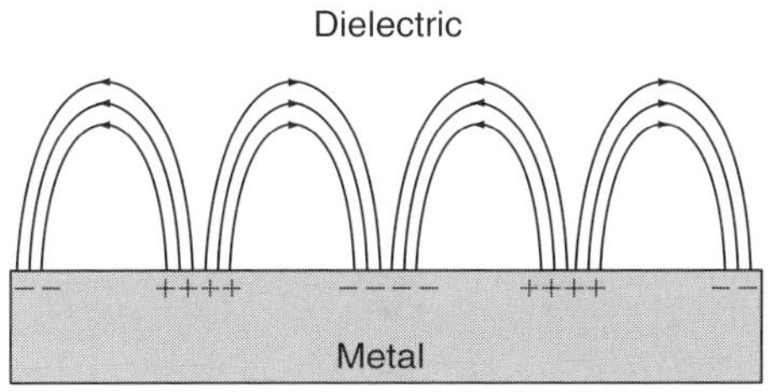

FIGURE 2.23 Surface plasmons: interaction of electrons and optical waves at a metal/dielectric interface.

$$n_r(F) = n_r - \frac{1}{2} r n_r^3 F \tag{2.32}$$

where r is the linear electro-optic coefficient (Pockels effect). Typical values of r lie in the range of 10^{-12} to 10^{-10} m V^{-1}. The refractive index may decrease or increase, depending on the direction of the electric field. If the material response is symmetric, i.e., invariant to the reversal of the electric field, the first derivative vanishes ($r = 0$) and the second-order Kerr effect dominates with

$$n_r(F) = n_r - \frac{1}{2} s n_r^3 F^2 \tag{2.33}$$

The Kerr coefficient s is typically in the range of 10^{-18} to 10^{-14} m^2 V^{-2}. The light velocity c/n_r changes with the applied field $F = V/d$ (V, bias; d, electrode distance). This is commonly described by the phase change

$$\Delta\phi = -\pi \frac{V}{V_\pi} \tag{2.34}$$

where V_π is the half-wave voltage that is required to shift the phase by 180° (π). In the linear case,

$$V_\pi = \frac{d}{L} \frac{\lambda}{r n_r^3} \tag{2.35}$$

where

L is the light travel distance

λ is the free-space wavelength

There are many types of electro-optic modulators [1,7,8]. We here use a popular intensity modulator as an example, which employs the principles of a Mach–Zehnder interferometer (Figure 2.24). It splits the input wave equally using a Y-junction waveguide. The refractive index in the upper branch is modified by applying the bias V. Consequently, a phase difference $\Delta\phi$ occurs when both parts of the wave are reunited at the output. Both waves cancel each other and the output power drops to zero if

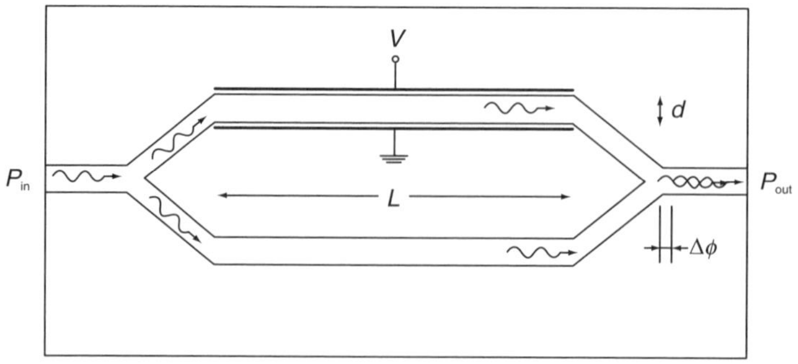

FIGURE 2.24 Mach–Zehnder modulator.

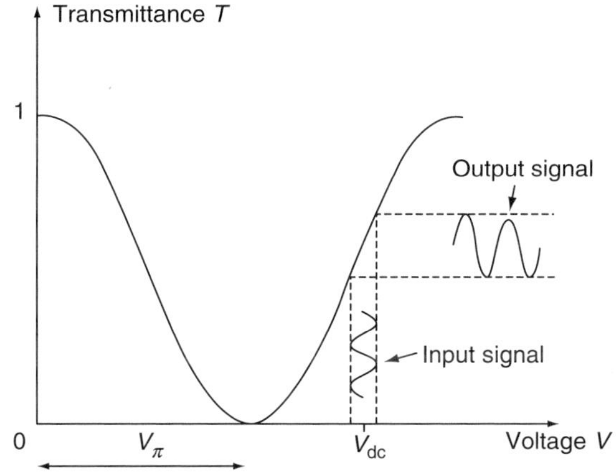

FIGURE 2.25 Modulator transmittance function $T(V)$.

the phase difference is exactly half a wavelength ($\Delta\phi = \pi$). Assuming that both branches are equally long, the ratio of output power P_{out} to input power P_{in} is given by the transmittance function

$$T(V) = \frac{P_{out}}{P_{in}} = \cos^2\left(\frac{\Delta\phi}{2}\right) = \cos^2\left(\frac{\pi}{2}\frac{V}{V_\pi}\right) \tag{2.36}$$

This periodical function is illustrated in Figure 2.25. In our simple example, full transmission is obtained at zero bias. Optical waveguide losses and coupling losses reduce the maximum transmittance of real devices. The transmission curve $T(V)$ may also be shifted along the voltage axis, due to built-in phase differences, e.g., from different lengths of both branches. Commonly, a built-in phase difference is chosen that keeps the operating voltage small. Variation of the input bias around the DC operation point V_{DC} creates a variation of the output power. This principle is used for electro-optic signal conversion in optical communication applications. On the other end, a photodetector can be employed for conversion of the optical signal back into the electrical domain.

2.7 SUMMARY

Semiconductor optoelectronic devices are traditionally based on absorption and generation of photons by electron–hole pair generation and recombination, respectively. As the photon generation can be spontaneous or stimulated, three basic types of devices are distinguished: photodetectors, LEDs, and lasers. Photodetectors transform light into electrical current (photocurrent). Photodiodes are operated at reverse bias; their performance is mainly characterized by the responsivity spectrum (Figure 2.4). Solar cells generate a forward bias (photovoltage) and electrical power; their main parameters are the open circuit voltage, the short circuit current, and the fill factor (Figure 2.7). The key effect in photoconductors is the electrical conductivity enhancement due to illumination; their photocurrent is proportional to the applied voltage (Figure 2.10). LEDs transform electrical current into noncoherent light (Figure 2.11); their efficiency is often limited by total internal reflection (Figure 2.12). Laser diodes are the most sophisticated optoelectronic devices, which require carrier inversion and optical feedback to reach the lasing threshold (Figures 2.14 through 2.16). Threshold current, slope efficiency, and modulation bandwidth are key laser parameters (Figures 2.17 and 2.18). Variations of the refractive index inside the device are often employed to confine and guide the optical wave within a small region of the size of the wavelength. Discrete waveguide modes are shown in Figure 2.22. The propagation velocity of the wave can be modified by applying an electric field that changes the refractive index of the waveguide. This principle is employed in electro-optic modulators to vary the intensity or the phase of an optical wave (Figures 2.24 and 2.25).

EXERCISE QUESTIONS

1. Solar cell under illumination of 500 W m^{-2} has a short circuit current of I_{sc} = 15 mA and an open circuit voltage of V_{oc} = 0.6 V. Calculate I_{sc} and V_{oc} for double the light intensity assuming a diode ideality factor of $m = 1$.
2. LED emission wavelength λ is sensitive to the temperature T, which is mainly related to the band gap shift $E_g(T)$. Consider an LED with an emission wavelength of 870 nm at 20°C that red-shifts by 2.8 nm for 10°C temperature increase. What is the band gap E_g and its temperature sensitivity $\Delta E_g/\Delta T$?
3. Consider a Mach–Zehnder modulator as in Figure 2.23 with a Pockels coefficient $r = 3.4 \times 10^{-12}$ m V^{-1} and a branch thickness d = 5 μm. For a maximum half-wave voltage of V_π = 10 V, what should be the minimum branch length L for modulator operation at 1.55 μm wavelength ($n_r = 2.2$)?

LIST OF ABBREVIATIONS

DC	direct current
DFB	distributed feedback
IR	infrared
IV	current–voltage
LED	light-emitting diode
LI	light–current
RC	resistance–capacitance
UV	ultraviolet
VCSEL	vertical-cavity surface-emitting laser

REFERENCES

1. Bhattacharya, P., *Semiconductor Optoelectronic Devices*, Prentice Hall, Upper Saddle River, 1997.
2. Donati, S., *Photodetectors: Devices, Circuits and Applications*, Prentice Hall, Upper Saddle River, 2000.
3. Ng, K.K., *Complete Guide to Semiconductor Devices*, John Wiley & Sons, New York, 2002.
4. Green, M.A., *Solar Cells: Operating Principles, Technology, and System Applications*, Prentice Hall, Englewood Cliffs, 1982.
5. Schubert, E.F., *Light-Emitting Diodes*, Cambridge University Press, Cambridge, 2003.
6. Chuang, S.L., *Physics of Optoelectronic Devices*, John Wiley & Sons, New York, 1995.
7. Coldren, L.A. and Corzine, S.W., *Diode Lasers and Photonic Integrated Circuits*, John Wiley & Sons, New York, 1995.
8. Saleh, B.E.A. and Teich, M.C., *Fundamentals of Photonics*, John Wiley & Sons, New York, 1991.
9. Rather, H., *Surface Plasmons*, Springer, Berlin, 1988.
10. Ozbay, E., Plasmonics: Merging photonics and electronics at nanoscale dimensions, *Science*, 311, 189–193, 2006.

3 Basic Electronic Structures and Charge Carrier Generation in Organic Optoelectronic Materials

Sam-Shajing Sun

CONTENTS

3.1 Introduction 48
3.2 Electron States, Orbitals, and Bands 48
  3.2.1 Electron States and Behavior in Materials 48
  3.2.2 Atomic Orbitals of Carbon Materials 55
  3.2.3 Molecular Orbitals of Carbon Materials 56
  3.2.4 Electronic Structures and States of Representative Organic Materials 59
3.3 Electronic Transitions and Carrier Generations 64
  3.3.1 Electron Transfers and Electronic Transitions 64
    3.3.1.1 Electron Transfers 64
    3.3.1.2 Electronic Transitions 67
  3.3.2 Carrier Generation Mechanisms 71
3.4 Analytical Techniques 82
  3.4.1 Determination of Frontier Orbital Levels 82
  3.4.2 Other Techniques 83
3.5 Summary 83
Exercise Questions 84
List of Abbreviations 85
Acknowledgment 85
References 86

Abstract This chapter briefly introduces some fundamental concepts and principles in a qualitative way (for students with diverse backgrounds at the college senior levels) on electronic structures, energy and charge carrier transfers of general electronic and optoelectronic materials, with emphasis on organic and polymeric conjugated materials. Specifically, the basic concepts and principles of quantum theory, electrons, orbitals, bands, ground and excited states, excitons, electron/energy transfers, and charge carrier generation mechanisms are briefly discussed. Both exciton and band models are presented to illustrate electronic transitions and carrier generations, including "chemical doping" and "photo doping" processes.

3.1 INTRODUCTION

Traditional or classic inorganic electronic and optoelectronic materials and devices have changed the way people live and think in the twentieth century. The development of classic and quantum mechanical theories, particularly the discovery and development of classic inorganic semiconductors and subsequent electronic/optoelectronic devices such as radios, telephones, televisions, integrated circuits and computers, has given birth to the human society of an "information age" and a "global village," where the information is processed and transferred at the level and speed of electrons and photons. Nonetheless, the demand and motivation for faster speed, larger capacity, smaller size, lighter weight, flexible shape, and lower cost devices have driven people to pay attention to the rapidly developing organic- and polymer-based electronic and optoelectronic materials, particularly since the discovery of conducting polymers in the 1970s [1–3]. The 2000 Nobel prize in chemistry (awarded to Alan G. MacDiarmid, Alan J. Heeger, and Hideki Shirakawa) manifested the progress and the potential of this field. There are similarities as well as differences between inorganic and organic semiconductors, and certain fundamental mechanisms of organic/polymeric optoelectronic processes are still under investigation and discussion. This chapter intends to introduce nonexpert readers to a brief account of the principles and concepts of both inorganic and organic semiconductors, and some basic concepts and principles on electronic structures and charge carrier generation mechanisms of mainly organic/polymeric semiconductors.

3.2 ELECTRON STATES, ORBITALS, AND BANDS

3.2.1 Electron States and Behavior in Materials

The electronic and optoelectronic properties of a material are directly correlated to the electronic structure (or electron configurations) of the material, and how the material is perturbed (or excited) by external forces such as electromagnetic radiations, heat, pressure (or mechanical force), and steady electric or magnetic fields. One fundamental question is where and how the electrons are located in the material and how they behave and respond to external perturbations. For instance, if many free or mobile electrons in a material are simultaneously moving coherently toward one direction driven by an external applied electric field, this would form an electric current. In this case, the average or summed velocity of all electrons (also called drift velocity of electrons) would be nonzero [4].

Materials are typically made out of molecules—the basic building block of a material. Each molecule, in turn, is made up of atoms. There are over 110 different kinds of atoms also called elements discovered so far, and they are positioned in the periodic table according to their electronic structure [5]. An atom is composed of a nucleus (containing neutrons and positively charged protons) and negatively charged electrons surrounding the nucleus. Note that certain materials such as diamond, silicon crystal, and metals are made up of a single type of atom directly. For a neutral atom, the number of protons in the nucleus is equal to the number of surrounding electrons, hence the materials appear neutral. A cation is a positively charged ion after a neutral atom loses electrons. Anions are negatively charged ions resulting after neutral atoms gain electrons.

According to modern quantum theory, an electron can be treated as both a particle and a wave, called wave–particle duality. And it can also be described by a space- and time-dependent wave function $\psi(x, y, z, t)$ or $\psi(\boldsymbol{r}, t)$ (also called eigenfunction) from the Schrödinger equations (or Schrödinger postulates) [5,6]

$$\hat{H}\psi(\boldsymbol{r},t) = i\hbar\frac{\partial\psi(\boldsymbol{r},t)}{\partial t} \tag{3.1}$$

where $\hat{H}$ is a Hamiltonian operator in the form of

$$\hat{H} = -\frac{\hbar^2}{2m}\nabla^2 + E_p(x, y, z, t) = -\frac{\hbar^2}{2m}\nabla^2 + E_p(r, t) \tag{3.2}$$

where

- E_p is the time- and space-dependent potential energy of the electron
- m is the physical, rest, or free mass of the electron ($m = 9.1 \times 10^{-31}$ kg, as compared to the effective or dynamic mass m^* of the electron that can be either smaller or bigger than m depending on the electron interaction with the materials)
- $\hbar$ ($=h/2\pi$) is the reduced Planck constant (also known as Dirac's constant, pronounced h-bar)
- h is the Planck's constant
- t is the time

Particle spatial position may be in Cartesian (x, y, z) or spherical ($\boldsymbol{r}$) = (r, θ, ϕ) coordinates. ∇^2 is the Laplacian operator or del-squared operator in the forms of

$$\nabla^2 = \frac{\partial^2}{\partial x^2} + \frac{\partial^2}{\partial y^2} + \frac{\partial^2}{\partial z^2} = \frac{1}{r^2}\left[\frac{\partial}{\partial r}\left(r^2\frac{\partial}{\partial r}\right) + \frac{1}{\sin\theta}\frac{\partial}{\partial\theta}\left(\sin\theta\frac{\partial}{\partial\theta}\right) + \frac{1}{\sin^2\theta}\frac{\partial^2}{\partial\phi^2}\right] \tag{3.3}$$

For most problems concerning electronic structures of the materials, the stationary or time-independent state wave function $\psi(\boldsymbol{r})$ would be sufficient ($\psi(\boldsymbol{r}, t) = \psi(\boldsymbol{r})\exp(-iEt/\hbar)$), and the Schrödinger equation can therefore be simplified as

$$\hat{H}\psi(\boldsymbol{r}) = E\psi(\boldsymbol{r}) \tag{3.4}$$

where E is the electron total energy (also called eigenvalue of ψ) expressed as

$$E(\boldsymbol{r}) = E_p(\boldsymbol{r}) + E_k(\boldsymbol{r}) \tag{3.5}$$

where $E_k(\boldsymbol{r})$ is the kinetic energy of the electron. The Hamiltonian operator now becomes

$$\hat{H} = -\frac{\hbar^2}{2m}\nabla^2 + E_p(\boldsymbol{r}) \tag{3.6}$$

In physics, the Schrödinger equation (proposed by the Austrian physicist Erwin Schrödinger in 1926) is of central importance in nonrelativistic quantum mechanics, playing a role for microscopic particles analogous to Newton's second law in classical mechanics for macroscopic particles. Microscopic particles include elementary particles, such as electrons, as well as systems of particles, such as atomic nuclei. Macroscopic particles vary in mass from small dust particles to heavy planets in the solar system. Though the Schrödinger equation cannot be derived, it can be shown to be consistent with experiments. The most valid test of a model is whether it faithfully describes the real world. The wave nature of the electron has been clearly shown in experiments like the electron diffraction experiments [5–7].

For a free electron in space without any boundary conditions, by solving the Schrödinger equation (Equation 3.1), for instance, a time-dependent wave function of the electron in one dimension (x) can be in a form of

$$\psi(x, t) = Ae^{i(kx - \omega t)} \tag{3.7}$$

where

A is a constant

$\boldsymbol{k}$ is called the electron wave vector that has the same direction as the electron momentum $\boldsymbol{p}$

The magnitude $|k|$ (called wave number) is related to the electron de Broglie wavelength λ by

$$|k| = \frac{2\pi}{\lambda} \tag{3.8}$$

Wave vector $\boldsymbol{k}$ in fact reflects traveling direction and momentum of an electron wave.

Also from the solutions of the Schrödinger equations, the total energy E of a free electron in space would be the same as its kinetic energy (since the potential energy is zero) and can be expressed as

$$E = \frac{\hbar^2}{2m}k^2 = \frac{p^2}{2m} \tag{3.9}$$

with

$$\boldsymbol{p} = m\boldsymbol{v} = \hbar\boldsymbol{k} \tag{3.10}$$

where $\boldsymbol{v}$ is the electron velocity.

Equation 3.9 indicates that for a free electron in space without any boundary conditions, its energy is continuous as shown in Figure 3.1a.

Unlike in classic mechanics where the exact position and momentum of a moving object at a specific time can be determined simultaneously and precisely, in quantum mechanics, the exact position and momentum (or speed) of an electron cannot be determined simultaneously and precisely, and this is called Heisenberg uncertainty principle that can be mathematically expressed as [5]

$$\Delta\boldsymbol{r}\,\Delta\boldsymbol{p} \geq \hbar/2 \tag{3.11}$$

where

$\Delta\boldsymbol{r}$ is the uncertainty of the position

$\Delta\boldsymbol{p}$ is the uncertainty of the momentum of the electron

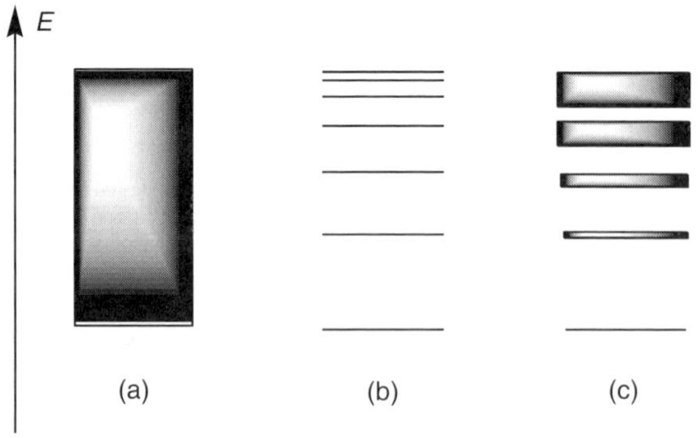

FIGURE 3.1 Schematic representation of electron energy levels/states in (a) electron in free space; (b) electron in a confined box; (c) electron in a periodic (such as crystalline) lattice.

Equation 3.11 implies that the product of the two uncertainties of an electron is at least $\hbar/2$.

Most importantly, the probability of finding an electron (or the electron probability density ρ) anywhere in the space can be expressed by [5]

$$\rho = \psi\psi^*(\mathrm{d}x\,\mathrm{d}y\,\mathrm{d}z) \tag{3.12}$$

where ψ^* represents the complex conjugate of ψ (if ψ contains imaginary part). Since the electron probability density in the whole space is unity, the normalization condition requires

$$\int_{-\infty}^{\infty}\int_{-\infty}^{\infty}\int_{-\infty}^{\infty}\psi\psi^*\,\mathrm{d}x\,\mathrm{d}y\,\mathrm{d}z = 1 \tag{3.13}$$

Electron wave functions are typically obtained by solving the Schrödinger equations with certain conditions such as specific boundary conditions and the general normalization condition mentioned above. Additionally, for electrons moving around the nuclei, the much heavier nuclei are treated (or approximated) as static when deriving the electron wave functions. This is called Born–Oppenheimer or adiabatic approximation [5–7].

For an electron confined in an one-dimension potential well of size L with infinitely high potential barrier walls ($U = \infty$ as shown in Figure 3.2a and c), like a particle confined in a box or a bound electron in an atom shown in Figure 3.3a, the electron wave function would become zero outside the potential well, i.e., $\psi(x) = 0$ for $x \le 0$ or $x \ge L$. By solving the time-independent Schrödinger equation, the electron wave function in one dimension will be in the form of

$$\psi_n(x) = A_n \sin\left(\frac{n\pi}{L}x\right) \tag{3.14}$$

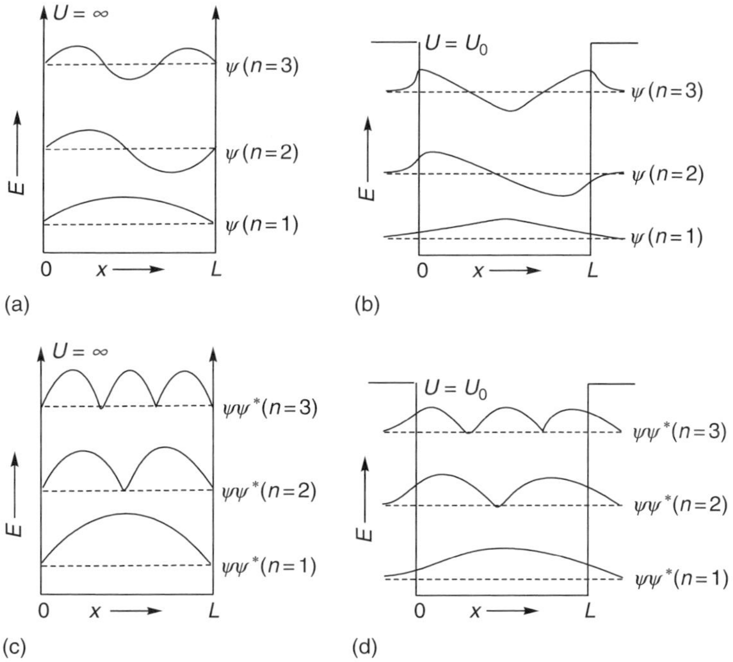

FIGURE 3.2 Scheme of electron wave functions confined in a one-dimensional (1-D) size L potential well with (a) infinite height potential walls; (b) finite height potential walls. Scheme of electron probability densities confined in a 1-D size L potential well with (c) infinite height potential walls; (d) finite height potential walls.

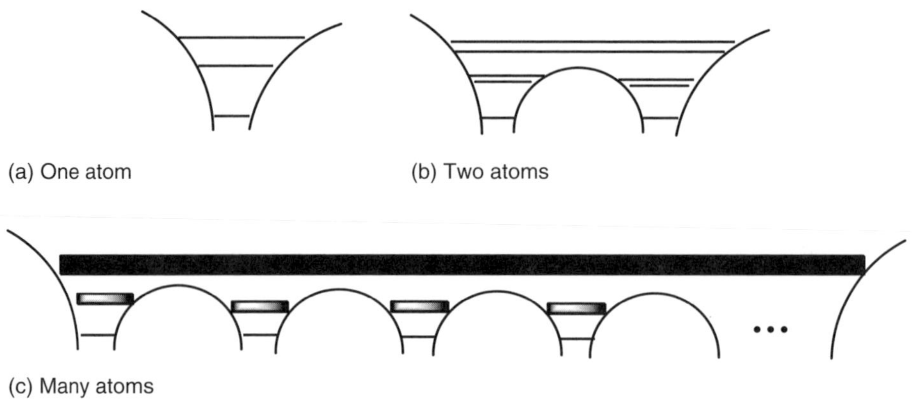

FIGURE 3.3 Schematic representation of atomic electronic potential surfaces and electron orbital levels in (a) one atom; (b) two atoms; (c) many atoms (such as crystalline) periodic structure.

where n is a nonzero integer. The wave functions with n = 1, 2, and 3 are schematically shown in Figure 3.2a, and the corresponding electron probability densities are shown in Figure 3.2c. The total energy E of such a confined electron will be in the general form

$$E_n = \frac{\hbar^2 \pi^2}{2mL^2} n^2 \tag{3.15}$$

Equation 3.15 reveals that for an electron completely confined in a box, only certain discrete energy levels (or orbitals) are allowed as shown in Figures 3.1b, 3.2a, and 3.3a. Since electrons in such an atom can only absorb or emit energies (or photons) among these discrete energy levels, therefore, only certain characteristic photon absorption and emission spectrum lines are allowed and shown in atomic spectra. This is the fundamental principle behind the atomic spectroscopy.

For an electron confined in a one-dimension potential well of size L but with finite height potential barrier walls ($U = U_0$, shown in Figure 3.2b and d), similar to a bound electron in a two-atom system as shown in Figure 3.3b, the solutions to the Schrödinger equation give wave functions with exponentially decaying penetration into the classically forbidden region outside the potential well [6,7]. The penetration of a potential barrier by an electron wave is called "tunneling." Tunneling is a quantum mechanical effect and has important applications such as tunnel diodes and tunnel electron microscope. Confining a particle to a smaller space requires larger confinement energy. Since the wave function penetration effectively "enlarges the box," the finite well energy levels are lower than those for the infinite well (Figure 3.2b and d).

For an electron in a two-atom molecule as shown in Figure 3.3b, while the lower energy electron orbitals near the nuclei may still be separated (or confined) by an interatomic electronic potential barrier, the higher atomic electron orbitals, particularly those orbitals with energy levels higher than the internuclei potential barrier, may overlap and couple to each other depending on their spatial orientation and form new interatomic orbitals called molecular orbitals (MOs) (to be discussed in more detail), so that an electron from either atom can stay in the interatomic molecular orbital and move among the two atoms. This electron is said to be delocalized among the two atoms, called electron delocalization.

For an electron in a periodically structured atomic or molecular lattice (such as in a closely packed atomic or molecular crystal) with periodic or alternating potential wells and barriers as shown in Figure 3.3c, the potential $E_p(\boldsymbol{r})$ can be represented as

$$E_p(\boldsymbol{r}) = E_p(\boldsymbol{r}+\boldsymbol{R}) \tag{3.16}$$

where $\boldsymbol{R}$ represents a spatial repeat unit. The solutions of the Schrödinger equation for the electron wave function ψ in such periodic potential is called a Bloch function, a Bloch wave, or a Bloch state in the form of [6,7]

$$\psi_{\boldsymbol{k}}(r) = u_{\boldsymbol{k}}(r)e^{ikr} \tag{3.17}$$

where

e^{ikr} is a typical plane wave envelope function

$u_k(r)$ is a periodic function that has the same periodicity as the potential in a form

$$u_{\boldsymbol{k}}(\boldsymbol{r}) = u_{\boldsymbol{k}}(\boldsymbol{r}+\boldsymbol{R}) \tag{3.18}$$

The concept of the Bloch function was developed by Felix Bloch in 1928 to describe the conduction of electrons in crystalline solids. Equation 3.17 (also called Bloch equation or Bloch's theorem) implies that for an electron in a periodic electronic potential lattice, the electron wave function is also in a periodic form. More generally, a Bloch wave description applies to any wave-like phenomenon in a periodic medium. For example, a periodic dielectric in electromagnetism leads to photonic crystals (i.e., materials which exhibit band gaps for photons), and a periodic acoustic medium leads to phononic crystals (i.e., materials which exhibit band gaps for phonons—quantized modes of lattice vibration).

From mathematical solutions of the Schrödinger equation (Equation 3.4) and the Bloch equation (Equation 3.17), a set of discrete energy states $E_n(\boldsymbol{k})$ can be obtained, where n is an integer ($n = 1, 2, 3, \ldots$) [6,7]. A plot of $E_n(\boldsymbol{k})$ versus $\boldsymbol{k}$ is schematically shown in Figure 3.4. As Figure 3.4 shows, at each n value, the electron energy (bold curves) can vary continuously along the electron wave vector $\boldsymbol{k}$ (also called $\boldsymbol{k}$-space) within a confined energy zone or band (between the two dashed lines) called "Brillouin zones" or "bands." The maximum vertical energy dispersion within each allowed electron bands (δE in Figure 3.4) is called bandwidth (BW), and the minimum forbidden energy region between the allowed bands (ΔE in Figure 3.4) is called band gap, energy gap, or stop band. It is a region where the electron is forbidden from propagating.

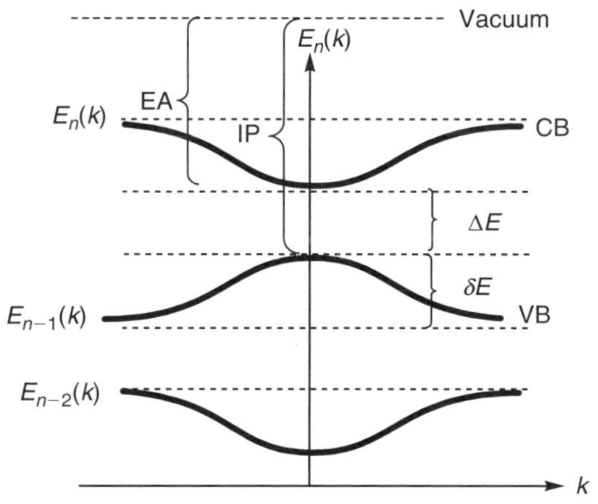

FIGURE 3.4 Schematic representations of electron energy $E_n(\boldsymbol{k})$ versus electron wave vector $\boldsymbol{k}$ in a periodic crystalline lattice where Bloch function is applicable. CB, conduction band; VB, valence band; EA, electron affinity; IP, ionization potential; ΔE, band gap; δE, bandwidth.

Typically, in the absence of any energy perturbations (i.e., at absolute zero temperature and without any external radiations), all electrons will be located at the lowest possible energy orbitals or bands. Thus, the highest electron completely occupied molecular orbital (i.e., the highest orbital that is filled with two electrons with opposite spin) is called highest occupied molecular orbital (HOMO), and the lowest electron unoccupied molecular orbital (i.e., the lowest orbital that is empty) is called lowest unoccupied molecular orbital (LUMO). Additionally, a single occupied molecular orbital (i.e., the orbital that has only one electron) is called SOMO. Additionally, from Bloch's theorem, HOMOs can overlap and couple to each other in a closely packed periodic structure to form HOMO bands, or stable valence bands (VB) if the bandwidths δE are substantial (i.e., above 0.1 eV) [8,9], and LUMOs (or SOMOs) can overlap and couple to each other in a closely packed periodic structure to form LUMO (or SOMO) bands, or stable conduction bands (CB) if the bandwidths δE are substantial (i.e., above 0.1 eV). Note that the effective orbital coupling and band formation require substantial matching and overlapping of orbital energy levels, orbital spacing (or distances), orbital orientations and geometries, etc. For traditional crystalline insulators and semiconductors, the band gap generally refers to the energy difference (ΔE) between the top of the valence band and the bottom of the conduction band (CB) as shown in Figure 3.4.

The ionization potential (IP) of a material is the energy necessary to bring the electrons from the uppermost occupied states, i.e., HOMO, or the valence band maximum (VBM), to just outside the surface (or the vacuum energy level, E_{VAC}) with zero kinetic energy (Figure 3.4). The electron affinity (EA) is defined as the energy difference between the LUMO or conduction band minimum (CBM) and the E_{VAC}. The Fermi level, E_F, of a material is defined as the average electron energy level of the materials, and is at HOMO level at absolute zero temperature for pure or undoped materials. The E_F level is typically located between the VB and CB levels depending on actual charge carrier densities and energy perturbations [4]. The work function is defined as the energy difference between the E_F, and the E_{VAC}.

In most chemical reactions, and in most electronic and optoelectronic processes, it is predominantly the valence electron transfers between the HOMO, the LUMO, the SOMO, the VB, and the CB. These orbitals and bands are therefore also called frontier orbitals/bands, and the information such as energy levels, shapes, orientations, and spatial distances of the frontier orbitals/bands is very crucial for the electronic and optoelectronic properties of the materials.

Because bands are solutions of the Bloch equation, in order for the materials to form or exhibit bands, it is crucial that the materials possess a periodic potential structure to satisfy the Bloch's theorem requirements. In most typical amorphous molecular or polymeric solids where the closely packed periodic potential structures are absent or very poor, it is not surprising that the bands are difficult to form or ever exist. However, in certain polycrystalline materials where both amorphous and crystalline domains coexist, bands may exist in the crystalline domains. Therefore, the band size (BS) may be defined as the average size of the actual periodic domain or path where Bloch function is applicable, and where electron transport is of coherent or tunneling type. In a classic single crystal semiconductor, since the Bloch function can be applicable to the whole crystal (except the boundary or edge regions of the crystal), the band size is therefore roughly the same as the crystal size. On the contrary, in the amorphous domains where the bands are poor or does not exist, charge transport follows incoherent or hopping mechanisms as will be discussed in detail in Chapter 4 of this book. The mean free path (MFP, l), or mean free distance (MFD) is defined in semiconductor physics as an average nonscattering (ballistic) electron transport length between two consecutive scattering centers expressed by

$$l = v\tau \tag{3.19}$$

where

v is the electron velocity

τ is the average electron transport relaxation time, also called mean free time (MFT), defined as the average time an electron travels between the two scattering centers [6,7]

3.2.2 Atomic Orbitals of Carbon Materials

By solving the Schrödinger equations for electrons surrounding a nucleus, similar to the case of an electron in a potential well, with either infinite or finite walls, it can be found that the electrons are located only in certain energetically and spatially discrete electron orbitals surrounding the nucleus, and the orbitals are subject to the confinement of the nuclear electronic potential wells (Figures 3.1 through 3.3). Each electron orbital is differentiated by its energy level, shape, and orientation as defined by the wave functions mentioned earlier. Such electron orbitals in atoms are called atomic orbitals (AOs). Also from the wave function solutions, each electron in an atom can be represented by a set of four unique quantum numbers [5,10]. These are the principal (or main shell) quantum number n ($n = 1, 2, 3, \ldots$) representing the main energy level or main shell of an electron; the orbital (also known as the azimuthal, angular, or subshell) quantum number l ($l = 0$ for s-type orbital, $l = 1$ for p-type orbital, $l = 2$ for d-type orbital, ..., $n-1$, etc.) representing the subshells or types of an electron orbital; the magnetic quantum number m_l ($m_l = 0, \pm 1, \ldots, \pm l$) representing the orientations (or angular momentum) of an orbital along a specified axis; and the spin quantum number m_s ($m_s = 1/2$ or $-1/2$) representing the spin orientation or spin angular momentum of an electron. Table 3.1 summarizes these quantum states and numbers of an electron in an atom.

From particle physics and statistical mechanics, since electrons have half integer spin quantum numbers, they belong to a family of elementary particles called fermions. The energy/state distribution of fermions in traditional semiconductors can be described by the Fermi–Dirac statistics. Fermions are in contrast to another major family of elementary particles called bosons that have integer spin quantum numbers, such as photons. The energy/state distribution of bosons can be described by Bose–Einstein statistics.

The shapes of the *s*-, *p*-atomic and sp^n ($n = 1, 2, 3$) hybrid atomic orbitals are schematically illustrated in Figure 3.5. In the figure, two different colors (dark and light) of the orbital lobes reflect the positive or negative phases of the electron wave function. The electron orbital shape roughly reflects the electron probability density. As Figure 3.5 shows, the *s* orbital is a spheric symmetric shape with the same phase in all directions, while the *p* (including sp^n) orbital has a dumbbell shape with two lobes of different phases and sizes.

According to Pauli exclusion principle, no two identical fermions may occupy the same quantum state simultaneously. For electrons in a single atom, the principle states that no two electrons can have the same four quantum numbers, i.e., every electron must be unique. Therefore, each electron orbital in an atom can only accommodate a maximum of two electrons with opposite spins. Also from Aufbau principle and Hund's rule, at ground state, the electrons will first occupy lowest energy state, and will first fill different orbitals with same single spin electron each before filling the same orbital with two electrons of opposite spins. Thus, the maximum number of electrons in each main shell is $2n^2$. While there are unlimited number of AOs surrounding a nucleus, there are, however, limited number of electrons in an atom. For instance, in a neutral carbon atom, there are six protons

TABLE 3.1
Quantum States and Numbers of an Electron in an Atom

Name	Symbol	Orbital Meaning	Range of Values	Value Examples
Principal quantum number	n	Main shell	$1 \le n$	$n = 1, 2, 3, \ldots$
Orbital quantum number (azimuthal quantum number)	l	Subshell	$0 \le l \le n - 1$	For $n = 3$: $l = 0$ (*s* orbital), 1 (*p* orbital), 2 (*d* orbital)
Magnetic quantum number	m_l	Energy shift	$-l \le m_l \le l$	For $l = 2$: $m_l = \pm 2, \pm 1, 0$
Spin quantum number	m_s	Spin	$-1/2$ or $1/2$	Always: $-1/2$ or $1/2$

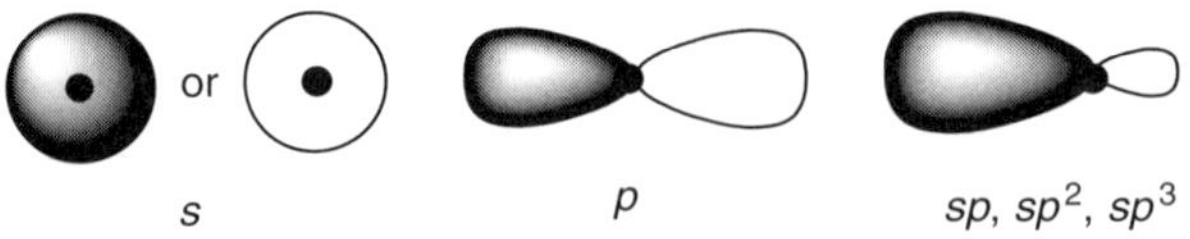

FIGURE 3.5 Shapes of some representative atomic electron orbitals. The dark dot in the middle represents the nucleus. The two colors of the orbital lobes represent the positive and negative phases of the electron wave functions. The orbital shapes roughly represent the electron probability density in space.

and six electrons, and the quantum numbers (or identity) of the six electrons in ground state can be represented by (n, l, m_l, m_s) as 1,0,0,1/2; 1,0,0,–1/2; 2,0,0,1/2; 2,0,0,–1/2; 2,1,0,1/2; 2,1,1 (or –1), 1/2; or can be simplified as $1s^22s^22p^2$. The silicon electron configuration can be represented as $1s^22s^22p^63s^23p^2$.

As stated earlier, in typical chemical reactions, including most electron transfer processes encountered in electronic and optoelectronic materials and devices, it is typically the electrons at the HOMOs, also called outer shell electrons (e.g., the $2s^22p^2$ electrons in carbon), that are engaged in electron transfer and chemical bonding processes. The outer shell electrons are also called valence shell electrons or simply valence electrons, as these electrons are involved in forming valence chemical bonds during most chemical reactions [10]. For instance, each of the four valence electrons of the carbon (2,0,0,1/2; 2,0,0,–1/2; 2,1,0,1/2; 2,1,1 [or –1],1/2) coupled with the one 1,0,0,1/2 electron of four hydrogen atoms and forms a methane molecule, CH_4, where the four valence chemical bonds (also called σ bonds, or σ molecular orbitals) are formed connecting the carbon and four hydrogen atoms. While each σ bonding orbital contains two electrons, one from the carbon, and one from the hydrogen, yet the σ antibonding orbital is empty. Since the four valence electrons ($2s^22p^2$) in carbon are different, while experimentally the four valence bonds in CH_4 appear to be the same, in order to account for this, Linus Pauling proposed a hybridized orbital theory. In chemistry, hybridisation or hybridization is the concept of mixing atomic orbitals to form new hybrid orbitals suitable for the qualitative description of atomic bonding properties. In this methane case, for instance, hybridization theory postulated that during the CH_4 bonding formation process, the carbon $2s$ orbital hybridizes with three $2p$ orbitals forming a four equivalent (same energy and shaped) sp^3 hybridized atomic orbitals (HAOs, see Figure 3.6a), where each HAO contains one electron, and that the four sp^3 hybridized orbitals are in a tetrahedral geometry with an angle of about 109.5° between any two orbital lobes (Figure 3.6e). Mathematically, the four hybridized sp^3 orbitals can be derived by different linear combinations of the one s and three p orbitals. Alternatively, the one carbon $2s$ orbital can hybridize with two $2p$ orbitals to form three equivalent sp^2 HAOs (Figure 3.6b) in a triangular plane geometry with an angle of 120° between any two orbital lobes (Figure 3.6d). Finally, the carbon $2s$ orbital can also hybridize with one carbon $2p$ orbital to form two equivalent sp hybridized orbitals in a linear (180°) shape (Figure 3.6c).

3.2.3 Molecular Orbitals of Carbon Materials

When two atoms are spatially very close to each other, and that substantial overlap (or coupling) of two AOs occurs, there is a possibility of two same electrons to coexist at the same location and at the same time, and this is prohibited from Pauli exclusion principle. Therefore, the two overlapped atomic orbitals split and form two new different energy (and shaped) orbitals called molecular orbitals. From mathematics, the linear combinations of any two atomic orbital wave functions can generate a pair of molecular orbital wave functions of different shapes and energy levels, one with energy level below the original atomic orbital level called bonding molecular orbital, and another with energy level above the original atomic orbital energy level called antibonding molecular orbital as shown in Figure 3.7 [5,10]. As shown in Figure 3.7a, when the two s atomic orbitals overlap in phase, a σ bonding molecular orbital is formed. When the two s atomic orbitals overlap or couple out of

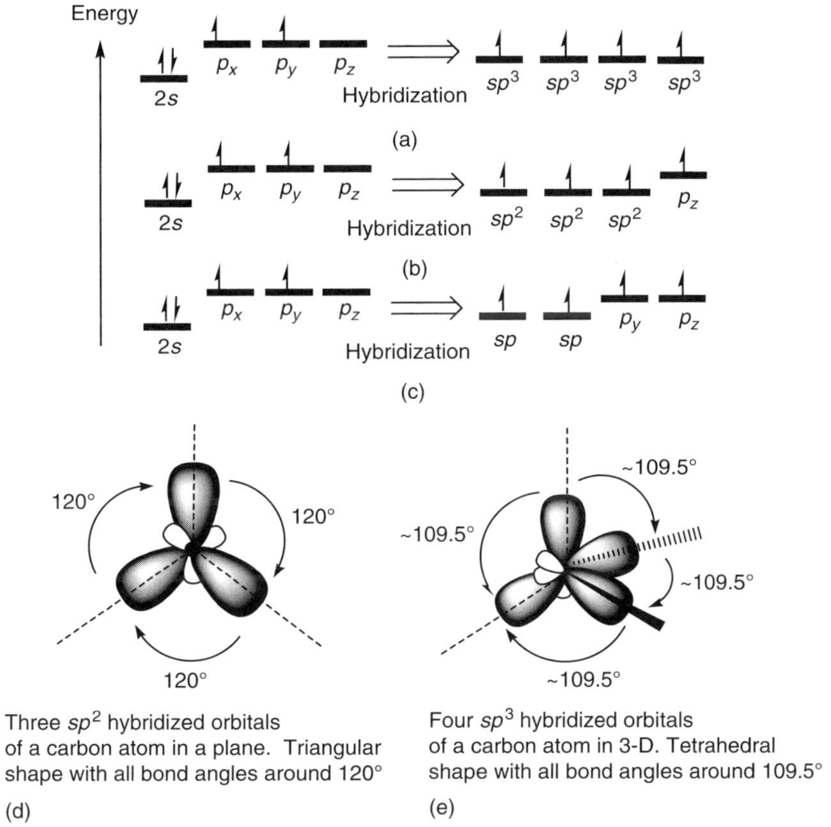

FIGURE 3.6 Scheme of AO hybridizations in (a) sp^3 hybridization; (b) sp^2 hybridization; (c) sp hybridization; (d) shape of sp^2 hybridized orbitals; (e) shape of sp^3 hybridized orbitals.

phase, a higher energy σ* antibonding type orbital is formed. The amount of energy decrease in σ bonding molecular orbital is the same as the amount of energy raise in σ* antibonding molecular orbital as shown in Figure 3.7b.

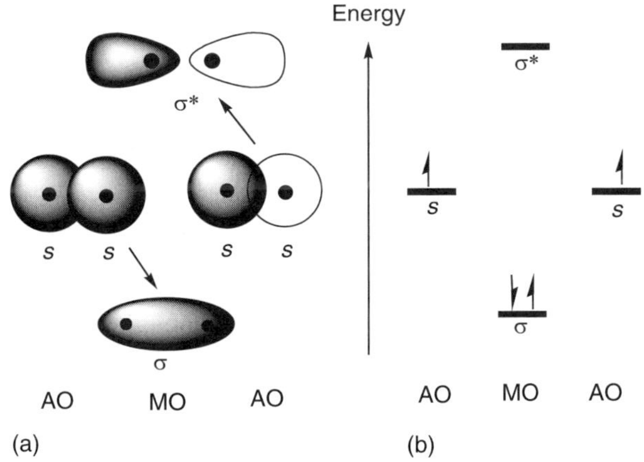

FIGURE 3.7 Scheme of a pair of σ-type MOs formed from the overlap of two s atomic orbitals in (a) orbital shape representation; (b) orbital energy level representation.

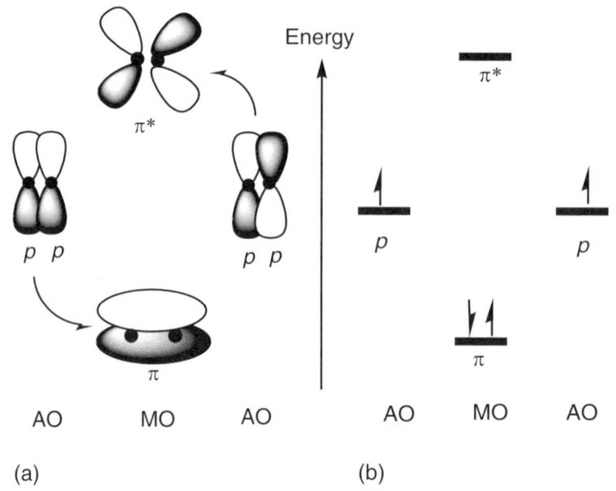

FIGURE 3.8 Scheme of a pair of π-type molecular orbitals formed from the overlap of two vertically aligned parallel p atomic orbitals in (a) orbital shape representation; (b) orbital energy level representation.

Likewise, as shown in Figure 3.8a, when the two *p* atomic orbitals overlap and couple in phase, a π bonding molecular orbital is formed. When the two *p* atomic orbitals overlap and couple out of phase, a higher energy π* antibonding type orbital is formed. The amount of energy decrease in π bonding molecular orbital is the same as the amount of energy raise in π* antibonding molecular orbital as shown in Figure 3.8b. Since the atomic valence electrons typically occupy only bonding molecular orbitals, the total energy of the molecular state would decrease in comparison to atomic state, the molecular states are therefore typically more stable than their atomic states.

Similarly, as shown in Figure 3.9a, when the lobes of two sp^n ($n = 1, 2, 3$) HAOs (or two *p* orbitals in horizontal orientation) overlap and couple in phase along the atomic center axis, a lower energy level σ bonding molecular orbital is formed. When such overlap is out of phase, a higher energy σ* type antibonding orbital is formed. The amount of energy decrease in σ bonding molecular orbital is the same as the amount of energy raise in σ* antibonding molecular orbital as shown in Figure 3.9b.

Thus, the potential problem of two same electrons coexisting at the same location is resolved. The molecular chemical structures of most carbon-based organic compounds are in fact formed

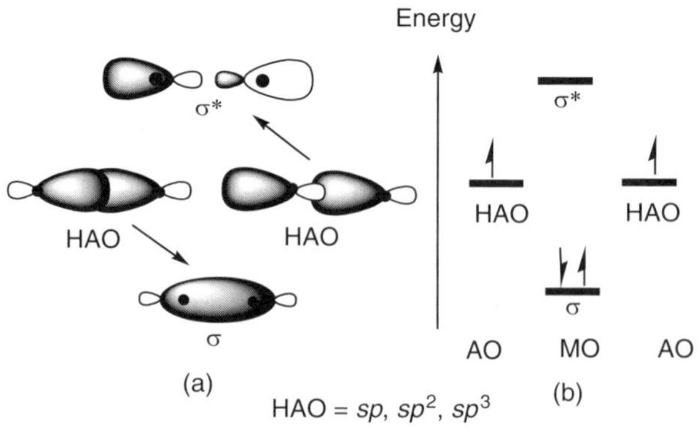

FIGURE 3.9 Scheme of a pair of σ-type molecular orbitals formed from the overlap of two horizontally aligned sp^n HAOs in (a) orbital shape representation; (b) orbital energy level representation.

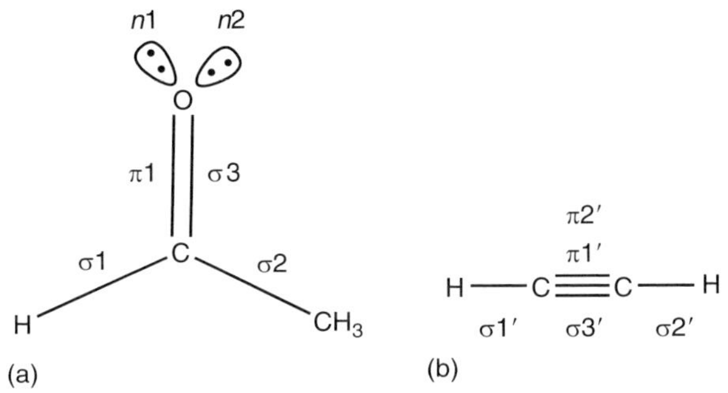

FIGURE 3.10 Schematic chemical structures of (a) ethanal (ethyl aldehyde); (b) ethyne (acetylene).

from carbon sp^3 HAOs coupling to each other forming relatively stable σ single bonds, as the energy required to excite an electron from σ to σ* level is much higher than typical thermal heating or UV–Vis level energies [10].

In some cases, certain valence electrons from atomic orbitals are not participating in molecular orbital formation or remain intact during molecular formation, such AOs in molecules are called nonbonding orbitals, and the electrons in it are called nonbonding electrons. In Figure 3.10, for instance, during the formation of ethanal molecule, the center carbon atom uses one of its three sp^2 HAOs to couple with the one 1*s* AO of the left hydrogen atom forming a σ1 molecular orbital (or σ1 single chemical bond). The center carbon atom uses its second sp^2 HAO to couple with one of the four sp^3 HAOs of the right side carbon forming the σ2 single bond. The remaining three sp^3 HAOs of the right side carbon each would couple with the 1*s* AO of three hydrogen and form three σ bonds. The center carbon atom uses its third sp^2 HAO to couple with one of the sp^2 HAOs of top oxygen forming the σ3 bond. The remaining p_z orbitals in both center carbon and top oxygen overlap and couple in direction normal to the triangular plane forming a π bond. Such two chemical bonds together (one σ and one π bond between the two atoms) is called a double bond and is typically represented by double lines as shown [10]. The two electron pairs remained in the two sp^2 HAOs of the top oxygen are called nonbonding or loan pair electrons, and their orbitals are also called nonbonding orbitals *n*1 and *n*2. During the formation of acetylene molecule as shown in Figure 3.10b, the two carbon atoms are in *sp* hybridization, each carbon uses one of its two *sp* HAOs couple with the 1s electron of a hydrogen atom forming σ1′ and σ2′ bonds. The remaining *sp* orbitals from each carbon then couple to each other forming the C–C σ3′ bond. Since each carbon now still has two *p* orbitals, and that the two *p* orbitals are orthagonal to each other in a plane normal to the C–C σ3′ bond (e.g., assuming that the C–C σ3′ bond is in *X*-axis or direction, the two *p* orbitals are called p_y and p_z orbitals). The coupling and overlaps of the p_y and p_z orbitals in the Y and Z directions formed two π bounds (π1′ and π2′) orthogonal to each other. Thus, a triple (alkyne) bound is formed, and ethyne (acetylene) molecule has a linear shape [10].

3.2.4 Electronic Structures and States of Representative Organic Materials

To illustrate the electronic structures of carbon-based materials, diamond and graphite are used here as examples. In the case of diamond as shown in Figure 3.11a, each carbon atom couples with four other carbon atoms via four sp^3 HAOs and forms four σ single chemical bonds in a tetrahedral three-dimensional (3-D) network. Due to the relatively large energy gap between the σ and σ* orbitals of the C–C bonds (i.e., energy gap of about 5.5 eV in diamond [7], much bigger than the typical energy perturbation at ambient condition such as room temperature and daylight of 0.02–2 eV), no electrons are expected to be excited and there are practically no free electrons at σ* conduction band and no free holes at σ valence band. This explains why diamond is an insulator in pure form and at ambient

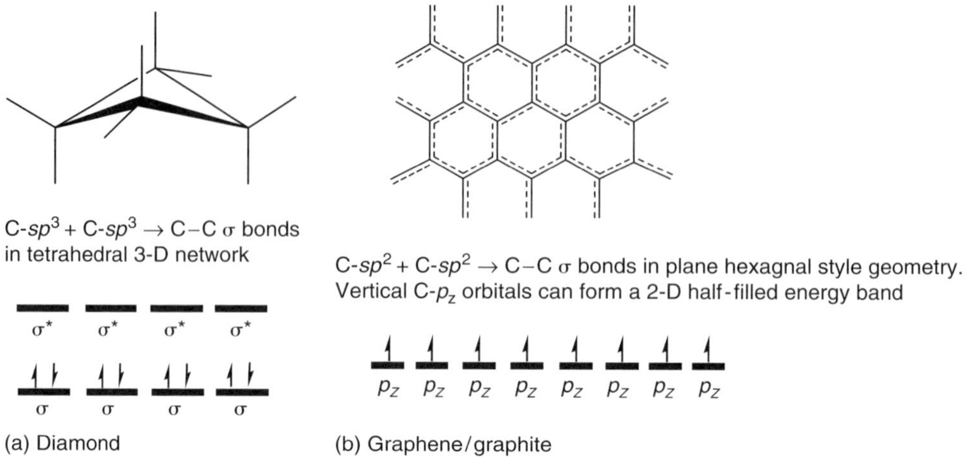

FIGURE 3.11 Schematic representation of atomic and molecular orbitals in (a) diamond; (b) graphene/graphite.

conditions. This is in contrast to silicon (or Germanium) crystals where the energy gap between the conduction band and the valence band is much smaller on the order of 1–2 eV, i.e., within daylight and ambient perturbation range, some electrons are excited from the valence band into the conduction band, so that semiconductors are formed. In the case of graphene (single layer graphite sheet, see Figure 3.11b), each carbon only couples with three other carbons to form σ bonds using three sp^2 HAOs, and thus a big carbon sheet or plane is formed. The remaining p_z orbitals on each carbon and normal to the carbon plane can overlap and couple with each other forming a two dimensional (2-D) big π electron band (bands). If there are no any defects, the electronic band size can be regarded as the size of the graphene sheet. Due to the fact that each p_z orbital contains only one electron, and up to two electrons can be in one p_z orbital, so the big π electron band is in fact theoretically half-filled metallic band. However, distortions, defects, and boundaries/edges of the carbon plane would break up such half-filled metallic band and yield energy gaps between conduction and valence band. In graphite, the interaction between graphene planes is rather weak, of the van der Waals type, and the overlap of wave functions on different planes is essentially nonexist. This explains why graphene or graphite is mainly conducting along the 2-D direction. Carbon nanotubes (CNTs) and fullerenes (C_{60} and derivatives) can be regarded to a certain degree as graphene sheets bent toward the shapes of tubes and spheres. In the case of fullerenes and CNTs, however, some carbon pentagons are necessary to mix with carbon hexagons in order to form a smooth sphere or tube surface. Since CNTs and fullerenes have half-filled π electron band in both curved and relatively flat geometries, both metallic (expected in relatively flat domains) and gaped semiconducting domains (expected in curved domains) are expected and indeed are the cases (see Chapter 8 of this book).

In ethene (ethylene) molecule as shown in Figure 3.12a, each of the two carbons uses three sp^2 HAOs to couple with two hydrogens and one other carbon forming three σ bonds. The two remaining p_z orbitals on the two carbon atoms normal to the ethylene plane overlap and couple to each other forming a π bond as shown in Figure 3.12b, and the π bonding orbital is schematically shown in Figure 3.12c. For conjugated polymer polyacetylene (PA) shown in Figure 3.12d, like in ethylene, each carbon on PA backbone uses three sp^2 HAOs forming three σ bonds with one hydrogen and two other carbons in a triangular manner as shown in Figure 3.12e. If all C–H atoms are in the same plane without any backbone distortion or twist, all p_z orbitals normal to such plane could then well overlap and couple to each other forming a long one-dimensional (1-D) and half-filled π electron band similar to graphene as depicted in Figure 3.12e and f. If so, then the PA should be metallic along the backbone direction. In reality, however, due to the well-known Peierls distortion mechanism in all

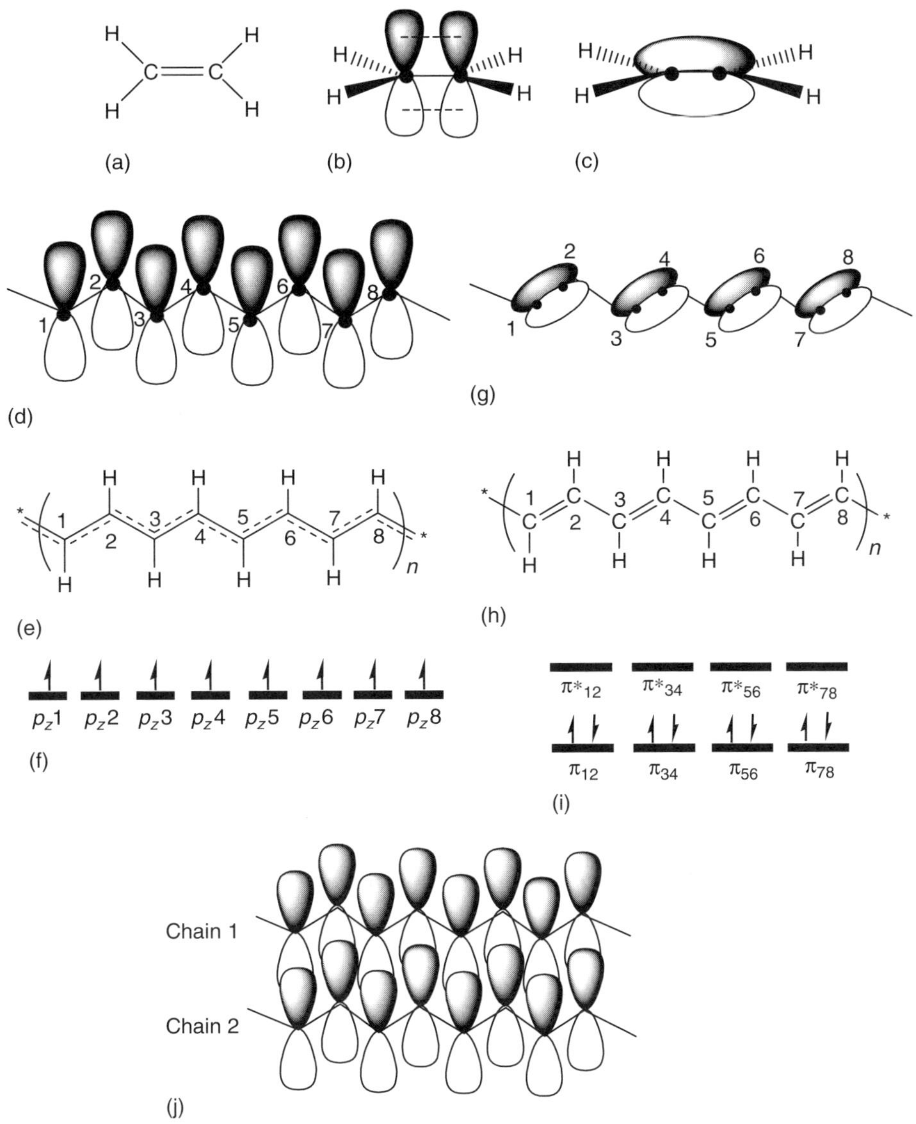

FIGURE 3.12 Schematic representation of C–C double π bonds, represented in (a) chemical structure of ethene (ethylene); (b) scheme of double bond formation by overlapping/coupling of two vertically aligned p_z orbitals at two adjacent carbon sites; (c) schematic shape of a π bonding orbital; (d) atomic p_z orbital representation of PA backbone; (e) chemical structure representation of PA assuming no 1-D Peierls distortion; (f) scheme of frontier orbitals and valence electron of PA assuming no Peierls distortion; (g) scheme of PA conjugated backbone showing π bonding orbitals; (h) chemical structure representation of PA assuming 1-D Peierls distortion applicable; (i) scheme of frontier orbitals and valence electron of PA assuming Peierls distortion; (j) atomic p_z orbital representation of two self-assembled and closely packed PA conjugated backbones.

1-D metals [6,7], every adjacent p_z orbitals are paired on PA backbone forming a single-double bond alternating conjugated polymer backbone shown in Figure 3.12g and h, where the π bonding orbitals (HOMOs) couple to each other forming a HOMO band (or valence band), and the π antibonding orbitals (LUMOs) overlap and couple forming a LUMO band (or conduction band). The difference of the single and double bond length is called the bond length alternation (BLA). The BLA is expected to become zero in a perfect or undistorted spatial conjugation situation such as in graphene 2-D sheet where the bandwidth is largest and band gap is smallest. The BLA would become largest in distorted

PA chain where the bandwidth is smallest, band gap is largest, and single and double bonds on the conjugated chain are well defined. The actual BLA would be between the two mentioned extreme cases depending on the extent of orbital overlaps and coupling, and those depending critically on factors such as PA backbone length (relevant to average polymer molecular weight, MW), backbone shape or geometry (e.g., trans- versus cis-, flat or straight versus twisted or bent), backbone packing pattern or morphologies, etc. For instance, Figure 3.12j depicts two PA chains in a straight and geometry matched close packing pattern. If all PA chains are packed as in Figure 3.12j, then the BLA and band gap would be small, and band size and bandwidth would be large.

In general, the more ordered and less distorted packing of PA backbones, the more well-defined solid-state structural periodicity and better coupling between electronic orbitals of adjacent carbon sites, the weaker lattice thermal vibrations or less electron–lattice (electron–phonon) vibronic coupling (i.e., at lower temperatures), the larger the bandwidth, and smaller the band gap (BG), the material's electronic properties are leaning more toward the "band regime", where the photo carrier generation may be the "primary mechanism" type and the carrier transport is dominated by coherent or band mechanism. On the contrary, the less ordered packing of PA backbones, the less defined solid-state structural periodicity and therefore the poorer electronic couplings between electronic orbitals of adjacent carbon sites, the more intense lattice thermal vibrations or stronger electron–phonon vibronic couplings (i.e., at higher temperatures), the smaller the bandwidth, the larger the band gap, the materials electronic properties are leaning more toward the "exciton regime", where the photo carrier generation would be the "secondary mechanism" type, and the carrier transport is dominated by incoherent (or hopping) mechanism [9]. Therefore, even for materials having the same molecular or chemical structures, different morphologies in solid state (i.e., well-ordered crystalline versus disordered amorphous) or even temperatures may render the materials in either the band-like transport regime or exciton-like hopping transport regimes. For instance, the conductivity (σ, in unit of $\Omega^{-1}\text{cm}^{-1}$ or S cm^{-1}) of doped polyacetylenes (PA) can vary from about 0.01 $\Omega^{-1}\text{cm}^{-1}$ (i.e., a semiconductor with AsF_5 doping, no stretching) all the way to about 100,000 $\Omega^{-1}\text{cm}^{-1}$ (i.e., a metal or conductor with iodine doping and stretching). Figure 3.27 depicts a conductivity chart of some key materials [1–3]. The conductivity (σ) of a material is defined as

$$\sigma = qN\mu \tag{3.20}$$

where

q is the basic electronic charge unit of the mobile charge carriers
N is the number density of mobile charge carriers (in unit of carrier numbers per volume such as cm^{-3})
μ is the charge carrier mobility (in unit of $\text{cm}^2\ \text{Vs}^{-1}$)

Carrier transport and modeling in organics are being reviewed in Chapter 4 as well as in Ref. [9].

While doping and defects/impurities in materials mainly affect the densities (N) of mobile charge carriers (to be elaborated in Section 3.3), the couplings and overlap of frontier orbitals (HOMOs or LUMOs) would affect the bandwidth and band gap, and that would critically affect the carrier transport mechanisms (either coherent or incoherent) and carrier mobility (μ). The dramatic conductivity differences of stretched and un-stretched PA films demonstrate the effects of molecular packing versus charge mobility. As a matter of fact, the estimated best charge mobility (μ) of doped polyacetylene is in the range of 100–200 $\text{cm}^2\text{Vs}^{-1}$, and this is in contrast to the measured charge mobility of 15,000 cm^2Vs in graphite, and the measured charge mobility of 80,000–100,000 cm^2Vs in single wall CNTs [9].

While polyacetylene can be regarded as a conjugated 1-D metal subject to Peierls distortion (though ordered and close solid-state molecular packing may minimize the distortion), a graphite sheet or graphene can be regarded as a conjugated 2-D metal, though it is also subject to distortions and boundaries of the plane. In the case of CNTs, due to a cylindrical 3-D shape, certain distortions of a plane in graphene 2-D sheet are in fact being restrained, and this may account for a better charge mobility of CNTs.

It was suggested at least 0.1 eV bandwidth is necessary for a band to be stable [8], i.e., those materials with bandwidth less than 0.1 eV could be treated as in charge carrier hopping transport regime. While most organic semiconductors have bandwidths less than 0.1 eV, certain organic crystals have been observed to have bandwidth up to 0.2 eV [8]. These are in contrary to classic inorganic crystalline semiconductors that typically exhibit bandwidths of several electron volts [6,7]. While band regime could exist in certain carefully grown and processed organic crystals at low temperatures, hopping regime is believed to be predominant in most organic or polymeric conjugated semiconductors, particularly solution processed polymers. It is also reasonable to expect both band domains and hopping domains coexist in a particular organic or polymeric semiconductor thin film due to coexistence of both crystalline and amorphous domains.

In addition to electronic orbitals and bands, the electron spin configurations or states are further represented by spin terms as listed in Table 3.2. Specifically, if there is a single unpaired electron, the spin quantum number (m_s) is one half, the sum (S) of spin quantum numbers is one half, the multiplicity ($2S + 1$) is 2, and the spin term is called Doublet. If all electrons are paired, the spin quantum numbers are positive and negative one halves in equal quantities, so the sum (S) of spin quantum numbers is zero, the multiplicity ($2S + 1$) is 1, and the spin term is called Singlet. If two unpaired electrons have the same spins (located in two different orbitals), the spin quantum numbers are two one-halves, so the sum (S) of spin quantum numbers is one, the multiplicity ($2S + 1$) is 3, and the spin term is called Triplet.

Certain molecular self-assembly or aggregates also form special electronic states. The H and J-aggregations are such examples. Specifically, J-aggregates (Figure 3.13a) are formed with the monomeric molecules arranged in one dimension such that the transition moment of the monomers is parallel and the angle between the transition moment and the line joining the molecular centers is zero (ideal case). The strong coupling of several self-assembled monomers results in a coherent excitation at redshifted wavelengths relative to the monomer. In addition, the spectrum gets narrower and the vibrational coupling to the molecular modes will be largely absent. H-aggregates (Figure 3.13b) are again a 1-D arrangement of strongly coupled monomers, but the transition moments of the monomers are perpendicular (ideal case) to the line of centers. In contrast to the side-by-side arrangement of molecules in J-aggregates, the arrangement in H-aggregates is face-to-face. The dipolar coupling between monomers in H-aggregates leads to a blueshift of the absorption band.

Experimentally, the absorptions (and emissions) for the H-aggregates would be blueshifted compared to non-aggregates, and the absorptions (and emissions) of the J-aggregates would be redshifted compared to non-aggregates. Typically, solid state or higher concentrations in solution at lower temperatures would favor aggregates formation, while dilute solutions in higher temperatures would favor non-aggregates (single molecular) state. Therefore, the typical aggregates component peaks should vary with these conditions.

Finally, excimers and exciplexes are examples of excited states aggregates [14]. An excimer (originally short for excited dimer) is a short-lived dimer composed of two identical molecules, at least

TABLE 3.2
Key Multiplicity States and Terms

Electron Pattern	Spin Quantum Numbers (m_s)	$S = \sum m_s$	Multiplicity $2S + 1$	Term
↑	1/2	1/2	2	Doublet
↑↓	1/2, −1/2	0	1	Singlet
↑↑	1/2, 1/2	1	3	Triplet

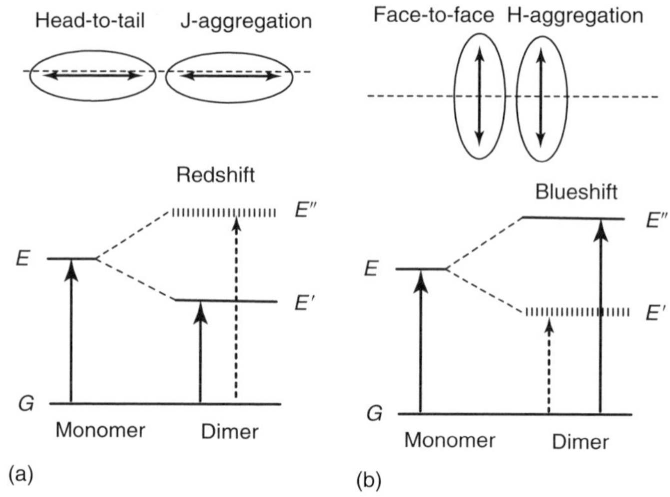

FIGURE 3.13 Molecular aggregates in (a) J-type aggregates; (b) H-type aggregates.

one of which is in an electronic excited state. Excimers are often diatomic and are formed between two atoms or molecules that would not bond if both were in the ground state. The lifetime of an excimer is very short, on the order of nanoseconds. An exciplex is similar except it is composed of two different molecules. Excimers or exciplexes are only formed when one of the dimer components is in the excited state. When the excimer/exciplex returns to the ground state, its components dissociate and often repel each other. The wavelength of an excimer/exciplex's emission is longer (smaller energy) than that of the excited monomer's emission. An excimer can thus be measured by fluorescent emissions.

3.3 ELECTRONIC TRANSITIONS AND CARRIER GENERATIONS

3.3.1 Electron Transfers and Electronic Transitions

3.3.1.1 Electron Transfers

In any electronic transitions involving an electron transfer process, e.g., an electron is transferred from an initial orbital to a final orbital (corresponding to a system electronic transition from an initial state to a final state) as occurred in many chemical and optoelectronic processes, the transition can be represented by an initial state function ψ_i and a final state function ψ_f. Assuming the transition is driven by a perturbation V, according to perturbation theory, the transition probability P_{if} can be expressed by [8]

$$P_{if} = \frac{1}{\hbar^2}\left|\langle \psi_i | V | \psi_f \rangle\right|^2 \left[\frac{\sin(\omega_{fi} t/2)}{\omega_{fi}/2}\right]^2 \tag{3.21}$$

where

t is the transition time
ω_{fi} is an angular frequency related to the transition energy $\hbar\,\omega_{fi}$ between the states i and f
$\langle \psi_i | V | \psi_f \rangle = V_{if}$ is called the electronic coupling matrix element of the electronic transition

If the final state has a continuum or distribution of states (e.g., different vibrational levels) represented by a density of state $\rho(E_f)$, the transition probability per unit time, or the transition rate constant (often shortened as transition rate or electron transfer rate) k_{if} can then be expressed by the Fermi's Golden Rule as

$$k_{if} = \frac{2\pi}{\hbar}|V_{if}|^2 \rho(E_f) \tag{3.22}$$

The expression for the rate obtained within the Frank–Condon approximation factorizes into an electronic and a nuclear vibrational contribution as [8]

$$k_{if} = \frac{2\pi}{\hbar}|V_{if}|^2 (\text{FCWD}) \tag{3.23}$$

where FCWD is the Frank–Condon-weighted density of states. When all vibrational modes are classical or at high temperatures regime (i.e., $\hbar\omega_i \ll kT$), the FCWD can be in the Arrhenius form of

$$\text{FCWD} = \frac{1}{\sqrt{4\pi\lambda kT}}\exp\left[-\frac{\Delta G^{\#}}{kT}\right] = \frac{1}{\sqrt{4\pi\lambda kT}}\exp\left[-\frac{(\Delta G^0 + \lambda)^2}{4\lambda kT}\right] \tag{3.24}$$

where

λ is the total lattice/nuclei and surrounding media reorganization energy cost of the transition
$\Delta G^{\#} = (\Delta G^0 + \lambda)^2/4\lambda$ is the transition activation energy
ΔG^0 is the standard Gibbs free energy change from the initial to final states

This leads to the semiclassical Marcus electron transfer equation

$$k_{if} = \frac{2\pi}{\hbar}|V_{if}|^2 \frac{1}{\sqrt{4\pi\lambda kT}}\exp\left[-\frac{(\Delta G^0 + \lambda)^2}{4\lambda kT}\right] \tag{3.25}$$

when certain lattice vibrational modes of the final state (countering the electron transfer) are much stronger compared to electron thermal energy (i.e., $\hbar\omega_i \gg kT$), the transition rate constant can then be better represented by Bixon and Jortner modified Marcus equation as

$$k_{if} = \frac{2\pi}{\hbar}|V_{if}|^2 \frac{1}{\sqrt{4\pi\lambda_s kT}}\sum_{n=0}^{\infty}\left\{e^{-S_i}\frac{S_i^n}{n!}\exp\left[-\frac{(\Delta G^0 + \lambda_s + n\hbar\omega_i)^2}{4\lambda_s kT}\right]\right\} \tag{3.26}$$

where the reorganization energy is divided into classic (or solvent) contribution (λ_s) and a strong lattice vibrational mode contribution ($\lambda_i = n\hbar\omega_i$). The Huang–Rhys factor $S_i = \lambda_i/\hbar\omega_i$ is a measure of the electron–phonon interactions.

In fact, any additional driving or counter-driving forces affecting the electron transfer should also be counted. These forces may include electromagnetic radiation, thermal or mechanical forces driving the electron transfer (e.g., "hot" initial state vibrational modes), electrical forces, magnetic forces, etc. Assuming ΔF to represent the sum of all those additional forces (driving force with negative signs and counter-driving forces with positive signs), the electron transfer rate constant in semiclassic form could then be approximated by

$$k_{if} = \frac{2\pi}{\hbar}|V_{if}|^2 \frac{1}{\sqrt{4\pi\lambda kT}}\exp\left[-\frac{(\Delta G^0 + \lambda + \Delta F)^2}{4\lambda kT}\right] \tag{3.27}$$

By definition, in any electron transfer reaction or process, the electron donor (D) is the entity that donates (gives out) the electron during such process, and likewise, the electron acceptor (A) is the entity that accepts (captures) the electron during such process. The profiles of a donor/acceptor (D/A) pair (such as the relative frontier orbital levels) could be different in different electron transfer processes (e.g., excited state or ground state electron transfers), and a donor may become an acceptor in different electron transfer processes. In analogy, in an energy (or exciton) transfer process, the energy donor (D) is the entity that donates (gives out) the exciton, and likewise, the energy acceptor (A) is the entity that accepts (captures) the exciton. The profiles of an energy transfer D/A pair (including the relative frontier orbital levels) may be different in different energy transfer processes, and a donor may be an acceptor in different energy transfer processes.

Figure 3.14a shows a simple and typical scheme where an electron is transferred from an electron donor (D) to an electron acceptor (A) as represented by a reaction $D/A \rightarrow D^+/A^-$, and the reaction is accompanied by a system (both donor and acceptor) standard Gibbs free energy reduction of ΔG^0 (driving force with a negative sign) and a reorganization energy cost of λ (counter-driving force with positive value) [11]. Q_r represents the nuclear/lattice geometry or coordinate of the initial or reactant state, $Q^{\#}$ represents the nuclear/lattice geometry or coordinate of transfer activation state, and Q_p represents the nuclear/lattice geometry or coordinate of the final or product state. $\Delta G^{\#}$ is the activation energy. Figure 3.14b shows a correlation between the electron transfer rate constant k versus the reaction standard Gibbs free energy based on Marcus theory. As the figure shows, when $-\Delta G^0 = \lambda$ (or $\Delta G^0 + \lambda = 0$), k reaches its maximum and $\Delta G^{\#} = 0$. On the left side of the maximum k value where k increases with $-\Delta G^0$, this is called normal region. However, on the right side of the maximum k value where k decreases with $-\Delta G^0$, this is called Marcus inverted region, as too much driving force

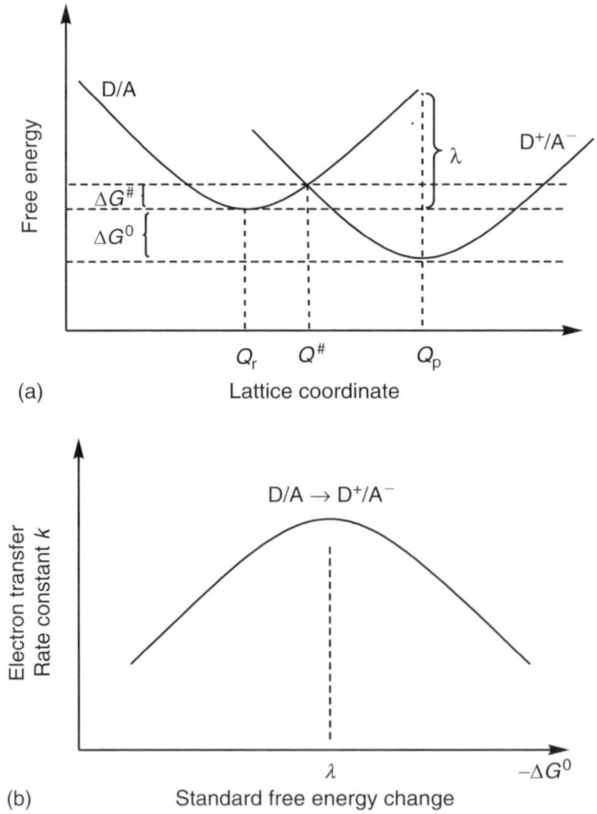

FIGURE 3.14 Electron transfer from a donor to an acceptor represented by (a) free energy surface versus lattice coordinate; (b) electron transfer rate constant versus standard free energy change.

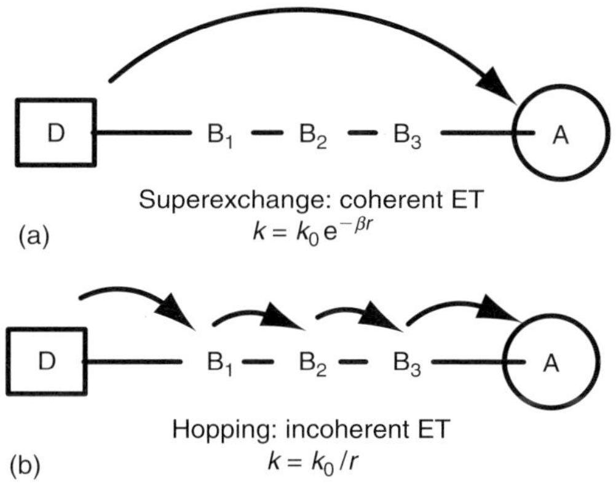

FIGURE 3.15 (a) Coherent electron transfer (super-exchange); (b) incoherent electron transfer (hopping).

would slow down the electron transfer rate, and this was somewhat unexpected in the early days [11]. In essence, Marcus theory predicts that electronic transition or electron transfer becomes fastest when all the driving and counter-driving forces are balanced.

In spatial domain, as shown in Figure 3.15, the electron transfer rate constant k correlates to the transfer distance r between the donor (D) and the acceptor (A) depending on the actual paths or mechanisms of the transfer process. For instance, in a coherent (or tunneling) type electron transfer, where the electronic coupling between the donor and acceptor is strong and the electron scattering is weak, the transfer rate constant is correlated to the transfer distance by

$$k = k_0 e^{-\beta r} \tag{3.28}$$

where the coefficient β varies in a range of 0.2–1.4 depending on the systems under consideration [12]. In an incoherent (or hopping) style electron transfer, where the electronic coupling between the donor and acceptor is relatively weak and the electron scattering is strong, the transfer rate is correlated to the transfer distance by

$$k = \frac{k_0}{r} \tag{3.29}$$

In the hopping case, the electron transfer rate can be approximated by Marcus electron transfer model [8,9], or approximated by Abbrium–Miller model (see Chapter 4).

Electron transfer rate is essential to many chemical, physical, and biological processes, particularly electronic and optoelectronic properties of many molecular materials systems.

3.3.1.2 Electronic Transitions

Using a simple diatomic molecular system as an example, the potential energy surfaces of ground (S_0) and first excited (S_1) states versus the diatomic nuclear distance can be schematically represented as top-distorted parabola (also called Morse potential) curves as shown in Figure 3.16 [5]. The horizontal lines (labeled as v_n, v'_n, v''_n, where n=0, 1, 2, ...) within each potential wells represent different vibrational energy levels. From quantum mechanics and by treating the diatomic molecular system as a harmonic oscillator (particularly near the bottom of the potential well), the vibrational energy levels E_v can be represented by

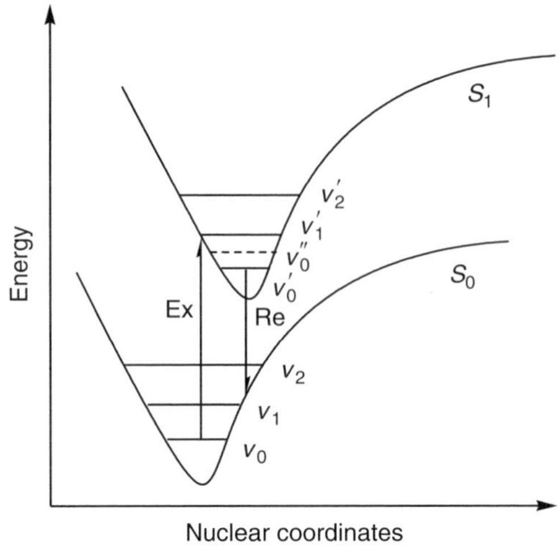

FIGURE 3.16 Scheme of an idealized two-level model showing electronic excitation (Ex), recombination (or relaxation, Re), intra-molecular vibrational modes (v_n and v'_n) and an inter-molecular or aggregate vibrational mode (v''_0).

$$E_v = \left(v + \frac{1}{2}\right) h v_o \quad (3.30)$$

where
v is the vibrational quantum numbers ($v = 0, 1, 2, \ldots$)
v_o is the fundamental frequency of the vibration between the two nuclei

Figure 3.16 illustrates that the potential energy minimum of ground (S_0) and first excited (S_1) states can be at different nuclear coordinates (e.g., different inter-nuclear distance in a simple diatomic system). Both the typical intra-molecular vibrational modes (represented by v_n and v'_n), and an inter-molecular vibrational mode (represented by v''_0, due to molecular aggregates at excited state) are shown. According to the Frank–Condon principle [5], an electronic transition typically proceed in a vertical transition manner (i.e., transitions $v_0 \rightarrow v'_1$ or $v'_0 \rightarrow v_1$ as shown in Figure 3.16), so that the nuclear coordinates essentially remain static during the initial ultra-fast electron transfer. If we apply Equation 3.27 to the case of a photo driven electron transfer (or transition), for instance, the photon driving force would be $\Delta F = -hv$. Assuming all nuclear vibrational modes are classical ($\hbar\omega_i \ll kT$), the rate (or intensity) of the transition would therefore be proportional to the energy factor exp $[-(\Delta G_0 + \lambda - hv)^2/4\lambda kT]$, which would yield a Gaussian bell shaped photo absorption peak, where the peak maximum corresponds to $hv = \Delta G^0 + \lambda$, and the low energy edge of the absorption peak could be used to estimate the potential energy gap ΔE of the system under certain approximations.

Once the molecule is at v'_1 level in S_1 state, it typically relaxes to the lower and more stable v'_0 level and geometry in S_1 state, called exciton vibronic relaxation, also called Kasha's rule. Relaxation/decay from excited S_1 state to ground S_0 state is also vertical, i.e., it may proceed between different vibrational levels if ground and excited states have different potential minimum (e.g., from v'_0 level in S_1 state to v_1 level in S_0 state shown in Figure 3.16). Excited state relaxation can be either radiative, i.e., via emitting photons, or nonradiative decay, i.e., via lattice thermal relaxation.

The excitation or absorption processes can be characterized by various absorption spectroscopic techniques (such as UV–Vis) in different electromagnetic energy ranges corresponding to the absorption energy, and the decay processes can also be studied by various emission spectroscopic techniques (such as fluorescence if the decay is radiative in UV–Vis).

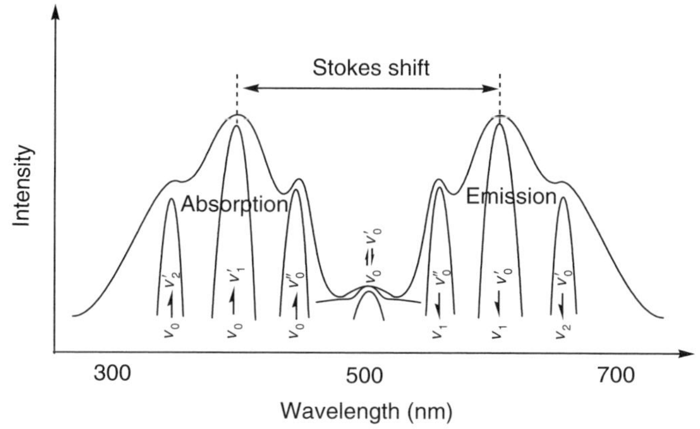

FIGURE 3.17 Scheme of idealized optical absorption and fluorescence emission bands of a two level model, where the sharp component peaks due to intra- or inter-molecular vibrational modes are also indicated. For instance, $v_0 \rightarrow v_1'$ denotes an excitation transition from v_0 level to v_1' level shown in Figure 3.16.

Stokes shift refers to the energy difference (or the redshift) of the emission spectrum peak versus the absorption spectrum peak (see Figure 3.17) [5]. In addition to vibrational relaxations that contribute to the Stokes shift, in solid states, several other factors can also contribute to the Stokes shift, for example, any lattice structural reorganizations (or electronic polarizations) after electronic transitions, aggregates (J- or H-type), or any excimers/exciplexes formations. Note if the potential energy surfaces of the ground and first excited states are the same or similar in shape, then the Stokes shift (which is in fact the sum of reorganization energies of both absorption and emission between S_0 and S_1 states) becomes about twice the magnitude of the reorganization energies in either $v_0 \rightarrow v_1'$ or $v_0' \rightarrow v_1$ transitions.

Figure 3.17 schematically depicts idealized absorption and emission mirror bands with component peaks representing either vibrational or aggregates modes. To determine if a component peak comes from vibrations or aggregations, spectroscopic concentration or heat-dependent experiments can be performed, as aggregate peaks are more intense at solid state or higher concentrations in solution and at lower temperatures.

In addition to the basic two-level electronic transitions described above, other major types of transitions are also often encountered. Those include transitions at higher excited state levels (also called super-excitations), intersystem crossing transitions, collisional quenchings, internal conversions, etc. All those transitions can be summarized in a Jablonski's diagram as shown in Figure 3.18 [5]. In this figure, S_0 represents the system ground state, S_1 represents the first excited state, S_2 represents the second excited state, and T_1 represents the first triplet state. 1a and 2a represent absorptions to S_1 and S_2 states; 1b and 2b represent vibrational and structural relaxations within S_1 and S_2 states; 1c and 2c represent singlet excited state emissions or fluorescence of S_1 and S_2 states; 1d represents an intersystem crossing from S_1 to T_1 state, i.e., a singlet exciton becomes a triplet exciton via an electron spin flip; 2d and 3d represent transitions from S_2 to S_1 states via either collision (2d) or internal conversion/structural relaxation (3d); and 1e represents radiative relaxation from the triplet exciton state T_1 to the singlet ground state S_0. Since this T_1–S_0 transition requires a spin flip coupled with an exciton relaxation, the process is somewhat hindered as compared to a singlet decay, therefore, such radiative decay (also called phosphorescence emission) is a much slower process compared to the singlet fluorescence emission in 1c and 2c. Alternatively, the S_1 exciton can relax to ground S_0 state via nonradiative mechanisms, such as thermal or structural relaxations. In these processes, heat instead of light would be released.

In general, the electronic transitions in organic compounds can be determined by UV/Vis spectroscopy provided that the transitions in the UV or visible range of the electromagnetic spectrum exist for this compound. For instance, electrons residing in the HOMO of a sigma bond can get excited to the LUMO of that bond. This process is written down as $\sigma \rightarrow \sigma^*$ transition. Likewise,

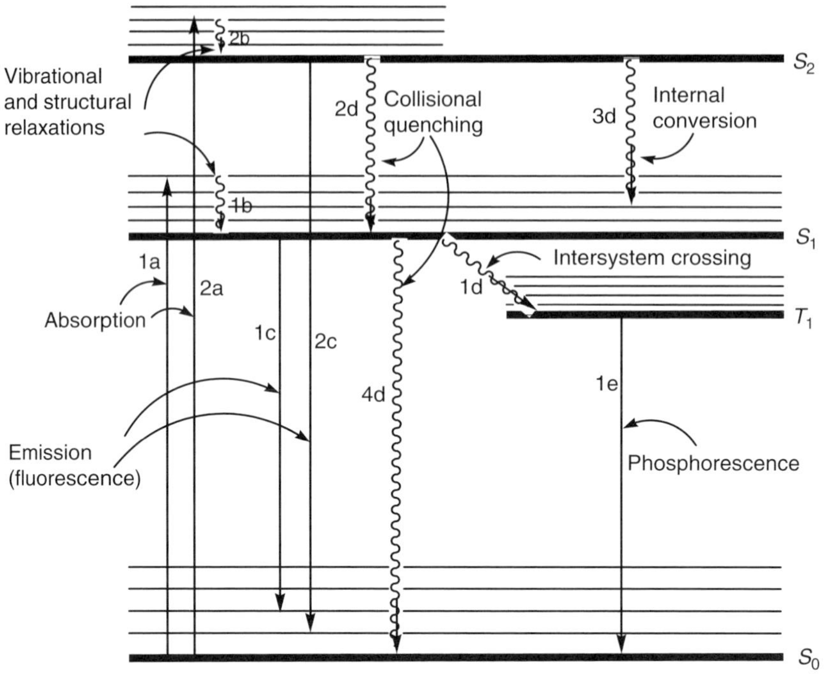

FIGURE 3.18 Jablonski's diagram depicting major transitions of a material in energy regime.

promotion of an electron from a π bonding orbital to an antibonding π* orbital is denoted as $\pi \rightarrow \pi^*$ transition. Auxochromes (a group of atoms such as –OH, $-NH_2$, and aldehyde groups attached to an organic chromophore that modifies the ability of that chromophore to absorb light) with loan pair electrons denoted as *n* have their own transitions, as do aromatic π bond transitions. The following molecular electronic transitions are quite common:

- $\sigma \rightarrow \sigma^*$
- $\pi \rightarrow \pi^*$
- $n \rightarrow \sigma^*$
- $n \rightarrow \pi^*$

In addition to these assignments, electronic transitions also have the so-called bands associated with them. The following bands are defined: the R-band from the German *radikalartig* or radical-like, the K-band from the German *Konjugierte* or conjugated, the B-band from benzoic, and the E-band from ethylenic (system devised by A. Burawoy in 1930). For example the absorption spectrum for ethane shows $\sigma \rightarrow \sigma^*$ transition at 135 nm and that of water shows $n \rightarrow \pi^*$ transition at 167 nm with an extinction coefficient of 7000. Benzene has three aromatic $\pi \rightarrow \pi^*$ transitions; two E-bands at 180 and 200 nm and one B-band at 255 nm with extinction coefficients of 60,000, 8,000, and 215, respectively. These absorptions are not narrow bands but are generally broad because the electronic transitions are superimposed on the other molecular energy states.

The electronic transitions of molecules in solution can depend strongly on the type of solvent with additional bathochromic shifts (a change of spectral band position in the absorption, reflectance, transmittance, or emission spectrum of a molecule to a longer wavelength or lower frequency, also called redshift) or hypsochromic shifts (a change of spectral band position in the absorption, reflectance, transmittance, or emission spectrum of a molecule to a shorter wavelength or higher frequency, also called blueshift). As briefly mentioned earlier, molecular aggregates in both solution and solid states would cause bathochromic or hypsochromic shifts depending on the nature and type of the aggregates.

3.3.2 Carrier Generation Mechanisms

When an energy matched photon hits a semiconductor, an exciton is usually first formed as shown in Figure 3.19. An exciton is a bound state of an electron and an imaginary particle called hole (a vacant site of an electron with one positive charge), and thus is also called a correlated electron–hole pair. Correlated means free electron spins are not detectable via traditional electron spin resonance (ESR) even if the excited single electron is at the LUMO orbital, and one remaining single electron is at HOMO orbital after the excitation. Exciton is a quasiparticle. A quasiparticle refers to a particle-like entity arising in certain systems of interacting particles. It can be thought of as a single particle moving through the system, surrounded by a cloud of other particles that are being pushed out of the way or dragged along by its motion, so that the entire entity moves along somewhat like a free particle. Excitons are integer spin particles thus obeying Bose statistics in the low-density limit. An exciton can diffuse in the material from one site to another called energy transfer process (to be described in detail later). Excitons can be treated in two limiting cases that depend on the properties of the material in question.

In a classic or traditional inorganic semiconductor as shown in Figure 3.19a, the dielectric constant is generally large, and as a result, charge screening effectively reduces the Coulomb interaction between electrons and holes of photo generated excitons. The result is a large-sized Wannier–Mott type exciton (typically >10 nm), much larger than the typical lattice spacing. As a result, the effect of the lattice potential can be incorporated into the effective mass of the electron and hole, and because of the lower effective masses and the screened Coulomb interaction, the exciton binding energy (E_B), i.e., the minimum energy required to dissociate an exciton into a free (or uncorrelated) electron and a free hole is relatively small (typically much less than 0.01 eV in Wannier–Mott type exciton), therefore, thermal energy of an electron ($E_T = kT = 0.025$ eV at room temperature, or thermal phonon energy $\hbar\omega$) would be sufficient to dissociate the Wannier–Mott type excitons. Thus, the freed electrons would be delocalized and traveling in the conduction band, and free holes would be delocalized and traveling in the valence band (VB). Both free electrons and holes (also called mobile charged carriers, or simply carriers) can be regarded as generated directly from photo excitations in the inorganic semiconductors. Free, uncorrelated, or mobile electrons and holes mean the electrons and holes can each move independently as individual particles and each can be detected by ESR. This is why free charge carrier photo generation in classic inorganic semiconductors is called primary photo carrier generation mechanism, and is also termed VB–CB transitions or simply band-to-band transitions [6,7].

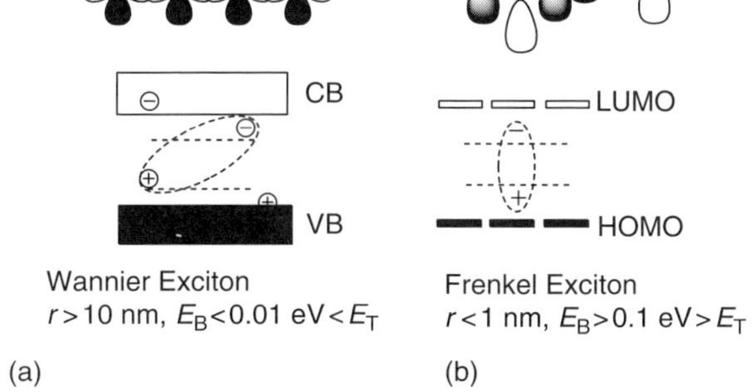

FIGURE 3.19 Schematic representation of (a) Wannier–Mott type excitons; (b) Frenkel-type excitons.

In most organic and polymeric semiconductors as shown in Figure 3.19b, the dielectric constant is generally much smaller than in inorganics, as a result, the Coulomb interaction between electron and hole of an exciton becomes very strong, therefore, a much smaller sized (mostly <1 nm) Frenkel-type exciton (also called exciton–polaron due to significant polarization and lattice distortion) is formed upon photo excitation [13–15]. The exciton orbital levels are within the LUMO/HOMO gap as shown in Figure 3.19b [13]. The exciton–polaron binding energy is relatively large (E_B is in the range of 0.1–1.5 eV [2,3,13–15]), i.e., thermal energy of electron at room temperature kT, or lattice thermal vibration/phonon energies ($\hbar\omega$) would not be sufficient to dissociate a Frenkel-type exciton, therefore, additional or secondary forces are needed in order to dissociate a Frenkel-type exciton into free electrons and holes [15]. This is why photo carriers in organics are generated mostly via a secondary photo carrier generation mechanism, also called photo doping mechanism referring a photo-induced charge separation [15].

In organics, free electron charge carriers are also termed electron–polarons or negative polarons; and free holes are also termed hole–polarons or positive polarons. Polarons are defined due to an electron or a hole in organics typically induces a relatively large local polarization and lattice distortion. The carrier together with the induced polarization is considered as one entity, which is called a polaron. Polaron is originally defined as a quasiparticle composed of a charge plus its accompanying polarization field. The resulting lattice polarization acts as a potential well that hinders the movements of the charge, thus decreasing its mobility. For instance, a slow moving electron in a dielectric crystal interacting with lattice ions through long-range forces will permanently be surrounded by a region of lattice polarization and deformation caused by the moving electron. Moving through the crystal, the electron carries the lattice distortion with it, thus one may speak of a cloud of phonons accompanying the electron. The polaron LUMO/HOMO levels are located between the pristine molecular LUMO/HOMO levels (see Figure 3.26) [13,14].

Finally, orbitals higher than LUMOs can also participate in the exciton formation (also called super-excitation), and this would lead to the formation of different types of excitons in the same material as demonstrated in some ultrafast two-photon absorption experiments.

The Bohr radius (r) or average size of an exciton in a semiconductor can be expressed by [6,15].

$$r = r_0 \varepsilon \left(\frac{m}{m^*} \right) \tag{3.31}$$

where

r_0 is the Bohr radius of hydrogen or atomic unit of length (about 0.053 nm)
ε is the dielectric constant of the materials
m and m^* are the rest (or free) and dynamic (or effective) mass of the electron, respectively

The concept of effective mass m^* is used because an electron would appear either heavier or lighter in a solid-state material depending on the interactions between the electron and the phonon, and it can be expressed by [6,7]

$$m^* = \frac{F}{\left(\frac{\mathrm{d}v}{\mathrm{d}t} \right)} = \frac{\hbar^2}{\left(\frac{\mathrm{d}^2 E}{\mathrm{d}k^2} \right)} \tag{3.32}$$

where

F is the net driving force
v is the velocity
E is the energy
k is the wave number of the electron

Equation 3.32 shows that the effective mass is inversely proportional to the curvature of the electron energy bands (E–k curve).

In general, the effective mass of the electron would become bigger with smaller charge delocalization or larger energy gap of the materials. For instance, the charge delocalization is typically smaller in organic semiconductors as compared to inorganic semiconductors, therefore, the electron effective mass is much bigger in organic semiconductors than in inorganic semiconductors. Additionally as mentioned above, the dielectric constants (ε) of organic semiconductors are generally smaller than inorganic semiconductors, and this results in poorer charge screening in organics. Both factors contribute to the average smaller exciton size of organic semiconductors as reflected in Equation 3.31. Finally, the Coulombic potential E_c is inversely proportional to both the exciton size and the dielectric constant of the material as represented by

$$E_c = -\frac{e^2}{4\pi\varepsilon\varepsilon_0 r} \tag{3.33}$$

A plot of E_c versus the exciton radius or size r is shown schematically in Figure 3.20, where the top curve (diamonds, assuming Frenkel exciton parameters E_c = 1 eV at 1 nm) represents a typical organic Frenkel-type exciton as occurred in most organic semiconductors with smaller dielectric constant, and the bottom curve (triangles, assuming Wannier exciton parameters E_c = 0.1 eV at 10 nm) represents a classic inorganic semiconductor exciton with larger dielectric constant. The middle horizontal dashed line represents the electron thermal energy at room temperature (kT = 0.025 eV). Figure 3.20 demonstrates that room temperature thermal energy is generally sufficient to overcome the exciton Coulombic binding energy of most classic inorganic semiconductors, while room temperature thermal energy is generally not sufficient to overcome the exciton Coulombic binding energy of most organic semiconductors, though exceptions may occur for highly crystalline or self-assembled organic materials with very large exciton sizes or larger dielectric constants.

Figure 3.21 shows ground state S_0, photo excited exciton state S_1, and exciton dissociated (or charge separated) state S_1' of a general molecular system in (a) frontier orbital representation and (b) free energy representation (Note the free energy can be represented by a single lattice coordinate, while the

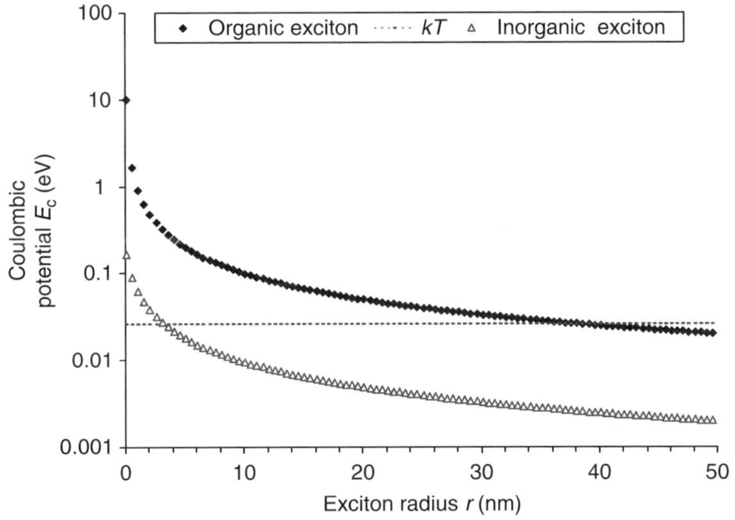

FIGURE 3.20 Scheme of Coulombic potential (in absolute value) for typical Frenkel-type excitons (diamonds) versus Wannier–Mott type excitons (triangles).

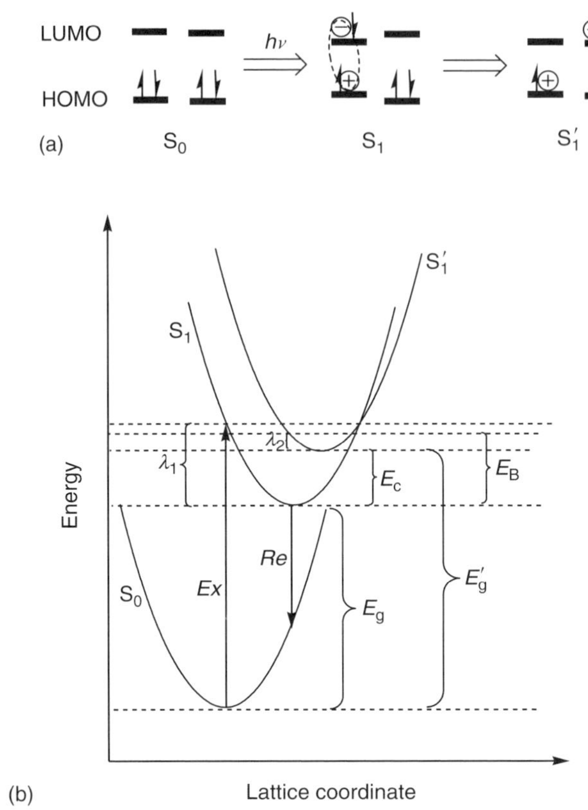

FIGURE 3.21 Scheme of photoelectric carrier generation processes in (a) frontier orbital representation; (b) free energy surface representation.

potential energy has multi-coordinates for a polyatomic molecular system. Also, the free energy change could approximate the potential energy change when the entropy change is small or negligible). Here, the main difference between S_1 and S_1' states is that, in S_1 or exciton state, the electron at LUMO is strongly correlated with the hole at HOMO, such that this exciton diffuses as one single quasiparticle. The S_1 state may also be an excimer or an exciplex as described earlier. While in S_1' or charge separated state, the electron at LUMO is regarded as uncorrelated to the hole at HOMO, so that electrons and holes can each diffuse separately as two independent or free charge carriers. Please note that in Figure 3.21a, the two schematic HOMO/LUMO pairs may not necessarily be immediately adjacent to each other, i.e., a certain spatial distance or several atomic/lattice sites may exist between the two charges.

Figure 3.21b schematically shows system free energy surfaces for the three states. In the figure, the energy gap, E_g (also called optical excitation energy gap) between S_0 and S_1 states can be experimentally estimated from the absorption band edge of typical UV–Vis spectrum. λ_1 is the reorganization energy of the $S_0 \rightarrow S_1$ transition and λ_2 is the reorganization energy during the $S_1 \rightarrow S_1'$ transition. Since the $S_1 \rightarrow S_1'$ transition is mainly exciton dissociation, the free energy gap between S_1 and S_1' states is mainly the exciton columbic binding energy E_c. Therefore, the electronic energy gap between S_0 and S_1' states is

$$E_g' = E_g + E_c \tag{3.34}$$

Based on semi-classical Marcus model, the magnitude of optimal driving force for the $S_1 \rightarrow S_1'$ transition appears to be $E_B = \lambda_2 + E_c$, where E_B approximates the total exciton binding energy. As mentioned earlier, for most organic or polymeric noncrystalline semiconductors, room temperature

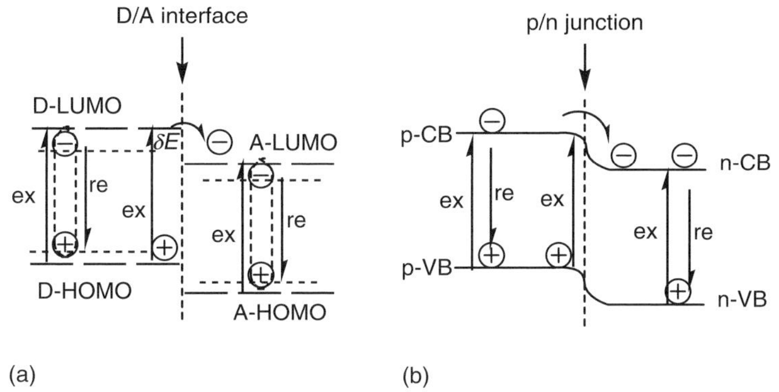

FIGURE 3.22 Scheme of electron (charge) transfers at (a) a D/A interface; (b) a p/n junction. ex, excitation; re, recombination.

thermal energy would not be sufficient to overcome E_B, additional force (or energy) is therefore needed in order to dissociate a Frenkel-type exciton into free charge carriers, the so-called secondary photo carrier generation mechanism [15]. The extra force or energy can be from externally applied electric fields, high temperature thermal energy, and in most cases from the frontier orbital level offsets (δE) between an electron donor and an electron acceptor as shown in Figure 3.22a [15]. As Figure 3.22a shows, the frontier orbital energy offset (δE—the energy difference between D-LUMO and A-LUMO) is in fact a key driving force to overcome E_B in order to incur the excited state electron transfer from the donor to the acceptor, as the D-LUMO → A-LUMO transition (exciton dissociation) is typically much faster than the D-LUMO → D-HOMO transition (exciton decay) [16]. This is somewhat analogous to a p/n junction as shown in Figure 3.22b where the electric field at the p/n junction facilitates the electron and hole separation, where electrons are pushed from the p-type semiconductor into the n-type semiconductor, and holes are pushed from n-type semiconductor into the p-type semiconductor [6,7]. However, while free charge carriers (electrons and holes) are generated in either p- or n-type traditional semiconductors, mainly neutral exciton–polarons are generated in organic donor or acceptor phases. Therefore, a donor/acceptor interface appears essential for photo charge carrier generation in typical organic optoelectronic such as photovoltaic devices [15,16].

The photo-induced charge separation due to the presence of a donor (D)/acceptor (A1) pair is also called photo doping process as schematically depicted in Figures 3.23a and 3.24. Specifically, in the photo doping situation (a), the A1-LUMO is close and slightly lower than the D-LUMO, the free

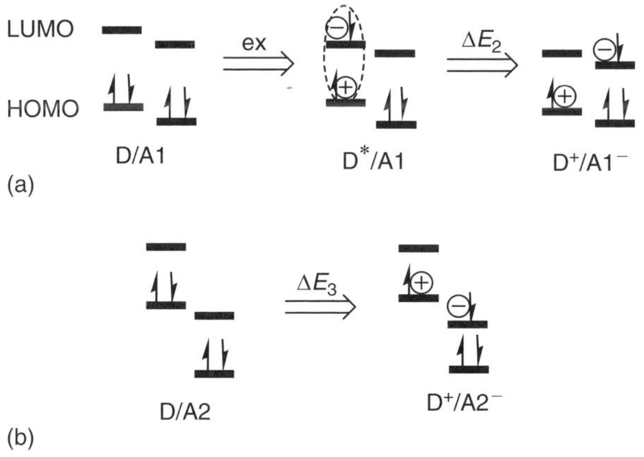

FIGURE 3.23 Scheme of frontier orbitals and electron transfers of (a) photo doping; (b) chemical doping.

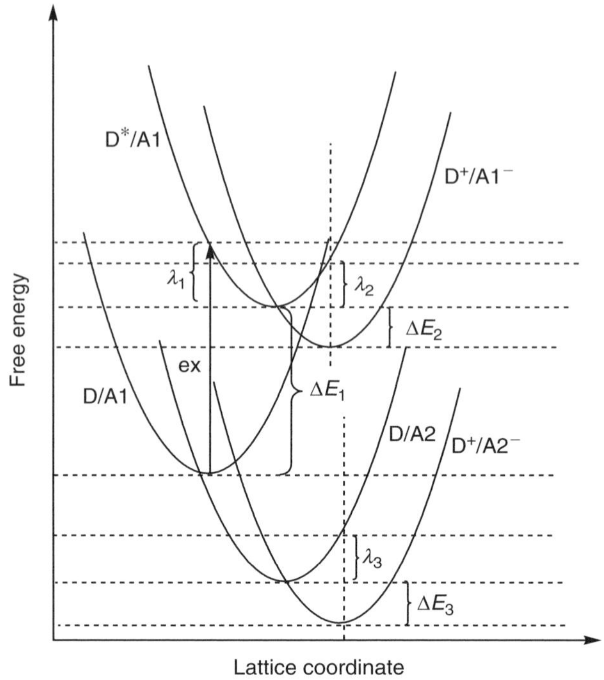

FIGURE 3.24 Scheme of free energy surfaces and electronic transitions of photo doping (D/A1 → D*/A1 → $D^+/A1^-$) and chemical doping (D/A2 → $D^+/A2^-$).

energy difference (ΔE_2) for the D-LUMO → A1-LUMO electron transfer could be optimum (e.g., $\Delta E_2 = \lambda_2$ shown in Figure 3.24), and that such transition can be much faster than the donor exciton relaxation (electron transfer from D-LUMO to D-HOMO) [16]. Note if the electronic coupling of D-HOMO/A-LUMO is stronger than D-HOMO/D-LUMO, then photo (or thermal) driven electron transfer from D-HOMO to A-LUMO may proceed directly to form an inter-molecular exciton (or exciplex). This occurs in many so called donor–acceptor type charge transfer chromophores or complexes.

In contrast, in the donor (D)/acceptor (A2) incurred chemical doping or ground state charge transfer process as schematically shown in Figures 3.23b and 3.24, the LUMO of the acceptor (A2) is typically close but weakly coupled to the HOMO of the donor (D), so the electron would transfer directly from the D-HOMO to the A2-LUMO [16,17]. For instance, in case the LUMO of A2 is lower than the HOMO of D, the free energy driving force ΔE_3 due to D-HOMO/A2-LUMO orbital offset could become optimum (e.g., $\Delta E_3 = \lambda_3$ as shown in Figure 3.24) to drive a ground state electron transfer at even absolute zero temperature [16]. At nonzero temperature, even if the LUMO of A2 is the same or slightly higher than the HOMO of D, electron transfer would still proceed due to thermal perturbation [6,16,17]. Though the transfer rate due to thermal perturbation may be very low due to weak driving forces, however, once such transfer occurs, and if the acceptor is a minority dopant dispersed in the majority donor phase as in Figure 3.25c, the left over hole site on D-HOMO can be filled immediately by an electron from an electron at nearby donor HOMO orbital, so the electron at the A2-LUMO is thus trapped, the freed or mobile hole (positive polaron) thus diffuses in the donor HOMO band forming a p-type semiconductor (or a conductor, depending on the carrier density and mobility). The acceptor dopant, or any acceptor type impurity or defect site can also be called the electron trap. If the donor is the minority dopant dispersed in the majority acceptor phase as in Figure 3.25b, then the freed and mobile electron (negative polaron) thus diffuses in the acceptor LUMO band, and the hole at the donor HOMO is trapped forming an n-type semiconductor (or a conductor depending on the density and mobility of the carriers). The donor dopant, or any donor

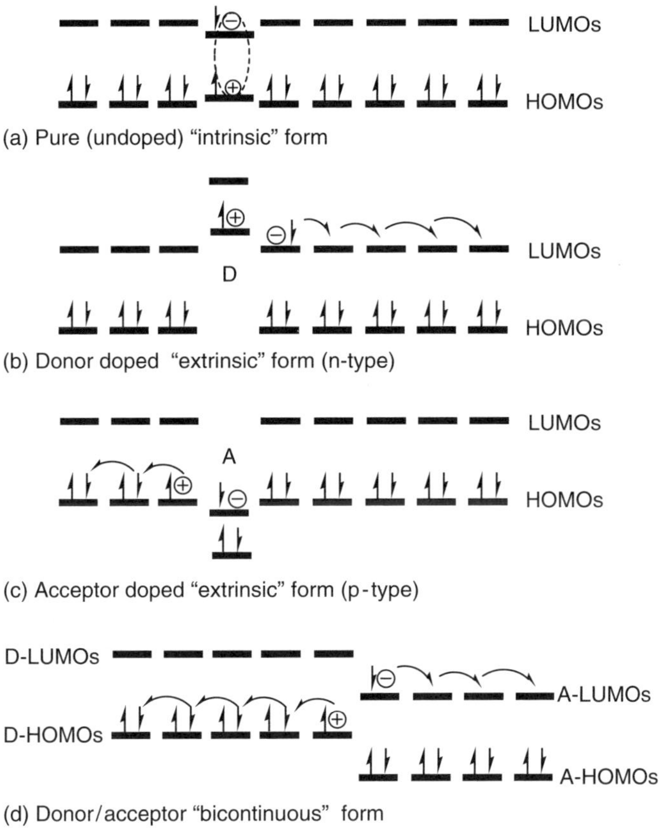

FIGURE 3.25 Frontier orbital and electron transfer schemes of organic/polymeric semiconductors and conductors in (a) pure (intrinsic) form; (b) donor (n-type) doped form; (c) acceptor (p-type) doped form; (d) donor/acceptor (p/n) junction form.

type impurity or structure defect, can therefore be called a hole trap. Certainly, the separated electrons and holes can still recombine as well. If the pair of recombining electron and hole is from the same original orbital or pair, it is called geminate pair and the recombination is called geminate recombination. If the pair of recombining electron and hole is not from the original pair, it is called nongeminate pair and the recombination is called nongeminate recombination [13].

Chemical doping can also be described in a classic way in certain systems [6,7]. For instance, when each atom (or nuclear site) has four valence electrons in its HOMOs in the majority phase, and the minority dopant has five valence electrons (donor type) in its HOMOs, four valence electrons from the dopant would couple with four valence electrons from the nearby four majority nuclei filling four new HOMO bonding orbitals. The fifth or the left over valence electron from the dopant can then be easily excited (via thermal or photo means) into the LUMO band of the majority phase forming an n-type semiconductor leaving a hole trapped at the dopant site. Alternatively, when each atom (or nuclear site) has four valence electrons in its HOMOs in the majority phase, and the minority dopant has three valence electrons (acceptor type) in its HOMOs, three valence electrons from the dopant would couple with four valence electrons from the nearby four majority nuclei filling four new HOMO bonding orbitals. The one new HOMO orbital that has only one electron (containing one vacant site for an electron) can be easily filled by an electron from the HOMO band of the majority phase forming a p-type semiconductor and leaving an electron trapped at the dopant site.

In general, n-type doping refers to a minority donor type dopant trapping a hole (or positive polaron) from the majority phase via donating an electron (or create a negative polaron) into the

LUMO band of the majority phase, and p-type doping generally refers to a minority acceptor type dopant trapping an electron from the majority phase leaving a mobile hole (or positive polaron) at the HOMO band of the majority phase.

In summary, compared to classic or traditional inorganic semiconductors, the frontier orbital schemes of intrinsic organic or polymeric semiconductor can be shown in Figure 3.25a, i.e., instead of uncorrelated free electrons and holes upon excitation (photo or thermal), an exciton (also called an exciton–polaron, or a correlated electron–hole pair) is typically generated in most organic or polymeric semiconductors. Though the exciton can diffuse in the material before decay (either radiatively or nonradiatively), exciton itself does not contribute to the conductivity of the materials, though the exciton contributes to the energy transport. However, like in traditional extrinsic semiconductors, an n-type of organic semiconductor (or conductor) could be generated upon a donor type of chemical doping as depicted in Figure 3.25b. Likewise, a p-type of organic semiconductor (or conductor) may be formed upon an acceptor type of chemical doping as depicted in Figure 3.25c. Finally, a donor and an acceptor can form a two-phase (binary phase, or bipolar type) material as shown in Figure 3.25d. In this case, the charge carriers generated at the D/A interface via either photo doping or chemical doping mechanisms can diffuse away from the interface in two separate phases due to chemical potential gradient [15]. This is somewhat similar to a traditional p/n junction semiconductor device. The main difference here is that the charge carriers are only generated at the D/A interface but not in D or A phase. This is a key mechanism of organic photovoltaics, i.e., photo doping in a donor/acceptor binary heterojunction system [15,16].

In addition to the free holes (positive polarons, the orbital levels are schematically shown in Figure 3.26a) and free electrons (negative polarons, Figure 3.26b), a number of other key charge (or energy) carriers in conjugated organic and polymeric materials include mobile positive bipolarons (Figure 3.26c), mobile negative bipolarons (Figure 3.26d), neutral solitons (Figure 3.26e), positive solitons (Figure 3.26f), negative solitons (Figure 3.26g), and energy carrier exciton–polarons (Figure 3.26h) [13,14]. Except the exciton–polaron (Figure 3.26h) that is an energy carrier but not a charge carrier, all other charged or neutral entities (Figure 3.26a through g) can be driven by an applied electric field and therefore called electric carriers, or simply carriers.

Specifically, while two polarons of opposite charge can couple and spin–correlate to each other forming an exciton–polaron (Figure 3.26h, and may subsequently decay to ground state by emitting a photon, a basic mechanism in organic light emitting materials/devices), two polarons of the same charge can also couple and spin–correlate to each other forming a relatively stable bipolaron entity (a quasiparticle) that can diffuse as a charge carrier containing either two positive (Figure 3.26c) or two

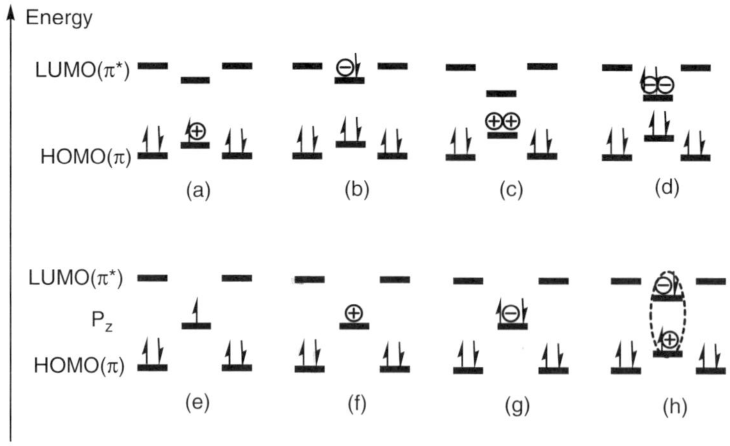

FIGURE 3.26 Orbital levels schemes of carriers: (a) positive (hole) polaron; (b) negative (electron) polaron; (c) positive (hole) bipolaron; (d) negative (electron) bipolaron; (e) neutral soliton; (f) positive (hole) soliton; (g) negative (electron) soliton; (h) exciton–polaron.

negative (Figure 3.26d) unit charges. When two polarons are close together, they can lower their energy by sharing the same distortions, which leads to an effective attraction between the two polarons. The correlated two unit charges typically share one orbital in a bipolaron, or remain in two separate orbitals forming a meta-stable polaron-pair. In either case, they are strongly coupled and correlated in a lower energy state. While a bipolaron is not responsive in the typical ESR single electron detection regime, a polaron-pair would exhibit a singlet-triplet splitting. Both bipolarons and polaron-pairs have integer spins and thus share some of the properties of bosons. Similar to a polaron, the frontier orbital levels of a bipolaron (and a polaron-pair) are also within the HOMO/LUMO gap of pristine material, but closer to the center than that of a polaron. Precise definition of a soliton is not easy and it involves substantial mathematics, however, a neutral soliton (Figure 3.26e) formation in polyacetylenes may be simplified as due to, for instance, a degenerate ground state structural distortion, or conjugation paring symmetry rearrangement in the polyacetylene backbone, so that one p_z electron of a backbone carbon atom somehow does not participate in the backbone conjugated π bands formation, thus generating one mobile soliton carrier (Figure 3.26e through g). The positive soliton (Figure 3.26f) can be regarded as the neutral soliton (Figure 3.26e) losses its electron, and the negative soliton (Figure 3.26g) can be regarded as the neutral soliton (Figure 3.26e) gained an electron with opposite spin.

Overall electrical conductivities of conjugated organic and polymeric materials are due to both density (N) and mobility (μ, Equation 3.17) of the mobile charge carriers as a result of the impurities, defects, and doping of the material.

Figure 3.27 shows electrical conductivity (σ) of some representative materials at room temperature (except poly-sulfur-nitride, which is a superconductor at 0.3 K). The resistivity (ρ, in unit of Ω cm) is the inverse of the conductivity ($\rho = 1/\sigma$). Depending on the magnitudes of carrier density (N) and carrier mobility (μ), from Figure 3.27, it can be seen that either insulator ($\sigma < 10^{-7}\ \Omega^{-1}\text{cm}^{-1}$), semiconductor ($10^{-7}\ \Omega^{-1}\text{cm}^{-1} \leq \sigma \leq 10^{2}\ \Omega^{-1}\text{cm}^{-1}$), or conductor ($\sigma > 10^{2}\ \Omega^{-1}\text{cm}^{-1}$) may be obtained from organic or polymeric conjugated materials.

In addition to the electron or charge transfers, energy transfer is also very common and critical in organic optoelectronic materials and devices. Energy transfer is essentially exciton transport or

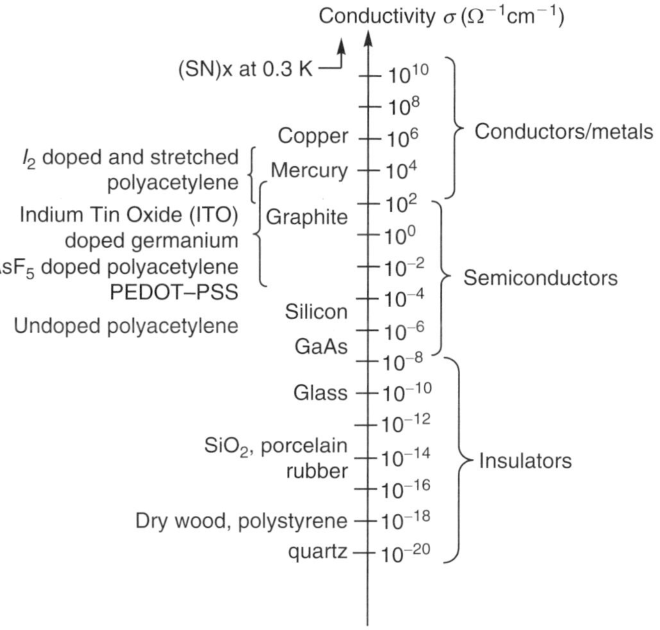

FIGURE 3.27 Room temperature conductivity (σ) of various materials.

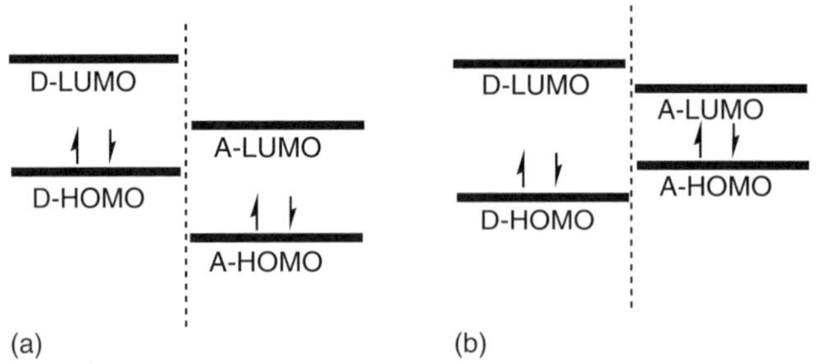

FIGURE 3.28 Frontier orbital energy level representations for a typical (a) electron transfer; (b) energy transfer between two different materials. D, donor; A, acceptor.

diffusion from one site to another. Figure 3.28 shows the general relative frontier orbital levels in the cases of (a) electron transfer and (b) energy transfer. As the figure demonstrates, when two sets of frontier orbitals are positioned in profile (a), electrons can be transferred from D-LUMO to A-LUMO in photo doping situation, or from D-HOMO directly to A-LUMO in chemical doping situation. Alternatively, if photo excitation occurs at acceptor site (i.e., A-HOMO → A-LUMO excitation), then the electron transfer from D-HOMO to the hole site at A-HOMO can also proceed (corresponding to the hole transfer from A-HOMO to D-HOMO) [16].

Among a number of energy (or exciton) transfer mechanisms, Förster and Dexter energy transfers are two well-known types of transfer protocols [16]. Förster energy transfer, also called Förster resonance energy transfer, describes an energy transfer mechanism between two fluorescent molecules as shown in Figure 3.29a. A fluorescent donor molecule is excited at its specific fluorescence excitation wavelength. By a long-range dipole–dipole coupling mechanism, this excited state is then nonradiatively transferred to a second molecule, the acceptor. The donor returns to the electronic ground state. When both molecules are fluorescent, the term fluorescence resonance energy transfer (FRET) is also used, although the energy is not actually transferred by fluorescence.

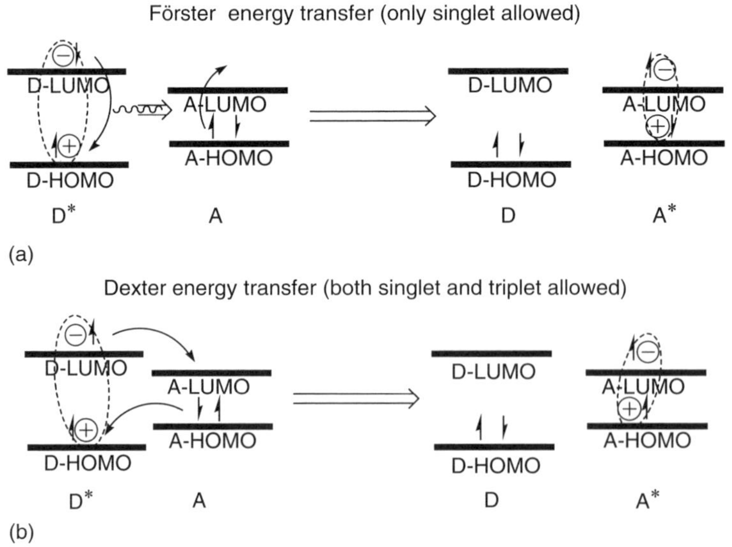

FIGURE 3.29 Scheme of (a) Förster energy transfer; (b) Dexter energy transfer.

The FRET efficiency is determined by three parameters:

1. Distance between the donor and the acceptor
2. Spectral overlap of the donor emission spectrum and the acceptor absorption spectrum
3. Relative orientation of the donor emission dipole moment and the acceptor absorption dipole moment

Due to the fact that the donor exciton relaxation energy is typically smaller than the absorbed photon energy (e.g., due to Stokes shift), the Förster energy transfer would result in gradual energy reduction or spectra redshift along the exciton diffusion or propagation direction. The most efficient Förster energy transfer occurs when the acceptor energy gap matches the donor emitted photon energy well, i.e., transfer coupling may be poor if the acceptor gap is too far away compared to the donor exciton energy. Also, the Förster energy transfer can occur between two remote sites of above 10 nm and is sensitive to molecular dipole orientation, i.e., those molecular dipoles aligned in parallel would have most effective energy transfer.

In contrast, in a Dexter energy transfer process as shown in Figure 3.29b, the electron near D-LUMO first transfers to the A-LUMO, and the hole near the D-HOMO at the same time transfers to A-HOMO, the electron at A-LUMO then relaxes with the hole at A-HOMO to form a new exciton, so it may be regarded as two separate charge transfer processes occurring simultaneously. Because of this, Dexter energy transfer can only proceed at close or adjacent sites (typically <1 nm), and it can proceed with both singlet and triplet exciton transfer, while Förster energy transfer can only proceed with singlet exciton transfer. Also, in the Dexter energy transfer, the A-HOMO and A-LUMO both are desirably located within the energy gap of the D-HOMO and D-LUMO (Figure 3.29b), while the relative positions of HOMO and LUMO are not critical in Förster energy transfer, as long as the energy gap of the acceptor matches the donor exciton emission. Even if the gap of the acceptor does not match the donor exciton emission well, Dexter energy transfer may occur as long as the energy offsets between the donor and acceptor are optimal for charge transfer.

When a donor and an acceptor entity are close to each other, predicting whether electron or energy transfer may occur will depend on the system free energy schemes as shown in Figure 3.30. For instance, if the free energy minimum of (D^+/A^-) is lower than both (D^*/A) and (D/A^*) as shown in Figure 3.30a, electron transfer from D^* to A would occur and (D^+/A^-) would be the energetically stable or measurable product. If the free energy minimum of (D^+/A^-) is lower than (D^*/A) but higher than the (D/A^*) as shown in Figure 3.30b, then energy transfer of (D^*/A) to (D/A^*) would dominate,

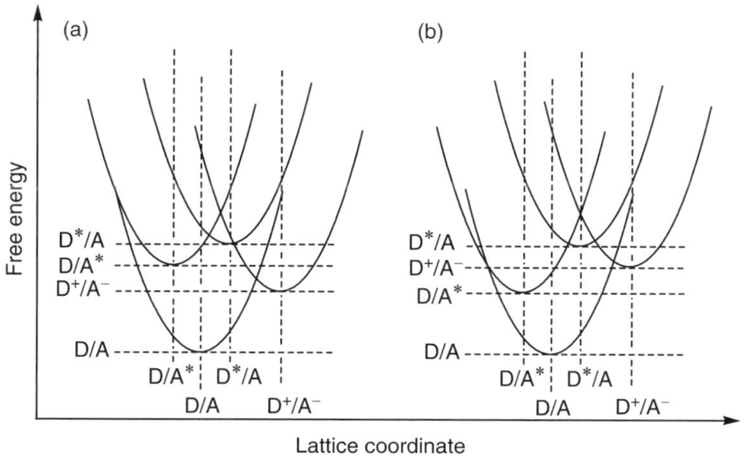

FIGURE 3.30 Free energy surface schemes of a D/A pair in (a) charge transfer; (b) energy transfer scenarios.

or (D/A*) would be the energetically stable or measurable product [8]. When there exist narrow energy gapped defect states (or a site) whose frontier orbital states are in the middle of the majority phase energy gap similar to an exciton–polaron, then an exciton can be trapped via energy transfer to this defect site. The defect site can therefore be called an exciton trap.

Due to the lifetime of the exciton, i.e., where one excited electron at LUMO decays back to its HOMO within the typical exciton lifetime of nano to pico second timescales, there is an average distance of the exciton diffusion, called the average exciton diffusion length (AEDL). The AEDL is highly sensitive to the molecular chemical structures, molecular packing pattern, or morphology of the materials as briefly reflected in both Förster and Dexter type of energy transfers discussed above. For instance, the AEDL can be in the range from 5 nm in amorphous polyphenylenevinylenes (PPV) thin film to above 100 nm in some organic molecular single crystals [2,3,13–15]. AEDL is very critical for any light harvesting such as solar cell applications [15].

3.4 ANALYTICAL TECHNIQUES

3.4.1 Determination of Frontier Orbital Levels

The frontier orbital levels can be determined or estimated using a number of spectroscopic techniques, including x-ray photoemission spectroscopy (XPS), ultraviolet photoemission spectroscopy (UPS), and inverse photoemission spectroscopy (IPES) (see Chapter 20). The UPS and IPES measurements probe directly the occupied and unoccupied electronic structure in the molecular interface region. Alternatively, the molecular frontier orbital levels can also be estimated from electrochemical analysis such as cyclic voltammetry (CV) in combination with UV–Vis optical absorption spectra. As shown in Figure 3.31, the Fermi level of a standard or reference electrode (typically silver or platinum) is typically located between the LUMO and HOMO of most organic conjugated materials. In order for a working electrode (correlated to the standard electrode) to grab an electron from the HOMO of the molecule, a positive potential (oxidation potential, in reference to the standard or reference electrode) needs to be applied to the working electrode and gradually lower the Fermi level of the working electrode until it reaches the HOMO level of the organic molecule where an electron can then transfer from the HOMO of the molecule onto the Fermi state of the electrode. This electron transfer would exhibit the first oxidation peak E_{ox} in the CV measurement. When the working electrode is applied a negative potential (reductive potential, in reference to the standard or reference electrode) and gradually increase its Fermi level until it reaches the LUMO level of the molecule, an electron can then transfer from the working electrode onto the LUMO orbital of the molecule. This corresponds to the first reduction potential peak E_{red}. As Figure 3.31 shows, the LUMO/HOMO levels can be easily obtained if the Fermi level (or work function, the energy difference between the Fermi level and the vacuum level) of the reference electrode is known. However, since the electrode work function would change due to different environmental and measurement conditions, such as solvent, electrolyte, concentration, and temperature, a reference standard compound (such as Ferrocene) whose HOMO level is known and assumed to be stable is typically used to calibrate the reference electrode work function. Once the reductive and the oxidative potentials of both the sample and the Ferrocene (X) are measured, the sample LUMO and HOMO levels can then be calculated using equation

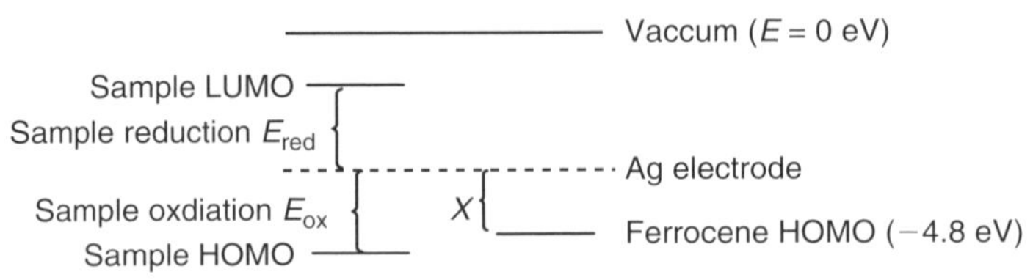

FIGURE 3.31 Energy level scheme of CV.

$$\text{HOMO/LUMO level} = [-E_{ox/red} - (4.8 - X)]\ \text{eV} \tag{3.35}$$

where

X is the Ferrocene oxidation onset (a positive value) versus reference electrode

$E_{ox/red}$ is either a HOMO positive oxidation onset E_{ox} (typically for donors) or a LUMO negative reduction onset E_{red} (typically for acceptors). The Ferrocene HOMO is assumed to be 4.8 eV below vacuum level.

In reality, however, since the solvent used may also be oxidized (or reduced) before the sample, quite often, only one potential peak (either the oxidation or the reduction) but not both can be measured, so once one orbital level is known, the other one can be estimated from the UV–Vis absorption band edge that approximates the HOMO–LUMO gap.

3.4.2 Other Techniques

There are a variety of experimental techniques that have been used to study or analyze the electronic and optical properties of organic and polymeric materials, including, but not limited to, basic techniques such as voltage–current measurements with or without lights, absorption, and emission spectroscopy in both ground and excited states and in either CW or pulsed modes, ESR spectroscopy with or without lights and at different temperatures, x-ray and small angle neutron scattering techniques, thin film characterization tools, various electron microscopies, etc. The readers are referred to relevant chapters of this and other textbooks for more details.

3.5 SUMMARY

The electronic, optoelectronic, and photonic properties of a material are correlated directly to the electronic structures and electron behaviors in the material. Specifically, all materials are composed of molecules and atoms, and there are over 110 different kinds of atoms called elements discovered so far. Each atom is composed of a nucleus (containing neutrons and positively charged protons) and negatively charged electrons (member of fermions) located in electron orbitals surrounding the nucleus. The number of electrons is the same as the number of protons in a neutral atom, and the number of protons determines the kind of element. Electron orbitals have different shapes, orientations, and discrete energy levels, and electrons can transfer from one orbital to another. When atoms combine to form molecules, certain atomic orbitals are overlapping or coupling to form molecular orbitals or chemical bonds. Two atomic orbitals typically couple (both spatially and energetically) and form two molecular orbitals, one with lower energy called bonding molecular orbital and one with higher energy called antibonding molecular orbital. Like many elementary particles, the electron exhibits a character of both particle and wave called particle–wave duality. The behavior of each electron can be described and represented by a wave function with four unique quantum numbers obtained from solutions of the Schrödinger equation. When solving the electron wave functions, Born–Oppenheimer or adiabatic approximation assumes the relatively heavy nuclei remain static during the ultra-fast electron movements. While the electron wave function represents the electron orbital, the square of the wave function at any spatial point represents the probability density of the electron at that point. Unlike classic large objects where the exact position and momentum of the objects can be determined simultaneously and precisely from principles of classic mechanics, the exact position and momentum of an electron cannot be determined simultaneously and precisely (Heisenberg uncertainty principle). Since each electron in an atom or a spatial point must be unique (Pauli exclusion principle), each electron orbital can therefore accommodate a maximum of two electrons with opposite spins. A correlated electron pair typically refers to a pair of two electrons in two separate orbitals but strongly interacting or correlating to each other and acting like one quasiparticle, these mainly include exciton–polarons, polaron-pairs, exciplexes, etc. Electron can transfer from one orbital to another (called electronic transitions) either intra-atomic or interatomic, provided there is a vacant site at the destination orbital. Franck–Condon principle dictates that during the initial ultra-fast electronic transition, the nuclei coordinates essentially remain static. The rate constant of the electron transfer can be described by Fermi's Golden Rule or by Marcus equation. One key feature of Marcus equation is that, in addition to a spatial electronic orbital coupling matrix element, the rate of the electron transfer is also dependent on an Gaussian exponential energy match term, and the rate would become fastest when all driving and counter-driving forces are balanced. Electrons can transfer from higher to lower energy orbitals without the need of an external driving force, or transfer to same or higher lying orbitals with external driving forces. These forces include, but not limited to, thermal forces, electrical forces, magnetic forces, energetic radiations, etc. Electron transport in a material (i.e., from one site to a distant remote site) via incoherent interorbital electron transfers is called hopping transport. In a material with periodic and closely packed atomic structure or electronic potential, the Bloch theorem dictates that HOMOs would couple to each other forming a valence band (VB), and LUMOs would couple to each other forming a conduction band. Electron transport in a band is coherent and much more smooth and faster than hopping. The bandwidth, band gap, and band size all depend on the interatomic and intermolecular electron orbital overlaps and couplings and therefore, the atomic/molecular packing assembly or morphology of the materials. Most electronic, optoelectronic, and photonic processes in materials are mainly due to the valence electron transfers between the frontier orbitals, i.e., HOMOs, LUMOs, and SOMOs, or frontier bands such as CB and VB. The electronic conductivity of a material is dependent on both the charge carrier density and mobility. Primary carrier generations (i.e., via VB–CB transitions like in classic inorganic semiconductors) in organic semiconductors are rare, and this is mainly due to the exciton binding energies of most organic semiconductors are much larger than room temperature thermal energy, so that charge carriers in most organic and polymeric materials (including positive or negative polarons; positive or negative bipolarons; and positive, negative, or neutral solitons) are typically generated by a secondary carrier generation process, i.e., via either chemical or photo doping processes. The principle of a photo doping is that an orbital energy offset (δE of either D-LUMO/A-LUMO or D-HOMO/A-HOMO) between a donor/acceptor binary pair constitutes a key driving force for an excited state interatomic electron transfer. The principle of a chemical doping is that an orbital energy offset (δE of D-HOMO/A-LUMO) or thermal perturbation constitutes a key driving force for a ground state interatomic electron transfer.

EXERCISE QUESTIONS

1. Why an electron can be treated as both a particle and a wave? What is the evidence to support the electron duality?
2. What are the main differences of an electron orbital versus the orbitals of the planets surrounding the Sun?
3. What is **k**-space? Why electron energy is plotted in **k**-space instead of the regular 3-D space?
4. What key assumptions or materials criteria are used to derive the electronic band model?
5. What are frontier orbitals and their relationships with electron bands?
6. What are the key differences of a graphene sheet versus a linear conjugated polymer in terms of electronic structures and properties?
7. Why charge transport or mobility is sensitive to the molecular packing or materials morphology?
8. How does thermal energy affect electron transfer processes?
9. How Fermi's Golden Rule and Marcus electron transfer model relate to each other? Are the two models applicable for intra-atomic interorbital electron transfers?
10. Why do many excitons favorably incur charge dissociation instead of the exciton decay at a donor/acceptor interface?
11. What are the common driving and counter-driving forces for electron transfers?

12. What are the relationships of Stokes shift versus the reorganization energies? In what conditions can Stokes shift be used to estimate the reorganization energies?
13. Why do triplet excitons have longer lifetime compared to singlet excitons?
14. What are the key differences of excitons in typical organic semiconductors compared to excitons in typical inorganic semiconductors?
15. Why do most inorganic semiconductors fit a primary photo carrier generation model, while most organic semiconductors fit a secondary photo carrier generation model?

LIST OF ABBREVIATIONS

A	acceptor
AEDL	average exciton diffusion length
AO	atomic orbital
BG	band gap
BLA	bond length alternation
BS	band size
BW	bandwidth
CB	conduction band
CBM	conduction band minimum
CV	cyclic voltammetry
D	donor
EA	electron affinity
E_g	energy gap
ESR	electron spin resonance
ET	electron transfer
FCWD	Frank–Condon-weighted density of states
FRET	Förster resonance energy transfer
HAO	hybrid atomic orbitals
HOMO	highest occupied molecular orbital
IP	ionization potential
IPES	inverse photoemission spectroscopy
LUMO	lowest unoccupied molecular orbital
MFD	mean free distance
MFP	mean free path
MFT	mean free time
MO	molecular orbital
MW	molecular weight
PA	polyacetylene
PPV	polyphenylenevinylene
SOMO	singly occupied molecular orbital
UPS	ultraviolet photoemission spectroscopy
VB	valence band
VBM	valence band maximum
XPS	x-ray photoemission spectroscopy

ACKNOWLEDGMENT

The author wishes to acknowledge and thank a number of funding agencies (particularly NASA, DoD, and NSF) for partial support of the research/educational activities related to the subject of this chapter. The author particularly wishes to thank Professor Henry A. Rowe (Chemistry), Professor

Igor V. Bondarev (Physics), and Dr. Demetrio Filho (Physics) for manuscript review, and Dr. Natasha Kirova for suggestions on bipolaron/polaron-pairs.

REFERENCES

1. Chiang, C.K., Druy, M.A., Gau, S.C., Heeger, A.J., Louis, E.J., MacDiarmid, A.G., Park, Y.W., and Shirakawa, H., Synthesis of highly conducting films of derivatives of polyacetylene, (CH)x, *J. Am. Chem. Soc.*, 100, 1013–1015, 1978.
2. Skotheim, T.A. and Reynolds, J.R., eds., *Handbook of Conducting Polymers*, 3rd edition, CRC Press: Boca Raton, Florida, 2007.
3. Nalwa, H.S., ed., *Handbook of Organic Electronics and Photonics*, American Scientific Publishers: Los Angeles, California, 2008.
4. Kwok, H.L., *Electronic Materials*, PWS: Boston, Massachusetts, 1997.
5. Laidler, K.J., Meiser, J.H., and Sanctuary, B.C., *Physical Chemistry*, 4th edition, Houghton Mifflin: Boston, Massachusetts, 2003.
6. Pierret, R.F., *Advanced Semiconductor Fundamentals*, 2nd edition, Prentice Hall/Pearson Education: Upper Saddle River, New Jersey, 2003.
7. Sze, S.M. and Ng, K.K., *Physics of Semiconductor Devices*, 3rd edition, Wiley Interscience: New York, 2006.
8. Brédas, J-L., Beljonne, D., Coropceanu, V., and Cornil, J., Charge-transfer and energy-transfer processes in π-conjugated oligomers and polymers: A molecular picture, *Chem. Rev.*, 104, 4971–5003, 2004.
9. Coropceanu, V., Cornil, J., Filho, D., Olivier, Y., Silbey, R., and Brédas, J-L., Charge transport in organic semiconductors, *Chem. Rev.*, 107, 926–952, 2007.
10. Ege, S., *Organic Chemistry*, 5th edition, Houghton Mifflin: Boston, Massachusetts, 2004.
11. Jortner, J. and Bixon, M., eds., *Electron Transfer—From Isolated Molecules to Biomolecules, Part 1, Advances in Chemical Physics,* Vol. 106, John Wiley and Sons, Inc: New York, 1999.
12. Pourtois, G., Beljonne, D., Cornil, J., Ratner, M., and Brédas, J-L., Photoinduced electron transfer processes along molecular wires based on phenylenevinylene oligomers: A quantum-chemical insight, *J. Am. Chem. Soc.*, 124, 4436–4447, 2002.
13. Arkhipov, V. and Bässler, H., Exciton dissociation in conjugated polymers, in *Handbook of Luminescence, Display Materials, and Devices*, Vol. 1, eds. Nalwa, H. and Rohwer, L., American Scientific Publishers: Los Angeles, California, 2003, Chapter 5, pp. 279–342.
14. Barford, W., *Electronic and Optical Properties of Conjugated Polymers*, Clarendon Press: Oxford, United Kingdom, 2005.
15. Sun, S. and Sariciftci, N., eds., *Organic Photovoltaics: Mechanisms, Materials and Devices*, CRC Press/Taylor & Francis: Boca Raton, Florida, 2005.
16. Sun, S., Organic and polymeric solar cells, in *Handbook of Organic Electronics and Photonics*, ed., Nalwa, S.H., American Scientific Publishers: Los Angeles, California, 2008, Vol. 3, Chapter 7, pp. 313–350.
17. Walzer, K., Maennig, B., Pfeiffer, M., and Leo, K., Highly efficient organic devices based on electrically doped transport layers, *Chem. Rev.*, 107, 1233–1271, 2007.

4 Charge Transport in Conducting Polymers

Vladimir N. Prigodin and Arthur J. Epstein

CONTENTS

4.1 Introduction 88
4.2 Overview of Three-Dimensional Charge Transport 88
4.2.1 Band Transport in Metals 89
4.2.1.1 Drude Metal 89
4.2.1.2 Drude Response of Quantum Particles 92
4.2.2 Hopping Conductivity in Insulators 95
4.2.2.1 Master Equation 95
4.2.2.2 Variable Range Hopping in Isotropic Systems 98
4.3 Electronic Polymers as Low-Dimensional Systems 103
4.3.1 Interchain Interaction 103
4.3.2 Variable Range Hopping Conductivity in Low Dimensions 104
4.3.2.1 Hopping in Single Chain 104
4.3.2.2 Quasi-1D System 106
4.3.2.3 Quasi-1D Fractal 107
4.3.2.4 Dielectric Properties of Poorly Conducting Polymers 111
4.3.3 Metallic State of Electronic Polymers 115
4.3.3.1 Metallic Response of Highly Conducting Polymers 115
4.3.3.2 Model of Resonant Transport 117
4.3.3.3 Average vs. Typical Transmission 120
4.3.3.4 Discussion 121
Exercise Questions 125
References 125

Abstract The electrical properties of conjugated polymers such as doped polyaniline and doped polypyrrole vary widely depending on their synthesis and processing route, doping level, etc. The variety demonstrates itself in the strong frequency and temperature dependencies of conductivity. These dependencies reflect the low dimensionality of a conducting network in the polymers. The network is formed by fast intrachain diffusion and rare interchain hops of charge carriers. Depending on polymer morphology the network is modeled by a quasi-one-dimensional system of randomly coupled chains or a fractal chain with dimensionality $1 + s$. We briefly discuss the experimental results for dielectric constant and conductivity in poorly conducting polymers. To describe the properties of doped polymers in the metallic state, we use the models of chain-linked granular metal.

4.1 INTRODUCTION

In the undoped state, the conjugated polymers are semiconductors with the Peierls–Mott energy gap determined by the internal structure of polymer repeat [1]. Their room temperature conductivities due to residual charge carriers are similar, ranging typically from 10^{-5} S/cm for undoped *trans*-polyacetylene [2–4] to 10^{-10} S/cm for undoped *cis*-polyacetylene [5] and the undoped emeraldine base form of polyaniline [6,7]. Their conductivities as a function of frequency and temperature vary widely, reflecting a variety of different charge conduction processes [3,4,7].

In highly doped polymers (we consider mostly these cases), there is a finite density of states at the Fermi level and these doped polymers were anticipated to be regular conductors. However, roughly depending on their room temperature conductivities (σ_{RT}) at low temperatures the heavily doped polymers are insulators, metals, or are in the critical regime. Poorly conducting, heavily doped samples (typically 1 dopant per 10–20 carbon atoms in the conjugated chain) with σ_{RT} of the order of or less than a few Siemens per centimeter have behavior that can be classified as "dielectric" [8–10], i.e., their conductivity decreases with decreasing temperature in some exponential manner [6]. In doped polymers with moderate σ_{RT} (of the order of tens of Siemens per centimeter) the conductivity continues to decrease with decreasing temperature [11]. However, the decay now follows a power law in a high temperature interval and, therefore these samples are near the insulator–metal transition (IMT). For many samples with σ_{RT} of the order of or >100 S/cm, the temperature and frequency dependencies of conductivity are close to being metallic and their conductivity is finite down to millikelvin temperatures [12,13].

An early point of view of the nature and mechanism of transport in doped polymers [14,15] assumed that the polymer structure can be ignored and the polymers behave similar to amorphous semiconductors and dirty metals. Following this analogy, at low doping levels charge carriers are bound by the fluctuations of the background potential and the charge transport is represented by phonon-assisted hops over localized states. This type of conduction is described by the Mott variable range hopping (VRH) model [16]. In more ordered samples of heavily doped polymers the localization radius increases and the transition from the disorder-induced Anderson insulator to metallic state happens. This IMT can be described by the Mott–Anderson model [17].

The above point of view prevailed since the earlier studies of conducting polymers [18,19]. With significant advances in conductivity of doped polymers, an alternative view on the charge transport in polymers, which exploits the chain morphology of polymers by considering them as low-dimensional structures, started to gain more attention [8,9,20,21]. The comprehensive experimental data obtained for the more conducting polymers does not follow the usual 3D models. There were observed effects which cannot be explained within conventional 3D models of charge transport. Before proceeding with the discussion of the details of charge transport in polymers, we remind the reader of the basic models of charge transport in conventional 3D solids (Section 4.2.2). Then, we describe how the charge transport is modified in low-dimensional systems and in parallel, we discuss the experimental data for organic materials and how the data for conducting polymers are consistent with these modifications.

4.2 OVERVIEW OF THREE-DIMENSIONAL CHARGE TRANSPORT

Electronic states in solids with a regular structure are grouped into bands [22]. At ideal structural order the band wave functions are spread over the whole sample and describe electrons that propagate over the sample with a fixed velocity. Disorder (impurities, irregularities, etc.) violates the ideal order and can lead to localization of a portion of the band states. The wave functions of localized states are concentrated in restricted spacial volumes and correspond to electrons trapped by disorder-induced fluctuations of the background potential. According to Mott and Anderson [23] in the presence of disorder the band states are split into extended and localized states as is shown in Figure 4.1. The boundary energy, E_c, separating localized and extended states is termed as the mobility edge.

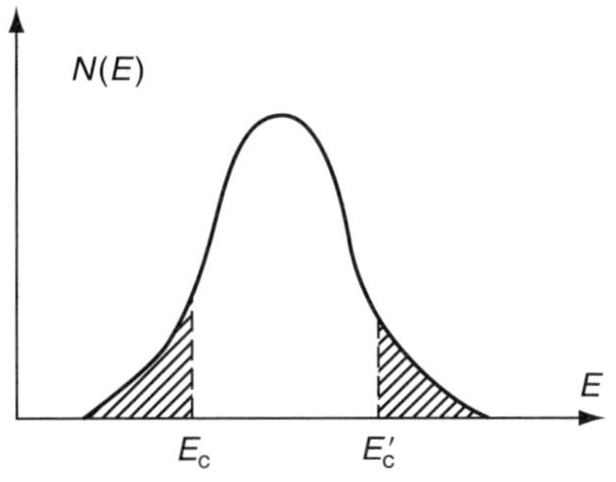

FIGURE 4.1 Density of band states in disordered solid. Dashed areas correspond to localized states and blank area means delocalized states. The boundary energies E_c and E_c' separating localized and extended states represent the mobility edges.

Filling electronic states with electrons is governed by the Pauli exclusion principle: one electron per state. As a result, at $T = 0$ all states below the Fermi energy E_F are occupied, while the states above E_F are empty. The position of E_F is determined by the electron concentration. For the case $E_F > E_c$ the system is an electric conductor while for $E_F < E_c$, all available electrons are localized and the system is an insulator, at least at $T = 0$. However, at nonzero temperatures electrons interacting with phonons can leave localized states and there is nonzero phonon-assisted conductivity in the insulators [16,24,25]. When E_F crosses E_c, the IMTs occur and the conductivity demonstrates critical behavior as a function of temperature, frequency, doping, etc. [17].

Thus, there are two fundamental mechanisms of charge carrier transport in solids: band transport which is realized in metals, and hopping transport which takes place in dielectrics with localized charge carriers. These two types exhibit entirely different temperature, frequency, and electric field dependencies. Here, we briefly describe the basic electric response of both conventional 3D conductors and insulators. As already discussed, organic materials due to their morphology have reduced dimensionality [26]. How hopping and band transport occur in low-dimensional systems, and the comparison of these laws with experiments, are discussed in the following sections.

4.2.1 Band Transport in Metals

4.2.1.1 Drude Metal

The behavior of band electrons mimics that of free particles [27], therefore it is sensible to consider first the electric response of free, classical charged particles in an external electric field, **F**, in the presence of random scattering. During the time interval between scattering the electron motion is described by Newton's law

$$m\frac{\mathrm{d}}{\mathrm{d}t}\mathbf{v} = e\mathbf{F}. \tag{4.1}$$

Therefore, the electron velocity evolves with time as $\mathbf{v}(t) = \mathbf{v}_0 + (e/m)\mathbf{F}t$, where $\mathbf{v}_0$ is the velocity of the electron after its last collision at $t = 0$. This evolution holds so long as the electron does not collide with another random center. We define the mean time of free path, τ, as a parameter that characterizes the frequency of random scatterings. The probability density to avoid scattering starting with $t = 0$ up to time t, is given by $(1/\tau)\exp[-t/\tau]$, and the average velocity of the electron between collisions is

$$\int \mathbf{v}(t)\exp[-t/\tau]\,\mathrm{d}t/\tau = \mathbf{v}_0 + e(\tau/m)\mathbf{F}. \tag{4.2}$$

The direction of initial velocity $\mathbf{v}_0$ is random. Therefore, for a large number of electrons with all possible $\mathbf{v}_0$, the nonzero flow of electrons is due only to the second term in Equation 4.2.

As a result, the mobility of free electrons, defined as the result of averaging over all electrons, $\langle \mathbf{v} \rangle = \mu\mathbf{F}$, is $\mu = e\tau/m$. The conductivity, written as $\sigma = e\mu n$, is then given by the equation

$$\sigma = \frac{e^2 n\tau}{m}, \tag{4.3}$$

where n is the electron concentration. The conductivity can also be represented in the following form:

$$\sigma = \frac{\omega_\mathrm{p}^2\tau}{4\pi}; \quad \omega_\mathrm{p}^2 = 4\pi\frac{e^2 n}{m}, \tag{4.4}$$

where instead of n the plasma frequency, ω_p, is introduced.

The generalization of the present consideration for a time-varying electric field enables us to get the frequency-dependent conductivity in the form

$$\sigma(\omega) = e^2\frac{n\tau}{m}\frac{1}{1 - i\omega\tau}. \tag{4.5}$$

Equations 4.3 through 4.5 represent the well-known Drude formula for the electronic conductivity in terms of the mean free path time. The Drude formula [22,27] enables us to outline the basic features of electric response in metals [28].

In general, scattering of electrons is produced by any deviation from periodicity of the background electric potential. It includes the impurities as well as the thermal vibrations of the host lattice. So the total scattering rate can be represented by two contributions

$$1/\tau = 1/\tau_\mathrm{imp} + 1/\tau_\mathrm{ph}, \tag{4.6}$$

where

τ_imp is the scattering time due to static disorder
τ_ph is the scattering time by lattice vibrations

The phonon scattering rate is temperature-dependent and this dependence can be approximated as $1/\tau_\mathrm{ph}(T) \sim T^\gamma$ at low temperatures and $1/\tau_\mathrm{ph}(T) \sim T$ at high temperatures. Therefore, the conductivity decreases with increasing temperature as a result of more intensive phonon scattering [22]. There is a characteristic temperature above which the phonon scattering dominates over the impurity one. Below this characteristic temperature the phonon scattering rate becomes weaker than the impurity one, and the conductivity approaches a residual conductivity with decreasing temperature. This zero-temperature conductivity entirely is determined by the impurity scattering τ_imp.

Considering frequency dependence, the Drude conductivity, Equation 4.5 can be split into the real and imaginary components. The real part

$$\Re\sigma(\omega) = \frac{\sigma_{dc}}{1+(\omega\tau)^2} \tag{4.7}$$

describes the absorption of electric field by the electronic system. $\Re\sigma(\omega)$ decreases with increasing frequency and this behavior is a characteristic feature of free electron response. For band transport the absorption of the external field at $\omega \gg 1/\tau$ is limited by the rate at which the electronic subsystem transfers its momentum to the environment. With increasing frequency of the external field, the electron motion becomes spatially restricted and therefore the scattering rate of electrons, as well as absorption, decreases with increasing frequency [28].

The imaginary part of ac-conductivity is related to the polarization properties of the electronic system. The dielectric constant $\varepsilon(\omega)$ is determined by the equation

$$\varepsilon(\omega) - 1 = -4\pi\frac{\Im\sigma(\omega)}{\omega} = \frac{-\omega_p^2\tau^2}{1+(\omega\tau)^2}, \tag{4.8}$$

and is negative at low frequency because of $\omega_p\tau \gg 1$. This fact means the electric current oscillations in time lag the electric field modulations or in other words, the displacement current is out-phase with the external electric field. In other words, the negative dielectric constant reflects inertia of free particles. With increasing frequency, the dielectric constant (Equation 4.8) increases taking into account that $\omega_p\tau \gg 1$ the dielectric constant crosses zero at $\omega = \omega_p$.

For comparison, we also present the ac-response of bound electrons. We assume that the electrons are bound near the fixed centers by a harmonic potential so that the electron displacement is described by the equation

$$\frac{d^2}{dt^2}\mathbf{x} = -\frac{1}{\tau}\frac{d}{dt}\mathbf{x} - \omega_0^2\mathbf{x} + \frac{e\mathbf{F}(t)}{m}, \tag{4.9}$$

where

τ is the relaxation time

ω_0 is the frequency of harmonic oscillations of electrons near the fixed centers

Assuming the electric field $\mathbf{F}(t)$ varies as $\Re F_\omega \exp(i\omega t)$, we find that the electron displacement is $\Re x_\omega \exp(i\omega t)$, where

$$x_\omega = -\frac{eF_\omega}{m}\frac{1}{\omega^2 - \omega_0^2 + i\omega/\tau}. \tag{4.10}$$

The contribution of each electron to the current is equal to $\Re J_\omega \exp(i\omega t)$, where $J_\omega = i\omega e x_\omega$ and the conductivity is

$$\sigma = e^2\frac{n\tau}{m}\frac{i\omega}{(\omega^2 - \omega_0^2)\tau + i\omega}. \tag{4.11}$$

In the limit of free electrons ($\omega \gg \omega_0$), Equation 4.11 reproduces the Drude formula (Equation 4.7). In the opposite case, $\omega \ll \omega_0$, the conductivity is

$$\sigma = \frac{e^2 n}{k}(-i\omega). \tag{4.12}$$

Here, $k = m\omega_0^2$ is the spring constant so that $\alpha = e^2/k$ is the polarization of a bound electron. The result (Equation 4.12) corresponds to a dielectric with the following dielectric constant in the limit of zero frequency:

$$\varepsilon(\omega = 0) = 1 + 4\pi\alpha n, \tag{4.13}$$

where n is the concentration of bound electrons.

The dc-transport in insulators occurs if electron transitions happen between different localized states. Such transitions are possible at nonzero temperatures due to thermal phonons that can match the energy deficit in electron transfers. This type of conductivity is considered in the Section 4.2.2.

4.2.1.2 Drude Response of Quantum Particles

We briefly outlined how to come to the Drude formula from first principles. In general, the electron dynamics obeys the Schrödinger equation,

$$i\hbar\frac{\partial}{\partial t}\Psi = H\Psi, \tag{4.14}$$

where the Hamiltonian, H, includes all possible interactions as well as external fields. The turning point in the study of the Schrödinger equation is the conjecture that the internal interactions of electrons provide a spacial scale such as the phase breaking length l_φ. The electron motion on lengths which are shorter than l_φ is quantum-mechanical or coherent while the electron motion over larger scales than l_φ can be described as classical [17,29]. In other words, l_φ is the shortest length within which the electron maintains memory about the phase of the wave function or the phase interference between different electronic processes is essential within spacial scale l_φ. Beyond the distance l_φ, accumulated uncertainty in the phase of the electron wave function exceeds 2π. Therefore, description of electronic motion over the scales larger than l_φ is given in terms of only modulus of wave function, i.e., electron density.

The phase variable of electron wave function may be washed out by interaction with phonons because collisions with phonons introduce unpredictable and nonreproducible phase shifts. The other sources of phase breaking are electron–electron interactions and the magnetic impurities, which transfer the electron into another energy and spin state. The magnetic field leads to an additional change of phase which depends on the direction of electron motion. Therefore, the application of magnetic fields suppresses in part the phase interference. It is important to stress that the scattering by static impurities does not destroy the phase variable and, namely the interference of impurity scattering leads to the phenomenon of weak localization.

Introducing l_φ enables one to eventually replace the Schrödinger equation on length scales larger than l_φ with the Boltzmann equation [27]

$$-\frac{\partial}{\partial t}f = \mathbf{v}(\mathbf{k})\nabla_{\mathbf{r}}f + (e\mathbf{F}/m)\nabla_{\mathbf{k}}f + \hat{S}f, \tag{4.15}$$

where

$f = f(\mathbf{r},\mathbf{k})$ is the distribution function of electrons with momentum $\mathbf{k}$ at a point $\mathbf{r}$

$\mathbf{v} = \nabla_{\mathbf{k}}E(\mathbf{k})$ and $E(\mathbf{k})$ are the electron dispersion

$\hat{S}$ is the collision operator that describes the transformation of the distribution function by the scattering processes

Equation 4.15 can be simplified if the external electric field is weak so that the function f weakly deviates from the equilibrium distribution f_0 given by the Fermi function

$$f_0(E) = \frac{1}{1 + \exp[(E - \zeta^*)/k_B T]}, \tag{4.16}$$

where $E = E(\mathbf{k})$ and ζ^* are the chemical potential for electrons determined by the electron concentration. The collision operator $\hat{S}$ in Equation 4.15 is responsible for the evolution of the system to the equilibrium state. The τ-approximation, used to describe this evolution, is given by the equation

$$\hat{S}f = \frac{1}{\tau}(f - f_0). \tag{4.17}$$

As a result, the Boltzmann equation (Equation 4.15) is reduced to the usual conservation law for a number of particles

$$-e\frac{\partial}{\partial t}n = \nabla \mathbf{J}, \tag{4.18}$$

with the drift–diffusion representation for the current

$$\mathbf{J} = (\sigma/e)\nabla\zeta; \quad \zeta = \zeta^* - e\varphi, \tag{4.19}$$

where

- ζ is the electrochemical potential
- φ is the electrical potential, $\mathbf{F} = -\nabla\varphi$
- n is the electron concentration and σ is the conductivity, which are related to each other with the Einstein equation

$$\sigma = en(\zeta^*)\mu = e^2 N(\zeta^*)D; \quad N(\zeta^*) = \frac{\partial}{\partial \zeta^*}n(\zeta^*); \quad n(\zeta^*) = \int \frac{d^3\mathbf{k}}{(2\pi)^3} f_0(E(\mathbf{k})), \tag{4.20}$$

where

- μ is the mobility
- D is the diffusion coefficient
- $N(\zeta^*)$ is the thermodynamic density of states

Equations 4.18 and 4.19 should be added by the Poisson equation

$$\varepsilon_0 \Delta\varphi = 4\pi(en + \rho), \tag{4.21}$$

with the appropriate boundary conditions for finding the electrical potential. In Equation 4.21, ρ is the immobile charge of the host lattice. Equation 4.21 treats electron–electron Coulomb interaction in a self-consistent manner. Equations 4.18 through 4.21 are used for simulation of electronic devices. Usually the electrical current is assumed to be fixed and it is required to find the voltage drop across a sample.

In application to metals the above parameters are specified to be $\zeta^* = E_F >> k_BT$ and

$$\sigma = 2e^2N(E_F)D; \quad D = \tau v_F^2/d; \quad N(E_F) = \int \frac{d^3\mathbf{k}}{(2\pi)^3}\delta(E_F - E(\mathbf{k})). \tag{4.22}$$

Turning to the Poisson equation (Equation 4.21) the screening length λ for metals can be found to be

$$1/\lambda^2 = (e^2/\varepsilon_0)2N(E_F). \tag{4.23}$$

Taking that into account, for metals $N(E_F) \sim n/E_F$, $n \sim 1/a^3$, and $e^2/(\varepsilon_0 a) \sim E_F$ we find that the screening length λ is on the order of lattice constant a. Thus any fluctuation of charge in metals is screened at the atomic scale, and according to Equation 4.21 the electric field inside the bulk remains constant.

Equations 4.19 through 4.23 point out that in metals only electrons at the Fermi level participate in charge transport, therefore the scattering time and velocity in the Drude equation (Equations 4.3 through 4.5) refer only to the Fermi electrons. Then the formula for the plasma frequency in Equation 4.4 is

$$\omega_p^2 = 8\pi e^2N(E_F)v_F^2/d. \tag{4.24}$$

So far the electron scattering is described in terms solely of scattering time. To get a microscopic expression, scattering time requires the specification of scattering. For elastic scattering by impurities, the collision operator in Equation 4.15 reads [27]

$$\hat{S}f = c_{imp}\int W(\mathbf{k},\mathbf{k}')[f(\mathbf{k}) - f(\mathbf{k}')]\,d\mathbf{k}', \tag{4.25}$$

where

c_{imp} is the concentration of impurities

$W(\mathbf{k},\mathbf{k}')$ is the probability of scattering for an electron with momentum $\mathbf{k}$ into the state with momentum $\mathbf{k}'$

For scattering by an impurity placed at point $\mathbf{r}_0$ with potential $V(\mathbf{r} - \mathbf{r}_0)$, W is given by the Fermi golden rule

$$W(\mathbf{k},\mathbf{k}') = \frac{2\pi}{\hbar}|V(\mathbf{k},\mathbf{k}')|^2\,\delta(E(\mathbf{k}) - E(\mathbf{k}')). \tag{4.26}$$

Here $V(\mathbf{q})$ is the Fourier transform of the scattering potential. For an isotropic electronic gas, scattering changes only the direction of electron propagation, therefore the scattering time is found to be

$$1/\tau = \pi N(E_F)c_{imp}\int d\theta\,|V(\theta)|^2(1 - \cos(\theta)), \tag{4.27}$$

where θ is the angle between $\mathbf{k}$ and $\mathbf{k}'$ in Equation 4.26.

Thus the above consideration proves that the Drude model for a metal is adequate provided that the de Broglie length of the electrons, λ, remains shorter than the mean free path, $l = v_F\tau$. In this case, we may speak about electrons as free particles which experience random scattering [29]. At $\lambda \le l$, such a description no longer makes sense and such electrons are expected to be localized. The above criterion for localization means the mobility edge E_c is on the order of $\hbar/\tau$. As the mobility edge E_c approaches the Fermi level the conductivity tends to be zero as $\sigma \approx (E_F - E_c)^\nu$, where ν is the scaling index [17]. When the Fermi energy is below E_c the system remains dielectric and its conductivity is zero at zero temperature. At nonzero temperatures electrons at the Fermi level, interacting with thermal phonons, can jump from one localized state to another and thereby provide the charge transport. This type of conductivity is discussed in Section 4.2.2.

4.2.2 Hopping Conductivity in Insulators

4.2.2.1 Master Equation

Hopping is a general mechanism of charge transport in disordered materials with localized charge carriers. A representative of such materials is a doped semiconductor where electrons hop among impurity states whose levels are inside the energy gap. Another example is different types of glassy materials such as amorphous semiconductors and dirty metals. The localization of carriers in this case is provided by large fluctuations of the background potential. In general, the calculation of hopping conductivity is a difficult problem [16,24,25,30], because localized states over which the electrons hop are randomly distributed in space and their energy levels are also random.

For polymers, hopping represents one of the basic mechanisms of charge transport because of their irregular structure, and as a consequence of these irregularities the localization of charge carriers occurs. Also due to the softness of the polymer backbone, the charge carriers induce the local polarization that also favors localization of charge carriers. In crystalline molecule materials, the polaronic effect is a factor that limits the charge transport [31]. At high enough temperatures and low electric fields the motion of electrons represents the activated hops among polaronic states while at low temperature and strong fields the band description for polarons is adequate [25]. Our consideration is concentrated on noncrystalline organic materials, where transport of electrons or polarons is controlled by disorder.

We start the consideration of hopping with a formulation of the problem. Let us suppose all the electron states to be localized. The transitions between the states (sites) are realized by phonon-assisted hops. The dynamics of this kind of charge transport is governed by the balance equation (master equation)

$$\frac{\mathrm{d}}{\mathrm{d}t}P_i = \sum_g \left[(1-P_i)w_{i,i+g}P_{i+g} - (1-P_{i+g})w_{i+g,i}P_i\right], \tag{4.28}$$

where i numerates the sites and $P_i(t)$ is the probability to find the electron at time t at a site i. The transition rate $w_{i+g,i}$ for a transfer of electron from site i to site $i+g$ per unit of time depends on the distance between the two sites as well as the distance on their energies, $w_{i+g,i} = w(|\mathbf{r}_{i+g} - \mathbf{r}_i|; E_{i+g}, E_i)$. The explicit dependence can be established by the Fermi golden rule. Multiphonon or polaronic effects may be important in the calculation of the hopping rate and were considered in a number of works [25,32–34]. For the purpose of our analysis, it is enough to consider common properties of hopping rates.

Phonons match the energy deficit for electron transitions and eventually lead to the equilibrium occupation of electronic states. The temperature dependence of transition rates should satisfy the requirement so that at the equilibrium occupation the right part of Equation 4.28 is identical to zero, i.e.,

$$\frac{w_{i+g,i}}{w_{i,i+g}} = \frac{(1-P_i^0)P_{i+g}^0}{(1-P_{i+g}^0)P_i^0}, \tag{4.29}$$

where P_i^0 is the equilibrium occupation function for a site i determined by its chemical potential ζ_i^* and energy E_i

$$P_i^0 = \frac{1}{1+\exp\left[\frac{E_i - z_i^*}{k_B T}\right]}. \tag{4.30}$$

Therefore, the principle of detail equilibrium (Equation 4.29) dictates

$$\frac{w_{i+g,i}}{w_{i,i+g}} = \exp\left[-\frac{E_{i+g} - E_i}{k_B T}\right]. \tag{4.31}$$

An external electric field **F** can be taken into account by shifting the site energy $E_i \rightarrow E_i - e(\mathbf{Er_i})$ with e being the charge of the carrier. According to Equation 4.31, in the presence of an electric field the relationship between transfer rates is modified to read

$$\frac{w_{i+g,i}(F)}{w_{i,i+g}(F)} = \exp\left[-\frac{E_{i+g} - E_i}{k_B T}\right] \exp\left[\frac{e\mathbf{F}(\mathbf{r}_{i+g} - \mathbf{r}_i)}{k_B T}\right]. \tag{4.32}$$

In this way, i.e., via charge transfer rates, the electric field enters into the master equation (Equation 4.28).

Applying an external electric field to the equilibrium system leads to the redistribution of its charge. At weak external field, the redistribution can be described by a shift $\delta\zeta_i^*$ of the local chemical potential ζ_i^* in the corresponding equilibrium function. Assuming that these field-induced shifts remain small, it can be written as

$$P_i = P_i^0(\zeta_i^* + \delta\zeta_i^*) = P_i^0(\zeta_i^*) + \frac{dP_i^0(\zeta_i^*)}{d\zeta_i^*}\delta\zeta_i^*, \tag{4.33}$$

where $dP_i^0(\zeta_i^*)/d\zeta_i^*$ is the local thermodynamic density of states at site i equal to

$$\frac{dP_i^0(\zeta_i^*)}{d\zeta_i^*} = \frac{(1 - P_i^0(\zeta_i^*))P_i^0(\zeta_i^*)}{k_B T}. \tag{4.34}$$

Treating Equation 4.28 within a linearized approximation with respect to $\delta\zeta_i^*$ and to external electric field F we come to the analog of the drift–diffusion equation 4.19 for the hopping transport

$$\frac{dP_i}{dt} = \sum_g \hat{w}_{i,i+g} \frac{\zeta_i - \zeta_{i+g}}{k_B T}. \tag{4.35}$$

Here $\hat{w}_{i,i+g} = \hat{w}_{i+g,i}$ is the hopping frequency that determines the equilibrium current from a site $i + g$ to a site i therefore

$$\hat{w}_{i,i+g} = (1 - P_i^0)w_{i,i+g}P_{i+g}^0 = (1 - P_{i+g}^0)w_{i+g,i}P_i^0 = \hat{w}_{i+g,i}. \tag{4.36}$$

In Equation 4.35 ζ_i represents the electrochemical potential

$$\zeta_i = \delta\zeta_i^* - e\varphi_i; \quad \varphi_i = -(\mathbf{Fr}_i). \tag{4.37}$$

In the stationary case, the left part of Equation 4.35 is zero and, in fact, Equation 4.35 represents Kirchhoff's law for an electrical circuit, where the conductance between sites i and $i+g$ is given by

$$G_{i+g,i} = e^2 \frac{\hat{w}_{i+g,i}}{k_B T}. \tag{4.38}$$

This type of description of the master equation in terms of an electric circuit scheme is known as the Miller–Abrahams network [35].

In the nonstationary case, the left part of Equation 4.35 is not zero and represents the rate of charge accumulation at site i. In an equivalent electric circuit scheme, the charge accumulation can be described by capacitors. As a result, in the ω-representation, the equivalent electric scheme for Equation 4.35 reads (see Figure 4.2)

$$i\omega C_i(u_i - \varphi_i) = \sum_g G_{i+g,i}\left[u_{i+g} - u_i\right], \tag{4.39}$$

where C_i is the capacitance at a site i

$$C_i = e^2 \frac{dP_i^0(\zeta_i^*)}{d\zeta_i^*} = \frac{e^2}{k_B T} P_i^0(1 - P_i^0). \tag{4.40}$$

In Equation 4.39 $u_i = \zeta_i/e$ is the electrochemical potential in units of electric potential. In Equation 4.37, φ is the electric potential determined by, e.g., the self-consistent Poisson equation (Equation 4.21).

Equations 4.28 and 4.39 should be completed by boundary and initial conditions. In numerical simulation, the current through the sample is fixed and the potential drop across the sample is calculated. Analysis of Equations 4.28 and 4.39 requires the spatial distribution of hopping sites

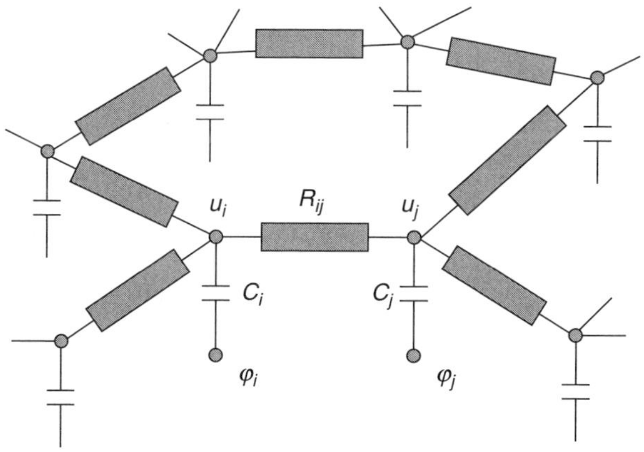

FIGURE 4.2 Equivalent electric circuit scheme for hopping network. The parameters are defined in the text.

and their energy. For disordered systems the assumption about random distribution of the localized states in space and over energy is used. In the case of a degenerate electron gas it leads to the so-called VRH considered in the Section 4.2.2.2.

Concluding this part, we present the exact solution of Equation 4.39 for the network of two coupled sites 1 and 2. Defining the current as

$$I = \frac{\mathrm{d}P}{\mathrm{d}t}; \quad P = e(P_1 r_1 + P_2 r_2), \tag{4.41}$$

where P is the dipole moment. Inserting Equation 4.39 into Equation 4.41, we find that $I(t) = \Re I_\omega \exp(-i\omega t)$, at the applied voltage $V_1 - V_2 = \Re V_\omega \exp(-i\omega t)$, where

$$I_\omega = G_\omega V_\omega; \quad G_\omega = -i\omega a \frac{G}{G/C - i\omega}. \tag{4.42}$$

Here $a = r_1 - r_2$, $1/C = 1/C_1 + 1/C_2$, and $G = G_{12}$.

G_ω in Equation 4.42 is the frequency-dependent conductance of the coupled pair of sites. The real part of G_ω determines the absorption coefficient and is

$$\Re G_\omega = \omega a \frac{G\omega}{(G/C)^2 = \omega^2}. \tag{4.43}$$

As a function of intersite hopping frequency, or G, the absorption coefficient exhibits the maximum at $G/C \sim \omega$ and at the maximum the absorption coefficient is

$$\Re G_\omega = \omega C a/2. \tag{4.44}$$

The maximum is due to the fact that for pairs with hopping frequency $G/C < \omega$, the electronic occupations are not able to follow the variation of external electric field and therefore their rate of absorption diminishes with decreasing hopping frequency. For pairs with higher frequency $G/C > \omega$, the rate of absorption is high, but such pairs reach equilibrium very fast and stay in their state for the majority of the time; therefore their absorption decreases with increasing hopping frequency.

The absorption within two site clusters is known as Debye losses [36]. As we discuss in the next section, the Debye losses prevail at the high frequency when only the transitions between the close states are realized.

4.2.2.2 Variable Range Hopping in Isotropic Systems

For large dimensionality ($d > 1$) the Miller–Abrahams network represents the connection of random conductances in parallel. Therefore, the observable conductance is determined by the average conductance and, thus, the problem is reduced to finding optimal intersite hopping frequency. For a degenerate electron gas with $E_F \gg k_B T$ the hopping rate w from one localized state j to the other i per unit time, including Fermi occupation numbers p_i, is approximated by the formula [37]:

$$\hat{w} = w_{ij} p_j (1 - p_i) = \nu_{\text{ph}} \exp\left(-2 f_{ij}\right); \tag{4.45}$$

$$f_{ij} = \frac{\left|E_j - E_i\right| + \left|E_j - E_{\text{F}}\right| + \left|E_i - E_{\text{F}}\right|}{4k_{\text{B}}T} + \frac{\left|\mathbf{r}_j - \mathbf{r}_i\right|}{\xi},$$

where

ξ is the localization radius

ν_{ph} is the attempt-frequency, which is of the order of the characteristic phonon frequency, $\nu_{\text{ph}} = 10^{12}\ \text{s}^{-1}$

The optimal hopping frequency is given by a result of the trade-off between the spatial overlap and the activation energy. As is shown by Mott, the optimal hopping frequency, w_{h}, and hopping distance, r_{h} are [16]

$$w_{\text{h}} = \nu_{\text{ph}} \exp\left[-\left(\frac{T_0}{T}\right)^{\frac{1}{d+1}}\right]; \quad r_{\text{h}} = \xi \left(\frac{T_0}{T}\right)^{\frac{1}{d+1}}, \tag{4.46}$$

where T_0 is given by the equation

$$k_{\text{B}} T_0 = B_d [N(E_{\text{F}}) \xi^d]^{-1}, \tag{4.47}$$

with $N(E_{\text{F}})$ being the density of states at the Fermi level per one spin projection and B_d is a numerical constant that depends on dimension d, for example $B_3 = 21.2$ [24]. The conductivity is related to the above parameters w_{h} and r_{h} by the Einstein relation

$$\sigma = \frac{e^2 n}{k_{\text{B}} T} D; \quad D = r_{\text{h}}^2 w_{\text{h}}, \tag{4.48}$$

where

D is the diffusion coefficient

n is the concentration of electrons

For a degenerate electron gas $n = 2N(E_{\text{F}})k_{\text{B}}T$. Collecting all these together, we come to the Mott law [16,25]

$$\frac{\sigma_{\text{dc}}(T)}{\sigma_0} = \frac{\nu_{\text{ph}}}{k_{\text{B}} T_0} x^2 e^{-x}; \quad x = \left(\frac{T_0}{T}\right)^{\frac{1}{d+1}}, \tag{4.49}$$

where $\sigma_0 = e^2 \xi^{2-d}$.

It is useful to present the expression for the Mott conductivity in the form

$$\sigma_{\text{dc}}(T) = \nu_{\text{ph}} (\varepsilon_{\text{s}} - 1) x^2 e^{-x}, \tag{4.50}$$

where $\nu_{\text{ph}} = 10^{12}(1/s) \approx 1\ \text{S/cm}$ and we introduced the static relative dielectric constant, ε_{s}, as (see below)

$$\varepsilon_0(\varepsilon_s - 1) = 2e^2 N(E_F)\xi^2, \tag{4.51}$$

where ε_0 is the permittivity of free space. We also note that at high temperatures ($x \approx 1$) the hopping conductivity is

$$\sigma_\infty = \sigma_0 \frac{\nu_{ph}}{k_B T_0} = 2e^2 N(E_F) D_\infty = \nu_{ph}(\varepsilon_s - 1), \tag{4.52}$$

where $D_\infty = \xi^2 \nu_{ph}$ is the diffusion coefficient corresponding to hops over distance ξ with frequency ν_{ph}.

The frequency below which the regime of dc-conductivity (Equations 4.49 through 4.51) realized is much less than the optimal hopping frequency w_h from Equation 4.46. With increasing frequency the conductivity increases, and at $\omega > w_h$ the frequency-dependent part of conductivity exceeds $\sigma_{dc}(T)$. As we considered in the previous section, the main contribution to $\sigma_{ac}(\omega)$ is provided by electron transitions between the pairs of states, the hopping frequency between which is equal to the frequency of an applied electric field, ω. The absorption of electric field is most effective for these pairs (Debye losses [36]). On the basis of these qualitative arguments we can write the ac-conductivity in the form

$$\Re\,\sigma_{ac}(\omega) = 2e^2 N(E_F) D_\omega z_\omega; \quad D_\omega = \omega r_\omega^2, \tag{4.53}$$

where r_ω is the size of the pair, whose internal hoping frequency is equal to the frequency of the applied field, i.e., according to Equation 4.46,

$$r_\omega = \frac{\xi}{2} \ln \frac{\nu_{ph}}{\omega}. \tag{4.54}$$

The factor z_ω in Equation 4.53 is the probability for an electron at a given site to be in proximity to another site to form the appropriate pair. Taking into account Equation 4.54 we can write

$$z_\omega = N(E_F) k_B T r_\omega^d = \frac{T}{T_0} \ln^d \frac{\nu_{ph}}{\omega}. \tag{4.55}$$

Collecting all together we have

$$\frac{\Re\,\sigma_{ac}(\omega)}{\sigma_0} = \frac{\omega T}{k_B T_0^2} \ln^{d+2} \frac{\nu_{ph}}{\omega}. \tag{4.56}$$

The present equation is the Mott–Austin formula for the Debye losses of degenerate electrons in the localized phase [16]. The extra power of the logarithm in Equation 4.56 in comparison with the standard expression [16,25] for configuration disorder is due to the energy variable in the VRH problem.

Equation 4.56 can also be written as

$$\frac{\Re\,\sigma_{ac}(\omega)}{\sigma_\infty} = \frac{\omega}{\nu_{ph}} \frac{T}{T_0} \ln^{d+2} \frac{\nu_{ph}}{\omega}. \tag{4.57}$$

The proportionality of the absorptive part of conductivity to ω reflects the rate of absorption or relaxation by selected pairs. The proportionality to T is related to the number of electrons participating in optical transitions for a degenerate electron gas at $\omega < k_B T$. For $w > k_B T$ the factor T should be replaced by ω and the formula 4.57 reads

$$\frac{\Re\,\sigma_{ac}(\omega)}{\sigma_\infty} = \frac{\omega^2}{\nu_{ph} k_B T_0} \ln^{d+2} \frac{\nu_{ph}}{\omega}. \tag{4.58}$$

The logarithmic power $d+2$ in Equations 4.56 through 4.58 originates from the square of the matrix element of the dipole moment r_ω^2 and the phase volume, including the energy component.

In general, between the low-frequency region with conductivity close to its static value and the region of pair hops (Equation 4.56), there exists an intermediate frequency range, where transport is effectively performed within finite, but large size clusters [25]. However, the formula (Equation 4.56) matches with $\sigma_{dc}(T)$ from Equations 4.49 and 4.52 at $\omega \approx \omega_h$. This matching means that the frequency interval for the multihopping within isolated clusters (cluster approximation) is narrow enough for isotropic VRH. In fact, the conductivity is entirely described by the dc-regime (Equation 4.49) and pair hops (Equation 4.57).

We now turn to the imaginary part of conductivity in the regime of VRH. First of all, we recall that in the absence of hopping ($T = 0$) the conductivity is imaginary and is given by the equation

$$\sigma(\omega \to 0) = -i\omega 2e^2 N(E_F)\xi^2. \tag{4.59}$$

Hence the dielectric constant $\varepsilon(\omega)$ defined as

$$\varepsilon_0(\varepsilon(\omega) - 1) = -\frac{4\pi}{\omega} \Im\,\sigma(\omega) \tag{4.60}$$

in the static limit is given by Equation 4.51. This value ε_s is the result of polarization of individual localized states of spatial size ξ. Also the static dielectric constant can be represented in the form

$$\varepsilon_s - 1 = (\xi/\lambda)^2; \quad 1/\lambda^2 = (e^2/\varepsilon_0)2N(E_F), \tag{4.61}$$

where λ is the Debye radius [27]. For the system under consideration, λ is of the atomic scale, and at the same time ξ is expected to be larger than the Bohr radius; therefore, $\varepsilon_s \gg 1$.

At finite temperature, additional nearby states are polarized due to phonon-assisted charge exchange. In the regime of pair hops the states whose separation is less than r_w are polarized. From the above arguments that lead to Equation 4.56, we can write

$$\frac{\varepsilon(\omega) - 1}{\varepsilon_s - 1} = 1 + \frac{T}{T_0} \ln^{d+3} \frac{\nu_{ph}}{\omega}. \tag{4.62}$$

The pair approximation is restricted to the frequency region $w_h < \omega < \nu_{ph}$.

At $\omega \ll w_h$ the frequency-dependent part of the response is described by the hydrodynamic approximation. The system is considered to be a conductive medium with some inhomogeneous inclusions [25]. These inhomogeneities produce internal dynamic polarization. It suggests that for ac-conductivity there is the regular low-frequency expansion

$$\sigma(\omega) = \sigma_{dc}(T)\left[1 - \frac{i\omega}{w_h} + \cdots\right], \tag{4.63}$$

where w_h is given by Equation 4.46. From here at $\omega \ll w_h$ we find

$$\frac{\varepsilon(\omega) - 1}{\varepsilon_s - 1} = \left(\frac{T_0}{T}\right)^{\frac{2}{d+1}}. \tag{4.64}$$

Equation 4.64 is matched to Equation 4.62 in the region of frequencies $\omega \approx w_h$.

As it follows from Equations 4.62 to 4.64, the dielectric constant of isotropic dielectric systems as a function of temperature has its maximum at $T = T_m$, where

$$T_m = \frac{T_0}{\ln^{d+1} \frac{\nu_{ph}}{\omega}}. \tag{4.65}$$

The initial increase of dielectric constant with temperature (Equation 4.62), corresponds to the increase of the regions, which become accessible for electrons due to thermal activation and which are polarized by an applied field. However, at higher temperatures the separate parts of space become less isolated leading to leakage of charge between polarization regions. As a result, the internal polarization decreases with increasing temperature, Equation 4.64, reflecting the increase of electron mobility. Schematically, the temperature dependence of the dielectric constant for isotropic VRH is shown in Figure 4.3.

As a function of frequency, the dielectric constant remains frequency independent for $\omega \lesssim w_h$ and decreases with increasing frequency at $\omega \lesssim w_h$. We note that the value of ε in the limit $T \to 0$ and $\omega \to 0$ strongly depends on the order of limits. When temperature first goes to zero and then $\omega \to 0$, the dielectric constant tends to ε_s. Otherwise the dielectric constant diverges.

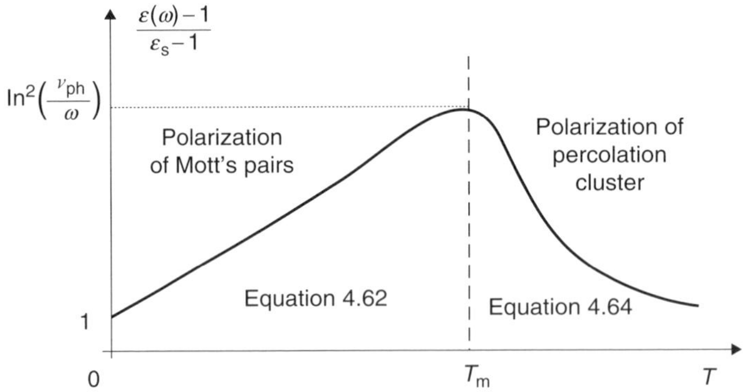

FIGURE 4.3 The temperature dependence of dielectric constant for isotropic variable range hopping ($d = 2,3$) with characteristic maximum. Initial linear increase of dielectric constant is due to polarization of closely located states. The decrease at high temperature reflects the decrease of charge inhomogeneity in the conducting state due to the increase of charge mobility with increasing temperature. The position of the maximum depends on the frequency (Equation 4.65). (Reprinted from Prigodin, V.N., Samukhin, A.N., and Epstein, A.J., *Synth. Met.*, 141, 155, 2004. Copyright 2004. With permission from Elsevier.)

4.3 ELECTRONIC POLYMERS AS LOW-DIMENSIONAL SYSTEMS

4.3.1 INTERCHAIN INTERACTION

As it was mentioned an early point of view of the nature and mechanism of transport in doped polymers [14,15] assumed that the polymers can be considered similar to amorphous semiconductors and dirty metals. With significant advances in conductivity of doped polymers, approach to the charge transport in polymers, which exploits the chain structures of polymers, was developed [8,9,20,21]. In contrast to amorphous semiconductors where single atoms have limited spatial correlation, in polymers the structural units are chains, which form disordered media. The neighboring atoms along chains are close to each other and arranged in a regular way, which may extend to hundreds of repeat units [38]. Therefore, electrons principally hop or tunnel along chains because of better overlapping between atoms of the same chain and due to their regular ordering. However, single noninteracting chains themselves cannot provide the macroscopic conductivity because of quantum localization caused by even weak disorder [23] and because of finite chain length. Therefore, transport properties of polymers are strongly controlled by interchain coupling [39,40].

Two models of interchain linkage were suggested for polymers. In the quasi-1D system of randomly interacting chains the interchain coupling is assumed to be by chaotically distributed links [41]. These can be due to direct overlap of orbitals of adjacent chains or bridges between the chains provided by dopants [42]. The interchain charge transfer is characterized by the interchain hopping frequency that is proportional to the concentration of links and to the rate of interchain hopping at individual bridges. As long as the effective interchain hopping rate remains substantially less than the intrachain one the system has specific 1D properties [43]. The system is 3D when the rate of interchain hops becomes comparable with intrachain hopping rate.

Another way to go beyond a 1D system is a fractal chain of dimensionality $d = 1 + s$ (see Figure 4.4). The fractal chain suggests a self-similar organization of chain links [44]. In a cube of size L, the chains form a set of bundles disconnected from each other. If for large enough L the number of chains in the maximum bundle is proportional to L^s, then the system can be identified as $(1 + s)$-dimensional. Obviously $s = 0$ for purely 1D systems (sets of uncoupled chains) and $s = 2$ for 3D systems.

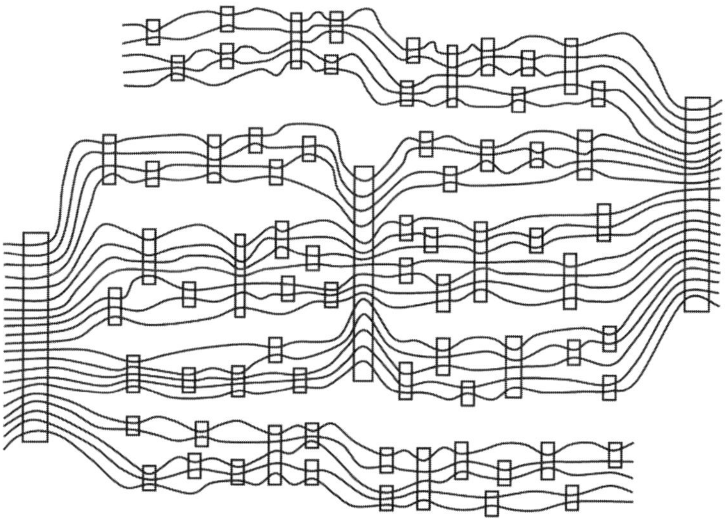

FIGURE 4.4 Schematic of chain fractal.

For both quasi-1D and fractal types of chain coupling the system is low dimensional and its transport preserves specific 1D features. The phenomenon of quantum localization by weak disorder in these systems was studied in Refs. [39,40,45].

In this section, we begun by considering the dielectric phase in this system at finite temperatures when there are thermally activated electronic transitions between localized states. We show that for the VRH mechanism of charge transport, the electric properties of quasi-1D and fractal chains are unique and differ from those of 3D networks [26]. We apply our results to explain the dielectric properties of poorly doped conducting polymers.

Later we discuss the metallic phase (finite conductivity at $T = 0$) in highly conducting polymers. It is pointed out that the electric properties of the metallic phase cannot be explained in terms of conventional Drude model used for 3D metals. We demonstrate that the adequate description can be obtained by taking into account the strong inhomogeneity of a conducting polymer. The model of chain-linked granular metal is consistent with experimental results.

4.3.2 Variable Range Hopping Conductivity in Low Dimensions

4.3.2.1 Hopping in Single Chain

In contrast to the 3D case, the VRH conductivity in the 1D system is determined not by the optimal hops but by the most difficult hops [37]. These hard hops are due to energy barriers that cannot be bypassed in the 1D case. The height of these barriers is of the order $3T_0/2$, and, hence, the frequency of overbarrier hopping is estimated to be [41]

$$\omega_1(T) = \nu_{\text{ph}} \exp\left(-\frac{3T_0}{2T}\right). \tag{4.66}$$

The barriers may be considered as having fixed heights at varying widths. These barriers can be overcome by tunneling through the barrier. With decreasing temperature, tunneling becomes preferable over activation, and, therefore the effective concentration $c(T)$ of barriers decreases as

$$c(T) = \frac{1}{\xi} \exp\left(-\frac{T_0}{2T}\right). \tag{4.67}$$

As a result, the dc-diffusion coefficient obeys the Arrhenius law

$$D = \frac{\omega_1}{c^2} = \xi^2 \nu_{\text{ph}} \exp\left(-\frac{T_0}{2T}\right), \tag{4.68}$$

and the dc-conductivity is written in the form

$$\sigma_{\text{dc}}(T) = \sigma_\infty \exp\left(-\frac{T_0}{2T}\right), \tag{4.69}$$

which strongly differs from the dependence that follows from Mott's argument, Equation 4.49 with $d = 1$.

Turning to ac-conductivity for 1D VRH, the essential point is that the hydrodynamic approximation is no longer valid. Besides the regular low-frequency expansion (Equation 4.63), there are irregular corrections in ω to the static conductivity [46]

$$\sigma_{ac}(\omega \lesssim w_1) = \sigma_{dc}(T)\left[1 - \sqrt{\frac{i\omega}{\omega_1}} + \cdots\right], \tag{4.70}$$

where ω_1 is given by Equation 4.66. It leads to the divergence of the dielectric constant at $\omega \to 0$

$$\frac{\varepsilon(\omega \to 0) - 1}{\varepsilon_s - 1} = \sqrt{\frac{v_{ph}}{\omega}} \exp\left(\frac{T_0}{4T}\right). \tag{4.71}$$

This singularity is a consequence of the slow spreading of the charge in 1D-disordered systems [46,25]. The behavior (Equation 4.71) holds as long as $\omega \lesssim \omega_1$, or $T \gtrsim T_{1\omega}$, where

$$T_{1\omega} = \frac{3T_0}{2\ln\frac{v_{ph}}{\omega}}. \tag{4.72}$$

At frequency $\omega \gtrsim \omega_1$ or $T \lesssim T_{1\omega}$ the electron motion is restricted to hops inside chain segments separated by energy barriers and the ac-conductivity is found to be [41]

$$\frac{\sigma_{ac}(\omega \gtrsim \omega_1)}{\sigma_0} = -\frac{i\omega}{T_0}\ln^2\frac{v_{ph}}{\omega}\left(\frac{v_{ph}}{i\omega}\right)^{2\gamma}. \tag{4.73}$$

The first two factors in Equation 4.73 represent the contribution to the conductivity due to the polarization of a chain cluster and the third factor is related to the concentration of clusters. The exponent γ in Equation 4.73 depends almost linearly on temperature and weakly, logarithmically on frequency

$$\gamma = \frac{p-1}{p+1}; \quad p^2 = 1 + 2\frac{T}{T_0}\ln\frac{v_{ph}}{\omega}, \tag{4.74}$$

so that γ reaches maximum 1/3 at $T = T_{1\omega}$.

Thus, the conductivity and dielectric constant of the 1D VRH system in the regime of multihopping or the cluster approximation ($\omega \gtrsim \omega_1$) are

$$\frac{\Re\,\sigma_{ac}(\omega)}{\sigma_0} \approx \frac{\omega}{T_0}\left(\frac{v_{ph}}{\omega}\right)^{2\gamma}\ln^2\frac{v_{ph}}{\omega}; \tag{4.75}$$

$$\frac{\varepsilon(\omega) - 1}{\varepsilon_s - 1} = \left(\frac{v_{ph}}{\omega}\right)^{2\gamma}\ln^2\frac{v_{ph}}{\omega}. \tag{4.76}$$

Equations 4.75 and 4.76 hold as long as $\omega \lesssim \omega_2$, where ω_2 is given by the equation

$$\omega_2(T) = v_{ph}\exp\left[-\left(\frac{T_0}{T}\right)^{1/2}\right]. \tag{4.77}$$

The characteristic frequency ω_2 represents the typical Mott hopping frequency. Above ω_2 the 1D conductivity is determined by the pair hops between close sites, and the conductivity is given by the Mott–Austin formula (Equation 4.56) with $d = 1$.

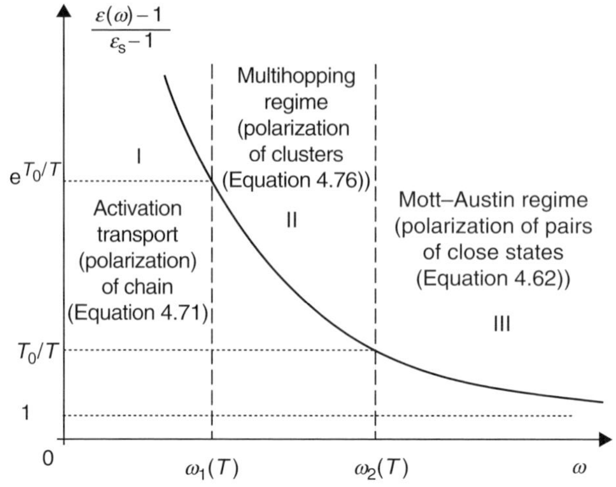

FIGURE 4.5 The frequency dependence of dielectric constant of 1D electronic system in the regime of variable range hopping. Three different regimes of ac-response can be easily identified.

The requirement $\omega \lesssim \omega_2$ is equivalent to $T \gtrsim T_{2\omega}$, where

$$T_{2\omega} = \frac{T_0}{\ln^2 \frac{\nu_{ph}}{\omega}}. \tag{4.78}$$

At $T = T_{2\omega}$, $\omega_2(T) = \omega$.

The frequency and temperature dependencies of the dielectric constant are shown in Figures 4.5 and 4.6.

4.3.2.2 Quasi-1D System

We assume that electrons can hop between the localized states belonging to different chains. The interchain hops happen at those places where the chains approach close to each other or through

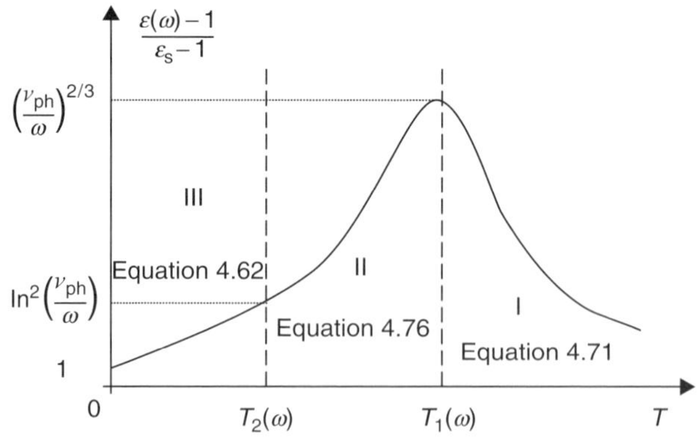

FIGURE 4.6 The temperature dependence of dielectric constant for the 1D system disordered chain. The initial linear increase describing the polarization of close states transforms into a strong exponential temperature dependence describing the polarization of extended chain cluster whose size is controlled by interchain hopping. At high temperatures the dielectric constant asymptotes to a constant value.

dopant atoms between the chains. If the localized states are located right at the point of closest approach and energies of the states are close to each other the interchain hopping rate is given by

$$w_{\perp}^{0} = \nu_{\text{ph}} \exp\left[-2r_{\perp}/\xi_{\perp}\right], \tag{4.79}$$

where

$r_{\perp}$ is the shortest distance between the chains

$\xi_{\perp}$ is the interchain localization radius, which is assumed to be much less than that along the chains, ξ

Assuming now that the energies of the localized states on interacting chains are uncorrelated as well as their distances from the location of closest approach and optimizing over these random variables, one can find that the effective interchain hopping rate is

$$\frac{w_{\perp}(T)}{w_{\perp}^{0}} = (\rho\xi)\sqrt{\frac{T_0}{T}} \exp\left[-\sqrt{\frac{T_0}{T}}\right], \tag{4.80}$$

where ρ is the linear concentration of interchain bridges. Essential for derivation of Equation 4.80 is the assumption that spatial overlap between the states is determined by only the distance between the states along the chains. Also we assumed that the interchain hopping paths occur in parallel to each other in contrast to intrachain hops, therefore effective hopping is determined by the optimal frequency of interchain hops.

Interchain hops are taken into account by adding $w_{\perp}(T)$ to $-i\omega$ in the above 1D Equations 4.70 through 4.73. As a result, the low temperature dc-conductivity of the quasi-1D system is given by the quasi-1D Mott's law

$$\frac{\sigma_{\text{dc}}(T)}{\sigma_0} = \frac{w_{\perp}^{0}}{T_0} \exp\left[-\sqrt{\frac{T_0}{T}}\right]. \tag{4.81}$$

The frequency dispersion of ac-conductivity arises only at $\omega > w_{\perp}(T)$ and follows the 1D dependency, Equations 4.73 through 4.75.

Introducing interchain hopping makes the static dielectric constant finite in a quasi-1D system. In the low-frequency regime $\omega < w_{\perp}(T)$ the dielectric constant now is

$$\frac{\varepsilon(\omega < w_{\perp}(T))}{\varepsilon_0} = \left(\frac{\nu_{\text{ph}}}{w_{\perp}^{0}}\right)^{\sqrt{\frac{T}{T_0}}}. \tag{4.82}$$

At higher frequencies $\omega > w_{\perp}(T)$ the dielectric constant shows the 1D behavior, Equation 4.62 with $d = 1$ and Equation 4.76, shown in Figure 4.5.

4.3.2.3 Quasi-1D Fractal

The fractal chain of $(1 + s)$ dimensionality was defined in the Introduction. For a fractal system a chain inside a cube of side L is connected with L^{s} other chains. Electron hops along chains are governed by Equation 4.46. At high temperatures $T \geq T_1$, where

$$T_1 = sT_0, \tag{4.83}$$

electron hops are mainly between the states of the same chain; therefore, the dc- and ac-conductivities are one-dimensional, Equations 4.73 and 4.75 (see Figure 4.5).

At $T \approx T_1$ the length of hops is comparable with the distance along the chains between the interchain links which is inversely proportional to *s*. For $T \lesssim T_1$, the preferable hops are between the states on different chains. As a result, the temperature dependence of static conductivity is given by the quasi-1D Mott law [44]

$$\frac{\sigma_{dc}(T)}{\sigma_0} = \frac{w_t(T)}{k_B T}; \quad \frac{w_t(T)}{\nu_{ph}} = \exp\left[-\sqrt{\frac{T_0}{sT}}\right]. \tag{4.84}$$

The increase in characteristic temperature by $1/s$ in the Mott law (Equation 4.84) can be understood if we recall that, in general, T_0 should be inversely proportional to the number of neighboring chains, i.e., *s*.

The regime of dc-conductivity takes place for $\omega < \omega_h$, where

$$\omega_h(T) = \left(\frac{4T}{sT_0}\right)^{2/s} w_t(T). \tag{4.85}$$

At $\omega \gtrsim \omega_h$ the conductivity is determined by electron motion within the finite-size clusters consisting of a large number of chains. With increasing frequency, the transverse size of the clusters decreases and at $\omega \approx \omega_m$, where

$$\omega_m(T) = w_t \exp\left(\sqrt{\frac{sT_0}{4T}}\right), \tag{4.86}$$

the cluster reduces to one chain. At $\omega \gtrsim \omega_m(T)$, the frequency dependence of conductivity has pure 1D behavior (Equations 4.62 and 4.76).

The overall frequency dependence of conductivity in the interval $\omega_h(T) < \omega_m < (T)$ was calculated in Ref. [44]. The results can be represented in the following universal form:

$$y \ln y = x, \tag{4.87}$$

where

$$y = \left[\frac{\sigma_{ac}(\omega)}{\sigma_{ac}(T)}\right]^{s/2} \tag{4.88}$$

and

$$x = \sqrt{\frac{T_2}{T}}\left(\frac{-i\omega}{w_t}\right)^{s/2}; \quad T_2 = \frac{s^3 T_0}{16}. \tag{4.89}$$

The ratio ω/w_t in Equation 4.89 can also be written as

$$\frac{\omega}{w_t} = \frac{\omega}{k_B T}\frac{\sigma_0}{\sigma_{dc}(T)}. \tag{4.90}$$

The principal temperature dependence of $\sigma_{dc}(T)$ is exponential, therefore we also can approximate the variable x in Equation 4.78 as

$$x \approx \left(\frac{-i\omega}{\sigma_{dc}(T)}\right)^{s/2}. \tag{4.91}$$

Equation 4.87 with Equations 4.88 and 4.91 at $s = 2$ coincide with that obtained for different 3D hopping models [47–49].

Equation 4.87 suggests the following presentation of the complex admittance of the fractal

$$\frac{\sigma_{ac}(\omega)}{\sigma_{dc}(T)} = \exp\left[\frac{2}{s}z - i\frac{\pi}{2}\frac{z}{1+z}\right], \tag{4.92}$$

where the variable z is related to the frequency by the equation

$$ze^{z} = \left(\frac{\omega}{\omega_1}\right)^{s/2}. \tag{4.93}$$

In other words, the admittance is the universal function of angle θ of dielectric losses ($\theta = \tan^{-1}(\Im\sigma(\omega)/\Re\sigma(\omega))$)

$$\frac{\sigma_{ac}(\omega)}{\sigma_{dc}(T)} = \exp\left[\frac{2}{s}\frac{\theta}{\pi/2-\theta} - i\theta\right], \tag{4.94}$$

where θ varies $0 \le \theta \le \pi/2$.

Equations 4.87 through 4.89 allow us to establish the frequency dependence in the explicit form. For $\omega \ll \omega_c(T)$, where

$$\omega_c(T) = \left(\frac{T}{T_2}\right)^{1/2} w_t(T), \tag{4.95}$$

the conductivity is

$$\frac{\sigma_{ac}(\omega)}{\sigma_{dc}} = \exp\left[\sqrt{\frac{sT_0}{4T}}\left(\frac{\omega}{w_t}\right)^{s/2}\right] \times \left[1 - i\frac{\pi}{2}\left(\frac{\omega}{\omega_c(T)}\right)^{s/2}\right] \tag{4.96}$$

We note that the conductivity remains purely absorptive and can be rewritten as

$$\frac{\sigma_{ac}(\omega)}{\sigma_0} = \frac{\nu_{ph}}{k_B T}\exp\left[-\sqrt{\frac{T_0}{sT}}\left(1 - s\left(\frac{\omega}{w_t}\right)^{s/2}\right)\right]. \tag{4.97}$$

Equation 4.97 demonstrates that at finite frequency the temperature dependence of conductivity starts to deviate from the 1D Mott law.

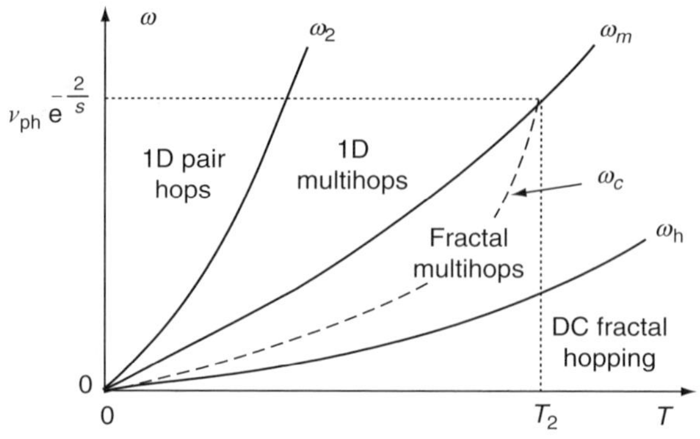

FIGURE 4.7 The diagram of hopping regimes for the fractal chain. The different boundary frequencies are defined in the text by Equations 4.77, 4.86, 4.95, and 4.85.

For $\omega \gg \omega_c(T)$ the conductivity reads

$$\frac{\sigma_{ac}(\omega)}{\sigma_0} = \frac{-4i\omega}{T_0}\left(\frac{sT_0}{4T\ln^2 \dfrac{i\omega}{\omega_1(T)}}\right)^{1/s}. \tag{4.98}$$

At the same time $\omega \ll \omega_m(T)$ the behavior (Equation 4.98) is realized only at $T < T_2$. Figure 4.7 gives the diagram of different hopping regimes of the chain fractal.

Equations 4.92 through 4.93 allow us to find the temperature dependence of dielectric constant at fixed frequency ω. Low temperature behavior is one-dimensional as long as the hopping distance is less than the distance between interchain links. The dielectric constant sharply increases with increasing temperature at $\omega_c(T) = \omega$ or at $T = T_c$, where

$$T_c = \frac{T_0}{4s\ln^2(\nu_{ph}/\omega)} \tag{4.99}$$

and $T_c < T_2$. For $s \ll 1$ the increase of dielectric constant approximately is described by the scaling formula

$$\frac{\varepsilon(\omega)-1}{\varepsilon_s - 1} = \left(\frac{s^2 T_c}{T_c - T}\right)^{2/s}, \tag{4.100}$$

and happens when polarization clusters extend to include many chains.

Above $T > T_c$ the conductivity is provided by percolation over the infinite cluster and the dielectric polarization arises around irregularities of the infinite cluster. According to Equation 4.96 the dielectric constant approximately follows the law

$$\frac{\varepsilon(\omega)-1}{\varepsilon_s - 1} = \frac{\sigma_{dc}(T)}{\omega}. \tag{4.101}$$

Table 4.1 summarizes the results for VRH in the systems with different dimensionality.

TABLE 4.1
Electrical Response of Localized Electrons in Systems of Different Dimensionality D

Response\D	1D	Quasi-1D	(1 + s)-Fractal	d = 2,3
$\frac{\sigma_{dc}(T)}{\sigma_0}$	$\exp\left[-\frac{T_0}{T}\right]$	$\exp\left[-\left(\frac{T_0}{T}\right)^{\frac{1}{2}}\right]$	$\exp\left[-\left(\frac{T_0}{sT}\right)^{\frac{1}{2}}\right]$	$\exp\left[-\left(\frac{T_0}{T}\right)^{\frac{1}{d+1}}\right]$
$\frac{\mathrm{Re}\,\sigma_{ac}(\omega)}{\sigma_0}$	$\frac{T}{T_0}\left\lvert\frac{\omega}{\nu_{ph}}\right\rvert^{1-\frac{T}{T_0}}$	$\frac{T}{T_0}\left\lvert\frac{\omega}{\nu_{ph}}\right\rvert^{1-\frac{T}{T_0}}$	$\exp\left[-\left(\frac{sT_0}{T}\right)^{\frac{1}{2}}\left\lvert\frac{\omega}{\nu_{ph}}\right\rvert^{\frac{s}{2}}\right]$	$\frac{T\omega}{T_2^0}\left\lvert\ln\frac{\nu_{ph}}{\omega}\right\rvert^{d+2}$
$\frac{\varepsilon(\omega)-1}{\varepsilon_s-1}$	$\exp\left[\frac{T}{T_0}\ln^2\frac{\nu_{ph}}{\omega}\right]$	$\exp\left[\left(\frac{T}{T_0}\right)^{\frac{1}{2}}\ln\frac{\nu_{ph}}{\omega_\perp^0}\right]$	$\exp\left[\left(\frac{4T}{sT_0}\right)^{\frac{1}{2}}\ln\frac{\nu_{ph}}{\omega}\right]$	$\frac{T}{T_0}\left\lvert\ln\frac{\nu_{ph}}{\omega}\right\rvert^{d+3}$
References	[37]	[10,41]	[44]	[16]

Note: Here $\sigma_0 = (\varepsilon_s - 1)T_0$, $\varepsilon_s - 1 = 2\,(e^2/\varepsilon_0)N(E_F)\xi^2$, $T_0 = (N(E_F)\xi^D)^{-1}$ and $N(E_F)$ the density of state at the Fermi level, ξ is the localization radius, ν_{ph} is the phonon frequency $10^{12}\ s^{-1}$, and $w_\perp^0$ is the interchain hopping frequency in a quasi-1D system of weakly coupled chains.

4.3.2.4 Dielectric Properties of Poorly Conducting Polymers

There are experimental data for poorly conducting polymers, which are difficult to reconcile with the 3D picture of isotropic hopping. The temperature-dependent conductivity often follows the form [50]: $\sigma_{dc} \propto \exp[-(T_0/T)^{1/2}]$. For a variable range hopping (VRH) mechanism the temperature dependence of hopping conductivity was initially derived [16,24,25] to be: $\sigma_{dc} \propto \exp[-(T_0/T)^{1/(d+1)}]$, where d is the system's dimensionality. For $d = 1$ this formula reproduces the observable dependence of $\sigma_{dc} \propto \exp[-(T_0/T)^{1/2}]$. However, the above approach fails for 1D VRH. The correct analysis of 1D VRH [37] yields the Arrhenius law, $\sigma_{dc} \propto \exp[-(T_0/T)]$ with T_0 set by the highest barrier that occurs in the chain.

The 1D VRH was generalized for a quasi-1D system [41] to include weak hops between the nearest-neighbor chains, thereby avoiding the highest barriers. This approximation results in a quasi-1D VRH law, $\sigma_{dc} \propto \exp[-(T_0/T)^{1/2}]$. For the $(1 + s)$ fractal network the VRH conductivity also obeys a quasi-1D Mott law: $\sigma_{dc} \propto \exp[-(T_1/T)^{1/2}]$ but the characteristic temperature T_1 is greater than T_0 for 1D chain by a factor $1/s$ [44].

Experimental measurements of microwave frequency conductivity and dielectric constant in poorly conducting doped samples [9,50–52] revealed that both are strongly dependent upon temperature, most probably according to the same quasi-1D Mott's law, i.e., $\exp[-(T_0/T)^{1/2}]$. In contrast, the usual theory of hopping transport predicts only a very weak power law temperature dependence for the frequency-dependent conductivity and the dielectric constant in 2D and 3D systems [25]. However, the above low-dimensional models can explain the strong temperature and frequency dependence for conductivity and dielectric constant observed for many poorly conducting polymers.

Variable range hopping is a basic mechanism of charge transport in disordered solids with localized carriers (for discussion on experimental measurements of conductivity and different theoretical models see, e.g., Refs. [48,49]). Their dc-conductivity follows the 3D Mott law (Equation 4.46), and their ac-response in a wide frequency range is described by the relation: $\sigma_{ac} = aT(-i\omega)^{\alpha}$, where α is close to but less than unity. The ac-behavior is well described with the Mott–Austin pair approximation (Equation 4.56).

Variable range hopping also can be identified by a study of the dielectric constant. The dielectric constant characterizes the internal elastic polarization by an external electric field. At a fixed frequency the 3D VRH dielectric constant as a function of temperature has a maximum at intermediate frequencies, where ac- and dc-conductivity are matched. The initial linear increase of dielectric constant with increasing temperature arises from polarization of isolated pairs of more remote states (Equation 4.62). Decrease of dielectric constant with increasing temperature, Equation 4.64, is due to suppression of internal polarization by increasing charge mobility.

As we discussed in the Introduction, the polymers are dielectrics at light doping and in the heavily doped case in the presence of strong disorder or weak interchain coupling. The conductivity in such polymers is provided by phonon-assisted hopping. In very poorly conducting polymers ($\sigma_{RT} \sim 10^{-4}$ S/cm or below), either the localization radius along the chain is comparable with interchain spacing or the concentration of carriers is small. In this case the intrachain and interchain hopping frequencies are comparable and the chain structure becomes unimportant since 3D hopping dominates. The experimental data for conductivity can then be fit with the 3D Mott law. We also point out that for lightly doped polymers the localized electrons produce a strong polarization of the lattice and, therefore, the electronic hops are multiphonon processes that can involve electron hopping among soliton [4,53] or polaron states [54], or electron hopping among bipolaron sites [6,55].

In more highly conducting polymers (but still insulators at low temperatures) the electrical properties are not consistent with the above conventional 3D variable range hopping [52,56,57]. The conductivity often is better fit with the quasi-1D Mott law (Figure 4.8) and the dielectric constant as well as ac-conductivity have strong temperature dependence (Figure 4.9). We suggested that these properties have a 1D or chain origin. The electrons hop more easily along chains due to well-overlapping electronic states of the same chains. The macroscopic properties in this case are strongly determined by the interchain linkages.

Electron micrographs show that in many of these materials polymeric chains are organized into macrofibrils [58], which in its turn may be distinctly seen to be subdivided into smaller ones [59]

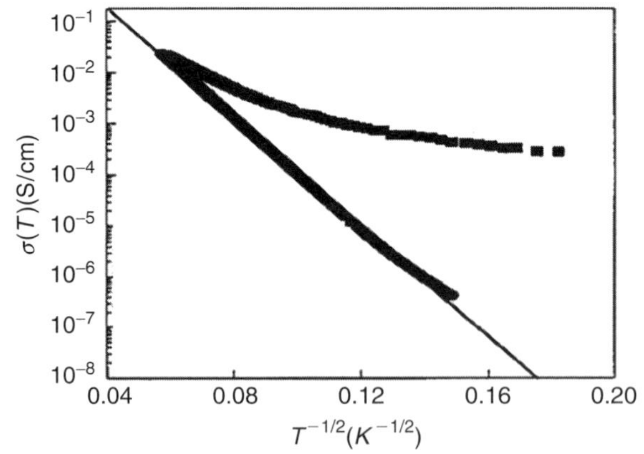

FIGURE 4.8 The temperature dependence of dc-conductivity (•) and microwave conductivity (6.5 GHz) of sulfonated polyaniline (▪). (Reprinted from Lee, W.-P., Du, G., Long, S.M., Epstein, A.J., Shimizu, S., Saitoh, T., and Uzawa, M., *Synth. Met.*, 84, 807, 1997. Copyright 1997. With permission from Elsevier.) The σ_{dc} (*T*) closely follows the quasi-1D Mott's law. It is noted that caution must be taken to review a large range of conductivity variation in identification of the temperature dependence.

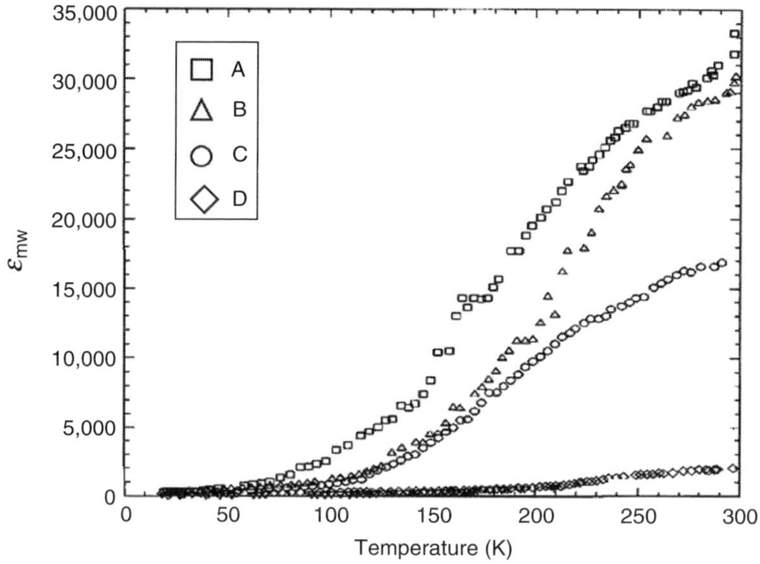

FIGURE 4.9 Temperature-dependent of microwave frequency (6.5 GHz) dielectric constant ε_{mw} (T) in the parallel direction for polyaniline films that have been stretched to provide partial orientation of the polymer chains. (Reprinted from Joo, J., Long, S.M., Pouget, J.P., Oh, E.J., MacDiarmid, A.G., and Epstein, A.J., *Phys. Rev. B*, 57, 9567, 1998. With permission. Copyright 1998 by the American Physical Society.) A and B are the most highly crystalline samples and C and D are the least crystalline samples.

and so on. A variety of nonoriented and oriented nanofiber materials can be prepared using polyaniline. In nonfibrillar forms of conducting polymers, e.g., polyaniline, x-ray data reveal the existence of highly ordered crystalline regions associated with metallic properties [8,60,61]. For example, a network of stretched polyaniline may be thought of as being constructed from long polymer chains randomly coupled by metallic islands of various sizes [55]. The volume fraction of metallic islands can be small. The model to describe this system is the fractal chain of $(1+s)$ dimension. Less structured or more homogeneous polymers are more adequately modeled as quasi-1D systems of randomly linked chains.

We have found that the properties of the $(1+s)$ fractal strongly differ from the 3D regular hopping system. The conductivity is given by the quasi-1D Mott law and the dielectric constant increases with temperature more rapidly than the linear increase with temperature observed for the 3D case. The main increase of dielectric constant with temperature takes place when the hopping distance at a specific frequency becomes comparable with the interchain distance (Equation 4.100).

The dielectric constant as a function of temperature does show a maximum and the position of the maximum shifts with frequency at low temperatures. There is a strong frequency dependence of the conductivity in the region of extremely low frequencies. The model of randomly linked chains has a very similar behavior. In Figure 4.10, we present data for a dielectric constant of poorly conducting polyaniline [62], which has a maximum as a function of temperature that supports the VRH hopping picture in the quasi-1D systems.

The data of Ref. [63] for the conductivity of carbon-black/polymer composites are presented in Figure 4.11. All data for different temperatures, frequencies, and samples are scaled onto a single master curve. The intermediate power-law behavior with very low exponent $b = 0.4$ is well identified. This behavior may be associated with the internal fractal structure whose transverse dimensionality is $s = 2b = 0.8$. The whole fractal dimensionality $1+s = 1.8$ coincides with the dimension found from SEM micrographs at low filler concentrations [64].

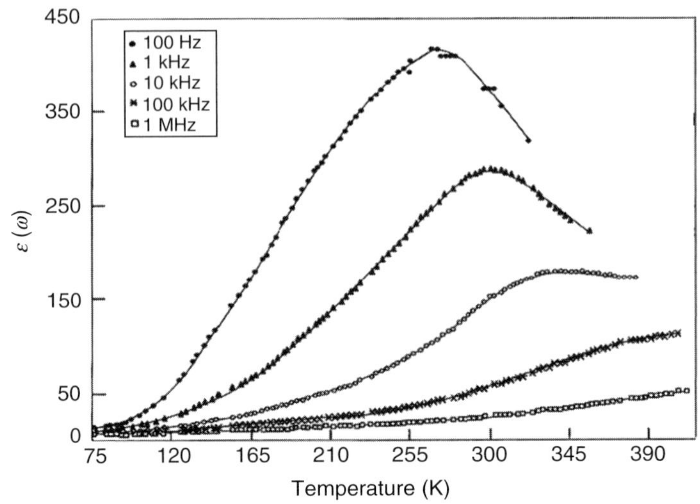

FIGURE 4.10 Dielectric constant plotted as a function of temperature for five fixed frequencies for doped polyaniline. (Reprinted from Singh, R., Arora, V., Tandon, R.P., Mansingh, A., and Chandra, S., *Synth. Met.*, 104, 137, 1999. Copyright 1999. With permission from Elsevier.)

In summary, we have shown that the observable strong temperature and frequency dependence of conductivity and dielectric constant of doped polymers with localized charge carriers can be described within the low-dimensional variable range hopping models. The peculiarities of low-dimensional hopping transport result from the fact that the disordered hopping system with dimensionality close to one behaves as weakly interacting clusters. The states within each cluster are well coupled and clusters are well isolated. The low-frequency electric properties of such a network are strongly

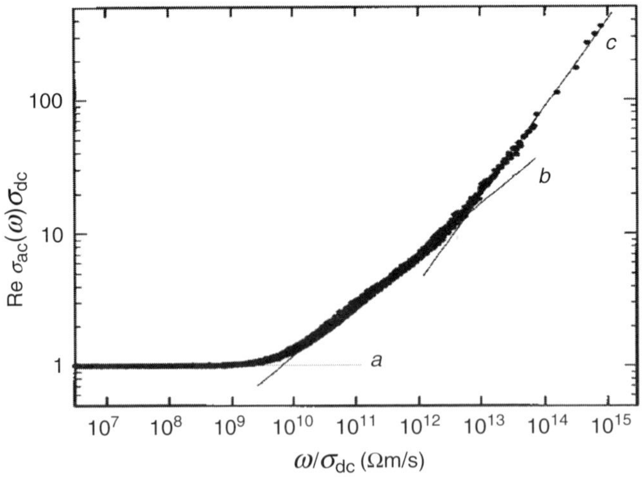

FIGURE 4.11 Scaled conductivity vs. scaled frequency for a conducting carbon-black (CB12)/nonconducting thermo-set polymer composite with carbon-black volume fraction p of 1×10^{-3} ($p = pc/3$). The data taken at about 20 different temperatures between 300 and 4.2 K. (Reprinted from Brom, H.B., Reedijk, J.A., Martens, H.C.F., Adriaanse, L.J., de Jongh, L.J., and Michels, M.A.J., *Phys. Stat. Sol.*, 205, 103, 1998. Copyright 1998 by the Wiley. With permission.) The three slopes are $a = 0$ (dc-regime), $b = 0.4$ (quasi-fractal with $s = 0.8$), and $c = 0.7$ (pair hops).

controlled by the weak charge transfer between clusters. Therefore, small changes of the intercluster connection with temperature or frequency cause large variations in macroscopic conductivity.

4.3.3 Metallic State of Electronic Polymers

4.3.3.1 Metallic Response of Highly Conducting Polymers

When first reported in 1977, the electrical conductivity of doped polymers had modest values at room temperature and decreased to zero with cooling [65]. Synthesis and processing efforts over the following 20 years eventually were successful in preparing heavily doped polyacetylene, polyaniline, and polypyrrole with a finite conductivity at very low temperatures ~10 mK [11,12]. However, the behavior of metallic polymers cannot be understood within the model of a conventional metal with fully delocalized band electrons. The metallic conductivity of these doped polymers decreases with decreasing temperature and a finite residual conductivity is within a decade of the room temperature value (see inset, Figure 4.12).

Though the decrease of the dc-conductivity for metallic polymers with decreasing temperature can be accounted for within the band model by effects of localization caused by disorder, the experimental optical and low-frequency conductivity, and dielectric constant in the metallic state of doped polymers in principle cannot be accounted for by band models with homogeneous disorder. Therefore, it was proposed that in highly conducting polymers there is a new mechanism of charge transport, resonance quantum tunneling among metallic domains [28].

The proposal about a new mechanism of transport is based on puzzles that are revealed in the frequency dependence of conductivity of highly conducting polymers through the optical and low-frequency measurements of conductivity (see Figures 4.12 and 4.13). Experiments [8,12,14] (Figure 4.12) show that the high-frequency ($\omega > 0.1$ eV) conductivity and dielectric constant generally follow a Drude law with the number of electrons ~10^{21} cm^{-3} corresponding to the total density of conduction electrons and conventional scattering time ~10^{-15} s in both the metallic and dielectric phases. At decreasing frequency the polymers in the dielectric phase progressively display insulator properties and ε becomes positive for frequency $\omega \sim 0.1$ eV, signaling that charge carriers are now localized. Microwave frequency (~6.6 GHz) dielectric constant experiments [12] yield localization lengths ~5 nm, depending on sample, dopants, and preparation conditions.

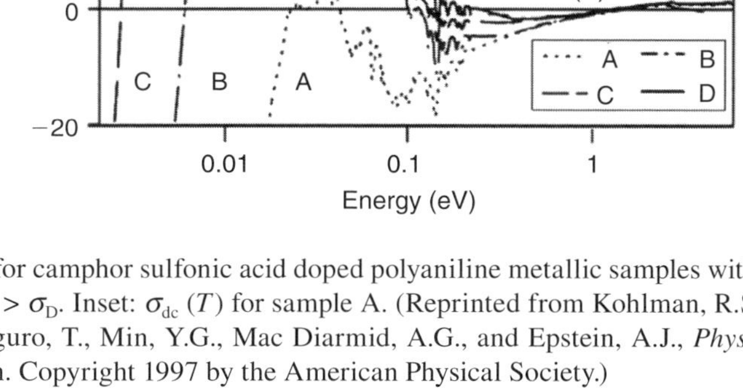

FIGURE 4.12 $\varepsilon(\omega)$ for camphor sulfonic acid doped polyaniline metallic samples with conductivities in the order of $\sigma_A > \sigma_B > \sigma_C > \sigma_D$. Inset: $\sigma_{dc}(T)$ for sample A. (Reprinted from Kohlman, R.S., Zibold, A., Tanner, D.B., Ihas, G.G., Ishiguro, T., Min, Y.G., Mac Diarmid, A.G., and Epstein, A.J., *Phys. Rev. Lett.*, 78, 3915, 1997. With permission. Copyright 1997 by the American Physical Society.)

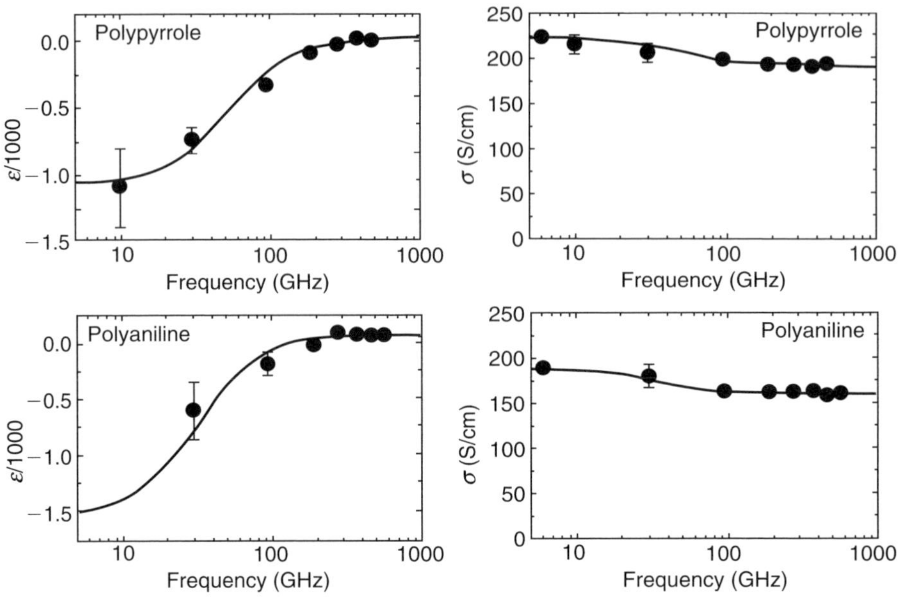

FIGURE 4.13 $\varepsilon(\omega)$ and $\sigma(\omega)$ for metallic-doped polymers. (Reprinted from Martens, H.C.F., Reedijak, J.A., Brom, H.B., de Leuw, D.M., and Menon, R., *Phys. Rev. B*, 63, 073203, 2001. With permission. Copyright 2001 by the American Physical Society.)

A puzzling feature of the metallic phase in polymers is that ε is similar to that of dielectric samples with decreasing frequency, also changing sign from negative to positive at approximately the same frequency ~0.1 eV. However, for metallic samples ε changes again to negative at yet lower frequencies, $\omega \sim 0.01$ eV, indicating that free electronic motion is presented [8,12]. The parameters of this low-frequency coherent phase are quite anomalous. From the Drude model, the relaxation time is found to be very long $\tau \sim 10^{-13}$ s; also, the new plasma frequency below which ε is again negative is very small ~0.01 eV [8,12].

Recently, this second zero crossing of dielectric constant at low frequency and the conclusion about a long relaxation time and a small plasma frequency were confirmed with radio frequency conductivity (see Figure 4.13) [66]. The results of Ref. [66] are very important because they were obtained by direct measurement of conductivity. The early frequency dependence of conductivity and dielectric constant was derived from the reflectance coefficient by using the Kramers–Kroning procedure that requires interpolation on reflectance in the limit $\omega \to 0$.

These experimental findings for low-frequency electromagnetic response are in contrast with the Anderson IMT [17] in which electronic behavior is controlled by homogeneous disorder. In the dielectric phase, electrons are bound by fluctuations of the random potential. On the metallic side of the transition free carriers have short scattering times. In the metallic phase near the transition ε is positive because the disorder causes dynamic polarization due to slowing diffusion by localization effects. When approaching the IMT transition the localization effects increase and ε diverges (dielectric catastrophe [67]).

The small plasma frequency and very long relaxation time of the metallic state in doped polymers can be explained [8,12,66] assuming that the conductivity is provided by a small fraction ~0.1% of the total carriers with long scattering time $>10^{-13}$ s. However, it is difficult to reconcile this conclusion with the behavior for high frequencies, which supports that the scattering time is usual $\sim 10^{-15}$ s and all available electrons participate in conduction. To account for these anomalies the possible presence of a collective mode, as in a charge density wave conductor, or superconductor, was suggested in Ref. [68].

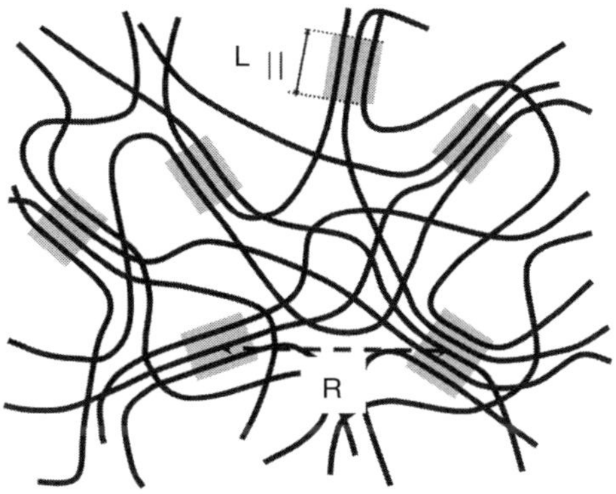

FIGURE 4.14 Schematic view on structure of polyaniline and polypyrrole. The lines represent polymer chains. The dashed squares mark the regions where polymer chains demonstrate the crystalline order. (Reprinted from Prigodin, V.N. and Epstein, A.J., *Synth. Met.*, 125, 43, 2001. Copyright 2001. With permission from Elsevier.)

4.3.3.2 Model of Resonant Transport

In explanation of the low-frequency anomaly in doped polymers in the metallic phase, the polymer chain morphology is very important. These materials are strongly inhomogeneous [60] with crystalline regions within which polymer chains are well ordered (Figure 4.14). When the IMT is approached delocalization first occurs inside these regions. Outside the crystalline regions the chain order is poor and the electronic wave functions are strongly localized. Therefore, the crystalline domains can be considered as nanoscale metallic dots embedded in an amorphous poorly conducting medium [10,69]. The metallic grains remain always spatially separated by amorphous regions, and, therefore, direct tunneling between grains is exponentially suppressed. The intergrain tunneling is possible through intermediate localized states in the disordered portion with strong contribution from resonance states whose energy is close to the Fermi level (Figure 4.15). The dynamics of resonance tunneling can account for the frequency-dependent anomalies in the conductivity and dielectric constant of the metallic phase of these doped polymers.

We assume that grains have $N_\perp$ chains densely packed over $N_\parallel$ repeat units yielding $N_\perp \times N_\parallel$ units cells in each grain. Neglecting intergrain coupling leaves the electronic levels inside grains quantized with mean level spacing $\Delta E = [\hat{N}(E_F)(N_\perp N_\parallel)]^{-1}$, where $\hat{N}(E_F)$ is the density of states per unit cell. For metallic-doped polypyrrole PPy(PF_6) [60]: $N_\perp = 3\times 8$, $N_\parallel = 7$, $\hat{N}(E_F) = 0.8$ states/(eV × ring), and $\Delta E = 7.4$ meV. For doped metallic polyaniline PAN(HCl) [8], $N_\perp = 9 \times 12$, $N_\parallel = 7$, $\hat{N}(E_F) = 1.1$ states/(eV × 2 rings), and $\Delta E = 1$ meV.

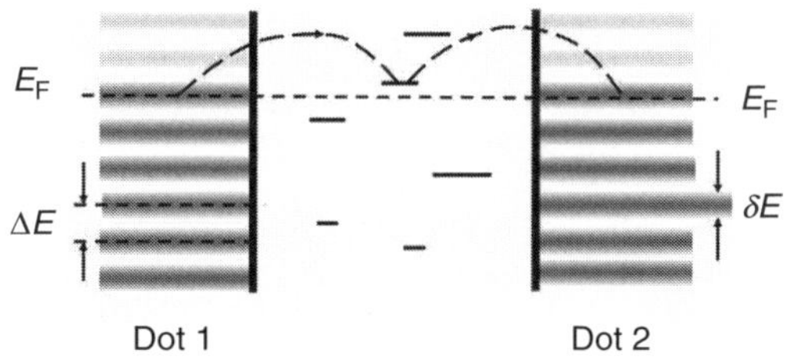

FIGURE 4.15 The electrical coupling between metallic grains is provided by resonance tunneling through localized states in the amorphous region. (Reprinted from Prigodin, V.N. and Epstein, A.J., *Synth. Met.*, 125, 43, 2001. Copyright 2001. With permission from Elsevier.)

We assume that the electronic states are delocalized inside grains and the electron's dynamics is diffusive with diffusion coefficient $D = v_F^2\tau/3$. This diffusive behavior is restricted to times smaller than the time, τ_T, for a charge carrier to cross the grain (Thouless time), $\tau_T = L_{\|}^2/D$, where $L_{\|} \sim 5$ nm is taken for the size of a typical grain [60]. At frequency $\omega\tau_T \gg 1$ the system should show bulk metal behavior and $\sigma(\omega)$ is given by the Drude formula

$$\sigma(\omega) = \frac{b}{4\pi}\frac{\Omega_p^2\tau}{1-i\omega\tau}; \quad \Omega_p^2 = 4\pi e^2 n/m, \tag{4.102}$$

where

Ω_p is the unscreened plasma frequency

b is the degree of crystallinity (usually, $b \sim 50\%$)

The Drude dielectric constant is

$$\varepsilon(\omega) = 1 - b\frac{(\Omega_p\tau)^2}{1+(\omega\tau)^2}, \tag{4.103}$$

and is negative for $\omega \ll \Omega_p$; $\varepsilon(w) \sim -b(\Omega/\omega)^2$.

Optical data [8] show that at high frequencies $\gg 0.1$ eV the system indeed is metal-like (Figure 4.12), which we attribute to the above metallic island response (Equations 4.102 and 4.103) (Figure 4.16). For PAN(CSA) metallic samples (Figure 4.12) the corresponding parameters are $\Omega_p \sim 2$ eV, $\tau \sim 10^{-15}$ s, and τ_T is estimated to be $\sim 5 \times 10^{-14}$ s. We note that the condition for applicability of the grain model, $\tau \ll \tau_T \ll 1/\Delta E \sim 10^{-13}$ s, is fulfilled.

The high-frequency Drude response (Equations 4.102 and 4.103) transforms with decreasing frequency into dielectric behavior at a frequency $1/\tau_T$. For $\omega\tau_T \ll 1$ electrons follow an external field and the conductivity is purely capacitive:

$$\sigma(\omega) = -i\omega e^2 bN(E_F)L_{\|}^2, \tag{4.104}$$

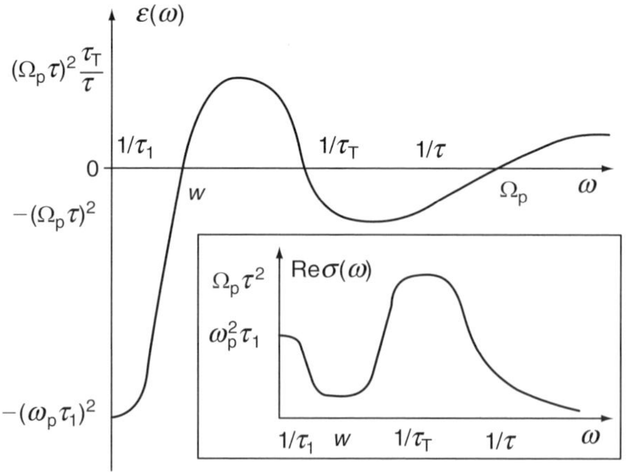

FIGURE 4.16 $\varepsilon(\omega)$ for the chain-linked granular model in the metallic phase. Inset: Re$\sigma(\omega)$. (Reprinted from Prigodin, V.N. and Epstein, A.J., *Synth. Met.*, 125, 43, 2001. Copyright 2001. With permission from Elsevier.)

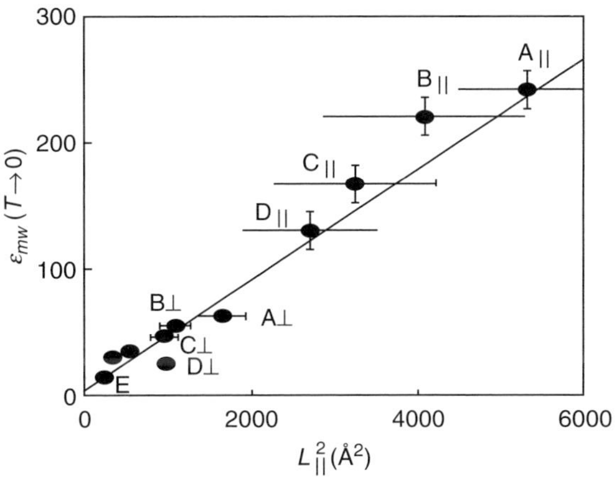

FIGURE 4.17 $\varepsilon_{mw}(T \to 0)$ vs. the square of crystalline coherence length $L_{\|}^2$ for various polyaniline samples. (Reprinted from Joo, J., Long, S.M., Pouget, J.P., Oh, E.J., MacDiarmid, A.G., and Epstein, A.J., *Phys. Rev. B*, 57, 9567, 1998. With permission. Copyright 1998 by the American Physical Society.)

with positive dielectric constant given by the polarization of grains:

$$\varepsilon(\omega \leq \tau_T) = b(\Omega_{\mathrm{p}} L_{\|} / v_{\mathrm{F}})^2. \tag{4.105}$$

The behavior (Equations 4.104 and 4.105) is in good agreement with indistinguishable experimental results (see Figure 4.17) for both dielectrics near the IMT and conductive phases at high and intermediate frequencies (0.001–0.1 eV) (Figures 4.12 and 4.13). Whether the low-frequency behavior is metallic or dielectric depends on the intergrain coupling.

Each grain is coupled to other grains by $2N_\perp$ independent chains through amorphous media. For simplicity, we assume that the two nearest grains are electrically connected by $N_\perp/z$ chains, where z is the number of nearest neighboring grains. In the metallic phase the intergrain coupling leads to broadening of quantized levels in the grains, $\delta E = 2N_\perp g\Delta E$, where g is the transmission coefficient between grains through a single chain. The IMT occurs when $\delta E \sim \Delta E$ (Thouless criteria of IMT [70]) and the critical chain-link coupling g_{c} satisfies

$$2N_\perp g_{\mathrm{c}} = 1. \tag{4.106}$$

For PAN(HCl) and similar PAN(CSA) this yields $g_{\mathrm{c}} \sim 10^{-2}$.

If $g < g_{\mathrm{c}}$ the system is a dielectric and the behavior (104) is retained for all $\omega\tau_T \ll 1$. However, on the metallic side ($g > g_{\mathrm{c}}$) electrons are delocalized over network of grains and their low-frequency motion is a random hopping among the grains. The hopping between grains is a quantum process and can be described by the mean transition frequency W. Introducing the mean distance between the centers of neighboring grains, R ($b \sim (L_{\|}/R)^3$), the corresponding diffusion coefficient D_3 and the macroscopic conductivity of the network are

$$D_3 = R^2 W; \quad \sigma(\omega \sim 0) = be^2 N(E_{\mathrm{F}}) D_3. \tag{4.107}$$

Approaching the IMT from the metallic side, W tends to 0 as [71]: $W = (\Delta E/(2z)) \exp[-2\pi (g_{\mathrm{c}}/(g-g_{\mathrm{c}}))^{1/2}]$.

In the metallic phase the hopping frequency, W, is related to the above model parameters as

$$W = \delta E/(2z) = (N_{\perp}/z)g\Delta E, \tag{4.108}$$

and the whole system can be represented as a network of random conductors. The nodes represent the grains where randomization of electronic motion happens. Combining all together the conductivity (Equation 4.107) also can be written in the simple form

$$\sigma(0) = (e^2 g)(N_{\perp}/z)(1/R). \tag{4.109}$$

Here, the first brackets represent the conductance of a single intergrain chain link, the second factor in brackets is a number of chains connecting neighboring grains, and R in the last brackets mimics the period of the grain network.

4.3.3.3 Average vs. Typical Transmission

Thus the problem of transport in the metallic phase far from the IMT is reduced to study of the average transmission coefficient g in Equation 4.109 for a chain of finite length. For direct tunneling between grains $g = \exp[-2L/\xi]$, where L is the length of the chain connecting neighboring grains and ξ is the localization length. For 50% crystallinity, L is of the same order as a grain size $L_{||}$, i.e., L ~ 5 nm. For weak localization, $\xi = 4l$, where $l = v_F\tau$ is the mean free path. For PAN(HCl) parameters assuming $v_F \sim 3\times10^7$ cm/s and $\tau \sim 10^{-15}$ s, it follows $\xi \sim 1.2$ nm. Therefore, g ~ 10^{-4} and direct tunneling is essentially suppressed.

However, the transmission coefficient is unity for resonance tunneling (Figure 4.15). The probability of finding a resonance state is proportional to the width of the resonance level γ. For constant ξ the resonance state needs to be in the center of the chain; therefore, $\gamma \sim (1/\tau)\exp[-L/\xi]$. As a result, the average transmission coefficient is determined by the probability of finding the resonance state at the center, i.e., $\langle g\rangle = (\gamma\tau) \sim \exp[-L/\xi]$. For PAN(HCl), we have $\langle g\rangle \sim 10^{-2}$ which is close to the critical value for IMT $g_c = 1/(2N_{\perp}) \sim 10^{-2}$. Thus, the probability to find a resonance state is small ~10^{-2}, therefore resonance tunneling is not taken into account. However, as the number of interconnecting chains for a given grain with others becomes large ~100, resonance coupling occurs between grains.

A principal difference between direct and resonance tunneling is the time for tunneling. Direct tunneling that occurs in a conventional granular metal is an almost instantaneous process, i.e., its characteristic time is the scattering time τ. Resonance tunneling that is anticipated to be in the metallic polymers shows a delay determined by the level width γ. The frequency-dependent transmission coefficient $g = g(\omega)$ for resonance tunneling is given by a generalization of the Bright–Wigner formula [72,73]

$$g(\omega) = [1 - i\omega/\gamma]^{-1}. \tag{4.110}$$

With Equations 4.109 and 4.110, we find that in the region $\omega \ll W$, $\sigma(\omega)$ can be written in the standard Drude form

$$\sigma(\omega) = (1/(4\pi))\omega_p^2\tau_1/(1 - i\omega\tau_1), \tag{4.111}$$

with ω_p being the plasma frequency determined by the frequency $\langle W\rangle$ of intergrain hops

$$(\omega_p/\Omega_p)^2 = (b\Delta E\tau)(\langle g\rangle)^2(R/l)^2, \tag{4.112}$$

and the relaxation time $\tau_l = (1/\gamma) = \tau<g> \sim \tau \exp[L/\xi]$ determined by the Wigner transmission time [73].

4.3.3.4 Discussion

The overall frequency dependence of the dielectric function and conductivity in the metallic state of polymers are shown in Figure 4.16. The dependencies (Equations 4.102 and 4.103; Equations 4.111 and 4.112) (Figure 4.16) agree with experimental observations (Figures 4.12 and 4.13). In the metallic phase at low frequencies (<0.01 eV) the studies [8,66] show that the polymers behave again like a Drude metal with anomalous long relaxation times and with very small plasma frequency (Figure 4.12). Microwave experiment and optical data yield $\tau_l \sim (10^{-13} - 10^{-12})$ s. Being compared with (Equations 4.111 and 4.112) the very small plasma frequency effectively corresponds to the decrease of electron concentration by a factor $(\omega_p/\Omega_p)^2 \sim 10^{-3}$ for R/l ~ 30, $\Delta E\tau \sim 10^{-2}$ and the increase of mean time of free path by a factor $\tau_l/\tau = 1/\langle g \rangle \sim 10^2$ to be $\tau_l \sim (10^{-13} - 10^{-12})$ s. Further, $\omega_p \sim 1/\tau_l$, which was observed in Ref. [66,74] (see Figure 4.18).

At finite temperatures, besides the resonance tunneling there is the contribution to conductivity due to phonon-assisted hopping. These tunneling and hopping channels connect between grains in parallel, therefore the total conductivity can be presented as

$$\sigma(T) = \sigma_r + \sigma_h. \tag{4.113}$$

The resonance conductivity σ_r in the metallic phase prevails only at low temperatures. σ_r provides the above frequency dependence but it has a weak temperature dependence. With increasing temperature, phonons increase the resonant transmission due to broadening resonance levels. As a result, the low-frequency part of electromagnetic response is shifted with increasing temperature to a range of higher frequencies as is experimentally observed [8] (see Figure 4.19).

The basic temperature dependence of total dc-conductivity is formed by the hopping term σ_h in Equation 4.113. Because electrons hop over localized states along single chains linking neighboring grains, $\sigma_h(T)$ follows the quasi-1D Mott variable range hopping

$$\sigma_h(T) = \sigma_0 \exp[-(T_1/T)^{1/2}]. \tag{4.114}$$

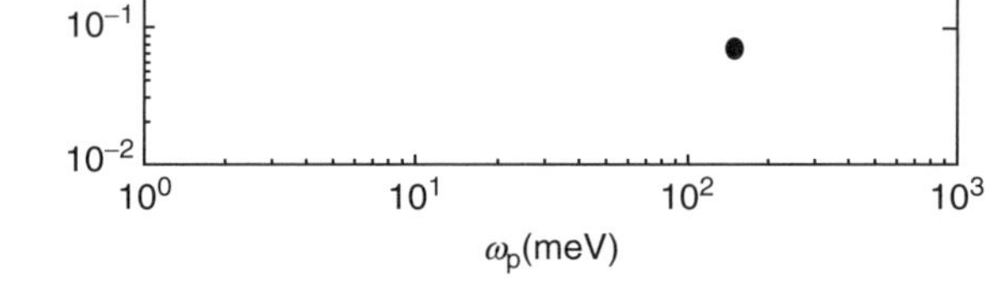

FIGURE 4.18 Experimental correlation between low-frequency relaxation time τ_l (intergrain tunneling time) and low plasma frequency ω_p (intergrain hopping frequency). (Reprinted from Martens, H.C.F. and Brom, H.B., *Phys. Rev. B*, 70, 241201, 2004. With permission. Copyright 2004 by the American Physical Society.)

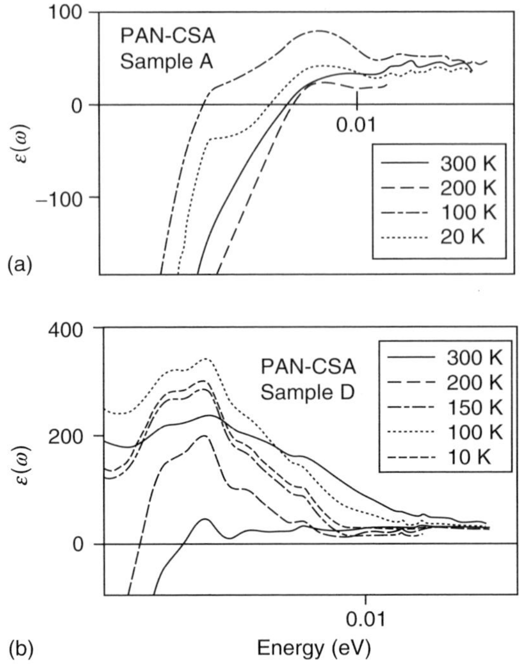

FIGURE 4.19 Dielectric response $\varepsilon(\omega)$ for (a) metallic PAN-CSA sample A and (b) insulating PAN-CSA sample D. The value of abscissa at the left hand axis is 0.002 eV. (Reprinted from Kohlman, R.S. and Epstein, A.J., *Handbook of Conducting Polymers*, T.A. Skotheim, R.L. Elsenbaumer, J.R. Reynolds (eds.), Marcel Dekker, New York, 1998, With permission.)

The temperature scale T_1 is determined by the characteristics of the electronic states of single chains

$$T_1 = C_1/[N_1\xi]. \tag{4.115}$$

We recall that here N_1 is the 1D Fermi density of states and ξ is the localization radius along a chain and $C_1 = 16$ is the numerical factor.

Using the quasi-1D VRH (Equation 4.114) for the present granular model is justified because there are a few chains connecting neighboring grains and the intergrain conductivity is determined by the most conducting one given by Mott's law. Equations 4.113 through 4.115 demonstrate very good agreement with experimental data [52]. It should be emphasized that to identify the exponent in the Mott VRH law the variation of conductivity with temperature should extend over a few orders of magnitude. In Ref. [52] the change of conductivity with temperature exceeds 8 decades and this enables the authors to conclude that the conductivity does follow the quasi-1D VRH law.

Recently the chain-linked granular model was used to describe the electric field effect in conducting polymers. It was observed that doped conducting polymer may be used as the active element for the transistor structure [75–77]. An example is shown in Figure 4.20. The doped conducting polymer PEDOT–PSSA is most often used in transistors, but the field effect also is observed for other conducting polymers, including polyaniline, polypyrrole, and their copolymers. Polyvinylphenol (PVP) was used as the insulator between the active channel and the gate.

I–V characteristics of this completely conducting polymer-based transistor are shown in Figure 4.20 at several gate voltage. At floating gate potential V_G, the source–drain voltage I_{SD} linearly increases with small source–drain voltage V_{SD} and the corresponding room temperature conductivity is 26 S/cm. As it is shown in Figure 4.20, I_{SD} decreases with positive gate voltage and the cutoff

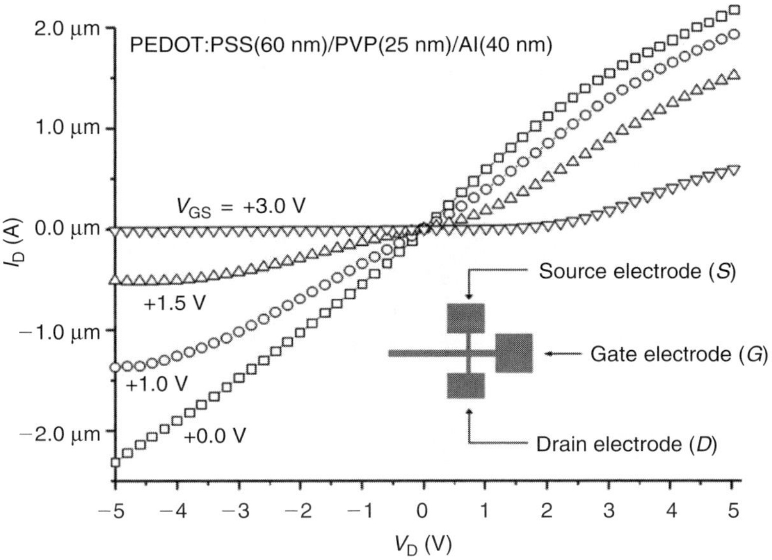

FIGURE 4.20 The drain–source current as the function of drain voltage of a thin active channel PEDOT:PSS/PVP film. (Reprinted from Prigodin, V.N., Hsu, F.C., Kim, Y.M., Park, J.H., Waldmann, O., and Epstein, A.J., *Synth. Met.*, 153, 157, 2003. Copyright 2003. With permission from Elsevier.) Note PEDOT = poly(ethel enedioxytheophene), the dielectric layer is PVP (PVP = poly(inylphenol)).

voltage for I_{SD} is estimated to be 3 eV. The ratio I_{on}/I_{off} is 100 for V_G = 3 V and reaches the higher value of 10^4 for higher gate voltage.

The field effect straightforwardly is explained within the chain-linked granular model. The main effect of applied positive gate voltage is injection of cations (H^+, Na^+, etc.) from the gate area into the polymer. Due to large free space between grains ions easily enter and are attracted by the negative charge of counterions. A hopping path for holes between grains is provided by localized states formed by the negative charge of counterions (see Figure 4.21).

Thus the density of hopping states in Equation 4.115 is related to the density n_d of charged counterions or dopants $N_1 \sim n_d/W$ where W is the characteristic level spread. Compensation of couterions with cations decreases n_d as n_d $(1 - q)$, where q is the degree of compensation. Then T_1 in Equation 4.114 decreases with compensation

$$T_1 = T_0/(1-\alpha q), \tag{4.116}$$

with $\alpha \gg 1$ being a material-dependent constant. Due to the exponential dependence of conductivity on T_1, small variations of q lead to large changes of conductivity. A comparison of the dependence (Equation 4.116) with experimental results (see Figure 4.22) shows very good agreement [78].

Critical concentration of ions to suppress the conductivity can be estimated from the following arguments. If we assume 50% crystallinity and grain size ~10 nm, then the intergrain hopping distance includes ~10 intermediate sites along the chain link. To interrupt this connection it is enough to introduce only one ion, i.e., approximately 5% dedoping may produce the appreciable effect as it is experimentally observed.

Concluding, metallic-doped polymers (polyaniline and polypyrrole) have an electromagnetic response that, when analyzed within the standard theory of metals, is provided by an extremely small fraction of the total number of available electrons ~0.1% (in contrast to ~100% for common metals) but with anomalous long scattering time ~10^{-3} s (~100 times longer than for common metals).

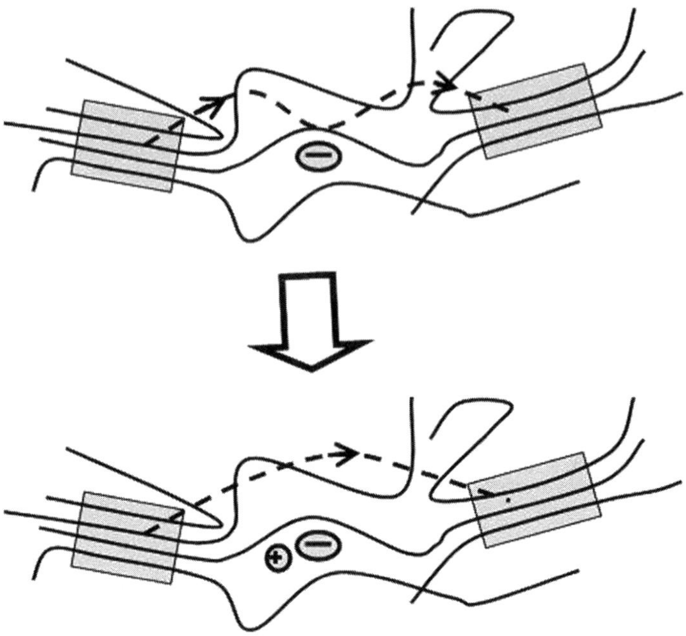

FIGURE 4.21 Ionic suppression of intergrain hopping. The compensation by cations of negative charge of acceptors removes the nearby localized state that provides for holes easy hopping between grains. (Reprinted from Prigodin, V.N., Hsu, F.C., Kim, Y.M., Park, J.H., Waldmann, O., and Epstein, A.J., *Synth. Met.*, 153, 157, 2003. Copyright 2003. With permission from Elsevier.)

It is shown that a network of metallic grains (the polymer's crystalline domains) connected by resonance quantum tunneling through localized states in the surrounding disordered medium produces this behavior. The small fraction of electrons is assigned to the low density of resonance states and the long scattering time is related to the narrow width of energy levels in resonance. The observation of electric field effect conducting polymers strongly supports the present model.

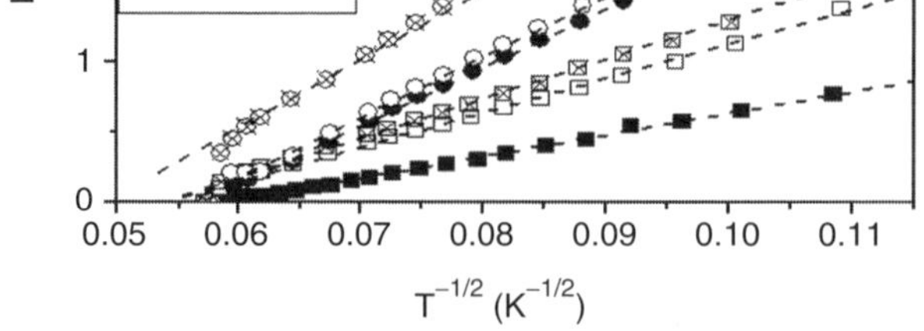

FIGURE 4.22 Quasi-1D Mott's law fit for temperature-dependent resistance at different conducting states R_f/R_0 induced by gate voltage applied at room temperature. (Reprinted from Hsu, F.-C., Prigodin, V.N., and Epstein, A.J., *Phys. Rev. B*, 74, 235219, 2006. With permission. Copyright 2006 by the American Physical Society.)

Thus, for charge motion, electronic polymers are neither typical of conventional 3D crystalline systems, nor 3D amorphous materials, but are a novel third state of matter with extended local order. As a result, the electron transport involves a broad hierarchy of relaxation times. It manifests itself in a very rich frequency dependence of conductivity and dielectric constant and a wide variety of dependencies on the history of a sample. The above model introduces three basic electronic processes characterized by the timescales: microscopic scattering time due to intrachain and intragrain disorder, mesoscopic time of diffusive spreading over the grains, and almost macroscopic time for resonance tunneling between the grains along the chain links. We identify signatures of these processes in the frequency dependence of conductivity.

EXERCISE QUESTIONS

1. Conductivity of metallic sample is $\sigma = 5 \times 10^5$ S/cm and the electron concentration is $n = 5 \times 10^{22}$ cm^{-3}. Calculate the scattering time, the mean free path, and the mobility by using the free electron model.
2. Concentration of localized electrons is $n = 10^{22}$ cm^{-3}. The localization radius is $\xi = 5$ nm. Calculate the value of scale T_0 in the temperature dependence of Mott 3D variable range hopping conductivity and the hopping distance at temperature, $T = 10$ mK.
3. Why does the dielectric constant increase more rapidly in a low-dimension hopping system than in the 3D isotropic one?
4. What is the difference in behavior of the resonance metal and the Drude metal as a function of relaxation time?

REFERENCES

1. A.J. Heeger, S. Kivelson, J.R. Schriffer, and W.P. Su, *Rev. Mod. Phys.* 60, 781, 1988.
2. A.J. Epstein, H.W. Gibson, P.M. Chaikin, W.G. Clark, and G. Grüner, *Phys. Rev. Lett.* 45, 1730, 1980.
3. A.J. Epstein, H. Rommelmann, R. Bigelow, H.W. Gibson, D.M. Hoffman, and D.B. Tanner, *Phys. Rev. Lett.* 50, 1866, 1983.
4. A.J. Epstein, H. Rommelmann, M. Abkowitz, and H.W. Gibson, *Phys. Rev. Lett.* 47, 1549, 1981.
5. A.J. Epstein, H. Rommelmann, and H.W. Gibson, *Phys. Rev. B* 31, 2502, 1985.
6. F. Zuo, M. Angelopoulos, A.G. MacDiarmid, and A.J. Epstein, *Phys. Rev. B* 39, 3570, 1989.
7. J.M. Ginder, A.J. Epstein, and A.G. MacDiarmid, *Synth. Met.* 29, 395, 1989.
8. R.S. Kohlman and A.J. Epstein, in T.A. Skotheim, R.L. Elsenbaumer, J.R. Reynolds (eds.), *Handbook of Conducting Polymers*, 2nd edition, Marcel Dekker, New York, 1998, pp. 85–121.
9. Z.H. Wang, H.H.S. Javadi, A. Ray, A.G. MacDiarmid, and A.J. Epstein, *Phys. Rev. B* 42, 5411, 1990.
10. J. Joo, Z. Oblakowski, G. Du, J.P. Pouget, E.J. Oh, J.M. Weisinger, Y. Min, A.G. MacDiarmid, and A.J. Epstein, *Phys. Rev. B* 49, 2977, 1994.
11. T. Ishiguro, H. Kaneko, Y. Nogami, H. Ishimoto, H. Nishiyama, J. Tsukamoto, A. Takahashi, M. Yhamaura, T. Hagiwara, and K. Sato, *Phys. Rev. Lett.* 69, 660, 1992.
12. R.S. Kohlman, A. Zibold, D.B. Tanner, G.G. Ihas, T. Ishiguro, Y.G. Min, A.G. Mac Diarmid, and A.J. Epstein, *Phys. Rev. Lett.* 78, 3915, 1997.
13. E.S. Choi, G.T. Kim, D.S. Suh, D.C. Kim, J.G. Park, and Y.W. Park, *Synth. Met.* 100, 3, 1999.
14. R. Menon, C.O. Yoon, D. Moses, and A.J. Heeger, in T.A. Skotheim, R.L. Elsenbaumer, J.R. Reynolds (eds.), *Handbook of Conducting Polymers*, 2nd edition, Marcel Dekker, New, York, 1998, p. 27.
15. K. Lee, A.J. Heeger, and Y. Cao, *Phys. Rev. B* 48, 14, 884, 1993.
16. N.F. Mott and E. Davis, *Electronic Processes in Non-Crystalline Materials*, Clarendon Press, Oxford, 1979.
17. P.A. Lee and T.R. Ramakrishnan, *Rev. Mod. Phys.* 57, 287, 1985.
18. J. Chroboczek and S. Summerfield, *J. Phys., Paris, Colloq. C* 3, 517, 1983.
19. K. Ehinger, S. Summerfield, W. Bauhofer, and S. Roth, *J. Phys. C* 17, 3753, 1984.
20. A.B. Kaiser, C.-J. Liu, P.W. Gilberd, B. Chapman, N.T. Kemp, B. Wessling, A.C. Partridge, W.T. Smith, and J. Shapiro, *Synth. Met.* 84, 699, 1997.
21. B. Beau, J.P. Travers, and E. Banka, *Synth. Met.* 101, 272, 1999.
22. N.W. Ashcroft and N.D. Mermin, *Solid State Physics*, Sauders College Publishers, New York, 1976.

23. N.F. Mott, *Metal-Insulator Transitions*, Taylor & Francis, London, 1990.
24. B.I. Shklovskii and A.L. Efros, *Electronic Properties of Doped Semiconductors*, Springer-Verlag, Berlin, 1984.
25. H. Böttger and V.V. Bryksin, *Hopping Conduction in Solids*, Akademie-Verlag, Berlin, 1985.
26. V.N. Prigodin, A.N. Samukhin, and A.J. Epstein, *Synth. Met.* 141, 155–164, 2004.
27. J.M. Ziman, *Principles of the Theory of Solids*, Cambridge University Press, 1964.
28. V.N. Prigodin and A.J. Epstein, *Synth. Met.* 125, 43–53, 2001; V.N. Prigodin and A.J. Epstein, *Europhys. Lett.* 60, 750, 2002.
29. B.L. Altshuler and B.D. Simons, in E. Akkermans, G. Montabaux, J.-L. Pichard, and J. Zinn-Justin (eds.), *Mesoscopic Quantum Physics*, North-Holland, Amsterdam, 1996.
30. D. Stauffer and A. Aharoni, *Introduction to the Percolation Theory*, Taylor & Francis, 1991.
31. S. Stafström, in T.A. Skotheim and J.R. Reynolds (eds.), *Conjugated Polymers: Theory, Synthesis, Properties, and Characterization (Handbook of Conducting Polymers, 3rd edition)*, Taylor & Francis, Boca Raton, 2007, pp. 2-1–2-21.
32. Z.S. Soos and J.M. Sin, *J. Chem. Phys.* 114, 3330, 2001.
33. J.L. Bredas, D. Betjonne, V. Uoropceanu, and J. Cornil, *Chem. Rev.* 104, 4971, 2004.
34. J.A. Freire, *Phys. Rev. B* 72, 125112, 2005.
35. A. Miller and E. Abrahams, *Phys. Rev.* 120, 745, 1960.
36. P. Debye, *Polar Molecules*, Chapter. 5, Chemical Catalog, New York, 1929.
37. J. Kurkijärvi, *Phys. Rev. B* 8, 922, 1973.
38. M. Laridjani and A.J. Epstein, *Eur. Phys. J. B* 7, 585, 1999.
39. V.N. Prigodin and Y.A. Firsov, *JETP Lett.* 38, 284, 1983.
40. N. Dupuis, *Phys. Rev. B* 56, 3086, 1997.
41. E.P. Nakhmedov, V.N. Prigodin, and A.N. Samukhin, *Fiz. Tverd. Tela (Leningrad)* 31, 31, 1989; *Sov. Phys. Solid State* 31, 368, 1989.
42. A. Yamashiro, A. Ikawa, and H. Fukutome, *Synth. Met.* 65, 233, 1994.
43. V.N. Prigodin and A.N. Samukhin, *Sov. Phys. Solid State* 26, 817, 1984.
44. A.N. Samukhin, V.N. Prigodin, L. Jastrabik, and A.J. Epstein, *Phys. Rev. B* 58, 11354, 1998.
45. B. Shapiro, *Phy. Rev. Lett.* 48, 823, 1982.
46. V.N. Prigodin and A.N. Samukhin, *Sov. Phys. Solid State* 25, 991, 1983.
47. V.V. Bryksin, *Fiz. Tverd. Tela (Leningrad)* 22, 2441, 1980; *Sov. Phys. Solid State* 22, 1421, 1980.
48. J.C. Dyre and T. Schroder, *Rev. Mod. Phys.* 72, 873, 2000.
49. J.R. MacDonald, *Phys. Rev. B* 49, 9428, 1994.
50. F. Zuo, M. Angelopoulos, A.G. MacDiarmid, and A.J. Epstein, *Phys. Rev. B* 36, 3475, 1987.
51. J. Joo, V.N. Prigodin, Y.G. Min, A.G. MacDiarmid, and A.J. Epstein, *Phys. Rev. B* 50, 12226, 1994.
52. J. Joo, S.M. Long, J.P. Pouget, E.J. Oh, A.G. MacDiarmid, and A.J. Epstein, *Phys. Rev. B* 57, 9567, 1998.
53. S. Kivelson, *Phys. Rev. Lett.* 46, 1344, 1981.
54. A.J. Epstein, in T.A. Skotheim (ed.), *Handbook of Conducting Polymers*, Marcel Dekker, New York and Basel, 1986, p. 1041.
55. H.H.S. Javadi, K.R. Cromack, A.G. MacDiarmid, and A.J. Epstein, *Phys. Rev. B* 39, 3579, 1989.
56. Z.H. Wang, C. Li, E.M. Scherr, A.G. MacDiarmid, and A.J. Epstein, *Phys. Rev. Lett.* 66, 1745, 1991.
57. W.-P. Lee, G. Du, S.M. Long, A.J. Epstein, S. Shimizu, T. Saitoh, and M. Uzawa, *Synth. Met.* 84, 807–808, 1997.
58. J. Tsukamoto, *Adv. Phys.* 41, 509, 1992.
59. K. Araya, T. Micoh, T. Narahara, K. Akagi, and H. Shirikawa, *Synth. Meth.* 17, 247, 1987.
60. J. Joo, Z. Oblakowski, G. Du, J.P. Pouget, E.J. Oh, J.M. Wiesinger, Y. Min, A.G. MacDiarmid, and A.J. Epstein, *Phys. Rev. B* 49, 2977, 1994; J.P. Pouget, Z. Oblakowski, Y. Nogami, P.A. Albouy, M. Laridjani, E.J. Oh, Y. Min, A.G. MacDiarmid, J. Tsukamoto, T. Ishiguro, and A.J. Epstein, *Synth. Meth.* 65, 131, 1994.
61. J.P. Travers, B. Sixou, D. Berner, A. Wolter, P. Rannou, B. Beau, B. Pepen-Donat, C. Barther, M. Guglielmi, N. Mermolliod, B. Gilles, D. Djurado, A.J. Attias, and M. Vautrin, *Synth. Meth.* 101, 359, 1999.
62. R. Singh, V. Arora, R.P. Tandon, A. Mansingh, and S. Chandra, *Synth. Met.* 104, 137–144, 1999.
63. H.B. Brom, J.A. Reedijk, H.C.F. Martens, L.J. Adriaanse, L.J. de Jongh, and M.A.J. Michels, *Phys. Stat. Sol.* 205, 103, 1998.
64. K.-M. Jager and D.H. McQueen, *Polymer* 42, 9575, 2001.
65. C.K. Chiang, C.R. Fincher Jr., Y.W. Park, A.J. Heeger, H. Shirakawa, E.L. Louis, S.C. Gau, and A.G. MacDiarmid, *Phys. Rev. Lett.* 39, 1098, 1977.
66. H.C.F. Martens, J.A. Reedijk, H.B. Brom, D.M. de Leuw, and R. Menon, *Phys. Rev. B* 63, 073203, 2001.

67. N.F. Mott and M. Kaveh, *Adv. Phys.* 34, 329, 1985.
68. K. Lee, M. Reghu, E.L. Yuh, N.S. Sariciftci, and A.J. Heeger, *Synth. Met.* 68, 287, 1995.
69. B. Bean and J.P. Travers, *Synth. Met.* 65, 101, 1999.
70. D.J. Thouless, *Phys. Rev. Lett.* 39, 1167, 1987.
71. K.B. Efetov and O. Viehweger, *Phys. Rev. B* 45, 11546, 1992.
72. M. Büttiker, A. Prêtre, and H. Thomas, *Phys. Rev. Lett.* 70, 4114, 1993.
73. J. Bolton, C.J. Lambert, V.I. Falko, V.N. Prigodin, and A.J. Epstein, *Phys. Rev. B* 60, 10569, 1999.
74. H.C.F. Martens and H.B. Brom, *Phys. Rev. B* 70, 241201, 2004.
75. J. Liu, N.J. Pinto, and A.G. MacDiarmid, *J. Appl. Phys.* 92, 6033, 2002.
76. A.J. Epstein, F.C. Hsu, N.R. Chiou, and V.N. Prigodin, *Curr. Appl. Phys.* 2, 339, 2002.
77. H. Okuzaki, M. Ishiharaa, and S. Ashizawa, *Synth. Met.* 137, 947, 2003.
78. F.-C. Hsu, V.N. Prigodin, and A.J. Epstein, *Phys. Rev. B* 74, 235219, 2006.

5 Major Classes of Organic Small Molecules for Electronics and Optoelectronics

Xianle Meng, Weihong Zhu, and He Tian

CONTENTS

5.1 Perylene Bisimides ... 131
5.1.1 Perylene Bisimide Derivatives for Solar Cells ... 134
5.1.2 Perylene Bisimide Derivatives for Organic Field-Effect Transistors ... 134
5.1.3 Perylene Bisimide Derivatives for Organic Light-Emitting Diodes ... 136
5.2 Phthalocyanines ... 137
5.2.1 Phthalocyanines for Organic Light-Emitting Diodes ... 137
5.2.2 Phthalocyanines for Solar Cells ... 138
5.2.3 Phthalocyanines for Nonlinear Optical Materials ... 139
5.2.4 Phthalocyanines for Organic Field-Effect Transistors ... 141
5.3 Porphyrins ... 141
5.3.1 Porphyrins for Solar Cells ... 144
5.3.2 Porphyrins for Organic Light-Emitting Diodes ... 146
5.3.3 Porphyrins for Nonlinear Optical Materials ... 146
5.3.4 Porphyrins for Optical Limiting Devices ... 147
5.4 Cyanines ... 147
5.4.1 Cyanines for Photovoltaic Devices ... 148
5.4.2 Cyanines for Nonlinear Optical Materials ... 150
5.5 Carbazoles and Triphenylamine Derivatives ... 151
5.6 Oxadiazoles ... 155
5.7 Tetrathiafulvalenes ... 157
5.8 Polycyclic Aromatic Hydrocarbons (Oligoacenes) ... 160
5.9 Small Molecule Complexes ... 164
5.10 Summary ... 166
Exercise Questions ... 166
List of Abbreviations ... 167
References ... 167

Abstract This chapter describes major classes of organic small molecules, such as perylene bisimides, phthalocyanines, porphyrins, cyanines, carbazoles and triphenylamine derivatives, oxadiazoles, tetrathiafulvalenes (TTFs), polycyclic aromatic hydrocarbons, and small molecule complexes used in electronics and optoelectronics. Latest examples of their applications are also included.

Over the past few decades, enormous progresses have been made in the field of organic optoelectronic devices. A first generation of visible organic light-emitting diodes (OLEDs) has been commercialized, organic solar cells have achieved high conversion efficiency, and thousands of papers on organic field-effect transistors (OFETs) and nonlinear optical materials have been published. Compared to inorganic materials, organic molecular materials display important roles in applications because of their low cost, easiness of device fabrication, high efficiency, and richness in molecular modification. The main properties of major class of organic small molecules are outlined in Table 5.1. Among a variety of well-studied small molecules, some of them are versatile

TABLE 5.1
Summary of Major Classes of Organic Small Molecules

Classes	Property	Application Fields	Parent or Basic Structures
Perylene bisimides	n-Type semiconductor	Solar cell OLED OFET	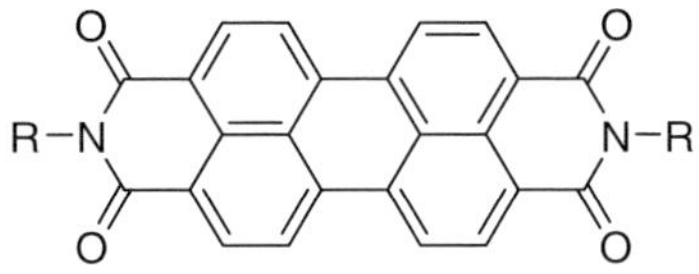
Phthalocyanines	p-Type semiconductor	OLED Solar cell NLO	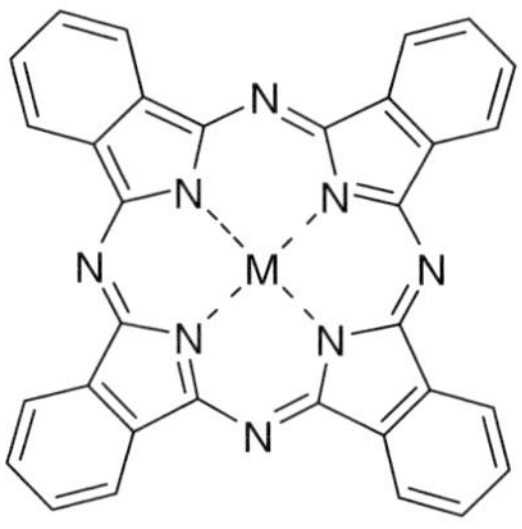 M = 2H or Metal
Porphyrins	p-Type semiconductor	Solar cell OLED NLO	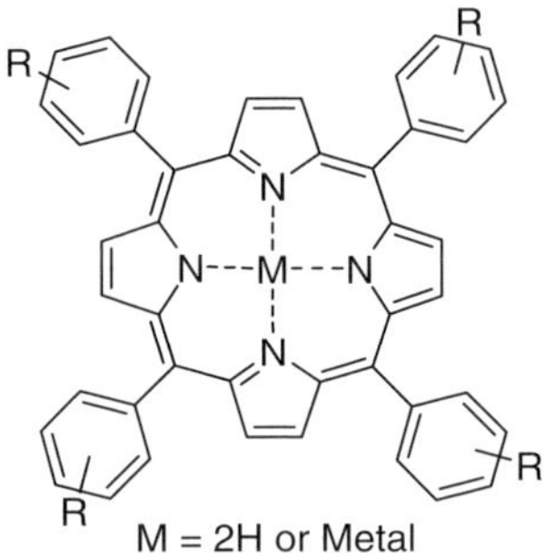 M = 2H or Metal
Cyanines	\	Solar cell NLO	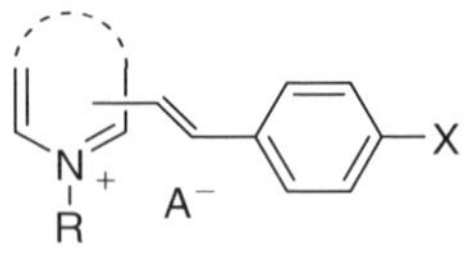
Carbazoles and triphenylamine derivatives	Hole transport	OLED Hole transport	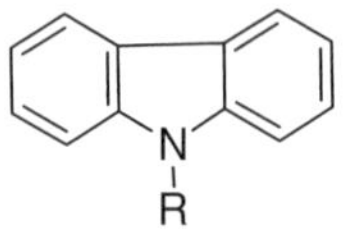 Carbazoles 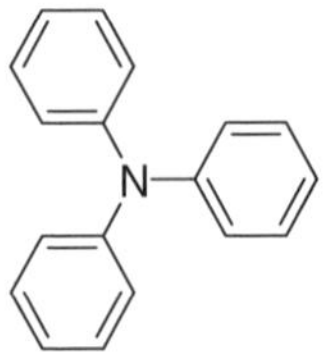 Triphenylamine

TABLE 5.1 (continued)
Summary of Major Classes of Organic Small Molecules

Classes	Property	Application Fields	Parent or Basic Structures
Oxadiazoles	Electron transport	OLED Electron transport	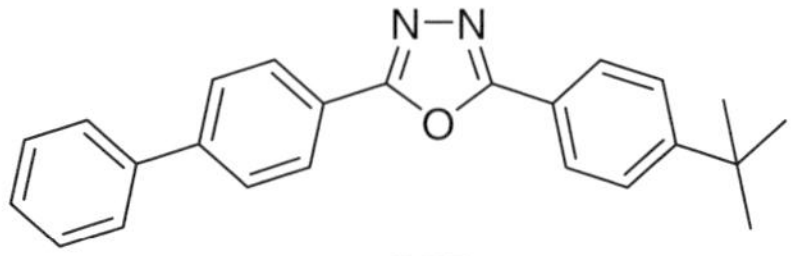 PBD
Tetrathiafulvalenes	Electron donor	Synthetic metal Solar cell	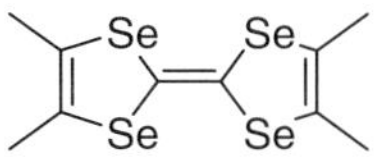 TMTSF 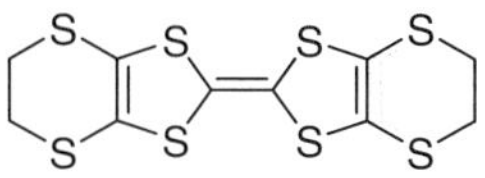 ET
Thiophene oligomers	Electron donor	OFET Solar cell	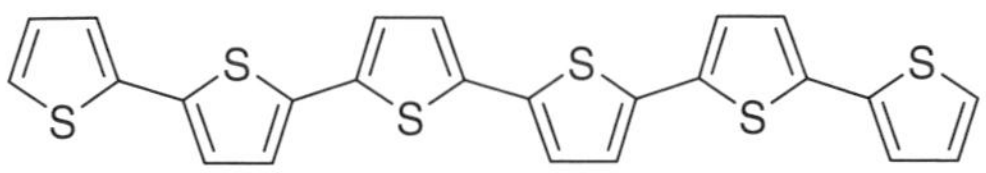
Polycyclic aromatic hydrocarbons	p-Type semiconductor	OFET OLED	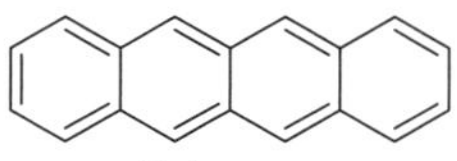 Tetracene 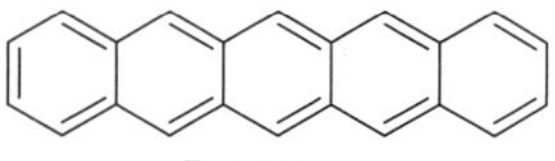 Pentacene 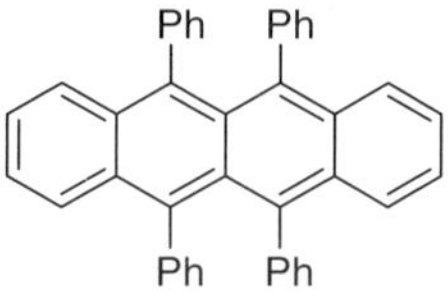 Rubrene
Complexes		OLED Solar cell	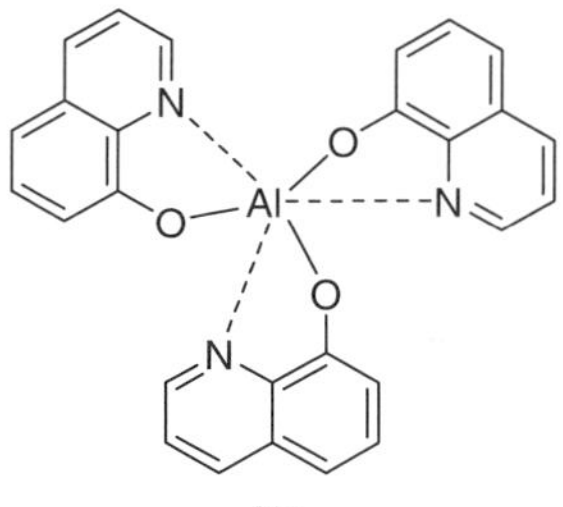 Alq$_3$ 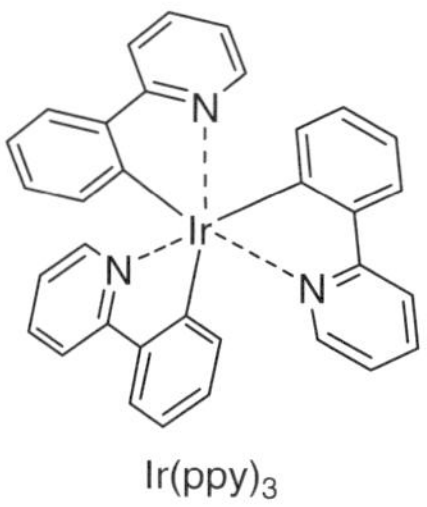 Ir(ppy)$_3$

in that they could be applied in many aspects of optoelectronics, such as phthalocyanines, porphyrins, and perylene bisimides. Some of them are mainly studied in one area, for example, the research on pentacene mainly lies in OFET. In this chapter, we mainly discuss the widely developed organic molecular materials for their applications in optoelectronics.

5.1 PERYLENE BISIMIDES

Vat dyes based on imides of perylene-3,4,9,10-tetracarboxylic acids (perylene bisimides) are widely applicable due to their outstanding chemical, thermal, and photochemical inertness [1]. Most preparations of perylene bisimides are outlined in Scheme 5.1. Although insoluble and high-melting perylene bisimides (**2**) could be easily obtained by the reaction of perylene tetracarboxylic acid bisanhydride (**1**) with a variety of aromatic and aliphatic amines, most topics of current interests

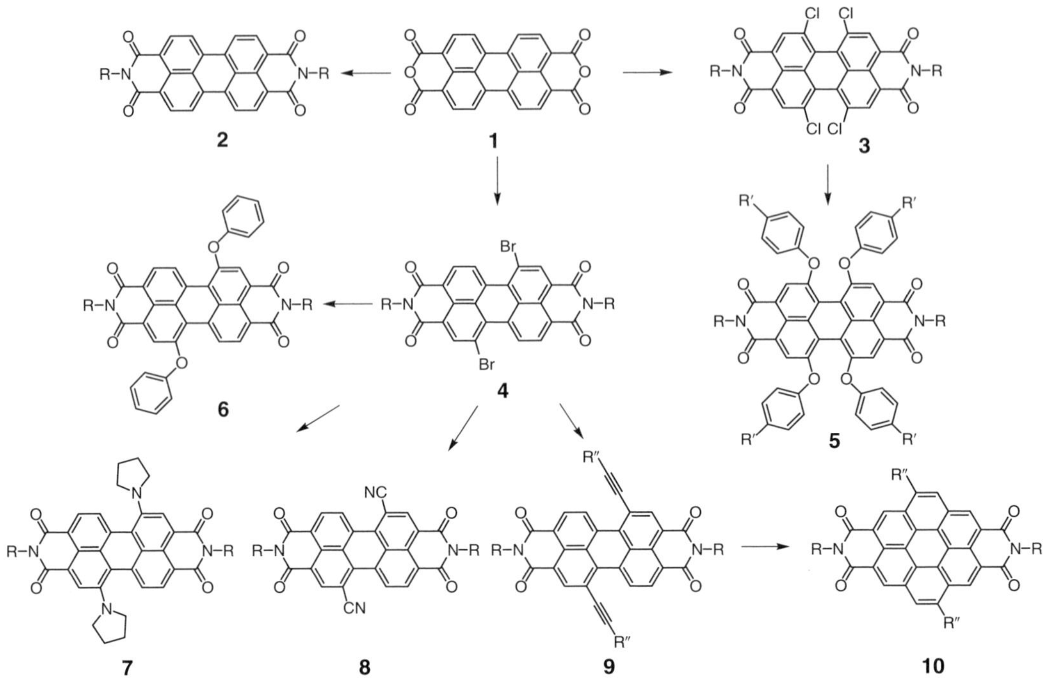

SCHEME 5.1 Synthetic routes to perylene bisimide derivatives.

require better soluble ones. Generally, there are two methods to improve the solubility of perylene bisimides (Scheme 5.1) [1,2]: (1) introducing solubilizing substituents at the imide nitrogen [3] and (2) incorporating substitute groups at the bay-area via the replacement of chlorine or bromine (**3**–**10**).

Figure 5.1 shows typical absorption and emission spectra of perylene bisimide **11** [4]. Generally, there are three peaks in the wide range of 400–550 nm, which corresponds to the transitions of $0 \rightarrow 2$, $1 \rightarrow 2$, and $0 \rightarrow 1$. Notably, the emission and absorption spectra are well mirrored, giving strong evidence of Franck–Condon principle. It has been proven that the imide substituent has a negligible influence on the absorption and emission properties of perylene bisimides because of the nodes of

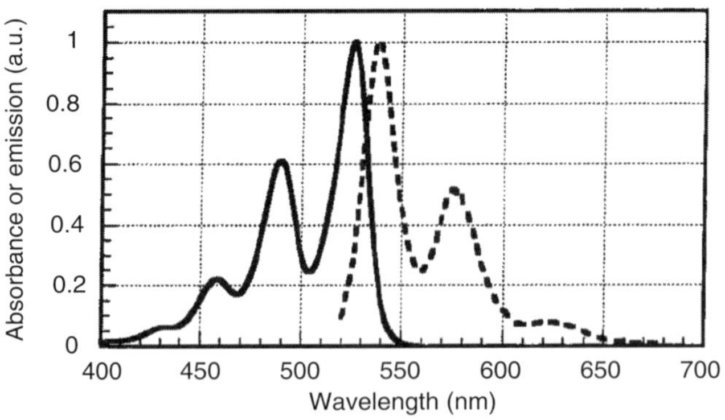

FIGURE 5.1 Normalized absorption (solid) and emission spectra (dash) of perylene bisimide **11** excited at 489 nm. (From Liu, S. et al., *J. Phys. Chem. B*, 106, 1307, 2002.)

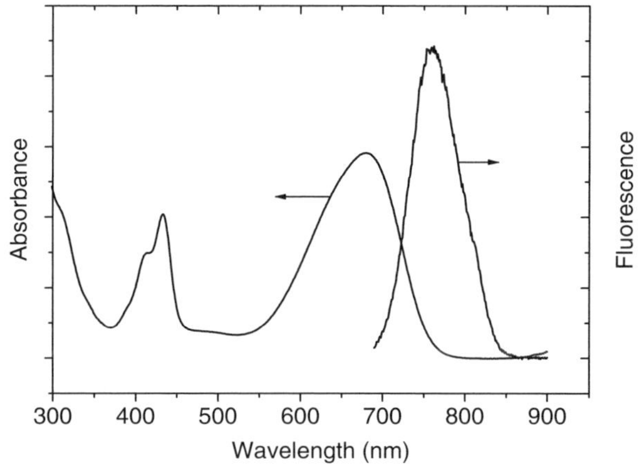

FIGURE 5.2 Absorption and fluorescence spectra of **12** excited at 645 nm in CH_2Cl_2 solution (10^{-5} M). (From Fan, L.Q., Xu, Y.P., and Tian, H., *Tetrahedron Lett.*, 46, 4443, 2005.)

the HOMO and LUMO orbitals at the imide nitrogens (Figure 5.1). However, the substitution at bay positions usually has pronounced effect on the absorption and emission properties of perylene bisimides [5,6]. Especially, the absorption and fluorescence peaks of **12** containing two electron donors (piperidine substituents) are shifted dramatically. Actually, the emission of perylene bisimide **12** bathochromically shifts to the infrared region with the peak at 759 nm (Figure 5.2) [6].

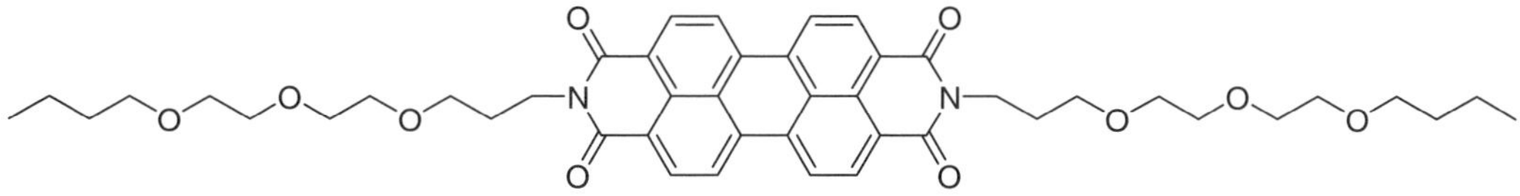

11

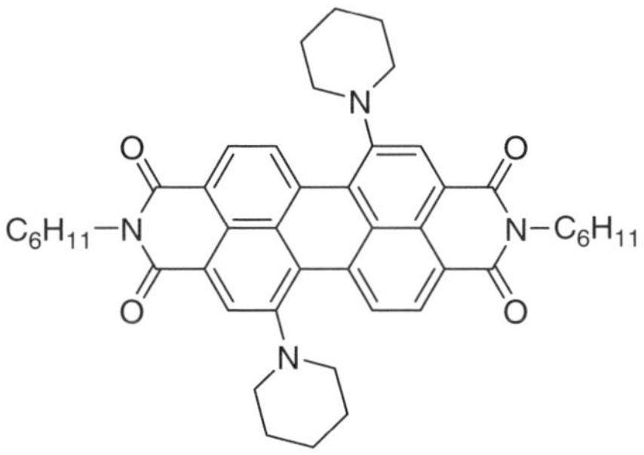

12

Up to now, more intensive studies have been done in the field of electronic materials, among which perylene bisimides are one of the best n-type semiconductors due to the high electron affinity. Organic semiconductors can conduct charges due to partial delocalization or charge hopping through molecules that are coupled by relatively weak van der Waals forces. Study on compound **13** shows that the electron mobility is much larger than its hole mobility [7]. The electron-deficient character of perylene bisimides is also a prerequisite for their high photochemical stability because photooxidation as the major destructive mechanism for dyestuffs is disfavored. On the other hand, the majority of these dyes are strong reductants in their photoexcited state [8]. This property has been widely applied to establish long-lived charge-separated states in photo-induced electron transfer cascades.

5.1.1 Perylene Bisimide Derivatives for Solar Cells

Perylene bisimides have large molar absorption coefficients, good electron-accepting properties, and possible generation of a highly conducting direction along the π–π stacking axis. These characters enable perylene derivatives to be good candidates for photovoltaic devices [9]. In 1986, Tang [10] first used 3,4,9,10-perylenetetracarboxylic bis-benzimidazole (**14**) as an electron acceptor for two-layer organic solar cell devices. And heterojunction of organic solar cell with **14** penetrated by polymer was reported to have the power conversion efficiency up to 1.9% [11]. Soluble perylene bisimide can bring down the cost to fabricate devices. Very high solubility has been realized by Langhals's so-called swallow-tail-substituents as in **15** and **16**. However, its highest incident photocurrent conversion efficiency (IPCE) of 19% and the power conversion efficiency of 0.18% achieved were not so satisfied. And **16** was reported to render the high open circuit voltage V_{OC} (up to 0.71 V) in photovoltaic device [9].

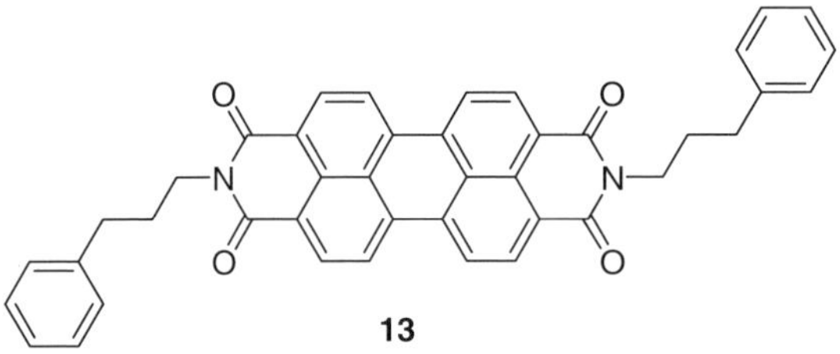

13

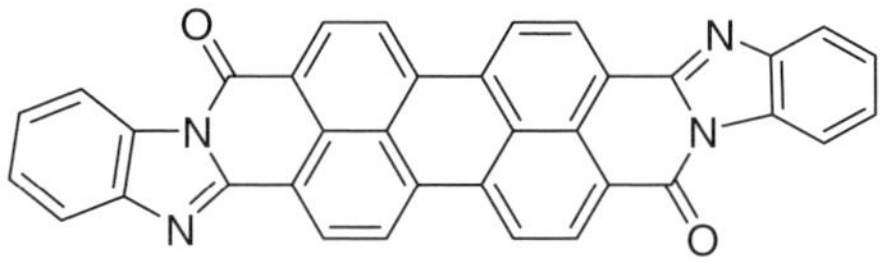

14

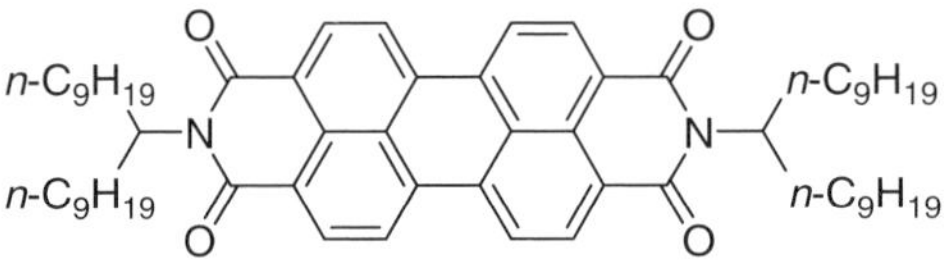

15

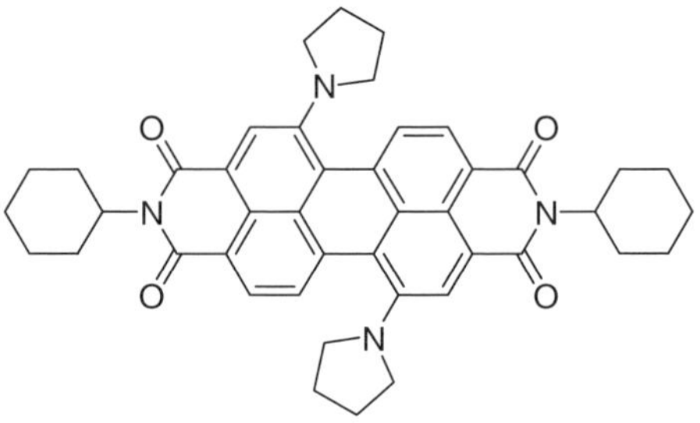

16

Since most photosensitizers of the Grätzel solar cells are based on Ru complexes, the research on other photosensitizers has been expanded. Perylene bisimides **17–20** are examples of photosensitizers studied in Grätzel solar cells. In the chemical structures of **17–20**, the incorporation of carboxylic or phenolic groups is aimed to increase the interaction between TiO_2 or SnO_2 and photosensitizers. **17** and **18** have four substituents on the bay-region of the parent core. A wide spectral photo-response on SnO_2 in visible region was observed with a maximum IPCE of 24% [12]. The film of **19** sensitized on TiO_2 yielded an IPCE of approximately 40% in the wavelength region of 440–530 nm, while that for perylene derivative **20** was approximately 14% at the wavelength region from 460 to 510 nm [13]. Syntheses of novel multichromophoric soluble perylene derivatives for wide spectral response for SnO_2 nanoporous electrode were also studied by Tian's group [14].

A perylene bisimide-fullerene dyad **21** in which a perylene moiety was attached to C_{60} through a pyrrolidine ring was fabricated into photovoltaic device. The photophysical properties and device performances of the dyad indicated an effective photo-induced energy transfer from the perylene moiety to C_{60} compared to the negligible electron transfer [15].

5.1.2 Perylene Bisimide Derivatives for Organic Field-Effect Transistors

Perylene derivatives as a well-defined n-type semiconductor were also widely studied in OFETs. The OFET based on **22** displays mobility as high as 0.10 $cm^2 \cdot V^{-1} \cdot s^{-1}$, threshold voltage of approximately

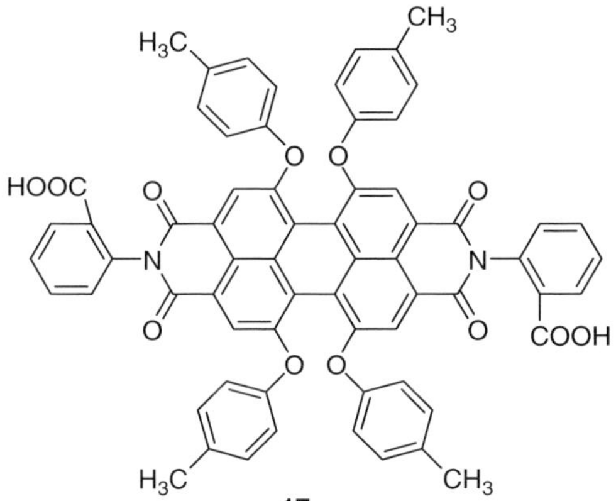

17

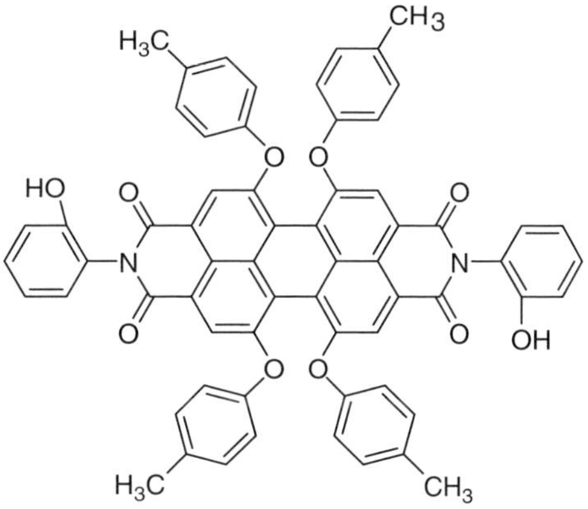

18

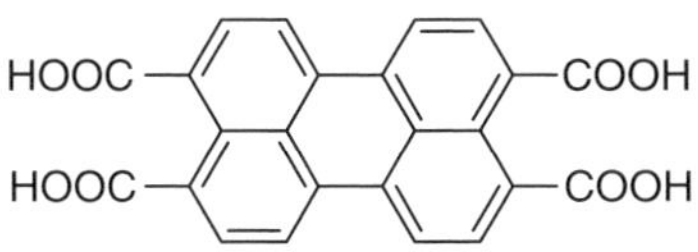

19

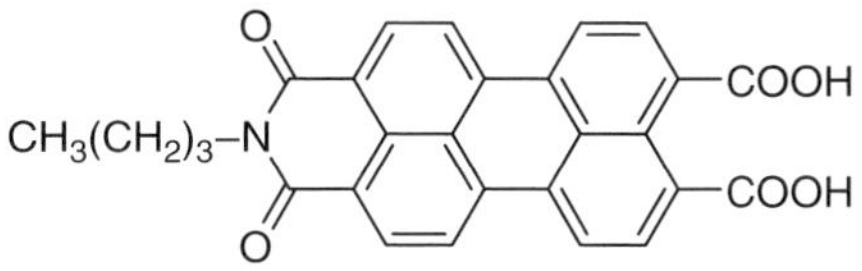

20

15 V, and I_{on}/I_{off} (+100 V/0 V) as high as 10^5, while **23** exhibits mobility as high as 0.64 $cm^2 \cdot V^{-1} \cdot s^{-1}$, threshold voltages between 20 and 30 V, and I_{on}/I_{off} (+100 V/–60 V) as high as 10^4 [16]. OFETs of **24–26** had a saturation electron mobility as high as 1.7 $cm^2 \cdot V^{-1} \cdot s^{-1}$ with on-to-off current ratios of 10^7, which could be competitive with pentacene p-channel OFETs reported [17–19]. A perylene bisimide (**27**) with strong electron-withdrawing fluorine atoms was quite interesting. Its electron mobility was measured as 1.29×10^{-5} $cm^2 \cdot V^{-1} \cdot s^{-1}$ with a spin-coated **27** film as transport layer [20].

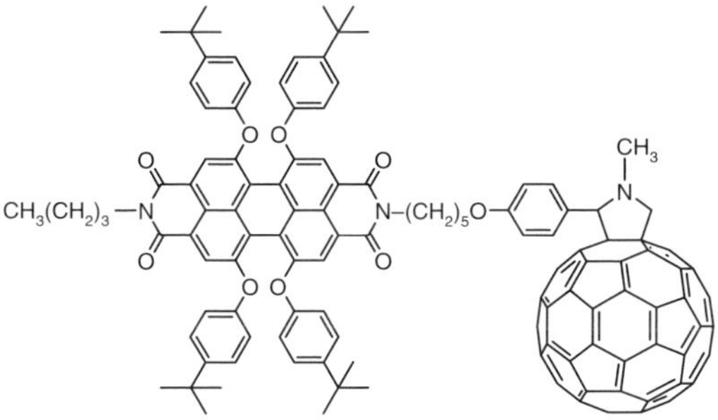

21

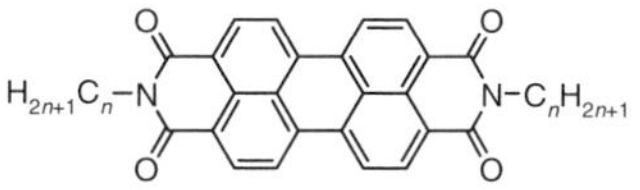

24: $n = 5$
25: $n = 6$
26: $n = 7$

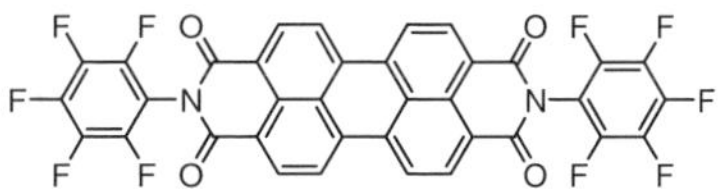

27

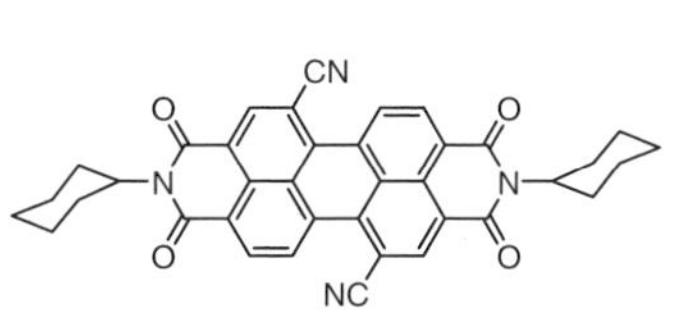

22

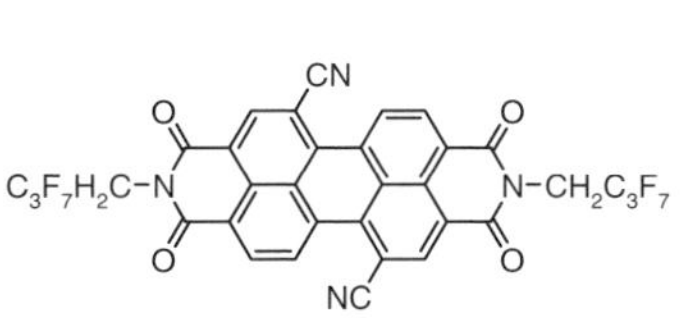

23

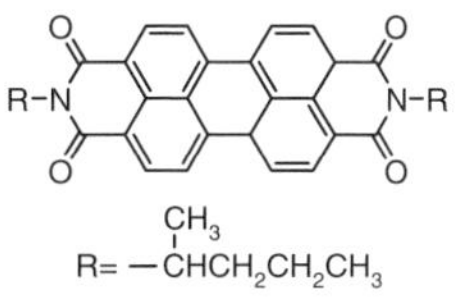

28

Recently, a single molecule n-type transistor based on perylene bisimides (Figure 5.3) was prepared, in which the current through the molecule can be reversibly controlled with a gate electrode over nearly three orders of magnitude at room temperature [21].

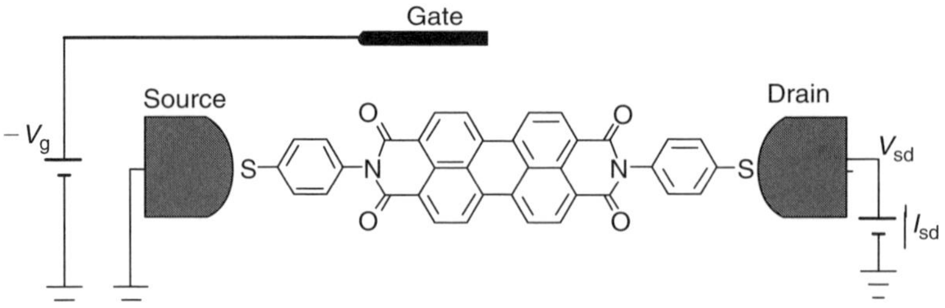

FIGURE 5.3 Schematic diagram of a single molecule transistor with an electrochemical gate based on perylene bisimide. (From Xu, B. et al., *J. Am. Chem. Soc.*, 127, 2386, 2005.)

5.1.3 Perylene Bisimide Derivatives for Organic Light-Emitting Diodes

An OLED with a double-heterostructure of indium tin oxide (ITO) substrate/aromatic diamine (TPD)/8-hydroxyquinoline aluminum (Alq_3)/**28**/Mg/Ag was fabricated by successive vapor deposition. The emitting color varied from the orange dominated by the red emission of **28** to greenish-yellow with a strong contribution of the green emission of Alq_3 as voltage increased, which is attributed to the electric field–induced quenching of excited states and voltage evolution of the recombination zone in **28** [22].

A starburst-type red-emitter based on perylene bisimide derivative has also been synthesized, into which an electron-transporting emitter unit of Alq_3 is incorporated [23]. In the system, the fluorescence of Alq_3 unit was significantly quenched due to the effective singlet energy transfer between two chromophore units in their excited states. A red electroluminescence (EL) emission originated from characteristic perylene bisimide moiety was observed. A turn-on voltage as low as 6 V, and the maximum luminescence of 2200 $cd \cdot m^{-2}$ at the driving voltage of 15 V (0.61 $cd \cdot A$) have been elaborated. More recently, dendron-functionalized perylene bisimides (**29** and **30**) with carrier-transporting ability were also applied as red electroluminescent materials [24]. The dendrimer designs are on the basis of following considerations: (1) dendron functionalization to incorporate carbazole or oxadiazole units to realize the carrier-injection adjustment; (2) tuning or improving solubility, functionality, and glass transition temperature (T_g) via well-defined dendrons; and (3) avoiding the luminescence quenching with the help of high site-isolation of dendrons to enhance core luminescence.

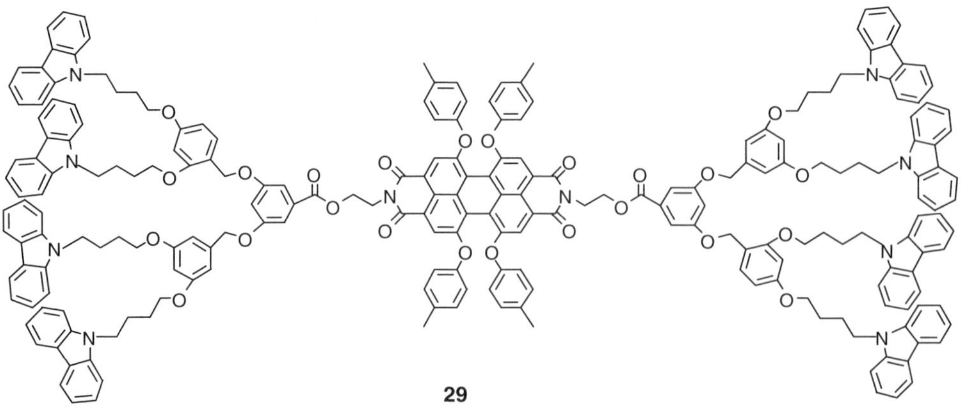

29

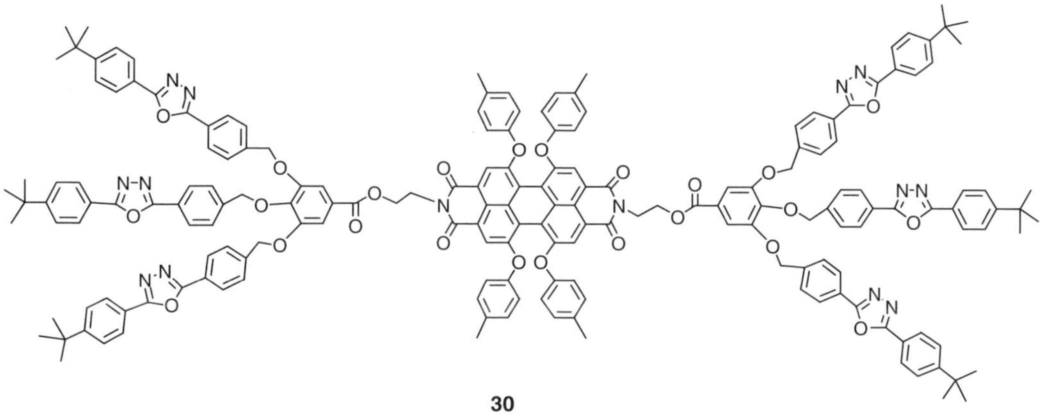

30

In conclusion, the broad visible absorption of perylene bisimides in the region of 400–550 nm and high fluorescent quantum yield with near unity are making them potential candidates for photosensitizers of solar cell, and EL emitters. Perylene monoimide and bisimide dyes have been proved to be the best fluorophores nowadays available for single molecule spectroscopy. Moreover, the assembly of perylene bisimide into supramolecular architectures through hydrogen-bonding, metal–ion coordination, and π–π stacking has been well studied by Würthner's group [8].

5.2 PHTHALOCYANINES

Phthalocyanine (Pc) is the tetraaza analog of tetrabenzoporphyrin. Its Cu(II) complex as well as benzene-substituted derivatives are extremely important pigments and dyes known for their brilliant turquoise shades and high lightfastness [1], which exhibit very strong characteristic optical absorption in the visible region, the so-called Q band. The syntheses of metal-free phthalocyanine **35** and metal complex **36** could follow different approaches with various starting materials of **31–34** (Scheme 5.2); among them isoindoline-1,3-diimine **34** is the most key intermediate. Functionalized phthalocyanines can be obtained either by direct substitution of the core molecule or by tailored synthesis from more-complicated starting materials [1]. Phthalocyanine pigments are practically insoluble in most organic solvents at ambient temperature, but they can often be recrystallized in small amounts from high-boiling solvents.

Phthalocyanine is a very stable π-conjugated macrocyclic ligand that can form metal complexes with almost all metal elements. Also, in most cases, the planarity of the ligand is retained. A variety of phthalocyanine molecules are thermally stable, and can thus be sublimed without decomposition. In contrast to Ca(II), Ba(II), Mg(II), or Cd(II) complexes, the corresponding Cu(II), Zn(II), Fe(II), Co(II), and Pt(II) complexes of phthalocyanine sublime in vacuo at 550°C–600°C. Therefore, the preparation of relative thin films by vacuum evaporation is feasible. Phthalocyanines have been widely studied in the field of optoelectronics such as OLEDs [25,26], photovoltaic devices [10,27–29], nonlinear optical (NLO) materials [30–37], and OFETs [38,39]. Aside from these applications, phthalocyanines were also utilized in the areas of optical recording [40] and molecule conductors [41].

5.2.1 PHTHALOCYANINES FOR ORGANIC LIGHT-EMITTING DIODES

The well-known dye pigment of Cu(II) phthalocyanine (CuPc) is chemically and thermally very stable, and has a number of interesting properties: being an organic semiconductor, easily forming a

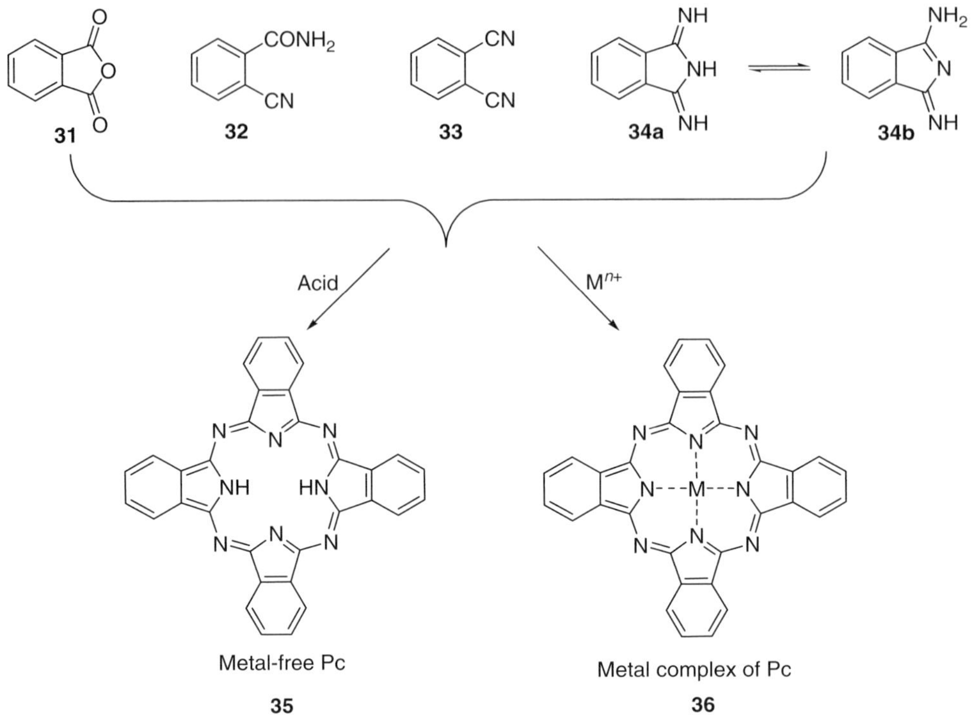

SCHEME 5.2 Synthetic routes of phthalocyanine and metal phthalocyanine.

compact and smooth thin films, and exhibiting photoconductivity and catalysis [25]. It has been widely utilized in OLEDs as a buffer layer between hole-transporting layer (HTL) and ITO anode (i.e., hole injection electrode). The buffer layer can improve the performance of OLEDs from several aspects such as suppressing noisy leakage current, reducing the operating voltage, and enhancing the thermal stability and quantum efficiency. However, the mechanism or function of CuPc as a buffer layer has still not fully been understood. Study by x-ray photoelectron spectroscopy on the interface of CuPc and ITO film shows that the oxygen in ITO film may diffuse into the organic semiconductor, and cause the alteration of electronic states in organic materials. The thickness of CuPc layer was critical and optimized. The HOMO level of CuPc is well matching between the ITO Fermi level and NPB HOMO level. The HOMO level and barrier gap of CuPc do not change for thicknesses up to 16 nm [25].

5.2.2 Phthalocyanines for Solar Cells

Stabilization of charge-separated states was observed in phthalocyanine–fullerene ensembles through supramolecular donor–acceptor interactions [27]. Actually, phthalocyanines are one of the most intensively studied dyes being as organic p-type semiconductors. A two-layer thin-film organic photovoltaic cell was first fabricated from CuPc and an n-type of perylene tetracarboxy diimide [10]. Organic/inorganic hybrid solar cells were also constructed from TiO_2 and zinc phthalocyanine (ZnPc). Study shows that the formed radical anions of oxygen drift toward the interface with the TiO_2 in this type of solar cell, resulting in a decrease of photocurrent [28]. Phthalocyanines **37** and **38** were anchored onto nanostructured TiO_2 films via the substituted ester groups. When the nanostructured TiO_2 electrode was treated with **38** in alkaline solution, a monochromatic IPCE of 4.3% was achieved at 690 nm [29].

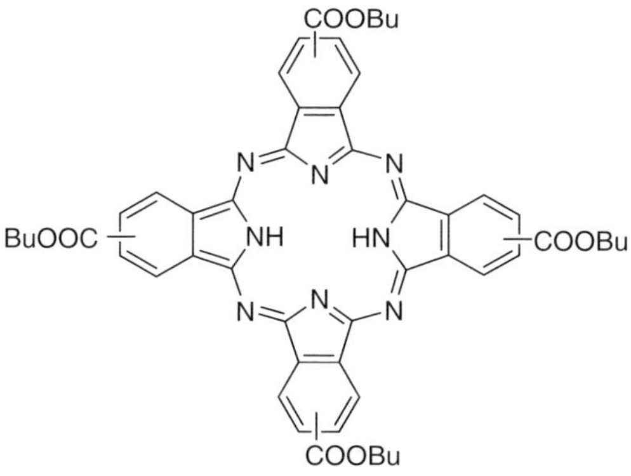

PcBu
37

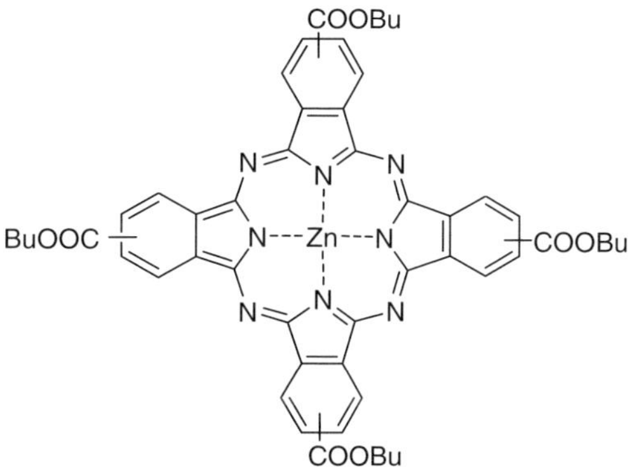

ZnPcBu
38

5.2.3 Phthalocyanines for Nonlinear Optical Materials

Strong nonlinearities in organic molecules usually arise from highly delocalized π-electron systems. Phthalocyanines, with their extensive two-dimensional 18 π-electron system, fulfill this requirement and have been, indeed, intensively investigated as NLO materials. These exhibit other additional advantages, namely, exceptional stability, versatility, and processability features. The architectural flexibility of phthalocyanines was well exemplified by a large number of metallic complexes, as well as by a large variety of substituents that can be attached to the phthalocyanine core [30]. Furthermore, some of the four isoindole units can be completely replaced by other heterocyclic moieties, giving rise to different phthalocyanine analogues. All these chemical variations can alter the electronic structure of macrocyclic core, thus allowing the fine-tuning of nonlinear response [30].

It has been developed to yield second-harmonic generation (SHG) by incorporating the proper donor and acceptor substituents to phthalocyanines since a decade ago. Three phthalocyanines of "push–pull" type **39–41** have an unusually strong dipole moment. The values of quadratic hyperpolarizability are quite significant and superior to those found for similar push–pull compounds lacking of a triple bond in the donor–acceptor path [31,32].

Langmuir–Blodgett (LB) films of a metal-free phthalocyanine, 1,8-naphthalimide-tri-*tert*-butylphthalocyanine **42**, were fabricated. The asymmetrically substituted phthalocyanine not only possesses good solubility in common organic solvents, but also possesses an ideal LB film forming ability. The SHG of a three-layer LB film of **42** is approximately 1.76 relative to a Y-cut quartz wedge $I_{2\omega}^{p\rightarrow p}$. The values of second-order susceptibility ($\chi^{(2)}$) and the molecular hyperpolarizability (β) from the multilayers are 8.32×10^{-9} and 5.97×10^{-30} esu, respectively [33].

Studies on the role of the central metal atom show that transition metals seems to strongly enhance the off-resonant $\chi^{(3)}$, especially for cobalt phthalocyanine. An ultrathin composite film containing both polyoxometalate anion $[PMo_{12}O_{40}]^{3-}$ (PMo_{12}) and a planar binuclear phthalocyanine, bi-CoPc **43**, was prepared by the electrostatic layer-by-layer self-assembly method, which showed a third-order nonlinear optical response $\chi^{(3)}$ of 4.21×10^{-12} esu [34].

Optical limiting (OL) is a nonlinear effect, a decrease in the transmittance of NLO materials under high-intensity illumination. Thus, the transmission of an optical limiter is high at normal light intensity and low for intense beams. The most useful one is the protection of optical elements and sensors against damage by exposure to sudden high-intensity light [30]. The optical limiting with Pcs was first reported for PcAlCl (**44**) [35]. It was found that phthalocyanines of PcAlCl, PcTiO, PcVO, and PcInCl possess larger NLO absorption coefficients than that of nickel and copper derivatives [36]. Axially modified metal phthalocyanines and naphthalocyanines were studied for optical

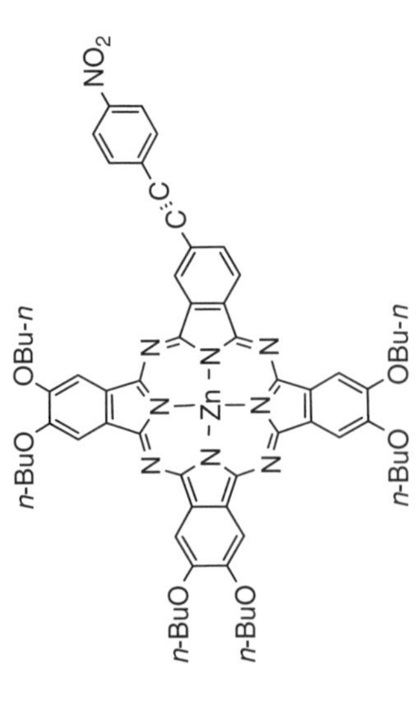

39

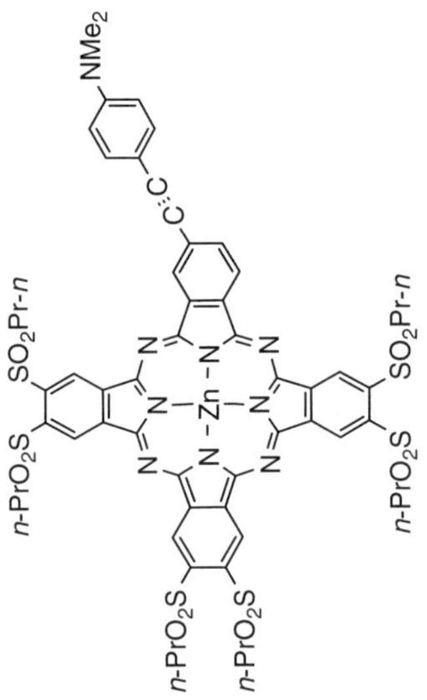

40

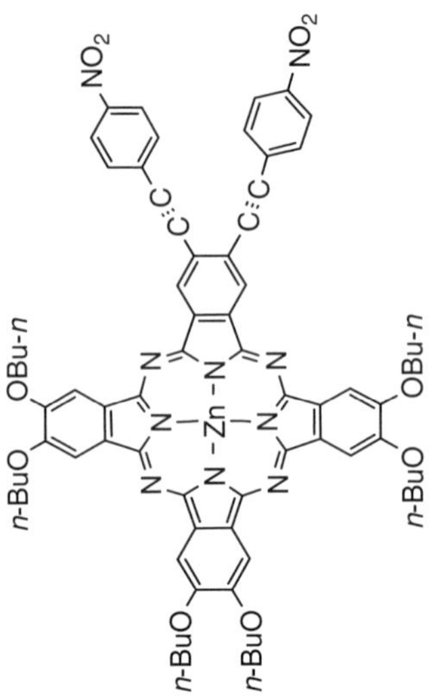

41

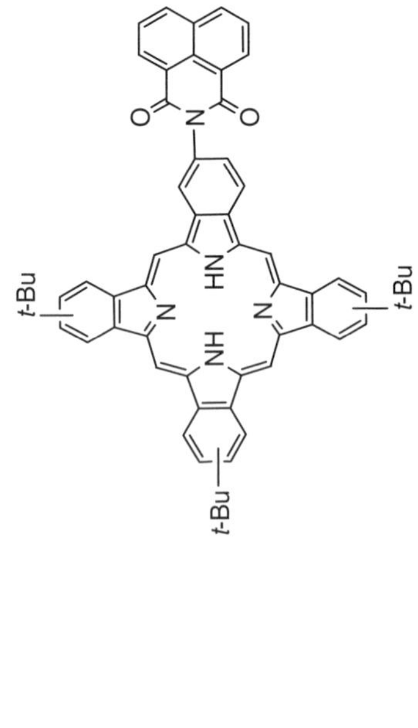

42

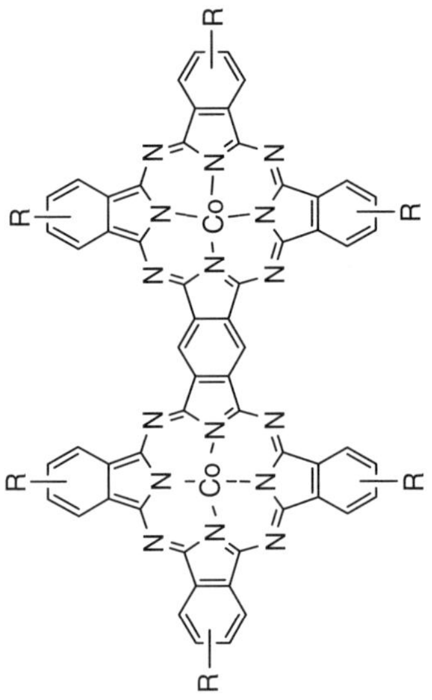

43 $R{=}SO_2NH_4$

limiting [35,36]. The optical limiting of axially modified In and Ga phthalocyanines **45–49** are very effective, exhibiting a range of saturation densities and absorption cross-section ratios [37].

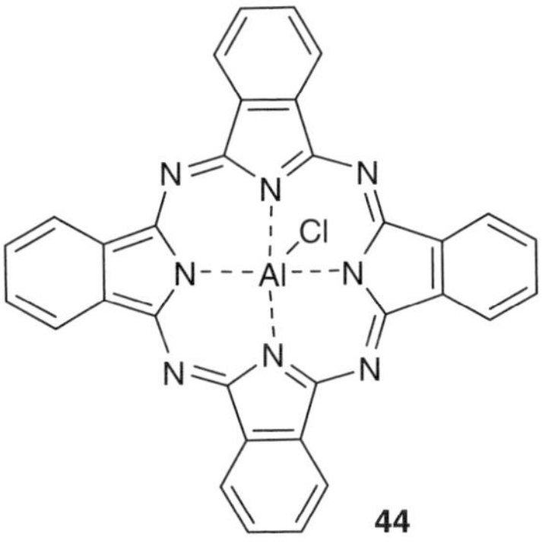

44

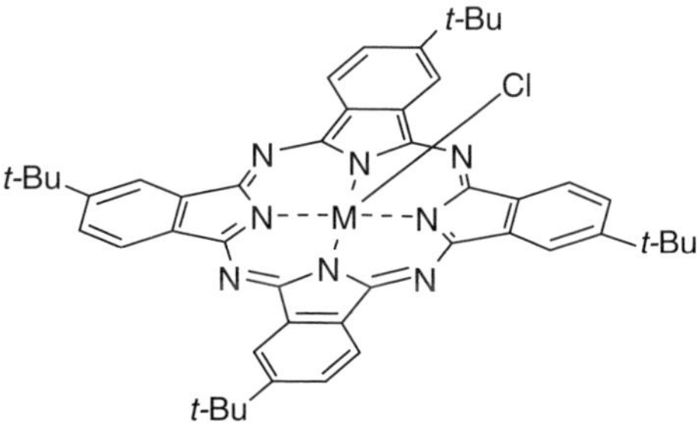

45 M = Ga^{3+}
46 M = In^{3+}

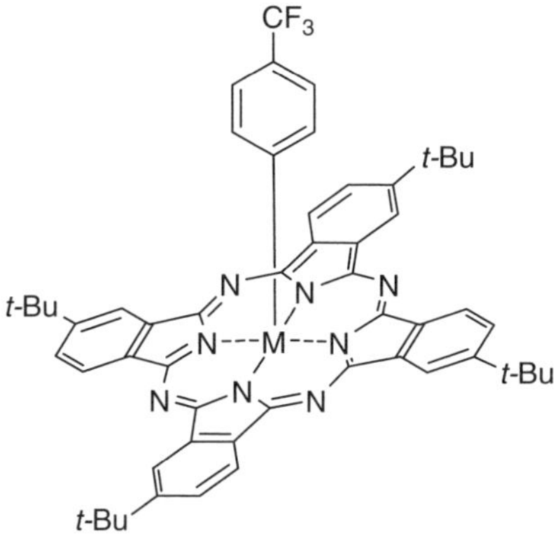

47 M = Ga^{3+}
48 M = In^{3+}

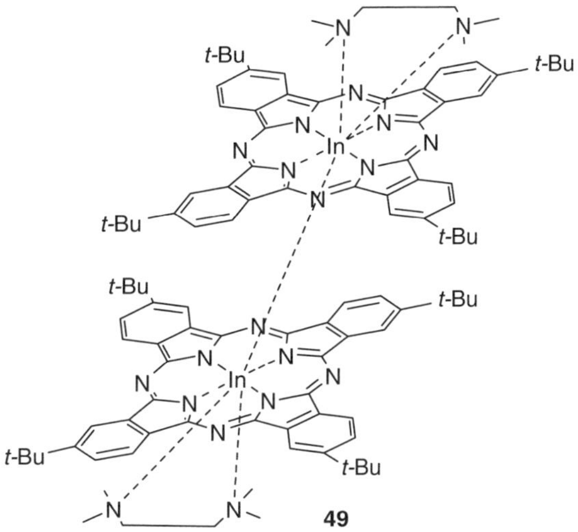

49

5.2.4 PHTHALOCYANINES FOR ORGANIC FIELD-EFFECT TRANSISTORS

Phthalocyanines are excellent candidates for building blocks because of their inherent symmetry, which makes rational designs of supramolecular structures easier to achieve. The most investigated members in this family are CuPc and CoPc. The mobilities of the vacuum deposited Pc thin films are found in the range of 10^{-5} to 10^{-3} $cm^2 \cdot V^{-1} \cdot s^{-1}$. More recently, a single crystal CuPc-based OFET have been reported with mobility as high as 1.0 $cm^2 \cdot V^{-1} \cdot s^{-1}$, the highest reported so far for Pc-based OFETs [38].

The LB films based on amphiphilic tris(phthalocyaninato) rare earth triple-decker complexes **50** and **51** (Figure 5.4) showed unexpectedly high OFET performance with carrier mobilities reaching 0.24–0.60 $cm^2 \cdot V^{-1} \cdot s^{-1}$, which is among the highest mobilities achieved so far for LB film-based OFETs [39].

5.3 PORPHYRINS

The parent structure of porphyrin (**52**) and corresponding metalloporphyrin (**53**) contain a skeleton of four pyrrole rings that are bridged in their α,α′-positions by four methine groups. The substituents on the positions of 2, 3, 7, 8, 12, 13, 17, and 18 are so-called β-porphyrins; the substituents on the positions of 5, 10, 15, and 20 are so-called *meso*-porphyrins [1]. Actually, natural porphyrins are of

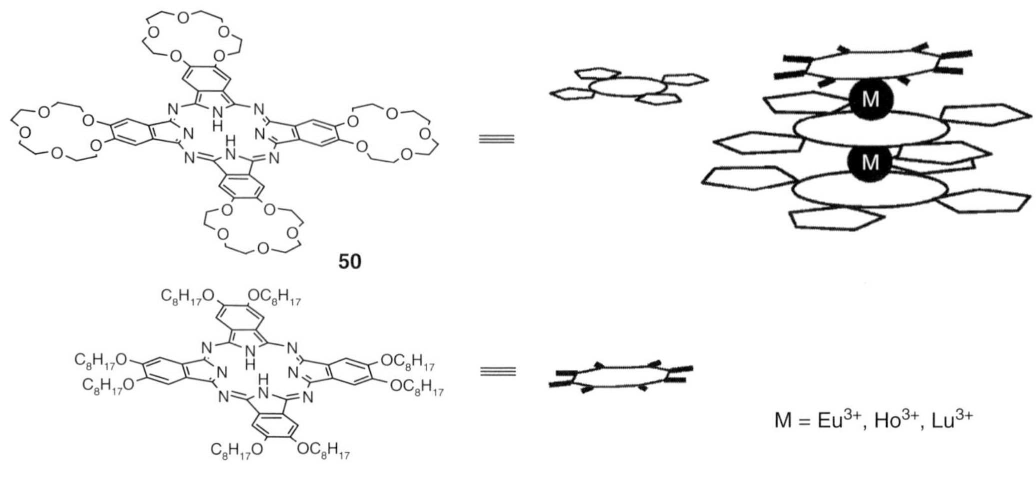

FIGURE 5.4 Schematic molecular structures of the triple-decker compounds. (From Chen, Y. et al., *J. Am. Chem. Soc.*, 127, 15700, 2005.)

enormous importance for almost all lives on Earth. Hemin **54** is the chromophore of hemoglobin in the blood of vertebrates. Chlorophylls **55**, which have striking similarities to porphyrins, are responsible for the green color of leaves and their photosynthesis.

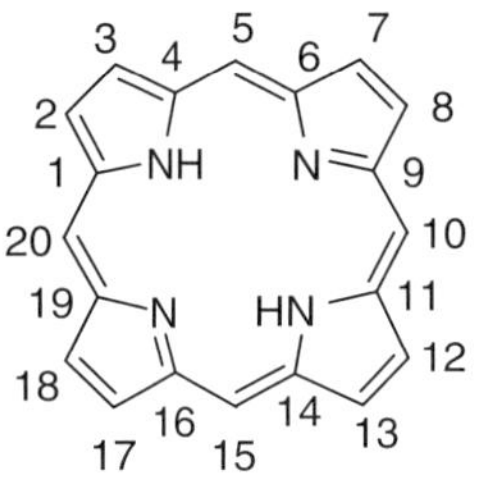

52

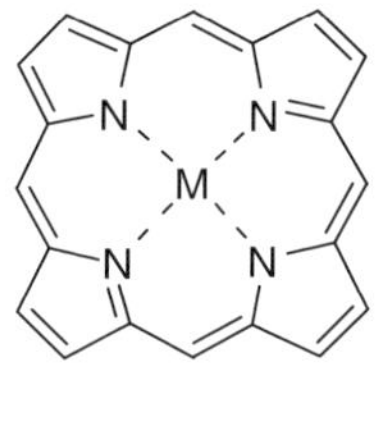

53

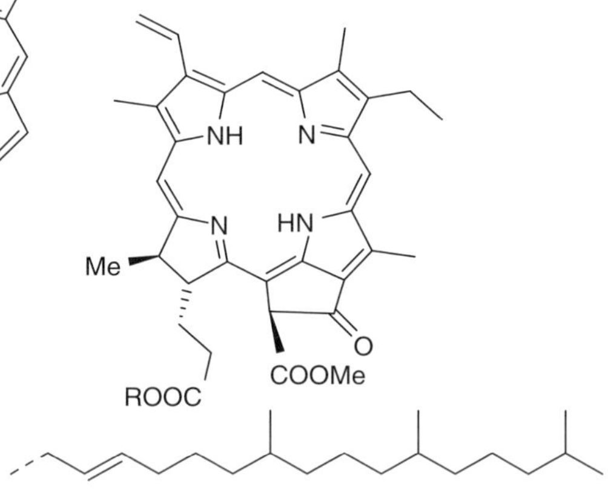

R = Phytyl residue

55

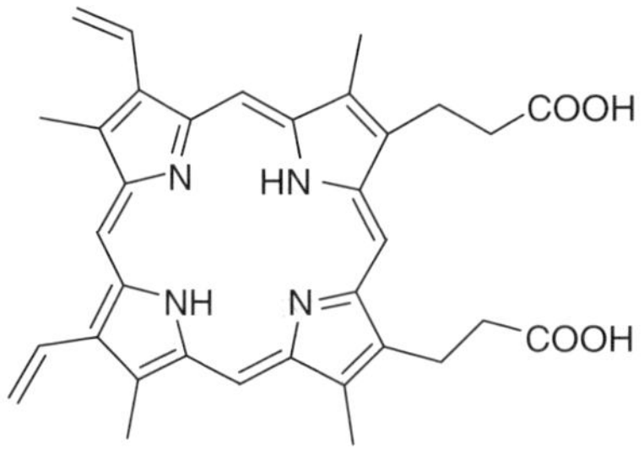

54

As shown in Scheme 5.3, aryl-substituted *meso*-porphyrins can be prepared via Adler and Longo's method by allowing substituted benzaldehyde and pyrrole to react for 30 min in refluxing propionic acid (141°C) open to the air [42]. In 1987, Lindsey presented another method [43] complementary to the Adler–Longo's procedure, allowing small quantities of porphyrins to be prepared from sensitive aldehydes in 30%–40% yield without difficult purification problems. Treatment of pyrrole and the

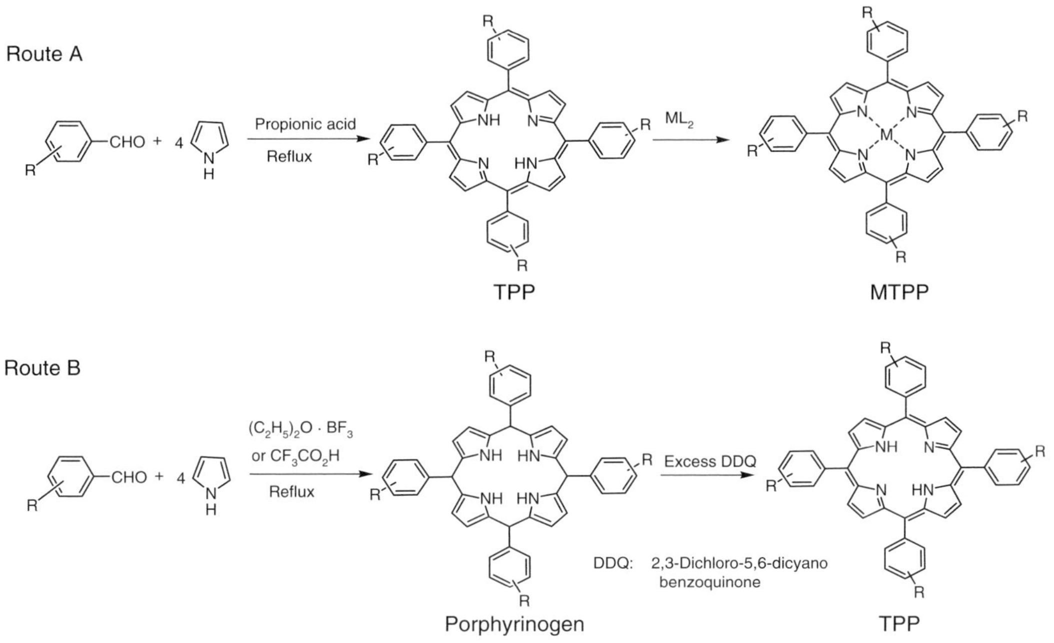

SCHEME 5.3 Synthetic routes for aryl-substituted *meso*-porphyrin: (A) Adler–Longo's method; (B) Lindsey's method.

desired benzaldehyde react reversibly at room temperature with trace acid catalysis to form the cyclic tetraphenylporphyrinogen at thermodynamic equilibrium. An oxidant is then added to irreversibly convert the porphyrinogen to porphyrin. However, Lindsey's method is very sensitive to the concentration. The reaction at high dilution or high concentration affords a negligible yield of the cyclic porphyrinogen. Lindsey's method is also extended to the preparation of *meso*-tetraalkylporphyrins and one hybrid porphyrin containing both aryl and alkyl substituents [43]. Also from a purely curiosity-driven point of view, macromolecules containing an extended fully conjugated π-electron network, such as expanded porphyrins and their heterologs, were also prepared and might provide insight into questions of aromaticity [44].

The porphyrins lie at the focal point formed from divergent fields of research, including solar energy conversion, catalysis, spectroscopy, and the development of organic metals. The *meso*-tetraphenylporphyrins (TPP) have been used in a wide variety of model studies. Porphyrins possess good chromophoric activity at the visible region of the spectrum and good electron-donating properties [1]. Table 5.2 lists the comparison of electronic absorption data of macrocyclic copper(II)

TABLE 5.2
Comparison of Electronic Absorption Data of Macrocyclic Copper(II) Complexes

Copper(II) Complexes	Naphthalocyanine	Phthalocyanine	Porphyrin	Tetrazaporphyrin
Q band λ_{max} (logε)	765 (5.23)	678 (5.34)	540 (2.04)	578 (4.98)
B band λ_{max} (logε)	—	350 (4.76)	417 (44.7)	334 (4.57)

Source: From Chen, Z.M., Xia, Z., and Wu, Y.Q., *Prog. Chem.*, 16, 820, 2004. With permission.

complexes [45]. Comparing with Q band of phthalocyanine, porphyrin has a blue shift by about 100 nm. Notably, for porphyrin, the molar coefficient of B band is much larger than that of Q band. Metal porphyrin complexes, which possess good optical properties at 400–650 nm, have been considered as novel high-density recordable optical disk storage materials [45].

5.3.1 Porphyrins for Solar Cells

The design of molecular systems capable of performing light-induced charge separation has been a subject of intensive research. Specifically, the interest in organic photovoltaics has risen. An essential question is how to create efficient charge separation in an organic photovoltaic system. One promising way is to use electron donor–acceptor molecules (bipolarity) to create a primary charge separation [46]. Fullerenes, which have excellent electron-accepting properties, linked with porphyrin units have attracted much attention [47,48]. Self-assembled monolayer of porphyrin–fullerene dyads **56** and **57** on ITO surfaces was used to fabricate photovoltaic devices. The system with porphyrin–fullerene dyad possesses the quantum yield of photocurrent generation 30 times larger than that of the corresponding system without C_{60}. This is ascribed to an efficient photo-induced electron transfer from the porphyrin singlet excited state to the C_{60} moiety [48].

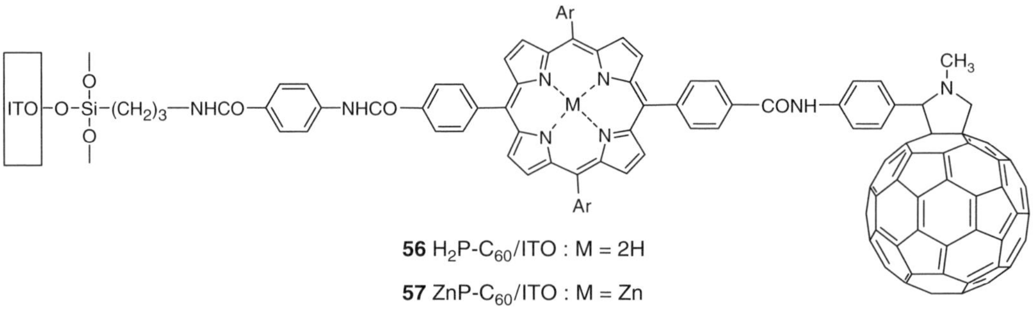

56 H_2P-C_{60}/ITO : M = 2H

57 ZnP-C_{60}/ITO : M = Zn

Würthner et al. utilized a self-assembly method to construct a side-to-face supramolecular array of chromophores, where a pyridyl-substituted perylene bisimide dye axially binds to two ruthenium porphyrin fragments. Wavelength-dependent electron and energy transfer pathways were observed [49]. Photochromic moieties were employed to control photo-induced electron transfer in triads [50–52]. Triad dimethyldihydropyrene–porphyrin–fullerene (DHP-P-C_{60}) **58** was a typical system, which can be reversibly converted between DHP form **58a** and cyclophanediene (CPD) form **58b** under light irradiation. Excitation of porphyrin moiety in **58a** resulted in a final DHP$^{\bullet+}$-P-$C_{60}^{\bullet-}$ state that has a lifetime of 2 μs. This is comparable to the short-lived (<10 ns) CPD-P$^{\bullet+}$-$C_{60}^{\bullet-}$ state generated from excitation of porphyrin moiety of **58b**. Many cycles may be performed without substantial degradation. Thus, light can be tuned in switching long-lived photo-induced charge separation on or off [52].

Organic solar cells using quaternary self-organization of porphyrins **59** and **60** and fullerene units by clusterization with gold nanoparticles on nanostructured SnO_2 electrodes were prepared (Figure 5.5). The film of the composite clusters with gold nanoparticle exhibits an IPCE as high as 54% and broad photocurrent action spectra (up to 1000 nm). The power conversion efficiency of such composite electrode reaches as high as 1.5%, which is 45 times higher than the system consisting of the both single components of porphyrin and fullerene [53].

Photovoltaic properties of five metalloporphyrins **61–65** adsorbed on TiO_2 were studied. Results showed that Zn(II) diamagnetic metalloporphyrins had very high IPCE conversion efficiencies compared to those observed for the Cu(II) paramagnetic metalloporphyrins. In addition, porphyrins with a phosphonate anchoring group show lower efficiencies than those with a carboxylate anchoring group. For porphyrin **62**, the IPCE is 75% for the Soret band and 50%–60% for the Q-band peaks [54].

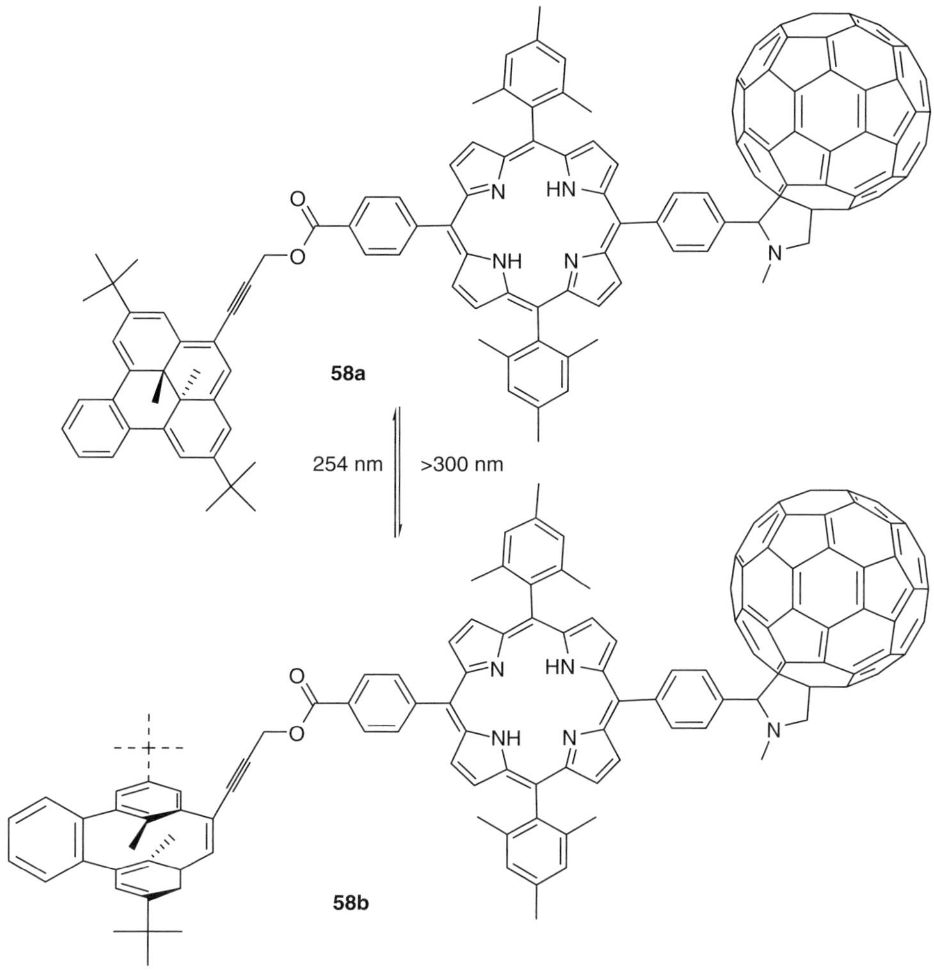

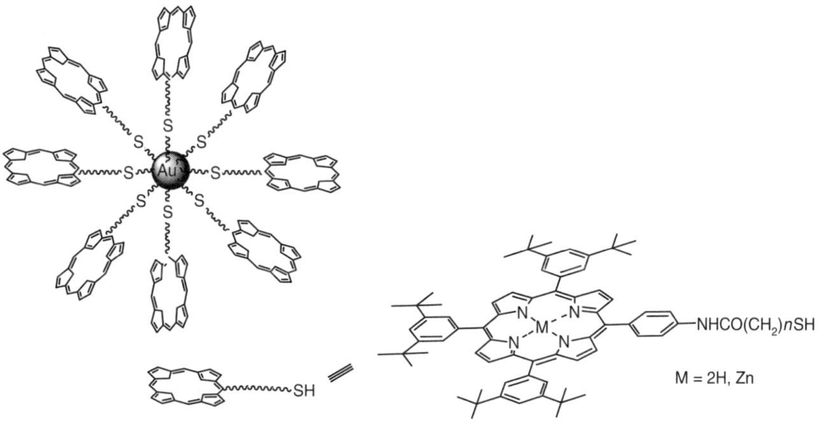

59 M = 2H : H_2Pc*n*MPC; (*n* = 5, 11, 15)

60 M = Zn : ZnPc*n*MPC; (*n* = 11, 15)

FIGURE 5.5 Schematic diagram of self-organization of porphyrins **59** and **60** by clusterization with gold nanoparticle.

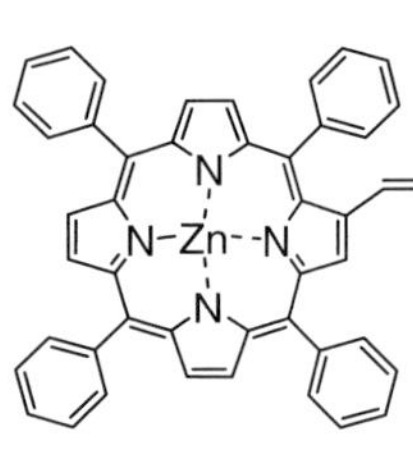

61

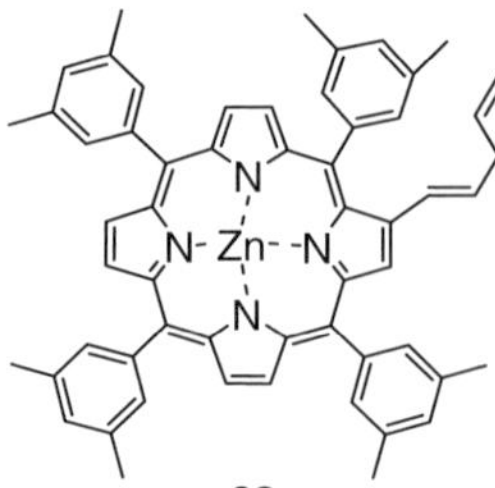

62

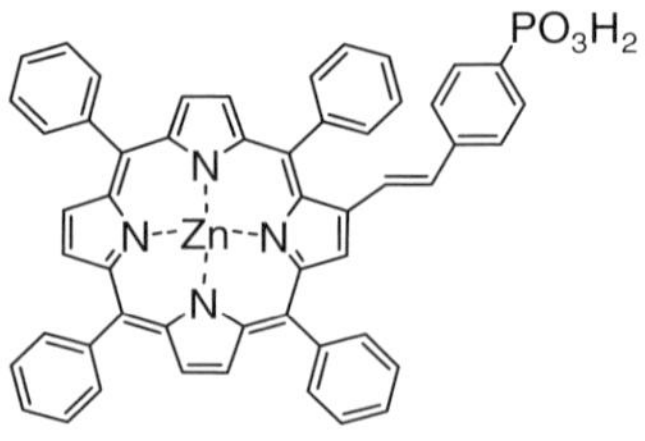

63

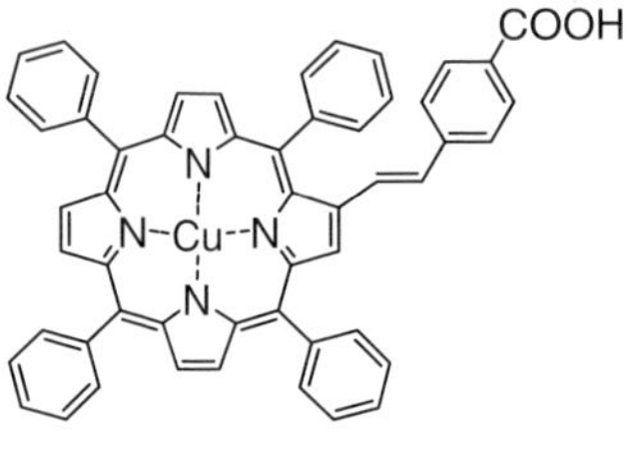

64

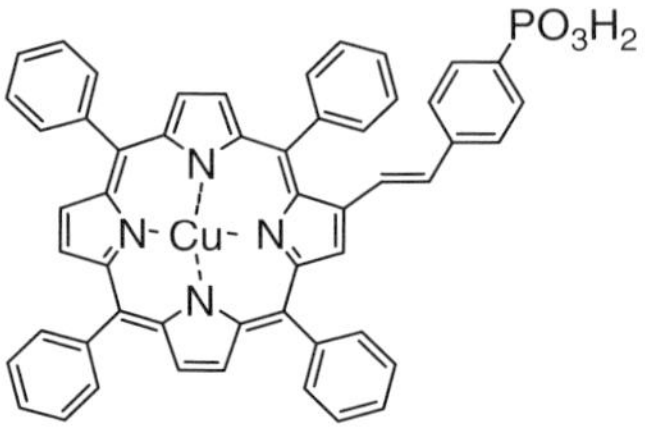

65

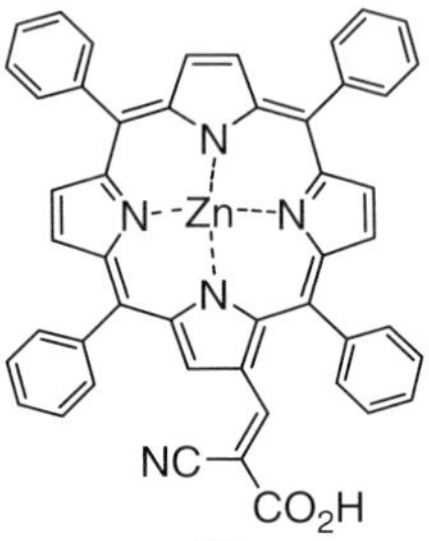

66

Another Zn(II) porphyrins **66** sensitized TiO_2 solar cells using an I^-/I^{3-} electrolyte produced high photovoltaic properties with a IPCE of 85% under standard AM 1.5 sunlight, a short circuit photocurrent density of 13.0 ± 0.5 mA·cm^{-2}, and an open circuit voltage of 610 ± 50 mV [55].

5.3.2 Porphyrins for Organic Light-Emitting Diodes

The platinum(II) porphyrins **67** and **68** exhibited strong phosphorescence in the red with narrow line widths, which has been used an electrophosphorescent emitter. When these were doped into aluminum(III) tris(8-hydroxyquinolate) (Alq_3) in the electron-transporting and light-emitting layer of OLEDs, bright saturated red emission with high efficiency at low-to-moderate current density has been achieved [56]. Interestingly, the photoexcited phosphorescent emission of Pt(II) porphyrin **69** increases by more than a 200-fold in a polystyrene film on nanotextured silver surfaces, coincident with a reduction in the triplet state lifetime by a factor of 5. Such large enhancement in luminescence quantum yields can be interpreted by the increased radiative rates due to interactions between the molecules and the electron plasma in nearby silver nanoparticles [57].

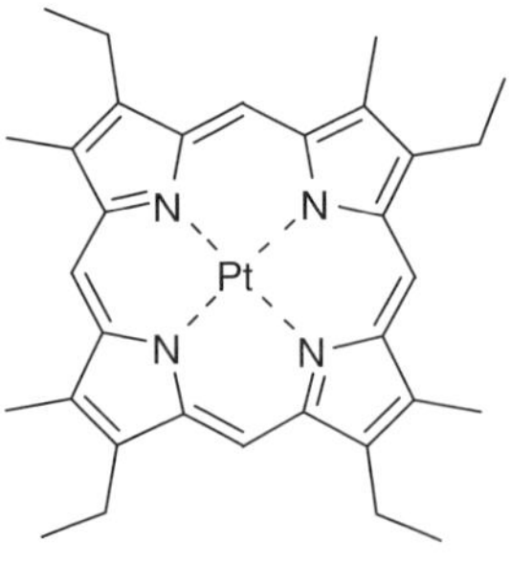

67

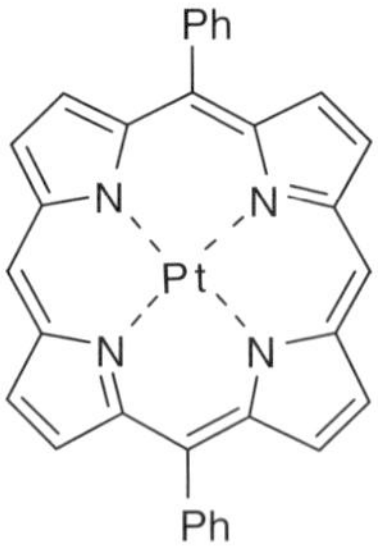

68

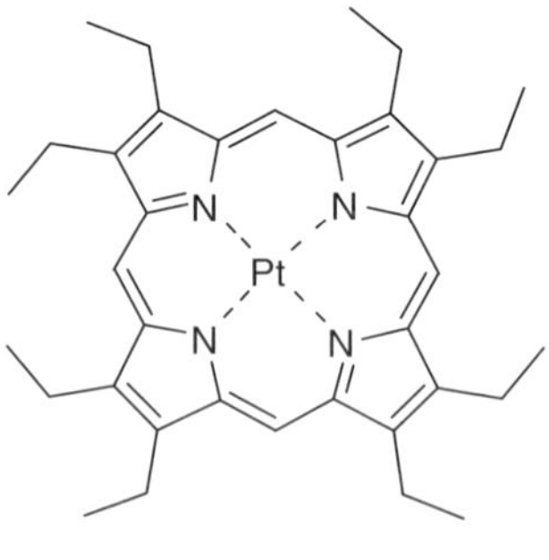

69

5.3.3 Porphyrins for Nonlinear Optical Materials

Earlier studies on porphyrin in NLO area were carried out by Suslick and coworkers [58] at the beginning of the 1990s. They examined the NLO properties of a set of *meso*-substituted porphyrins with both 4′-(dimethylamino)-phenyl and 4″-nitrophenyl moieties. These push–pull porphyrins were measured by the EFISH technique and showed β values of 10–30 × 10^{-30} esu [30,58].

Porphyrin derivates **70** bearing donor and acceptor are essentially coplanar, enabling effective electronic interactions within the molecule. Particularly, the copper derivative **70a** possesses exceptionally high resonance-enhanced β values, near 5000 × 10^{-30} esu, obtained by the Hyper-Rayleigh scattering (HRS) technique [59,60].

The effect of the metallic ions and axial ligands on the molecular hyperpolarizability of porphyrins has been investigated. Z-scan experiments over different transition metal complexes exhibit a significant increase for divalent ions with decreasing d-shell occupancy. In the case of tetravalent metal centers, a dramatic enhancement of the nonlinear parameter was found for electronegative axial ligands, for example, iodine [30]. An ultrathin nanoscopic multilayer film has been fabricated through the electrostatic layer-by-layer self-assembly of negatively charged porphyrin **71** and oppositely charged polyethylenimine. The nonlinear optical properties of ultrathin film were studied by Z-scan technique with laser duration of 8 ns at a wavelength of 532 nm. The nonlinear absorption coefficient and refractive index of the self-assembly ultrathin film are -9.7×10^{-5} $m \cdot W^{-1}$ and -7.56×10^{-12} $m^2 \cdot W^{-1}$, respectively, which are clearly better than that of **71** in solution [61].

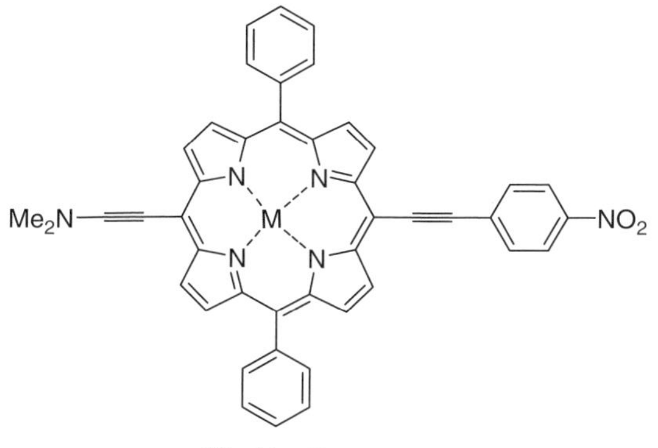

70a M = Cu
70b M = Zn

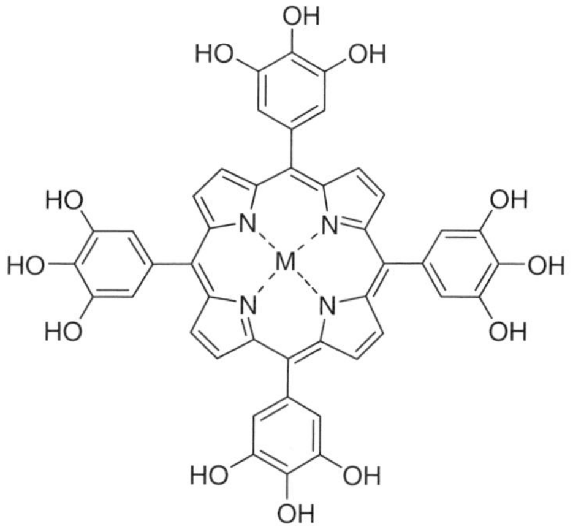

71

5.3.4 Porphyrins for Optical Limiting Devices

A series of *meso*-alkynyl porphyrin group III and IV metal complexes **72** were studied as strong candidates for use in optical limiting devices. The heavier metal complexes of In(III), Tl(III), and Pb(II) offer the most sensitive response at around 530 nm, and $\sigma^{T}_{ex}/\sigma_{gr}$ values in the range of 45–48 was recorded. These values are the largest reported for macrocyclic dyes. The In(III) and Tl(III) complexes exhibit strong nonlinear absorption across their transmission window from 480–620 nm, have relatively long triplet lifetimes ($\tau_T \approx 800$ ns) in air saturated solutions, and a strong nonlinear response is predicted from the picosecond to microsecond timescale [30,62].

5.4 CYANINES

The name cyanine was originally referred to the Williams dyes, but it has been later extended to the compounds shown in Figure 5.6, where the nitrogen atom at the end of the conjugated chain usually

$$R_2\ddot{N}{-}(CH{=}CH)_r CH{=}\overset{+}{N}R_2' \rightleftharpoons R_2\overset{+}{N}{=}CH-(CH{=}CH)_r{-}\ddot{N}R_2'$$

FIGURE 5.6 General structure of cyanine compounds.

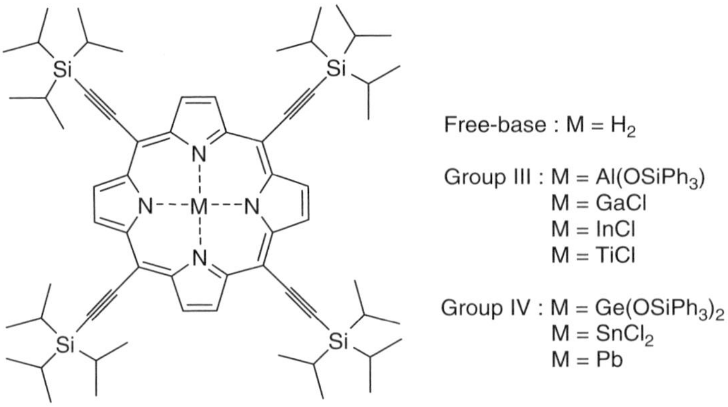

72

(but not all) is one of the members of a heterocycle. A general method for the synthesis of stilbazolium dyes (hemicyanines) is involved in condensation of a 2/4-methyl quaternary salt of the base with substituted benzaldehyde in the presence of a suitable base (Scheme 5.4). Dyes of cyanine family usually display bright color owing to their narrow absorption band. Among these categories of dyes, the cyanine, merocyanine, and squaraine dyes are substances with a variety of colors but are not widely used for dyeing purpose, as they are decolorized by light and acid. These have been employed extensively as spectral sensitizers for silver halide photography and other large band gap semiconductor materials, in optical disks as recording media, in industrial paints, for trapping of solar energy, as laser materials, in light-harvesting systems of photosynthesis, as photorefractive materials, as antitumor agents, and as probes for biological systems [63]. Here, we mainly discuss the application of cyanines in photovoltaic device and NLO materials.

5.4.1 Cyanines for Photovoltaic Devices

Cyanine dyes have very large absorption extinction coefficients ($\sim 10^5\ M^{-1} \cdot cm^{-1}$) and are easily processed from solution. And cyanine dyes have the possibility to form J- and H-aggregates with highly delocalized excitons when attached to TiO_2 film [63–65]. For all these reasons, cyanine dyes possess ideal characteristics to be used in fully organic photovoltaic devices, especially in sensitized dyes on nanocrystalline TiO_2 [65].

The absorption peaks of cyanine monomers in solution are very sharp. However, the absorption spectra shows that when sensitized on TiO_2, compounds **73** and **74** form two absorption peaks, which are corresponding to H-aggregate and monomer absorptions on TiO_2 surface. Incorporation of carboxylic groups results in performance in a sensitized solar cell comparable with a control ruthenium complex [64].

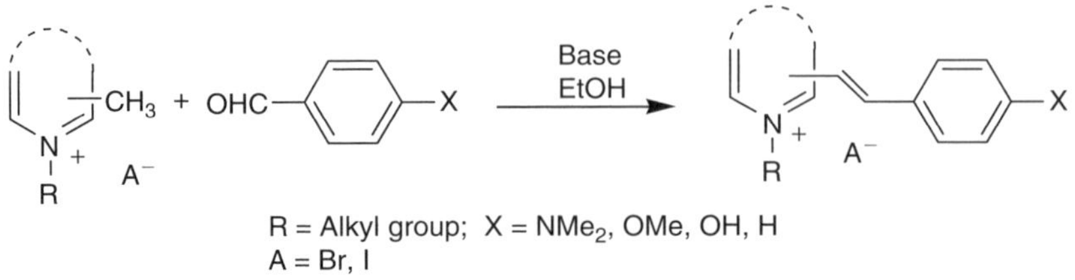

SCHEME 5.4 General method for the synthesis of cyanines.

Although cyanines could cover a broad spectrum in visible area through aggregation, the absorption is still far from the requirements as photosensitizers for solar cell. Combination of several dyes on the TiO_2 surface could fill this gap. For instance, the mixtures of cyanines **75** and **76** could be employed to co-sensitize the solar cell in a much wider spectrum, resulting in a high photoelectric conversion yield of 3.4% [66].

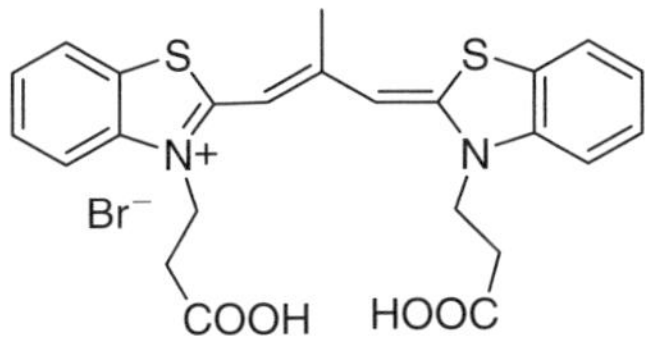

73

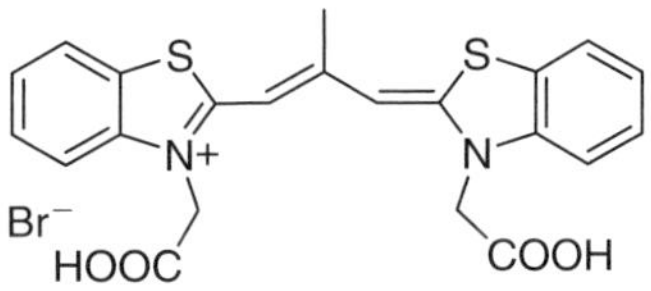

74

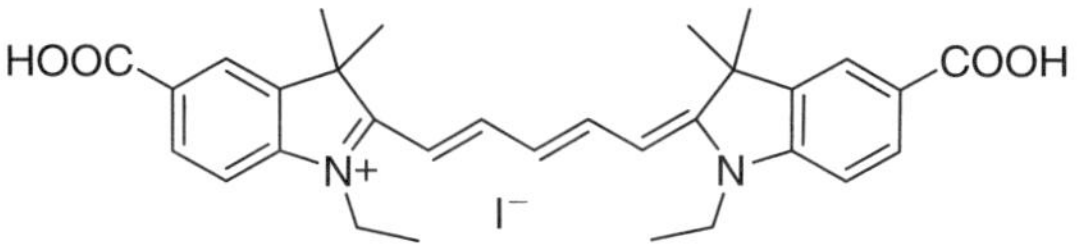

75

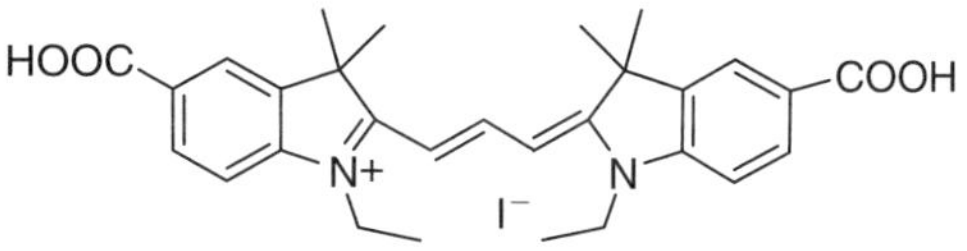

76

Upon adsorption on a TiO_2 electrode, all absorption spectra of four hemicyanine dyes **77–80** become broad compared with their respective spectra in ethanol solution. Hemicyanine **78** or **80** bearing two carboxyl groups exhibits a broader absorption spectrum on TiO_2 film, about two (or three) times more adsorption capability than that of **77** or **79**. Clearly, the light-to-electricity conversion property is dependent on the number of carboxylic groups. Among the four hemicyanine dyes, hemicyanine **80** generated the highest photoelectric conversion yield of 4.9%, with a short circuit photocurrent of 21.4 $mA \cdot cm^{-2}$, an open circuit voltage of 424 mV, and a fill factor of 0.49 under irradiation with 90.0 $mW \cdot cm^{-2}$ white light from a Xe lamp [67]. Overall conversion efficiency is 2.12% for **81a** sensitized nanocrystalline TiO_2 solar cell, which is much larger than **81b** (0.85%) and **81c** (0.51%). It might be attributed to the large steric hindrance, thus decreasing their adsorption on the electrode [68].

The heterojunction photovoltaic devices using cyanine perchlorate **82** as donor and C_{60} as acceptor were also studied [69]. Also, single and double layer photovoltaic devices were fabricated with two cyanine–fullerene dyads **83** and **84**. The fluorescence of the cyanine unit was quenched efficiently owing to the efficient photo-induced electron transfer from cyanine unit to C_{60} unit. However, the light conversion efficiency of the double layer device only reaches 0.1% under white light irradiation (3.1 $mW \cdot cm^{-2}$) [70].

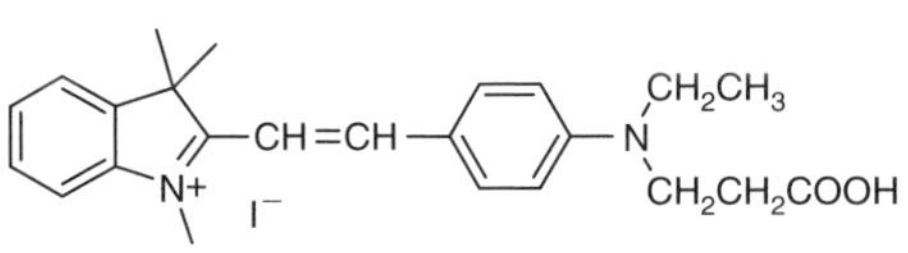

77

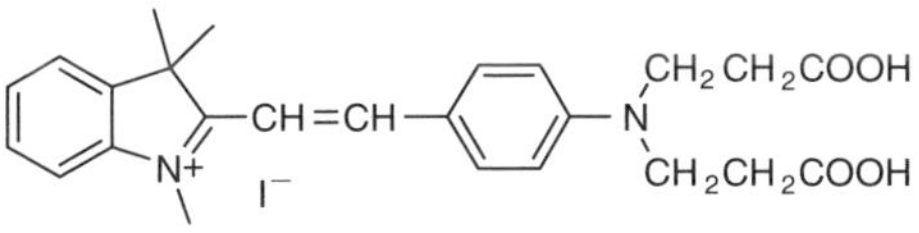

78

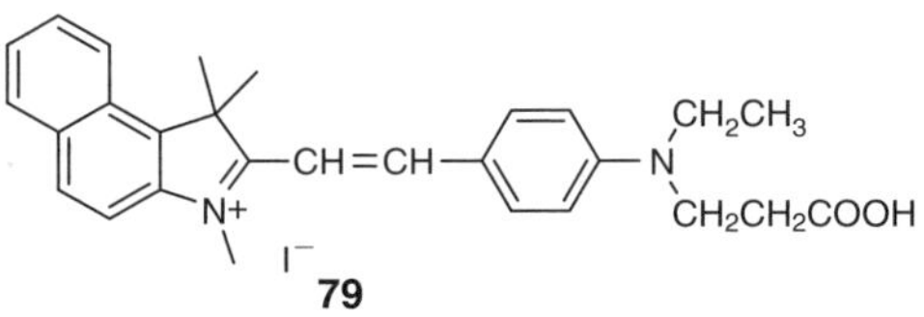

79

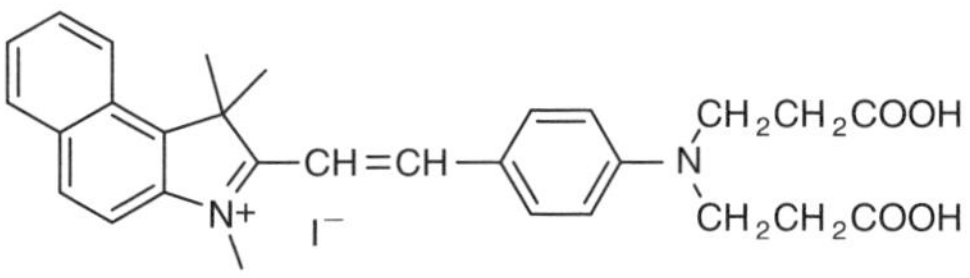

80

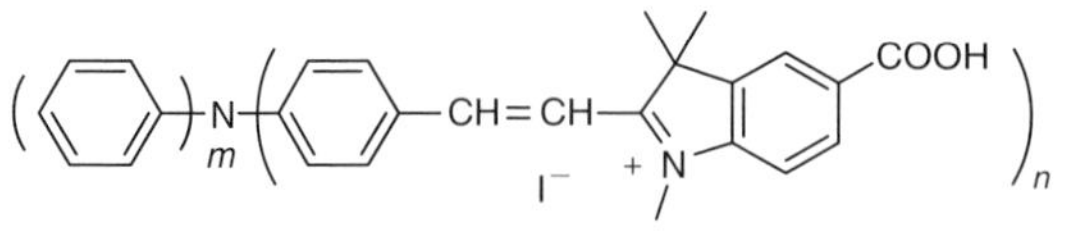

81a $m = 2, n = 1$
81b $m = 1, n = 2$
81c $m = 0, n = 3$

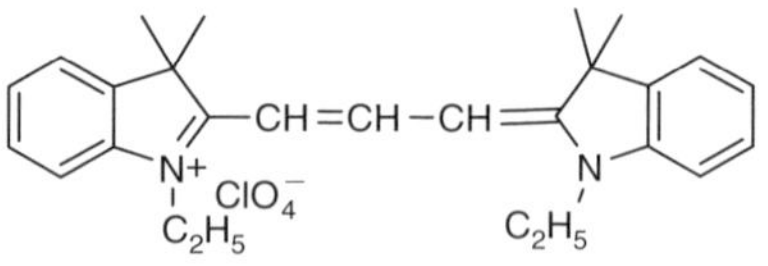

82

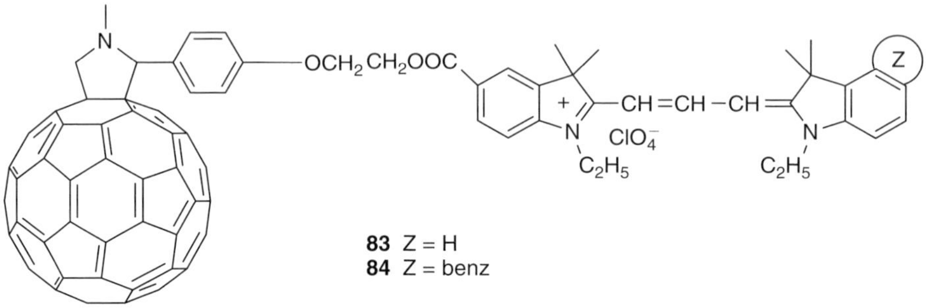

83 Z = H
84 Z = benz

5.4.2 Cyanines for Nonlinear Optical Materials

Many hemicyanines, merocyanines, and azacyanines show high NLO properties owing to their wide transparent range, high NLO coefficiency, large molecular hyperpolarizability, and extremely short response time. They are also used in optical signal processing such as amplification, frequency conversion (i.e., SHG), modulation, laser technology, telecommunication, data storage, and optical switches [63]. The effect of electron-withdrawing substituents on the first-order hyperpolarizability β of conjugated systems has received a great deal of attention in recent years. The substituted cyanine stems from the large and negative γ values of the parent molecule. The cyanines with an odd number of carbons are interesting because of their large and negative γ, which grow superlinearly with chain length. A given substituent produces opposite effects at alternating positions along the molecular backbone (even and odd carbons) [71,72].

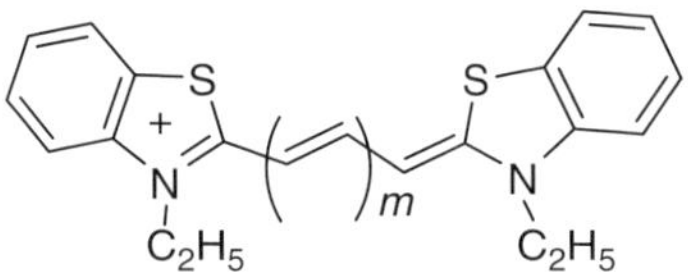

85a $m = 0$
85b $m = 1$
85c $m = 2$
85d $m = 3$

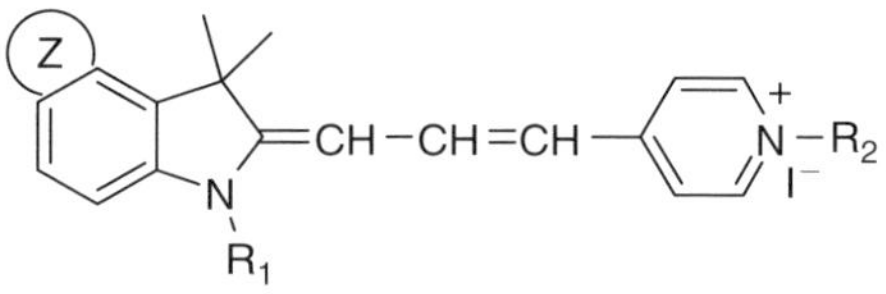

86a Z = H, R_1 = Me, R_2 = Me
86b Z = H, R_1 = Me, R_2 = Bu
86c Z = benz, R_1 = Me, R_2 = Me
86d Z = benz, R_1 = Bu, R^2 = Me

Hyper-Rayleigh scattering spectroscopy studies on dithiacarbocyanine **85a–d** suggest significant bond-length alternation in the ground state. Semiempirical molecular orbital calculations (at the AM1 level) indicated that isomerization locks in the bond-length alternation, enabling isomerization about a low-order bond. According to the generally accepted relaxation mechanism for S_1 cyanines, bond-length alternation must likewise be prerequisite to S_1–S_0 internal conversion [73].

Methine cyanine dyes with unsymmetrical structures using the Z-scan technique showed that cyanines **86a–d** have good nonlinear absorption properties. The molecular second-order NLO (β) coefficients of **86b** and **86c** reach 1.10×10^{-10} and 3.51×10^{-10} cm·mW^{-1}, respectively [74].

Merocyanine dyes have some of the highest known molecular second-order NLO (β) coefficients. Their ionic resonance forms have an extremely large dipole moment, which may reach 50 D, while

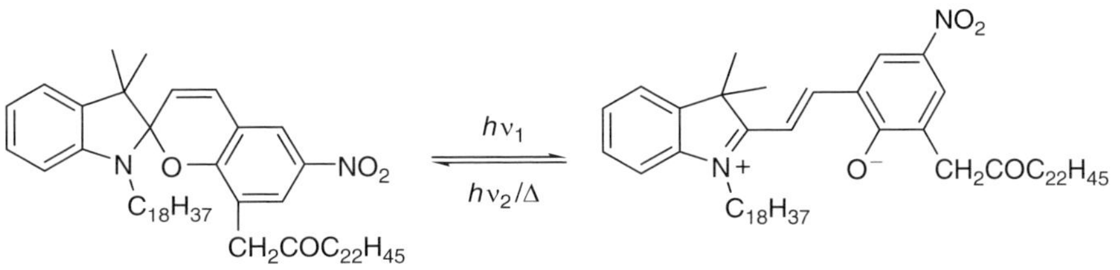

87a **87b**

SCHEME 5.5 Photochromic conversion of a spiropyran with NLO effect.

the quinoid form has a much smaller dipole moment. The photochromic conversion from spiropyran to merocyanine is accompanied by a major change in the conjugation of the π electrons, thus resulting in great effect on NLO properties [75]. Actually, both of the second- and third-order NLO properties have been observed in the photochromic system of spiropyran and merocyanine. The photochromic transformation from spiropyran **87a** to merocyanine **87b** caused a large light-switchable change in the third-order nonlinearity, in the same way as the second-order nonlinearity (Scheme 5.5) [76].

5.5 CARBAZOLES AND TRIPHENYLAMINE DERIVATIVES

The derivatives of carbazole and triphenylamine have been paid increasing attention because of their unique optical properties and strong hole-transporting ability in optoelectronic devices. By now, one of typical hole-transporting units is triphenylamine unit, its time-of-flight hole mobility reaches 10^{-3} $cm^2 \cdot V^{-1} \cdot s^{-1}$ at 10^5 $V \cdot cm^{-1}$ for amorphous glassy films at room temperature [77].

Numerous failure mechanisms have been proposed for the degradation in OLED performance over time. The presence of moisture and oxygen is believed to enhance the formation of dark spots, an effect that can be reduced by encapsulation. It has also been proposed that expansion of the HTL in a thermally stressed OLED can induce strain-driven failure. This effect can be reduced using HTL materials with high glass transition temperatures (T_g). As shown in Figure 5.7, Forrest et al. studied several hole-transporting materials with high T_g, including the most commonly used TPD (T_g 65°C) and α-NPD or NPB (T_g 95°C) [78]. The materials have the same biphenyl backbone as TPD and α-NPD, but utilize different amine substituents, indicating that T_g can be increased by introducing different aryl substituent groups on the nitrogen atom of each amine group. Compared with TPD, PPD exhibits very high T_g reaching 152°C when the 9-phenanthyl group is attached on the nitrogen atom. From Table 5.3, all these derivatives of triphenylamine and carbazole have high ionization potential, resulting in a high hole mobility, also an ability to block electron injection from the electron-transporting layer (ETL) to the HTL.

Molecular asymmetry is thought to form high T_g glasses. Amorphous materials can form stable glasses with high T_g, which possess excellent processability, transparency, isotropic properties, and, most important, the absence of grain boundaries. Amorphous thin films with high T_g in OLEDs are less vulnerable to the Joule heating, which accelerates the formation of crystalline boundaries. The correlation between molecular structure and the glass-forming property, T_g, and stability of the glassy state has been well discussed by Shirota [79].

On the basis of the simple idea that nonplanar molecular structure might prevent easy packing of molecules and hence ready crystallization, Shirota [79] proposed a concept of starburst molecules to design several families of amorphous materials with high hole mobility, for examples, 4,4′,4″-tris (diphenylamino)triphenylamine (TDATA), 1,3,5-tris(diphenylamino)benzene (TDAB), 2,4,6-tris (diphenylamino)-1,3,5-triazine (TDATz), 1,3,5-tris(4-diphenylaminophenyl)benzene (TDAPB), and various derivatives of each family. These families have been found to readily form amorphous glasses above room temperature, except for the parent compounds TDAB, TDATz, and *p*-MTDATz, either by rapidly cooling with liquid nitrogen or by slowly cooling on standing in air [79].

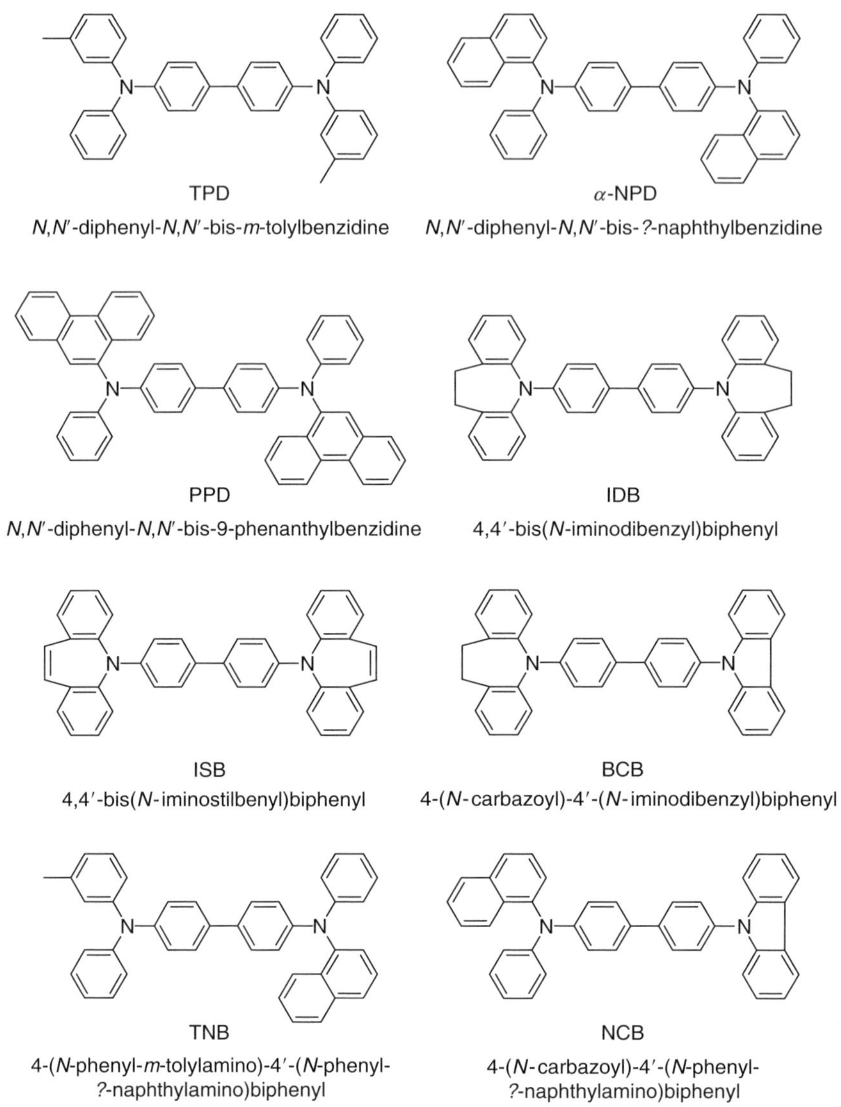

FIGURE 5.7 Several typical hole-transporting materials with high T_g, derived from the same biphenyl backbone as TPD and α-NPD, but utilize different amine substituents. (From O'Brien, D.F. et al., *Adv. Mater.*, 10, 1108, 1998. With permission.)

TABLE 5.3
Physical Properties of Typical Hole-Transporting Materials Derived from Carbazole and Triphenylamine Unit

Physical Properties	PPD	ISB	IDB	α-NPD	TPD	BCB	NCB	TNB
T_g (°C)	152	115	110	95	65	110	109	85
Electron affinity (eV)	—	2.06	1.95	2.60	2.30	2.13	2.56	2.60
Ionization potential (eV)	—	5.46	5.30	5.70	5.50	5.53	5.78	5.58

Source: From O'Brien, D.F. et al., *Adv. Mater.*, 10, 1108, 1998. With permission

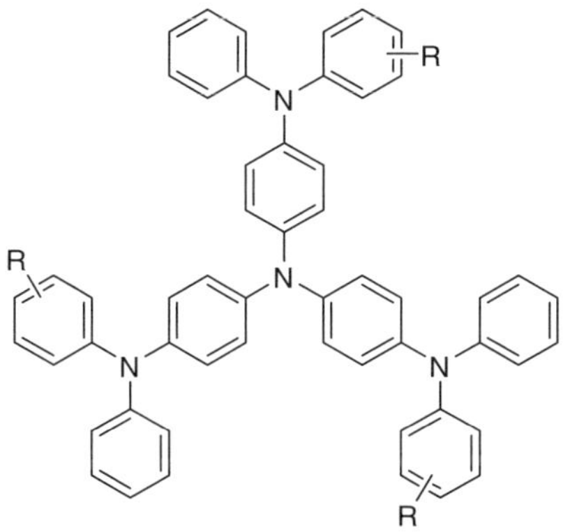

TDATA: R = H
o-MTDATA: R = o-Me
m-MTDATA: R = m-Me
p-MTDATA: R = p-Me

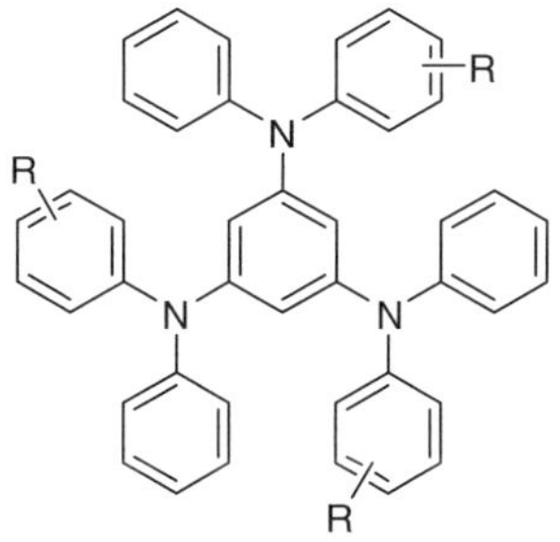

TDAB: R = H
o-MTDAB: R = o-Me
m-MTDAB: R = m-Me
p-MTDAB: R = p-Me

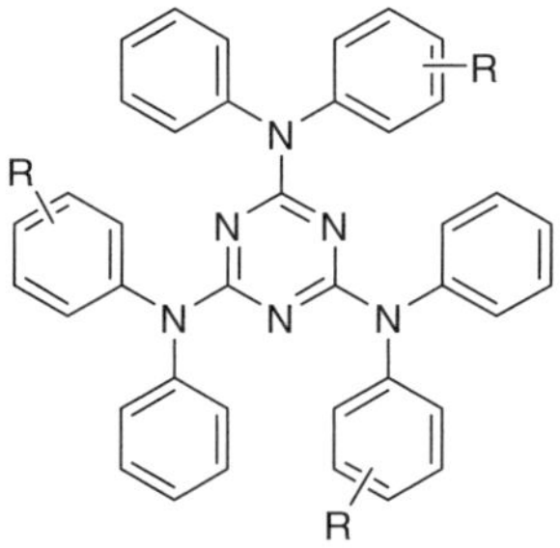

TDATz: R = H
o-MTDATz: R = o-Me
m-MTDATz: R = m-Me
p-MTDATz: R = p-Me

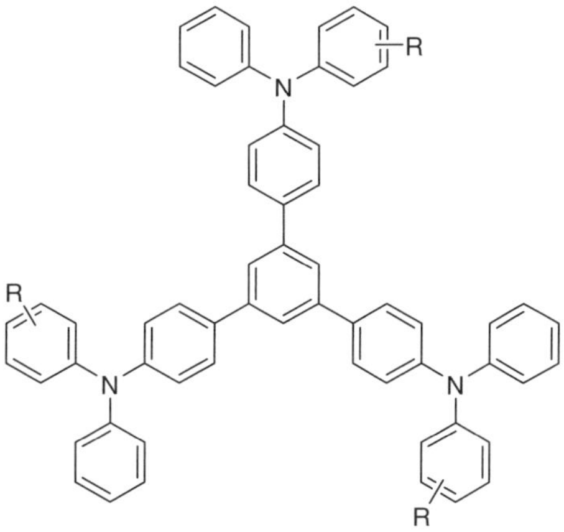

TDAPB: R = H
o-MTDAPB: R = o-Me
m-MTDAPB: R = m-Me
p-MTDAPB: R = p-Me

Combination of both triphenylamine and carbazole units were developed into a starburst **88** with a high T_g temperature of 132°C [80]. The introduction of **88** into the standard NPB/Alq_3 OLED as the hole injecting and transporting layer dramatically enhanced the device efficiency to 5.7 $cd \cdot A^{-1}$ and 2.2 $lm \cdot W^{-1}$, which are a factor of two higher than those of the standard OLED without the layer of carbazole derivative **88**.

Carbazole derivative **89** with a high T_g of 164°C is substantially higher than that of widely used NPB. Furthermore, **89** shows excellent hole-transporting properties in OLEDs, and its performance is practically equivalent to NPB [81]. A branched carbazole derivative 1,3,5-tris (2-(9-ethylcarbazoyl-3)ethylene)benzene **90** was constructed by using 1,3,5-tris(ethylene)benzene as core, which is comparable to NPB in terms of HOMO/LUMO energy levels. However, it is superior to NPB from the view of its higher T_g (130°C) and easiness of synthesis. The latter features suggest that **90** can be a potential alternative material to NPB and TPD especially for high-temperature applications in OLEDs and other organic electronic devices [82]. The triphenylamine-based hole-transporting material **91** also exhibits a high T_g of 167°C, indicating that the twisted fluorene unit is very beneficial to the amorphous and thermal properties [83]. A series of building blocks of thiophene–Ar–NPh_2 were assembled into star-shaped amines utilizing 3,6-disubstituted carbazole and 1,3,5-trisubstituted benzene as the central units. Use of sterically

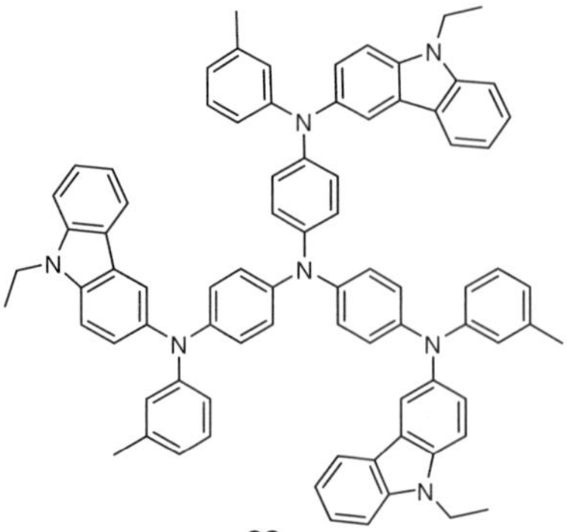
88

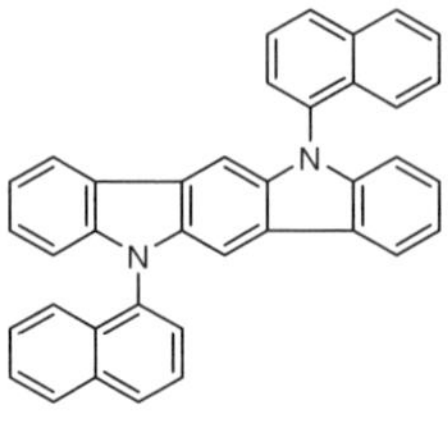
89

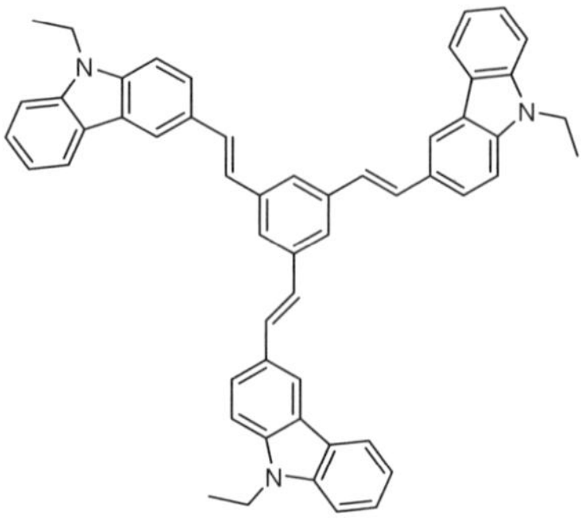
90

more demanding spacers is expected to inhibit intermolecular interactions that could give rise to exciplex emission [84].

Star-shaped derivatives (**92–94**) with a triphenylamine core and carbazole or fluorene sidearms have been tested as organic semiconductors in solution processed OFETs. Mobilities of $3 \times 10^{-4}\ cm^2 \cdot V^{-1} \cdot s^{-1}$, high on/off ratios of up to 10^5, and low threshold voltages were obtained. These new materials show very small hysteresis and an exceptionally high stability under ambient conditions [85].

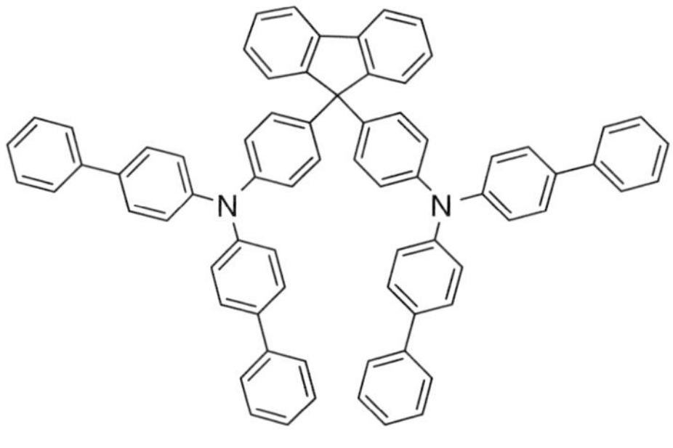
91

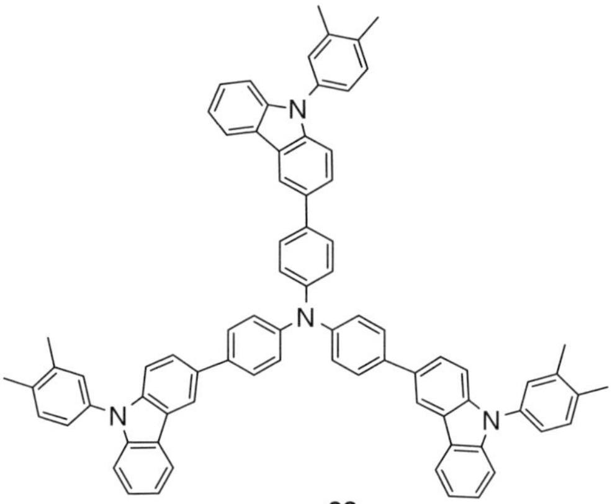
92

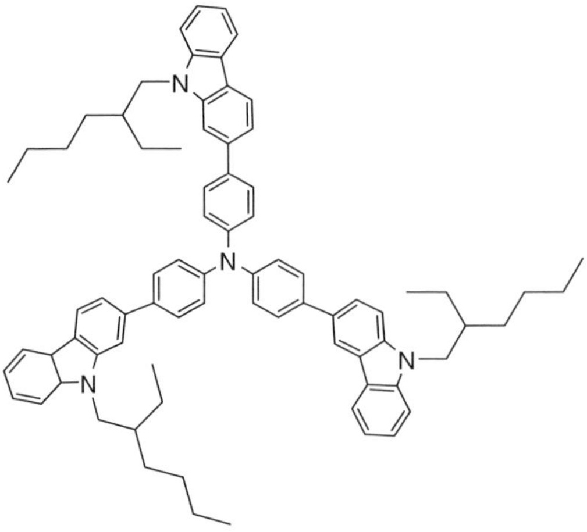
93

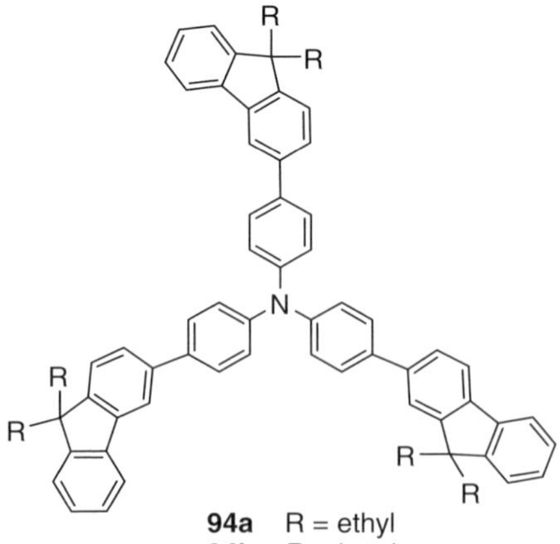
94a R = ethyl
94b R = butyl
94c R = hexyl

In conclusion, the strategy of branched and star-shaped configuration is widely utilized for designing triarylamine- or carbazole-based hole-transporting materials. For specific application in optoelectronic device, two factors such as amorphous state and high T_g have to be considered. Moreover, because of the inherent reductive quenching, most triarylamines are either nonluminescent or only weakly emitting. Additionally, larger star-shaped triarylamines easily form exciplexes when contacted with the electron-transporting or -emitting layers [84]. In addition, due to the high electron-donating and low oxidation potential properties of triphenylamine, the fluorescence quenching induced by possible through-bond photo-induced electron transfer (PIET) should be also desirably considered [86].

5.6 OXADIAZOLES

Oxadiazoles are among the most widely investigated electron-transporting materials for OLEDs. The oxadiazole molecule of 2-(4-biphenyl)-5-(4-*tert*-butylphenyl)-1,3,4-oxadiazole (PBD), with an electron affinity of 2.16 eV and an ionization potential of 6.06 eV, was first used as an electron transport material in a bilayer OLED based on a triphenylamine derivative as the emissive material [87]. Since the initial studies of PBD, 2,5-diaryl-1,3,4-oxadiazoles have been widely exploited as electron-transporting and hole-blocking (ETHB) materials in EL devices due to their electron-deficient nature and thermal stability.

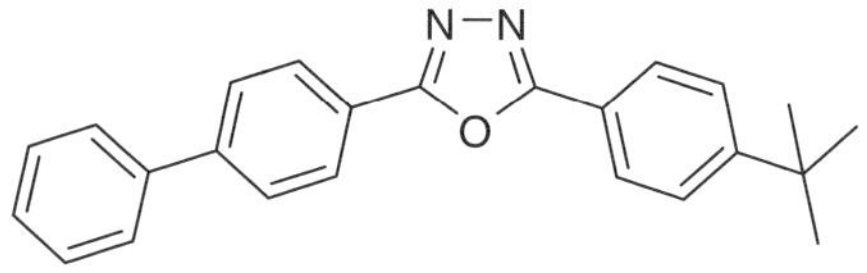

PBD

However, the vacuum-evaporated amorphous PBD thin films (T_g ~ 60°C) crystallize over time due to Joule heating during device operation. To overcome this problem, PBD was dispersed in a poly(methyl methacrylate) (PMMA) matrix that can be spin coated. External quantum efficiencies of 2%–4% were achieved by blending PBD in PPV-based electroluminescent polymers at 20 wt% concentrations [88]. Improvements by factors of 8–10 in external quantum efficiencies of PPV-based LEDs have also been achieved by using a PBD/PMMA blend as the electron transport materials [89].

Several tetraphenylmethane-based oxadiazoles (**95a–c**) with higher T_g were developed in an effort to increase the stability of the amorphous films. Since π-conjugation is limited to one arm of the four-armed compound, the optical and electrochemical properties of **95a–c** are very similar to those of PBD, except that **95c** has a slightly higher electron affinity of 2.26 eV due to the peripheral CF_3 groups. However, the thermal properties of **95a–c** are significantly enhanced compared to those of PBD, with T_g of 97°C–175°C and T_d of 405°C–499°C. In terms of device performance, the external quantum efficiency was comparable at ~0.7% when using **95c** or PBD as the electron-transporting materials and Alq_3 as the emissive layer. However, the devices with **95c** are expected to be more stable owing to its superior thermal properties; **95c** crystallizes above 200°C compared to 70°C–90°C for PBD. Since compounds **95a–c** are soluble in most organic solvents, uniform thin films for devices can be made by spin coating, rather than by vacuum evaporation [90].

Hole-transporting and electron-transporting properties are opposing functions associated with different structural features and often used independently. However, attempts have also been made to incorporate all the above functions in the same molecule or polymer [91] in an effort to attain a maximum performance. It is believed that this approach will make device fabrication simpler and cost effective. Owing to the complexity of dopant, the attention on dyad and triad compounds were concentrated in which both electron- and hole-transporting units were connected directly with the

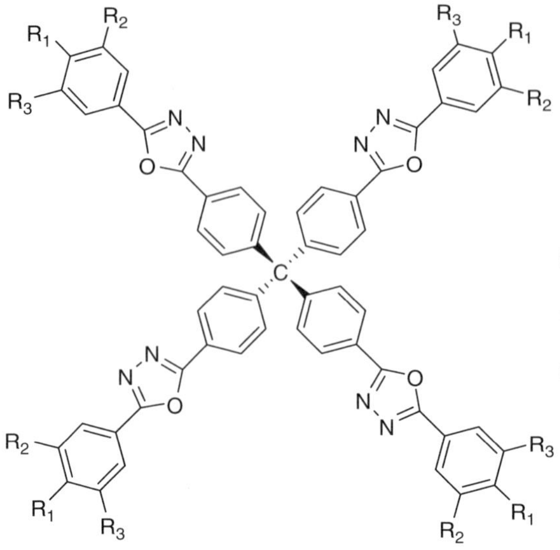

	R_1	R_2	R_3
95a	H	$C(CH_3)_3$	$C(CH_3)_3$
95b	OCH_3	OCH_3	H
95c	H	CF_3	H

emitting unit via covalent bonds. However, one aspect should be considered is that the injection of electrons and holes must be balanced [92].

An electron-transporting moiety (oxadiazole) and a hole-transporting moiety (carbazole) were combined in compound **96**, and a three-layer device was fabricated with a configuration of ITO/TPD/**96**/Alq_3/$Mg_{0.9}Ag_{0.1}$/Ag, which exhibited a blue emission peak at ~470 nm ($x = 0.14$, $y = 0.19$) with a maximum luminance of 26200 $cd \cdot m^{-2}$ at a driving voltage of 15 V and a maximum luminous efficiency of 2.25 $lm \cdot W^{-1}$ [93].

Metal complexes of oxadiazole were also studied. Devices with the concise configuration of ITO/TPD/**97**/Al showed bright blue EL emission centered at 468 nm with a maximum luminance of 2900 $cd \cdot m^{-2}$. A current efficiency of 3.9 $cd \cdot A^{-1}$ and power efficiency of 1.1 $lm \cdot W^{-1}$ were obtained. The efficiency of an ITO/NPB/Alq_3/Al device increased considerably when **97** was inserted between Alq_3 and aluminum. The impressive EL efficiency may be attributed to the following: (1) the oxadiazole segment in **97** has excellent electron-transporting ability and (2) the lithium salt is favorable for electron injection [94].

The second-order nonlinear optical properties of oxadiazoles (**98** and **99**) based on push–pull chromophores were characterized by the EFISH technique. The maximum observed $\mu_g\beta$ is 900 and 1250×10^{-48} esu for **98** and **99**, respectively, in conditions far from resonance enhancement (fundamental laser wavelength 1.907 μm) [95].

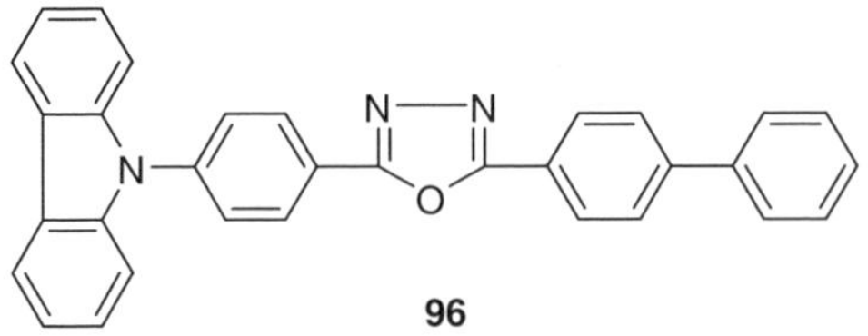

96

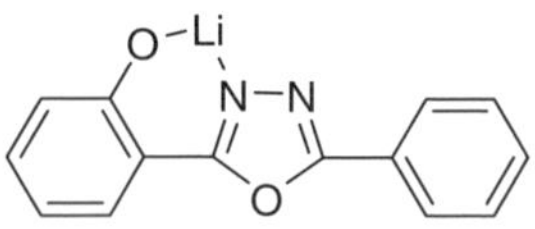

97

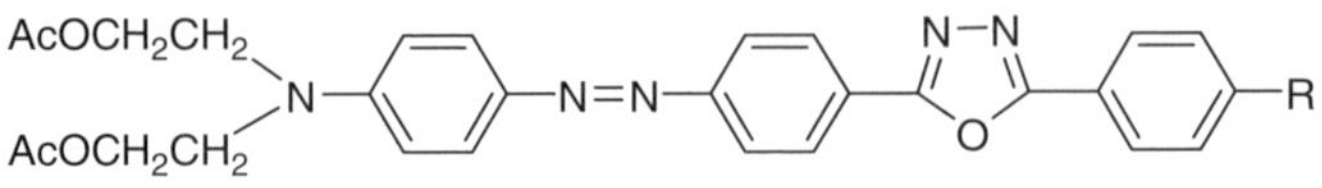

98 R = NO_2

99 R = CH = CH–  NO_2

5.7 TETRATHIAFULVALENES

Tetrathiafulvalenes is a family of sulfur-based electron donor containing 2-(1,3-dithiol-2-ylidene)-1,3-dithiole (TTF) as backbone. TTFs were first synthesized by Wudl group in 1970 [96]. In 1973, Ferraris et al. [97] synthesized the first organic synthetic metal TTF-TCNQ, a charge-transfer (CT) salt of TTF and tetracyanoquinodimethane (TCNQ), with the room-temperature conductivity of 500 $S \cdot cm^{-1}$. TTF-TCNQ is the prototype of CT complexes where HOMO and LUMO bands of the open-shell donors and acceptors, respectively, contribute to the conduction. Up to now, the synthesis, properties, and applications of TTF derivatives have been shed much light and well reviewed [98,99].

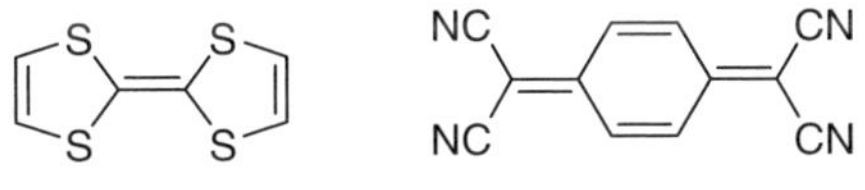

TTFs are good electron donors that form stable open-shell species by transferring a HOMO π-electron. On the other hand, TTFs have planar configurations that promote their stacking as a consequence of the π–π orbital overlap. Depending on the packing pattern in the solid, the intermolecular electronic interaction, described by the transfer integrals, varies and, as a consequence, the bulk electronic and magnetic properties of the molecular solid can be modified. In addition, it is synthetically possible to introduce a large number of substituents in the 2, 3, 6, and 7 positions of the TTF core [99]. Exactly, the field of TTF chemistry is still active, making use of the electron-donor and charge-transfer properties for different applications such as electrochemical switches, sensors, surface modification agents, and so forth.

Extensive studies on chemical modifications of the TTF skeleton have been made to develop new molecular conductors with metallic conductivity and superconductivity. The syntheses of TTF were diversible, and here just a few typical routes were listed in Scheme 5.6. Route 1 shows the coupling reaction between 1,3-dichalcogenole-2-chalcogenones using trialkyl phosphite, e.g., $(MeO)_3P$

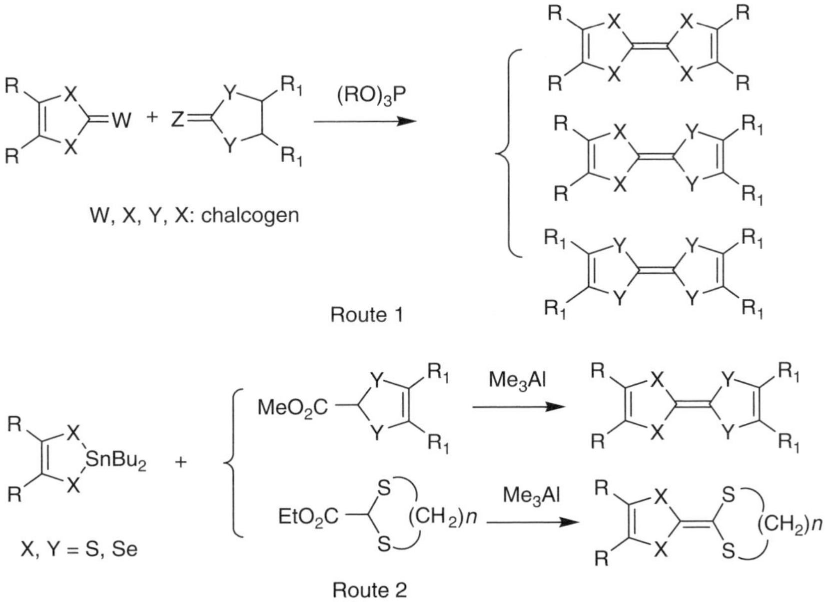

SCHEME 5.6 Synthetic routes for TTFs.

or $(EtO)_3P$, is a most convenient pathway to construct not only tetrachalcogenafulvalene (TCF) derivatives but also π-electron donors containing 1,3-dichalcogenol-2-ylidene moieties. And route 2 is recognized as a versatile nonphosphite methodology for constructing unsymmetrical TTFs without the formation of symmetrical by-products.

In 1970s, the emphasis on CT complexes was to synthesize new derivatives of TTF and TCNQ for the breakthrough of superconductivity. Research on the development of synthetic metals and superconductors was stimulated by TCNQ and was subsequently accelerated by the discovery of superconductivity in the CT salts of tetramethyltetraselenafulvalene (TMTSF) and bis(ethylenedithio)-tetrathiafulvalene (ET). Simultaneously, continuous interest in this field has been sustained by synthetic organic chemists, resulting in the preparation of a huge number of π-electron donors for molecular conductors [100].

Most organic superconductors based on TTFs have 2:1 composition as represented by $(TMTSF)_2X$ and $(ET)_2X$. Since X is an anion with a negative charge (−1), the donor molecule has a half positive charge (+½). Because the HOMO orbital of a donor molecule can contain up to two electrons (or holes), the +½ charge corresponds to quarter filling; three-quarters of the HOMO energy band is occupied by electrons, and the remaining one-quarter is empty. The quarter filling seems to be an important requisite of the existing theory of organic superconductors, in particular of the universal phase diagrams of quasi-one-dimensional $(TMTSF)_2X$ series and two-dimensional κ-phase $(ET)_2X$ salts [101].

Organic molecular crystals of $(ET)_2X$ are low-dimensional electronic systems showing a wide range of physical properties from semiconductors to superconductors. One of the member of this family, β-$(ET)_2I_3$, was the first quasi-two-dimensional (2D) organic superconductor at ambient pressure. It was found that this compound can exist in low-temperature (β_L) and high-temperature (β_H) superconducting states. Another member of the $(ET)_2X$ family, β′-$(ET)_2ICl_2$, has become "intriguingly famous." This compound being a Mott insulator transforms to a superconductor with a T_c of 14.2 K under very high pressure [102]. An attachment of two methyl groups to a ET molecule succeeded in introducing moderate negative chemical pressure and dimerization to afford a new superconductor, β-$(m\text{-DMBEDT})_2PF_6$, with a T_c of 4.3 K (onset) under 4.0 kbar [103].

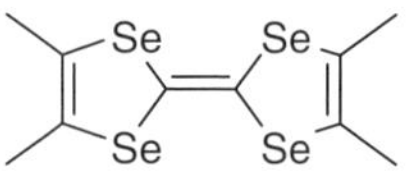
TMTSF

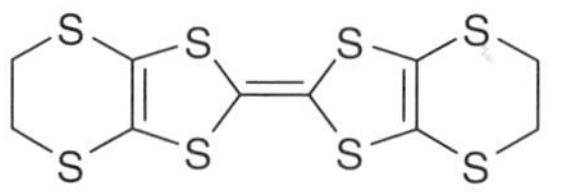
ET

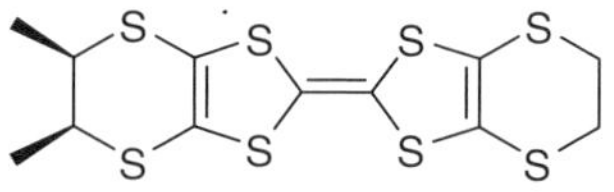
m-DMBEDT

Clear observation of field-effect hole mobility in single crystals of a TTF derivative with thiophene substituents was reported, the organic semiconductor dithiophene–tetrathiafulvalene (DT-TTF) grown by drop casting, a very simple method. The electrical measurements performed on the DT-TTF crystals are typical of a p-type semiconductor; as a more negative voltage is applied, more holes are induced in the semiconductor with an increase of conductivity. The maximum mobility was achieved at $1.4\ cm^2 \cdot V^{-1} \cdot s^{-1}$. The isostructural asymmetric donor ETT-TTF is also a good semiconductor to form OFETs. Nevertheless, the mobility of ETT-TTF achieved on a single crystal is an order of magnitude lower due to its disordered structure. The same case was also observed with ETEDT-TTF [99].

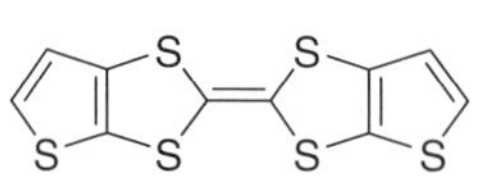
DT-TTF

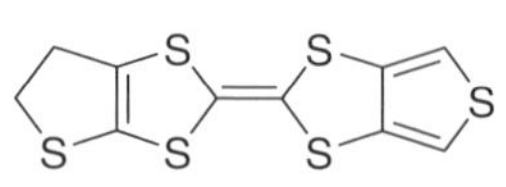
ETT-TTF

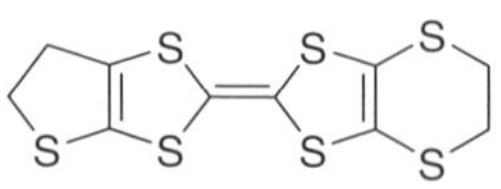
ETEDT-TTF

Because of the strong electron-donating properties of TTFs, their thin films are labile to be oxidized, resulting in poor FET performances in thin films. Four OFET devices were fabricated with TTF derivatives of dibenzo TTF **100**, dinaphtho TTF **101**, dipyrazino TTF **102**, and diquinoxalino TTF **103** [104]. Incorporation of aromatic rings to the TTF skeleton was favorable to enhance the intermolecular interactions, leading to excellent p-type FET performances. In addition, a π-stacking structure was constructed by using electron-accepting quinoxaline rings, which was also useful to enhance the stability of FET devices to oxygen [104].

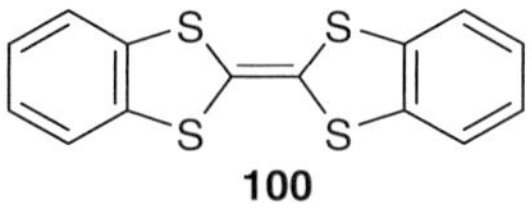

100

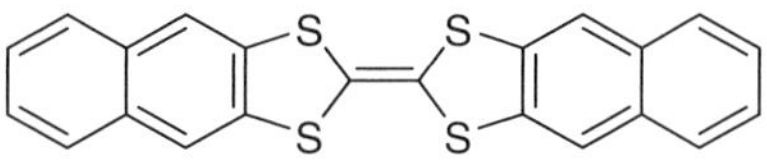

101

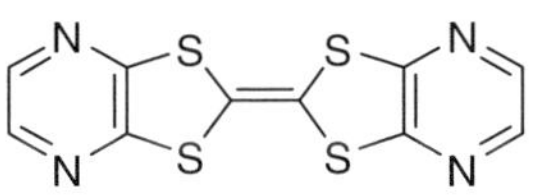

102

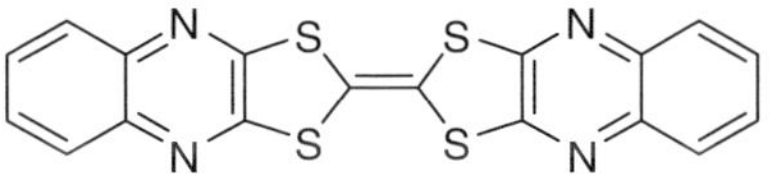

103

Photo-induced charge separation based on TTFs and other components in dyads and triads was also widely investigated. Three isomers of C_{60}-TTF dyads (C_{60}-*X*-TTF, *X* = *o*, *m*, and *p*) have been studied by changing the linking positions (ortho, meta, and para) at a phenyl group that is attached to the methano-[60] fullerene. These dyads show clear intramolecular charge-transfer absorption bands in the steady-state absorption spectra, indicative of an intramolecular charge-transfer interaction between the C_{60} and TTF moieties in the ground state. The increase in intensity of charge-transfer absorption bands followed the order of C_{60}-*o*-TTF > C_{60}-*m*-TTF > C_{60}-*p*-TTF, which can be reasonably explained by the optimized molecular configurations calculated at the ab initio level. Appreciable charge-separated species were generated for C_{60}-*p*-TTF. The lifetimes of charge-transfer and charge-separated states are increased in the order of C_{60}-*o*-TTF < C_{60}-*m*-TTF < C_{60}-*p*-TTF. Overall, the ground and excited states are controlled by the difference in proximity between the C_{60} and TTF moieties, that is, depending on the linking positions [105].

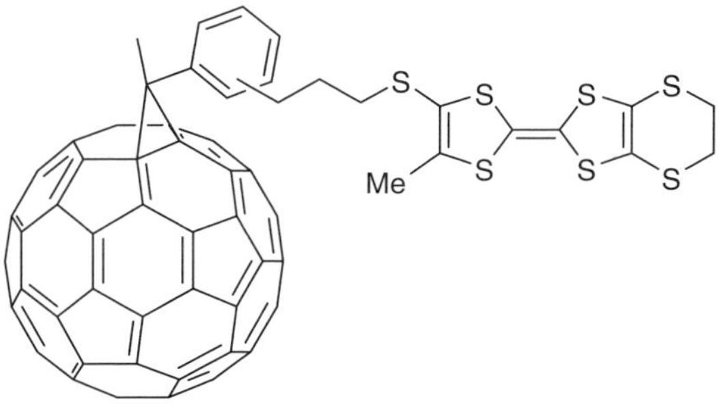

C_{60}-*X*-TTF (*X* = *o*, *m*, and *p*)

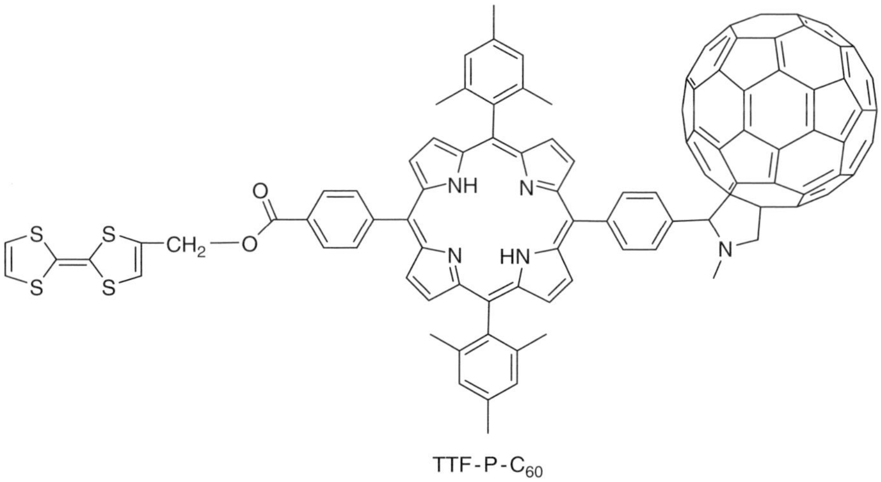

TTF-P-C_{60}

Photo-induced electron transfer has been observed in a molecular triad TTF-P-C_{60}, consisting of a porphyrin (P) covalently linked to a TTF and a C_{60} cage, in the different phases of the commercially available liquid crystal E-7 and in a glass of 2-methyltetrahydrofuran. The $TTF^{\bullet+}$-P-$C_{60}^{\bullet-}$ state is formed from the TTF-^{1}P-C_{60} singlet state via an initial TTF-$P^{\bullet+}$-$C_{60}^{\bullet-}$ charge-separated state. Long-lived charge separation (~8 μs) for the singlet-born radical pair is observed in the 2-methyltetrahydrofuran glass at cryogenic temperatures. In the nematic phase of E-7, a high degree of ordering in the liquid crystal is achieved by the molecular triad. In this phase, both singlet- and triplet-initiated electron transfer routes are concurrently active. At room temperature in the presence of the external magnetic field, the triplet-born radical pair $^{T}(TTF^{\bullet+}$-P-$C_{60}^{\bullet-})$ has a lifetime of ~7 μs, while that of the singlet-born radical pair $^{S}(TTF^{\bullet+}$-P-$C_{60}^{\bullet-})$ is much shorter (<1 μs). The difference in lifetimes is ascribed to spin dynamic effects in the magnetic field [106].

The electrochemistry, and photophysical properties of a highly soluble C_{60}-TTF-C_{60} triad containing two C_{60} units was also studied. Although no significant electronic interaction was found between the electroactive species in the ground state, an efficient intramolecular electron transfer occurs in the excited state resulting in the formation of a long-lived charge-separated state. These data reveal that electroactive C_{60} dimers are appealing ensembles and potentially useful in photovoltaic devices [107].

Zhu et al. presented another example of a fluorescence switch (compound 104) based on a supramolecular dyad with TTF and porphyrin units, in which a TTF derivative bearing a pyridyl group was axially connected to (tetraphenylporphyrinato)zinc(II) (ZnTPP) through metal coordination of pyridyl group. Such metal coordination method provides a more accessible strategy because it is not severely limited by the synthetic route with respect to other covalently linked TTF hybrids [108].

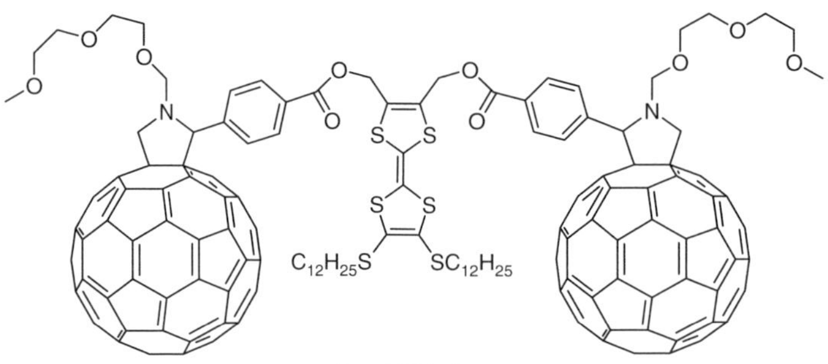

C_{60}-TTF-C_{60}

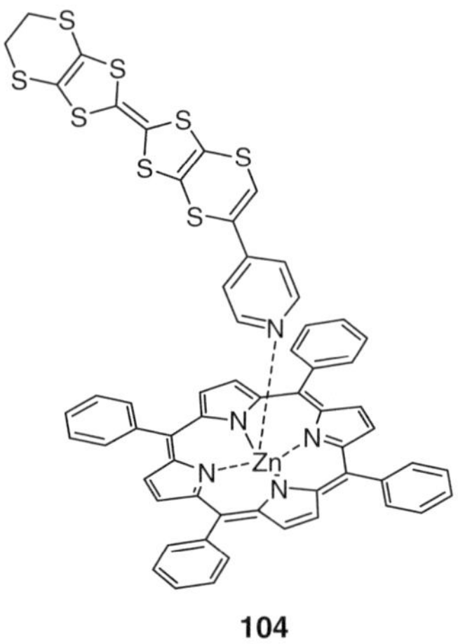

104

5.8 POLYCYCLIC AROMATIC HYDROCARBONS (OLIGOACENES)

Linear polycyclic aromatic hydrocarbons are composed of laterally fused benzene rings, so-called linear acenes. They are currently among the most widely studied organic π-functional materials due to their importance as organic semiconducting materials. Some typical representatives are anthracene, tetracene, and pentacene. Tetracene and pentacene are among the most promising molecular conductors for OFETs. Up to now, the chemistry, applications, and especially controlling the HOMO–LUMO gap of oligoacenes have been well reviewed by Wudl et al. [96].

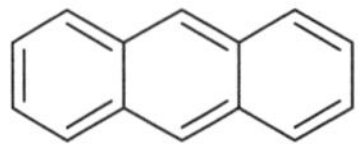

Anthracene

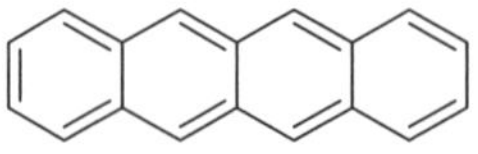

Tetracene

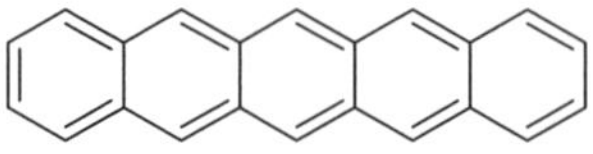

Pentacene

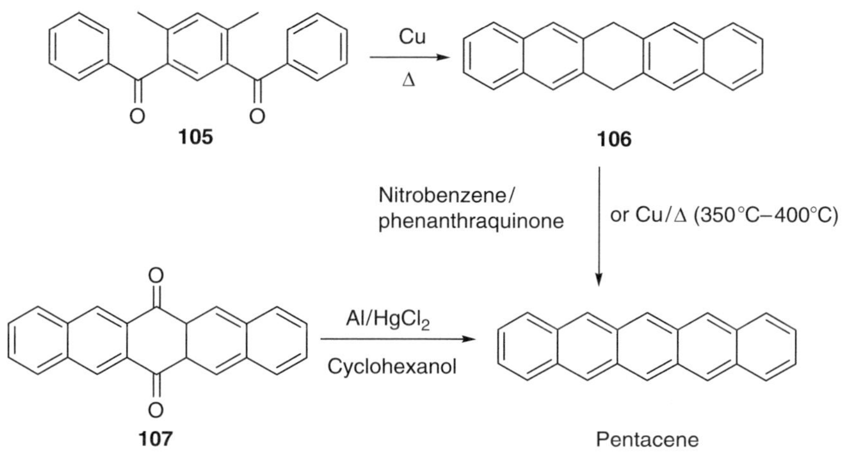

SCHEME 5.7 Synthetic routes for pentacene.

Despite large strides in sophistication of theoretical treatments over the last 30 years, the electronic structure, stability, aromaticity, and, most importantly, band gap (or HOMO–LUMO gap) in oligoacenes and polyacenes are still subjects of controversy. Study shows that increasing the number of rings in polycyclic aromatics results in successive reduction of the band gap (reflected in the UV spectra), reduction of the ionization potential, and an increase in proton and electron affinities, which is also reflected in decreased stability [98].

The lower homologues, from benzene to anthracene, can be extracted from coal, while higher homologues, such as pentacene or hexacene, can be obtained only by multistep synthesis. Pentacene was synthesized from the cyclization of **105** and dehydrogenation of dihydropentacene (**106**), and a more convenience method by reduction of **107** using aluminum–cyclohexanol mixture was mostly employed (Scheme 5.7) [109].

Cyclic voltammograms of both pentacene and tetramethylpentacene **108** showed quasi-reversible oxidation and reduction in hot 1,2-dichlorobenzene with n-Bu_4NPF_6 electrolyte. The oxidation potentials and reduction potentials ($E^{1/2}$ vs Ag wire, Fc/Fc^+ shows 0.47 V under these conditions) are +0.73 V, −1.66 V for pentacene, and +0.58 V, −1.66 V for **108**, respectively. The HOMO–LUMO gaps, determined from the $E^{1/2}$ of oxidation and reduction potentials, were 2.4 and 2.3 eV for pentacene and **108**, respectively. An OTFT fabricated with compound **108** with a mobility of 0.30 $cm^2 \cdot V^{-1} \cdot s^{-1}$ renders it the first pentacene derivative with mobility similar to that of the parent pentacene [98,110].

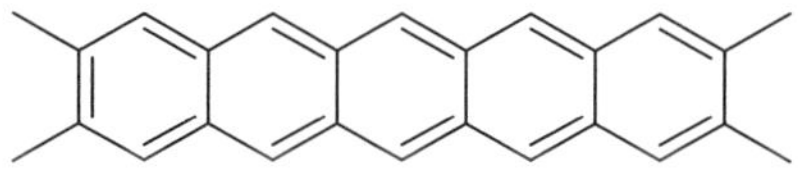

108

Actually, pentacene is among the most promising molecular conductors based on acenes and semiconductors for OFETs. OFETs are very similar to commonly used silicon-based TFT transistors, which contain amorphous silicon as the semiconductor material (charge carrier mobility about 0.5 $cm^2 \cdot V^{-1} \cdot s^{-1}$). The field-effect mobility in pentacene-based TFTs has been improving to equal that of hydrogenated amorphous silicon (0.5 $cm^2 \cdot V^{-1} \cdot s^{-1}$) and to supersede it up to 3 $cm^2 \cdot V^{-1} \cdot s^{-1}$. Pentacene shows the high hole mobility because its linear fused ring structure with extended π-conjugation and rigid planarity is suitable for intermolecular π–π overlap and face-to-edge interactions in the solid state [98].

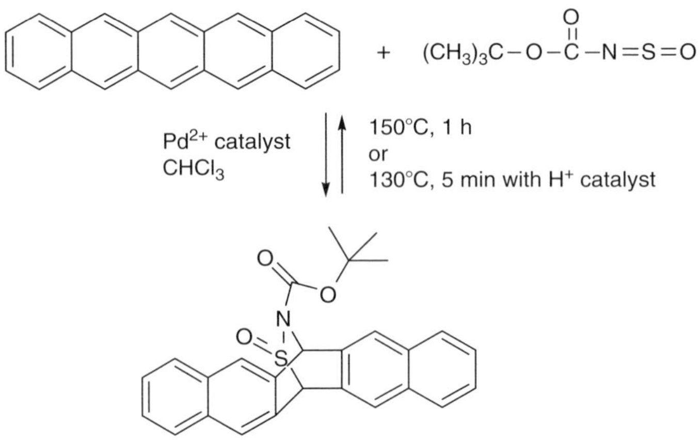

109

SCHEME 5.8 Conversion between pentene and its precursor.

However, the poor solubility of pentacene is limited to the application. This might be solved by using its precursors. As a case illustrated in Scheme 5.8, soluble pentacene precursor **109** is succeeded in incorporating an acid-labile moiety into the molecular structure, which enables its conversion back to pentacene initiated by ultraviolet light in the presence of a photoacid-generating compound (Scheme 5.8). In the saturation regime, the mobility could also be reached at 0.13 $cm^2 \cdot V^{-1} \cdot s^{-1}$ with an I_{on}/I_{off} ratio of 3×10^5 [111].

Anthracene is not active in OFETs; however, rather high mobilities (0.18 $cm^2 \cdot V^{-1} \cdot s^{-1}$) were achieved using oligoanthracene **110** [112]. OFETs of optically transparent freestanding single crystals of unsubstituted tetracene have been fabricated, which exhibit effective hole channel mobility μ_{eff} up to 0.15 $cm^2 \cdot V^{-1} \cdot s^{-1}$, and on/off ratios up to 2×10^7 [113,114].

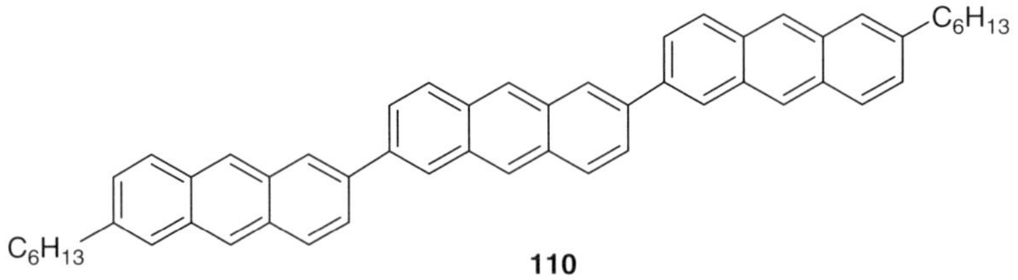

110

Another oligoacene should be mentioned. Rubrene (5,6,11,12-tetraphenylnaphthacene) has been frequently used in improving the efficiency of the red-doping emitters of DCJTB in OLEDs. To solve the doping problem of concentration quenching, Hamada et al. [115] reported an innovative and quite successful solution using rubrene as "assist dopant". This might be attributed to the well-aligned energy level of Alq_3-rubrene-DCM fluorophore that facilitates the Förster resonance energy transfer. Rubrene can fill in the gap of energy transfer between Alq_3 host and DCM-type red dopants. Now, rubrene can be synthesized through a method described by Dodge (Scheme 5.9) [116].

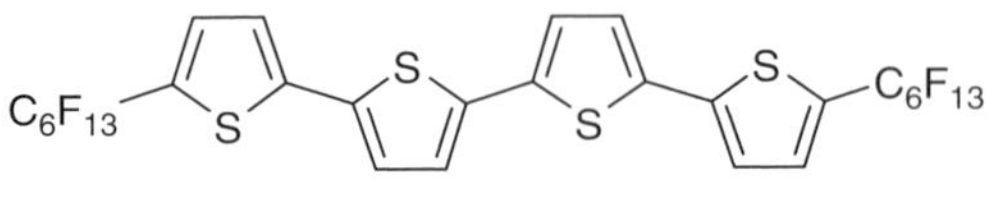

DF-4T

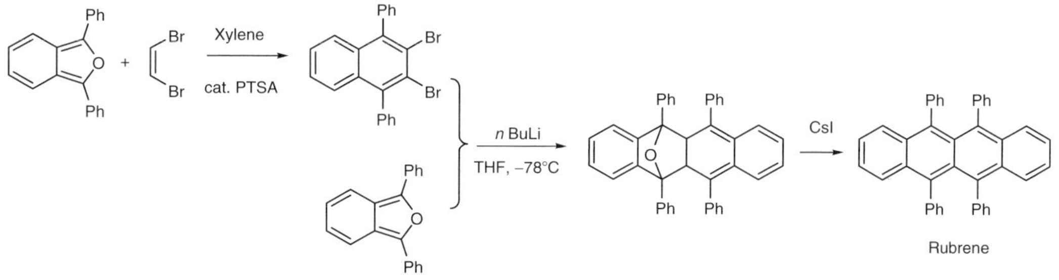

SCHEME 5.9 Synthetic route of rubrene described by Dodge.

The transistors based on rubrene were constructed by laminating a monolithic elastomeric transistor stamp against the surface of a crystal with the organic crystal placed on top of the source and drain electrodes. Strong field-effect modulation of the channel conductance was observed, with on/off ratios as high as 10^6. Rubrene-based OFETs with charge carrier mobilities as high as 15 $cm^2 \cdot V^{-1} \cdot s^{-1}$ were realized by using this method with respect to that of 3–15 $cm^2 \cdot V^{-1} \cdot s^{-1}$ obtained for different rubrene crystals [117]. Single crystal of rubrene grown by a vapor-phase process shows excellent crystallinity and very small rocking curve width. FETs based on pure rubrene single crystal with colloidal graphite electrodes demonstrate a maximal mobility of 15 $cm^2 \cdot V^{-1} \cdot s^{-1}$ with strong anisotropy [118].

Another p-type typical semiconductor for OTFT was thiophene oligomer. The most typical DHα-6t was widely studied as OTFT and its mobility was approximately 0.03–0.06 $cm^2 \cdot V^{-1} \cdot s^{-1}$ [119,120]. Synthetic routes of thiophene oligomer were shown in Scheme 5.10 [121].

Marks and his coworkers investigated the OFET fabricated by a series of n-type semiconductors which has fluorocarbon chains at the thiophene termini. The α,ω-diperfluorohexyl-4T (DF-4T) exhibits a mobility of 0.22 $cm^2 \cdot V^{-1} \cdot s^{-1}$ and an I_{on}:I_{off} ratio of 10^6, one of the highest so far reported for an n-type organic semiconductor [122].

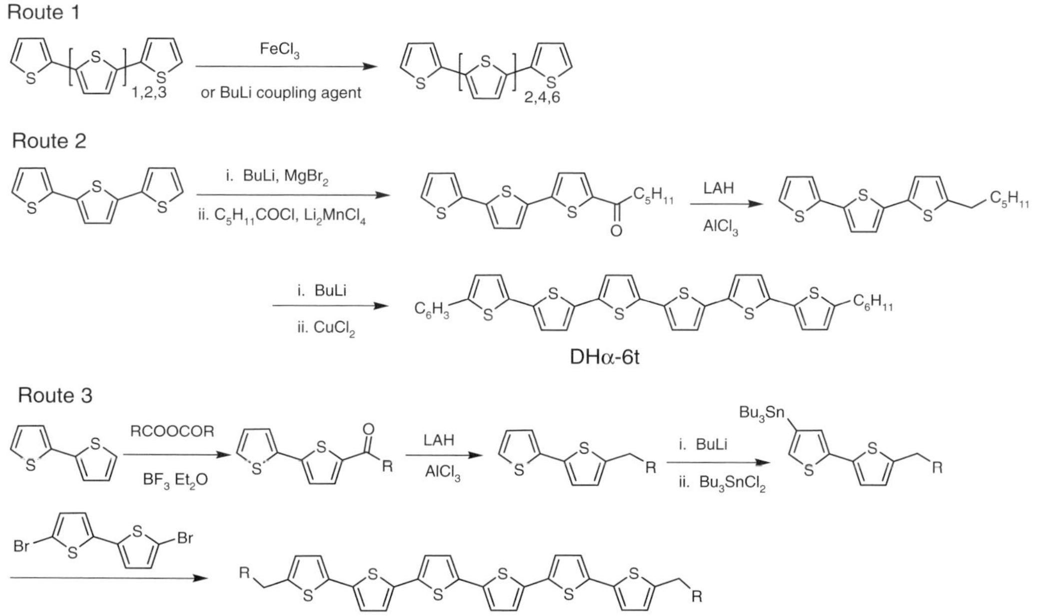

SCHEME 5.10 Synthetic routes of thiophene oligomers.

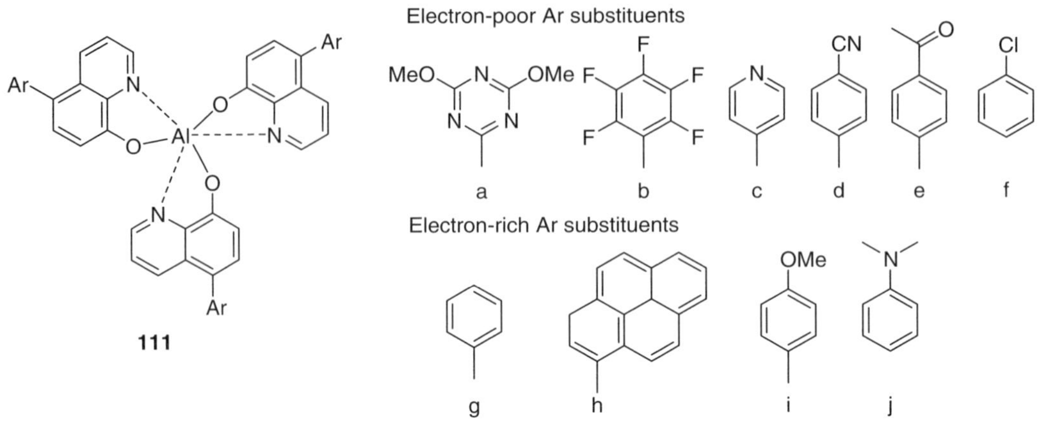

111

5.9 SMALL MOLECULE COMPLEXES

Small molecule complexes also play important roles in the electronic and optoelectronic application. The most famous small molecule complexes for OLED are 8-hydroquinoline aluminum (Alq_3) and Ir complex. Alq_3 is still one of the widely used fundamental materials in this area due to its excellent thermal stability, high fluorescent efficiency, and relatively good electron mobility since Tang and VanSlyke [123] first utilized it as luminophore for OLED device. Because the efficiency and lifetime of current blue and red OLEDs are not satisfactory, a lot of researchers are focusing on the molecular tailoring and growth of a new crystalline phase of Alq_3 to shift its emission from the green to the blue or red region. Most modifications are taken on the C5 position of quinolinolate ligand. Anzenbacher and his coworkers [124] had developed a series of 5-substitued 8-hydroquiline aluminum **111a–j** via Suzuki coupling reaction. The electronic nature of aryl substituents affects the emission color and fluorescence quantum yield. Electron-donating groups cause a blue shift in the complex emission. The maximum wavelength of the luminescence spectrum range could be tuned from 490 to 612 nm.

Tian and his coworkers [125] had incorporated hole-transporting carbazole unit to Alq_3, thus obtaining a functional complex **112** that shows yellow luminescence with high quantum yield. Electron-withdrawing atom F was also introduced into Alq_3 to form a series compound **113**. With fluorination at C-5 the emission is red-shifted with a tremendously decreased intensity, fluorination at C-6 causes a blue shift with a significantly increased intensity, and fluorination at C-7 has a minor effect on both the color and intensity of Alq_3's emission [126].

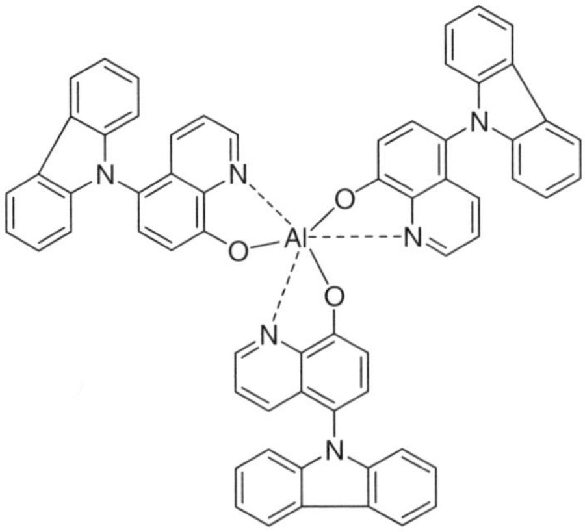

112

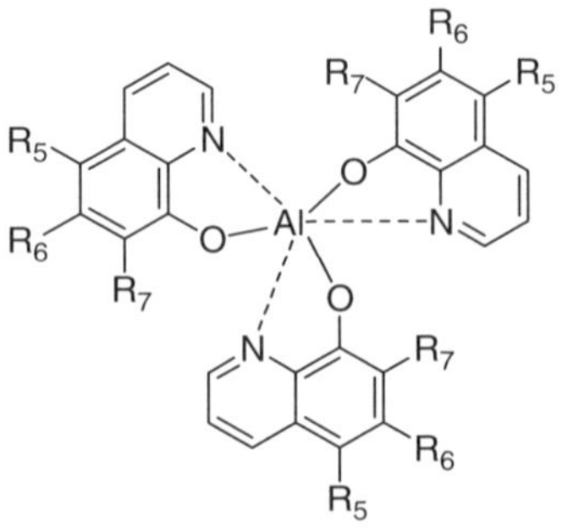

113

Both theoretical predictions and experimental measurements give a singlet/triplet ratio for these excitons of 1 to 3. The singlet/triplet ratio thus implies a limitation of 25% for the internal quantum efficiency of OLEDs based on fluorescence. However, electrophosphorescence could significantly improve the efficiency of OLED. The phosphorescence of Ir complexes has been widely studied due to its high efficiency and easiness to tune the luminescence wavelength. In the case of Ir(ppy)$_3$ doping system, the maximum external quantum efficiency could be reached to 19.2%, and keeping over 15% even at high current densities of 10–20 mA · cm^2 [127,128].

Up to now, the luminescence color of Ir complex can cover the whole visible spectrum via tuning ligands. For examples, PhqIr and FIrpic can be utilized as the electrophosphorescent luminophore with high external quantum efficiency for red and blue light [129,130], respectively. Moreover, a broad range color tuning has been realized by controlling the relative energy level (LX3) of the ancillary ligand in **114a–h** [131].

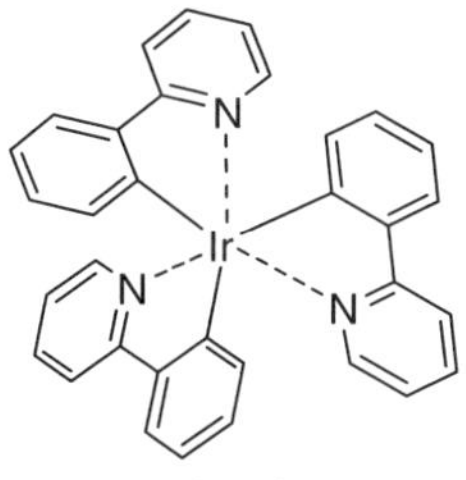

Ir(ppy)$_3$

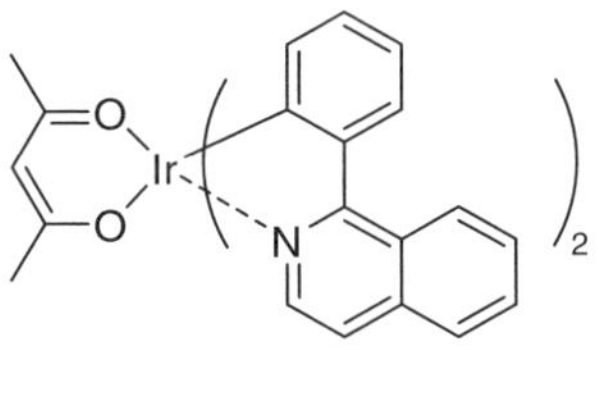

PhqIr

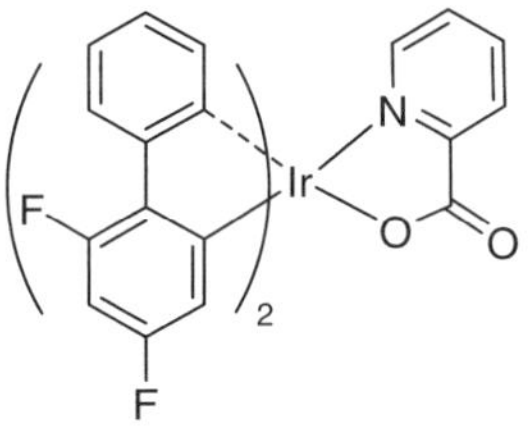

FIrpic

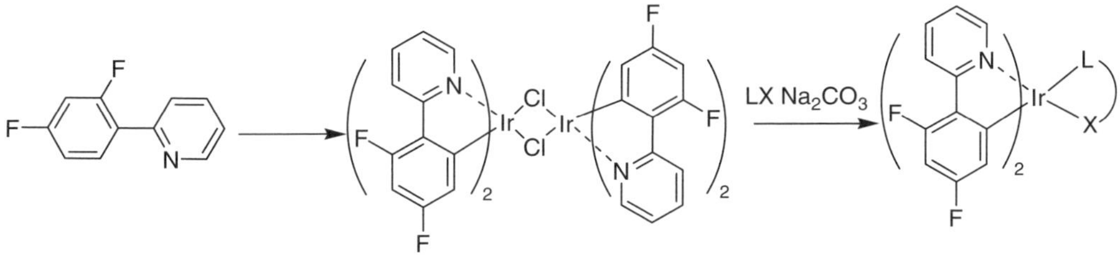

114

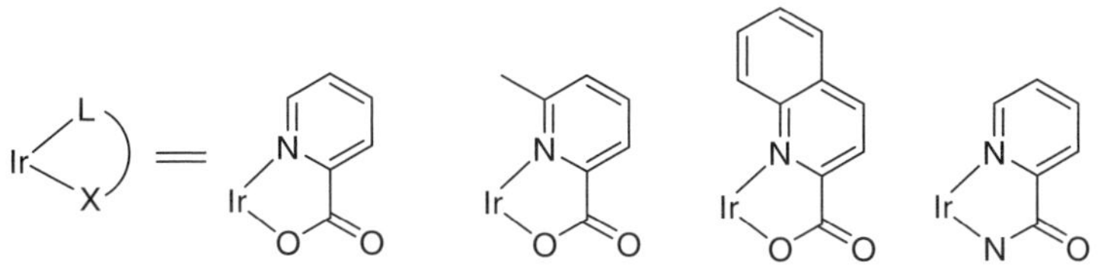

a b c d

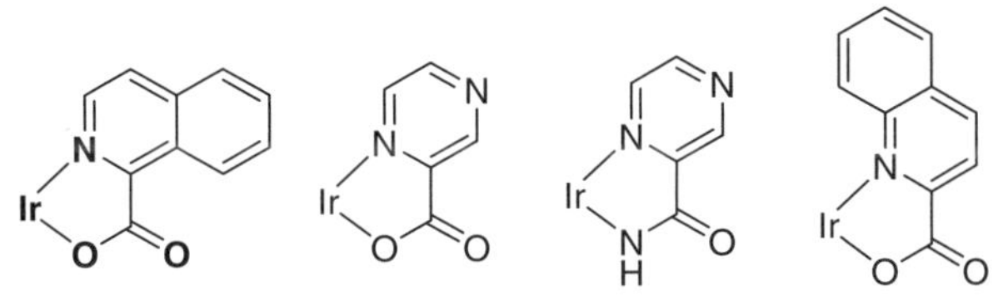

e f g h

Ruthenium complex is worthy of acclaim in Grätzel-type solar cells. Since an overall conversion efficiency of 10.4% was reported by Grätzel and his coworkers, expanding research had been focus on nanocrystalline TiO_2 solar cells with Ru complexes as sensitizers [132]. Among them, the so-called black dye (**115**) has attracted more and more attention. The carboxylic groups grafted on the bipyridine ligand can anchor the complex to nanocrystalline TiO_2 films. As well established, it achieves very efficient sensitization over the whole visible range extending into the near-IR region up to 920 nm, yielding high incident photon-to-current efficiencies (IPCE). A short circuit photocurrent density and the open circuit voltage were 20.5 $mA \cdot cm^{-2}$ (AM 1.5) and 0.72 V, respectively. Recently, 4- and 4,4′-oligophenylenevinylene-functionalized Ru(II)-bipyridine sensitizers were used in dye-sensitized TiO2 solar cells (DSSCs) to study their sensitizing properties [133].

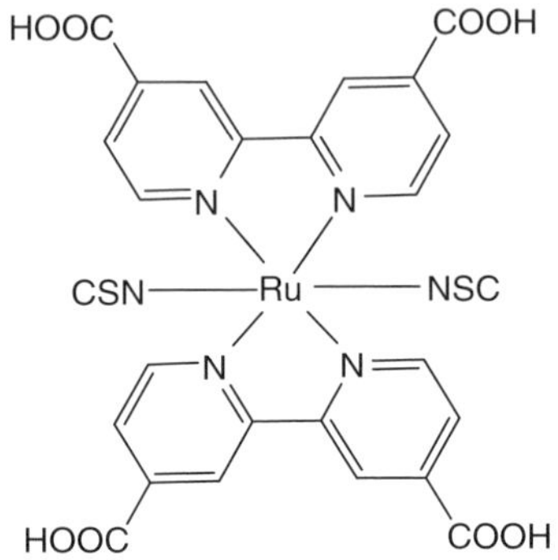

115

5.10 SUMMARY

The advantages of organic functional materials over inorganic ones are their easiness of processing and the tunability of their properties through a simple chemical modification. Moreover, the incorporation of functional groups has endowed the molecular materials with unique and interesting optoelectronic properties. In this chapter, major class of molecular materials for electronic and optoelectronics is described. Especially, latest applications of organic small molecules in OLEDs, photovoltaic devices, thin-film transistors are concerned. However, up to now, the fragility of many molecular materials is still a major limitation toward their application in real-world electronic devices. The increase of stability should become one of the major objectives in the future search for optoelectronic materials.

EXERCISE QUESTIONS

1. Most of luminescent materials are highly emissive in their dilute solutions but become weakly luminescent when fabricated into thin films. An exception of 1,1,2,3,4,5-hexaphenylsilole has been observed, which exhibit very strong luminescence upon aggregation (aggregation-induced emission, AIE). Would you please explain the phenomenon?

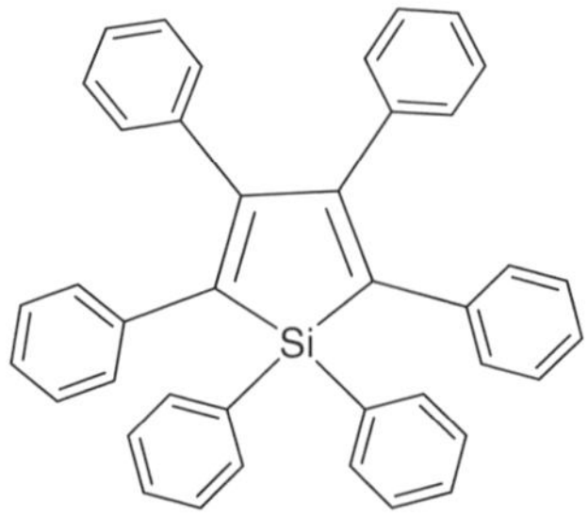

1,1,2,3,4,5-hexaphenylsilole

2. Please predict the properties of the following *β*-porphyrin according to your knowledge.

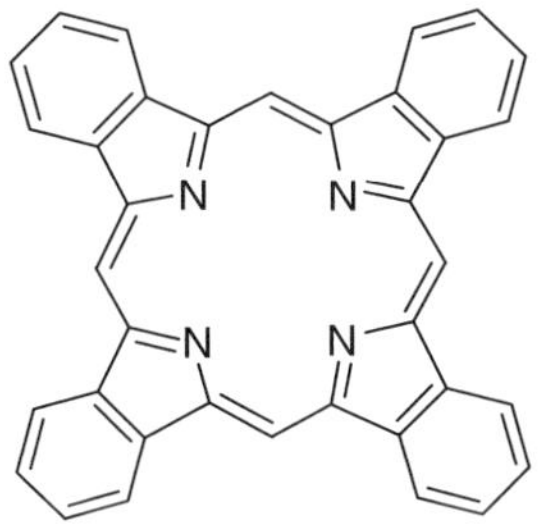

LIST OF ABBREVIATIONS

TTF	tetrathiafulvalenes
OFET	organic field-effect transistors
OLED	organic light-emitting diodes
NLO	nonlinear optical
IPCE	incident photocurrent conversion efficiency
ITO	indium tin oxide
EL	electroluminescence
SHG	second-harmonic generation
HTL	hole-transporting layer
LB	Langmuir–Blodgett
OL	Optical limiting
Alq_3	aluminum(III) tris(8-hydroxyquinolate)
T_g	glass transition temperatures
ETL	electron-transporting layer
ETHB	electron-transporting and hole-blocking

REFERENCES

1. Zollinger, H., *Color Chemistry*, 3rd ed., VCH, Weinheim, 2003.
2. Lukáč, I. and Langhals, H., Darstellung und fluoreszenzverhalten von 2,3,4,4a,10a,11,12,13-Octahydro-1,4a,10a,14-tetraazaviolanthron-derivaten, *Chem. Ber.*, 116, 3524, 1983.
3. Langhals, H. et al., A novel fluorescent dye with strong, anisotropic solid-state fluorescence, small stokes shift, and high photostability, *Angew. Chem. Int. Ed.*, 44, 2427–2428, 2005.
4. Liu, S. et al., Self-organizing liquid crystal perylene bisimide thin films: Spectroscopy, crystallinity, and molecular orientation, *J. Phys. Chem. B*, 106, 1307–1315, 2002.
5. Ahrens, M.J. et al., Bis(*n*-octylamino)perylene-3,4:9,10-bis(dicarboximide)s and their radical cations: Synthesis, electrochemistry, and ENDOR spectroscopy, *J. Org. Chem.*, 71, 2107–2114, 2006.
6. Fan, L.Q., Xu, Y.P., and Tian, H., 1,6-Disubstituted perylene bisimides: Concise synthesis and characterization as near-infrared fluorescent dyes, *Tetrahedron Lett.*, 46, 4443–4447, 2005.
7. Kim, J.Y. et al., Mobility of electrons and holes in an n-type organic semiconductor perylene bisimide thin film, *Curr. App. Phys.*, 5, 615–618, 2005.
8. Würthner, F., Perylene bisimide dyes as versatile building blocks for functional supramolecular architectures, *Chem. Comm.*, 1564–1579, 2004.
9. Nakamura, J., Yokoe, C., and Murata, K., Efficient organic solar cells by penetration of conjugated polymers into perylene pigments, *J. Appl. Phys.*, 96, 6878–6883, 2004.
10. Tang, C.W., Two-layer organic photovoltaic cell, *Appl. Phys. Lett.*, 48, 183–185, 1986.
11. Shin, W.S. et al., Effects of functional groups at perylene bisimide derivatives on organic photovoltaic device application, *J. Mater. Chem.*, 16, 384–390, 2006.
12. Tian, H. et al., Wide spectral photosensitization for SnO_2 nanoporous electrode with soluble perylene derivatives and cyanine dyes, *Synth. Met.*, 121, 1557–1558, 2001.

13. Wang, S. et al., Dye sensitization of nanocrystalline TiO_2 by perylene derivatives, *Synth. Met.*, 128, 299–304, 2002.
14. Tian, H. et al., Synthesis of novel multi-chromophoric soluble perylene derivatives and their photosensitizing properties with wide spectral response for SnO_2 nanoporous electrode, *J. Mater. Chem.*, 10, 2708–2715, 2000.
15. Hua, J. et al., Novel soluble and thermally-stable fullerene dyad containing perylene, *J. Mater. Chem.*, 14, 1849–1853, 2004.
16. Jones, B.A. et al., High-mobility air-stable *n*-type semiconductors with processing versatility: Dicyanoperylene-3,4,9,10-bis(dicarboximides), *Angew. Chem. Int. Ed.*, 43, 6363–6366, 2004.
17. Malenfant, P.R.L. et al., *N*-type organic thin-film transistor with high field-effect mobility based on a N,N′-dialkyl-3,4,9,10-perylene tetracarboxylic diimide derivative, *Appl. Phys. Lett.*, 80, 2517–2519, 2002.
18. Chesterfield, R.J. et al., Variable temperature film and contact resistance measurements on operating n-channel organic thin film transistors, *J. Appl. Phys.*, 95, 6396–6405, 2004.
19. Chesterfield, R.J. et al., Organic thin film transistors based on N-alkyl perylene bisimides: Charge transport kinetics as a function of gate voltage and temperature, *J. Phys. Chem. B*, 108, 19281–19292, 2004.
20. Chen, H. et al., A novel organic *n*-type material: Fluorinated perylene bisimide, *Sol. Energy Mater. Sol. Cells*, 87, 521–527, 2005.
21. Xu, B. et al., Large Gate modulation in the current of a room temperature single molecule transistor, *J. Am. Chem. Soc.*, 127, 2386–2387, 2005.
22. Kalinowski, J. et al., Voltage-tunable-color multilayer organic light emitting diode, *Appl. Phys. Lett.*, 68, 2317–2319, 1996.
23. Fan, L.Q. et al., Novel red-light emitting metal complex based on asymmetric perylene bisimide and 8-hydroxyquinoline dyads, *Synth. Met.*, 145, 203–210, 2004.
24. Pan, J.F. et al., Dendron-functionalized perylene diimides with carrier-transporting ability for red luminescent materials, *Polymer*, 46, 7658–7669, 2005.
25. Tadayyon, S.M. et al., CuPc buffer layer role in OLED performance: A study of the interfacial band energies, *Org. Electronics*, 5, 157–166, 2004.
26. Vestweber, H. and Rie, W., Highly efficient and stable organic light-emitting diodes, *Synth. Met.*, 91, 181–185, 1997.
27. de la Escosura, A. et al., Stabilization of charge-separated states in phthalocyanine–fullerene ensembles through supramolecular donor–acceptor interactions, *J. Am. Chem. Soc.*, 128, 4112–4118, 2006.
28. Huisman, C.L., Goossens, A., and Schoonman, J., Photodoping of zinc phthalocyanine: Formation, mobility, and influence of oxygen radicals in phthalocyanine-based solar cells, *J. Phys. Chem. B*, 106, 10578–10584, 2002.
29. He, J. et al., Phthalocyanine-sensitized nanostructured TiO_2 electrodes prepared by a novel anchoring method, *Langmuir*, 17, 2743–2747, 2001.
30. de la Torre, G. et al., Role of structural factors in the nonlinear optical properties of phthalocyanines and related compounds, *Chem. Rev.*, 104, 3723–3750, 2004.
31. Maya, E.M. et al., Synthesis of novel push–pull unsymmetrically substituted alkynyl phthalocyanines, *J. Org. Chem.*, 65, 2733–2739, 2000.
32. Maya, E.M. et al., Novel push–pull phthalocyanines as targets for second-order nonlinear applications, *J. Phys. Chem. A*, 107, 2110–2117, 2003.
33. Liu, Y.Q., Hu, W.P., Xu, Y., Liu, S.G., and Zhu, D.B., Characterization, second-harmonic generation, and gas-sensitive properties of Langmuir–Blodgett films of 1,8-naphthalimide-tri-*tert*-butylphthalocyanine, *J. Phys. Chem. B.*, 104, 11859–11863, 2000.
34. Xu, L. et al., Preparation and nonlinear optical properties of ultrathin composite films containing both a polyoxometalate anion and a binuclear phthalocyanine, *New J. Chem.*, 26, 782–786, 2002.
35. Chen, Y. et al., Axially modified gallium phthalocyanines and naphthalocyanines for optical limiting, *Chem. Soc. Rev.*, 34, 517–529, 2005.
36. Calvete, M., Yang, G.Y., and Hanack, M., Porphyrins and phthalocyanines as materials for optical limiting, *Synth. Met.*, 141, 231–243, 2004.
37. Chen, Y. et al., Strong optical limiting of soluble axially substituted gallium and indium phthalocyanines, *Adv. Mater.*, 15, 899–902, 2003.
38. Zeis, R., Siegrist, T., and Kloc, C., Single-crystal field-effect transistors based on copper phthalocyanine, *Appl. Phys. Lett.*, 86, 22103, 2005.
39. Chen, Y. et al., High Performance organic field-effect transistors based on amphiphilic tris (phthalocyaninato) rare earth triple-decker complexes, *J. Am. Chem. Soc.*, 127, 15700–15701, 2005.

40. Kuder, J.E., Organic active layer materials for optical recording, *J. Imag. Sci.*, 32, 51–56, 1988.
41. Inabe, T. and Tajima, H., Phthalocyanines-versatile components of molecular conductors, *Chem. Rev.*, 104, 5503–5533, 2004.
42. Adler, A.D. et al., A simplified synthesis for *meso*-tetraphenylporphyrin, *J. Org. Chem.*, 32, 476, 1967.
43. Lindsey, J.S., et al., Rothemund and Adler-Longo reactions revisited: Synthesis of tetraphenylporphyrins under equilibrium conditions, *J. Org. Chem.*, 52, 827–836, 1987.
44. Jasat, A. and Dolphin, D., Expanded porphyrins and their heterologs, *Chem. Rev.*, 6 2267–2340, 1997.
45. Chen, Z.M., Xia, Z., and Wu, Y.Q., High density recordable optical disk storage materials: Metal porphyrin complexes, *Prog. Chem.*, 16, 820–828, 2004.
46. Vuorinen, T. et al., Photoinduced electron transfer in Langmuir–Blodgett monolayers of porphyrin–fullerene dyads, *Langmuir*, 21, 5383–5390, 2005.
47. Lu, F.S. et al., Synthesis and chemical properties of conjugated polyacetylenes having pendant fullerene and/or porphyrin units, *Macromolecules*, 37, 7444–7450, 2004.
48. Yamada, H. et al., Photovoltaic properties of self-assembled monolayers of porphyrins and porphyrin–fullerene dyads on ITO and gold surfaces, *J. Am. Chem. Soc.*, 125, 9129–9139, 2003.
49. Prodi, A., Chiorboli, C., Scandola, F., Iengo, E., Alessio, E., Dobrawa, R., and Würthner, F., Wavelength-dependent electron and energy transfer pathways in a side-to-face ruthenium porphyrin/perylene bisimide assembly, *J. Am. Chem. Soc.*, 127, 1454–1462, 2005.
50. Liddell, P.A. et al., Photonic switching of photoinduced electron transfer in a dithienylethene–porphyrin–fullerene triad molecule, *J. Am. Chem. Soc.*, 124, 7668–7669, 2002.
51. Terazono, Y. et al., Photonic control of photoinduced electron transfer via switching of redox potentials in a photochromic moiety, *J. Phys. Chem. B*, 108, 1812–1814, 2004.
52. Liddell, P.A. et al., Photonic switching of photoinduced electron transfer in a dihydropyrene–porphyrin–fullerene molecular triad, *J. Am. Chem. Soc.*, 126, 4803–4811, 2004.
53. Hasobe, T. et al., Photovoltaic cells using composite nanoclusters of porphyrins and fullerenes with gold nanoparticles, *J. Am. Chem. Soc.*, 127, 1216–1228, 2005.
54. Nazeeruddin, Md.K., Humphry-Baker, R., Officer, D.L., Campbell, W.M., Burrell, A.K., and Gratzel, M., Application of metalloporphyrins in nanocrystalline dye-sensitized solar cells for conversion of sunlight into electricity, *Langmuir*, 20, 6514–6517, 2004.
55. Wang, Q., Campbell, W.M., Bonfantani, E.E., Jolley, K.W., Officer, D.L., Walsh, P.J., Gordon, K., Humphry-Baker, R., Nazeeruddin, M.K., and Gratzel, M., Efficient light harvesting by using green Zn-porphyrin-sensitized nanocrystalline TiO_2 films, *J. Phys. Chem. B*, 109, 15397–15409, 2005.
56. Kwong, R.C. et al., Efficient, saturated red organic light emitting devices based on phosphorescent platinum(II) porphyrins, *Chem. Mater.*, 11, 3709–3713, 1999.
57. Pan, S. and Rothberg, L.J., Enhancement of platinum octaethyl porphyrin phosphorescence near nanotextured silver surfaces, *J. Am. Chem. Soc.*, 127, 6087–6094, 2005.
58. Suslick, K.S. et al., Push–pull porphyrins as nonlinear optical materials, *J. Am. Chem. Soc.*, 114, 6928–6930, 1992.
59. LeCours, S.M., et al., Push–pull arylethynyl porphyrins: New chromophores that exhibit large molecular first-order hyperpolarizabilities, *J. Am. Chem. Soc.*, 118, 1497, 1996.
60. Priyadarshy, S., Therien, M.J., and Beratan, D.N., Acetylenyl-linked, porphyrin-bridged, donor–acceptor molecules: A theoretical analysis of the molecular first hyperpolarizability in highly conjugated push–pull chromophore structures, *J. Am. Chem. Soc.*, 118, 1504–1510, 1996.
61. Jiang, L. et al., Third-order nonlinear optical properties of an ultrathin film containing a porphroupyrin derivative, *J. Phys. Chem. B*, 109, 6311–6315, 2005.
62. Krivokapic, A. et al., *Meso*-tetra-alkynyl porphyrins for optical limiting-a survey of g III and IV metal complexes, *Adv. Mater.*, 13, 652–656, 2001.
63. Mishra, A. et al., Cyanines during the 1990s: A review, *Chem. Rev.*, 100, 1973–2011, 2000.
64. Higgins, D.A., Reid, P.J., and Barbara, P.F., Structure and exciton dynamics in J-aggregates studied by polarization-dependent near-field scanning optical microscopy, *J. Phys. Chem.*, 100, 1174–1180, 1996.
65. Meng, F.S. and Tian, H., Solar cells based on cyanine and polymethine dyes, in *Organic Photovoltaics: Mechanisms, Materials, and Devices*, Sun, S.S. and Sariciftci, N.S., Eds., Taylor & Francis, Boca Raton, 2005, Chap. 13.
66. Guo, M. et al., Photoelectrochemical studies of nanocrystalline TiO_2 co-sensitized by novel cyanine dyes, *Sol. Energy Mater. Sol. Cells*, 88, 23–35, 2005.
67. Yao, Q. et al., Photoelectric conversion properties of four novel carboxylated hemicyanine dyes on TiO_2 electrode, *J. Mater. Chem*, 13, 1048–1053, 2003.

68. Meng, F.S. et al., Novel cyanine dyes with multi-carboxyl groups and their sensitization on nanocrystalline TiO_2 electrode, *Synth. Met.*, 137, 1543–1544, 2003.
69. Nüesch, F. et al., Interface modification to optimize charge separation in cyanine heterojunction photovoltaic devices, *Sol. Energy Mater. Sol. Cells*, 87, 817–824, 2005.
70. Meng, F.S. et al., Synthesis of novel cyanine–fullerene dyads for photovoltaic devices, *J. Mater. Chem.*, 15, 979–986, 2005.
71. Raos, G. and Zoppo, D.M., Substituent effects on the second-order hyperpolarisability of cyanine cations, *J. Mol. Struct.: Theochem.*, 589–590, 439–445, 2002.
72. Zoppo, D.M., Bianco, A., and Zerbi, G., Nonlinear optical properties of cyanine systems, *Synth. Met.*, 124, 183–184, 2001.
73. Marks, A.F., Noah, A.K., and Sahyun, M.R.V., Bond-length alternation in symmetrical cyanine dyes, *J. Photochem. Photobiol. A: Chem.*, 139, 143–149, 2001.
74. Wang, J. et al., Syntheses and nonlinear absorption of novel unsymmetrical cyanines, *Dyes Pigments*, 57, 171–179, 2003.
75. Berkovic, G., Krongauz, V., and Weiss, V., Spiropyrans and spirooxazines for memories and switches, *Chem. Rev.*, 100, 1741–1753, 2000.
76. Tamaoki, N. et al., Photoreversible optical nonlinearities of polymeric films containing spiropyran with long alkyl chains, *Appl. Phys. Lett.*, 69, 1188–1190, 1996.
77. Stolka, M., Janus, J.F., and Pai, D.M., Hole transport in solid solutions of a diamine in polycarbonate, *J. Phys. Chem.*, 88, 4707–4714, 1984.
78. O'Brien, D.F. et al., Hole transporting materials with high glass transition temperatures for use in organic light-emitting devices, *Adv. Mater.*, 10, 1108–1112, 1998.
79. Shirota, Y., Organic materials for electronic and optoelectronic devices, *J. Mater. Chem.*, 10, 1–25, 2000.
80. Li, J. et al., Novel starburst molecule as a hole injecting and transporting material for organic light-emitting devices, *Chem. Mater.*, 17, 615–619, 2005.
81. Hu, N. et al., 5,11-Dihydro-5,11-di-1-naphthylindolo[3,2-b]carbazole: Atropisomerism in a novel hole-transport molecule for organic light-emitting diodes, *J. Am. Chem. Soc.*, 121, 5097–5098, 1999.
82. Li, J. et al., A high T_g carbazole-based hole-transporting material for organic light-emitting devices, *Chem. Mater.*, 17, 1208–1212, 2005.
83. Ko, C.W. and Tao, Y.T., 9,9-Bis{4-di-(*p*-biphenyl)aminophenyl}fluorine: A high T_g and efficient hole-transporting material for electroluminescent devices, *Synth. Met.*, 126, 37–41, 2002.
84. Thomas, K.R.J. et al., New star-shaped luminescent triarylamines: Synthesis, thermal, photophysical, and electroluminescent characteristics, *Chem. Mater.*, 14, 1354–1361, 2002.
85. Sonntag, M. et al., Novel star-shaped triphenylamine-based molecular glasses and their use in OFETs, *Chem. Mater.*, 17, 3031–3039, 2005.
86. Zhu, W.H. et al., Singlet energy transfer and photoinduced electron transfer in star-shaped naphthalimide derivatives based on triphenylamine, *Bull. Chem. Soc. Jpn.*, 78, 1362–1367, 2005.
87. Adachi, C., Tsutsui, T., and Saito, S., Organic electroluminescent device having a hole conductor as an emitting layer, *Appl. Phys. Lett.*, 55, 1489–1491, 1989.
88. Cao, Y. et al., Improved quantum efficiency for electroluminescence in semiconducting polymers, *Nature*, 397, 414–417, 1999.
89. Brown, A.R. et al., Poly(*p*-phenylenevinylene) light-emitting diodes: Enhanced electroluminescent efficiency through charge carrier confinement, *Appl. Phys. Lett.*, 61, 2793, 1992.
90. Yeh, H.C. et al., Synthesis, properties, and applications of tetraphenylmethane-based molecular materials for light-emitting devices, *Chem. Mater.*, 13, 2788–2796, 2001.
91. Tian, H., Zhu, W., and Elschner, A., Single-layer electroluminescence device made with novel copolymers containing electron- and hole-transporting moieties, *Synth. Met.*, 111–112, 481–483, 2000.
92. Tian, H. and Yang, S., Intramolecular triplet energy transfer in multi-chromophoric dyes and its influence on the photostability, *J. Photochem. Photobiol. C: Photochem. Rev.*, 3, 67–76, 2002.
93. Guan, M. et al., High-performance blue electroluminescent devices based on 2-(4-biphenylyl)-5-(4-carbazole-9-yl)phenyl-1,3,4-oxadiazole, *Chem. Commun.*, 2708–2709, 2003.
94. Liang, F. et al., A hydroxyphenyloxadiazole lithium complex as a highly efficient blue emitter and interface material in organic light-emitting diodes, *J. Mater. Chem.*, 13, 2922–2926, 2003.
95. Carella, A. et al., Synthesis and second order nonlinear optical properties of new chromophores containing 1,3,4-oxadiazole and thiophene rings, *J. Chem. Soc., Perkin Trans.*, 2, 1791–1795, 2002.
96. Wudl, F., Smith, G.M., and Hufnagel, E.J., Bis-1,3-dithiolium chloride: An unusually stable organic radical cation, *J. Chem. Soc. Chem. Comm.*, 1453–1455, 1970.

97. Ferraris, J. et al., Electron transfer in a new highly conducting donor–acceptor complex, *J. Am. Chem. Soc.*, 95, 948–949, 1973.
98. Bendikov, M., Wudl, F., and Perepichka, D.F., Tetrathiafulvalenes, oligoacenes, and their buckminsterfullerene derivatives: The brick and mortar of organic electronics, *Chem. Rev.*, 104, 4891–4945, 2004.
99. Rovira, C., Bis(ethylenethio)tetrathiafulvalene (BET-TTF) and related dissymmetrical electron donors: From the molecule to functional molecular materials and devices (OFETs), *Chem. Rev.*, 104, 5289–5317, 2004.
100. Yamada, J. et al., New trends in the synthesis of π-electron donors for molecular conductors and superconductors, *Chem. Rev.*, 104, 5057–5083, 2004.
101. Mori, T., Organic conductors with unusual band fillings, *Chem. Rev.*, 104, 4947–4969, 2004.
102. Shibaeva, R.P. and Yagubskii, E.B., Molecular conductors and superconductors based on trihalides of BEDT-TTF and some of its analogues, *Chem. Rev.*, 104, 5347–5378, 2004.
103. Kimura, S. et al., A new organic superconductor β-(*meso*-DMBEDT-TTF)$_2$PF$_6$, *Chem. Commun.*, 2454–2455, 2004.
104. Naraso, N. et al., High-performance organic field-effect transistors based on π-extended tetrathiafulvalene derivatives, *J. Am. Chem. Soc.*, 127, 10142–10143, 2005.
105. Nishikawa, H. et al., Photophysical study of new methanofullerene-TTF dyads: An obvious intramolecular charge transfer in the ground states, *J. Phys. Chem. A*, 108, 1881–1890, 2004.
106. Valentin, M.D. et al., Photoinduced long-lived charge separation in a tetrathiafulvalene–porphyrin–fullerene triad detected by time-resolved electron paramagnetic resonance, *J. Phys. Chem. B*, 109, 14401–14409, 2005.
107. Segura, J.L. et al., A new photoactive and highly soluble C_{60}-TTF-C_{60} dimer: Charge separation and recombination, *Org. Lett.*, 2, 4021–4024, 2000.
108. Xiao, X.W., Xu, W., Zhang, D,Q., Xu, H., Lu, H.Y., and Zhu, D.B., A new fluorescence-switch based on supermolecular dyad with (tetraphenylporphyrinato) zinc(II) and tetrathiafulvalene units, *J. Mater. Chem.*, 15, 2557–2561, 2005.
109. Gooding's, E.P., Matched, D.A., and Owen, G., Synthesis, structure, and electrical properties of naphthacene, pentacene, and hexacene sulphides. *J. Chem. Soc., Perkin Trans. I*, 11, 1310–1314, 1972.
110. Meng, H. et al., Tetramethylpentacene: Remarkable absence of steric effect on field effect mobility, *Adv. Mater.*, 15, 1090–1093, 2003.
111. Weidkamp, K.P., Afzali, A., Tromp, R.M., and Hamers, R.J., Pentacene organic transistors and ring oscillators on glass and on flexible polymeric substrates, *J. Am. Chem. Soc.* 2004, 126, 12740–12741.
112. Ito, K. et al., Oligo(2,6-anthrylene)s: Acene-oligomer approach for organic field-effect transistors, *Angew. Chem. Int. Ed.*, 42, 1159–1162, 2003.
113. Butko, V.Y., Chi, X., and Ramirez, A.P., Free-standing tetracene single crystal field effect transistor, *Solid State Commun.*, 128, 431–434, 2003.
114. de Boer, R.W.I., Klapwijk, T.M., and Morpurgo, A.F., Field-effect transistors on tetracene single crystals, *Appl. Phys. Lett.*, 83, 4345–4347, 2003.
115. Hamada, Y. et al., Red organic light-emitting diodes using an emitting assist dopant, *Appl. Phys. Lett.*, 75, 1682–1684, 1999.
116. Dodge, J.A., Bain, J.D., and Richard, A., Chambedin regioselective synthesis of substituted rubrenes, *J. Org. Chem.*, 55, 4190–4198, 1990.
117. Sundar, V.C. et al., Elastomeric transistor stamps: Reversible probing of charge transport in organic crystals, *Science*, 303, 1644–1646, 2004.
118. Zeis, R. et al., Field effect studies on rubrene and impurities of rubrene, *Chem. Mater.* 18, 244–248, 2006.
119. Gamier, F. et al., Molecular engineering of organic semiconductors: Design of self-assembly properties in conjugated thiophene oligomers, *J. Am. Chem. Soc.*, 115, 8716–8721, 1993.
120. Tian, H. et al., Novel thiophene-aryl co-oligomers for organic thin film transistors, *J. Mater. Chem.*, 15, 3026–3033, 2005.
121. Katz, H.E. et al., Synthetic chemistry for ultrapure, processable, and high-mobility organic transistor semiconductors, *Acc. Chem. Res.*, 34, 359–369, 2001.
122. Facchetti, A. et al., Building blocks for n-type molecular and polymeric electronics. perfluoroalkyl-versus alkyl-functionalized oligothiophenes (nT; n = 2–6). systematics of thin film microstructure, semiconductor performance, and modeling of majority charge injection in field-Effect transistors, *J. Am. Chem. Soc.*, 126, 13859–13874, 2004.
123. Tang, C.W. and VanSlyke, S.A., Organic electroluminescence diode, *Appl. Phys. Lett.*, 51, 913–915, 1987.

124. Pohl, R., Montes, V.A., Shinar, J., and Anzenbacher Jr. P., Red–green–blue emission from tris(5-aryl-8-quinolinolate) Al(III) complexes, *J. Org. Chem.*, 69, 1723–1725, 2004.
125. Xie, J., Ning, Z.J., and Tian, H., A soluble 5-carbazolium-8-hydroxyquinoline Al(III) complex as a dipolar luminescent material, *Tetrahedron Lett.*, 46, 8559–8562, 2005.
126. Shi, Y.W. et al., Fluorinated Alq_3 derivatives with tunable optical properties, *Chem. Commun.*, 1941–1943, 2006.
127. Dedeian, K. et al., A new synthetic route to the preparation of a series of strong photoreducing agents: Fac tris-*ortho*-metalated complexes of Iridium(III) with substituted 2-phenylpyridine, *Inorg. Chem.*, 30, 1685–1687, 1991.
128. Ikai, M. et al., Highly efficient phosphorescence from organic light-emitting devices with an exciton-block layer, *Appl. Phys. Lett.*, 79, 156–158, 2001.
129. Jiang, C. et al., High-efficiency, Saturated red-phosphorescent polymer light-emitting diodes based on conjugated and non-conjugated polymers doped with an Ir complex, *Adv. Mater.*, 16, 537–541, 2004.
130. Yeh, S.J. *et al.*, New dopant and host materials for blue-light-emitting phosphorescent organic electroluminescent devices, *Adv. Mater.*, 17, 285–289, 2005.
131. You, Y. and Park, S.Y., Inter-ligand energy transfer and related emission change in the cyclometalated heteroleptic Iridium complex: Facile and efficient color tuning over the whole visible range by the ancillary ligand structure, *J. Am. Chem. Soc.*, 127, 12438–12439, 2005.
132. Nazeeruddin, M.K., Kay, A., Rodicio, I., Humphry-Baker, R., Mueller, E., Liska, P., Vlachopoulos, N., and Graetzel, M., Conversion of light to electricity by cis-Xzbis(2,2′-bipyridyl-4,4′-dicarboxylate)ruthe nium(11) charge-transfer sensitizers (X = C1-, Br-, I-, CN-, and SCN-) on nanocrystalline TiO_2 electrodes, *J. Am. Chem. Soc.*, 115, 6382–6390, 1993.
133. Jang, S.R. et al., Oligophenylenevinylene-functionalized Ru(II)-bipyridine sensitizers for efficient dye-sensitized nanocrystalline TiO_2 solar ells, *Chem. Mater.*, 18, 5604–5608, 2006.

6 Major Classes of Conjugated Polymers and Synthetic Strategies

Yongfang Li and Jianhui Hou

CONTENTS

6.1 Introduction 174
6.2 Doped Conjugated Polymers: Conducting Polymers 175
6.2.1 Properties of Conducting Polymers 175
6.2.1.1 Doping Properties 177
6.2.1.2 Conductivity 178
6.2.1.3 Absorption Spectra 178
6.2.1.4 Solubility 180
6.2.1.5 Electrochemical Properties 181
6.2.2 Electrochemical Preparation of Conducting Polymers 182
6.2.2.1 Electrochemical Preparation of Polypyrrole 183
6.2.2.2 Mechanism of the Electropolymerization of Pyrrole 184
6.2.2.3 Electrochemical Preparation of Polyaniline and Polythiophene 185
6.2.3 Chemical Catalytic or Oxidative Polymerization 185
6.2.3.1 Polyacetylene and Its Derivatives 185
6.2.3.2 Oxidative Polymerization of Pyrrole and Thiophene 186
6.2.3.3 Preparation of Soluble Conducting Polyaniline 187
6.3 Basic Properties of Neutral Conjugated Polymers 187
6.3.1 Electronic Energy Levels 187
6.3.2 Optical Properties 189
6.3.3 Effect of Side Chain Substitution on the Properties of Conjugated Polymers 191
6.3.4 Charge Carrier Mobility 192
6.4 Synthesis of Soluble Polythiophene Derivatives 193
6.4.1 Synthesis of Poly(3-alkylthiophene)s 193
6.4.1.1 Regioregularity of P3ATs 193
6.4.1.2 Synthesis of P3ATs by Chemical Oxidation Method 194
6.4.1.3 Synthesis of P3ATs by the McCullough and Grignard Metathesis Method 195
6.4.1.4 Synthesis of P3ATs by the Rieke Method 196
6.4.1.5 Synthesis of HT–HT P3ATs by the Stille and Suzuki Coupling Reaction 196
6.4.1.6 Ni(0) Catalytic Dehalogenation Method for P3ATs 197
6.4.2 Synthesis of Phenyl Substituted Polythiophene Derivatives 198
6.4.3 Synthesis of Polythiophenes with Conjugated Side Chains 198
6.4.4 Synthesis of Poly(3-alkoxythiophene)s 199

6.5 Synthesis of Poly(*p*-phenylene vinylene) and Its Derivatives 200
6.5.1 Synthesis of PPV by the Wessling Precursor Method 200
6.5.2 Synthesis of PPV and Its Derivatives by the Gilch Method 201
6.5.3 Synthesis of PPVs by the Heck Coupling Reaction 202
6.5.4 Synthesis of PPV and Its Derivatives by the Wittig Reaction 202
6.5.5 Synthesis of CN-PPVs by the Knoevenagel Condensation 203
6.6 Synthesis of Poly(*p*-phenylene), Polyfluorene, and Their Derivatives 203
6.6.1 Synthesis of Poly(*p*-phenylene) and Its Derivatives 203
6.6.2 Synthesis of PF and PPP by Ni(0)-Mediated Polymerization 204
6.6.3 Synthesis of PF by the Suzuki Reaction 204
6.7 Synthesis of Poly(aryleneethynylene)s 204
6.7.1 Synthesis of Poly(aryleneethynylene)s by Pd(0) Catalytic Method 204
6.7.2 Synthesis of PPEs by Alkyne Metathesis 205
6.8 Summary 206
Exercise Questions 207
List of Abbreviations 207
References 208

Abstract This chapter introduces the molecular structures, basic properties, main applications, and the synthetic methods of the major classes of conjugated polymers. The conjugated polymers are the polymers with conjugated main chains, including polyacetylene (PA), polypyrrole (PPy), polyaniline (PAn), polythiophene (PT), poly(*p*-phenylene vinylene) (PPV), poly (*p*-phenylene) (PPP), polyfluorene (PF), poly(thienylene vinylene) (PTV), etc. According to their steady state, research interests, and applications, they can be divided into doped conducting polymers and neutral conjugated polymers. The former include PA, PPy, PAn, PT, and PTV; and the latter include PPV, PT, PF, and PPP. For doped conducting polymers, the doping structures, doping properties, conductivity, absorption spectra, solubility, electrochemical properties, and their electrochemical preparation are described. For neutral conjugated polymers, the electronic energy level, optical properties, effect of side chain substitution, and various synthetic routes for the preparation of soluble conjugated polymers are elucidated.

6.1 INTRODUCTION

Since the discovery of conducting polyacetylene in 1977 by Shirakawa, MacDiarmid, and Heeger et al. [1,2], conjugated polymers have attracted great attention all over the world [3,4], due to their novel properties and many potential applications as conducting polymers in the fields of electrode materials for batteries, electrochromic displays, modified electrode and biosensors, transparent conducting materials, etc., and as intrinsic semiconducting polymers in the fields of polymer light-emitting diodes (PLEDs), field-effect transistors (FET), polymer solar cells (PSCs), etc. The most important characteristics of the conjugated polymers are their molecular structures with conjugated main chains and the doping properties of the conjugated main chains. The conjugated polymers can be oxidized to their p-doped state or be reduced to their n-doped state, and they become conducting polymers after the doping, with conductivity increasing by a order of 6–9. Figure 6.1 shows the major classes of conjugated polymers well studied so far. Among the conjugated polymers, polyacetylene (PA), polypyrrole (PPy), polyaniline (PAn), polythiophene (PT), and poly(thienylene vinylene) (PTV) are intensively studied as conducting polymers, while the soluble derivatives of PT, poly(*p*-phenylene vinylene) (PPV), poly(*p*-phenylene) (PPP), polyfluorene (PF), and poly(phenylene ethynylene) (PPE) attracted interests as intrinsic semiconducting polymers in their neutral state for the applications in PLEDs, FETs, PSCs, etc. Table 6.1 lists the main characteristics of the most popular conjugated polymers.

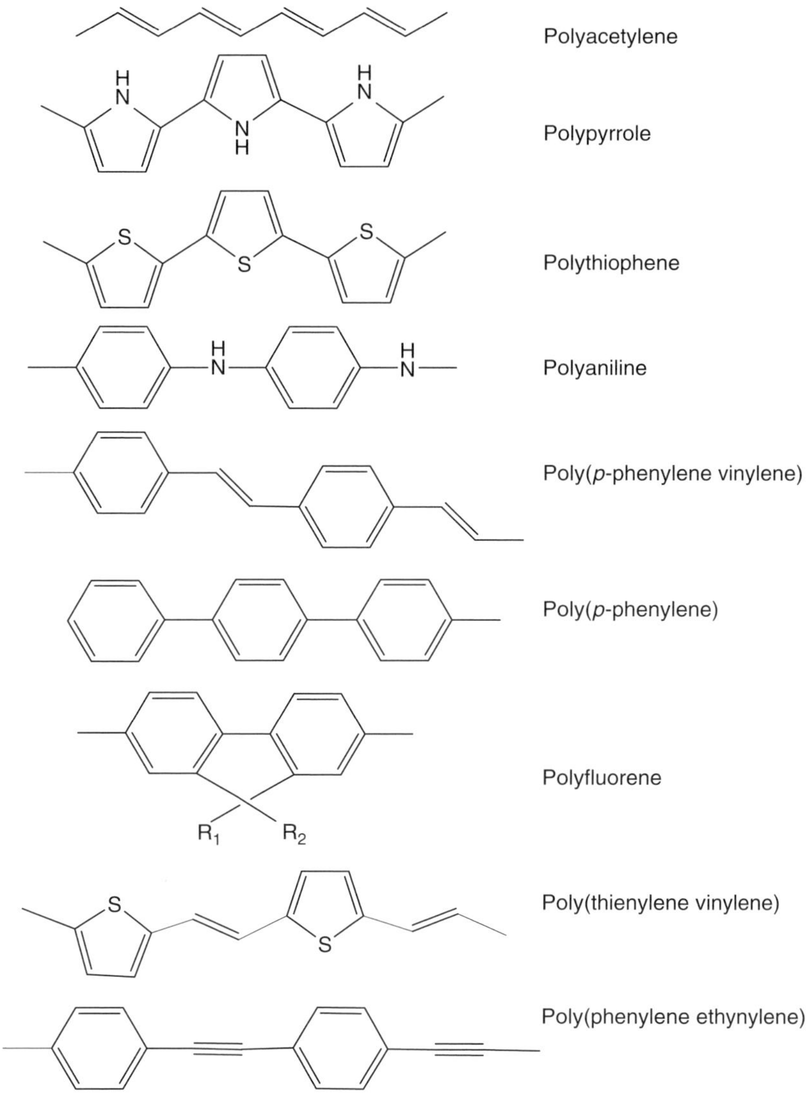

FIGURE 6.1 Molecular structures of the major classes of conjugated polymers.

6.2 DOPED CONJUGATED POLYMERS: CONDUCTING POLYMERS

As mentioned above, conducting polymers are the conjugated polymers in their doped state. In fact, all the conducting polymers studied till now are in their p-doped state, because their n-doped state is either not found (e.g., polypyrrole and polyaniline) or unstable (e.g., polythiophene). Generally, almost all the conjugated polymers can be doped into conducting polymers, but only a few of them were studied as conducting polymers because the other conjugated polymers are unstable in their doped state. The stable doped (p-doped) conducting polymers mainly include PPy, PAn, and PT.

6.2.1 Properties of Conducting Polymers

Doping properties, conductivity, absorption spectra, solubility, and electrochemical properties are the mostly concerned properties of the conducting polymers.

TABLE 6.1
Main Characteristics of the Most Popular Conjugated Polymers

Conjugated Polymers	Steady State	Synthetic Method	State of the Sample	Solubility	Applications
Polyacetylene	Unstable	Chemical polymerization	Film	Insoluble	Theoretical studies for the nature of the charge carriers and charge transportation in conducting polymers
Polypyrrole	p-Doped conducting state	Electrochemical polymerization	Film	Insoluble	Conducting polymer films, electrode materials for batteries, modified electrode and biosensors, capacitors, electrochromics
Polyaniline	p-Doped conducting state	Electrochemical polymerization	Loosely bounded layer on electrode	Insoluble	Electrode materials for biosensors, batteries and electrochromics, modified electrode
Polyaniline	p-Doped conducting state	Chemical oxidation polymerization	Powder	Soluble in *m*-cresol with camphor sulphonic acid or dodecylbenzene sulphonic acid as dopant	Electrode materials, anti-corrosion, microwave absorption material, anti-statics, modification layer on ITO for the application in PLEDs
Polythiophene	p-Doped conducting state	Electrochemical polymerization	Film	Insoluble	Electrochromics
Poly(3-hexylthiophene) (P3HT) or poly(3-alkylthiophene) (P3AT)	Neutral state	Chemical polymerization	Powder	Soluble in THF, xylene, chlorobenzene, etc.	Donor material in PSCs, red polymer in PLEDs, semiconductor in FETs
Poly(ethylenedioxythiophene): polystyrene sulfonate (PEDOT: PSS)	p-Doped conducting state	Chemical polymerization	Colloid aqueous solution	Soluble in water	Transparent conducting electrode, coating layer for unistatic painting, modification layer on ITO electrode used in PLEDs and PSCs, capacitors, etc.
Poly(2-methoxy-5-(2′-ethylhexyloxy)-1,4-phenylene vinylene) (MEH-PPV)	Neutral state	Chemical polymerization	Powder	Soluble in THF, xylene, chlorobenzene, etc.	Orange luminescent polymer in PLEDs, donor material in PSCs
Poly(9,9-dioxylfluorene) (PFO)	Neutral state	Chemical polymerization	Powder	Soluble in THF, xylene, chlorobenzene, etc.	Blue luminescent polymer in PLEDs, host material for phosphorescent PLEDs and for white light PLEDs

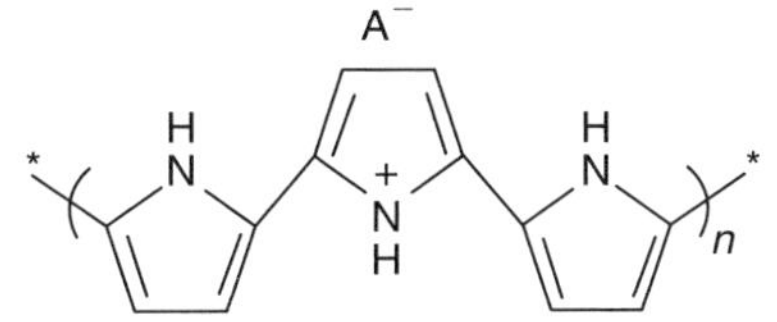

FIGURE 6.2 p-Doping structure of polypyrrole (A^- represents counter anion).

6.2.1.1 Doping Properties

Doping in conducting polymers is different from that in inorganic semiconductors. Inorganic semiconductors possess ordered crystal structure, and their doping is realized by replacement of few crystalline atoms with other atoms with one more or less valence electron. While conducting polymers are in amorphous structure, their doping needs the oxidation or reduction of their conjugated main chains to input holes or electrons, with the intercalation of counter ions for keeping the electro-neutrality. Figure 6.2 shows the p-doping structure of polypyrrole. The dopant concentration in conducting polymers can reach $10^{21}/cm^3$, which is much higher (several orders higher) than that in inorganic semiconductors. Doping degree of a conducting polymer is the doping charge numbers per monomer unit in the conjugated main chains, for example, the doping degree of polypyrrole is 0.33 (Figure 6.2). The doping of conducting polymers can be realized by charge transfer of donors or acceptors (chemical doping), or by electrochemical oxidation or reduction (electrochemical doping).

Chemical doping includes p-doping (oxidation doping) or n-doping (reduction doping). In the p-doping, the oxidant (acceptor), such as I_2, Br_2, and AsF_5, is used as the dopant. After the p-doping, the main chains of the conjugated polymer were oxidized to give out electrons and the dopants get electrons to become counter anions. The reaction of the p-doping can be expressed as

$$\mathrm{CP} + (3/2)\,\mathrm{I_2} \rightarrow \mathrm{CP^+(I_3^-)} \tag{6.1}$$

where CP represents conjugated polymer molecules.

In n-doping, the reductant (donor) such as Na vapor and $Na^+ (C_{10}H_8)^-$ is used as the dopant. After the n-doping, the main chains of the conjugated polymer were reduced to get electrons and the dopants lose electrons to become counter cations. The reaction equation of the n-doping can be expressed as

$$\mathrm{CP + Na \rightarrow CP^-(Na^+)} \quad \text{or} \quad \mathrm{CP + Na^+(C_{10}H_8)^- \rightarrow CP^-(Na^+) + C_{10}H_8} \tag{6.2}$$

The doping of some conducting polymers (such as polyaniline) can be realized by proton acid doping (Figure 6.3). In fact, the proton acid doping is a kind of chemical p-doping. In Figure 6.3, the proton in the acid molecule combines with an N-atom on the main chain of PAn and transfers its positive charge to the PAn main chain, while the anion of the acid molecule becomes a counter anion of the doped PAn. The proton acid doping property of PAn provides a convenient way for preparing soluble conducting PAn, which is discussed later in this chapter.

Electrochemical doping is performed by the electrochemical oxidation or reduction reactions. For electrochemical p-doping, the conjugated main chains are oxidized and the solution anions will dope into the polymer as counter anions:

$$\mathrm{CP - e^- + A^- \rightleftarrows CP^+(A^-)} \tag{6.3}$$

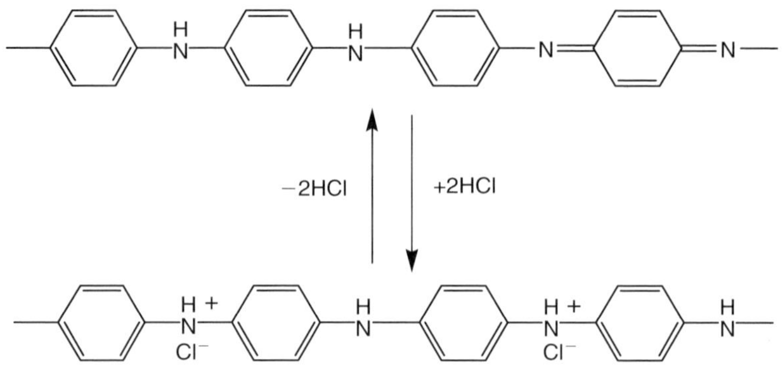

FIGURE 6.3 Proton acid doping of polyaniline (the upper molecular structure of PAn is called emeraldine PAn).

For electrochemical n-doping, the conjugated main chains are reduced and the solution cations will dope into the polymer as counter cations:

$$CP + e^- + M^+ \rightleftarrows CP^-(M^+) \tag{6.4}$$

Almost all the conjugated polymers can be doped by electrochemical doping. But the stability of the doped state depends on the value of the doping potentials. For the p-doping, if the doping potential is low, such as less than 0.5 V versus saturated calomel electrode (SCE), the doped conducting state will be very stable; but if the doping potential is high, such as more than 1.0 V versus SCE, the doped conducting state will be unstable while the neutral state of the conjugated polymer will be stable.

6.2.1.2 Conductivity

Figure 6.4 shows the conductivity range of the p-doped conducting polymers and compares their conductivities with the traditional electronic materials. The highest conductivity of conducting polymers reached till now is 10^5 S/cm for the polyacetylene film after orientation by drawing. Generally, the conductivity of the doped conducting polymers is in the range of 10^{-3} to 10^3 S/cm. Usually, the conductivity of the conducting polymers shows the temperature dependence of semiconductors, and it meets the Mott variable range hoping (VRH) model:

$$\sigma(T) = \sigma_0 \exp[-(T_0/T)^{1/(n+1)}] \tag{6.5}$$

where

σ_0 is a prefactor that is independent or weakly dependent on temperature

n is the space dimension number of the material, $n = 1, 2, 3$ indicates it is one-dimension, two-dimension, or three-dimension VRH transportation

6.2.1.3 Absorption Spectra

An important characteristic of the doped conducting polymers is a strong absorption peak in the near infrared region, while for the neutral conjugated polymer there is no infrared absorption and its absorption peak appears in the UV–visible (UV–vis) region. Hence, the absorption spectrum in the visible–near infrared (Vis–NIR) region becomes a valuable method to distinguish the doping or

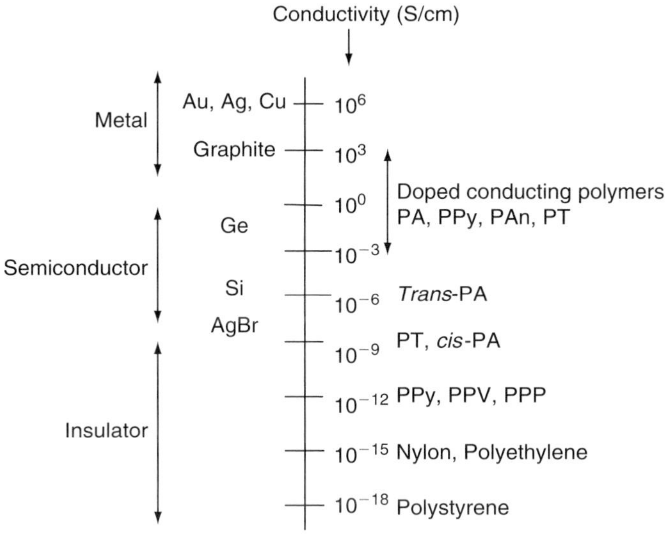

FIGURE 6.4 Comparison of the conductivity of conducting polymers with other electronic materials.

dedoping states of the conjugated polymers. The Vis–NIR absorption spectra of polypyrrole can be used as a representative of the conducting polymers (Figure 6.5). The absorption peak of the neutral PPy is located at ca. 400–410 nm, which is corresponding to the π–π* absorption of the conjugated main chains. After doping, there appears a strong absorption peak at a wavelength longer than 800 nm, and the visible absorption peak redshifted and weakened.

The structure of PAn is more complicated than PPy and PT because there are benzene and quinone structures in its main chains of the emeraldine PAn (see Figure 6.3), so the absorption spectrum of PAn shows some difference from that of PPy. Figure 6.6 shows the Vis–NIR absorption spectra of PAn. For the emeraldine form of PAn, there are two absorption peaks at 320 and 630 nm, respectively. The absorption peak at 320 nm corresponds to the π–π* absorption of the main chains of the emeraldine PAn, and the absorption peak at 630 nm comes from the transition from the highest occupied molecular orbital (HOMO) of the benzene ring to the lowest unoccupied molecular orbital (LUMO) of the quinone ring. After proton acid doping, there appears a strong absorption at ca. 950 nm, with the disappearance of the 630 nm absorption and the weakness of the 320 nm absorption.

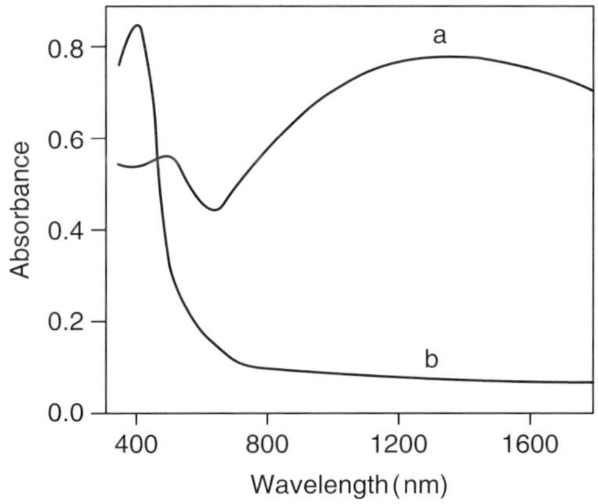

FIGURE 6.5 Absorption spectra of PPy: (a) p-doped state and (b) neutral state.

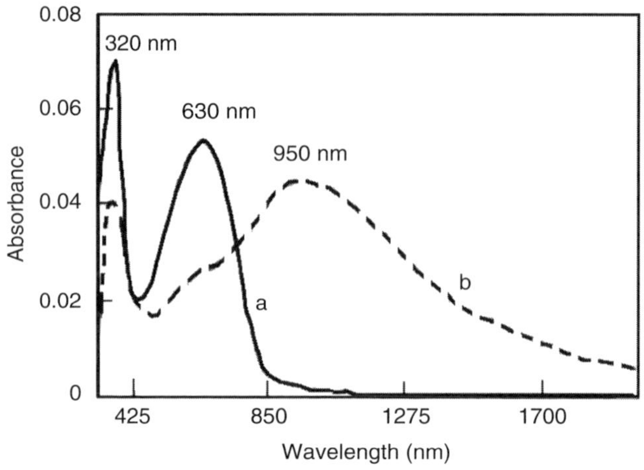

FIGURE 6.6 Absorption spectra of PAn: (a) emeraldine PAn and (b) proton acid doped PAn.

6.2.1.4 Solubility

Conducting PA, PPy, and PT without flexible side chains are insoluble and infusible, which makes the processing of the conducting polymers difficult. Substitution with flexible side chains could improve the solubility, but it often results in the decrease of the conductivity of the conducting polymers. For conducting PPy, introducing long alkyl or alkoxy side chain at the 3- and 4-positions or *N*-position of its pyrrole ring makes it soluble in organic solvents, but the conductivity of the PPy with the substituents reduced dramatically (Table 6.2). Probably, the long substituents distort the main chains of the conducting polymers, resulting in the decrease of its conjugation and conductivity. The flexible side chains make the neutral conjugated polymer soluble in some organic solvents, but sometimes it becomes insoluble after doping.

An alternative and more effective way to solve the solubility problem is to use the counter anion induction method proposed by Cao et al. in 1992 [5]. They doped PAn in *m*-cresol with functionalized proton acid (such as camphor sulphonic acid or dodecylbenzene sulphonic acid) as dopant, and obtained the clear green conducting PAn solution. The conductivity of the PAn film cast from the PAn solution reached 200 S/cm, which is much higher than that of the PAn prepared by common methods. Table 6.3 lists the solubility and conductivity of the doped PAn with functionalized proton acid H^+ (SO_3^--R).

TABLE 6.2
Influence of the Alkyl Substituents on the Properties of PPy(BF_4^-) Films

Substituents	Doping Degree	Density (g/cm^3)	Conductivity (S/cm)
—	0.25–0.32	1.48	30–100
3-Methyl	0.25	1.36	4
3,4-Dimethyl	—	—	10
N-methyl	0.23–0.29	1.46	0.001
N-ethyl	0.20	1.36	0.002
N-propyl	0.20	1.28	0.001
N-*n*-butyl	0.11	1.24	0.0001

TABLE 6.3
Solubility and Conductivity of the Doped PAn with Functionalized Proton Acid H^+ (SO_3^--R)

R in the Dopant of H^+ (SO_3^--R)	Conductivity (S/cm)		Solubility in				
	Powder	Film	Xylene	CH_3Cl	*m*-Cresol	Formic Acid	DMSO
C_6H_{13}	10		B	B			
C_8H_{17}	19		B	B		B	
(L,D)camphor	1.8	100–400		A	A	A	B
4-Dodecylbenzene	26.4	100–250	A	A	A	B	
o-Anisidine-5-	0.0077				B	B	B
p-Chlorobenzene	7.3				B	B	B
4-Nitrotoluene-2-	0.057				B	B	B
Dinonylnaphthalene	0.000018		B	A	A		

Note: A, very soluble at room temperature; B, soluble at room temperature.

6.2.1.5 Electrochemical Properties

The electrochemical properties are one among the most important properties of conducting polymers, because many applications of conducting polymers (such as batteries, electrochromic displays, modified electrodes, and enzyme electrode) are related to the electrochemical properties. Because the p-doped state of the conducting polymers is their steady state, the electrochemical studies are usually carried out for the reduction (dedoping) of the doped conducting polymers and redoping of the dedoped conjugated polymers. Due to space limitations, the electrochemical properties of only PPy and PAn are introduced here.

The dedoping and redoping of conducting PPy can be expressed as follows:

$$PPy^+(A^-) + e^- \underset{p^-\ \text{Doping}}{\overset{\text{Dedoping}}{\rightleftharpoons}} PPy^0 + A^- \tag{6.6}$$

where

PPy^+ (A^-) represents the p-doped PPy with A^- counter anions
PPy^0 represents the dedoped neutral PPy

The redox potential of PPy is ca. −0.4 to −0.2 V versus SCE. Hence the electrochemical studies of PPy films can be performed in aqueous solutions. The mechanism and reversibility of the redox processes of PPy in aqueous solution depend on the nature of anions and pH value of the aqueous solution. In an acidic $NaNO_3$ aqueous solution with pH 2–5, the cyclic voltammogram of $PPy(NO_3^-)$ film shows a couple of reversible reduction (p-dedoping) and reoxidation (p-doping) peaks in the potential range of 0.3 to −0.8 V versus SCE [6]. The reduction accompanies with the dedoping of the NO_3^- counter anions, and the oxidation occurs with the redoping of the solution NO_3^- anions. If the supporting electrolyte salt of $NaNO_3$ is changed to TsONa (sodium tosylate) in the aqueous solution, the reoxidation will be difficult and irreversible because of the difficulty of the doping of TsO^- anions due to its large size.

If the counter anion in PPy is the large-sized TsO^-, etc., the difficulty of the counter-anion dedoping results in negative shift of the reduction potential of the PPy film. And in this case, the

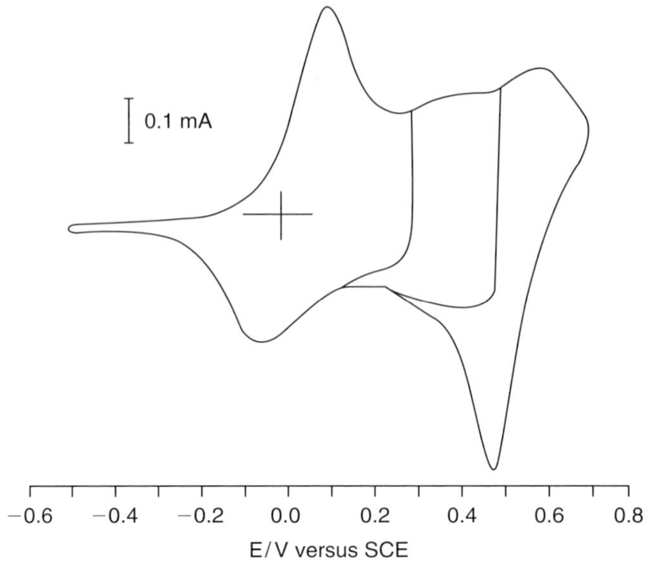

FIGURE 6.7 Cyclic voltammograms of PAn(NO_3^-) in pH 1.5, 1 mol/L $NaNO_3$ aqueous solution, v = 40 mV/s.

reduction mechanism is not the dedoping of the counter anions, but the doping of the solution cations, as shown in the following equation:

$$PP_y^+(A^-) + e^- + M^+ \rightleftharpoons PP_y^0(M^+A^-) \tag{6.7}$$

In the cyclic voltammetry of PPy, the potential range cannot exceed 0.5 V versus SCE, because an irreversible overoxidation reaction of PPy will occur at a potential higher than 0.5 V versus SCE. Overoxidation results in the degradation of the PPy main chains and the loss of its conductivity and electrochemical activity [7].

The electrochemical properties of PAn are also related to the preparation conditions and the pH values of the aqueous solution. Due to the instability (proton acid dedoping) of the doped PAn in a neutral or basic solution, PAn will lose its electrochemical activity in the neutral or basic solutions. Therefore, the electrochemical studies of PAn are usually performed in an acidic solution. Figure 6.7 shows the cyclic voltammograms of PAn(NO_3^-) in pH 1.5, 1 mol/L $NaNO_3$ aqueous solution [8]. There are three couples of redox peaks in the potential range of –0.5 to 0.7 V versus SCE. The redox processes between –0.4 and 0.28 V are corresponding to the p-doping/dedoping of the neutral PAn, which is similar to the redox processes of PPy in Equation 6.6. At a potential over 0.3 V, the p-doped PAn is oxidized further and its partial benzene ring is oxidized into quinone structure with the dedoping of counter anions. As the potential exceeds 0.8 V, there also occurs the overoxidation of PAn that results in the loss of its electrochemical activity. The intermediate couple of weak redox peaks in Figure 6.7 comes from some overoxidation product produced during the electrochemical preparation of PAn at a higher oxidation potential [9].

6.2.2 Electrochemical Preparation of Conducting Polymers

Electrochemical polymerization is a main method for the preparation of conducting polymer films, especially for the preparation of PPy. The electrochemical approach has the advantage of one-step production of conducting polymer films onto a metal electrode surface. Moreover, the properties of the conducting polymer films produced electrochemically can be modulated easily by changing the

TABLE 6.4
Oxidative Polymerization Potential of Monomers and the Conductivity of the As-Prepared Conducting Polymers

Monomer	Conductivity of the As-Prepared Polymers (S/cm)	Polymerization Potential (V versus SCE)
Pyrrole	30–200	0.7
Aniline	1–20	0.8
Indole	0.005–0.01	0.9
Thiophene	10–100	1.6

counter anions in the electrolyte solutions, and the thickness of the conducting polymer films can be precisely controlled by controlling the charge amount of the electropolymerization. In addition, electropolymerization is also very important in the preparation of polymer modified electrode, enzyme electrode immobilized by conducting polymers, polymer electrode for electrochromic displays, etc.

The electrochemical preparation of conducting polymers is performed by the oxidative polymerization of their corresponding monomers in an electrolyte solution. Three methods, i.e., constant current, constant potential, and potential scanning in a potential range, are commonly utilized. The anode, on which the conducting polymer is deposited, should be inert metal such as Pt, Au, carbon, stainless steel, or indium tin oxide (ITO) conducting glass. There are many factors influencing the electropolymerization processes, such as solvents, supporting electrolyte salts, concentration of the monomers, and pH value of the electrolyte solutions, as well as polymerization potential, current, temperature, etc. Polymerization potential of the monomers is the key factor to affect the electropolymerization. Table 6.4 lists the polymerization potentials of several important monomers for the synthesis of conducting polymers. The lower the oxidation polymerization potential is, the easier the electropolymerization of the monomers will be. Obviously, the electropolymerization of pyrrole is the easiest among the monomers.

6.2.2.1 Electrochemical Preparation of Polypyrrole

The oxidative polymerization potential of pyrrole is ca. 0.7 V versus SCE, so pyrrole can be electropolymerized in both organic and aqueous solutions. Controlling the electropolymerization conditions is crucial for the preparation of high-quality PPy films. The conductivity (σ) of the as-prepared PPy films strongly depends on the nature of the solution anions, which changes from the order of 10^{-1} S/cm to the order of 10^2 S/cm for different counter anions. Generally, the anions of strong acids and the surfactant anions are favorable for the preparation of high-quality PPy films. In the aqueous solutions, flexible PPy films with σ higher than 100 S/cm can be produced with the surfactant anions such as tosylate and benzene sulphonate. The concentration of the electrolyte anions also plays an important role. Too low concentration of the anions will lead to poor PPy films. The concentration of the anions should be no lower than 0.1 M. And the concentration of pyrrole monomer is usually 0.1 M. The effect of solvent on the electropolymerization was found to be related to the donor number (DN) of the solvent [10]. The lower the DN value of the solvent, the higher the conductivity of the as-prepared PPy films (Table 6.5). Water is a special solvent in comparison with organic solvent, its acidity can be regulated by changing pH values. The optimizing pH value of the aqueous solutions for pyrrole electropolymerization is between pH 2 and 5.5. In a basic aqueous solution, conducting PPy cannot be produced. In addition to the solvent and supporting electrolyte, small amount of additives in the solutions can sometimes influence the electropolymerization. Surfactants have been proven to be the effective additives to improve the quality of the PPy films. With nonionic surfactant nonylphenol polyethyleneoxy (10) as an additive in the sodium tosylate (TsONa) aqueous solution, the tensile strength of the as-prepared PPy film reached 127 MPa, which is five times higher than that of the PPy

TABLE 6.5
Effect of Solvent on the Conductivity of the As-Prepared PPy Films

		Conductivity of the As-Prepared PPy Films (S/cm)		
Solvents	**Donor Number**	**PPy(BF_4^-)**	**PPy(ClO_4^-)**	**PPy(TsO^-)**
DMSO	29.8	7×10^{-6}		
DMF	26.6	1×10^{-4}	5×10^{-4}	0.008
TBP	23.7		1	
TMP	23.0	1.0	20	0.09
THF	20.0		31	
H_2O	18.0	8.4	34	79
PC	15.1	67	55	90
CH_3NO_2	2.7	69	56	

film produced from the solution without the surfactant additive [11]. The surfactant molecules may adsorb on the surface of the electrode where PPy will deposit and change the interface structure between the electrode and the solution, which benefits the deposition of high-quality PPy films.

The potential control is also important for the electropolymerization. The potential for the pyrrole polymerization should be controlled at no higher than 0.75 V versus SCE, usually at 0.65–0.70 V versus SCE. The potential over 0.8 V versus SCE will result in overoxidation degradation of the deposited PPy films. Temperature is another factor that influences the electropolymerization. Usually, high-quality PPy films with longer conjugation chains, less defects, and higher conductivity were obtained at lower temperatures (lower than 20°C). At higher temperature, defect structures of PPy are easily formed, which results in decrease of its conductivity.

6.2.2.2 Mechanism of the Electropolymerization of Pyrrole

The electrochemical polymerization processes of pyrrole can be explained by a mechanism of anion-participated cation-radical polymerization [12]:

$$(\text{H–N–H}) \underset{k_{-1}}{\overset{k_1}{\rightleftarrows}} (\text{H–N–H})_{\text{ad}} \tag{6.8}$$

$$(\text{H–N–H})_{\text{ad}} - \text{e} + \text{A}^- \xrightarrow{k_2} \text{A}^-(\text{H–N–H})_{\text{ad}}^{+\bullet} \tag{6.9}$$

$$2\text{A}^-(\text{H–N–H})_{\text{ad}}^{+\bullet} \xrightarrow{k_3} (\text{H–N–N–H}) + 2\text{A}^- + 2\text{H}^+ \tag{6.10}$$

$$(\text{H–N–N–H}) - \text{e} + \text{A}^- \xrightarrow{k_4} \text{A}^-(\text{H–N–N–H})_{\text{ad}}^{+\bullet} \tag{6.11}$$

$$\text{A}^-(\text{H–N–N–H})_{\text{ad}} + \text{A}^-(\text{H–N–H})_{\text{ad}}^{+\bullet} \xrightarrow{k_5} (\text{H–N–N–N–H}) + 2\text{A}^- + 2\text{H}^+ \tag{6.12}$$

$$(\text{H–N–N–N–H}) - \text{e} + \text{A}^- \xrightarrow{k_6} (\text{H–N–N–N–H})^+\text{A}^- \tag{6.13}$$

In Equations 6.8 through 6.13, H–N–H represents pyrrole molecule and two H molecules represent two α-H of pyrrole. First, pyrrole molecules adsorb on the electrode surface and reach adsorption equilibrium (Equation 6.8), then the adsorbed pyrrole molecule is oxidized into cation-radical under the polymerization potential and at the same time the cation-radical combines with a solution anion to form an ion-pair of the cation-radical and the anion (Equation 6.9). Two ion-pairs couple together to form a pyrrole dimer by removing two protons and two anions (Equation 6.10). The polymerization of pyrrole is carried out as so on, and the polypyrrole chain is formed finally. The polypyrrole molecules will be oxidized to their p-doped state immediately during the electropolymerization, since the p-doping potential of PPy is much lower than that of the electropolymerization.

6.2.2.3 Electrochemical Preparation of Polyaniline and Polythiophene

The electrochemical preparation processes of conducting PAn and PT are similar to that of PPy, the main difference between them is that for PAn, the electrochemical polymerization must be performed in a strong acidic solution, and for PT, it has to be carried out in an organic solution due to the higher polymerization potential of thiophene monomer (see Table 6.4).

For the electrochemical polymerization of aniline, common electrolyte solution used is 0.1 M aniline in 1 M H_2SO_4, HCl, $HClO_4$, or HBF_4 aqueous solutions. Here the strong acidic solution is used because the doped PAn produced in the electropolymerization is stable only in the low pH value solutions. The conductivity of the as-prepared PAn is usually in the order of 10^{-1}–10^{1} S/cm. Overoxidation is a baffling situation met in the electropolymerization of aniline. At the polymerization potential (ca. 0.8 V versus SCE) of aniline, the polyaniline deposited on the electrode could be overoxidized, which leads to partial dissolution of the film. Hence, cyclic potential scan in a potential range (e.g., –0.15 to 0.78 V versus SCE) is usually preferred for the preparation of a high-quality PAn product.

For the electrochemical polymerization of thiophene, a common electrolyte solution is 0.1 M $(Bu)_4NClO_4$ + 0.2 M thiophene in CH_3CN. The electropolymerization can be performed at 1.6 V versus SCE under an inert atmosphere. The conductivity of the as-prepared PT film is in the range of 10–100 S/cm. The high oxidative polymerization potential limits the selection of solvents and solution anions in the electrolyte solutions and causes some problems such as overoxidation degradation of the as-prepared PT films. Therefore, efforts have devoted to decreasing the polymerization potential. One approach is to carry out the electropolymerization from thiophene oligomers. The polymerization potential of bithiophene is ca. 1.20 V versus SCE, 0.45 V decreased in comparison with that of thiophene monomer. Another approach is to perform the electropolymerization of thiophene in boron trifluoride/ethylene ether (BFEE) solution. In the BFEE solution, the electropolymerization potential of thiophene is decreased to ca. 1.2 V, and high-quality PT film can be obtained [13].

Poly(3,4-ethylenedioxythiophene) (PEDOT) has drawn much attention recently because of its applications as modifying layer on ITO electrode in PLEDs and PSCs, transparent electrode materials, antistatic painting, and capacitors. PEDOT can also be prepared by the electropolymerization of its monomer 3,4-ethylenedioxythiophene (EDOT). The polymerization potential of EDOT is 1.49 V versus SCE, which is lower than that of thiophene. By using the dimer of EDOT as a monomer, the polymerization potential can be decreased further to 0.84 V versus SCE.

The mechanism of the electropolymerization of aniline and thiophene is similar to that of pyrrole as mentioned above. It should follow the same anion-participated cation-radical polymerization mechanism.

6.2.3 Chemical Catalytic or Oxidative Polymerization

6.2.3.1 Polyacetylene and Its Derivatives

Polyacetylene is the first conducting polymer, and possesses the simplest conjugated molecular structure among the conjugation polymers. The studies on PA mainly focused on the synthesis of PA films, the conductivity of the doped PA films, and the theoretical aspect related to the nature of the charge carriers and the charge transportation processes in the conducting polymers. The application of PA is very limited due to the instability of PA in both neutral and doped states. Among the studies of PA, the synthesis of PA film is of special importance, since the discovery of conducting polymers benefited from Shirakawa's synthetic route toward the formation of PA films. The following synthetic conditions are crucial for producing PA films: (1) homogenous catalytic system (such as $Ti(OBu)_4$-$AlEt_3$), (2) higher catalyst concentration, (3) lower chain transfer rate (to get higher molecular weight), and (4) proper solvent. A typical polymerization method is as follows: under a strict inert (high purity N_2) atmosphere, 1 ml toluene, 0.14 ml (0.004 mol) $Ti(OBu)_4$, and 0.22 ml (0.016 mol) $AlEt_3$ were added into a flask. After allowing to react for 30 min, decrease the temperature to −78°C and draw vacuum of the flask. Turning the flask to make the reaction solution attached to the wall of the flask homogeneously. Pass acetylene gas rapidly and a red polyacetylene film will be formed immediately after passing the acetylene gas. The thickness of the PA film can be modulated by controlling the catalyst concentration, the pressure of the acetylene gas, and the polymerization time. Generally, with the acetylene pressure of 79,980 Pa and the catalyst concentration mentioned above, the PA film with a thickness of 0.1 mm can be produced at −78°C after reacting for 1–4 h. The polymerization reaction is terminated by withdrawing the unreacted acetylene gas. The PA film is washed with toluene for several times till the PA film becomes colorless and transparent. The PA films should be kept in vacuum for further use.

Polyacetylene without its substituents is unstable in air, so at present PA is interested only for theoretical studies of conducting polymers such as the nature of the charge carriers and the charge transportation processes. But the polyacetylene derivatives with chromophore pendants find basic research interests and applications in PLEDs [14]. The existence of two hydrogen atoms in its repeat unit offers ample opportunity to decorate the backbone with pendants. Many efforts have been devoted to the decoration of polyacetylene main chains [14]. The main difficulty in the polymerization of the functional acetylenes is the incompatibility of the polar groups of the functional acetylenes with the metathesis catalysts used for the polymerization. The Rh-based insertion catalysts were successfully used in the synthesis of the substituted PAs, because of their tolerance of different function groups. A series of Rh complexes Rh(diene)L were prepared by Tang et al. [15]. A catalyst system that can initiate acetylene polymerizations in water and air was developed, giving stereoregular polymers in high yields (see Scheme 6.1) [15].

6.2.3.2 Oxidative Polymerization of Pyrrole and Thiophene

Conducting polypyrrole can also be prepared by oxidative polymerization of pyrrole in an aqueous solution with oxidant. But the product of the chemical polymerization is in the form of a powder. A typical polymerization method is as follows: 100 ml of 0.2 M dodecylbenzenesulfonic acid (DBSA) and 0.2 M pyrrole aqueous solution are vigorously stirred to get an emulsion system. And 0.26 M of $FeCl_3$ predissolved in de-ionized water is added to the emulsion at 0°C. After 24 h, large amount of

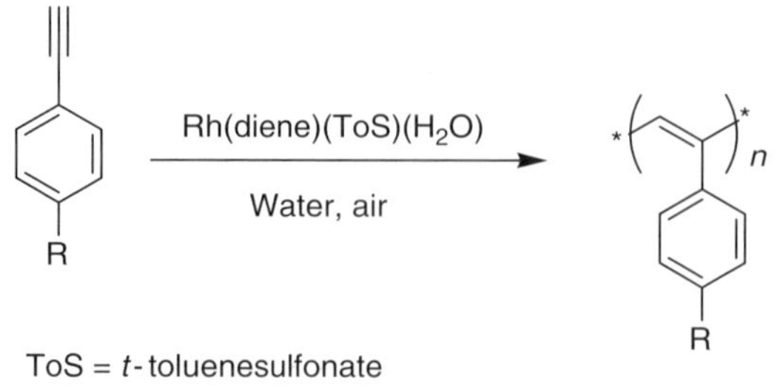

SCHEME 6.1 Synthesis of polyacetylene derivatives with functional pendants.

excess methanol was poured into the solution to terminate the reaction. Then, the resulting polypyrrole precipitate was vacuum-filtered and washed copiously with distilled water, methanol, and acetone for several times. Finally, it was dried under dynamic vacuum for at least 40 h at room temperature.

Conducting polythiophene powder can also be prepared with the similar oxidative polymerization route from bithiophene as that of polypyrrole mentioned above. The powders of the conducting polypyrrole and polythiophene prepared by the chemical oxidation method are insoluble and infusible.

6.2.3.3 Preparation of Soluble Conducting Polyaniline

Polyaniline is the most important conducting polymer and has realized mass production, with the advantages of low cost and easy preparation. And most importantly, PAn can be made soluble in water by counter anion-induction method. Therefore, PAn is promising for many applications as a conducting polymer. The preparation processes of conducting polyaniline powders and solutions are as follows:

1. Preparation of HCl doped PAn: The HCl doped PAn can be prepared using MacDiarmid's method [16]. 2 ml of (0.022 mol) aniline is dissolved in 120 ml, 1 M HCl solution, and is then cooled to 0°C–5°C in ice water bath. Forty ml of 0.5 M $NH_4S_2O_4$ and 1 M HCl solution are added into the aniline solution dropwise in 1 h with magnetic stirring at 0°C–5°C, then keep stirring to react for 7 h. The precipitate produced in the reaction is removed by filtration, washed repeatedly with 1 M HCI, and dried under dynamic vacuum for 48 h. The PAn powder so obtained is in HCl-doped state (PAn–HCl).
2. Preparation of emeraldine base PAn: The PAn–HCl powder is put in 0.1 M ammonia water and stirred for 3 h at room temperature. The deposits are filtered, washed by de-ionized water till the water is neutral, and dried in vacuum for 48 h, then the emeraldine PAn is obtained. The emeraldine PAn can be dissolved in *N*-methylpyrrolidinone (NMP), and the insulating emeraldine PAn film can be cast from the solution on a glass. The emeraldine PAn film can be doped into conducting PAn film by dipping it into a proton acid (such as HCl) solution.
3. Preparation of conducting PAn solution [5]: A typical preparation process (functionalized proton acid doping) is as follows: 0.91 g (0.01 mol) emeraldine PAn powder and 4.646 g (0.02 mol) camphor sulphonic acid (CSA) are dissolved in 100 ml *m*-cresol, ultrasonic treatment for 1 h, magnetic stirring at room temperature for 24 h. Then a clear green conducting PAn solution can be obtained. The conducting PAn film can be cast from the solution on the surface of a glass.

6.3 BASIC PROPERTIES OF NEUTRAL CONJUGATED POLYMERS

Since Burroughes et al. [17] found the electroluminescent phenomenon of PPV in 1990, neutral conjugated polymers have attracted much attention all over the world, due to their promising applications as semiconductor polymers in the optoelectronic devices such as PLEDs, polymer FETs, and PSCs. The conjugated polymers used in PLEDs are also called luminescent polymers, whereas those used in PSCs could be called as photovoltaic polymers. The most important properties of the conjugated polymers concerned for the application in the polymer optoelectronic devices include the electronic energy levels, optical properties (absorption and photoluminescence), charge carrier mobility, etc.

6.3.1 Electronic Energy Levels

In the PLEDs, the conjugated polymers (luminescent polymers) are sandwiched between the ITO anode and a metal cathode (see Chapter 13 for more details). The operation voltage and the charge injection energy barrier of the PLED devices depend on the work function of the electrode materials and the electronic energy levels of the HOMO and LUMO of the conjugated polymers. The color of the electroluminescence depends on the energy gap of the luminescent polymers. In PSCs, the exciton charge separation, open circuit voltage, and the charge collection on the electrodes are also tightly related to the HOMO and LUMO energy levels of the conjugated polymers. The solar light harvest ability (absorption spectrum) of the conjugated polymers relies on their energy gap and electronic energy levels. Hence, the electronic energy level is a key parameter of the conjugated polymers.

The photon energies of the visible light from blue to red are 3.1–1.8 eV corresponding to the wavelength of 400–700 nm, hence, the E_g of the luminescent polymers should locate in the range of 1.8–3.1 eV. For absorbing the solar light, the E_g of the photovoltaic polymers should also be in the range of 1.8–3.1 eV.

The factors influencing the electronic energy levels of the conjugated polymers are as follows [18]:

1. For the conjugated polymers composed of aromatic rings connected by single bond between them, the deviation of adjacent aromatic rings from a molecular plane will result in the increase of E_g.
2. Electron-donating or electron-accepting ability of substituents affects the HOMO and LUMO energy levels and the E_g of the conjugated polymers.
3. Interchain interactions between the conjugated main chains will decrease the E_g of the conjugated polymers at solid state.

E_g values of the conjugated polymers can be calculated from the long wavelength absorption edge of their absorption spectra. If the absorption edge is λ_{edge} (nm), then $E_g = 1240/\lambda_{edge}$ (eV). The E_g value calculated from the absorption edge is sometimes called the optical E_g^{opt}, for distinguishing with the E_g value obtained from electrochemical measurement which will be discussed in the following.

The HOMO and LUMO energy levels of the conjugated polymers can be estimated from their onset oxidation (p-doping) potential and onset reduction (n-doping) potential, respectively [19]. The onset potentials can be obtained from the cyclic voltammograms of the conjugated polymers, as shown in Figure 6.8. In the positive potential region, there is a couple of reversible redox peaks corresponding to the p-doping/dedoping of the conjugated polymer, and the onset oxidation potential can be obtained from the oxidation peak (it is 0.10 V versus Ag/Ag^+ in the case of poly(3-hexylthiophene) [P3HT] as shown in Figure 6.8). In the negative potential region, there is also a couple of reversible redox peaks corresponding to the n-doping/dedoping of the conjugated polymer, and the onset reduction potential can be obtained from the reduction peak (it is −2.10 V versus Ag/Ag^+ in the case of P3HT in Figure 6.8). If we use φ_{ox} and φ_{red} to represent the onset oxidation and

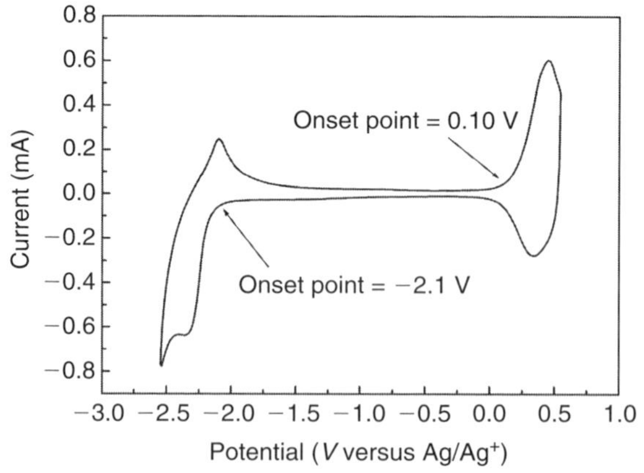

FIGURE 6.8 Cyclic voltammogram of poly(3-hexylthiophene) film on Pt electrode in 0.1 M Bu_4NPF_6 acetonitrile solution.

reduction potentials, respectively, then the HOMO and LUMO energy levels of the conjugated polymers can be calculated from the following equation:

$$\text{HOMO} = -(\varphi_{\text{ox}} + C)\ (\text{eV}); \quad \text{LUMO} = -(\varphi_{\text{red}} + C)\ (\text{eV}) \tag{6.14}$$

where
the unit of the potential is V versus RE (reference electrode)
C is a constant related to the reference electrode used

If the RE is SCE, ferrocene/ferrocium (Fc/Fc^+), Ag/Ag^+, then C takes the value of 4.4, 4.8, 4.71, respectively. If silver wire is used as the reference electrode, the electrode potential should be calibrated with the Fc/Fc^+ reference electrode. From the HOMO and LUMO energy levels, E_g can be calculated from the difference of LUMO and HOMO or from the difference of the onset oxidation and reduction potentials:

$$E_g = \text{LUMO} - \text{HOMO} = e(\varphi_{\text{ox}} - \varphi_{\text{red}})\ (\text{eV}) \tag{6.15}$$

The E_g value calculated from the electrochemical measurement is sometimes called as the electrochemical E_g^{EC}, as compared to the optical E_g^{opt}.

6.3.2 Optical Properties

Absorption and photoluminescence (PL) spectra are of primary importance for the neutral conjugated polymers. Figure 6.9 shows the absorption spectra of poly(9,9-bioctylfluorene) (PFO), Poly(2-methoxy-5-(2′-ethylhexyloxy)-1,4-phenylene vinylene) (MEH-PPV), P3HT, and poly(3-hexylthienylene vinylene) (P3HTV). It can be seen that the absorption spectra of the conjugated polymers composed of thiophene rings (P3HT, P3HTV) are redshifted in comparison with that composed of benzene rings (PFO, PPV). And the insertion of vinylene units between the benzene rings and between the thiophene rings also results in the redshift of the absorption spectra. From the viewpoint of solar light utilization in PSCs, obviously, MEH-PPV and P3HT are much better than PFO.

Figure 6.10 shows the PL spectra of PFO, MEH-PPV, and P3HT. From the PL spectra of the conjugated polymers, the luminescence color of the polymer can be determined. The PL peak of PFO is at ca. 420 nm, indicating PFO is a blue emitter. While MEH-PPV is an orange emitter with its PL peak at ca. 575 nm and P3HT is a red emitter with its PL peak at ca. 650 nm. In addition, it can be seen from Figure 6.10 that the PL spectra are quite broad and there is a shoulder peak at the

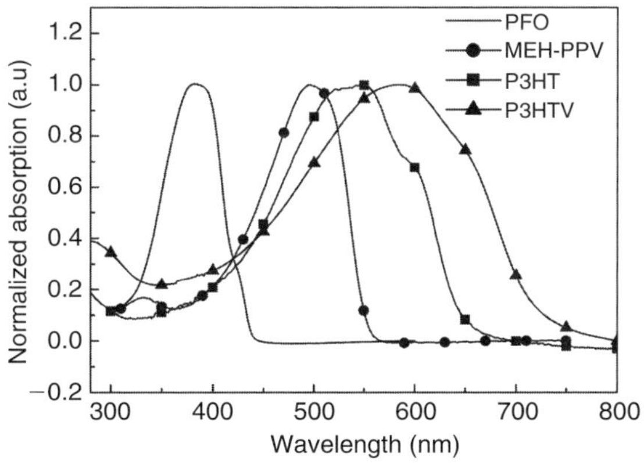

FIGURE 6.9 Absorption spectra of PFO, MEH-PPV, P3HT, and P3HTV.

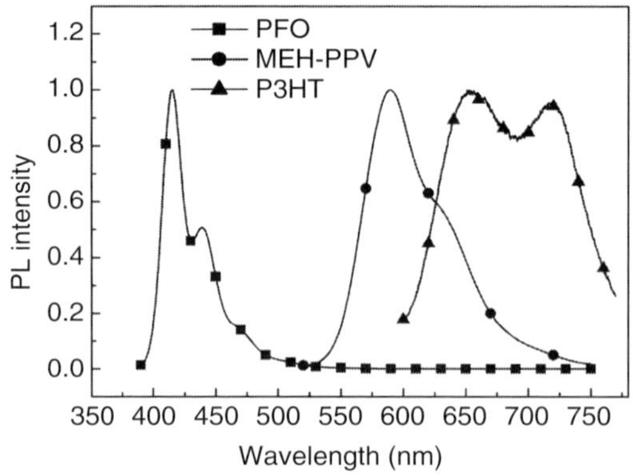

FIGURE 6.10 Photoluminescence spectra of PFO, MEH-PPV, and P3HT films.

longer wavelength direction for each PL spectrum of the conjugated polymers. The longer wavelength shoulder peak may come from the interchain interactions of the conjugated polymers.

In comparison with the absorption spectra in Figure 6.9, it can be seen that the PL spectrum in Figure 6.10 is redshifted obviously. For more clear comparison of the absorption and PL spectra, Figure 6.11 shows the absorption and PL spectra of MEH-PPV together. The redshift of the PL spectrum is a general phenomenon for the conjugated polymers, and the peak wavelength difference between the PL spectrum and the absorption spectrum is called Stokes shift. The reason of the Stokes shift can be analyzed from the following aspects: different conjugation length of the polymer main chains exists in the luminescent polymers, and the excitons will relax to the longest conjugated chains that possess the lowest energy. Moreover, the relaxation of the excitons to the lowest energy excited state occurs *vir* inter- or intramolecular energy transfer, or through vibronic relaxation, before they release energy by luminescence. Thus the PL spectra are redshifted in comparison with the absorption spectra of the conjugated polymers. Furthermore, the formation of excimers explained later in the text also results in the redshift of the PL spectrum.

Fluorescent properties of the conjugated polymers are tightly related to the distribution of their conjugation length. The absorption and PL spectra of the conjugated polymers are quite broad, which results from the nonuniformity of the length of the conjugated main chains. The polymer chains with different conjugation lengths show shifted absorption spectra (longer conjugation length shows redshifted absorption, while shorter conjugation length will show blueshifted absorption), so that the broader absorption spectra appeared. Polymer structure regularity (e.g., regioregularity, *cis–trans* ratio of C=C bonds) and different degrees of chain packing in the solid samples are additional reason for spectrum broadening.

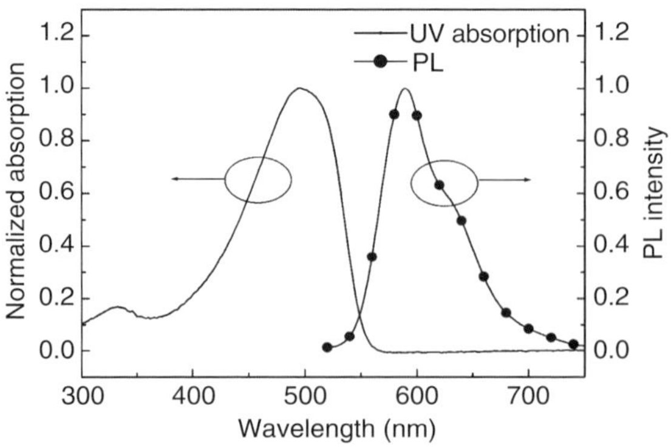

FIGURE 6.11 Absorption and PL spectra of MEH-PPV film.

The interchain interactions of the conjugated polymers also influence their absorption and fluorescent spectra. There are two kinds of the interchain interactions that affect the optical properties:

1. Excimer: An excimer is a dimer formed by an excited conjugated molecular chain and its adjacent ground state conjugated chain of the same polymer. Its stability comes from the overlap of the π–π* orbitals of the two conjugated molecular chains. The formation of the excimer needs the two conjugated molecular chains close to each other to the distance that is short enough to form the π–π* interaction of the two molecules but long enough for avoiding the ground state interaction to form the aggregate. For the common conjugated polymers, the interchain distance of the excimers is 0.3–0.4 nm. The excimer does not affect the absorption spectra of the luminescent polymers, but it makes a long wavelength peak appear in the fluorescent spectra of the conjugated polymers, which results in broadening of the fluorescent spectra. In addition, the existence of the excimers often results in decrease of the fluorescent quantum efficiency.
2. Aggregate: When the distance between the conjugated chains of a luminescent polymer is short enough to make the electronic wavefunctions of the conjugated chains at both ground and exited states that overlap over two or more conjugated chains, the wavefunction-overlapped conjugated chains form the aggregate. The difference between the aggregate and the excimer is that in aggregates there is strong interaction between the ground state conjugated chains, which makes a new absorption band appear at longer wavelength region in the solid film, in comparison with its dilute solution. The interchain aggregates make both absorption and fluorescent spectra broaden.

6.3.3 Effect of Side Chain Substitution on the Properties of Conjugated Polymers

The processability and film-forming properties of the conjugated polymers are crucial for the fabrication of high-performance PLEDs and PSCs. The conjugated polymers of PPV, PT, PPP, etc., without substituted side chains are insoluble and infusible, which limits their applications. The strong interchain interactions are responsible for the insolubility. To solve the problem, alkyl or alkoxy soft side chains are widely used as substituent groups on the conjugated main chains, which could reduce the interchain interactions and induce solubility. Figure 6.12 shows the molecular structures of some of the side-chain substituted conjugated polymers.

The substituent groups not only improve the solubility, but also regulate the electronic energy levels (HOMO, LUMO, and E_g) and improve the absorption spectra of the conjugated polymers. For instance, MEH-PPV with alkoxy side chains and BuEH-PPV with alkyl side chains are both soluble in organic solvents such as toluene and xylene. The emission light color of MEH-PPV is orange–red that is a little redshifted in comparison with that of PPV. While the color of emission light of BuEH-PPV is green, a little blueshifted in comparison with that of PPV. The PL efficiencies of both polymers increased obviously in comparison with that of PPV. Similarly, the PT and PPP derivatives with the alkyl or alkoxy side chains are also soluble in the organic solvents. In addition to the alkyl and alkoxy side chains, alkyl-phenylene side chains are also utilized in the PPV derivatives (such as BP-PPV in Figure 6.12) for getting stable and highly efficient green luminescent polymers.

Bredas et al. [20] studied the effect of the electron-donating and electron-accepting ability of the substituents on the electronic energy levels of PPV, by quantum chemistry calculation (VEH method). They calculated the energy gap (E_g), ionization potential (IP), and electron affinity (EA) of the PPV derivatives. IP and EA of a conjugated polymer are the absolute values of its HOMO and LUMO, respectively. The calculated E_g values are basically the same as those obtained from the absorption spectra, which indicates that the calculated parameters of the electronic structure are quite reliable. It can be seen that PPV derivatives with the electron-donating substituents of alkoxy possess smaller IP and EA than those of PPV, and the decrease of IP is more than that of EA.

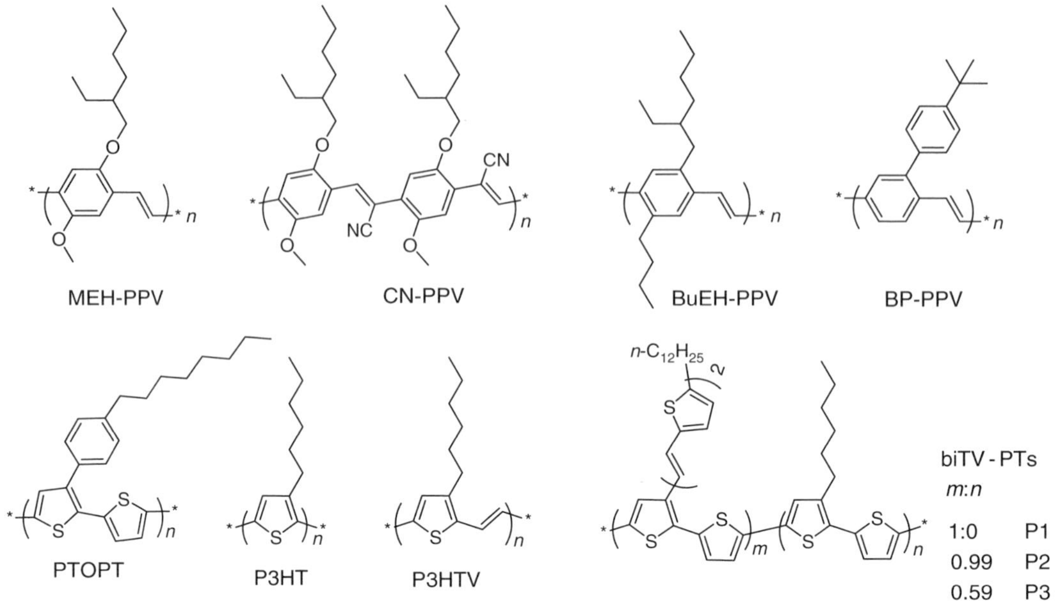

FIGURE 6.12 Molecular structures of some conjugated polymer derivatives with side chains.

The PPV derivative with the electron-withdrawing substituent of -CN makes the IP and EA increased, and the increase of EA is more than that of IP. Interestingly, both the substitutions of the electron-donating groups and the electron-withdrawing groups make the E_g of PPV reduced.

The side chain substitution can also be used to regulate absorption spectra of the conjugated polymers. Recently, the authors synthesized a series of polythiophene derivatives with conjugated side chains for the application in PSCs. It was found that the polymers possess two absorption peaks in the UV–vis region, the one at 300–400 nm belongs to the absorption of the conjugated side chains and the other at 500–600 nm belongs to the absorption of the conjugated polythiophene main chains. By prolonging the conjugation length of the conjugated side chains, the UV absorption of the side chains can be moved to visible region. And further by controlling the ratio of the units with the conjugated side chains in the polythiophene main chains, the new polythiophene derivatives with broad absorption band and strong absorbance in the visible region were obtained [21]. A bithienylenevinylene substituted polythiophene shows a broad and strong absorption band covering the range of 380–650 nm (P3 in Figure 6.13, the molecular structure of P3 is shown in Figure 6.12), and the photovoltaic properties of the polymer are greatly enhanced [22].

6.3.4 Charge Carrier Mobility

Charge carrier mobility is an important parameter for the applications of the conjugated polymers in FET and PSC. But divergent values were reported for the conjugated polymers, due to the different measurement methods and the effect of morphology and aggregation state on the values of the mobility. For MEH-PPV film, the hole mobility is in the order of 10^{-7} cm^2/V s measured by space charge limited current (SCLC) method [23]. For regioregular poly(3-hexylthiophene) (P3HT), the hole mobility values reported are from 10^{-2} to 10^{-3} cm^2/V s (measured from FET) to 10^{-5} to 10^{-6} cm^2/V s (measured by SCLC) for a spin-coated polymer film. Generally, the hole mobility of the conjugated polymer films is in the order of 10^{-5} to 10^{-7} cm^2/V s, which is much lower (6–8 orders lower) than that of the silicon materials used in inorganic solar cells. Therefore, design and synthesis of high hole mobility conjugated polymers is a hard task for the applications of the conjugated

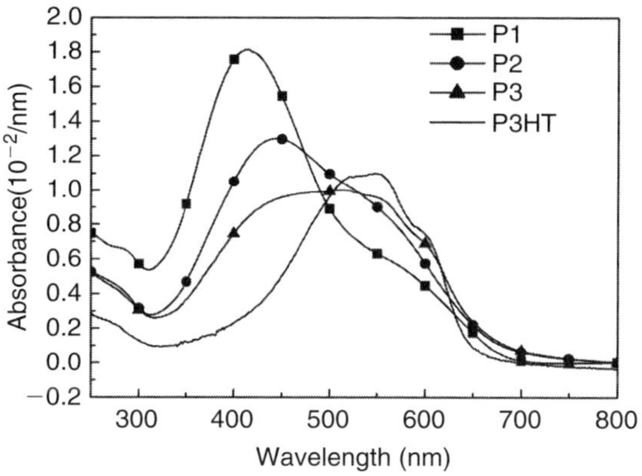

FIGURE 6.13 Absorption spectra of P3HT and the bithienylenevinylene substituted polythiophene film.

polymers in the PSCs and in FETs. The low mobility of the conjugated polymers is limited by the amorphous structure of the conjugated polymer films and by the hopping transportation between polymer chains.

Recently, the authors designed and synthesized a cross-linked polythiophene with conjugated vinylene–phenylene–vinylene linking bridges. The hole mobility of the polymer with 5% conjugated linking bridges reached 10^{-3} cm^2/V s [24], two orders increased in comparison with that of the linear conjugated polythiophenes. This result indicates that the cross-linking by conjugated bridges could be an effective approach toward the higher mobility of the conjugated polymers.

6.4 SYNTHESIS OF SOLUBLE POLYTHIOPHENE DERIVATIVES

6.4.1 Synthesis of Poly(3-alkylthiophene)s

Poly(3-alkylthiophene)s (P3ATs), especially poly(3-hexylthiophene) (P3HT), are the most important soluble polythiophene derivatives for the application in PSCs and FETs. So here we introduce the detailed synthetic strategies for the preparation of P3ATs.

6.4.1.1 Regioregularity of P3ATs

An important concept in the P3ATs is the regioregularity. Generally, the polymerization of thiophene is carried out at 2- and 5-positions of thiophene ring (see Scheme 6.2). In P3ATs, the repeated units are asymmetric, there exist three coupling forms when two thiophene rings are coupled together between the 2- and 5-positions: (1) head-to-tail (HT) coupling, (2) head-to-head (HH) coupling, and (3) tail-to-tail (TT) coupling: head is the 2-position of the repeating unit, and tail is the 5-position. This leads to a mixture of four chemically distinct triad regioisomers when 3-substituted (or asymmetric) thiophene monomers are employed (Scheme 6.2) [25]. The HT–HT structure of P3AT are denoted as regioregular P3AT, the other three structures are denoted as regioirregular or regiorandom P3AT, and the HT–HT isomer proportion in the polymers is named as regioregularity of the P3ATs. Regioregular poly(3-substituted thiophene) can easily access low-energy planar conformation, leading to highly conjugated polymers, better absorption, and higher mobility of the polymers. Decrease of the regioregularity will lead to some torsion of the polythiophene main chains and result in shorter length of effective conjugation and greater band gap. Hence the regioregularity is an important factor in the characterization of poly(3-substituted thiophene).

^{1}H NMR and ^{13}C NMR are the main methods used for determining the regioregularity of P3ATs [26]. In a regioregular P3AT (HT–HT coupling ≈100%), the proton on position 4 of thiophene ring

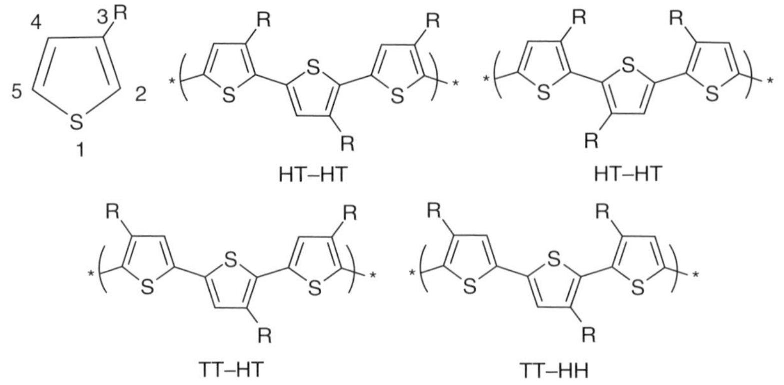

SCHEME 6.2 Monomer position number and possible regiochemical coupling in poly(3-substituted thiophene)s.

should exhibit a neat peak at $\delta = 6.98$ ppm. There are four chemically distinct triad regioisomers in regioirregular P3ATs, as shown in Scheme 6.2. In ^{1}H NMR spectra, TT–HT isomer has a peak at $\delta = 7.00$ ppm, HH–TT isomer has a peak at $\delta = 7.05$ ppm, and HT–HH isomer has a peak at $\delta = 7.02$ ppm. So, with the integral area of the peaks, the relative ratio of the HT–HT couplings to the non-HT–HT couplings can be determined. In the ^{13}C NMR spectrum, regioregular P3AT exhibits four resonance peaks at aromatic region ($\delta = 128.5$, 130.5, 134.0, and 140.0 ppm), but regioirregular P3AT shows many resonance peaks from 120 to 150 ppm. So, the degree of the structural regioregularity can also be determined from the ^{13}C NMR spectrum.

6.4.1.2 Synthesis of P3ATs by Chemical Oxidation Method

Many polythiophene derivatives can be easily synthesized by chemical oxidation method as shown in Scheme 6.3 [27]. In typical oxidation polymerization of PTs, thiophene monomers, such as thiophene, 3-alkylthiophenes, and 3-phenylthiophenes, are dissolved in chloroform. Under the protection of inert gas, the excess oxidant ($FeCl_3$, $MoCl_5$, or $RuCl_3$) is added (see Scheme 6.3). The polymerization could be carried out for several hours. The $FeCl_3$ oxidation method has been used for the synthesis of P3ATs and poly(3-phenylthiophene)s. The molecular weight of the polymers prepared from this method is in the range of 30–300 K, and the polydispersity ranges from 1.5 to 5.0. The regioregularity of the P3ATs is 70%–80%, but for poly(3-phenyl thiophene)s (P3PTs) the regioregularity can reach 90%–95% by this method.

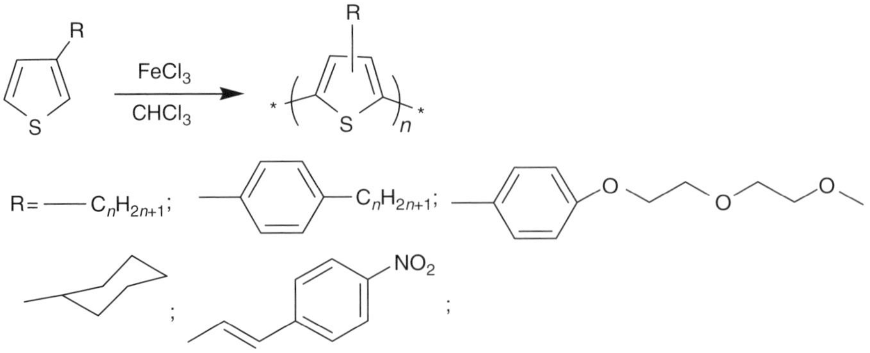

SCHEME 6.3 Preparation of polythiophene derivatives by $FeCl_3$ oxidation.

Very high molecular weight has been reported in the synthesis of P3ATs using the $FeCl_3$ method by bubbling dry air through the reaction mixture during the polymerization [28]. After isolation and dedoping of the polymer with concentrated ammonia solutions, and washing, the P3ATs with molecular weight of 68–175 K (refractive index), 77–146 K (UV–vis), and 204–398 K (light scattering) were obtained.

The poor reproducibility of the $FeCl_3$ oxidation reaction is one of the major problems. As reported by Pomerantz et al. [28], the polymerization of 3-octylthiophene with $FeCl_3$ was repeated under identical reaction conditions for five times. The molecular weights of the five samples of poly(3-octylthiophene) showed molecular weights ranging from 54 to 122 K (UV–vis) with PDIs ranging from 1.6 to 2.7. Additionally, from identical preparation conditions, the polymer samples were found to contain different levels of Fe impurities. For example, the percentage of Fe impurity in poly(3-octylthiophene) from $FeCl_3$ method ranges from 9.6 to 0.15 mol%. Despite the shortcomings, the $FeCl_3$ method is a well-established method to polymerize thiophene derivatives and continues to be the most widely used and straightforward method to prepare PT and its derivatives.

6.4.1.3 Synthesis of P3ATs by the McCullough and Grignard Metathesis Method

In 1992, McCullough reported the first synthetic method of head-to-tail coupled P3AT (HT–HT P3AT) with nearly 100% regioregularity [29]. In this method (see Scheme 6.4a), the monomer, 2-bromo-5-(bromomagnesio)-3-alkylthiophene, is obtained from 2-bromo-5-alkylthiophene at cryogenic temperature, and then is polymerized with catalytic amounts of $Ni(dppp)Cl_2$ (dppp = diphenylphosphinopropane). HT–HT P3ATs are obtained with the yields of 44%–69% and the molecular weight of $M_n = 20–40$ K and polydispersity indice (PDI) ≈ 1.4. The resulting HT–HT P3AT is precipitated in MeOH, washed with MeOH and hexane sequentially, and then recovered by Soxhlet extraction with chloroform. In the reaction, cryogenic temperature is used because of the tendency of the substituted lithiated aryl and heteroaryl halides to scramble by metal–halogen exchange at elevated temperature, which results in the formation of the mixtures of isomeric products.

Recently, a new method was developed to synthesize HT–HT P3ATs (see Scheme 6.4b). Treatment of a 2,5-dibromo-3-alkylthiophene with 1 equiv. of an alkyl Grignard reagent results in a magnesiumbromine exchange reaction, which is also referred to Grignard metathesis (GRIM) [30]. This reaction proceeds with a moderate degree of regioselectivity producing two isomers **1** and **2** (Scheme 6.4b) with a molar ratio of 85:15. As shown in Scheme 6.4b, the active monomer to be polymerized is simply prepared by the treatment of 2,5-dibromo-3-alkylthiophene with a cheap and easy available Grignard reagent solution. Treatment of the mixture of **1** and **2** with a catalytic amount of $Ni(dppp)Cl_2$ affords analytically pure, highly regioregular P3ATs. The molecular weight of the polymer obtained from this method is $M_n = 20–60$ K with a DPI = 1.5–3.0, and the yield of the polymerization can reach 78%. This method can be performed at room temperature, the reaction setup is cost-effective and the polymerization is quick and easy. This reaction can also be used to polymerize

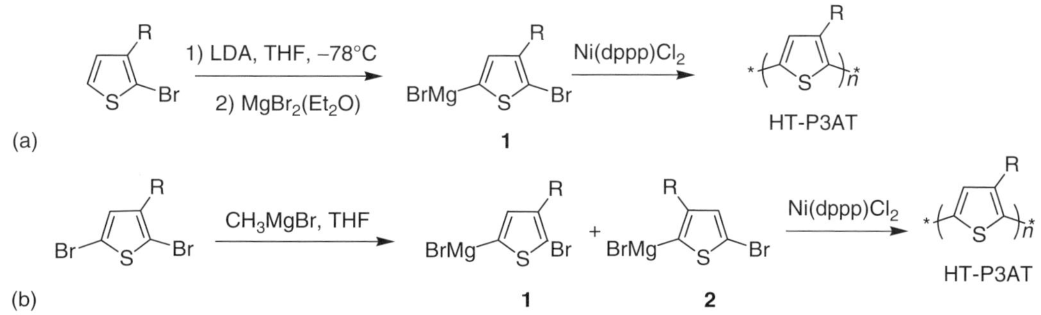

SCHEME 6.4 Synthesis of HT–HT P3ATs by the (a) McCullough and (b) GRIM method.

other thiophene monomers bearing esters, ethers, alkyl bromides, styryl, and alkoxy groups. But the thiophene monomers with the groups, which can react with Grignard reagent such as nitro, cyano, carbonyl, and hydroxyl groups, cannot be polymerized with this method.

6.4.1.4 Synthesis of P3ATs by the Rieke Method

The Rieke method, as shown in Scheme 6.5, is the second approach to prepare the regioregular P3AT [31]. In this method, a 2,5-dibromo-3-alkylthiophene is added to a solution of highly reactive "Rieke zinc" (Zn*). A mixture of **3** and **4** is obtained, and the ratio between these two isomers is dependent upon the reaction temperature. This metalated intermediate undergoes regioselective polymerization in a Negishi cross-coupling reaction to yield the desired HT–HT P3AT (Scheme 6.5). Interestingly, the percentage of HT–HT couplings is altered dramatically by the use of nickel and palladium catalysts. The regioselective control was rationalized on the basis of the steric congestion at the reductive elimination step in the catalytic cycle. Although the ratio of **3** to **4** is about 90:10, when Ni(dppe)Cl_2 or Ni(dppp)Cl_2 is used as catalyst, the HT–HT couplings are more than 98.5%. Other catalysts with less bulky, labile ligands like PPh_3 combined with larger metal centers like Pd lead to a regiorandom sample of P3AT.

Since there is no risk of metal–halogen exchange in **3** and **4**, 2-bromo-5-bromozinico-3-alkylthiophene is prepared with the similar method as 2-bromo-5-(bromomagnesio)-3-alkylthiophene, except that $ZnCl_2$ is used instead of $MgBr_2 \cdot Et_2O$. The addition of a Ni cross-coupling catalyst leads to the formation of a regioregular P3AT with ~100% HT–HT couplings. The molecular weight and the PDI of the P3ATs from the Rieke method are similar to that from the GRIM method. However, organic zinc reagents are tolerant of more functional groups than Grignard reagent, PTs with the substituents of nitryl, cyano, and carbonyl groups can be prepared by the Rieke method.

6.4.1.5 Synthesis of HT–HT P3ATs by the Stille and Suzuki Coupling Reaction

The Stille [32] and Suzuki [33] coupling methods are two convenient approaches to synthesize HT–HT P3ATs (Scheme 6.6). Although by the GRIM and Rieke method, nearly 100% HT–HT couplings can be obtained, the active organometals in these methods are generated in situ and used without any purification, and the high reactivity of the organometallic groups may be a limitation when additional reactive functional groups are present on the polymers. Thus, in the case of existing reactive

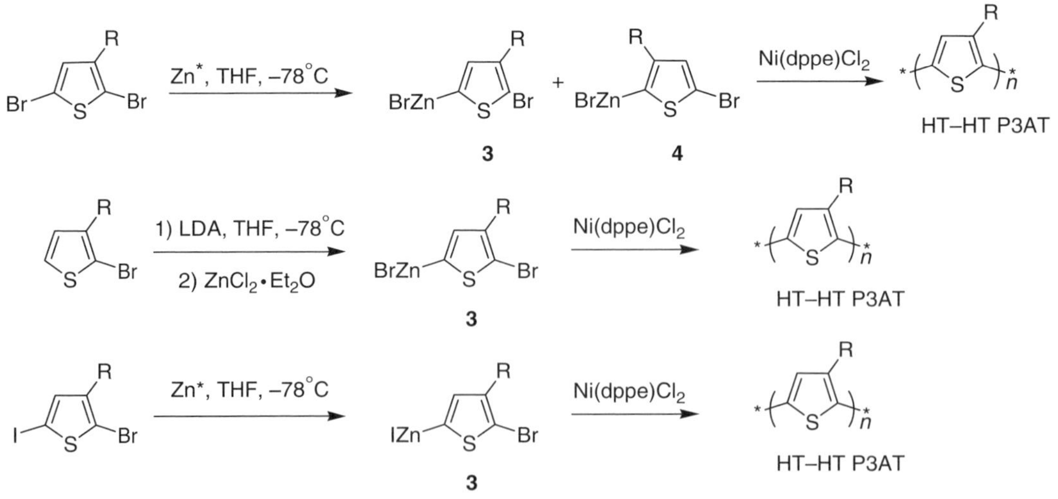

SCHEME 6.5 Synthesis of HT–HT P3AT by the Rieke method.

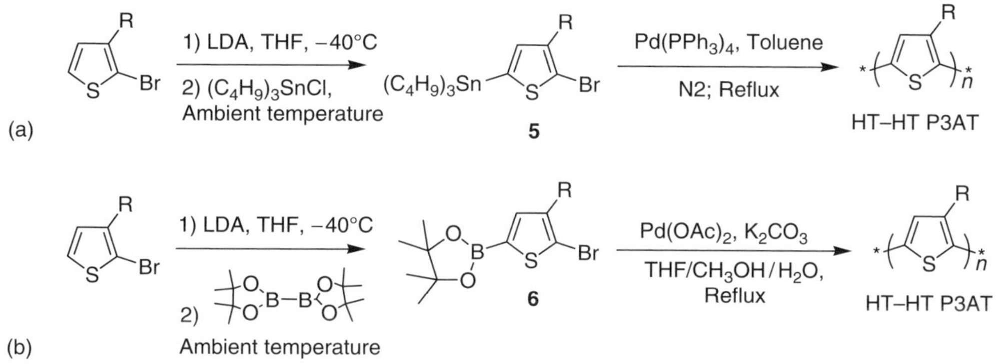

SCHEME 6.6 Synthesis of HT–HT PTs by the (a) Stille reaction and (b) Suzuki reaction.

functional groups, alternative method using compatible reactive groups (which allows direct synthesis of the functionalized soluble and regioregular P3ATs) should be employed. In the Stille and Suzuki methods, organostannic and organoboronic esters are used as monomers, respectively, and the precursors of the polymers can be separated and purified conveniently. As shown in Scheme 6.6, organostannic, **5**, and organoboronic, **6**, are prepared by similar method as **1** (Scheme 6.4) and **3** (Scheme 6.5). The two monomers, **5** and **6**, are polymerized with catalytic amount of $Pd(PPh_3)_4$ and $Pd(OAC)_2$, respectively, and the inert atmosphere is the most crucial condition for the polymerization. By the Suzuki method, poly(3-octylthiophene) with HT–HT couplings >96% was obtained, and the polymer has a molecular weight of $M_n = 40$ K and DPI = 1.5 with a yield of 55%. Due to their compatibility with a large number of organic functional groups, these two methods exhibit advantage in the synthesis of multifunctional polythiophenes.

6.4.1.6 Ni(0) Catalytic Dehalogenation Method for P3ATs

P3AT can also be synthesized from 2,5-diiodo-3-alkylthiophene and zero-valence nickel catalysts (Scheme 6.7) [34]. The polycondensation polymerization of 2,5-dihalo-3-alkythiophenes has shown that a good yield of P3AT (60%–95%) can be obtained from 2,5-dibromothiophene, $Ni(cod)_2$ (cod = cyclooctadiene), and triphenylphosphine (PPh_3) at 60°C–80°C in dimethylformamide (DMF) (Scheme 6.4). Diiodothiophenes are found to be more active monomers than dibromothiophenes. This type of organometallic coupling polymerization proceeds with predominantly HH type of couplings. This would give a P3AT with mainly HH–TT couplings. This is interpreted as selective oxidative addition of Ni to the less sterically hindered 5-position on the alkylthiophene. Molecular weight by this method is $M_n = 7.4$ K, PDI = 4.

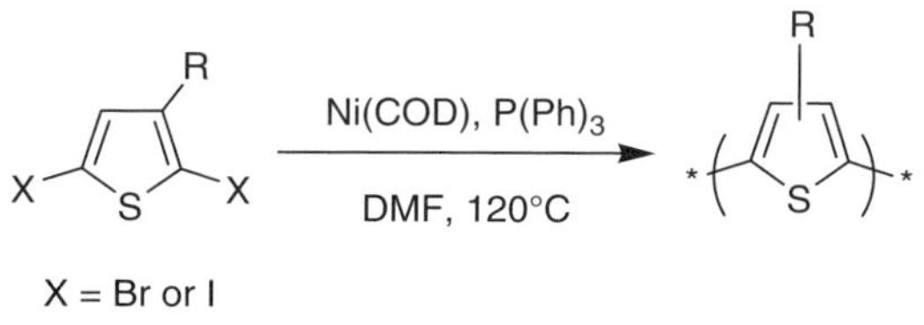

SCHEME 6.7 Synthesis of P3ATs by Ni(0) catalytic dehalogenation reaction.

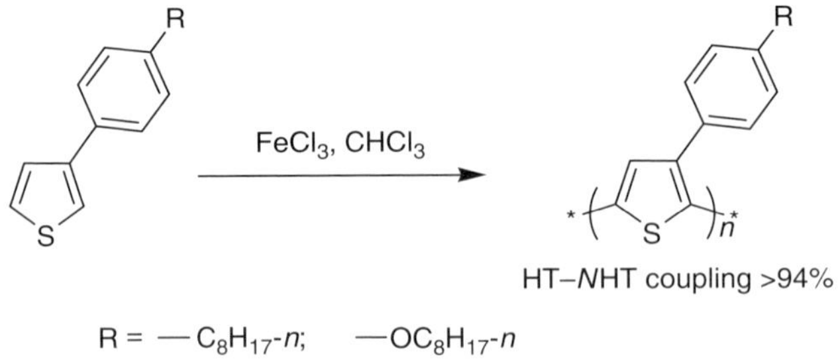

SCHEME 6.8 Synthesis of nearly regioregular P3PTs by oxidative polymerization.

6.4.2 Synthesis of Phenyl Substituted Polythiophene Derivatives

Regiorandom poly(3-phenyl-thiophene)s (P3PTs) were synthesized by Ueda et al. [35], with the polycondensation polymerization of 2,5-dihalothiophenes using Ni(0) catalytic dehalogenation reaction. Nearly regioregular (94% HT–HT) poly(3-(4-octylphenyl)-thiophene) has been prepared using $FeCl_3$ to oxidatively polymerize 3-(4-octylphenyl)thiophene (Scheme 6.8).

6.4.3 Synthesis of Polythiophenes with Conjugated Side Chains

The polythiophene derivatives with conjugated side chains (such as phenylene vinylene and thienylene vinylene) are interesting polymers for regulating the absorption spectra of the conjugated polymers. So the synthesis method of the side-chain conjugated PTs is introduced here. Phenylene-vinylene substituted PTs was first synthesized by $FeCl_3$ oxidation method, as show in Scheme 6.9 [36]. The polymer **7** was obtained as an insoluble solid. To improve the processibility of compound **7**, 3-alkylthiophene was added before the oxidative polymerization, and the polymer **8** was obtained as a soluble solid. Recently, by the GRIM method and Stille method, the authors synthesized several soluble polythiophene derivatives with the conjugated side chains of phenylene vinylene and thienylene vinylene, as shown in Scheme 6.10 [21,22]. When the conjugated side chain is **9** or **11**

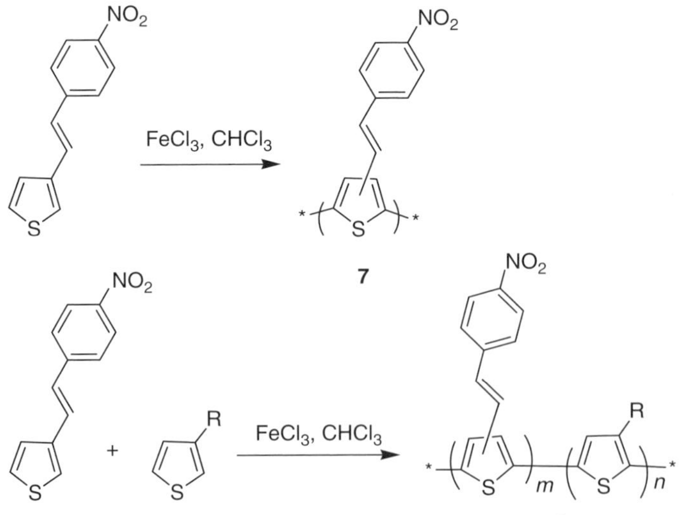

SCHEME 6.9 Synthesis of phenylene-vinylene substituted PTs by $FeCl_3$ method.

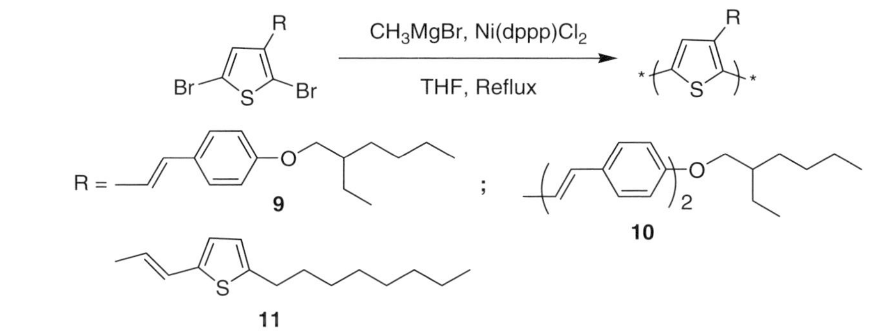

SCHEME 6.10 Synthesis of phenylene-vinylene and thienylene-vinylene substituted PTs by the GRIM method.

(see Scheme 6.10), molecular weight of the polymers is about $M_n = 29$ K with PDI = 2.3. While, when the conjugated side chain is **10**, the molecular weight of the polymer is only about $M_n = 4.3$ K, the big steric hindrance of the side chain is the main reason for the low polymerization degree. The Stille coupling reaction can also be used to prepare polythiophenes with the conjugated side chains. In Scheme 6.11, the coupling between 2,5-bis(tributylstannyl)-thiophene and 2,5-dibromothiphenes produces the polythiophene derivatives with $M_n = 21–52$ K and PDI = 1.6–2.3.

6.4.4 Synthesis of Poly(3-alkoxythiophene)s

If an alkoxy group is attached to thiophene ring directly, the band gap of the polymer can be reduced effectively, and the oxidative potential of the polymer will decrease by a substantial amount. So the conducting state of the polymer is stabilized [37]. Scheme 6.12 shows the molecular structures of some poly(3-alkoxythiophene)s, where polymer **16** can be used as molecular recognition units for chemical sensing or for self-assembly of the polymer. Generally, the poly(3-alkyloxythiophene)s can be synthesized by the same method as that for the synthesis of P3ATs. With $FeCl_3$ oxidative method, the polythiophene derivatives are usually region random, while for polymer **19**, 95% regio-regular HT–HT coupling can be obtained.

Very stable conducting polymer **18**, PEDOT, can also be prepared by $FeCl_3$ oxidative method. The solubility of the PEDOT from the $FeCl_3$ method is very sensitive to the amount of $FeCl_3$ used in

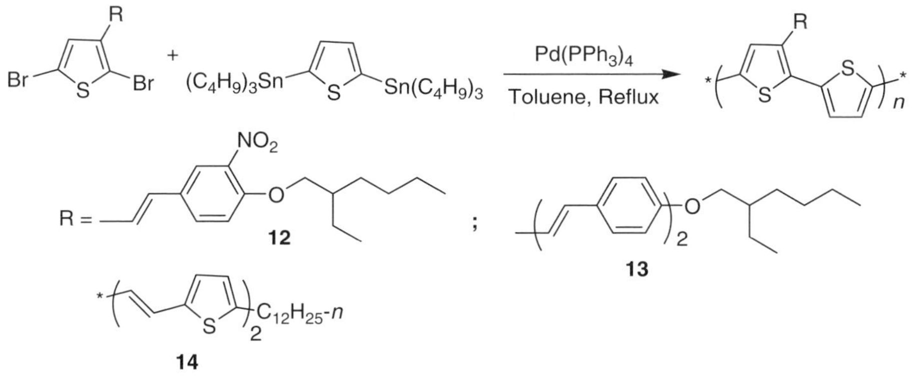

SCHEME 6.11 Synthesis of phenylene-vinylene and thienylene-vinylene substituted PTs by the Stille coupling method.

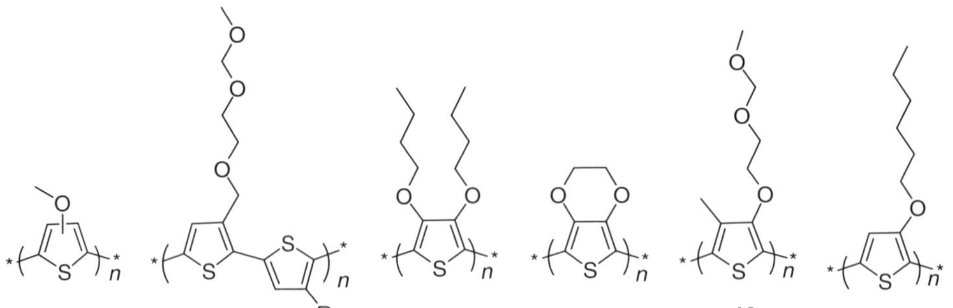

SCHEME 6.12 Structure of some poly(alkoxythiophene)s.

the polymerization. By controlling the ratio of $FeCl_3$ to the monomer EDOT, the solubility of the resulting polymer can be improved.

By GRIM method, from 2,5-dibromo-3-hexyloxythiophene, a regioregular polymer **20**, can be prepared with good yield, and the solubility, molecular weight, and HT–HT coupling content are similar to those of poly(3-hexylthiophene) prepared by GRIM method.

6.5 SYNTHESIS OF POLY(*p*-PHENYLENE VINYLENE) AND ITS DERIVATIVES

6.5.1 Synthesis of PPV by the Wessling Precursor Method

Poly(*p*-phenylene vinylene) and its derivatives can be synthesized via Wessling precursor route [38], as shown in Scheme 6.13. At the first step, a soluble PPV precursor is synthesized by the reaction of bis(sulfonium halide) salts of *p*-xylene with base (NaOH) in water or alcohol solution. For the application in PLEDs, the precursor solution is spin-cast on ITO substrate, then PPV film is formed after thermal treatment of the precursor film at 180°C–300°C under vacuum. Vanderzande et al. [39] and Mullen et al. [40] improved the Wessling method by replacing chloride with bromide in the starting chemicals, and the temperature of the heat treatment is decreased to 100°C for the decomposition of the precursor to PPV.

There are some drawbacks with the Wessling route. Defects and impurities in the polymer could exist due to the oxidation of the precursor polymer, the residual precursor moieties, and the undesired side reactions during the thermal conversion (see Scheme 6.14). (Significant improvement on reducing the side reactions during the thermo-conversion was realized by the use of cycloalkylene sulfonium groups instead of dialkylene sulfonium groups.) In addition, there is a problem of the possible reaction with volatile corrosive elimination products during the thermal treatment, and the liberation of volatile corrosive elimination products could be harmful to other components of PLEDs when this method is used in the fabrication of PLEDs (for example, ITO layer used as the

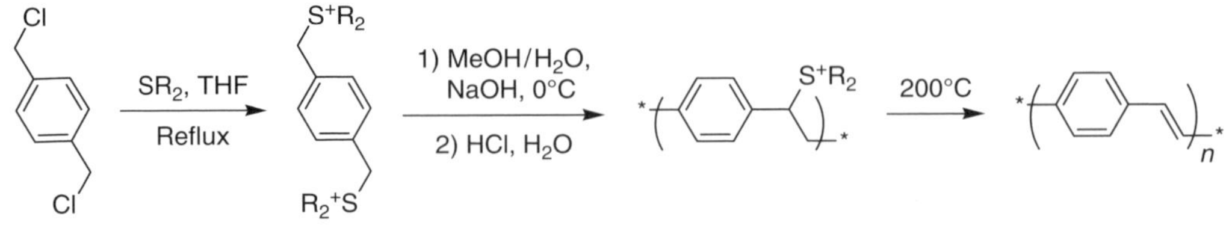

SCHEME 6.13 Synthesis of PPV by the Wessling precursor route.

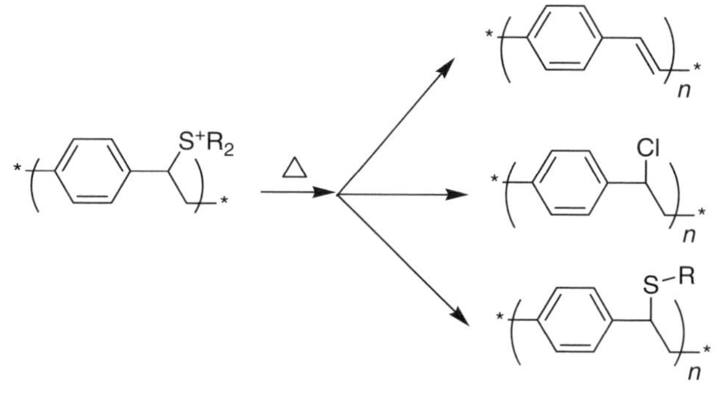

SCHEME 6.14 Possible products and defects in the Wessling precursor method.

anode is corrosion sensitive). Therefore, the Wessling method is mainly used for the preparation of the insoluble unsubstituted PPV, although it can also be used to prepare soluble PPV derivatives, since the method is complicated for the posttreatment.

6.5.2 Synthesis of PPV and Its Derivatives by the Gilch Method

Poly(*p*-phenylene vinylene) and its derivatives can be prepared conveniently by the Gilch method [41] in which 1,4-bis(chloromethyl) (or bromomethyl) arenes are treated with potassium *t*-butoxide in a non-hydroxylic solvent like tetrahydrofuran (THF). As shown in Scheme 6.15, a precursor of **21** is formed in situ at first, and then it is converted to PPV, **22**, by base-induced abstraction of hydrohalogenides. The temperature of the reaction, the concentration of the monomer and the base, and the speed of the base addition are all crucial conditions for getting high molecular weight and narrower PDI of the polymer. The molecular weight can also be controlled by using a benzylchloride derivative as a chain terminator. Many PPV derivatives, such as MEH-PPV, MDMO-PPV, and other PPVs with bulky groups on side chains (see Scheme 6.15), have been prepared with great molecular weight and high purity by the Gilch method, making it to be the most popular method for PPV preparation.

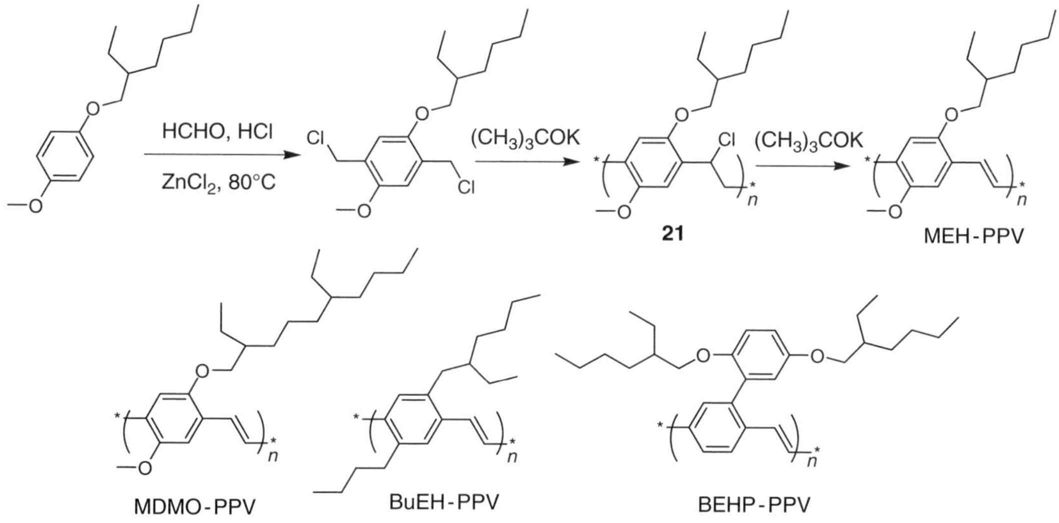

SCHEME 6.15 Synthesis of MEH-PPV by the Gilch method and the PPVs that have been synthesized by this method.

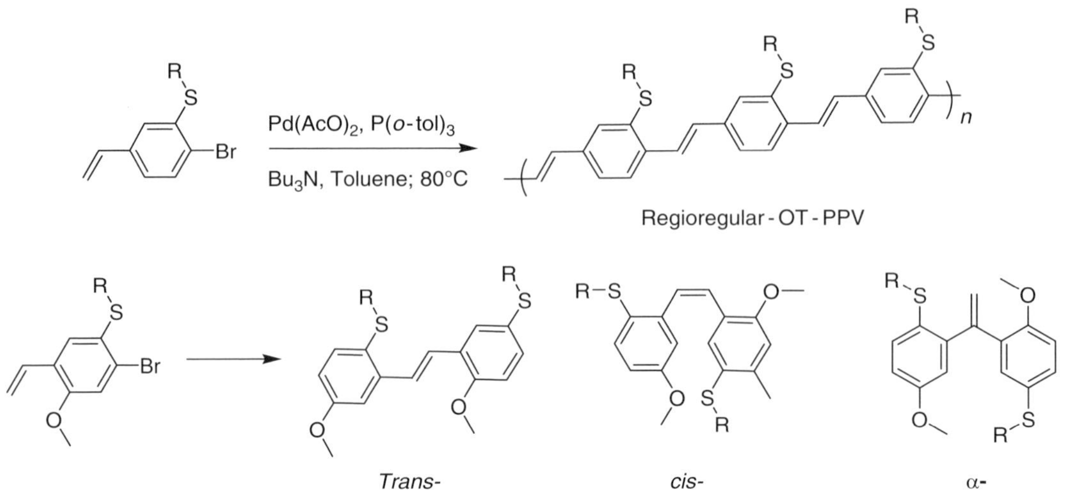

SCHEME 6.16 Synthesis of the alkylthio-substituted PPV by the Heck coupling reaction.

6.5.3 Synthesis of PPVs by the Heck Coupling Reaction

By the Heck reaction [42], organic halides and vinylbenzene compounds are coupled to generate a carbon–carbon bond under the catalysis of Pd(0). The vinylbenzene compounds can be easily synthesized by the Wittig reaction with high yield from benzaldehyde, and the organic halides can also be prepared easily. Many functional groups, such as aldehyde, ester, nitro (-NO_2), cyano (-CN), hydroxy, and carboxy, have no obviously effect on the coupling reaction, and a lot of solvents could be used for the reaction, such as DMF, DMA, NMP, toluene, and even water. Therefore, this reaction is widely used for the preparation of PPVs, and it has been proved to be superior for the synthesis of multifunctional substituted PPVs.

In the Heck coupling reaction, the reactivity order of the halides is I > Br > Cl. Because of weak reactivity, chlorides are seldom used in the polymerization. $Pd(OAc)_2$ is used as catalyst, and $P(Ph_3)$ or $As(Ph_3)$ is used as ligand. The electron-pushing group on the benzene halides, such as alkoxy group, can reduce the reactivity of halogen. Therefore, higher reaction temperature is often used to accelerate the coupling reaction when there is an electron-pushing group. However, an increase of reaction temperature may lead to the formation of α structure, a defect in polymer main chain (Scheme 6.16). Commonly, the Heck reaction is performed between 1,4-divinylbenzene and aromatic dibromide, and regiorandom polymer is synthesized, but when the vinyl and the halide groups are in one asymmetry molecule, a regioregular alkylthio-substituted PPV can be obtained, as shown in Scheme 6.16 [43]. The molecular weight of poly(octylthio-phenylene vinylene) (OT-PPV) is about 28 K with PDI = 2.2.

6.5.4 Synthesis of PPV and Its Derivatives by the Wittig Reaction

The synthesis of PPV and its derivatives by the Wittig reaction [44] are carried out by two steps, as shown in the upper part of Scheme 6.17. Under inert atmosphere, bis(triphenyl phosphine) salt is suspended in tetrahydrofuran or toluene, and dealt with strong alkali such as *n*-butyl lithium, *t*-butoxy sodium, and other alkoxy sodium. Then *p*-phthalaldehyde or aryl dialdehyde is added. After refluxing for one to several hours, PPVs can be obtained with a good yield.

As a similar method to Wittig reaction, Hornor–Emmons reaction can also be used to prepare copolymers of PPV with poly(thienylene vinylene) [45], as shown in the lower part of Scheme 6.17. In this reaction, bis(phosphonate ester) compound is mixed with aryl dialdehyde directly instead of

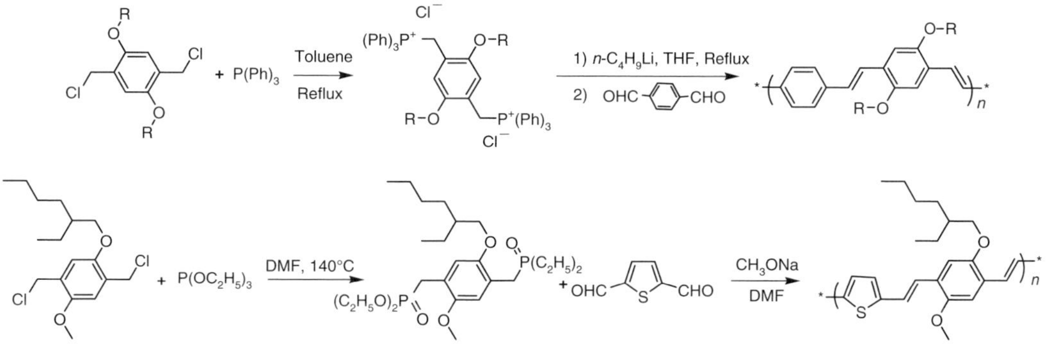

SCHEME 6.17 Synthesis of PPV and its derivatives by the Wittig and Hornor–Emmons reactions.

bis(triphenyl phosphine) salt. Sodium methoxide or potassium hydroxide can be used as a base, and dimethyl formamide, *N*-methyl pyrrolidone, or tetrahydrofuran can be used as a solvent. Hornor–Emmons reaction can be carried out by one step under ambient temperature and atmosphere with a good yield.

6.5.5 Synthesis of CN-PPVs by the Knoevenagel Condensation

Cyano-substituted PPV (CN-PPV) is important, as a polymer acceptor, in the applications of polymer optoelectronic devices. CN-PPV can be synthesized by a Knoevenagel condensation between equimolar amounts of a terephthaldehyde and a 1,4-diacetonnitrile-benzene derivatives [46], as shown in Scheme 6.18. The condensation reaction takes place upon addition of excess potassium *t*-butoxide or tetrabutylammonium hydroxide in THF/*t*-butanol mixture at 50°C. 1,4-Diacetonnitrile-benzene derivatives can be synthesized from 1,4-dichloromethyl-benzene derivatives by nucleophilic reaction, and sodium cyanide or zinc cyanide can be used as nucleophilic reagent. In Knoevenagel condensation reaction, tetrabutylammonium hydroxide is used as catalyst, while toluene, DMF, and THF are used as solvents (THF is usually added to increase solubility of the product and to avoid premature precipitation). The molecular weight of the polymer is $M_n = 20$ K with PDI = 2.0.

6.6 SYNTHESIS OF POLY(*p*-PHENYLENE), POLYFLUORENE, AND THEIR DERIVATIVES

6.6.1 Synthesis of Poly(*p*-phenylene) and Its Derivatives

Poly(*p*-phenylene) is the first conjugated polymer showing blue electroluminescence. It can be prepared following a precursor route, as shown in Scheme 6.19 [47]. A film of the precursor polymer is

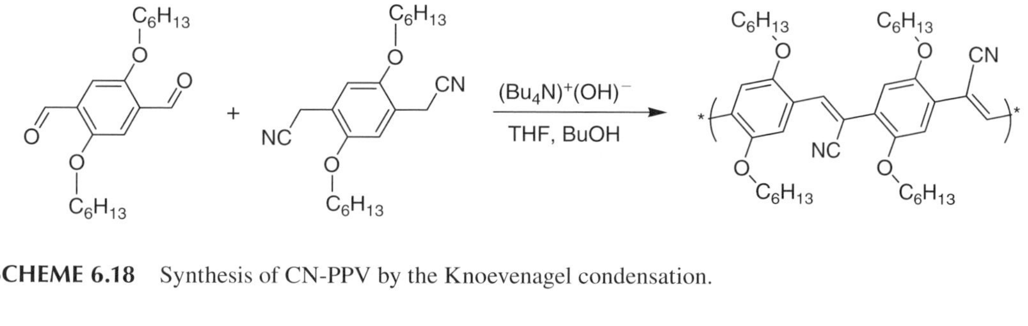

SCHEME 6.18 Synthesis of CN-PPV by the Knoevenagel condensation.

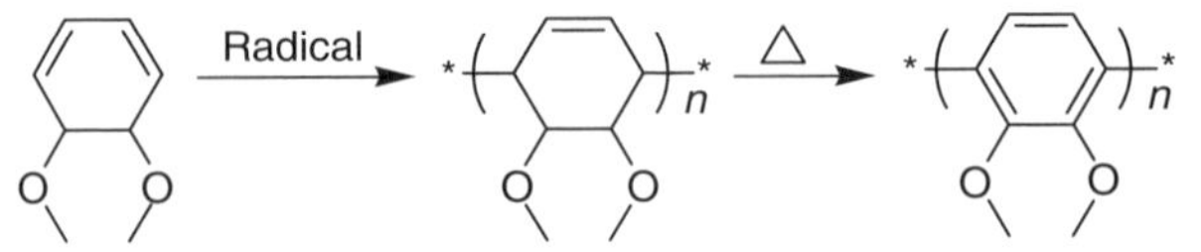

SCHEME 6.19 Synthesis of PPP by precursor route.

spin-coated from a precursor solution on an ITO electrode and converted to PPP by thermo-conversion in vacuum at 340°C. There are several problems with this method. The PPPs prepared by this method are of low molecular weight and make poor quality films.

6.6.2 Synthesis of PF and PPP by Ni(0)-Mediated Polymerization

Pei and Yang [48] successfully synthesized high molecular weight polyfluorene homo- and copolymers by a Ni(0)-mediated polymerization of 9,9-disubstituted-2,7-dibromofluorene monomers (Yamamoto polymerization) [49] (Scheme 6.20).

6.6.3 Synthesis of PF by the Suzuki Reaction

Another important synthetic route for the preparation of PFs is via the Suzuki reaction (Scheme 6.21). The reaction between diboronic functionalized aromatic units and diiodo (or dibromo) aromatic derivatives is performed with a catalytic amount of $Pd(PPh_3)_4$ and K_2CO_3. Since PPP and its derivatives have similar structure as PFs, they can also be synthesized by Suzuki polymerization.

6.7 SYNTHESIS OF POLY(ARYLENEETHYNYLENE)S

6.7.1 Synthesis of Poly(aryleneethynylene)s by Pd(0) Catalytic Method

Poly(aryleneethynylene)s (PAEs), including Poly(phenyleneethynylene)s (PPE), poly(thienyleneeth ynylene)s (PTE) and other heterocyclic poly(aryleneethynylene)s, are synthesized mainly by the Pd-catalyzed coupling of terminal alkynes to aromatic bromides or iodides in amine solvents [50], as shown in Scheme 6.22. Both bromo- and iodoaromatic compounds can be used in this reaction. In comparison with aryl bromides, the iodides react much faster, and the reaction can be carried out under a lower temperature. As a consequence, the polymerization can be performed under mild conditions when iodides are used, so that problems including cross-linking and formation of defects can

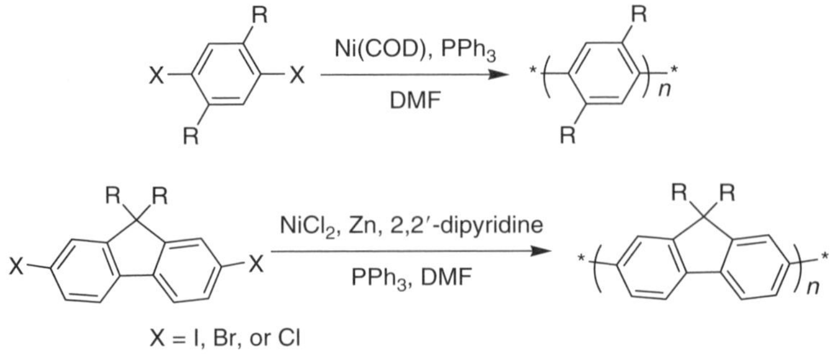

SCHEME 6.20 Synthesis of PPP and PF by Ni(0) catalytic dehalogenation reaction.

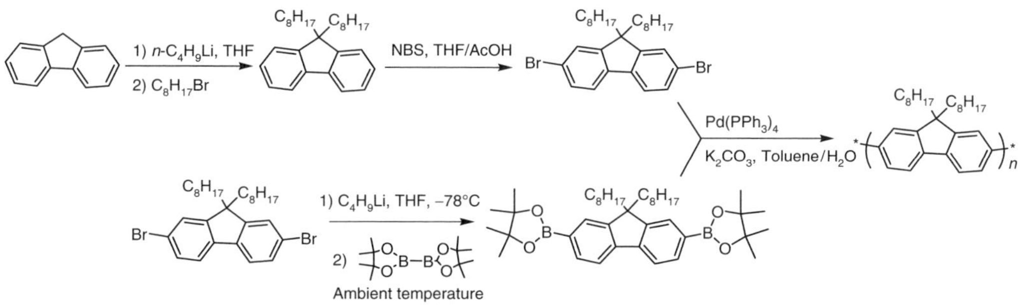

SCHEME 6.21 Synthesis of PFs by the Suzuki coupling reaction.

be minimized. Therefore, if available, iodoarenes are the preferred substance for the reaction. The reaction is also dramatically influenced by the substituent on aryl halides. Electron-withdrawing group, such as nitro, on the halide improves the rate and the yield of the coupling reaction, while the electron-pushing group, such as alkoxy, reduces the reaction. The electron-withdrawing substituents on *ortho-* or *para*-position are more efficient than that on the *meta*-position. Most frequently 0.1–5 mol% $(Ph_3P)_2PdCl_2$ and varying amounts of CuI are used in both organic and polymer-forming reactions. In the cases where the haloarenes are sufficiently active (typically iodoarenes), much smaller amounts (0.1%–0.3%) should be sufficient. An organic amine, such as tributylamine, triethylamine, and piperidine mixed with toluene or THF, is used as solvent. Suzuki coupling as that of synthesis of PFs and PTs (Scheme 6.23) can also be used to prepare PPEs.

There are several shortcomings of the Pd methodology. (1) Even under strict exclusion of air, diyne defects exist with the use of Pd(0) catalysts. (2) Dialkoxy- and dialkyldiiodobenzenes give, even under optimum conditions, PPEs with degree of polymerization (DP) less than 100 according to GPC which itself overestimates the DP. (3) Formation of the defects of dehalogenation and phosphonium salt (substitution of the iodide end group by triarylphosphine, leading to phosphonium-substituted chains). (4) The Palladium- and phosphorus-containing catalyst residues need to be removed and purified after the polymerization. Therefore, it is necessary to find an alternative non-Pd-based synthetic route of PAEs.

6.7.2 Synthesis of PPEs by Alkyne Metathesis

In 1997, Weiss et al. [51] reported the first use of Schrock's tungsten-carbyne [52] for the preparation of some PPEs including 2,5-dihexyl-PPE. The dipropynylated monomer **22** (Scheme 6.24) should be treated with the tungsten-carbyne in a carefully dried solvent under strict exclusion of air and water at elevated temperature for 12–16 h. Since the Schrock catalyst $(t\text{BuO})_3\text{WC-}t\text{Bu}$ is very active, it has to be synthesized just before use, and it is very sensitive to air and particularly sensitive to water.

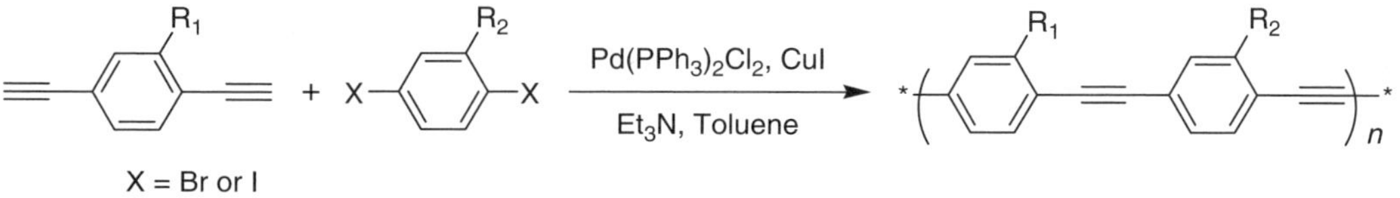

SCHEME 6.22 Synthesis of PPE by the Heck coupling reaction.

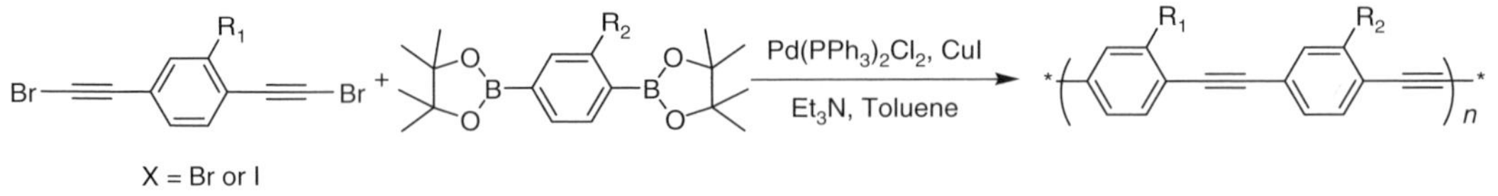

SCHEME 6.23 Synthesis of PPE by the Suzuki coupling reaction.

Bunz et al. optimized the reaction conditions of the alkyne metathesis utilizing $Mo(CO)_6$ and 4-chlorophenol by increasing the reaction temperature from 105°C to 130°C–150°C (see Scheme 6.25). And the reaction was performed with a purge of nitrogen to remove the formed 2-butyne. By this method, they obtained a polymer with a degree of polymerization of 180 (GPC). It was indicated that this reaction is competitive to the Pd-catalyzed coupling reaction.

6.8 SUMMARY

Conjugated polymers are the key materials for the organic/polymer electrical, electrochemical, and optoelectronic applications. The main classes of conjugated polymers include PA, PPy, PAn, PT, PPV, PPP, PF, PTV, PPE, and their derivatives. For most of the conjugated polymers, they can be p-doped (oxidized) at positive potential region, and can be n-doped (reduced) in the negative potential region. After doping, their conductivity can increase by 6–9 orders, and the doped conducting polymers show a strong absorption peak in the near infrared region. For conducting polymers of PPy, PAn, and PT, the reversible electrochemical properties are very important for their applications as electrode materials in batteries, electrochromic displays, modified electrodes, and biosensors. Electrochemical oxidation polymerization is the main method for the preparation of PPy films and conducting polymer electrodes. The electronic energy levels and solubility of the conjugated polymers can be modulated with substituents of functional groups. The electronic energy levels (HOMO, LUMO, and E_g) of the conjugated polymers, which are important parameters for the applications of the intrinsic conjugated polymers in optoelectronic devices, can be estimated from the absorption spectra (absorption edge) and the electrochemical measurement (onset p-doping and n-doping potentials). Flexible alkyl or alkoxy substituents improve the solubility greatly. The electron-donating substituents raise the HOMO and the electron-withdrawing groups drop the LUMO of the conjugated polymers. The absorption spectra redshift and E_g reduces from PPP, PF, PPV, PT to PTV. As luminescent polymers, PPP and PF emit blue light, PPV emits green–orange light, and PT emits red light. The synthesis of the soluble neutral conjugated polymers is usually carried out by catalytic polymerization with suitable catalysts. Various synthesis strategies can be used to prepare the conjugated polymers, as described in Sections 6.4–6.7.

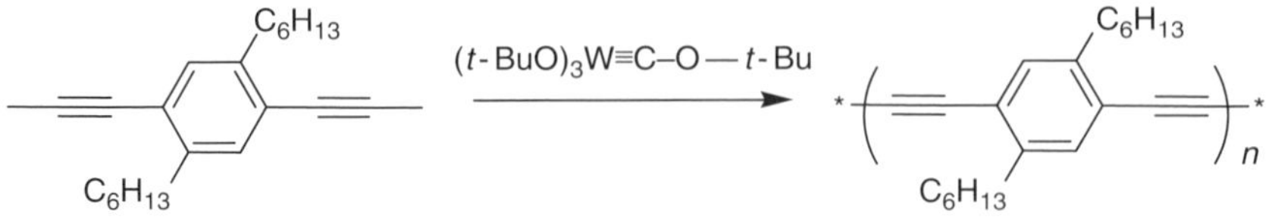

22

SCHEME 6.24 Synthesis of PPEs by alkyne metathesis.

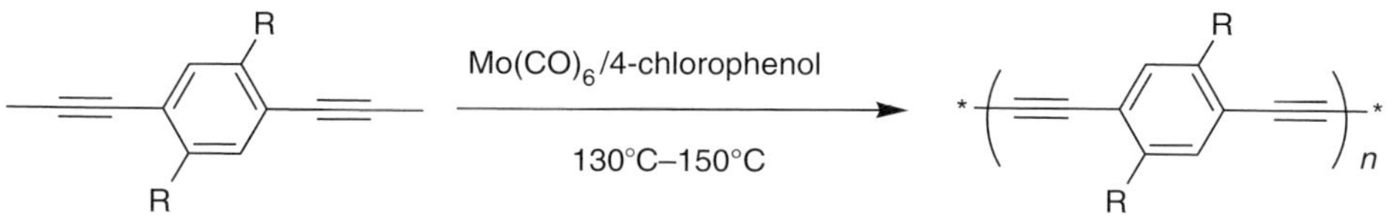

SCHEME 6.25 Synthesis of PPEs by alkyne metathesis at a higher temperature.

EXERCISE QUESTIONS

1. Provide the names of the major conjugated polymers you know.
2. Could the electrochemical polymerization of pyrrole be performed in an aqueous solution? And could the electrochemical polymerization of thiophene be performed in an aqueous solution? Why?
3. How does p-doping change the absorption spectra of the conjugated polymers?
4. How to prepare a soluble conducting polyaniline solution?
5. Why could the HOMO/LUMO energy levels of the conjugated polymers be estimated by electrochemical measurement? Please write the equations for the calculation.
6. How do the substituent groups on the conjugated polymers influence the properties of the polymers?

LIST OF ABBREVIATIONS

BFEE	boron trifluoride/ethylene ether
CSA	camphor sulphonic acid
DBSA	dodecylbenzenesulfonic acid
DMF	dimethylformamide
DN	donor number
EA	electron affinity
FET	field-effect transistor
GRIM	Grignard metathesis
HOMO	highest occupied molecular orbital
IP	ionization potential
ITO	indium tin oxide
LUMO	lowest unoccupied molecular orbital
MEH-PPV	poly(2-methoxy-5-(2′-ethylhexyloxy)-1,4-phenylene vinylene)
NMP	*N*-methylpyrrolidinone
PA	polyacetylene
PAn	polyaniline
PEDOT:PSS	poly(ethylenedioxythiophene):polystyrene sulfonate
PF	polyfluorene
PFO	poly(9,9-dioxylfluorene)
P3AT	poly(3-alkylthiophene)
P3HT	poly(3-hexylthiophene)
PL	photoluminescence
PLED	polymer light-emitting diode
PPE	poly(phenylene ethynylene)
PPP	poly(*p*-phenylene)
PPV	poly(*p*-phenylene vinylene)

PPy	polypyrrole
PSC	polymer solar cell
PTV	poly(thienylene vinylene)
PT	polythiophene
SCE	saturated calomel electrode
SCLC	space charge limited current
THF	tetrahydrofuran

REFERENCES

1. Shirakawa, H., Louis, E.L., MacDiarmid, A.G., Chiang, C.K., Heeger, A.J., *J. Chem. Soc., Chem. Commun.*, 1977, 578.
2. Chiang, C.K., Fincher, C.R., Jr., Park, Y.W., Heeger, A.J., Sharakawa, H., Louis, E.J., Gau, S.C., MacDiarmid, A.G., *Phys. Rev. Lett.*, 1977, 39: 1098.
3. Skotheim, T.A., Elsenbaumer, R.L., Reynolds, J.R., Ed., *Handbook of Conducting Polymers*, 2nd edition, Marcel Dekker, New York, 1998.
4. Farchioni, R., Grosso, G., Eds., *Organic Electronic Materials—Conjugated Polymers and Low Molecular Weight Organic Solids*, Springer, Berlin, 2001.
5. Cao, Y., Smith, P., Heeger, A.J., *Synth. Met.*, 1992, 48: 91.
6. Li, Y.F., Qian, R.Y., *Synth. Meth.*, 1989, 28: C127; Li, Y.F., Qian, R.Y., *J. Electroanal. Chem.*, 1993, 362: 267.
7. Li, Y.F., Qian, R.Y., *Electrochim. Acta*, 2000, 45: 1727.
8. Li, Y.F., Yan, B.Z., Yang, J., Cao, Y., Qian, R.Y., *Synth. Met.*, 1988, 25: 79.
9. Genies, E.M., Lapkowski, M., Penneau, J.F., *J. Electroanal. Chem.*, 1988, 249: 97.
10. Ouyang, J.Y., Li, Y.F., *Polymer*, 1997, 38: 1971.
11. Ouyang, J.Y., Li, Y.F., *Polymer*, 1997, 38: 3997.
12. Li, Y. F., *J. Electroanal. Chem.*, 1997, 433: 181.
13. Shi, G.Q., Jin, S., Xue, G., Li, C., *Science*, 1995, 276: 994.
14. Lam, J.Y., Tang, B.Z., *Acc. Chem. Res.*, 2005, 38: 745–754.
15. Tang, B.Z., Poon, W.H., Leung, W.H., Peng, H., *Macromolecules*, 1997, 30: 2209–2212.
16. Chaing, J.C., MacDiarmid, A.G., *Synth. Met.*, 1986, 13: 193.
17. Burroughes, J.H., Bradley, D.D.C., Brown, A.R., Marks, R.N., Mackay, K., Friend, R.H., Burns, P.L., Holmes, A.B., *Nature*, 1990, 347: 539.
18. Roncali, J., *Chem. Rev.*, 1997, 97: 173–205.
19. Li, Y.F., Cao, Y., Gao, J., Wang, D.L., Yu, G., Heeger, A.J., *Synth. Met.*, 1999, 99: 243.
20. Cornil, J., dos Santos, D.A., Beljonne, D., Bredas, J.L., *J. Phys. Chem.*, 1995, 99: 5604.
21. Hou, J.H., Huo, L.J., He, C., Yang, C.H., Li, Y.F., *Macromolecules*, 2006, 39: 594–603; Hou, J.H., Yang, C.H., He, C., Li, Y.F., *Chem. Commun.*, 2006, 871–873.
22. Hou, J.H., Tan, Z.A., Yan, Y., He, Y.J., Yang, C.H., Li, Y.F., *J. Am. Chem. Soc.*, 2006, 128: 4911–4916.
23. Malliaras, G.G., Salem, J.R., Brock, P.J., Scott, C., *Phys. Rev. B*, 1998, 58: 13411.
24. Zhou, E.J., Tan, Z.A., Yang, C.H., Li, Y.F., *Macromol. Rapid Commun.*, 2006, 27: 793–798.
25. Sato, M., Morii, H., *Polym. Commun.*, 1991, 32: 42.
26. McCullough, R.D., *Adv. Mater.*, 1998, 10: 93.
27. Sugimoto, R., Takeda, S., Gu, H.B., Yoshino, K., *Chem. Express*, 1986, 1: 635.
28. Pomerantz, M., Tseng, J.J., Zhu, H., Sproull, S.J., Reynolds, J.R., Uitz, R., Arnott, H.J., Haider, H.I., *Synth. Met.*, 1991, 41–43: 825.
29. McCullough, R.D., Lowe, R.D., *Chem. Commun.*, 1992, 70.
30. Robert, S., Loewe, P., Ewbank, C., Liu, J.S., Zhai, L., McCullough, R.D., *Macromolecules*, 2001, 34: 4324–4333.
31. Chen, T.-A., Rieke, R.D., *Synth. Met.*, 1993, 60: 175.
32. Iraqi, A., Barker, G.W., *J. Mater. Chem.*, 1998, 8: 25.
33. Guillerez, S., Bidan, G., *Synth. Met.*, 1998, 93: 123.
34. McCullough, R.D., Jayaraman, M., *Chem. Commun.*, 1995, 135.
35. Ueda, M., Miyaji, Y., Ito, T., Oba, Y., Sone, T., *Macromolecules*, 1991, 24: 2694.
36. Greenwald, Y., Cohen, G., Poplawski, J., Ehrenfreund, E., Speiser, S., Davidov, D., *J. Am. Chem. Soc.*, 1996, 118: 2980.
37. Tanaka, S., Sato, M.A., Kaeriyama, K., *Synth. Met.*, 1988, 25: 277.

38. Wessling, R.A., *J. Polym. Sci., Polym. Symp.*, 1985, 72: 55.
39. Beerden, A., Vanderzande, D., Gelan, J., *Synth. Met.*, 1992, 52: 387.
40. Garay, R.O., Baier, U., Bubesk, C., Mullen, K., *Adv. Mater.*, 1993, 5: 561.
41. Gilch, H.G., Wheelwright, W.L., *J. Polym. Sci. A, Polym. Chem.*, 1996, 34: 13.
42. Beletskaya, I.P., Cheprakov, A.V., *Chem. Rev.*, 2000, 100: 3009–3066.
43. Hou, J.H., Fan, B.H., Huo, L.J., He, C., Yang, C.H., Li, Y.F., *J. Poly. Sci. A, Polymer Chem.*, 2006, 44: 1279.
44. Lahti, P.M., Moderelli, D.A., Denton, F.R., Lenz, R.W., Karasz, F.E., *J. Am. Chem. Soc.*, 1988, 110: 7258.
45. Hou, J.H., Yang, C.H., Qiao, J., Li, Y.F., *Synth. Met.*, 2005, 150: 297.
46. Kim, D.J., Kim, S.H., Zyung, T., Kim, J.J., Cho, I., Choi, S.K., *Macromolecules*, 1996, 29: 3657.
47. Ballard, D.G.H., Courtis, A., Shrley, L.M., Taylor, S.C., *Macromolecules*, 1989, 21: 294.
48. Pei, Q., Yang, Y., *J. Am. Chem. Soc.*, 1996, 118: 7416.
49. Yamamoto, T., Hayashi, Y., Yamamoto, A., *Bull. Chem. Soc. Jpn.*, 1978, 51: 209.
50. Bunz, U.H.F., *Chem. Rev.*, 2000, 100, 1605.
51. Weiss, K., Michel, A., Auth, E.M., Bunz, U.H.F., Mangel, T., Mullen, K., *Angew. Chem.*, 1997, 36: 506.
52. Schrock, R.R., Clark, D.N., Sancho, J., Wengrovius, J.H., Pederson, S.F., *Organometallics*, 1982, 1: 1645.

7 Low Energy Gap, Conducting, and Transparent Polymers

Arvind Kumar, Yogesh Ner, and Gregory A. Sotzing

CONTENTS

7.1 Introduction 211
7.2 Energy Gap in Conjugated Polymers 213
7.3 Different Classes of Low Energy Gap Polymers 216
  7.3.1 Fused Aromatics 216
  7.3.2 Electron-Donating Substituents 219
  7.3.3 Electron-Withdrawing Substituents 221
  7.3.4 Push–Pull Donor/Acceptors 221
  7.3.5 Vinylenes and Methines 225
7.4 Processability of Low Energy Gap Polymers 228
7.5 Summary 230
Exercise Questions 232
List of Abbreviations 233
Acknowledgment 233
References 233

Abstract This chapter discusses the electrical and optical properties of polymers having band gap energy (E_g) of 1.8 eV or lower and optically transparent in their conducting state. The factors affecting the E_g of polymer are discussed in detail with the characterization techniques used to determine E_g. Low energy gap polymers (LEGPs) are primarily categorized into five groups, and several examples are cited herein for a better understanding of the structure–property relationship. Some problems related to the processability of LEGPs are mentioned, and approaches to improve the processability are discussed with examples from the literature. This chapter does not cover the synthesis of monomers and polymers or the fundamentals of optoelectronic devices as they are discussed in the previous chapters of this book.

7.1 INTRODUCTION

π-Conjugated polymers (CPs) generally consist of linear chains having delocalized electrons along alternating single and double bonds. CPs have been dubbed as the fourth generation of polymeric materials due to their wide range of applications, including as active materials in optoelectronic devices such as organic light-emitting diodes (OLEDs), transistors, volatile organic gas sensors, photovoltaics, electrochromics, energy storage batteries, nonlinear optics, electrostatic discharge coatings, and protective coatings for corrosion. In the vast field of CPs, low energy gap polymers (LEGPs) are of great importance due to their electrical and optical properties, with the ultimate aim of making CPs with zero energy gap (E_g) that should exhibit metallic or semimetallic properties

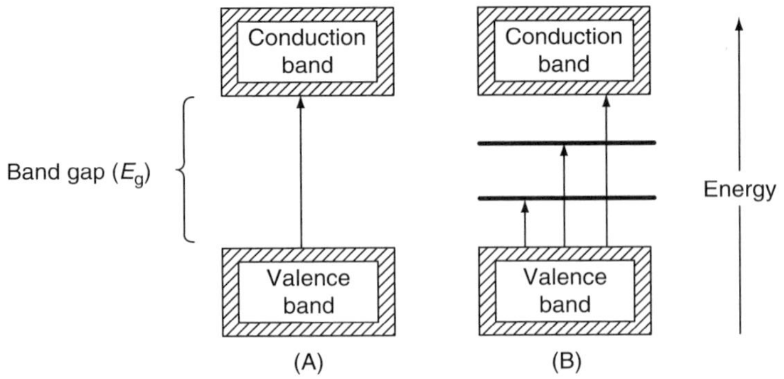

FIGURE 7.1 E_g in (A) neutral and (B) doped conducting polymer. (From Pomerantz, M., *Handbook of Conducting Polymers*, Skotheim, T.A., Elsenbaumer, R.L, Reynolds, J.R., Eds., Marcel Dekker, Inc., New York, 1998. With permission.)

without the need for doping. Generally, conjugated polymers are insulating in their neutral state but when doped by oxidation or reduction they behave as semiconductors. There are some reports on the metallic nature of some of these materials. CPs in their doped semiconducting or metallic forms are referred as conductive polymers. The electro-optical properties of CPs depend on the energy difference between their highest occupied molecular orbital (HOMO) or valence band and lowest unoccupied molecular orbital (LUMO) or conduction band, which is defined as the energy or band gap (E_g) (Figure 7.1). Shirakawa, McDiarmid, and Heeger [1] discovered the conducting properties of polyacetylene (PA) upon iodine doping, and triggered the research toward new materials and applications for CPs to replace inorganic materials to make inexpensive, flexible, and light weight devices. Since PAs have energy gaps ranging from 1.5 to 1.7 eV, LEGPs are defined as materials with energy gaps of 1.7 eV or less where 1 eV is the energy of a photon at the wavelength of 1240 nm. Some researchers have chosen 1.5 eV as an arbitrary cutoff [2]. LEGPs are generally expected to be optically transparent in the visible (vis) region in their conducting state, i.e., no absorption at 800 nm (1.55 eV) or lower, and that may be the other reason to define polymers with E_g's of 1.5 eV or lower as LEGPs. In this chapter, we extend this limit to 1.8 eV so as to include those polymers having absorption maxima (λ_{max}) at 600 nm or higher in their neutral state and show good optical transparency in their conductive state. A perfect example of this is the commercially available and highly popular optically transparent conducting polymer, poly(3,4-ethylenedioxythiophene) (PEDOT). Conductivity of CPs has been reported to increase with a decrease in E_g, provided that there are similar morphologies between the two materials being compared. One explanation is the higher population of charge carriers generated thermally within the conduction band. In general, LEGPs have lower oxidation and reduction potentials resulting in higher stabilities of both the p-doped and n-doped states.

Electrical and optical properties of CPs generally depend on their redox states, and also on the extent of doping. Upon doping, CPs convert to their conducting form and a redshift in the absorption maximum is observed resulting in a different color. This forms the basis of electrochromic devices (Figure 7.2). For example, poly(thiophene)s are yellow–orange in their neutral state and blue in their conducting state. Colors of conjugated polymers are, in general, complementary to the visible wavelengths absorbed.

LEGPs generally show a maximum absorption at around 600 nm or longer wavelength in their neutral state with an onset of absorption (λ_{onset}) in near-infrared (NIR) region. In the doped/conductive state, LEGPs show maximum absorption in the NIR or IR region with little or no absorbance in the visible region due to the redshift in the absorption spectra as a result of a structural ordering of the conjugated polymer backbone, removal of electrons, and a longer conjugation length of the more planarized quinoidal backbone (Figure 7.2). Thus, LEGPs can mainly be used for two different

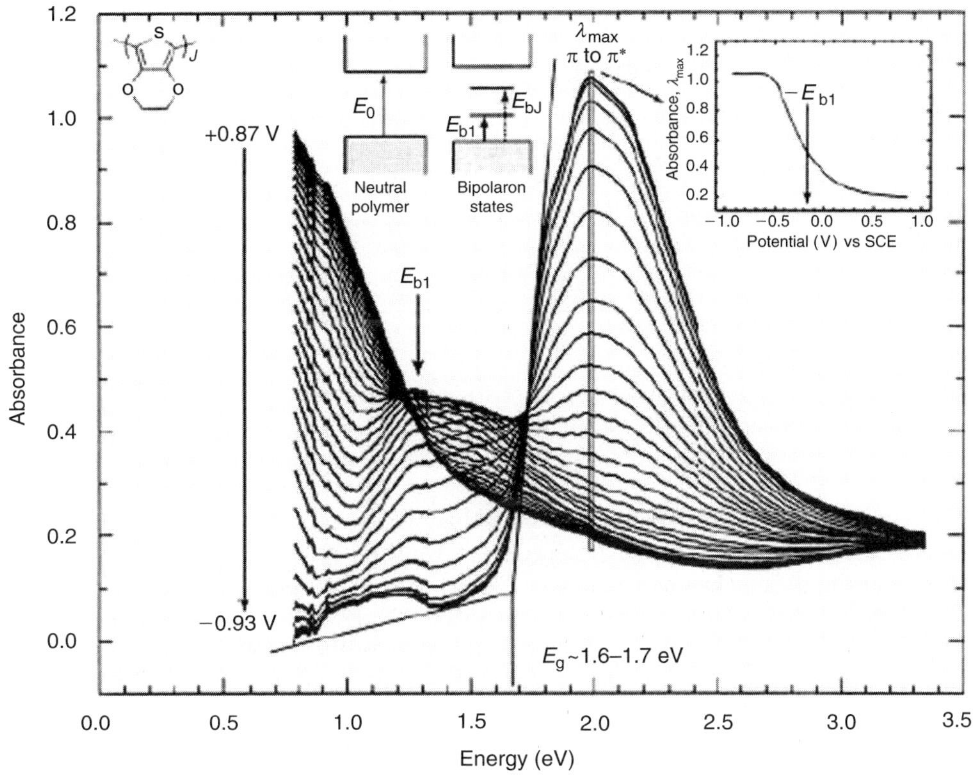

FIGURE 7.2 Spectroelectrochemistry of PEDOT. Film was deposited from propylene carbonate (0.1 M) tetrabutylammonium perchlorate (TBAP), (~300 mM) EDOT and switched in acetonitrile (0.1 M) TBAP. Inset shows absorbance versus potential. Band gap is determined by extrapolating the onset of the π to π* absorbance to the background absorbance. The E_{b1} transition is allowed and is visible at intermediate doping levels. (From Groenendaal, L., Zotti, G., Aubert, P.-H., Waybright, S.M., Reynolds, J.R., *Adv. Mater.*, 15, 855, 2003. With permission.)

types of optoelectronic applications: electrochromic devices involving switching between colored and transparent states on applying a potential (provided large photopic contrasts can be achieved), and as optically transparent conductors involving only the stable, doped state [51].

7.2 ENERGY GAP IN CONJUGATED POLYMERS

The highest energy π orbital and lowest energy π* orbital within organic compounds are designated as the HOMO and LUMO, respectively. The energy gap (E_g), defined as the energy difference between the HOMO and LUMO, decreases with an increase in conjugation. Upon incorporation of additional repeat units into a CP backbone, discrete π and π* orbitals begin to stack upon each other that are very close in energy forming bands of π and π* orbitals. These bands are also referred to as the valence and conduction bands, respectively. The width of the band increases with an increase in the number of repeating units resulting in a lowering of E_g. The width of the band also depends on the extent of π orbital overlap or delocalization of π electrons, and the breadth of the band generally increases with an increase in π orbital overlap. It can be envisioned that a very low E_g polymer can be achieved by having a large number of repeat units with unobstructed π orbital overlap. The E_g of CPs corresponding to the π to π* transition and the formation of bands of molecular orbitals are shown in Figure 7.3 [52].

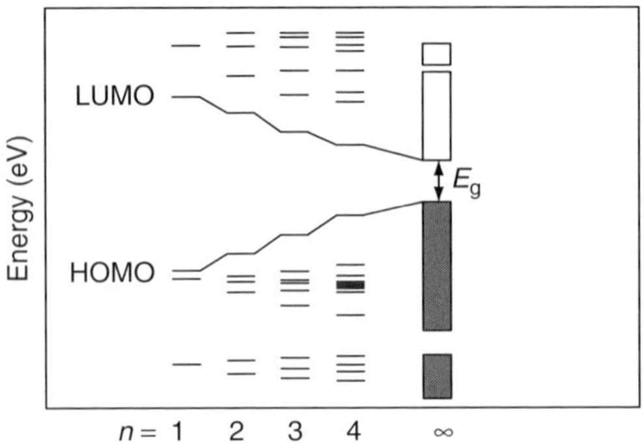

FIGURE 7.3 Energy band formation during polymerization of conjugated monomer. (From Ajayaghosh, A., *Chem. Soc. Rev.*, 32, 181, 2003. With permission.)

The oxidation of CPs and formation of hole charge carriers involve the removal of electrons from the valence band, resulting in the p-doped form, while the reduction process involves the addition of electrons to the conduction band, resulting in the n-doped form. The onset of the oxidation and reduction process is representative of the energies of the HOMO and the LUMO.

There are two common methods for the determination of the energies of the energy gap. One method is electrochemistry, by which the potentials for electron removal from the HOMO and electron injection into the LUMO are determined. The energy difference between these two electron transfer events is the energy gap. The electrochemical method assumes negligible intermolecular interactions [3–5].

$$E_g \sim (E_{ox} - E_{red}) + (S^+ + S^-)\left\{1 - \frac{[1-(1/\varepsilon_2)]}{[1-(1/\varepsilon_1)]}\right\} \tag{7.1}$$

where

E_{ox} and E_{red} are the polymer oxidation and reduction potential onsets, respectively
S^+ and S^- are the solvation energy of the positive or negative ions minus the solvation energy of the neutral molecule, respectively
ε_1 and ε_2 are the dielectric constants of the solution and the solid, respectively

A second common method for determining E_g is through obtaining an electromagnetic spectrum of the neutral conjugated polymer, in which the π to π* transition generally lies within the lower-energy NIR region or higher-energy visible region. Through this technique, the E_g is defined as the onset for the π to π* transition. Where the spectra are not reported, it is useful to report both the onset and the maximum for the absorption. In the case of narrow or sharp onset of absorbance peak, the wavelength at which it deviates from the absorbance baseline (at the longer wavelength side) is generally taken as E_g. While in the case of broad onset of absorption peak, energy gap is determined by extrapolating the onset of the π to π* absorbance to the background absorbance as shown in Figure 7.2.

The E_g for CPs depends on several parameters encompassing both intra- and intermolecular considerations. Intramolecular parameters such as the nature of the substituents and stereochemistry, and intermolecular parameters such as interchain interactions may alter E_g and can be expressed as the following equation:

$$E_g = E^{\delta r} + E^{\theta} + E^{Res} + E^{Sub} + E^{int} \tag{7.2}$$

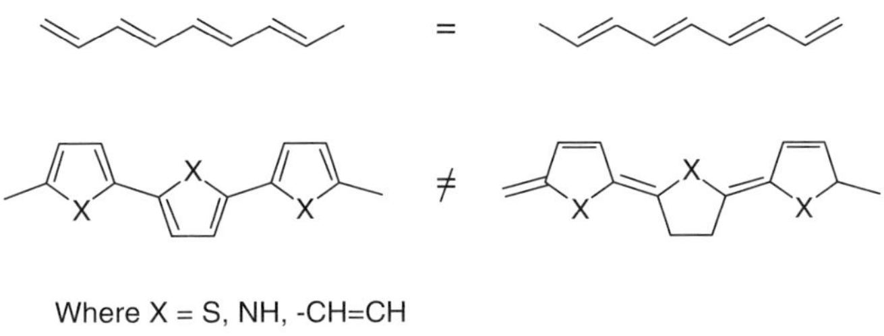

FIGURE 7.4 Aromatic and quinonoidal forms of conjugated polymers.

where $E^{\delta r}$, E^{θ}, E^{Res}, E^{Sub}, and E^{int} represent the energy related to bond length alternation (BLA), coplanarity deviation, resonance, substituent, and interchain interactions, respectively.

$E^{\delta r}$ is the energy corresponding to BLA, which is related to the difference between single and double bond length. BLA is also defined quantitatively as δr, which is a measure of the relative degree of aromatic and quinonoid character in CPs. $\delta r > 0$ or $\delta r < 0$ indicates a quinonoid or aromatic structure, respectively. The quinonoid form generally has a lower E_g than the aromatic form [6]. BLA is absent in poly(acetylene) due to two degenerate ground states, while polyaromatic polymers such as poly(thiophene), poly(pyrrole), and poly(*p*-phenylene) have nondegenerate forms, aromatic and quinonoid, resulting in BLA as shown in Figure 7.4 [7].

E^{θ} represents the energy related to the structure deviation from coplanarity, and may arise from the interannular rotations of aromatic rings through the single bond connecting them. Besides following the $(4n + 2)\pi$ electron rule, coplanarity is a prerequisite for a molecule to be aromatic (which means that there is a delocalization of π electrons). As the conductivity of CPs arise from the delocalization of π electrons, any deviation from planar structure leads to a lower conjugation length and higher E_g. Rotational distortion limits the effective conjugation length in oligothiophene, thereby increasing the E_g by E^{θ} [8–10].

E^{Res} represents the energy related to the resonance energy of the monomer in the case of poly(aromatic)s. The delocalization of π electrons along the chain is in competition with their confinement within the aromatic ring, owing to the resonance [11] with the former being responsible for higher conjugation. Thus, higher resonance stabilization energy of aromatic monomers (E^{Res}) decreases the delocalization of π electrons along the chain, resulting in a higher E_g.

E^{Sub} is the energy related to the inductive and mesomeric effect of the substituents attached directly to the monomer ring. Electron-withdrawing substituents reduce the E_g by decreasing the LUMO level energy whereas electron-donating substituents increase the energy of the HOMO level resulting in E_g reduction (Figure 7.5). As substitution generally involves simple chemistry, this is a very popular method among chemists to modify or tune the E_g.

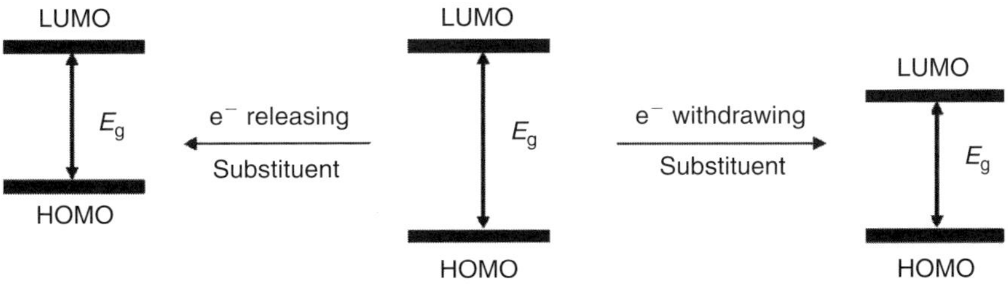

FIGURE 7.5 Effect of electron-donating or electron-withdrawing substituents on E_g of CPs.

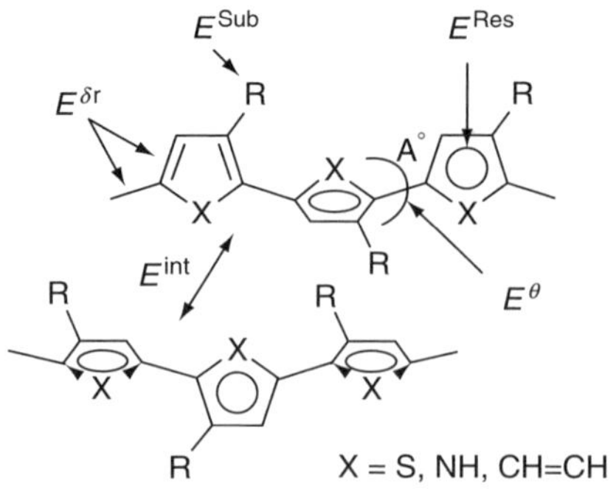

FIGURE 7.6 Factors affecting the E_g of conducting polymer. (From Roncali, J., *Chem. Rev.*, 97, 173, 1997. With permission.)

All the four terms above describe the factors affecting the E_g of a single conducting polymer chain. However, in reality, it is also important to consider the interchain interactions to define the E_g. The last term in Equation 7.2, E^{int}, accounts for the energy related to the interchain interaction in the solid state. Figure 7.6 represents both inter- and intra-chain factors responsible for determining E_g of the conducting polymer [53].

In summary, LEGPs can be achieved by attaching electron-withdrawing or electron-releasing groups to an aromatic ring or by decreasing BLA, coplanarity deviation, and π-electron delocalization within the ring.

7.3 DIFFERENT CLASSES OF LOW ENERGY GAP POLYMERS

Numerous LEGPs have been prepared, and their properties, including E_g, have been studied both experimentally and theoretically. LEGPs can be classified on the basis of several parameters, but in this section, they are categorized on the basis of the approach or technique used to control the E_g from a molecular basis. In some cases only one factor predominates in controlling the E_g, while in other cases two or more factors can be involved. The optical and electrical properties of different types of LEGPs and their structures are discussed in this section.

7.3.1 Fused Aromatics

Wudl et al. [12] reported the first LEGP, poly(isothianaphthene) (PITN) or poly(benzo[c]thiophene), synthesized from isothianaphthene (ITN) having the fused benzene and thiophene rings depicted in Figure 7.7. PITN (**1**) was reported to have an E_g of 1.0–1.2 eV while unsubstituted and alkyl-substituted poly(thiophene)s have an E_g of ~2.1 eV. The low E_g of PITN can be attributed to the stability of its quinonoid form (**2**) by the aromatic benzene ring. It has been reported that there is a competition between the two rings to maintain their aromaticity, and benzene has a higher probability to maintain aromatic character at the expense of the thiophene ring, since benzene has a higher aromatic resonance energy (1.56 eV) compared to that of thiophene (1.26 eV). Thus, structure **2** is more stable

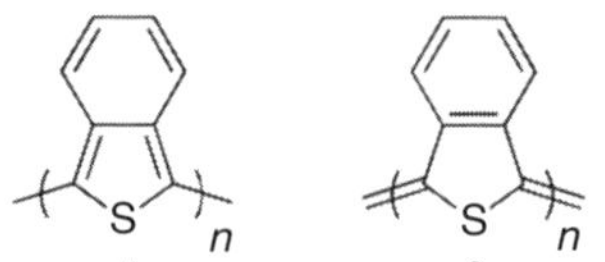

1 **2**

FIGURE 7.7 Aromatic (**1**) and quinonoidal (**2**) form of PITN.

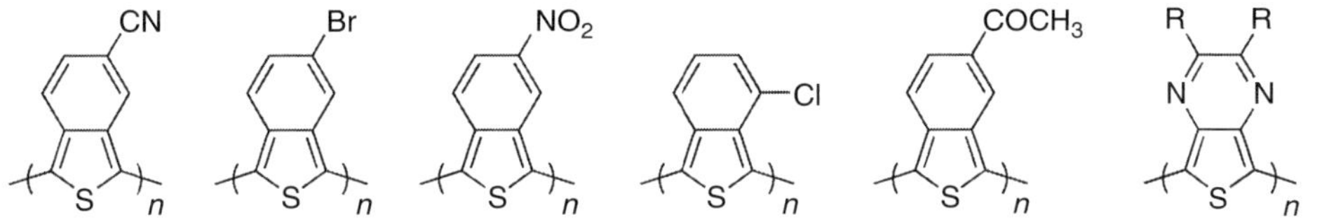

3 [65%] **4** [60%] **5** [55%] **6** [60%] **7** [50%] **8a**: R = H **8b**: R = C_6H_1

FIGURE 7.8 Substituted PITN and poly(thieno[3,4-*b*]pyrazine).

than **1**, leading to the stability of the quinonoidal form. The conductivity for the chemically or electrochemically prepared p-doped PITN was reported to be 50 S/cm [13].

Substituted ITN was expected to exhibit a low E_g due to either a decrease in the LUMO energy level or an increase in the HOMO energy level; however, the E_g obtained from optical data lies in the range of 1.0–1.3 eV, which is comparable to the unsubstituted PITN. Although E_g was not found to decrease by substitution, optical transparency improved. Substituted PITN showed an optical transparency in the range of 50%–65% as compared to 40% obtained in the case of PITN. The percentage transparency values, along with the structures, are shown in Figure 7.8 [14].

Another LEGP based on a fused ring system is poly(thieno[3,4-*b*]pyrazine) (**8**) and its derivatives, in which carbon atoms directly attached to the thiophene ring of ITN are substituted for nitrogen atoms. Polymer **8b** showed an optical absorption maximum at 875 nm in solution and at 915 nm in solution cast films with an E_g of 0.95 eV (1178 nm) (Figure 7.9) [15]. The lower E_g of **8** as compared to PITN may be due to less steric interactions between neighboring monomer units, as there is no hydrogen atom attached to the nitrogen.

Another approach to lower the E_g involves monomers containing two units, one of which prefers the aromatic character and the other prefers the quinoidal structure. The starting monomer with both units has higher conjugation, and is expected to produce polymers with higher conjugation length and lower E_g as compared with the polymers obtained from monomers having only aromatic unit. The other advantage is the lower oxidation potential of monomers that have more conjugation, which helps to deter side reactions of cationic radicals and/or overoxidation of the resultant polymer. This approach is expected to produce CPs with fewer α–β linkages as the starting monomer contains exclusive α–α linkages.

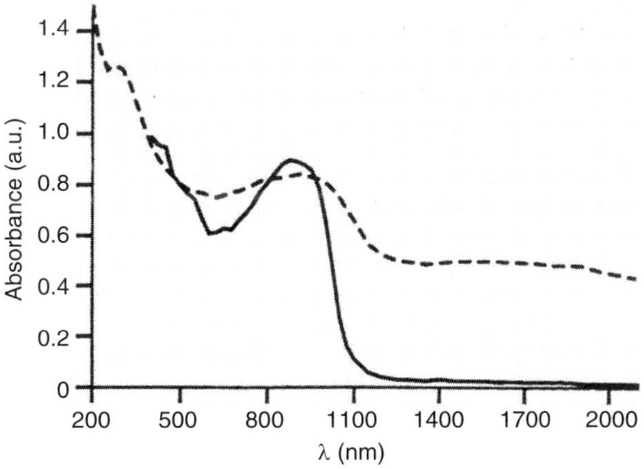

FIGURE 7.9 Optical spectrum of **8b**. (From Pomerantz, M., Chaloner-Gill, B., Harding, L.O., Tseng, J.J., Pomerantz, W.J., *J. Chem. Soc., Chem. Commun.*, 22, 1672, 1992. With permission.)

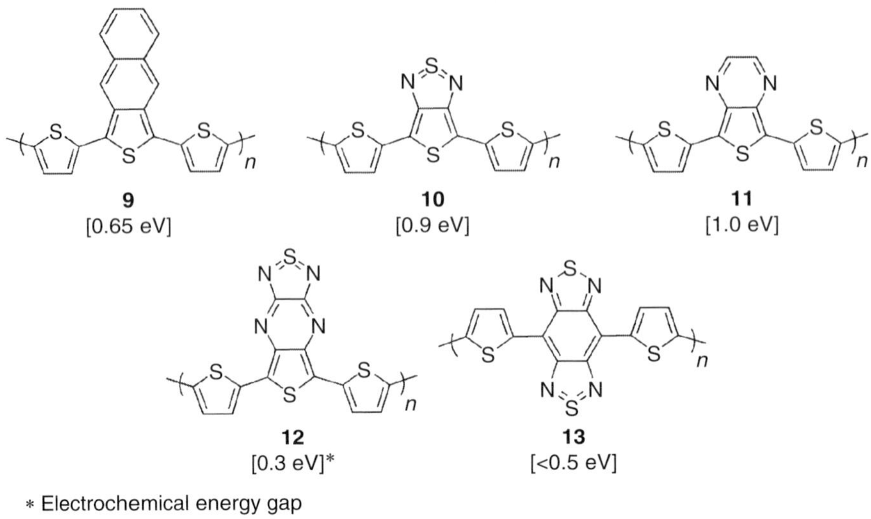

FIGURE 7.10 Some examples of low E_g polymers obtained from aromatic-quinoidal systems.

Some examples following this approach, with their E_g values, are shown in Figure 7.10. Polymers **9** [16,17] and **10** [18] have reported E_g values lower than 1.0 eV while polymer **11** [19] had E_g of 1.0 eV. The precursors for **13** showed an absorption maximum at 700 nm and reported to produce polymer **13** with an E_g of 0.5 eV [20]. Polymers **12** [21] and **13** were reported to be among the lowest E_g polymers to date.

Replacing the central thiophene ring in structures **10** and **11** with a benzene ring yielded the higher E_g conjugated polymers **14** and **15** (Figure 7.11). The higher energy gap may be attributed to more contribution to the orbital coefficient by the electron-donating sulfur atom, since the coupling positions are part of a 5-membered thiophene ring [22].

E_g was found to increase with the replacement of one or both of the fused thiadiazole rings of **13** with a pyrazine ring as shown for **16** and **17** (Figure 7.12) [20]. This could be attributed to the short intermolecular S–N contacts, which contribute in controlling the structural order in the polymer [23].

Thieno[3,4-*b*]thiopene (T34bT) and thieno[3,4-*b*]furan (T34bF) have also been explored as LEGPs and their p-doped polymers have been looked at as optically transparent conductors. The beauty in such systems is the simplicity of the monomer; yet, the monomer has three positions alpha to the heteroatom that are available for oxidative coupling. It is this aspect that sets these systems apart from ITN, in that the polymers from T34bT and T34bF could be highly branched or cross-linked (Scheme 7.1). Thus, potentially poly(thieno[3,4-*b*]thiophene) (**20**) [24,25] and poly(thieno[3,4-*b*]furan) (**21**) [26] could exhibit cross conjugation via coupling through all of its α positions in a hypothetical branched or networked structure as shown in Scheme 7.1. As α positions in the polymer are predicted to be far separated from each other, steric interactions will not be significant and repeat

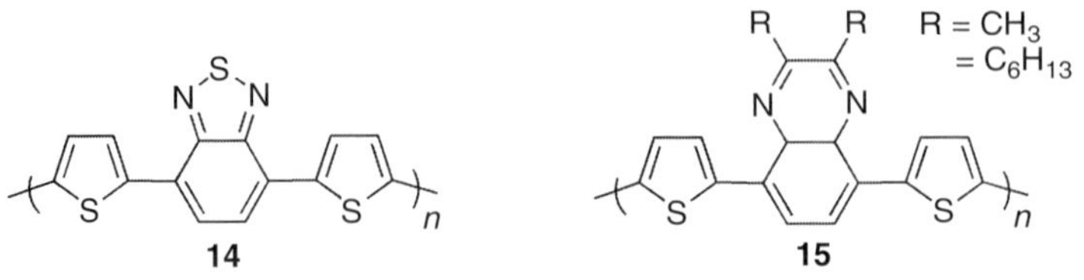

FIGURE 7.11 LEGPs of fused ring polymer with phenyl ring as a central unit.

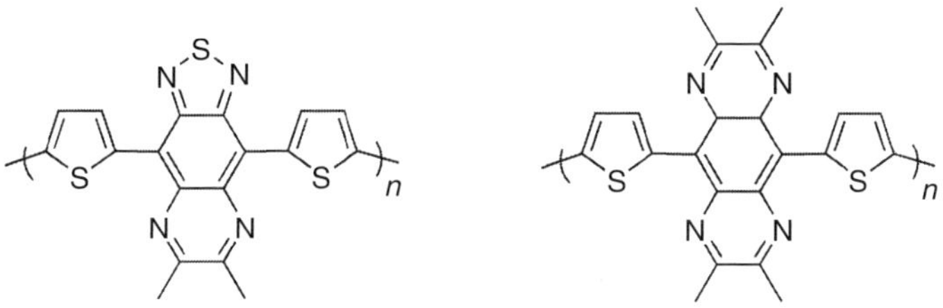

16 [0.7 eV]

17 [0.9 eV]

FIGURE 7.12 LEGPs with different central quinoidal units.

units could exist in coplanarity. The neutral CP **20** has a λ_{max} at 846 nm with an E_g of 0.85 eV in the neutral state while a redshift of λ_{max} in NIR region was observed in the oxidized/conductive state as shown in Figure 7.13. High optical transparency in the visible region in conductive state makes this polymer a promising candidate for optically transparent conductor applications. Polymer **21** showed an energy gap of 1.03 eV with a λ_{max} at 720 nm and also exhibits high optical transparency in the oxidized/conductive state.

7.3.2 Electron-Donating Substituents

Polymers synthesized from aromatic monomers having electron-donating groups are an important class of LEGPs, and their energy gap can be controlled by attaching different groups of varying electron-donating strength. The most commonly employed electron-donating groups are alkoxy and amino. Almost all the classes of LEGPs have been studied for electron-releasing substituent effects but this section mainly focusses on poly(3,4-alkylenedioxythiophene) (**22**) and related systems (Figure 7.14). In 3,4-alkylenedioxythiophene, 3 and 4 positions of thiophene ring are substituted with electron-donating alkoxy connecting groups, which reduce the E_g of polymer by increasing the HOMO energy level. The polymers obtained from parent thiophene showed a high energy gap of 2.1 eV while poly(3,4-alkylenedioxythiophene)s have an energy gap of 1.5–1.7 eV. Another advantage of using these monomers is that both the β positions to sulfur are substituted and hence polymerization

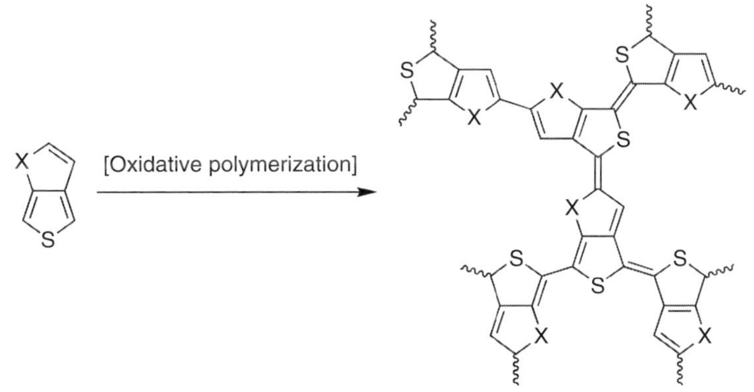

X = S **18** **20**
X = O **19** **21**

SCHEME 7.1 Synthesis of cross conjugated polymer network **20** and **21** from monomers **18** and **19**.

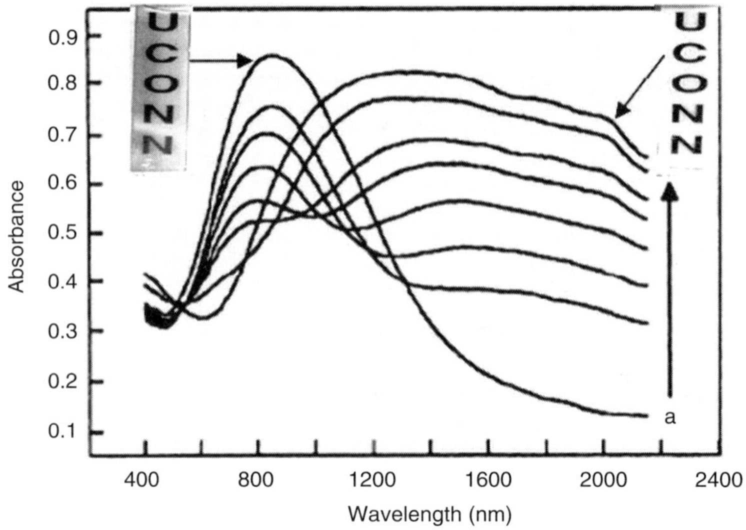

FIGURE 7.13 UV–vis–NIR spectrum of a 0.1 μm thick poly(thieno[3,4-*b*]thiophene) (**20**) film on indium tin oxide (ITO) glass at different potentials. The film was first electrochemically reduced at −0.8 V, dipped into a 0.1 M TBAP/acetonitrile containing 0.2 vol% hydrazine for full reduction, and then placed into a 0.1 M TBAP/acetonitrile, and the UV–vis–NIR spectrum was taken (a). Poly(T34bT) was then sequentially oxidized to (b) −0.4, (c) −0.3, (d) −0.2, (e) −0.1, (f) 0.0, (g) 0.15, and (h) 0.4 V vs. Ag/Ag+ reference electrode (0.47 V vs. NHE). Two inset pictures with "UCONN" are 0.8 μm thick poly(T34bT) (**20**) film coated on an ITO-coated glass slide. (From Sotzing, G.A. and Lee, K., *Macromolecules*, 35, 7281, 2002. With permission.)

proceeds exclusively through the α positions to produce linear, defect free, polymers. The same advantages can be expected from substitute derivatives of 3,4-alkylenedioxythiophene (**24–26**).

Poly(3,4-ethylenedioxythiophene) (PEDOT) (**23**) has a maximum absorption at 580–610 nm and is transmissive sky blue in the oxidized p-doped form. Optical transmissitivity of PEDOT is reported to be over 75% throughout the visible region, which is comparable to indium tin oxide (ITO), the most commonly used optical transparent electrode (Figure 7.15) [27]. Polymers obtained from dibenzyl substituted ProDOT (**29**) [28] have been reported to have an optical contrast of 89%, one of the highest contrasts reported for conjugated polymers.

A brief mention about polymers prepared from 3,4-alkyenedioxypyrrole is required, even though they have higher band gaps compared to 3,4-alkyenedioxythiophene, resulting from the shift of both HOMO and LUMO levels. These polymers are of special interest as they showed a high optical contrast between their neutral and oxidized states, and are highly transmissive in their oxidized/conductive state. Poly(propylenedioxypyrrole) (PProDOP) (**30**) shows a high optical contrast at 534 nm, very close to the wavelength (550 nm) at which human eyes are most sensitive, and it also shows an optical transmittance of 70% throughout the visible spectrum (Figure 7.16) [29]. This system demonstrates that a conjugated polymer need not be of low band gap to generate a p-doped conductive polymer that is highly optically transmissive.

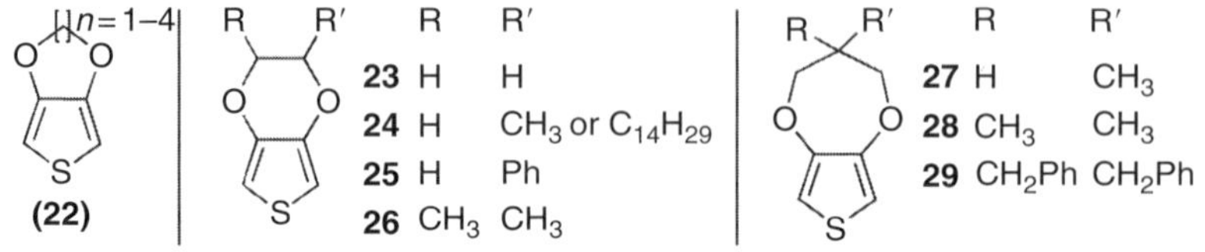

FIGURE 7.14 Some examples of substituted and unsubstituted 3,4-alkylenedioxythiophene.

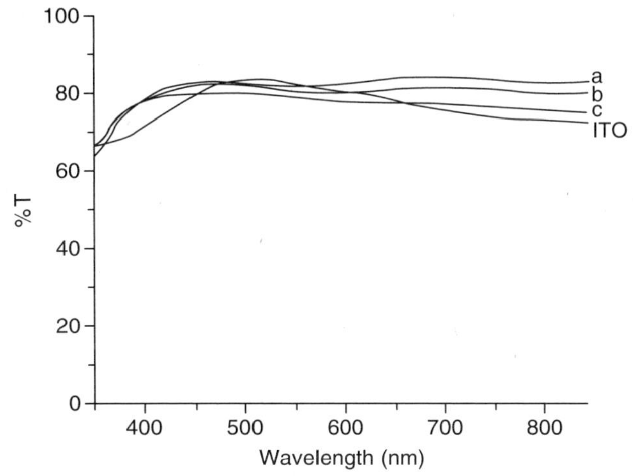

FIGURE 7.15 Percent transmittance (%T) of the PEDOT–PSS-coated transparent film electrodes in the visible region with (a) one layer, (b) two layers, and (c) three layers. Electrodes with three layers yield a surface resistivity of 600 V/square with an average %T value of 77% through the visible spectrum. %T spectrum of an indium tin oxide (ITO) electrode (bold line) is also shown for comparison. (From Argun, A.A., Cirpan, A., and Reynolds, J.R., *Adv. Mater.*, 15, 1338, 2003. With permission.)

7.3.3 Electron-Withdrawing Substituents

Another approach to reduce the E_g is to lower the LUMO energy level through the use of electron-withdrawing groups. The most commonly employed electron-withdrawing groups are cyano, nitro, and carbonyl groups. The effect of electron-withdrawing groups on the energy gap and other properties of conducting polymers have also been studied for almost all classes of LEGPs.

Cyclopenta[2,1-*b*:3′,4′-*b*′]dithiophene-containing polymers having different electron-withdrawing groups have been prepared, and the E_g of these polymers were found to decrease with an increase in strength of the electron-withdrawing substituent (Figure 7.17). In this class of LEGPs in many cases, besides the electron-withdrawing effect, a resonance form of the repeating unit contributes to the lowering of the E_g by decreasing the aromaticity of the monomer, e.g., **31A** and **31B** [30]. Poly(**32**) [31] exhibited a lower E_g than poly(**31**), which may be attributed to the stronger electron-withdrawing effect of dicyanomethylene group as compared to a carbonyl group. The lower E_g of poly(**32**) and poly(**34**) [32] when compared to poly(**31**) implies that the electron-withdrawing groups at the bridged sp^2 carbons significantly reduce the E_g of the polymer. Poly(**33**) showed a considerably smaller potential difference between oxidation and reduction than that of poly(**31**), possibly due to the minimal influence of the dioxolane on the HOMO energy level relative to the effect of ketone or cyano groups [33]. The optical E_g of poly(**33**) was reported to be 1.10–1.20 eV, similar to that of poly(**31**), while poly(bithiophene) (PBT) without any electron-withdrawing group showed an E_g of ~2.1 eV (Figure 7.18). Lower band gap of poly(**33**) can also be explained on the basis of the enhanced planarity and possible decrease of BLA caused by the rigidification of the π-conjugated system.

7.3.4 Push–Pull Donor/Acceptors

Low energy gap polymers having both donors and acceptors (DA) in direct conjugation with each other have been termed push–pull systems. DA systems are of interest because these utilize the benefit of both electron acceptor and electron donor groups for reducing the E_g by lowering LUMO energy level and by raising HOMO energy level (Figure 7.5). The resultant E_g may be just the summation of two effects or a synergistic effect of both groups. The significantly strong communication between

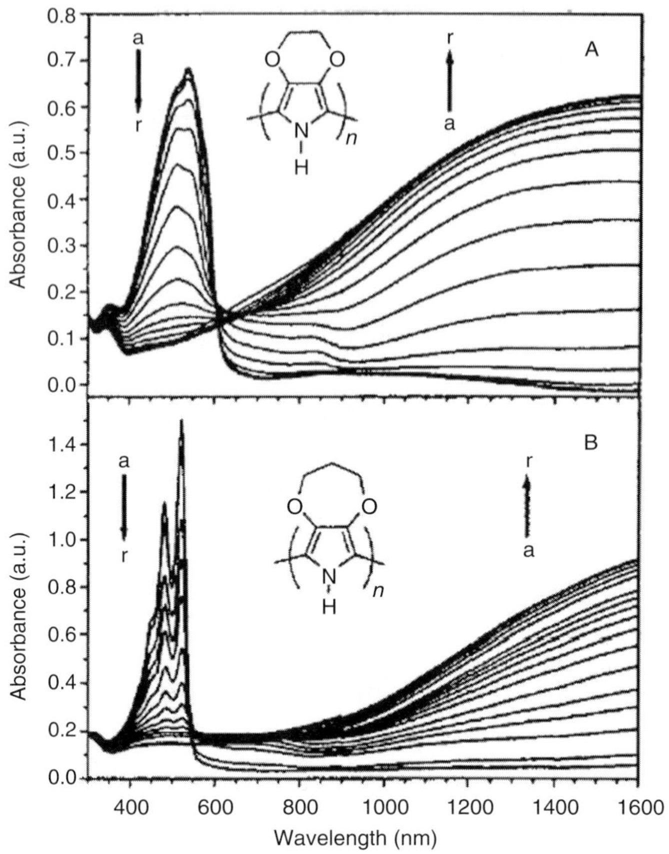

FIGURE 7.16 (A) Spectroelectrochemistry of poly(ethylenedioxypyrrole) (PEDOP) in 0.1 M $LiClO_4$/propylene carbonate at applied potentials of (a) −1.33 V, (b) −1.23 V, (c) −1.13 V, (d) −1.03 V, (e) −0.98 V, (f) −0.93 V, (g) −0.88 V, (h) −0.83 V, (i) −0.78 V, (j) −0.73 V, (k) −0.68 V, (l) −0.63 V, (m) −0.53 V, (n) −0.43 V, (o) −0.33 V, (p) −0.23 V, (q) −0.13 V, and (r) +0.07V vs. Fc/Fc$^+$. (B) Spectroelectrochemistry of poly(propylenedioxypyr role) (PProDOP) (**30**) in 0.1 M $LiClO_4$/PC at applied potentials of (a) −1.13 V, (b) −1.03 V, (c) −0.93 V, (d) −0.88 V, (e) −0.83 V, (f) −0.73 V, (g) −0.62 V, (h) −0.58 V, (i) −0.53 V, (j) −0.48 V, (k) −0.43 V, (l) −0.38 V, (m) −0.33 V, (n) −0.23 V, (o) −0.13 V, (p) −0.03 V, (q) +0.07 V, and (r) +0.17V vs. Fc/Fc$^+$. (From Schottland, P., Zong, K., Gaupp, C.L., Thompson, B.C., Thomas, C.A., Giurgiu, I., Hickman, R., Abboud, K.A., and Reynolds, J.R., *Macromolecules*, 33, 7051, 2000. With permission.)

donor and acceptor moieties increases the double bond character or the quinoidal structure of the conjugated polymer, resulting in a low E_g. To prepare a very low E_g polymer, the HOMO level of the donor group and LUMO level of the acceptor group should be close in energy, as shown in Figure 7.19. During polymerization, the HOMO and LUMO energy levels of the monomer will be converted into the valence and conduction bands, respectively as shown in Figure 7.3, resulting in the further reduction of E_g. However, the deviation from coplanarity due to steric interactions between neighboring aromatic units or the diminishing size of the atomic orbitals (AOs) at the coupling position may increase the E_g. Thus, it is imperative that DA be chosen so as to reduce such steric interactions. The most popular push–pull LEGPs consist of alkoxy and cyano substituents [52].

For DA systems, the most commonly used electron-withdrawing groups are nitro, cyano, pyrazine, and the derivatives of these groups. The commonly used electron-donating groups are amino, thiophene, furan, pyrrole, selenophene, EDOT, and their derivatives. Aromatics such as furan, thiophene, and pyrrole have substantial electron density, themselves, and can also be considered as donors. Some

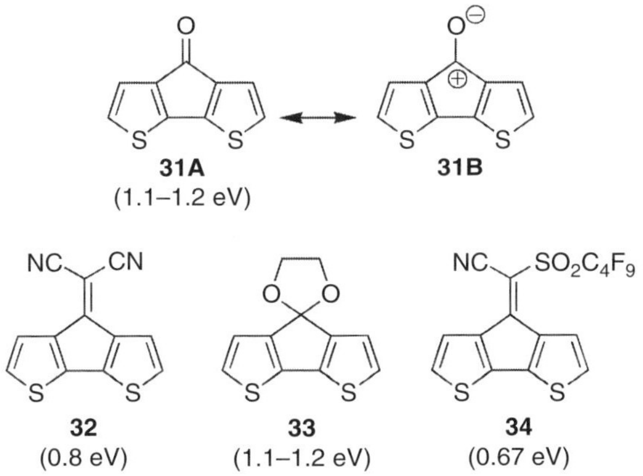

FIGURE 7.17 Electron-withdrawing substituents attached to cyclopenta[2,1-*b*:3′,4′-*b*′]dithiophene system (**31–34**). (From Ferraris, J.P. and Lambert, T.M., *J. Chem. Soc., Chem. Commun.*, 1268, 1991; Ferraris, J.P., Henderson, C., Torres, D., and Meaker, D., *Symth. Met.*, 72, 147, 1995; Brisset H., Thobie-Gautier, C., Gorgues, A., Jubault, M., and Roncali, J., *J. Chem. Soc., Chem. Commun.*, 1305, 1994; Akoudad, S. and Roncali, J., *Chem. Commun.*, 2081, 1998.)

examples of LEGPs obtained using DA systems are shown in Figure 7.20. Polymer **37** showed an E_g of 0.36 eV with a maximum absorption at 0.86 eV (1430 nm) as shown in Figure 7.21 [34].

Polymer **35** synthesized by Zang and Tour [35] showed an E_g of 1.1 and 1.4 eV in solid state and solution, respectively. The lower E_g in the solid state is a result of the rigidity (coplanarity) in the polymer backbone due to the contribution of mesomeric structure, **35b** (Figure 7.22).

All the above examples employed high E_g monomers such as thiophene, pyrrole, and their derivatives to prepare LEGPs using the DA approach. Polymers with even lower E_g are expected if lower

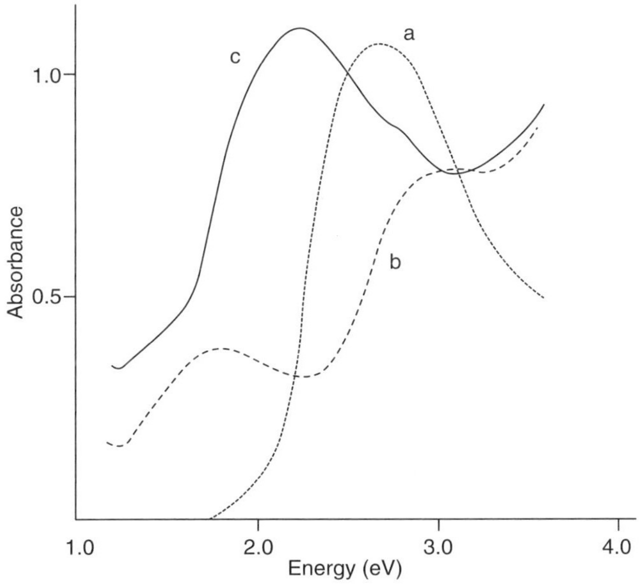

FIGURE 7.18 Electronic absorption spectra of undoped polymers on ITO: (a) dotted **PBT**; (b) poly(**31**); (c) poly(**33**). (From Brisset, H., Thobie-Gautier, C., Gorgues, A., Jubault, M., Roncali, J., *J. Chem. Soc., Chem. Commun.*, 1305, 1994. With permission.)

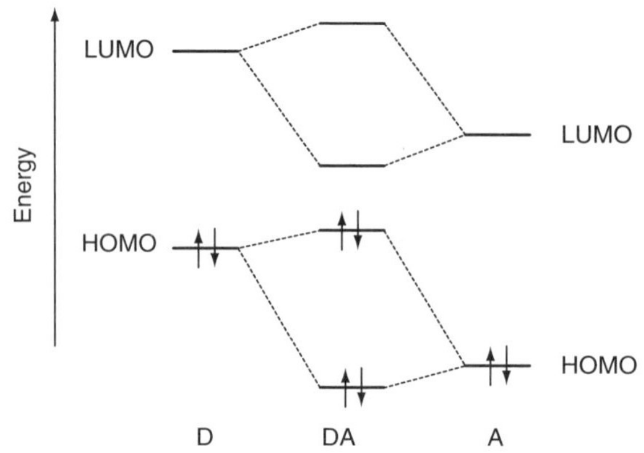

FIGURE 7.19 Molecular orbital interaction in donor (D) and acceptor (A) moieties leading to a DA monomer with an unusually low HOMO–LUMO energy separation. (From Ajayaghosh, A., *Chem. Soc. Rev.*, 32, 181, 2003. With permission.)

E_g monomers are used, such as organic dyes, which are known to absorb in visible to NIR region. Squaraine dyes are known to be the best for this purpose, and are obtained by condensation reaction between electron rich aromatic and heterocyclic molecules such as pyrrole, azulenes, phenols, and *N*,*N*-dialkylanilines with 3,4-dihydroxy-3-cyclobutene-1,2-dione (squaric acid) [36]. Polysquaraines (**46**) and polycroconaines (**47**) of Figure 7.23 were found to have E_g's in the range of 1.0–1.5 eV, with **47** having the lower E_g due to the stronger donor–acceptor interactions.

Incorporation of electron-donating groups into the structure resulted in polymer **48** with an E_g between 0.7–1.1 eV as shown in Figure 7.24 [37]. **48g** showed a maximum absorption in NIR region

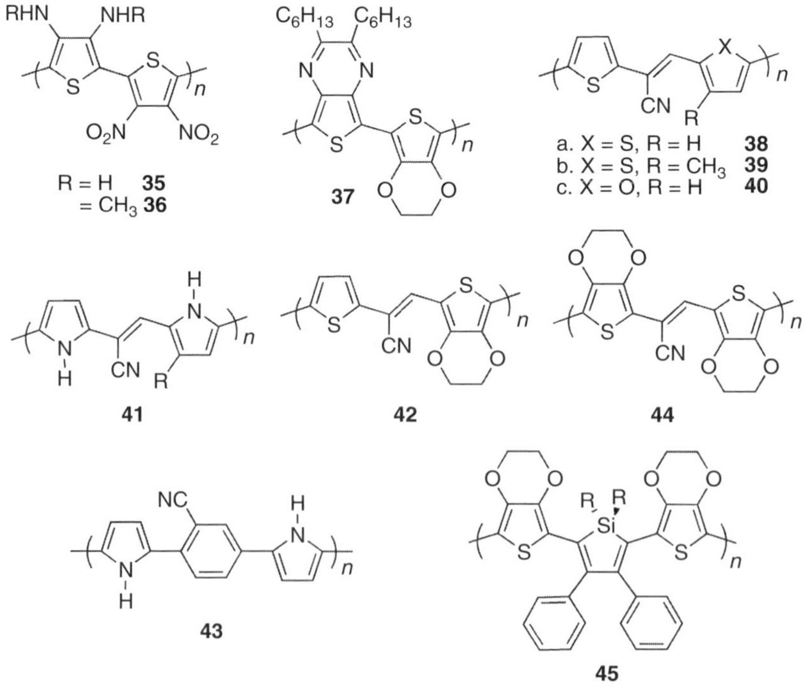

FIGURE 7.20 Examples of LEGPs using the push–pull strategy.

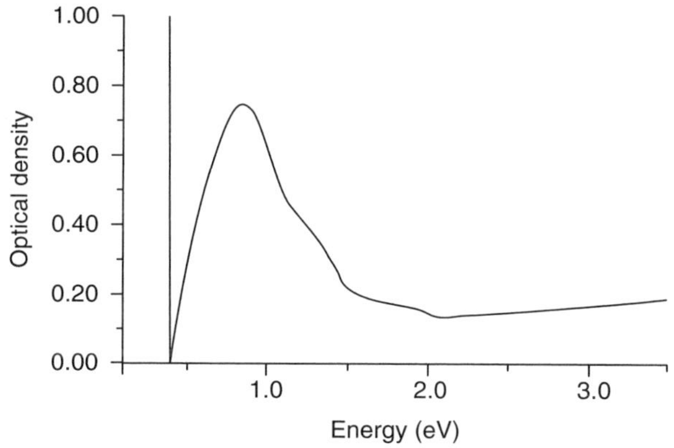

FIGURE 7.21 Optical spectrum of neutral poly(**37**) on ITO (the straight line at 0.40 eV is due to ITO absorption). (From Akoudad, S. and Roncali, J., *Chem. Commun.*, 2081, 1998. With permission.)

at 868 nm with a shoulder at 975 nm in solution while the λ_{max} was observed at 881 nm with a shoulder at 1021 nm in the solid state as shown in Figure 7.25. The E_g for **48g** was found to be 1.09 eV (1140 nm) and 1.02 eV (1215 nm) in solution and solid state, respectively. The electrical conductivity of these polymers was reported to be in the range of 10^{-4} to 10^{-7} S/cm.

7.3.5 Vinylenes and Methines

Another interesting class of LEGPs is poly(arylenevinylenes) (**49**). Poly(thienylenevinylene)s (PTV) (**50–53**), as shown in Figure 7.26, are of special interest as they are reported to have low E_g. The vinylene unit serves the purpose of reducing steric interactions between neighboring thiophene repeat units, thereby increasing the effective conjugation length and thereby lowering the E_g. Another reason for its low E_g could be the prevention or limitation of rotational disorder due to the presence of olefinic linkages with defined configurations. For example, poly(3,4-dibutoxythienylenevinylene) (**52**) is reported to have λ_{max} at 700 nm with an E_g of 1.20 eV, as shown in Figure 7.27 [38]. The electron-donating butoxy groups further aided in reducing the E_g.

Two EDOT units separated by a vinyl group (**53**) is an interesting model compound for poly(arylenevinylenes), and electropolymerization of **53** resulted in a polymer having an E_g of 1.4 eV for the same reasons [39].

Poly(heteroarylenemethines) (**54**) are polymers having both aromatic and quinoidal moieties in different proportions and are expected to have low E_g's since the quinoidal forms of PT possess lower E_g's when compared with their aromatic forms [40,41]. Polymer (**56**), with the incorporation of a bithiophene ring ($y = 1$), showed an absorption maximum at 580–650 nm with an E_g in the range of 1.14–1.45 eV depending on the substituents of the phenyl ring (Figure 7.28) [42]. Electron-donating groups at the para position were reported to reduce the E_g, while the electron-withdrawing substituents

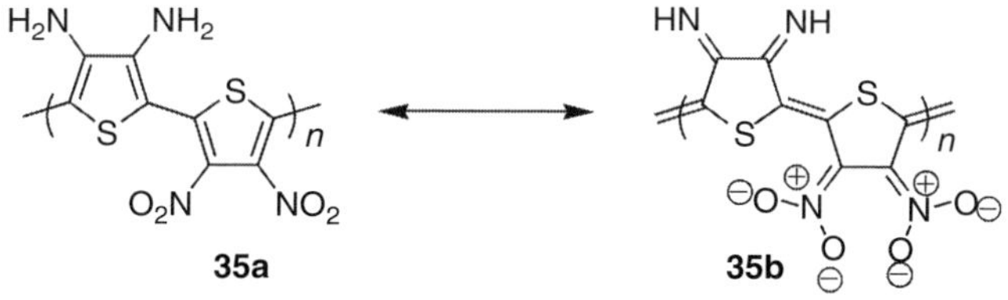

FIGURE 7.22 Resonance form of polymer **35**. (From Zhang, Q.T. and Tour, J.M., *J. Am. Chem. Soc.*, 120, 5355, 1998.)

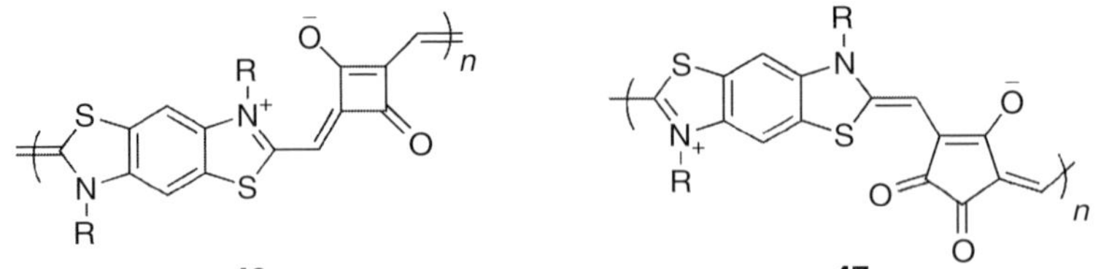

46 47

FIGURE 7.23 Structure of polysquaraines (**46**) and polycroconaines (**47**). (From Sirringhaus, H., Kawase, T., Friend, R.H., Shimoda, T., Inbasekaran, M., Wu. W., and Woo, E.P., *Science*, 290, 2123, 2000. With permission; Lee, B. Seshadri, V., and Sotzing, G.A., *Langmuir*, 21, 10797, 2005.)

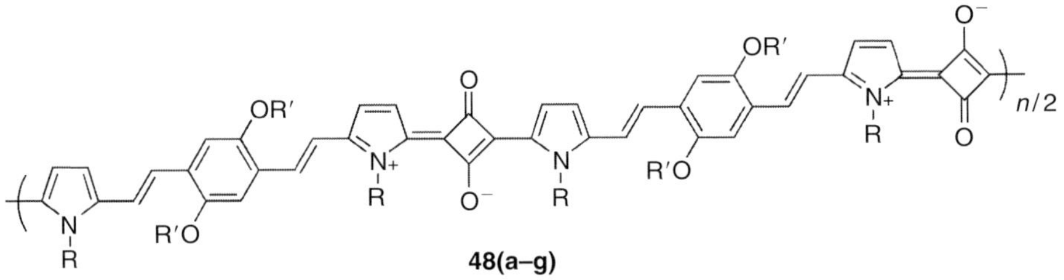

48(a–g)

a, R = CH_3	R′ = C_4H_9	**d**, R = $C_{12}H_{25}$	R′ = CH_3
b, R = CH_3	R′ = C_8H_{17}	**e**, R = $C_{12}H_{25}$	R′ = C_4H_9
c, R = CH_3	R′ = $C_{12}H_{25}$	**f**, R = $C_{12}H_{25}$	R′ = C_8H_{17}
		g, R = $C_{12}H_{25}$	R′ = $C_{12}H_{25}$

FIGURE 7.24 LEGPs based on polysquaraines dyes, [**48(a–g)**]. (From Meng, H., Perepichka, D.F., and Wudl, F. *Chem. Int. Ed.*, 42, 658, 2003. With permission; Meng, H., Perepichka, D.F., Bendikov, M., Wudl, F., Pan, G.Z., Yu, W., Dong, W., and Brown, S., *J. Am. Chem. Soc.*, 125, 15151, 2003.)

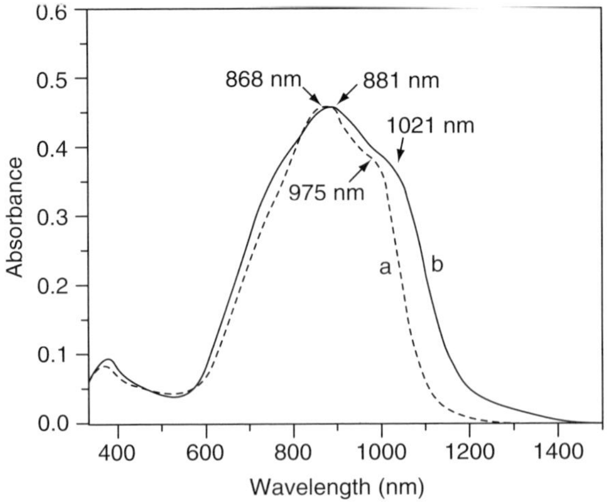

FIGURE 7.25 Comparison of the solution (a) and solid state (b) absorption spectra of polysquaraine **48g**. (From Eldo, J. and Ajayghosh, A., *Chem. Mater.*, 14, 410, 2002. With permission.)

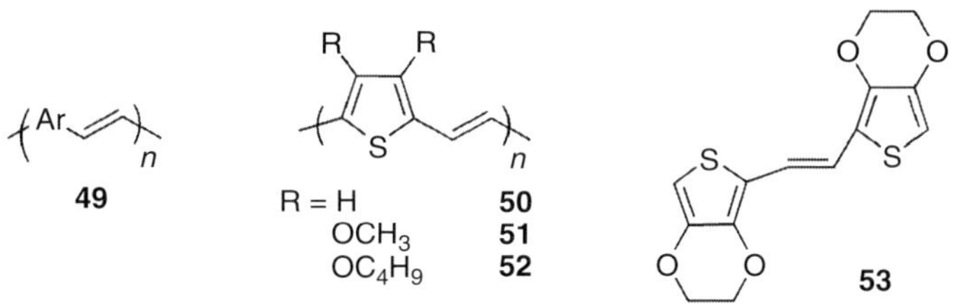

FIGURE 7.26 Examples of poly(arylenevinylenes) and dithienylethylenes.

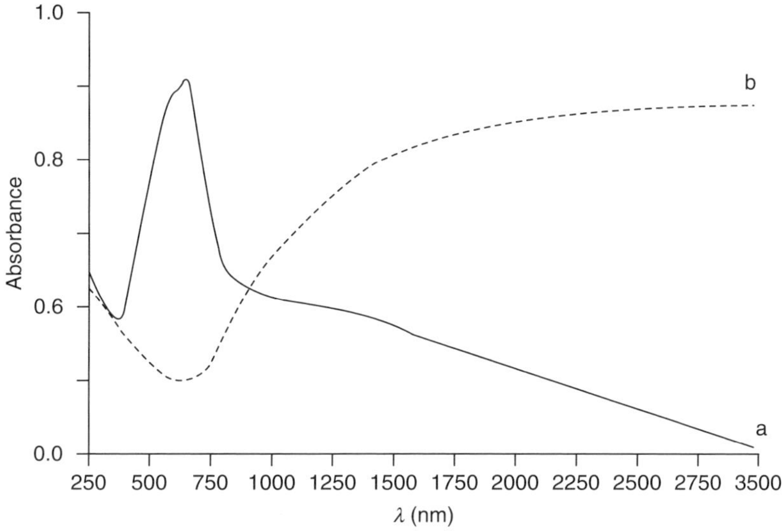

FIGURE 7.27 Electronic absorption spectra of poly(3,4-dibutoxythienylenevinylene) (**52**) on ITO: (a) neutral and (b) doped. (From Cheng, H. and Elsenbaumer, R.L., *J. Chem. Soc., Chem. Commun.*, 1451, 1995. With permission.)

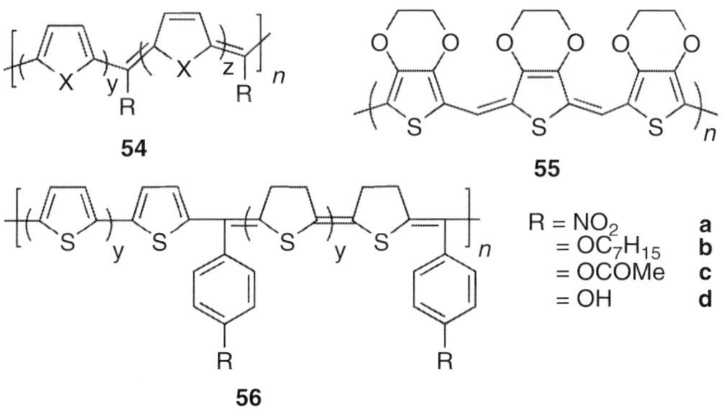

FIGURE 7.28 Poly(heteroarylenemethine)-based LEGPs.

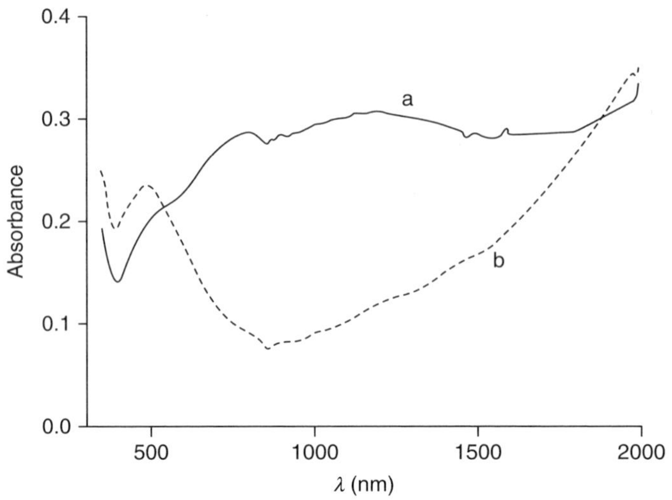

FIGURE 7.29 Vis–NIR spectrum of (a) doped and (b) neutral forms of nonclassical conjugated polymer based on EDOT. (From Benincori, T., Rizzo, S., Sannicolò, F., Schiavon, G. Zecchin, S., and Zotti, G., *Macromolecules*, 36, 5114, 2003. With permission.)

increased the E_g of the polymer. The E_g of polymer (**55**) was found to be 1.0 eV electrochemically but the neutral polymer showed an absorption maximum at 500 nm with an onset of absorption at ~800–900 nm (Figure 7.29) [43].

7.4 PROCESSABILITY OF LOW ENERGY GAP POLYMERS

Conjugated polymers generally have low solution viscosities due to their rigid backbone, and, thus, it is more difficult to process them compared to conventional polymers, such as polystyrene, polymethylmethacrylate, polyvinyl acetate, etc. For many CPs, the best mode of generation is to deposit them onto the conducting substrate via electrochemical polymerization from a monomer solution. However, this is not a cost-effective method because yields are low and disposal of the electrolyte bath solution is required after only a few depositions. The most common approach to increase the processability of conjugated polymers is the substitution of long alkyl group on the monomer unit to prepare a polymer that is soluble in organic solvents, and thus can be processed from the polymer solution via spin, spray, and dip coating, and ink jetting. Regioregular poly(3-alkylthiophene)s (P3ATs) are the most promising conjugated polymers under this class.

Another approach involves the polymerization of monomer using a chemical oxidant in the presence of water-soluble polymeric electrolyte, which will act as a template. Such a polymeric template helps to stabilize the dispersion of CPs in solution. The most successful example of this method is the PEDOT–poly(styrene sulfonate) (PEDOT–PSS) (**57**) (Figure 7.30) dispersion, which is a commercially successful conducting polymer. PEDOT–PSS dispersions consist mainly of 1.3 wt% polymer (with PEDOT to PSS ratio of 1:2.5 by weight) in water, which is produced by adding sodium persulfate to a solution containing EDOT and PSS [44]. A thin transparent film of PEDOT–PSS can be obtained by spray, dip, or spin coating of the dispersion, and has been demonstrated to be used in many applications, like OLEDs, photovoltaics, and field-effect transistors, to name a few. Different grades of PEDOT–PSS dispersions are commercially available.

Conductivity of PEDOT–PSS films have a broad range of conductivities, from 10^{-5} up to 900 S/cm, depending on solvents, additives, cure temperatures, and time. The conductivity of the polymer film was increased from 0.05–0.075 to 12–19 S/cm by adding a saturated polyol, like sorbitol, to the dispersion, and then casting a film from the resulting solution [45].

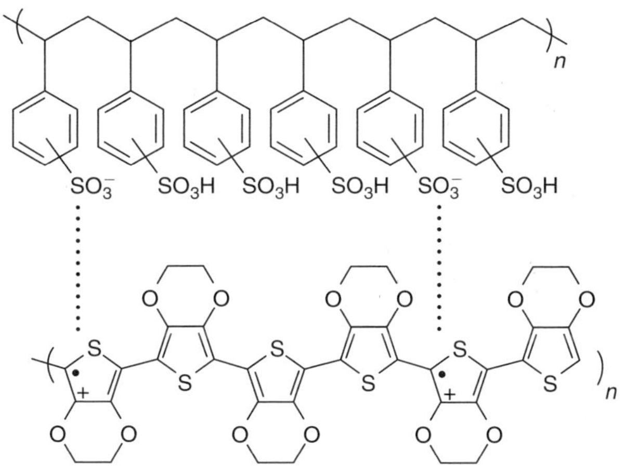

FIGURE 7.30 Chemical structure of PEDOT–PSS blends (Baytron P). (From Groenendaal, L., Jonas, F., Freitag, D., Pielartzik, H., Reynolds, J.R., *Adv. Mater.*, 12, 481, 2000. With permission.)

Sirringhaus et al. [46] demonstrated a fast and inexpensive way to pattern PEDOT–PSS dispersions via ink jet printing (IJP), and the resulting patterned surface was used for making an all-polymer transistor circuit (Figure 7.31).

A similar approach has been used to make stable poly(thieno[3,4-*b*]thiophene) (PT34bT) (**20**) dispersions in water using PSS as a template [47]. The conductivity of the film obtained by spin coating PT34bT-PSS dispersions was reported to be 10^{-1} to 10^{-4} S/cm, and the film showed great optical transparency in the oxidized/conducting state. This indicates that the template approach can be applied to other LEGPs to make them water processable.

Wudl and coworkers [48] have recently reported a facile way to synthesize highly conducting PEDOT in the solid state without using any solvent. The process involves the sublimation of 2,5-dibromo-3,4-ethylenedioxythiophene (DBEDOT) and simultaneous polymerization to PEDOT, as shown in Figure 7.32. The proposed mechanism for solid-state polymerization (SSP) is shown in Figure 7.33.

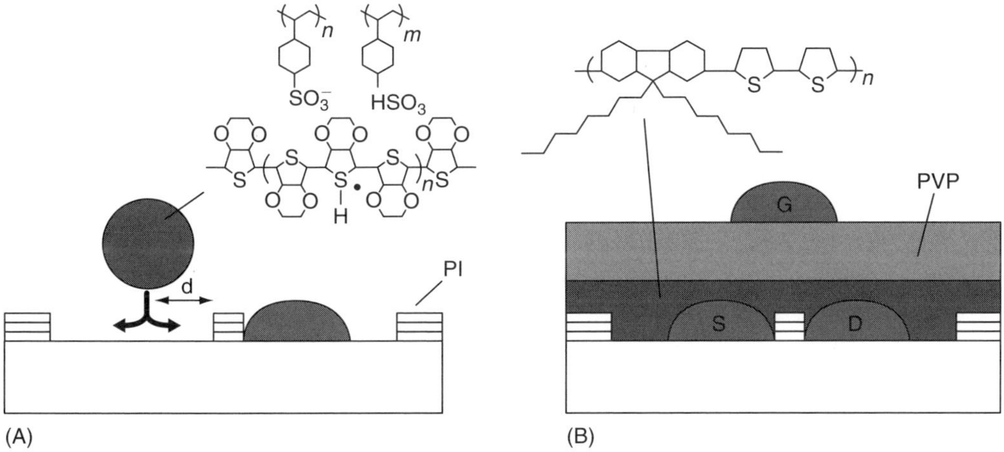

FIGURE 7.31 (A) Schematic diagram of high-resolution ink jet printing (IJP) onto a pre-patterned substrate. (B) Schematic diagram of the top-gate IJP TFT configuration with an F8T2 semiconducting layer (S, source; D, drain; and G, gate). (From Sirringhaus, H., Kawase, T., Friend, R.H., Shimoda, T., Inbasekaran, M., Wu, W., and Woo, E.P., *Science*, 290, 2123, 2000. With permission.)

FIGURE 7.32 Photographs of crystals of DBEDOT (left) and PEDOT (right), formed on heating the crystals of the monomer at 60°C for 8 h. (From Meng, H., Perepichka, D.F., and Wudl, F., *Angew, Chem. Int. Ed.*, 42, 658, 2003. With permission.)

As SSP does not involve any solvent, PEDOT films were patterned onto flexible plastic substrates, thereby demonstrating their potential application in flexible plastic electronics (Figure 7.34).

Defieuw et al. [49] polymerized 1,3-dihydrobenzo[c]thiophene in the presence of concentrated sulfuric acid and dispersed the resulting polymer in water using λ-carrageenan as the surfactant. A transparent film was made on a PET substrate. It possessed an optical density of 0.02 and had surface resistivity as low as 8.5 Ω/cm^2. The neutral state of PITN was found to be stable in air but the doped/conductive form was unstable in air and rapidly dedoped. Our group has also reported the sulfonation of insoluble PT34bT to make a water-processable LEGP [50]. Besides providing processability, the optical properties of PT34bT were reported to be controlled by the degree of sulfonation (Figure 7.35A), resulting in an optical band gap that can be tuned through the use of a polycation. The optical properties of polymer were also found to vary with the number of bilayers obtained in the layer-by-layer (LBL) method of assembly, as shown in Figure 7.35B.

7.5 SUMMARY

LEGPs having energy gaps of 1.5 eV or lower are generally optically transparent in the visible region in their conducting state making them a promising candidate for optoelectronic devices. The E_g of the polymer depends on several parameters, and can be reduced by minimizing the BLA along the polymer chain, increasing the planarity of the system for effective conjugation, decreasing the aromaticity of the monomer unit for more delocalization of π electrons along the chain, decreasing the LUMO energy level, increasing the HOMO energy level, and improving interchain interactions.

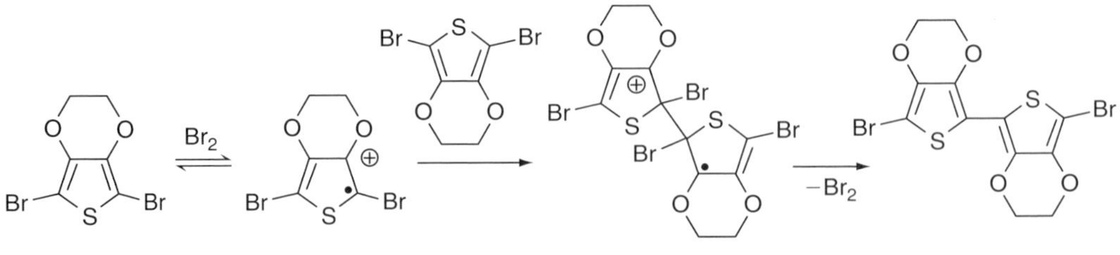

FIGURE 7.33 Proposed mechanism of the initiation of SSP of DBEDOT. (From Meng, H., Perepichka, D.F., Bendikov, M., Wudl, F., Pan, G.Z., Yu, W., Dong, W., and Brown, S., *J. Am. Chem. Soc.*, 125, 15151, 2003. With permission.)

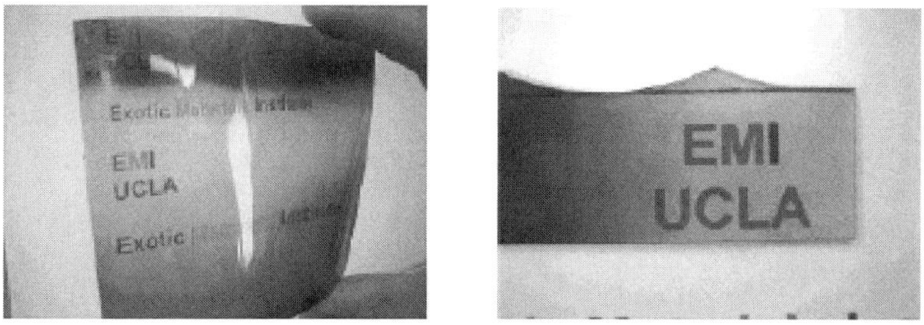

FIGURE 7.34 Semitransparent conducting films of PEDOT on a plastic substrate (left) and a glass slide (right), prepared by in situ SSP of vacuum deposited DBEDOT. (From Meng, H., Perepichka, D.F., Bendikov, M., Wudl, F., Pan, G.Z., Yu, W., Dong, W., and Brown, S., *J. Am. Chem. Soc.*, 125, 15151, 2003. With permission.)

On the basis of the strategy used to achieve a low E_g, LEGPs are grouped into four classes: fused ring systems, aromatic monomers (which have only electron-donating or electron-withdrawing groups), donor–acceptor or push–pull systems consisting of strong electron DA, and vinylenes/methines that decrease steric interactions between neighboring repeat units. As LEGPs are generally insoluble and infusible, different approaches have been attempted to make them processable, such as substituting long alkyl chains to make them soluble in organic media, polymer dispersion in water in the presence of a polyelectrolyte, SSP, and post-derivatization (e.g., sulfonation of the conducting polymer backbone). It should not be assumed that a low band gap polymer translates to a more highly transparent conductive polymer upon p-doping when compared to a high band gap polymer. A long-extant problem in the field of optically transparent conductive polymers is the motion of the isosbestic point when going from a neutral conjugated polymer to the p-doped conductive polymer, as indicated in Figures 7.2 and 7.13. Once the problem of the moving isosbestic point has been circumvented for the p-doping of LEGPs, the preparation of very highly transparent optical conductors will become a reality.

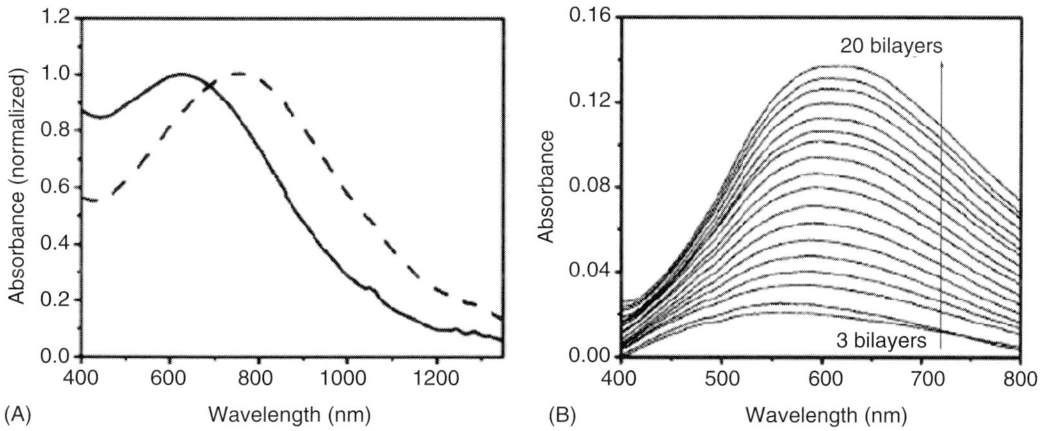

FIGURE 7.35 (A) Vis–NIR spectra of neutral 56-SPoT (dashed line) and neutral 65-SPoT (solid line) in water. The prefix 56 and 65 indicated the % of sulfonation in PT34bT. (B) Optical spectra of an LBL film of 56-SPoT/PEI as a function of bilayer number. (From Lee, B., Seshadri, V., Palko, H., Sotzing, G.A., *Adv. Mater.*, 17, 1792, 2005. With permission.)

EXERCISE QUESTIONS

1. Define the energy gap in a conjugated polymer and in LEGPs. List their importance in applications.
2. Why is the onset for absorption of a spectral response taken as the energy gap of a conjugated polymer?
3. Why does a conjugated polymer absorb at longer wavelengths than simple organic molecules? Explain the redshift in the absorption spectrum upon p-doping of the neutral conjugated polymer?
4. Identify each group as being electron withdrawing or electron donating, as it pertains to these groups being directly attached to a conjugated polymer.

$$-OCH_3, >C{=}O, -CH_3, -CF_3, -CH_2CH_3, -NH_2, -NO_2$$

5. Give the aromatic and quinoidal structures for four low energy gap conjugated polymers having nondegenerate ground states.
6. Arrange the following monomers in ascending order in terms of conjugation length and energy gap, respectively.

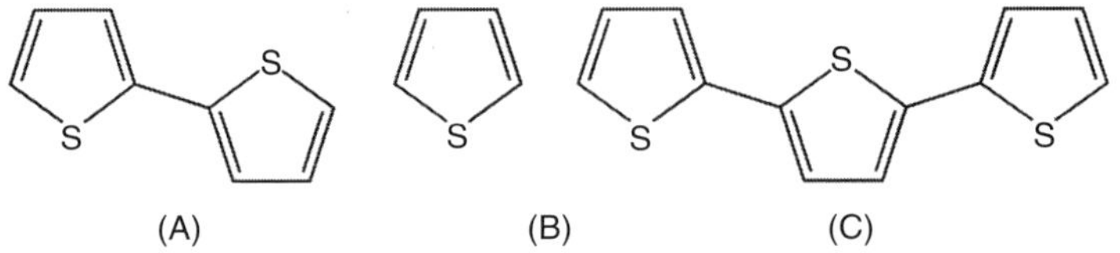

SCHEME 7.2

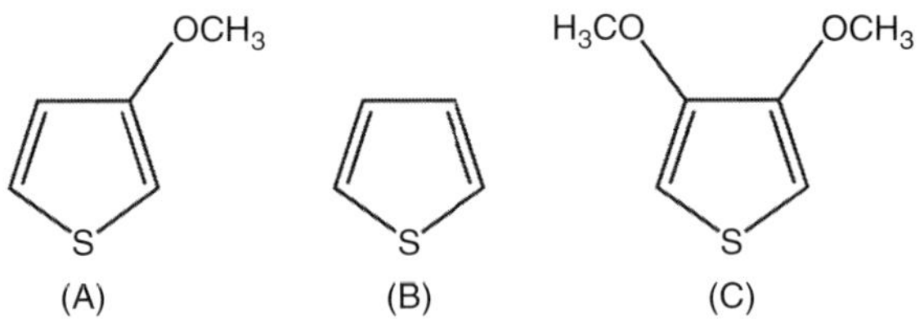

SCHEME 7.3

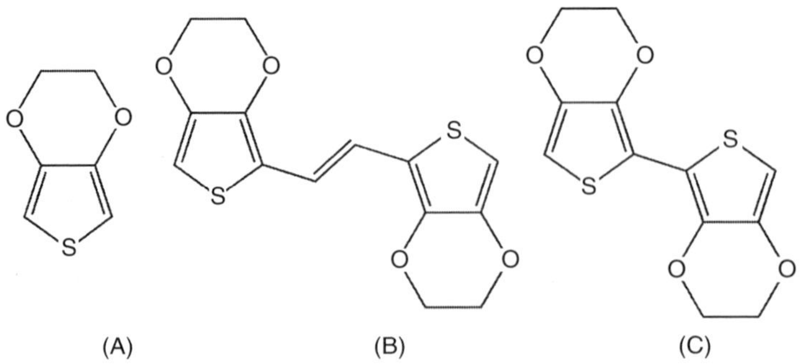

SCHEME 7.4

LIST OF ABBREVIATIONS

AOs	atomic orbitals
BLA	bond length alternation
CPs	Π-Conjugated polymers
DA	donors and acceptors
DBEDOT	2,5-dibromo-3,4-ethylenedioxythiophene
E_g	gap energy
HOMO	highest occupied molecular orbital
IJP	ink jet printing
ITO	indium tin oxide
LBL	layer-by-layer
LEGPs	low energy gap polymers
LUMO	lowest unoccupied molecular orbital
NIR	near-infrared
OLEDs	organic light-emitting diodes
P3AT	poly(3-alkylthiophene)
PA	polyacetylene
PBT	poly(bithiophene)
PEDOP	poly(ethylenedioxypyrrole)
PEDOT	poly(3,4-ethylenedioxythiophene)
PITN	poly(isothianaphthene)
PProDOP	poly(propylenedioxypyrrole)
ProDOT	propylenedioxythiophene
PSS	poly(styrene sulfonate)
PT	poly(thiophene)
PTV	poly(thienylenevinylene)
SSP	solid-state polymerization
T34bF	thieno[3,4-b]furan
T34bT	thieno[3,4-b]thiopene
TBAP	tetrabutylammonium perchlorate

ACKNOWLEDGMENT

We would like to thank the National Science Foundation (NSF CAREER CHE-0349121) for support.

REFERENCES

1. Ito, T., Shirakawa, H., and Ikeda, S., Simultaneous polymerization and formation of polyacetylene film on the surface of a concentrated soluble Ziegler-type catalyst solution, *J. Polym. Sci. Chem. Ed.*, **1974**, *12*, 11–20; Chiang, C.K., Park, Y.W., Heeger, A.J., Shirakawa, H., Louis, E.J., and MacDiarmid, A.G., Electrical conductivity in doped polyacetylene, *Phys. Rev. Lett.*, **1977**, *39*, 1098–1101.
2. Pomerantz, M., Low band gap conducting polymers, in *Handbook of Conducting Polymers*, Skotheim, T.A., Elsenbaumer, R.L., Reynolds, J.R., Eds., Marcel Dekker, Inc., New York, **1998**, Chap. 11.
3. Parker, V.D., Energetics of electrode reactions. II. The relationship between redox potentials, ionization potentials, electron affinities, and solvation energies of aromatic hydrocarbons, *J. Am. Chem. Soc.*, **1976**, *98*, 98–103.
4. Lyons, L.E., Energy gaps in organic semiconductors derived from electrochemical data, *Aust. J. Chem.*, **1980**, *33*, 1717–1725.
5. Loutfy, R.O. and Cheng, Y.C., Investigation of energy levels due to transition metal impurities in metal-free phthalocyanine, *J. Chem. Phys.*, **1980**, *73*, 2902–2910.
6. Bredas, J.L., Relationship between band gap and bond length alternation in organic conjugated polymers, *J. Chem. Phys.*, **1985**, *82*, 3808–3811.

7. Lee, Y.-S. and Kertesz, M., The effect of heteroatomic substitutions on the band gap of polyacetylene and polyparaphenylene derivatives, *J. Chem. Phys.*, **1988**, *88*, 2609–2617.
8. Ten Hoeve, W., Wynberg, H., Havinga, E.E., and Meijer, E.W., Substituted 2,2′:5′,2″:5″,2‴:5‴,2⁗:5⁗,2′′′′′:5′′′′′,2′′′′′′:5′′′′′′,2′′′′′′′:5′′′′′′′,2′′′′′′′′:2′′′′′′′′′,2′′′′′′′′′′–undecithiophenes, the longest characterized oligothiophenes, *J. Am. Chem. Soc.*, **1991**, *113*, 5887–5889.
9. Guay, J., Kasai, P., Diaz, A., Wu, R., Tour, J.M., and Dao, L.H., Chain-length dependence of electrochemical and electronic properties of neutral and oxidized soluble. α,α-coupled thiophene oligomers, *Chem. Mater.*, **1992**, *4*, 1097–1105.
10. Sato, M. and Hiroi, M., Synthesis and properties of hexyl-substituted oligothiophenes, *Synth. Met.*, **1995**, *71*, 2085–2086.
11. Hernandez, V., Castiglioni, C., Del Zopo, M., and Zerbi, G., Confinement potential and π-electron delocalization in polyconjugated organic materials, *Phys. Rev. B*, **1994**, *50*, 9815–9823.
12. Wudl, F., Kobayashi, M., and Heeger, A.J., Poly(isothianaphthene), *J. Org. Chem.*, **1984**, *49*, 3382–3384.
13. Kenoue, Y., Wudl, F., and Heeger, A.J., A novel substituted poly(isothianaphthene), *Synth. Met.*, **1991**, *40*, 1–12.
14. Eiji, F., Polyisothianaphthene Polymer and Conductive Material, Japanese Patent Publication Number: 02-252727, **1990**.
15. Pomerantz, M., Chaloner-Gill, B., Harding, L.O., Tseng, J.J., and Pomerantz, W.J., Poly(2,3-dihexylthieno-[3,4-b]pyrazine). A new processable low band-gap polyheterocycle, *J. Chem. Soc., Chem. Commun.*, **1992**, *22*, 1672–1673.
16. Lakshmikantham, M.V., Lorcy, D., Scordilis-Kelley, C., Wu, X.L., Parakka, J.P., Metzger, R.M., and Cava, M.P., Poly(naphtho[2,3-c]thiophene-alt-bithiophene): A novel low band gap polymer, *Adv. Mater.*, **1993**, *5*, 723–726.
17. Metzger, R.M., Wang, P., Wu, X. -L., Tormos, G.V., Lorcy, D., Shcherbakova, I., Lakshmikantham, M.V., and Cava, M.P., Langmuir–Blodgett superconductors based on C60, new donor-acceptor systems, and a new conducting polymer , *Synth. Met.*, **1995**, *70*, 1435–1438.
18. Tanaka, S. and Yamashita, Y., Synthesis of a narrow band gap heterocyclic polymer: Poly-4,6-di(2-thienyl)thieno[3,4-c][1,2,5]thiadiazole, *Synth. Met.*, **1993**, *55*, 1251–1254.
19. Kitamura, C., Tanaka, S., and Yamashita, Y., Synthesis of new narrow bandgap polymers based on 5,7-di(2-thienyl)thieno[3,4-b]pyrazine and its derivatives, *J. Chem. Soc., Chem. Commun.*, **1994**, 1585–1586.
20. Karikomi, M., Kitamura, C., Tanaka, S., and Yamashita, Y., New narrow-bandgap polymer composed of benzobis(1,2,5-thiadiazole) and thiophenes, *J. Am. Chem. Soc.*, **1995**, *117*, 6791–6792.
21. Tanaka, S., and Yamashita, Y., Syntheses of narrow band gap heterocyclic copolymers of aromatic-donor and quinonoid-acceptor units, *Synth. Met.*, **1995**, *69*, 599–600.
22. Kitamura, C., Tanaka, S., and Yamashita, Y., Design of narrow-bandgap polymers. syntheses and properties of monomers and polymers containing aromatic-donor and *o*-quinoid-acceptor units *Chem. Mater.*, **1996**, *8*, 570–578.
23. Suzuki, T., Fujii, H., Yamashita, Y., Kabuto, C., Tanaka, S., Harasawa, H., Mukai, T., and Miyashi, T., Clathrate formation and molecular recognition by novel chalcogen-cyano interactions in tetracyanoquinodimethanes fused with thiadiazole and selenadiazole rings, *J. Am. Chem. Soc.*, **1992**, *114*, 3034–3043.
24. Lee, K., and Sotzing, G.A., Poly(thieno[3,4-b]thiophene). A new stable low band gap conducting polymer, *Macromolecules*, **2001,** *34*, 5746–5747.
25. Sotzing, G.A. and Lee, K., Poly(thieno[3,4-b]thiophene): A p- and n-dopable polythiophene exhibiting high optical transparency in the semiconducting state, *Macromolecules*, **2002**, *35*, 7281–7286.
26. Kumar, A., Buyukmumcu, Z., and Sotzing, G.A. Poly(thieno[3,4-b]furan). a new low band gap conjugated polymer, *Macromolecules*, **2006**, *39*, 2723–2725.
27. Argun, A.A., Cirpan, A., and Reynolds, J.R., The first truly all-polymer electrochromic devices, *Adv. Mater.*, **2003**, *15*, 1338–1341.
28. Krishnamoorthy, K., Ambade, A.V., Kanungo, M., Contractor A.Q., and Kumar, A., Rational design of an electrochromic polymer with high contrast in the visible region: Dibenzyl substituted poly(3,4-propylenedioxy thiophene), *J. Mater. Chem.*, **2001**, *11*, 2909–2911; Krishnamoorthy, K., Ambade, A.V., Mishra, S.P., Kanungo M., Contractor, A.Q., and Kumar, A., Dendronized electrochromic polymer based on poly(3,4-ethylenedioxy thiophene), *Polymer*, **2002**, *43*, 6465–6470.
29. Schottland, P., Zong, K., Gaupp, C.L., Thompson, B.C., Thomas, C.A., Giurgiu, I., Hickman, R., Abboud, K.A., and Reynolds J.R., Poly(3,4-alkylenedioxypyrrole)s: Highly stable electronically conducting and electrochromic polymers, *Macromolecules*, **2000**, *33*, 7051–7061.

30. Lambert, T.M. and Ferraris, J.P., Narrow band gap polymers: Polycyclopenta[2,1-b;3,4-b′]dithiophen-4-one, *J. Chem. Soc., Chem. Commun.*, **1991**, 752–754.
31. Ferraris, J.P. and Lambert, T.M., Narrow bandgap polymers: Poly-4-dicyanomethylene-4H-cyclopenta[2,1-b;3,4-b′]dithiophene (PCDM), *J. Chem. Soc., Chem. Commun.*, **1991**, 1268–1270.
32. Ferraris, J.P., Henderson, C., Torres, D., and Meeker, D., Synthesis, spectroelectrochemistry and application in electrochromic devices of an n- and p-dopable conducting polymer, *Synth. Met.*, **1995**, *72*, 147–152.
33. Brisset, H., Thobie-Gautier, C., Gorgues, A., Jubault, M., and Roncali, J., Novel narrow bandgap polymers from sp3 carbon-bridged bithienyls: Poly(4,4-ethylenedioxy-4H-cyclopenta[2,1-b;3,4-b′] dithiophene), *J. Chem. Soc., Chem. Commun.*, **1994**, 1305.
34. Akoudad, S. and Roncali, J., Electrogenerated poly(thiophenes) with extremely narrow bandgap and high stability under n-doping cycling, *Chem. Commun.*, **1998**, 2081–2082.
35. Zhang, Q.T. and Tour, J.M., Alternating donor/acceptor repeat units in polythiophenes. Intramolecular charge transfer for reducing band gaps in fully substituted conjugated polymers, *J. Am. Chem. Soc.*, **1998**, *120*, 5355–5362.
36. Schmidt, A.H., Reaktionen von quadratsäure und quadratsäure-derivaten, *Synthesis*, **1980**, 961–994 and references cited therein.
37. Ajayaghosh, A. and Eldo, J., A novel approach toward low optical band gap polysquaraines, *Org. Lett.*, **2001**, *3*, 2595–2598; Eldo, J. and Ajayghosh, A., New low band gap polymers: Control of optical and electronic properties in near infrared absorbing π-conjugated polysquaraines, *Chem. Mater.*, **2002**, *14*, 410–418.
38. Cheng, H. and Elsenbaumer, R.L., New precursors and polymerization route for the preparation of high molecular mass poly(3,4-dialkoxy-2,5-thienylenevinylene)s: Low band gap conductive polymers, *J. Chem. Soc., Chem. Commun.*, **1995**, *1980*, 1451–1452.
39. Sotzing, G.A. and Reynolds, J.R., Poly[trans-bis(3,4-ethylenedioxythiophene)vinylene]: A low band-gap polymer with rapid redox switching capabilities between conducting transmissive and insulating absorptive states, *J. Chem. Soc., Chem. Commun.*, **1995**, 703–704.
40. Brédas, J.L., Relationship between band gap and bond length alternation in organic conjugated polymers, *J. Chem. Phys.*, **1985**, *82*, 3808–3811.
41. Brédas, J.L., Bipolarons in doped conjugated polymers: A critical comparison between theoretical results and experimental data, *Mol. Cryst. Liq. Cryst.*, **1985**, *118*, 49–56.
42. Chen, W.-C. and Jenekhe, S.A., Small bandgap conducting polymers based on conjugated poly(heteroarylene methines). 1. Precursor poly(heteroarylene methylenes), *Macromolecules* **1995**, *28*, 454–464; Chen, W.C. and Jenekhe, S.A., Small bandgap conducting polymers based on conjugated poly(heteroarylene methines). 2. Synthesis, structure, and properties, *Macromolecules*, **1995**, *28*, 465–480.
43. Benincori, T., Rizzo, S., Sannicolò, F., Schiavon, G., Zecchin, S., and Zotti, G., An electrochemically prepared small-bandgap poly(biheteroarylidenemethine): Poly{bi[(3,4-ethylenedioxy) thienylene]methine}, *Macromolecules*, **2003**, *36*, 5114–5118.
44. Bayer A.G, *Eur. Patent*, **1991**, *440*, 957; Gevaert, A., *Eur. Patent*, **1993**, *564*, 911; Jonas, F., Krafft, W., and Muys, B., *Macromol. Symp.*, **1995**, *100*, 169; Groenendaal, L., Jonas, F., Freitag, D., Pielartzik, H., Reynolds, J.R., Poly (3,4-ethylenedioxythiophene) and its derivatives: Past, present, and future, *Adv. Mater.*, *12*, 481–494, 2000.
45. Jonsson, S.K.M., Birgerson, J., Crispin, X., Greczynski, G., Osikowicz, W., Denier van der Gon, A.W., Salaneck, W.R., and Fahlman, M., The effects of solvents on the morphology and sheet resistance in poly(3,4-ethylenedioxythiophene)–polystyrenesulfonic acid (PEDOT–PSS) films, *Synth. Met.*, **2003**, *139*, 1–10.
46. Sirringhaus, H., Kawase, T., Friend, R.H., Shimoda, T., Inbasekaran, M., Wu, W., and Woo, E.P., High-resolution inkjet printing of all-polymer transistor circuits, *Science*, **2000**, *290*, 2123–2126.
47. Lee, B., Seshadri, V., and Sotzing, G.A., Poly(thieno[3,4-*b*]thiophene)-Poly(styrene sulfonate): A low band gap, water dispersible conjugated polymer, *Langmuir*, **2005**, *21*, 10797–10802.
48. Meng, H., Perepichka, D.F., and Wudl, F., Facile solid-state synthesis of highly conducting poly (ethylenedioxythiophene), *Angew, Chem. Int. Ed.*, **2003**, *42*, 658–661; Meng, H., Perepichka, D.F., Bendikov, M., Wudl, F., Pan, G.Z., Yu, W., Dong, W., and Brown, S., Solid-state synthesis of a conducting polythiophene via an unprecedented heterocyclic coupling reaction, *J. Am. Chem. Soc.*, **2003**, *125*, 15151–15162.
49. Defieuw, G., Samijn, R., Hoogmartens, I., Vanderzande, D., and Gelan, J., Antistatic polymer layers based on poly(isothianaphthene) applied from aqueous compositions, *Synth. Met.*, **2003**, *57*, 3702–3706.
50. Lee, B., Seshadri, V., Palko, H., and Sotzing, G.A., Ring-sulfonated poly(thieno thiophene), *Adv. Mater.*, **2005**, *17*, 1792–1795.

51. Groenendaal, L., Zotti, G., Aubert, P.-H., Waybright, S.M., and Reynolds, J.R., Electrochemistry of poly (3,4-alkylenedioxythiophene) derivatives, *Adv. Mater.*, **2003,** *15*, 855.
52. Ajayaghosh, A., Donor–acceptor type low band gap polymers: Polysquaraines and related systems, *Chem. Soc. Rev.*, **2003**, *32*, 181.
53. Roncali, J., Synthetic principles for bandgap control in linear π-conjugated systems, *Chem. Rev.*, **1997**, *97*, 173.

8 Conjugated Polymers, Fullerene C_{60}, and Carbon Nanotubes for Optoelectronic Devices

Liangti Qu, Liming Dai, and Sam-Shajing Sun

CONTENTS

8.1 Introduction 238
8.2 Organic Optoelectronic Materials 238
  8.2.1 Conjugated Polymers 238
  8.2.2 Fullerene C_{60} and Carbon Nanotubes 239
8.3 Electronic Devices 241
  8.3.1 Schottky Barrier Diodes 241
  8.3.2 Field-Effect Transistors 244
8.4 Optoelectronic Devices for Converting Electrons into Photons 248
  8.4.1 Light-Emitting Polymer Displays 248
  8.4.2 C_{60} Light Bulbs 252
  8.4.3 Carbon Nanotube Displays 253
8.5 Optoelectronic Devices for Converting Photons into Electrons 255
  8.5.1 C_{60}-Containing Polymer Photovoltaic Cells 255
  8.5.2 Nanotube-Containing Polymer Photovoltaic Cells 257
8.6 Concluding Remarks 259
Exercise Questions 259
List of Abbreviations 259
Acknowledgment 259
References 260

Abstract A brief summary of recent developments in the field of organic optoelectronic devices is given, with a focus on conjugated polymers, fullerene C_{60}, and carbon nanotubes. The device design, performance, and underlying principles are discussed. Examples include Schottky barrier diodes, field-effect transistors (FETs), photovoltaic cells, light-emitting diodes (LEDs), and field-emitting displays to cover optoelectronic devices for both the current and light generation.

8.1 INTRODUCTION

Optoelectronic devices based on semiconductors may be divided into two categories depending on whether or not a current or light is produced. The former includes thin film transistors [1,2], photoconductors [3], and photovoltaic cells [4], while the latter encompasses light-emitting diodes (LEDs) [5] and optically/electrically pumped lasers [6]. Since the discovery of the first transistor in the 1950s [7], optoelectronic industries have been dominated by inorganic semiconductors (e.g., silicon, germanium, gallium arsenide, and gallium phosphide) [8]. Until very recently, organic materials have been only used as passive components, such as insulators or photoresists for fabricating silicon chip circuitry in the microelectronic industry. This is because the relatively high level of structural disorder and band gap energy of organic semiconducting materials often leads to much lower charge carrier mobilities than those of inorganic semiconductors. As a consequence, it is difficult, though not impossible, for organic semiconductors to achieve the switching frequencies required for certain microelectronic devices. However, the recent advancements in organic polymer LEDs, fullerene C_{60}-sensitized photovoltaic cells, and carbon nanotube nanoelectronics [9] have attracted considerable attention for the commercial development of optoelectronic devices based on active organic materials.

Fabricating organic devices is often less expensive than fabricating devices made of inorganic semiconductors, as films of organic materials can be more easily deposited over a large area. Besides, organic materials are often flexible and light. Two major classes of organic materials are currently under extensive investigation for applications in organic optoelectronic devices: (1) solution-cast high molecular weight films and (2) small organic molecular coatings formed by sublimation. This chapter focuses on conjugated polymers, fullerene C_{60}, and carbon nanotubes (Figure 8.1) for optoelectronic device applications.

8.2 ORGANIC OPTOELECTRONIC MATERIALS

8.2.1 Conjugated Polymers

Polymers have been traditionally used as electrically insulating materials. After all, metal cables are coated in plastic to insulate them. The visit of MacDiarmid to Shirakawa at Tokyo Institute of Technology in 1974 and, later, Shirakawa to MacDiarmid and Heeger at University of Pennsylvania in

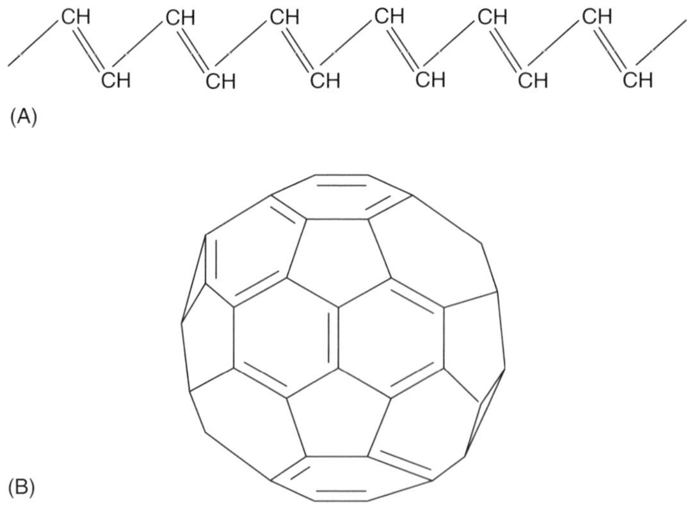

FIGURE 8.1 Conjugated structures with different dimensionalities: (A) polyacetylene; (B) buckminsterfullerene C_{60};

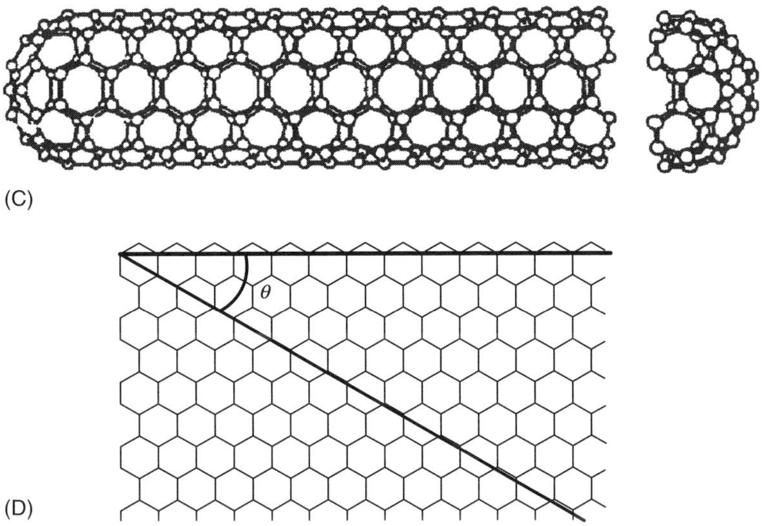

FIGURE 8.1 (continued) (C) [5,5] armchair single-walled carbon nanotubes (SWNTs); (D) schematically illustrating how the graphene sheet is rolled into a nanotube cylinder.

1976 led to the discovery of conducting polyacetylene—a prototype conjugated conducting polymers (Figure 8.1A) [10]. This finding opened up the important new field of polymers for electronic applications and was recognized by the 2000 Nobel Prize in Chemistry [11]. The subsequent discovery of the electroluminescent light emission from conjugated poly(*p*-phenylene vinylene), by Friend's group at Cavendish Laboratory in 1990 [12], revealed the significance for the use of conjugated polymers in optoelectronic devices. Various conjugated polymers can now be synthesized to show the processing advantages of plastics and the optoelectronic properties of inorganic semiconductors or metals; with a conductivity even up to 1.7×10^5 S/cm—comparing favorably with the value of 106 S/cm for copper or silver [13] (Table 8.1).

Just like metals have high conductivity due to the free movement of electrons through their structure, for organic materials to be electronically conductive, they must possess not only charge carriers but also an orbital system that allows the charge carriers move [9]. The conjugated structure can meet the second requirement through a continuous overlapping of π orbitals. Most organic materials, however, do not have intrinsic charge carriers. The required charge carriers may be provided by partial oxidation (p-doping, Figure 8.2A) of the organic (macro) molecules with electron acceptors (e.g., I_2, AsF_5) or by partial reduction (n-doping, Figure 8.2B) with electron donors (e.g., Na, K). Through such a chemical doping process, charged defects are introduced, which could then be available as the charge carriers.

8.2.2 Fullerene C_{60} and Carbon Nanotubes

Sometimes history repeats itself. The visit made by Kroto in 1985 to Smalley and Curl at Rice University led also to a Nobel prize-winning discovery of buckminsterfullerene C_{60}—a conjugated molecule with a soccer-ball-like structure consisting of 12 pentagons and 20 hexagons facing symmetrically (Figure 8.1B) [14]. Although Ōsawa [15] had theoretically predicted the structure of C_{60} in 1970 and Huffman [16] might have produced the C_{60} molecules in his graphite smoke as early as 1973, it was the 1985 discovery that generated a great deal of interest and created an entirely new branch of carbon chemistry [16]. Fullerenes consist of 20 hexagonal and 12 pentagonal rings as the basis of an icosahedral symmetry closed-cage structure. Each carbon atom is bonded to three others

TABLE 8.1
Some Conjugated Conducting Polymers

Polymer (Date Conductivity Discovered)	Structure	π–π* Gap (eV)	Conductivity[a] (S/cm)
Polyacetylene and analogues			
Polyacetylene (1977)	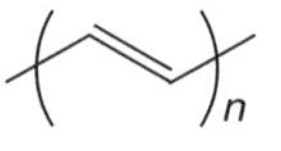	1.5	10^3–1.7×10^5
Polypyrrole (1979)	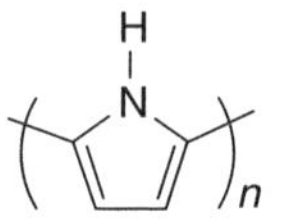	3.1	10^2–7.5×10^3
Polythiophene (1981)	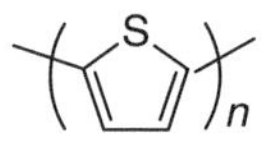	2.0	10–10^3
Polyphenylene and analogues			
Poly(paraphenylene) (1979)	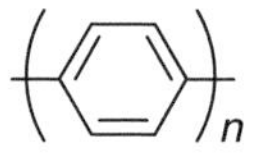	3.0	10^2–10^3
Poly(*p*-phenylene vinylene) (1979)	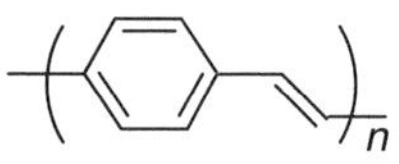	2.5	3–5×10^3
Polyaniline (1980)	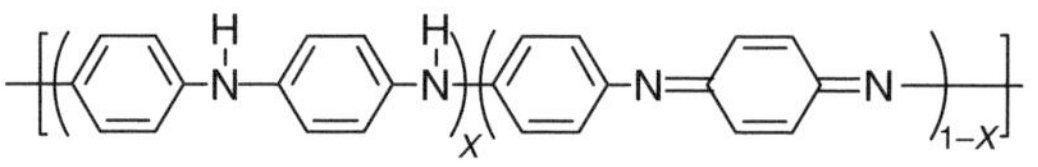	3.2	30–200

[a] The range of conductivities listed is from that originally found to the highest values obtained to date.

and is sp^2 hybridized. The C_{60} molecule has two bond lengths—the 6:6 ring bonds can be considered double bonds and are shorter than the 6:5 bonds. C_{60} is not superaromatic as it tends to avoid double bonds in the pentagonal rings, resulting in poor electron delocalization. (The curvature of the C_{60} surface causes a hybridization of the atomic 2s and 2p levels into the π and σ orbitals, which have hybridization between planar [sp^2] and tetrahedral [sp^3]. The electrons in the σ orbitals participate in the C–C covalent bonding, while the π orbitals protrude from the C_{60} surface with asymmetric lobes outside and inside the carbon framework. This hybridization causes the extremely high electronegativity

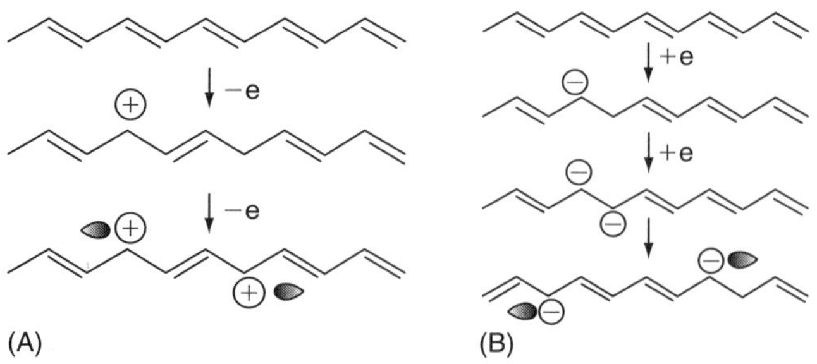

FIGURE 8.2 A schematic description of the doping of *trans*-polyacetylene chain: (A) p-doping and (B) n-doping.

of C_{60}, i.e., the electron affinity is ~2.7 eV.) As a result, C_{60} behaves like an electron-deficient alkene, and reacts readily with electron-rich species. The geodesic and electronic bonding factors in the structure account for the stability of the molecule. In theory, an infinite number of fullerenes can exist, their structure based on pentagonal and hexagonal rings constructed according to rules for making icosahedra.

The subsequent discovery of carbon nanotubes by Iijima [17] in 1991 opened up a new era in material science and nanotechnology [18]. Carbon nanotubes may be viewed as a graphite sheet (graphene) that is rolled up into a nanoscale tube form (single-walled carbon nanotubes [SWNTs]) (Figure 8.1C) or with additional graphene tubes around the core of an SWNT (multiwalled carbon nanotubes [MWNTs]) [18]. An ideal nanotube can be thought of as a hexagonal network of carbon atoms (like a graphene) that has been rolled up to make a seamless cylinder (Figure 8.1D). Just a nanometer across, the cylinder can be tens of microns long with each end being "capped" by half of a fullerene molecule. SWNTs can be thought of as the fundamental cylindrical structure, forming the building blocks of both MWNTs and the ordered arrays of SWNTs. Carbon nanotubes come in a variety of diameters and lengths. Depending on the growth process, the length of the tubes can be from approximately 100 nm to several millimeters. Diameters vary from 1 to ~20 nm. Another parameter describing carbon nanotubes is their chiral angle. This angle is specified by how the graphene sheet is rolled into a cylinder (Figure 8.1D). By convention, a nanotube with its axis collinear with the horizontal ($\theta = 0$) line in Figure 8.1D is called a zigzag nanotube. This name derives from the appearance of the half fullerene molecule that can cap the end of the tube. Being one giant molecule, carbon nanotubes have unusual mechanical and electrical properties. The conductivity of SWNTs can vary from semiconductive to metallic depending on the chiral angle of the tube and its diameter. In fact, the measured best charge mobility (μ) of carbon nanotubes is in the range of 80,000–100,000 $cm^2/V\,s$, and these are in contrast to a measured best charge mobility of 15,000 $cm^2/V\,s$ in graphite, and estimated best charge mobility of about 100–200 $cm^2/V\,s$ in doped PA (Chapter 3).

The mechanical properties of nanotubes are also unusual. Numerical simulations predict the Young's modulus of SWNTs to be in excess of a terapascal. In view of the large number of well-defined carbon–carbon single and double bonds in their molecular structure, fullerenes and carbon nanotube are polymeric in essence. While conjugated polymers have widely been regarded as quasi-one-dimensional semiconductors [9], fullerenes and carbon nanotubes can be considered as a quantum dot and quantum wires, respectively [18]. Having a conjugated all-carbon structure with unusual molecular symmetries, fullerenes and carbon nanotubes have also been shown to possess some similar optoelectronic properties to conjugated polymers attractive for various applications, including photovoltaic cells and field-emitting displays. Compared to typical π electron conjugated polymers, C_{60} and carbon nanotubes are relatively very stable and incur little photooxidative degradation. This is possibly due to their relatively inert chemical reactivity.

8.3 ELECTRONIC DEVICES

8.3.1 Schottky Barrier Diodes

The Schottky diode (named after German physicist Walter H. Schottky) is a semiconductor diode with a low forward voltage drop and a very fast switching action desirable for switch mode power supplies, discharge-protection for solar cells connected to lead-acid batteries, prevention of transistor saturation, and many others. Instead of a semiconductor–semiconductor junction, a Schottky diode uses a metal–semiconductor junction as a Schottky barrier. During normal operation, Schottky diode allows majority carriers (e.g., electrons for an n-doped semiconductor) to be quickly injected into the conduction band of the metal contact on the other side of the diode to become free moving electrons. This minimizes the possible slow random recombination of n- and p-type carriers and causes conduction faster than an ordinary p–n rectifier diode, and hence faster and smaller circuits

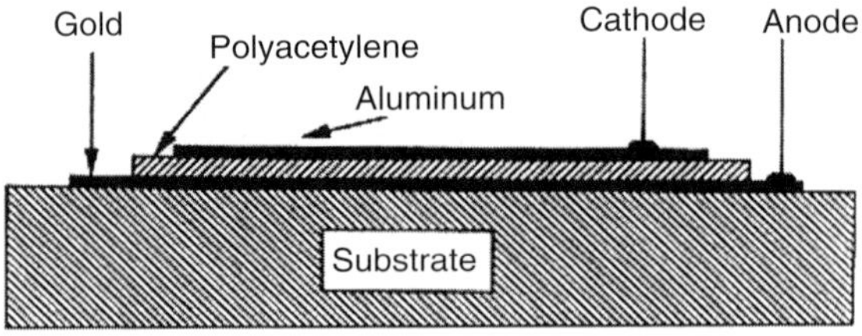

FIGURE 8.3 A typical Schottky barrier diode structure with a layer of polyacetylene (500–1000 nm) sandwiched between a thin layer of aluminum, chromium, or indium (20 nm) and gold (20 nm) supported by a flat glass substrate. (From Burroughes, J.H., Jones, C.A., and Friend, R.H., *Nature*, 335, 137, 1988. With permission.)

that can even operate at higher frequencies. Figure 8.3 shows a typical Schottky barrier diode structure with polyacetylene sandwiched between two metal contact layers, in which one of the metal layers (i.e., Al) forms the Schottky barrier with polyacetylene and the other (gold) provides an ohmic contact.

Although the fabrication of this type of Schottky barrier diode is straightforward, it provides a very useful structure for studying the electrical and electro-optical properties of semiconducting materials, including conducting polymers. For a p-type semiconductor, the ideal Schottky barrier is formed with a metal of a work function (Φ_m) lower than that of the semiconductor (Φ_s). As shown in Figure 8.3, Φ_s is given by

$$\Phi_s = \chi + \xi \tag{8.1}$$

where χ and ξ are the electron affinity (energy difference between the bottom of the conduction band and the vacuum level) and the energy gap between the Fermi level and bottom of the conduction band, respectively. The formation of the Schottky junction between the semiconductor and the low-work-function metal leads to the equalization of the Fermi energies, at zero bias, by forming a dipole charge layer with positive charge density in the metal and a region of negative charge density (i.e., the depletion layer) at the interface. As the charge concentration in a metal is normally much higher than the charge concentration in a semiconductor, the semiconductor only undergoes the band bending. This sets up a barrier (i.e., Schottky barrier, Φ_b), which is given by $\chi + E_g - \Phi_m$, for the flow of holes toward the metal contact.

In addition to the $J(V)$ characteristics of the junction (Figure 8.4), detailed information on the nature of the junction may be obtained from the complex impedance. The capacitance of the junction in the depletion regime at reverse bias depends strongly on the variation in the width of the depletion regime with bias and is given by

$$C/A = (q\varepsilon_o\varepsilon_r N_a/2)^{1/2}\left[\frac{1}{V_{d0}+V}\right]^{1/2} \tag{8.2}$$

where

A is the area of the junction
q is the electronic charge
$V_{d0} = \Phi_b - (E_F - E_V)$ is the diffusion voltage at zero bias
N_a is the acceptor dopant concentration
ε_r and ε_o are the relative dielectric constant and the dielectric constant in vacuum, respectively

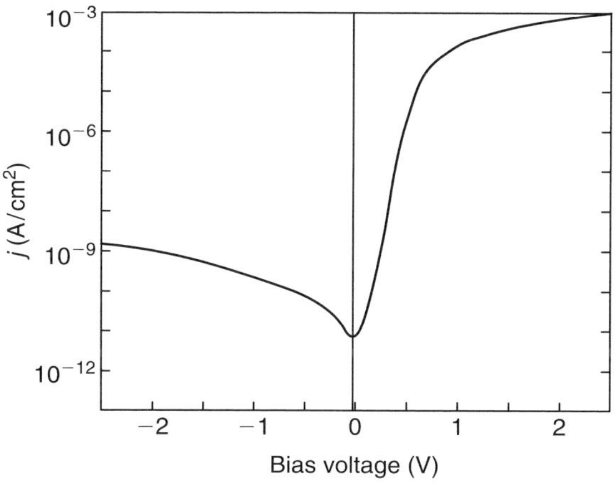

FIGURE 8.4 Forward and reverse characteristics for a Schottky diode with the Al/polyacetylene/Au structure, which gives a very good diode characteristic with a forward to reverse ratio as high as 5×10^{10} at a bias of +1.5 V. (From Burroughes, J.H., Jones, C.A., and Friend, R.H., *Nature*, 335, 137, 1988. With permission.)

As can be seen from Equation 8.2, the slope of $1/C^2$ versus bias voltage should allow direct measurement of N_a and V_{d0}. From these data, information on the dependence of the depletion width (w) with bias voltage can be obtained by

$$w = \left[2\varepsilon_o \varepsilon_r \frac{V_{d0} + V}{qN_a} \right]^{1/2} \tag{8.3}$$

The increase in width of the depletion regime with increasing reverse bias has resulted from the movement of the extrinsic charge carriers to leave the depletion region with a space charge density (qN_a) due to the acceptors. The increase in absorbance (δ_α) by an increase in the reverse bias voltage is then related with the increase in the depletion width (δ_w) by

$$\delta_\alpha = -\frac{\Delta T}{T} = \delta_w N_a \sigma \tag{8.4}$$

where

ΔT is the fractional change in the transmission (T)
σ is the optical cross-section of the soliton

Therefore, the Schottky diodes can also be used for optical measurement since thin metal contacts (e.g., the 20 nm thick Al and Au films) allow adequate optical transmission through the device. As shown in Figure 8.5, it is possible to obtain the high sensitivity spectrum of the change in optical transmission with bias voltage.

The plot of $[T(V) - T(0)]/T(0)$ obtained by taking the bias voltage between 0 and a large negative value (−35 V) gives a positive signal for a bleaching induced by the reverse bias condition. The bleaching has been attributed to the loss of soliton levels associated with the extrinsic charge carriers that were swept out of the semiconductor to form the depletion layer. The oscillations at higher energies were considered to be due to modulation of interference fringes [19].

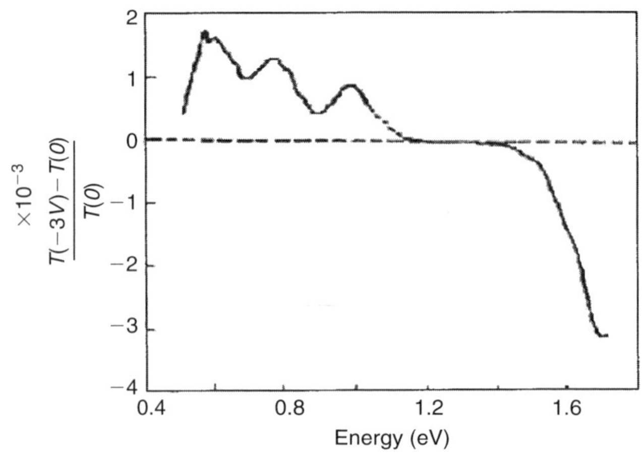

FIGURE 8.5 Bias voltage modulated optical transmission for an Al/polyacetylene Schottky diode of thickness 900 nm. $T(V)$ is the optical transmission at bias voltage V (–3 V with respect to the blocking contact). (From Burroughes, J.H., Jones, C.A., and Friend, R.H., *Nature*, 335, 137, 1988. With permission.)

One of the major advantages in the use of Schottky diode structures for such optical measurements is that it is straightforward to obtain a quantitative measurement of the optical change and of the number of charges that are responsible for it. For example, the capacitance versus bias voltage (not shown) for the device corresponding to Figure 8.5 shows the expected behavior in reverse bias up to –35 V with a value for the acceptor concentration of $N_a = 2.1 \times 10^{16}$ cm^{-3}. Then, the change in width of the depletion layer (w) with the change in bias voltage from 0 to –35 V is calculated to be 470 nm and the optical cross-section per injected charge at the peak of the "mid-gap" absorption band (σ) to be 1.6×10^{16} cm^2 (Equation 8.4).

Single-walled carbon nanotubes are attractive materials in both fundamental science and technology due to their unique electrical, mechanical, and chemical properties [18]. In particular, SWNTs are promising as active materials for building electronic devices, including diodes. Figure 8.6A shows a scheme of SWNT-Schottky diode using Ti and Pt for the Schottky barrier and Ohmic contact, respectively [20].

Figure 8.6B shows a typical atomic force microscope (AFM) image for the nanotube Schottky diode. The metal layer thicknesses were Ti ~20 nm; Pt ~17 nm; Au ~100 nm, and typical lengths of the SWNTs between contact pads ranged from 1.7 to 2.5 μm. This device exhibits rectification as shown in Figure 8.7.

Similarly, fullerenes have also been used for the fabrication of certain diode devices. For example, diode characteristics have been demonstrated by sandwiching two semiconducting layers of C_{60} intercalated with a porous conducting polyaniline network between two metal electrodes [21].

8.3.2 Field-Effect Transistors

The addition of an insulating layer between the semiconductor and one of the metal layers of the Schottky diode structure and the replacement of the top electrode by a source and drain contacts onto the insulator layer gives the metal–insulator–semiconductor field-effect transistor (MISFET), as exemplified by Figure 8.8 for a typical MISFET device fabricated on a silicon wafer.

In field-effect transistors (FETs), the flow of carriers between the source and drain contacts is controlled by the application of an electric field onto the gate electrode. Current flow along the main conduction path of the semiconductor in a field-effect device is dominated by the majority carriers,

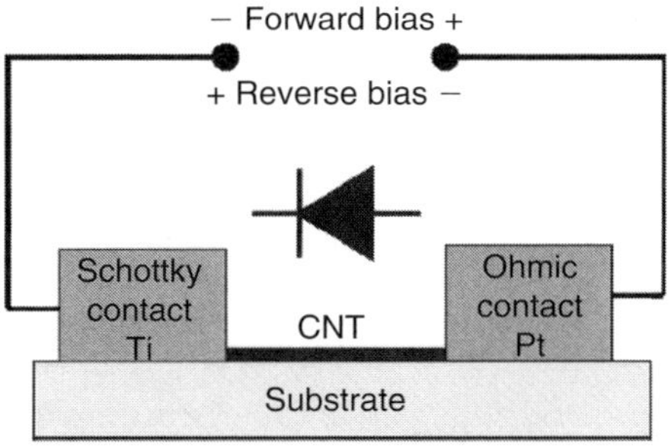

(A)

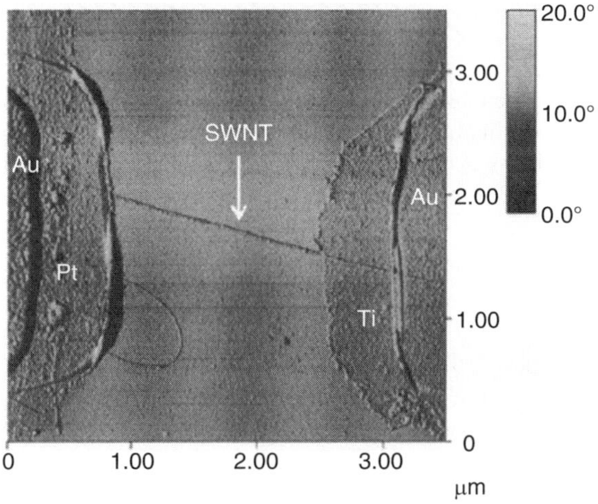

(B)

FIGURE 8.6 (A) Schematic representation of the SWNT-Schottky diode with Ti and Pt contacts; (B) a typical AFM phase plot image of the SWNT-Schottky diode. (From Manohara, H.M., Wong, E.W., Schlecht, E., Hunt, B.D., and Siegel, P.H., *Nano Lett.*, 5, 1469, 2005. With permission.)

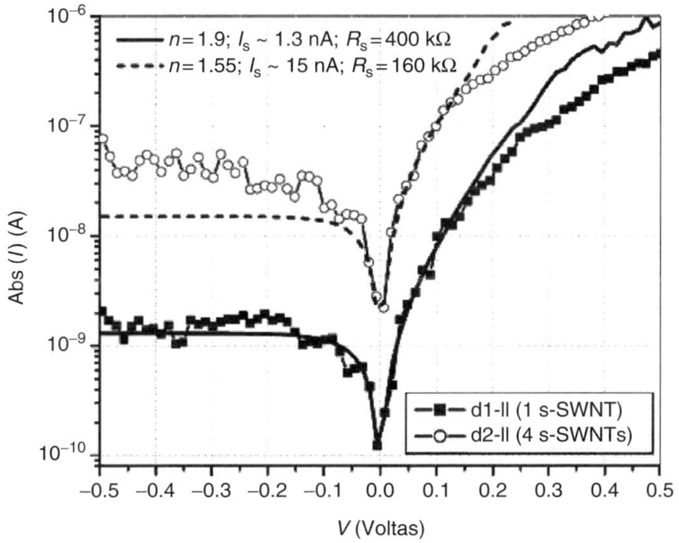

FIGURE 8.7 Ideality curve fits in low bias range for d1-II (with single s-SWNT) and d2-II (with four s-SWNTs) Schottky diodes. The curve shows the absolute magnitude of the current plotted against the corresponding voltage. (From Manohara, H.M., Wong, E.W., Schlecht, E., Hunt, B.D., and Siegel, P.H., *Nano Lett.*, 5, 1469, 2005. With permission.)

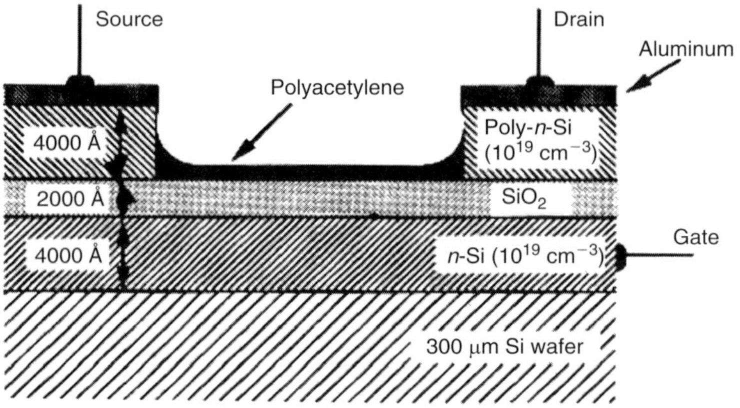

FIGURE 8.8 A MISFET device fabricated on a silicon wafer substrate, using polyacetylene as the semiconductor, silicon dioxide as the insulator layer, heavily n-doped silicon as the bottom metallic layer (the gate), and the source and drain contacts. Aluminum is used as a blocking contact. (From Burroughes, J.H., Jones, C.A., and Friend, R.H., *Nature*, 335, 137, 1988. With permission.)

leading to their performance being relatively unaffected by external conditions (e.g., ambient temperature changes and radiation degradation). The variation of the channel conductance with gate voltage (V_{gs}) for constant drain–source voltage (V_{ds}) defines the transconductance (g_m):

$$g_m = \frac{\Delta I_{ds}}{\Delta V_{gs}} \tag{8.5}$$

For low values of V_{ds}, g_m gives a direct measure of the carrier mobility through the following equation:

$$g_m = \frac{w}{l}(\mu C_i V_{ds}) \tag{8.6}$$

where
- w is the depletion width
- l is the mean free path of the carriers in the semiconductor (which is estimated to be 3 nm for polyacetylene)
- μ is the charge carrier mobility
- C_i is the gate capacitance of the insulator layer [22]

Furthermore, the saturation current (I_{sat}) for the channel at high values of V_{ds} is given by

$$I_{sat} = \frac{\mu C_i w}{2l(V_{gs} - V_{th})^{1/2}} \tag{8.7}$$

where V_{th} is the threshold voltage at which the accumulation channel forms.

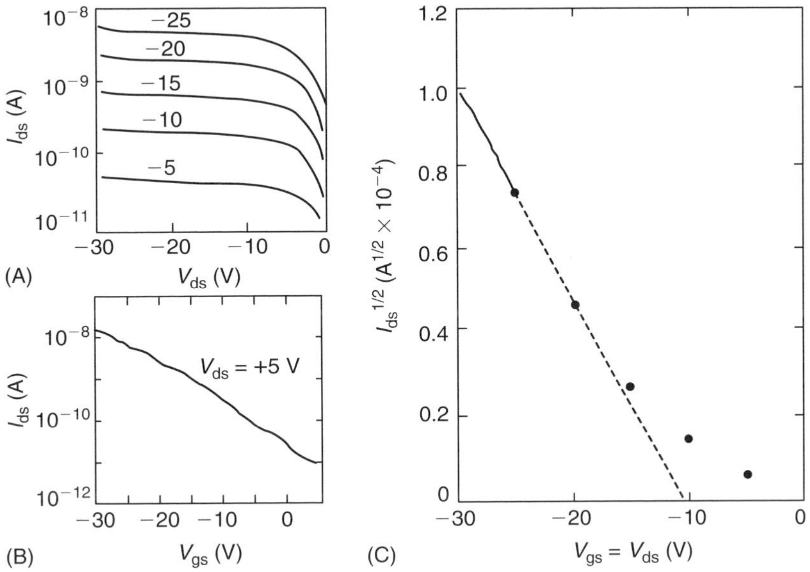

FIGURE 8.9 Characteristics of MISFET transistor with Au source and drain contacts. (A) I_{ds} versus V_{ds} for various negative values of V_{gs}; (B) I_{ds} versus V_{gs} for $V_{ds} = +5$ V; (C) $(I_{ds})^{1/2}$ versus $V_{ds} = V_{gs}$. (From Burroughes, J.H. and Friend, R.H., *Conjugated Polymers*, Brédas, J.L. and Silbey, R. (Eds.), Kluwer Academic Publishers, Dordrecht, 1991. With permission.)

While Figure 8.9A and B shows the plots of I_{ds} against V_{ds} at various negative values of V_{gs} (Figure 8.9A) and the variation of I_{ds}, with V_{gs} for $V_{ds} = +5$ V (Figure 8.9B), the plot of $(I_{ds})^{1/2}$ against $V_{ds} = V_{gs}$ is given in Figure 8.9C. From the slope of the linear region found at the higher values of V_{ds} in Figure 8.9C (the broken line), the values of V_{th} and μ are estimated to be −12 V and 9×10^{-6} cm²/V s, respectively.

Recently, Dekker and coworkers have reported molecular FETs based on a semiconducting SWNT [23]. Avouris and coworkers have independently demonstrated the use of structurally deformed MWNTs in nanotube FETs [24]. A cross-section view of the nanotube FETs, which consist of either an individual SWNT or MWNT bridging two electrodes deposited on gate oxide film supported by a doped Si wafer back gate, is schematically shown in Figure 8.10. Electron beam lithography was used to define the Au electrodes. The source–drain current I through the nanotubes was measured at room temperature as a function of the bias voltage V_{SD} and the gate voltage V_G.

Figure 8.11A shows the I–V_{SD} characteristics for a device consisting of a single SWNT with a diameter of 1.6 nm at different values of the gate voltage. As can be seen, the I–V_{SD} curve is linear with a resistance of $R = 2.9$ MΩ at $V_G = 0$ V and the I–V_{SD} curves remain linear for $V_G < 0$ V. For $V_G \gg 0$ V, the I–V_{SD} curves become increasingly nonlinear up to a point where the current becomes undetectable,

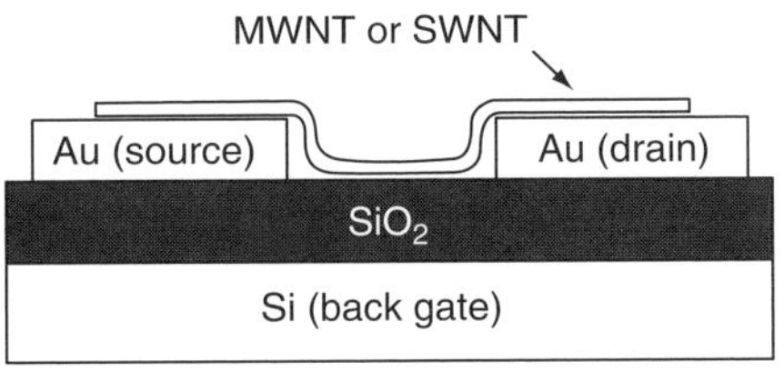

FIGURE 8.10 A cross-section view of the nanotube FET device shows schematically a single carbon nanotube bridging the gap between two gold electrodes. The silicon substrate is used as back gate. (From Martel, R., Schmidt, T., Shea, H.R., Hertel, T., and Avouris, P., *Appl. Phys. Lett.*, 73, 2447, 1998. With permission.)

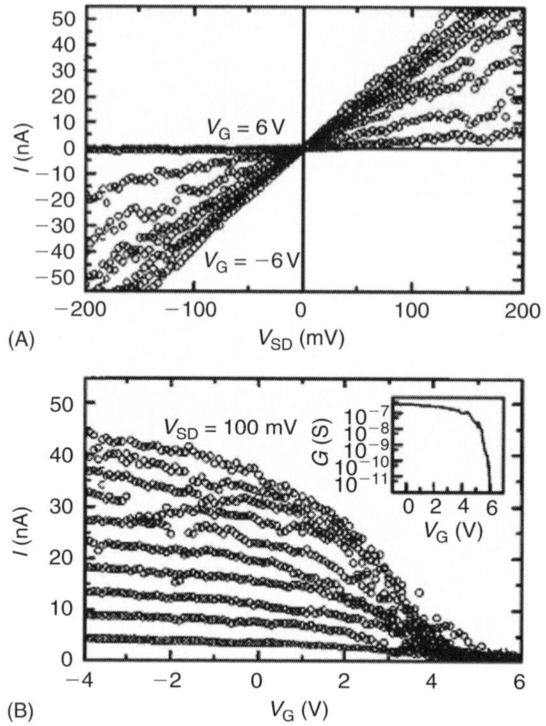

FIGURE 8.11 Output and transfer characteristics for the SWNT FET schematically shown in Figure 8.10 (A) I–V_{SD} curves measured for V_G = –6, 0, 1, 2, 3, 4, 5, and 6 V; (B) I–V_G curves for V_{SD} = 10–100 mV in steps of 10 mV. Inset shows that the gate modulates the conductance by five orders of magnitude (V_{SD} = 10 mV). (From Martel, R., Schmidt, T., Shea, H.R., Hertel, T., and Avouris, P., *Appl. Phys. Lett.*, 73, 2447, 1998. With permission.)

indicating a controllable transition between a quasi-metallic and an insulating state of the nanotube. The transfer characteristics I–V_G of the nanotube device for different source–drain voltages are given in Figure 8.11B, which shows that the source–drain current decreases strongly with increasing gate voltage. The inverse dependence of the source–drain current on the positive gate voltage indicates not only that the nanotube device operates as a FET but also that positive carriers (holes) are dominate carriers transporting through the semiconducting SWNT.

Haddon and coworkers [25] reported the first n-channel FET based a thin film of C_{60} in 1995. These authors first deposited chromium and gold pads on SiO_2/Si using lithographic techniques to give the source (S) and drain (D) electrodes, and then wired the substrates to an ultrahigh vacuum feedthrough for depositing C_{60} thin films through a shadow mask (Figure 8.12A). Figure 8.12B shows I–V_{SD} characteristics for the n-channel transistor with an 800 Å thick film of C_{60}. As expected, a drain current (I_D) increases with drain–source voltage (V_{ds}) before gradually leveling off to approach the saturation current at a given gate voltage (V_G), and the drain current increases rapidly with increasing gate voltage (Figure 8.12B).

8.4 OPTOELECTRONIC DEVICES FOR CONVERTING ELECTRONS INTO PHOTONS

8.4.1 Light-Emitting Polymer Displays

Since the discovery of electroluminescent light emission from poly(p-phenylene vinylene) (PPV), in 1990 by Friend's group [12], various other electroluminescent conjugated polymers have been synthesized [26–28]. Figure 8.13 schematically shows a typical structure representative of the

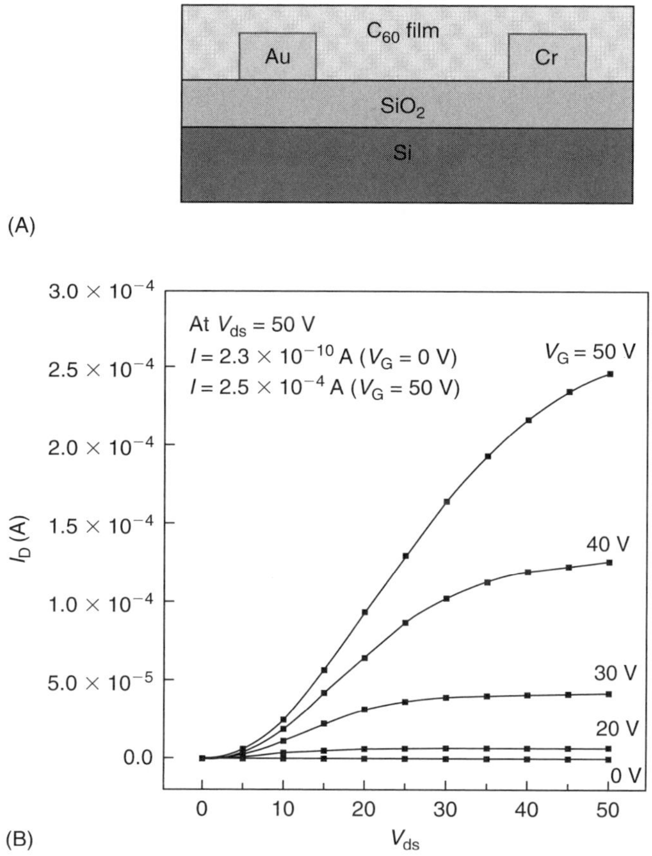

FIGURE 8.12 (A) A schematic representation of the C_{60} thin film n-channel FET and (B) drain current (I_D) versus drain–source voltage (V_{ds}) under various gate voltages (V_G) for the C_{60} thin film transistor. (From Haddon, R.C., Perel, A.S., Morris, R.C., Palstra, T.T.M., Hebard, A.F., and Fleming, R.M., *Appl. Phys. Lett.*, 67, 121, 1995. With permission.)

simplest single-layer LED devices using PPV as the light-emitting polymer. Among the two electrodes, at least one should be transparent or semitransparent. In practice, a transparent indium tin oxide (ITO)-coated glass is often used as the anode and a layer of low-work-function metal (e.g., aluminum, calcium) acts as the cathode.

Poly(*p*-phenylene vinylene) prepared by the sulfonium precursor route has an energy gap between the π and π^* states of about 2.5 eV (see Table 8.1). Photoexcitation of an electron from the highest occupied molecular orbital (HOMO) to the lowest unoccupied molecular orbital (LUMO) creates a singlet exciton (Figure 8.14A), which produces luminescence at a longer wavelength (the Stokes shift) through radioactive decay. Polymer LEDs work under a similar general principle [29]. Upon the application of an electrical voltage onto a LED device, electrons from the low-work-function cathode

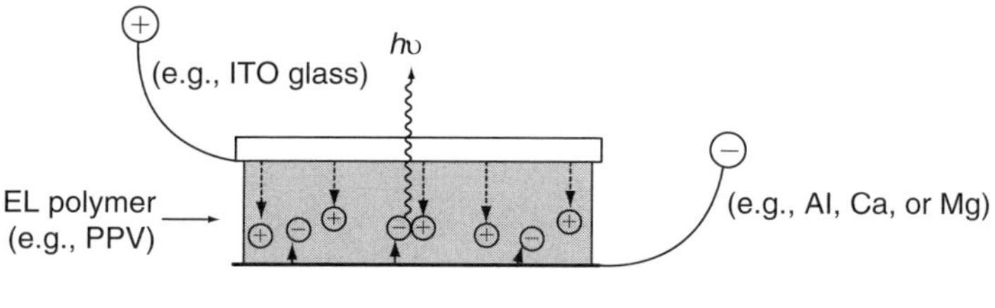

FIGURE 8.13 Schematic representation of a single-layer polymer LED.

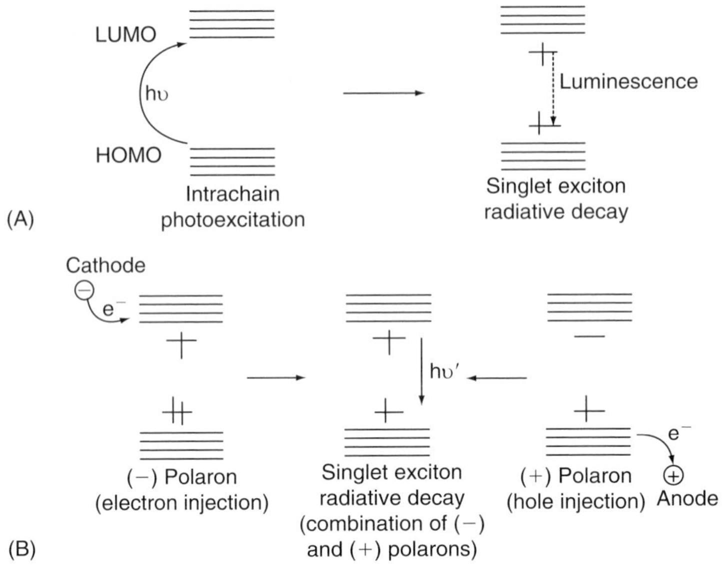

FIGURE 8.14 The working principle for (A) photoluminescence (PL) and (B) electroluminescence (EL). (From Dai, L., *Polym. Adv. Technol.*, 10, 357, 1999. With permission.)

(e.g., Al in Figure 8.13) are injected into the LUMO of the PPV layer, leading to the formation of negatively charged polarons, whereas holes from the high-work-function anode (i.e., ITO in Figure 8.13) are injected into the HOMO, producing positively charged polarons. These negatively and positively charged polarons migrate under the influence of the applied electric field and combine in the band gap of the PPV layer (Figure 8.14B), resulting in the formation of the same singlet exciton as is produced by the photoexcitation. The singlet exciton thus produced can also decay radiatively with the emission of light. The wavelength (and hence the color) of the photons thus produced depends on the energy gap of the organic light-emitting material, coupled with a Stokes shift (Figure 8.15). To maintain a low operation voltage, a very thin polymer layer (~100 nm) is usually used.

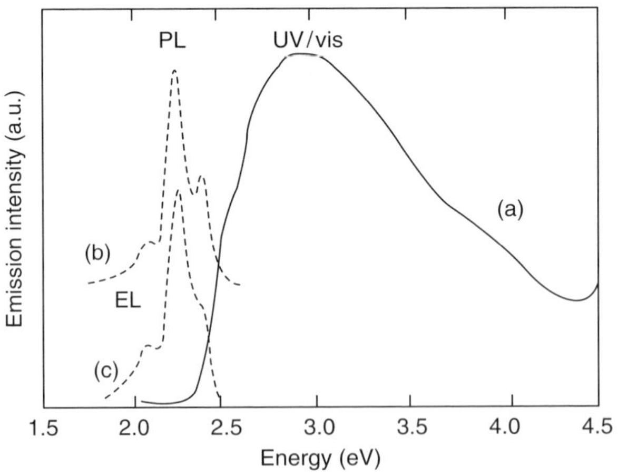

FIGURE 8.15 The absorption (curve a) and emission (curves b and c) spectra of PPV. The similarity of the emission spectra generated by photoexcitation (Figure 8.14A) and by charge injection (Figure 8.14B) suggests that the same excited state is responsible for both PL and EL. (From Brown, A.R., Bradley, D.D.C., Burroughes, J.H., Friend, R.H., Greenham, N.C., Burn, P.L., Holmes, A.B., and Kraft, A., *Appl. Phys. Lett.*, 61, 2793, 1992. With permission.)

Although the same working principle applies to both photoluminescence (PL) and electroluminescence (EL), the theoretical maximum EL efficiency may only reach one-quarter of the best PL efficiency (21% for PPV) as electron–hole capture is expected to result in the loss of 75% of the electron–hole pairs to triplet excitons, which do not decay radiatively with high efficiency [5]. Recently, however, there have been arguments that the EL efficiency could be higher than this limit due to the high quantum yield of spatially indirect excitons (i.e., bound polaron pairs) [30]. The introduction of species that allow efficient triplet luminescence has also been shown to be an attractive approach for improving EL efficiency [31]. For comparison purposes, it is worthwhile to mention here that the first LED device with the Al/PPV/ITO structure reported by Burroughes [12] had an efficiency of only 0.05%. A few years later, this group obtained a very high internal efficiency (~4%) from both a single active layer device based on the PPV-dimethoxy-substituted PPV copolymer [27,28] and a PPV/cyano-substituted PPV bilayer device [32,33]. Although the synthesis of EL copolymers with alternating conjugated and nonconjugated segments has since been demonstrated to be a useful method for improving the EL efficiency through exciton confinement [34]; the work functions of the cathode and anode need to be matched with the corresponding LUMO and HOMO levels of the EL polymer in a particular LED to achieve efficient emission.

The internal quantum efficiency, η_{int}, defined as the ratio of the number of photons produced within an EL device to the number of electrons flowing through the external circuit [5], is given by

$$\eta_{int} = g r_{st} q \tag{8.8}$$

where

g is the number of excitons formed per electron flowing through the external circuit
r_{st} is the fraction of excitons formed as singlets
q is the efficiency of radiative decay of these singlet excitons

Therefore, to improve the EL efficiencies, various issues concerning the formation of excitons and their decay need to be considered. The process of exciton formation, in turn, involves carrier injection, carrier transport, and carrier combination. As we shall see later, the polymer–electrode interface has an important impact on the carrier injection and carrier transport.

Balanced charge injection and transport is a prerequisite for LEDs with high quantum efficiency [5]. Balanced injection may be achieved using two polymer–electrode interfaces with equal barriers, which must be small for low operating voltages [5]. The work functions for some of the commonly used electrode materials are given in Table 8.2.

The schematic energy level diagram (Figure 8.16) shows the ionization potential (IP) and electron affinity (EA) of the light-emitting polymer, the work function of the anode (Φ_a) and cathode (Φ_c), and the barriers to injection of electrons (ΔE_e) and holes (ΔE_h). By taking the IP and the band gap (i.e., LUMO–HOMO) of PPV as 5.1 and 2.5 eV, respectively, we should have

$$\text{EA} = 5.1 - 2.5 = 2.6\ \text{eV} \tag{8.9}$$

$$\Delta E_e = \Phi_c - 2.6\ \text{(eV)} \tag{8.10}$$

and

$$\Delta E_h = \text{IP} - \Phi_a \tag{8.11}$$

TABLE 8.2
Work Function of Electrode Materials and Barriers to Injection of Electrons and Holes

Electrode	Work Function (eV)[a]	Barrier to Injection (eV)
Al	4.2	1.6
Ca	2.9	0.3
Mg	3.7	1.1
Au	5.3	2.7
In	4.1	1.5
Cu	4.7	2.1
Cr	4.5	1.9
ITO	4.6	0.5

[a] See Ref. [35].

It can be seen that the Ca/PPV/ITO provides a better match in the barriers for electron and hole injection than Al/PPV/ITO. Consequently, the quantum efficiency of a PPV/ITO-based LED was significantly improved when calcium was used in place of aluminum [36]. However, metals of a low work function are generally oxygen-reactive. The use of highly reactive metals with a low work-function (e.g., calcium) often requires careful encapsulation, which makes this method difficult, if not impossible, for practical applications. To circumvent this difficulty, modifications of the electrodes [37] and the charge injection characteristics [38] by surface and interface control of the polymer–metal (electrode) or the polymer–polymer interfaces have been exploited as alternative avenues for improving the EL efficiencies.

8.4.2 C_{60} Light Bulbs

Although the fluorescence of C_{60} and its monoadducts is very weak, Hutchison and coworkers [39] have reported that a C_{60} adduct, T_h-hexapyrrolidine (THP) (Figure 8.17A) has a strong yellow–green

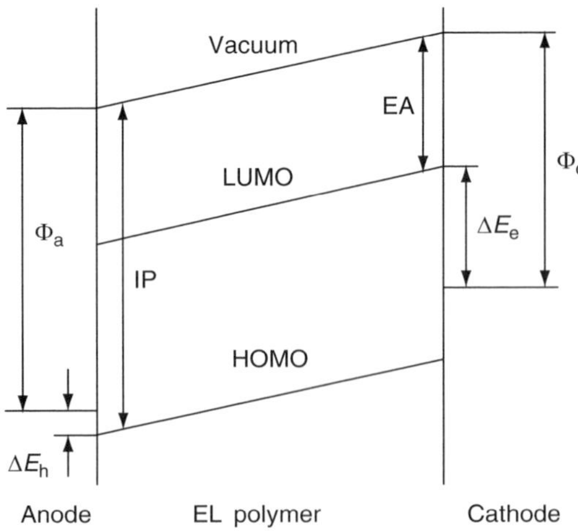

FIGURE 8.16 A schematic energy level diagram.

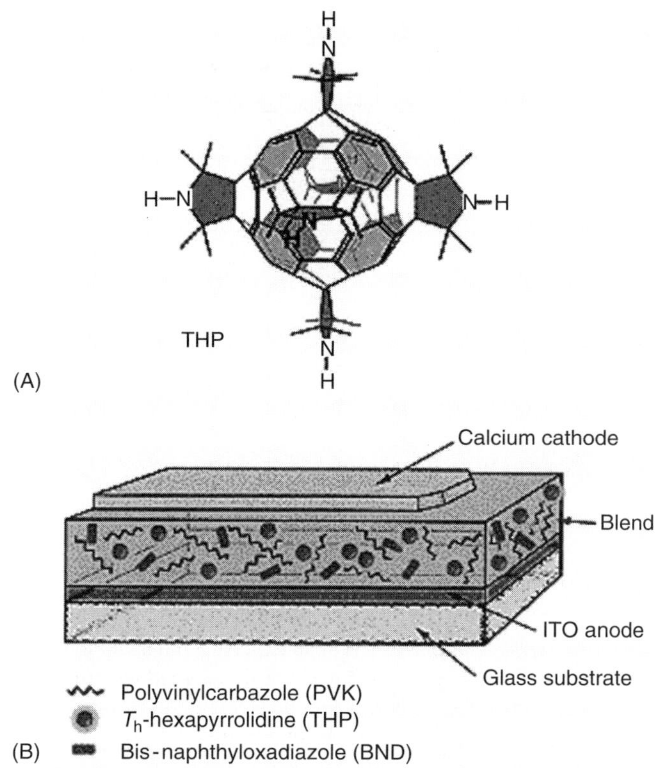

FIGURE 8.17 (A) THP and (B) device configuration. (From Hutchison, K., Gao, J., Schick, G., Rubin, Y., and Wudl, F., *J. Am. Chem. Soc.*, 121, 5611, 1999. With permission.)

fluorescence due to a higher excitation energy and larger single-triplet gap caused by a reduction of conjugation within the fullerene core. Largely contradicting the conventional belief that C_{60} is an efficient luminescence quencher [40] and exhibits electroluminescence with apparent decomposition, as it is slightly n-doped C_{60} [41], Wudl and coworkers [39] have successfully used THP as an emissive material by blending it with a hole-transporting poly(9-vinylcarbazole), PVK, and an electron-transporting 2,5-bis-(4-naphthyl)-1,3,4-oxadiazole (BND) to fabricate single-layer LEDs. They found that the Ca/PVK-THP-BND/ITO device emitted strong white light, leading to the first bucky light bulb.

8.4.3 Carbon Nanotube Displays

The interesting physicochemical properties, together with the unique molecular symmetries, have led to diverse applications for carbon nanotubes, including as new electron field emitters in panel displays [18]. The carbon nanotube electron emitters work on a similar principle to a conventional cathode ray tube, but their small size could lead to thinner, more flexible, and energy-efficient display screens with a higher resolution. Figure 8.18A shows a schematic drawing of a typical experimental set up used by de Heer et al. [42] to study the nanotube field emissions. As can be seen, the aligned carbon nanotube film is separated from a fine copper grid cover by a $d = 30$ μm thick perforated mica sheet with a 1 mm diameter hole. Upon application of a potential V between the grid and the nanotube film, an average field $E_0 = V/d$ is induced which, in turn, causes the extraction of

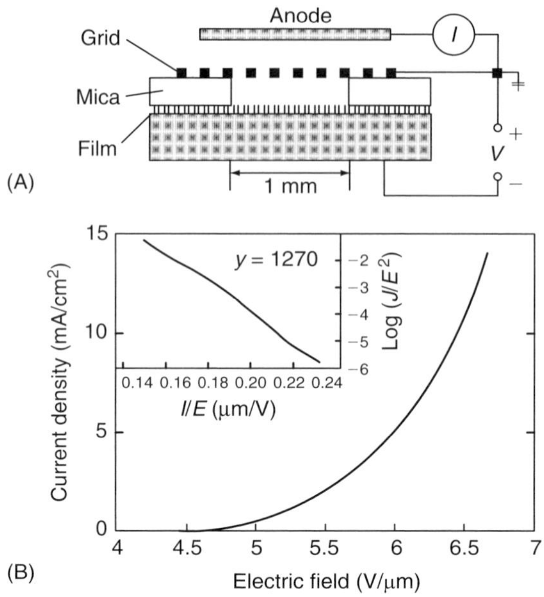

FIGURE 8.18 (A) A schematic representation of a typical experimental setup for studying the field emission from aligned carbon nanotube films and (B) the *I–V* curve for the aligned nanotube field emitter. Inset shows the Fowler–Nordheim plot of the emission current. (From de Heer, W.A., Bonard, J.M., Fauth, K., Chatelain, A., Forro, L., and Ugarte, D., *Adv. Mater.*, 9, 87, 1997. With permission.)

electrons from the nanotubes. The emitted electrons then pass through the grid, focusing on a fluorescent screen for picture displays or on an anode plate for measuring the current as a function of the applied field.

Figure 8.18B reproduces the *I–V* curve recorded from the nanotube field emitter shown in Figure 8.18A. By plotting the log of I/E^2 against $1/E$, a straight line was obtained (inset of Figure 8.18B), indicating that the *I–V* data thus measured were found to fit the Fowler–Nordheim (FN) equation [43]:

$$I = a E_{\mathrm{eff}}^2 \exp(-b/E_{\mathrm{eff}}) \tag{8.12}$$

where

a and b are constants

E_{eff} is the electric field strength at the emitting point

The field enhancement factor γ (i.e., $\gamma = E_{\mathrm{eff}}/E_0$, the ratio of the electric field at the emitting point to the applied field) was deduced from the slope to be 1270 in this particular case. Apart from γ, two other important parameters used to characterize field emitters are the turn-on field, E_{to}, defined as the electric field required to emit a current $I = 10\ \mu\mathrm{A/cm^2}$, and the threshold field, E_{thr}, corresponding to $I = 10\ \mathrm{mA/cm^2}$.

The field-emission study on SWNTs carried out by de Heer and coworkers [44] showed some promising results with E_{to} and E_{thr} in the range of 1.5–4.5 and 3.9–7.8 V/μm, respectively. These values, like those for most MWNTs, are far lower than corresponding values for other film emitters. Field emissions from an epoxy-carbon nanotube composite have also been reported [45]. More interestingly, Saito et al. [46] have constructed an electron tube lighting element equipped

(A)

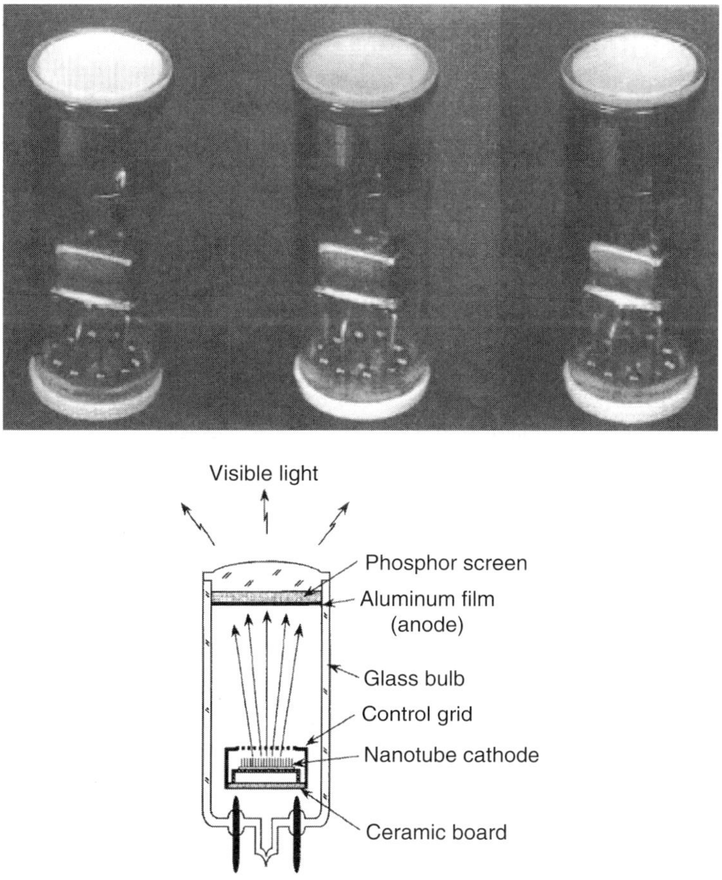

(B)

FIGURE 8.19 (A) The cathode ray lighting elements equipped with a carbon nanotube field-emitting cathode, displaying the three primary colors: green (ZnS:Cu, Al), red (Y_2O_3:Eu), and blue (ZnS:Ag) and (B) a schematic presentation of the carbon nanotube field-emission lamps, in which electrons are extracted from the cathode by applying a voltage to the control grid. The electrons are accelerated toward the phosphor screen, which consequently emits light. The color of the light depends on the choice of the phosphor. (From Saito, Y., Uemura, S., and Hamaguchi, K., *Jpn. J. Appl. Phys.*, 37, L346, 1998. With permission.)

with MWNT field emitters as a cathode (Figure 8.19A and B). In this study, stable electron emission, bright luminance, and long life suitable for various practical applications have been demonstrated. More recently, some prototypes of carbon nanotube field-emission flat panel displays were reported, notably by Samsung Advanced Institute of Technology in Korea and Ise Electronics Corp. in Japan [47].

8.5 OPTOELECTRONIC DEVICES FOR CONVERTING PHOTONS INTO ELECTRONS

8.5.1 C_{60}-Containing Polymer Photovoltaic Cells

The interaction of buckminsterfullerenes with light has attracted considerable interest among scientists in the exploration of applications related to photophysical, photochemical, and photo-induced charge transfer properties of fullerenes. The photo-induced charge transfer properties of fullerenes are largely determined by their electronic structures. Although a HOMO–LUMO energy gap as high as 4.9 eV has been determined for isolated C_{60} molecules [48,49], it was found to be diminished in solution or in the condensed solid state due to intermolecular electronic screening [50]. In fact, the minimum energy required to create a separated electron and hole in a C_{60} film was found to be of the order of 2.3–2.6 eV [51], which is well within the wavelength of UV/vis light.

The photo-induced charge transfer of fullerenes is of importance for the development of polymeric photovoltaic cells, which can be used to store light energy as electron relays for producing electricity. The photovoltaic effect involves the generation of electrons and holes in a semiconducting device under illumination, and subsequent charge collection at opposite electrodes. Inorganic semiconductors, such as silicon, gallium arsenide, and sulfide salts, have been widely used in conventional photovoltaic cells in which free electrons and holes were produced directly upon photon absorption.

As mentioned earlier, a major problem associated with inorganic semiconductor-based photovoltaic cells is the high cost for the production of the semiconductors. Therefore, organic dyes and conjugated polymeric semiconductors have received considerable attention in the search for novel photovoltaic cells. Since the fabrication of devices from inorganic semiconductors or organic dyes often involves relatively expensive techniques of vacuum evaporation or vapor deposition, photovoltaic cells based on soluble conjugated polymers, which can form thin films even over large areas by solution processing methods, become most attractive. However, the energy conversion efficiency that can be achieved with conjugated polymers is relatively low. Unlike their inorganic counterparts, photon absorption by conjugated polymers at room temperature typically creates tightly bound electron–hole pairs (i.e., Frenkel type excitons). Charge collection, therefore, requires dissociation of the excitons, a process which is known to be favorable at the interface between semiconducting materials with different ionization potentials and electron affinities [52]. Accordingly, devices with a single photoresponsive polymer layer often have small quantum yields and the typical energy conversion efficiencies of $10^{-3}\%$ to $10^{-2}\%$ for pure conjugated polymer-based photovoltaic cells are too low for them to be used in practical applications.

The observation of photovoltaic effects arising from the photo-induced charge transfer at the interface between conjugated polymers as donors and C_{60} film as acceptors [54] suggests interesting opportunities for improving energy conversion efficiencies of photovoltaic cells based on conjugated polymers. Indeed, increased quantum yields have been obtained by the addition of C_{60} to form heterojunctions with conjugated polymers, such as PPV, poly[2-methoxy-5-(2′-ethyl-hexyloxy)-*p*-phenylene vinylene] (MEH-PPV) [54,55] and platinum-polyyne [56]. In these conjugated polymer-C_{60} systems, excitons generated in either layer diffuse toward the interface between the layers. Although the photo-induced charge transfer between the excited C_{60} acceptor and a conducting polymer donor can occur very rapidly on a sub-picosecond timescale [57], with a quantum efficiency of close to unity for charge separation from donor to acceptor, the conversion efficiency of a bilayer heterojunction device is still limited [54] by several other factors. Firstly, since the efficient charge separation occurs only at the heterojunction interface, the overall conversion efficiency is diminished by the limited effective interfacial area available in the layer structure. Secondly, because the exciton diffusion range is typically at least a factor of 10 smaller than the optical absorption depth [52], the photoexcitations produced far from the interface recombine before diffusing to the heterojunction. Finally, the conversion efficiency is also limited by the carrier collection efficiency. To overcome these deficiencies, interpenetrating networks consisting of two semiconducting polymers (also called bulk heterojunction) have been developed as ideal photovoltaic materials for a high-efficiency photovoltaic conversion [53] (Figure 8.20). The interpenetrating network structure provides the spatially distributed interfaces necessary for both an efficient photogeneration and a facile collection of the electrons and holes [53]. As expected, significantly improved conversion efficiencies have been reported for photovoltaic cells based on interpenetrating network composites consisting of MEH-PPV and C_{60} derivatives [53]. As a matter of fact, a poly(3-hexylthiophene): [6,6]-phenyl-C61-butyric acid methyl ester (P3HT/PCBM) composite based solar cell has achieved a best power conversion efficiency of about 5% at AM 1.5 G standard sun test condition [60,61].

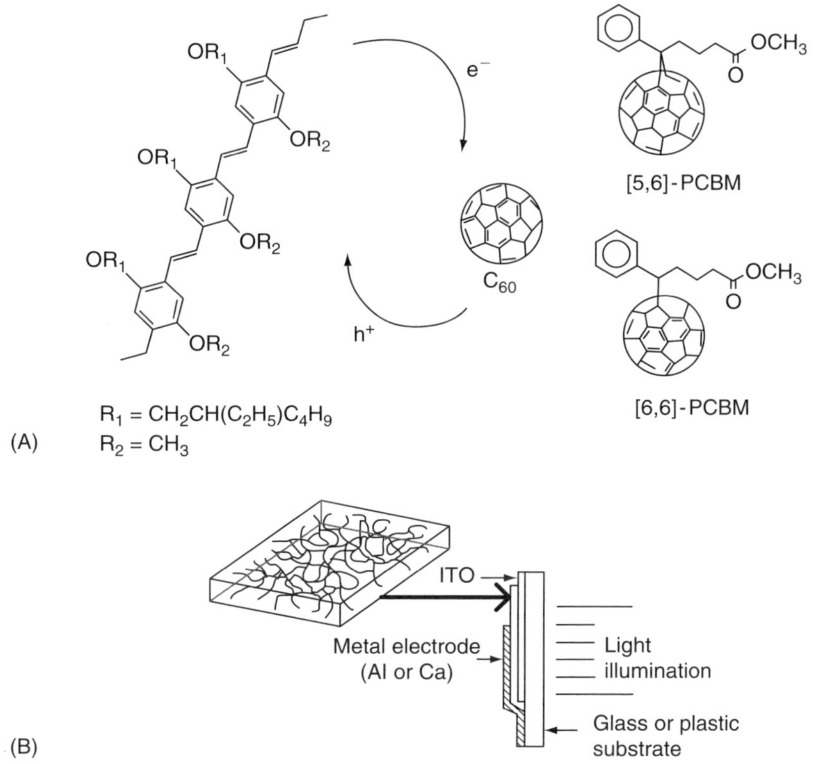

FIGURE 8.20 Schematic illustrations of (A) charge transfer between C_{60} derivatives and MEH-PPV and (B) the interpenetrating conjugated polymer-C_{60} (donor–acceptor) network and the photovoltaic cell. (From Yu, G., Gao, J., Hummelen, J.C., Wudl, F., and Heeger, A.J., *Science*, 270, 1789, 1995. With permission.)

8.5.2 Nanotube-Containing Polymer Photovoltaic Cells

Although solar cells contain the bulk heterojunction layer comprised of buckminsterfullerene (C_{60}) dispersed in a photovoltaic polymer is finding some success [58], there are associated limitations that hinder a real competitiveness with silicon-based solar cells. Two major problems are the instability of the polymers and poor charge transport in the C_{60} network. Encapsulation of the devices in air and moisture-free packaging can alleviate the problems with polymer stability but the charge transport issue remains at the forefront of the challenges with these solar cells. Charge transport is an inhibiting feature in C_{60}-based systems due to the spherical shape of these molecules and the fact that a conductive network must be formed by the successive touching of the C_{60} molecules throughout the heterojunction. The use of carbon nanotubes, instead of C_{60}, should be a viable approach to solve the charge transport problem and a significant advancement for the development of polymer photovoltaic cells of better performance. Because fullerenes have boosted the efficiency of polymer photovoltaic cells, therefore, carbon nanotubes have also been used to improve the device performance of polymer photovoltaic cells. In particular, Kymakis and Amaratunga [59] have spin-cast composite films of poly(3-octylthiophene), P3OT, and SWNTs on ITO-coated quartz substrates. They found that diodes with the Al/P3OT–SWNT/ITO structure (Figure 8.21A) showed photovoltaic behavior with an open circuit voltage of 0.7–0.9 V even at a low nanotube concentration (<1%). As shown in Figure 8.21B, the short circuit current is increased by two orders of magnitude compared with the pristine polymer diodes, with an increased fill factor from 0.3 to 0.4 for the nanotube–polymer cell. The improved device performance has been attributed to the good electronic properties of carbon nanotubes and their large surface areas.

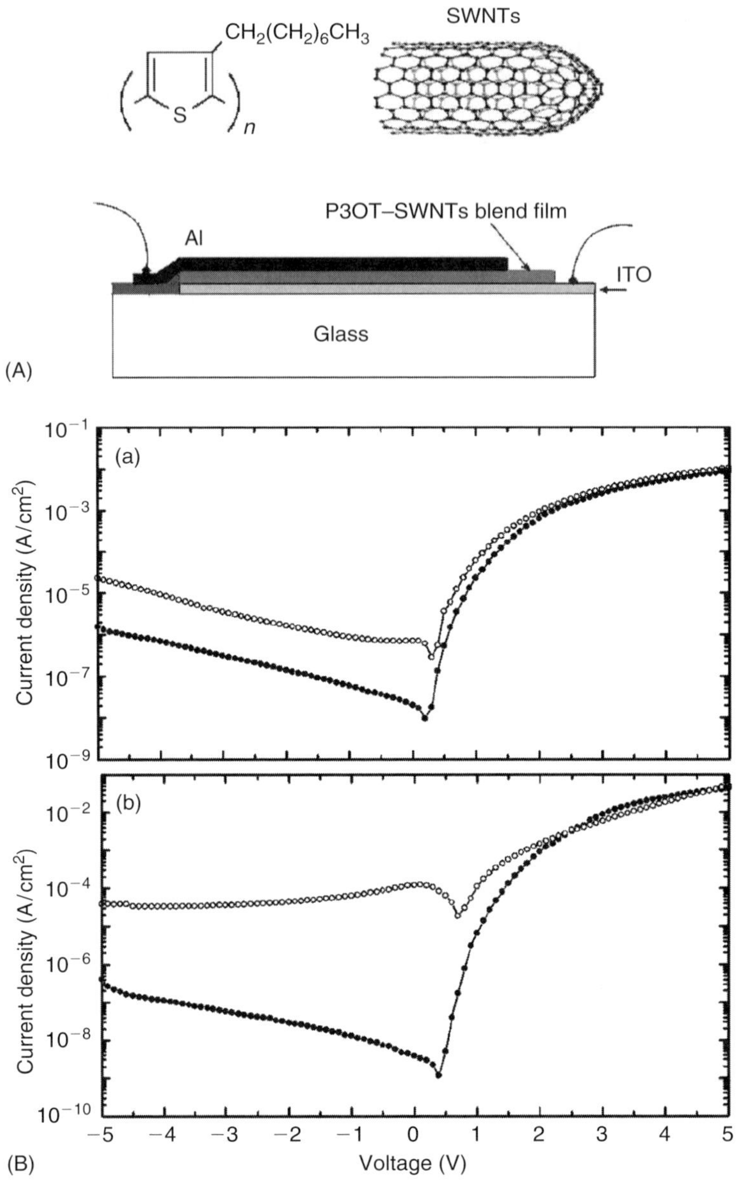

FIGURE 8.21 (A) A typical device architecture of the P3OT/SWNT photovoltaic cell. (B) (a) *I–V* characteristic of an ITO/P3OT/Al device in dark (filled circles) and under illumination (open circles); (b) the same data for an ITO/P3OT–SWNT/Al device. (From Kymakis, E. and Amaratunga, G.A.J., *Appl. Phys. Lett.*, 80, 112, 2002.)

Recently, photoconversion efficiencies of up to 5.2% and 9.3% have also been reported for flexible polymer solar cells containing C_{60} fullerene [60,61] and carbon nanotube [62]. Comparing with C_{60}, the long cylindrically shaped carbon nanotubes can also provide higher electron mobility with a lower percolation threshold, facilitating the subsequent charge transport to the electrode. In fact, carbon nanotubes are known as an extremely electron-conductive semiconductor [63]. The higher carrier mobility allows thicker devices to harvest more photons without sacrificing internal quantum efficiency. Further, improved light absorption in the UV and red region can be achieved upon the addition of the dye molecules [63]. Carbon nanotubes can also provide good mechanical stability and high thermal conductivity [18]. Clearly, a promising potential for future research exists in this area.

8.6 CONCLUDING REMARKS

Owing to the substantial π-electron delocalization, conjugated conducting polymers, fullerenes, and carbon nanotubes have been shown to possess interesting optoelectronic properties attractive for a wide range of potential applications, including FETs, LEDs, field-emitting flat panel displays, and organic photovoltaic cells. We have briefly reviewed recent progresses in the development of some most promising organic optoelectronic devices involving conjugated conducting polymers, fullerene C_{60}, and carbon nanotubes as active materials. The optoelectronic properties for each of the materials were discussed, along with their device principles and performance. The continuing arrival of new organic optoelectronic materials, together with the rapid development in the device design and fabrication technologies, has been shown to greatly improve the performance of these organic optoelectronic devices. Continued research and development in this field of organic optoelectronics should hasten the ongoing race to the technology marketplace.

EXERCISE QUESTIONS

1. What is C_{60}? What is PCBM? How many hexagons and pentagons are there in C_{60}?
2. What is a carbon nanotube? What are the main differences between a carbon nanotube and C_{60}? What are the main differences between an SWNT and an MWNT?
3. What advantages a carbon nanotube has when compared with C_{60} for solar cell applications?
4. Read the Nobel lectures on the discovery of fullerenes:
   a. Curl, R.F. 1997 Dawn of the fullerenes: Conjecture and experiment, *Angew. Chem. Int. Ed.* 36, 1566.
   b. Kroto, H. 1997 Symmetry, space, stars, and C_{60}, *Angew. Chem. Int. Ed.* 36, 1578.
   c. Smalley, R.E. 1997 Discovery the fullerenes, *Angew. Chem. Int. Ed.* 36, 1594.
5. Find out all you can about carbon nanotubes.

LIST OF ABBREVIATIONS

AFM	atomic force microscope
BND	2,5-bis-(4-naphthyl)-1,3,4-oxadiazole
FETs	field-effect transistors
HOMO	the highest occupied molecular orbital
ITO	transparent indium tin oxide
LEDs	light-emitting diodes
LUMO	the lowest unoccupied molecular orbital
MEH-PPV	poly[2-methoxy-5-(2′-ethylhexyloxy)-*p*-phenylene vinylene]
MISFET	metal–insulator–semiconductor field-effect transistor
MWNTs	multiwalled carbon nanotubes
PVK	poly(9-vinylcarbazole)
P3OT	poly(3-octylthiophene)
PPV	poly(*p*-phenylene vinylene)
SWNTs	single-walled carbon nanotubes
THP	T_h-hexapyrrolidine

ACKNOWLEDGMENT

We thank the authors whose work was cited and apologize to the authors of papers not cited here due to the space limitation. We are grateful for financial support from the AFRL/ML-HBCU (05-S555-0006-C5).

REFERENCES

1. Horowitz, G. 1998 *Adv. Mater.* 10, 365.
2. Lovinger, A.J. and Rothberg, L.J. 1996 *J. Mater. Res.* 11, 1581.
3. Mylnikov, V.S. 1994 *Photoconducting Polymers,* Advances in Polymer Science, Vol. 115, Berlin: Springer-Verlag.
4. Luque, A. and Hegedus, S. 2003 *Handbook of Photovoltaic Science and Engineering*, Hoboken, NJ: Wiley.
5. Greenham, N.C. and Friend, R.H. 1995 *Solid State Phys.* 49, 1.
6. Hide, F., Díaz-García, M.A., Schwartz, B.J., and Heeger, A.J. 1997 *Acc. Chem. Res.* 30, 430.
7. http://www.cedmagic.com/history/transistor-1947.html.
8. Quillec, M. 1996 *Materials for Optoelectronics*, Boston: Kluwer Academic Publishers.
9. Dai, L. 2004 *Intelligent Macromolecules for Smart Devices: From Materials Synthesis to Device Applications*, Berlin: Springer-Verlag.
10. Shirakawa, H., Louis, E.J., MacDiarmid, A.G., Chiang, C.K., and Heeger, A.J. 1977 *Chem. Commun.*, 16, 578.
11. Jones, N. 2000 *New Sci. Oct.* 21, 14.
12. Burroughes, J.H., Bradley, D.C.C., Brown, A.R., Mackey, M.K., Friend, R.H., and Burn, P.L. 1990 *Nature* 347, 539.
13. Naarmann, H. and Theophilou, N. 1987 *Synth. Met.* 22, 1.
14. Kroto, H.W., Heath, J.R., O'Brien, S.C., Curl, R.F., and Smalley, R.E. 1985 *Nature* 318, 162.
15. Ōsawa, E. 1970 *Kagaku (Japanese)* 25, 854.
16. Huffman, D. 1991 *Phys. Today* 11, 22.
17. Iijima, S. 1991 *Nature* 354, 56.
18. Dai, L. (Ed.) 2006 *Carbon Nanotechnology: Recent Developments in Chemistry, Physics, Materials Science and Device Applications*, Amsterdam: Elsevier.
19. Burroughes, J.H. and Friend, R.H. 1991 in *Conjugated Polymers*, Brédas, J.L. and Silbey, R. (Eds.), Dordrecht: Kluwer Academic Publishers.
20. Manohara, H.M., Wong, E.W., Schlecht, E., Hunt, B.D., and Siegel, P.H. 2005 *Nano Lett.* 5, 1469.
21. McElvain, J., Keshavarz, M., Wang, H., Wudl, F., and Heeger, A.J. 1997 *J. Appl. Phys.* 81, 6468.
22. Rhoderick, E.H. and Williams, R.H. 1988, *Metal-Semiconductor Contacts*, Oxford: Oxford University Press.
23. Tans, S.J., Verschueren, A.R.M., and Dekker, C. 1998 *Nature* 393, 49.
24. Martel, R., Schmidt, T., Shea, H.R., Hertel, T., and Avouris, P. 1998 *Appl. Phys. Lett.* 73, 2447.
25. Haddon, R.C., Perel, A.S., Morris, R.C., Palstra, T.T.M., Hebard, A.F., and Fleming, R.M. 1995 *Appl. Phys. Lett.* 67, 121.
26. Bradley, D.C. 1992 *Adv. Mater.* 4, 756, and references therein.
27. Burn, P.L., Bradley, D.D.C., Friend, R.H., Halliday, D.A., Holmes, A.B., Jackson, R.W., and Kraft, A. 1992 *J. Chem. Soc., Perkin Trans.* 1, 3225.
28. Burn, P.L., Kraft, A., Baigent, D.R., Bradley, D.D.C., Brown, A.R., Friend, R.H., and Gymer, R.W. 1992 *Nature* 356, 47.
29. Dai, L. 1999 *Polym. Adv. Technol.* 10, 357.
30. Graupner, W., Leising, G., Lanzani, G., Nisoli, M., De Silvestri, S., and Scherf, U. 1996 *Phys. Rev. Lett.* 76, 847.
31. Hertel, D., Setayesh, S., Nothofer, H.G., Scherf, U., Müllen, K., and Bässler, H. 2001 *Adv. Mater.* 13, 65.
32. Brown, A.R., Bradley, D.D.C., Burroughes, J.H., Friend, R.H., Greenham, N.C., Burn, P.L., Kraft, A., and Holmes, A.B. 1992 *Appl. Phys. Lett.* 61, 2793.
33. Greenham, N.C., Moratti, S.C., Bradley, D.D.C., Friend, R.H., and Holmes, A.B. 1993 *Nature* 365, 628.
34. Hiberer, A., van Hutten, P.F., Wildeman, J., and Hadziioannou, G. 1997 *Macromol. Chem. Phys.* 198, 2211, and references cited therein.
35. Lide, D.R. (Ed.) 1996 *CRC Handbook of Chemistry and Physics*, 77th edn., Boca Raton: CRC Press.
36. Nakayama, T., Itoh, Y., and Kakuta, A. 1993 *Appl. Phys. Lett.* 63, 594.
37. Bradley, D.D.C. and Tsutsui, T. (Eds.) 1995 *Organic Electroluminescence*, Cambridge: Cambridge University Press.
38. Salaneck, W.R. and Bredas, J.L. 1996 *Adv. Mater.* 8, 48.
39. Hutchison, K., Gao, J., Schick, G., Rubin, Y., and Wudl, F. 1999 *J. Am. Chem. Soc.* 121, 5611.
40. Sariciftci, N.S., Braun, D., Zhang, C., Srdranov, V., Heeger, A.J., and Wudl, F. 1992 *Science* 258, 1474.
41. Palstra, T.T.M., Haddon, R.C., and Lyons, K.B. 1997 *Carbon* 35, 1825.

42. de Heer, W.A., Bonard, J.M., Fauth, K., Chatelain, A., Forro, L., and Ugarte, D. 1997 *Adv. Mater.* 9, 87.
43. Fowler, R.H. and Nordheim, L.W. 1928 *Proc. Royal. Soc. A* (London) 119, 173.
44. Bonard, J.M., Salvetat, J.P., Stockli, T., de Heer, W.A., Forro, L., and Chatelain, A. 1998 *Appl. Phys. Lett.* 73, 918.
45. Collins, P.G. and Zettl, A. 1996 *Appl. Phys. Lett.* 69, 1969.
46. Saito, Y., Uemura, S., and Hamaguchi, K. 1998 *Jpn. J. Appl. Phys.* 37, L346.
47. de Heer, W.A. and Martel, R. 2000 *Phys. World* June, 49.
48. Skotheim, T.A. (Ed.) 1986 *Handbook of Conducting Polymers*, New York: Marcel Dekker.
49. Lichtenberger, D.L., Nebesney, K.W., Ray, C.D., Huffman, D.R., and Lamb, L.D. 1991 *Chem. Phys. Lett.* 176, 203.
50. Hammond, G.S. and Kuck, V.J. (Eds.) 1992 *Fullerenes: Synthesis, Properties, and Chemistry of Large Carbon Clusters*, ACS Symposium Series 481, Washington DC: American Chemical Society.
51. Lof, R.W., van Veenendaal, M.A., Koopmans, B., Jonkrnan, H.T., and Sawatzky, G.A. 1992 *Phys. Rev. Lett.* 68, 3924.
52. Tang, C.W. 1986 *Appl. Phys. Lett.* 48, 183.
53. Yu, G. Gao, J., Hummelen, J.C., Wudl, F., and Heeger, A.J., 1995 *Science* 270, 1789.
54. Kraabel, B., Hummelen, J.C., Vacar, D., Moses, D., Sariciftci, N.S., Heeger, A.J., and Wudl, F. 1996 *J. Chem. Phys.* 104, 4267.
55. Yu, G., Pakbaz, V., and Heeger, A.J. 1994 *Appl. Phys. Lett.* 64, 3422.
56. Kohler, A., Wittmann, H.F., Friend, R.H., Khan, M.S., and Lewis, J. 1996 *Synth. Met.* 77, 147.
57. Kamat, P.V. and Asmus, K.D. 1996 *Electrochem. Soc. Interface* Spring, 22.
58. Cho, Y. -J., Ahn, T.K., Song, H., Kim, K.S., Lee, C.Y., Seo, W.S., Lee, K., Kim, S.K., Kim, D., and Park, J.T. 2005 *J. Am. Chem. Soc.* 127, 2380.
59. Kymakis, E. and Amaratunga, G.A.J. 2002 *Appl. Phys. Lett.* 80, 112.
60. Reyes-Reyes, M., Kim, K., Dewald, J., Lopez-Sandoval, R., Avadhanula, A., Curran, S., and Carroll, D.L. 2005 *Org. Lett.* 7, 5749.
61. Li, G., Shrotriya, V., Huang, J.S., Yao, Y., Moriarty, T., Emery, K., and Yang, Y. 2005 *Nat. Mater.* 4, 864.
62. Rahman, G.M.A., Guldi, D.M., Cagnoli, R., Mucci, A., Schenetti, L., Vaccari, L., and Prato, M. 2005 *J. Am. Chem. Soc.* 127, 10051.
63. Jin, M.H.-C. and Dai, L. 2005 Vertically aligned carbon nanotubes for organic photovoltaic devices, in *Organic Photovoltaics: Mechanisms, Materials and Devices*, Sun, S. and Sariciftci, N.S. (Eds.), Boca Raton: CRC Press.

9 Introduction of Organic Superconducting Materials

Hatsumi Mori

CONTENTS

9.1 Basic Concepts of Organic Superconductors 263
9.1.1 Characteristics of Organic Superconductors 263
9.1.2 History of Organic Semiconductors, Organic Metals, and Organic Superconductors 271
9.2 Crystal Growth of Organic Superconductors 274
9.2.1 Electrocrystallization Method 274
9.3 Organic Superconductors 276
9.3.1 Pseudo-One-Dimensional Organic Superconductors: TMTSF Salts 276
9.3.2 Two-Dimensional Organic Superconductors: BEDT-TTF Salts 279
9.3.2.1 β-Type BEDT-TTF Salts 279
9.3.2.2 κ-Type BEDT-TTF Salts: Competition between Antiferromagnetic and SC States 280
9.3.2.3 β′-Type BEDT-TTF Salts: Competition between Antiferromagnetic and SC States 280
9.3.3 Three-Dimensional Molecular Superconductors 282
9.4 Summary 283
Exercise Questions 284
List of Abbreviations 284
References 285

Abstract Since the first discovery of an organic superconductor in 1980, $TMTSF_2PF_6$ (TMTSF = tetramethyltetraselenafulvalene) with the superconducting (SC) transition temperature (T_c) = 0.9 K under 12 kbar, approximately 144 organic superconductors have been found. Among them, the highest T_c of the two-dimensional (2D) sulfur-based organic superconductors is 14.2 K under 82 kbar for β′-$(BEDT\text{-}TTF)_2ICl_2$ (BEDT-TTF = bis(ethylenedithio)tetrathiafu lvalene) and 3D C_{60}-based superconductor is 33 K for Cs_2RbC_{60}. In this chapter, the development, crystal growth, and the characteristic features of pseudo-1D, 2D, and 3D molecular superconductors are introduced.

9.1 BASIC CONCEPTS OF ORGANIC SUPERCONDUCTORS

9.1.1 Characteristics of Organic Superconductors

In contrast to inorganic metals, one-component metal-free organic materials could not show metallic conductivity at an ambient pressure, since the molecules have a closed shell with small transfer integrals and a large on-site Coulomb repulsion energy. Therefore, the necessary conditions to synthesize organic conductors are the following:

- To generate charge carriers, the charge transfer (CT) complexes composed of donors and acceptors to produce hole and electron carriers, donors and anions with holes, or cations and acceptors with electrons should be prepared (Figure 9.1a). The degree of charge transfer (γ) is in the range of $0.5 \leq \gamma < 1$.
- To construct conducting path, the segregated donor and acceptor columns should be formed (Figure 9.1b).

The organic superconductors [1–6] basically possess the framework of TTF (tetrathiafulvalene) molecule [7]. Since TTF is a 7 π system (Figure 9.1a), this molecule is easily oxidized and the oxidized one is stable as 6 π Hückel rule. The molecular structures of donors and acceptors constituting organic superconductors and the list of organic superconductors are shown in Figure 9.2, and Tables 9.1 and 9.2, respectively. The organic molecules are flat and construct a stacking column. Along the stacking direction, the most conductive path is formed and the kinetic energy of carriers is proportional to the overlap of π orbitals of molecules that extended perpendicular to a molecular plane. Owing to a flat molecule, conductivity is anisotropic and the conducting path is 1D or 2D (Figure 9.1c).

Most organic superconductors are cation radical charge transfer complexes composed of donors (D) and closed shell anions (X^-) like $(D^{\gamma+})_m(X^-)_n$, the so-called hole superconductors. There also exists some anion radical complexes constituting closed shell cations (tetramethylammonium [TMA^+])

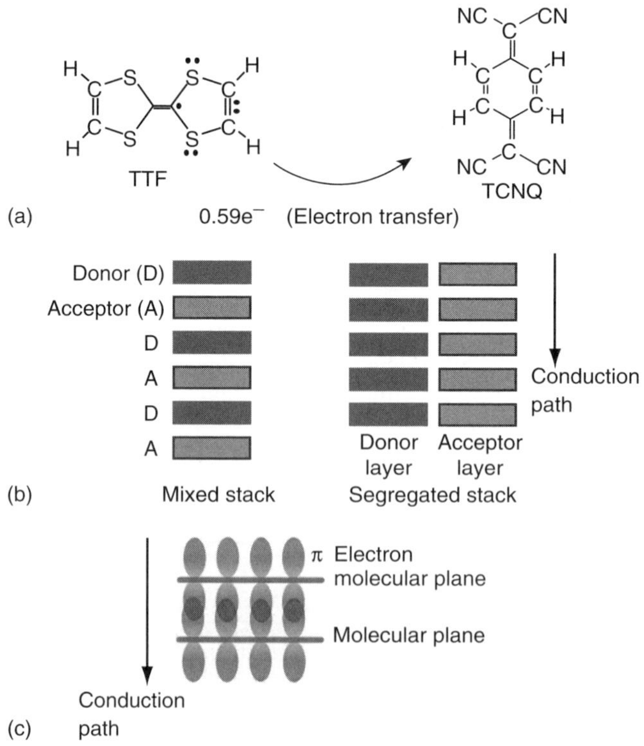

FIGURE 9.1 (a) CT from an organic donor of TTF to an acceptor of TCNQ (tetracyanoquinodimethane) by $0.59e^-$, (b) mixed stack and segregated stack, and (c) a stack of π electron molecular plane.

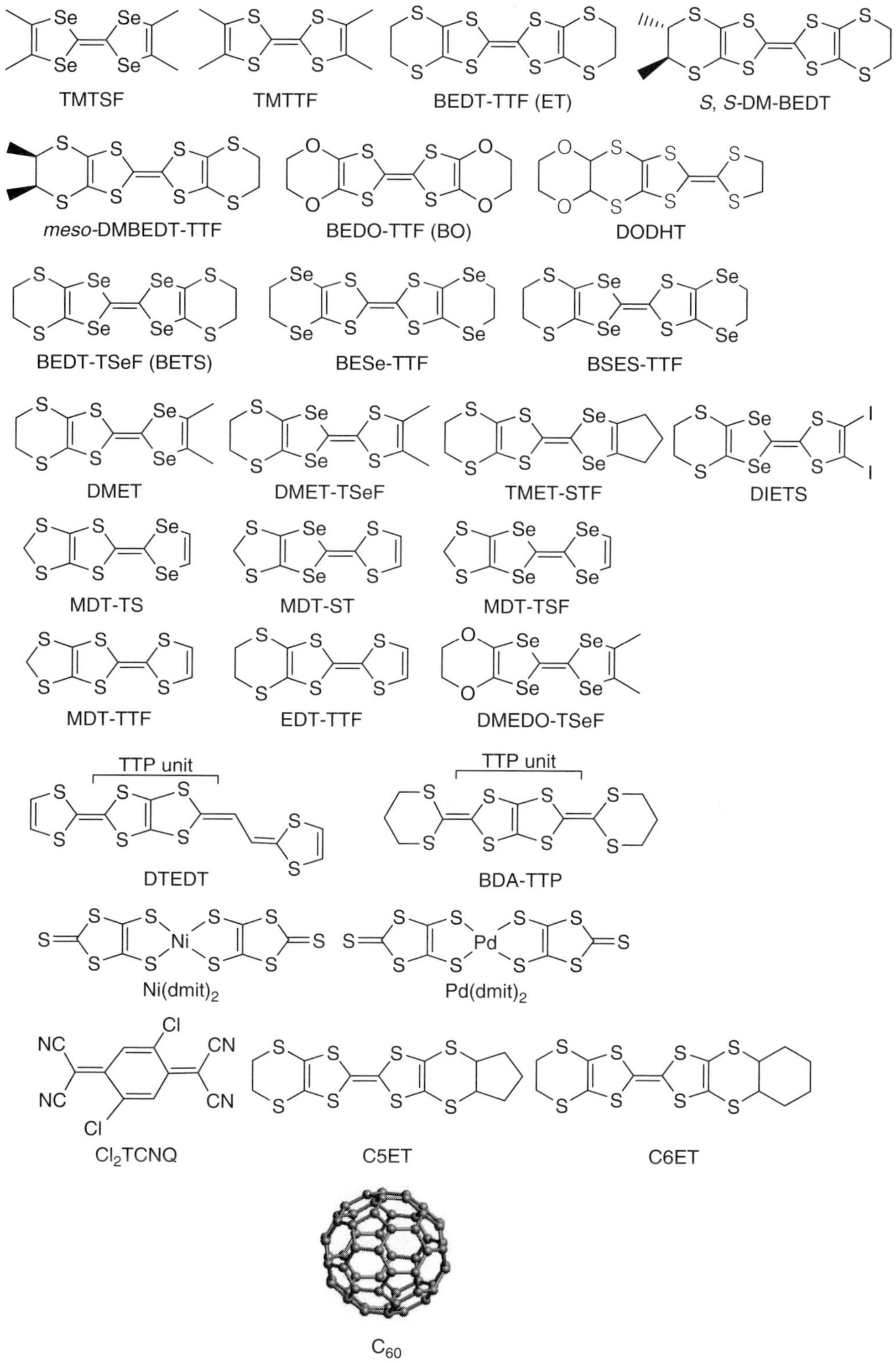

FIGURE 9.2 Molecular structures of donors and acceptors constituting organic superconductors and related materials.

TABLE 9.1
List of Molecular Superconductors

Compound	T_c	P_c	T_{MI}	Electronic Property
TMTSF Complexes				
$(TMTSF)_2PF_6$	0.9	12	12	$T_{SDW} = 12$ K
$(TMTSF)_2AsF_6$	1.1	12	12	$T_{SDW} = 12$ K
$(TMTSF)_2SbF_6$	0.38	10.5	17	$T_{SDW} = 17$ K
$(TMTSF)_2TaF_6$	1.35	11	11	$T_{SDW} = 11$ K
$(TMTSF)_2ClO_4$	1.4	0		$T_{DO} = 24$ K
$(TMTSF)_2ReO_4$	1.3	9.5	180	$T_{DO} = 180$ K
$(TMTSF)_2FSO_3$	2.1	6.5	86	$T_{DO} = 86$ K
$(TMTSF)_2NbF_6$	1.26	12	12	$T_{SDW} = 12$ K
TMTTF Complexes				
$(TMTTF)_2PF_6$	1.8	54	250	$T_{CO} = 70$ K, $T_{SP} = 15$ K
$(TMTTF)_2Br$	0.8	26	100	$T_N = 15$ K
$(TMTTF)_2SbF_6$	2.6	61	154	$T_{CO} = 154$ K, $T_N = 8$ K
$(TMTTF)_2BF_4$	1.38	33.5	210	$T_{DO} = 40$ K
BEDT-TTF Complexes				
β″-$(BEDT\text{-}TTF)_3Cl_2(H_2O)_2$	2	16		
β-$(BEDT\text{-}TTF)_2ReO_4$	2	4	81	$T_{DO} = 81$ K
α-$(BEDT\text{-}TTF)_2I_3$	3.3	0		I_2 doping
	7.2	2		Uniaxial strain
β-$(BEDT\text{-}TTF)_2I_3$ low T_c	1.5	0		$T_{incommensurate} = 175$ K
β-$(BEDT\text{-}TTF)_2I_3$ high T_c	8.1	0		
γ-$(BEDT\text{-}TTF)_2I_{2.5}$	2.5	0		
θ-$(BEDT\text{-}TTF)_2I_3$	3.6	0		
κ-$(BEDT\text{-}TTF)_2I_3$	3.6	0		
β-$(BEDT\text{-}TTF)_2IBr_2$	2.7	0		
β-$(BEDT\text{-}TTF)_2AuI_2$	4.9	0		
κ-$(BEDT\text{-}TTF)_2Cu(NCS)_2$	10.4	0		
κ-$(BEDT\text{-}TTF)_4Hg_{2.89}Cl_8$	1.8	>12	<9	
κ-$(BEDT\text{-}TTF)_4Hg_{2.89}Br_8$	4	0		
α-$(BEDT\text{-}TTF)_2NH_4Hg(SCN)_4$	0.8	0		
κ-$(BEDT\text{-}TTF)_2Cu[N(CN)_2]Br$	11.6	0		
κ-$(BEDT\text{-}TTF)_2Cu[N(CN)_2]Cl$	12.8	0.3		
κ-$(BEDT\text{-}TTF)_2Cu[N(CN)_2]I$	8	10		
κ-$(BEDT\text{-}TTF)_2Ag(CN)_2H_2O$	5	0		
κ-$(BEDT\text{-}TTF)_2Cu_2(CN)_3$	2.8	1.5		
κ′-$(BEDT\text{-}TTF)_2Cu_2(CN)_3$	3.8	0		
β″-$(BEDT\text{-}TTF)_4Pt(CN)_4H_2O$	2	6.5		
κ-$(BEDT\text{-}TTF)_2Cu[N(CN)_2]CN$	10.7	0		
β″-$(BEDT\text{-}TTF)_4Pd(CN)_4H_2O$	1.2	7		
$κ_H$-$(BEDT\text{-}TTF)_2Cu(CF_3)_4(TCE)_x$	9.2	0		
$κ_L$-$(BEDT\text{-}TTF)_2Cu(CF_3)_4(TCE)$	4	0		
$κ_H$-$(BEDT\text{-}TTF)_2Ag(CF_3)_4(TCE)$	11.1	0		
$κ_L$-$(BEDT\text{-}TTF)_2Ag(CF_3)_4(TCE)$	2.4	0		

TABLE 9.1 (continued)
List of Molecular Superconductors

Compound	T_c	P_c	T_{MI}	Electronic Property
κ_H-(BEDT-TTF)$_2$Au(CF$_3$)$_4$(TCE)	10.5	0		
κ_L-(BEDT-TTF)$_2$Au(CF$_3$)$_4$(TCE)	2.1	0		
κ_L-(BEDT-TTF)$_2$Cu(CF$_3$)$_4$– (1-bromo-1,2-dichloroethane)	3.5	0		
κ_L-(BEDT-TTF)$_2$Cu(CF$_3$)$_4$– (2-bromo-1,1,-dichloroethane)	4.9	0		
κ_L-(BEDT-TTF)$_2$Cu(CF$_3$)$_4$– (1,2-dibromo-1-chloroethane)	5.5	0		
κ_L-(BEDT-TTF)$_2$Cu(CF$_3$)$_4$– (1,1,2-tribromoethane)	5.2	0		
κ_L-(BEDT-TTF)$_2$Ag(CF$_3$)$_4$– (1-bromo-1,2-dichloroethane)	3.8	0		
κ_L-(BEDT-TTF)$_2$Ag(CF$_3$)$_4$– (1-bromo-1,2-dichloroethane)	7.3	0		
κ_H-(BEDT-TTF)$_2$Ag(CF$_3$)$_4$– (2-bromo-1,1,-dichloroethane)	10.2	0		
κ_L-(BEDT-TTF)$_2$Ag(CF$_3$)$_4$– (2-bromo-1,1,-dichloroethane)	4.1	0		
κ_L-(BEDT-TTF)$_2$Ag(CF$_3$)$_4$– (1,2-dibromo-1-chloroethane)	4.5	0		
κ_H-(BEDT-TTF)$_2$Ag(CF$_3$)$_4$– (1,1,2-tribromoethane)	7.2	0		
κ_L-(BEDT-TTF)$_2$Ag(CF$_3$)$_4$– (1,1,2-tribromoethane)	4.8	0		
κ_L-(BEDT-TTF)$_2$Au(CF$_3$)$_4$– (1-bromo-1,2-dichloroethane)	3.2	0		
κ_L-(BEDT-TTF)$_2$Au(CF$_3$)$_4$– (2-bromo-1,1,-dichloroethane)	5	0		
κ_L-(BEDT-TTF)$_2$Au(CF$_3$)$_4$– (1,2-dibromo-1-chloroethane)	5	0		
κ_L-(BEDT-TTF)$_2$Au(CF$_3$)$_4$– (1,1,2-tribromoethane)	5.8	0		
β''-(BEDT-TTF)$_4$(H$_3$O)Fe(C$_2$O$_4$)$_3$PhCN	8	0		
β''-(BEDT-TTF)$_4$(H$_3$O)Cr(C$_2$O$_4$)$_3$PhCN	6	0		
β''-(BEDT-TTF)$_4$AFe(C$_2$O$_4$)$_3$PhNO$_2$[A = H$_3$O or NH$_4$]	6.2	0		
β''-(BEDT-TTF)$_4$ACr(C$_2$O$_4$)$_3$PhNO$_2$[A = H$_3$O or NH$_4$]	5.8	0		
β''-(BEDT-TTF)$_4$(H$_3$O)Fe(C$_2$O$_4$)$_3$PhBr	4	0		
β''-(BEDT-TTF)$_4$(H$_3$O)Cr(C$_2$O$_4$)$_3$PhBr	1.5	0		
β''-(BEDT-TTF)$_4$(H$_3$O)Ga(C$_2$O$_4$)$_3$PhNO$_2$	7.5	0		
β''-(BEDT-TTF)$_4$(H$_3$O)Ga(C$_2$O$_4$)$_3$Pyridine	1.5	0		
β''-(BEDT-TTF)$_2$SF$_5$CH$_2$SF$_2$SO$_3$	5.2	0		
β'-(BEDT-TTF)$_2$ICl$_2$	14.2	82		
β'-(BEDT-TTF)$_2$IBrCl	8.5	86		
α-(BEDT-TTF)$_2$KHg(SCN)$_4$	1	3		Uniaxial strain
***S,S*-DM-BEDT Complex**				
κ-(*S,S*-DM-BEDT)$_2$ClO$_4$	2.6	5.8		

(continued)

TABLE 9.1 (continued)
List of Molecular Superconductors

Compound	T_c	P_c	T_{MI}	Electronic Property
BEDO-TTF Complexes				
β″-$(BEDO\text{-}TTF)_3Cu_2(NCS)_3$	1.06	0		
β″-$(BEDO\text{-}TTF)_2ReO_4$	0.9	0		
BEDT-TSeF Complexes				
λ-$(BEDT\text{-}TSeF)_2GaCl_4$	5.5	0		
$(BEDT\text{-}TSeF)(Cl_2TCNQ)$	1.3	3.5		
κ-$(BEDT\text{-}TSeF)_2FeBr_4$	1.1	0		$T_N = 2.5$ K
λ-$(BEDT\text{-}TSeF)_2FeCl_4$	1.8	3		
	0.1			17 T
κ-$(BEDT\text{-}TSeF)_2FeCl_4$	0.1			$T_N = 0.45$ K
κ-$(BEDT\text{-}TSeF)_2GaBr_4$	1	0		
κ-$(BEDT\text{-}TSeF)_2TlCl_4$	2.5	0		
BESe-TTF Complex				
κ-$(BESe\text{-}TTF)_2Cu[N(CN)_2]Br$	1.5	7.5		
ESET-TTF Complex				
κ-$(ESET\text{-}TTF)_2Cu[N(CN)_2]Br$	4.7	3.2		
DMET-TSeF Complexes				
$(DMET\text{-}TSeF)_2AuI_2$	0.58	0		
$(DMET\text{-}TSeF)_2I_3$	0.5	0		
DMET Complexes				
β-$(DMET)_2Au(CN)_2$	0.8	5	25	$T_{SDW} = 24$ K
β-$(DMET)_2AuCl_2$	0.83	0		
β-$(DMET)_2AuBr_2$	1	1.5		
κ-$(DMET)_2AuBr_2$	1.9	0		
β-$(DMET)_2AuI_2$	0.55	5	20	$T_{SDW} = 16$ K
β-$(DMET)_2I_3$	0.47	0		
β-$(DMET)_2IBr_2$	0.59	0		
β-$(DMET)_2CuCl_2$	0.8	0		
MDT-TTF Complex				
κ-$(MDT\text{-}TTF)_2AuI_2$	3.5	0		
DTEDT Complex				
$(DTEDT)_3[Au(CN)_2]$	4	0		
TMET-STF Complex				
$(TMET\text{-}STF)_2BF_4$	4.1	0		
MDT-TSF Complexes				
β-$(MDT\text{-}TSF)(AuI_2)_{0.44}$	4.5	0		
β-$(MDT\text{-}TSF)I_{1.27}$	4.6	0		
β-$(MDT\text{-}TSF)I_{1.19}Br_{0.08}$	5	0		

TABLE 9.1 (continued)
List of Molecular Superconductors

Compound	T_c	P_c	T_{MI}	Electronic Property
MDT-ST Complexes				
β-(MDT-ST)$I_{1.27}$	3.6	0		
β-(MDT-ST)$I_{1.27-\delta}Br_{\delta}(\delta \sim 0.2)$	3.2	0		
MDT-TS Complex				
β-(MDT-TS)$(AuI_2)_{0.441}$	4.7	11.4		
MDSe-TSF Complex				
κ-$(MDSe\text{-}TSF)_2Br$	4	0		
EDT-TTF Complex				
β-$(EDT\text{-}TTF)_4Hg_{3-\delta}I_8$	8.1	0		$\delta = 0.1–0.2$
BDA-TTP Complexes				
β-$(BDA\text{-}TTP)_2SbF_6$	6.9	0		
β-$(BDA\text{-}TTP)_2AsF_6$	5.9	0		
β-$(BDA\text{-}TTP)_2PF_6$	5.9	0		
β-$(BDA\text{-}TTP)_2GaCl_4$	3.1	7.6		
β-$(BDA\text{-}TTP)_2FeCl_4$	2	6.3		
β-$(BDA\text{-}TTP)_2I_3$	7.6	10.3		
BODHT Complexes				
β″-$(DODHT)_2PF_6$	3.1	16.5		
β″-$(DODHT)_2AsF_6$	3.3	16.5		
β″-$(DODHT)_2BF_4 \cdot H_2O$	3.2	15.5		
DIETS Complex				
θ-$(DIETS)_2[Au(CN)_4]$	8.6	10		
***meso*-DMBEDT-TTF Complexes**				
β-(*meso*-DMBEDT-TTF)$_2PF_6$	4.3	4		
β-(*meso*-DMBEDT-TTF)$_2AsF_6$	4.3	3.8		
DMEDO-TSeF Complex				
κ-$(DMEDO\text{-}TSeF)_2[Au(CN)_4]$ (THF)	4.8	3.0		
dmit Complexes				
$TTF[Ni(dmit)_2]_2$	1.6	7		
$Me_4N[Ni(dmit)_2]_2$	5	7	100	
$(EDT\text{-}TTF)[Ni(dmit)_2]$	1.3	0		
C_{60} Complexes				
α-$TTF[Pd(dmit)_2]_2$	1.7	22	220	
α′-$TTF[Pd(dmit)_2]_2$	5.93	24	220	
β-$Me_4N[Pd(dmit)_2]_2$	6.2	6.5		
β′-$Me_4As[Pd(dmit)_2]_2$	4	7		
β′-$Me_4Sb[Pd(dmit)_2]_2$	3	10		Uniaxial strain
	8.4	4.5		

(*continued*)

TABLE 9.1 (continued)
List of Molecular Superconductors

Compound	T_c	P_c	T_{MI}	Electronic Property
$Me_2Et_2N[Pd(dmit)_2]_2$	4	2.4		
β′-$Me_2Et_2P[Pd(dmit)_2]_2$	4	6.9		Uniaxial strain
	6.1	4.5		
β′-$Me_2Et_2As[Pd(dmit)_2]_2$	5.5	8		
Na_2KC_{60}	2.5			
Na_2RbC_{60}	2.5			
Na_2CsC_{60}	12			
K_3C_{60}	19			
K_2RbC_{60}	21.8			
K_2CsC_{60}	24.5			
KRb_2C_{60}	26.4			
Rb_3C_{60}	28, 30			
Rb_2CsC_{60}	31			
$RbCs_2C_{60}$	33			
Li_2CsC_{60}	12			
$KRbCsC_{60}$	29			
$(NH_3)_xNaK_2C_{60}$	11			
$(NH_3)_xNaRb_2C_{60}$	13			
$(NH_3)_xK_3C_{60}$	8.5			
$(NH_3)\ K_3C_{60}$	28	14.8		
Cs_3C_{60}	40			
$Na_xN_yC_{60}$	14.7			
Ca_5C_{60}	8.9	5		
Sr_4C_{60}	4.4			
Ba_4C_{60}	6.7			
$Yb_{2.75}C_{60}$	6			
Sm_xC_{60}	8			
La_xC_{60}	12.5			
$K_3Ba_3C_{60}$	5.6			
$Rb_3Ba_3C_{60}$	3 ~ 5			

Source: From Mori, H., *J. Phys. Soc. Jpn.*, 75, 051003, 2006. With permission.

T_c, SC transition temperature; P_c, applied pressure, in which SC state appears; T_{MI}, metal–insulator transition temperature; T_{SDW}, SDW transition temperature; T_{DO}, anion disorder–order temperature; T_{CO}, charge-ordered temperature; T_N, Neel temperature; and the Greek symbol denotes the donor arrangement as shown in Figure 9.7.

and acceptors (A) like $(TMA^+)_m(A^{\gamma-})_n$ in $M(dmit)_2$ system [M = Pd, Ni], namely electron superconductors. No SC complex like $(D^{\gamma+})_m(A^{\gamma-})_n$ with both hole and electron carriers has been found. Purely organic superconductors are $(TMTSF)_2FSO_3$, β″-$(BEDT\text{-}TTF)_2SF_5CH_2CF_2SO_3$, and $(BEDT\text{-}TSF)_2(Cl_2TCNQ)$.

Organic conductors are very sensitive to pressure and temperature owing to its softness. With applying pressure or lowering temperatures, the intermolecular interaction increases owing to a

TABLE 9.2
List of Abbreviated Organic Donors

TMTSF	Tetramethyltetraselenafulvalene
TMTTF	Tetramethyltetrathiafulvalene
BEDT-TTF	bis(ethylenedithio)tetrathiafulvalene
S,S-DM-BEDT	*S,S*-dimethy-bis(ethylenedithio)tetrathiafulvalene
meso-DMBEDT-TTF	*meso*-dimethy-bis(ethylenedithio)tetrathiafulvalene
BEDO-TTF (BO)	bis(ethylenedioxy)tetrathiafulvalene
DODHT	(1,4-dioxane-2,3-diyldithio)dihydrotetrathiafulvalene
BEDT-TSeF (BETS)	bis(ethylenedithio)tetraselenafulvalene
BESe-TTF	bis(ethylenediselena)tetrathiafulvalene
ESET-TTF	ethylenediselena(ethylenedithio)tetrathiafulvalene
DMET	dimethyl(ethylenedithio)diselenadithiafulvalene
DMET-TSeF	dimethyl(ethylenedithio)tetraselenafulvalene
TMET-STF	trimethylene(ethylenedithio)deselenadithiafulvalene
DIETS	diiodo(ethylenedithio)deselenadithiafulvalene
MDT-TS	Methylenedithiadiselenafulvalene
MDT-ST	Methylenedithiodiselenadithiafulvalene
MDT-TSF	Methylenedithiotetraselenafulvalene
MDSe-TSF	Methylenediselenatetraselenafulvalene
MDT-TTF	Methylenedithiotetrathiafulvalene
EDT-TTF	Ethylenedithiotetrathiafulvalene
DMEDO-TSeF	Dimethylethylenedioxytetraselenafulvalene
DTEDT	2-(1,3-dithiol-2-ylidene)-5-(2-ethanediylidene-1,3-dithiole)-1,3,4,6-tetrathiapentalene
BDA-TTP	2,5-bis(1,3-dithian-2-ylidene)-1,3,4,6-tetrathiapentalene
$M(dmit)_2$[M = Ni, Pd]	Metal(1,3-dithiol-2-thione-4,5-dihiolate)$_2$
Cl2TCNQ	Dichlorotetracyanoquinodimethane
C5ET	(*cis*-1,2-cyclopentylene-1,2-dithio)ethylenedithiotetrathiafulvalene
C6ET	(*cis*-1,2-cyclohexylene-1,2-dithio)ethylenedithiotetrathiafulvalene

decrease of lattice, so that conductivity increases. Moreover, as for a half of organic superconductors, the SC state is observed only in a pressurized condition by suppression of a magnetic order or lattice distortion. Once SC state appears, T_c decreases with an increase of pressure by $\Delta T_c \sim -1$ K kbar^{-1}.

9.1.2 History of Organic Semiconductors, Organic Metals, and Organic Superconductors

Organic compounds are utilized as insulating materials as an insulating cover made of the organic polymer PVC (poly[vinylchloride]) around Cu wires. The intermolecular charge transfer theory proposed by R.S. Mulliken supports the birth of an organic conductor; by the hybridization of the highest occupied molecular orbital (HOMO) of a donor (D) and the lowest unoccupied molecular orbital (LOMO) of an acceptor (A), a more stable bonding orbital (Φ_1) and a more unstable antibonding orbital (Φ_2) are obtained (Figure 9.3).

$$\Phi_1 = a\,\Phi\,(D^0A^0) + b\,\Phi\,(D^+A^-)$$

$$\Phi_2 = a^*\,\Phi\,(D^0A^0) - b^*\,\Phi\,(D^+A^-)$$

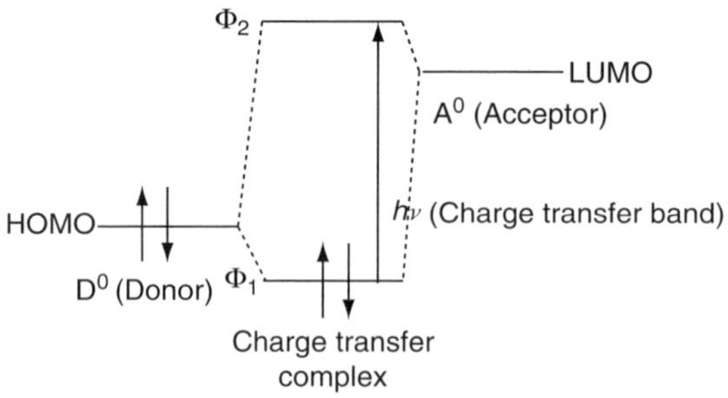

FIGURE 9.3 Charge transfer from HOMO (the highest occupied molecular orbital) to LUMO (the lowest unoccupied molecular orbital) and charge transfer band.

The optical absorption spectra between Φ_1 and Φ_2 are CT bands. The electron transfer from donors to acceptors produces hole and electron carriers when a CT complex is formed. Among CT complexes, the first organic semiconductor, perylene-Br_2 (1 Ωcm, E_a = 0.055 eV), was reported by Japanese chemists in 1954 [8]. The history of organic semiconductors, metals, and superconductors is shown in Table 9.3. In the 1960s and 1970s, the novel organic acceptor TCNQ (tetracyanoquinodi methane) and the organic donor TTF (tetrathiafulvalene) [7], constituting the main skeleton of organic superconductors, were synthesized, and relatively high-conducting TCNQ or TTF complexes were studied, respectively. The redox potentials of TTF and TCNQ are $E_{1/2}{}^1$ = +0.37 (TTF–e^- → $TTF^{+\bullet}$)

TABLE 9.3
History of Organic Semiconductors, Metals, and Superconductors

1950s Development of organic semiconductors

1952 R.S. Mulliken: charge transfer theory (1966 Nobel prize for chemistry)

1954 Discovery of organic semiconductor: perylene-Br_2 [8] (1 Ω cm, E_a = 0.055 eV)

1957 Bardeen–Cooper–Schrieffer (BCS) theory (1972 Nobel prize for physics)

1960s Development of organic conductors

1960 Synthesis of organic acceptor TCNQ (tetracyanoquinodimethane) and good conductivity of TCNQ charge transfer salts (σ_{RT} ~ 100 S cm^{-1})

1964 W. A. Little's model: possibility of T_c ~ 1000 K organic superconductor

1970s Development of organic metals

1970 Synthesis of an organic donor TTF

1971 Synthesis of polyacetylene film

1973 Discovery of the first stable organic metal: TTF·TCNQ

1977 High conductivity of doping to polyacetylene film (2000 Nobel prize for chemistry)

1980s Development of organic superconductors

1980 Discovery of the first organic superconductor: a pseudo-one-dimensional superconductors, $(TMTSF)_2PF_6$ (T_c = 0.9 K under 12 kbar) [9]

1988 Discovery of the first organic superconductor with T_c over 10 K: two-dimensional superconductors, κ-(BEDT-TTF)$_2$Cu(NCS)$_2$ (T_c = 10.4 K) [12]

1991 Discovery of molecular superconductor: three-dimensional superconductors, K_3C_{60} with T_c = 18 K [11]

2001 Magnetic field-induced superconductor λ-$BETS_2FeCl_4$

2003 Discovery of the highest T_c = 14.2 K (82 kbar) for β′-(BEDT-TTF)$_2$ICl$_2$ [10]

Source: From Mori, H., *J. Phys. Soc. Jpn.*, 75, 051003, 2006. With permission.

and $E_{1/2}^2$ = +0.76 V ($TTF^{+\bullet}$–e^- → TTF^{2+}); and $E_{1/2}^1$ = +0.22 (TCNQ + e^- → $TCNQ^{-\bullet}$) and $E_{1/2}^2$ = −0.34 V ($TCNQ^{-\bullet}$ + e^- → $TCNQ^{2-}$) (versus SCE [saturated calomel electrode], CH_3CN, 0.1 M TBA•BF_4, 20°C–22°C, 10~20 mVs^{-1}). In 1973, the first stable organic metal down to 59 K for TTF·TCNQ was discovered. Its electronic instability below 59 K coupled with a lattice distortion owing to the one-dimensionality induces the metal–insulator Peierls transition, which attracted physicists. Chemists continued important work to increase the dimensionality of organic conductors to suppress 1D Peierls instability. By the introduction of larger Se orbitals to increase the intercolumnar interaction, the first organic superconductor $(TMTSF)_2PF_6$ was discovered in 1980 [9]. This salt is a pseudo-1D conductor, and the magnetic order of spin density wave (SDW) transition occurs at 12 K. As shown in Figure 9.4d, SDW state is a magnetic order, where the density of the spin magnetic moment is modulated spatially in a periodic way. By applying an external pressure to suppress the SDW magnetic order, superconductivity was observed at T_c = 0.9 K under 12 kbar. The pseudo-1D TMTSF salts belong to the first generation of organic superconductors. The second generation is 2D BEDT-TTF superconductors. In particular, the highest T_c = 14.2 K (82 kbar) among sulfur-based organic superconductors is recorded for β′-$(BEDT\text{-}TTF)_2ICl_2$ [10], in which the SC state is obtained under high pressure to suppress the antiferromagnetic (AF) insulating state. As shown in Figure 9.4b, AF state is a magnetic order, where up and down spins are arranged alternately and the total spontaneous magnetic moment is zero. The β′-system has a dimerized structure, so that the effective half-filled band gives a large on-site coulomb repulsion energy (U) and the external high pressure affords a large bandwidth (W). When U/W becomes ~1, the SC state appears. In 1991, the 3D molecular superconductor K_3C_{60} with T_c = 18 K [11] was discovered. The highest T_c = 33 K among molecular superconductors is recorded for Cs_2RbC_{60}. In this book, the preparation, and chemical and physical properties of pseudo-1D, 2D, and 3D molecular superconductors are described.

In 2000, the third-generation λ-$(BEDT\text{-}TSeF)_2FeCl_4$ (BEDT-TSeF = bis(ethylenedithio)tetraselenafulvalene, abbreviated BETS) was discovered to be a magnetic field-induced organic superconductor. Usually magnetic field destabilizes superconductivity, but in this system, magnetic field induces the superconductivity. This material is composed of organic conductive BETS layers and inorganic

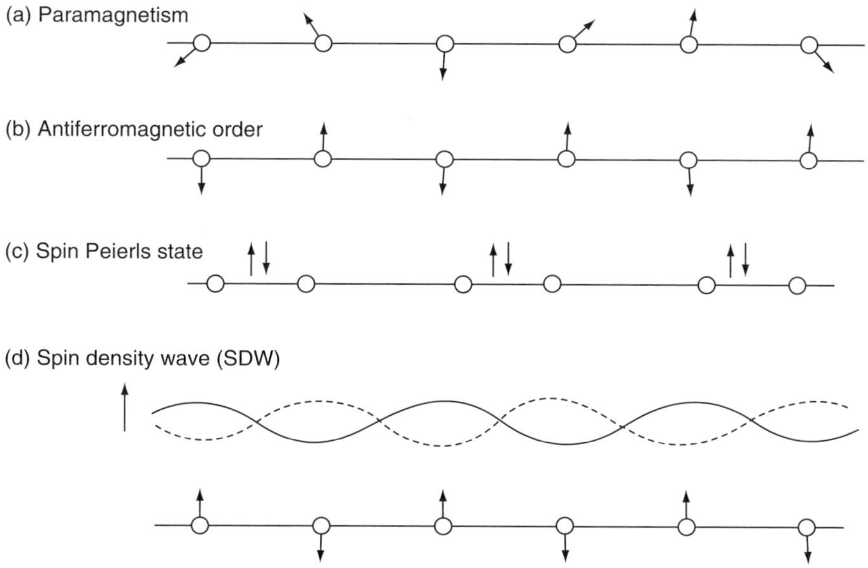

FIGURE 9.4 (a) Paramagnetism, (b) antiferromagnetic order, (c) spin Peierls state, and (d) spin density wave state.

$Fe(III)Cl_4^-$ ($S = 5/2$) anion sheets. In the paramagnetic phase, the localized 3d moments of Fe^{3+} are aligned along the external field (H). Owing to the strong negative exchange interaction J between the 3d and π spins, the π spins experience a strong internal magnetic field (H_{int}) created by the 3d moments, whose direction is antiparallel to H. Therefore, the resulting field experienced by the π spins approaches zero when $H \sim H_{int}$. Under this condition, superconductivity appears in the magnetic field of 17 T, which is qualitatively understood by the Fischer theory based on the Jaccarino–Peter effect.

9.2 CRYSTAL GROWTH OF ORGANIC SUPERCONDUCTORS

It is important to prepare high-quality, large single crystals to study chemical and physical properties of organic superconductors. Generally, the growth of organic crystals is carried out in a gas phase (gas sublimation or gas reaction method), liquid phase (condensed, slow cooling, liquid sublimation, or electrocrystallization method), or melted phase (normal freezing or zone melting method). The single crystal growth of organic superconductors has been performed mainly in a liquid phase using an electrocrystallization method for TTF-based superconductors and a gas reaction method for the C_{60} system.

9.2.1 Electrocrystallization Method

Most of the organic superconductors are prepared by the galvanostatic method. For example, single crystals of the organic superconductor κ-(BEDT-TTF)$_2$Cu(NCS)$_2$ (Figure 9.5a) are prepared electrochemically using BEDT-TTF as a donor and CuSCN, KSCN, and 18-crown-6 ether as counter anions in an organic solvent under an inert gas. The next reaction might proceed in the anode.

$$\text{BEDT-TTF} \xrightarrow{-e^-} \text{BEDT-TTF}^+$$
$$\text{BEDT-TTF}^+ + \text{BEDT-TTF} + \text{Cu(NCS)}_2^- \longrightarrow \kappa\text{-(BEDT-TTF)}_2\text{Cu(NCS)}_2$$

The details of the preparation are as follows:

1. *Purification of Starting Materials*: This process is very important to obtain high-quality crystals. Purification of CuSCN is done by dissolving CuSCN (4 g) and KSCN (64 g) in hot water. After cooling, water is added to the filtrate and the obtained white CuSCN powder is filtered off. This purification process is repeated thrice. KSCN and 18-crown-6 ether are recrystallized in ethanol and acetonitrile, respectively, and are dried completely. The solvent, 1,1,2-trichloroethane, is purified by stirring 1,1,2-trichloroethane (1500 g) and conc. H_2SO_4 (100 mL) for one night, and after removal of H_2SO_4, 1,1,2-trichloroethane is washed with water, aq. $NaHCO_3$, water, saturated aq. NaCl, and is then dried with $CaCl_2$ overnight. After filtration with basic Alumina, 1,1,2-trichloroethane is distilled off. The freshly distilled solvent is then filtered through basic alumina just before the preparation of single crystals to remove off the decomposed chloride compounds present in 1,1,2-trichloroethane.
2. *Crystal Growth by Electrocrystallization Method*: This method is carried out in a pilex glass cell (Figure 9.5b and c). Into the pilex glass cell purged by inert gas (Figure 9.5b), BEDT-TTF (30 mg), CuSCN (70 mg), KSCN (120 mg), 18-crown-6 ether (210 mg), and spinner are added in an anode side. By flowing an inert gas, the filtered 1,1,2-trichloroethane (90 mL) and distilled ethanol (10 mL) are added. The donor and counter anions are dissolved by the stirring and microwave method, and the undissolved chemicals remain in the bottom of the cell. After setting, Pt electrode is washed with carbon paste and aqua regia and is

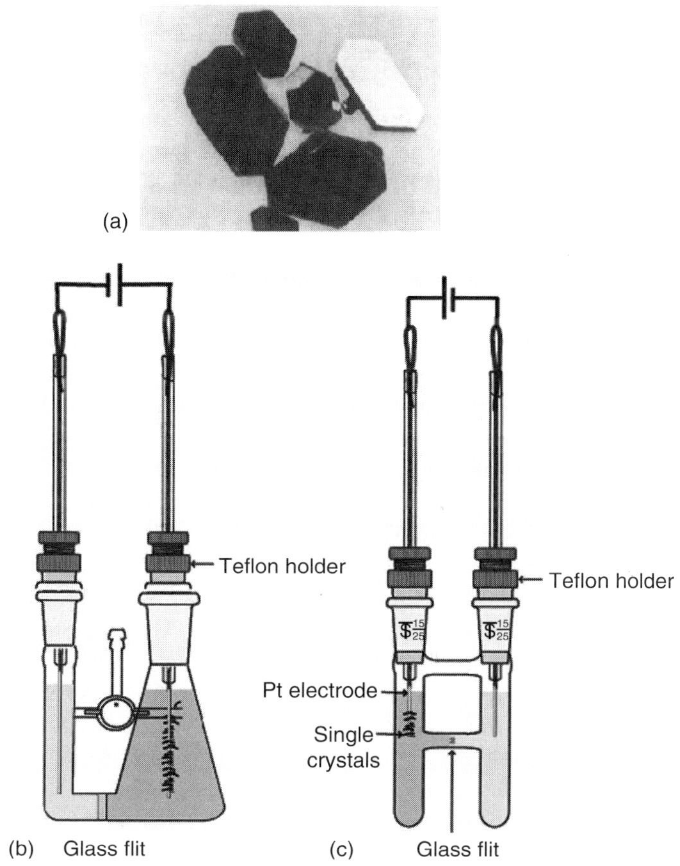

FIGURE 9.5 (a) Single crystals of κ-$(BEDT\text{-}TTF)_2Cu(NCS)_2$ and electrocrystallization of (b) 100 mL cell and (c) 20 mL cell.

burned, a constant current (0.5 μA) is applied for several weeks, so as to grow black plate crystals on the Pt electrode. The obtained crystals are washed with methanol and dried.

In this electrocrystallization method, the quality and phase of crystals depend upon the condition of glass cells, electrode, solvent, temperature, and current. There exist two-type of glass cells (100 mL, Figure 9.5b; and 18 mL, Figure 9.5c), where a glass flit divides an anode and a cathode to prevent the mixture of oxidized or reduced materials. The stable Pt electrode (1–2 mmϕ) is mainly used as well as Au, Ni, or W electrodes. As solvent, 1,1,2-trichloroethane, chlorobenzene, 1,2-dichloroethane, 1,2-dichloromethane, tetrahydrofuran, acetonitrile, benzonitrile, nitrobenzene, 1,1,1-trichloroethane, or dimethylformamide are used. In order to increase the concentration of counter anions in an organic solvent, 5%–10% vol ethanol or methanol is used. As counter anions to dissolve organic solvent, tetrabuthylammonium (TBA) salt, K^+ (18-crown-6) salt, tetraphenylphosphonium (TPP) salt, tetraphenylarsonium (TPA) salt, bis(triphenylphosphoranylidene)ammonium (PPN) salt, etc. are used. As for pseudo-1D organic superconductor, $TMTSF_2ClO_4$, single crystals are obtained by the electrocrystallization method by using TMTSF and TBA•ClO_4 in 1,1,2-trichloroethane under constant current of 0.25 μA.

The electrocrystallization is performed mainly by a constant current method of 0.25~50 μA or a regulated current method to apply current proportional to crystal area, and a constant electric potential method to use a reference electrode like SCE.

9.3 ORGANIC SUPERCONDUCTORS

9.3.1 Pseudo-One-Dimensional Organic Superconductors: TMTSF Salts

The TMTSF molecule and most of its complexes were synthesized by K. Bechgaard, and are present in the so-called Bechgaard salts. He has introduced the larger orbital Se atoms instead of S atoms into the TTF skeleton to increase the intercolumnar interaction and discovered the first organic superconductor $TMTSF_2PF_6$ at 0.9 K under 12 kbar [9].

All Bechgaard salts are 2:1 stoichiometry $(TMTSF^{+0.5})_2X^{-1}$ and their crystal structures are isostructural. Although the TMTSF molecules stack along the *a*-axis, the distance between Se atoms is similar in the intracolumnar direction [3.874(2)–4.133(2) Å] and intercolumnar direction [3.879(1)–3.959(2) Å] (Figure 9.6a) [13]. Therefore, the obtained transfer integrals of t_a, t_b, and t_c are 0.25, 0.025, and 0.0015 eV, and the calculated band structure by the tight-binding approximation is pseudo-1D; the Fermi surface is open along the *a*-axis and deeply warped along the b-direction (Figure 9.6b) [14]. TMTSF salts with an octahedral anion, $TMTSF_2X$ (X = PF_6, AsF_6, SbF_6, TaF_6, NbF_6), show metallic behavior and the distinct metal–insulator (MI) transitions at 11–17 K, below which the SDW state is observed by electron spin resonance (ESR), nuclear magnetic resonance (NMR), and magnetic susceptibility measurements (Figures 9.4d and 9.6c). The incommensurate wave vector (0.5*a**, 0.24*b**, –0.06*c**) of the PF_6 salt is consistent with the nesting vector of a pair of pseudo-1D Fermi surfaces. The suppression of the SDW state with pressure by increasing the intercolumnar interactions has induced the superconductivities of $TMTSF_2X$ (X = PF_6, AsF_6, SbF_6, TaF_6, NbF_6) at 0.9 K (12 kbar), 1.1 K (12 kbar), 0.38 K (10.5 kbar), 1.35 K (11 kbar), and 1.26 K (12 kbar), respectively.

The TMTSF salts with a nonsymmetrical tetrahedral anion, $TMTSF_2X$ (X = ClO_4, ReO_4, FSO_3), exhibit the anion orientational disorder–order transitions at 24, 180, and 86 K, respectively. Because the structural transition of the ClO_4 salt under a slow cooling is (*a**, *b**/2, *c**) without the lattice distortion along the *a** donor stacking axis, superconductivity has been observed at T_c = 1.4 K at ambient pressure. The rapid cooling, however, has induced the SDW state below 5 K with a randomly orientated ClO_4 anion. On the other hand, the ReO_4 salt shows the metal–insulator transition caused by the (*a**/2, *b**/2, *c**/2) anion ordering. Under 9.5 kbar by the suppression of the structural modulation along the stacking axis, (*a**, *b**/2, *c**/2), superconductivity has been observed at 1.3 K. Influenced by the anion ordering, the specific transport behavior was observed: the metal–insulator transition at approximately 100 K, and the reentrant metallic behavior below 70 K, and then, the resistivity drop at 1.3 K. As for the FSO_3 salt, the suppression of the anion disorder has also afforded the highest T_c, 2.1 K, among the TMTSF salts. Recently, the unique phase diagram of FSO_3 salt under pressure has been proposed, reflected by the dipole moment of the FSO_3 anion.

TMTSF salts are type II superconductors. The type II superconductors have the lower critical field (H_{c1}) and the upper critical field (H_{c2}). Below H_{c1}, the applied magnetic field is excluded, namely Meissner effect. Between H_{c1} and H_{c2}, the magnetic fields enter the superconductors, so that SC and normal regions coexist. And above H_{c2}, the SC state is destabilized and normal state appears. On the other hand, in the type I superconductor, the Meissner state transfers to the normal state at the critical field (H_c). The upper critical field (H_{c2}) and the obtained coherence length of $(TMTSF)_2ClO_4$ are listed in Table 9.4. The jump of specific heat at 1.22 K was observed with γ = 10.5 mJ mol^{-1} K^{-2}, β = 11.4 mJ mol^{-1} K^{-4}, ΔC = 21.4 mJ mol^{-1} K^{-1}, which induces $\Delta C/\gamma T_c = 1.67$, which is close to 1.43 based upon Bardeen–Cooper–Schrieffer (BCS) theory. The BCS theory is proposed by J. Bardeen, L.N. Cooper, and J.R. Schrieffer, in which superconductivity is Bose–Einstein condensation of electron pairs, namely Cooper pairs. They also calculated the superconducting (SC) gap at 0 K (Δ_0) as $2\,|\Delta_0| = 3.53\,k_BT_c$ (k_B; Boltzmann constant). The tunnel spectroscopy, which is a method to investigate the electronic state of junction metal electrode or insulators by current–voltage dependence, affords $2\,\Delta = 0.44$ mJ for $(TMTSF)_2ClO_4$. This value is consistent to the BCS theory. The anomalous symmetry of superconductivity for $TMTSF_2ClO_4$ has been observed with no coherence peak at T_c by the NQR relaxation measurement, suggesting an anisotropic singlet state or a triplet state. The SC

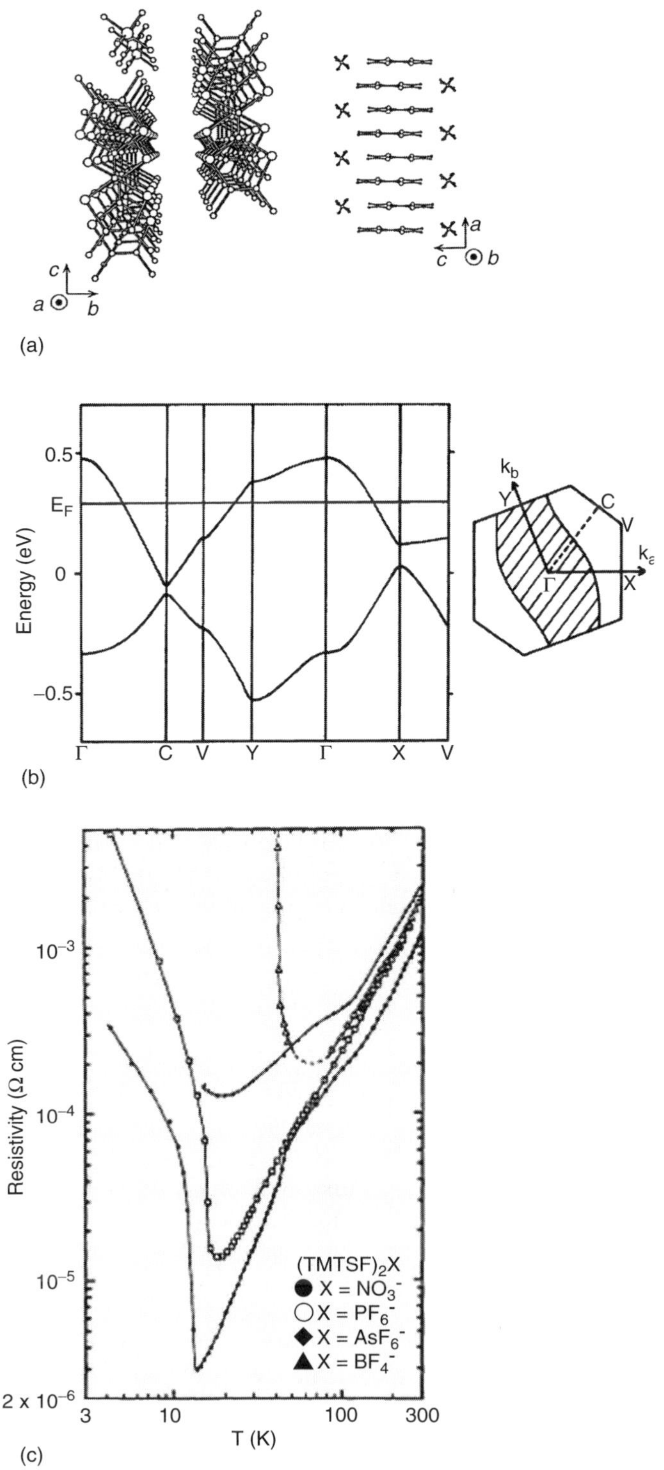

FIGURE 9.6 (a) Crystal structure. (From Bechgaard, K. et al., *Solid State Commun.*, 33, 1119–1125, 1980. With permission.) (b) Fermi surface. (From Mori, T. et al., *Bull. Chem. Soc. Jpn.*, 57, 627–633, 1984. With permission.) (c) Metal–insulator transition of $TMTSF_2X$ (X = NO_3, PF_6, AsF_6, and BF_4). (From Mori, H., *J. Phys. Soc. Jpn.*, 75, 2006. With permission.)

(continued)

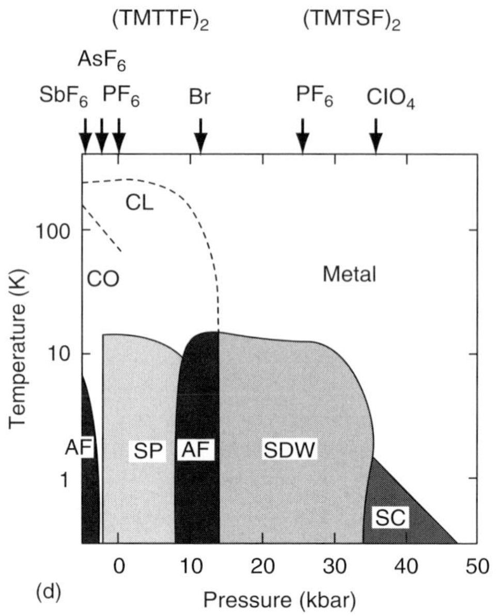

FIGURE 9.6 (continued) (d) Electronic phase diagram of $(TMTTF)_2X$ and $(TMTSF)_2X$. (From Mori, H., *J. Phys. Soc. Jpn.*, 75, 2006. With permission.)

state probed by ^{77}Se NMR Knight shift and the anomalously high critical fields near the SC/SDW phase boundary suggest the triplet superconductivity state.

The electronic phase diagram including sulfur-substituted TMTTF complexes are proposed (Figure 9.6d). The TMTTF complexes are more 1D than TMTSF salts, so that the electronic state changes from metallic, charge-ordered state, to Spin Peierls (SP) or AF state by lowering temperatures. As shown in Figure 9.4c, SP state is nonmagnetic state, where the second-order phase transition

TABLE 9.4
List of SC Parameters

	$(TMTSF)_2ClO_4$	β_L-ET_2I_3	$\kappa ET_2Cu(NCS)_2$	K_3C_{60}
T_c (K)	1.4 K	1.5 K	10.4 K	18 K
H_{c1} (mT)	0.02 (a, 0.05 K)	0.005 (a, 0.1 K)	0.07 (//, 5 K)	13 (0 K)
	0.10 (b, 0.05 K)	0.009 (b′, 0.1 K)	20 (⊥, 1.5 K)	
	1.0 (c, 0.05 K)	0.036(*c**, 0.5 K)		
H_{c2} (T)	2.8 (a, 0 K)	2.09 (a, 0 K)	24.5 (//, 0.5 K)	49 (0 K)
	2.1 (b, 0 K)	2.48 (b′, 0 K)		
	0.16 (c, 0 K)	0.081 (c, 0 K)	5.5 (⊥, 0.5 K)	
ξ (Å)	706 (a, 0 K)	633 (a, 0 K)	29 (//)	26
	335 (b, 0 K)	608 (b′, 0 K)	3.1 (⊥)	
	20.3 (c, 0 K)	29 (*c**, 0 K)		
dT_c/dP (K $kbar^{-1}$)	−0.08	−0.3	−3	−0.78
γ(mJ mol^{-1} K^{-2})	10.5	24	34	45
β (mJ mol^{-1} K^{-4})	11.4	19	10	
ΔC (mJ mol^{-1} K^{-1})	21.4		730	1200
2Δ (meV)	0.44	1	4.2	4.4

makes 1D Heisenberg AF spin system to the spin singlet state. The unified phase diagram contains rich electronic states; by applying physical and chemical pressure, the ground state transforms from AF, SP, AF, SDW, to SC state. Actually, the superconductivity of $TMTTF_2PF_6$ is obtained at 1.8 K under high pressure of 54 kbar, which comfirms the phase diagram.

The pseudo 1D superconductor of a TMTSF system is attractive to research not only unusual SC mechanism, but also physical phenomena like field-induced SDW state, quantum hall effect, etc.

9.3.2 Two-Dimensional Organic Superconductors: BEDT-TTF Salts

To increase bandwidth and dimensionality by larger intermolecular interactions, the salts of the BEDT-TTF molecule with eight sulfur atoms have been studied. Among them, β″-(BEDT-TTF)$_2$ClO$_4$(TCE)$_{0.5}$ (TCE = 1,1,2-trichloroethane) was found to be 2D as σ_c/σ_a (2 K) = 0.4, and to exhibit metallic behavior down to T_{min} = 15 K. Owing to the nonplanar conformation of the BEDT-TTF molecule with an outer ethylenedithio-ring by 0.3 ~ 0.6 Å from the least square plane and a thermal vibration, a diagonal donor stacking, as well as a face-to-face one, induces two-dimensionality. Consequently, a variety of donor arrangements is observed in the BEDT-TTF salts. The Greek symbol (a) α-type, (b) θ-type, (c) κ-type, (d) λ-type, (e) β-type, (f) β′-type, and (g) β″-type in Figure 9.7 denotes the arrangement of BEDT-TTF molecule, which determines the electronic structure of BEDT-TTF complexes.

As shown in Table 9.1, a BEDT-TTF molecule affords about a half of organic superconductors (49 kinds/total 144 kinds) and the highest T_c = 14 K under 82 kbar for β′-(BEDT-TTF)$_2$ICl$_2$. Therefore, BEDT-TTF system has been intensively studied so far by chemists and physicists.

9.3.2.1 β-Type BEDT-TTF Salts

The β-type donor arrangement is similar to that of TMTSF complexes, not only a strong dimerize structure in the column, but also a large side-by-side intercolumnar interaction affords 2D electronic state. The tight-binding band calculation based on the extended Hückel method for β-(BEDT-TTF)$_2$I$_3$ was carried out to elucidate the 2D Fermi surface, which has been proved quantitatively by the Schubnikov–de Haas oscillation of the magnetoresistance [22].

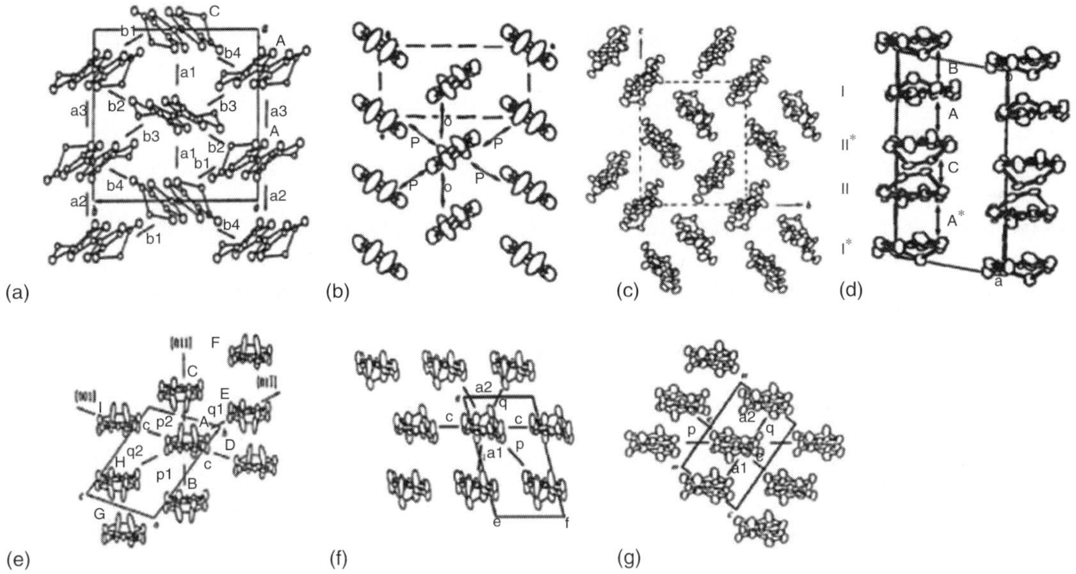

FIGURE 9.7 Molecular arrangements of (a) α-type, (b) θ-type, (c) κ-type, (d) λ-type, (e) β-type, (f) β′-type, and (g) β″-type BEDT-TTF salts.

β-(BEDT-TTF)$_2$X [X = I$_3$, AuI$_2$, IBr$_2$] are ambient pressure superconductors. The lower T_c of β$_L$-(BEDT-TTF)$_2$I$_3$, 1.5 K, increases suddenly to the higher T_c of 7.5 K under about 1 kbar by the suppression of a long-range order of a lattice instability induced by a conformational disorder of an ethylenedithio group. The released pressure below 125 K affords the higher T_c (= 8.1 K) phase, β$_H$-(BEDT-TTF)$_2$I$_3$. By applying hydrostatic pressure, the T_c of β-(BEDT-TTF)$_2$X [X = I$_3$, AuI$_2$, IBr$_2$] decrease by 0.8–1.2 K kbar^{-1}, which might be related to a decrease of density of state with an increase of bandwidth.

9.3.2.2 κ-Type BEDT-TTF Salts: Competition between Antiferromagnetic and SC States

Because the T_c of β-(BEDT-TTF)$_2$X [X = I$_3$, AuI$_2$, IBr$_2$] decrease by applying pressure, a negative chemical pressure is applied by utilizing longer linear anions so as to increase T_c. A 2D superconductor κ-(BEDT-TTF)$_2$Cu(NCS)$_2$ with T_c = 10.4 K was prepared [12]. The crystal structure analysis shows that two BEDT-TTF molecules make a dimer, which is arranged in a checkerboard manner in the 2D plane (Figure 9.7c). Owing to the dimerized structure, an effective half-filled band structure is found to be a strongly correlated system.

The temperature dependences of electrical resistivity for κ-(BEDT-TTF)$_2$X [X = Cu(NCS)$_2$, Cu(CN)[N(CN)$_2$], Cu[N(CN)$_2$]Cl, and Cu[N(CN)$_2$]Br] are shown in Figure 9.8c [17]. The κ-(BEDT-TTF)$_2$Cu[N(CN)$_2$]Cl is a Mott insulator. Mott transition is one of the metal–insulator transitions, where the insulating state is induced by electron–electron repulsion Coulomb energy. As for κ-(BEDT-TTF)$_2$Cu(NCS)$_2$, the nonmonotonic temperature dependence is considered to reflect the Mott-boundary; the temperature dependence is metallic down to 270 K, and increases again with the resistivity maximum around 100 K, and decreases again with the SC transition at 10.4 K. The room temperature resistivity along the *b*-axis is 0.05–0.1 Ω cm (σ_b = 10–20 S cm^{-1}) and nearly isotropic in the 2D bc plane: (σ_a: σ_b: σ_c = 1/600: 1: 1.2).

The temperature dependences of magnetic susceptibility for κ-(BEDT-TTF)$_2$X are elucidated in Figure 9.8d [17]. At high temperature, all the salts show temperature-insensitive behavior with 4×10^{-4} emu mol^{-1} and decrease at low temperatures. The abrupt increase of susceptibility at 30 K for κ-(BEDT-TTF)$_2$Cu[N(CN)$_2$]Cl is due to canting of the spins that undergo AF transition.

Not only the resistivity but also reflectance spectra exhibit 2D of electronic state for κ-(BEDT-TTF)$_2$X. As shown in Figure 9.9 [18], the optical spectra indicate the 2D Drude-like metallic spectra in the bc plane. The electronic states of a series of κ-type BEDT-TTF salts are elucidated in the phase diagram as shown in Figure 9.10. In the vicinity of a SC state, there exists a Mott–Hubbard AF insulating state, where the Coulomb repulsion energy between carriers is stronger than kinetic energy of carriers. By changing *W*/*U* (*W*: bandwidth, *U*: on-site Coulomb repulsion energy) or applying pressure, an AF state is transformed to a SC and a metallic state [16,18–21].

The SC electrons have wave and particle duality. The symmetry of the wave function for SC electrons (Cooper pairs) belongs to the common s-wave with isotropic SC gap at the Fermi level (E_F) or unconventional p- or d-wave with nodal gapless point or line at E_F. As for κ-(BEDT-TTF)$_2$X [X = Cu(NCS)$_2$ and Cu[N(CN)$_2$Br]], there exists controversial experimental results to probe normal s-wave or non s-wave symmetry.

9.3.2.3 β′-Type BEDT-TTF Salts: Competition between Antiferromagnetic and SC States

β′-(BEDT-TTF)$_2$X (X = ICl$_2$, IBrCl, AuCl$_2$) are strongly dimerized systems with an effective half-filled band structure, which is a Mott insulator. With decreasing temperature, the AF orders of the X = ICl$_2$, IBrCl, and AuCl$_2$ salts are observed at 22, 19.5, and 28 K, respectively. Under high pressure, the AF order is suppressed, and the superconductivities of the X = ICl$_2$ and IBrCl salts have been observed at 14 K under 82 kbar with the highest T_c among sulfur-based organic superconductors, and 8.5 K under 86 kbar, respectively. The large *U* owing to a strong dimerization and the large *W* owing to the applied high pressure might afford the highest T_c.

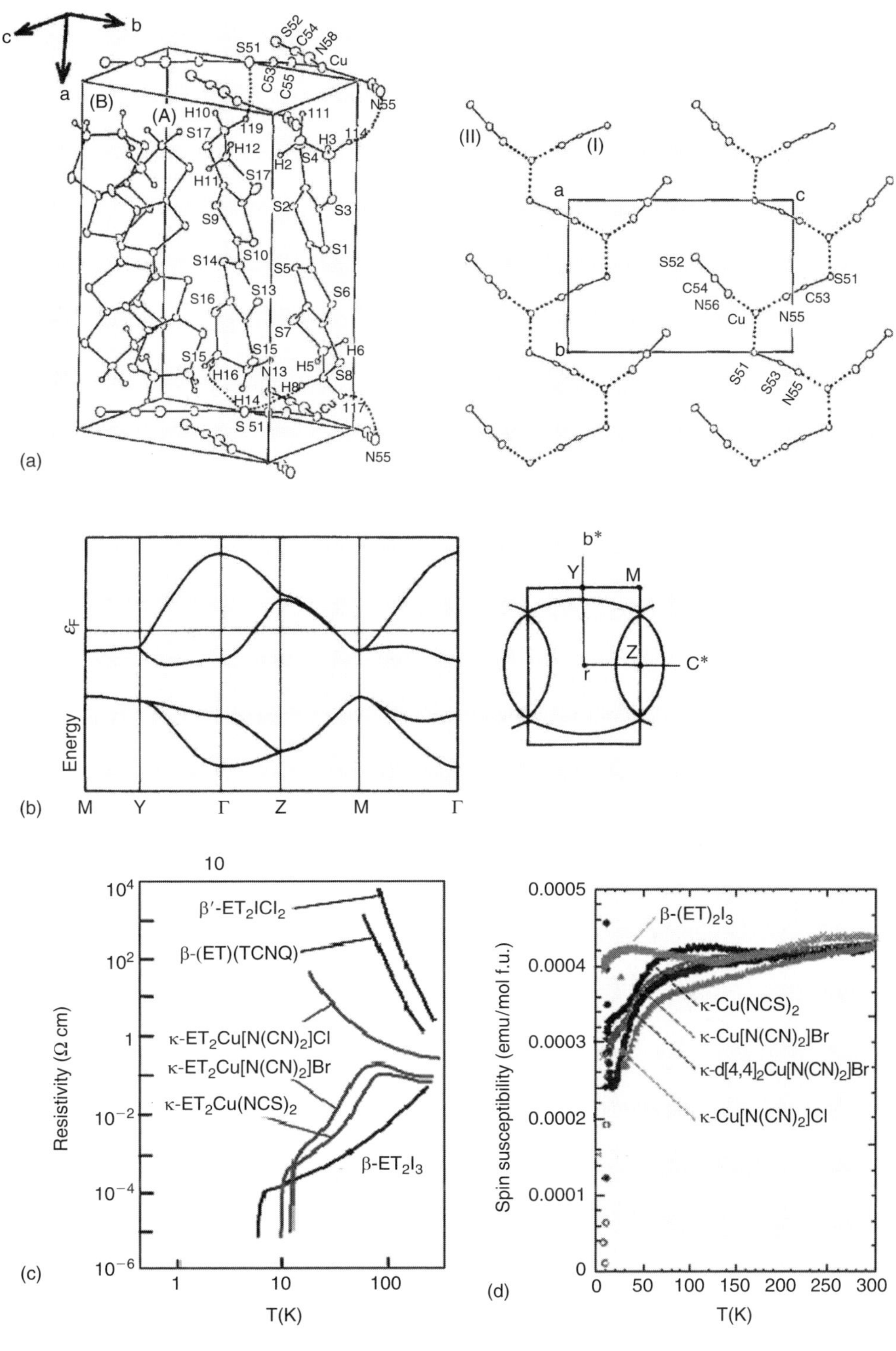

FIGURE 9.8 (a) Crystal structure. (From Urayama, H. et al., *Synth. Metals*, 27, A393, 1988. With permission.) (b) Band structure of κ-(BEDT-TTF)$_2$Cu(NCS)$_2$. (From Oshima, K., et al., *Phys. Rev. B*, 38, 938, 1988. With permission.) (c) Electrical resistivities. (From Kanoda, K., *J. Phys. Soc. Jpn.*, 75, 051007, 2006. With permission.) (d) Magnetic susceptibilities for a series of κ-type BEDT-TTF salts. (From Kanoda, K., *J. Phys. Soc. Jpn.*, 75, 051007, 2006. With permission.)

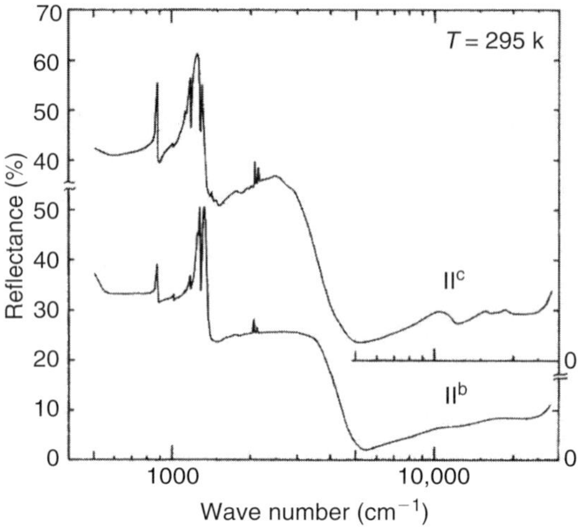

FIGURE 9.9 Reflectance spectra of κ-(BEDT-TTF)$_2$Cu(NCS)$_2$. (From Sugano, T. et al., *Phys. Rev. B*, 37, 9100, 1988. With permission.)

9.3.3 Three-Dimensional Molecular Superconductors

In 1985, H.W. Kroto, R.E. Smally, and R.F. Curl found C_{60} molecule with high I_h symmetry. In 1990, W. Kratschmer and D.R. Huffman performed in high-yield preparation of C_{60}. In 1991, A.F. Hebard, M.J. Rosseinsky, and R.C. Haddon discovered the superconductivity of K_3C_{60} [11].

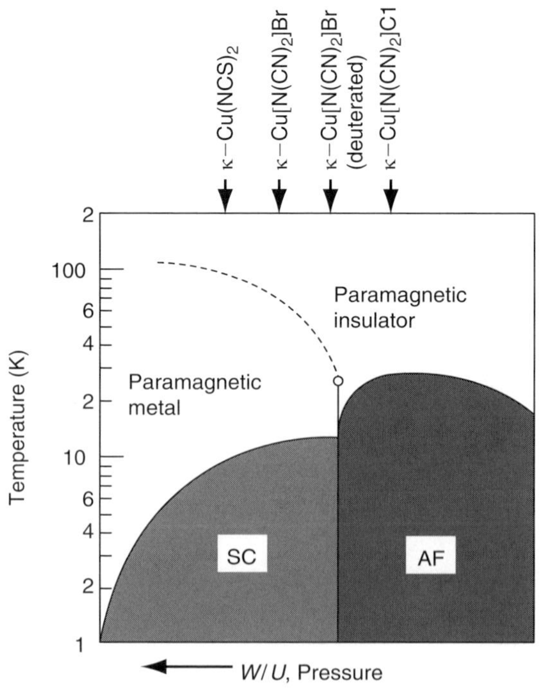

FIGURE 9.10 Electronic phase diagram of κ-type BEDT-TTF salts. SC: superconducting; AF: antiferromagnetic states. (From Mori, H., *J. Phys. Soc. Jpn.*, 75, 051003, 2006. With permission.)

The HOMO of C_{60} is calculated fivefold degenerate states with h_u symmetry and the LUMO is threefold degenerate states with t_{1u} symmetry. The ionization potential is small, so that C_{60} is a weak acceptor as $E_1 = -0.38$ V, $E_2 = -0.79$ V, $E_3 = -1.28$ V in dichlorobenzene versus SCE. C_{60} crystal belongs to face-centered cubic (fcc) system with $a = 14.17$ Å, K_3C_{60} is also fcc with $a = 14.24$ Å, and K_4C_{60} and K_6C_{60} are body-centered orthorhombic and body-centered cubic, respectively. Only K_3C_{60} shows superconductivity and the relative C_{60} complexes, AC_{60} (A = Na_2K, Na_2Rb, Na_2Cs, K_2Rb, K_2Cs, KRb_2, Rb_3, Rb_2Cs, $RbCs_2$, and Li_2Cs), are also fcc and indicate SC, where T_c is proportional to lattice constant a. The relationship is related to the narrower bandwidth and the larger density of state at the Fermi level for larger volume of C_{60} complexes.

The SC parameters of K_3C_{60} are summarized in Table 9.4. The magnetization data indicates $H_{c2}(0) = 49$ T, $\xi = 26$ Å, $H_{c1}(0) = 132$ mT, $\lambda_L = 2400$ Å, $Jc(1T) = 1.2 \times 10^5$ A/cm^2. The pressure dependence of K_3C_{60} is $dT_c/dP = -0.63$ K kbar^{-1}.

The measurements of μSR and NMR relaxation rate with Hebel–Slichter coherence peak just below T_c due to the divergence of density of states for quasiparticles indicates that K_3C_{60} is a normal s-wave superconductor. The isotope effect is the method to investigate how the SC transition temperature (T_c) changes by the substitution of the isotope atoms in superconductors. This effect indicates that T_c is proportional to ω_D (phonone frequency), which depends upon $M^{-\alpha}$ (M: molecular weight). The observed α of $K_3{}^{13}C_{60}$ is 0.3, which is close to the normal isotope effect 0.5.

9.4 SUMMARY

Usually, organic materials are utilized as insulators. By introducing the concept of charge transfer (CT) (Figures 9.1 and 9.3), carriers are introduced, so that the first conducting organic material, Perylene-Br_2, was found in 1954. The development of a superior organic donor TTF and an acceptor TCNQ made the first stable organic CT salt TTF•TCNQ metallic down to 59 K in 1973. In 1980, the first organic superconductor $TMTSF_2PF_6$ ($T_c = 0.9$ K under 12 kbar) was discovered. This pseudo-1D superconductor (Figure 9.6b) is obtained by the suppression of the SDW state (Figure 9.4d) under an ambient pressure. The TMTSF salts with nonsymmetrical tetrahedral anion, $TMTSF_2X$ (X = ReO_4, FSO_3), exhibit the anion orientational disorder–order transition, which induces the metal–insulator transition (Figure 9.6c). $TMTSF_2ClO_4$ is the first ambient pressure organic superconductor without the lattice distortion along the donor stacking axis. The universal phase diagram of $TMTTF_2X$ and $TMTSF_2X$ elucidates that the versatile ground states change from AF (antiferromagnetic order), SP (spin Peierls state), AF, SDW, to SC with applying pressure (Figures 9.4 and 9.6d). TMTSF salts are type II superconductors. The anomalous triplet SC symmetry for $TMTSF_2ClO_4$ has been observed by NMR and magnetotransport measurements.

To increase bandwidth and dimensionality by larger intermolecular interactions, the salts of the BEDT-TTF molecule with eight sulfur atoms have been developed. Owing to the nonplanar conformation of the BEDT-TTF molecule with an outer ethylenedithio-ring, a diagonal donor stacking as well as a face-to-face one induces 2D. The variety of 2D donor arrangements is shown in Figure 9.7. A half of organic superconductors (49 kinds/total 144 kinds) are BEDT-TTF CT salts and the highest $T_c = 14$ K under 82 kbar is recorded for β′-(BEDT-TTF)$_2$ICl$_2$. Especially, κ-type BEDT-TTF salts afford several 10 K superconductors. In the κ-type donor arrangement, two donor molecules make a dimer, which is arranged as the checkerboard manner in the 2D plane and an effective half-filled band structure is found to be a strongly correlated system. The electronic phase diagram of κ-type salts elucidates that the SC state is next to the AF state. As for the Mott insulator κ-(BEDT-TTF)$_2$Cu[N(CN)$_2$Cl], the superconductor $T_c = 12.8$ K was obtained by the suppression of the AF order. The symmetry of the Cooper pairs in κ-(BEDT-TTF)$_2$X [X = Cu(NCS)$_2$ and Cu[N(CN)$_2$Br]] has been controversial to probe s-wave or non-s-wave symmetry.

The 3D molecular superconductor is the AC_{60} system (A = Na_2K, Na_2Rb, Na_2Cs, K_2Rb, K_2Cs, KRb_2, Rb_3, Rb_2Cs, $RbCs_2$, Li_2Cs), where T_c is proportional to the fcc lattice constant a. The relationship is related to the narrower bandwidth and the larger density of state at the Fermi level for larger volume of C_{60} complexes. The C^{13} isotope complex indicates that the AC_{60} system follows the normal BCS theory.

EXERCISE QUESTIONS

1. The organic superconductors κ-$(BEDT\text{-}TTF)_2Cu(NCS)_2$ (9.5 mg) is prepared electrochemically by utilizing BEDT-TTF ($S_8C_{10}H_8$, 30 mg), KSCN (120 mg), CuSCN (70 mg), and 18-crown-6 ether ($C_{12}O_6H_{24}$, 210 mg). Calculate the obtained yield of κ-$(BEDT\text{-}TTF)_2Cu(NCS)_2$ crystals.

   Atomic weight: Cu = 63.5, K = 39.1, S = 32, O = 16, N = 14, C = 12, H = 1.

2. Explain the reason why T_c's of molecular superconductors decrease by applying pressure.

LIST OF ABBREVIATIONS

2D	two-dimensional
3D	three-dimensional
BDA-TTP	2,5-bis(1,3-dithian-2-ylidene)-1,3,4,6-tetrathiapentalene
BEDO-TTF (BO)	bis(ethylenedioxy)tetrathiafulvalene
BEDT-TTF	bis(ethylenedithio)tetrathiafulvalene
BEDT-TSeF (BETS)	bis(ethylenedithio)tetraselenafulvalene
BESe-TTF	bis(ethylenediselena)tetrathiafulvalene
C5ET(*cis*-1,2-cyclopentylene-1,2-dithio)	ethylenedithiotetrathiafulvalene
C6ET (*cis*-1,2-cyclohexylene-1,2-dithio)	ethylenedithiotetrathiafulvalene
Cl_2TCNQ	dichlorotetracyanoquinodimethane
DIETS	diiodo(ethylenedithio)deselenadithiafulvalene
DMEDO-TSeF	dimethylethylenedioxytetraselenafulvalene
DMET	dimethyl(ethylenedithio)diselenadithiafulvalene
DMET-TSeF	dimethyl(ethylenedithio)tetraselenafulvalene
DTEDT	2-(1,3-dithiol-2-ylidene)-5-(2-ethanediylidene-1, 3-dithiole)-1,3,4,6-tetrathiapentalene
DODHT	(1,4-dioxane-2,3-diyldithio)dihydrotetrathiafulvalene
EDT-TTF	ethylenedithiotetrathiafulvalene
ESET-TTF	ethylenediselena(ethylenedithio)tetrathiafulvalene
ESR	electron spin resonance
H_{C1}	lower critical field
H_{C2}	upper critical field
HOMO	the highest occupied molecular orbital
LUMO	the lowest unoccupied molecular orbital
meso-DMBEDT-TTF	*meso*-dimethy-bis(ethylenedithio)tetrathiafulvalene
MDT-TS	methylenedithiadiselenafulvalene
MDT-ST	Methylenedithiodiselenadithiafulvalene
MDT-TSF	Methylenedithiotetraselenafulvalene
MDSe-TSF	methylenediselenatetraselenafulvalene
MDT-TTF	methylenedithiotetrathiafulvalene
$M(dmit)_2$ [M = Ni, Pd]	metal(1,3-dithiol-2-thione-4,5-dihiolate)$_2$
NMR	nuclear magnetic resonance
SC	superconducting
S,S-DM-BEDT	*S,S*-dimethy-bis(ethylenedithio)tetrathiafulvalene
T_c	superconducting transition temperature

T_{CO}	charge-ordered temperature
T_{DO}	anion disorder–order temperature
T_{MI}	Metal–insulator transition temperature
T_N	Neel temperature
T_{SDW}	spin density wave transition temperature
TMTSF	tetramethyltetraselenafulvalene
TMTTF	tetramethyltetrathiafulvalene
TMET-STF	trimethylene(ethylenedithio)deselenadithiafulvalene

REFERENCES

1. Ishiguro, T., Yamaji, K., and Saito, G., *Organic Superconductors*, 2nd edition, Springer, Heidelberg, NY, 1998.
2. Williams, J.M. et al., *Organic Superconductors (Including Fullerenes: Synthesis, Structure, Properties, and Theory)*, Prentice Hall, Englewood Cliffs, NJ, 1992.
3. Batail, P. et al., Molecular conductors, *Chem. Rev.*, 104, 4887–5781, 2004 (for recent reviews).
4. Mori, H., Organic conductors, *J. Phys. Soc. Jpn.*, 75, 051001–051016, 2006 (for recent reviews).
5. Fukuyama, H. et al., Materials viewpoint of organic superconductors, *J. Phys. Soc. Jpn.*, 75, 051003, 2006.
6. Ishiguro, T. et al., *Advances in Synthetic Metals, Twenty Years of Progress in Science and Technology*, Bernier, P., Lefrant, S., and Bidan, G., Eds., Elsevier, Amsterdam, The Netherlands, 1999 (for recent reviews).
7. Sugimoto, T. et al., *TTF Chemistry: Fundamentals and Applications of Tetrathiafulvalene*, Yamada, J. and Sugimoto, T., Eds., Kodansya, Tokyo, Springer, Berlin, 2004.
8. Akamatu, H., Inokuchi, H., and Matsunaga, Y., Electrical conductivity of the perylene-bromine complex, *Nature*, 173, 168, 1954.
9. Jerome, D. et al., Superconductivity in a synthetic organic conductor $(TMTSF)_2PF_6$, *J. Phys. Lett.*, 41, L95-L98, 1980.
10. Taniguchi, H. et al., Superconductivity at 14.2 K in layered organics under extreme pressure, *J. Phys. Soc. Jpn.*, 72, 468–471, 2003.
11. Hebard, A.F. et al., Superconductivity at 18 K in potassium-doped C60, *Nature*, 350, 600–601, 1991.
12. Urayama, H. et al., A new ambient pressure organic superconductor based on BEDT-TTF with T_c higher than 10 K (T_c = 10.4 K), *Chem. Lett.*, 55–56, 1988.
13. Bechgaard, K. et al., The properties of five highly conducting salts: $(TMTSF)_2X$, $X = PF_6^-$, AsF_6^-, SbF_6^-, BF_4^- and NO_3^-, derived from tetramethyltetraselenafulvalene (TMTSF), *Solid State Commun.*, 33, 1119–1125, 1980.
14. Mori, T. et al., Intermolecular interaction of TTF and BEDT-TTF in organic metals, *Bull. Chem. Soc. Jpn.*, 57, 627–633, 1984.
15. Urayama, H. et al., Crystal and electronic structures and physical properties of T_c = 10.4 K superconductor, $(BEDT\text{-}TTF)_2Cu(NCS)_2$, *Synth. Metals*, 27, A393–A400, 1988.
16. Oshima, K. et al., Shubnikov-de Haas effect and the Fermi surface in an ambient-pressure organic superconductor $[bis(ethylenedithiolo)tetrathiafulvalene]_2Cu(NCS)_2$, *Phys. Rev. B*, 38, 938–941, 1988.
17. Kanoda, K., Metal–insulator transition in $\kappa\text{-}(ET)_2X$ and $(DCNQI)_2M$: Two contrasting manifestation of electron correlation, *J. Phys. Soc. Jpn.*, 75, 051007, 2006.
18. Sugano, T. et al., In-plane quasi-isotropic organic superconductor di[bis(ethylenedithiolo)tetrathiafulvalne]bis(isothiocyanato)cuprate(I), $(BEDT\text{-}TTF)_2[Cu(NCS)_2]$: Polarized reflectance spectra, *Phys. Rev. B*, 37, 9100–9102, 1988.
19. Kino, H. and Fukuyama, H., Electronic states of conducting organic $\kappa\text{-}(BEDT\text{-}TTF)_2X$, *J. Phys. Soc. Jpn.*, 64, 2726–2729, 1995.
20. MacKenzie, R., Similarity between organic and cuprate superconductors, *Science*, 278, 820–821, 1997.
21. Lang, M. and Mueller, J., Organic superconductors, in *The Physics of Superconductors: Cuprate, Heavy-Fermion, and Organic Superconductors*, Bennemann, K.H. and Ketterson, J.B., Eds., Springer-Verlag, Heidelberg, Germany 2004.
22. Wosnitza, J., SC properties of quasi-two-dimensional organic metals, *Physica C*, 317–318, 98–107, 1999.

10 Molecular Semiconductors for Organic Field-Effect Transistors

Antonio Facchetti

CONTENTS

10.1 Introduction 287
10.2 Organic Field-Effect Transistors: Device Structure and Operation 288
10.3 Material Requirements for High OFET Charge Transport Efficiency 290
 10.3.1 Semiconductor Molecular Design Requirements 291
 10.3.2 Frontier Molecular Orbital Energies 291
 10.3.3 Crystal Structure 294
 10.3.4 Film Morphology and Microstructure 296
 10.3.5 Dielectric and Contact Materials 297
10.4 Molecular Semiconductor Families for OFETs 300
 10.4.1 Acenes and Oligophenylenes 300
 10.4.2 Thiophene-Based Molecules 302
 10.4.3 Two-Dimensional Fused Rings 307
10.5 Single-Crystal Transistors 311
Exercise Questions 314
List of Abbreviations 314
Acknowledgments 314
References 314

Abstract π-Conjugated organic molecules are capable of transporting charge and interact efficiently with light similar to inorganic materials. Therefore, these systems can act as semiconductors in optoelectronic devices. However, organic chemistry offers the advantage of tailoring material functional properties via modifications of the molecular core units, opening new exciting possibilities for inexpensive device manufacture. This chapter reviews fundamental aspects behind the structural design/realization of molecular semiconductors for organic field-effect transistors (OFET). An introduction to the basic principles of OFET and the state-of-the-art organic molecular semiconductors for organic transistors is presented.

10.1 INTRODUCTION

Organic semiconductors have been studied since the 1940s, and the initial industrial application of these compounds was xerography, a process exploiting their photoconductive properties [1,2]. One class of materials receiving significant attention is that of π-conjugated small molecules, mainly oligomeric structures. These systems exhibit a common feature of having π-conjugated bonds, giving rise to delocalized, filled, and empty π-orbitals, which greatly impact their optical and electrical properties. In contrast to inorganic semiconductors, the solid-state structure of organic semiconductors is based on weak van der Waals and dipole–dipole interactions between neighboring molecules, imparting within them the properties of both semiconductors and insulators. Modern interest in organic-based electronic devices stems directly from the realization that a linear π-conjugated system (doped polyacetylene) was found in the 1970s to exhibit metallic conductivity. Some of the earliest examples of organic compounds finding application in electronic components include efficient organic light-emitting diodes [3] and organic field-effect transistors (OFETs) [4]. However, the potential of active electronic devices such as solar cells, light emitters, and thin-film transistors remained unfulfilled for decades because organic materials have often proved to be environmentally unstable and far less efficient than inorganic materials.

The pioneering studies on organic materials addressed both fundamental questions of the semiconducting properties and also the great potential presented by a new generation of electro-optic materials and devices. However, the molecular design or engineering of selected organic molecules was not a major activity in this field until the late 1980s. New synthetic tools and better understanding of charge transport in organic molecules resulted in the realization of new classes of conjugated materials with enhanced semiconducting properties. In particular, for transistor applications, the question was how to tune/improve device characteristics on the basis of designed molecular building blocks. At present, a number of organic semiconductor classes have been discovered, the principal of which are based on π-conjugated small molecules and polymers. These results generated an exciting library of organic semiconductors, which allowed the organic electronic solid-state community to draw important relations between molecular structure and structural organization in the solid state, film morphology, and ultimately, charge transport efficiency. In addition, new chemical–physical phenomena can be revealed by systematically studying properly designed sets of these systems.

This chapter provides the fundamental chemical aspects behind structural design, realization, and charge transport properties of various organic molecular semiconductors for organic field-effect transistor (OFET) applications [5], starting from OFET structure and operational principles and ending with a summary of the main molecular semiconductor families. State-of-the-art transistor performance reached by molecular semiconductors both as polycrystalline thin films and single crystals is presented here.

10.2 ORGANIC FIELD-EFFECT TRANSISTORS: DEVICE STRUCTURE AND OPERATION

Inorganic materials are at the core of the modern semiconductor industry, and are the current constituents of metal-oxide-semiconductor field-effect transistors (MOSFETs)—the crucial building blocks of electronic circuits [6]. According to Moore's law, these devices have proven to be reliable, and highly efficient, with performance increasing regularly [7]. Instead of competing with conventional silicon/gallium arsenite (GaAs) technologies, organic FETs fabricated with organic materials used as semiconductors (OFETs) may find niche applications in low-performance radio-frequency technologies [8], sensors [9,10], and light emission [11], and in integrated optoelectronic devices, such as pixel drive and switching elements used in displays [12,13]. Organic semiconductors have been intensively investigated since they offer numerous straightforward attractions: vapor phase or solution film deposition and good compatibility with various substrates, including flexible plastics [14–16], and broad opportunities for rational structural tailoring [17–19]. This trend is driven by the demand for inexpensive, large-area, flexible, and lightweight devices and the possibility of processing these materials at far lower temperatures compared to conventional silicon technologies.

The most common and easier to realize OFET device configuration is that of a bottom-gate thin-film transistor, in which a thin film of the organic semiconductor is deposited on top of a dielectric with an underlying gate (G) electrode (Figure 10.1). Charge-injecting source–drain (S–D) electrodes providing the contacts are defined either on top of the organic film (top-contact configuration) or on the surface of the FET substrate before deposition of the semiconductor film (bottom-contact

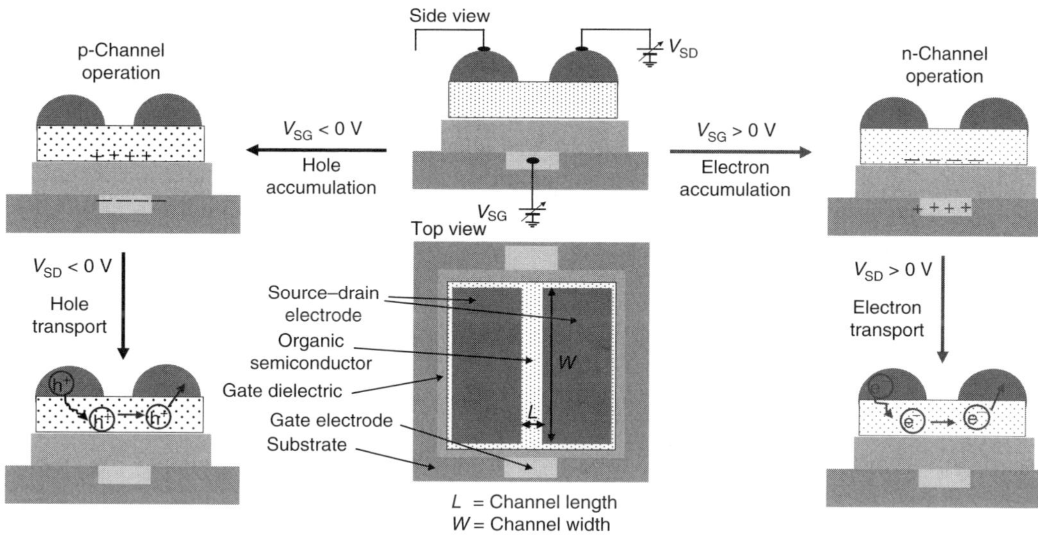

FIGURE 10.1 Structure of OFET device with highlighted material component (center), p-channel (right), and n-channel (left) operation.

configuration). In this electronic device, the current between source and drain electrodes is minimal when no voltage is applied between the gate and source electrodes, and the device is in "off" state. When voltage is applied to the gate, electrons or holes can be induced into the semiconductor at the interface with the dielectric layer and the source–drain current increases (Figure 10.1). This is called the "on" state of the transistor. The basic relationships describing the OFET drain current are given in Equations 10.1 and 10.2

$$(I_{SD})_{lin} = (W/L)\mu C_i(V_{SG} - V_T - V_{SD}/2)V_{SD} \tag{10.1}$$

$$(I_{SD})_{sat} = (W/2L)\mu C_i(V_{SG} - V_T)^2 \tag{10.2}$$

where

- μ is the field-effect carrier mobility of the semiconductor
- W is the channel width
- L is the channel length
- C_i is the capacitance per unit area of the gate dielectric
- V_T is the threshold voltage
- V_{SD} is the source–drain voltage
- V_{SG} is the source-gate voltage

On increasing the magnitudes of V_{SD} and V_{SG}, a linear current regime (Equation 10.1) is initially observed at low drain voltages ($V_{SD} < V_{SG}$), followed by a saturation regime (Equation 10.2) when the drain voltage exceeds the gate voltage. Note that organic FETs normally operate in the accumulation mode, where the increase in the magnitude of V_{SG} enhances channel conductivity, in contrast to conventional inorganic transistors. Figure 10.2 collects examples of OFETs I–V plots from which the carrier mobility can be calculated. The output plot is obtained by plotting I_{SD} versus V_{SD} and the transfer plot is obtained when I_{SD} is plotted against V_{SG}. From the latter plot the device current on:

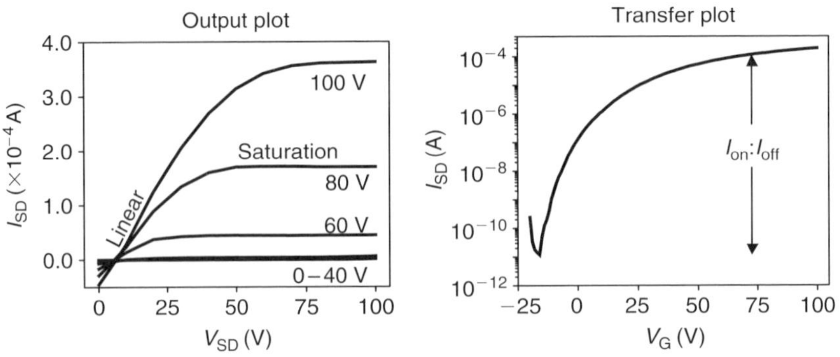

FIGURE 10.2 Example of output and transfer plots for n-channel semiconductors.

off ratios (I_{on}:I_{off}) can be easily calculated, which represent the parameters describing how efficiently the FET operates as an electronic valve. Other important OFET parameters are the threshold voltage (V_T) and the subthreshold swing (S).

When a molecule-based FET is active upon the application of negative V_{SG} and V_{SD}, the organic semiconductor is said to be p-channel (or p-type) since holes are the majority charge carriers (Figure 10.1). On the other hand, when a (positive) source–drain current is observed upon the application of positive V_{SG} and V_{SD}, the semiconductor is n-channel (or n-type) since electrons are mobile (Figure 10.1). In few cases, OFETs operate for both V_{SG} and V_{SD} polarities and the semiconductor is said to be ambipolar. Note the fundamental difference between n/p-organic and -inorganic semiconductors; the former is based on the gate voltage sign at which they are active, whereas the latter is based on the mobile carrier type due to the (chemical) doping process. Therefore, it is important to stress that the categorization of an organic semiconductor as p- or n-type has no absolute meaning, but it is strongly related to the FET device structure/material combination on which the transport characteristics are measured.

10.3 MATERIAL REQUIREMENTS FOR HIGH OFET CHARGE TRANSPORT EFFICIENCY

Efficient FET charge transport and therefore high device performance can be achieved by the use of proper materials and material combinations to fabricate the transistor. Source, drain, and gate contacts must exhibit large conductivity and ensure ohmic contacts to enhance device speed. The gate dielectric must exhibit large dielectric strength to ensure charge carrier accumulation in the FET channel upon application of the gate field. As far as the most important FET material is concerned, the organic semiconductors, it must satisfy general criteria relating both injection and current-carrying characteristics, in particular: (i) highest occupied molecular orbital (HOMO) and lowest unoccupied molecular orbital (LUMO) energies of the individual molecules (perturbed by their placement in a crystal lattice) must be at levels where holes/electrons can be incorporated at accessible applied electric fields, (ii) crystal structure of the material must provide sufficient overlap of frontier orbitals to allow efficient charge migration between neighboring molecules, (iii) solid should be extremely pure since impurities act as charge carrier traps, (iv) molecules should be preferentially orient with the long molecular axes approximately parallel to the FET substrate normal since the most efficient charge transport occurs along the direction of intermolecular π–π stacking, and (v) crystalline domains of the semiconductor must cover the area between source and drain contacts uniformly, hence the film should posses a single crystal-like morphology. All of these material aspects are described in the following sections.

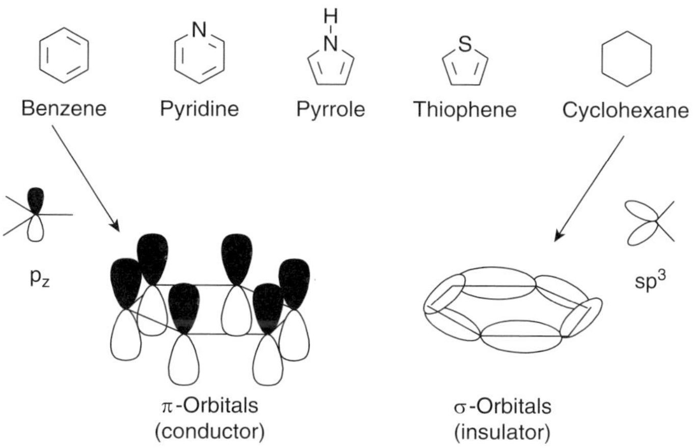

FIGURE 10.3 Chemical structure of common (hetero)aromatic rings and cyclohexane generating conducting and insulating molecular orbitals, respectively.

10.3.1 Semiconductor Molecular Design Requirements

Early candidate organic materials used in electronic devices largely included π-conjugated small molecules and polymers based on structural modifications of very simple aromatic building blocks such as ethylene, acetylene, benzene [20], thiophene [21], pyrrole [22], and aniline [23] (Figure 10.3). These molecules exhibit similar electronic structure in which the p_z orbitals of the C and hetero (S, N) atoms overlap generating molecular orbitals delocalized along the entire molecule/ring. The electronic structure generated from the overlap of carbon p_z orbitals is called π-conjugated or conductive in contrast to sp^3 orbital overlap giving rise to σ-localized or nonconjugated (or insulating) electronic structures (e.g., benzene versus cyclohexane).

The filled (or bonding) π-molecular orbital with the highest energy is called HOMO whereas the empty (or antibonding) molecular orbital with the lowest energy is called LUMO. When some of these aromatic rings are fused/linked or mixed to afford larger molecular (or polymeric) chemical structures, their HOMO and LUMO orbitals interact generating new conjugated molecular orbitals (larger HOMOs and LUMOs) extending along the length of the entire structure (Figure 10.4).

The solid-state intermolecular interactions between π-molecular orbitals (π–π interactions) mainly account for their unique optical and electrical properties. One of the key structural prerequisite to achieve extensive intramolecular π-conjugation, hence fully delocalized HOMO and LUMO, and good intermolecular π–π interactions is molecular planarity (Figure 10.5). Therefore, all efficient charge transporting organic materials have a planar or nearly planar molecular core. Other factors affecting the electronic behavior of π-conjugated electronic materials are (i) bond length alternation caused by an intrinsic electronic instability (Perierls distortion) [24], (ii) deviation from molecular planarity due to core substitution, which limits efficient intermolecular π-orbital overlap, (iii) localized carrier trapping arising from aromatic stabilization and functional groups, (iv) electronic and steric influences of heteroatoms and pendant substituents, and (v) intermolecular electronic communication (π–π interaction) in the solid state.

10.3.2 Frontier Molecular Orbital Energies

The energies of the frontier molecular orbitals (FMOs, hence HOMO and LUMO) are one of the key parameters affecting majority of charge transport in the OFET channel as well as the efficiency of charge injection (contact resistance) from the source and drain contacts. An efficient way to tune

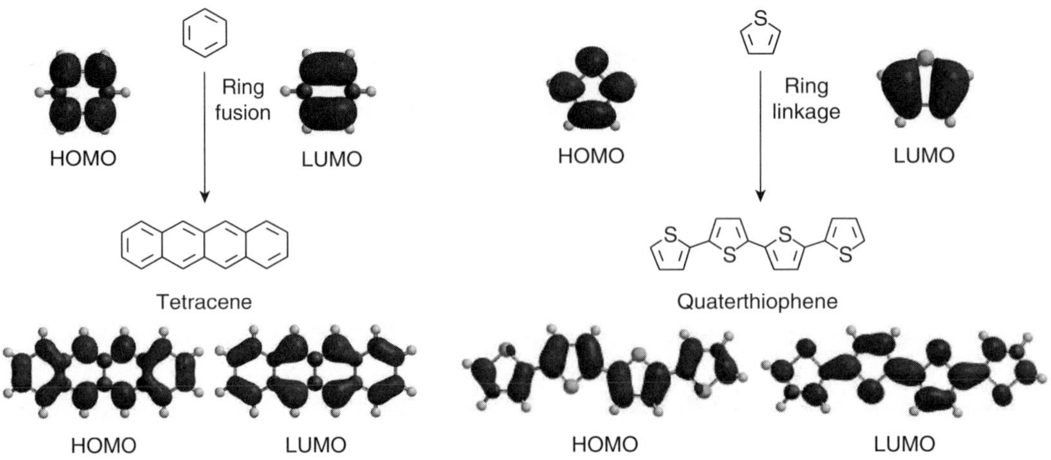

FIGURE 10.4 The highest occupied molecular orbitals (HOMOs) and the lowest unoccupied molecular orbitals (LUMOs) of benzene (left) and thiophene (right) and their corresponding molecular semiconductors tetracene and quaterthiophene, respectively.

HOMO and LUMO energies is via substitution of the molecular core with heteroatoms (e.g., S, Si, N, or O) and with proper chemical functionalities. Organic substituents are divided into electron-withdrawing (e.g., –F, $-C_nF_{2n+1}$, –CN, –COR) and electron-donating (e.g., $-C_nH_{2n+1}$, –OR, NR_2). As an example, the variations of the FMO energies for oligothiophenes (*n*Ts) upon substitutions with *n*-hexyl (electron-donating) and *n*-perfluorohexyl (electron-withdrawing) substituents are described (for structures see Figure 10.6). An approximate but common and extremely easy way to estimate HOMO and LUMO energies is using electrochemical measurements (cyclic voltammetry experiments) usually in combination with optical spectroscopy. Note that these measurements provide good information on energy variations rather than the absolute FMO energy values. Furthermore molecular orbital computations are also a valuable tool to estimate molecular orbital energy variations upon core substitution, and a useful guide for the molecular design of optimized structure before undertaking their synthesis.

Figure 10.7 shows the experimental (from electrochemistry) and computed HOMO and LUMO energies for various unsubstituted/substituted oligothiophenes as a function of the core length *n* (number of thiophene ring constituting the core). This work was performed at Northwestern University [25].

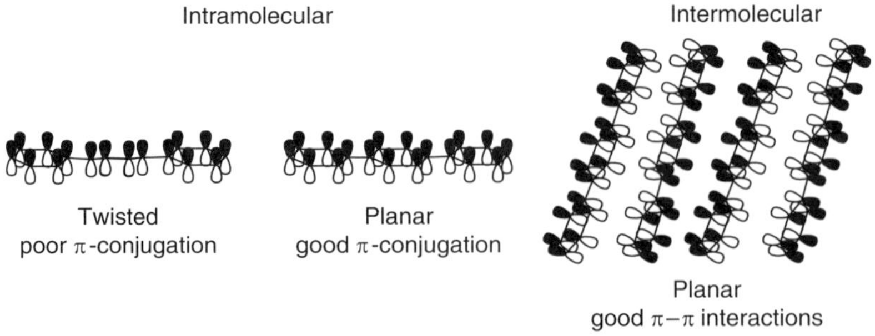

FIGURE 10.5 Schematic representation of twisted and planar conformations affecting intramolecular π-conjugation and intermolecular packing.

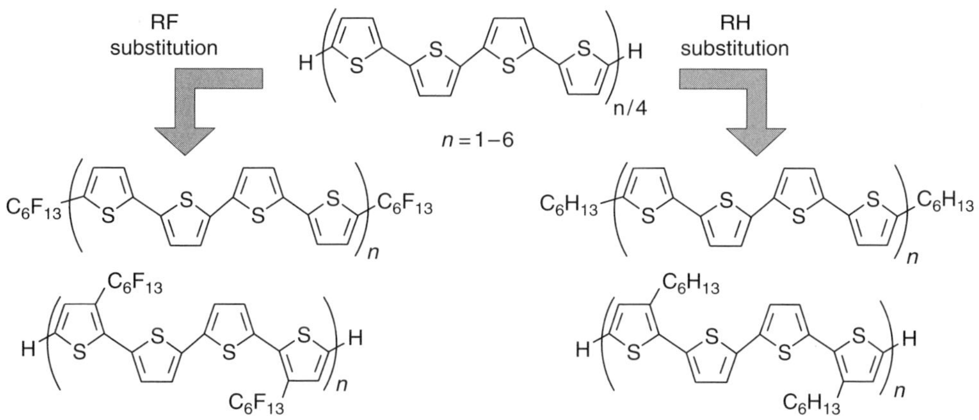

FIGURE 10.6 Chemical structure of oligothiophenes (*n*Ts) and the corresponding α, ω-disubstituted derivatives with perfluorohexyl (DFH-*n*Ts) and hexyl (DH-*n*Ts) chains.

From these plots it is evident that oligothiophene substitution with *n*-hexyl groups raises both HOMO and LUMO energies versus the unsubstituted analogues, while *n*-perfluorohexyl groups lowers both groups of energy levels (e.g., ΔE + 0.11 eV for the dihexyl-**6T** (n = 6) HOMO and LUMO versus **6T**, and –0.41 and –0.38 eV for the diperfluorohexyl-**6T** HOMO and LUMO, respectively, versus **6T**). Particularly instructive here are the linear energetic trends revealed by plotting the experimental/computed orbital energies versus the number of thiophene units in the core (Figure 10.7). The HOMO and LUMO energy plots of fluorocarbon-substituted and alkyl-substituted *n*Ts are located below and above those of the unsubstituted oligothiophenes *n*Ts, respectively. This result is in agreement with the established donating/accepting properties of hexyl/perfluorohexyl substituents. Important information on the molecular HOMO and LUMO energy levels can be drawn from an accurate comparison of the experimental trends with those predicted by theoretical calculations. While HOMO and LUMO eigenvalues from density functional theory (DFT) methods cannot be formally taken as either ionization potentials or electron affinities [26], previous work has shown that B3LYP-derived eigenvalues compare favorably with experimental ionization potentials and electron affinities. The present systematic agreement between computed HOMO and LUMO

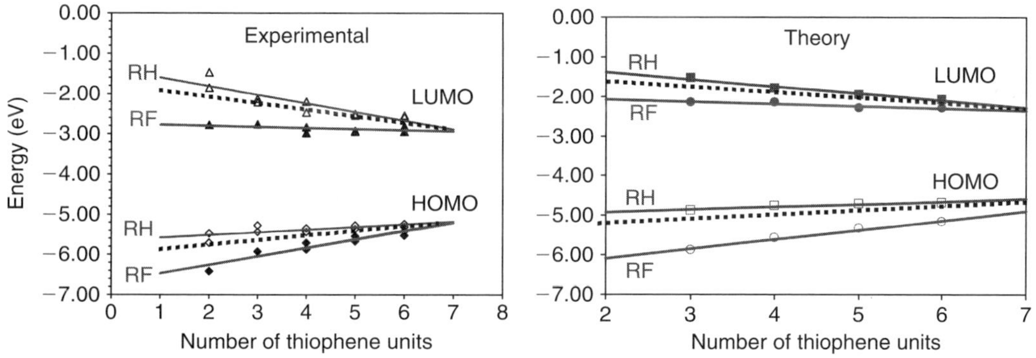

FIGURE 10.7 Experimental and theoretical HOMO and LUMO energy plots versus the number of core thiophene units (n) for DFH-*n*Ts and DH-*n*Ts.

energies and experimental trends is extremely good, confirming the validity of these methods. The disparity between computed and electrochemical HOMO/LUMO absolute energy values derives from the difference in environments, and previous work has shown that solvation effects such as polarizability and cavity radius can be used to linearly adjust (via a Kamlet–Taft relationship) [27] computed ionization potentials and electron affinities to compare with electrochemical data [28].

10.3.3 Crystal Structure

Since the charge transport properties of organic semiconductors depend intimately upon their molecular and crystal structure, when possible these have been investigated intensely. This possibility of easy access to molecular crystal structure represents an advantage of molecular versus polymeric semiconductors. The knowledge and details of the molecular structural parameters and intermolecular packing characteristics can be used for additional molecular design tuning. The intermolecular order adopted by the individual molecules in the solid state is one of the most important parameter affecting the electro-optic properties of organic materials. In general, good electronic performance requires strong electronic coupling (π–π intermolecular interactions) between adjacent molecules. For most organic materials this strong intermolecular overlap is achieved via two common solid-state packing motifs. In the "herringbone" arrangement aromatic edge-to-face interaction dominates. Alternatively, the molecules can adopt a coplanar or cofacial arrangement and stack typically with some degree of displacement along the long and short axes of the molecules (Figure 10.8).

As representative examples of these two packing motifs, the crystal structure of sexithiophene (**6T**) and pentacene (**P5**) (herringbone, Figure 10.9), and mixed fluoroarene-thiophene oligomers (**FTnFs**) (coplanar, Figure 10.10) will be described. The first single-crystal study of **6T** was conducted by Garnier et al. [29]. Macroscopic single crystals were obtained by sublimation under reduced pressure under Ar. In this low-temperature structure, the molecules were found to be strictly planar. Another unit cell proposed later is for a high-temperature polymorph of **6T** obtained via the melt-growth process with a temperature gradient [30]. The latter structure shows some differences from the low-temperature phase, especially a significant reduction in the interlayer spacing. However, also for this polymorph the **6T** molecules are planar with the inter-ring dihedral angle not exceeding 2°. Furthermore, all of these structures have in common the herringbone packing, which is usual among all oligothiophenes (nTs, n = 2–7). The herringbone angle ranges from 55° to 67°. The herringbone packing is important in determining the intermolecular interactions and transport properties of **6T**. For the low-temperature form, the C–S distances are 0.391 and 0.33 nm and other distances (C–C and S–S) are also greater than the sum of the corresponding van der Waals radii. However, for the high-temperature form (which has the densest packing) intermolecular C–S distances are considerably shorter [31], 0.368–0.385 nm (side-to-side) and 0.352 nm (end-to-end), leading to strong C–S interactions.

Like sexithiophene and shorter fused acenes, pentacene crystallizes in the classic acene herringbone motif [32], although a number of polymorphs have been found [33]. Note that the polymorphic nature of many organic semiconductors complicates transport studies, since band structure calculations

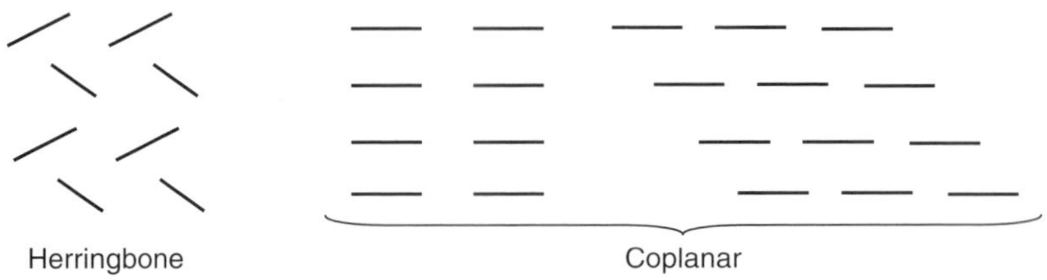

FIGURE 10.8 Schematic representation of herringbone and coplanar molecular packings.

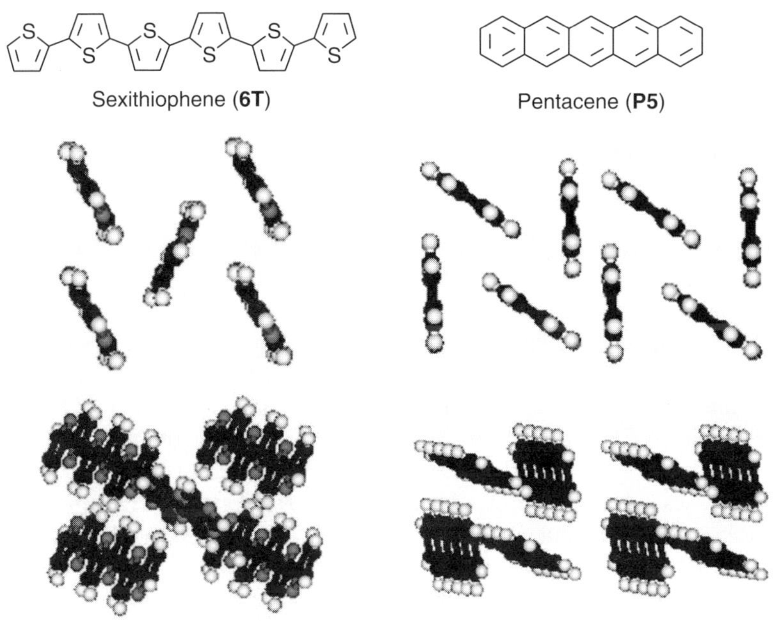

FIGURE 10.9 Crystal structure of sexithiophene and pentacene.

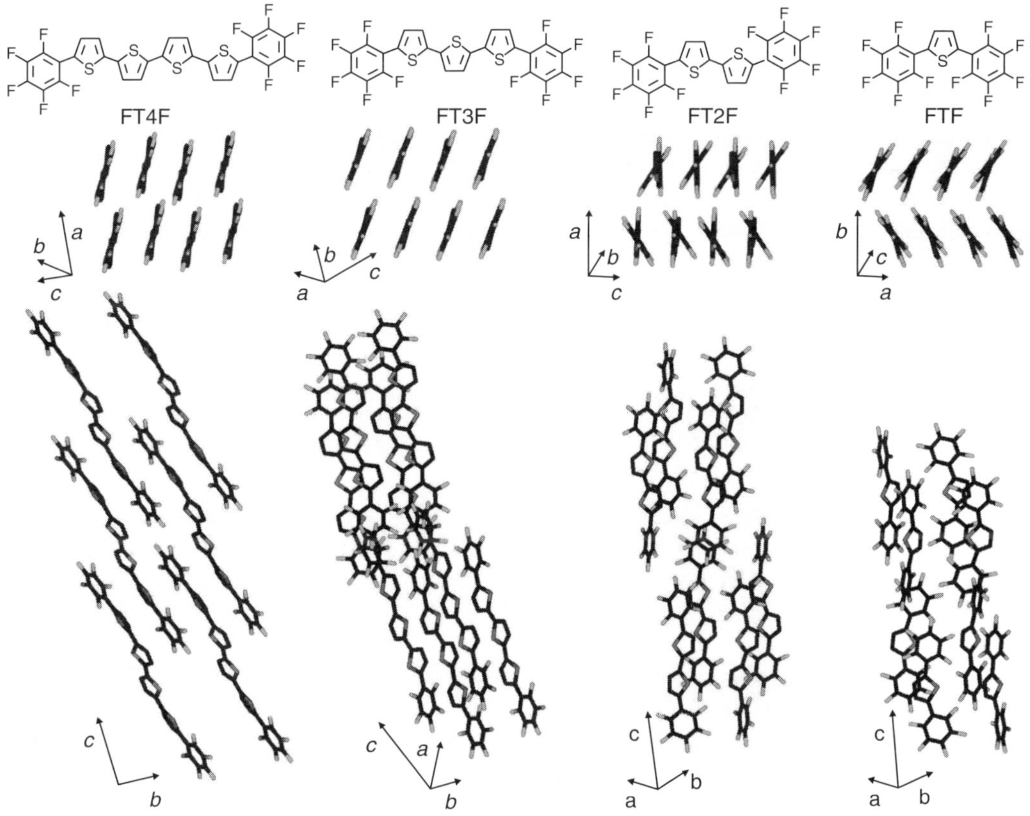

FIGURE 10.10 Crystal structure of mixed perfluoroarene-thiophene oligomers (FTnFs).

(a way to predict how efficient is the transport of holes and electrons in the molecular crystal valence and conducting bands, respectively) show significant electronic differences between polymorphic forms [34]. The crystal structure of pentacene was first reported by Campbell et al. [35] for crystals grown from solution. They observed a triclinic crystal structure using 0*kl* and *h*0*l* Weissenberg films. The molecules are detected along the *c* axis with a *d*(001)-spacing of 14.5 Å. A much smaller *d*-spacing of 14.1 Å was recently reported by Holmes et al. [36], also for crystals grown from solution.

On the other hand, the crystal structures of fluoroarene-thiophenes **FTnF** ($n = 1$–4) exhibit the common feature of close cofacial packing of the whole molecule or molecular fragments between the electron-rich (thiophene) and electron-deficient (perfluoroarene) subunits (Figure 10.10) [37]. The molecular structures of the longer perfluoroarene end-substituted oligomers **FT4F** and **FT3F** reveal substantially planar cores with a maximum core torsion of ~17° (between the two innermost thiophene rings) and 8° (between the perfluorophenyl and outer thiophene rings), respectively. Radically different is the molecular structure of the shorter oligomers, **FT2F** and **FTF**, where one of the perfluorophenyl groups is twisted with respect to the remaining part of the molecule (quite planar) at an angle ~36° and ~22°, respectively. These angles are substantially greater than in typical unsubstituted oligothiophenes (~5°) and comparable to the ~30° value required to achieve sufficient intramolecular π-overlap for an efficient conduction band structure [38]. While **FT4F** molecules exhibit uncommon "syn" conformation between outer pairs of thiophene rings, the end-substituted systems **FT3F** and **FT2F** display typical all-trans conformation of their thiophene cores. Molecular packing is also quite different between longer and shorter oligomers. The large core dimension of molecule **FT4F** allows each perfluorophenyl ring to interact with two of the four core thiophenes resulting in parallel layers stacking along the *b* axis for the whole molecule via displacement of each molecular layer by one-half a molecule with respect to the other. The minimum interlayer distance between cofacial molecules in adjacent planes is 3.20 (F_4–C_6'), comparable to or smaller than the sum of F–C (3.15–3.30 Å), S–C (3.45–3.50 Å), and C–C (3.30–3.40 Å) van der Waals radii [39]. On the other hand, for the shorter oligothiophenes **FT2F** and **FTF**, only one of the two perfluorophenyl rings can stack face-to-face with the electron-rich dithiophene molecular portion, with the consequence that the packing of part of the molecule is π–π face-to-face and that of the remaining perfluorophenyl group is herringbone-like, as found in the crystal structure of perfluorobenzene [40]. A more complex packing is observed for molecule **FT3F**. In this case the core planarity yields molecular layers extending along the ab plane in which the molecules within the layer are face-to-face stacked whereas those of alternating layers intermingle with their end perfluoroarene groups in a herringbone fashion. The minimum intermolecular distances between cofacial molecular fragments/molecules are 3.49 Å (C_2–S_1') for **FTF**, 3.72 Å (C_{17}–S_1') for **FT2F**, and 3.32 Å (F_2–S_3') for **FT3F**. There values are much greater than that observed for **FT4F** (3.20 Å), probably because of the reduced number of closely π–π stacked rings.

10.3.4 Film Morphology and Microstructure

Obtaining the crystal structure of single-crystalline molecular semiconductors is a powerful tool to access the solid-state bulk packing characteristics. However, single crystals do not provide a complete picture of how the molecules behave when processed into a thin film for OFET fabrication. Deposition conditions and interactions with the surface play a critical role in determining the crystal structure, crystallite/molecular orientation, film continuity, and crystalline grain size within the film. It is well known that in polycrystalline organic films, the molecules can orient differently with respect to the substrate surface [41,42], and their grain size and microstructure is strongly affected by substrate surface chemical characteristics [43] and film growth temperature [44]. For instance, vapor-deposited pentacene molecules orient either parallel or perpendicular to the substrate surface depending on whether the substrate is a metal (Au, Cu) or an insulator. Substrate effects are even more pronounced for solution-processed semiconductor films [45]. Therefore, investigation of the organic semiconductor thin-film microstructure and morphology versus film deposition parameters and substrate characteristics allows better understanding of important structure–device property relationships.

Film morphology and grain size distribution within polycrystalline films can be investigated using a variety of microscopy techniques such as scanning electron microscopy (SEM) and transmission electron microscopy (TEM). These techniques consent to investigate larger scale morphology and grain boundary size and can also be used to obtain film diffraction patterns. Atomic force microscopy (AFM) and scanning tunneling microscopy (STM) can also be used to image the film morphology. In addition, AFM can image nanometer-scale vertical features such as step heights within the films, providing useful molecular geometrical parameters. Note that to be an active and efficient charge transporter, the film must cover continuously the region between the source and drain electrodes. Examples of film morphologies observed using AFM and SEM are shown in Figure 10.11. In the AFM image (Figure 10.11), the pentacene crystal steps extending a few micrometers are clearly visible, and the SEM image shows an example of solution-processed film of a fluorocarbon-substituted 4T (**DFH-4T**). It can be noted that the large and flat crystals extending several microns are interrupted by deep fractures.

Additional structural information within the film such as the texture and preferential molecular orientation relative to the substrate surface can be gathered from x-ray diffraction (XRD) [46] and grazing-incidence x-ray diffraction (GIXD) [47]. In general, molecular semiconductors orient with their long axis nearly perpendicular to the substrate and form layered planes parallel to the substrate. Figure 10.12 shows the XRD pattern of three perfluorohexyl-substituted oligothiophene films fabricated by vapor deposition (DFH-*n*T, $n = 4$–6) [48]. All of the films are highly crystalline as demonstrated by the large number of sharp Bragg XRD reflections. These films are composed of molecular layers, each of thickness d, which can be measured by applying the Bragg equation. Recently, near-edge x-ray absorption fine structure (NEXAFS) spectroscopy has also found utility in characterizing thin films of organic semiconductors. NEXAFS is depth-sensitive and can be used to determine chemical bond orientation as well as chemical identity [49].

10.3.5 Dielectric and Contact Materials

Considering the OFET structures and the implications of Equations 10.1 and 10.2, it becomes immediately evident that to achieve good OFET performance, the organic semiconductor is not the only critical component. It is also very important to incorporate suitable gate dielectric and contact materials. Regarding the former material, dielectrics for OFETs must satisfy requirements applicable

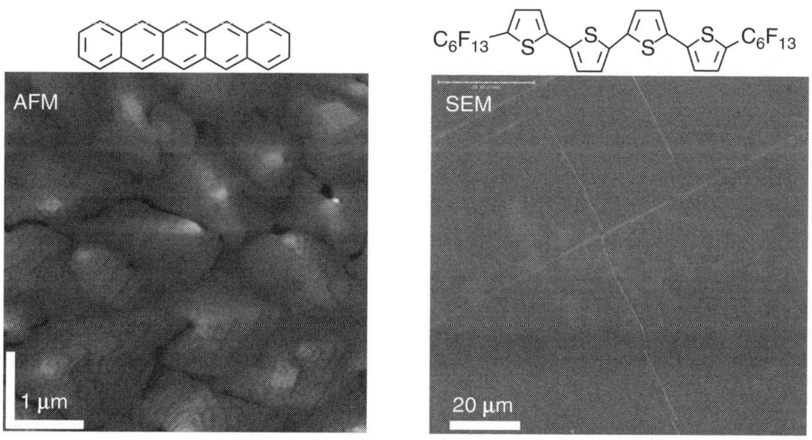

FIGURE 10.11 AFM and SEM images of pentacene and DFH-4T films, respectively.

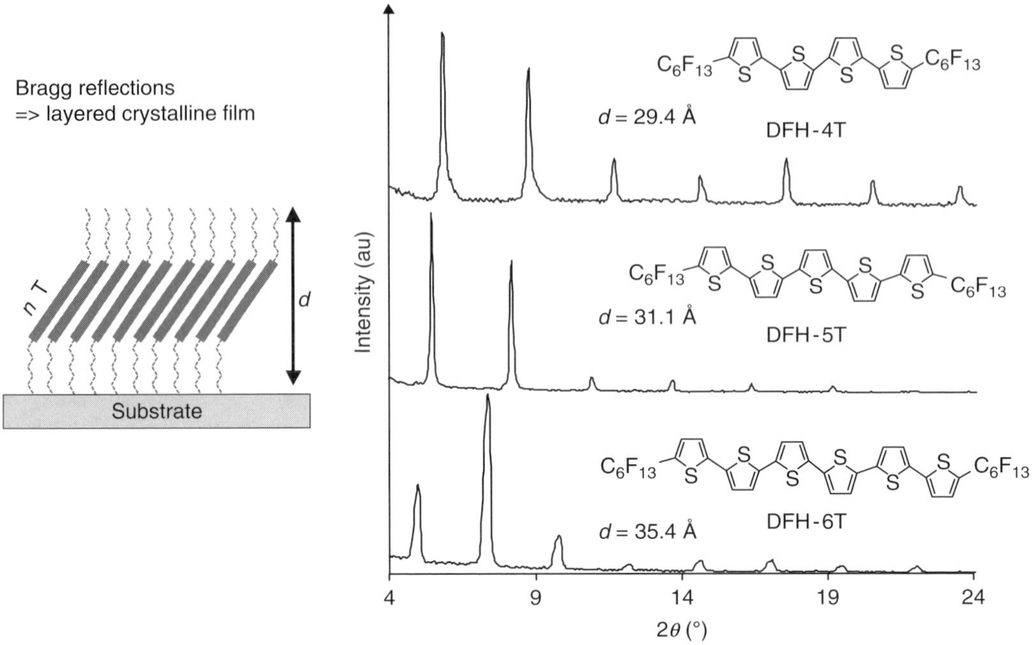

FIGURE 10.12 Schematic representation (left) of molecular orientation of DFH-*n*T (*n* = 4, 5, and 6) molecules on the SiO_2 surface. XRD plots of DFH-*n*T films vapor deposited on Si-SiO_2 substrates.

to gate inorganic transistor insulators as well [50]. The crucial parameters are the maximum possible electric displacement D_{max} the gate insulator can sustain

$$D_{max} = \varepsilon_0 k E_B \quad (10.3)$$

where
- k is the dielectric constant
- E_B is the dielectric breakdown field
- ε_0 is the vacuum permettivity

and the capacitance per unit area is given by

$$C_i = \varepsilon_0 (k/d) \quad (10.4)$$

where d is the insulator thickness

The capacitance magnitude is governed not only by the k value but also by the thickness (d) for which a pinhole-free film can be achieved, and thus may reflect the deposition procedure as well as intrinsic material properties.

Silicon dioxide (SiO_2) is the dielectric mostly used in OFETs. One motivation to search for SiO_2 alternatives is to significantly reduce the OFET operating voltage. In fact, while the carrier mobilities of organic semiconductors have now approached/surpassed those of amorphous Si, this has generally been achieved at very large source–drain/source-gate biases, typically >30–50 V. OFET operation employing such large biases will incur excessive power consumption. Therefore, according to Equations 10.1 and 10.2, a viable approach to substantially increase the drain current while operating

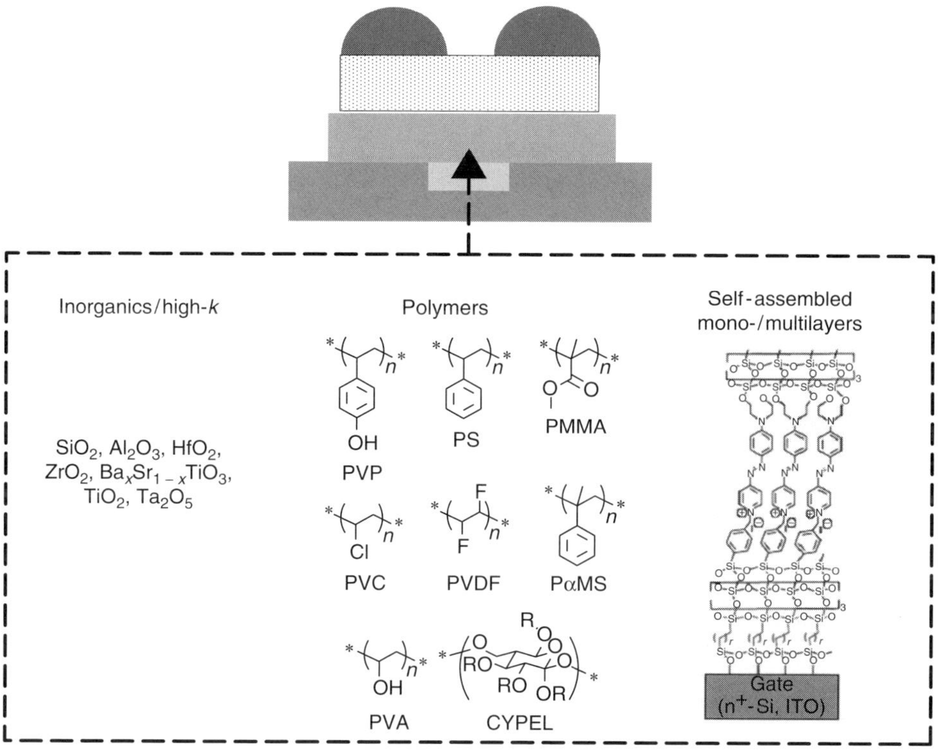

FIGURE 10.13 Gate insulator families used on organic TFTs.

at low biases is to increase the capacitance of the dielectric. The second motivation is that OFET gate dielectric must fulfill demands specific to organic electronics, which include low-cost/low-capital investment manufacture of organic electronic circuits. OFET gate insulators should ideally be compatible with flexible substrates and be processable/printable from solution, but should also be insoluble in the solvent used for deposition of the organic semiconductor (in the case of bottom-gate structures), and should be compatible with the gate line deposition process (in the case of top-gate configurations). The confluence of these two important goals has stimulated search of new dielectrics [51]. The three major dielectric classes used in OFETs are those of high-*k* inorganics, polymers, and self-assembled mono- and multilayer small molecules (Figure 10.13).

Regarding the contact materials, Au is the most used metal contact. However, Au is expensive and although recently progress in discovering printable Au-based inks has been achieved, alternative materials have been tested including conducting polymers (e.g., polyaniline, polypyrrole, and PEDOT:PSS) [52] and Ag-based particles/inks (Figure 10.14) [53]. As a general rule charge injection in organic semiconductors can be achieved by matching the work function of the electrode to either the HOMO for hole injection or LUMO for electron injection. p-Type organic semiconductors typically have HOMO levels between −4.9 and −5.5 eV, resulting in ohmic contact with high work-function metals such as gold (5.1 eV) and platinum (5.6 eV) [54]. To achieve correct OFET device operation, contact resistance must be as small as possible and must not dominate the bulk (or channel) resistance of the semiconductor. However, in many cases contact resistance and barrier height to charge injection can be significant, reducing on currents appreciably [55]. n-Type materials typically have LUMO levels between −3 and −4 eV and should have better contact with low work-function metals such as aluminum and calcium. Unfortunately, these metals are highly reactive and degrade rapidly with air exposure. It has been found that gold typically forms good top contacts with organic n-type materials, even though it has a much higher work function [56].

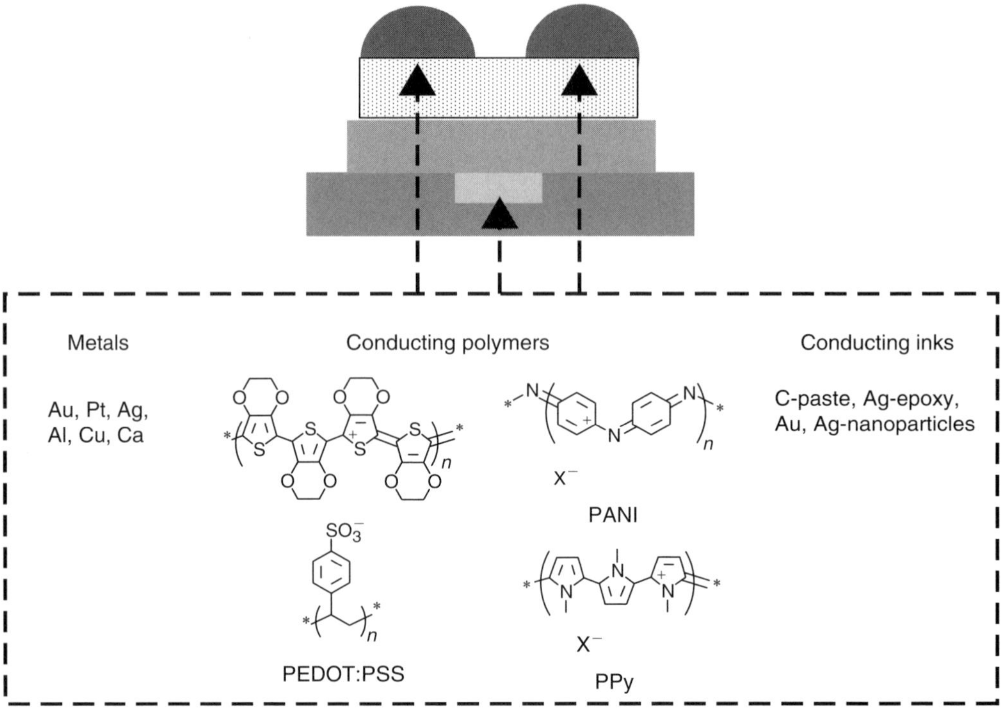

FIGURE 10.14 Conducting materials used in organic TFTs.

10.4 MOLECULAR SEMICONDUCTOR FAMILIES FOR OFETs

At present, many classes of organic semiconductors have been discovered, ranging from small molecules based on (hetero)aromatic rings, conjugated oligomers and polymers, dendrimers, and hybrid organic–inorganic structures (Figure 10.15). Recent review articles provide details of these structures and their up-to-date performance in OFETs [57,58]. In the following sections representative examples of molecular semiconductor families are described.

10.4.1 Acenes and Oligophenylenes

Initial experiments on the transport properties of organic semiconductors were performed on simple aromatic hydrocarbons such as anthracene and tetracene [59]. These systems belong to the acene family and, as for phenylenes, are formed by fusion/connection of benzene rings (Figure 10.16).

To date, the largest polyacene synthesized is heptacene (Figure 10.17) [60]. However, during the last decade, pentacene has become one of the most extensively studied organic semiconductors for FETs [61]. It is commercially available, and exhibits the highest acene mobility yet achieved for a polycrystalline film. Jackson et al. reported a large mobility of 1.5 cm^2/V s, on/off current ratio $>10^8$, near zero threshold voltage, and a subthreshold slope <1.6 V per decade [62]. Recently, mobility of 6 cm^2/V s was reported by Kelley et al. [63]. High-performance pentacene devices have also been obtained employing polymeric gate dielectric layers (mainly polyvinylphenol) with carrier mobilities as large as 3 cm^2/V s, subthreshold swings as low as 1.2 V per decade, and on/off current ratio of about 10^5 [64]. High-quality pentacene single crystal can be prepared from the vapor phase in a stream of hydrogen or forming flowing gas [65]. Iodine-doped pentacene is a p-type semiconductor [66], while alkaline metal-doped pentacene is an n-type semiconductor [67]. However, pentacene is

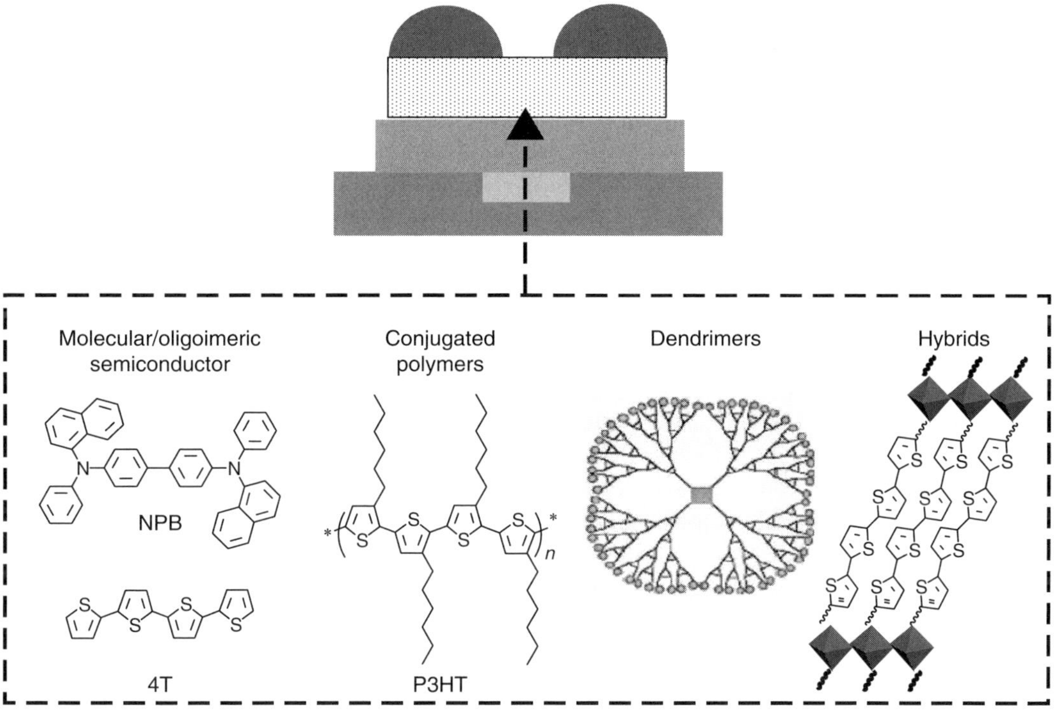

FIGURE 10.15 Semiconducting families used in organic TFTs.

plagued with disadvantages of oxidative instability and intense absorption in the visible spectrum [68]. These properties significantly limit the application of pentacene in practical electronic devices requiring ambient stability and optical transparency (i.e., in displays). Furthermore, unsubstituted oligoacenes are not compatible with the increasingly popular, printing-based device fabrication methods given their limited solubilities.

To circumvent such problems, various strategies for solution-processable oligomeric materials have been developed. One of these consists of preparing a soluble precursor molecule that is not semiconducting but can be converted to its semiconducting, insoluble form upon heating. In the first approach, the pentacene precursor is the corresponding tetrachlorobenzene adduct, which can be prepared through a multistep synthesis (Figure 10.18). This **P5** precursor is soluble in dichloromethane and forms continuous, amorphous films when spun onto substrates. The conversion to pentacene is accomplished by heating the films at 140°C–220°C in vacuum. Tetrachlorobenzene is eliminated in the conversion process. This approach has been realized for pentacene, with initial reported mobilities 0.01–0.03 cm²/V s [69]. In a more recent study, a mobility of 0.2 cm²/V s was achieved by treating the Si/SiO_2 substrate with hexamethyldisilazane (HMDS) before spin coating the precursor, and by optimizing the conversion conditions [70].

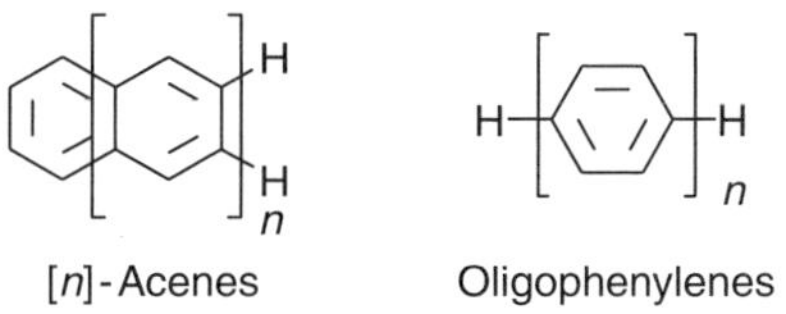

FIGURE 10.16 General structure of [*n*]-acenes and oligophenylenes.

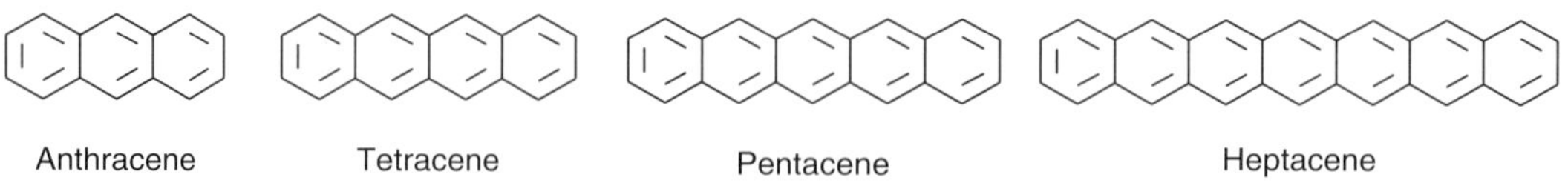

FIGURE 10.17 Chemical structure of anthracene, tetracene, pentacene, and heptacene.

In an alternative, elegant approach, the synthesis of a pentacene precursor was achieved by the Lewis acid-catalyzed Diels–Alder reaction of pentacene with heterodienophiles such as *N*-sulfinyl acetamide (Figure 10.18) [71]. This precursor is very soluble in chlorinated solvents, and conversion to pentacene is achieved by moderate heating. A record mobility of 0.89 cm²/V s was measured.

Solubilization of the pentacene core can also be achieved by core substitution with alkyl groups in proper positions. For instance, the tetramethylpentacene (**Me$_4$-P5**) has been synthesized and characterized by the Bao and Wudl groups (Figure 10.19) [72]. Other interesting soluble pentacene derivatives were introduced by Anthony et al. [73,74]. FET data strongly depend on the single-crystal packing characteristics with the greatest FET performance (μ up to 1 cm²/V s) for those systems, such as **P5-TIPSA**, exhibiting close π–π core stacking. Perfluorination of pentacene to afford perfluoropentacene (**F-P5**) has also been achieved resulting in an n-type semiconductor for OFETs [75]. Perfluoropentacene is a planar molecule and adopts a herringbone structure as observed for pentacene. **F-P5**-based OFETs exhibit an electron mobility of ~0.1 cm²/V s. The authors also reported complementary inverter circuits fabricated by combining **P5** and **F-P5** OFETs.

FET devices have also been fabricated using thermally evaporated films of the oligophenylenes *p-n*P (Figure 10.20). The mobility of these systems ranges from 10^{-2} cm²/V s for ***p*-4T** to 0.1 cm²/V s for ***p*-6T** with on/off current ratio from 10^5 to 10^6 [76]. While perfluorinated acenes larger than anthracene are unknown, perfluorinated oligophenylenes *p*-nFP have been reported as promising n-type semiconductors for OLEDs [77]. However, no FET data were reported. The synthetic pathway to these systems, for instance ***p*-5FP**, involves the reaction of 1,4-dibromoperfluorobenzene with an excess of 2,3,5,6-tetrafluorophenylcopper to afford trimer *T*, which was then brominated and finally subjected to reaction with pentafluorophenylcopper to afford the final product (Figure 10.21).

10.4.2 Thiophene-Based Molecules

The thiophene ring (see Figure 10.3) is the most common building block used in semiconducting materials. All thiophene-based oligomers and mixing thiophene with ethene, benzene, and other

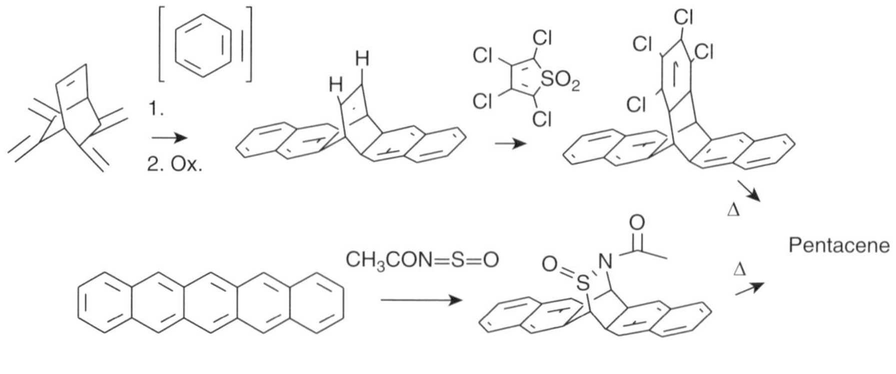

FIGURE 10.18 Synthetic routs to soluble pentacene precursors and back to pentacene upon heating.

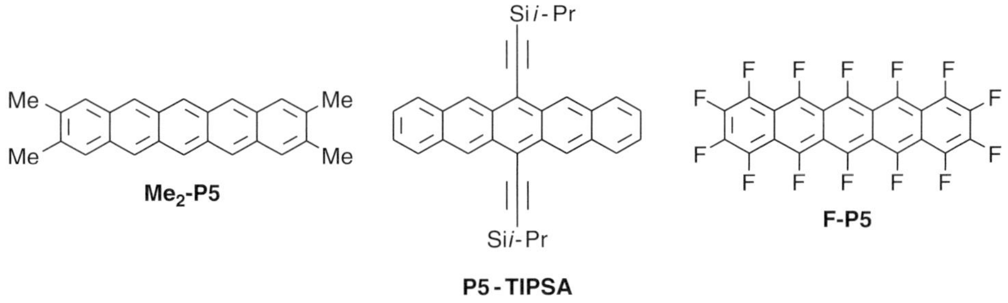

FIGURE 10.19 Chemical structure of various pentacene derivatives.

heterocycles have generated exciting libraries of compounds (Figure 10.22). Unsubstituted (*n*Ts) and alkyl-substituted (e.g., DH-*n*Ts) thiophene oligomers have been extensively investigated for OFETs applications. FETs based on polycrystalline, vapor-deposited sexithiophene (**6T**) [78] and α,ω-dihexyl-sexithiophene films were first systematically studied by Garnier et al. [79] and played a very important role in the evolution of the field of organic transistors. These studies demonstrated that relatively high mobilities (0.002 and 0.05 cm^2/V s for **6T** and **DH-6T**, respectively) are attainable from polycrystalline organic films and delineated the strategies that should be followed to increase FET performance. Employing rigid, rod-like molecules, such as thiophene oligomers, with large π-conjugation lengths extending along the long axis of the molecule, and close molecular packing of the molecules along at least one of the short molecular axes (π-stacking) are two important requirements for high carrier mobility. In addition, α,ω-alkyl substitution enhances performance since it induces a higher degree of film self-organization. However, very recently Hajlaoui et al. reported field-effect mobility in a vacuum-deposited film of unsubstituted **8T** as high as 0.33 cm^2/V s [80].

The even-ring number members of this family can be synthesized by oxidative coupling of the corresponding monomers with either $FeCl_3$ or $BuLi/CuCl_2$ [81]. However, these methods were found to produce difficult-to-eliminate by-products, which depress FET performance. Alternative synthetic pathways to thiophene oligomers, in particular 6T derivatives, were introduced by Katz et al. using ferric acetylacetonate as oxidizing reagent [82]. The compound **DH-6T** can be prepared by Stille coupling [83]. Unsubstituted quinquethiophene and end-substituted quater-, quinque-, and sexithiophene display sufficient solubility in organic solvents to allow fabrication of field-effect devices by solution processing techniques. Initial studies of solution-processed oligothiophene transistors gave mobilities of $\sim 5 \times 10^{-5}$ cm^2/V s for quinquethiophene and α,ω-diethylquaterthiophene (**DE-4T**) [84]. More recently, mobilities in the range of 0.01–0.1 cm^2/V s have been achieved using films of solution-processed substituted oligothiophenes [85]. The mobility was found to depend strongly on film morphology, which can be controlled by processing conditions, such as solution concentration, substrate temperature during casting, solvent choice, and environmental conditions during film drying. Barbarella et al. described the synthesis, crystal structure, and electrical measurements on the di- and

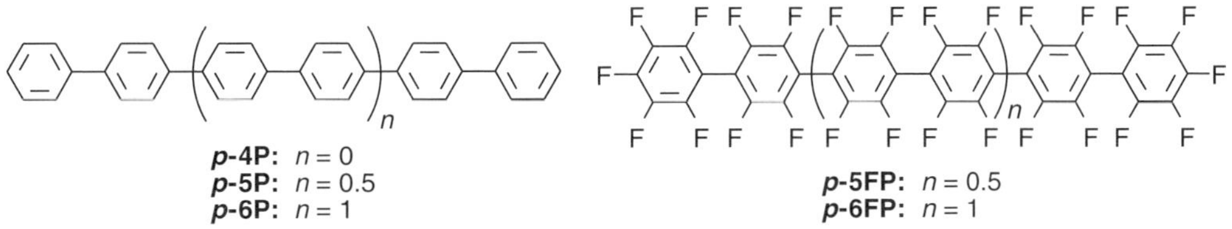

FIGURE 10.20 Chemical structure of various oligophenylene derivatives.

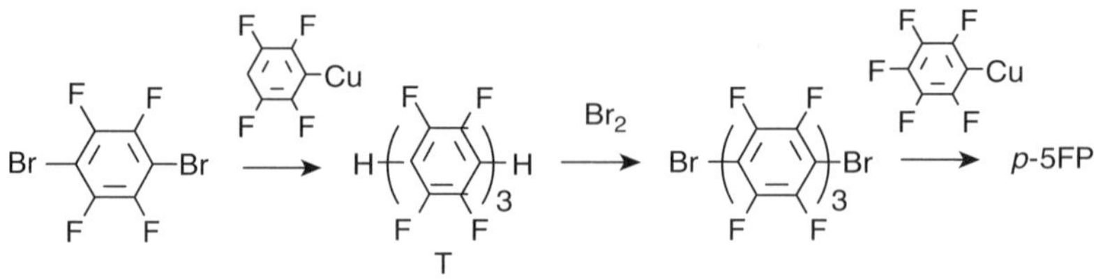

FIGURE 10.21 Synthesis of *p*-5FP.

tetramethyl-substituted 6Ts [86]. Compound **Me_2-6T** was found to exhibit high mobility (0.02 cm^2/V s) as an evaporated film and good solid-state order in packing. More recently, Facchetti et al. found that the carrier mobility of regiochemically pure **isoDH-6T** approaches 0.1 cm^2/V s [87].

A problem associated with using electron-rich thiophene-based semiconductors is their sensitivity to atmospheric doping, which diminishes the current on/off ratio and alters the threshold voltage. A strategy to enhance environmental stability is to decrease the availability of these electrons, hence reduce the HOMO energy. An effective way to reduce HOMO energies involves using shorter thiophene oligomers than **6T**. Another means of improving environmental stability without shortening the molecule is to incorporate less oxidizable heteroaromatics such as thiazole (**BHT4Z2**) [88] and azines (e.g., pyrimidine, **DH-TPmT2PmT**) (Figure 10.22) [89]. Compound **BHT4Z2** was prepared starting from thiazole, which was subjected to dimerization and iodination to afford 5,5′-diiodo-2,2′-bithiazole. Subsequent Stille coupling of 5-hexylbithiophene with the diiodide intermediate produces **DHT4Z2** in good yields. This system was shown to exhibit a higher on/off current ratio versus the all-thiophene analogue **DH-6T**, and its threshold voltage is markedly less susceptible to drift.

Recent results proved that the reduction of HOMO energies are also obtained by replacing some thiophene rings with phenyl rings (Figure 10.23) [90]. Mixed thiophene-phenylene oligomers **dH-PPTPP** and **dH-PTTP** were prepared by Stille coupling, and display high mobilities as evaporated (up to 0.09 cm^2/V s) and solution-cast (up to 0.03 cm^2/V s) films on both n-Si/SiO_2 and ITO/glass resin substrates [91]. In addition, since reduction of the HOMO energy is accompanied by a shift of the threshold voltage toward the accumulation regime, simple reversible nonvolatile memory elements have been fabricated by solution-only techniques. Finally, Ichikawa et al. showed that epitaxially

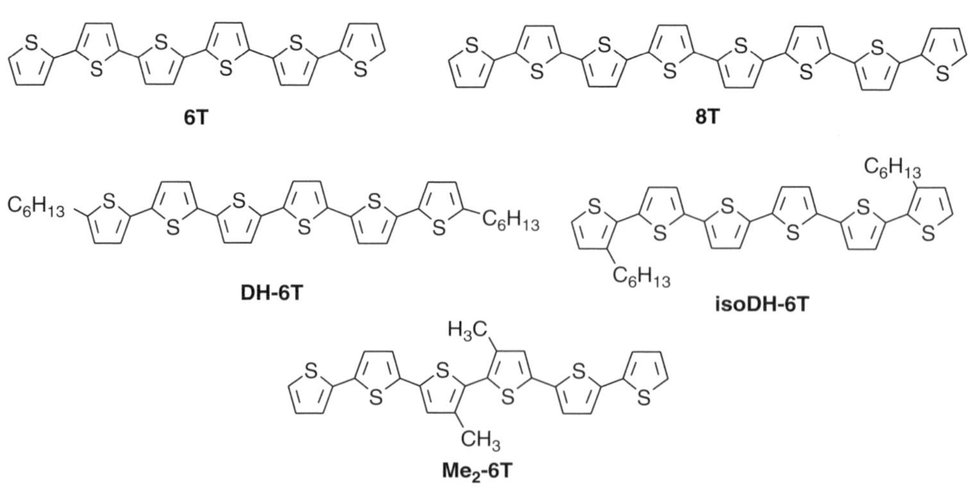

FIGURE 10.22 Chemical structure of various sexithiophene derivatives.

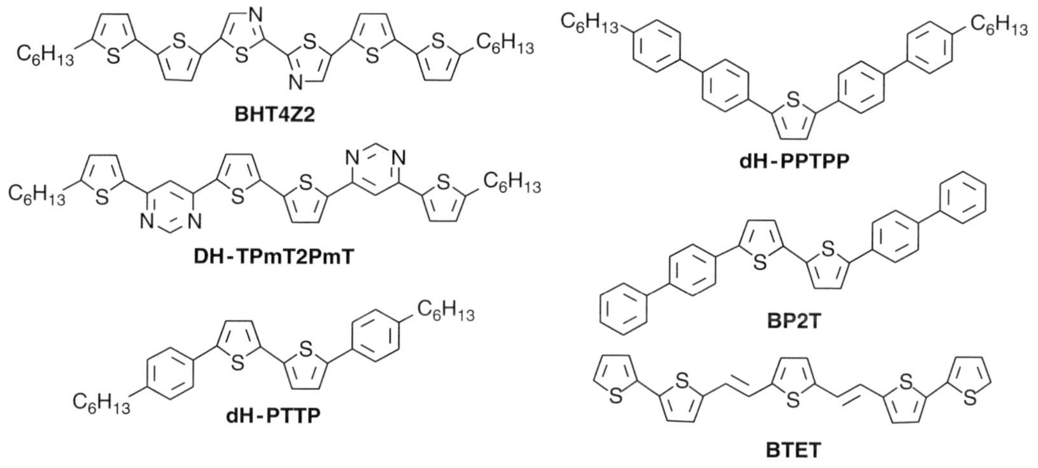

FIGURE 10.23 Chemical structure of various thiophene-based small molecules.

grown thiophene-phenylene system **BP2T** on KCl single-crystal substrates exhibits a hole mobility (0.29–0.66 cm²/V s) [92] much larger than that of **BP2T** vacuum-deposited directly on SiO_2 at room temperature and close to that of high-quality oligothiophene single crystals. In contrast to phenylene incorporation, interposition of a double bond between thiophene rings was found to decrease compound stability, although one thienylenevinylene oligomer, all-*trans*-2,5-bis(2-(2,2′-bithien-5-yl) ethenyl)thiophene (**BTET**), synthesized using Wittig chemistry, has a measured mobility of 0.008–0.012 cm²/V s as an evaporated film and 0.0014 cm²/V s as a spin-coated film [93].

High-performance environmentally stable thiophene-fluorene derivatives have been synthesized by a Suzuki coupling reaction of dibromodithiophene with the corresponding pinacolato boronic ester-substituted fluorene and 2-hexylfluorene (Figure 10.24). Mobilities as high as 0.11 cm²/V s and an on/off current ratio of 10^5 have been reached for evaporated films [94].

A successful approach to new organic semiconductors has been to combine the molecular shape of pentacene, which leads to a favorable crystal packing geometry and orientation, with thiophene end groups that should increase stability and also provide points of attachment for solubilizing substituents. Thus, anthradithiophene derivatives (Figure 10.25) were prepared and characterized for the first time by Katz et al. [95]. These products were obtained as a mixture of "syn" and "anti" isomers, and separation was not possible. ATD field-effect mobility is an order of magnitude lower than that of pentacene, about 0.1 cm²/V s, but the on/off current ratio is higher versus zero gate. A highly ordered thin-film morphology is observed and is consistent with the electrical characteristics. Hexyl-, dodecyl-, and octadecyl-disubstituted derivatives were also found to demonstrate modulation typical of FET devices, with the first two exhibiting higher mobilities (~0.15 cm²/V s) than the parent compound, and with increased solubility. The third material still has significant activity (mobility ~0.06 cm²/V s), even though it consists mostly of nonconjugated C12 chains.

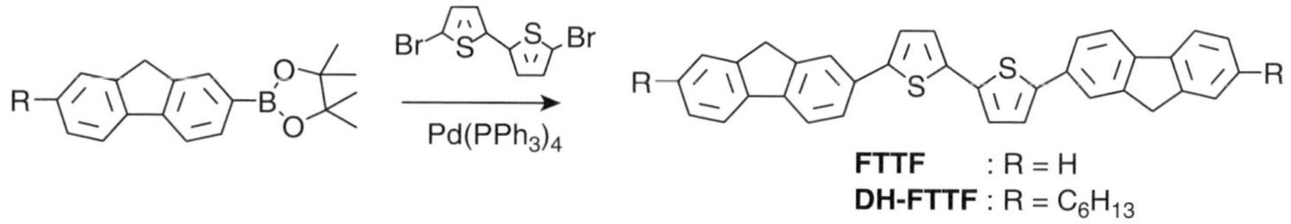

FIGURE 10.24 Synthesis and chemical structure of fluorine-based molucules.

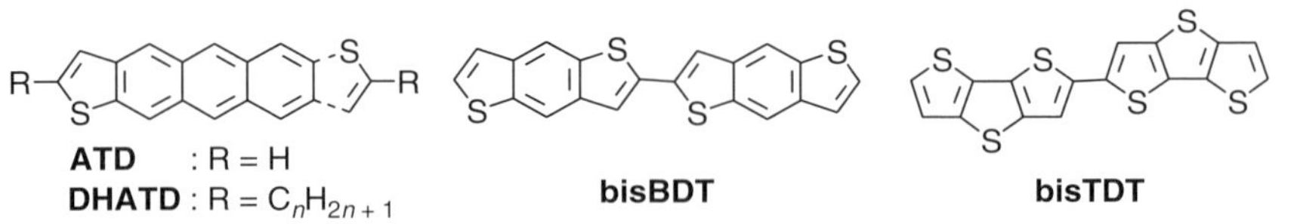

FIGURE 10.25 Chemical structure of ATD, BDT, and TDT derivatives.

The performance of thiophene-based semiconductors was further explored using as core units benzodithiophene (BDT) and thienodithiophene (TDT). Benzodithiophene can be synthesized exclusively as the anti isomer through a four-step synthesis starting from 2,3-dibromothiophene, which was reacted with thiophene-2-carboxyaldehyde to afford the coupled alcohol. Reduction of the methine bridge with $LiAlH_4/AlCl_3$, formylation, and final cyclization affords BDT. Appropriate dimerization and coupling reactions yielded different BDT-based compounds. The building block TDT was prepared starting from 3-bromothiophene, which was lithiated and reacted with bis(phenylsulphonyl)sulfide to afford 3,3′-dithienylsulfide. $CuCl_2$-promoted coupling afforded the annelated final product. Dimerization of unsubstituted BDT and TDT to **bisBDT** and **bisTDT**, respectively, was achieved using the corresponding monolithiated species and $Fe(acac)_3$. Other oxidizing agents were found to give either low reaction yields or by-products. Compounds **bisBDT** and **bisTDT** were purified by vacuum sublimation and exhibit remarkable thermal stability. The mobility of **bisBDT** [96] was found to be 0.04 cm^2/V s when vapor deposited on a substrate at 100°C. The corresponding all-thiophene dimer (**bisTDT**) [97] exhibits a mobility of 0.05 cm^2/V s and very high on/off current ratio of 10^8. The crystal structure of **bisTDT** reveals a completely coplanar conformation with a unique compressed π–π stacking, considerably different from the herringbone motif found in α6T and pentacene. Very recently, extension to other benzo[1,2-b:4,5-b′]dichalcogenophenes (X = Se, Te other than S) [98] and the utilization of larger π-cores resulted in derivatives exhibiting FET mobilities of 0.17–2.0 cm^2/V s [99].

The energy levels of thiophene-based oligomers can also be substantially tuned by substitution with electron-withdrawing groups to provide broad families of electron-transporting (n-channel) semiconductors. In a series of papers, Marks et al. described the synthesis and characterization of α,ω-perfluorohexyl-substituted thiophene oligomers (see Figure 10.6 for general formula) [25,100,101]. For instance, DFH-4T (Figure 10.26) exhibits electron mobilities for vacuum-deposited films as high as 0.24 cm^2/V s and on/off current ratios of 10^8 (in ~20 mTorr vacuum) [48]. As for other classes of fluorinated materials, the mechanism of electron carrier stabilization is not completely clear and probably reflects an interplay of other effects besides reduction of LUMO energies. Following the same substitution strategy, mixed phenylene-thiophene [102] and phenylene-thiazole (e.g., **CF-PTZ** on OTS-treated substrates) [103] derivatives functionalized with perfluoroalkyl groups have been investigated and have been found to exhibit n-type conductivity (Figure 10.26).

Other important molecular n-channel semiconductors were realized following similar synthetic strategies (Figure 10.27). Hence, Frisbie et al. reported on the properties of a quinodimethane terthiophene (**QM3T**) [104]. This compound was found to behave as n-channel material in air and

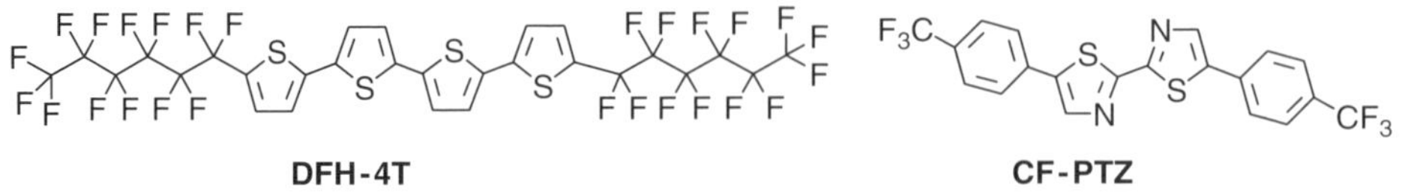

FIGURE 10.26 Chemical structure of DFH-4T and CF-PTZ.

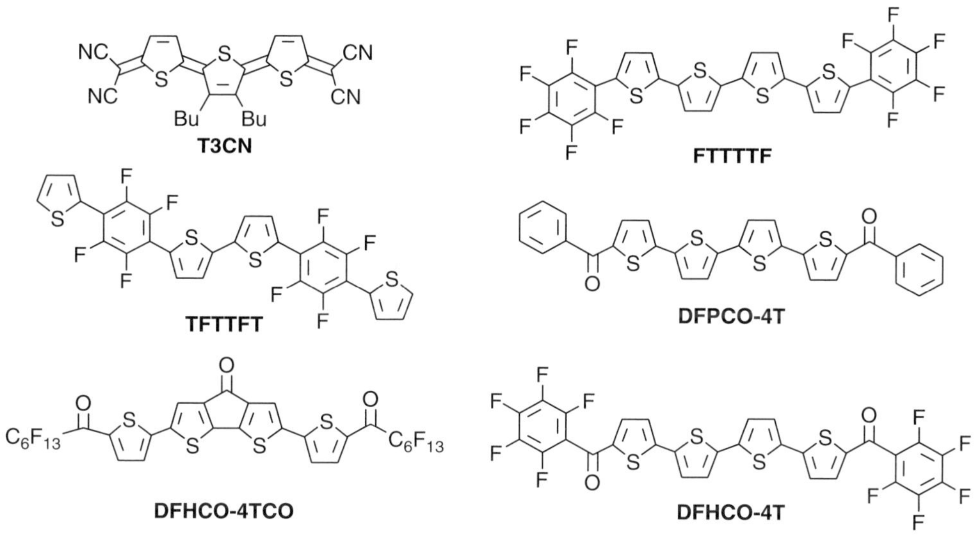

FIGURE 10.27 Chemical structure of various thiophene-based n-channel semiconductors.

exhibited a mobility of 0.5 and 0.002 cm^2/V s for vapor- and solution-deposited films, respectively. An extended series with even greater performance and ambipolar behavior has also been reported [105,106]. Very recently, high-performance tetrathiofulvalene (TTF)-based FETs have been investigated, mainly as p-type semiconductors [107,108]. However, Yamashita and coworkers [109] reported a TTF series where depending on core substitution and length both high p-(up to 0.6 cm^2/V s) and n-type (up to 0.1 cm^2/V s) carrier mobilities were measured. Recently, The Marks Group also designed and synthesized a new family of mixed perfluorophenyl-thiophene (F-T) oligomers to address the role of regiochemical substitution and π-core extension on the properties of the corresponding thin films [110]. These oligomers exhibit unique packing characteristics as shown in Figure 10.10. For the first time, both n-(for **FTTTTF**, up to ~0.4 cm^2/V s) and p-type (for **TFTTFT**, up to ~0.1 cm^2/V s) transport have been achieved within the same series by manipulating relative positions of the perfluoroarene building blocks. Proper combination of LUMO energy reduction, versus the corresponding fluorine-free analogues, and component regiochemistry results in n-type activity. Shorter oligomers were found to be solution processable and form highly ordered films exhibiting single crystal-like morphology (Figure 10.10) [37]. Recently, our group realized new oligothiophenes containing/substituted with carbonyl groups. Within these structures one compound is an air-stable n-channel oligothiophene (**DFHCO-4TCO**, mobility up to 0.01 cm^2/V s) [111], whereas other systems exhibit very high electron mobility both as vapor-deposited (**DFHCO-4T**, mobility up to 0.6 cm^2/V s) and as solution-cast (**DFPCO-4T**, mobility up to 0.24 cm^2/V s) [112] films.

10.4.3 Two-Dimensional Fused Rings

Perylene is one of the simplest two-dimensional (2D) fused aromatic rings, a class of compounds derived from the fusion of aromatics and heterocycles along ring bonds in two directions (Figure 10.28). Perylene conductivity measurements were reported in the early 1960s by Hideo et al. [113]. Furthermore, the electron and hole mobilities of perylene single crystals were measured using conventional time-of-flight methods and were found to be 0.017 and 0.02 cm^2/V s, respectively [114]. FET devices based on perylene have been recently studied and mobilities were quite low [115].

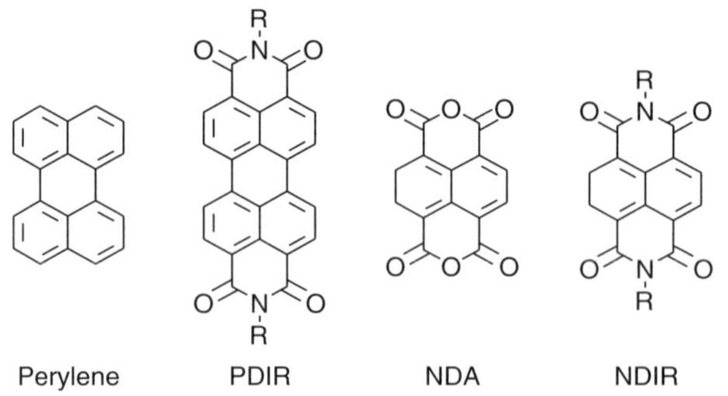

FIGURE 10.28 Chemical structure of perylene and naphthalenedianhydride (NDA) and corresponding *N*-alkyl-tetracarboxydiimides.

Having the same perylene backbone are perylene derivatives. For instance, Malenfant et al. reported a record mobility of 0.6 cm²/V s and on/off current ratio >10^5 for **PDI8** (R = C_8H_{17}) [116]. Frisbie et al. studied the microstructural and electrical properties of **PDI5** (R = C_5H_{11}) in detail [117]. Very recently, a record mobility of 2.1 cm²/V s was reported for a *N,N′*-ditridecyl-3,4,9,10-perylene-tetracarboxylic diimide (**PDI13**, R = $C_{13}H_{27}$) derivative after thermal annealing at 140°C [118]. Other interesting 2D molecules are naphthalene carbodiimide derivatives (NDIs). They are commonly synthesized from NDA (naphthalene carbodianhydride) and amines [119] and are purified by sublimation. These compounds have been studied extensively by Katz et al. as efficient electron-transport materials [120]. The unsubstituted system exhibits quite low thin-film mobilities, in the range of 10^{-5} and 10^{-4} cm²/V s for NDA and **NDIH** (R = H), respectively [121]. On the other hand, for alkyl-substituted NDIs, mobilities of 0.16 cm²/V s (R = C_8H_{17}), 0.01 cm²/V s (R = $C_{12}H_{25}$), and 0.005 cm²/V s (R = $C_{18}H_{37}$) have been measured under vacuum. The incorporation of fluoroalkyl groups on the side chains greatly stabilizes NDI solids with respect to electron transport in air [122].

Very recently, Marks and Wasielewski have reported the FET properties of new core-cyanated PDI [123] and NDI [124] derivatives exhibiting very large electron transport in air (Figure 10.29). For instance, **PDIF-CN$_2$** and **NDI8-CN$_2$** exhibit electron mobilities of ~0.6 and ~0.1 cm²/V s, respectively, in air. Complementary circuits employing **PDI8-CN$_2$**, in combination with pentacene as the p-type counterpart, have also been demonstrated [125].

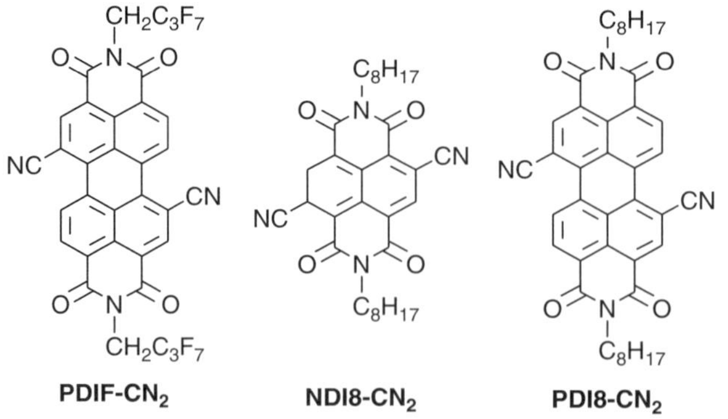

FIGURE 10.29 Chemical structure of various core-cyanated PDI and NDI derivatives.

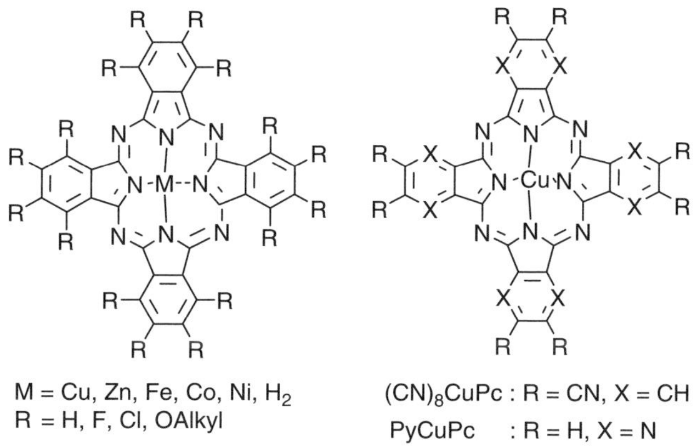

FIGURE 10.30 Chemical structure of various metallophthalocyanines.

Another type of 2D fused ring system intensively investigated is that of phthalocyanine (Pc, Figure 10.30). Pcs are probably the first reported organic semiconductors [126]. The Pc molecule has the structure of a molecular cage, into which various metals can be introduced. Most of the unsubstituted and metal-coordinated phthalocyanines are commercially available, thermally stable up to 400°C, and readily form films by vacuum deposition.

They are commonly used as organic semiconductors in numerous applications, such as solar cells, light-emitting diodes, and nonlinear optical materials [127]. The field effect was reported in a Pc as early as 1970 [128] and OFETs were fabricated in 1988 [129]. Their field-effect mobility ranges between 0.0001 and 0.01 cm^2/V s [130]. The most common and best performing phthalocyanine for p-channel transistors is copper phthalocyanine (**CuPc**; R = H, M = Ni, Figure 10.30). Mobilities as high as 0.02 cm^2/V s and an on/off current ratios $>10^5$ have been reported [131]. The rare-earth Lu and Tm bis(phthalocyanines) were also shown to have relatively high p-type mobilities (ca. 0.015 cm^2/V s) [132]. In addition, the same paper reports n-type behavior for these systems under vacuum. Soluble phthalocyanines can be produced by functionalizing with alkyl and alkoxyl side chains. Short-chain substituted Pcs have good solubility in various organic solvents, such as chloroform and toluene, and are also sufficiently volatile for vacuum sublimation. However, the crystallinity of these compounds is usually very poor. On the other hand, most of the other substituted phthalocyanines with longer linear alkyl and alkoxy side chains (e.g., hexyl or higher) are not sufficiently volatile for vacuum sublimation. Therefore, their purification is far more difficult, and chromatographic separation sometimes degrades them. Relatively low mobilities were measured for octa-octoxy copper phthalocyanine films cast from solution, despite the high apparent crystallinity of the film. Very recently, OFETs based on **CuPc-NiPc** composites exhibit mobility as high as 0.05 cm^2/V s [133]. In addition, the same paper reports n-type behavior for these systems under vacuum. Finally, a recent paper describes the OFET properties of **CuPc** nanorods exhibiting p-type mobilities as high as 0.1 cm^2/V s and on/off current ratio of 100 [134]. Probably one of the most spectacular example of using **CuPc** in combination with pentacene and PTCDI (an n-type semiconductor) is in the application of artificial skin [135]. Like acenes and thiophene oligomers, the LUMO energy of phthalocyanine derivatives can be reduced by substitution with electron-withdrawing groups to give n-channel semiconductors [136]. Several phthalocyanine derivatives with fluoro, chloro, CN, and azine substituents can be readily synthesized [137]. The perfluorinated metallophthalocyanines are easily purified by vacuum sublimation, and highly ordered semiconductor films can be prepared [138]. Again, the best transistor performance (mobility up to 0.03 cm^2/V s) was reported for the copper complex.

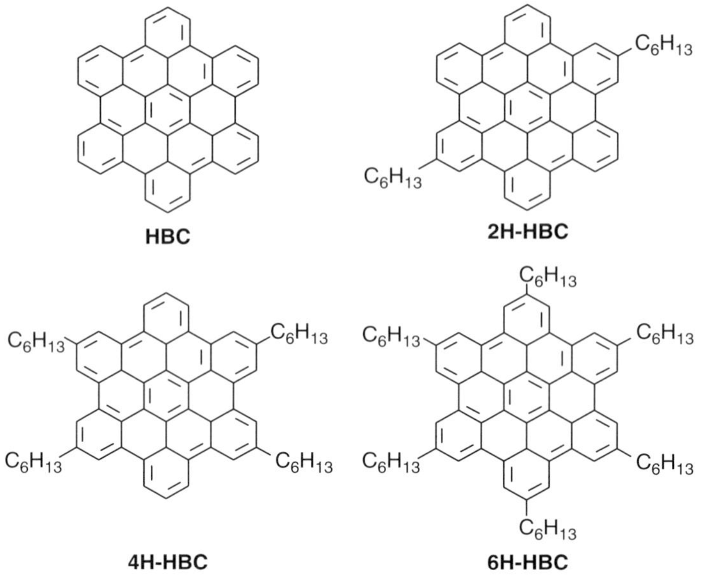

FIGURE 10.31 Chemical structure of various hexabenzocoronene derivatives.

Another interesting 2D organic semiconductor is hexabenzocoronene (**HBC**, Figure 10.31). HBC-based molecules' possibility of forming columnar stack of discotic liquid crystalline materials has been recognized as a structure that allows facile carrier transport along the stack [139]. These systems exhibit intrinsic mobilities of up to 1 cm^2/V s using a pulse-radiolysis time-resolved microwave conductivity technique, however, FET mobilities of only ~10^{-3} cm^2/V s have been initially measured [140]. Mori et al. studied HBC and di-(**2H-HBC**), tetra-(**4H-HBC**), and hexa-alkyl (**6H-HBC**) substituted derivatives and have demonstrated mobilities of ~0.01 cm^2/V s for vapor-deposited **4H-HBC** films. This good performance could be explained by the self-assembly in a 2D conductor of the HBC derivatives, i.e., the dihexyl and tetrahexyl derivatives, in contrast to the self-assembly in the 1D conductor of the hexahexyl derivatives and low self-assembling property of the unsubstituted HBC [141].

More recently, FETs with highly ordered active layers were fabricated by zone-casting techniques using **HBC-C12** as the semiconductor deposited on hydrophobic substrates to form columnar structures (Figure 10.32) [142]. AFM images showed that the disks form columns that

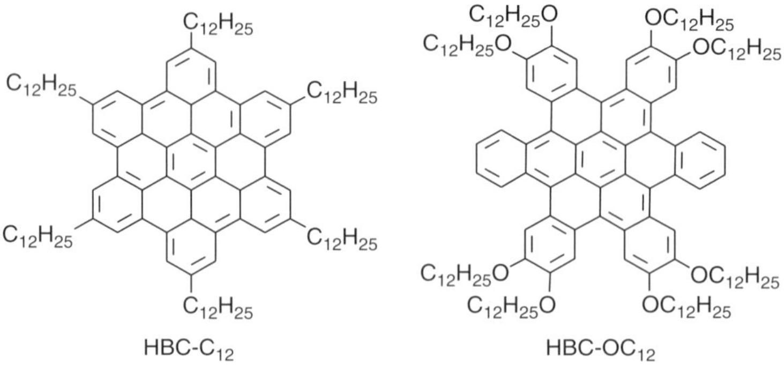

FIGURE 10.32 Chemical structure of HBC-C_{12} and HBC-OC_{12}.

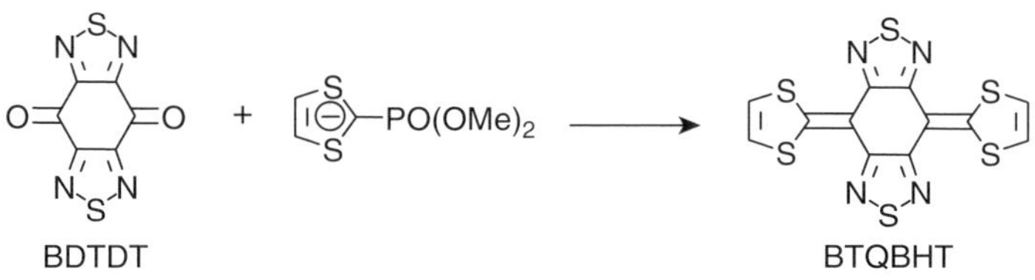

FIGURE 10.33 Synthesis and structure of BTQBHT.

possess long-range order on the scale of square centimeters. The high out-of-plane order was verified by XRD which shows out-of-plane periodicity of 1.98 nm. FETs were fabricated on Si-SiO_2 gate dielectric substrates by direct zone-casting of **6D-HBC** films on which Au source–drain contacts were vacuum deposited with a shadow mask. These devices exhibit hole mobilities as high as 0.01 cm²/V s and I_{on}:I_{off} > 10^4. Xiao et al. showed that **HBC-OC12** organizes into molecular stacks which then organize into cables or fibers [143]. These cables are ordered both on the molecular scale and on the supramolecular scale having orthorhombic unit cell. OFET devices fabricated with individual fiber with Au/Cr electrodes exhibit hole-transporting characteristics with carrier mobility ~0.02 cm²/V s.

Finally, another interesting material is the thiadiazole derivative **BTQBT** (Figure 10.33). This compound was synthesized by Wittig–Horner reaction of benzothiadiazole-dione BDTDT with the carbanion generated from 2-(dimethoxyphosphinyl)-1,3-dithiole in THF [144]. Earlier studies showed that BTQBT exhibits a large Hall mobility of 4 cm²/V s [145]. FETs were fabricated on Si (gate)/SiN_x (dielectric) substrates in both configurations, and **BTQBT** was deposited by organic molecular-beam deposition. Mobilities of 0.044 cm²/V s were reported. Unfortunately, due to the relatively high conductivity of the **BTQBT** films, current on/off ratios of only ~10^3 were achieved [146]. Very recently, the hole mobility and on/off ratio of **BTQBT** films under ultrahigh vacuum conditions reached 0.2 cm²/V s and 10^8, respectively, indicating great potential for further improvements [147].

10.5 SINGLE-CRYSTAL TRANSISTORS

The use of a single crystal of organic material as the semiconductor for OFETs consents to underscore intrinsic material properties independently of grain boundary effects and different crystallite phases/orientations characteristic of polycrystalline films. Note that crystals with near-perfect order over very large size can now be obtained by both vapor and solution-phase methods (vide infra). While impurities and lattice disorder and defects may still ultimately limit material performance [148], the single-crystal FET (SCFET) structure offers, in principle, the possibility to investigate the best possible transport properties achievable for a particular molecule. These devices are also a useful tool for examining intrinsic physical properties of organic semiconductor materials and a test bed to discover new phenomena. Furthermore, newly developed patterning techniques for single crystals—by both vapor [149] and solution [150] methods—suggest that SCFETs may again mature, this time into a viable method for hands-free circuit fabrication.

Most of SCFETs have been fabricated using single crystals grown from the vapor phase due to the poor solubility of the majority of high-mobility organic semiconductors. Figure 10.34 shows the structure of some of the best molecular semiconductors for SCFETs reported to date.

Single crystals of rod-like planar organic molecules often grow as very large (centimeter size) but thin (micrometer size) plates making them very fragile and difficult to handle. Therefore, SCFET fabrication requires great expertise to achieve good device yields. As for conventional OFETs, SCFETs can be fabricated with different device geometries. The top-contact SCFET structure requires the fabrication of the device directly on top of the organic single crystal. As opposed to conventional

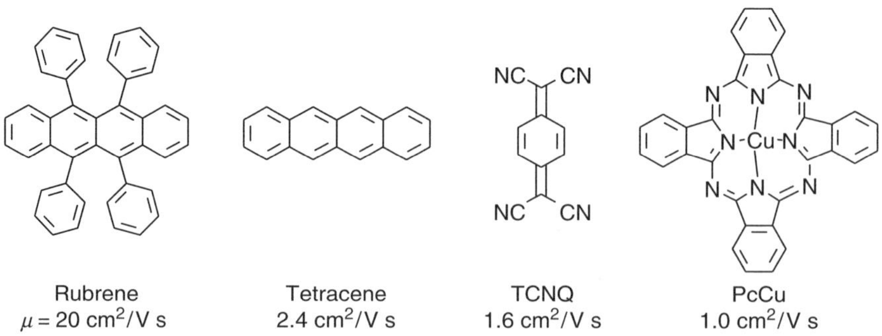

FIGURE 10.34 Chemical structure of various molecular semiconductors investigated in single-crystal TFTs and their maximum carrier mobilities.

top-contact organic FETs, however, the top-contact SCFET is most often a top-gate structure with the semiconductor crystal acting as the substrate on which the dielectric and then the gate conductor are deposited (Figure 10.35). These fabrication steps on top of the thin organic plate have been associated to the formation of defects at the critical semiconductor/electrode and semiconductor/dielectric interfaces. Furthermore, the initial needs to apply the electrodes by hands and the limitations of shadow-mask feature sizes necessitate large crystals and limit device density. The deposition of a dielectric layer on top of this structure has proven even more problematic, with first attempts at sputtering oxides onto the crystal surface irreparably damaging it [151]. This problem was addressed with the introduction of a conformal evaporated parylene dielectric film (parylene is a polymeric insulator, see Figure 10.35) [152]. This technique has resulted in high mobility being recorded in a variety of organic and inorganic single crystals and substantial increase of the device yields [153]. The electrostatic bonding technique, in which a thin single crystal adheres to the dielectric surface by simply being brought into contact, has been used to fabricate SCFETs on thermally grown SiO_2 substrates as well as sputtered metal-oxide dielectrics [154]. As with thin-film devices on bare oxides, leakage, hysteresis, and otherwise nonideal transistor behavior is generally observed. While each of these techniques has achieved improved device performance on conventional dielectrics, replacement of the metal-oxide dielectric with inert, nonpolar dielectrics has resulted in the highest field-effect mobilities and more consistent electrical characteristics [155].

Fabrication of bottom-contact SCFETs from solution has primarily been performed by drop-cast formation of crystals on substrates with a large array of devices, whereby a number of crystals formed in solution will bridge the preformed electrodes and form functioning devices. This technique was effective to demonstrate excellent mobilities for dithiophene-tetrathiofulvalene (**DT-TTF**) [156] and for other molecular semiconducting families [157]. Despite the fact that the methods described above have made tremendous progress in the past few years, they all present the same

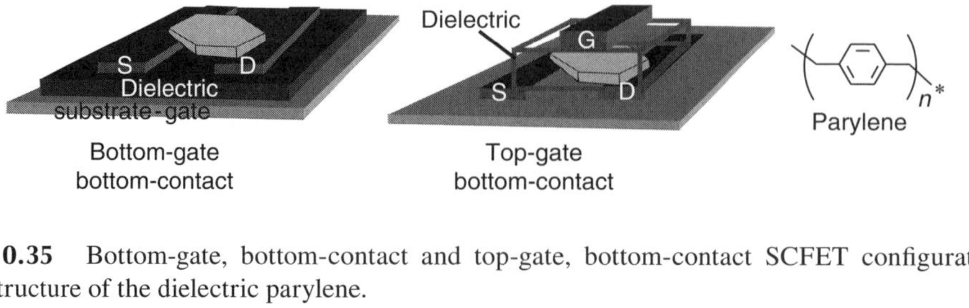

FIGURE 10.35 Bottom-gate, bottom-contact and top-gate, bottom-contact SCFET configurations and chemical structure of the dielectric parylene.

limitation due to the manual placement of fragile single crystals, which must be hand-selected for quality and consistency. Furthermore, using these techniques it is difficult to consistently obtain high-quality devices and extract meaningful data limiting their use for fundamental study. Furthermore, from a manufacturing viewpoint, the required precise location and orientation of the single crystal into the FET device structure are major obstacles for their potential industrial application.

An alternative approach would be to grow the single crystal directly on the predefined FET structure, hence it is necessary to control crystal nucleation and growth on a predetermined substrate location. While selective nucleation of crystalline inorganics has been demonstrated previously [158], analogous selectivity had not been reported for organics until recently. First effort toward selective crystal growth was performed for anthracene on self-assembled monolayer (SAM)-modified Au substrates. SAM of various thiol-based functionalities were first deposited to the Au surface by microcontact printing; the vacant regions were then backfilled with a different SAM molecule. An anthracene solution was then either drop-cast or applied by dip-coating. This afforded not only selective nucleation of many crystals in domains ranging from 100 to 300 μm in size but also an explicit comparison of nucleation density between the different functionalities. While no devices were reported, a follow-up study by the same group resulted in arrays of functioning transistors. More recently, tall pillars of an organosilane film were deposited via microcontact printing onto an oxide substrate (the FET dielectric) with pattern features as small as 5 μm. When patterned substrates were subjected to a stream of sublimed semiconductor in a sealed ampoule, crystals selectively nucleated on the treated regions. When the patterns are aligned with arrays of electrodes, crystal growth results in a large number of functioning transistors, as demonstrated with rubrene, pentacene, tetracene, C_{60}, tetracyanoquinodimethane (**TCNQ**), and **FCuPc** [159].

The ability of SCFETs to probe the intrinsic properties of semiconductor materials makes them ideal for the screening of novel materials, as it provides an upper limit to the performance attainable by thin films of the same material. While many electrical and stability criteria must be met for the ultimate incorporation of organic semiconductors into commercial applications, the main motivation for investigating SCFETs has been to increase the device speed. Field-effect mobilities reported for a number of organic semiconductor crystals, tested with SCFETs, are also listed in Figure 10.34. The performance of crystalline semiconductors is better than their polycrystalline counterparts. As SCFET performance and consistency improve with their optimization, so does the size and quality of the database of structures and performance measures. At the current rate of material development, the hope is that it will be possible to draw meaningful structure/property relationships not only within families of materials, which has been done with some success [160], but also across those with widely varying functionalities.

SCFETs based on rubrene are the best performing organic field-effect devices reported to date. Since the discovery of rubrene as an excellent hole-transport material, its use in the exploration of transport has been ubiquitous. Its high mobility, stability, and ease of growth have made it a favorite material for fundamental studies of charge transport along different crystallographic directions. In principle, identification of crystal axes by matching crystallographic data to crystal shape consents to assess relationship between molecular orientation, proximity, and the efficiency of charge transport [161]. Indeed, Sunder et al. demonstrated that the introduction of an elastomer dielectric allows repeated removal and relamination of thick rubrene single crystals enabling probing of FET transport in multiple crystal axes. Similar results were also reported by Podzorov et al. [162] with the use of the air-gap technique and by Zeis et al. [163] using a top-contact, top-gate configuration with a parylene dielectric. These results, demonstrate both the great promise and remaining challenges of this approach. Note that while the data from different groups working on rubrene SCFETs show similar trends, offering additional data for comparison with rubrene's molecular packing motif, the difference between the absolute value data sets illustrates the variation even between well-documented methods, and between crystals. Pentacene SCFETs have also been investigated following similar approaches [164].

In conclusion, SCFETs have developed tremendously only the last few years, from a simple tool for a deeper understanding of organic material performance into a possible industrial process. While a much deeper understanding of transport in molecular semiconductors has been harvested, many important issues need to be further explored. For example, while mobilities for single-crystal organics are, with little exception, better than their polycrystalline analogues, wide variation has been reported for other characteristics, such as threshold voltage and current on/off ratios. Furthermore, device performance and transport characteristics do not follow predicted trends between families of crystalline materials, most notably the polyacenes. Consequently, more studies are necessary to underscore the importance of crystal morphology at critical device interfaces.

EXERCISE QUESTIONS

1. Write the structure of pentacene, sexithiophene, rubrene, and cupperphthalocyanine. Which of these compound is an hydrocarbon?
2. Name the most important molecular orbitals of organic semiconductors?
3. What is the technique used to determine the crystal structure of an organic single-crystal compound? If the same compound is deposited as a thin polycrystalline film, how is the degree of texturing (crystallinity) investigated?
4. Draw the chemical structure of three molecules belonging to the family of 2D fused rings.
5. What does SCFET stand for? Why is this so important? What are the most common SCFET device configurations?

LIST OF ABBREVIATIONS

6T	sexithiophene
AFM	atomic force microscopy
BDT	benzodithiophene
FET	field-effect transistor
FMOs	frontier molecular orbitals
GaAs	gallium arsenite
GIXD	grazing-incident x-ray diffraction
HOMO	highest occupied molecular orbital
LUMO	lowest occupied molecular orbital
MOSFET	metal-oxide field-effect transistor
OFET	organic field-effect transistor
P5	pentacene
SCFET	single-crystal field-effect transistor
SEM	scanning electron microscopy
TDT	thienodithiophene
TEM	tunnelling electron microscopy
TTF	tetrathiofulvalene
XRD	x-ray diffraction

ACKNOWLEDGMENTS

The author would like to thank Polyera Corporation for financial support and professor Tobin J. Marks for his helpful discussion.

REFERENCES

1. Pope, M. and Swenberg, C.E. *Electronic Processes in Organic Crystals and Polymers*, Oxford University Press, Oxford, 1999.
2. Inokuchi, H. *Org. Electron.* 2006, *7*, 62.

3. Tang, C.W. et al. *Appl. Phys. Lett.* 1987, *51*, 913; Burroughs, J. H. et al. *Nature* 1990, *347*, 539.
4. Ebisawa, F., Kurokawa, T., and Nara, S. *J. Appl. Phys.* 1983, *54*, 3255.
5. Klauk, H. *Organic Electronics: Materials, Manufacturing, and Applications*, Wiley, Weinheim, Germany, 2006.
6. Sze, S.M. *Semiconductor Devices: Physics and Technology*, Wiley, New York, 1985, pp. 6–7, 216–218, 507–510.
7. Moore, G.E. *IEEE IEDM Tech. Dig.* 1975, *21*, 11.
8. Subramanian, V. et al. *IEEE Trans. Comp. Pack. Tech.* 2005, *28*, 742.
9. Torsi, L. and Dodabalapur, A. *Anal. Chem.* 2005, *77*, 380A.
10. Katz, H.E. et al. *J. Appl. Phys.* 2002, *91*, 1572.
11. Muccini, M. *Nature Mater.* 2006, *5*, 605.
12. Kitamura, M. et al. *Japn. J. Appl. Phys.* 2003, *42*, 2483.
13. Huitema, H.E.A. et al. *Adv. Mater.* 2002, *14*, 1201.
14. Wang, Z. et al. *Adv. Mater.* 2003, *15*, 1009.
15. Halik, M. et al. *Adv. Mater.* 2002, *14*, 1717.
16. Sirringhaus, H. et al. *Science* 2000, *290*, 2123.
17. Yamaguchi, S. et al. *Pure Appl. Chem.* 2006, *78*, 721.
18. Kunugi, Y. et al. *Chem. Mater.* 2003, *15*, 6.
19. Dimitrakopoulos, C.D. and Malenfant, P.R.L. *Adv. Mater.* 2002, *14*, 99.
20. Kovacic, P. and Jones, M.B. *Chem. Rev.* 1987, *87*, 357.
21. Chung, T.C. et al. *Phys. Rev. B* 1984, *30*, 702.
22. Diaz, A.F. et al. *J. Chem. Soc. Chem. Commun.* 1979, *14*, 635; Diaz, A.F. and Lacroix, J.C. *New J. Chem.* 1988, *12*, 171.
23. Ginder, J.M. et al. *Solid State Comm.* 1997, *63*, 97.
24. Peierls, R.E. *Quantum Theory of Solids*, Oxford Press, London, 1956.
25. Facchetti, A. et al. *J. Am. Chem. Soc.* 2004, *126*, 13480.
26. Godby, R.W., Schluter, M., and Sham, L.J. *Phys. Rev. B* 1988, *37*, 10159; Stowasser, R. and Hoffmann, R., *J. Am. Chem. Soc.* 1999, *121*, 3414.
27. Kamlet, M.J. et al. *J. Org. Chem.* 1983, *48*, 2877.
28. Jonsson, M. et al. *J. Chem. Soc., Perkin Trans.* 1999, *2*, 425.
29. Horowitz, G. et al. *Chem. Mater.* 1995, *7*, 1337.
30. Laudise, R.A. et al. *J. Cryst. Growth* 1995, *152*, 241.
31. Siegrist, T.R. et al. *J. Mater. Res.* 1995, *10*, 2170.
32. Anthony, J.E. *Chem. Rev.* 2006, *106*, 5028.
33. Mattheus, C.C. et al. *Acta Crystallogr.* 2001, *C57*, 939.
34. Siegrist, T. et al. *Angew. Chem. Int. Ed.* 2001, *40*, 1732.
35. Campbell, R.B., Robertson, J.M., and Trotter, J. *Acta Cryst.* 1961, *14*, 705.
36. Holmes, D. et al. *Chem. Eur. J.* 1999, *5*, 3399.
37. Yoon, M.-H. et al. *J. Am. Chem. Soc.* 2006, *128*, 5792.
38. McCullough, R.D. *Adv. Mater.* 1998, *10*, 93; Brédas, J.-L. *J. Chem. Phys.* 1985, *82*, 3809.
39. Huheey, J.E., Keiter, E.A., and Keiter, R.L. *Inorganic Chemistry*, Harper Collins: New York, 1993, p. 292.
40. Bertolucci, M.D. and Marsh, R.E. *J. Appl. Cryst.* 1974, *7*, 87.
41. Peisert, H., Knupfer, M., and Fink, J. *Recent Res. Dev. Appl. Phys.* 2002, *5*, 129.
42. Chen, X.L. et al. *Chem. Mater.* 2001, *13*, 1341; Yasuda, T. et al. *Chem. Mater.* 2005, *17*, 264.
43. Cicoira, F. et al. *Adv. Funct. Mater.* 2005, *15*, 375; Kim, D.H. et al. *Adv. Funct. Mater.* 2005, *15*, 77; Verlaak, S. et al. *Phys. Rev. B* 2003, *68*, 195409.
44. Dinelli, F. et al. *Synth. Methods* 2004, *146*, 373; Hajlaoui, M.E. et al. *Synth. Methods* 2002, *129*, 215; Noh, Y.Y. et al. *Adv. Mater.* 2003, *15*, 699.
45. Chang, P.C. et al. *Chem. Mater.* 2004, *16*, 4783.
46. Azumi, R. et al. *Bull. Chem. Soc. Jpn.* 2003, *76*, 1561; Moret, M. et al. *J. Mater. Chem.* 2005, *15*, 2444.
47. Fritz, S.E. et al. *J. Am. Chem. Soc.* 2004, *126*, 4084; Merlo, J.A. et al. *J. Am. Chem. Soc.* 2005, *127*, 3997; Moulin, J.F. et al. *Nucl. Instrum. Methods Phys. Res., Sect. B* 2006, *246*, 122; Yang, H. et al. *J. Am. Chem. Soc.* 2005, *127*, 11542.
48. Facchetti, A. et al. *Adv. Mater.* 2003, *15*, 33.
49. DeLongchamp, D.M. et al. *Adv. Mater.* 2005, *17*, 2340; Murphy, A.R. et al. *Chem. Mater.* 2005, *17*, 6033; Pattison, L.R. et al. *Macromolecules* 2006, *39*, 2225; DeLongchamp, D.M. et al. *Chem. Mater.* 2005, *17*, 5610.
50. Sze, S.M. *Semiconductor Devices: Physics and Technology*, 2nd edition, Wiley, New York, 1985.

51. Facchetti, A. et al. *Adv. Mater.* 2005, *17*, 1705; Veres, J. et al. *Chem. Mater.* 2004, *16*, 4543.
52. Gelinck, G.H., Geuns, T.C.T., and de Leeuw, D.M. *Appl. Phys. Lett.* 2000, *77*, 1487; Drury, C.J. et al. *Appl. Phys. Lett.* 1998, *73*, 108; Sirringhaus, H. et al. *Science* 2000, *290*, 2123.
53. Wu, Y., Li, Y., and Ong, B.S. *J. Am. Chem. Soc.* 2007, *129*, 1862; Li, Y., Wu, Y., and Ong, B.S. *J. Am. Chem. Soc.* 2005, *127*, 3266; Wu, Y., Li, Y., and Ong, B.S. *J. Am. Chem. Soc.* 2006, *128*, 4202; Gray, C. et al. *Proc. SPIE* 2001, *4466*, 89; Tate, J. et al. *Langmuir* 2000, *16*, 6054.
54. Fichou, D. *Handbook of Oligo- and Polythiophenes*, Wiley-VCH, New York, 1998.
55. Shen, Y. et al. *Chem. Phys. Chem.* 2004, *5*, 16.
56. Newman, C.R. et al. *Chem. Mater.* 2004, *16*, 4436.
57. Facchetti, A. *Mater. Today* 2007, *10*, 28.
58. Murphy, A.R. and Frechet, J.M.J. *Chem. Rev.* 2007, *107*, 1066; Locklin, J. et al. *Polym. Rev.* 2006, *46*, 79; Anthopoulos, T. D. et al. *Adv. Mat.* 2006, *18*, 1900; Chabinye, M. and Loo, Y.L. *J. Macromol. Sci. Polym. Rev.* 2006, *46*, 1; Tulevski, G. S. et al. *J. Am. Chem. Soc.* 2006, *128*, 1788; Muccini, M. *Nature Mater.* 2006, *5*, 605; Sirringhaus, H., *Adv. Mat.* 2005, *17*, 2411; Dimitrakopoulos, C.D. and Malenfant P.R.L. *Adv. Mater.* 2002, *14*, 99.
59. Karl, N. et al. *Synth. Methods* 1991, *42*, 2473; Bree, A., Carswell, D.J., and Lyons, L.E. *J. Chem. Soc.* 1955, 1728; Carswell, D.J., Ferguson, J., and Lyons, L.E. *Nature* 1954, *173*, 736; Carswell, D.J. *J. Chem. Phys.* 1953, *21*, 1890.
60. Bailey, W.J. and Liaio, C.W. *J. Am. Chem. Soc.* 1955, *77*, 992.
61. Rang, Z. et al. *Appl. Phys. Lett.* 2001, *79*, 2731; Klauk, H. et al. *IEEE Trans. Electron Devices* 1999, *46*, 1258; Dimitrakopoulos, C.D., Brown, A.R., and Pomp, A. *J. Appl. Phys.* 1996, *80*, 2501; Laquindanum, J.G. et al. *Chem. Mater.* 1996, *8*, 2542; Horowitz, G. et al. *Synth. Methods* 1991, *41–43*, 1127.
62. Lin, Y.-Y. et al. *IEEE Trans. Electron Device Lett.* 1997, *18*, 606.
63. Kelley, T.W. et al. *Mater. Res. Soc. Symp. Proc.* 2003, *771*, 169.
64. Klauk, H. et al. *J. Appl. Phys.* 2002, *92*, 5259.
65. Laudise, R.A. et al. *J. Cryst. Growth* 1998, *187*, 449; Kloc, C. et al. *J. Cryst. Growth* 1997, *182*, 416.
66. Minakata, T., Imai, H., and Ozaki, M. *J. Appl. Phys.* 1992, *72*, 4178.
67. Minakata, T., Ozaki, M., and Imai, H. *J. Appl. Phys.* 1993, *74*, 1079
68. Laquindanum, J.G., Katz, H.E., and Lovinger, J. *J. Am. Chem. Soc.* 1998, *120*, 664.
69. Brown, A.R. et al. *Synth. Methods* 1997, *88*, 37; Brown, A.R. et al. *J. Appl. Phys.* 1996, *79*, 2136; Brown, A.R. et al. *Science* 1995, *270*, 972.
70. Herwig, P.T. and Müllen, K. *Adv. Mater.* 1999, *11*, 480.
71. Afzali, A., Dimitrakopoulos, C.D., and Breen, T.L. *J. Am. Chem. Soc.* 2002, *124*, 8812.
72. Meng, H. *Adv. Mater.* 2003, *15*, 1090.
73. Anthony, J.E. et al. *J. Am. Chem. Soc.* 2001, *123*, 9482.
74. Payne, M. *J. Am. Chem. Soc.* 2005, *127*, 4986.
75. Sakamoto, Y., Suzuki, T., Kobayashi, M., Gao, Y., Fukai, Y., Inoue, Y., Sato, F., and Tokito, S. *J. Am. Chem. Soc.* 2004, *126*, 8138.
76. Chen, L.X. et al. *Chem. Mater.* 2001, *13*, 1341; Gundlach, D.J. et al. *Appl. Phys. Lett.* 1997, *71*, 3853.
77. Sakamoto, Y. et al. *J. Am. Chem. Soc.* 2000, *122*, 1832; Heidenhain, S.B. et al. *J. Am. Chem. Soc.* 2000, *122*, 10240.
78. Horowitz, G. et al. *Synth. Methods* 1992, *51*, 419; Horowitz, G. et al. *Solid State Commun.* 1989, *72*, 381.
79. Garnier, F. et al. *J. Am. Chem. Soc.* 1993, *115*, 8716.
80. Hajlaoui, M.E. et al. *Synth. Methods* 2002, *129*, 215.
81. Fichou, D. et al. *Adv. Mater.* 1997, *9*, 75; Hajlaoui, R. et al. *Adv. Mater.* 1997, *9*, 557; Fichou, D., Horowitz, G.G., and Garnier, F. Eur. Pat. Appl. EP, 402, 269, 1990; FR Appl. 89/7, 610, 1989.
82. Torsi, L. et al. *Phys. Rev. B: Condens. Matter Mater. Phys.* 1998, *57*, 2271; Dodabalapur, A., Torsi, L., and Katz, H.E. *Science* 1995, *268*, 270; Torsi, L. et al. *Science* 1996, *272*, 1462; Katz, H.E., Torsi, L., and Dodabalapur, A. *Chem. Mater.* 1995, *7*, 2235.
83. Li, W. et al. *Chem. Mater.* 1999, *11*, 458; Katz, H.E., Laquindanum, J.G., and Lovinger, A.J. *Chem. Mater.* 1998, *10*, 633.
84. Akimichi, H. et al. *Appl. Phys. Lett.* 1991, *58*, 1500.
85. Katz, H.E. et al. *Nature* 2000, *404*, 478; Katz, H.E. et al. *Synth. Methods* 1999, *102*, 897; Katz, H.E., Laquindanum, J.G., and Lovinger, A.J. *Chem. Mater.* 1998, *10*, 633; Garnier, F. et al. *Chem. Mater.* 1998, *10*, 3334.
86. Barbarella, G. et al. *J. Am. Chem. Soc.* 1999, *121*, 8920.
87. Facchetti, A. et al. *J. Am. Chem. Soc.* 2004, *126*, 13859.

88. Li, W. et al. *Chem. Mater.* 1999, *11*, 458.
89. Facchetti, A. et al. Unpublished results.
90. Katz, H.E. et al. *Appl. Phys.* 2002, *91*, 1572; Hong, X.M. et al. *Chem. Mater.* 2001, *13*, 4686.
91. Mushrush, M. et al. *J. Am. Chem. Soc.* 2003, *125*, 9414.
92. Ichikawa, M. et al. *Adv. Mater.* 2002, *14*, 1272.
93. Dimitrakopoulos, C.D. et al. *Synth. Methods* 1997, *89*, 193.
94. Meng, H. et al. *J. Am. Chem. Soc.* 2001, *123*, 9214.
95. Laquindanum, J.G., Katz, H.E., and Lovinger, A.J. *J. Am. Chem. Soc.* 1998, *120*, 664.
96. Laquindanum, J.G. et al. *Adv. Mater.* 1997, *9*, 36.
97. Li, X.-C. et al. *J. Am. Chem. Soc.* 1998, *120*, 2206.
98. Takimiya, K. et al. *J. Am. Chem. Soc.* 2004, *126*, 5084.
99. Takimiya, K. et al. *J. Am. Chem. Soc.* 2006, *128*, 12604.
100. Facchetti, A. and Marks, T.J. *Polym. Prepr.* 2002, *43*, 734.
101. Facchetti, A. et al. *Angew. Chem. Int. Ed.* 2000, *39*, 4547.
102. Facchetti, A. et al. *Chem. Mater.* 2004, *16*, 4715.
103. Ando, S. et al. *J. Am. Chem. Soc.* 2005, *127*, 14996.
104. Pappenfus, T.M. et al. *J. Am. Chem. Soc.* 2002, *124*, 4184.
105. Casado, J. et al. *J. Am. Chem. Soc.* 2002, *124*, 12380.
106. Chesterfield, R.J. *Adv. Mater.* 2003, *15*, 1278.
107. Mas-Torrent, M. et al. *J. Am. Chem. Soc.* 2004, *126*, 984.
108. Gao, X. et al. *Chem. Commun.* 2006, 2750.
109. Naraso, N. et al. *J. Am. Chem. Soc.* 2006, *128*, 9598.
110. Facchetti, A. et al. *Angew. Chem. Int. Ed.* 2003, *42*, 3900.
111. Yoon, M.-H. et al. *J. Am. Chem. Soc.* 2005, *127*, 1348.
112. Letizia, J. et al. *J. Am. Chem. Soc.* 2005, *127*, 13476.
113. Uchida, T. and Akamatsu, H. *Bull. Chem. Soc. Jpn.* 1961, *34*, 1015; Sano, M. and Akamatzu, H. *Bull. Chem. Soc. Jpn.* 1961, *34*, 1569.
114. Maruyama, Y. et al. *Mol. Cryst. Liquid Cryst.* 1973, *20*, 373.
115. Suga, T. et al. *Synth. Methods* 1999, *102*, 1050.
116. Malenfant, P.R.L. et al. *Appl. Phys. Let.* 2002, *80*, 2517.
117. Chesterfield, R. et al. *J. Appl. Phys.* 2004, *95*, 6396.
118. Tatemichi, S. et al. *Appl. Phys. Lett.* 2006, *89*, 112108/1.
119. Rademacher, A., Maerkle, S., and Langhals, H. *Chem. Ber.* 1982, *115*, 2927.
120. Katz, H.E. et al. *J. Am. Chem. Soc.* 2000, *122*, 7787.
121. Laquindanum, L.G. et al. *J. Am. Chem. Soc.* 1996, *118*, 11331.
122. Katz, H.E. et al. *Nature* 2000, *404*, 478.
123. Jones, B. et al. *Angew. Chem. Int. Ed.* 2004, *43*, 6363.
124. Jones, B. et al. *Chem. Mater.* 2007, *19*, 2703.
125. Yoo, B. et al. *IEEE Electr. Dev. Lett.* 2006, *27*, 737.
126. Eley, D.D. et al. *Trans. Faraday Soc.* 1953, *49*, 79; Eley, D.D. *Nature* 1948, *162*, 819.
127. Moser, F. H. *Phthalocyanines*, CRC Press, London, 1983.
128. Barbe, D.F. and Westgate, C.R. *J. Phys. Chem. Solids* 1970, *31*, 2679.
129. Madru, R. et al. *Chem. Phys. Lett.* 1988, *145*, 343.
130. Bao, Z., Lovinger, A.J., and Dodabalapur, A. *Appl. Phys. Lett.* 1996, *69*, 3066; Clarisse, C. and Riou, M.T. *J. Appl. Phys.* 1991, *69*, 3324.
131. Bao, Z., Lovinger, A.J., and Dodabalapur, A. *Adv. Mater.* 1997, *9*, 42; Bao, Z., Lovinger, A.J., and Dodabalapur, A. *Appl. Phys. Lett.* 1996, *69*, 3066.
132. Guilland, G. et al. *Chem. Phys. Lett.* 1990, *167*, 503.
133. Zhang, J. et al. *Adv. Mater.* 2005, *17*, 1191.
134. Tang, Q. et al. *Adv. Mater.* 2006, *18*, 65.
135. Someya, T. *Proc. Nat. Acad. Sci. USA* 2005, *102*, 12321.
136. Schlettwein, D. et al. *Chem. Mater.* 1994, *6*, 3.
137. Schlettwein, D., Wöhrle, D., and Jaeger, N.I. *J. Electrochem. Soc.* 1989, *136*, 2882; Jones, J.G. and Twigg, M.V. *Inorg. Chem.* 1969, *8*, 2018.
138. Bao, Z., Lovinger, A.J., and Brown, J. *J. Am. Chem. Soc.* 1998, *120*, 207.
139. van de Craats, M. et al. *Adv. Mater.* 1999, *11*, 1469.
140. van de Craats, M. et al. *Adv. Mater,.* 2003, *15*, 495.
141. Mori, T. et al. *J. Appl. Phys.* 2005, *97*, 066102/1.

142. Pisula, W. et al. *Adv. Mater.* 2005, *17*, 684.
143. Xiao, S. et al. *J. Am. Chem. Soc.* 2006, *128*, 10700.
144. Yamashita, Y. et al. *J. Org. Chem.* 1992, *57*, 5517.
145. Iameda, K. et al. *J. Mater. Chem.* 1992, *2*, 115.
146. Xue, J. and Forrest, S. R. *Appl. Phys. Lett.* 2001, *79*, 3714.
147. Takada, M. et al. *Jpn. J. Appl. Phys.* 2002, *41*, L4-L6.
148. Roberson, L.B. et al. *J. Am. Chem. Soc.* 2005, *127*, 3069; Zeis, R. et al. *Chem. Mater.* 2006, *18*, 244.
149. Briseno, A.L. et al. *Nature* 2006, *444*, 913; Briseno, A.L. et al. *Adv. Mater.* 2006, *18*, 2320.
150. Briseno, A.L. et al. *J. Am. Chem. Soc.* 2005, *127*, 12164.
151. de Boer, R.W.I. et al. *Phys. Status Solidi A* 2004, *201*, 1302.
152. Podzorov, V. et al. *Appl. Phys. Lett.* 2003, *82*, 1739.
153. Podzorov, V. et al. *Appl. Phys. Lett.* 2003, *83*, 3504; Moon, H. et al. *J. Am. Chem. Soc.* 2004, *126*, 15322; Kotani, M. et al. *Chem. Phys.* 2006, *325*, 160; Yamada, K. et al. *Appl. Phys. Lett.* 2006, *88*, 122110; Butko, V.Y. et al. *Phys. Rev. B* 2005, *72*, 081312.
154. de Boer, R.W.I. et al. *Appl. Phys. Lett.* 2003, *83*, 4345; Goldmann, C. et al. *J. Appl. Phys.* 2004, *96*, 2080; Stassen, A.F. et al. *Appl. Phys. Lett.* 2004, *85*, 3899.
155. Chua, L.-L. et al. *Nature* 2005, *434*, 194; Zaumseil, J. et al. *J. Appl. Phys.* 2003, *93*, 6117.
156. Mas-Torrent, M. et al. *J. Am. Chem. Soc.* 2004, *126*, 984.
157. Mas-Torrent, M. et al. *J. Am. Chem. Soc.* 2004, *126*, 8546; Mas-Torrent, M. et al. *Appl. Phys. Lett.* 2005, *86*, 012110.
158. Aizenberg, J. et al. *Nature* 1999, *398*, 495.
159. Briseno, A.L. et al. *Nature* 2006, *444*, 913.
160. Anthony, J.E. et al. *J. Am. Chem. Soc.* 2001, *123*, 9482; Curtis, M.D. et al. *J. Am. Chem. Soc.* 2004, *126*, 4318.
161. Sundar, V.C. et al. *Science* 2004, *303*, 1644.
162. Podzorov, V. et al. *Phys. Rev. Lett.* 2004, *93*, 086602.
163. Zeis, R. et al. *Chem. Mater.* 2006, *18*, 244.
164. Lee, J.Y. et al. *Appl. Phys. Lett.* 2006, *88*, 252106.

11 Polymer Field-Effect Transistors

Henrik G.O. Sandberg

CONTENTS

11.1 Introduction 320
11.2 Electronic Transport in Conjugated Polymers and in Polymer Field-Effect Transistors 321
11.2.1 Metal–Insulator–Semiconductor Diode and Transistor Device Structure 323
11.2.1.1 Field Effect 324
11.2.2 Characterization of Polymer Transistors 326
11.2.3 Solution Processing 329
11.2.3.1 Molecular Order in Conjugated Polymers 330
11.2.3.2 Thin-Film Preparation 330
11.3 Materials for Polymer Field-Effect Transistors 332
11.3.1 Semiconductor Layer 332
11.3.1.1 Thiophene Derivative Polymers 334
11.3.1.2 Fluorene-Based Conjugated Polymers 334
11.3.1.3 Poly(phenylene vinylene) 335
11.3.2 Other Device Layers 335
11.3.2.1 Insulators 336
11.3.2.2 Conducting and Electrode Materials 338
11.4 Polymer Field-Effect Transistor Processing and Manufacturing; Future Potential Applications 340
11.4.1 Low-Cost Printable Electronics 341
11.4.1.1 Ink-Jet Printing 342
11.5 Summary 343
Exercise Questions 344
List of Abbreviations 344
Acknowledgment 345
Refereneces 345

Abstract Printable electronics is a key application area for conjugated polymers. Much emphasis is placed on developing a transistor that could be built merely from solution-processable materials, show stable electronic performance, and offer a sufficient lifetime for any chosen application. Conjugated polymers offer one of the most versatile material alternatives for printed transistors. Solubility in common solvents is an important advantage held by polymers compared with their molecular counterparts. Therefore, the focus of polymer transistors research is on devices that may be manufactured at low cost by solution processing, e.g., printing. The semiconductor layer is of specific relevance when transistors are developed, though the importance of the insulating and conducting materials should not be neglected.

Furthermore, the interfaces between layers play an important role in device performance. Stability is one of the key issues when processing is scaled up both in device fabrication volumes and in circuit integration. This chapter gives a general introduction to polymer transistors. The basic operating principle for the devices and currently the most essential materials are discussed, and a future outlook with regard to processing and applications, in particular, are presented. The chapter should be read as a complement to Chapter 10 with a special emphasis on polymers, and should be regarded more as a review of the field.

The research field of organic electronics has largely concentrated on studying basic electronic components. Organic transistors [1–3], transistor circuits [4–6], and optoelectronic devices including light-emitting diodes [7,8] and photovoltaic cells [9,10] are active electronic components that use organic materials, such as polymers, and organic molecules as a semiconducting material. Other standard electronic components, such as passive components using organic materials like resistors and capacitors, are less complicated, and, thus, receive less attention. Although passive components are relevant in electronic circuits, these are not useful as research model platforms when compared with active devices such as the transistor.

The transistor is a versatile model system for basic research on material properties of conjugated polymers. Insulating organic materials have been used widely in traditional electronics, for example in capacitors, but play a rather passive role in applications. When investigated in the context of organic electronics, these materials are often studied as insulator layers in thin-film transistors (TFTs) [11,12]. Transistors are the main building blocks of modern integrated electronics and have, thus, been essential for the transition from traditional silicon-based electronics to organic electronics. This chapter discusses polymer organic transistors. The main focus is on devices using a conjugated polymer as the semiconducting material in a TFT of a metal–insulator–semiconductor field-effect transistor (MISFET) type. Progress has been fast in this area, and recently even large integrated optoelectronic devices and circuits made of flexible materials have emerged [13,14]. Characterization of transistor devices is briefly discussed here, but the main characterization tools and methods described in the previous chapter on molecular transistors are also valid for polymer devices.

11.1 INTRODUCTION

The first transistor made was a bipolar device developed at Bell Laboratories in 1947 [15,16]. At that time, semiconductivity was just about being discovered and the fundamental properties of semiconducting materials were studied in general, finally resulting in the great electronics revolution. The transistor device was discovered to be a practical tool for estimating a number of material properties, such as the evaluation of charge carrier mobility in the semiconductor in the case of polymer electronics in the conjugated polymer. Furthermore, it was established that the transistor device was a very useful electronic component, for example, in amplifiers and later in digital applications, replacing the bulky and expensive fragile components that had been used earlier. The computers and cell phones we use today are all results of the same research process.

Since the first observation of polyacetylene (PA) being a semiconductor material with a conductivity of 10^{-4} to 10^{-5} S/cm [17], the field of polymer electronics has grown very rapidly. Today it forms a well-established field of research. PA was a powder material and not very practical or easy to handle. General interest in research in this field was greatly increased in 1971, when freestanding films of semiconducting polyacetylene were synthesized [18]. Soon thereafter, many other conjugated polymers with interesting properties were discovered. The synthesis and properties of conducting polythiophene, one of the most popular conjugated polymers [19], were demonstrated 10 years later in 1980 [20]. Although photocurrent was observed in oligothiophene thin films by Kuhn et al. as early as in 1974 [21], it was only after the discovery of the charge transport properties of α-sexithiophene [22,23] that it was suggested that small molecules could be superior to polymers due to the improved control of defects and film quality in these materials. Presently, both conducting polymers and small conjugated molecules are subjects of intense research, and the technology is reaching the industrial stage in many different areas, for example in the field of integrated optoelectronic devices [4,5,14,24–26]. The drawback of molecular materials has been their poor solubility in common solvents, and polymeric materials have so far thus been the stronger candidate for printable electronics. Printable electronics is one of the main motivations for the development of polymer transistors and is also the focus of this chapter. However, soluble oligomers with good electronic properties have recently been developed, and subsequently, the distinction between polymeric and molecular electronics may soon be disappearing.

11.2 ELECTRONIC TRANSPORT IN CONJUGATED POLYMERS AND IN POLYMER FIELD-EFFECT TRANSISTORS

A simple device structure useful for estimating the basic properties of any material (such as electrical conductivity, σ, and the dielectric constant, ε) is a sandwich structure where contacts of size $a \times b$ are applied to both surfaces of a slab of thickness d of the material under study. Polymer films are normally disordered materials and can in general be regarded as amorphous bulk materials. If the dimensions a and b are much larger than the length d this is essentially a plate capacitor giving the dielectric value, ε_r, from the following equation:

$$C = \varepsilon_r \varepsilon_0 \frac{A}{d} \tag{11.1}$$

where

A is the contact area $A = a\,b$

C is the capacitance

C can be directly measured and will be frequency dependent. These devices, as schematically shown in Figure 11.1, are normally operated in DC mode or at a very low frequency, and C can alternatively be estimated from a quasi-static estimate for the capacitance.

When a voltage, V, is applied over the capacitor, the device is charged, and for a known capacitance, C, the charge on either side of the insulator layer is the product

$$Q = CV \tag{11.2}$$

In a field-effect transistor (FET), this charge is accumulated in the semiconductor in the transistor channel, a mechanism that can also be referred to as the field effect.

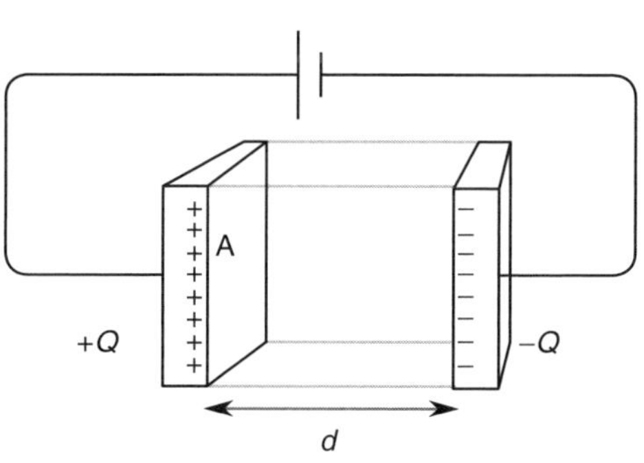

FIGURE 11.1 The capacitor structure. When a voltage is applied, charge Q is accumulated on the electrodes at the interface with the insulating material.

A simple two-terminal device is sufficient for estimating some parameters while other parameters might require the three-terminal transistor structures to be evaluated. The most common device structure of the polymer transistor is the TFT where the various materials of the device are applied as consecutive layers. An additive manufacturing process is well suited for these devices, and established printing techniques are modified to allow for fast and low-cost manufacturing. In general, the device characterization methods and formulas used for MISFET-type devices may be applied for polymer transistors as a first approximation. However, organic transistors are normally operated in the accumulation regime. This transistor design was first introduced as early as in 1960 when a thermally oxidized silicon substrate was used [27] but is still one of the most commonly used platforms for characterizing organic transistors. Molecular and polymer transistors essentially differ in one respect: molecular materials can be obtained in a crystalline morphology while the molecular order in polymer films is largely amorphous. Even so, the transport properties of polymer transistors may in exceptional special cases be isotropic. Despite this, for basic characterization, the semiconductor is regarded as a disordered material.

The alternated single and double bonds in the carbon chain are characteristic of a conjugated polymer or molecule [28, Section 3]. See for example the structure of the materials discussed in the following sections. This structure in the polymer backbone provides delocalized charge carriers in overlapping electronic π-orbitals. In their pure state, conjugated polymers are best described as electrical insulators [28]. Their conductivity, σ, is proportional to the charge carrier mobility, μ, and the free charge carrier concentration, n, according to

$$\sigma = en\mu \tag{11.3}$$

where e is the unit electronic charge. Conjugated polymers have relatively large band gaps, which result in low carrier concentration and low intrinsic conductivity, thus making the materials potentially suitable for field-effect devices. The relative carrier concentration induced by an electric field is large resulting in good modulation of the conductivity.

Different mechanisms for electrical transport in conjugated organic materials have been proposed. The most distinct differences between the suggested mechanisms relate to band type transport, on the one hand, and hopping transport, on the other [29, Chapter 1]. In reality, the complex transport mechanism most probably represents a situation somewhere in between the two alternatives mentioned above. The identification of the correct description of the transport model is a currently ongoing task in this field. Pure band-like transport is only identified in very pure molecular material with an extreme high degree of order between the individual molecules. Band transport is generally not observed in bulk material of polymers. Concerning small molecules, the band transport model is only valid for small crystalline grains or crystallites within an amorphous matrix of material. Further information about crystallization of polymers and related topics can be found in sources such as Ref. [30]. Single crystals of organic conjugated material can only be obtained by using sophisticated film preparation methods and pure and rod-like molecules such as pentacene and sexithiophene.

Transport in polymers is often described as a variable range hopping (VRH) process [31,32]. VRH has been extensively modeled and shows good agreement with experimental results. Characteristic of a hopping process is that due to thermally activated mechanisms, the conductivity rises with temperature. Stimulated by thermal activation, the charge carrier acquires additional energy for tunneling to the next hopping site. Polymer films represent disordered materials despite possible microscopic orientation, and the physical mechanism of the electronic transport involved will not be covered here in detail. On a macroscopic scale, there are regions of semicrystalline domains with good transport along the chains and between the adjacent chains. Good π-interactions are required for efficient transport between the adjacent chains [33,34]. The molecular packing of the conjugated polymer chains determines the π-interaction and thus the charge transport in the material. The conjugated polymer backbone may be interrupted, for example, by an impurity giving rise to the charge

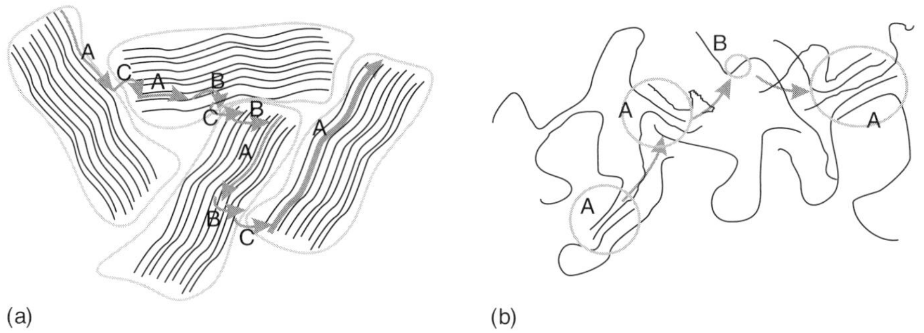

FIGURE 11.2 (a) The hopping transport mechanism principle in a conductivity network. Intra-chain transport is indicated with A, B represents interchain transport, and C indicates transport between different ordered domains. The arrows show the path of the charge carrier through the material. (b) Hopping transport through ordered domains A and single hopping sites B in a disordered material with small amount of ordered domains in an amorphous matrix, which is typically the case in a polymeric material.

hopping to the neighboring chain. At the edge of the domain, the charge will have to traverse a longer distance to the next domain. The three transport mechanisms are schematically depicted in Figure 11.2a.

The hopping transport will have to overcome large amorphous regions in a less ordered material. The charge uses single hopping sites to reach the next ordered domain, as is schematically shown in Figure 11.2b. The current through the material is then largely dependent on the availability of suitable isolated hopping sites. The current can then be easily modulated by small changes in these sites and may give rise to a sharp transition between an insulating and a conducting state in the material. This is a concept brought forward, for example, by Epstein and coworkers to explain the apparent field effect in some devices using highly doped conjugated polymers as active material [35–37].

Polymer FETs differ from their molecular counterparts, as there are many crystalline molecular materials available, while there are very few polymers showing ordered and directional morphology. Only a few reports have been published where some degree of crystallinity is observed in semiconducting polymer films, and, in general, the standard characterization methods for organic field-effect transistors (OFETs) give reasonable results for a given system. Specialized methods, usually derived from the standard characterization methods, have been proposed to observe particular properties of a device or material, but the conclusions reached have generally been applicable only for that particular system and for that particular way of processing the films and layers. The device processing of polymer FETs influence the behavior of the measured sample to a large extent. As a result, for a comprehensive characterization of a polymer FET, both the materials and the processing must be investigated as equally important aspects of the system. For example, electrode material conductivity, contact resistance between conductor and semiconductor polymers, and interface quality at the semiconductor–insulator bilayer [38] are of great importance for the results of a practical characterization of a polymer transistor.

11.2.1 Metal–Insulator–Semiconductor Diode and Transistor Device Structure

A transistor is in its most general form a variable resistor. In the history of organic electronics, the concept has been used in a flexible manner, and the distinction between transistors, FETs, and other field-dependent devices such as chemical resistors [39] and electrochemical transistors [40,41] is often rather vague. The organic MISFET device geometry is a layered structure where an insulator (the gate dielectric) is sandwiched between a metal (or other gate electrode material) and a semiconductor. In organic electronics, a conjugated polymer (or oligomer/molecule) naturally acts as the

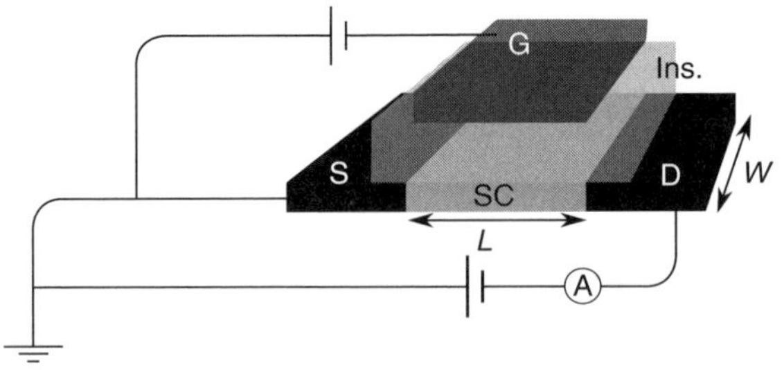

FIGURE 11.3 Typical device structure of an organic TFT (in the so-called top-gate configuration, the gate electrode layer is applied last) in a measurement circuit, the output characteristics are measured as the drain current as a function of drain voltage at different gate voltages. In a real measurement, the current is measured on all three terminals of the device. The transistor channel region dimensions are marked W (width) and L (length). Usually, the width is much larger than the length of the channel region.

semiconductor. Source and drain electrodes in contact with the semiconductor define the transistor channel area where the conductivity is modulated by the gate potential. A typical device structure is shown in Figure 11.3 where also the measurement circuitry is schematically drawn.

The current is measured on all three terminals (gate, source, and drain) and plotted against drain or gate voltage to give the device characteristics. From these I–V curves, material and device parameters can be estimated. The standard theory behind the I–V characteristics can be found in basic semiconductor theory [42–44], to mention a few examples.

11.2.1.1 Field Effect

The particular phenomenon where the conductivity of a semiconductor is modulated by an electric field applied normal to the surface of the semiconductor is known as the field effect [43]. Early field-effect devices were proposed in the 1920s and 1930s, for example, by Lilienfeld; but because the modern semiconductor materials were not available at that time, the concept remained theoretical. The MISFET in the 1960s signified the emergence of modern electronics. Two of the most significant developments of the transistor device were the invention of the dynamic random access memory (DRAM) and the charge coupled device (CCD). The DRAM cell was first demonstrated by Dennard in 1968 [45] and is now likely to have been produced in larger numbers than any other man-made objects.

To understand the field effect in a TFT, one must consider the metal–insulator–semiconductor (MIS) capacitor. Originally, its structure was formed by a metal with an insulating oxide layer in contact with a semiconductor called the metal–oxide–semiconductor (MOS) structure. In the MIS capacitor, the main simplifying assumptions are that (1) the metallic gate is considered an equipotential region; (2) the oxide is a perfect insulator with no current flowing through the structure; (3) the semiconductor is uniformly doped and sufficiently thick, and field-free bulk semiconductor is reached before the back contact; (4) there are no interfacial charges at the insulator–semiconductor interface or within the insulator, and (5) the back contact with the semiconductor is ohmic. The structure is schematically drawn in Figure 11.4.

When a circuit is formed by these three layers, the fermi energies of the layers are aligned in an energy diagram and different operational regions occur by biasing the device, as is seen in Figure 11.5. One can argue in the most qualitative way that the band-bending that appears in the figure provides the charge carriers that are modulated in a FET device. In a perfect device there is some band-bending even at zero gate potential due to a built-in field in the device (unless the fermi levels are identical). A small potential is required to reach flat-band conditions where there is no field within the semiconductor and the carrier concentration is the same at the interface as in the semiconductor bulk. When a negative bias is applied to the gate, accumulation of charges appears at the

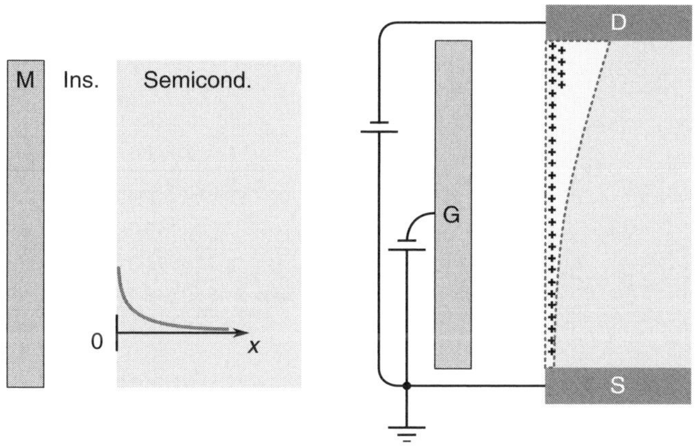

FIGURE 11.4 Schematic drawing of a MOS capacitor and the corresponding FET circuit used for modeling the MISFET structure. The area marked "M" is the metallic gate electrode and the gray area is the semiconductor separated by an insulating layer. The *x*-axis shows the distance from the insulator–semiconductor interface into the bulk of the semiconductor and the gray curve indicates a induced charge carrier distribution (left). The lightened area of the semiconductor (right) schematically shows the accumulated charge carrier (+) region.

insulator–semiconductor interface (for a p-type semiconductor). This is observed as an enhanced conductivity in a FET. If, on the other hand, a positive bias is applied, the same region is depleted of charges until at some point opposite charges are accumulated and the device reaches inversion.

Normally, organic FETs are operated in an accumulation mode and the source–drain current is modulated by applying negative bias to the gate electrode. The threshold voltage is related to the flat-band condition and the doping level of the semiconductor. Conjugated polymers normally require a positive gate voltage (depletion) to turn off the device.

Accumulation of charges in the FET channel region increases the concentration of charge carriers and a larger current may flow through the device. The behavior of the MIS capacitor is the basis of all FET operation (for further details, see for example Ref. [42]). The extension of the accumulation layer into the semiconductor layer is of some interest for TFT devices in which ultrathin semiconductor films are used. It appears, however, that even a monolayer of conjugated polymer material will not change the FET characteristics drastically [46], which means that condition 3 above is always fulfilled. The other four conditions are in varying degrees seen as discrepancies in measured polymer FETs.

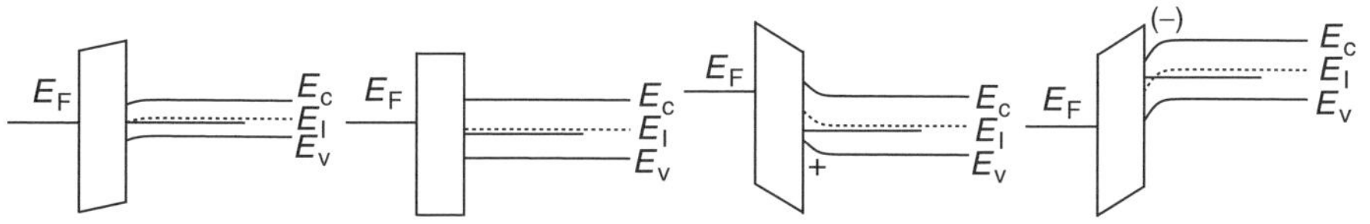

FIGURE 11.5 Different biasing regions for a p-type device showing the energy diagrams at $V_G = 0$ V, under flat-band, in accumulation and under depletion conditions (from left to right). Flat-band condition corresponds to a bias resulting in no band-bending in the semiconductor. In an ideal device, flat-band appears at $V_G = 0$ V. For details see, for example, Refs. [42,43]. The energy level indices are F = fermi, C = conduction, V = valence, and I = intrinsic.

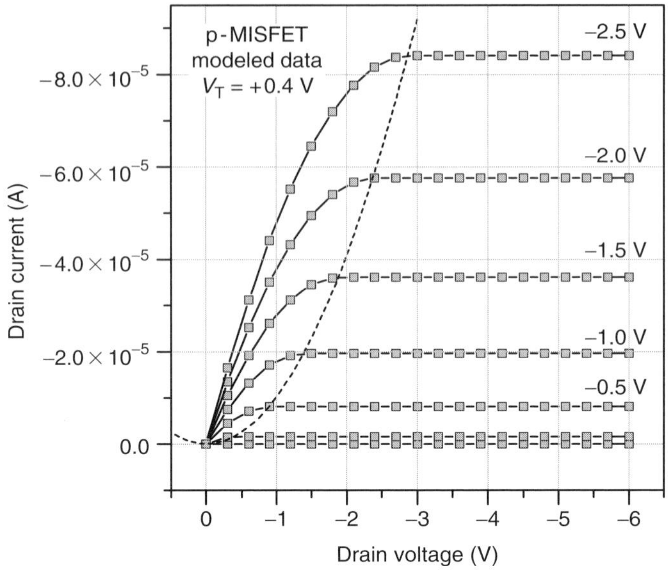

FIGURE 11.6 Typical output characteristics for a p-type MISFET.

11.2.2 Characterization of Polymer Transistors

Typical output curves of a FET are schematically drawn in Figures 11.6 and 11.7. The characteristics are ideal *I–V* curves generated using the SPICE model (Simulation Program with Integrated Circuits Emphasis), a general purpose analogue circuit simulator that was originally developed at the Electronics Research Laboratory of the University of California, Berkeley, in 1975. In a MISFET, the current through the transistor channel (between source and drain electrodes) can be controlled

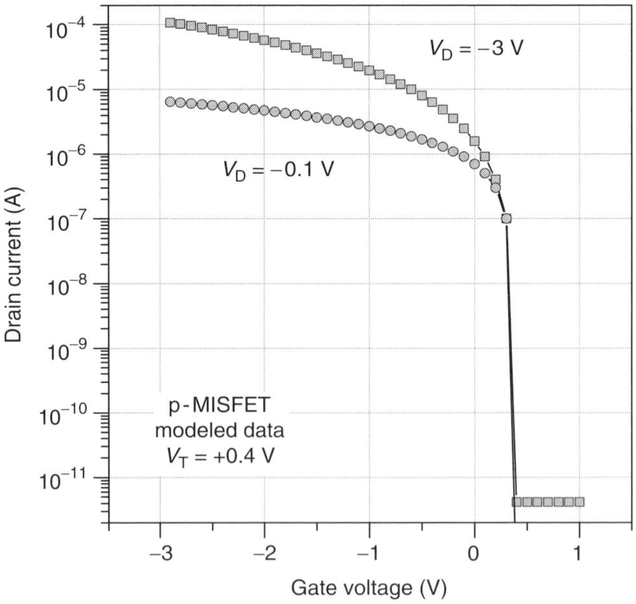

FIGURE 11.7 Typical transfer curve for a p-type MISFET.

by the voltage connected to the gate electrode. Figure 11.6 shows several source–drain sweeps at different gate voltages. The linear region is where the drain current is strongly dependent on the drain voltage. The region where the drain current is flat is known as the saturated region. In the saturated region, the source–drain current is, thus, a function of only the gate voltage.

The main equations used for device characterization express the drain current through the device at various measurement conditions. The expression for the drain current in the linear region is

$$I_{\mathrm{D}} = \frac{W}{L}\mu C_{\mathrm{i}}(V_{\mathrm{G}} - V_{\mathrm{T}})V_{\mathrm{D}} \tag{11.4}$$

and in the saturated region

$$I_{\mathrm{D}} = \frac{W}{2L}\mu C_{\mathrm{i}}(V_{\mathrm{G}} - V_{\mathrm{T}})^2 \tag{11.5}$$

where

- W and L are the channel dimensions (width and length, respectively)
- C_{i} is the gate insulator capacitance per unit area
- V_{T} is the threshold voltage
- μ is the charge carrier mobility for the semiconductor in the transistor channel area
- V_{G} and V_{D} are the voltages at the gate and drain contacts, respectively
- Source voltage, V_{S}, is connected to ground in the measurement circuit

The transition between the saturated and the linear regions appears approximately where $V_{\mathrm{D}} = V_{\mathrm{G}} - V_{\mathrm{T}}$. This is indicated with a dotted line in Figure 11.6 and can be expressed using Equation 11.5 if $V_{\mathrm{G}} - V_{\mathrm{T}}$ is substituted with V_{D}.

On the basis of these *I–V* curves, device parameters can be estimated using Equations 11.4 and 11.5. Furthermore, the transconductance

$$g_{\mathrm{m}} = \left.\frac{\partial I_{\mathrm{D}}}{\partial V_{\mathrm{G}}}\right|_{V_{\mathrm{D}} \to 0} = \frac{W}{L}\mu C_{\mathrm{i}} V_{\mathrm{D}} \tag{11.6}$$

is useful for obtaining accurate estimates of the mobility. It ought to be noted that the threshold voltage, which sometimes drifts in experimental devices, is not directly included as a parameter in the expression [42]. Traditional MISFET-devices have a constant charge carrier mobility, and estimates made from the linear and the saturated regions should give the same result. For organic devices, however, this is often not the case. Modified equations have been calculated for the case of amorphous silicon, where the mobility is highly dependent on the gate voltage due to traps [47]. The situation is quite similar in the case of organic devices.

An important property of the transistor is the on/off ratio, which determines both switching efficiency and the leakage in the OFF state, the latter being something that ought to be minimized for digital applications and which increases the power consumption of the device. A simple definition of the on/off ratio is the drain current in the saturation regime at a normal (maximum) operation gate voltage divided by the current at $V_{\mathrm{G}} = 0$ V.

Thin-film transistors using organic materials are essentially MISFET devices working in accumulation mode, and the standard theory holds for estimating most device and material properties, even if there are certain aspects that require specific attention. In organic semiconductors, the charge carrier mobility is usually field and concentration dependent. This will give a larger estimate for mobility if estimated from the saturated region. It is generally considered that the most accurate value of charge carrier mobility is obtained from the low field region (the linear region). Furthermore, in practical experiments, the saturation is frequently less than perfect, and a clear transition point in the output curve (at the point of intersect with the dotted line as in Figure 11.6) may prove difficult to define. It should be borne in mind that in FET devices the charge carrier mobility is only estimated for a certain device configuration and by using specified electric fields. Another common issue is contact resistance between the semiconductor and the source–drain electrodes. This is often visible as bending of the *I–V* curves close to origin in the output characteristics. The contact resistance may be estimated by measuring devices of different channel length while assuming that the channel resistance is linearly proportional to the channel length. Effects related to the source–drain contacts grow in importance when device manufacture is scaled up. This is important for polymer field-effect transistors (PFETs) since their most significant application prospective lies in the field of printable electronics where large quantities of devices/circuits are to be manufactured with low cost. Another way of estimating the contact resistance on a single device, without the need to make identical transistors with different channel length, is to use the gate conductance, g_D, measurement.

$$g_D = \left.\frac{\partial I_{DS}}{\partial V_D}\right|_{V_G=\text{const.}} = \frac{W}{L}\mu C_i(V_G - V_T) \tag{11.7}$$

The gate conductance is empirically evaluated from the measured data in the linear regime where the induced charge carrier distribution in the channel is fairly uniform. Several gate sweeps are performed at increasing drain voltages and the gate conductance is estimated from the measured data at different gate voltage until a linear channel conductance is found. This line will cross the *x*-axis (drain voltage) at an offset voltage, $V_{D\,offset}$. The voltage at the drain contact in Equation 11.4, V_D, is reduced with this offset voltage to compensate for the error caused by the imperfect contacts. Another less-traditional measurement method that can be useful for the characterization of OFETs is to measure the response of the device using the carrier extraction using a linearly increasing voltage (CELIV) technique [48]. This is a simple and direct measurement and gives a quasi-static estimate for the capacitance. The method is developed for general transport measurements in conjugated polymers and similar materials but can also be utilized to estimate the capacitance of a thin-film type sample.

Organic field-effect transistors are mainly characterized by measuring the *I–V* curves of the device and by using the equations indicated above. With the source contact connected to ground, the voltage at one of the other contacts (drain or gate) is set, while a voltage sweep is performed at the remaining third contact. During this process, the current is measured at all three contacts, thus giving the typical *I–V* curves of a MISFET-type transistor, as demonstrated in Figures 11.6 and 11.7. In addition to the variables applied in the standard theory, the currents through the three contacts of the organic FET are determined by various other variables. The quality of the device and the materials used are reflected in the characteristics but the device geometry also contributes to the results. Figure 11.8 demonstrates a typical PFET output curve.

Organic devices are often divided into two groups based on whether the semiconducting material is a conjugated polymer or an oligomer and whether all used materials are organic. The latter are usually referred to as all-polymer devices. The behavior of all-polymer transistors is, furthermore, influenced by other factors, including imperfect insulator layers, nonohmic contacts to the electrodes, and chemically or physically ill-defined substrate surfaces. Naturally, there are also clear advantages with all-polymer devices; one of the most important advantages from a practical point of view is that all layers of the device are solution processable and easily implemented in a mass-production context. Other possible advantages include good mechanical and chemical interaction between different materials, and by suitable material choice and printing ink tuning, also excellent matching of the electronic properties of the different materials in the device can be reached. One of the most important properties of organic materials in general is the possibility of tuning material properties through an almost infinite number of synthetic methods and treatment alternatives.

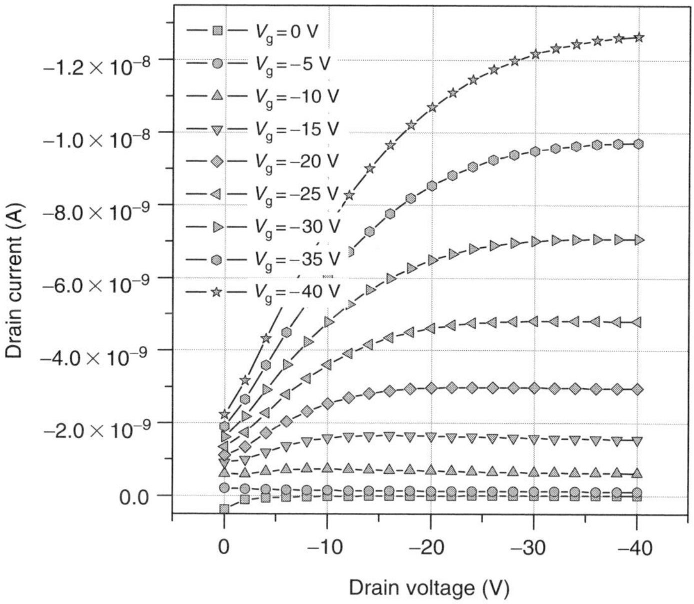

FIGURE 11.8 Typical polymer FET output curves. The following are the typical discrepancies shown: (a) the transition from the linear to the saturated regime is very smooth, (b) there is an offset between the *I–V* curves at low field proportional to the gate leakage, and (c) the *I–V* curves tails off toward origin, testimony of nonideal contacts between semiconductor and source–drain contacts.

The current through a MISFET is determined by the mobile charge carriers generated by accumulation of charge carriers to the FET channel region close to the gate dielectric-organic semiconductor interface. Organic semiconductors are usually unintentionally doped, for example, by impurities present in the environment or by residues from the material synthesis. Consequently, there will always be competition between the bulk conductivity of the slightly doped semiconductor film and the field-effect conductivity due to the accumulated charge. Since the bulk conductivity of the semiconductor degrades the performance of the FET by providing alternative pathways for the source–drain current, the total thickness of the semiconductor film should ideally match the thickness of the accumulation layer, thus optimizing the on/off ratio. Indeed, traditional silicon-based transistors should not be as fast as they currently are based only on material properties. The development of the technology has amounted to very fast silicon-based transistors by reducing the semiconductor layer thickness to a fraction of the total accumulation/depletion depth for that material.

It is possible to distinguish between a two-dimensional and a three-dimensional transport mechanism in the device by studying the conductivity of the device as a function of temperature [49]. It has been shown that by minimizing the thickness of the semiconductor, the transport mechanism will get the characteristics of two-dimensional transport. Traditional OFETs have been characterized as a function of semiconductor thickness to establish the minimum layer thickness required. On the basis of the thickness of the semiconductor layer, it is found that it corresponds to the expected thickness of a single monolayer of polymer on the surface of the sample. It is, thus, concluded that the field-effect transport in a FET can be described as two-dimensional transport and, furthermore, that a single layer of material is sufficient to construct a FET.

11.2.3 Solution Processing

Where a silicon wafer is used both as support substrate and as working gate electrode simultaneously, the gate dielectric is a thermally grown SiO_2 film with a thickness in the range of 200–300 nm. The typical electrode material is gold, evaporated through a shadow mask in vacuum. The substrates are normally cleaned in an ultrasonic bath involving three steps: first in deionized ultrapure water, then in acetone, and finally in isopropanol. Some use a more rigorous cleaning process similar to the Radio Corporation of America (RCA) or the piranha cleaning process. Various surface modification techniques may be used to modify, for example, the wettability and the surface energy of the substrate. For example, a self-assembled monolayer of a silylating agent passivates a silicon oxide surface and alters its surface energy making it hydrophobic.

Most polymeric materials are soluble in common organic solvents and may be deposited by solution processing. Solubility in organic solvents is, in the case of the semiconductors, ensured by the presence of side chains or by end chains covalently bound to the conjugated main chain. Solution processability may involve specialized techniques, such as laser assisted liftoff for patterning [50]. However, in most cases established wet processing methods are utilized, such as spin coating and different printing techniques similar to those applied in the graphics industry.

11.2.3.1 Molecular Order in Conjugated Polymers

Charge travels rapidly along a perfect conjugated polymer chain. The conjugation length is, however, limited in any conjugated polymer, and the current is mainly controlled by bottlenecks. For example, a border region, a break in the conjugation of the polymer backbone, or deep traps formed by impurities may constitute such bottlenecks. In hopping type transport, the charge carrier makes hops both in energy and in position. The distribution of hopping sites is crucial.

For example, regioregular poly(3-hexylthiophene) (RR-P3HT) is, in the best case, ordered in microcrystalline domains where there is good π–π interaction between different polymer chains. If the bulk of the film is well ordered, charge will travel efficiently in the direction of the π-stacking while transport, for example, in the direction parallel to the thiophene rings is less efficient. It has been established that both the molecular weight of the material and the degree of head-to-tail coupling in the chain are determined by the molecular order in the resulting thin film [51]. The highest charge carrier mobility is measured for devices where the RR-P3HT is ordered with the plane of the polymer chains positioning edge-on onto the substrate. Since the charge transport in most materials is not isotropic, control of the microscopic order in the films is desired and knowledge of the packing properties of the material is highly relevant when designing new polymers or methods for transistor applications. However, for polymers and other solution-processed materials, the film morphology is close to isotropic, especially when deposited, for example, by a printing process. Variations between individual devices or samples in measured properties such as current level and calculated mobilities are more related to nonuniform impurity levels in the semiconductor and general doping level of the material or layer thickness variations. Although highly ordered conjugated polymer thin films are reported, highly specialized techniques are normally required that are generally unsuitable for production of OFETs [52,53]. These techniques are suitable for material research but are currently not well suited for the manufacture of transistor and circuit.

11.2.3.2 Thin-Film Preparation

Thin films of solution-processable material can be deposited in a variety of ways. The most straightforward and simplest way of depositing material is by drop casting. The solution is applied onto the substrate, which is then left in a horizontal position until the solvent has evaporated and a solid film of the material has formed on the surface. This is a rather slow process (depending on the solvent), which leaves a considerable amount of material on the substrate and thereby usually yields relatively thick films. The process is close to equilibrium during the whole drying process, so the material may tend to aggregate in solution while the concentration in the deposited drop of solution rises.

FIGURE 11.9 Schematic drawing of the spin coating process. The material is deposited in solution and the substrate is then rotated to flatten the drop of liquid while it dries out.

The choice of solvent for a specific material is a tool used to determine the time a material is allowed to reorganize into its optimal molecular state. The drying behavior of a polyalkylthiophene solution involves an increase of the internal order of the film in the very last stages of drying [54].

Spin coating is a considerably less time-consuming method where the solution is deposited on a rotating substrate, forcing most of the material to fly off the substrate leaving an even solution layer with a thickness that depends on the spinning speed, the solution concentration, and the interaction between the substrate and the solution (Figure 11.9). Spinning the sample will result in spin coating that gives repeatable film thickness control and allows for the preparation of thin films down to almost nanometer scale. Owing to the rapid film forming and drying process this is a nonequilibrium process and the resulting film is in a largely disordered state.

To yield a film thickness of approximately 10–40 nm of a conjugated material, a solution concentration of about 2–6 mg/mL is spin coated at around 1000–3000 rpm. The insulators are normally required as thicker films up to around 1 μm, and typically a solution of concentration ~100 mg/mL is deposited at a spinning speed of 1000–2000 rpm. However, the concentrations and spinning speeds required vary greatly depending on, for example, the material and solvent used.

Various surface treatment methods and sample conditioning steps may be applied to reach the desired film morphology. Annealing of polymer films initially serves as a way of ensuring that the film is perfectly dry and free of solvent, but the annealing also allows the polymer chains to adjust to each other. In addition, an increased molecular order may be obtained in the film. Another important property that determines the resulting film morphology is the surface energy of the substrate or the underlying film or substrate. The surface energy affects the wetting and coverage of the applied solution on the substrate. The surface energy may be tuned by applying self-assembled monolayers, for example, hexamethyl-disilazane (HMDS), or by applying plasma treatment.

Various printing techniques have been proposed for polymer electronics. Conjugated polymers are suitable for printed electronics since polymers are fairly easily modified to increase their solubility in common solvents. Printing techniques are important for the processing of polymer FETs. Even though ink-jet printing has been the initial method of choice, various printing techniques known from the graphics industry are applied also for printing electronic materials, and from a manufacturing point of view may prove more suitable than ink-jet printing.

11.3 MATERIALS FOR POLYMER FIELD-EFFECT TRANSISTORS

Research on transistors based on conjugated polymers is performed both to characterize the material itself and to develop more efficient devices. Material performance, on the one hand, and device design, on the other hand, need to be improved to optimize OFETs in applications. Polymers are most likely to be applied in low-cost devices where manufacturing costs must be kept at a minimum; this implies that the materials need to be stable in room atmosphere and easily processable. The device structure used for research is a standard MISFET-type structure, which may not be ideal for actual applications of the technology. Completely new device concepts may be required to reach performance and stability objectives.

With regard to the performance and device properties of organic FETs in thin-film formation, specific attention needs to be given to film morphology and interface interactions between different solution-processed layers. Device properties depend, for example, on material composition, film morphology, interlayer interactions, and impurity control.

Poly(3-hexylthiophene) (P3HT) is one of the most studied and most promising polymers for electronic applications, especially in its regioregular form [52]. The material is, however, sensitive to air and humidity and, for this reason, may not be suitable for large-scale applications even though less air sensitive grades have recently been developed. To minimize the effect of the environment, production devices will be encapsulated, which reduces the significance of the air sensitivity of the semiconductor material. To reduce the requirements on the encapsulation, the material properties must be optimized. A wide spectrum of derivatives of alkylthiophenes has been developed and new improved variants are frequently reported. Other promising and widely used materials are polyfluorenes and their derivatives.

The semiconductor material, the conjugated polymer in this case, is the central material when discussing polymer FETs, but optimized conductors and insulators are also required for all-polymer devices. Frequently used polymer conductor materials are polyaniline (PANI) derivatives or a highly doped polythiophene derivative, PEDOT:PSS. There are also inorganic materials such as metal nanoparticle dispersions that may be used for conductors in practical applications.

Insulator materials are widely available as most polymers are electrical insulators. For FET applications, an insulator with a high dielectric constant as the field effect is linearly dependent on the dielectric constant, ε. The mechanical properties of the insulator material and its interface properties in relation to the conjugated polymer are more important for practical applications.

11.3.1 SEMICONDUCTOR LAYER

Amorphous silicon technology has determined the level of performance that is targeted by organic systems. This is mainly due to the vast range of applications implementing amorphous silicon despite the complicated processing procedure and the subsequently high manufacturing cost involved. The benchmark position for organic transistors has been claimed by the small organic molecule pentacene. Pentacene has shown the best performance in transistor applications so far with excellent charge carrier mobility values and relatively good stability. The material, however, is not soluble in common organic solvents, even though soluble precursors have recently been presented [55,56]. Pentacene is usually deposited by vacuum evaporation or vapor deposition techniques.

Sexithiophene (T6) is another high-performance molecular material. Its performance resembles that of pentacene but sexithiophene is more sensitive to the environmental factors such as humidity and air. Like pentacene, T6 is not solution processable. Nevertheless, it bears the great advantage that the molecule can be synthesized with different side chains or can be attached to other molecular or polymer blocks, which makes it easily soluble in common organic solvents. The added material will influence the electronic properties of the material. Therefore, finding a clean and efficient synthesis of an optimized T6 derivative has received much attention.

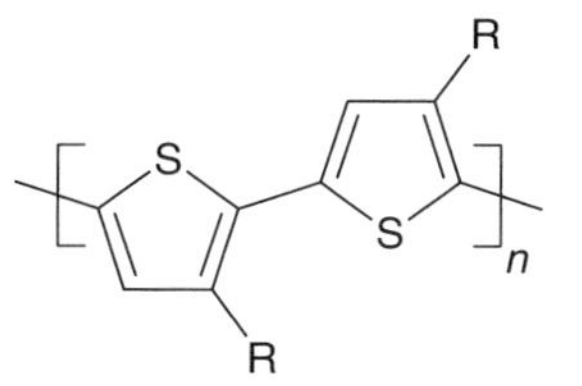

FIGURE 11.10 Chemical structure of RR-P3HT where the side chain R is a regular hexyl chain.

The morphology of the material is of a great importance and affects the electronic transport [57]. Pentacene and T6, together with many other small molecules, form polycrystalline films with a quality that depends on the material deposition technique applied. This is different from polymers, which usually form amorphous films due to the bulkiness of the polymer chains and their solubility-enhancing side chains. Some polymers also self-organize and form semicrystalline films due to the π-stacking properties of the conjugated backbone, which potentially forms good transport layers [58]. RR-P3HT (see chemical structure in Figure 11.10), used here in the work of Hassenkam et al., is one of the most extensively studied systems. The packing of the individual conjugated blocks is an ideal one with extensive π–π interaction, which gives large charge delocalization and efficient interchain transport. The transport is usually limited by the grain boundaries between crystalline grains, as in the case of a polycrystalline film or by disordered regions with a low concentration of available hopping sites.

The ordering of RR-P3HT films also depends on the molecular weight of the material and the degree of regioregularity, i.e., the percentage of head-to-tail coupling of the polymer chains [51]. Two different orientations of the crystalline domains are identified in the polymer film. A larger charge carrier mobility is recorded for the sample with the π-stacking in the plane of the substrate; hence good transport in the direction of the transistor channel.

Both small molecules and polymers are represented when the best field-effect mobilities reported per year are presented [59] (see the table in this review from 2002). Reported hole mobility has increased 10-fold around every 2 years, exceeding the hole mobility for amorphous silicon in the late 1990s.

Although solution-processed polyacetylene was reported already in 1983 [60], the first successful polymer material has been polythiophene with the first reports appearing in 1986 and with a record hole mobility on the order of 10^{-5} cm^2/Vs the same year [61]. Naturally, even the performance of polyacetylene increased during the early years [62]. Thiophene-based polymers have largely dominated the development of p-type conjugated polymers for transistor applications [63,64], (for a few examples), reaching 0.1 cm^2/Vs by the turn of the millennium [65].

The only competing polymeric material at that time was poly(thienylene vinylene) (PTV); however, with only very few reports [66]. The material can be produced as a precursor polymer that is applied in solution and converted to the conjugated polymer PTV (see chemical structure in Figure 11.11) by heating the material to around 150°C [66–68]. RR-P3HT, however, remains the

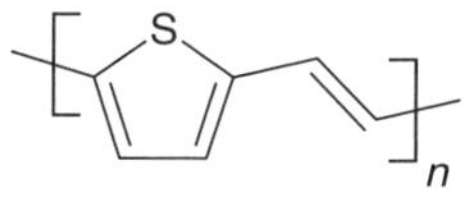

FIGURE 11.11 The chemical structure of PTV. (From Sze, S.M., *Physics of Semiconductor Devices,* 2nd ed., John Wiley & Son, New York, 1981; Pierret, R.F., *Field Effect Devices*, 2nd ed., Addison-Wesley Publishing Company, Reading, MA 1990. With permission.)

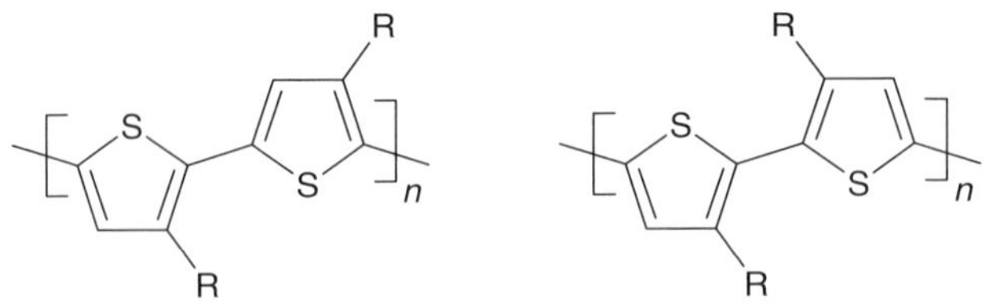

FIGURE 11.12 The chemical structure of RR-P3HT with the side chain always attached to the same position on the thiophene ring (left). In a regiorandom poly(3-hexylthiophene), the side chains are randomly attached to each other in a head-to-head or head-to-tail configuration.

favored material for high-performance polymer semiconductor layers, as the high mobility results for PTV have proved hard to reproduce.

11.3.1.1 Thiophene Derivative Polymers

Poly(alkyl thiophene) (PAT) has been produced with a range of alkyl chains, each derivative having specific solubility and morphology properties. The most successful of the PAT derivatives was the one with hexyl chains attached to the thiophene rings. PAT was first successfully synthesized with the alkyl chains attached randomly to the backbone, a mixture of the conformations shown in Figure 11.12. Later, synthetic routes were developed that enabled the polymer to achieve regularity in the polymer chains and, consequently, higher performance as in the case of regioregular PAT. A high regioregularity in the polymer allows the individual chains to pack more tightly with a higher degree of π-overlap between the conjugated backbones, resulting in higher field-effect mobility in PFETs [51,69,70].

Derivatives of PAT are sought after by synthetic methods, aiming at eliminating the problematic sensitivity to air and humidity. P3HT is easily oxidized in contact with air or humidity, which degrades the device performance. There are, however, recent reports of pure regioregular P3HT showing remarkable stability even in room atmosphere [71].

A wide range of different PAT derivatives have been demonstrated showing that very small changes to the side chain and the polymer structure may alter some properties of the resulting material drastically. The development of stable PAT derivatives is always a compromise between environmental stability, processability, and device/material performance, both electrically and structurally. The objective is the development of a material that can be easily and safely deposited in, for example, a printing facility; is suitably flexible while still being durable; shows stable performance even without or with little encapsulation; and still maintains the good electronic properties of the best reported RR-P3HT results.

11.3.1.2 Fluorene-Based Conjugated Polymers

Fluorene-based monomers offer a number of advantages for applications when used as semiconductor material in devices. They maintain a relatively high degree of delocalization due to their chemical structure even when the solubility of the material is optimized. Various fluorene-based polymers and copolymers have been developed for use in LEDs and transistors [72–74]. As for the PAT-based materials, one of the greatest challenges is to find the optimal compromise between stability and electrical performance. One method to increase the hole mobility of polymers is to blend the conjugated polymer with fullerene-based particles [75]. Combining fluorene and thiophene blocks in a copolymer has provided several promising materials [5,76–78].

Poly(9,9-dioctylfluorene-*co*-bithiophene) (F8T2) (see chemical structure in Figure 11.13) is one very popular fluorine-based copolymer that shows good performance in PFETs [5,79].

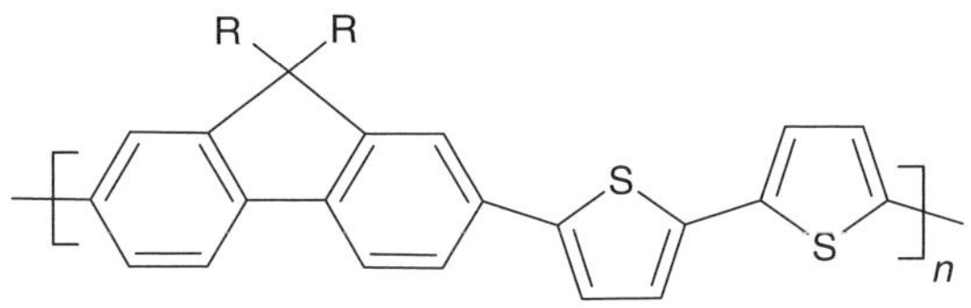

FIGURE 11.13 Chemical structure of F8T2. The side chains R = C_8H_{15}.

The structural order is important for PFET applications, and a higher order in the morphology of the F8T2 films can also be induced by using, for example, aligned polyimide substrates [80,81]. With the transition from an amorphous to an ordered phase, an almost 10-fold increase in hole mobility is observed in the direction of the polymer backbone. This material is a liquid crystalline material that can be deposited with printing techniques such as ink-jet and is therefore promising for plastic electronic applications [82]. Fluorene-based transistors may also emit light directly from the transistor channel with the light intensity controlled by the gate potential [83].

11.3.1.3 Poly(phenylene vinylene)

Poly(*p*-phenylene vinylene) (PPV) is similar in structure to PTV, with the thiophene ring being substituted with a benzene ring (see chemical structure in Figure 11.14). The material properties are further modified by adding one or several side chains or groups to the 2 and 5 positions of the benzene ring [84–88].

Poly(*p*-phenylene vinylene) derivatives were investigated in parallel to, for example, the PATs and fluorine-based materials in search of suitable materials for FETs and LEDs. An always important question with regard to the overall charge transport for all conjugated polymers is the optimization of inter- and intra-chain transport, something that has been intensely studied with various techniques [89], (an example where PPV is used). These results indicate that considerable improvement in the performance characteristics of devices could be reached presuming that better control of material purity and structural order can be obtained. The field-effect mobility, however, is generally orders of magnitude lower than that of the best PATs, often reported in the range of 10^{-6} to 10^{-5} cm^2/Vs [86], but it can also be optimized by suitable polymer modification, thin-film conditioning, and annealing, reaching around 10^{-2} cm^2/Vs [90–92]. Furthermore, a discrepancy is often observed between the charge carrier mobility in FETs (lateral transport) and LEDs (vertical transport) [84,93]. A proposed explanation is that the charge carrier density is orders of magnitude lower in LEDs than in FETs and that the field-effect mobility is carrier density dependent [94].

11.3.2 Other Device Layers

Critical for the overall performance of a PFET are the choice of gate insulator [95,96], and the contact resistance at the interface between the semiconductor and the material used for source and drain electrodes [97].

As a dielectric layer in OFETs, silicon oxide has been widely used for characterizing the semiconductor layer, as it is a very well-known material taken from the conventional Si-based FET

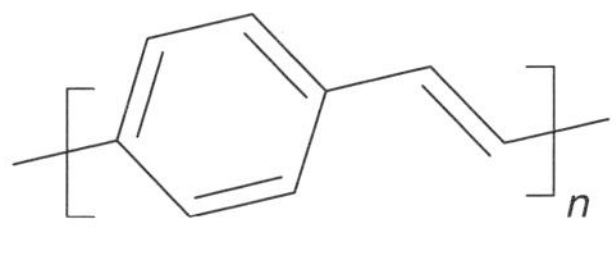

FIGURE 11.14 The chemical structure of poly(*p*-phenylene vinylene).

technology. The performance of the material is considered electrically superior to most polymeric materials. Using doped Si wafers as substrate is a simple and well-defined system for basic material research. The substrate functions simultaneously as the gate electrode and as a mechanical support, while a thermal oxide layer acts as the gate insulator. The system, however, is not a particularly good platform for actual devices and circuit manufacture where cost and processing complexity are of great importance. A transition to plastic substrates using polymeric insulator and conductor materials is desired.

11.3.2.1 Insulators

Most plastics are electrical insulators. Even though there are many polymer insulator materials available that can be easily processed and handled in solution on common solvents, the requirements for an insulator with good performance in a PFET application are quite stringent [96,98]. This is also evident from the equations used for characterizing the devices. The maximum possible electric displacement, D_{max}, that the dielectric can sustain is one of the crucial parameters and depends on the dielectric constant and the dielectric breakdown field, E_b as

$$D_{max} = \varepsilon_0 \varepsilon_r E_b \tag{11.8}$$

As seen from Equation 11.1, the minimum thickness for which a pinhole-free film may be produced is crucial for the manufacture of insulators with an optimized capacitance. One of the most promising methods for making ultrathin pinhole-free films is using self-assembly methods. This can be taken further by designed self-assembling building blocks incorporating three-dimensional cross-linking and π-electron constituents enabling precise solution phase fabrication of extremely thin layers on the order of 2.3–5.5 nm. These layers are nano-structurally ordered pinhole-free multilayer dielectrics with high-capacitance, high *k* value, and low leakage [99].

Cross-linked polymers are very promising, although certain problems relating to multilayer processing and to dielectric properties of the material persist. Robust materials are required for practical FETs, and cross-linking is currently the best option for low-cost transistors [100]. Cross-linking also allows the material to be manufactured in thinner films, thus offering the potential of enhanced field-effect mobility in FETs [100].

Poly(vinyl phenol) (PVP), also called polyhydroxystyrene (PHS), is one of the most widely used polymer insulators in organic transistors (see the chemical structure of PVP in Figure 11.15). Other materials that are quite popular are polystyrene (PS) on which PVP is based by adding a hydroxyl group to the aromatic ring; poly(vinyl alcohol) (PVA); poly(methyl methacrylate) (PMMA); and various derivatives of these or block-*co*-polymers [38,98]. These are rather brittle materials and not ideal for flexible devices. The introduction of the hydroxyl group for PVP gives the material some flexibility as well as provides reactive groups for cross-linking of the material. The hydroxyl group also gives the material a slight hygroscopicity, which is one of the reasons for hysteresis in

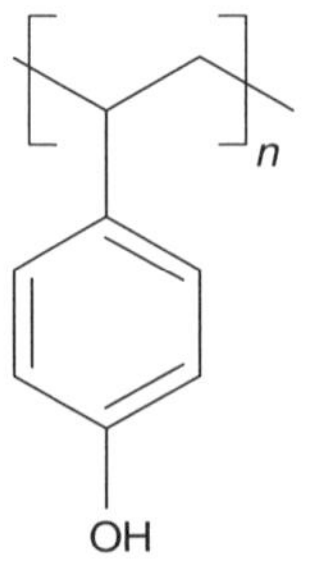

FIGURE 11.15 Structure of PVP.

devices using this material. To benefit from the advantages of both materials, Yoon et al. used a block-*co*-polymer with both PVP and PS [100].

One of the earliest curable insulators with excellent dielectric properties was polyimide (PI) made from a solution-processable precursor. However, the temperatures required for the polymerization of the PI precursor are on the order of several hundred degrees centigrade, which is too high for inclusion in an all-polymer process with several layers of sensitive polymer films. Many potential substrate materials and semiconductor polymers would melt at the temperatures required for the polymerization of PI. However, recently, certain PI precursors have become available that can be polymerized at lower temperatures of around 50°C–100°C [11]. There is a tendency toward ultraviolet (UV) curable materials since this eliminates many of the obstacles relating to the heating of these soft materials to elevated temperatures. Especially in the case of all-polymer devices where several layers of polymer film are laminated in a very tight structure, a room temperature process is to be preferred. Some layers may be very thin and the interface area of the thin-film stack is of crucial importance for the device operation. However, the effect that UV light may have on conjugated polymers remains unresolved and much research is still required in this area.

Self-assembly techniques for the manufacture of ultrathin layers have been demonstrated by Yoon et al. as one means of lowering the operation voltage for organic transistors [99].

Poly(vinyl phenol) is a PS derivative with a hydroxyl group attached to each monomer unit. This adds the property of relatively easily cross-linking of the material, a property that has been shown to be of great importance for the practical application of the material for field-effect devices and especially multilayer structures [101,102].

Thermal cross-linking has been the method of choice to achieve these structures. Acrylate added readily provides cross-linking when baked at temperatures up to 100°C–200°C. The transistor performance, however, was decreased as acrylate was added to the material, using pentacene as semiconductor [103].

A more practical approach could be based on photo-curable polymers since solution processing of the transistors requires close to room temperature processing. However, hysteresis appears in the device characteristics and it has been shown that this is because of the amount of hydroxyl groups in the insulator polymer material [104]. Lee et al. propose that this hysteresis may be caused by mobile ion charges or charge trapping in the interface. They demonstrate the effect of the hydroxyl groups by synthesizing polymers with 20%, 50%, 70%, and 100% hydroxyl group concentration on the polymer units [103]. The polymer derivatives were all similarly hygroscopic giving the conclusion that the main cause for hysteresis is the hydroxyl group content in the starting material [104]. This is an important point since PVP is one of the most popular materials used as insulator polymer to date for OFETs, and PVP represents the 100% hydroxyl group content. By decreasing the hydroxyl group content to 20%, the hysteresis was decreased to around 0.5 V from several volts, demonstrating that not only material optimization is necessary for OFET applications but also suitable materials can, indeed, be found by relatively straightforward synthetic routes. Another consequence is that due to the hydroscopicity of the PVP transistor, devices may also show field-dependent effects that are due to mobile ions in the insulator. This results in an apparent very high charge carrier mobility, but still with a slow switching speed when measured in room atmosphere [105,106].

In a standard MISFET device, the performance of the device may be increased by using the so-called high-k materials, insulators with a high dielectric constant ε. However, even if the number of charge carriers in the channel is increased and, thus, a better modulation of the current should be measured, this may in organic devices result in energy disorder at the interface due to a high dipole interaction. There are several reports where good performance has been achieved by using materials with a lower dielectric constant and where the purity of the material is emphasized, thus decreasing the disorder effect [107–109]. Benzocyclobutene (BCB) and its derivatives are currently the most commonly used materials in this context, as these may be produced with high purity and yield extremely thin pinhole-free films. In this case, the electronic performance is increased by the low thickness of the insulator layer and decreased interference by disorder. It has even been shown by using BCB as the gate dielectric that if the purity of the materials and the process are maintained, both p-type and n-type transistors may be made from RR-P3HT, a material where normally only p-type conduction may be seen [108]. The conclusion has been that electrons are very easily trapped by hydroxyl groups at the insulator–semiconductor interface when using PVP as the gate insulator, which is why only p-type transistors have been made using polymers with hydroxyl groups. Furthermore, BCB is a cross-linkable polymer allowing chemically and mechanically robust insulator films. BCB is even used as insulator material in amorphous silicon FETs where it outperforms silicon nitride as the gate insulator [110].

11.3.2.2 Conducting and Electrode Materials

Metal electrodes have traditionally been used for OFETs since the material is well defined and stable. However, charge carrier injection has often been a performance-limiting factor, and polymer conductors may show a significantly lower built-in injection barrier. In a pentacene transistor presented by Koch et al., PEDOT:PSS showed a much lower injection barrier than gold even though the work function for the materials is roughly the same [111].

Polyacetylene (PA) is commonly known as the prototype conducting polymer. Since the discovery in 1977 that the polymer could be p- or n-doped to the metallic state either chemically or electronically, the field of conducting polymers has continued to develop very rapidly and many other conducting polymers and their derivatives have been discovered. The main doping mechanisms in the early conducting polymers are briefly discussed by McDiarmid in his Nobel lecture [112].

The most popular polymeric conductors are poly(3,4-ethylenedioxythiophene) (PEDOT) doped with poly(styrene sulfonic acid) (PSS, hereafter referred to as PEDOT:PSS) and PANI. These are polymers that are additionally doped to increase the conductivity. These materials have been extensively used as an antistatic coating, for example, as packaging materials for electronic components. This is partly due to their easy processing capacity and because their electrical conductivity is almost ideal for this particular application. A thin layer of polymer is virtually transparent and colorless; prevents electrostatic discharges, for example, during plastic film handling; and reduces dust buildup on the coated material.

Polymer contacts have been shown to exhibit smaller hole injection barriers to organic semiconductors compared with those for metallic contacts due to a decrease in interfacial dipoles [111]. The decrease in the injection barrier has been shown to improve the electrical performance of both light-emitting diodes and organic transistors [113].

Despite these developments, metals may still prove to be the best electrode material for low-cost printed electronics as metal nanoparticle-based printing inks are developed, providing very stable electrode materials with excellent surface properties (flatness) and superior conductivity. As an example see commercial silver nanoparticle ink-jet ink from Cabot Inc. Various routes have recently been presented for developing metal-based inks for ink-jet printing targeted at low-cost electronics [114].

PEDOT:PSS (see chemical structure in Figure 11.16) has already for some years been commercially available from Starck GmbH (Bayer) as Baytron P (see http://www.baytron.com/index.php?page_id=1179, visited July 14, 2006). This material was originally developed for organic light-emitting diodes to be used as a modification layer and hole transport material for the transparent anode, for example, indium tin oxide. One of the challenges of this material is that it is applied as a dispersion, not a solution. This is, naturally, also a great advantage: the material is available as an aqueous dispersion and therefore available for various printing techniques and contains no harmful organic solvents that would limit its use; it fulfills most of the requirements of low-cost printed electronics. Although the solid content of the dispersion is a few percent, the material will contract upon drying if the layer is not sufficiently thin [115]. The material is also hygroscopic and thus requires shielding of the final device, or at least handing in moisture-free environment until further processing is complete, which may constitute a cost-increasing manufacturing step.

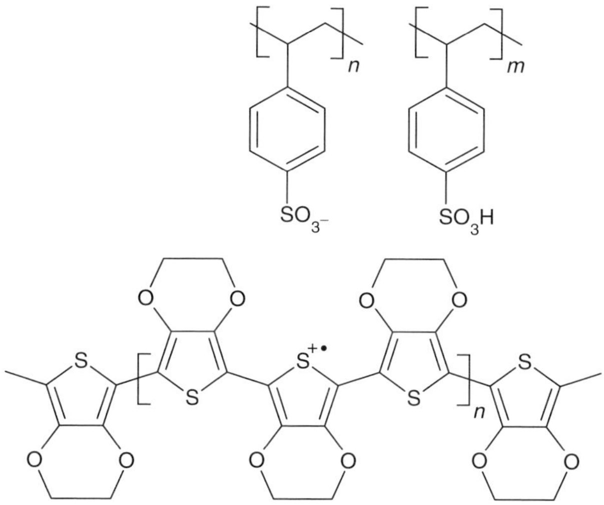

FIGURE 11.16 Structure of PEDOT:PSS. The acid is a strong dopant for the conjugated polymer resulting in a highly conductive material.

A general advantage of polymer electronics is that the materials may be directly printed using common printing techniques already available, such as screen printing, gravure, flexo, and ink-jet printing. Because of the patterning requirements of the source and drain electrodes, printing techniques have in most cases been more rarely used for depositing this electrode layer in a polymer FET. Nevertheless, PEDOT:PSS may be directly printed using, for example, ink-jet methods [5,116,117].

PEDOT:PSS may also be used as the active switching layer in organic transistors. These are not traditional field-effect devices, however, but electrochemical transistors [40,118]. These devices have a simple structure with the same material working both as electrodes and the semiconductor. The only material required in addition to PEDOT:PSS is an electrolyte that functions as an equivalent to the insulator layer in a FET. An explanation to the device mechanism that gives rise to current saturation in this switching device similar to that of a FET is proposed to be de-doping of the area close to the drain contact [119]. Similar field-effect devices prepared completely from an insulator and PEDOT:PSS or polypyrrole (PPy) doped with Cl^- (PPy:Cl^-) are also reported. These devices have threshold turnoff V_G's positive about +15 V (varying with polymer, geometry, and preparation conditions). The device is in the ON (conductive) state when the gate voltage V_G is applied and the current ratio I_{on}/I_{off} can exceed 10^4 at room temperature [36,37].

Polyaniline remained a widely studied material for a long time, since it has several stable oxidation states and the conductivity of the material can be controlled over a very large conductivity range from around 10^{-12} S/cm (typical for a glass insulator) up to around 100 S/cm, the conductivity of mercury metal (silver and copper conductivity is around 10^5 to 10^6 S/cm). The conductivity of PANI can be further controlled by mixing (doping) with carbon nanotubes [120]. In Figure 11.17, the structure of PANI is shown, with the different oxidation configurations indicated as blocks (indexed n and m). Leucoemeraldine with $n = 1$, $m = 0$ is the fully reduced state. Pernigraniline is the fully oxidized state ($n = 0$, $m = 1$) with imine links replacing amine links. The emeraldine ($n = m = 0.5$) form of PANI, often referred to as emeraldine base (EB), is either neutral or only partially reduced or oxidized. Emeraldine base is regarded as the most useful form of PANI due to its high stability at room temperature, compared to the easily oxidized leucoemeraldine and the easily degraded pernigraniline. Additionally, the emeraldine base PANI can function as a semiconductor when doped by a protic acid. Doping emeraldine base with acid (dopant) results in a conductive emeraldine salt (ES).

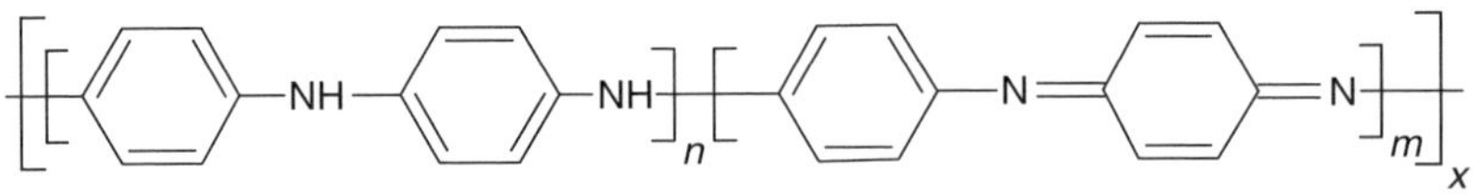

FIGURE 11.17 Chemical structure of PANI. The different stable oxidation states have different proportions of n and m.

PANI is commercially available in large quantities as bulk raw material or as ink solutions or dispersions for screen printing, gravure printing, or spray coating. Currently, the most successful commercial application for PANI is as a semitransparent EMI (electromagnetic interference) shield coating.

11.4 POLYMER FIELD-EFFECT TRANSISTOR PROCESSING AND MANUFACTURING; FUTURE POTENTIAL APPLICATIONS

The devices used for conjugated polymer material characterization are simple TFT structures, as presented in Section 11.2.1. It is formed by a four-layer sandwich structure in either the top-gate or bottom-gate configuration. Regardless of whether the gate electrode layer or the source–drain electrode layer is applied as the first layer on the substrate, the device performance should be the same. However, characterization of devices always gives results that follow no simple rule or expression but are a mixture of device and material properties [121]. Improved properties of devices are obtained when the different layers are patterned, which is why printing techniques are beneficial compared to the coating methods. By patterning the semiconductor such that the material is deposited only in the channel area, the risk of leakage through pinholes or around edges of the device may be avoided. Furthermore, cross-talk between different transistors may be avoided. Switching speed due to capacitive electrode overlap may be decreased, for example, thus driving the development and research on processing techniques in parallel with materials and device characterization.

Polymer field-effect transistor devices may also have properties not traditionally associated with FETs. This may open new application areas for PFET devices that are not limited by the still inferior stability and electrical performance of PFETs as compared to traditional MISFETs or amorphous silicon devices. Such properties may be the above-mentioned ionic effects in PVP-based transistors in air [105], purely electrochemical transistors [119,122,123], or chemically sensitive FETs (CHEMFETs) [40,124]. The commonly observed instability of PFETs points at the material suitability for sensor applications if only the response of the device is stable and reproducible [125]. These special applications may often utilize the same materials that are used for normal OFETs such as RR-P3HT [126].

The basic device structure presented earlier is sufficient for material characterization, but various novel device structures and applications have been proposed in parallel with new materials and device characteristics. Many of these are even on the boundaries of today's field-effect devices altogether. Morana et al. present a double gate device allowing both more flexibility in controlling the device and higher performance capacity [127]. Since many of the applications driving the development of OFETs are related to display technology, various device structures and geometries have been proposed that combine an organic transistor and the light-emitting component. To simplify the processing, it is desirable to use the same organic semiconductor for both devices [84]. There have even been reports of transistors that emit light from the transistor channel area [83]. Since electronics today is mostly based on digital logic, various types of logic gates optimized for OFETs have been developed, and materials have been developed to show both p- and n-type transport, which is a necessity for simple logic gate structures [128].

For plastic electronics to be truly organic they must also incorporate polymer substrates and electrode materials. The processing of traditional electronics involves metal and semiconductor deposition in layers that are insulated from each other by insulator (normally oxide) layers. In the early days of conjugated polymer/oligomer research, well-defined inorganic materials were used as substrates, electrodes, and as gate insulator layers. The most common sample configuration has been a highly doped silicon wafer with a top layer of Si^+/SiO_2 comprising a gate electrode and gate dielectric. Metal source and drain electrodes were deposited on top of this silicon wafer. The active conjugated semiconductor was deposited by a solution process through spin coating or solution coating. In the case of oligomers, thermal evaporation was usually employed due to the poor solubility properties of the material. Because the inorganic materials are well known and their electronic behavior predictable, the device structure obtained by this method is useful for studying the intrinsic properties of the conjugated materials.

Inorganic materials have been gradually replaced by organic alternatives and the effect of this transition has been studied in detail elsewhere [121]. The transition to plastic substrates introduced flexible device structures, but requires more stringent selection of other materials comprising the device. Naturally, for example, brittle polymers are not suitable for flexible structures. Furthermore, the use of polymer conductors/electrodes and polymer insulators alters the device properties, as the electronic characteristics are also determined by interaction between the different layers of the device, on the one hand, and by the properties of the material, on the other.

Complicated integrated circuits have been built by using rather standard semiconductor-manufacturing techniques that involve photo-lithographic steps and batch processing [129]. The demonstrated devices consist of hundreds of field-effect transistors and hundreds of vertical interconnects and are operational even if the sample is sharply bent. Many other similar devices have been introduced including active pixels [122,130] and small organic displays [13,24]. Recently, much more complex circuits have been demonstrated; the technology approaches the state of more traditional electronics technologies, even when using novel manufacturing techniques suitable for polymer solution processing [14].

The application areas for polymer FETs may be divided into two groups. First, since the electronic performance of conjugated polymers approaches the performance of amorphous silicon, applications where amorphous silicon is currently used are targeted by PFET technology. In this area, polymers might provide simplified processing and consequently savings in relation to production cost. Second, PFETs target entirely new applications where electronics has not been viable until now, either due to processing or economic reasons. Printed electronics based on PFETs already begin to have an impact in the field of packaging, advertising, and related application areas.

11.4.1 Low-Cost Printable Electronics

Today, rather complicated electronics are being incorporated even in single use items such as labels and information tags. This creates a need to develop new methods for the fabrication of low-cost electronic circuits. There are visions of a remotely readable and writable circuit attached, for example, to every packet of fresh food where data such as manufacture date and best before dates can be stored. Printed radio-frequency identification (RFID) tags have already reached the commercial stage.

High end electronics rely on consumers' demand for high-quality products, even at a high cost. The concept of single use electronics manufactured in a production or packaging line environment, for example in a dairy factory, can only be realized, if the technology can be developed for standard printing machinery or similar technology. The fact that polymers and organic molecules can be easily handled in solution makes them ideal for this kind of applications. Moreover, the techniques for making a solution suitable for printing are well known. Thus, new printing techniques are being developed while old printing methods are being modified to be suitable for printing conjugated polymers [131]. There have been reports of printed displays and electronic integrated circuits with good performance [13,14,26].

Printing conjugated material from solution will normally yield an amorphous film. The molecular ordering is determined, first, by the self-ordering properties of the material, and second, by the time needed for this self-organizing process to occur. Plastic films such as poly(ethylene terephthalate) (PET) or poly(ethylene naphthalate) (PEN) are normally used as substrate material for printed circuits. Sheets of paper covered with suitable surface layers have also been investigated as a means of mechanical support. A great challenge is the transition from individual transistor to reliable integrated circuits comprising hundreds or thousands of transistors and other components [132]. Other critical aspects are the quality and the uniformity of the printed source–drain electrode gap, the transistor channel, and the flatness of the printed material. Printing techniques have currently not quite reached the required specifications for reliable uniform devices.

One of the main driving forces in the field of organic electronics is the potentially inexpensive and simple device-manufacturing techniques available for solution-processed materials. Manufacturing of large quantities of devices is most efficiently performed through a reel-to-reel process utilizing the standard printing techniques available. The first all-printed organic transistor, which used an ITO-covered substrate as gate, was made using screen printing techniques [133]. There have been numerous attempts to debut printed organic transistors, first concentrating on single elements of the device and then on complete devices that have been manufactured using methods like stamping, μ-contact printing, screen printing, soft lithography, or ink-jet printing [134], just to mention a few. Ink-jet printing differs from most other printing techniques that use templates for patterning, as it is a 100% noncontact technique that allows high resolution deposition of the material [5], and has initially been the most popular and most successful technique for printed electronic devices. The technique as such, however, is still not capable of making narrow well-defined transistor channels with high-quality electrode edges. It is, on the other hand, suitable for gate electrode deposition, and minimizes the overlap between gate and source–drain electrode pairs, thus reducing gate leakage currents and the risk of electrical breakdown due to shorting through the insulator film. The critical dimension in a FET is the channel length, which is defined by the distance between the drain and source electrodes. Faster switching speeds and higher output currents can be achieved by reducing this channel length. Bao investigated various printing techniques and compared their suitability for FET manufacture and observed that without substrate surface treatment the resolution for screen printing, ink-jet printing, and thermal laser printing in on the order of 75, 25, and 5 μm, respectively [135].

When making structures where several layers of solution-deposited material are applied in sequence, the solubility properties of the different materials often determine the choice of device structure or the actual polymers that may be used. Naturally, the next layer should not dissolve the underlying film. Materials with cross-linking properties and annealing procedures have been used to reduce the mixing of the different materials in the device, to increase the quality of all-polymer devices, and order to make the samples more stable in room environment [136].

11.4.1.1 Ink-Jet Printing

Ink-jet printing has been established as a technique for organic electronics due to the economical use of material, precise patterning by an additive noncontact process, and a fairly straightforward way to control material deposition amount and speed [137,138]. All-solution-processed FETs have been prepared in ambient air, and direct ink-jet printing of complete transistor circuits has been demonstrated, including via-hole interconnections based on solution-processed polymer conductors, insulators, and self-organizing semiconductors [5,139]. Moreover, a flexible electrophoretic display has been demonstrated by laminating an ink-jet printed active-matrix backplane with an electrophoretic device [140].

There are numerous alternative solution processing techniques available besides ink-jet, even though most of these methods are not based on noncontact deposition [141]. Roll-to-roll techniques related to a stamping type process include coating and printing processes such as roll-coating and doctor-blading as well as gravure or flexographic printing [142,143]. Modified versions of standard graphic printers may be utilized, or specialized techniques relying on, for example, capillary forces and solvent absorbing elastomers may be developed [144].

Most processing methods are combinatory techniques involving steps utilizing various printing technologies together with other wet processing techniques [145,146]. A high device yield, uniformity, and the resolution required for thin-film electronic applications can be achieved by using a substrate that contains a surface energy pattern to control the flow and spreading of droplets when using, for example, the ink-jet technique [147–149]. Ink-jet technique may also be utilized for additively depositing the mask layout for standard wet lithography as an alternative to a photo-lithographic mask [150].

The substrate material is also crucial for determining the total cost of the technology in use. Most flexible organic electronics demonstrators are prepared on flexible plastic substrates but also paper has been used [40]. Keeping the applications in advertising and packaging in mind, paper would seem to constitute the ultimate substrate material, though its surface roughness and inhomogeneous chemical properties still offer great challenges. The field of PFETs is maturing rapidly and is gradually evolving from being a research environment to a production technology.

11.5 SUMMARY

One of the most interesting application areas for conjugated polymers is in the field of printable electronics. New areas for applications of electronics are emerging, for example, in advertising and packaging, and low-cost technologies for making electronic devices and circuits are required to meet the challenges presented by these industries. A key component in this context is a transistor that can be built from only solution-processable materials. Solution processing allows the use of established printing technology for device manufacture at low cost and with very high throughput.

The semiconductor layer in a transistor, the conjugated polymer thin film, forms the core when studying and developing polymer transistors. Conjugated polymers with sufficiently high charge carrier mobility and environmental stability for some practical applications have been developed and are beginning to provide some competition for amorphous silicon. However, even if the conjugated polymer possesses excellent properties, the device may not perform well, as both the insulating and conducting materials used in the device play a significant role in its performance. For example, interface properties between layers and parasitic resistance in the conductors are properties in the printed circuit that may prove to be the critical limiting factors of the whole system.

Furthermore, the transistor that would sufficiently meet the criteria should show stable performance and a sufficient lifetime for the purpose of the relevant applications. It is widely reported that charge carrier mobility is dependent on both charge carrier concentration and electric field, and the simple equation for conductivity (Equation 11.3) easily gets rather more complicated, i.e., the conductivity in the channel region of a TFT is not in reality uniform but graded according to the applied voltages and depends on intentional or unintentional doping levels. Drift in the threshold voltage is also a common problem in reported polymer transistors. Stability of the device is one of the key issues when device processing is scaled up both in device fabrication volumes and circuit integration. Hysteresis or drift in the current through the transistor due to material instability or processing inaccuracy is partly responsible for preventing large-scale integration and application of polymer transistors. Another important issue is the absolute electronic performance of the material. Is the field-effect charge carrier mobility, for example, too low for practical applications and is the maximum current modulation of the device acceptable?

Conjugated polymers offer one of the most versatile material alternatives for printable transistors. Solubility of all materials in common solvents is a key advantage for polymer transistors when compared to most of their molecular counterparts discussed in Chapter 10. This property has made printable electronics one of the most interesting application areas for conjugated polymers. One of the most widely used conjugated polymers for transistor applications are derivatives of polythiophene, poly(phenylene vinylene), and fluorene-based materials. Acting as insulators, PVP has acted as the main reference for some time and several high-performance insulators have been reported based on this material. Cross-linking of the insulator thin film is increasingly used as a tool to improve the chemical and mechanical stability of the insulator layer. A chemically stable middle layer is crucial for the successful solution processing of multilayer devices. As electrodes, there are currently two commonly used commercial alternative materials, one based on a doped polythiophene,

PEDOT, and the other on PANI. Some of the most important forces driving the development of the synthesis of polymers for the purposes of electronics are the need for pure materials; materials that are stable in air; and materials that are easily turned into printing inks and are preferably even cross-linkable for robustness and chemical integrity.

This chapter gives an introduction to the field of polymer transistors and printable transistors, as well as the currently most central materials used in this field. The basic operating principle for the devices is discussed, and should be read as a complement to the theory in the previous chapter on molecular transistors. Most characterization methods and result interpretation models presented in the chapter on molecular transistors may also be applied for polymer FETs, and therefore this chapter is deliberately given the character of a broader review of the field of polymer FETs rather than giving an in-depth iteration of fundamental OFET characterization methods. The chapter does not elaborate on the fine details of charge transport in conjugated polymers and the technicalities of the printing processes, but is to be regarded as a general introduction to the field.

EXERCISE QUESTIONS

1. Polymer transistors compete with molecular devices especially in the area of processability of the materials. Polymers are generally soluble in organic solvents and can be applied by solution processing. However, materials must be chosen carefully, both semiconductor and electrode materials, to avoid inefficient contacts to the semiconductor. How can the contact resistance be estimated from the standard *I–V* characteristics of the PFET, or how can the effect of the contact resistance be eliminated in device characterization?
2. Low-cost, preferably printed integrated circuits are proposed as potential applications for organic electronics and polymer FETs. Modern electronics is largely based on logic circuits based on transistors. A single transistor may be connected as an amplifier, with a gain value directly dependent on the transistor properties. How can the gain of a transistor be estimated directly from the transistor *I–V* characteristics? For theoretical background, see also the previous chapter about molecular FETs and the references to this chapter.

LIST OF ABBREVIATIONS

BCB	benzocyclobutene
CCD	charge coupled device
CELIV	carrier extraction using a linearly increasing voltage
CHEMFET	chemically sensitive FET
DRAM	dynamic random access memory
EB	emeraldine base
EMI	electromagnetic interference
ES	emeraldine salt
F8T2	poly(9,9-dioctylfluorene-*co*-bithiophene)
FET	field-effect transistor
HMDS	hexamethyl-disilazane
ITO	indium tin oxide
LED	light-emitting diode
MIS	metal–insulator–semiconductor
MISFET	metal–insulator–semiconductor field-effect transistor
MOS	metal–oxide–semiconductor
OFET	organic field-effect transistor
P3HT	poly(3-hexylthiophene)
PA	polyacetylene
PANI	polyaniline

PAT	poly(alkyl thiophene)
PEDOT	poly(3,4-ethylenedioxythiophene)
PEDOT:PSS	poly(styrene sulfonic acid) doped poly(3,4-ethylenedioxythiophene)
PEN	poly(ethylene naphthalate)
PET	poly(ethylene terephthalate)
PFET	polymer FET
PHS	polyhydroxystyrene
PI	polyimide
PMMA	polymethylmetacrylate
PPV	poly(*p*-phenylene vinylene)
PPy	polypyrrole
PS	polystyrene
PSS	poly(styrene sulfonic acid)
PTV	poly(thienylene vinylene)
PVA	poly(vinyl alcohol)
PVP	poly(vinyl phenol)
RCA	Radio Corporation of America
RFID	radio-frequency identification
RR-P3HT	regioregular poly(3-hexylthiophene)
SPICE	Simulation Program with Integrated Circuits Emphasis
T6	sexithiophene
TFT	thin-film transistor
UV	ultraviolet
VRH	variable range hopping

ACKNOWLEDGMENT

The author would like to thank Dr. Tomas Bäcklund for numerous excellent suggestions and comments on the manuscript.

REFERENCES

1. Garnier, F., Hajlaoui, R., Yassar, A. et al., All-polymer field-effect transistor realized by printing techniques, *Science*, 265, 1684 1994.
2. Dodabalapur, A., Torsi, L., and Katz, H.E., Organic transistors: Two-dimensional transport and improved electrical transport, *Science*, 268, 270 1995.
3. Horowitz, G., Organic field-effect transistors, *Adv. Mater.*, 10, 365–377 1998.
4. Matters, M., de Leeuw, D.M., Vissenberg, M.J.M.C. et al., Organic field-effect transistors and all-polymer integrated circuits, *Opt. Mater.*, 12, 189 1999.
5. Sirringhaus, H., Kawase, T., Friend, R. et al., High-resolution inkjet printing of all-polymer transistor circuits, *Science*, 290, 2123–2126 2000.
6. Gelinck, G., Geuns, T., and de Leeuw, D., High-performance all-polymer integrated circuit, *Appl. Phys. Lett.*, 77, 1487 2000.
7. Burroughes, J.H., Bradley, D.D.C., Brown, A.R. et al., Light-emitting diodes based on conjugated polymers, *Nature*, 347, 539–541 1990.
8. Friend, R., Gymer, R., Holmes, A. et al., Electroluminiscence in conjugated polymers, *Nature*, 397, 121 1999.
9. Noma, N., Tsuzuki, T., and Shirota, Y., Alpha-thiophene octamer as a new class of photoactive material for photoelectrical conversion, *Adv. Mater.*, 7, 647–648 1995.
10. Videlot, C., El Kassmi, A., and Fichou, D., Photovoltaic properties of octithiophene-based schottky and p/n junction cells: Influence of molecular orientation, *Sol. Energy Mater. Sol. Cells*, 63, 69–82 2000.
11. Liu, Y., Cui, T.H., and Varahramyan, K., All-polymer capacitor fabricated with inkjet printing technique, *Solid-State Electron.*, 47, 1543–1548 2003.
12. Halik, M., Hagen, K., Zschieschang, U. et al., Polymer gate dielectrics and conducting polymer contacts for high-performance organic thin-film transistors, *Adv. Mater.*, 14, 1717 2002.

13. Gelinck, G., Huitema, H., Veenendaal, E.V. et al., Flexible active-matrix displays and shift registers based on solution-processed organic transistors, *Nat. Mater.*, 3, 106–110 2004.
14. Knobloch, A., Manuelli, A., Bernds, A. et al., Fully printed integrated circuits from solution processable polymers, *J. Appl.Phys.*, 96, 2286–2291 2004.
15. Bardeen, J. and Brattain, W.H., The transistor, a semi-conductor triode, *Phys. Rev.*, 74, 230–231 1948.
16. Shockley, W., Bardeen, J., and Brattain, W.H., The electronic theory of the transistor, *Science*, 108, 678–679 1948.
17. Watson, W.H., McMordie, W.C., and Lands, L.G., Polymerization of alkynes by Ziegler-type catalyst, *J. Polym. Sci.*, 55, 137 1961.
18. Shirakawa, H. and Ikeda, S., Infrared spectra of poly(acetylene), *Polym. J.*, 2, 231 1971.
19. Fichou, D., *Handbook of Oligo- and Polythiophenes*. Weinheim: Wiley-VCH 1999.
20. Lin, J.W.P. and Dudek, L.P., Synthesis and properties of poly(2,5-thienylene), *J. Polym. Sci. Polym. Chem.*, 18, 2869–2873 1980.
21. Schoeler, U., Tews, K.H., and Kuhn, H., Potential model of dye molecule from measurements of photocurrent in monolayer assemblies, *J. Chem. Phys.*, 61, 5009–5016 1974.
22. Fichou, D., Horowitz, G., Nishikitani, Y. et al., Conjugated oligomers for molecular electronics: Schottky diodes on vacuum evaporated films of alpha-sexithienyl, *Chemtronics*, 3, 176 1988.
23. Horowitz, G., Fichou, D., and Garnier, F., Alpha-sexithienyl—a p-type and n-type dopable molecular semiconductor, *Solid State Commun.*, 70, 385–388 1989.
24. Huitema, H.E.A., Gelinck, G.H., van der Putten, J.B.P.H. et al., Plastic transistors in active-matrix displays, *Nature*, 414, 599 2001.
25. Sheraw, C.D., Zhou, L., Huang, J.R. et al., Organic thin-film transistor-driven polymer-dispersed liquid crystal displays on flexible polymeric substrates, *Appl. Phys. Lett.*, 80, 1088 2002.
26. Rogers, J.A. and Bao, Z., Printed plastic electronics and paperlike displays, *J. Polym. Sci. Polym. Chem.*, 40, 3327–3334 2002.
27. Kahng, D. and Atalla, M.M., Silicon-silicon dioxide field induced surface devices. IRE Solid State Device Research Conference, Pittsburgh 1960.
28. Kroschwitz, J.I., Ed., *Electrical and Electronic Properties of Polymers: A State-of-the-Art Compendium*. Encyclopedia reprint series, New York: Wiley 1988.
29. Kao, K.C. and Hwang, W., *Electrical Transport in Solids*, Vol. 14 of *International Series in the Science of the Solid State*. Headington Hill Hall, Oxford: Pergamon Press Ltd. 1981.
30. Gedde, U.W., *Polymer Physics*. Chapman & Hall 1995.
31. Mott, N.F. and Davis, E.A., *Electronic Processes in Non-Crystalline Materials*, 2nd ed. Oxford: Oxford University Press, 1979.
32. Mott, N.F., *Conduction in Non-Crystalline Materials*. Oxford: Clarendon Press, 1993.
33. Garnier, F., Horowitz, G., Fichou, D. et al., Molecular order in organic-based field-effect transistors, *Synth. Met.*, 81, 163–171 1996.
34. Gundlach, D.J., Lin, Y.Y., Jackson, T.N. et al., Pentacene organic thin-film transistors—molecular ordering and mobility, *IEEE Electr. Dev. Lett.*, 18, 87–89 1997.
35. Prigodin, V.N. and Epstein, A.J., Nature of insulator-metal transition an dnovel mechanism of charge transport in the metallic state of highly doped electronic polymers, *Synth. Met.*, 125, 43 2002.
36. Epstein, A.J., Hsu, F.C., Chiou, N.R. et al., Electric-field induced ion-leveraged metal–insulator transition in conducting polymer-based field effect devices, *Curr. Appl. Phys.*, 2, 339–343 2002.
37. Epstein, A.J., Hsu, F.C., Chiou, N.R. et al., Doped conducting polymer-based field effect devices, *Synth. Met.*, 137, 859–861 2003.
38. Unni, K.N., Dabos-Seignon, S., and Nunzi, J.M., Influence of the polymer dielectric characteristics on the performance of a quaterthiophene organic field-effect transistor, *J. Mater. Sci.*, 41, 317–322 2006.
39. Jones, E.T.T., Chyan, O.M., and Wrighton, M.S., Preparation and characterization of molecule-based transistors with a 50-nm source–drain separation with use of shadow deposition techniques—toward faster, more sensitive molecule-based devices, *J. Am. Chem. Soc.*, 109, 5526–5528 1987.
40. Nilsson, D., Kugler, T., Svensson, P.O. et al., An all-organic sensor-transistor based on a novel electrochemical transducer concept printed electrochemical sensors on paper, *Sens. Actuators, B*, 86, 193–197 2002.
41. Chen, M.X., Nilsson, D., Kugler, T. et al., Electric current rectification by an all-organic electrochemical device, *Appl. Phys. Lett.*, 81, 2011–2013 2002.
42. Sze, S.M., *Physics of Semiconductor Devices,* 2nd ed. New York: John Wiley & Son, 1981.
43. Pierret, R.F., *Field Effect Devices*, Vol. IV of *Modular Series on Solid State Devices*, 2nd ed. Reading, MA: Addison-Wesley Publishing Company, 1990.

44. Tecklenburg, R., Paasch, G., and Scheinert, S., Theory of organic field effect transistors, *Adv. Mater. Opt. Electron.*, 8, 285 1998.
45. Dennard, R.H., Field-effect transistor memory, US Patent 3 387 286 1968.
46. Sandberg, H., Frey, G., Shkunov, M. et al., Ultrathin regioregular poly(3-hexyl thiophene) field-effect transistors, *Langmuir*, 18, 10176–10182 2002.
47. Shur, M., Hack, M., and Shaw, J.G., A new analytic model for amorphous silicon thin-film transistors, *J. Appl. Phys.*, 66, 3371 1989.
48. Juska, G., Arlauskas, K., Viliunas, M. et al., Extraction current transients: New method of study of charge transport in microcrystalline silicon, *Phys. Rev. Lett.*, 84, 4946–4949 2000.
49. Hamilton, E.M., Variable range hopping in a nonuniform density of states, *Philos. Mag.*, 26, 1043 1972.
50. Kim, S.J., Ahn, T., Sun, M.C. et al., Low-leakage polymeric thin-film transistors fabricated by laser assisted lift-off technique, *Jpn. J. Appl. Phys. Part 2*, 44, 1109–1111 2005.
51. Sirringhaus, H., Brown, P.J., Friend, R.H. et al., Two-dimensional charge transport in self-organized, highmobility conjugated polymers, *Nature*, 401, 685–688 1999.
52. Sirringhaus, H., Tessler, N., Thomas, D.S. et al., *High-Mobility Conjugated Polymer Field-Effect Transistors*, Vol. 39 of Advances in Solid State Physics, p. 101. Braunschweig: Vieweg, 1999.
53. Mena-Osteriz, E., Meyer, A., Langeveld-Voss, B.M.W. et al., Two-dimensional crystals of poly (3-alkylthiophene)s: Direct visualization of polymer folds in submolecular resolution, *Angew. Chem. Int. Edit.*, 39, 2680 2000.
54. Breiby, D., Samulesen, E., and Konovalov, O., The drying behaviour of conjugated polymer solutions, *Synth. Met.*, 139, 361–369 2003.
55. Afzali, A., Breen, T., and Kagan, C., An efficient synthesis of symmetrical oligothiophenes: Synthesis and transport properties of a soluble sexithiophene derivative, *Chem. Mater.*, 14, 1742 2002.
56. Afzali, A., Dimitrakopoulos, C.D., and Breen, T.L., High-performance, solution-processed organic thin film transistors from a novel pentacene precursor, *J. Am. Chem. Soc.*, 124, 8812–8813 2002.
57. Fichou, D., Structural order in conjugated oligothiophenes and its implications on opto-electronic devices, *J. Mater. Chem.*, 10, 571–588 2000.
58. Hassenkam, T., Greve, D., and Björnholm, T., Direct visualization of the nanoscale morphology of conducting polythiophene monolayers studied by electrostatic force microscopy, *Adv. Mater.*, 13, 631 2001.
59. Dimitrakopoulos, C. and Malenfant, P., Organic thin film transistors for large area electronics, *Adv. Mater.*, 14, 99 2002.
60. Ebisawa, F., Kurokawa, T., and Nara, S., Electrical properties of polyacetylene/polysiloxane interface, *J. Appl. Phys.*, 54, 3255–9 1983.
61. Tsumura, A., Koezuka, H., and Ando, T., Macromolecular electronic device: Field-effect transistor with a polythiophene thin film, *Appl. Phys. Lett.*, 49, 1210–12 1986.
62. Burroughes, J.H., Jones, C.A., and Friend, R.H., New semiconductor-device physics in polymer diodes and transistors, *Nature*, 335, 137–141 1988.
63. Assadi, A., Svensson, C., Willander, M. et al., Field-effect mobility or poly(3-hexylthiophene), *Appl. Phys. Lett.*, 53, 195 1988.
64. Paloheimo, J., Kuivalainen, P., Stubb, H. et al., Molecular field-effect transistor using conducting polymer Langmuir-Blodgett films, *Appl. Phys. Lett.*, 56, 1157 1990.
65. Sirringhaus, H., Tessler, N., and Friend, R.H., Integrated optoelectronic devices based on conjugated polymers, *Science*, 280, 1741 1998.
66. Fuchigami, H., Tsumura, A., and Koezuka, H., Polythienylenevinylene thin-film transistor with high carrier mobility, *Appl. Phys. Lett.*, 63, 1372–4 1993.
67. Murata, H., Tokito, S., Tsutsui, T. et al., Preparation of high-quality poly(2,5-thienylene-vinylene) films, *Synth. Met.*, 36, 95–102 1990.
68. Brown, A., Jarrett, C., de Leeuw, D. et al., Field-effect transistors made from solution-processed organic semiconductors, *Synth. Met.*, 88, 37–55 1997.
69. Bao, Z., Dodabalapur, A., and Lovinger, A.J., Soluble and processable regioregular poly(3-hexylthiophene) for thin film field-effect-transistor applications with high mobility, *Appl. Phys. Lett.*, 69, 4108 1996.
70. Sirringhaus, H., Tessler, N., and Friend, R.H., Integrated, high-mobility polymer field-effect transistors driving polymer light-emitting diodes, *Synth. Met.*, 102, 857–860 1999.
71. Ficker, J., Ullman, A., Fix, W. et al., Stability of polythiophene-based transistors and circuits, *J. Appl. Phys.*, 94, 2638 2003.
72. Bernius, M., Inbasekaran, M., Woo, E. et al., Fluorene-based polymers-preparation and applications, *J. Mater. Sci.: Mater. Electron.*, 11, 111–116 2000.

73. Rawcliffe, R., Bradley, D.D., and Campbell, A.J., Comparison between bulk and held-effect mobility in polyfluorene copolymer field-effect transistors, *Proc. SPIE*, 5217, 25–34 2003.
74. Heeney, M., Bailey, C., Giles, M. et al., Alkylidene fluorene liquid crystalline semiconducting polymers for organic field effect transistor devices, *Macromolecules*, 37, 5250–5256 2004.
75. Andersson, L.M. and Inganäs, O., Acceptor influence on hole mobility in fullerene blends with alternating copolymers of fluorene, *Appl. Phys. Lett.*, 88, 082103 2006.
76. Surin, M., Sonar, P., Grimsdale, A.C. et al., Supramolecular organization in fluorene/indenofluorene-oligothiophene alternating conjugated copolymers, *Adv. Funct. Mater.*, 15, 1426–1434 2005.
77. Chen, M., Crispin, X., Perzon, E. et al., High carrier mobility in low band gap polymer-based field-effect transistors, *Appl. Phys. Lett.*, 87, 252105 2005.
78. Porzio, W., Destri, S., Giovanella, U. et al., Fluorenone-thiophene derivative for organic field effect transistors: A combined structural, morphological and electrical study, *Thin Solid Films*, 492, 212–220 2005.
79. Sirringhaus, H., Wilson, R.J., Friend, R.H. et al., Mobility enhancement in conjugated polymer field-effect transistors through chain alignment in a liquid-crystalline phase, *Appl. Phys. Lett.*, 77, 406–408 2000.
80. Kinder, L., Kanicki, J., Swensen, J. et al., Structural ordering in F8T2 polyfluorene thin-film transistors, *Proc. SPIE*, 5217, 35–42 2003.
81. Kinder, L., Kanicki, J., and Petroff, P., Structural ordering and enhanced carrier mobility in organic polymer thin film transistors, *Synth. Met.*, 146, 181–185 2004.
82. Newsome, C.J., Kawase, T., Shimoda, T. et al., The phase behaviour of polymer semiconductor films and its influence on the mobility in FET devices, *Proc. SPIE*, 5217, 16–24 2003.
83. Ahles, M., Hepp, A., Schmechel, R. et al., Light emission from a polymer transistor, *Appl. Phys. Lett.*, 84, 428–430 2004.
84. Tzeng, K., Meng, H., Tzeng, M. et al., One-polymer active pixel, *Appl. Phys. Lett.*, 84, 619–621 2004.
85. Tanase, C., Blom, P., Meijer, E. et al., Hole transport in polymeric field-effect transistors and light-emitting diodes, *Proc. SPIE*, 5217, 80–86 2003.
86. Geens, W., Shaheen, S.E., Wessling, B. et al., Dependence of field-effect hole mobility of PPV-based polymer films on the spin-casting solvent, *Org. Electron.*, 3, 105–110 2002.
87. Scheinert, S., Paasch, G., Pohlmann, S. et al., Field effect in organic devices with solution-doped arylamino-poly-(phenylene-vinylene), *Solid-State Electron.*, 44, 845–853 2000.
88. Jorgensen, M., Sommer-Larsen, P., Norrman, K. et al., Vacuum UV of a PPV type polymer and possible PTFE induced orientation, *Synth. Met.*, 142, 121–125 2004.
89. Hoofman, R.J., de Haas, M.P., Siebbeles, L.D. et al., Highly mobile electrons and holes on isolated chains of the semiconducting polymer poly(phenylene vinylene), *Nature*, 392, 54–56 1998.
90. Van Breemen, A.J., Herwig, P.T., Chlon, C.H. et al., High-performance solution-processable poly (p-phenylene vinylene)s for air-stable organic field-effect transistors, *Adv. Funct. Mater.*, 15, 872–876 2005.
91. Tanase, C., Wildeman, J., Blom, P. et al., Optimization of the charge transport in poly(phenylene vinylene) derivatives by processing and chemical modification, *J. Appl. Phys.*, 97, 123703 2005.
92. Tanase, C., Blom, P.W.M., Mulder, M. et al., Enhanced hole transport in poly (*p*-phenylene vinylene) planar metal-polymer-metal devices, *J. Appl. Phys.*, 99, 103702 2006.
93. Tanase, C., Blom, P., De Leeuw, D. et al., Charge carrier density dependence of the hole mobility in poly(*p*-phenylene vinylene), *Phys. Status Solidi A*, 201, 1236–1245 2004.
94. Tanase, C., Meijer, E.J., Blom, P.W.M. et al., Unification of the hole transport in polymeric field-effect transistors and light-emitting diodes, *Phys. Rev. Lett.*, 91, 216601 2003.
95. Bao, Z.N., Kuck, V., Rogers, J.A. et al., Silsesquioxane resins as high-performance solution processible dielectric materials for organic transistor applications, *Adv. Funct. Mater.*, 12, 526–531 2002.
96. Facchetti, A., Yoon, M.H., and Marks, T.J., Gate dielectrics for organic field-effect transistors: New opportunities for organic electronics, *Adv. Mater.*, 17, 1705–1725 2005.
97. Necliudov, P., Shur, M., Gundlach, D. et al., Contact resistance extraction in pentacene thin film transistors, *Solid State Electron.*, 47, 259–262 2003.
98. Peng, X., Horowitz, G., Fichou, D. et al., All-organic thin-film transistors made of alpha-sexithienyl semiconducting and various polymeric insulating layers, *Appl. Phys. Lett.*, 57, 2013–15 1990.
99. Yoon, M.H., Facchetti, A., and Marks, T., σ-π molecular dielectric multilayers for low-voltage organic thin-film transistors, *Proc. Natl. Acad. Sci. U.S.A.*, 102, 4678–4682 2005.
100. Yoon, M.H., Yan, H., Facchetti, A. et al., Low-voltage organic field-effect transistors and inverters enabled by ultrathin cross-linked polymers as gate dielectrics, *J. Am. Chem. Soc.*, 127, 10388–10395 2005.

101. Klauk, H., Halik, M., Zschieschang, U. et al., High-mobility polymer gate dielectric pentacene thin film transistors, *J. Appl. Phys.*, 92, 5259–5263 2002.
102. Hwang, D., Park, J.H., Lee, J. et al., Improving resistance to gate bias stress in pentacene tfts with optimally cured polymer dielectric layers, *J. Electrochem. Soc.*, 153, 23–26 2006.
103. Lee, S., Koo, B., Park, J.G. et al., Development of high-performance organic thin-film transistors for large-area displays, *MRS Bull.*, 31, 455–459 2006.
104. Lee, S., Koo, B., Shin, J. et al., Effects of hydroxyl groups in polymeric dielectrics on organic transistor performance, *Appl. Phys. Lett.*, 88, 162109 2006.
105. Sandberg, H.G.O., Bäcklund, T.G., Österbacka, R. et al., High-performance all-polymer transistor utilizing a hygroscopic insulator, *Adv. Mater.*, 16, 1112 2004.
106. Bäcklund, T., Sandberg, H., Österbacka, R. et al., Current modulation of a hygroscopic insulator organic field-effect transistor, *Appl. Phys. Lett.*, 85, 3887 2004.
107. Sheraw, C., Gundlach, D., and Jackson, T., Spin-on polymer gate dielectric for high performance organic thin film transistors, *Proc. MRS*, 558, 403–408 2000.
108. Chua, L.L., Zaumseil, J., Chang, J.F. et al., General observation of n-type field-effect behaviour in organic semiconductors, *Nature*, 434, 194–199 2005.
109. Marjanovic, N., Singh, T., Dennler, G. et al., Photoresponse of organic field-effect transistors based on conjugated polymer/fullerene blends, *Org. Electron.*, 7, 188–194 2006.
110. Won, S.H., Hur, J.H., Lee, C.B. et al., Hydrogenated amorphous silicon thin-film transistor on plastic with an organic gate insulator, *IEEE Electr. Dev. Lett.*, 25, 132–134 2004.
111. Kahn, A., Koch, N., and Gao, W., Electronic structure and electrical properties of interfaces between metals and π-conjugated molecular films, *J. Polym. Sci., Part B: Polym. Phys.*, 41, 2529–2548 2003.
112. MacDiarmid, A., Synthetic metals: A novel role for organic polymers, *Synth. Met.*, 125, 11–22 2001.
113. Lee, K.S., Blanchet, G.B., Gao, F. et al., Direct patterning of conductive water-soluble polyaniline for thin-film organic electronics, *Appl. Phys. Lett.*, 86, 074102 2005.
114. Xue, F., Liu, Z., Su, Y. et al., Inkjet printed silver source/drain electrodes for low-cost polymer thin film transistors, *Microelectron. Eng.*, 83, 298–302 2006.
115. Bäcklund, T.G., Sandberg, H.G.O., Österbacka, R. et al., A novel method to orient semiconducting polymer films, *Adv. Funct. Mater.*, 15, 1095–1099 2005.
116. Sirringhaus, H., Kawase, T., and Friend, R.H., High-resolution ink-jet printing of all-polymer transistor circuits, *MRS Bull.*, 26, 539–543 2001.
117. Lim, J.A., Cho, J.H., Park, Y.D. et al., Solvent effect of inkjet printed source/drain electrodes on electrical properties of polymer thin-film transistors, *Appl. Phys. Lett.*, 88, 082102 2006.
118. Nilsson, D., Chen, M.X., Kugler, T. et al., Bi-stable and dynamic current modulation in electrochemical organic transistors, *Adv. Mater.*, 14, 51–54 2002.
119. Robinson, N.D., Svensson, P.O., Nilsson, D. et al., On the current saturation observed in electrochemical polymer transistors, *J. Electrochem. Soc.*, 153, 39–44 2006.
120. Lefenfeld, M., Blanchet, G., and Rogers, J., High-performance contacts in plastic transistors and logic gates that use printed electrodes of DNNSA-PANI doped with single-walled carbon nanotubes, *Adv. Mater.*, 15, 1188–1191 2003.
121. Bäcklund, T., Sandberg, H., Österbacka, R. et al., Towards all-polymer field-effect transistors with solution processable materials, *Synth. Met.*, 148, 87–91 2004.
122. Andersson, P., Nilsson, D., Svensson, P.O. et al., Active matrix displays based on all-organic electrochemical smart pixels printed on paper, *Adv. Mater.*, 14, 1460–1464 2002.
123. Andersson, P., Nilsson, D., Svensson, P.O. et al., Organic electrochemical smart pixels, *Proc. MRS*, 736, 289–294 2002.
124. Janata, J. and Josowicz, M., Conducting polymers in electronic chemical sensors, *Nat. Mater.*, 2, 19–24 2003.
125. Bartic, C. and Borghs, G., Organic thin-film transistors as transducers for (bio) analytical applications, *Anal. Bioanal. Chem.*, 384, 354–365 2006.
126. Wang, L., Fine, D., Sharma, D. et al., Nanoscale organic and polymeric field-effect transistors as chemical sensors, *Anal. Bioanal. Chem.*, 384, 310–321 2006.
127. Morana, M., Bret, G., and Brabec, C., Double-gate organic field-effect transistor, *Appl. Phys. Lett.*, 87, 153511 2005.
128. Hayashi, Y., Kanamori, H., Yamada, I. et al., Facile fabrication method for pn -type and ambipolar transport polyphenylenevinylene-based thin-film field-effect transistors by blending c60 fullerene, *Appl. Phys. Lett.*, 86, 052104 2005.
129. Drury, C.J., Mutsaers, C.M.J., Hart, C.M. et al., Low-cost all-polymer integrated circuits, *Appl. Phys. Lett.*, 73, 108–110 1998.

130. Dodabalapur, A., Bao, Z., Makhija, A. et al., Organic smart pixels, *Appl. Phys. Lett.*, 73, 142–144 1998.
131. Katz, H.E., Recent advances in semiconductor performance and printing processes for organic transistor-based electronics, *Chem. Mater.*, 16, 4748–4756 2004.
132. Clemens, W., Fix, W., Ficker, J. et al., From polymer transistors toward printed electronics, *J. Mater. Res.*, 19, 1963–1973 2004.
133. Bao, Z.N., Feng, Y., Dodabalapur, A. et al., High-performance plastic transistors fabricated by printing techniques, *Chem. Mater.*, 9, 1299 1997.
134. de Gans, B., Duineveld, P., and Schubert, U., Inkjet printing of polymers: State of the art and future developments, *Adv. Mater.*, 16, 203–213 2004.
135. Bao, Z., Conducting polymers: Fine printing, *Nat. Mater.*, 3, 137–138 2004.
136. Huisman, B.H., Valeton, J.J.P., Nijssen, W. et al., Oligothiophene-based networks applied for field-effect transistors, *Adv. Mater.*, 15, 2002–2005 2003.
137. Paul, K.E., Wong, W.S., Ready, S.E. et al., Additive jet printing of polymer thin-film transistors, *Appl. Phys. Lett.*, 83, 2070–2072 2003.
138. Shimoda, T. and Kawase, T., All-polymer thin film transistor fabricated by high-resolution ink-jet printing, Digest of Technical Papers—IEEE International Solid-State Circuits Conference, 47, 285–287 2003.
139. Kawase, T., Sirringhaus, H., Friend, R.H. et al., All-polymer thin film transistors fabricated by inkjet printing, *Proc. SPIE*, 4466, 80–88 2001.
140. Kawase, T., Moriya, S., Newsome, C.J. et al., Inkjet printing of polymeric field-effect transistors and its applications, *Jpn. J. Appl. Phys. Part 1*, 44, 3649–3658 2005.
141. Zhang, J., Brazis, P., Chowdhuri, A.R. et al., Investigation of using contact and non-contact printing technologies for organic transistor fabrication, *Proc. MRS*, 725, 155–160 2002.
142. Knobloch, A., Bernds, A., and Clemens, W., An approach towards the printing of polymer circuits, *Proc. MRS*, 736, 277–281 2002.
143. Zielke, D., Hubler, A.C., Hahn, U. et al., Polymer-based organic field-effect transistor using offset printed source/drain structures, *Appl. Phys. Lett.*, 87, 123508 2005.
144. Salleo, A., Wong, W.S., Chabinyc, M.L. et al., Polymer thin-film transistor arrays patterned by stamping, *Adv. Funct. Mater.*, 15, 1105–1110 2005.
145. Chang, Y.T., Liou, C.H., and Wen, C.B., Inkjet revolution, *IEEE Circuits Devices Magazine*, 21, 8–11 2005.
146. Schellekens, J., Burdinski, D., Saalmink, M. et al., Wave printing (ii): Polymer MISFETs using micro-contact printing, *Proc. MRS*, EXS, 21–23 2004.
147. Burns, S.E., Cain, P., Mills, J. et al., Inkjet printing of polymer thin-film transistor circuits, *MRS Bull.*, 28, 829–834 2003.
148. Kawase, T., Shimoda, T., Newsome, C. et al., Inkjet printing of polymer thin film transistors, *Thin Solid Films*, 438–439, 279–287 2003.
149. Wang, J., Gu, J., Zenhausern, F. et al., Low-cost fabrication of submicron all polymer field effect transistors, *Appl. Phys. Lett.*, 88, 133502 2006.
150. Arias, A., Ready, S., Lujan, R. et al., All jet-printed polymer thin-film transistor active-matrix backplanes, *Appl. Phys. Lett.*, 85, 3304–3306 2004.

12 Organic Molecular Light-Emitting Materials and Devices

Franky So and Jianmin Shi

CONTENTS

12.1 Introduction ... 352
12.1.1 Organic Light-Emitting Diodes ... 352
12.1.2 Operation Mechanism of OLEDs ... 352
12.2 Organic Molecular Light-Emitting Device Structures ... 353
12.2.1 Double-Layer OLED Devices ... 354
12.2.2 Multilayer OLED Devices ... 355
12.2.3 Stacked OLED Devices ... 355
12.3 Organic Molecular Light-Emitting Materials ... 356
12.3.1 Basic Physical Property Requirements for Organic Molecular Light-Emitting Materials ... 356
12.3.2 Anodes and Cathodes ... 357
12.3.3 Hole Injection Materials ... 358
12.3.4 Hole Transport Materials ... 360
12.3.5 Electron Transport Materials ... 361
12.3.6 Electron Blocking Materials and Hole Blocking Materials ... 361
12.3.7 Light-Emitting Luminescent Materials ... 363
12.3.7.1 Light-Emitting Host Materials ... 365
12.3.7.2 Light-Emitting Dopants ... 366
12.4 Summary and Conclusion ... 371
Exercise Questions ... 372
Bibliography ... 372

Abstract Organic light-emitting devices (OLEDs) based on molecular materials have been one of the most important organic optoelectronic devices. These devices have now been widely used for commercially mobile displays as well as for flat screen TVs. Because of the rapid progress made in device efficiency and lifetime, OLEDs are now even considered for solid state lighting applications to replace conventional incandescent light bulbs. In this chapter, we will first review the basic device physics, the operation principle, and different device architectures to achieve high efficiency and high brightness. We will then present a detail description of the materials used for OLED fabrication. Specifically, the requirements for electrode materials, carrier transport and injection materials, emitting hosts and dopants will be discussed in detail.

12.1 INTRODUCTION

12.1.1 ORGANIC LIGHT-EMITTING DIODES

Light-emitting diodes (LEDs) are optoelectronic devices, which generate light when they are electrically biased in the forward direction. The early commercial LEDs devices, in 1960s, were based on inorganic semiconductors such as gallium arsenide phosphide (GaAsP) as an emitter and the efficiencies were very low. After 40 years of development, the efficiencies of inorganic LEDs have been significantly improved and they are used in a wide range of applications such as telecommunications, indicator lights, and more recently in solid-state lighting. For flat panel displays, the applications of LEDs have been limited. High-resolution pixelated LED arrays are very expensive to fabricate and the application of LEDs in displays has been limited to bill board displays where individual LEDs are manually mounted on the display boards.

Light-emitting diodes made with organic materials are called organic light-emitting diodes (OLEDs). Before the invention of OLEDs, organic-based devices could be operated only in electroluminescence mode. The first organic electroluminescence device was demonstrated in the 1950s, and very high operating voltages were required. These devices were made with anthracene single crystals doped with tetracene (a blue-emitting fluorescence dye) sandwiching between two electrodes. Very high drive voltages were required and the efficiencies were very low. In the 1980s, a major breakthrough was made. Low-voltage OLEDS were demonstrated. In contrast to the first electroluminescent devices, the new OLED devices were based on a multilayer structure and they consisted of a transparent anode, a hole transporting layer, an electron/emitting layer, and a cathode. During operation, electrons and holes are injected from a cathode and an anode, respectively, and recombination of electrons and holes leads to efficient light generation. The operation principle of OLEDs is similar to that of LEDs.

12.1.2 OPERATION MECHANISM OF OLEDs

When an electric field is applied to the electroluminescent layer of an OLED,

1. Holes are injected from anode into the electroluminescence layer, the ground-state molecule A will be excited and a radical cation is formed:

$$A \xrightarrow{-e} A^{\overset{\oplus}{\cdot}}$$

2. Holes are transported toward the cathode:

$$A^{\overset{\oplus}{\cdot}} + A \longrightarrow A + A^{\overset{\oplus}{\cdot}}$$

3. Electrons are injected from the cathode into the electroluminescence layer and a radical anion A^- is formed:

$$A \xrightarrow{+e} A^{\overset{\ominus}{\cdot}}$$

4. Electrons are transported toward the anode:

$$A^{\overset{\ominus}{\cdot}} + A \longrightarrow A + A^{\overset{\ominus}{\cdot}}$$

5. Electrons and holes recombine to form excitons A^*:

$$A^{\oplus}\cdot + A^{\ominus}\cdot \longrightarrow A^* + A$$

6. Excitons relax to ground state and emit a photon:

$$A^* \longrightarrow A + h\nu$$

Schematic representation of the operation of an OLED is shown in Figure 12.1. Efficient hole injection from an anode and efficient electron injection from a cathode are key to obtaining high-efficiency devices. In inorganic semiconductors, carrier injection is achieved by heavy doping the semiconductors (n- or p-type) at the contacts to allow tunneling of carriers through the barriers. In organic semiconductors, the strategy to obtain efficient hole and electron injection is to match the work function of an electrode with either the highest occupied molecular orbital (HOMO) level of an organic semiconductor for hole injection or the lowest unoccupied molecular orbital (LUMO) level for electron injection. While this strategy works for hole injection because of the good match of many metal work functions with the HOMO energies of organic materials, highly reactive low work function metals are required for electron injection contacts. This is problematic because low work function metals are highly reactive and prone to corrosion due to moisture and oxygen.

12.2 ORGANIC MOLECULAR LIGHT-EMITTING DEVICE STRUCTURES

In order to maximize the OLED device efficiency, balance of electron and hole transport is important. Light generation requires recombination of both types of carriers and dominance of electron or hole current will lead to nonradiative recombination. To achieve balance of carrier transport, it requires both balanced injection and balanced transport of both types of carriers. Transport of carriers is determined by carrier mobility, a parameter describing how fast carriers are moving in a material upon application of an electric field. Ideally, both types of carriers are injected efficiently into the electroluminescent layer and equal number of electrons and holes recombine in the middle of the emitting layer to give off light.

However, in most cases carrier transport in organic materials is highly imbalance, namely, the carrier mobilities of electrons and holes are very different. In many organic materials, they either preferentially transport electrons or holes. Imbalance transport will result in electron–hole recombination at either the

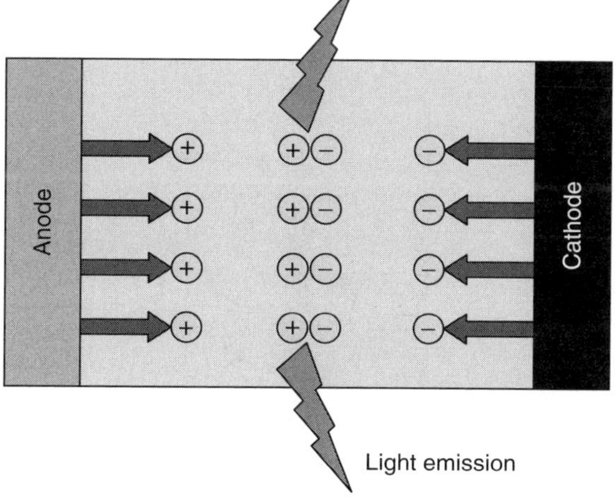

FIGURE 12.1 Operation of an OLED.

cathode or anode interface, leading to quenching (or nonradiative recombination) and significantly reduction in overall device efficiency. In addition to transport, charge injection also plays an important role in determining the charge balance. Efficient hole injection from an anode requires good matching of its function to the HOMO level of the organic layer and efficient electron injection from the cathode requires good matching of its work function to the LUMO energy of the organic layer. If this requirement is not met, this will cause low carrier injection efficiencies and high drive voltages.

12.2.1 Double-Layer OLED Devices

In the 1980s, Tang and Van Slyke at the Eastman Kodak Company made a breakthrough in OLEDs. For the first time, they fabricated an OLED with two organic layers sandwiched between two electrodes. The OLED device consisted of a transparent indium-tin-oxide (ITO) on glass as a transparent anode, a diamine layer as a hole transporting layer, a tris(hydroxyquinoline aluminum) (Alq_3) as an electron transporting as well as an emitting layer, and a magnesium–silver alloy layer as a cathode. The ITO layer has a work function close to the HOMO energy of the diamine layer for efficient hole injection while the magnesium–silver alloy has a work function close to the LUMO level of Alq_3 for efficient electron injection. Compared to the first organic electroluminescence devices made in the 1960s, the double-layer device was more efficient and operated at a much lower voltage. The improved device performance was due to three key factors. First, work functions of the cathode and anode used in the devices match the LUMO of the electron transporting layer and the HOMO of the hole transporting layer, respectively. This is important for efficient carrier injection. Second, diamine has high hole mobility for hole transport, while Alq_3 has a relatively high electron mobility for electron transport. This way, a more balanced carrier transport is achieved, resulting in high device efficiency. Third, the total layer thickness was kept below 200 nm to reduce the operating voltage. In the original anthracene electroluminescence devices, bulk crystals were used, resulting in operating voltages larger than 100 V. By reducing layer thickness to less than 200 nm, the device operating voltage was also reduced to less than 10 V.

Figure 12.2 shows the structure of a double-layer OLED. In this device, holes are injected from the ITO electrode through the diamine layer to the Alq_3 layer while electrons are injected from the magnesium–silver cathode into the Alq_3 layer. Since the electron mobility in Alq_3 is much larger than the hole mobility, electrons and holes meet at the diamine/Alq_3 interface and recombine to give off green light.

The combination of the device architecture and the properly chosen organic materials and electrodes enabled the fabrication of low voltage OLEDs, which allows this technology to be used for commercial products. The concept of separating the transport of holes and electrons in two or more layers served as a platform for more sophisticated OLED device and material design.

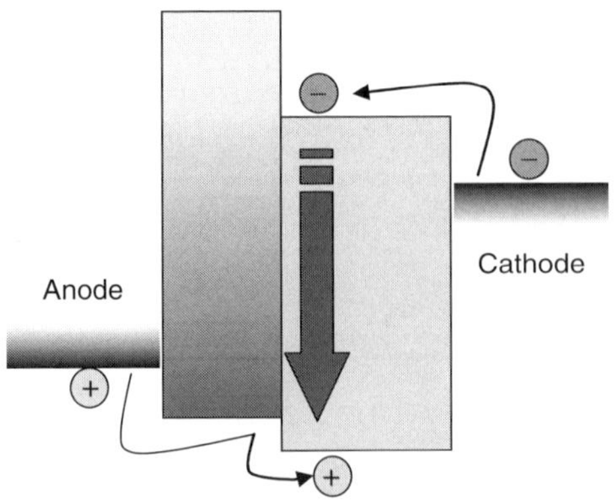

FIGURE 12.2 Double-layer OLED.

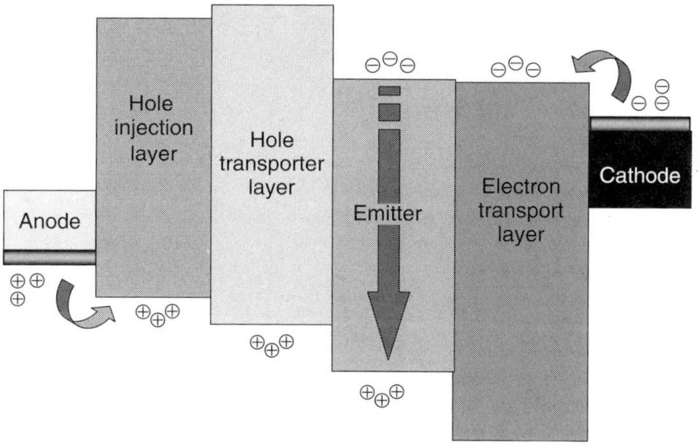

FIGURE 12.3 Multilayer OLED device structure.

12.2.2 Multilayer OLED Devices

The original two-layer OLED device served as a basic building block for more sophisticated devices with optimum device performance. Depending on the emitting materials used in the device, a device consisting of two organic layers might not give optimum device performance and more layers are necessary. To facilitate carrier injection, carrier injection layers with proper energy alignment with the injection electrodes are necessary. Specifically, an electron injection layer (EIL) with its LUMO energy matching the work function of the cathode is needed, while a hole injection layer (HIL) with its HOMO energy matching the work function of the transparent anode is needed. To transport the injected carriers from the injection layer to the emitting layer, electron and hole transport layers are needed. Ideally, the electron transport layer should have high electron mobility and the hole transport layer should have a high hole mobility. In addition, these transport layers have a large energy gap in such a way that they provide a path for one type of carrier transport, while acting as a blocker for the other type. For example, the hole transport layer should allow the transport of holes while simultaneously blocking the transport of electrons. This can be achieved by choosing a wide gap hole transport material with its HOMO-level matching the HOMO level of the HIL. Since it is a wide gap material, its LUMO energy is small and it provides an energy barrier for transport of electrons. Today, an OLED device consisting of five layers or more is common. In some cases, a single transport layer serves the purpose of both carrier transport and injection layers, hence the total number of layers is reduced. A multilayer OLED device consists of an anode, a HIL, a hole transporting layer, an emitting layer, an electron transporting layer, and a cathode (Figure 12.3).

The light-emitting color can be tuned depending on the emitter materials used in the devices. During operation, holes from the anode are injected into the emitting layer through the hole transporting layer, and electrons from the cathode are injected into the emitting layer through electron transport layer. Subsequently, electrons and holes recombine to form excitons in the emitting layer and the excitons return to ground state to give off light.

12.2.3 Stacked OLED Devices

In recent years, because of the progress made in device efficiencies, OLED devices are considered for solid-state lighting applications. For light applications, the brightness required is many times larger than that of a typical flat panel display. Under such conditions, OLED devices need to be operated at very high current densities. There are two problems associated with OLED devices

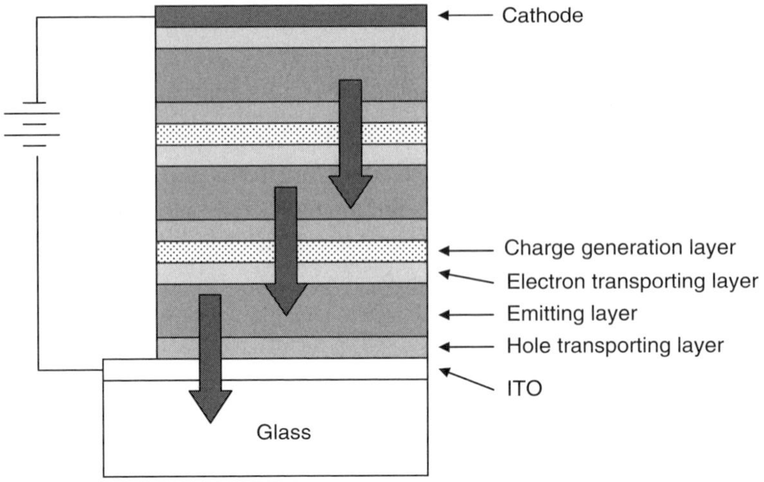

FIGURE 12.4 A three-cell stack OLED.

operating at high current densities. First, devices operating at high current densities degrade at a rate much higher than that for devices operating at low current densities. Second, as the current density increases, the device operating voltage will also increase, leading to substantial decrease in power efficiency. To alleviate these problems, multiple OLED devices can be fabricated within one stack. This is called stacked OLEDs or tandem OLED cells. Essentially, the device simply consists of multiple OLEDs connected in series within one stack. Compared to a single cell device, each cell in the stacked OLED only needs to operate at a fraction of the voltage and current density level required. Each OLED cell in the stacked OLED device independently emits light of different color or same color. For lighting applications where white light emission is required, red, green, and blue cells are fabricated within one stack resulting in white light emission. Since the stack consists of several OLED cells, the total stack operating voltage will be higher depending on the number of cells in the stack. A typical stacked OLED is shown in Figure 12.4. The key to this device architecture is to have a transparent cathode in a single cell. Early stacked OLEDs were made with a semitransparent cathode. However, even an ultrathin metal cathode has very limited optical transmission and using thin metal cathode is not practical. To get around the problem, transparent metal oxides are used to supply electrons and holes to two opposite cells, and they are called charge generation layer. It has been found that some of the metal oxides can be effective injection holes into the HIL of one OLED cell while it can also be effective injection electrons into the EIL of another OLED cell. With this configuration, the device brightness and the operating lifetime will be substantially improved. For example, OLEDs emitting at 5000 nits can have lifetime exceeding several thousand hours.

12.3 ORGANIC MOLECULAR LIGHT-EMITTING MATERIALS

12.3.1 Basic Physical Property Requirements for Organic Molecular Light-Emitting Materials

The multilayer OLED devices are made with small molecular materials and they are fabricated by vacuum vapor deposition process. Most small molecule organic materials used for OLED applications have sublimation or evaporation temperatures ranging from 200°C to 500°C. Typical molecular OLEDs are fabricated in a multisource vacuum evaporator as shown in Figure 12.5.

In order to be used for OLED fabrication process, there are two requirements for the organic materials:

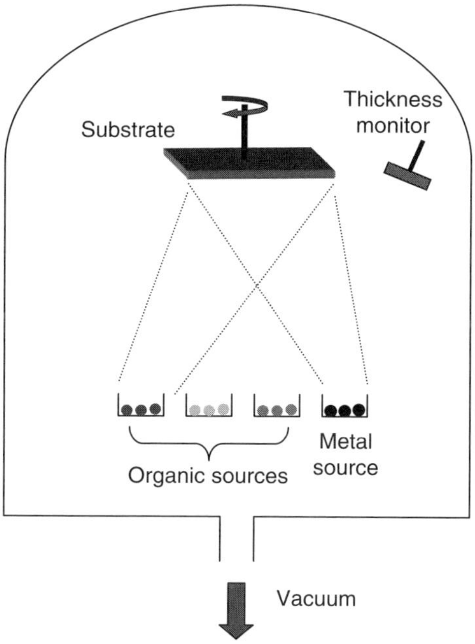

FIGURE 12.5 Multisource evaporator for OLED fabrication.

1. They should have good thermal stability and can be sublimed under vacuum without decomposition. Materials decompose before reaching the sublimation temperature cannot be used for OLED fabrication.
2. These organic materials should form uniform thin films by vacuum vapor deposition process. In most cases, amorphous materials will be desirable for OLED fabrication because amorphous materials tend to form homogenous thin films. In some cases, hole injection materials such as copper phthalocyanine are polycrystalline with very fine grain size. These polycrystalline films do have good enough film uniformity for device fabrication.

12.3.2 Anodes and Cathodes

In most cases, the anode in OLEDs is typically a high work function transparent conducting material. The most commonly used transparent thin-film anode is ITO. Thin films of ITO can be deposited by radio frequency sputtering or electron-beam evaporation on transparent substrates, such as glass or plastics. In addition to ITO, many other conducting metal oxides such as $GaInO_3$ and ZnInSnO have also been used as anode material. Conducting polymers such as polyaniline and poly(3,4-ethylenedioxythiophene) poly(styrenesulfonate) (PEDOT:PSS) have also been used as a hole injecting electrode. These materials tend to have better hole injection properties compared to bare ITO. In most cases, these materials are deposited on the top of ITO as a HIL rather than a stand-alone electrode. Successful development of the organic transparent conductors will enable the realization of many areas in organic electronics and flexible electronics.

The cathode or the electron injecting electrode is generally a thin film of low work function metal or metallic alloy. Low work function is required to facilitate efficient electron injection into the LUMO of the EIL. The cathode material used in the Tang's devices was an alloy of magnesium and silver. Later on, other low work function metals such as lithium and cesium have been used for the cathode. Recently, composite cathodes consisting of alkaline halides, such as LiF and CsF, as an interface layer with aluminum electrode have also been used for OLED cathodes and these cathodes are most popular among OLED researchers. The commonly used cathodes in OLED devices are listed in Table 12.1.

TABLE 12.1
Low Work Function Metals in OLED Devices

Electrodes	Metal or Alloys	Work Function (eV)
Anode	ITO	4.6
Cathode	Al	4.28
Cathode	Ag	4.26
Cathode	In	4.12
Cathode	Mg	3.66
Cathode	Ca	2.87
Cathode	Li	2.90
Cathode	Cs	2.14

For efficient carrier injection, the work function of the injection electrode has to match the HOMO or the LUMO level of the transport layer. In addition to matching the electrode work function, the electrode/organic interfaces play a very important role in determining the device performance. For example, surface treatment of ITO can strongly affect its surface work function, and hence has a strong effect on the device operating voltage and efficiency.

Most OLED devices are made with the top metal electrode and light is emitted via the bottom transparent electrode. In some active matrix OLED displays where each OLED pixel is driven with a thin-film transistor, and it is desirable to have light emission via the top electrode and a top transparent electrode. Such devices are called top-emitting OLEDs. Top-emitting OLEDs are used in active matrix displays or microdisplays where a large pixel fill factor is required and a top-emitting OLED allows fabrication of an entire OLED on the top of a thin-film transistor. The most common way to fabricate a transparent cathode is to deposit a very thin layer of cathode metal (<10 nm) and overcoat it with a thin film of ITO. Combining a transparent cathode and a transparent anode, a total transparent OLED can also be realized. In this case, light emits from both the anode and the cathode sides of the device.

12.3.3 Hole Injection Materials

In many OLED devices, matching the work function of the injection electrode and the HOMO or LUMO levels of the transport layer is not possible. For example, the work function of ITO is between 4.6 and 4.8 eV, while the HOMO level of a common hole transporting amine is about 5.5 eV. Hole injection across a 0.7–0.9 eV barrier is not efficient. To improve the carrier injection efficiency, a buffer HIL can be inserted between the electrode and the carrier transport layers.

For example, the HOMO level of HIL should be between the anode work function and the hole transport layer HOMO level. Likewise, the LUMO level of the EIL should be between the cathode work function and the electron transport layer LUMO level.

Another approach to improve the charge injection efficiency is to dope the charge transport layer with either a p- or an n-dopant. This is a common approach in inorganic semiconductor devices where ohmic contacts are made by heavily doping the contact region with p- or n-dopant. Heavily doped transport layers serve two purposes. First, doping reduces the width of the contact barrier and hence carriers can easily be injected by tunneling. Second, doping can substantially increase the conductivity of the transport layer and hence the operating voltage can be reduced. A device with a p-doped hole transport layer and an n-doped electron transport layer is called a PIN device.

Table 12.2 lists the commonly used hole- and electron-injecting materials in OLED devices using ITO as anode and Mg/Ag as cathode.

TABLE 12.2
Hole Injection Materials

Materials	Processes
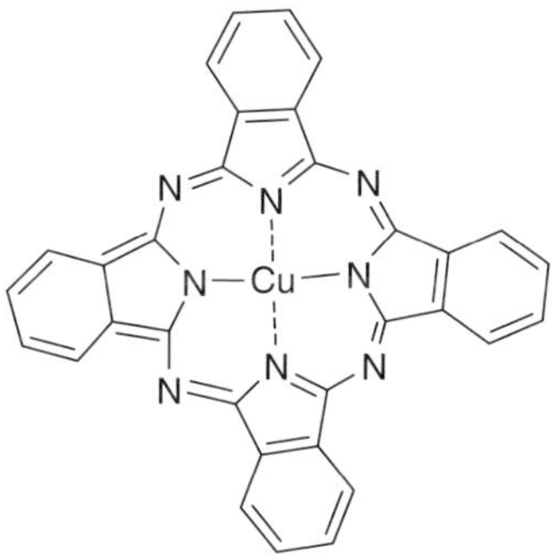 Copper phthalocyanine (CuPc)	Vacuum vapor deposition
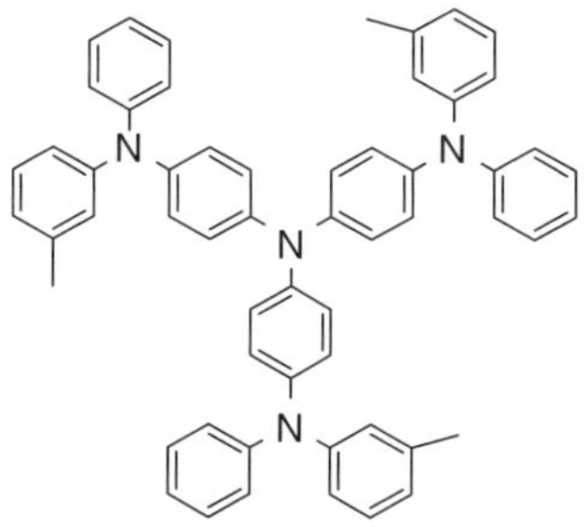 4,4′,4″-tris[3-methylphenyl(phenyl)amino]triphenylamine	Vacuum vapor deposition
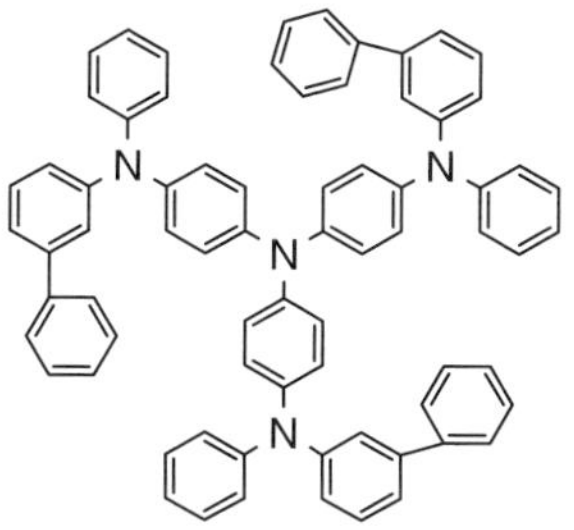 4,4′,4″-tris[biphenyl-3-yl-(phenyl)amino]triphenylamine	Vacuum vapor deposition
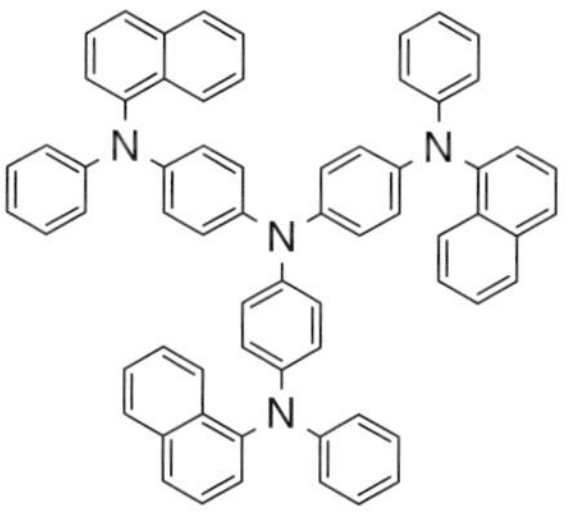 4,4′,4″-tris[1-naphthyl(phenyl)amino]triphenylamine	Vacuum vapor deposition

(continued)

TABLE 12.2 (continued)
Hole Injection Materials

Materials	Processes
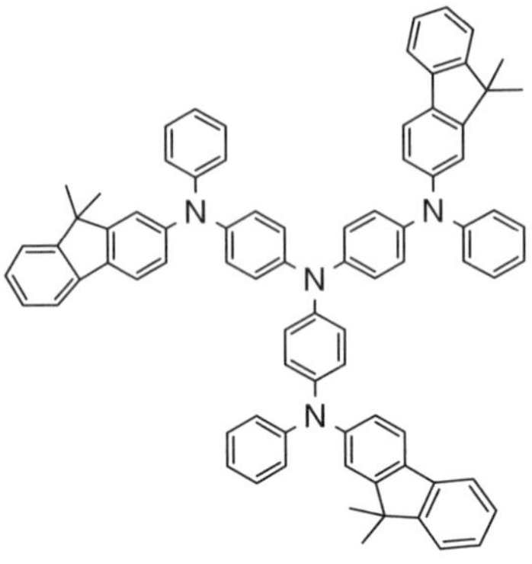 4,4′,4″-tris[9,9-dimethyl-2-fluorenyl(phenyl)amino]triphenylamine	Vacuum vapor deposition
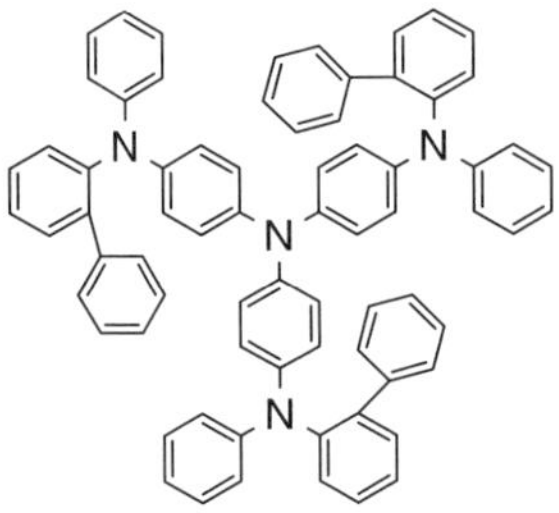 4,4′,4″-tris[biphenyl-2-yl-(phenyl)amino]triphenylamine	Vacuum vapor deposition
CFx polymer	CHF_3 gas plasma
Poly(3,4ethylenedioxythiophene)Poly(styrenesulfonate) (PEDOT/PSS)	Solution process

12.3.4 Hole Transport Materials

The hole transport layer in an OLED device serves two purposes. First, it provides a path for hole-injected materials from the electrode to be injected into the emitting layer. Second, it also serves as an electron blocker to confine electrons within the emitting layer. Numerous materials have been developed for hole transport layers in OLED devices. Most of these hole transport materials, which were originally developed for charge transport layer in xerography, are arylamine derivatives. These arylamine derivatives used in xerography have high hole mobilities, typically on the order of 10^{-4} cm^2/Vs. In xerography, they are usually blended with a polymeric binder and thin films of the polymer blend are used as the charge transport layers in a photoconductive system.

However, in addition to high mobility, the following physical and chemical properties are important for hole transport materials used in OLED devices:

1. They form uniform pinhole free glassy films by vacuum deposition. Film uniformity is important as it affects the carrier transport properties of the films.
2. They should have low ionization potential. The low ionization potential enables efficient hole injection from the HIL to the hole transport layer.
3. They should have wide optical band gaps. The wide optical band gap serves the following purposes:

i. Wide optical band gap hole transport materials have small LUMO energy levels and serve as electron blocking layers, thus confining the injected carriers within the emitting layer and leading to radiative recombination.
ii. Wide optical band gap hole transport materials can be used as a host for light-emitting layer. This approach has been used to generate white light emission from two light-emitting layers in OLED devices.

4. They are expected to have a high glass transition temperature (T_g). Specifically, it is desirable to have a T_g higher than 100°C. While high glass transition temperature might not be necessary for high electroluminescence (EL) efficiency and good operational stability under low temperature environment, it is critical for OLED devices operated under elevated temperature conditions.

Table 12.3 provides a list of commonly used hole transport materials in OLED devices.

12.3.5 Electron Transport Materials

In an OLED device, the electron transport layer is between the light-emitting layer and the cathode. It is responsible for transporting the injected electrons from the cathode to the emitting layer. It also serves as a hole blocker to confine holes in the emitting layer. In contrast to hole transporting materials, very little is known about the electron transporting properties of organic materials before the invention of OLEDs. So far, most of the high mobility electron transport materials have been developed exclusively for OLED devices. In addition of high electron mobility, the following physical and chemical properties are important for electron transport materials:

1. Similar to hole transport materials, they should form uniform pinhole free glassy film.
2. They have to have high electron affinity. The high electron affinity enables efficient electron injection from the cathode to the electron transport layer.
3. They should have a wide optical band gap. The wide optical band gap property provides the following characteristics for OLED devices:
   i. Minimize optical loss due to light absorption of electron transport layer.
   ii. Wide optical band gap electron transport materials have deep HOMO energy level and it can block holes from entering into the electron transport layer, thus effectively confine carriers within the emitting layer.
4. They are expected to have a high T_g for temperature stability.

Some common electron transport materials used in OLED devices are listed in Table 12.4.

Among these electron transport materials, the most important one was tris(8-hydroxyquinolinato) aluminum, an aluminum chelate referred as Alq_3. Alq_3 has electron mobility about 10^{-6} cm^2/Vs with a HOMO energy level range from 5.7 to 5.9 eV and a LUMO energy level range from 3.0 to 3.2 eV depending on the specific measurement techniques. This is important because its LUMO level is not too shallow and metals such as magnesium or magnesium alloys work effectively as a cathode in Alq_3-based OLED devices. Alq_3 was found to be not only a good electron transport material but also a good host material for green and red emitters.

12.3.6 Electron Blocking Materials and Hole Blocking Materials

In heterostructure OLEDs, carrier confinement is important to achieve high efficiency. This is done by using wide gap hole and electron transport layers sandwiching the emitting layer. These carrier transport layers allow transport of electrons (or holes) and simultaneously block transport of holes (or electrons).

The following are the commonly used hole blocking and electron blocking materials in OLED devices (Table 12.5).

TABLE 12.3
Hole Transport Materials

Materials	HOMO (eV)	LUMO (eV)	T_g (°C)
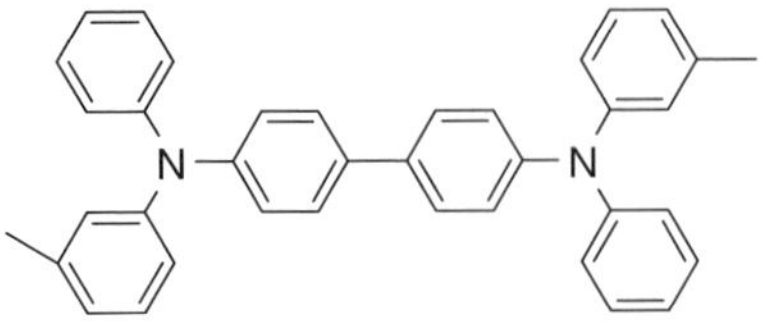 *N,N'*-bis(3-methylphenyl)-*N,N'*-diphenyl-[1,1'-biphenyl]-4,4'-diamine	5.5	2.4	62
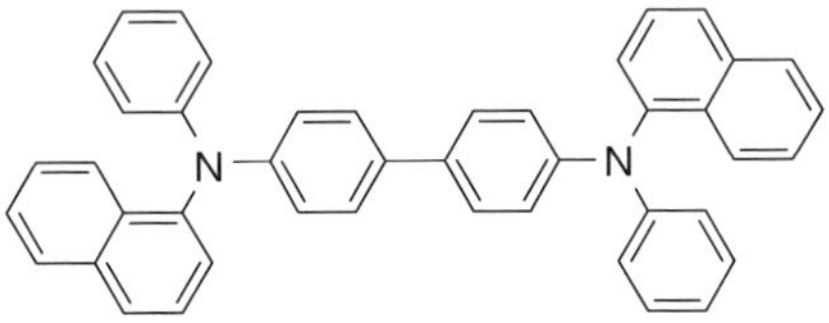 *N,N'*-bis(biphenyl-3-yl)-*N,N'*-diphenyl-[1,1'-biphenyl]-4,4'-diamine	5.5	2.4	98
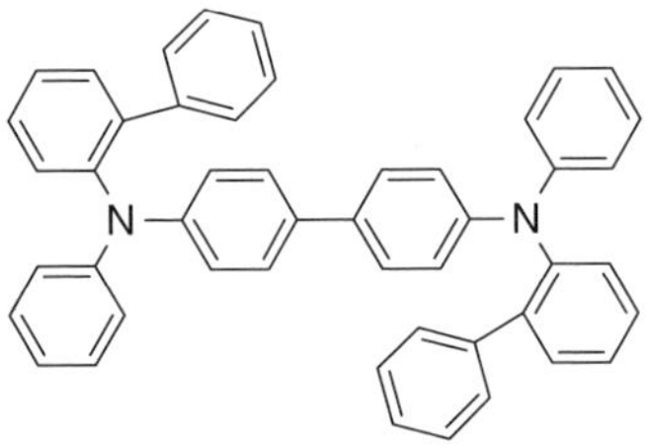 *N,N'*-bis(biphenyl-2-yl)-*N,N'*-diphenyl-[1,1'-biphenyl]-4,4'-diamine	5.5	2.4	75
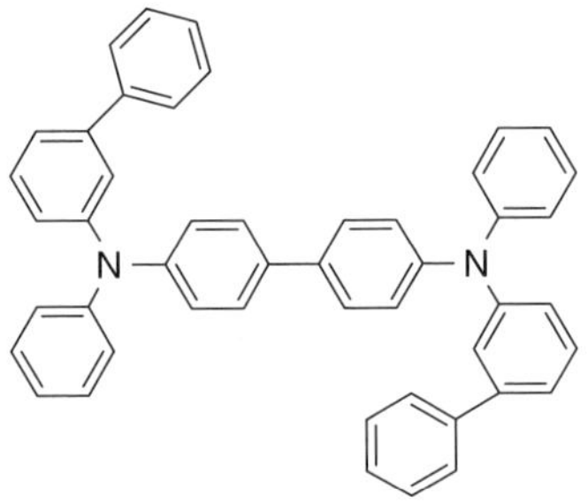 *N,N'*-bis(1-naphthyl)-*N,N'*-diphenyl-[1,1'-biphenyl]-4,4'-diamine	5.5	2.4	81
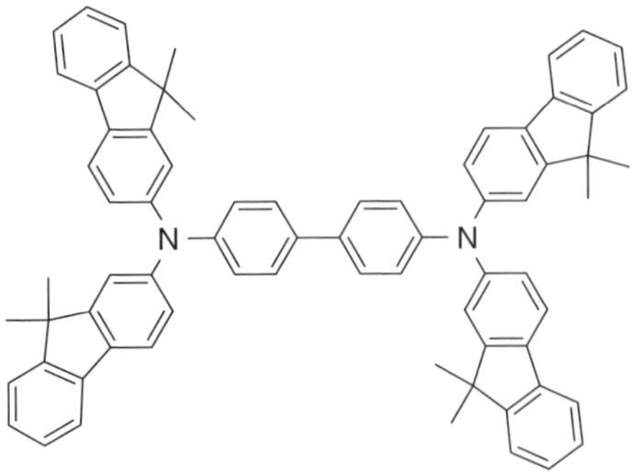 *N,N'*-bis(9,9-dimethyl-2-fluorenyl)-*N,N'*-diphenyl-[1,1'-biphenyl]-4,4'-diamine	5.4	2.4	165

TABLE 12.4
Typical Electron Transport Materials

Electron Transport Materials	HOMO (eV)	LUMO (eV)
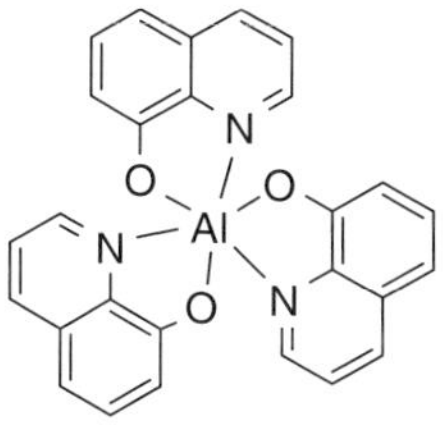 Tris(8-hydroxyquinolinato)aluminum	5.8	3.0
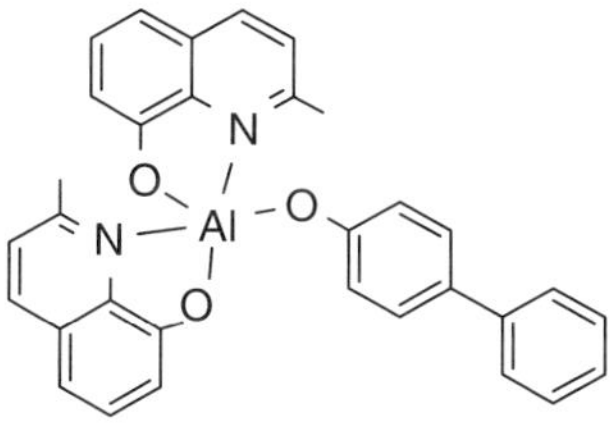 Bis(8-hydroxy-2-methylquinolinato)-(4-phenylphenolato)aluminum	5.7	3.2
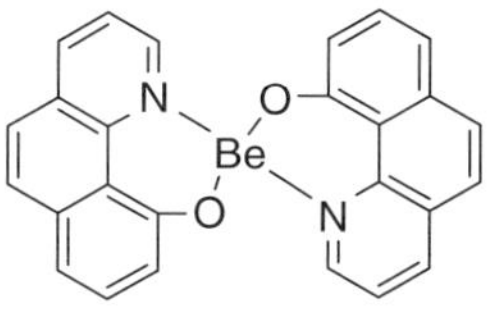 Bis(10-hydroxybenzo quinolinato)beryllium	5.5	2.9
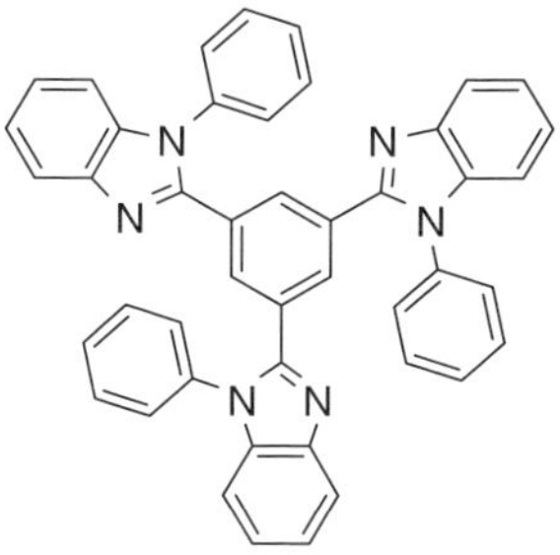 1,3,5-Tris(*N*-phenylbenzimidazol-2-yl)benzene	6.2	2.7
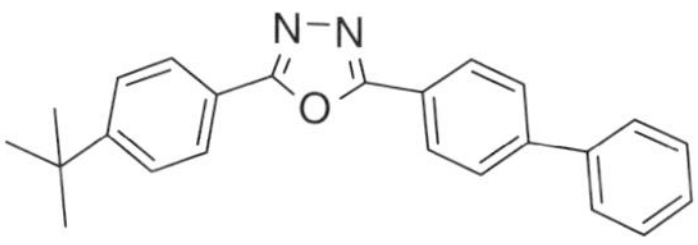 2-(4′-tert-Butylphenyl)-5-(4″-biphenylyl)-1,3,4-oxadiazole	6.2	2.3

12.3.7 Light-Emitting Luminescent Materials

Many isolated organic dye molecules exhibit very high fluorescence quantum yields. For example, when low concentrations of dye molecules are present in solutions, fluorescence yield can be close to 100%. However, when they exist in a condense phase, aggregates of organic fluorescence quantum yield can be drastically reduced due to formation of exciplexes.

Because of this reason, fluorescence dye molecules are doped into a host matrix in an OLED device. Doping can be achieved by co-evaporating the dopant with the host and typical doping

TABLE 12.5
Electron Blocking and Hole Blocking Materials

Materials	Function	HOMO (eV)	LUMO (eV)
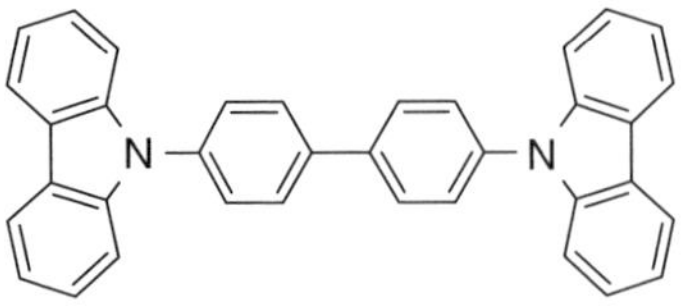	Electron blocking	5.8	3.5
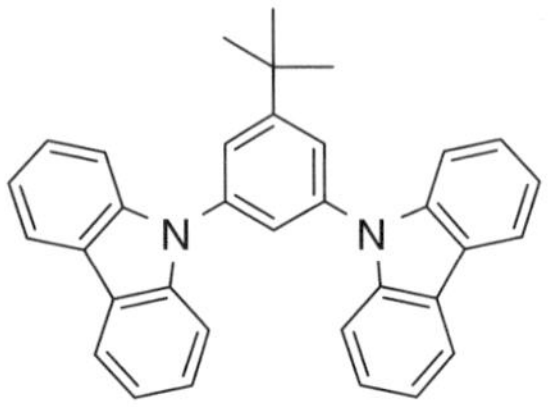	Electron blocking	5.9	2.6
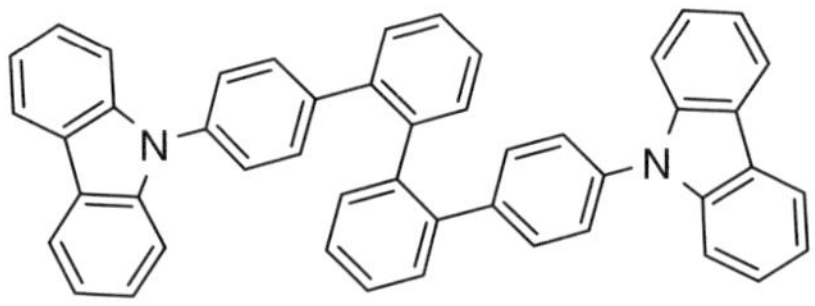	Electron blocking	6.1	2.6
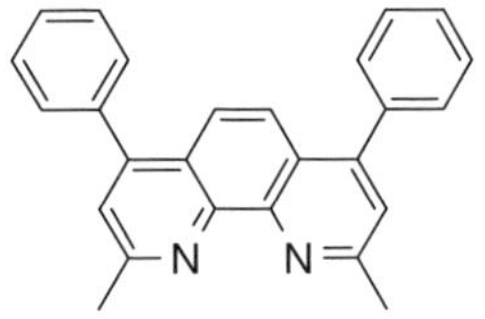	Hole blocking	6.5	3.0
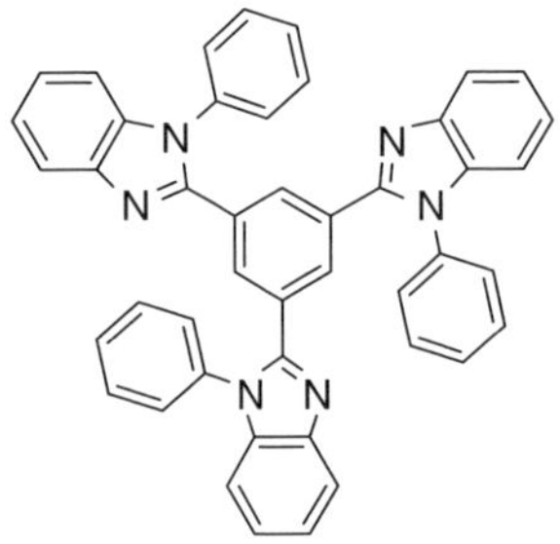	Hole blocking	6.2	2.7
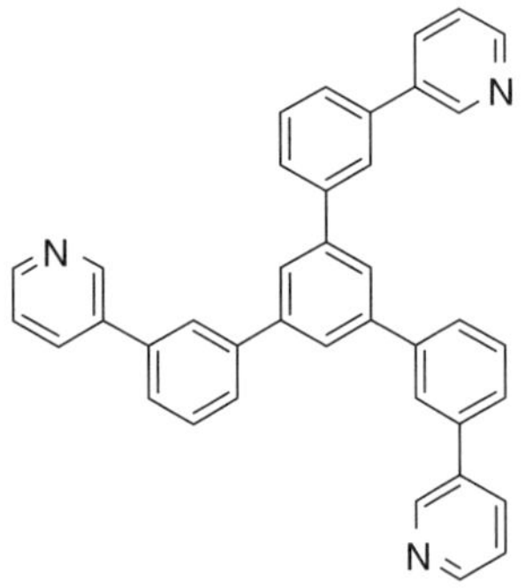	Hole blocking	6.7	2.7

concentrations of the fluorescent dye vary from a fraction of a percent to a few percents. Emission from dopant molecule can be obtained by two possible pathways. The first pathway is that electrons and holes recombine at host molecule sites and form excited states at the host molecules. The excited state (or exciton) energies are then transferred to the dopant molecules and excitons are then formed at the dopant molecular sites. This process is called Förster energy transfer. This way, emission is due to excited states of dopant molecules and independent of the host molecules. This process allows tuning of the emission wavelength without changing the host materials. Another pathway is that electrons and holes are directly injected into the dopant molecules and recombination occurs at the dopant molecular sites. Molecular doping enables the same host materials used with different color dopant molecules. For example, Alq_3 is a green emitter. It is also commonly used as a host material for green, yellow, orange, and red emitter dopants. This greatly simplifies the molecular design.

12.3.7.1 Light-Emitting Host Materials

Light-emitting host materials should be able to transport electrons and holes. In addition, the following properties are also required for light-emitting host materials:

1. Good thin-film forming properties
2. High luminescence efficiency
3. Wide optical band gap compared to emitter dopants
4. Minimum interaction of its ground state and excited state with carrier transport layers

Many electron transport materials are effective light-emitting host materials provided that they are not the luminescent quenchers. For example, electron transport materials such as Alq_3 and some derivative of *N*-substituted imidzole have been used as host materials in OLED devices.

Some hole transport materials can also be used for light-emitting host materials. For example, *N,N′*-Bis(naphthalen-1-yl)-*N,N′*-bis(phenyl)benzidine (NPB), one of the most commonly used hole transport materials can be doped with rubrene to emit yellow light. In the case of phosphorescent devices, carbazole derivative 4,4′-Bis(carbazol-9-yl)biphenyl (CBP) can be doped with iridium complexes to emit green or red light (Table 12.6).

TABLE 12.6
Green and Red Fluorescent Host Materials

Host Materials	Dopant Emitting Colors
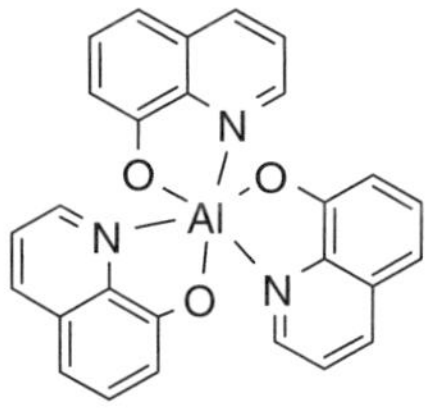 Alq_3	Green and red
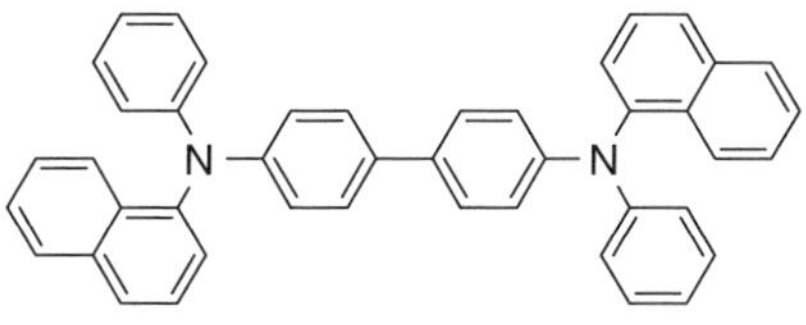 NPB	Green and red

TABLE 12.7
Blue Light-Emitting Host Materials

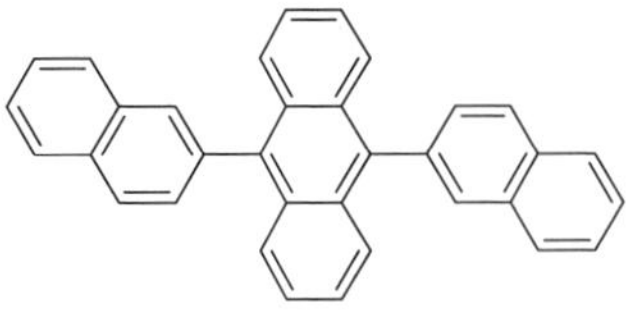

9,10-Bis(2-naphthyl)anthracene

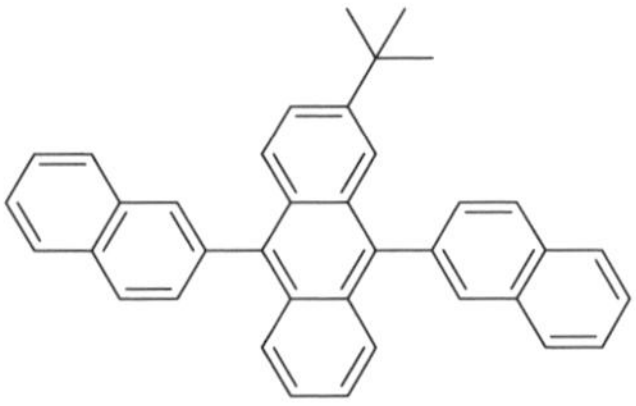

2-*tert*-Butyl-9,10-bis(2-naphthyl)anthracene

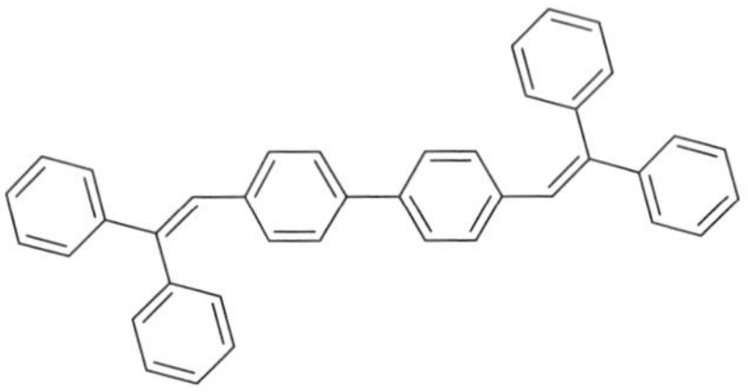

4,4′-Bis(2,2-diphenylethenyl)-1,1′-biphenyl

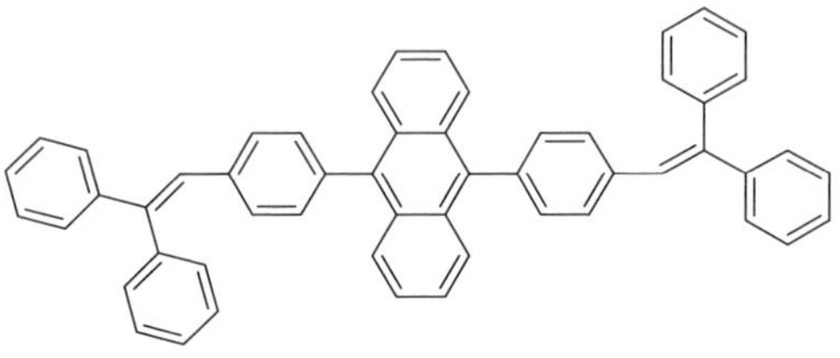

9,10-Bis[4-(2,2-diphenylethenyl)phenyl]anthracene

High electroluminescence efficiency requires efficient electron and hole injection into the host materials. To achieve efficient carrier injection, the host materials should have large electron affinity and low ionization potential. These requirements have two drawbacks: first, these materials usually do not have enough wide energy gap for blue-emitting dopants. Second, if the light-emitting hosts are polar molecules, then it has the possibility to form charge complexes at the interfaces with its neighboring carrier transporting layers. It is desirable to have nonpolar wide energy gap materials as the light-emitting hosts. The example of this class of materials is typically aromatic hydrocarbons, such as anthracene derivatives. These organic materials have small polarity and small Stoke shift due to its rigid molecular structure, and typically efficient blue emission can be achieved if they are doped with blue fluorescent emitters (Table 12.7).

12.3.7.2 Light-Emitting Dopants

Light-emitting dopant materials play an important role in the performance of the OLED device. Light-emitting dopant materials affect device efficiency and operational stability. The device emission color can be tuned to emit light covering the entire visible spectrum. The following are the requirements for light-emitting dopant materials:

1. Light-emitting dopant materials should have high luminescent efficiencies.
2. The absorption spectrum of light-emitting dopant materials should have a strong overlap with the emission spectrum of the light-emitting host materials to allow efficient energy transfer.
3. Light-emitting dopant materials should have relatively narrow emission bandwidth to maintain chromatic purity.

12.3.7.2.1 Fluorescent Dopant Materials

Relatively, high-efficiency green-emitting dopant molecules are more readily available. One of the most common high-efficiency green-fluorescent molecules used is the coumarin derivatives. Among all color emitters, green emitters have the best luminous efficiency and operational stability. Some of these coumarin derivatives with their emission wavelength in doped Alq_3 are summarized as follows (Table 12.8).

TABLE 12.8
Green-Fluorescent Dopants

Dopants	Emission Wavelengths (nm)
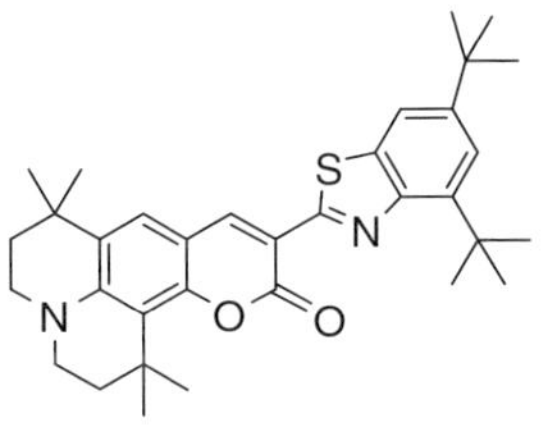 C-545TB	545
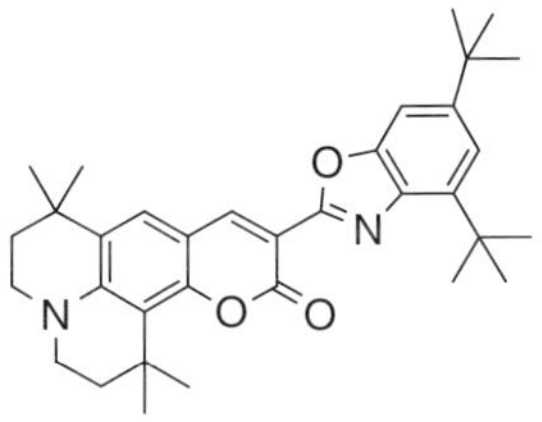 C-525TB	525
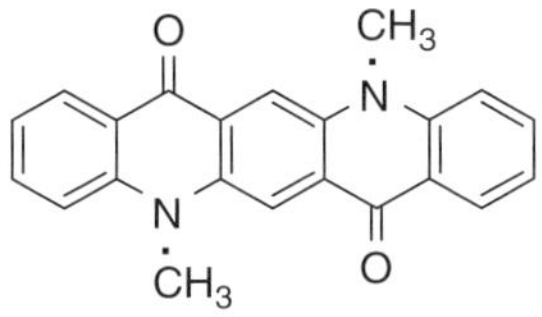 DMQA	540
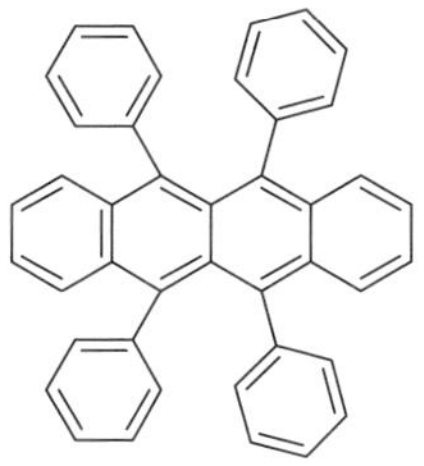 Rubrene	560

While saturated green-fluorescent materials are readily available for OLEDs, saturated red emitters are not very common. Many fluorescent dyes have peak wavelength between 630 and 650 nm, they usually accompany with a long tail extending into the yellow region of the spectrum, leading to nonsaturated pale orange emission. The most useful red fluorescent light-emitting dopant materials for OLED devices are derivatives of 4-dicyanomethylene-6-dialkylaminostyryl-4H-pyran. It has been found that the emission wavelength and EL efficiency of this class of fluorescent molecules in OLED devices are highly concentration dependent. Increasing the concentration of the light-emitting dopant molecules in Alq_3 light-emitting host material will redshift the EL emission spectrum, leading to a more saturated red color. However, the EL efficiency was significantly reduced due to dopant material concentration quenching. This problem has been partially solved by modifying the molecular structure by introducing more bulky group on pyran ring to minimize concentration quenching. Some of the red compounds and their emission wavelengths are shown in Table 12.9.

It is most challenging to achieve high efficient blue light-emitting OLED devices with high operational stability. The reason is that high energy excitons in large gap materials are more reactive

TABLE 12.9
Red Fluorescent Dopants

Red Fluorescent Dopants	Emission Wavelengths (nm)
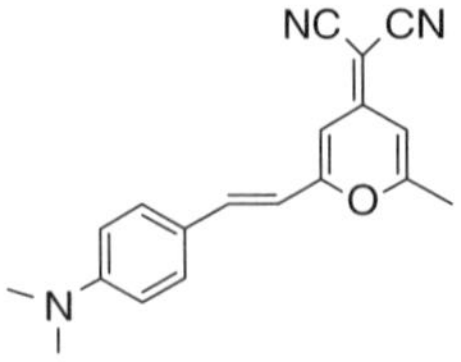	585
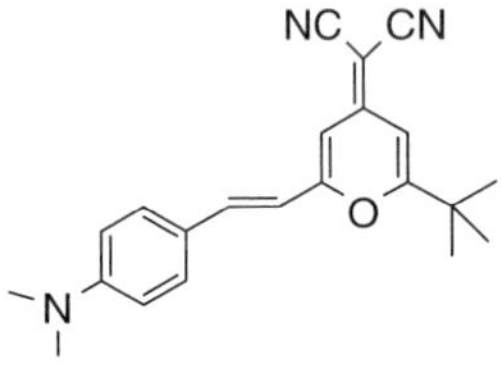	583
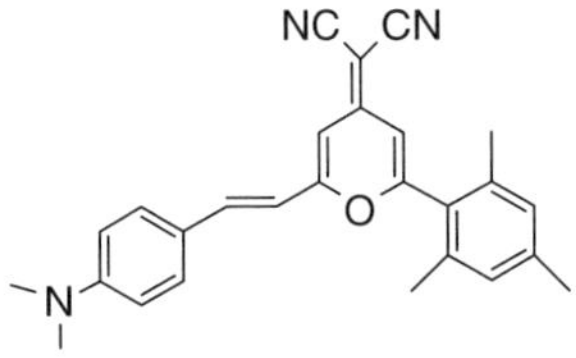	597
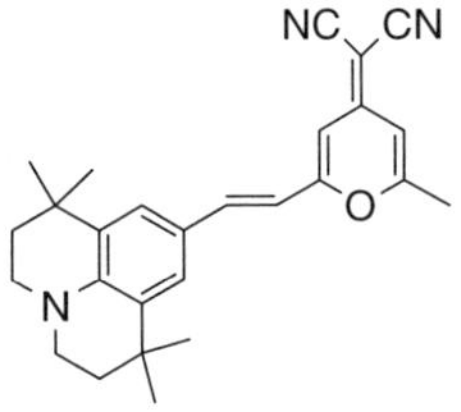	618
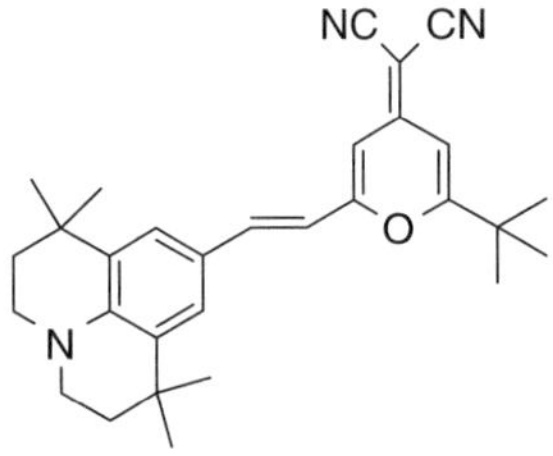	615
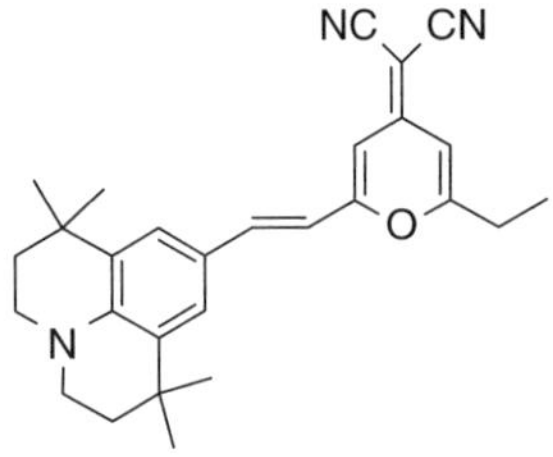	622
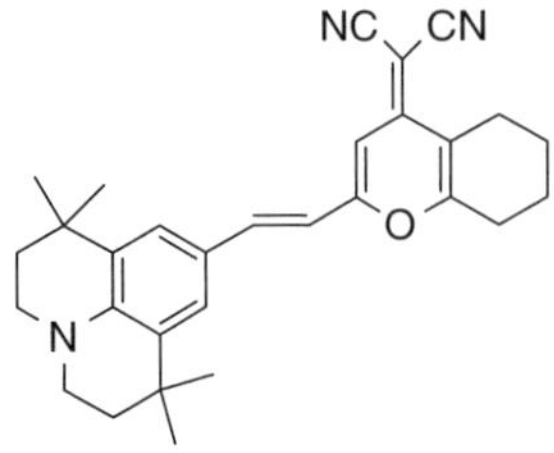	619

TABLE 12.9 (continued)
Red Fluorescent Dopants

Red Fluorescent Dopants	Emission Wavelengths (nm)
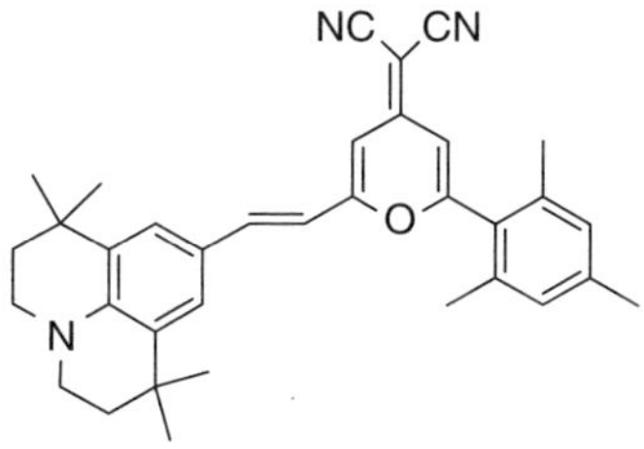	636

with oxygen than low-energy excitons. They also require large band gap host materials to maintain efficient exothermic energy transfer from the host to the blue dopants. Wide gap materials have either a small LUMO level or a large HOMO level, and carrier injection into the wide gap host materials is also very challenging. It usually designs the host materials that act as a relative inert matrix to allow for direct exciton formation on the dopant sites. For example, anthracene derivatives used as light-emitting blue host materials demonstrated improved EL efficiency and operational stability in OLED devices (Table 12.10).

TABLE 12.10
Blue Fluorescent Dopants

Blue Fluorescent Dopants	Emission Wavelengths (nm)
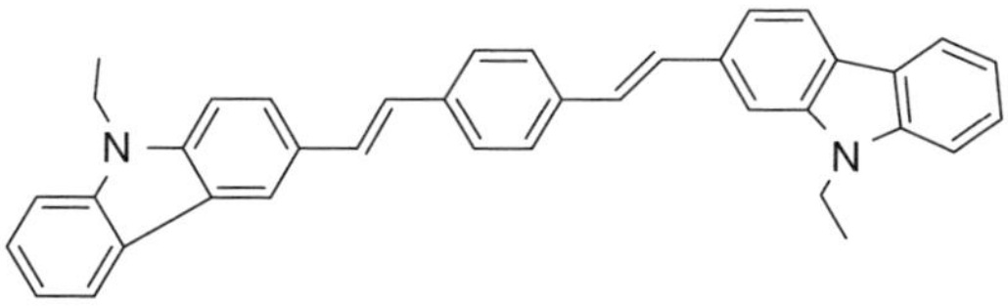	476
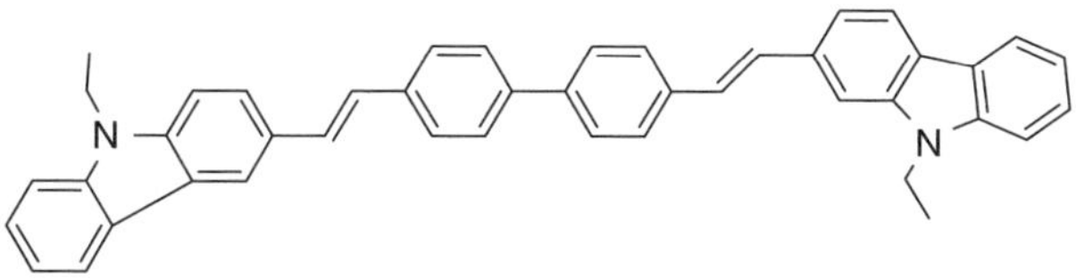	468
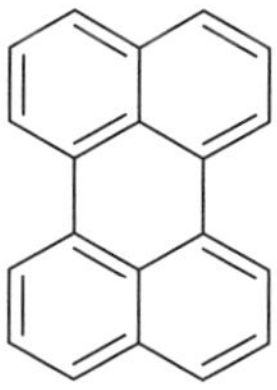	460
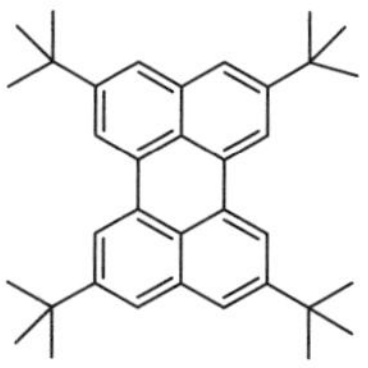	465

12.3.7.2.2 Phosphorescent Dopant Materials

In organic molecules, the excited states can exist either in singlet state or triplet state. According to the spin statistics, 25% of the excited states are singlet and 75% of the excited states are triplet. Since triplet states are not radiative, the maximum fluorescence efficiency of an organic molecule is about 25%. In some cases where a heavy metal atom is used, due to the strong spin–orbit interaction between the heavy metal atom and its ligands, the singlet and the triplet states form mixed states. As a result, both singlet and triplet states can be used to harvest light emission and in principle 100% efficiency is possible. However, most phosphorescent organic molecules (or triplet emitters) only have high phosphorescent quantum yield at extremely low temperature. Only a few classes of rare-earth metal organic complexes show a high phosphorescent quantum yield above room temperature. Among them, iridium chelates is one of the most important class of high-efficiency phosphorescent light-emitting dopant materials (Table 12.11).

TABLE 12.11
Phosphorescent-Emitting Dopants

Phosphorescent Dopants	Emission Wavelengths (nm)	Phosphorescent Dopants	Emission Wavelengths (nm)
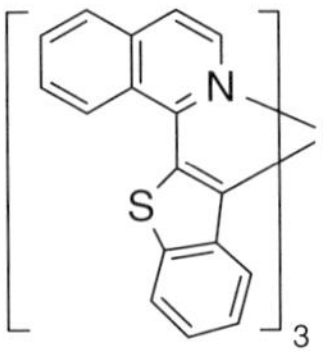	688	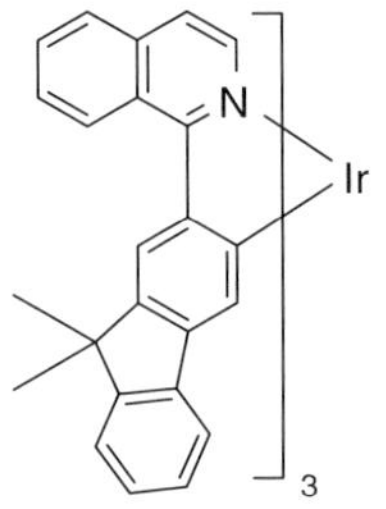	652
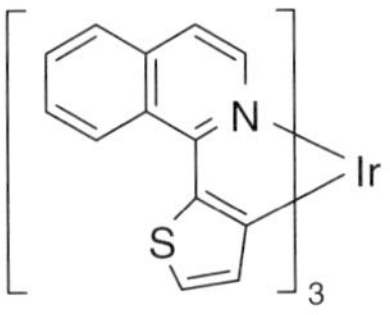	644	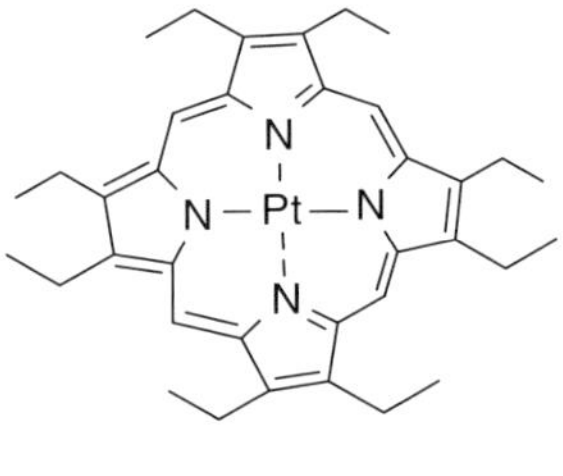	635
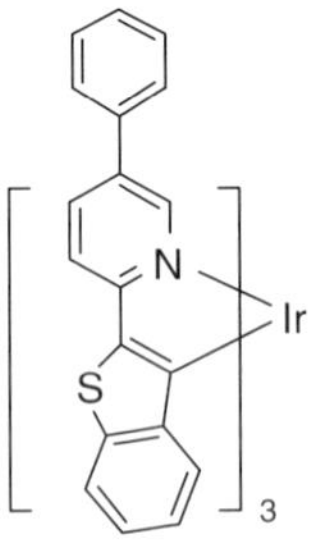	627	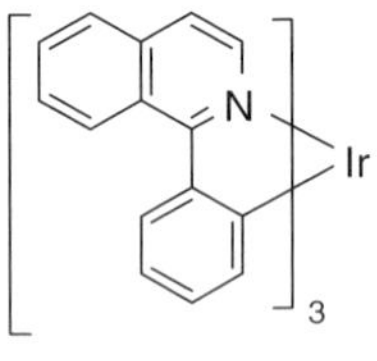	623
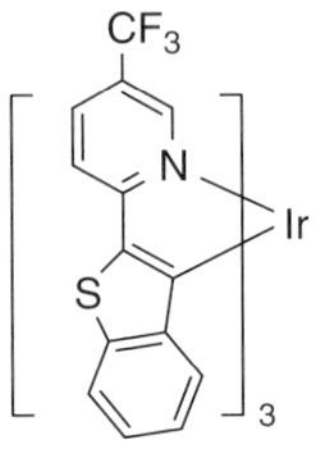	617	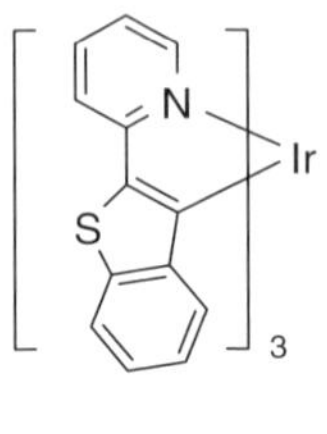	596

TABLE 12.11 (continued)
Phosphorescent-Emitting Dopants

Phosphorescent Dopants	Emission Wavelengths (nm)	Phosphorescent Dopants	Emission Wavelengths (nm)
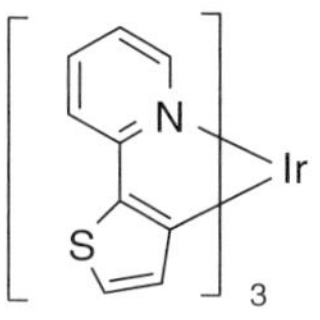	550	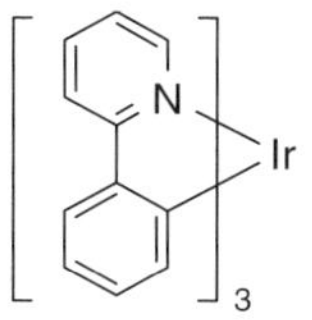	514
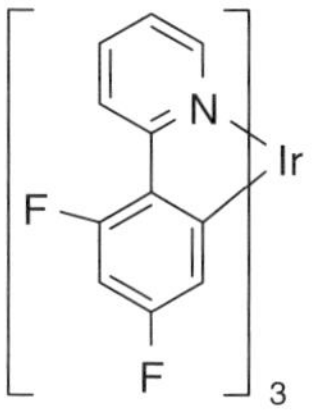	688	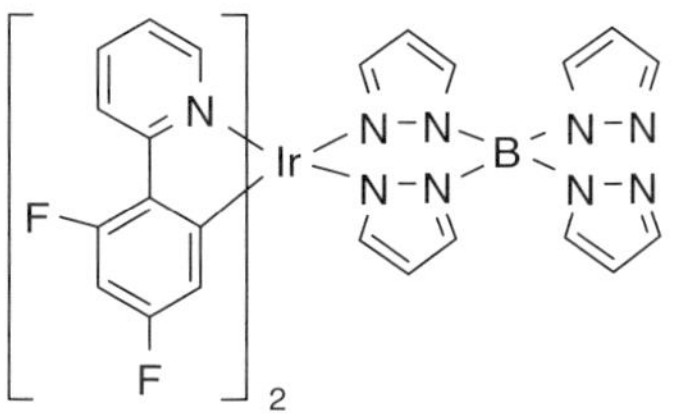	652
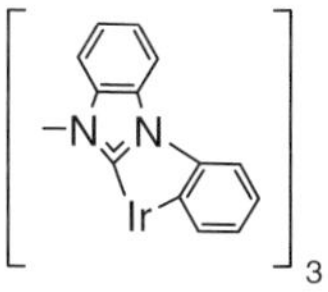	389		

Similar to aluminum chalets, iridium complexes can be designed by chelating with different organic legends to emit blue, green, and red light in the visible range. The following examples of iridium complexes are used in OLED devices to produced very high EL efficiency.

The major challenge in phosphorescent system is the design of phosphorescent emitter and host molecules. The phosphorescence states usually have longer lifetime than the fluorescence states. In order to minimize the temperature dependence and nonradiative decay, the excited state of the phosphorescence molecule is preferred to have short lifetime and rigid molecular structure. In designing fluorescence dopant/host system, only singlet states need to be considered. However, since emission of phosphorescent molecules is coming from triplet states, triplet state energies need to be considered. For example, it is important that both the host and the carrier blocking molecules have higher triplet energies than the dopant molecules. Otherwise, triplet quenching will result and this presents an even greater challenge for blue-emitting phosphorescent emitters due to narrower choice of wide gap host materials.

12.4 SUMMARY AND CONCLUSION

OLEDs have started entering into commercial products in the last few years. Most of the applications have been used in mobile displays because of its thin-form factor, high contrast, and low power consumption. Going forward, OLEDs are also considered for large size TV displays and solid-state lighting. The commercialization of OLED technology for these applications requires major development in manufacturing technology. In addition to display and lighting applications, OLEDs have also been considered for chemical and biological sensing applications as well. Because of their form factor, OLEDs offer many advantages over inorganic LED light sources for these applications.

EXERCISE QUESTIONS

1. What are the key factors determining the performance of an OLED?
2. Compare the performance of fluorescent OLEDs and phosphorescent OLEDs.
3. What are the advantages of the tandem OLED structure compared to the single cell structure?
4. Why is it more challenging to make phosphorescent blue-emitting devices compared to blue fluorescent devices?
5. What are the major differences in device operation between small molecule OLEDs and PLEDs?

BIBLIOGRAPHY

1. C.W. Tang and S.A. VanSlyke, *Appl. Phys. Lett.*, 1987, 51, 913.
2. W. Brutting, S. Berleb, and A. Muckl, *Org. Electron.*, 2001, 2, 1.
3. U. Mitschke and P. Bauerle, *J. Mater. Chem.*, 2000, 10, 1471.
4. J. Kalinowski, *Organic Light-Emitting Diodes: Principles, Characteristics and Processes*, Marcel Dekker, New York, 2004.
5. K. Mullen and U. Scherf, *Organic Light Emitting Devices: Synthesis, Properties and Applications*, Wiley-VCH, New York, 2006.
6. J. Shinar, *Organic Light-Emitting Devices*, Springer-Verlag, New York, 2004.
7. S.M. Sze, *Physics of Semiconductor Devices*, Wiley, New York, 1981.

13 Polymer Light-Emitting Diodes: Devices and Materials

Xiong Gong and Shu Wang

CONTENTS

13.1 Introduction 374
13.2 Polymer Light-Emitting Diodes Fabricated from Conjugated Polymers 374
 13.2.1 Devices Architecture 374
 13.2.2 Device Fabrication 375
13.3 Devices Physics of Polymer Light-Emitting Diodes 376
 13.3.1 Elementary Microscopic Process of Polymer Light-Emitting Diodes 376
  13.3.1.1 Injection 376
  13.3.1.2 Carrier Transport 376
  13.3.1.3 Carrier Recombination 376
  13.3.1.4 Photon Emission 377
  13.3.1.5 Photon Extraction 377
 13.3.2 Carrier Transport in Polymer Light-Emitting Diodes 377
 13.3.3 Electrical Characteristic of Polymer Light-Emitting Diodes 378
  13.3.3.1 Current–Voltage Characteristics 378
  13.3.3.2 Space-Charge-Limited Currents 380
  13.3.3.3 Injection-Limited Currents 380
  13.3.3.4 Diffusion-Controlled Currents 381
 13.3.4 Fowler–Nordheim Tunneling in Conjugated Polymer Metal–Insulator–Metal Diodes 381
  13.3.4.1 Single Carrier Devices 382
  13.3.4.2 Light-Emitting Diode Operating Voltage and Efficiency 385
  13.3.4.3 Limits of the Model 386
 13.3.5 Approaches to Improved Carrier Injection 386
13.4 Materials for Polymer Light-Emitting Diodes 388
 13.4.1 Conjugated Polymers for Polymer Light-Emitting Diodes 388
  13.4.1.1 Poly(*p*-phenylene vinylene)s 388
  13.4.1.2 Polyphenylenes 390
  13.4.1.3 Polyfluorenes 391
  13.4.1.4 Polythiophenes 392
 13.4.2 Anode and Cathode 392
  13.4.2.1 Anodes 392
  13.4.2.2 Cathodes 392
 13.4.3 Hole-Injection/Transporting Materials 393
  13.4.3.1 Hole-Injection Materials 393
  13.4.3.2 Hole-Transporting Materials 393
 13.4.4 Electron-Transporting Materials 393

13.5 Accurate Measurement of Polymer Light-Emitting Diodes Device Parameters 394
13.6 Summary 398
Exercise Questions 398
List of Abbreviations 398
Acknowledgments 399
Bibliography 399

Abstract This chapter introduces basic principles and materials of polymer light-emitting diodes (PLEDs). Such devices transform electricity from conjugated polymers into light. The chapter begins with the view of the architecture and fabrications of PLEDs followed by description of devices physics including elementary microscopic processes, carriers transport, electrical characteristics, Fowler–Nordheim tunneling in conjugated polymer metal–insulator–metal (MIM) diodes, and approaches to improve carrier injection. Different materials used for fabrication of PLEDs are highlighted and accurate measurement of PLEDs device parameters is described in this chapter.

13.1 INTRODUCTION

Since the discovery of the metallic properties of doped polyacetylene in 1977, remarkable progress has been made in synthesizing conjugated polymers with unique properties: the electronic and optical properties of metals and semiconductors in combination with the processing advantages and mechanical properties of polymers. These materials are under development for use in electronic and optical applications, including polymer light-emitting diodes (PLEDs), photodetectors, photovoltaic cells, sensors, field-effect transistors, and lasers.

Research carried out over the last decade has demonstrated the commercial opportunities associated with implementation of PLEDs in passive and active matrix displays. As a result, the development of PLEDs that show efficient, stable blue, green, and red emission is an active ongoing research effort in laboratories in Europe, Asia, and the United States.

The purpose of this chapter is to introduce concepts and progress in the field of PLEDs and polymer laser. This chapter is organized as follows: Section 13.2 describes PLEDs fabricated from conjugated polymers. Following the description of device physics of PLEDs in Section 13.3, we discuss the materials used for PLEDs in Section 13.4. Section 13.5 describes the method for accurate measurement of PLEDs device parameters. The recommended references and further reading materials follow Section 13.6 that summarize the remaining challenges in PLEDs.

13.2 POLYMER LIGHT-EMITTING DIODES FABRICATED FROM CONJUGATED POLYMERS

13.2.1 Devices Architecture

Figure 13.1 shows the basic and simplified architecture of a PLED, which is largely built-up of conjugated polymers. Under the action of a driving voltage of a few or a couple of volts, electrons are injected from a metal cathode with a low work-function into the electronic state corresponding to the lowest unoccupied molecular orbital (LUMO) of a conjugated polymer; holes are injected from a bilayer anode (PEDOT:PSS [poly(3,4-ethylene dioxythiophene: poly[styrene sulfonic acid] on indium tin oxide [ITO]) into the electronic state corresponding to the highest occupied molecular orbital (HOMO) of a conjugated polymer. Both electrons and holes, coming from the different electrodes, move from opposite directions toward the recombination zone where they can combine to form excitons. This leads to a population of excited states of the emissive polymers that subsequently emits light.

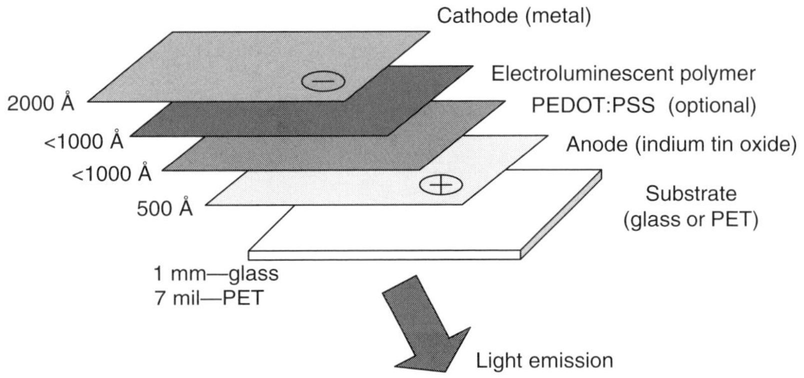

FIGURE 13.1 Devices architecture of single-layer PLEDs.

13.2.2 Device Fabrication

Single-layer PLEDs were prepared according to the following procedure: the ITO-coated glass substrate was first cleaned with detergent, then ultrasonicated in acetone and isopropyl, and subsequently dried in an oven for at least 6 h. Medium conducting PEDOT:PSS, for example, Baytron P 4083, was spin-casted (4000 rpm) with thickness ~40 nm from aqueous solution after passing a 0.45 μm filter. The substrate was dried for 10 min at 160°C in air, and then moved into a glove box for spin casting the emissive layer. Subsequently, the device was pumped down in vacuum ($<10^{-7}$ torr), and a ~5 nm Ba or Ca with 200 nm Al or Ag encapsulation film was deposited on top of the active layer.

For the demonstration device in Figure 13.2, polyaniline was chosen as the anode material because it is flexible, conducts current, and is transparent to visible light (there is no ITO layer in the device of Figure 13.2). More generally, however, a bilayer electrode comprising a thin layer of metallic polymer cast onto ITO is used for the anode; the ITO carries the current and the metallic polymer layer serves to planarize the ITO and to improve the injection of holes into the luminescent conjugated polymer. The emissive layer of the display shown in Figure 13.2 was formed by spin casting a layer of poly(*p*-phenylene vinylene) (PPV) derivatives over the polyaniline. The top electrodes that define the seven-segment display were formed by evaporating calcium through a patterned shadow mask. Because of the low conductivity of undoped emissive polymers, it was not necessary to pattern the polymer or the bottom electrode to prevent current spreading between neighboring pixels.

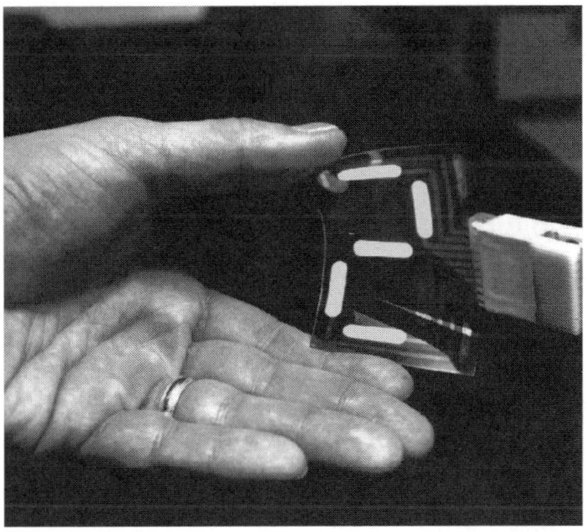

FIGURE 13.2 Photo of a thin film flexible seven-segment display fabricated from conjugated polymers.

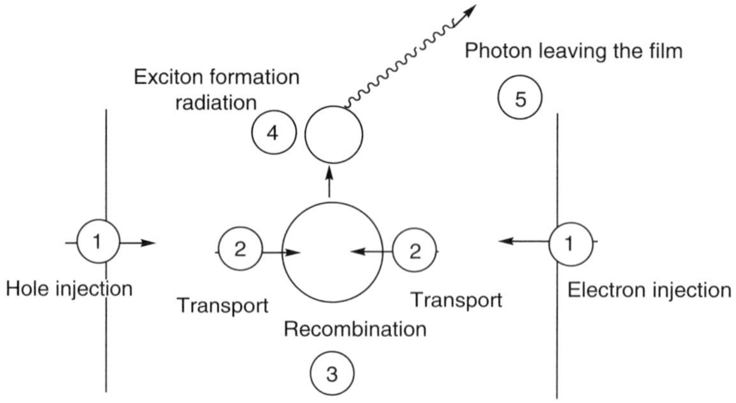

FIGURE 13.3 Schematic drawing of the five elementary steps of PLEDs.

13.3 DEVICES PHYSICS OF POLYMER LIGHT-EMITTING DIODES

13.3.1 Elementary Microscopic Process of Polymer Light-Emitting Diodes

It is convenient to consider the overall operation of PLEDs as a sequence of five steps: charge injection at the electrodes, transport through the structure, recombination into a neutral excited state, emission of a photon, and transport of the photon out of the diode. These steps are shown in Figure 13.3.

13.3.1.1 Injection

The equilibrium concentrations of carriers in the polymer materials used are very small; the forbidden gap is large, and shallow donors or acceptors are absent. The carriers present in the operating PLEDs must be injected from electrodes. Injection may either proceed over the interface barrier by thermionic emission or through tunneling. The injection barrier is given by the equations $\Delta E_e = W_c - A$ and $\Delta E_h = I - W_a$ for electrons and holes, respectively; W_a is the work function of the anode, W_c is the work function of the cathode, I is the ionization potential, and A is the electron affinity of conjugated polymers in the solid state.

13.3.1.2 Carrier Transport

Once an electron and a hole have been injected, they must be transported through the film to the point where they will recombine. In PLEDs, carrier transport is believed to take place by hopping between localized states. These localized states are usually loosely referred to as polarons and bipolarons. The details of carrier transport are described in Section of 13.3.2.

13.3.1.3 Carrier Recombination

The dielectric constant of conjugated polymers is small, typically 3–4. Coulomb screening is inefficient and the attractive interaction between an electron and a hole is $>kT$, large, and typically 15 nm at room temperature. Therefore, it is believed that the carrier recombination in PLEDs is Langevin recombination. The recombination rate constant is given by the equation

$$\gamma_r = \frac{e(\mu^+ + \mu^-)}{\varepsilon\varepsilon_0} \tag{13.1}$$

and is independent of any applied field. This has been well established experimentally in the case of molecular crystals, and is quite different from the case of semiconductors.

13.3.1.4 Photon Emission

Almost all conjugated polymers used in PLEDs are composed of light elements, mainly C and H; thus spin-orbit coupling effects are inefficient. Neutral excited states are well classified as singlet ($S = 0$) and triplets ($S = 1$), according to their spins and multiplicity. Higher spin states are generally at too high energies to play any role here. The singlet–triplet energy splitting is typically 1 eV. If the quantum yield for singlet emission, the fluorescence, is η_s, and for triplet emission, the phosphorescence, is η_t, then the electroluminescence quantum yield is proportional to

$$\frac{\eta_s + 3\eta_t}{4} \tag{13.2}$$

13.3.1.5 Photon Extraction

The emitting layer of conjugated polymers being generally disordered and nontextured, the emission is, therefore, isotropic inside the film, leading to an approximately Lambertian emission pattern outside the diode. This is a luminance that is independent of the angle of view, and is indeed advantageous for display applications. An accurate measurement of PLEDs device performance is discussed in Section 13.5.

13.3.2 Carrier Transport in Polymer Light-Emitting Diodes

Most of the organic electroluminescence materials, small molecules, and conjugated polymers are low conductance materials. The h^+ mobility in these materials is typically 10^{-7} to 10^{-3} cm^2/V s, and the e^- mobility is typically lower by a factor of 10–100. However, it is now clear that the low mobility is due to the disorder in the amorphous or polycrystalline materials. The application of an external field causes injection of h^+'s from the ITO and of an e^-'s from the cathode. The injection from the metallic injection is usually less efficient than from the ITO. The asymmetry in carrier injection leads to an imbalance in the concentrations of the injected carriers that reduce the device's efficiency.

Unlike inorganic semiconductors, the transport and the injection properties in PLEDs are determined by intersite hopping of charge carrier between localized states as well as hopping from delocalized states in the metal to localized states in the organic layer. The actual transition rate from one site to another depends on their energy difference and on the distance between the sites. The energy states involved in the hopping transport of h^+'s and e^-'s from narrowbands are around the HOMO and LUMO levels. The widths of these bands are determined by the intermolecular interactions and by the level of disorder.

The transport in PLEDs has been extensively studied by time-of-flight (TOF), and analysis of the DC current–voltage characteristics. The universal dependence of charge carrier mobility on the electric field is

$$\mu(E,T) = \mu(0,T)\exp(\gamma\sqrt{E}) \tag{13.3}$$

where

$\mu(0,T)$ is the low-field mobility

γ is an empirically determined coefficient, which is observed for the vast majority of materials

Two models, polaron and disorder, were applied for explanation of carrier hopping in PLEDs. The models based on polaron formation assume that a localized carrier interacts strongly with molecular vibrations of the host and neighboring molecules; so, significant relaxation of the local molecule structure occurs around carrier. This carrier can move to an adjacent molecule only by carrying that relaxation along with it. Clearly, that relaxation or stabilization lowers the energy of the negative carrier below the LUMO level and the energy of the positive carrier above the HOMO level. The experimental evidence for polaronic relaxation in PPV and other conjugated polymers are extensive.

Although experimental evidence for polaronic relaxation is extensive, other experiments render the polarons models problematic due to the following reasons: (1) the use of the Arrhenius relation to describe the temperature dependence of the mobility leads to prefactor mobilities well in excess of unity, and (2) the polaron models cannot account for the dispersive transport observed at low temperature.

In the disorder models, it is assumed that the coupling of a charge carrier to molecular modes is weak, and the activation energy reflects the static disorder of the hopping sites. In polaron models, it is assumed that the energetic disorder energy is small compared to the deformation energy. The fundamental difference between disorder and polaron models is related to the difference in energy of hopping sites due to disorder and the change in molecular conformation upon addition or removal of a charge at a given site. Generally, despite the better agreement between the disorder-based models and transport measurements, it is widely believed that the charge carrier exists as polarons rather than free e^-'s and h^+'s.

13.3.3 Electrical Characteristic of Polymer Light-Emitting Diodes

13.3.3.1 Current–Voltage Characteristics

The current–voltage characteristics of PLEDs, measured over many orders of magnitude in applied field, exhibit several well-pronounced regimes. An example presented in Figure 13.4 shows the

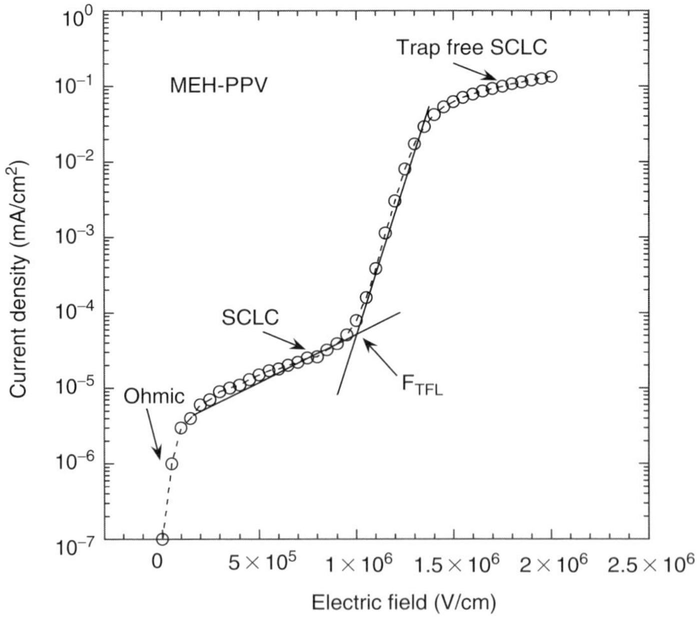

FIGURE 13.4 Unipolar current versus electric field for 80 nm thickness MEH-PPV with device configuration of ITO/MEH-PPV/Ca.

low-field ohmic conduction and space-charge-limited-current (SCLC) in the presence of shallow traps followed by the free-trap conduction in the upper limit of applied field.

Although having superficially similar device characteristics to conventional diodes made from inorganic semiconductors, polymeric devices differ in many details. For Schottky-barrier or p–n junction-type devices, the operational voltage is typically equal to, or less than, the inorganic semiconductor band gap. In contrast, polymer devices typically begin to generate light at voltage much greater than their band gap. Reducing the thickness lowers the device operating voltage. Figure 13.5 clearly demonstrates that the *I–V* characteristics depend not on the voltage but instead on the electric field strength. It has also pointed out clearly a tunneling model for carrier injection in which one, or

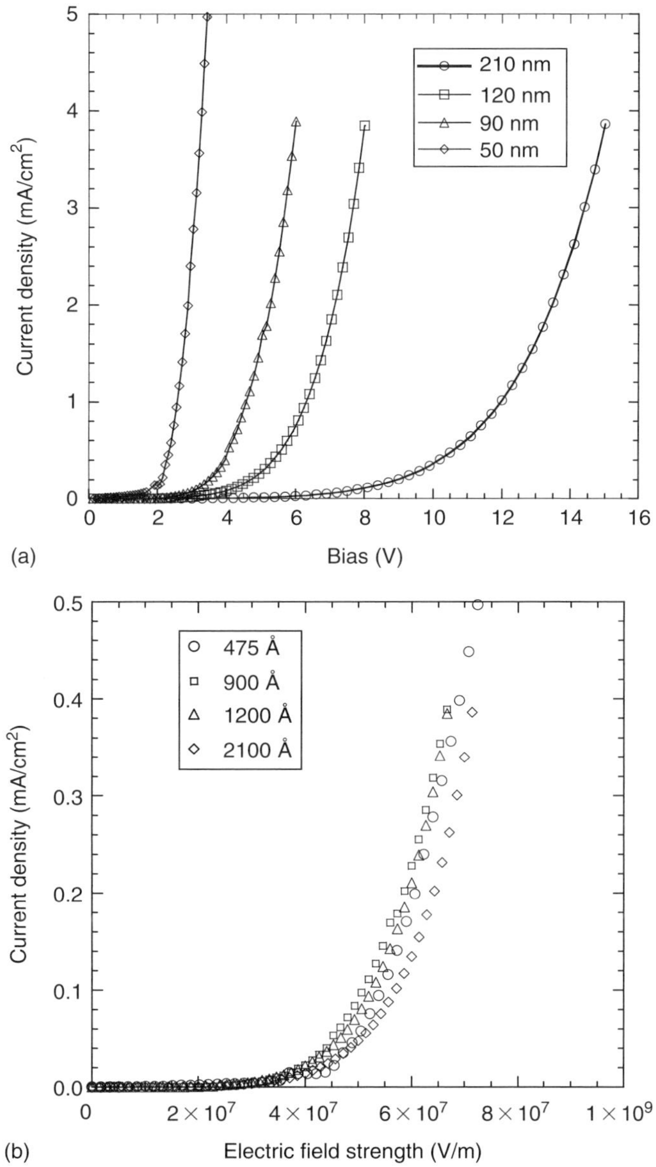

FIGURE 13.5 (a) Thickness dependence of the *I–V* characteristics in an ITO/MEH-PPV/Ca device (b) electric field versus current for the above devices.

both, of the carriers is field emitted through a barrier nature at the electrode/polymer interface. The details in tunneling model are described in Section 13.3.4.

13.3.3.2 Space-Charge-Limited Currents

Organic solids are usually insulators. Unlike in inorganic semiconductors, impurities normally act as traps for charge carriers rather than as sources of charge carriers. Exceptions from that rule are conjugated polymers. For example, PPV fabricated via a special precursor route turned out to be p-doped with doping concentrations in the order of 10^{17} cm^{-3}. In that case, a Schottky-type depletion zone can be established near a metal contact. However, in vast majority of cases, the concentration of impurities is small enough not to perturb the electric field distribution inside a solid-state sample. In such cases, the dark electrical conduction is very low, and the solids are considered as good insulators. Such solids can be made to conduct a relatively large current if the contacts permit the introduction of an excess of free carriers in them. If the carrier enters through a surface boundary, the process is referred to as charge injection. The charge injected conduction is governed by charge injection barriers at the electrode contacts and charge transport properties of materials. Depending on the charge injection efficiency and mobility of charge carrier, the current is either SCLC or injection-limited current (ILC).

For a perfectly ordered or disordered insulating material, or those containing very shallow traps ($\Delta E \ll kT$), the SCL current in a sample of thickness d obeys Child's law:

$$j_{\text{SCL}} = \frac{9}{8}\varepsilon_0 \varepsilon \mu \frac{F^2}{d} \tag{13.4}$$

In the presence of discrete traps

$$j_{\text{SCL}} = \frac{9}{8}\varepsilon_0 \varepsilon \Theta \mu \frac{F^2}{d} \tag{13.5}$$

where

μ is the microscopic mobility of the carriers
ε is the dielectric constant
ε_0 is the permittivity in vacuum
Θ is the fraction of free (n_f) to trapped (n_t) space-charge

If local traps are distributed in energy (E), they will be filled from bottom to top as electric fields, F, increase. The quasi-fermi level will scan the distribution shifting toward the transport band, and $\Theta \approx n_f/n_t$ will become a function of F.

13.3.3.3 Injection-Limited Currents

The current becomes limited by injection when the average charge density in the sample approaches to the charge density at the injecting contact. The injecting contact can no longer act as a reservoir and thus ceases to be ohmic. The current from such an electrode will saturate at sufficiently high voltage. On the other hand, very high electric fields can make some contacts ohmic by causing a strong injection via tunneling or other mechanisms superlinear with electric field. Although the average charge density in the samples is comparable with the charge density at the contact, both of them should be much smaller than the capacitor charge related to unit volume. ILC will be observed only for relatively low currents at high electric fields with high mobility and large value dielectric permittivity materials formed into high chemical and structural perfection thin layers. These are often the met features of thin organic films sandwiched between metals or semiconductors with moderate work-functions.

13.3.3.4 Diffusion-Controlled Currents

Charge carrier injection from a metallic electrode is said to be diffusion controlled if space-charge effects can be neglected and the diffusion current is comparable or exceeding the drift current flow.

13.3.4 Fowler–Nordheim Tunneling in Conjugated Polymer Metal–Insulator–Metal Diodes

The operating mechanism of PLEDs is quite different from conventional p–n junction LEDs. In a PLED, a pure undoped film of luminescent conjugated polymer is sandwiched between a high work-function metal anode and a low work-function metal cathode. The charge carrier concentration in such pure semiconducting films is sufficiently low (~10^{14} to 10^{15} cm^{-3}) such that any residual carriers introduced by impurities, etc., are swept out by the built-in field that arises from the difference in work functions of the two electrodes. The depletion depth of pristine PPV is ~250 μm, which is much larger than the thickness of the polymer layer in an LED (typically <100 nm). Consequently, the electronic structure of the LED can be approximated by the rigid band model displayed in Figure 13.6. The built-in field causes the uniform slope in the energies of the states in the bulk of the conjugated polymer; there is negligible band bending.

When a positive bias is applied to the LED, the fermi level of the cathode is raised relative to that of the anode, as shown in Figure 13.6c. Thus, the thickness of the barrier is a function of the applied voltage; the barrier thickness decreases as the voltage is increased. Carriers tunnel through the barrier primarily by Fowler–Nordheim field emission tunneling from the anode and cathode into the π-band (holes from the anode), and the π*-band (electrons from the cathode), of the conjugated polymer. Thermionic emission over the barriers can also play a role if the barriers are small and the

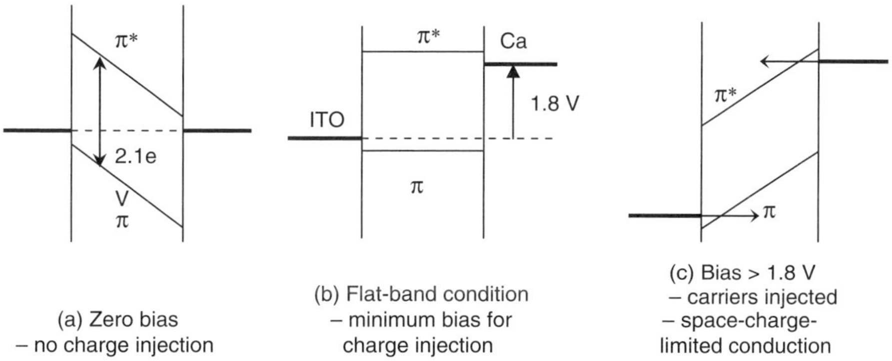

FIGURE 13.6 The electronic structure of the LED. (a) Zero bias with a common fermi level across the device (note that within the semiconducting polymer layer, the fermi level moves across the energy gap on going from the anode to the cathode). (b) Flat band condition occurs when the applied voltage equals the difference in the work functions of the anode and the cathode. This is the minimum voltage required for injection of electrons and holes. Ideally, the electroluminescent emission should turn on at this voltage. (c) Forward bias; carriers are injected through the triangular barriers at the anode (holes) and cathode (electrons) and meet within the polymer film where they radiatively recombine (electroluminescence).

temperature is relatively high. Since the rate of injection by Fowler–Nordheim tunneling is determined by the strength of the electric field, it is important for the polymer layer to be thin so that high electric fields can be obtained at low voltages.

To optimize the performance of PLEDs, it is important to minimize the barriers for charge injection by choosing electrodes with work functions that are well matched to the bands of the polymer. ITO, polyaniline, polypyrrole, and PEDOT are the most commonly used anode materials. These also have the important property of being transparent and therefore allow the emitted light to escape from the device. Calcium, barium, and magnesium are commonly chosen as the cathode because of their low work-functions. Unfortunately, low work-function metals are highly reactive. PLEDs must therefore be hermetically sealed for long life. Improved electron injection from stable metals such as aluminum can be achieved by coating the electrode with a polar self-assembled monolayer. The dipole layer effectively shifts the electrode work-function.

If the electrodes are well matched to the bands of the polymer, then the barrier for charge injection is small and the current that passes through the LED is not limited by injection. Instead, the hole current is space-charge limited and the electron current is trap limited. Space-charge limiting arises because the space-charge that builds up near the anode due to the population of holes screens the field between the two electrodes and thereby limits the current. The traps that limit electron transport originate from defects that have energy levels just below the conduction band (due to disorder in the polymer).

Once electrons and holes have been injected into the polymer, they must encounter each other and recombine radiatively to give off light. In this context, the low mobility of the charge carriers (polarons) in semiconducting polymers is helpful since the slow drift of the charge carriers across the thickness of the semiconducting polymer will allow enough time for the carriers to meet and recombine radiatively.

There are several factors that determine the efficiency of an LED. Maximum efficiency can only be achieved through balanced electron and hole currents. If one carrier type is injected much more efficiently and drifts in the applied electric field with higher mobility than the other, then many of the majority carriers will traverse the entire polymer layer without recombining with a minority carrier. As shown in Figure 13.6, this problem can be minimized by carefully choosing appropriate electrodes so that the fermi level of the anode is close in energy to the top of the π-band and the fermi level of the cathode is close in energy to the bottom of the π^*-band. With such well-matched electrodes, both carriers are injected efficiently.

As demonstrated in Figure 13.7, diodes fabricated as described above show excellent rectification ratios and strong electroluminescent light emission in forward bias. Light emission turns on close to the flat band condition; i.e., when the applied voltage is greater than the difference between the work functions of the two electrodes (see Figure 13.6). The current at voltages below 2 V in Figure 13.7b is residual leakage current resulting from microshorts and imperfections.

Parker observed that the turn-on voltages for devices with different thicknesses of the conjugated polymer layer are roughly the same and equal to the voltage required to reach the flat band condition. As shown in Figure 13.8, however, the current density is a strong function of the thickness.

Understanding the operating mechanism of polymer LEDs is complicated by the two-carrier nature of these diodes. To study the details of the carrier injection, single carrier devices were fabricated in which the current density of one of the carriers was reduced to negligible levels.

13.3.4.1 Single Carrier Devices

"Hole-only" devices were fabricated by replacing the low work-function Ca cathode (2.9 eV) with higher work-function metals such as In (4.2 eV), Al (4.4 eV), Ag (4.6 eV), Cu (4.7 eV), or Au (5.2 eV); see Figure 13.9.

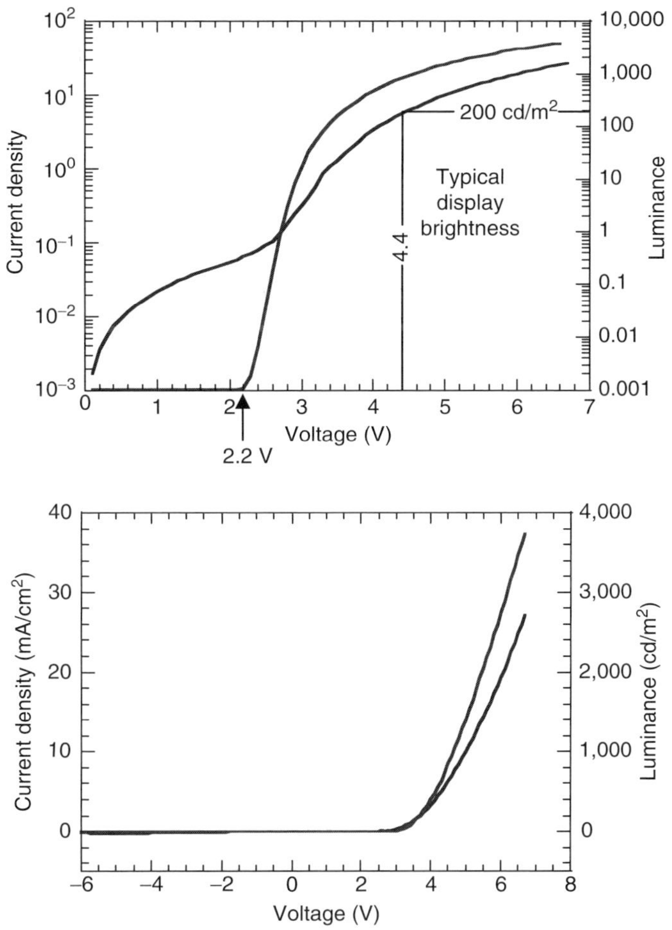

FIGURE 13.7 Current density and luminance versus voltage for devices fabricated with MEH-PPV using polyaniline as the anode and calcium as the cathode.

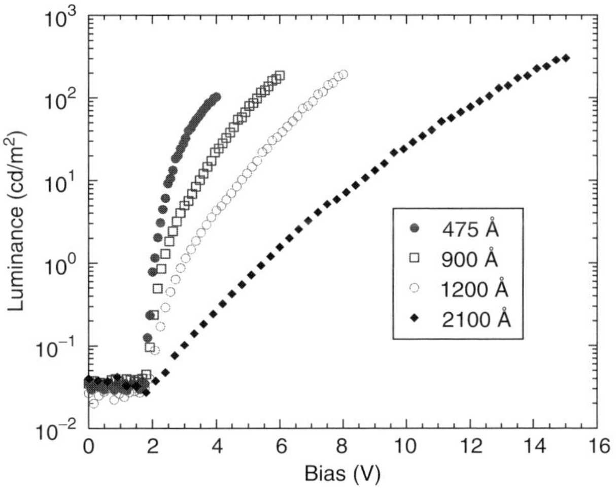

FIGURE 13.8 Luminance versus voltage for ITO/MEH-PPV/Ca LEDs with semiconducting polymer layers of different thicknesses. Note that turn-on voltage is independent of thickness and equal to that needed to reach the flat band condition (see Figure 13.6b).

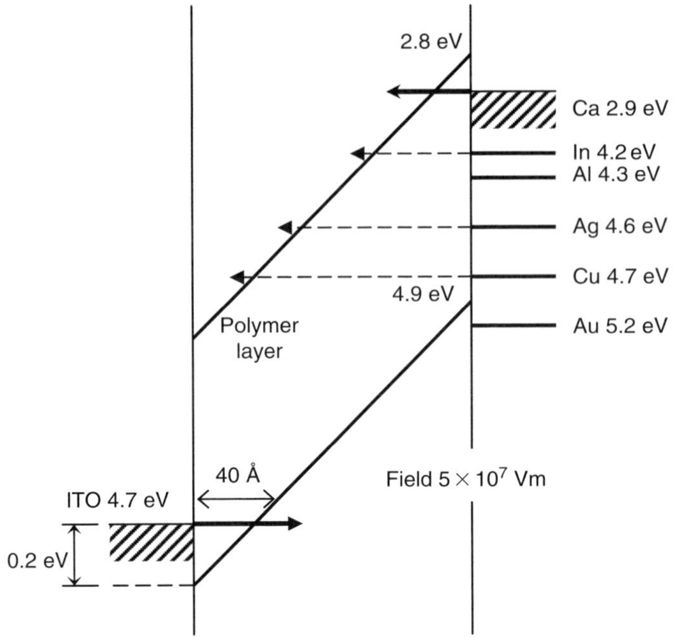

FIGURE 13.9 Band diagram (in forward bias) indicating the magnitudes of the triangular tunneling barrier for various cathode materials.

Increasing the work function of the cathode increases the offset between the fermi energy of the cathode and the bottom of the π*-band of poly[2-methoxy-5-(2′-ethylhexyloxy)-*p*-phenylenevinylene] (MEH-PPV) at 2.8 eV. This reduces the number of injected electrons to levels at which the injected holes dominate. Despite the fact that the work function of the cathode increases by more than 1 eV (from In to Au), the *I–V* curves of the devices made with In, Al, Ag, Cu, and Au cathodes are almost identical, indicating that electron injection has been shut off.

Fowler–Nordheim tunneling theory predicts that the tunneling current is an exponential function of $1/F$

$$I \propto F^2 \exp(-\kappa/F) \tag{13.6}$$

where

- I is the current
- F is the electric field strength
- κ is a parameter that depends on the barrier shape

For a triangular barrier,

$$\kappa = 8\pi(2m^*)^{1/2}\varphi^{3/2}/3qh \tag{13.7}$$

where

- φ is the barrier height
- m^* is the effective mass of holes in the semiconducting polymer
- q is the electron charge
- h is the Planck's constant

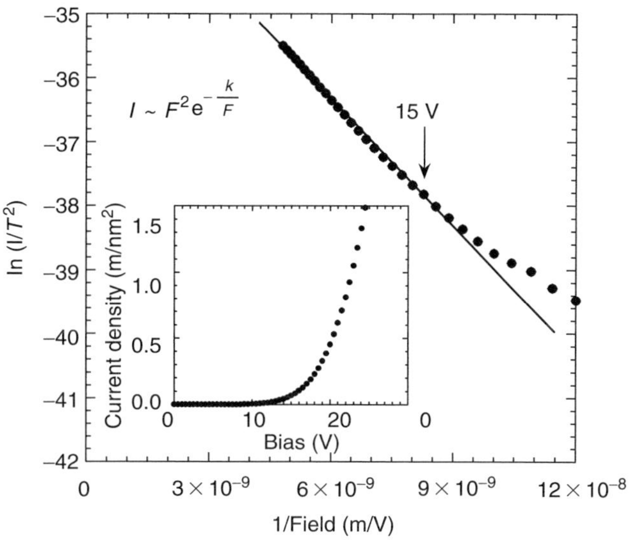

FIGURE 13.10 Fowler–Nordheim plot of $\ln[I/F^2]$ versus $1/F$ for a hole-only device fabricated using MEH-PPV with ITO as the anode and Au as the cathode.

The I–V data from a hole-only device fabricated from MEH-PPV with Au as the cathode and ITO as the anode are plotted as I/F^2 versus $1/F$ in Figure 13.10. As predicted, the plot is close to linear particularly at high fields. The literal assumption of tunneling through a triangular barrier appears to be an excellent approximation. The deviation from linearity at lower fields probably indicates an additional contribution to the current from thermionic emission.

Assuming that the electric field is constant across the semiconducting polymer and that $m^* = m$ (the free electron mass), the calculated barrier heights are found to be 0.2–0.3 eV for all the hole-only devices indicated in Figure 13.9. The fact that the infrared barrier height is unchanged despite the large variation in cathode work-function indicates that the barrier must be at the ITO/polymer interface. Indeed, $\varphi = 0.2$–0.3 eV is in agreement with the energy diagram in Figure 13.9.

A similar approach can be taken to study electron injection. Replacing ITO with a lower work-function metal (e.g., Nd or Mg) yields devices in which the carriers are almost exclusively electrons. A similar analysis of the data from a range of electron-only devices indicates that electrons tunnel into the π^*-band of MEH-PPV at 4.9 eV through a triangular barrier at the polymer/cathode interface. With Ca as the cathode, this barrier is 0.1 eV, again in agreement with the energy diagram in Figure 13.9.

13.3.4.2 Light-Emitting Diode Operating Voltage and Efficiency

The operating voltage is very sensitive to the barrier height. This sensitivity is predicted by the Fowler–Nordheim tunneling model. Equations 13.6 and 13.7 indicate that for the same tunneling current, the ratio $\varphi^{3/2}/V$ must be the same. Thus, increasing the barrier height from 0.1 eV (PANI) to 0.6 eV (Cr) should increase the operating voltage by a factor of 11. The experimentally determined increase was a factor of 9.

It is obvious that the device efficiency, η, must also be very sensitive to the barrier height, since the efficiency is limited by the minority carrier density. As suggested by Equations 13.6 and 13.7, Figure 13.11 plots $\ln(\eta)$ versus $\varphi^{3/2}$. The excellent agreement between the theory and the data confirms the use of the Fowler–Nordheim tunneling model for describing the carrier injection into the band structure of the conjugated polymer.

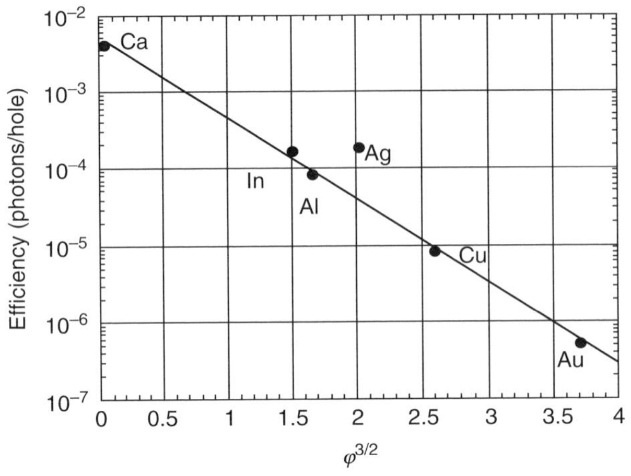

FIGURE 13.11 Semilog plot of quantum efficiency (photons/electron) versus $\varphi^{3/2}$.

There is evidence of interface modification when metal films are deposited onto conjugated polymers. However, since these interactions involve specific chemical reactions between the metal atoms and certain groups on the polymer, it is not clear how such chemical reactions could give rise to the systematic variation in the device efficiency with work function that is evident in Figure 13.11.

13.3.4.3 Limits of the Model

When the barrier heights are small, the simple comparison of work functions for the prediction of barrier heights is inappropriate; thermionic emission becomes important for barriers less than a few tenths of an electron volt. Moreover, the barrier will not be perfectly triangular as the model assumes, but will be somewhat ill-defined as a result of disorder-induced band-tailing near the polymer band edges. In addition, the evaporated electrode materials are likely to be full of defects (possibly even amorphous) and are, therefore, not expected to have precisely the well-defined work-function values listed in reference books. Other factors, such as the mobility of each charge species and space-charge limiting effects, may also have increasing influence over the performance for low barrier-height devices. Nevertheless, it is clear that an energy band picture for the conjugated polymer supplemented by Fowler–Nordheim tunneling theory provides an excellent starting point for understanding the operating mechanism of LEDs fabricated from conjugated polymers.

13.3.5 Approaches to Improved Carrier Injection

It is not always possible to find electrode materials that are well matched to the electronic structure of an electroluminescent polymer. For example, for large band-gap blue-emitting polymers such as the polyfluorenes, the energy barrier at the cathode is typically too large for efficient injection of electrons via Fowler–Nordheim tunneling. Similarly, for more electronegative, stable conjugated polymers, the barrier at the anode will be too large for efficient hole injection. In such cases, hole or electron blocking layers can be added to improve the balance of electron and hole currents. Figure 13.12 shows the electronic structure of an LED with a hole blocking layer. The blocking layer creates a barrier at the interface of two polymers that blocks the flow of the majority carrier. As the density of the majority carrier increases at the blocking interface, the electric field at the minority

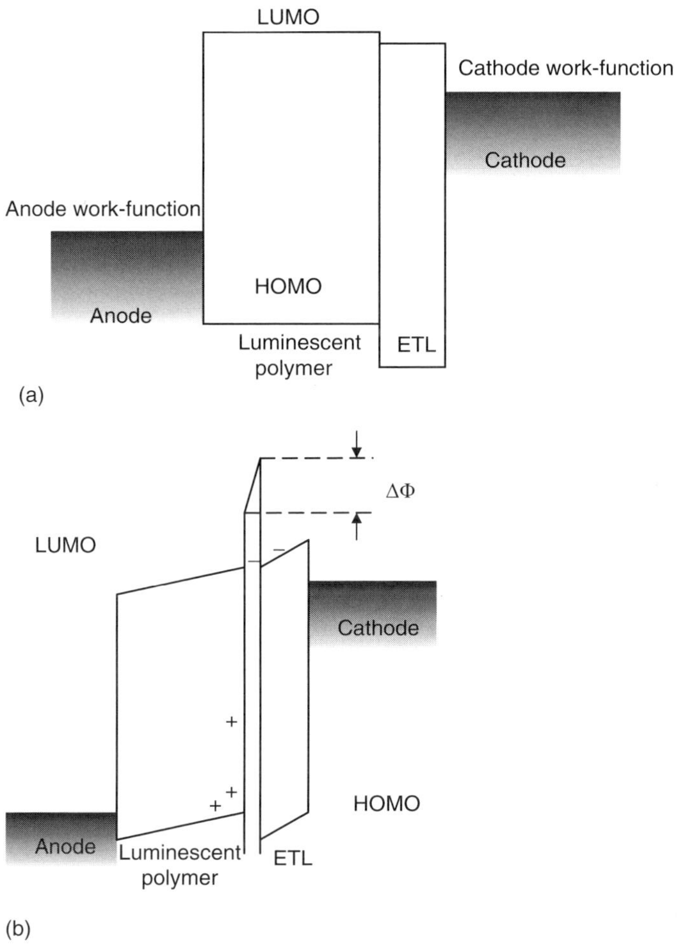

FIGURE 13.12 (a) Flat band diagram of the electronic structure of a LED with a hole blocking layer. (b) Electronic structure of an LED with a hole blocking layer in forward bias. The electric field is primarily across the electron-transport layer (ETL); the field in the luminescent polymer is nearly zero.

carrier injecting electrode increases, thereby enhancing the minority carrier injection. As a result, the electron and hole currents will be more nearly balanced. The analogous diagram can be drawn for a device with an electron blocking layer.

Improved injection can also be achieved through the addition of defects and impurities. End-capping with electron-accepting groups or hole-accepting groups is a useful example. As shown in Figure 13.13, such groups lead to the formation of localized states near the band edges. Thus, in forward bias an electron can tunnel into the π*-band in two steps: first from the cathode into a localized state on an electron-accepting site, and then from that site into the π*-band. Since the tunneling probability decreases exponentially as $\varphi^{3/2}$, the two-step injection can be more efficient than direct tunneling through the barrier into the π*-band. As sketched in Figure 13.13 (dotted arrow), a two-step process involving tunneling followed by thermal excitation from the impurity acceptor state into the band can also help.

Electron-accepting or hole-accepting groups can also be incorporated into the chain (copolymers) by the addition of appropriate small molecules into the film. In the latter approach, however, clustering and phase separation are likely problems.

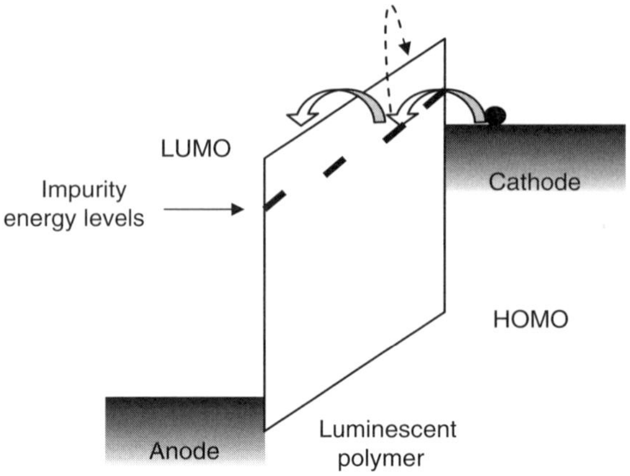

FIGURE 13.13 End-capping with electron-accepting groups or hole-accepting groups leads to the formation of localized states near the band edges; electron-accepting states are at energies just below the π*-band, and hole-accepting states are at energies just above the π-band. Alternatively, such acceptor units can be inserted into the main chain as dopants. Small molecule acceptors can also be used.

13.4 MATERIALS FOR POLYMER LIGHT-EMITTING DIODES

13.4.1 Conjugated Polymers for Polymer Light-Emitting Diodes

The design of luminescent materials for use in PLED devices is as critical to device performance as the process of constructing the device itself. Processability, purity, thermal and oxidative stability, color of emission, luminance efficiency, balance of charge carrier mobility, and others are among many important material properties required for a system to be viable in commercial PLED devices applications. In the last two decades, several types of conjugated polymers with different molecular structures have been developed and applied for fabrication of PLEDs. Among the conjugated polymers poly(*p*-phenylene vinylene)s (PPVs), polyphenylenes (PPPs), polyfluorenes (PFs), and polythiophenes (PTs), the PPVs and the PFs have been widely studied and emerged as the leading candidates for PLEDs applications because of their tunable emission frequencies and relatively high photoluminescence yields.

Semiconducting polymers have been synthesized with different molecular structures and with an associated wide range of energy gaps. Consequently, luminescent semiconducting polymers can be obtained with emission colors that span the full range of the visible spectrum. Side chain functionalization of the same main chain (e.g., with alkoxy or alkyl groups) can be used to shift the color of the emitted light over a substantial portion of the visible spectrum. The use of synthesis to create homopolymers with different molecular structures and to create copolymers is the band-gap engineering methodology for semiconducting polymers. Figure 13.14 shows a few examples that demonstrate the range of different band gaps and emission colors that can be obtained through band-gap engineering. The use of copolymers is particularly interesting. By using block-copolymers, well-defined quantum well structures can be created.

13.4.1.1 Poly(*p*-phenylene vinylene)s

Poly(*p*-phenylene vinylene) is an insoluble, intractable, and infusible conjugated polymers with green–yellow emission. The core molecular structure of PPV is shown in Figure 13.15. The problem of PPV for making PLEDs is that any direct synthesis will produce a solid. An alternative solution is to use a solution-processable precursor polymer, which is the first cast from solution and then

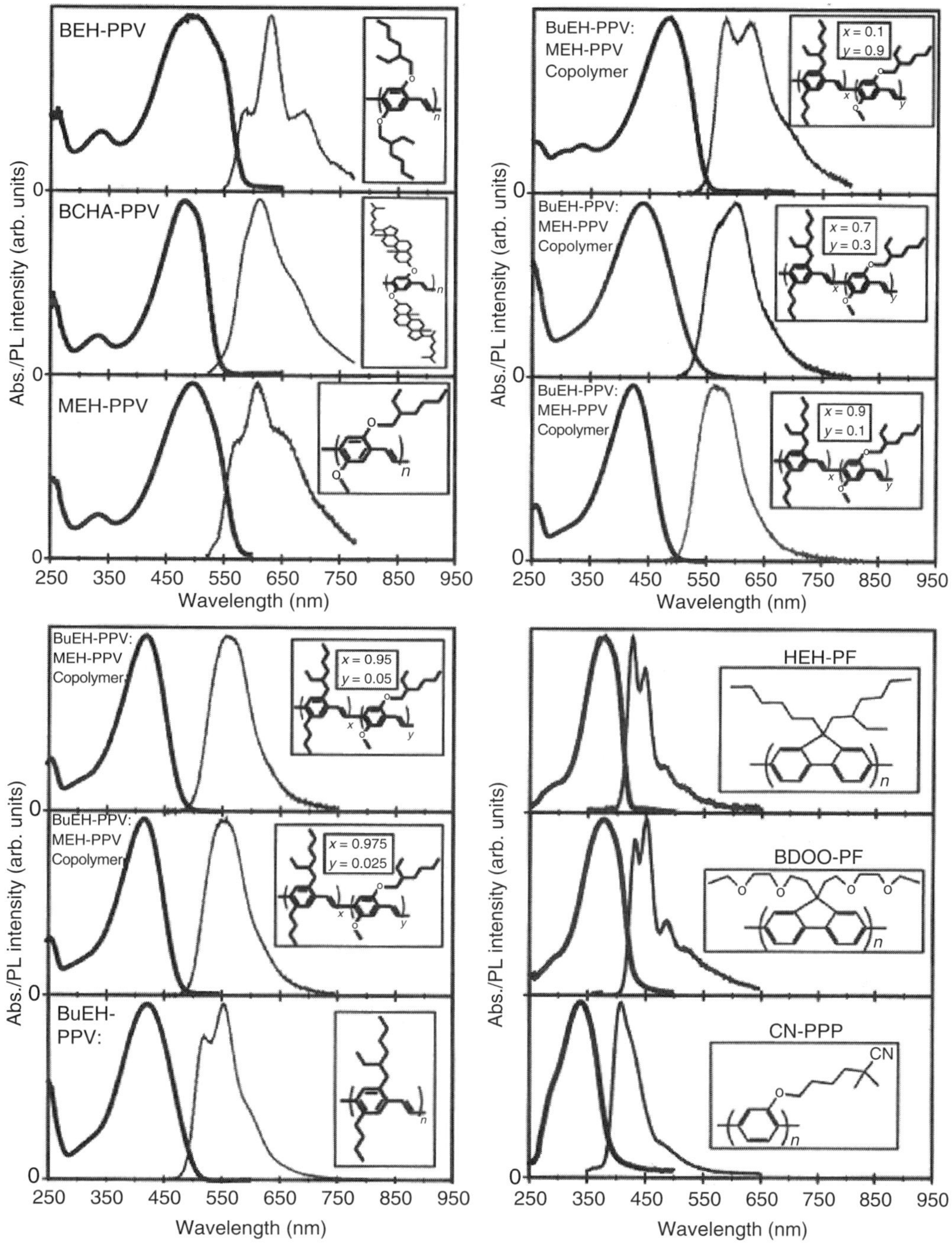

FIGURE 13.14 Absorption and emission spectra of a number of conjugated polymers with different molecular structures. The emission colors span the visible spectrum.

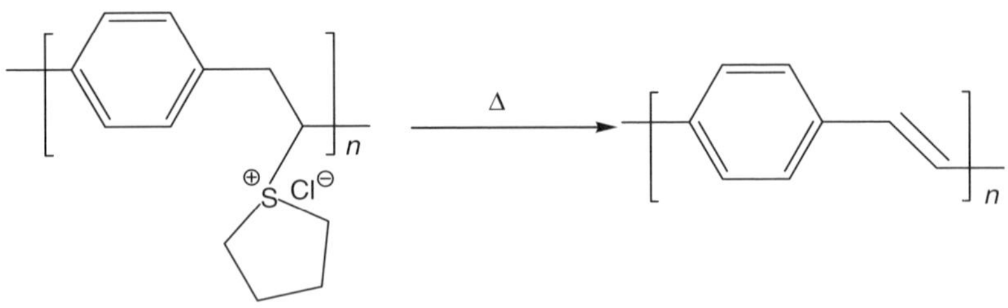

FIGURE 13.15 Synthesis of PPV through a precursor polymer.

is subsequently converted into PPV by thermal treatment. The synthesis of PPV through a precursor polymer is shown in Figure 13.15. The conversion temperature can be reduced to 100°C, from over 200°C, by using bromide salt, instead of the chloride counterpart, or by vacuum treatment.

Since 1991, Fred Wudl and his colleagues demonstrated the synthesis of soluble MEH-PPV, and many PPVs derivatives have been reported. These include alkyl- and alkoxy-substituted, cyano-substituted, silyl-substituted, and phenyl-substituted PPVs. The representative molecular structures of PPV derivatives are shown in Figure 13.16.

13.4.1.2 Polyphenylenes

Polyphenylenes (PPPs) have received considerable attraction due to their blue color emission. Like PPV, the unsubstituted PPP is insoluble in any solvent. Soluble PPPs can be designed by incorporation of alkyl-, aryl-, or alkoxy-solublizing groups into their side chains. The representative molecular structures of PPP derivatives are shown in Figure 13.17.

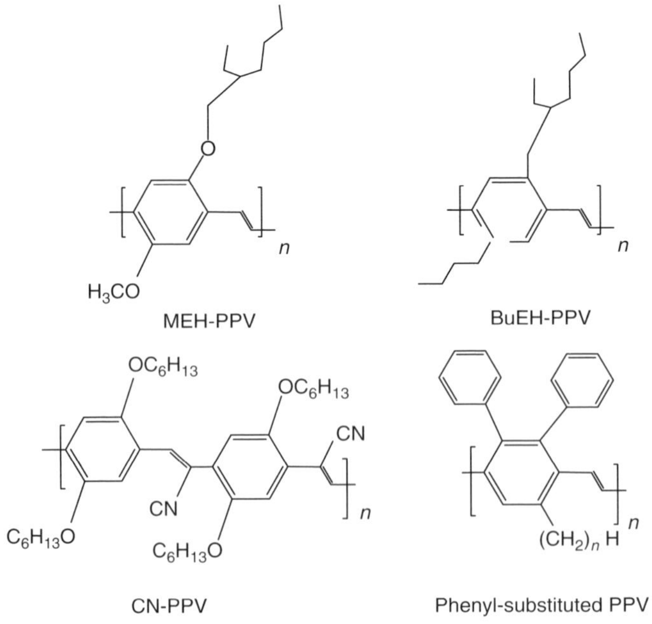

FIGURE 13.16 Molecular structures of representative PPV derivatives.

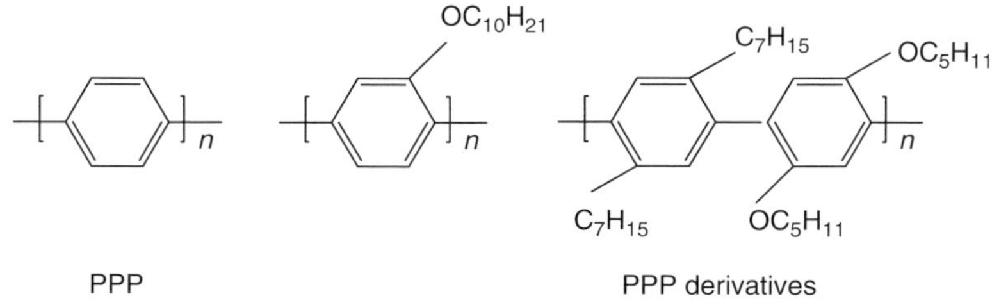

FIGURE 13.17 Representative molecular structures of PPP derivatives.

13.4.1.3 Polyfluorenes

Polyfluorenes (PFs) with substitutes at C-9 are soluble in conventional organic solvents. PFs and their copolymers have evolved as a major class of emitting materials for PLEDs. Solid-state PL yield of up to 80% have been reported, and PLEDs fabricated from PFs and their copolymers exhibit blue, green, and red emissions. The molecular structures of PFs and their copolymers are shown in Figure 13.18.

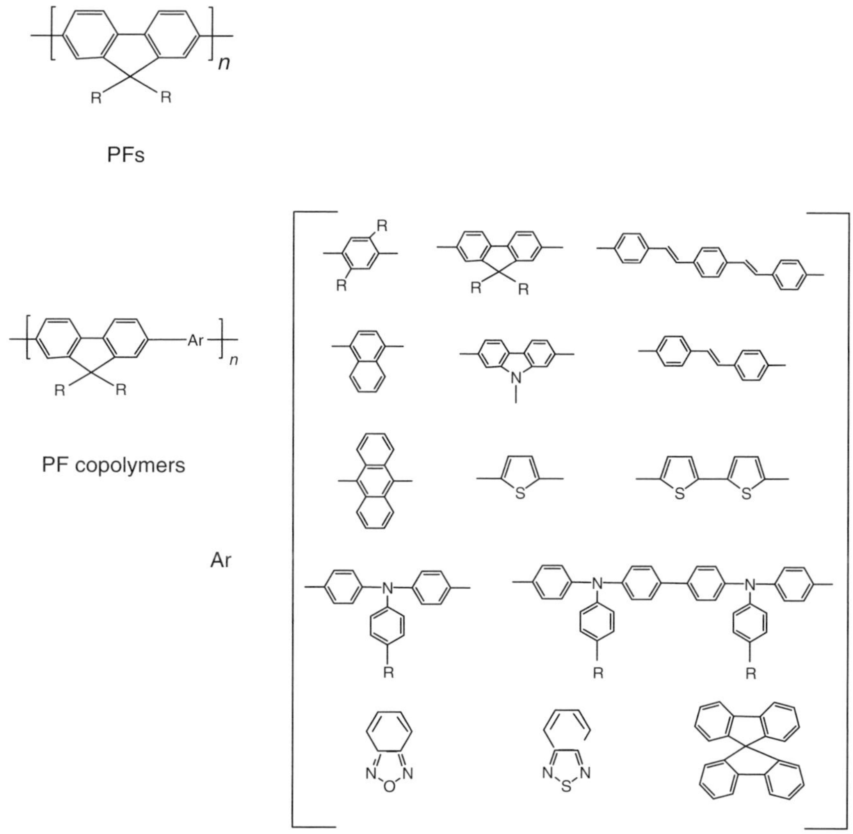

FIGURE 13.18 Molecular structures of PFs and examples of monomers, Ar, used with PF copolymer.

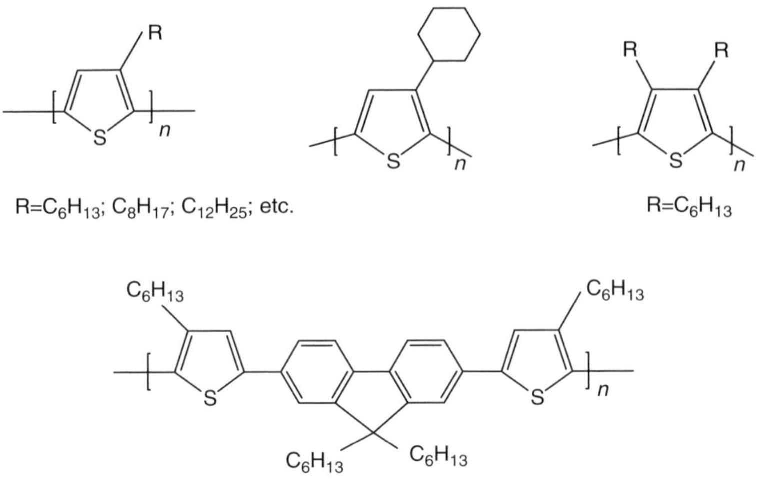

FIGURE 13.19 Representative molecular structures of PTs and their derivatives.

13.4.1.4 Polythiophenes

Soluble poly(3-alkythiophene)s with alkyl chain length varying from 6 to 22 carbon atoms have been reported. The emission intensity increases with increasing side chain length, and emission wavelength is blue-shifted with cyclohexyl group substituents or alkyl groups substitutions in both 3-and 4-positions. Since the PTs have low PL quantum yield, the PTs-based PLEDs are made from their derivatives. Figure 13.19 shows the representative molecular structures of PTs and PT derivatives.

13.4.2 Anode and Cathode

13.4.2.1 Anodes

Relatively few materials have been explored as anodes for PLEDs. An ITO is widely used as the anode in PLEDs because it has the virtue of optical transparency, but it is not a well-controlled material. Several alternative materials have been recently examined as anodes. These materials include doped ITO (Ge-doped ITO and fluorine-doped tin oxide [FTO]), doped zinc oxide (Al-doped zinc oxide [AZO]), and transparent conductive oxides (TCO).

13.4.2.2 Cathodes

Different cathode metals have been employed in device architecture with varying degree of success. The most obvious change to the choice of electron-injecting media is to vary the electronic work-function and observe performance. Table 13.1 summarized metallic electronic work-functions and ionization potentials.

The attempt to use Ca, K, and Li for effective cathode materials revealed that they exhibit poor corrosion resistance and high chemical reactivity with the organic medium. Thus, a variety of low work-function metal alloys such as Mg–Ag and Al–Li, and metal compounds liking alkali metal compounds/Al, are used for cathodes.

LiF and CsF exhibit pronounced differences from those reactive alkali metal compounds, as it is thermodynamically stable with respect to either Al or organic/polymeric materials. Devices with the bilayer electrode showed significantly better *I–V* characteristics and higher electroluminescence efficiencies than that with a standard Mg, Ag, Al cathode. Therefore, LiF/Al and CsF/Al have been applied to both PLEDs and organic light-emitting diodes (OLEDs) to form an effective electron injector.

TABLE 13.1
Electronic Properties of Typical Electrode Metals

Element	Ionization Potential (eV)	Preferred Work Function (eV)
Cs	3.89	2.14
K	4.34	2.30
Ba	5.21	2.70
Na	5.14	2.75
Ca	6.11	2.87
Li	5.39	2.90
Mg	7.65	3.66
In	5.79	4.12
Ag	7.58	4.30
Al	5.99	4.28
Nb	6.88	4.30
Cr	6.77	4.50
Cu	7.73	4.65
Si	8.15	4.85
Au	9.23	5.10

13.4.3 Hole-Injection/Transporting Materials

13.4.3.1 Hole-Injection Materials

Besides surface treatments of ITO, the other approach to enhance hole injection is to insert a nanometer thick layer with its HOMO between the ITO fermi level and the HOMO of the hole-transporting layer (HTL). This layer is termed as the hole-injection layer (HIL). HIL creates a ladder-type energy structure, which has shown to improve hole injection. A variety of materials have been shown to act as a HIL: copper phtalocyanine (CuPC); alkaline halogen such as LiF, amorphous carbon, and platinum; and conductive polymers such as PEDOT:PSS, starburst polyamines, and polyaniline.

13.4.3.2 Hole-Transporting Materials

The hole-transport layer in PLEDs is to facilitate hole injection from the anode into the conjugated polymer layer, accepting holes, and transporting injected holes to the emitting layer. HTL also functions to block the electrons from escaping from the emitting layer to the anode. Therefore, hole-transporting materials should fulfill the requirements of energy level matching for the injection of holes from the anode. They should posses electron-donating properties, and their anodic oxidation processes should be reversible to form stable cation radicals. They also should form homogenous thin films with both morphological and thermal stability. Table 13.2 summarizes widely used hole-transporting materials and their physical properties. Recently, water-soluble conjugated polymers used as the HTL have been reported.

13.4.4 Electron-Transporting Materials

The electron-transport layer (ETL) in PLEDs is to facilitate electron injection from the cathode into the conjugated polymer layer, accepting electron, and transporting injected electron to the emitting layer. ETL also functions to block the holes from escaping from the emitting layer to the cathode. Therefore, electron-transporting materials should fulfill the requirements of energy level matching for the injection of electron from the cathode. Their electron mobility should be desirably high. They also should form homogenous thin films with both morphological and thermal stability. Table 13.2 summarizes

TABLE 13.2
HTL/ETL Materials and Their Glass-Transition Temperatures and Oxidation/Reduction Potentials

Materials	T_g (°C)	$E_{1/2}^{ox}/E_{1/2}^{red}$	Functions
m-MTDATA	75	0.06	HTL
1-TNATA	113	0.08	HTL
t-Bu-TBATA	203	0.09	HTL
o-PTDATA	93	0.06	HTL
TFATA	131	0.08	HTL
TCTA	151	0.69	HTL
p-MTDAPB	110	0.64	HTL
TFAPB	150	0.61	HTL
TPTE	140		HTL
p-BPD	102	0.50	HTL
TPD	60	0.48	HTL
A-NPD	100	0.51	HTL
CBP		0.72	HTL
OXD-7			ETL
TAZ			ETL
TPOB	137	−2.10 ($E_{1/2}^{red}$)	ETL
Spiro-PBD	163		ETL

Note: Vs. Ag/Ag^+.

some electron-transporting materials and their physical properties. Recently, water-soluble conjugated polymers used as the ETL have also been reported.

13.5 ACCURATE MEASUREMENT OF POLYMER LIGHT-EMITTING DIODES DEVICE PARAMETERS

A prime concern of the PLED community is the level of uncertainty and inconsistency of results found in published scientific literature. Here, we summarize an accurate method for measurement of PLED optical properties.

In addition to transmittance, absorbance, and color temperature, light is described by the following:

Flux the total luminous power (lm)
Intensity the angular concentration of flux (cd)
Illuminance the surface density of incident flux (lx or lm/m^2)
Luminance the intensity emitted per unit area (cd/m^2)

Table 13.3 summarizes the various quantities, their units, and the related measurement techniques.

Because the major application of the PLEDs is for displays, the response of the human eye, described by the photopic luminosity function, must be taken into account. The photopic luminosity function is shown in Figure 13.20. By using the photopic luminosity function, the radiance (W/[sr m^2]) is converted into the luminance (cd/m^2 or lm/[sr m^2]). Therefore, photometry is used to measure the forward viewing luminance at the surface of the PLEDs.

TABLE 13.3
Quantity and Unit of Measurement

Technique of Measurement	Type of Measurement	Quantity	Unit of Measurement
Photometry	Total flux	Total luminous flux	lm
Radiometry	Total flux	Total radiant flux	W
Spectroradiometry	Total flux	Total spectral flux	W/nm
Photometry	Angular intensity	Luminous intensity	cd = lm/sr
Radiometry	Angular intensity	Radiant intensity	W/sr
Spectroradiometry	Angular intensity	Spectroradiometric intensity	W/(sr nm)
Photometry	At a surface	Illuminance	lx = lm/m^2
Radiometry	At a surface	Irradiance	W/m^2
Spectroradiometry	At a surface	Spectral irradiance	W/(m^2 nm)
Photometry	At a surface	Luminance	cd/m^2
Photometry	At a surface	Radiance	W/(sr m^2)
Spectroradiometry	At a surface	Spectral radiance	W/(sr m^2 nm)

The luminous intensity (luminance) can be determined by measuring the flux in any given solid angle, Ω (the ratio of the size of aperture divided by the square of the distance between the light and the aperture). Consider a flat emitting surface, each point of which emits light equally in all directions; i.e., a Lambertian source. The PLEDs is a Lambertian source if the luminous intensity follows the cosine law when measured with a small area detector placed far away from the surface (very small Ω, see below).We will assume that for the PLEDs, deviations from Lambertian emission are relatively small.

To measure the luminous intensity, one must first choose a reference direction for the measurements, and then determine the solid angle to be used in the measurement. For display applications, the reference direction should be chosen as the forward viewing direction (along the direction perpendicular to the surface of the PLEDs). The luminous intensity is defined as the emission in cd/m^2 from the emitting surface.

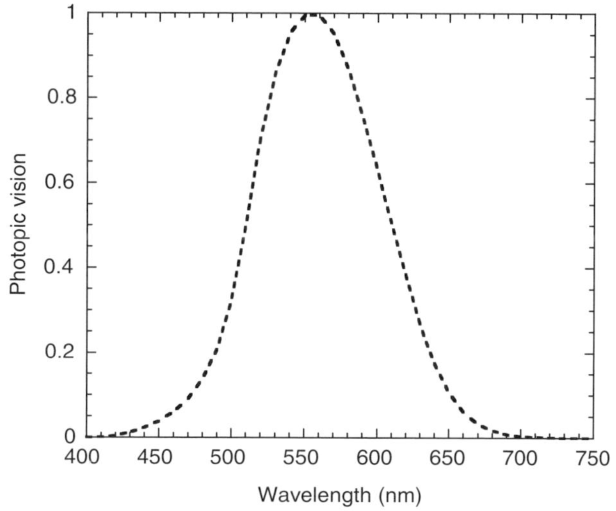

FIGURE 13.20 The photopic luminosity function.

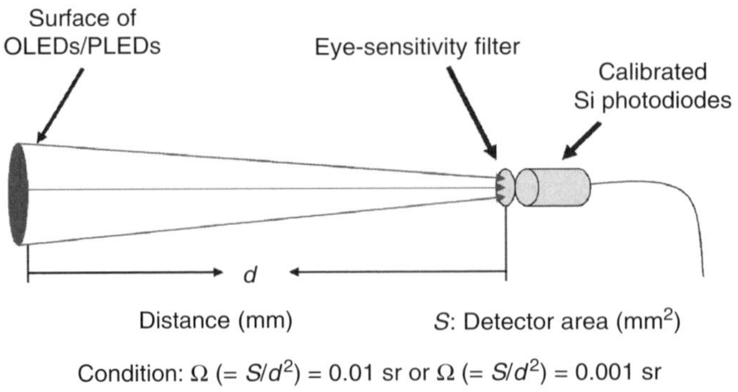

FIGURE 13.21 Configuration for measuring the luminous intensity of PLEDs.

An accurate and convenient configuration for measuring luminous efficiency from the PLEDs is shown in Figure 13.21. Because the luminous efficiency strongly depends on the PLEDs emission spectrum (even for constant quantum efficiency) an eye-sensitivity filter is mounted directly onto the surface of the calibrated photodiode (see Figure 13.21).

Assuming a Lambertian intensity profile and a disc-shaped source with radius r, the light intensity on a point detector placed a distance d away (see Figure 13.21) can be expressed as follows:

$$I = I_o\left(\frac{r^2}{r^2 + d^2}\right) \tag{13.8}$$

For a detector with a finite area, this expression can be used provided the detector subtends a sufficiently small solid angle (see Figure 13.21). The condition recommended by Commission Internationale de I'Eclairage (CIE) is that $\Omega \leq 0.01$ sr. Under these conditions, the luminous intensity (Λ, in candela) can be obtained by comparing the signal from the PLEDs with that obtained from a calibrated lamp. The brightness is then given by $L = \Lambda/\pi r^2$.

The data from a set of such measurements are shown in Figure 13.22 (for these data, $r = 2$ mm). As expected, the signal falls off as d^{-2}. Note that even for $d/r = 1$, there is no serious error from the finite area of the detector.

Once the luminance, L (cd/m^2), is accurately measured, the luminous efficiency, LE (cd/A), luminous power efficiency, PE (lm/W), and external quantum efficiency, η_{ext} (the ratio of the number of photons emitted by the PLEDs into the viewing direction of the number of electrons injected) can be determined using the following expressions:

$$\begin{aligned} \text{LE(cd/A)} &= \frac{L}{j} \\ \text{PE(lm/W)} &= \frac{\pi * L}{jV} \\ \eta_{ext}(\%) &= \frac{5.0 \times 10^3}{(h\nu)\Phi(\lambda)}\text{LE} \end{aligned} \tag{13.9}$$

where

j is the current density
V is the applied voltage
$h\nu$ is the photon energy (in eV) of the emission
$\Phi(\lambda)$ is the photopic luminosity function (Figure 13.20)

Another option is to place the PLEDs into an integrating sphere containing a calibrated detector, and then to measure the total output from the device. In an integrating sphere, all the emitted photons are collected, including those that are guided to the edge of the substrate. Note, however, that photons emitted from the edge of the substrate are not useful in display applications. Thus, the integrating sphere approach must be used with caution. If the edges of the PLEDs are sealed to absorb the waveguided and scattered light, the external quantum efficiency can be accurately obtained. Once the external quantum efficiency is measured, *L*, LE, and PE can be calculated using the above equations. The values obtained by using the integrating sphere are consistent with corresponding values obtained from the configuration shown in Figure 13.22.

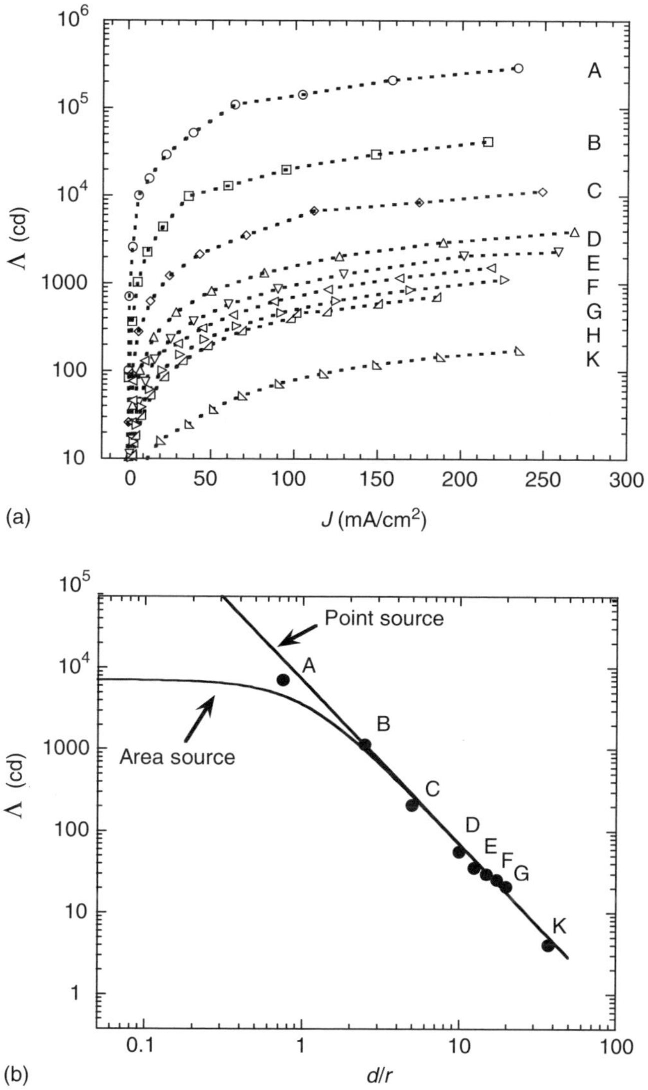

FIGURE 13.22 (a) The luminous intensity (Λ, cd) versus current density (*J*, mA/cm^2) measured at the different distances (*d*, mm) between the surface of the PLEDs and the calibrated photodiodes. (A) $d = 3$ mm; (B) $d = 10$ mm; (C) $d = 20$ mm; (D) $d = 40$ mm; (E) $d = 50$ mm; (F) $d = 60$ mm; (G) $d = 70$ mm; (H) $d = 80$ mm; and (K) $d = 150$ mm. The PLEDs are made from MEH-PPV. The surface area of calibrated photodiodes is 4×4 mm^2.

The condition for measurement of PLEDs carried out in our laboratories at UC Santa Barbara are the following: the surface area (S) of the calibrated photodiode with eye-sensitivity filter is 4 × 4 mm; the distance (d) between the photodiode and the PLEDs is 4 cm; i.e., $\Omega = 0.01$ sr ($\Omega = S/d^2$).

Note that as indicated by Optronic Laboratories, the response of a diffuser/fiber combination is far from uniform or ideal. Not only does the fiber/diffuser combination fail to provide the correct measurement area but also the nonuniform spatial response makes the result highly sensitive to alignment. Therefore, the fiber/diffuser combination should be avoided if the PLEDs intensity is to be correctly measured.

13.6 SUMMARY

Research on PLEDs has achieved significant progress in the last decade. However, there are some challenges that limit the widespread applications of PLEDs in commercial products. Among the most important challenges are achieving high materials purity, low-cost, high brightness, and long operational lifetimes. For emissive devices, impurities may result in exciton quenching, and hence decrease the emission efficiency, and decrease operational stability. In addition to the impurities of materials, the challenges preventing the widespread manufacturing and the use of the white PLEDs for solid-state lighting are manufacturing cost, efficiency, low out-coupling efficiency, large-area devices, and low operational stability.

EXERCISE QUESTIONS

1. A PLEDs pixel emits 6000 cd/m^2 of brightness blue light peaked at 475 nm at current density of 12 mA/cm^2 and 8 V. Calculate external LE, external PE, and external quantum efficiency.
2. The measurement from electrochemistry turned out that the LUMO and HOMO of a conjugated polymer are −3.2 eV and −5.8 eV, respectively. The band gap of this conjugated polymer is 2.6 eV. Calculate ionization potential and electron affinity of this conjugated polymer.
3. What are the key factors to achieve high performance PLEDs?
4. Why apply Fowler–Nordheim tunneling model to describe conjugated polymer MIM diodes?
5. Why PLEDs have high reversed leakage current compared with OLEDs (small molecular LEDs)?

LIST OF ABBREVIATIONS

PLEDs	polymer light-emitting diodes
LUMO	lowest unoccupied molecular orbital
HOMO	highest occupied molecular orbital
ITO	indium tin oxide
W_a	the work functions of the anode
W_c	the work functions of the cathode
I	ionization potential of conjugated polymers in the solid state
A	electron affinity of conjugated polymers in the solid state
T	temperature
η_s	quantum yield for singlet emission, the fluorescence
η_t	quantum yield for triplet emission, the phosphorescence
TOF	time-of-flight
$\mu(E, T)$	charge carrier mobility
$\mu(0, T)$	low-field mobility
γ	empirically determined coefficient
SCLC	space-charge-limited-current

ILC	injection limited current
ε	the dielectric constant
ε_0	the permittivity in vacuum
Θ	the fraction of free (n_f) to trapped (n_t) space charge
F	electric fields
MIM	metal-insulator-metal
φ	the barrier height
m^*	the effective mass of holes in the semiconducting polymer
q	electron charge
h	Planck's constant
PEDOT	poly(3,4-ethylene dioxythiophene)
PSS	poly(styrene sulfonic acid)
PPV	poly(p-phenylene vinylene)
MEH-PPV	poly[2-methoxy-5-(2′-ethylhexyloxy)-*p*-phenylenevinylene]
PPP	polyphenylenes
PF	polyfluorenes
PT	polythiophenes
CuPC	copper phtalocyanine
FTO	fluorine-doped tin oxide
AZO	al-doped zinc oxide
TCO	transparent conductive oxides
OLEDs	organic light-emitting diodes
HIL	hole-transporting layer
HTL	hole-transport layer
ETL	electron-transport layer
T_g	glass transition temperature
$E_{1/2}^{ox}$	oxidization potential
$E_{1/2}^{red}$	reduction potential
Ω	solid angle
L	luminance, L (cd/m^2)
LE	luminous efficiency
PE	power efficiency
η_{ext}	external quantum efficiency
j	current density
V	applied voltage
$h\nu$	photon energy of the emission
$\Phi(\lambda)$	photopic luminosity function
CIE	Commission Internationale de l'Eclairage

ACKNOWLEDGMENTS

We gratefully acknowledge K.C. Wong Education Foundation, Hong Kong, for financial support.

BIBLIOGRAPHY

1. A.J. Heeger, *Angew. Chem. Int. Ed.*, **2001**, 40, 2591; *Solid State Comm.*, **1998**, 107, 673; *Rev. Modern Phys.*, **2001**, 73, 681.
2. D. Braun and A.J. Heeger, *Appl. Phys. Lett.*, **1991**, 58, 1982.
3. G. Yu, J. Wang, J. McElvain, and A.J. Heeger, *Adv. Mater.*, **1998**, 10, 1431.
4. P.S. Heeger and A.J. Heeger, *Proc. Nat. Aca. Sci. USA*, **1999**, 96, 12219.
5. C.D. Dimitrakopoulos and D.J. Mascaro, *IBM J. Res & Dev.*, **2001**, 45, 11.
6. M.D. McGehee and A.J. Heeger, *Adv. Mat.*, **2000**, 12, 1655.

7. U. Scherf, E.J.W. List, *Adv. Mater.*, **2002**, 14, 477.
8. S. Setayesh, D. Marsitzky and K. Müllen, *Macromolecules*, **2000**, 33, 2016.
9. U. Scherf, *J. Mater. Chem.*, **1999**, 9, 1853.
10. J. Kido, H. Shionoya, and K. Nagai, *Appl. Phys. Lett.*, **1995**, 67, 2281.
11. C. Zhang and A.J. Heeger, *J. Appl. Phys.*, **1998**, 84, 1579.
12. Y. Cao, G.M. Klavetter, N. Colaneri, and A.J. Heeger, *Nature*, **1992**, 357, 477.
13. J.C. Whitaker, *Video Display Engineering*, McGraw-Hill Companies Inc. **2001**, p. 52.
14. N.C. Greenham, R.H. Friend, D.D.C. Bradley, *Adv. Mater.*, **1994**, 6, 491.
15. A.D. Ryer, *Light Measurement Handbook*, International Light Inc., **1998**.
16. H.A.E. Keitz, *Light Calculations and Measurements*, 2nd edition, Macmillan and Co. Ltd., **1971**.
17. I.D. Parker, *J. Appl. Phys.*, **1994**, 75, 1656.
18. S.M. Sze, *Physics of Semiconductor Devices*, Wiley, New York, **1981**.
19. X. Gong, S. Wang, G.C. Bazan, and A.J. Heeger, *Adv. Mater.*, **2005**, 17, 2053.
20. K. Muellen and U. Scherf, *Organic Light-Emitting Devices*, Wiley-Vch, **2005**, p. 153.

14 Organic and Polymeric Photovoltaic Materials and Devices

Sam-Shajing Sun and Cheng Zhang

CONTENTS

14.1 Introduction ... 401
 14.1.1 Solar Energy and Photoelectric Conversion ... 401
 14.1.2 Organic Photovoltaics versus Inorganic Photovoltaics ... 403
14.2 Evolution and Optimizations of Organic Solar Cells in the Spatial Domain ... 405
 14.2.1 Single-Material Organic Solar Cells ... 405
 14.2.2 Donor/Acceptor Bilayer Heterojunction Organic Solar Cells ... 405
 14.2.3 Donor/Acceptor Blend Bulk Heterojunction Organic Solar Cells ... 407
 14.2.4 Bicontinuous-Ordered Nanostructure Solar Cells ... 409
14.3 Optimizations of Organic Solar Cells in the Energy Domain ... 412
14.4 Organic Photovoltaic Cell Fabrication and Characterization ... 414
14.5 Summary ... 418
Exercise Questions ... 418
List of Abbreviations ... 418
Acknowledgments ... 419
References ... 419

Abstract This chapter introduces basic concepts, principles, and key developments of organic photovoltaic (PV) materials and devices for students and nonexperts. Specifically, this chapter starts from background of solar energy to basic concepts and principles of PV processes in both inorganic and organic materials, followed by a presentation of several key types of organic and polymeric PV materials and solar cells from single phase Schottky cell, to donor/acceptor (D/A) bilayer heterojunction Tang cell, to D/A blend bulk heterojunction cell, to D/A bicontinuous-ordered nanostructured (BONS) cell, and to multilayer/multi-gap tandem cell. The chapter also covers certain ongoing optimization approaches for improving the organic PV materials and solar cells.

14.1 INTRODUCTION

14.1.1 Solar Energy and Photoelectric Conversion

Energy and environment have become the top two concerns and challenges of mankind nowadays, and these two issues are closely correlated to each other. As estimated, over 80% of energy sources of the human society are obtained from burning of fossil fuels such as coal, oil, and natural gas [1–3]. However, carbon dioxide and other toxic gases released from fossil fuel burning are causing many potential environmental problems such as the greenhouse effect. In addition, fossil fuel deposits on the earth are not unlimited. As a result, new technologies making use of alternative renewable and clean energy sources have become major research objectives and challenges of scientists in the twenty-first century.

Solar energy is perhaps the biggest renewable and clean energy source on the earth. The amount of incoming solar radiation per unit area on the outer surface of earth's atmosphere in a plane perpendicular to the rays is roughly 1366 W/m^2 (or 1.366 Sun, the solar constant, measured by satellite) [1]. While traveling through the atmosphere, the incoming solar radiation is partially reflected and absorbed. The degree of solar intensity reduction is dependent on the thickness of the air (air mass, AM) and therefore, the zenith angle of the Sun at a specific point on the earth's surface. So, AM0 (air mass zero) sunlight is the solar radiation that reaches the earth's upper atmosphere and has a power density equal to the solar constant. AM1.0 sunlight is the solar radiation received on the ground when the Sun is directly overhead (0° zenith angle, the angle between irradiation from a space object and the gravitation vector). AM1.5 sunlight is the solar radiation received on the ground when the Sun is at 48.2° zenith angle and has a power density of approximately 1000 W/m^2 (one Sun) on a sunny day when the receiving surface is perpendicular to the solar ray. The actual amount of solar energy at ground level of a specific location changes over the time of the year and over different weather conditions. In North America, the average insolation at ground level during an entire year (including nights and periods of cloudy weather) lies between 125 and 375 W/m^2 (calculated from satellite data collected from 1991 to 1993).

In 2006, the world's electricity consumption rate is about 1.94×10^{12} W, and the world's primary energy consumption rate (including petroleum, natural gas, coal, and electric power) is about 1.57×10^{13} W. They are only 0.0011% and 0.0090% of solar energy reaching the earth (1.740×10^{17} W above the atmosphere) [1,2]. Although solar energy can be used in many ways, including solar lighting, solar water heating, conversion of sunlight into electricity is the most desired way since electricity can be easily transported over long distances and can satisfy almost every need for energy from industry to household. This conversion can be done in two main ways. One way is to use concentrated sunlight to heat up water or air to power a steam or a turbine engine to generate electricity. Another way is to use the PV effect and develop inexpensive PV devices, which are the foci of this chapter.

Photoelectric conversion was first observed in an experiment of selenium with light by William Grylls Adams and his student Richard Evans Day in the last half of the nineteenth century [1–3]. The first practical application of PV cells was on the U.S. satellite Vanguard (to power its communication with the earth for many years) in 1955, two years after the revolutionary development of doped silicon p–n junction PV cell at Bell Laboratories. Since its beginning, inorganic semiconductor materials have dominated the field in terms of energy conversion efficiency. The power conversion efficiency of single crystal silicon PV cell has reached 25%, and multi-junction cells consisting of materials of different band gaps, i.e., Ge (0.66 eV), GaAs (1.23 eV), and GaInP (1.85 eV), have shown efficiency of about 40% [2] (Spectrolab, Inc., 2005). Due to the high manufacturing costs, PV cells are mostly used on satellites/spaceships, in remote areas, such as on oil rigs, at homes on small islands, and on portable devices. As the price of PV cells continues to drop, even in urban areas, government and businesses find that for certain devices, such as call boxes on the highways, illuminators on bus stop shelters, it is more cost effective to use solar cells instead of connecting devices to the power grid. However, the cost of electricity from PV cells is still relatively much higher than the cost of electricity generated in an advanced coal facility (20 ¢/kW h versus 3.5 ¢/kW h in 2006), preventing the massive use of PV technologies [1,2].

The high cost of inorganic semiconducting PV cells is associated with fabrication steps including crystal growth, wafer production, high-temperature processing. Organic, especially polymeric, PV materials may allow us to develop low-cost PV technology due to the following advantages:

1. Low-cost solution-based thin-film fabrication via spin coating, spraying, or printing.
2. Flexibility which may allow low-cost and large area mass production and installation.
3. Low processing temperature allows easy integration of plastic cells with other products.
4. Tunability of optoelectronic and chemical properties of polymers via molecular design and engineering.

14.1.2 Organic Photovoltaics versus Inorganic Photovoltaics

In a donor–acceptor (D/A) binary-type organic PV cell, photoelectric conversion is accomplished by five consecutive steps [3,4]:

1. Absorption of photons and the formation of excitons (tightly bound electron–hole pairs)
2. Exciton diffusion to a donor–acceptor interface
3. Charge separation at the interface
4. Charge transport to the anode (holes) and cathode (electrons)
5. Charge collection by electrodes

Similar processes are involved in inorganic p–n junction PV cells (Figure 14.1), except that free charge carriers are generated upon absorption.

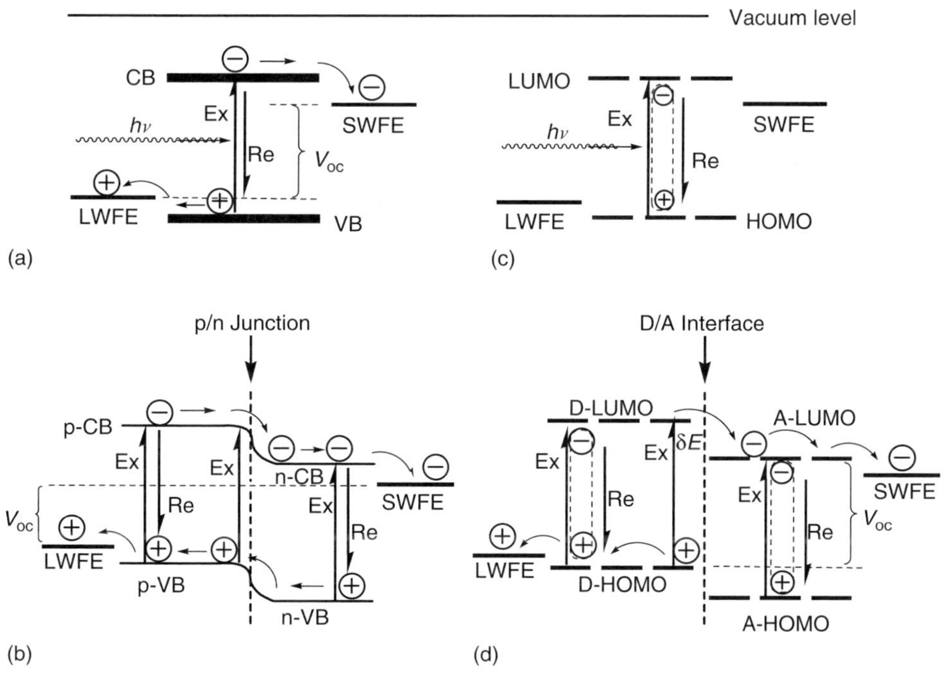

FIGURE 14.1 Frontier orbital/band energy level schemes of solar cells in (a) a single inorganic semiconductor cell (Fritz cell), (b) an inorganic p–n junction solar cell (Bell Lab cell), (c) an organic single-material solar cell (Schottky cell), and (d) an organic donor/acceptor heterojunction solar cell (Tang cell). *D,* electron donor; *A*, electron acceptor; Ex, excitation; Re, relaxation (or recombination); and δE, the energy level offset between donor-lowest unoccupied molecular orbital (D-LUMO) and acceptor-lowest unoccupied molecular orbital (A-LUMO).

The energy level schemes of both inorganic and organic solar cells are shown in Figure 14.1. Ex and Re denote excitation and recombination, respectively. In principle, materials will most effectively capture those photons that match the energy gaps of the materials. Commonly used inorganic semiconductors form conduction bands (CBs) and valence bands (VBs), where charge carriers are delocalized and easily transported, and the VB–CB energy gaps are typically between 1–2 eV (e.g., 1.1 eV for Si, 1.34 eV for GaAs). On the contrary, most organic semiconductors have optical energy gaps over 2.0 eV, larger than the energies of photons (1–2 eV) in the most intense region of the solar spectrum under AM1.5 condition. This means sunlight photon capture in organic semiconductors is generally poor.

In an inorganic single semiconductor cell as shown in Figure 14.1a, light absorption leads directly to the generation of free charge carriers including free electrons in the conduction band and free holes in the valence band. The free electrons then diffuse to and are eventually collected by the negative small work function electrode (SWFE). The free or mobile holes diffuse to and ultimately being collected by the positive large work function electrode (LWFE). The open circuit voltage (V_{oc}, i.e., the photovoltage when the external current is zero) is determined mainly by the work function difference of the two electrodes. However, electron/hole recombination is a major source of loss.

In the p–n junction inorganic solar cells (Figure 14.1b), since the photo-generated electrons and holes in either p- or n-type semiconductor can be further separated from each other by the electric field at the p–n junction, the chance of electron/hole recombination is greatly reduced. The electrons and holes can now diffuse in separate domains of the two electrodes, respectively; thus the PV effects are greatly enhanced over the single-material cells.

In contrast, in most organic and polymeric semiconductors (except a few highly crystalline systems such as pentacenes and rubrenes), the intermolecular electronic overlap or coupling of orbitals is very poor due to loose and random packing of molecules. Therefore, stable long-range delocalized conduction band and valence band cannot be formed. Instead, localized frontier orbitals (i.e., the highest occupied molecular orbital [HOMO], and the lowest unoccupied molecular orbital [LUMO], see Figure 14.1c and d) are formed. Due to lower degree of charge delocalization as well as weaker charge screening (smaller dielectric constants), photo excitations mainly generate a strongly bound electron–hole pair called exciton (see Figure 14.1c). These excitons have a very strong Coulombic attraction and thus a smaller Bohr radius (the average distance between the electron and the hole) [5]. Such small excitons, called Frenkel-type excitons—as compared to the large-sized Wannier-type excitons—are dominant in organic semiconductors [1,2]. These Frenkel-type excitons have binding energies (i.e., the minimum energy required to dissociate the exciton into free or uncorrelated electron and hole) typically over 0.1 eV, much greater than the thermal energy at 25°C ($kT = 0.026$ eV). Therefore, a secondary force is needed to dissociate the Frenkel-type excitons. The exciton can diffuse over a distance of about 5–50 nm in most organic semiconductors within its lifetime of pico to nanoseconds before it decays or relaxes back to ground state, and this does not contribute to the formation of free charge carriers. Thus, PV effects in most single-material organic semiconductor devices are typically undetectable or extremely weak. Because of this, a typical all-organic/polymer PV cells would therefore require an electron-rich molecule/polymer (donor) as well as an electron deficient molecule/polymer (acceptor), such that the frontier orbital level offsets (δE) (Figure 14.1d) between the LUMOs (or HOMOs) of donor and acceptor would constitute a potential driving force for dissociating the excitons. The donor is the one with a higher level of LUMO and HOMO when compared with an acceptor (Figure 14.1d). Once the excitons formed in the donor domain diffuse into the donor/acceptor interface, the energy offset (δE) at the interface will drive the electron to hop from LUMO of the donor to the LUMO of the acceptor at a timescale of picoseconds or femtoseconds, and leaving holes behind in the donor HOMO. In the case when an acceptor is the light absorber, a higher HOMO level in donor would enable the hole generated in acceptor's HOMO to hop into the donor's HOMO. All these charge transfer processes follow the Marcus electron transfer theory as elaborated elsewhere [6,7].

However, in a highly ordered and well-packed organic crystalline structure where frontier orbitals are spatially well overlapped between adjacent molecules, the formation of stable valance and conduction bands may become feasible. Recent experimental data on the charge mobility μ (up to over 20 cm^2/V s) of highly ordered crystalline rubrene may provide a direct evidence for the band formation in an organic crystal [8]. It should be noted that such band behavior is rarely observed for most organic semiconductors.

It is common to both inorganic and organic PV cells that photon absorption in areas far away from the p–n junction or D/A interface does not contribute significantly to the energy conversion due to charge carrier recombination in inorganic semiconductors, or exciton decay in organic semiconductors. This is particularly true for organic PV cells because organic molecules and polymers have very short average exciton diffusion lengths (AEDL), (i.e., the average distance excitons can travel before they are quenched via recombination) about 5–100 nm. Additionally, solution-processed polymer films usually have much shorter AEDL than organic crystals (processed from either high vacuum or solution). After excitons reach the D/A interfaces and dissociate into electrons and holes, the generated charge carriers generally transport to the two opposite electrodes via the thermal hopping mechanism. In addition to the density of photo-generated charge carriers, the charge mobility (μ, measured in cm^2/V s) also critically affects the conductivity (see Chapter 3), and therefore, the efficiency of the PV cell.

14.2 EVOLUTION AND OPTIMIZATIONS OF ORGANIC SOLAR CELLS IN THE SPATIAL DOMAIN

14.2.1 Single-Material Organic Solar Cells

The earliest photoconductivity observation in organic materials was reported in anthracene in 1907 [9]. Later, some organic dyes and biological materials, such as methylene blue, carotenes, chlorophylls, and related porphyrins or phthalocyanines were also found to exhibit photoconductive or very weak PV effects with a best power conversion efficiency η_p of about 0.05%. Conjugated polymers such as polyacetylene, polythiophenes, and poly-*p*-phenylenevinylenes (PPVs) were also investigated for PV effects, with best power conversion efficiency of 0.1% achieved under white light illumination [10]. These first-generation organic PV cells are composed of a single layer of organic semiconductor sandwiched between two metal electrodes of different work functions. A thin layer of Schottky barrier forms at the contact of the metal electrode with an SWFE and typical p-type of organic layer (Figure 14.2). The diffusion of electrons from active metal into p-type organic material generates a depletion region (depleted of holes). The resulting field causes LUMO and HOMO to rise up (appears to be a band bending toward the Schottky contact) [11]. This field can cause dissociation of excitons generated in or diffused into the depletion region. Thus, the generated electrons are collected by the SWFE and holes are transported across the film to the LWFE. Because of short diffusion length, most excitons generated outside the depletion region do not contribute to photocurrent. This explains the very low photoelectric efficiency of single-material organic PV cells.

14.2.2 Donor/Acceptor Bilayer Heterojunction Organic Solar Cells

The second-generation PV cells consist of a D/A bilayer as shown in Figure 14.3. This bilayer heterojunction-type cell was first demonstrated by C. Tang in 1986 with a surprisingly high power conversion efficiency (η_p) of 1% under AM2 illumination [12]. Since then, many other D/A bilayer systems have been investigated intensively, including, for instance, molecule-fullerene, polymer-fullerene, and polymer-polymer D–A bilayer systems, with a best 1% power conversion efficiency achieved in a PPV/C60 bilayer cell [13]. As shown in Figure 14.3, once a photo-generated Frenkel exciton in either donor or acceptor layer diffuses to the D/A interface, the electrons would transfer

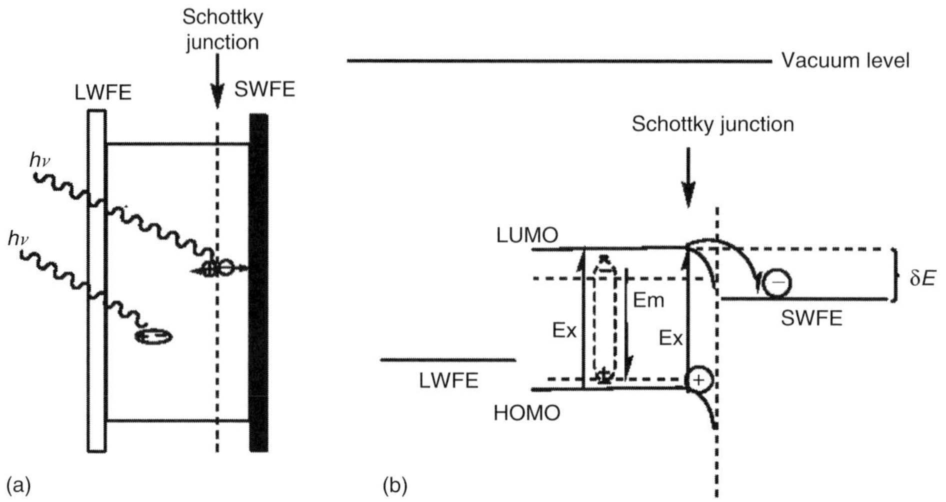

FIGURE 14.2 First-generation single-material organic PV cell or Schottky cell with (a) device structure and (b) energy diagram.

to or remain in the acceptor LUMO and the holes would transfer to or remain in the donor HOMO. The frontier orbital energy offsets between the donor and an acceptor are the key driving force for the exciton dissociation (or charge separation) [3]. Due to both the internal field across the electrodes and chemical potential driving forces [5], the electrons and holes then diffuse to their respective electrodes much quicker than in single-layered cells. The likelihood of carrier recombination is relatively small due to electrons and holes now diffusing in two separate layers or domains. However, the efficiency of this type of cell is limited due to short exciton diffusion length. Figures 14.4 and 14.5 exhibit the chemical structures of mostly used or studied organic donors and acceptors, and Figure 14.6 shows the frontier HOMO/LUMO levels of some commonly used conjugated polymers.

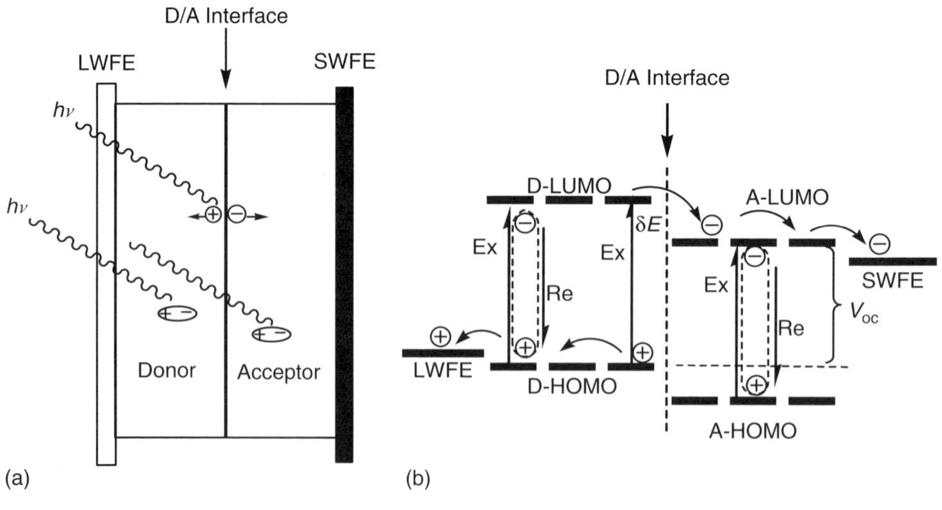

FIGURE 14.3 Second-generation D/A bilayer heterojunction organic PV cell or Tang cell with (a) device structure and (b) energy diagram.

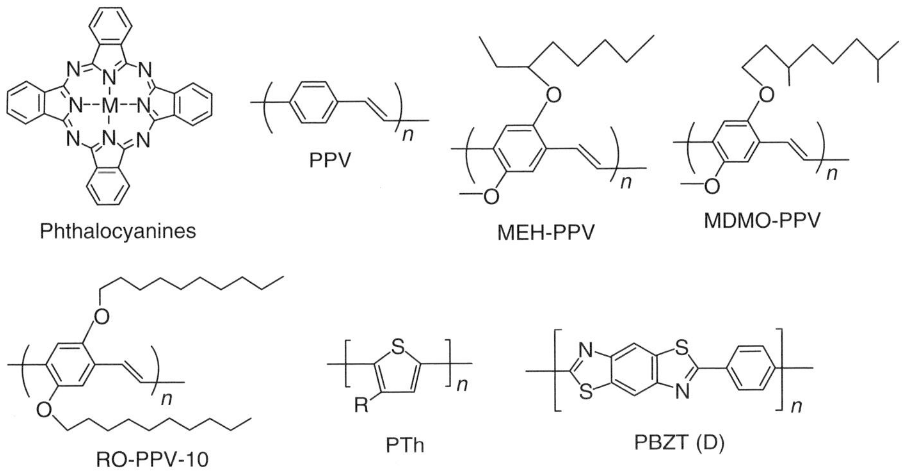

FIGURE 14.4 Representative organic/polymeric electron donors (p-type semiconductors). PPV, poly-*p*-phenylenevinylene; MEH-PPV, poly(2-methoxy-5-(2-ethylhexyloxy)-1,4-phenylene vinylene); MDMO-PPV, poly [2-methoxy-5-(3′,7′-dimethyloctyloxy)-1–4-phenylene vinylene]; PBZT, poly(*p*-phenylene benzobisthiazole); PTh, polythiophene; and RO-PPV-10, poly(2,5-didecyloxy-1,4-phenylene vinylene).

14.2.3 Donor/Acceptor Blend Bulk Heterojunction Organic Solar Cells

A logical way to enhance D/A binary PV cell efficiency is to maximize the D/A interface via intimately blending donor and acceptor materials to form bulk heterojunction, and that at least the dimension of photon-capturing domain is on the same order of AEDL (between 5 and 50 nm in most organic semiconductors). Efficiency indeed has been greatly improved in such bulk heterojunction cells (so called the third-generation organic solar cells). For instance, it was found that a cell with a D–i–A trilayer structure, where D represents a metal-free phthalocyanine, A represents a perylene tetracarboxylic diimide (PTCDI), and i represents a D/A co-deposited layer, had nearly doubled the

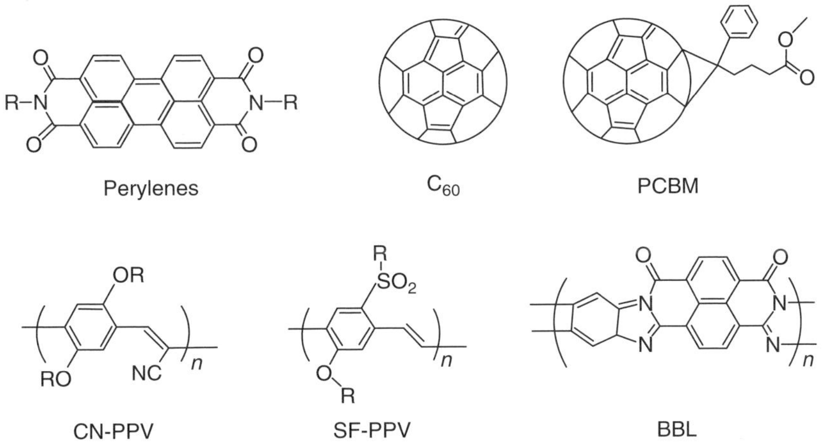

FIGURE 14.5 Representative organic/polymeric electron acceptors (n-type semiconductors). CN-PPV, poly(2,5,2′,5′-tetraalkoxy-7,8′-dicyanodi-*p*-phenylene vinylene); BBL, poly(benzamidazobenzo phenanthroline); PCBM, [6,6]-phenyl-C_{61}-butyric acid methyl ester; and SF-PPV, poly[2-alkoxy-5-alkanesulfonyl-1–4-phenylene vinylene].

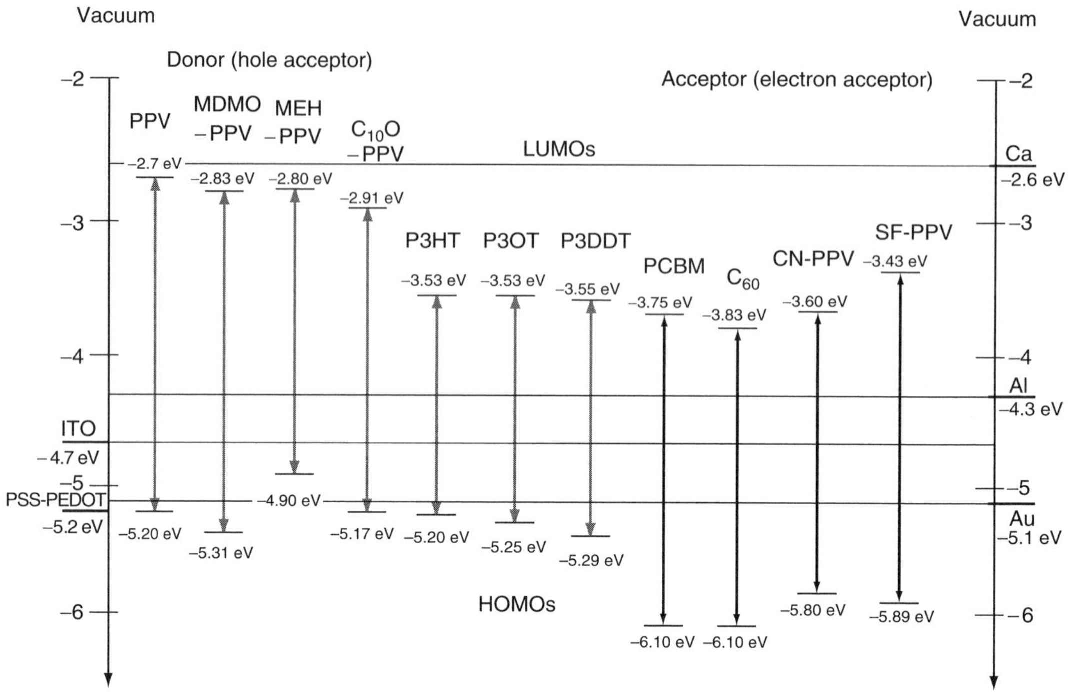

FIGURE 14.6 Frontier orbital levels of representative polymeric electron donors and acceptors; the work functions of several representative electrodes are also shown on the sides. P3HT, poly(3-hexylthiophene); P3OT, poly(3-octyl thiophene); and P3DDT, poly(3-dodecylthiophene).

photoelectric power conversion efficiency over that of corresponding D–A bilayer cell under similar conditions [14]. So far, numerous bulk heterojunction cells using PPV or polythiophene derivatives (e.g., MEH-PPV, MDMO-PPV, P3HT, or P3OT) as the donor, and CN-PPV or fullerene derivatives (e.g., PCBM) as the acceptor have been intensely studied [3,4], and near unity photoinduced charge separation (also called internal quantum efficiency), and between 1% and 6% photoelectric power conversion efficiencies have been reported under different conditions [15–17]. The higher efficiencies of D/A blend cells over D–A bilayered cells can be attributed mainly to the reduction of exciton loss due to a larger D/A interface, and also due to the reduction of photon loss as the films now can be made thicker to harvest more photons. However, carrier loss could become a major problem as phase continuity from any donor or acceptor domain to the anode or cathode can hardly be realized in a physical blend. In other words, charge carriers can be easily trapped in isolated domains and eventually quenched via charge recombination. Additionally, if the donor and acceptor are in direct contact with both electrodes, carrier recombination at the organic/electrodes interface would be severe, and carrier collection efficiency at electrodes would be poor. This problem may be minimized by using a D–i–A trilayer structure, where i layer can be prepared by co-deposition of D and A or by interdiffusion of D and A [18]. However, this trilayer structure does not guarantee good performance because the morphology of an i layer still needs to be optimized (Figure 14.7).

In D/A blend cells, since electron and hole conduction pathways may not be well separated, high degree of charge carrier recombination and low carrier mobility are expected. Due to the typical poor charge carrier mobility of organic semiconductors (orders of magnitudes lower than typical inorganic semiconductors), thickness of active materials in an organic PV cell must be very small (about 100–200 nm), much smaller than the thickness of a typical inorganic p–n junction cell (on

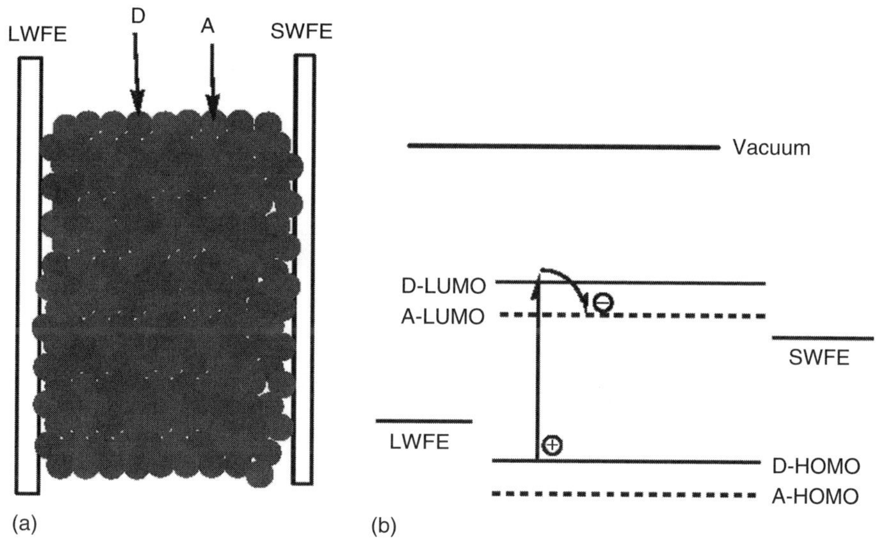

FIGURE 14.7 Third-generation D/A blend-type organic PV cells or bulk heterojunction cells with (a) device structure and (b) energy diagram.

microns scale). However, organic semiconductors have relatively strong absorption coefficients (10^4–10^5 cm^{-1}), enabling substantial absorption even in 100 nm thick devices.

In recent years, most device optimization efforts were focused on the morphology optimization of D/A blend cells. Different thermal annealing techniques have been used to improve the morphology. For example, it was found that slow evaporation of 1,2-dichlorobenzene solvent (for P3HT and PCBM) allows the formation of P3HT domains of better crystallinity and resulted in an increased hole mobility, or reduced series resistance, and balanced charge transport [19]. By changing solvent evaporation protocol from drying on a 70°C hot plate (fast) to drying at room temperature under the cover of a Petri dish (slow), the efficiency was increased from 1.4% to 4.4% [16]. A similar improvement on P3HT/PCBM PV cell (higher crystallinity, improved transport, and up to 5% power conversion efficiency) was realized by thermally annealing the film at 150°C for 30 min [17].

14.2.4 Bicontinuous-Ordered Nanostructure Solar Cells

Although significant improvement in power conversion efficiencies has been obtained in D/A blend systems (power conversion efficiency up to about 5%), the desired nanoscale phase separation, domain size uniformity, and phase bicontinuity can hardly be realized in simple physical D/A blends. Stability and reproducibility of blends are also issues becauses the morphology of a blend is strongly dependent on processing and operational conditions. A bicontinuous-ordered nanostructure (BONS) solar cell (Figure 14.8) with D/A separated phases appears to be ideal for minimizing both the exciton loss and the carrier loss [6,20,21]. A variety of ways to achieve BONS has recently been investigated. For inorganic or inorganic–organic hybrid cells, these include the use of vertically aligned carbon nanotubes, semiconducting nanorods, or nanoporous inorganic semiconductor thin films grown on the substrate as both the electron acceptor and the template for an electron donating polymer that could fill in the pores [3,21,22] (Figure 14.9). The main challenges are (1) to prepare uniformly distributed and vertically aligned tubes/rods on substrates coated with transparent electrode (indium tin oxide [ITO]), (2) to control diameter and spacing around 5–50 nm and length of over 100 nm for the nanorods/tubes assembly, and (3) to obtain good packing and alignment of conjugated polymer chains in the confined space between the rods or tubes.

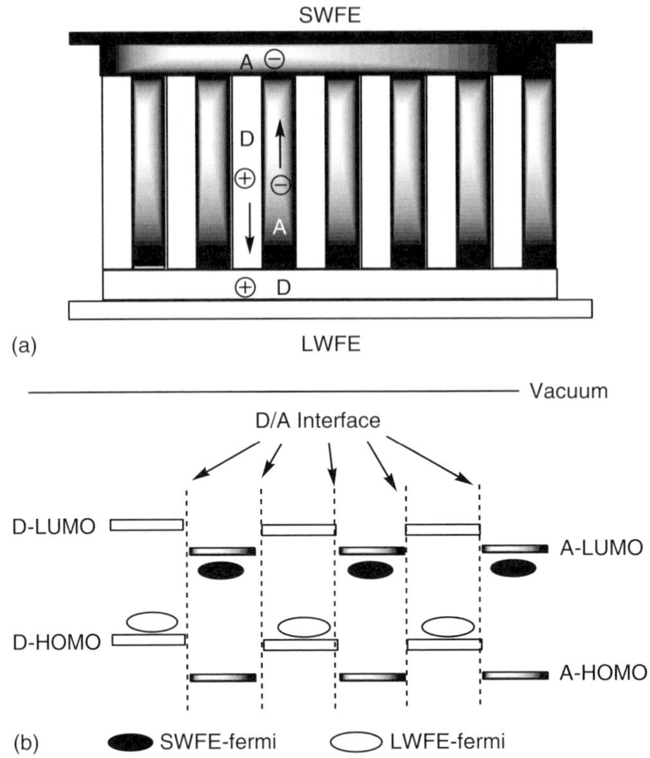

FIGURE 14.8 Fourth-generation D/A BONS-type PV cells with (a) device structure and (b) energy diagram. The vertically aligned donor or acceptor columns shown in (a) can be inorganic semiconductor nanorods, nanowires, carbon nanotubes, conjugated semiconducting polymers, or discotic-type stacked organic small molecules.

BONS can also be achieved via block copolymers. Figure 14.10 shows some representative nanostructures of flexible diblock copolymers as functions of volume fraction and temperature. Among them, the cylinder, gyroid, and lamellar structures appear satisfying the BONS requirement and appear ideal for use in PV cells. One advantage of block polymer is that its two-phase nanostructure

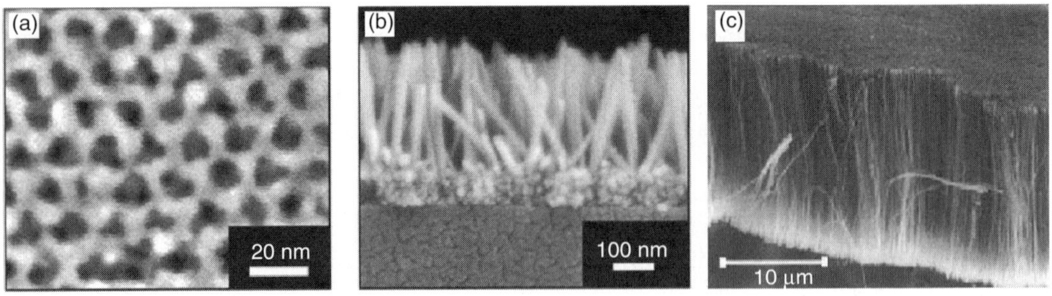

FIGURE 14.9 Scanning electron microscope images of (a) mesoporous titania (plane view). (From Coakley, K., Liu, Y., McGehee, M., Frindell, K., and Stucky, G., *Adv. Funct. Mater.*, 13, 301, 2003.) (b) A brush of zinc oxide nanowires (cross-section view), respectively. (Courtesy of S. Shaheen, *National Renewable Energy Lab*) (c) A typical transmission electron microscopic image of carbon nanotube coaxial nanowires produced by pyrolysis of FePc. (From Sun, S. and Sariciftci, N.S. (Eds.), *Organic Photovoltaics: Mechanisms, Materials and Devices*, CRC Press, Boca Raton, FL, 2005.)

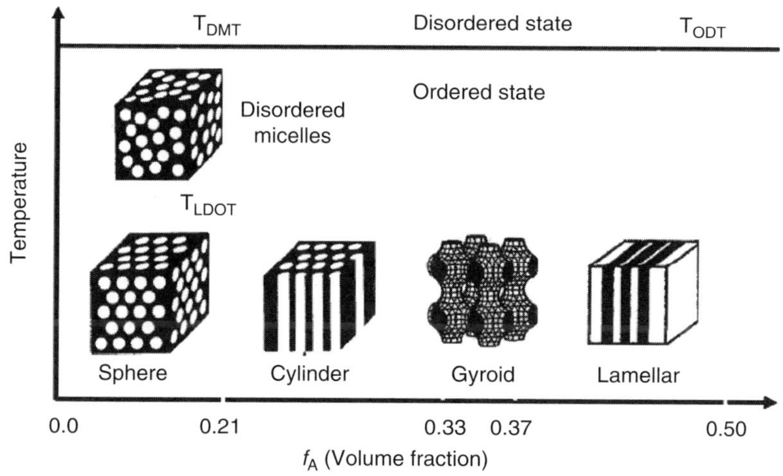

FIGURE 14.10 Some representative self-assembly structures of diblock copolymers.

is thermodynamically more stable compared to physical blends [23]. Figure 14.11 shows two representative D/A block copolymers developed for PV applications [24–26]. For instance, the PV properties of a –DBAB-block copolymer have been shown to be much better than the corresponding D/A blend, and XRD/AFM studies revealed the formation of lamellar domains in –DBAB-block copolymer thin film after simple thermal annealing [26]. However, the size of such lamella domains may be limited due to the dispersive nature of the donor and acceptor blocks in the block copolymer.

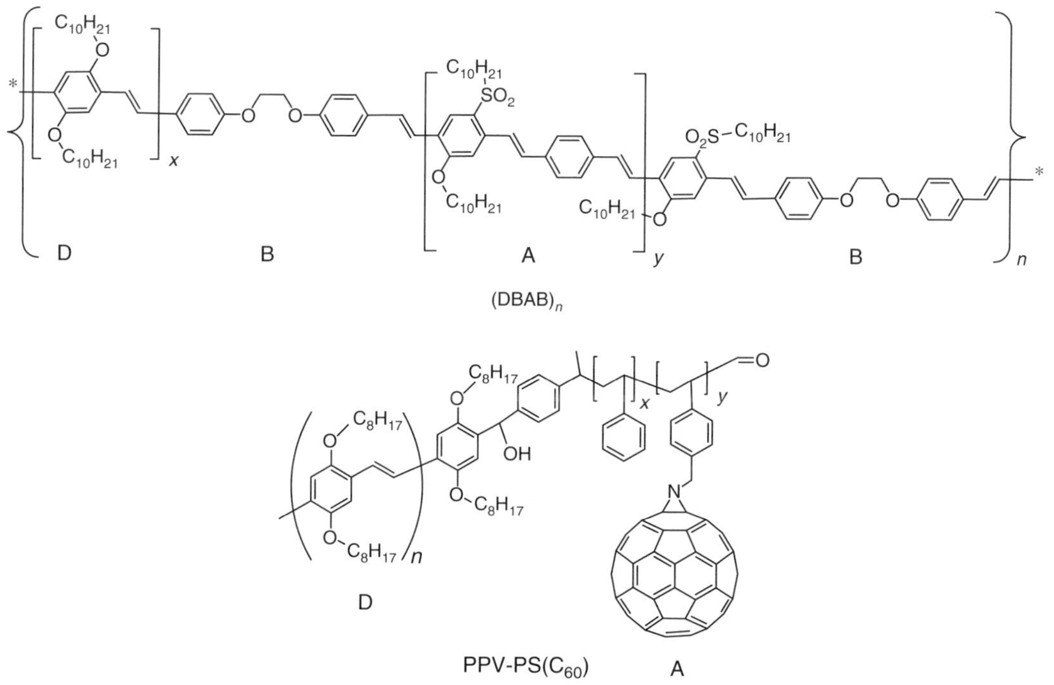

FIGURE 14.11 Two representative D/A block copolymer systems studied.

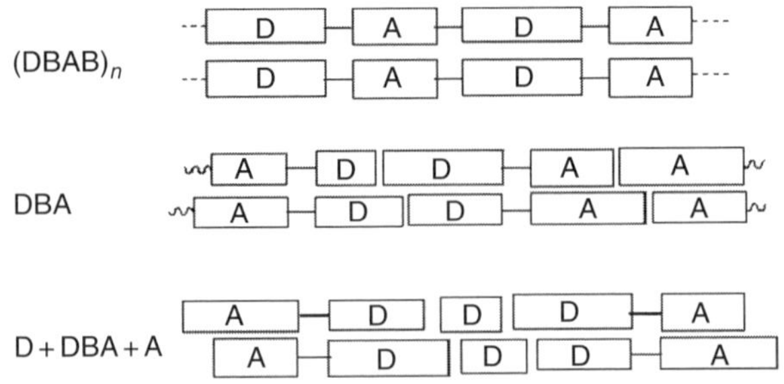

FIGURE 14.12 Different stacking possibilities of $(DBAB)_n$ and DBA block copolymers (see text for details).

To address this issue, a DBA-triblock copolymer (shown in Figure 14.12) has also been proposed [27]. As Figure 14.12 shows, DBA would allow more flexibility in self-assembling and is expected to enable large domain BONS formation from broad dispersive block copolymers. Furthermore, DBA can be blended with D and A to gain greater flexibility in polymer packing due to the higher mobility of the free donor and acceptor polymer blocks.

Discotic donor–acceptor-type molecules offer another opportunity to realize BONS. They can self-organize from solution (or vacuum) into columnar stacks in which one-dimensional hole, electron, or ambipolar transports with nanometer scale cross-section diameter may be achieved. While stacked crystalline structure has already been reported for single component discotic liquid crystal (LC) materials [28], columnar structures from discotic molecules containing both donor and acceptor and with proper stacking order have yet to be demonstrated.

14.3 OPTIMIZATIONS OF ORGANIC SOLAR CELLS IN THE ENERGY DOMAIN

As discussed in Section 14.1.2, photoelectric conversion in an organic solar cell involves five processes. To achieve high power conversion efficiency, all five steps need to be optimized. In the first step, absorption of sunlight should be maximized to minimize the photon loss. In the second and third steps, large D/A interface area and small domain sizes of donor and acceptor are essential to facilitate exciton diffusion and dissociation at the D/A interface (i.e., to minimize the exciton loss). In the fourth step (charge transport), bicontinuous morphology character of donor and acceptor is critical because each electron and hole need to have a continuous pathway to reach the respective electrodes, and good crystalline packing of molecules/polymers is the key to realization of higher mobility of charge carriers. In the fifth step, the carrier loss may be severe if the energy levels of the organic semiconductors and the electrodes are not matched. Therefore, the cell optimizations need to be conducted in both energy and spatial domains [6].

As Figure 14.13 shows, the sunlight spectrum is very broad, from ultraviolet (over 3 eV, or wavelength shorter than 400 nm) all the way to infrared (less than 1.0 eV, or wavelength longer than 1200 nm), with maximum photon flux between 1.0 and 2.0 eV (1240 and 620 nm) on the surface of the earth (for AM 1.5 standard solar spectrum, see Figure 14.13 [1–3]). A single material can only absorb photons of energy closely matching its optical excitation energy gap [6].

To capture as much sunlight as possible, a tandem cell structure shown in Figure 14.13 can be used [21]. A tandem PV cell is a stack of serially connected subcells with different energy gaps. For effective sunlight capture purpose, an ideal tandem cell may need to have energy gaps of the subcells from UV all the way to IR side of the sunlight spectrum in descending order. There should be a very thin, transparent conducting layer (TCL) between any two adjacent cells serving as electron–hole charge recombination sites. Since the cells are serially connected, photovoltages add up from each subcell. However, current density also needs to be balanced between the subcells to achieve the optimal performance. The tandem-type cells have been used in both inorganic solar cells with maximum power conversion efficiency of over 40% [2], and in organic solar cells with a maximum power

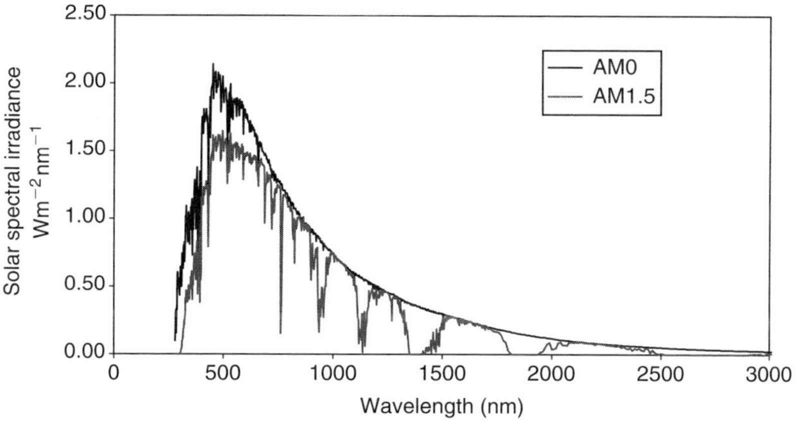

FIGURE 14.13 Standard solar spectral irradiance above the atmosphere (AM0, extraterrestrial, top curve) and under the reference air mass at 1.5 (AM1.5) conditions (ASTM G173-03, 2003, bottom curve. See rredc.nrel.gov/solar/spectra/am1.5 or www.astm.org/): Conditions for the terrestrial standard: An AM of 1.5 (solar zenith angle 48.19°s), a turbidity of 0.27, and the receiving surface is at 37° tilt toward the equator, facing the Sun (i.e., the surface normal points to the sun, at an elevation of 48.81° above the horizon). These conditions are representative of average conditions in the 48 contiguous states of the United States. In real life, a large range of atmospheric conditions can be encountered, resulting in more or less variations in atmospheric extinction.

conversion efficiency of over 5% [29] under one Sun-simulated AM1.5G solar illumination. Since most of the existing widely used conjugated semiconducting polymers generally have energy gaps over 2.0 eV, the current main challenge for developing polymer-based tandem cells would depend on the development of new low energy gap, processable, and stable semiconducting polymers (Figure 14.14).

In addition to optimization of photon capture via energy gap engineering, electronic transitions in the materials must also be optimized to maximize photoinduced charge separation and at the same

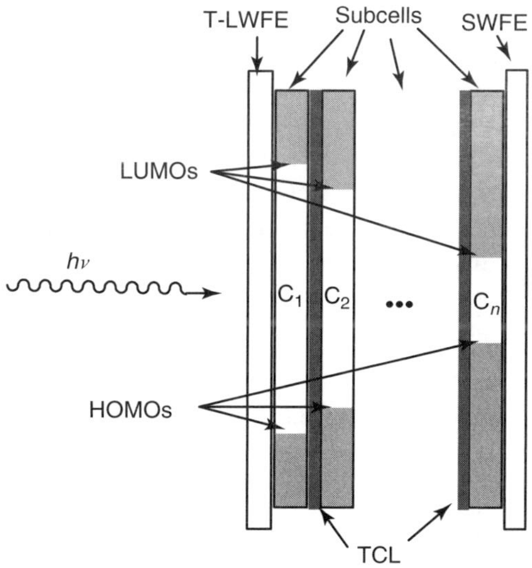

FIGURE 14.14 Scheme of a tandem style stacked subcells with excitation energy gaps grading from large to small along the light propagation direction. TCL, transparent conducting layer and T-LWFE, transparent large work function electrode.

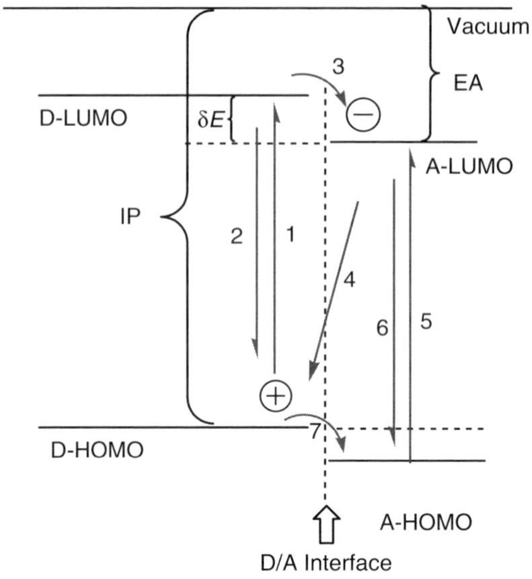

FIGURE 14.15 Scheme of molecular frontier orbitals and photoinduced electron transfer processes in a D/A binary heterojunction solar cell. δE denotes energy offset between D-LUMO and A-LUMO. IP denotes ionization potential of the donor, and EA denotes electron affinity of the acceptor.

time to minimize the charge recombination. Figure 14.15 shows all key electronic transitions of a D/A binary PV system. Specifically, transitions **1** and **5** denote photoexcitations in donor and acceptor, respectively; transitions **2** and **6** denote exciton intra-atomic recombination. Transitions **3** and **7** denote inter-atomic charge separations (exciton dissociations) at the donor/acceptor interface, and transition **4** denotes inter-atomic charge recombination at the D/A interface. As mentioned earlier, the frontier orbital (HOMOs or LUMOs) energy offset (δE) between the donor and the acceptor provides a key driving force for the exciton dissociations (transitions **3** and **7**). When the offset is too small, transitions **3** and **7** become inefficient because the driving force is weaker than the counter-driving forces. On the other hand, a too large offset would also slow down such an electron transfer based on Marcus electron transfer theory [30]. The same principle also accounts for the fact that the exciton recombination (transitions **2** and **6**) is generally much slower than the charge separation (transitions **3** and **7**) [7]. Too large offset of the LUMOs may lead to strong coupling between D-HOMO and A-LUMO and ground state electron transfer (photoinduced charge transfer exciplex formation or chemical doping) instead of photoinduced excited state electron transfer (photo doping) would likely occur [21], and this is undesirable for PV applications. Large offset also reduces the open circuit voltage [31]. For material systems in which both donor and acceptor absorb light, an optimal HOMO energy offset is also critical [7].

14.4 ORGANIC PHOTOVOLTAIC CELL FABRICATION AND CHARACTERIZATION

In a typical organic or polymeric PV cell device, as shown in Figure 14.16, the organic or polymeric semiconductor PV layer (called the active layer) is typically sandwiched between a transparent conducting electrode (TCE) (for e.g., ITO-coated glass as the LWFE, bottom) and a metal electrode (e.g., aluminum as the SWFE, top). It was also found that a poly(ethylene dioxythiophene):polystyrene sulfonic (PEDOT–PSS) acid layer, chemical structure shown in Figure 14.17, greatly facilitates the hole transfer between the active layer and the ITO electrode, and a thin LiF layer greatly facilitates

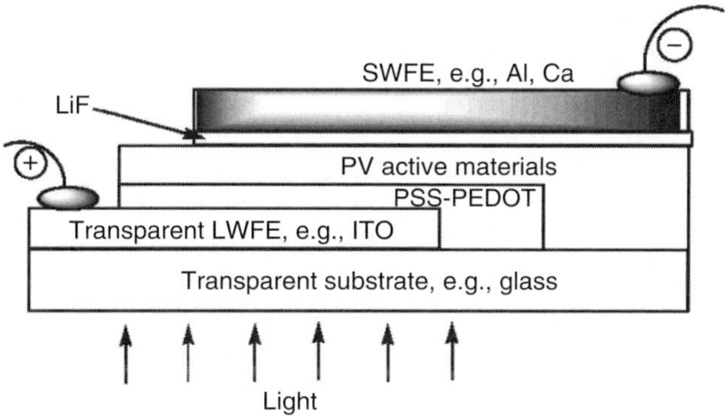

FIGURE 14.16 General scheme of an organic/polymer PV cell with buffer layers.

the electron transfer between the active layer and the metal electrode [3,4]. For small organic molecules, high vacuum (at least 10^{-6} torr) vapor deposition and occasionally solution crystallization protocols have been used to grow thin films on TCE substrates, followed by the vacuum deposition of metal electrode on top of the PV active layer. For polymers, solution spin coating (small devices), ink-jet printing, or roll-to-roll printing (large-sized sheets) protocols can be used. Solution processing generally offers advantage of low cost and convenience on large scale industrial productions.

The conductivity of PEDOT–PSS varies depending on processing conditions; it can reach 80 s/cm if produced by electropolymerization, and can be as low as 0.03 s/cm if produced by chemical polymerization [3]. PEDOT–PSS is commercially available in an aqueous dispersion, which can form uniform, transparent, conductive film by spin coating. The most important physical properties for its application in devices are the high work function (5.2 eV) and the smooth surface. The improvement of the device performance by the application of the PEDOT-PSS layer may be attributed to several factors. For instance, the PEDOT–PSS layer was believed to help smooth both the surface roughness and the conductivities of commercial ITO glass surfaces, which have been found to be very rough and have nonuniform conductivities [3]. It is also believed that the work function of PEDOT–PSS lies between the work function of the ITO (4.7 eV) and the HOMO levels of most organic donor materials. This intermediate level would facilitate the hole transport from the polymer to the ITO electrode. Possibly, PEDOT–PSS simply prevents an acceptor in direct contact with the ITO, so charge recombination at the ITO interface is minimized [21].

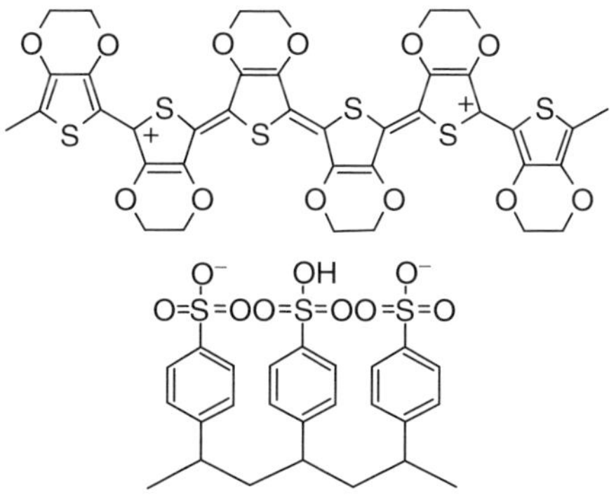

FIGURE 14.17 Chemical structure of PEDOT–PSS.

Insertion of a thin lithium fluoride (LiF) layer between the active layer and the SWFE metal electrode has been shown to improve electron injection in some organic light-emitting diodes (LEDs) [3,4]. The LiF layer is typically very thin (<1 nm), since thicker layers are detrimental to electron injection. While the exact causes were not clear, the improved device optoelectronic performance of incorporating LiF in some cases may be attributed to, for instance, the LiF has been found to modify work function of the metal electrode due to the dipolar nature of LiF layer. This would facilitate the electron transfer. The LiF layer may also prevent donor in direct contact with the metal, and prevents the chemical reactions between the organic active layer and the metal, and reduces the serial resistance of the interface between the metal and the active layer [3,4].

The maximum short-circuit photocurrent density J_{sc}, which occurs when the load is of zero resistance such as a metal, can be calculated from the photon flux F [32]

$$J_{sc} = q_e \int_0^{\lambda_c} \eta F d\lambda \tag{14.1}$$

where

q_e is the charge of an electron
λ_c is the absorption cutoff wavelength of the PV material
η is the internal quantum efficiency of photon–electron conversion

Under AM1 sunlight radiation ($F = 5 \times 10^{17}/\text{cm}^2\ \text{s}$), assuming all photons are absorbed and a unity photon–electron quantum efficiency ($\eta = 1$), an upper limit of 80 mA/cm^2 can be estimated for J_{sc} based on AM1 solar spectrum. In reality, only photons with energy matching the optical energy gap of the active material can be captured. Although short-circuit current may increase with decreasing materials energy gap due to more lower energy solar photons being captured, the open circuit voltage V_{oc} generally decreases with decreasing energy gap, so the power conversion efficiency for a single layer cell peaks at a certain energy gap (E_g).

In Figure 14.18, the current density–voltage (J–V) characteristics are shown for a general PV cell in the dark and under illumination. In the dark, there is almost no current flowing until the applied forward bias overcomes the countering open circuit voltage (V_{oc}). Under illumination, photoinduced electrons will transfer from the acceptor LUMO to the negative metal electrode, then through the load to the positive ITO electrode where they combine with holes coming out of the HOMO of donor material. This current can be suppressed by a forward bias (defined as an external

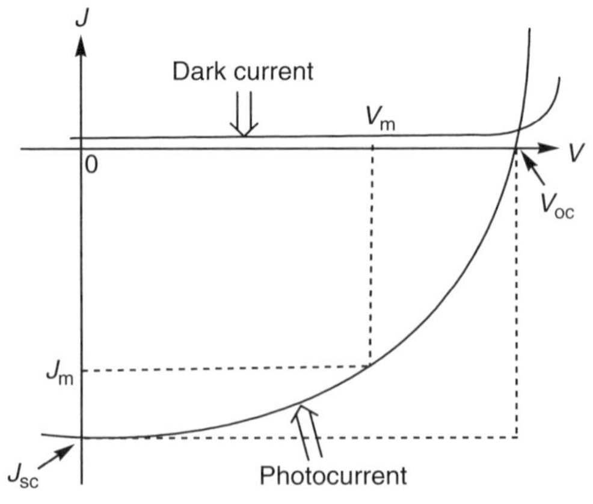

FIGURE 14.18 A schematic current J–V plot of a PV cell in the dark and under illumination.

negative potential on the metal electrode (−) and an external positive potential on the ITO electrode (+)) and is reduced to zero when the applied voltage reaches a value called open circuit voltage (V_{oc}). At a certain point of the J–V curve between zero bias and V_{oc} (with coordinates [J_m, V_m] shown in Figure 14.18), the product of the current J–V (i.e., the output power density) reaches maximum ($P_m = J_m V_m$). The ratio of P_m over $V_{oc} \times J_{sc}$ denotes the maximum power output fraction of the cell and is called the fill factor (FF),

$$\mathrm{FF} = \frac{J_m V_m}{J_{sc} V_{oc}} \tag{14.2}$$

With this, photoelectric power conversion efficiency (PCE or η_p) can be written as

$$\eta_p = \frac{P_m}{P_{in}} = \frac{J_m V_m}{P_{in}} = \frac{\mathrm{FF} \cdot J_{sc} V_{oc}}{P_{in}} \tag{14.3}$$

where P_{in} is the incident light power density.

The current–voltage (I–V) characteristics of a PV device can be described by

$$I = I_0 \left\{ \exp\left[\frac{q_e}{nkT}(U - IR_s) \right] - 1 \right\} + \frac{U - IR_s}{R_{sh}} - I_{ph}, \tag{14.4}$$

where

I_0 is the dark current
q_e is the elementary charge
n is the diode ideality factor (typically $1 < n < 2$)
U is the applied voltage
R_s is the series resistance
R_{sh} is the shunt (or parallel) resistance
I_{ph} is the photocurrent

The corresponding equivalent circuit is given in Figure 14.19. The series resistance is the sum of all series resistive contributions in device from bulk transport and interface charge transfer. High shunt resistance, which prevents leakage currents and low series resistance, is required to achieve a high fill factor. For a pinhole free organic active layer, the parallel (shunt) resistivity was approximately the same for many organic photodiodes (typically over 1 kΩ), but the total serial resistivity may vary depending on materials and device fabrications. In comparison to inorganic PV cells, organic solar cells typically have very large R_s due to low carrier density and poor charge mobility.

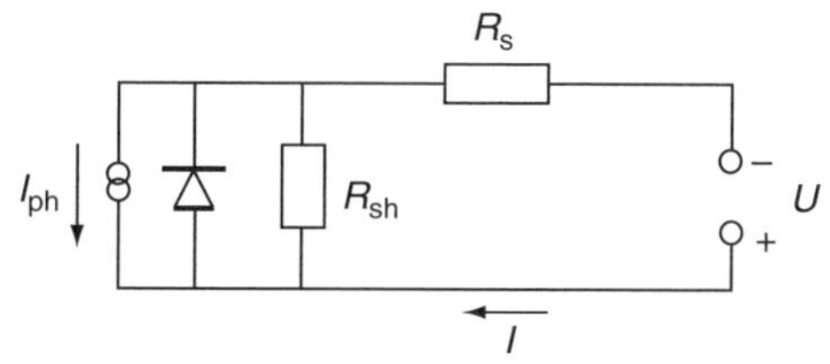

FIGURE 14.19 Equivalent circuit for a solar cell as described in Equation 14.4.

Because of the special charge generation mechanisms in organic devices, the efficiency-limiting factors are therefore distinct from those in conventional inorganic PV cells (e.g., silicon p–n junction PV cells). For instance, while the maximum photovoltage (V_{oc}) achievable in silicon cells is generally limited to the magnitude of the built-in potential of the electrodes, it is common to observe experimentally V_{oc} greater than the built-in potential of electrodes in organic-based PV devices [3,4]. It is believed the maximum V_{oc} of an organic D/A binary cell is closely correlated to the gap between the HOMO of the donor and the LUMO of the acceptor as shown in Figure 14.1d [31].

14.5 SUMMARY

Organic PV cells involve five critical steps: photon capture and exciton generation, exciton diffusion to D/A interface, exciton dissociation at D/A interface, carrier diffusion to the respective electrodes, and carrier collection by the respective electrodes. Organic D/A binary PV materials are different from inorganic semiconductor p–n junction PV in two main aspects. First, due to differences mainly in morphology and charge screening, photo-generated excitons in organics are mostly Frenkel-type tightly bound electron–hole pairs that cannot be dissociated by room temperature thermal energy. A secondary force such as the energy offset of a D/A pair would be needed for such exciton dissociation. Since excitons typically have finite lifetime and AEDL limit, this dictates that donor and acceptor domain sizes have to be on the same order as the AEDL. Second, charge carriers mobility of organic semiconductors are relatively low, which is a key reason for the low energy conversion efficiencies in organic/polymeric PV cells.

Optimizations in both spatial and energy/time domains should be pursued to achieve high efficiency in organic and polymeric PV devices. In the spatial domain, film-processing conditions must be optimized for bulk heterojunction or BONS materials to realize better packing of conjugated systems to improve mobility. New material systems such as D/A block copolymers, hybrid materials consisting of vertically aligned/grown carbon nanotubes/fibers or inorganic semiconductor oxide nanorods/pillars with polymer donors, and discotic liquid crystal materials are all under investigation. BONS material systems appear to have the desired or optimal nanostructures for PV applications. In the energy/time domain, energy gap-graded tandem-type cell structure can be used to maximize sunlight capture. D/A frontier orbital energy offsets also need to be optimized. With optimizations in both space and energy/time domains, it is expected that high efficiency organic photoelectric power conversion efficiency can be realized.

EXERCISE QUESTIONS

1. What are the major differences between an organic D/A heterojunction PV cell and an inorganic semiconductor p–n junction PV cell?
2. Why do organic bulk D/A heterojunction (physical blend) PV cells exhibit better PV properties than an organic D/A bilayer heterojunction PV cell?
3. What are the potential advantages of DBA block copolymers over D/A physical blends?
4. How do frontier orbital levels affect PV properties?
5. List strategies that can be used to improve efficiency of organic PV cells.

LIST OF ABBREVIATIONS

AM0	air mass zero
AM1.5	air mass at 1.5
BBL	poly(benzamidazobenzo phenanthroline)
BONS	bicontinuous-ordered nanostructure
CN-PPV	poly(2,5,2′,5′-tetraalkoxy-7,8′-dicyanodi-*p*-phenylenevinylene)
HOMO	highest occupied molecular orbital

IQE	internal quantum efficiency
ITO	indium tin oxide
LiF	lithium fluoride
LWFE	large work function electrode
LUMO	lowest unoccupied molecular orbital
MDMO-PPV	poly [2-methoxy-5-(3′,7′-dimethyloctyloxy)-1-4-phenylene vinylene]
MEH-PPV	poly(2-methoxy-5-(2-ethylhexyloxy)-1,4-phenylene vinylene)
P3DDT	poly(3-dodecylthiophene)
P3HT	poly(3-hexylthiophene)
P3OT	poly(3-octyl thiophene)
PBZT	poly(*p*-phenylene benzobisthiazole)
PCBM	[6,6]-phenyl-C_{61}-butyric acid methyl ester
PCE	power conversion efficiency
PEDOT–PSS	poly(ethylene dioxythiophene):polystyrene sulfonic acid
PPV	poly-1,4-phenylenevinylene, or, poly-*p*-phenylenevinylene
PTh	polythiophene
RO-PPV-10	poly(2,5-didecyloxy-1,4-phenylene vinylene)
SF-PPV	poly[2-alkoxy-5-alkanesulfonyl-1-4-phenylene vinylene]
SWFE	small work function electrode
TCE	transparent conducting electrode
TCL	transparent conducting layer
V_{oc}	open circuit voltage

ACKNOWLEDGMENTS

The authors would like to acknowledge the research/educational grant supports to subject related to this chapter from a number of funding agencies including NASA, DoD (MDA, AFOSR), the National Science Foundation, the Department of Education (Title III award).

REFERENCES

1. A. Luque and S. Hegedus (Eds.), *Handbook of Photovoltaic Science and Engineering*, Wiley, The Atrium, England, 2003.
2. L. Kazmerski, *J. Elect. Spec. Rel. Phen.*, 150, 105–135, 2006.
3. S. Sun and N.S. Sariciftci (Eds.), *Organic Photovoltaics: Mechanisms, Materials and Devices*, CRC Press, Boca Raton, FL, 2005.
4. H. Hoppe and N.S. Sariciftci, *J. Mater. Res.*, 19, 1924–1945, 2004.
5. B.A. Gregg, Coulomb forces in excitonic solar cells, in *Organic Photovoltaics: Mechanisms, Materials and Devices*, S. Sun and N.S. Sariciftci (Eds.), CRC Press, Boca Raton, FL, 2005, p. 139.
6. S. Sun and C. Bonner, Optimizations of organic solar cells in both space and energy/time domains, in *Organic Photovoltaics: Mechanisms, Materials and Devices*, S. Sun and N.S. Sariciftci (Eds.), CRC Press, Boca Raton, FL, Chapter 8, 2005, pp. 183–214.
7. S. Sun, *Mater. Sci. Eng. B*, 116(3), 251–256, 2005.
8. V. Podzorov, E. Menard, A. Borissov, V. Kiryukhin, J.A. Rogers, and M.E. Gershenson, *Phys. Rev. Lett.*, 93, 086602, 2004.
9. A. Pochettino, *Acad. Lincei Rendiconti*, 15, 355–363, 1906.
10. S. Karg, W. Riess, V. Dyakonov, and M. Schwoerer, *Synth. Metals*, 54, 427, 1993.
11. P. Peumans, A. Yakimov, and S.R. Forrest, *J. Appl. Phy.*, 93(7), 3693–3723, 2003.
12. C.W. Tang, *Appl. Phys. Lett.*, 48, 183–185, 1986.
13. J.J.M. Halls, K. Pichler, R.H. Friend, S.C. Moratti, and A.B. Holmes, *Appl. Phys. Lett.*, 68, 3120, 1996.
14. M. Hiramoto, H. Fujiwara, and M. Yokoyama, *Appl. Phys. Lett.*, 58, 1062–1064, 1991.
15. J. Xue, S. Uchida, B. Rand, and S.R. Forrest, *Appl. Phys. Lett.*, 86, 5757–5759, 2004.
16. G. Li, V. Shrotriya, J. Huang, Y. Yao, T. Moriarty, K. Emery, and Y. Yang, *Nature Materials*, 4, 864–868, 2005.

17. W. Ma, C. Yang, X. Gong, K. Lee, and A.J. Heeger, *Adv. Func. Mater.*, 15, 1617–1622, 2005.
18. M. Drees, R. Davis, and R. Heflin, Polymer-fullerene concentration gradient photovoltaic devices by thermally controlled interdiffusion, in *Organic Photovoltaics: Mechanisms, Materials and Devices*, S. Sun and N.S. Sariciftci (Eds.), CRC Press, Boca Raton, FL, 559, 2005.
19. Z. Bao and J. Locklin (Eds.), *Organic Field-Effect Transistors*, CRC Press, Boca Raton, FL, 2007.
20. S. Sun, *Sol. Ener. Mater. Sol. Cel.*, 79(2), 257, 2003.
21. S. Sun, Organic and polymeric solar cells, in *Handbook of Organic Electronics and Photonics*, S.H. Nalwa (Ed.), American Scientific Publishers, Los Angeles, CA, vol. 2, Chapter 23, 2006.
22. K. Coakley, Y. Liu, M. McGehee, K. Frindell, and G. Stucky, *Adv. Funct. Mater.*, 13(4), 301, 2003.
23. N. Hadjichristidis, S. Pispas, and G. Floudas (Eds.), *Block Copolymers: Synthetic Strategies, Physical Properties, and Applications*, John Wiley & Sons, New York, 2003.
24. B. de Boer, U. Stalmach, P.F. van Hutten, C. Melzer, V.V. Krasnikov, and G. Hadziioannou, *Polymer*, 42, 9097, 2001.
25. C. Zhang, S. Choi, J. Haliburton, R. Li, T. Cleveland, S. Sun, A. Ledbetter, and C. Bonner, *Macromolecules*, 39, 4317, 2006.
26. S. Sun, C. Zhang, S. Choi, A. Ledbetter, C. Bonner, M. Drees, and S. Sariciftci, *Appl. Phys. Lett.*, 90, 043117, 2007.
27. C. Zhang, T. Cleveland, and S. Sun, *Poly. Mater. Sci. Eng.*, 96, 242–243, 2007.
28. L. Schmidt-Mende, A. Fechtenkötter, K. Müllen, E. Moons, R.H. Friend, and J.D. MacKenzie, *Science*, 293, 1119, 2001.
29. J. Xue, S. Uchida, B. Rand, and S.R. Forrest, *Appl. Phys. Lett.*, 86, 5757, 2005.
30. V. Balzani (Ed.), *Electron Transfer in Chemistry*, Wiley-VCH, New York, 2002.
31. C.J. Brabec, A. Cravino, D. Meissner, N.S. Sariciftci, T. Fromherz, M. Minse, L. Sanchez, and J.C. Hummelen, *Adv. Funct. Mater.*, 11, 374–380, 2001.
32. K.K. Wong. *Complete Guide to Semiconductor Devices*, 2nd edn., John Wiley & Sons, New York, 2002, p. 516.

15 Organic Molecular Nonlinear Optical Materials and Devices

Mojca Jazbinsek and Peter Günter

CONTENTS

15.1 Nonlinear Optical Processes and Applications 422
15.1.1 Origins of the Nonlinear Optical Response 422
15.1.1.1 Nonlinear Wave Equation 423
15.1.1.2 Symmetry Properties and Dispersion of Nonlinear Optical Susceptibilities 424
15.1.1.3 Crystal Optics 424
15.1.2 Second-Order Nonlinear Optical Effects and Applications 426
15.1.2.1 Sum-Frequency Generation and Optical-Frequency Doubling 427
15.1.2.2 Difference-Frequency Generation, Optical Parametric Generation/Oscillation, and Optical Rectification 429
15.1.2.3 Conservation of Energy and Momentum (Phase Matching) 430
15.1.2.4 Linear Electro-optic Effect and Applications 433
15.1.3 Third-Order Nonlinear Optical Effects and Applications 437
15.1.3.1 Self-Phase Modulation 439
15.1.3.2 Self-Focusing and Spatial Kerr Solitons 439
15.1.3.3 Optical Phase Conjugation 440
15.2 Organic Nonlinear Optical Materials 441
15.2.1 Comparison of Organic and Inorganic Materials for Nonlinear and Electro-optics 441
15.2.2 Organic Molecules for Second-Order Nonlinear Optics 445
15.2.3 Microscopic and Macroscopic Nonlinearities of Organic Molecules 445
15.2.4 Basic Types of Organic Materials Exhibiting Macroscopic Second-Order Nonlinearity 449
15.3 Organic Nonlinear Optical Single Crystals 450
15.3.1 Noncentrosymmetric Crystalline Packing: Approaches 450
15.3.2 Examples of Molecular Crystals 456
15.3.3 Growth of Bulk and Thin-Film Organic Crystals 460
15.3.4 4-*N*,*N*-Dimethylamino-4′-*N*′-Methyl Stilbazolium Tosylate (DAST) 460
15.4 Summary 461
Exercise Questions 462
Acknowledgment 463
List of Abbreviations 464
References 464

Abstract This chapter introduces the basic principles of nonlinear optical effects in organic materials. Various types of nonlinear optical processes are included, together with an overview of most important applications. The origins of optical and nonlinear optical response in solid materials are discussed, and key differences between inorganic and organic materials are described. Basic molecular structures of various types of organic materials for second-order nonlinear optics and models to describe their response are introduced. A separate discussion is devoted to organic nonlinear optical single crystals with best stability characteristics among organic nonlinear optical materials.

15.1 NONLINEAR OPTICAL PROCESSES AND APPLICATIONS

Optical properties of illuminated materials may be modified as soon as the light intensity becomes high enough. The change of the optical properties influences in turn the light propagation itself so that interactions between light beams become possible. Owing to this nonlinear behavior of optical media a wide variety of phenomena occur with many useful practical applications. The field of nonlinear optics started to expand rapidly after the invention of the laser in 1960. Nowadays, most commonly used nonlinear optical effects in applications include second-harmonic generation (SHG), parametric amplification, optical modulation, optical switching, optical filtering, data storage, signal processing, and others. In this section, the origin and basic applications of nonlinearity in optical media are presented.

15.1.1 Origins of the Nonlinear Optical Response

A response of a dielectric material to an applied electric field $\boldsymbol{E}$ is described by an induced polarization density $\boldsymbol{P}$. The medium is said to be nonlinear if the relation between the induced polarization density $\boldsymbol{P}$ and the applied field $\boldsymbol{E}$ is nonlinear. According to a simple classical theory, the charged particles in a dielectric material are bound together with bonds that have some elasticity. When the electric field is applied, electric dipole moments are induced, which oscillate with the same frequency as the applied field, when the restraining forces are harmonic. For high enough external fields, the restraining elastic forces for charged particles become nonlinear functions of the displacement, and therefore the relation between the polarization density $\boldsymbol{P}$ and the applied field $\boldsymbol{E}$ becomes nonlinear.

Consider a case where the optical field illuminating the material is a monochromatic plane wave with a frequency ω. Its time variation at a certain position can be described as

$$\boldsymbol{E}(t) = \frac{1}{2}\boldsymbol{E}(\omega)\mathrm{e}^{-i\omega t} + \mathrm{cc}, \tag{15.1}$$

where cc denotes the complex conjugate, so that $\boldsymbol{E}(t)$ is a real quantity. The linear response of the material may be expressed with the induced polarization density in the form

$$\boldsymbol{P}(t) = \varepsilon_0 \int_{-\infty}^{\infty} \chi(t-t')\boldsymbol{E}(t')\mathrm{d}t'; \quad \chi(t-t') = 0 \quad \text{for } t < t', \tag{15.2}$$

which reflects the fact that the material response does not occur instantaneously. Here $\chi(t)$ is the time-dependent linear susceptibility and ε_0 is the electric constant (permittivity of vacuum). We can write

$$\boldsymbol{P}(t) = \frac{1}{2}\boldsymbol{P}(\omega)\mathrm{e}^{-i\omega t} + \mathrm{cc}, \tag{15.3}$$

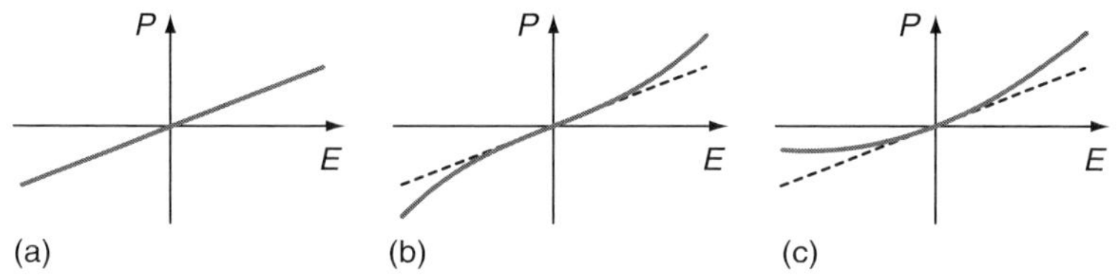

FIGURE 15.1 Induced polarization, P, in dielectrics for (a) linear medium, (b) nonlinear centrosymmetric medium, and (c) nonlinear noncentrosymmetric medium.

where

$$\boldsymbol{P}(\omega) = \varepsilon_0 \chi(\omega) \boldsymbol{E}(\omega) \tag{15.4}$$

and $\chi(\omega)$ is the Fourier transform of $\chi(t)$. The linear response is completely described by the frequency-dependent susceptibility $\chi(\omega)$ that is related to the dielectric constant $\varepsilon(\omega) = 1 + \chi(\omega)$.

The optical response of a nonlinear material can be described in terms of the linear polarization P^{L} and the nonlinear polarization P^{NL} induced by the electric field E.

$$P = P^{\mathrm{L}} + P^{\mathrm{NL}} = \varepsilon_0 \chi^{(1)} E + \varepsilon_0 \chi^{(2)} E^2 + \varepsilon_0 \chi^{(3)} E^3 + \cdots, \tag{15.5}$$

where the function that relates the dielectric response (polarizability) P to the electric field E is expanded in Taylor's series around $E = 0$ and $\chi^{(n)}$ presents the nth-order susceptibility. For symmetry reasons, the odd-order susceptibilities are present in any material, whereas the even-order ones only occur in noncentrosymmetric materials. The linear and nonlinear polarization responses are illustrated in Figure 15.1.

15.1.1.1 Nonlinear Wave Equation

To describe the propagation of light in a nonlinear medium, one has to solve the wave equation that can be derived from the Maxwell's equations for a homogeneous dielectric medium

$$\nabla^2 \boldsymbol{E}(\boldsymbol{r},t) - \frac{1}{c^2}\frac{\partial^2 \boldsymbol{E}(\boldsymbol{r},t)}{\partial t^2} = \mu_0 \frac{\partial^2 \boldsymbol{P}(\boldsymbol{r},t)}{\partial t^2}, \tag{15.6}$$

where

$c = 1/\sqrt{\varepsilon_0 \mu_0}$ is the speed of light in vacuum
μ_0 is the magnetic constant (permeability of vacuum)
$\boldsymbol{E}(\boldsymbol{r},t)$ is the electric field of the optical wave
$\boldsymbol{P}(\boldsymbol{r},t) = \varepsilon_0 \chi^{(1)} \boldsymbol{E}(\boldsymbol{r},t) + \boldsymbol{P}^{\mathrm{NL}}(\boldsymbol{r},t)$ is the induced material polarization

In general the optical field that fulfills the wave equation in a medium will contain several frequency components. The variation of the electric field amplitude in time t and space $\boldsymbol{r}$ can be written as

$$\boldsymbol{E}(\boldsymbol{r},t) = \frac{1}{2}\sum_{\boldsymbol{k},\omega}\left(\boldsymbol{E}(\boldsymbol{k},\omega)\mathrm{e}^{\mathrm{i}(\boldsymbol{kr}-\omega t)} + \mathrm{cc}\right), \tag{15.7}$$

where
$\boldsymbol{k}$ is the wave vector of the optical field component
ω is its frequency

For the case of a linear medium with $\boldsymbol{P}^{NL}=0$, inserting Equation 15.7 into the wave equation (Equation 15.6) gives $|\boldsymbol{k}(\omega)|=\omega/c \cdot n(\omega)$, where the refractive index $n(\omega)$ is defined with $n^2(\omega)=\varepsilon(\omega)=1+\chi^{(1)}(\omega)$.

In an anisotropic nonlinear medium, each component of the polarization density may be written as

$$P_i = \varepsilon_0 \chi_{ij}^{(1)} E_j + \varepsilon_0 \chi_{ijk}^{(2)} E_j E_k + \varepsilon_0 \chi_{ijkl}^{(3)} E_j E_k E_l + \cdots, \quad (15.8)$$

where the usual Einstein summation convention holds. Note that in a general dispersive medium, susceptibility tensor elements also depend on the frequency of the electric fields involved. By substituting Equation 15.7 into Equation 15.8, we get a distorted profile of the polarization $\boldsymbol{P}$.

15.1.1.2 Symmetry Properties and Dispersion of Nonlinear Optical Susceptibilities

Consider first nondispersive media, where the frequency dependences of the susceptibility elements can be omitted. Because the coefficient $\chi_{ijk}^{(2)}$ is a multiplier of the product $E_j E_k$, it must be invariant to the exchange of j and k. Similarly, $\chi_{ijkl}^{(3)}$ is invariant to any permutations of j, k, and l. In nondissipative media, where there is no energy loss, we can also exchange the first two indices i and j. This can be seen if we consider the free energy of the system, where all the included fields occur equally. In this case the tensors $\chi_{ij}^{(1)}$, $\chi_{ijk}^{(2)}$, and $\chi_{ijkl}^{(3)}$ are invariant to any permutations of their indices (the so-called Kleinman symmetry).

For dispersive media, the frequency dependence of the susceptibility tensors has to be included. For example, the second term in Equation 15.8 is then of the form

$$\varepsilon_0 \chi_{ijk}^{(2)}(-\omega_3, \omega_1, \omega_2) E_j(\omega_1) E_k(\omega_2), \quad (15.9)$$

if the polarization at frequency $\omega_3=\omega_1+\omega_2$ is produced. Because the product $E_j(\omega_1)E_k(\omega_2)$ is commutative, the indices j and k in $\chi_{ijk}^{(2)}$ can be exchanged, but only if the frequencies are exchanged as well

$$\chi_{ijk}^{(2)}(-\omega_3, \omega_1, \omega_2) = \chi_{ikj}^{(2)}(-\omega_3, \omega_2, \omega_1). \quad (15.10)$$

Similarly, the third-order susceptibility tensor $\chi_{ijkl}^{(3)}(-\omega_4; \omega_1, \omega_2, \omega_3)$ is invariant under all 3! permutations of pairs (j,ω_1), (k,ω_2), (l,ω_3). If all the optical frequencies occurring in the susceptibility tensor dependence are far from the transition frequencies of the nonlinear medium (no absorption), this permutation includes the additional pair $(i,-\omega_4)$.

Optical wavelengths are usually large compared to the dimensions of the polarizable units; therefore, the local field may in most cases be considered uniform and the spatial dispersion (the $\boldsymbol{k}$ dependence) of the susceptibility neglected [1].

15.1.1.3 Crystal Optics

Nonlinear optical crystals are typically highly anisotropic. This means that even in the linear regime, light propagation characteristics (like phase velocity) will strongly depend on the direction of propagation with respect to the crystal axes and light polarization state. The linear optical properties of an anisotropic medium are usually described by the so-called optical indicatrix or index ellipsoid

$$\varepsilon_{ij}^{-1} x_i x_j = 1, \quad (15.11)$$

where

ε_{ij}^{-1} is the inverse dielectric tensor at optical frequencies

$x_{i,j}$ are the coordinates in a chosen coordinate system

Since ε_{ij} is a symmetric tensor, it can be diagonalized, i.e., in a certain coordinate system, the so-called optical principal-axes system, it will only contain diagonal elements

$$\text{General system: } \varepsilon = \begin{pmatrix} \varepsilon_{11} & \varepsilon_{12} & \varepsilon_{13} \\ \varepsilon_{12} & \varepsilon_{22} & \varepsilon_{23} \\ \varepsilon_{13} & \varepsilon_{23} & \varepsilon_{33} \end{pmatrix}. \tag{15.12}$$

$$\text{Principal-axes system: } \varepsilon = \begin{pmatrix} \varepsilon_{11} & 0 & 0 \\ 0 & \varepsilon_{22} & 0 \\ 0 & 0 & \varepsilon_{33} \end{pmatrix} = \begin{pmatrix} n_1^2 & 0 & 0 \\ 0 & n_2^2 & 0 \\ 0 & 0 & n_3^2 \end{pmatrix}, \tag{15.13}$$

where the principal refractive indices n_1, n_2, and n_3 were defined. In the principal-axes system the index ellipsoid has therefore a simple form

$$\frac{x_1^2}{n_1^2} + \frac{x_2^2}{n_2^2} + \frac{x_3^2}{n_3^2} = 1. \tag{15.14}$$

The index ellipsoid may be used to determine the polarization and refractive indices of the two normal modes of a wave traveling in an arbitrary direction of wave propagation $\boldsymbol{k}$ (see Ref. [2]). This is accomplished by intersecting the index ellipsoid with a plane normal to the propagation direction $\boldsymbol{k}$ (see Figure 15.2). The resulting intersection is an ellipse whose major and minor axes directions present the polarization directions of the two normal modes, and whose major and minor axes half-lengths are equal to the corresponding refractive indices. The detailed derivation follows the solutions

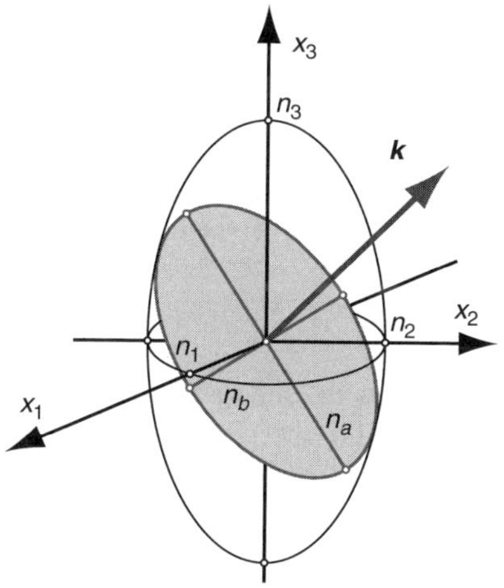

FIGURE 15.2 Optical indicatrix in the principal-axes system $x_1x_2x_3$. The refractive indices of the two normal modes for a wave traveling in the direction $\boldsymbol{k}$ are n_a and n_b. The polarizations of these two modes are parallel to the major and the minor axis of the intersection ellipse, respectively.

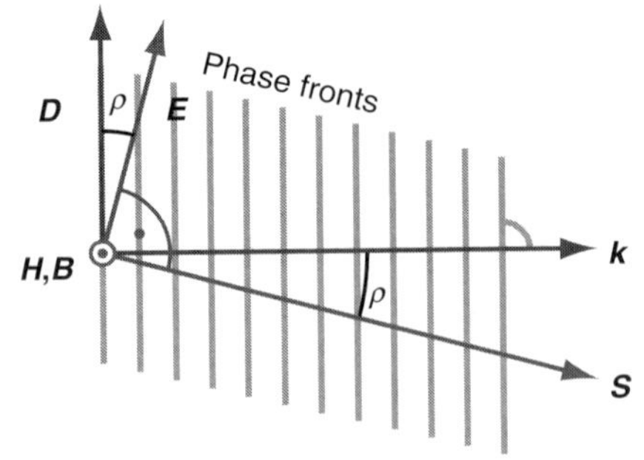

FIGURE 15.3 Directions of electromagnetic fields ***D***, ***E***, ***H***, and ***B***, phase propagation direction, ***k***, and energy propagation direction, ***S***, in an anisotropic medium. The vectors ***D***, ***E***, ***k***, and ***S*** lie in the plane to which ***H*** and ***B*** are normal. The angle between the energy and the phase propagation direction is denoted by ρ and equals the angle between ***E*** and ***D***.

of the Maxwell's equations and can be found in many textbooks on crystal optics. If the intersection is a circle, then the crystal direction parallel to the corresponding propagation direction ***k*** is called the optical axis. Crystals with $n_1=n_2=n_3$ are called optically isotropic, crystals with $n_1=n_2\neq n_3$ are optically uniaxial (since they contain one optical axis along x_3), and crystals with $n_1\neq n_2\neq n_3$ are optically biaxial (with two optical axes, none of which is parallel to one of the principal optical axes).

Since the dielectric tensor ε relates the electric field vector ***E*** and the electric displacement vector $\boldsymbol{D} = \boldsymbol{P} + \varepsilon_0\boldsymbol{E}$ as $\boldsymbol{D} = \varepsilon_0\varepsilon\boldsymbol{E}$, or in components

$$D_i = \varepsilon_0 \varepsilon_{ij} E_k, \tag{15.15}$$

the electric displacement vector ***D*** (and therefore the induced polarization ***P***) is in general not parallel to the electric field ***E*** in an anisotropic medium. From Maxwell's equations, for a nonmagnetic anisotropic medium, it follows that [2]

$$\boldsymbol{H} \perp \boldsymbol{D} \perp \boldsymbol{k}, \tag{15.16}$$

$$\boldsymbol{H} \perp \boldsymbol{E} \perp \boldsymbol{S}, \tag{15.17}$$

where

H is the magnetic field vector of the light wave
k is the wave vector along the propagation direction (normal to the phase fronts of the light wave)
$\boldsymbol{S} = \boldsymbol{E} \times \boldsymbol{H}$ is the Poynting vector along the energy propagation direction

The magnetic induction vector ***B*** is simply related to ***H*** as in vacuum $\boldsymbol{B} = \mu_0\boldsymbol{H}$. The mutual positions of the electromagnetic fields involved and energy/phase propagation directions are illustrated in Figure 15.3.

15.1.2 Second-Order Nonlinear Optical Effects and Applications

Based on the characteristics of the electric fields involved, several second-order nonlinear optical effects can be distinguished (see Figure 15.4). Those that are important have their own notation and will be introduced in the following sections.

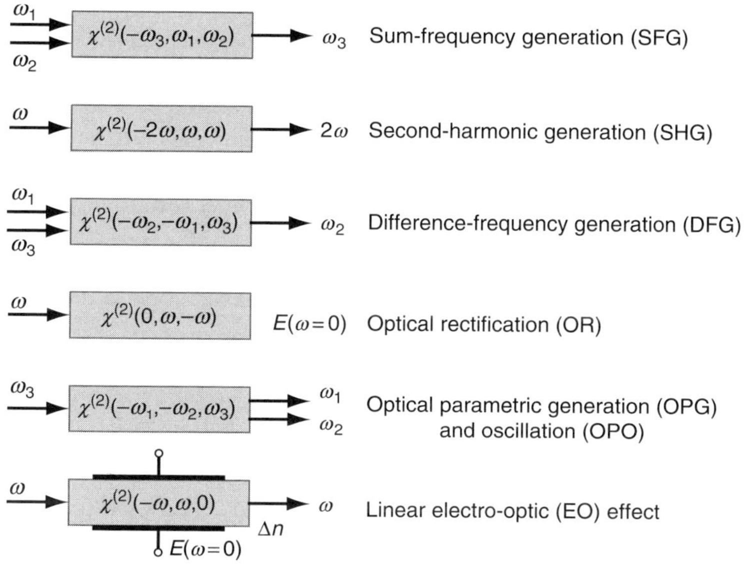

FIGURE 15.4 Schematic representation of important nonlinear optical effects of the second order, including the linear electro-optic effect.

15.1.2.1 Sum-Frequency Generation and Optical-Frequency Doubling

Sum-frequency generation is the mixing of two incident light waves of frequencies ω_1 and ω_2 creating a wave of $\omega_3 = \omega_1 + \omega_2$. This situation is represented by the nonlinear polarization

$$P_i^{\omega_3} = \varepsilon_0 \chi_{ijk}^{(2)}(-\omega_3, \omega_1, \omega_2) E_j^{\omega_1} E_k^{\omega_2}. \tag{15.18}$$

Optical-frequency doubling or SHG is a special case of sum-frequency generation. Only one light wave of frequency ω is incident and is "mixing with itself", thus generating a wave at twice the frequency. The nonlinear polarization for SHG can be also expressed by the use of the nonlinear optical coefficient d_{ijk}, which is often used for the nonlinear optical characterization of macroscopic samples

$$P_i^{2\omega} = \frac{1}{2}\varepsilon_0 \chi_{ijk}^{(2)}(-2\omega,\omega,\omega) E_j^{\omega} E_k^{\omega} = \varepsilon_0 d_{ijk}(-2\omega,\omega,\omega) E_j^{\omega} E_k^{\omega}. \tag{15.19}$$

Sum-frequency generation and SHG are standard techniques used to create a new coherent output from existing laser systems, especially to access the short wavelength range toward the ultraviolet region.

The efficiency for harmonic generation can be calculated using the nonlinear wave equation (Equation 15.6) with the nonlinear polarizations given above. As an example we consider here optical-frequency doubling of a plane monochromatic pump wave at frequency ω with a nonlinear polarization (Equation 15.19) generated inside the material. For simplicity, we consider that the interacting waves at ω and at 2ω are polarized in the same direction, and that they all propagate along the direction z. Therefore, we can consider scalar quantities and the electric field composed of two harmonic components

$$E(z,t) = \frac{1}{2}E_1(z)e^{i(k_1 z-\omega t)} + \frac{1}{2}E_2(z)e^{i(k_2 z-2\omega t)} + \text{cc}, \tag{15.20}$$

where $E_1(z)$ and $k_1 = n^{\omega}(\omega/c)$ are the amplitude and the wave vector of the pump wave at frequency ω, and $E_2(z)$ and $k_2 = n^{2\omega}(2\omega/c)$ are the amplitude and the wave vector of the generated second-harmonic wave at frequency 2ω. Here $n^{\omega} \equiv n(\omega)$ and $n^{2\omega} \equiv n(2\omega)$ denote the refractive indices of the fundamental and second-harmonic wave, which are usually not equal even for the same polarization of both waves due to dispersion. We will also consider the following common assumptions:

- *Undepleted pump approximation* that is valid if only a small part of the pump beam energy is transferred to the second-harmonic beam, i.e., $\mathrm{d}E_1/\mathrm{d}z = 0$.
- *Slowly varying envelope approximation* for the generated wave, saying that $|\mathrm{d}^2E_2/\mathrm{d}z^2| \ll |k_2\mathrm{d}E_2/\mathrm{d}z|$, i.e., the envelope $E_2(z)$ varies slowly in the z direction in comparison with the wavelength $\lambda_2 = 2\pi/k_2$.

Using the above assumptions and inserting the nonlinear polarization from Equation 15.19 gives $P^{NL}(z,t) = (1/2)\varepsilon_0 d_{SHG} E_1^2 \mathrm{e}^{i(2k_1z-2\omega t)} + \mathrm{cc}$, and the optical field (Equation 15.20) into the wave equation (Equation 15.6) results in

$$ik_2 \frac{\mathrm{d}E_2(z)}{\mathrm{d}z} \mathrm{e}^{i(k_2 z - 2\omega t)} = -\frac{1}{2}\frac{d_{SHG}E_1^2(2\omega)^2}{c^2}\mathrm{e}^{i(2k_1 z - 2\omega t)}. \quad (15.21)$$

Introducing a phase mismatch $\Delta k \equiv k_2 - 2k_1$ we can rewrite Equation 15.21 as

$$\mathrm{d}E_2 = i\omega \frac{1}{cn^{2\omega}} d_{SHG} E_1^2 \mathrm{e}^{-i\Delta k z}\mathrm{d}z \quad (15.22)$$

that can be easily integrated over the crystal length L from $z = 0$ to $z = L$ giving

$$E_2(L) = i\omega \frac{1}{cn^{2\omega}} d_{SHG} E_1^2 L \mathrm{e}^{-i\Delta kL/2} \operatorname{sinc}\left(\frac{\Delta kL}{2}\right). \quad (15.23)$$

We can now calculate the intensity $I^{2\omega}(L) = (1/2)\varepsilon_0 c n^{2\omega}|E_2(L)|^2$ of the generated second-harmonic wave for an interaction region of length L. The efficiency η of SHG is therefore

$$\eta = \frac{I^{2\omega}(L)}{I^{\omega}} = \frac{2\omega^2}{\varepsilon_0 c^3}\left(\frac{d_{SHG}^2}{n^{2\omega}(n^{\omega})^2}\right) I^{\omega} L^2 \operatorname{sinc}^2\left(\frac{\Delta kL}{2}\right). \quad (15.24)$$

We note the following:

- Efficiency of SHG is proportional to the pump intensity I^{ω}, at least as long as the undepleted pump approximation is valid. We can therefore increase the efficiency by increasing the pump beam intensity, which can be done e.g., by using high-Q resonators or using waveguiding structures.
- Considering the material properties, the efficiency scales with d^2/n^3, which is a useful figure of merit for SHG, i.e., materials with high nonlinear susceptibility and low refractive index will have a better figure of merit for SHG. Figures of merit of different organic and inorganic materials are discussed later (see Figure 15.21).
- For $\Delta k = 0$ the efficiency increases quadratically with the interaction length L.
- Nonzero phase mismatch parameter $\Delta k \neq 0$ will considerably decrease the efficiency. This effect is illustrated in Figure 15.5, where also the coherence length l_c is defined as

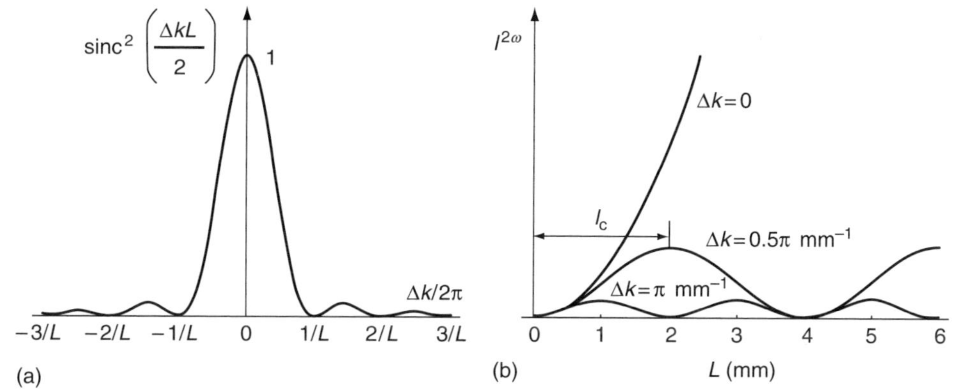

FIGURE 15.5 (a) Factor that reduces the efficiency of the SHG due to the phase mismatch Δk. (b) Second-harmonic intensity as a function of the crystal thickness for different values of the phase mismatch Δk.

$$l_c = \frac{\pi}{\Delta k} = \frac{\lambda}{4(n^{2\omega} - n^{\omega})} \tag{15.25}$$

and measures a useful material length for frequency doubling. If we do not carefully choose the conditions, l_c will be in the order of 1–100 μm, which will lead to a very small macroscopic efficiency η. We will discuss the possibilities to achieve the so-called phase-matching condition with $\Delta k = 0$ later on.

15.1.2.2 Difference-Frequency Generation, Optical Parametric Generation/Oscillation, and Optical Rectification

Difference-frequency generation (DFG) is characterized as the interaction of two input beams of frequencies ω_3 and ω_1 resulting in an optical field with the frequency $\omega_2 = \omega_3 - \omega_1$, i.e., the difference of the two. The nonlinear polarization for the DFG can be written as

$$P_i^{\omega_2} = \varepsilon_0 \chi_{ijk}^{(2)}(-\omega_2, \omega_3, -\omega_1) E_j^{\omega_3} (E_k^{\omega_1})^* . \tag{15.26}$$

The beam with the frequency ω_3 is typically the strongest and therefore referred to as the pump beam. Optical parametric generation (OPG) is a special case of DFG, where only the pump beam is incident on the nonlinear material generating two beams at the frequencies ω_1 and ω_2. These frequencies are selected based on the phase-matching condition. In order to enhance the efficiency of either process, the nonlinear medium can be placed inside a cavity with highly reflecting mirrors for the frequencies ω_1 and/or ω_2; in this case the process is referred to as the optical parametric oscillation (OPO).

In contrast to sum-frequency generation, DFG and OPO are well suited to achieve coherent light sources at longer wavelengths, i.e., the near and mid-infrared region. Another application is optical parametric amplification, where the strong pump beam at frequency ω_3 tranfers energy to amplify an optical signal at frequency ω_1. In this case the output at frequency ω_2 remains unused and is thus often referred to as idler. Finally OPG and OPO are of particular importance because they allow to turn a single frequency laser into a broadly tunable laser system by adjusting the phase-matching condition (Equation 15.29) using, e.g., angle or temperature tuning that are introduced in the following section.

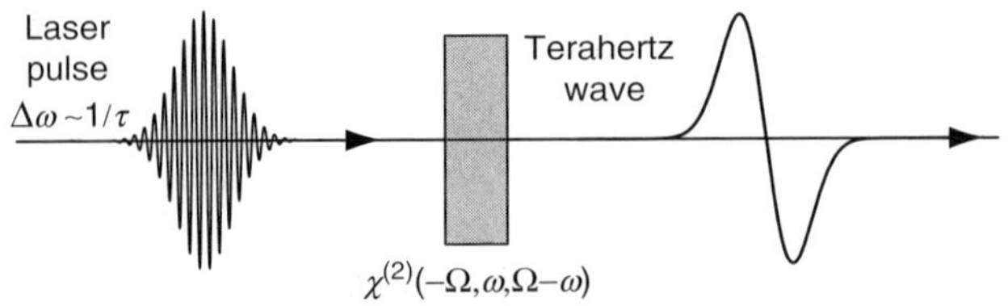

FIGURE 15.6 Schematic illustration of terahertz wave generation via optical rectification in a second-order nonlinear optical crystal. Initial ultrashort optical pulse ($\tau \sim 100$ fs) contains several frequency components that are mixing with each other in the crystal to generate a nonlinear polarization oscillating in the terahertz frequency range.

Optical rectification (OR) is a special case of DFG. Similarly as in the process of SHG, only one light wave of frequency ω is incident. In the case of optical rectification, this wave is "mixing with itself" to generate a static polarization in the material according to

$$P_i^0 = \frac{1}{2}\varepsilon_0 \chi_{ijk}^{(2)}(0,\omega,-\omega)E_j^{\omega}(E_k^{\omega})^*. \tag{15.27}$$

Optical rectification can be used for example to generate waves at terahertz (THz) frequencies using ultrashort laser pulses with a large bandwidth as pump waves. The frequency components of such pulses are differenced with each other to produce nonlinear polarization from zero to several terahertz (Figure 15.6).

15.1.2.3 Conservation of Energy and Momentum (Phase Matching)

All the nonlinear processes previously discussed have in common that they conserve energy, which was already implicitly assumed:

$$\omega_3 = \omega_1 + \omega_2 \text{(energy conservation)}. \tag{15.28}$$

Another common feature is that efficient nonlinear interaction can only occur if momentum is conserved. This requirement, also referred to as phase matching, is manifested in the following condition:

$$\boldsymbol{k}_{\omega_3} = \boldsymbol{k}_{\omega_1} + \boldsymbol{k}_{\omega_2} \text{(momentum conservation or phase matching)}. \tag{15.29}$$

For collinear parametric interactions, where all wave vectors $\boldsymbol{k}_i$ are parallel to one another, the phase-matching condition simplifies to

$$n^{\omega_3}\omega_3 = n^{\omega_1}\omega_1 + n^{\omega_2}\omega_2, \tag{15.30}$$

where n^{ω_i} are the refractive indices of the waves at frequency ω_i. The nonlinear polarization for general directions in the crystal can be written as

$$|\boldsymbol{P}^{\omega_3}| = 2\varepsilon_0 d_{\text{eff}} |\boldsymbol{E}^{\omega_1}||\boldsymbol{E}^{\omega_2}| \text{(sum-frequency generation)}, \tag{15.31}$$

$$d_{\text{eff}} = \sum_{ijk} d_{ijk}(\omega_3, \omega_1, \omega_2)\cos(\alpha_i^{\omega_3})\cos(\alpha_j^{\omega_1})\cos(\alpha_k^{\omega_2}), \tag{15.32}$$

where d_{eff} is the effective nonlinear optical coefficient and α_i^{ω} are the angles between the electric field vector $\boldsymbol{E}^{\omega}$ of the wave with frequency ω and the main axis i of the optical indicatrix. Note that in general the angle ρ between the wave vector and the Poynting vector (see Figure 15.3) has to be taken into account to calculate the electric field vectors and α_i^{ω}.

15.1.2.3.1 Phase-Matching Techniques

We will here illustrate some possibilities of phase matching for the example of SHG, which are however applicable also for other nonlinear optical processes. For SHG, the condition (Equation 15.30) simplifies to

$$n^{2\omega} = n^{\omega}, \tag{15.33}$$

which will lead to $\Delta k = 0$ and a quadratic increase of the second-harmonic intensity as a function of the interaction length (see Figure 15.5). Since most materials exhibit a positive dispersion of the refractive index and therefore $n^{2\omega} > n^{\omega}$, phase matching is most often achieved using birefringence, meaning that the polarizations of the interacting waves will not all be parallel. We distinguish type I and type II phase matching: by type I phase matching the nonlinear polarization is generated using two photons of the same linear polarization, while for type II phase matching it is generated using two photons of the orthogonal linear polarization. Figure 15.7 illustrates the two types of phase matching for the case of an optically uniaxial crystal with $n_e > n_o$, where $n_e = n_3$ is the so-called extraordinary refractive index and $n_o = n_1 = n_2$ is the ordinary index.

Due to dispersion, $n_o^{2\omega} > n_o^{\omega}$ and $n_e^{2\omega} > n_e^{\omega}$. Phase matching is possible if $n_e^{\omega} > n_o^{2\omega} > n_o^{\omega}$. The phase-matching angle $\theta_{\text{PM}}^{\text{I}}$ for the phase matching of type I is characterized by $n_o^{2\omega} = n_e^{\omega}(\theta_{\text{PM}}^{\text{I}})$, where $n_e^{\omega}(\theta_{\text{PM}}^{\text{I}})$ can be calculated considering Figure 15.2 as

$$\frac{1}{(n_e^{\omega}(\theta_{\text{PM}}^{\text{I}}))^2} = \frac{\cos^2(\theta_{\text{PM}}^{\text{I}})}{(n_o^{\omega})^2} + \frac{\sin^2(\theta_{\text{PM}}^{\text{I}})}{(n_e^{\omega})^2}. \tag{15.34}$$

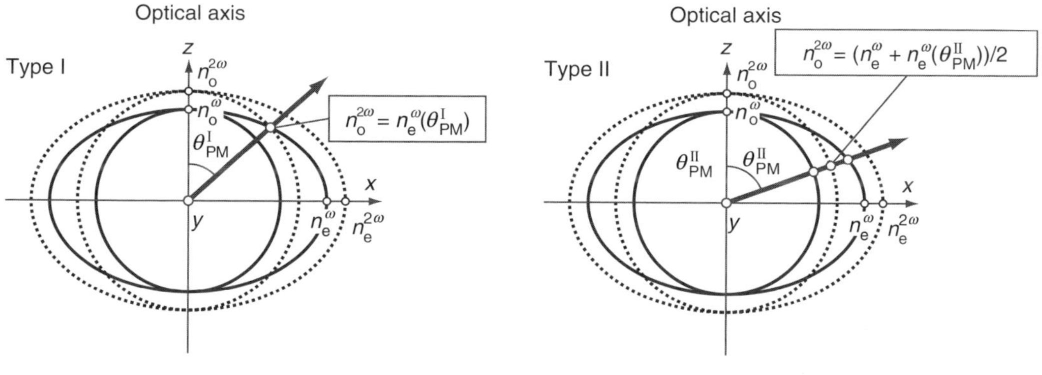

FIGURE 15.7 Type I (left) and type II (right) phase matching for a uniaxial crystal with $n_e > n_o$. Circles and ellipses are the so-called normal surfaces of the two orthogonal eigenwaves at ω (solid curves) and at 2ω (dotted curves), which give the refractive index as a function of the propagation direction (the direction of the wavevector k). The normal surfaces can be constructed from the optical indicatrix according to Figure 15.2.

For the phase matching of type II, the two photons at ω are polarized orthogonally and therefore they travel with a different phase velocity. Considering also the condition (Equation 15.30) for the case of SHG we obtain

$$n_o^{2\omega} = \frac{1}{2}(n_o^{\omega} + n_e^{\omega}(\theta_{PM}^{II})), \tag{15.35}$$

where $n_e^{\omega}(\theta_{PM}^{II})$ can be expressed analogously as $n_e^{\omega}(\theta_{PM}^{I})$ in Equation 15.34. We can tune the phase-matching condition for different fundamental frequencies ω by changing the propagation angle (angle tuning) or the temperature of the crystal through $n(T)$ (temperature tuning).

For the example above, the energy propagation directions of the interacting waves may not match, which will also reduce the efficiency. We distinguish critical and noncritical phase matching. Noncritical phase matching is possible only if the phase matching can be achieved for the propagation along the dielectric axis of the crystal, i.e., $\theta_{PM} = 90°$. In this case the so-called walk-off angle between the energy propagation direction of the fundamental and the second-harmonic beam will be zero, which will considerably increase the acceptance angle for the incoming fundamental beam.

Quasi-phase matching (QPM) is another possibility beside birefringence to increase the efficiency of the frequency conversion processes. For QPM case there is no restriction to propagate along a certain direction in the material that allows for $\Delta k = 0$. With QPM we can also employ the diagonal nonlinear optical coefficients, since the polarizations of the interacting waves may all be parallel. However, QPM requires a certain periodical modulation of the nonlinear optical susceptibility, which is technologically much more challenging. The basic principle of QPM for an example of SHG is as follows: let us consider a material, in which the effective nonlinear optical coefficient d_{eff} changes sign after each coherence length l_c as defined in Equation 15.25. This can be done by the so-called periodical poling of the material, which changes the polarization direction of the material units (domains) with a period of $2l_c$ (see Figure 15.8b). In a homogeneously poled material, after passing one coherence length l_c, a destructive interference of the newly generated second-harmonic light with the light generated from the beginning starts to decrease the $I^{2\omega}$ signal, until after $2l_c$ it drops to zero. As a result, we have a periodic modulation of $I^{2\omega}$ with the thickness (Figure 15.8a). Change of the d_{eff} sign after one l_c introduces a phase shift of π (since $-1 = \exp^{i\pi}$) to the newly generated light, which will result in a constructive interference with the signal generated in the previous domain. Thus we have a continuous increase of the $I^{2\omega}$ signal with

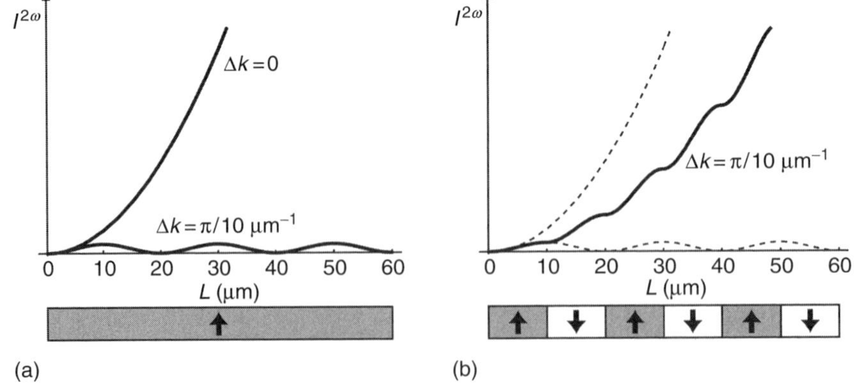

FIGURE 15.8 Intensity of the second-harmonic wave as a function of the propagation length, L, in (a) homogeneously poled material with $\Delta k = 0$ and $\Delta k = \pi/10\ \mu m^{-1}$, (b) periodically poled material with $\Delta k = \pi/10\ \mu m^{-1}$ and domain lengths of 10 μm.

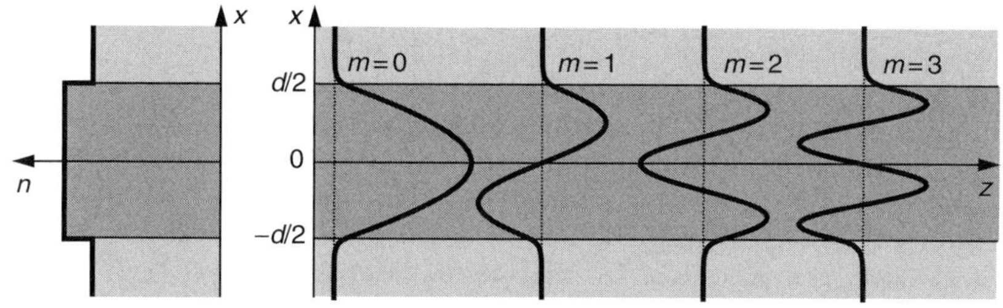

FIGURE 15.9 Optical field profiles of the first four waveguiding modes $m = 0, 1, 2, 3$ in a waveguide (right) with the refractive index profile that is shown on the left. The number of different modes supported by a waveguide depends on waveguide dimensions, refractive indices, and light wavelength.

the thickness as shown in Figure 15.8b. For a perfectly poled material with a period of $2l_c$ it can be shown that the average $I^{2\omega}$ signal increases with $((2/\pi)d_{eff}L)^2$, which can be compared with the increase in the homogeneously poled material with $\Delta k = 0$ that is proportional to $(d_{eff}L)^2$ as given in Equation 15.24.

Mode dispersion in optical waveguides is another possibility to achieve phase matching in nonlinear optical materials. Optical waveguides are basic units of integrated-optics elements allowing for guided-wave propagation in micrometer-size cross sections without experiencing diffraction. They consist of a core region with dimensions comparable to light wavelength, which has a higher refractive index than the surrounding cladding region, resulting in light confinement in the region of high refractive index (see Chapter 2). By choosing the refractive indices of the core and of the cladding region, we can tune the effective refractive index of the light propagating in a waveguide. Depending also on the dimensions of the core regions, several optical modes with different effective indices may propagate in a waveguide (see Figure 15.9).

We can now achieve phase matching $\Delta k = 0$ by matching the effective index of one mode (e.g., $m = 0$) at frequency ω with another mode (e.g., $m = 1$) at frequency 2ω.

Waveguides are very important elements for increasing the efficiency of the frequency conversion that scales with intensity I^{ω} as shown in Equation 15.24. Due to the strong light confinement into areas in the order of $A = 1\ \mu m^2$, we can achieve very high intensities $I^{\omega} = P^{\omega}/A$ even for low pump powers P^{ω}.

15.1.2.4 Linear Electro-optic Effect and Applications

Linear electro-optic effect or Pockels effect is defined as the deformation and rotation of the optical indicatrix if an electric field is applied to a noncentrosymmetric sample. The linear electro-optic effect can be also expressed using the nonlinear $\chi^{(2)}$ tensor as

$$P_i^{\omega} = 2\varepsilon_0 \chi_{ijk}^{(2)}(-\omega,\omega,0)E_j^{\omega}E_k^{0}. \tag{15.36}$$

The electro-optic effect is generally considered separately from nonlinear optical effects, because one of the fields involved E_k^0 is not an optical but a static electric field. As long as the frequency of the electric field applied is much smaller than that of the light field, phase matching is always fulfilled for this process. Typically the linear electro-optic effect is described in terms of the change of the optical indicatrix

$$\Delta\left(\frac{1}{n^2}\right)_{ij} = r_{ijk}E_k. \tag{15.37}$$

Equation 15.37 is also defining the electro-optic tensor r_{ijk}. For small changes, the linear refractive index change can be approximated by

$$\Delta n_i \cong -\frac{1}{2} n_i^3 r_{iik} E_k. \tag{15.38}$$

Inserting the above nonlinear polarization (Equation 15.36) into the wave equation (Equation 15.6) and assuming a small change of the refractive index, we can relate the electro-optic tensor coefficients defined above with the second-order susceptibility coefficients as

$$\chi_{ijk}^{(2)}(-\omega,\omega,0) = -\frac{1}{2} n_i^2 n_j^2 r_{ijk}(\omega). \tag{15.39}$$

The linear electro-optic effect is widely used in electro-optic modulators. These devices exploit the induced phase change of an optical wave, which can also be converted to a change in the intensity. Therefore, the phase or the optical intensity can be controlled by an electrical signal, a frequent task in telecommunications.

Electro-optic modulators are optical devices used to modulate a beam of light, which can be used to transmit information. Due to high frequencies of the optical waves, very high modulation bandwiths are possible with light as information carrier. The optical beam may be modulated in phase, frequency, polarization, amplitude, and direction. In the following paragraphs, we briefly describe the most common modulation geometries including integrated-optics and high-frequency modulators.

Using phase modulators the information is presented with variations of the phase of the carrier wave as shown in Figure 15.10. As we have seen above in Equation 15.38, the refractive index change of a biased crystal will depend on the effective electro-optic coefficient r (that will depend on the orientation of the crystal and light polarization) and the applied voltage $V(t)$ as

$$\Delta n = -\frac{1}{2} n^3 r \frac{V(t)}{d}, \tag{15.40}$$

where d is the thickness of the crystal. The phase of the wave after passing a crystal of length l will change with the voltage as

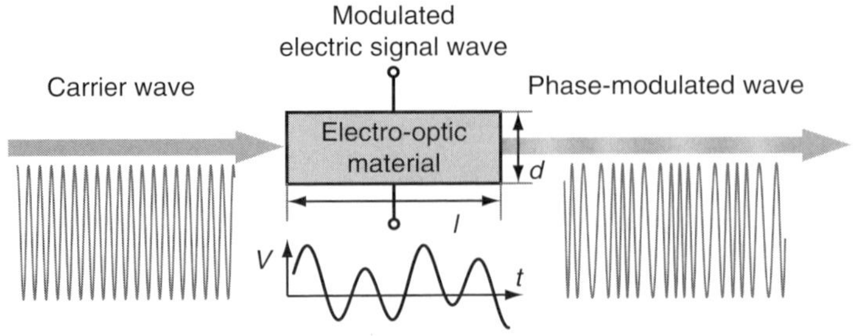

FIGURE 15.10 Phase modulator: the phase of the output beam is controlled by the voltage applied to the crystal.

$$\Delta\phi = k(V)l - k(0)l = \frac{2\pi}{\lambda}\Delta nl = -\frac{\pi n^3 r}{\lambda}V(t)\frac{l}{d}. \qquad (15.41)$$

In the case of simple sine modulation of the applied voltage $V(t) = V_m \sin \omega_m t$, the output optical wave $E(t)$ can be written as

$$E(t) = A\cos(\omega t - \Delta\phi) = A\cos\left(\omega t + \frac{\pi n^3 r}{\lambda}\frac{l}{d}V_m \sin\omega_m t\right) = A\cos(\omega t + \phi_m \sin\omega_m t), \qquad (15.42)$$

i.e., the output optical field is phase modulated with the modulation index

$$\phi_m = \frac{\pi n^3 r}{\lambda}\frac{l}{d}V_m. \qquad (15.43)$$

Amplitude (intensity) modulation can be achieved, for example, by putting a phase modulator in one branch of an interferometer, that is most commonly a Mach–Zehnder interferometer as shown in Figure 15.11a. A beam splitter is used to divide the laser light into two beams, one of which is passing a phase electro-optic modulator. The two beams are combined again with another beam splitter. If the incoming beam has an amplitude A and the beam splitters are dividing a wave into two equal parts, the outgoing field is the sum of the two interfering beams as

$$E_{out}(t) = E_1(t) + E_2(t) = \frac{1}{2}A\cos(\omega t) + \frac{1}{2}A\cos(\omega t - \Delta\phi) = A\cos\left(\frac{\Delta\phi}{2}\right)\cos\left(\omega t - \frac{\Delta\phi}{2}\right). \qquad (15.44)$$

The outgoing signal beam intensity is proportional to $E^2(t)$ averaged in time:

$$I_{out} = I_{in}\cos^2\left(\frac{\Delta\phi}{2}\right) = I_{in}\cos^2\left(\frac{\pi n^3 r}{\lambda}\frac{l}{d}\frac{V}{2}\right) = I_{in}\cos^2\left(\frac{\pi}{2}\frac{V}{V_\pi}\right), \qquad (15.45)$$

where I_{in} is the input beam intensity and the half-wave voltage V_π has been defined as

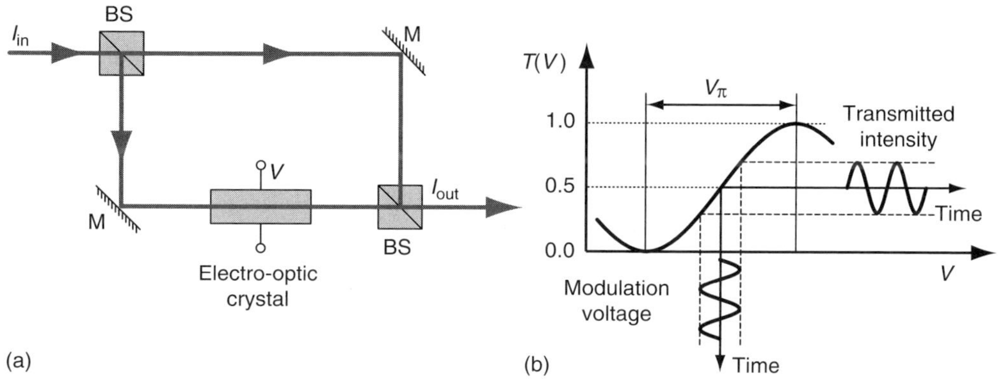

FIGURE 15.11 (a) Mach–Zehnder interferometer with a phase electro-optic modulator in one arm acts as an amplitude electro-optic modulator. BS: beam splitter; M: mirror. (b) Transmittance of the interferometer as a function of the applied voltage.

$$V_\pi = \left(\frac{\lambda}{n^3 r}\right)\left(\frac{d}{l}\right). \tag{15.46}$$

The resulting transmittance of the interferometer $T = I_{out}/I_{in}$ as a function of the applied voltage is shown in Figure 15.11b. We see that the half-wave voltage V_π is an applied voltage needed to change the transmittance from 0 to 1, i.e., to operate the device as an optical switch. By operating the device in a limited region around $T(V) = 0.5$ (that can be adjusted simply by changing the length of one arm of the interferometer), we obtain a linear intensity modulator. In this case the modulated transmitted light intensity is directly proportional to the modulation of the electric signal $V(t)$, as also illustrated in Figure 15.11b, as

$$T(\text{small mod. amplitude}) \approx \frac{1}{2}\left(1 + \frac{\pi}{V_\pi} V(t)\right). \tag{15.47}$$

Half-wave voltage V_π given by Equation 15.46 is an important parameter for device applications. Half-wave voltage for a certain modulator geometry and carrier wavelength will be low for large $n^3 r$, which is the material figure of merit for electro-optics. Figures of merit of some selected organic and inorganic materials are discussed later on (Table 15.2). On the other hand, half-wave voltage can be reduced by optimizing the geometry, i.e., long propagation distances l and short electrode distances d will greatly reduce the V_π. However, there exist limits for reducing the lateral dimension d connected to light diffraction: light focused on small spots in the range of square micrometers will be diffracted so that after few micrometers of propagation its lateral dimensions will double. The solution for this are integrated-optics modulators, in which light is confined to waveguiding cores with dimensions comparable to the light wavelength. Figure 15.12 shows the simplest schemes of integrated-optics phase and amplitude modulators.

Mach–Zehnder integrated-optics elements are CMOS compatible and allow for high-speed (>1 GHz) modulation, however the present devices based on $LiNbO_3$ are still having a relatively big size (lengths in the order of 1 cm) and a relatively high power consumption (several watts). Compact-size and low-power devices are the subject of present studies. A promising structure is a microring resonator that allows for long electrical and optical interaction lengths on a very small scale (in the order of 10–100 μm), allowing for low-power consumption and very large scale integration (VLSI).

The basic design of a microresonator is shown in Figure 15.13. Light guided in the input waveguide will be partially coupled into the microring. At resonator frequencies the intensity of

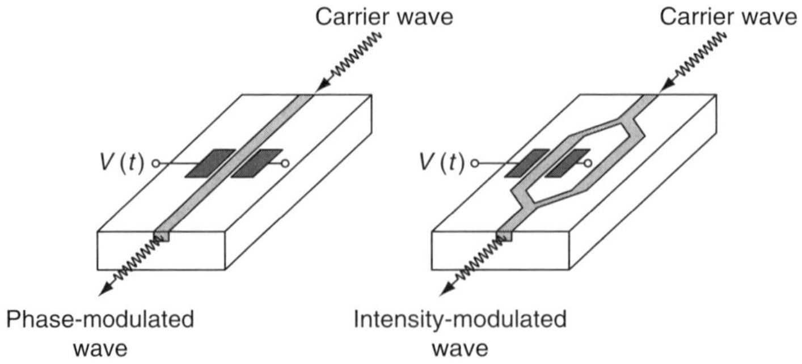

FIGURE 15.12 An integrated-optics phase modulator (left) and a Mach–Zehnder intensity modulator (right) based on the electro-optic effect.

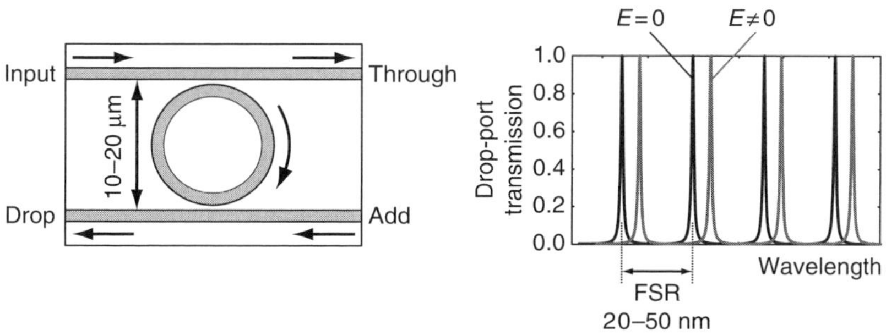

FIGURE 15.13 A basic scheme of a very large scale integrated (VLSI) microring device, viewed from the top (left). The device consists of two waveguides that are coupled by a microring waveguide. Light coming from the input-port waveguide will be transmitted through the through-port waveguide unless a resonance condition is met: in this case the intensity of light will increase dramatically inside the microring and will be coupled to a drop-port waveguide. Drop-port transmission is shown on the right as a function of light wavelength. The free-spectral range (FSR) is the wavelength separation between the adjacent transmission peaks. If the ring is made of an electro-optic material, we can tune its refractive index and therefore also the resonance condition by applying an external electric field.

the light in the ring will increase dramatically, and will be coupled out into the drop-port waveguide. The graph in Figure 15.13 presents a result of such a filtering by observing the light transmission through the drop port while scanning the wavelength. One can see sharp increases of the intensity when the resonance condition is satisfied. The resonance condition is satisfied, if after one round trip in the ring the light is in phase with the light coupled into the microring. The resonance condition is quite sensitive and depends on the geometry, wavelength, and the material refractive index. Changing the refractive index by means of the electro-optic effect, we can effectively tune the resonance condition. Based on microresonators, several types of high-speed photonic elements can be constructed, including optical switches, modulators, wavelength multiplexers, and filters.

When approaching high-frequency modulation (100 MHz and more), there are essentially two other parameters that can limit the efficiency and speed of the electro-optic response:

- Response time of the electric circuit $\tau = RC$ will limit the bandwidth of the device, since only a small fraction of the voltage will be applied to the crystal for modulation frequencies $\omega > 1/\tau$. Therefore, decreasing the size of the modulator (like in integrated-optics modulators) and using materials with low dielectric constants will remarkably increase the bandwidth due to a reduced capacitance C of the device.
- For high-speed modulation the applied voltage may vary considerably during the transit time of the optical field. This effect can be eliminated in the so-called traveling wave modulators, in which the modulation field is applied as a microwave traveling in the same direction as the optical field. The traveling time effect can be completely eliminated for the velocity-matched conditions, i.e., when optical and electric waves travel through the electro-optic material with the same speed.

15.1.3 Third-Order Nonlinear Optical Effects and Applications

In centrosymmetric media the second-order nonlinear term is absent, the dominant nonlinearity is of third order

$$P_i^{\mathrm{NL}} = \varepsilon_0 \chi_{ijkl}^{(3)} E_j E_k E_l \tag{15.48}$$

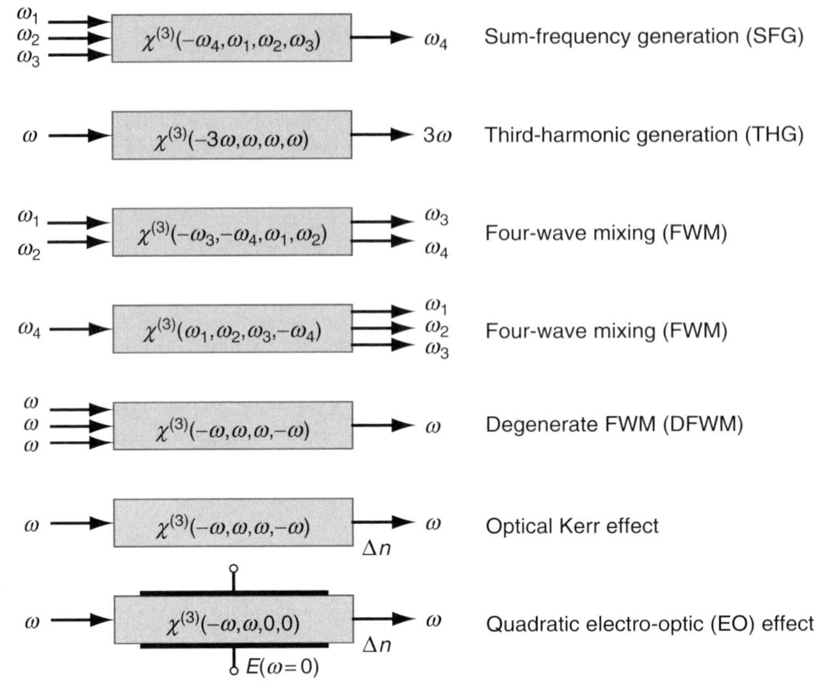

FIGURE 15.14 Schematic representation of important nonlinear optical effects of the third order, including the quadratic electro-optic effect.

and gives rise to third-harmonic generation and related mixing phenomena that are schematically illustrated in Figure 15.14. The third-order response of every medium to a monochromatic optical field at ω is a nonlinear polarization containing components at frequency 3ω and at frequency ω as

$$P_i^{3\omega} = \frac{1}{4}\varepsilon_0 \chi_{ijkl}^{(3)}(-3\omega,\omega,\omega,\omega)E_j^{\omega}E_k^{\omega}E_l^{\omega}, \tag{15.49}$$

$$P_i^{\omega} = \frac{3}{4}\varepsilon_0 \chi_{ijkl}^{(3)}(-\omega,-\omega,\omega,\omega)E_j^{*\omega}E_k^{\omega}E_l^{\omega}. \tag{15.50}$$

The first process is known as the third-harmonic generation. Similar to the process of SHG, the efficiency of generating the third-harmonic frequency will be negligible if the phase-matching condition is not satisfied. Note that the numerical factors 1/4 and 3/4 consider all possible permutations of the electric fields on the right-hand side that contribute to the induced polarization [1].

The second effect, the generation of the polarization component at frequency ω is equivalent to changing the effective refractive index according to

$$\Delta n(\omega) \cong \frac{3\chi^{(3)}(-\omega,-\omega,\omega,\omega)}{8n_0(\omega)}|E^{\omega}|^2, \tag{15.51}$$

where $n_0(\omega)$ is the unperturbed index at frequency ω and we omitted the tensor notation for simplicity. The relation (Equation 15.51) can be easily derived if we insert the nonlinear polarization

(Equation 15.50) into the wave equation for the propagation of an optical field inside a medium. The overall refractive index is therefore a linear function of the optical intensity $I \propto |E^{\omega}|^2$

$$n(I) = n_0 + n_2 I. \tag{15.52}$$

This effect is known as the optical Kerr effect. It is a self-induced effect, in which the phase velocity of the wave depends on the wave's own intensity. As a result, a wide variety of important processes occur, such as self-phase modulation and self-focusing effect.

15.1.3.1 Self-Phase Modulation

The result of the optical Kerr effect is the optical wave traveling in a third-order nonlinear medium, experiencing an additional phase shift $\Delta\phi^{NL}$, which is proportional to the optical intensity:

$$\phi = \frac{2\pi}{\lambda_0} n(I)L = \phi_0 + \Delta\phi^{NL}, \quad \Delta\phi^{NL} = \frac{2\pi}{\lambda_0} n_2 LI, \tag{15.53}$$

where
- ϕ is the total phase shift
- λ_0 is the light wavelength in vacuum
- L is the interacting length

Such self-phase modulation is useful in all-optical switching applications. An illustrative example for an optical switching device is shown in Figure 15.15, where a nonlinear material is placed in a Fabry–Perot interferometer. At low-incident intensity, the phase shift fails to satisfy the resonance condition for the interferometer. Therefore, the transmitted light intensity is blocked. As the intensity increases, the refractive index changes. When it reaches the value that satisfies the resonance condition, the transmitted beam is switched on. Device operation is controlled by the light beam itself.

15.1.3.2 Self-Focusing and Spatial Kerr Solitons

In reality, optical beams are not in the form of plane waves, but have certain transverse intensity dependence. The laser output signal has usually a Gaussian transverse intensity distribution. If such a beam is transmitted through a film of material with nonlinear refractive index, the phase difference at the center of the beam, where the intensity is the highest, is larger than at the edge of the beam. The resulting wavefront curvature is similar to the ordinary lens effect. The nonlinear material therefore acts as a lens with intensity-dependent focal length (Figure 15.16).

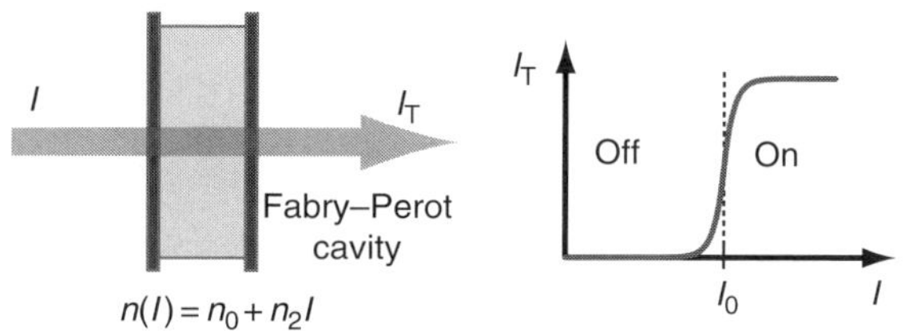

FIGURE 15.15 Optical switching behavior, exhibited by a third-order nonlinear material in a Fabry–Perot interferometer.

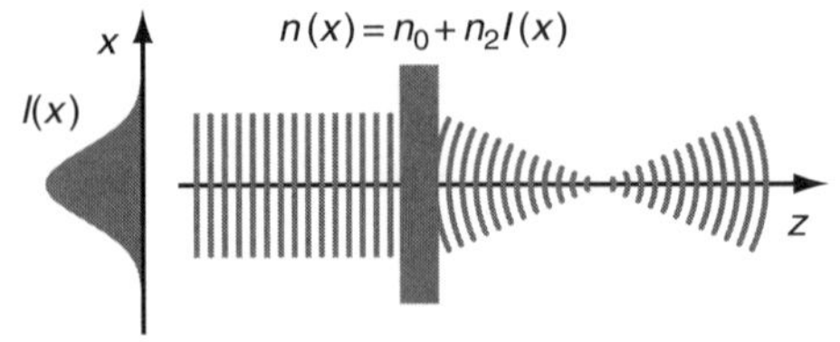

FIGURE 15.16 Self-focusing effect of a beam with transverse intensity dependence.

Consider now an intense optical beam traveling through a substantial thickness of a homogeneous medium instead of a thin sheet. In a linear medium, the beam of light will spread out due to diffraction as illustrated in Figure 15.17a. In a nonlinear medium, self-focusing effect counteracts diffraction. If the diffraction is exactly compensated by the nonlinear focusing effect, the beam propagates without changing its shape. Such self-guided beams are called spatial solitons (Figure 15.17b).

15.1.3.3 Optical Phase Conjugation

The response of a medium to a superposition of three monochromatic waves at frequencies ω_1, ω_3, and ω_4 is an example of four-wave mixing. A lot of different frequencies can be generated if frequency- and phase-matching conditions are satisfied. Widely used is the special case of degenerate four-wave mixing, where all four waves are of the same frequency $\omega_1 = \omega_2 = \omega_3 = \omega_4 = \omega$, but they propagate in different directions. If two waves (3 and 4, see Figure 15.18) are uniform plane waves traveling in opposite directions

$$E_3(\boldsymbol{r}) = A_3 \mathrm{e}^{i\boldsymbol{k}_3\cdot\boldsymbol{r}}, \quad E_4(\boldsymbol{r}) = A_4 \mathrm{e}^{i\boldsymbol{k}_4\cdot\boldsymbol{r}}, \quad \boldsymbol{k}_4 = -\boldsymbol{k}_3, \tag{15.54}$$

then the induced polarization density is

$$P_2(\boldsymbol{r}) = \frac{3}{2}\varepsilon_0 \chi^{(3)} A_3 A_4 E_1^*(\boldsymbol{r}). \tag{15.55}$$

Note that the numerical factor of 3/2 considers all possible permutations of the electric fields on the right-hand side that contribute to the induced polarization [1]. This polarization is the source emitting an optical wave

$$E_2(\boldsymbol{r}) \propto A_3 A_4 E_1^*(\boldsymbol{r}). \tag{15.56}$$

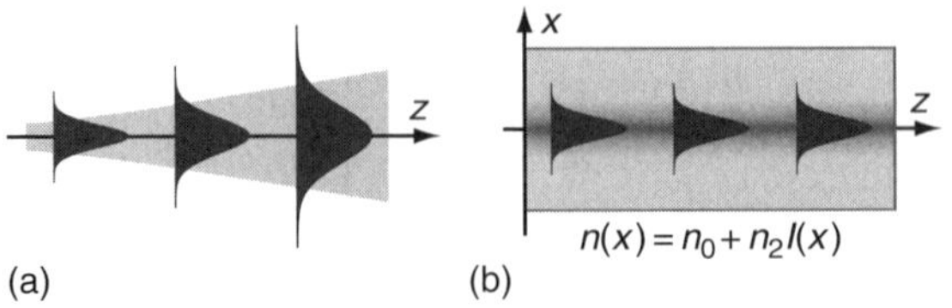

FIGURE 15.17 Comparison between (a) a Gaussian beam traveling in a linear medium and (b) a spatial soliton traveling in a nonlinear medium.

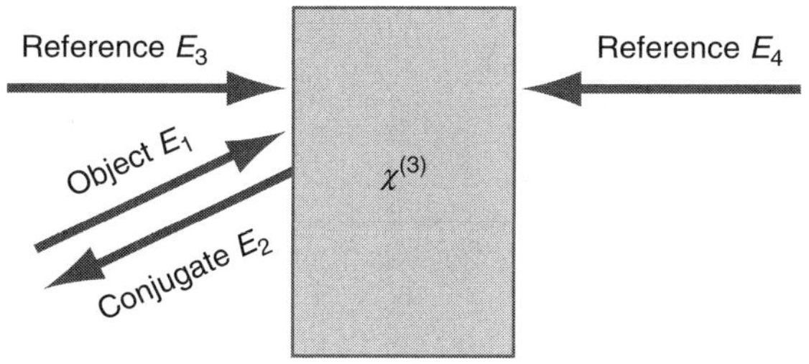

FIGURE 15.18 Degenerate four-wave mixing configuration, in which the generated wave 2 is the phase conjugate version of the wave 1.

In this case the generated wave is a conjugated version of the wave 1. The device can be used as a phase conjugate mirror (see Figure 15.19).

15.2 ORGANIC NONLINEAR OPTICAL MATERIALS

15.2.1 Comparison of Organic and Inorganic Materials for Nonlinear and Electro-optics

There is an essential difference between the optical response in organic and in inorganic materials considering the origin of the nonlinear as well as of the linear polarizability of a material. The electronic polarizability of molecular units presents the dominant contribution in organic materials. Inorganic materials are based on strong bonding between the lattice components (ions), which are acting as additional polarizable elements. However, the contributions of ions or lattice vibrations to the polarizability response are essential only for low frequencies of the applied electric field, since the dynamics of ions is in general much slower compared to electrons. Figure 15.20 schematically illustrates different lattice contributions and their frequency range for the linear response (dielectric constant ε) and for the nonlinear response (electro-optic coefficient r).

The "free" (also called unclamped or zero stress) electro-optic coefficient r^{T} in the low-frequency region contains three different contributions: from acoustic lattice vibrations (acoustic phonons) r^{a}, from optical lattice vibrations r^{o}, and from electrons r^{e}

$$r^{\mathrm{T}} = r^{\mathrm{a}} + r^{\mathrm{o}} + r^{\mathrm{e}} = r^{\mathrm{a}} + r^{\mathrm{S}}, \tag{15.57}$$

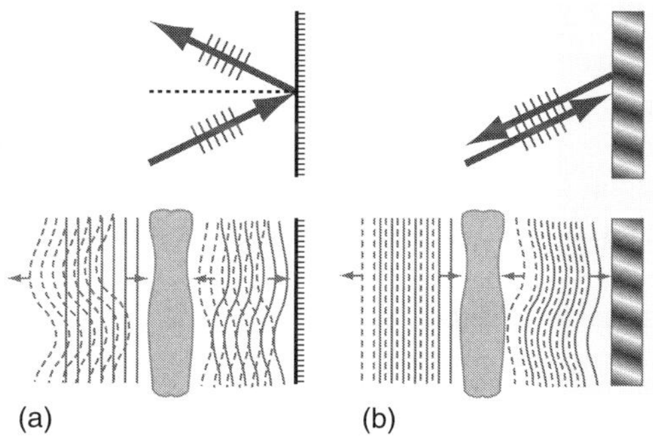

FIGURE 15.19 Comparison between (a) ordinary mirror and (b) phase conjugate mirror. Phase conjugate mirror reflects the incident wave back to itself, so that it propagates exactly in the opposite direction (above) and compensates distorted wave fronts (below).

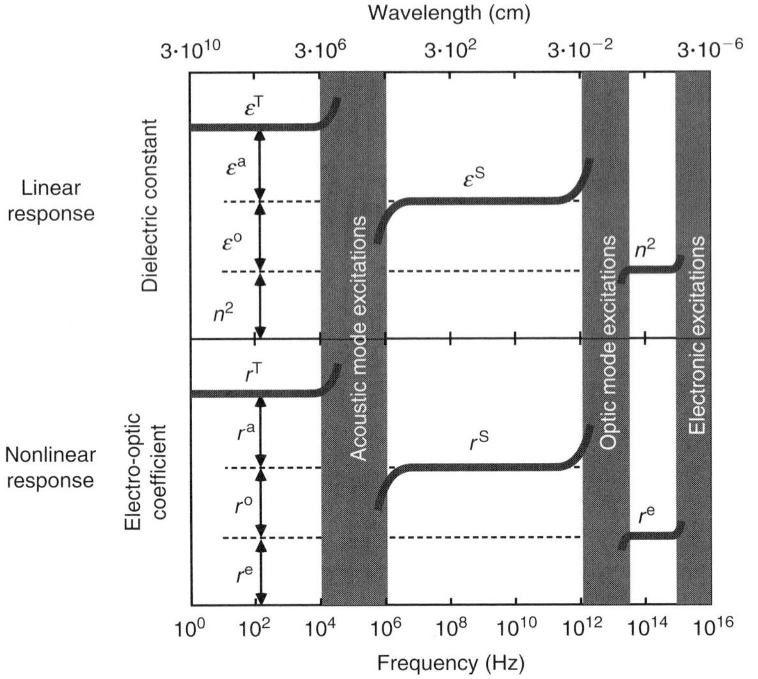

FIGURE 15.20 Schematics of the material linear and of the nonlinear optical response as a function of the frequency of the external electric field.

where $r^S = r^o + r^e$ is the clamped (zero strain) electro-optic coefficient. An analogous description is also valid for the linear response. Note that the dielectric constant, ε, can be related to the refractive index, n, and the electro-optic coefficient, r, to the nonlinear optical coefficient, d, only at optical frequencies, for which the response is purely of electronic origin

$$\varepsilon^e = n^2 \tag{15.58}$$

$$r^e = -\frac{4d}{n^4}, \tag{15.59}$$

where the last equality follows from Equation 15.19 and Equation 15.39.

If we compare the measured optical and nonlinear optical parameters and the frequency dependence in organic and in inorganic materials (see Table 15.1), we can mostly observe that in organic materials the electronic contribution is dominant, whereas in inorganic materials the greater part of the response comes from the acoustic and optical lattice vibration modes. This difference is of essential importance for the applications as discussed in the following.

Because of the trend to higher and higher data rates in optical communication technology there is an increasing need for new materials with large and fast responding nonlinearities, which can be used for all-optical or electro-optic switching and modulation. Today most of the commercialized products use inorganic materials such as $LiNbO_3$, since these materials are well understood, have excellent mechanical and chemical stability, and sufficiently large nonlinear optical coefficients for many applications. On the order hand, organic materials have several important advantages in terms of high-frequency electro-optic applications. They are naturally suited for traveling wave

TABLE 15.1
Comparison of Inorganic Crystals $LiNbO_3$ and $KNbO_3$, and an Organic Crystal DAST with Respect to the Contribution to the Linear and Nonlinear Optical Effects

	ε^T	n^2	r^T (pm/V)	r^S (pm/V)	r^e (pm/V)	d (pm/V)
$LiNbO_3$	28	4.5	30	30	5.8	−34
$KNbO_3$	44	4.5	63	34	3.7	−21
DAST	5.2	4.6	47	48	36	−290

Notes: The values for $LiNbO_3$ and $KNbO_3$ are given at 1.06 μm, and for DAST (4-*N,N*-dimethylamino-4′-*N*′-methyl stilbazolium tosylate) at 1.5 μm. ε^T is the unclamped dielectric constant; n is the refractive index; r^T, r^S, and r^e are the unclamped, clamped, and electronic electro-optic coefficients; and d is the nonlinear optical coefficient.

electro-optic modulators. Due to the low dielectric constant at low frequencies $\varepsilon \approx n^2$, the electric wave travels at about the same speed as the optical wave, which is not the case for most inorganic electro-optic materials with $\varepsilon \gg n^2$. This kind of phase matching is important when building high-frequency modulators. The low dielectric constant of organic materials will also increase the electro-optic modulator bandwidth that is limited by the response time of the electric circuit $\tau = RC$ that is proportional to ε. Another advantage of organic over inorganic materials is the almost constant electro-optic coefficient over an extremely wide frequency range. This property is essential when building broadband electro-optic modulators and field detectors. Table 15.2 lists selected inorganic and organic materials, their properties and figures of merit relevant to electro-optics.

TABLE 15.2
Selection of Electro-optic Materials and Their Parameters Relevant for the Electro-optic Response

	r (pm/V)	n	n^3r (pm/V)	ε	r/ε (pm/V)	λ (nm)
$LiNbO_3$	31.5	2.2	340	28	1.1	633
GaAs	1.2	3.5	51	13.2	0.09	1020
$KNbO_3$	63.3	2.2	650	44	1.4	633
	35[a]	2.2	350[a]	24[a]	1.5[a]	633
$Sn_2P_2S_6$	170	2.8	4000	230	0.74	1313
DAST[b]	92	2.5	1470	5.2	18	720
	53	2.2	530	5.2	10	1313
Polymers						
A-095.11[c]	20	1.66	92	2.8	7.1	1313
CLD-1[d]	130	1.65	584	3.5	37	1313

Note: λ is the corresponding wavelength, r is the electro-optic coefficient (unclamped, if not specified), n is the refractive index, and ε is the low-frequency dielectric constant (unclamped, if not specified); semiconductor GaAs, inorganic ferroelectric crystals $LiNbO_3$, $KNbO_3$, $Sn_2P_2S_6$, organic crystal DAST, and polymers A-095.11 and CLD-1.

[a] Clamped value.

[b] Organic crystal 4-*N,N*-dimethylamino-4′-*N*′-methyl stilbazolium tosylate.

[c] Polyimide side-chain polymer based on disperse red.

[d] Phenyltetraene chromophore in a guest–host polymer system.

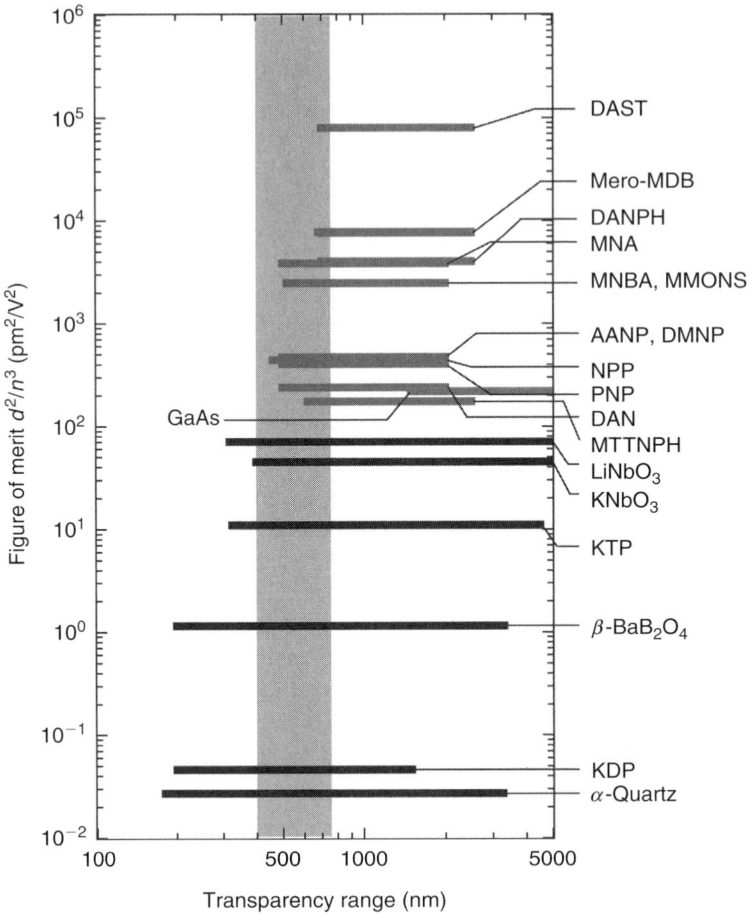

FIGURE 15.21 The figure of merit for frequency doubling, d^2/n^3, versus the transparency range of various organic and inorganic crystals. The shaded area indicates the visible spectral range. Inorganic crystals are indicated in black, organic crystals in grey. It is clearly seen that organic crystals are much superior in terms of the figure of merit.

The almost purely electronic response is also of advantage for terahertz generation via optical rectification in nonlinear optical materials. The high terahertz generation efficiency can only be obtained if the so-called velocity-matching between pump pulses at optical frequencies and the generated terahertz waves is achieved. This is possible in organic materials, because the relatively low dielectric constants allow for the matching between the terahertz phase velocity and the pump optical pulse group velocity [3].

Another advantage of organic nonlinear optical materials compared to inorganic materials are high nonlinear optical figures of merit. This is illustrated in Figure 15.21 that shows figures of merit for frequency conversion d^2/n^3 versus transparency range for various organic and inorganic crystals. One can clearly see that the nonlinear optical figures of merit of organic materials can be several orders of magnitude higher than in best inorganic materials.

In the following section, we describe the origin of the nonlinear optical response in organic materials from the microscopic point of view. We focus on organic acentric materials with a pronounced second-order nonlinearity, suitable for electro-optic applications, frequency conversion, and terahertz wave generation.

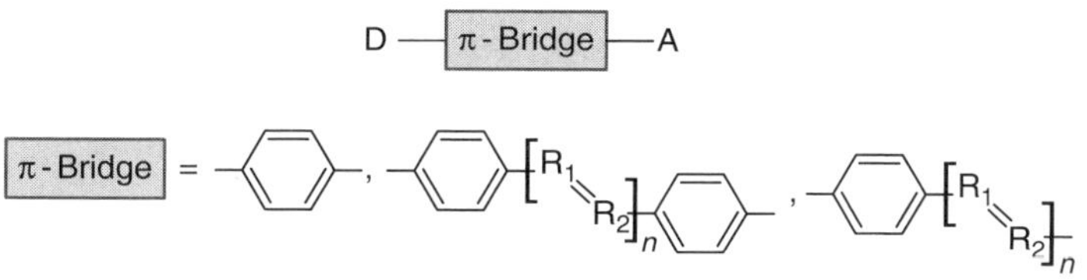

FIGURE 15.22 Typical dipolar organic molecules for second-order nonlinear effects. The electron donor (D) group is connected to the electron acceptor (A) group through a π-conjugated structure. The most common systems are those containing one benzene ring (benzene analogs) and those containing two benzene rings (stilbene analogs). R_1 and R_2 are usually carbon or nitrogen.

15.2.2 Organic Molecules for Second-Order Nonlinear Optics

The basic design of organic nonlinear optical molecules is based on π-conjugated systems. π-conjugated structures are regions of delocalized electronic charge distribution resulting from the overlap of π orbitals. This delocalization leads to a high mobility of the electron density. The electron distribution can be distorted by substituents at both sides of the π-conjugated system. The extent of the redistribution is measured by the dipole moment, the linear redistribution in response to an applied field by the (linear) polarizability α, and the asymmetry of this redistribution with respect to the polarity of the field by the first-order hyperpolarizability β. The optical nonlinearity of organic molecules can be increased either by increasing the conjugation length or by using appropriate electron donor and electron acceptor groups (Figure 15.22). Typical electron donor groups are, for example, CH_3, NH_2, OH, $N(CH_3)_2$, and OCH_3, and typical electron acceptor groups are CN, NO_2, NO, and CHO.

The addition of appropriate functionality at the ends of the π system can enhance the asymmetric electron distribution in either or both the ground state and excited state configurations (Figure 15.23). Another important aspect is the planarity of the molecule, which will affect the size of the π-electron system and the mobility of electrons; twist angles between rings can for example considerably reduce the charge-transfer contribution and thus molecular nonlinearity.

15.2.3 Microscopic and Macroscopic Nonlinearities of Organic Molecules

While Equation 15.8 governs nonlinear effects on a macroscopic scale, one can also consider this problem on the molecular level. The dipole moment of the molecule $\boldsymbol{p}$ consists of its ground state dipole moment μ_g and the induced contribution. The corresponding expansion

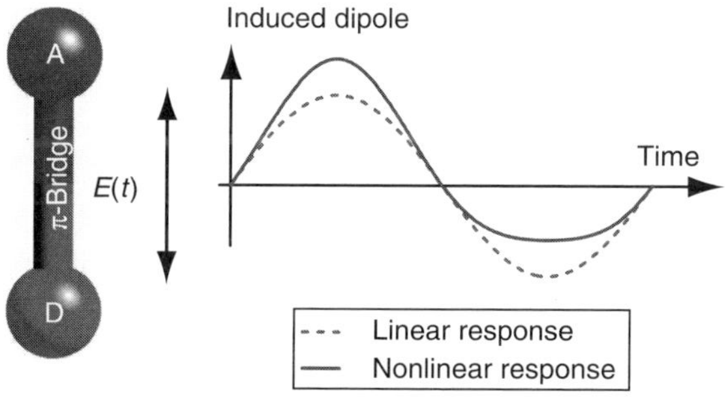

FIGURE 15.23 Simple picture of the physical mechanisms of the nonlinearity of donor–acceptor π-conjugated molecules. If we excite such molecules with an optical field, we induce an asymmetric electronic response of the polarization. This is due to the nature of the substituents: the electron cloud (i.e., the electronic response) favors the acceptor over the donor.

$$p_i = \mu_{g,i} + \varepsilon_0 \alpha_{ij} E_j + \varepsilon_0 \beta_{ijk} E_j E_k + \cdots \quad (15.60)$$

defines the molecular coefficients: linear polarizability, α_{ij}, and first-order hyperpolarizability, β_{ijk}. The electric field, $\boldsymbol{E}$, in Equation 15.60 is the local electric field at the place of the molecule.

The microscopic second-order nonlinear optical hyperpolarizability can be expressed by a simple two-level model for many molecules [4], which considers only two levels in the energy diagram of the molecule: the ground state, g, and the first-excited charge-transfer excitation state, e, for which one electron is transferred from the donor to the acceptor. This approximation holds quite well, since for most organic molecules higher excited states lie much higher in the energy diagram than the photon energy of the light used. Additionally, for many molecules with strong nonlinearity along a single charge-transfer axis, it is reasonable to consider only one dominant hyperpolarizability tensor component β_{zzz} along this axis. Using the two-level model for $\beta_{zzz}(-\omega_3,\omega_1,\omega_2)$ we obtain [4]

$$\beta_{zzz}(-\omega_3,\omega_1,\omega_2) = \frac{1}{2\varepsilon_0\hbar^2} \frac{\omega_{eg}^2(3\omega_{eg}^2 + \omega_1\omega_2 - \omega_3^2)}{(\omega_{eg}^2 - \omega_1^2)(\omega_{eg}^2 - \omega_2^2)(\omega_{eg}^2 - \omega_3^2)} \Delta\mu\mu_{eg}^2, \quad (15.61)$$

where

ω_{eg} is the resonant frequency of the transition, i.e., $\hbar\omega_{eg}$ is the energy difference between the ground state, g, and the excited state, e

$\Delta\mu = \mu_e - \mu_g$ is the difference between the excited and the ground state dipole moments

$\mu_{eg} = \langle e|\mu_z|g\rangle$ is the transition dipole moment between the excited and the ground state

ω_1, ω_2, ω_3 are the involved interacting light frequencies that are all below ω_{eg}

We can also introduce a dispersion-free hyperpolarizability β_0 that is extrapolated to infinite wavelengths (low frequencies), away from electronic resonances

$$\beta_0 = \frac{3}{2\varepsilon_0\hbar^2} \frac{\Delta\mu\mu_{eg}^2}{\omega_{eg}^2}. \quad (15.62)$$

This parameter is very useful when comparing the nonlinearity of molecules with different resonant frequencies ω_{eg}.

For the case of SHG, we then obtain the following dispersion relation:

$$\beta_{zzz}(-2\omega,\omega,\omega) = \frac{\omega_{eg}^4}{(\omega_{eg}^2 - 4\omega^2)(\omega_{eg}^2 - \omega^2)} \beta_0 \quad (15.63)$$

and for the linear electro-optic effect

$$\beta_{zzz}(-\omega,\omega,0) = \frac{\omega_{eg}^2(3\omega_{eg}^2 - \omega^2)}{3(\omega_{eg}^2 - \omega^2)^2} \beta_0. \quad (15.64)$$

The relation between the macroscopic and the molecular coefficients is nontrivial because of the interactions between neighboring molecules. However, most often the macroscopic second-order nonlinearities of organic materials can be well explained by the nonlinearities of the constituent molecules (oriented-gas model [5]). For example, the nonlinear optical and the electro-optic coefficients can be expressed assuming only one dominant tensor element β_{zzz}

$$
\begin{aligned}
d_{ijk}(-2\omega,\omega,\omega) &= \frac{1}{2}\chi_{ijk}^{(2)}(-2\omega,\omega,\omega) \\
&= \frac{1}{2}N\frac{1}{n(g)}f_i^{2\omega}f_j^{\omega}f_k^{\omega}\sum_{s}^{n(g)}\sum_{mnp}^{3}\cos(\theta_{im}^{s})\cos(\theta_{jn}^{s})\cos(\theta_{kp}^{s})\beta_{mnp}(-2\omega,\omega,\omega) \\
&= \frac{1}{2}N\frac{1}{n(g)}f_i^{2\omega}f_j^{\omega}f_k^{\omega}\sum_{s}^{n(g)}\cos(\theta_{iz}^{s})\cos(\theta_{jz}^{s})\cos(\theta_{kz}^{s})\beta_{zzz}(-2\omega,\omega,\omega)
\end{aligned}
\quad (15.65)
$$

and

$$
\begin{aligned}
r_{ijk}(-\omega,\omega,0) &= -\frac{2}{n_i^2(\omega)n_j^2(\omega)}\chi_{ijk}^{(2)}(-\omega,\omega,0) \\
&= -N\frac{1}{n(g)}\frac{2f_i^{\omega}f_j^{\omega}f_k^{0}}{n_i^2(\omega)n_j^2(\omega)}\sum_{s}^{n(g)}\sum_{mnp}^{3}\cos(\theta_{im}^{s})\cos(\theta_{jn}^{s})\cos(\theta_{kp}^{s})\beta_{mnp}(-\omega,\omega,0) \\
&= -N\frac{1}{n(g)}\frac{2f_i^{\omega}f_j^{\omega}f_k^{0}}{n_i^2(\omega)n_j^2(\omega)}\sum_{s}^{n(g)}\cos(\theta_{iz}^{s})\cos(\theta_{jz}^{s})\cos(\theta_{kz}^{s})\beta_{zzz}(-\omega,\omega,0),
\end{aligned}
\quad (15.66)
$$

where

θ_{im}^{s} is the angle between the dielectric axis i and the molecular axis m
N is the number of molecules per unit volume
$n(g)$ is the number of equivalent positions in the unit cell
s is a site in the unit cell
β_{mnp} is the molecular first-order hyperpolarizability
f_i^{ω} is the local field correction that is usually simplified using the Lorentz approximation

$$
f_i^{\omega} = \frac{n_i^2(\omega)+2}{3}. \quad (15.67)
$$

Note that Equation 15.66 gives only the electronic contribution to the electro-optic coefficient, which presents a major contribution to the electro-optic effect in organic materials.

Figure 15.24 schematically shows the molecules and projection angles θ_{kz} between the polar axis k and the molecular axis z. To maximize the diagonal electro-optic or nonlinear optical coefficient

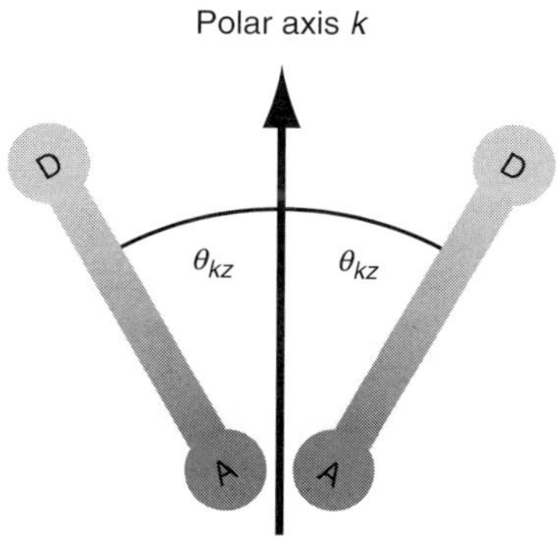

FIGURE 15.24 Nonlinear optical molecules in the oriented-gas model. For phase-matched nonlinear optical applications the optimal angle θ_{kz} between the polar axis of the crystal and the charge-transfer axes of the molecules is 54.7°, while for electro-optics the optimal $\theta_{kz}=0°$.

along the polar axis, r_{kkk} and d_{kkk} respectively, the projection angles θ_{kz} should be close to zero, i.e., the charge-transfer axes of the molecules should be close to parallel. For nonlinear optical applications such as frequency doubling, phase matching should also be considered. It was shown that for phase-matched applications the optimal angle between the charge-transfer axes of the molecules and the polar axis of the crystal is 54.7° for most crystallographic classes [5].

The relationships between the micro- and the macroscopic nonlinearities given above give us a basic idea about the optimized packing of molecules. However, they still do not allow to determine the macroscopic coefficients from the known microscopic nonlinearities precisely, since intermolecular interactions will reduce the molecular nonlinearity β in the solid state [6–8]. Although the relation between macroscopic nonlinearity and molecular structure including the intermolecular interactions is not yet completely understood, the mechanisms leading to large microscopic effects are well known. During the last decades, a large number of nonlinear optical molecules have been synthesized and investigated allowing scientists to gain insight into the chemistry and physics of optical hyperpolarizabilities. Improvement of the values of the hyperpolarizabilities by using new electron donor and acceptor groups led to many new nonlinear optical materials.

The molecular nonlinearity is strongly correlated to the transparency of the materials. Increasing the nonlinearity will shift the wavelength of maximum absorption λ_{eg} toward longer wavelengths. The first-order hyperpolarizability β_{ijk} depends on the strength of donor and acceptor substituents (Figures 15.22 and 15.23) as well as on the extension of the π-electron system (conjugation length). This behavior is illustrated in Figure 15.25 that shows the connection of the dispersion-free hyperpolarizability β_0 (calculated by extrapolation to infinite optical wavelengths) with the wavelength of maximum absorption λ_{eg} in solution. The shaded area schematically shows the experimentally determined range of available molecules. Theoretical calculations of the optimal nonresonant hyperpolarizabilities β_0 have shown that they depend only on the number of electrons in the molecule N and the resonant frequency of the transition ω_{eg} so that $\beta_0^{\mathrm{OPT}} \propto N^{3/2}\,\omega_{eg}^{-7/2}$ [9]. We have included here only those chromophores from which acentric crystals have been obtained. There exist many chromophores with even higher molecular nonlinearities that are

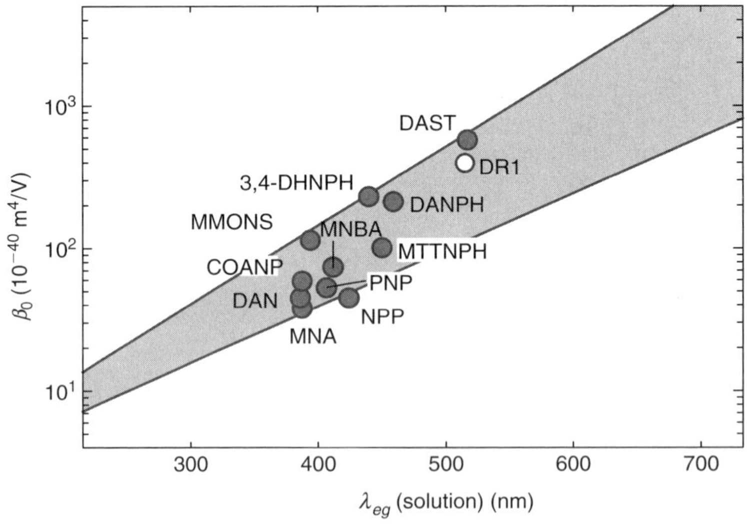

FIGURE 15.25 First-order hyperpolarizability, β_0, extrapolated to infinite wavelengths versus wavelength of maximum absorption for various molecules. The shaded area represents the range of values experimentally determined in various laboratories. For full names and molecular structures, see Table 15.3. Acentric crystals are available for all chromophores included except DR1 (disperse red 1).

mainly used in poled-polymer systems—for these we refer to Chapter 16. We also consider only the approach using dipolar molecules (Figure 15.22), which has led so far to molecular crystals that could be used for macroscopic investigations and device applications.

Using optimized hyperpolarizabilities, the upper limits of the electro-optic and the nonlinear optical coefficients were calculated [10] based on Equations 15.65 and 15.66 by neglecting all the intermolecular contributions (except for local field corrections). The calculations of upper limits with regard to electro-optic and nonlinear optical coefficients show that the macroscopic susceptibilities of crystalline materials based on highly extended π-conjugated donor–acceptor chromophores, e.g., donor–acceptor-disubstituted stilbenes, have by far not reached the upper limit yet. Much improvement is still possible if solid-state materials can be formed from molecules with highly increased nonlinearities and optimal packing.

15.2.4 Basic Types of Organic Materials Exhibiting Macroscopic Second-Order Nonlinearity

Optimizing the molecular nonlinear response by molecular engineering presents only one part in designing a nonlinear optical or electro-optic material. To show a macroscopic second-order nonlinearity, the arrangement of molecules plays an important role. If the molecules are arranged arbitrarily, as in a liquid, the molecular second-order nonlinearities are averaged to a zero macroscopic effect. There exist basically four kinds of materials showing a macroscopic second-order nonlinear optical response as illustrated in Figure 15.26:

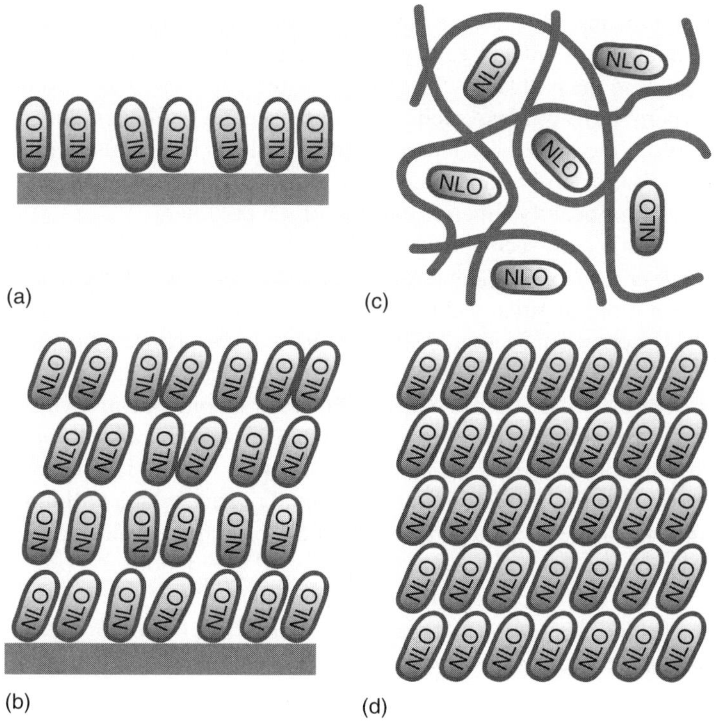

FIGURE 15.26 Schematic illustration of different types of macroscopic second-order nonlinear optical organic materials based on π-conjugated dipolar molecules: (a) Langmuir–Blodgett films, (b) molecular beam epitaxy (MBE), (c) poled polymers, and (d) single crystals.

(a) Thin films of highly ordered organic chromophores can be produced by the Langmuir–Blodgett technique or self-assembly of mono/multilayers (SAM). In the Langmuir–Blodgett technique a monolayer of molecules on a water surface is transferred layer-by-layer to a substrate. SAM technique is based on spontaneous assembly of the molecules by the immersion of an appropriate substrate into the solution.
(b) By molecular beam epitaxy (MBE) thin films of organic molecules can be produced. A beam of molecules is produced in ultrahigh vacuum using a Knudsen cell. This beam is directed onto a substrate where the molecules condensate and under certain circumstances a noncentrosymmetric ordering can be achieved.
(c) Nonlinear optical chromophores can be incorporated into a polymer matrix. By heating the polymer over the glass transition temperature and by applying a strong electric field at the same time, they can be partially oriented. By cooling down below the glass transition temperature, the polar arrangement of the chromophores gets frozen in, to obtain the so-called poled polymers. This is a very promising technique, since thin films can be easily produced and also large electro-optic effects can be reached (see Chapter 16).
(d) Growth of noncentrosymmetric organic crystals is the fourth possibility. With this technique the highest possible chromophore density and the best long-term orientational stability can be reached. A problem with this technique is the preferred antiparallel ordering of dipolar molecules, but there exist several techniques to overcome this preference and to enforce a polar crystallization, as discussed in the following section.

15.3 ORGANIC NONLINEAR OPTICAL SINGLE CRYSTALS

15.3.1 Noncentrosymmetric Crystalline Packing: Approaches

There have been significant advances in understanding and optimizing classical π-conjugated donor–acceptor chromophores with large first-order molecular hyperpolarizabilities in the area of organic nonlinear optics in the past [9,11–19]. However, there are only few chromophores with very large molecular hyperpolarizabilities, such as donor–acceptor stilbenes and tolanes, that form potentially useful crystalline materials. This interest in molecular crystals stems from the fact that the long-term orientational stability and photochemical stability, as well as the optical quality of molecular crystals are significantly superior to those of the polymers.

The basic requirement for any second-order nonlinear optical material is that it crystallizes in a noncentrosymmetric space group, so that the microscopic effects favorably add up to a macroscopic effect. The suitable packing of molecules that yields optimized macroscopic effects can be determined by the oriented-gas model, as discussed in Section 15.2.3 (see Figure 15.24). However, controlled crystallization of large organic molecules with desired optical properties is still a challenging topic. A major problem in achieving this is that most organic molecules will crystallize in a centrosymmetric space group, which is usually attributed to dipolar interaction forces that favor antiparallel chromophore alignment.

Crystal growth is the prototype of self-assembly in nature. The molecules pack into the crystalline lattice that corresponds to a minimum in the potential energy, which will in first approximation depend on the geometry of the molecules (close-packing principle). More precisely it will include electrostatic (Coulomb), van der Waals, and hydrogen bond interactions. A correct determination of the structure requires a more fundamental quantum mechanical approach that is computationally very difficult. The prediction of the crystal structure based on the π-conjugated donor–acceptor chromophores has not yet been made possible. However, there have been several approaches identified to obtain noncentrosymmetric nonlinear optical organic crystals that are listed below:

(a) Use of molecular asymmetry: Molecules tend to undergo shape simplification during crystal growth, which gives rise to dimers and then high-order aggregates in order to adapt a close packing in the solid state [20]. The high tendency of achiral molecules to crystallize centrosymmetrically could be due to such a close-packing driving force. Therefore, if the symmetry of the chromophores is reduced, dimerization and subsequent aggregation is no longer of advantage to the close packing and increases the probability of acentric crystallization. This symmetry reduction can be accomplished by either the introduction of molecular (structural) asymmetry or the incorporation of steric (bulky) substituents into the chromophore. These two approaches were widely and successfully applied to benzenoid chromophores. An introduction of a substituent at the 3-position of 4-nitrobenzylidene 4-donor-substituted aniline can induce a favorable noncentrosymmetric packing for large optical nonlinearities [21]. This led to the discovery of 4′-nitrobenzylidene-3-acetamino-4-methoxyaniline, MNBA (Figure 15.27a), which shows a large SHG efficiency that is 230

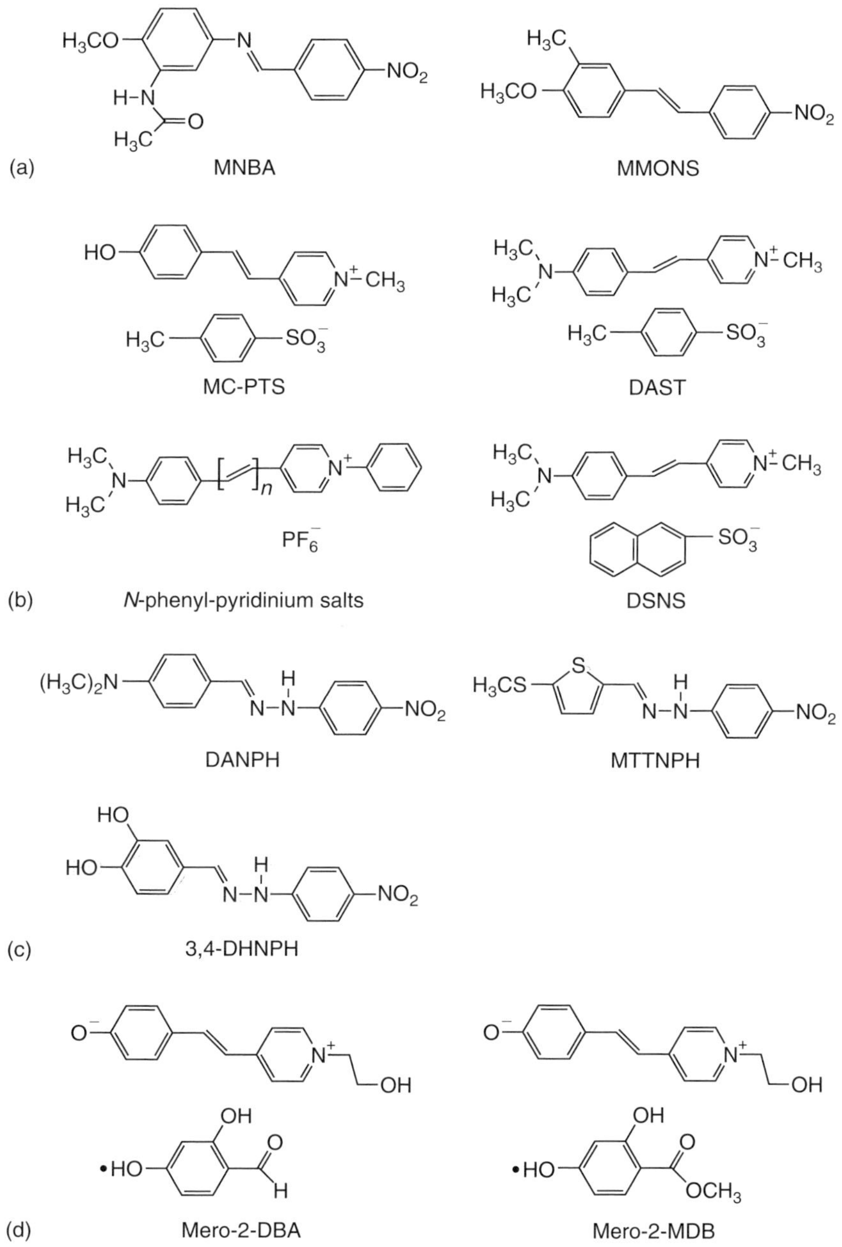

FIGURE 15.27 Engineering strategies for inducing a noncentrosymmetric packing of nonlinear chromophores and examples: (a) use of molecular asymmetry, (b) use of strong Coulomb interactions, (c) use of non-rod-shaped π-conjugated cores, and (d) supramolecular synthetic approach.

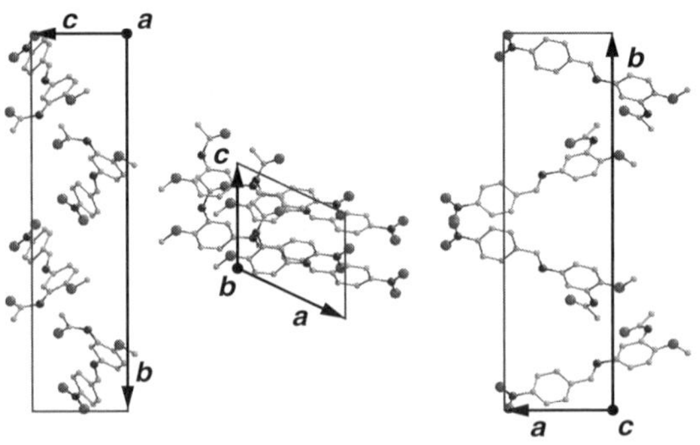

FIGURE 15.28 Crystal structure of MNBA with the point-group symmetry *m* projected along the crystallographic axes *a*, *b*, and *c*, respectively. Hydrogen atoms have been omitted for clarity. The polar axis points approximately along the $4\boldsymbol{a}+\boldsymbol{c}$ crystallographic vector in the mirror plane that is normal to the *b* axis (projection in the middle). The angle between the long axes of the molecules and the polar axis is about 18.7°.

times of that of urea standard. Another example using this approach is 3-methyl-4-methoxy-4′-nitrostilbene, MMONS, which shows an SHG powder efficiency of 1250 times that of urea [22].

The crystallographic structure of MNBA as determined by x-ray single crystal diffraction is shown in Figure 15.28. MNBA has a space-group symmetry *Cc* (point group *m*), which means that in addition to the translational symmetry that is common for all crystals, it also has a mirror symmetry plane (or, more precisely, a glide plane) that is in this case normal to the crystallographic *b* axis. This further means that all the physical properties should be symmetric under the mirror symmetry operation transforming *b* to –*b* [23]. Therefore, the angles between the *b* axis, and the *a* and the *c* axes should be exactly 90°. More importantly, we cannot have any asymmetry (for example a polar axis) that is needed for the second-order effects along the *b* axis. If we look at the projection along the *b* axis (Figure 15.28 in the middle), i.e., parallel to the mirror plane, we can see that the polar axis, pointing approximately from the acceptor to the donor group of the molecules in this projection, is approximately aligned along the [4,0,1] crystallographic vector. The projections along the *a* and the *c* axes in Figure 15.28 show that the long axes of the molecules make a certain angle to the polar crystal axis, which is about $\theta_{kz} \simeq 18.7°$, where θ_{kz} was defined in Figure 15.24. MNBA has a nonlinear optical coefficient d_{111} of about 130 pm/V at 1.06 μm [24].

(b) Use of strong Coulomb interactions can help to override weak dipole–dipole interactions to induce a noncentrosymmetric packing. The validity of this concept was proved for the case of 4-dimethylamino-*N*-methylstilbazolium salts. This led to the discovery of 4-dimethylamino-*N*-methylstilbazolium methylsulfate, DMSM, which shows an SHG efficiency of 220 times of that of urea [25].

The 4-toluenesulfonate anion was found to be an effective counterion to induce the noncentrosymmetric packing of stilbazolium chromophores that led to the development of 4-hydroxy-*N*-methylstilbazolium 4-toluene-sulfonate, MC-PTS (Figure 15.27b), which exhibits an SHG signal of 14 times of that of urea at 1.06 μm [26].

Marder and coworkers adopted the same strategy to perform an extensive investigation by means of varying the counterions of various stilbazolium chromophores including 2-*N*-methylstilbazolium and 4-*N*-methylstilbazolium cations [27]. They found that whereas rod-shaped 4-*N*-methylstilbazolium cations can often be forced to crystallize noncentrosymmetrically, this is not true for non-rod-shaped 2-*N*-methylstilbazolium cations. The 4-dimethylamino-*N*-methylstilbazolium 4-toluene-sulfonate, DAST, was shown to exhibit a very large powder SHG efficiency (1000 times urea) at 1.9 μm. Presently, DAST is still known as the best organic nonlinear optical crystal and its optical and nonlinear optical properties are presented in more detail later on. As in several other stilbazolium-based acentric crystals, a polar ionic sheet packing motif was evidenced in the crystal packing of DAST. However, by either incorporating a nonplanar or bulky donating group or replacing the phenyl ring with a heteroaromatic ring into the skeleton of 4-*N*-methylstilbazolium cations, the probability of getting acentric crystals goes down significantly. This suggests that the polar ionic sheet is very sensitive to the structural change of the stilbazolium cation. Changes in counter-anion have been identified as an effective molecular engineering strategy to develop new ionic organic crystals with noncentrosymmetric structure [28–30]. Using this approach, DSNS (4-*N*,*N*-dimethylamino-4-*N*-methyl-stilbazolium 2-naphthalenesulfonate), a promising DAST derivative with perfectly parallel chromophores, leading to a 50% higher nonlinearity in the crystalline powder than DAST at 1.9 μm, was developed [31].

Based on the favourable results with *N*-methyl stilbazolium chromophore of DAST, Coe et al. developed a new series of *N*-aryl pyridinium analogues with higher molecular polarizabilities [18,32–34]. Several new promising salts with comparable or even higher powder test efficiencies as DAST were thus obtained [33].

The crystallographic structure of a *N*-phenyl pyridinium salt with $n = 1$ (see Figure 15.27), trans-4-dimethylamino-*N*-phenyl-stilbazolium hexafluorophosphate (DAPSH), is shown in Figure 15.29. DAPSH also has a space-group symmetry *Cc* (point group *m*) with the mirror symmetry plane normal to the *b* crystallographic axis. We can see that the molecules align head-to-tail with an almost parallel stacking, making an angle of about $\theta_{kz} \simeq 15.5°$ with the polar crystal axis. DAPSH shows a powder test efficiency of about the same as DAST at 1.9 μm.

Figure 15.30 shows the crystallographic structure of DSNS with presently the highest powder SHG efficiency among the crystalline organic materials, 50% larger as for DAST. It has the space-group symmetry *P*1 (point group 1), which is a lowest possible symmetry, i.e., without any additional symmetry to the translation. This allows for the perfectly parallel packing of the stilbazolium chromophores with $\theta_{kz} = 0°$ that is ideal for maximizing the diagonal nonlinear optical and the electro-optic coefficient.

(c) Use of non-rod-shaped π-conjugated cores: In contrast to donor–acceptor disubstituted stilbene derivatives, hydrazone derivatives generally adopt a bent, non-rod-shaped conformation in the solid state because of the nonrigid nitrogen–nitrogen single bond

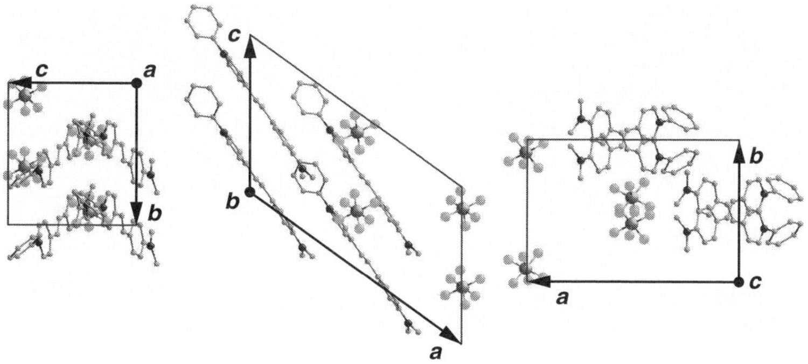

FIGURE 15.29 Crystal structure of DAPSH with the point-group symmetry *m* projected along the crystallographic axes *a*, *b*, and *c*, respectively. Hydrogen atoms have been omitted for clarity. The polar axis points approximately along the $\boldsymbol{a} - \boldsymbol{c}$ crystallographic vector in the mirror plane that is normal to the *b* axis (projection in the middle). The angle between the long axes of the molecules and the polar axis is about 15.5°.

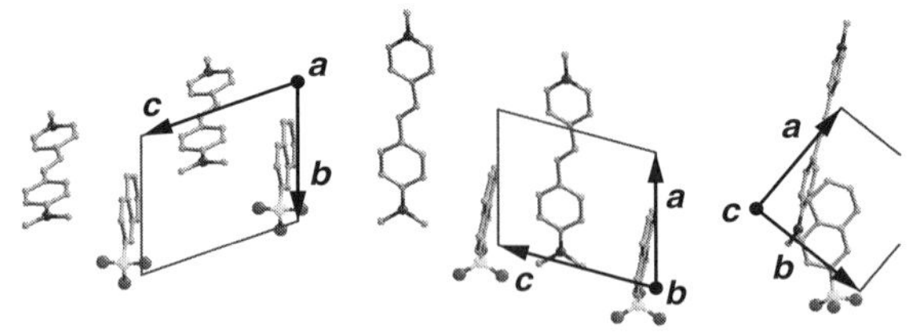

FIGURE 15.30 Crystal structure of DSNS with the point-group symmetry 1 showing molecules from two unit cells projected along the crystallographic axes *a*, *b*, and *c*, respectively. Hydrogen atoms have been omitted for clarity. The polar axis points approximately along the $2\boldsymbol{a}-\boldsymbol{b}$ crystallographic vector (see the projection on the right). There is only one chromophore in the unit cell, meaning that all chromophores are perfectly parallel, leading to the optimal packing for the electro-optic applications with $\theta_{kz}=0°$.

(–CH=N–NH–). Donor-substituted (hetero)-aromatic aldehyde-4-nitrophenylhydrazones show a relatively very high tendency to form a noncentrosymmetric packing [35,36]. Of particular importance is that the majority of these acentric crystals exhibit very strong SHG signals that are at least two orders of magnitudes greater than that of urea. Furthermore, most of the hydrazone crystals developed show very good crystallinity and high thermal stability. The flexibility of the hydrazone backbone poses a problem of polymorphism; however, with a proper control of the growth conditions such as careful choices of solvent and method of crystal growth, the desirable acentric bulk crystal phase can be selectively grown [37].

The best example in this class is 4-dimethylaminobenzaldehyde-4-nitrophenyl-hydrazone, DANPH (Figure 15.27c), which exhibits a very strong SHG signal that is comparable to that of DAST. Another potential candidate is 5-(methylthio)thiophenecarboxaldehyde-4-nitrophenylhydrazone, MTTNPH, which also shows the same order of powder efficiency as DANPH. A third example is 3,4-DHNPH with an excellent alignment of the chromophores in the crystal lattice and a molecular hyperpolarizabilitycomparable to DANPH [38].

Figure 15.31 shows a crystal packing diagram of MTTNPH with a space group $Pca2_1$ and point group *mm*2. This symmetry includes a 180° rotational symmetry axis along the *c* crystallographic axis and two mirror planes, one normal to the *a* and the other normal to the

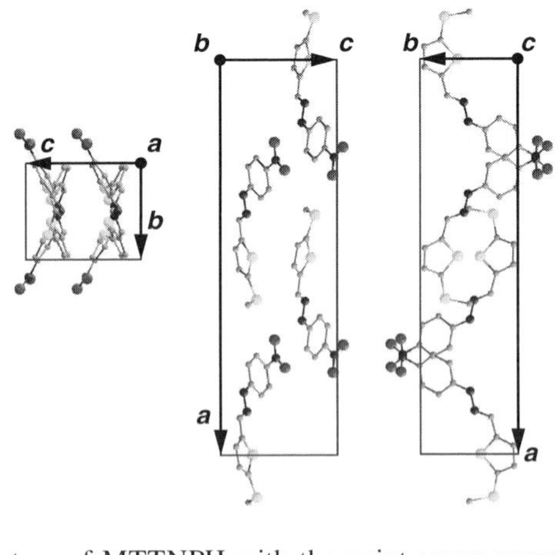

FIGURE 15.31 Crystal structure of MTTNPH with the point-group symmetry *mm*2 projected along the crystallographic axes *a*, *b*, and *c*, respectively. Hydrogen atoms have been omitted for clarity. The polar axis points along the ***c*** crystallographic vector.

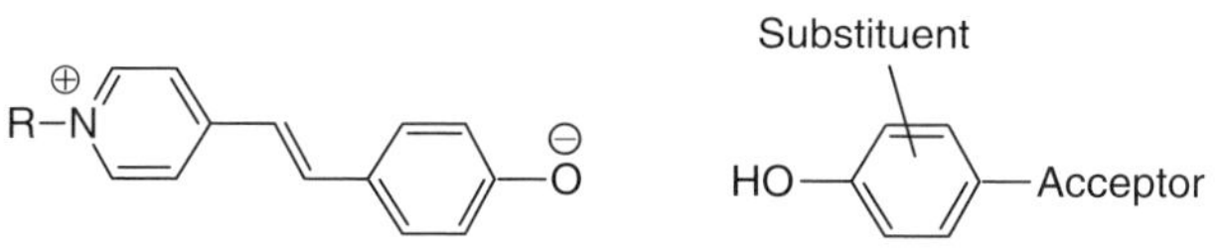

Mero-1: R = CH_3
Mero-2: R = CH_2CH_2OH

Class I phenolic derivatives

FIGURE 15.32 Chemical structures of Mero-1, Mero-2, and Class I phenolic derivatives.

b crystallographic axis. This implies that the crystallographic vectors should be perpendicular to each other and the polar axis should be parallel to the *c* axis following the symmetry considerations. The angle θ_{kz} between the polar crystal axis and molecular charge-transfer axes is about 55°, which is optimal for the phase-matched frequency conversion applications. For these applications the off-diagonal nonlinear optical coefficient d_{322}, which was measured to be around 32 pm/V at 1.3 μm [39], can be used.

(d) Supramolecular synthetic approach involves molecular or ionic aggregates or assemblies to favour the desirable crystallographic packing. This approach offers more design feasibility as one or both molecules can be tailor-made or can be modified to fit one another to acquire the desirable molecular properties in the solid state. Furthermore, the physical properties such as melting point and solubility as well as the crystal properties such as crystallinity and ease of crystal growth of the co-crystals can usually be improved compared to those of their starting components. Etter and coworkers first demonstrated the induction of a net dipole moment with a complimentary host–guest pair of 4-aminobenzoic acid and 3,5-dinitrobenzoic acid; however, the SHG signal generated by this co-crystal is in the order of the urea standard [40].

Co-crystals formed from the merocyanine dyes (Mero-1 and Mero-2) and the class I phenolic derivatives, in which the electron acceptor is para-related to the phenolic functionality together with a substituent either in the ortho- or meta-position (Figure 15.32), have shown the highest tendency of forming acentric to co-crystals. In addition, a large fraction of acentric co-crystals (25%) based on Mero-2 and the class I phenolic derivatives exhibit strong second-harmonic signals that are at least two orders of magnitudes larger than that of urea. Their packing motifs can be distinctively divided into two groups.

The type I co-crystal is generally characterized by anionic and cationic assemblies or arrays. An interesting example in this class is the co-crystal Mero-2-DBA (DBA = 2,4-dihydroxybenzaldehyde, see Figure 15.27d) [41]. Mero-2-DBA contains a water molecule and packs noncentrosymmetrically with space group *P*1 and point group 1. The anionic assembly is constructed by the co-aggregation of two DBA molecules in which one of the molecules gives up a proton and bonds to another by a hydrogen bond. Additionally, Mero-2 acquires a proton and co-aggregates in antiparallel fashion with another Mero-2 by a short hydrogen bond constituting a cationic assembly. An interesting fact is that, although the net dipole moment almost vanishes in this arrangement, the Mero-2-DBA co-crystal exhibits a large second-harmonic signal in the powder test. This can be explained by the asymmetric position of the hydrogen bonded proton between the two Mero-2 dyes, which results in a positive reinforcement of molecular hyperpolarizabilities within the cationic assembly, since Mero-2 has a negative sign of the hyperpolarizability and the protonated form of Mero-2 ([Mero-2-H]$^+$) has a positive sign of the hyperpolarizability (see Figure 15.33). As a consequence, the co-crystal Mero-2-DBA is a potential candidate for linear electro-optic effects because of its perfectly parallel alignment of molecular hyperpolarizabilities in the solid state.

Type II co-crystals are formed by linear molecular aggregates. One of the representative examples in this class is the co-crystal Mero-2-DAP that exhibits a very strong SHG signal

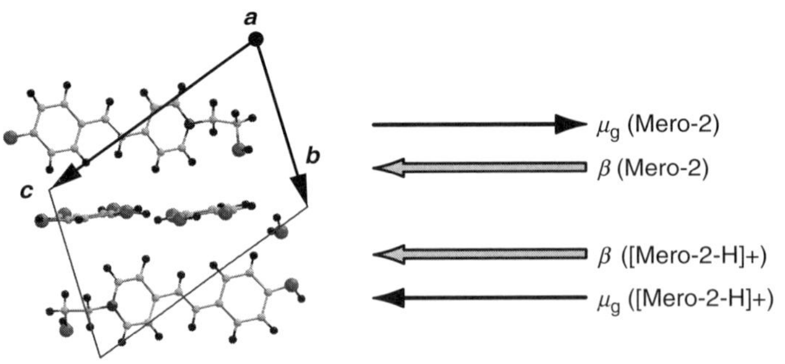

FIGURE 15.33 Crystal packing of Mero-2-DBA crystal projected along the *a* axis with the directions of the ground state dipole moments and the main directions of the first-order hyperpolarizabilities.

that is three orders of magnitudes larger than that of urea [42]. The molecular aggregate is assembled by the highly electronegative oxygen of Mero-2 and the acidic proton of the phenolic derivative through a short hydrogen bond. These rod-like aggregates connect laterally by hydrogen bonds resulting in a staircase-like polar chain. These polar chains align in a parallel fashion constituting a two-dimensional acentric layer, which is found to be the common and key feature of all the highly noncentrosymmetric co-crystals in this class. Since the charge-transfer axis of Mero-2 is inclined by an angle of about 70° to the polar direction of the crystal, this co-crystal is a candidate for phase-matched nonlinear optical effects. In addition, in this system the orientation of the merocyanine dye can be changed and tuned within the crystal lattice by careful selection of a guest molecule–phenolic derivative, provided that the linear molecular aggregate and the acentric layer packing motifs are maintained. Although Mero-2 exists only in an amorphous state by itself, both types of co-crystals formed show greatly improved crystalline and physical properties compared to its constituted components. Another interesting type II crystal shown in Figure 15.34 is Mero-2-MDB that is optimized for electro-optic applications due to the almost parallel alignment of the nonlinear optical chromophores. Mero-2-MDB has a space-group symmetry *Cc* (point group *m*) and a high nonlinear optical coefficient d_{111} of about 270 pm/V at 1.3 μm [43].

15.3.2 Examples of Molecular Crystals

Table 15.3 lists some examples of second-order nonlinear optical organic crystals. The crystals are organized with respect to the cutoff wavelength λ_c, where λ_c was taken from literature and not adjusted for a consistent definition. The values of the nonlinear optical coefficients *d* were adjusted to the same reference value, e.g., d_{11} (α-quartz) = 0.3 pm/V (λ =1064 nm) and = 0.28 pm/V (λ = 1907 nm).

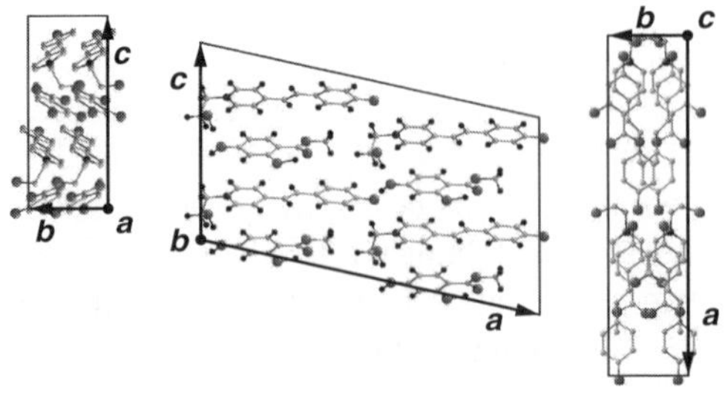

FIGURE 15.34 X-ray structure of a hydrogen-bond directed, acentric layer structure of the co-crystal Mero-2-MDB projected along the crystallographic axes *a*, *b* and *c*, respectively. Hydrogen atoms are only shown in the projection along the *b* axis.

TABLE 15.3
Examples of Molecular Crystals That Have Been Investigated for Their Nonlinear Optical or Electro-optic Response

Material	Point Group	λ_c(nm)	*d*, *r* (pm/V)	T_m(°C)
	Cutoff Wavelength $\lambda_c \leq 450$ nm			
Urea 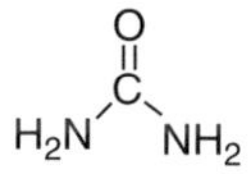	$\bar{4}2m$	200	d_{14}(480–640 nm) = 1.0 d_{14}(1064 nm) = 1.1 r_{41}(633 nm) = 1.9 r_{63}(633 nm) = 0.8	133–135
IAPU(isopropyl-4-acetylphenylurea) 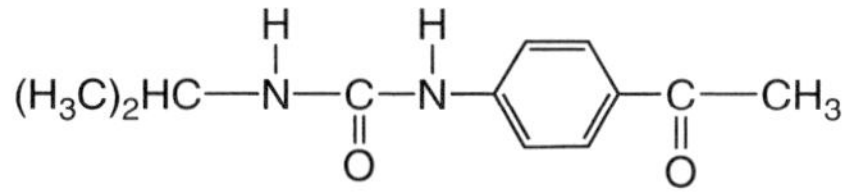	2	380	d_{22}(1064 nm) = 30.5	150
DMACB(dimethylaminocyanobiphenyl) 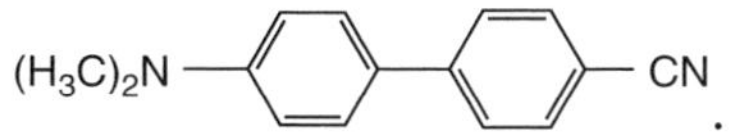	*m*	420	d_{11}(1064 nm) = 276 r_{11}(1064 nm) = 55	—
DMNP(3,5-dimethyl-1-(4-nitrophenyl)-pyrazole) 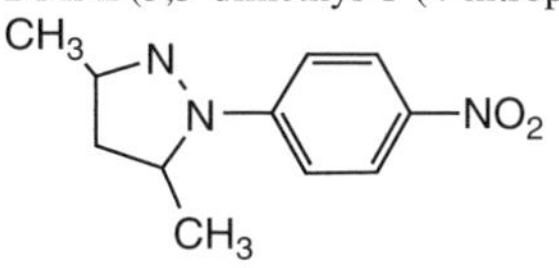	*mm*2	450	d_{21}(950 nm) = 90 d_{22}(950 nm) = 29	102.5
	Cutoff Wavelength 450 nm $\leq \lambda_c \leq$ 550 nm			
MNA(2-methyl-4-nitroaniline) 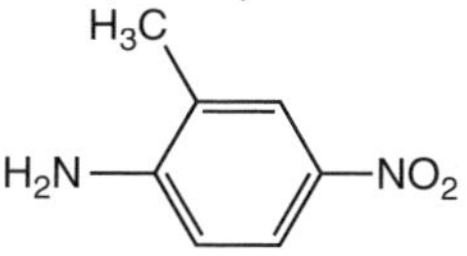	*m*	480	d_{11}(1064 nm) = 150 d_{12}(1064 nm) = 23 r_{11}(633 nm) = 67	133
DAN(4-(*N*,*N*-dimethylamino)-3-acetamidonitrobenzene) 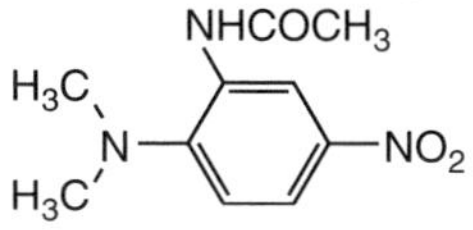	2	485	d_{23}(1064 nm) = 38 r_{32}(633 nm) = 13	165.7
mNA(meta-nitroaniline) 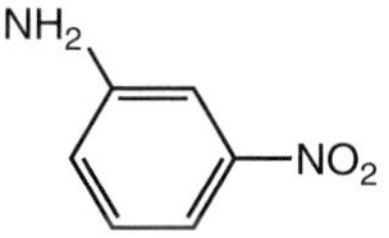	*mm*2	500	d_{31}(1064 nm) = 11.7 d_{33}(1064 nm) = 12.3 r_{33}(633 nm) = 16.7	112
POM(3-methyl-4-nitropyridine-1-oxide) 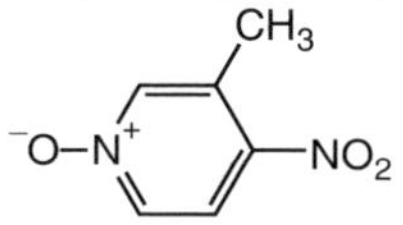	222	460	d_{14}(1064 nm) = 6.0 r_{41}(633 nm) = 3.6 r_{52}(633 nm) = 5.2 r_{63}(633 nm) = 2.6	136
NPP(*N*-(4-nitrophenyl)-(L)-prolinol) 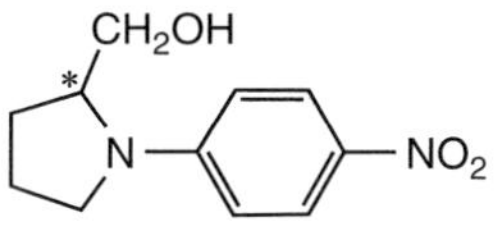	2	500	d_{21}(1064 nm) = 51	116

(continued)

TABLE 15.3 (continued)
Examples of Molecular Crystals That Have Been Investigated for Their Nonlinear Optical or Electro-optic Response

Material	Point Group	λ_c(nm)	*d*, *r* (pm/V)	T_m(°C)
PNP(2-(*N*-prolinol)-5-nitropyridine) 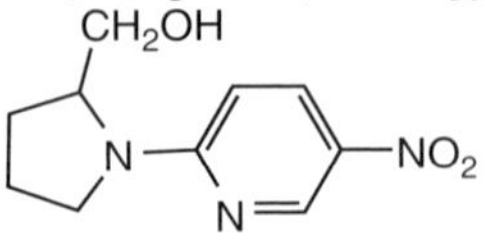	2	490	d_{22}(1064 nm) = 16.8 d_{21}(1064 nm) = 51 d_{22}(1064 nm) = 16.2 r_{22}(514 nm) = 28.3 r_{22}(633 nm) = 12.8	83
COANP(2-cyclooctylamino-5-nitropyridine) 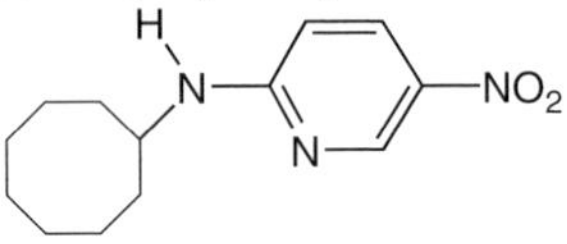	*mm*2	490	d_{32}(1064 nm) = 32 r_{33}(514 nm) = 28 r_{33}(633 nm) = 15	72.8
(-)MBANP((-)-2-(α-methylbenzylamino)-5-nitropyridine) 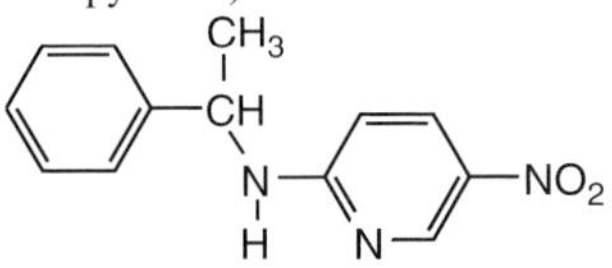	2	450	d_{22}(1064 nm) = 36 r_{eff}(488 nm) = 31.4 r_{eff}(514 nm) = 26.6 r_{eff}(633 nm) = 18.2	83
AANP(2-adamantylamino-5-nitropyridine) 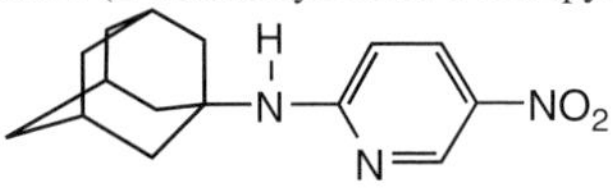	*mm*2	460	d_{31}(1064 nm) = 48 d_{33}(1064 nm) = 36	167
MMONS(3-methyl-4-methoxy-4′-nitrostilbene) 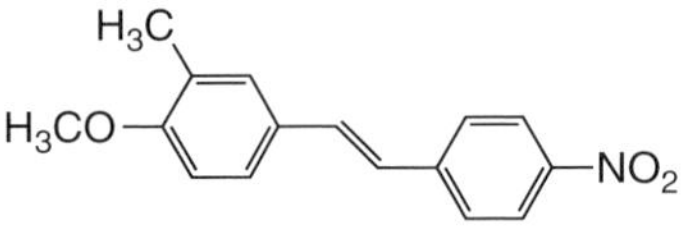	*mm*2	515	d_{33}(1064 nm) = 112 d_{24}(1064 nm) = 43 r_{33}(633 nm) = 40	—
MNBA(4′-nitrobenzylidene-3-acetamino-4-methoxyaniline) 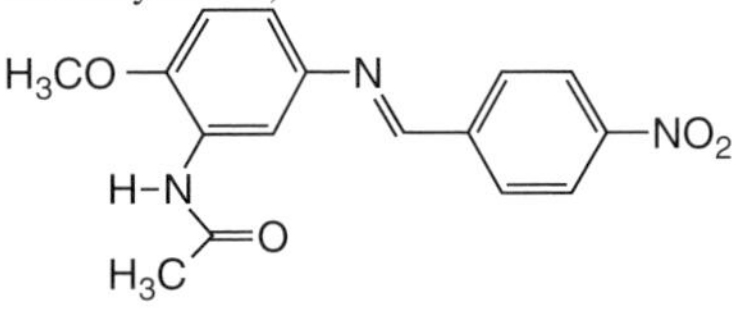	*m*	520	d_{11}(1064 nm) = 131 r_{11}(532 nm) = 50 r_{11}(633 nm) = 29	—
MC-PTS(4-hydroxy-*N*-methylstilbazolium-4-toluene sulfonate) 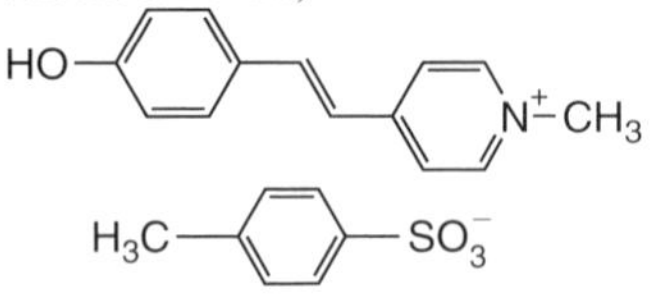	1	510	d_{11}(1064 nm) = 314	—
Cutoff Wavelength $\lambda_c \geq 550$ nm				
DR1(disperse red 1)DR1 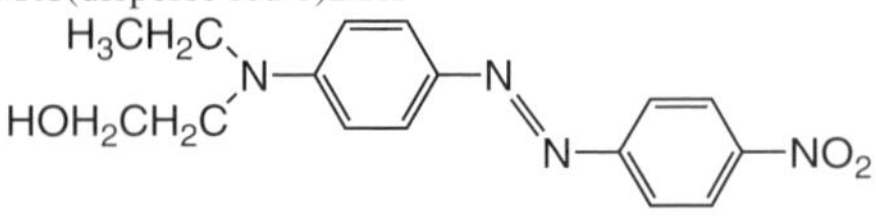	—	—	—	160–162

TABLE 15.3 (continued)
Examples of Molecular Crystals That Have Been Investigated for Their Nonlinear Optical or Electro-optic Response

Material	Point Group	λ_c(nm)	*d*, *r* (pm/V)	T_m(°C)
DAST(4′-dimethylamino-*N*-methyl-4-stilbazolium tosylate) 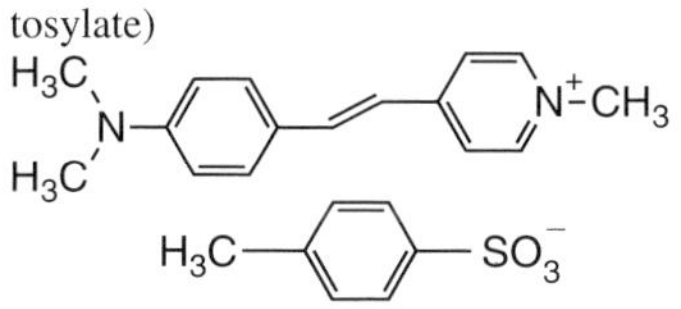	*m*	700	d_{11}(1318 nm) = 1010 d_{11}(1542 nm) = 290 d_{26}(1542 nm) = 39 r_{11}(720 nm) = 92 r_{11}(1313 nm) = 53 r_{11}(1535 nm) = 47	256
N-phenyl-pyridinium salts 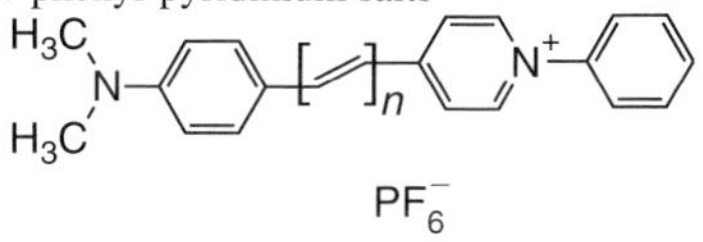	*m* (*n* = 1) *m* (*n* = 2)	~700 (*n* = 1) ~800 (*n* = 2)	Powder test: 470 times urea (*n* = 1, 1.9 μm) Powder test: 550 times urea (*n* = 2, 1.9 μm)	— —
Mero-2-MDB(4-2-[1[(2-hydroxyethyl)-4-pyridylidene]-ethylidene-cyclo-hexa-2,5-dien-1-one-methyl-2,4-dihydroxybenzoate) 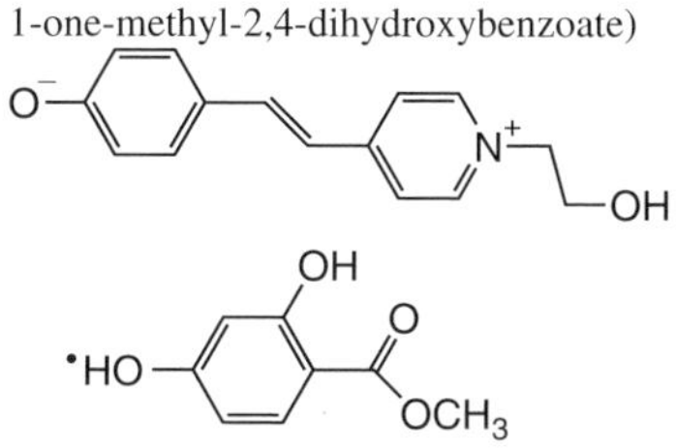	*m*	Phase II: 680 Phase I: 615	Phase II: d_{11}(1318 nm) = 267 r_{11}(1313 nm) = 34 Phase I: d_{11}(1318 nm) = 108 r_{11}(1313 nm) = 24	185
DANPH(4-dimethylaminobenzaldehyde-4-nitrophenylhydrazone) 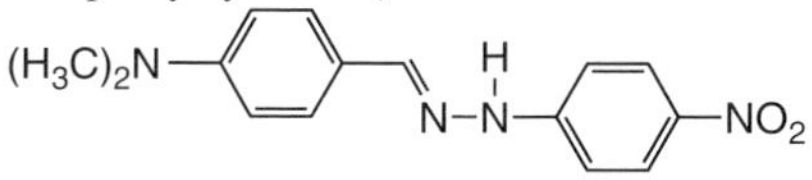	*m*	670	d_{12}(1542 nm) = 200 d_{11}(1542 nm) = 150	186
MTTNPH(5-(methylthio)-thiophenecarboxaldehyde-4-nitrophenylhydrazone) 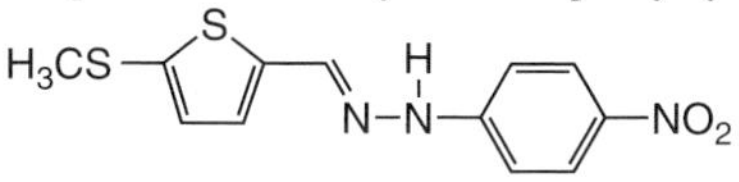	*mm*2	620	d_{32}(1313 nm) = 32	172
3,4-DHNPH(3,4-dihydroxybenzaldehyde-4-nitrophenyl-hydrazone) 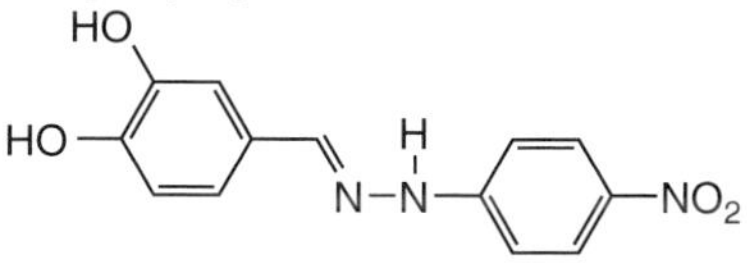	*m*	730	—	230

Sources: Bosshard, C., Bösch, M., Liakatas, I., Jäger, M. and Günter, P., in *Nonlinear Optical Effects and Materials*, P. Günter, ed., Springer Series in Optical Science, New York, 2000, p. 163; Nalwa, H.S. and Miyata, S., eds., *Nonlinear Optics of Organic Molecules and Polymers*, CRC Press, Boca Raton, Florida, 1997.

Note: λ_c is the cutoff wavelength in the bulk, *d* is the nonlinear optical coefficient, *r* is the electro-optic coefficient, and T_m is the melting point. The values of the nonlinear optical coefficients *d* were adjusted to the same reference value (d_{11} (α-quartz) = 0.3 pm/V (1064 nm) and 0.28 pm/V (1907 nm)). For compounds that the nonlinear optical coefficients are not yet known, results of the Kurtz and Perry powder test [45] are given. For more examples including the molecular hyperpolarizabilities see Refs. [17] and [44].

- Cutoff wavelength $\lambda_c \leq 450$ nm, "white" materials: urea, IAPU, DMACB, DMNP, …
- Cutoff wavelength 450 nm $\leq \lambda_c \leq$ 550 nm, "yellow" to "orange" materials: MNA, DAN, mNA, POM, NPP, PNP, COANP, (-)MBANP, AANP, MMONS, MNBA, MC-PTS, …
- Cutoff wavelength $\lambda_c \geq 550$ nm, "orange" to "red" materials: DR1, DAST, *N*-phenyl-pyridinium salts, Mero-2-MDB, DANPH, MTTNPH, 3,4-DHNPH, …

For more examples including the molecular hyperpolarizabilities see e.g., Refs. [17,44].

15.3.3 Growth of Bulk and Thin-Film Organic Crystals

Crystallization of organic nonlinear optical molecules is based on solution growth, melt growth, and vapor growth, depending on:

(i) production of either three-dimensional bulk, two-dimensional thin platelet, and one-dimensional fiber-like crystals;
(ii) melting temperature, the solubility, and the vapor pressure;
(iii) thermal stability and general chemical stability of melts, solutions, and solids;
(iv) purity of starting materials;
(v) formation of solution or solvent inclusions and other defects.

Since most of highly nonlinear organic molecules show melt degradation effects, solution growth techniques are most common and represent in many cases the only possibility of obtaining large bulk crystals. They are however limited to rather low growth rates (in many cases ~0.1 to 1 mm/day) and often suffer from solvent or solution inclusion problems. Melt growth techniques on the other hand do not possess these restrictions and exhibit relatively fast growth rates compared to solution growth techniques, they are therefore preferred for the molecules that exhibit high long-term thermal stability. The vapor growth method is rarely used for organic nonlinear optical crystals. The crystals grown from the vapor phase are of high purity and lattice perfection, however often with low growth rates in the range 0.1–0.01 mm/h.

15.3.4 4-*N*,*N*-Dimethylamino-4′-*N*′-Methyl Stilbazolium Tosylate (DAST)

4-*N*,*N*-dimethylamino-4′-*N*′-methyl stilbazolium tosylate (DAST) is the most well investigated and the only commercially available highly nonlinear optical organic crystal to date [46]. It is an organic salt based on strong Coulomb interactions between the positively charged nonlinear optical chromophore and the negatively charged counter-anion tosylate (see Figure 15.27b). The crystallographic structure of DAST is shown in Figure 15.35. DAST has a space-group symmetry *Cc* (point group *m*) with the mirror symmetry plane normal to the *b* crystallographic axis.

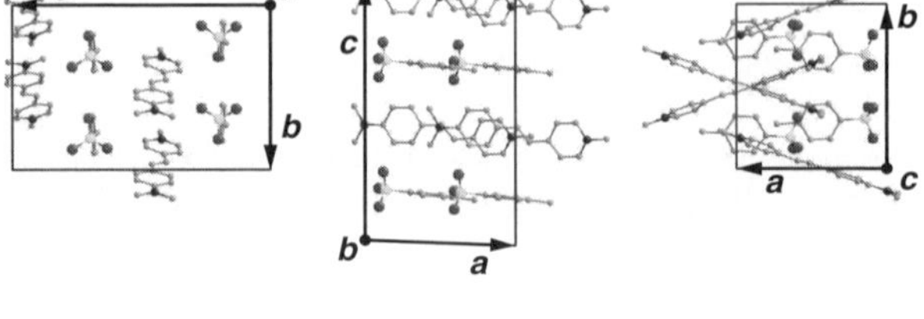

FIGURE 15.35 X-ray structure of the ionic DAST crystal with the point-group symmetry *m* showing molecules from one unit cell, projected along the crystallographic axes *a*, *b*, and *c*, respectively. Hydrogen atoms have been omitted for clarity. The mirror plane is the *ac* crystallographic plane (projection in the middle). The chromophores make an angle of 20° with respect to the polar *a* axis (right).

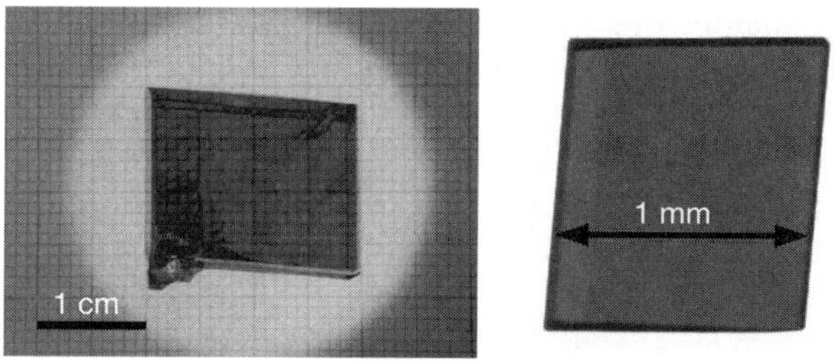

FIGURE 15.36 Bulk single crystal of DAST (left), grown from the solution by controlled temperature lowering technique (From Laveant, P., Medrano, C., Ruiz, B., and Gunter, P., *Chimia*, 57, 349, 2003; Pan, F., Wong, M.S., Bosshard, C., and Gunter, P., *Adv. Mater.*, 8, 592, 1996), and a single-crystalline thin film (right), grown by planar solution growth. (From Manetta, S., Ehrensperger, M., Bosshard, C., and Gunter, P., *C.R. Physique*, 3, 449, 2002.)

The reasons for the growing interest in obtaining high-quality DAST crystals are its extraordinarily high nonlinearities, the high second-order nonlinear optical and the electro-optical coefficients, being respectively 10 times and twice as large as those of the inorganic standard $LiNbO_3$. Optical and nonlinear optical properties of DAST crystals are reported in Refs. [47–50].

Bulk DAST crystal growth has been investigated by different groups [51–53]. Supersaturated methanol solutions are used and the temperature of the solution is carefully reduced. The growth starts from either spontaneous nucleation or seed introduction. A very high thermal stability of the solution (under 0.01°C over several weeks) is required for high-quality bulk samples. For integrated-optics applications, single crystalline thin films with a thickness in the range of 0.2–10 μm are needed. An overview of the presently available techniques for obtaining thin-film organic crystals and for the fabrication of organic crystalline waveguides is given in Ref. [54]. Thin films of DAST have been produced by mechanical methods, i.e., polishing down bulk crystals [55] and etching [56]; epitaxial growth methods [57,58]; solution capillary method [57,59]; planar solution growth methods [57]; and vapor growth method [60]. Figure 15.36 shows photos of bulk and thin-film single crystals of DAST grown from solution.

DAST crystals have been used to demonstrate prototype photonic devices for OPG [50], electro-optic modulation and electro-optic sampling [48,61,62], optical waveguides [58,63–65], and terahertz generation and detection [3,52,66–68].

15.4 SUMMARY

Nonlinear optical materials enable light to interact with light and the external electric fields. This allows for a variety of effects and applications in photonics, including the generation of new optical wavelengths and terahertz waves, light modulation and optical data processing, switching and steering of light beams among others. Efficient interactions between light waves and electric fields require specially designed materials. Organic nonlinear optical materials offer a large number of design possibilities with large and extremely fast nonlinearities compared to their inorganic alternatives. The research on the fundamental aspects of light-matter interactions in organic media, and new concepts and functionalities for photonic elements is strongly motivated by the need for the development of high transmission bandwidths and wavelength division multiplexing systems in telecommunication technologies.

The basic requirement for second-order nonlinear optical materials that allow for, e.g., SHG, terahertz wave generation, and electro-optic modulation and sampling, is the noncentrosymmetry of their structure. Therefore, even molecules with largest microscopic nonlinearities cannot operate efficiently on a macroscopic level, unless they are ordered into a noncentrosymmetric structure. The perfectly parallel arrangement of highly polar π-conjugated molecules offers the optimal configuration for maximizing the macroscopic electro-optic effect. Such a configuration is for most of the molecules, mainly due to their high polarity, nonoptimal and therefore special engineering techniques are required for achieving a macroscopic noncentrosymmetric order. In this chapter, an approach based on the self-assembly of the molecules to form polar single crystals is introduced.

The packing of the molecules into a crystal lattice is governed by very complex interactions between the molecules, and presently it is not possible to predict the crystal structure of the highly nonlinear chromophores based on the molecular composition. However, there have been several approaches identified that can lead to a higher probability of obtaining a noncentrosymmetric polar arrangement of π-conjugated molecules. These include the design of the molecules with high molecular asymmetry and the capability of building strong Coulomb or hydrogen intermolecular bonds. An optimized packing of the organic molecules into a highly ordered crystalline lattice can result in the highest possible chromophore density and the best long-term device stability among organic nonlinear optical materials.

Organic nonlinear optical DAST crystals are presently commercially available and are being used for terahertz generation and detection in several research laboratories. The efforts are going on for the production of integrated electro-optics devices made of organic crystals, due to their favorable electro-optic and stability properties that promise to meet the future requirements for ultrahigh transmission bandwidth photonic devices.

EXERCISE QUESTIONS

1. Optical Kerr effect is defined by Equation 15.52 as a light intensity I dependent refractive index change $\Delta n = n_2 I$ in a third-order nonlinear optical medium. Calculate n_2 using the wave Equation 15.6 and the expression (Equation 15.50) for the nonlinear polarization generated in a $\chi^{(3)}$ medium at frequency ω. For simplicity you can omit the tensor notation, i.e., consider a linearly polarized monochromatic optical plane wave propagating in the z direction $E(z,t) = \frac{1}{2}E^{\omega}e^{i(kz-\omega t)} + \text{cc}$. Light intensity is related to the electric field amplitude E^{ω} by $I^{\omega} = \frac{1}{2}\varepsilon_0 cn|E^{\omega}|^2$.

   Calculate the nonlinear refractive index n_2 for a material with $\chi^{(3)} = 10^{-18}$ m^2/V^2 and the linear refractive index $n_0 = 2.0$.
2. DAST crystal is used as a phase electro-optic modulator as shown in Figure 15.10 at $\lambda = 720$ nm employing its largest coefficient r_{111} (for the parameters see Table 15.3) with the corresponding dielectric constant of $\varepsilon_{11} = 5.2$. The crystal is $l = 1$ cm long and the lateral dimensions are $d = 3$ mm $\times$ $d' = 3$ mm (d is the distance between the electrodes and d' the length of the crystal perpendicular to l and d). Determine the half-wave voltage V_{π}, the transit time through the crystal and the electric capacitance C of the modulator. The voltage is applied using a source with 50 Ω resistance. Which factor limits the speed of the device? Compare the results with the same-size $LiNbO_3$ crystal with $r = 31$ pm/V, $n = 2.2$, and $\varepsilon = 28$.
3. Consider a single-mode microring resonator as shown in Figure 15.13 made of a DAST electro-optic crystal. The effective refractive index for the optical mode inside the ring is $n = 2.1$ with negligible dispersion in the vicinity of the telecommunication wavelength 1.55 μm, where the electro-optic coefficient is $r = 47$ pm/V. Calculate the free-spectral range (FSR) of a ring with the diameter of $d = 100$ μm. How much will the resonance shift, if we apply an electric field of 10 V/μm?
4. MTTNPH organic crystal (Figure 15.31) has the point-group symmetry $mm2$ that contains a 180° rotational symmetry axis along the c crystallographic axis and two mirror planes, one normal to the a and the other normal to the b crystallographic axis. We choose the coordinate system so that x_1 is parallel to the a crystallographic axis a, x_2 to the b axis and x_3 to the c axis. The $mm2$ symmetry means [23] that the crystal is optically biaxial, i.e., with three different principal refractive indices $n_1 \neq n_2 \neq n_3$, and that the following electro-optic coefficients are not zero: r_{113}, r_{223}, r_{333}, $r_{232} = r_{322}$, and $r_{131} = r_{311}$. All the other electro-optic coefficients should be zero due to symmetry reasons, i.e., invariance of the physical parameters under point-group symmetry transformations [23].

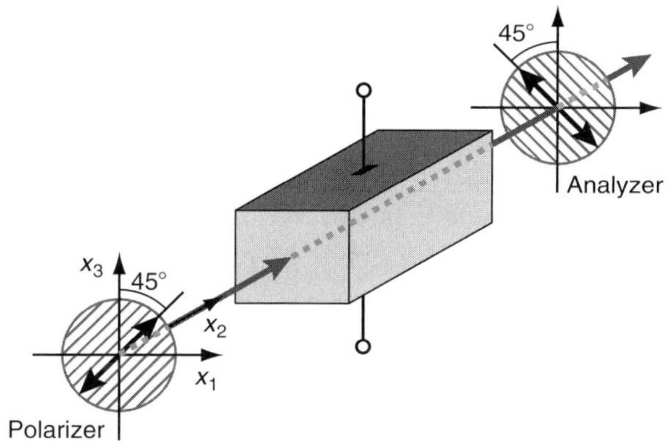

FIGURE 15.37 An amplitude electro-optic modulator using an electro-optic crystal between two crossed polarizers.

(a) We apply an electric field E_3 along the x_3 axis of a MTTNPH crystal, how much will the optical properties change due to the electro-optic effect?

(b) We have described an amplitude electro-optic modulator using an interferometer geometry (Figure 15.11). We can also build an amplitude modulator by placing an electro-optic crystal between two crossed polarizers (polarizer and analyzer) that are oriented at 45° with respect to the crystal axes as illustrated in Figure 15.37.
As an electro-optic crystal, we use MTTNPH and apply the electric field: along the $x_3 = c$ axis. Find an expression for the transmission through such a modulator as a function of the applied electric field.

5. (a) Find the relation between the nonlinear optical coefficient $d_{ijk}(-2\omega',\omega',\omega')$ for SHG at fundamental frequency ω' and the electronic part of the electro-optic coefficient $r^e_{ijk}(\omega)$ at frequency ω for an organic material with the dispersion relation for the diagonal molecular hyperpolarizability β_{zzz} following the two-level model (Equation 15.61) and the macroscopic nonlinearities following the oriented-gas model as given by Equations 15.65 and 15.66.

(b) Calculate the electro-optic coefficient r^e_{111} for DAST crystal at $\lambda = 1.55$ μm if you know the nonlinear coefficient for SHG $d_{111} = 210$ pm/V at $\lambda' = 1.9$ μm fundamental wavelength. The refractive index dispersion of DAST is given by the Sellmeier equation

$$n_1^2(\omega) = A + \frac{B}{\omega_{eg}^2 - \omega^2}, \tag{15.68}$$

where
$A = 2.675$
$B = 20.3706 \cdot 10^{30}\ \mathrm{s}^{-1}$

$\omega_{eg} = 3.521 \cdot 10^{15}\ \mathrm{s}^{-1}$ is the resonant frequency of the transition between the ground state, g, and the excited state, e ($\lambda_{eg} = 535.3$ nm). The corresponding dielectric constant of DAST is $\varepsilon_1 = 5.2$.

ACKNOWLEDGMENT

This work was supported by the Swiss National Science Foundation.

LIST OF ABBREVIATIONS

cc	complex conjugate
CMOS	complementary metal–oxide–semiconductor
DFG	difference-frequency generation
DFWM	degenerate four-wave mixing
EO	electro-optic, or electro-optics
FSR	free spectral range
FWM	four-wave mixing
NLO	nonlinear optical, or nonlinear optics
OPG	optical parametric generation
OPO	optical parametric oscillation
OR	optical rectification
QPM	quasi-phase matching
SFG	sum-frequency generation
SHG	second-harmonic generation
THG	third-harmonic generation
VLSI	very large scale integrated, or very large scale integration

REFERENCES

1. P.N. Butcher and D. Cotter, *The Elements of Nonlinear Optics* (Cambridge University Press, Cambridge, 1990).
2. B.E.A. Saleh and M.C. Teich, *Fundamentals of Photonics* (Wiley, New York, 1991).
3. A. Schneider, M. Neis, M. Stillhart, B. Ruiz, R. Khan, and P. Gunter, Generation of terahertz pulses through optical rectification in organic DAST crystals: Theory and experiment, *J. Opt. Soc. Am. B* 23, 1822, 2006.
4. J.L. Oudar and D.S. Chemla, Hyperpolarizabilities of nitroanilines and their relations to excited-state dipole-moment, *J. Chem. Phys.* 66, 2664, 1977.
5. J. Zyss and J.L. Oudar, Relations between microscopic and macroscopic lowest-order optical nonlinearities of molecular-crystals with one-dimensional or two-dimensional units, *Phys. Rev. A* 26, 2028, 1982.
6. L.R. Dalton, W.H. Steier, B.H. Robinson, C. Zhang, A. Ren, S. Garner, A.T. Chen, T. Londergan, L. Irwin, B. Carlson, L. Fifield, G. Phelan, C. Kincaid, J. Amend, and A. Jen, From molecules to opto-chips: Organic electro-optic materials, *J. Mater. Chem.* 9, 1905, 1999.
7. C. Bosshard, R. Spreiter, and P. Gunter, Microscopic nonlinearities of two-component organic crystals, *J. Opt. Soc. Am. B* 18, 1620, 2001.
8. I. Liakatas, C. Cai, M. Bosch, M. Jager, C. Bosshard, P. Gunter, C. Zhang, and L.R. Dalton, Importance of intermolecular interactions in the nonlinear optical properties of poled polymers, *Appl. Phys. Lett.* 76, 1368, 2000.
9. M.G. Kuzyk, Physical limits on electronic nonlinear molecular susceptibilities, *Phys. Rev. Lett.* 85, 1218, 2000.
10. C. Bosshard, K. Sutter, R. Schlesser, and P. Gunter, Electrooptic effects in molecular crystals, *J. Opt. Soc. Am. B* 10, 867, 1993.
11. J. Zyss, *Molecular Nonlinear Optics: Materials, Physics, Devices* (Academic Press, Boston, 1994).
12. J. Zyss and J.F. Nicoud, Status and perspectives for molecular nonlinear optics: From crystals to polymers and fundamentals to applications, *Curr. Opin. Solid State Mater. Sci.* 1, 533, 1996.
13. C. Bosshard, K. Sutter, P. Prêtre, J. Hulliger, M. Flörsheimer, P. Kaatz, and P. Günter, *Organic Nonlinear Optical Materials* (Gordon and Breach Science, Amsterdam, 1995).
14. C. Bosshard and P. Günter, in *Nonlinear Optics of Organic Molecules and Polymers*, edited by H.S. Nalwa and S. Miyata (CRC Press, Boca Raton, FL, 1997, p. 391).
15. D.M. Burland, Optical nonlinearities in chemistry, *Chem. Rev.* 94, 1, 1994.
16. S.R. Marder, L.T. Cheng, B.G. Tiemann, A.C. Friedli, M. Blancharddesce, J.W. Perry, and J. Skindhoj, Large 1st hyperpolarizabilities in push-pull polyenes by tuning of the bond-length alternation and aromaticity, *Science* 263, 511, 1994.
17. C. Bosshard, M. Bösch, I. Liakatas, M. Jäger, and P. Günter, in *Nonlinear Optical Effects and Materials*, Vol. 72, edited by P. Günter (Springer Series in Optical Science, Berlin, Heidelberg, New York, 2000, p. 163).
18. K. Clays and B.J. Coe, Design strategies versus limiting theory for engineering large second-order nonlinear optical polarizabilities in charged organic molecules, *Chem. Mater.* 15, 642, 2003.

19. L.R. Dalton, Rational design of organic electro-optic materials, *J. Phys. Condens. Matter.* 15, R897, 2003.
20. A.I. Kitaigorodskii, *Molecular Crystals and Molecules* (Academic Press, New York, 1973).
21. T. Tsunekawa, T. Gotoh, and M. Iwamoto, New organic nonlinear optical-crystals of benzylidene-aniline derivative, *Chem. Phys. Lett.* 166, 353, 1990.
22. L.T. Cheng, W. Tam, S.H. Stevenson, G.R. Meredith, G. Rikken, and S.R. Marder, Experimental investigations of organic molecular nonlinear optical polarizabilities. 1. Methods and results on benzene and stilbene derivatives, *J. Phys. Chem.* 95, 10631, 1991.
23. J.F. Nye, *Physical Properties of Crystals* (Clarendon, Oxford, 2000).
24. G. Knopfle, C. Bosshard, R. Schlesser, and P. Gunter, Optical, nonlinear-optical, and electrooptical properties of 4′-nitrobenzylidene-3-acetamino-4-methoxyaniline (MNBA) crystals, *IEEE J. Quant. Electron.* 30, 1303, 1994.
25. G.R. Meredith, *in Nonlinear Optical Properties of Organic and Polymeric Materials*, Vol. 233, edited by D.J. Williams, ACS Symposium Series (Washington, DC, 1983, p. 27).
26. S. Okada, A. Masaki, H. Matsuda, H. Nakanishi, M. Kato, R. Muramatsu, and M. Otsuka, Synthesis and crystal-structure of a novel organic ion-complex crystal for 2nd-order nonlinear optics, *Jpn. J. Appl. Phys.* 1(29), 1112, 1990.
27. S.R. Marder, J.W. Perry, and W.P. Schaefer, Synthesis of organic salts with large 2nd-order optical nonlinearities, *Science* 245, 626, 1989.
28. S.R. Marder, J.W. Perry, and C.P. Yakymyshyn, Organic salts with large 2nd-order optical nonlinearities, *Chem. Mater.* 6, 1137, 1994.
29. S. Okada, K. Nogi, A.K. Tsuji, X.M. Duan, H. Oikawa, H. Matsuda, and H. Nakanishi, Ethyl-substituted stilbazolium derivatives for second-order nonlinear optics, *Jpn. J. Appl. Phys.* 42, 668, 2003.
30. Z. Yang, S. Aravazhi, A. Schneider, P. Seiler, M. Jazbinsek, and P. Gunter, Single crystals of stilbazolium derivatives for second-order nonlinear optics, *Adv. Funct. Mater.* 15, 1072, 2005.
31. B. Ruiz, Z. Yang, V. Gramlich, M. Jazbinsek, and P. Gunter, Synthesis and crystal structure of a new stilbazolium salt with large second-order optical nonlinearity, *J. Mater. Chem.* 16, 2839, 2006.
32. B.J. Coe, J.A. Harris, I. Asselberghs, K. Clays, G. Olbrechts, A. Persoons, J.T. Hupp, R.C. Johnson, S.J. Coles, M.B. Hursthouse, and K. Nakatani, Quadratic nonlinear optical properties of *N*-aryl stilbazolium dyes, *Adv. Funct. Mater.* 12, 110, 2002.
33. B.J. Coe, J.A. Harris, I. Asselberghs, K. Wostyn, K. Clays, A. Persoons, B.S. Brunschwig, S.J. Coles, T. Gelbrich, M.E. Light, M.B. Hursthouse, and K. Nakatani, Quadratic optical nonlinearities of *N*-methyl and *N*-aryl pyridinium salts, *Adv. Funct. Mater.* 13, 347, 2003.
34. B.J. Coe, D. Beljonne, H. Vogel, J. Garin, and J. Orduna, Theoretical analyses of the effects on the linear and quadratic nonlinear optical properties of *N*-arylation of pyridinium groups in stilbazolium dyes, *J. Phys. Chem. A* 109, 10, 052, 2005.
35. M.S. Wong, U. Meier, F. Pan, V. Gramlich, C. Bosshard, and P. Gunter, Five-membered heteroaromatic hydrazone derivatives for second-order nonlinear optics, *Adv. Mater.* 8, 416, 1996.
36. M.S. Wong, C. Bosshard, F. Pan, and P. Gunter, Non-classical donor–acceptor chromophores for second order nonlinear optics, *Adv. Mater.* 8, 677, 1996.
37. F. Pan, C. Bosshard, M.S. Wong, C. Serbutoviez, K. Schenk, V. Gramlich, and P. Gunter, Selective growth of polymorphs: An investigation of the organic nonlinear optical crystal 5-nitro-2-thiophenecarboxaldehyde-4-methylphenylhydrazone, *Chem. Mater.* 9, 1328, 1997.
38. I. Liakatas, M.S. Wong, V. Gramlich, C. Bosshard, and P. Gunter, Novel, highly nonlinear optical molecular crystals based on multidonor-substituted 4-nitrophenylhydrazones, *Adv. Mater.* 10, 777, 1998.
39. F. Pan, M.S. Wong, M. Bosch, C. Bosshard, U. Meier, and P. Gunter, A highly efficient organic second-order nonlinear optical crystal based an a donor–acceptor substituted 4-nitrophenylhydrazone, *Appl. Phys. Lett.* 71, 2064, 1997.
40. M.C. Etter and G.M. Frankenbach, Hydrogen-bond directed cocrystallization as a tool for designing acentric organic solids, *Chem. Mater.* 1, 10, 1989.
41. F. Pan, M.S. Wong, V. Gramlich, C. Bosshard, and P. Gunter, A novel and perfectly aligned highly electro-optic organic cocrystal of a merocyanine dye and 2,4-dihydroxybenzaldehyde, *J. Am. Chem. Soc.* 118, 6315, 1996.
42. M.S. Wong, F. Pan, V. Gramlich, C. Bosshard, and P. Gunter, Self-assembly of an acentric co-crystal of a highly hyperpolarizable merocyanine dye with optimized alignment for nonlinear optics, *Adv. Mater.* 9, 554, 1997.
43. M.S. Wong, F. Pan, M. Bosch, R. Spreiter, C. Bosshard, P. Gunter, and V. Gramlich, Novel electro-optic molecular cocrystals with ideal chromophoric orientation and large second-order optical nonlinearities, *J. Opt. Soc. Am. B* 15, 426, 1998.
44. H.S. Nalwa and S. Miyata, Eds., *Nonlinear Optics of Organic Molecules and Polymers* (CRC Press, Boca Raton, FL, 1997).

45. S.K. Kurtz and T.T. Perry, A powder technique for evaluation of nonlinear optical materials, *J. Appl. Phys.* 39, 3798, 1968.
46. P. Laveant, C. Medrano, B. Ruiz, and P. Gunter, Rainbow photonics—growth of nonlinear optical DAST crystals, *Chimia* 57, 349, 2003.
47. L. Mutter, P. Dittrich, M. Jazbinsek, and P. Gunter, Growth and planar structuring of DAST crystals for optical applications, *J. Nonlinear Opt. Phys.* 13, 559, 2004.
48. F. Pan, G. Knopfle, C. Bosshard, S. Follonier, R. Spreiter, M.S. Wong, and P. Gunter, Electro-optic properties of the organic salt 4-*N*,*N*-dimethylamino-4′-*N*′-methyl-stilbazolium tosylate, *Appl. Phys. Lett.* 69, 13, 1996.
49. R. Spreiter, C. Bosshard, F. Pan, and P. Gunter, High-frequency response and acoustic phonon contribution of the linear electro-optic effect in DAST, *Opt. Lett.* 22, 564, 1997.
50. U. Meier, M. Bosch, C. Bosshard, F. Pan, and P. Gunter, Parametric interactions in the organic salt 4-*N*, *N*-dimethylamino-4′-*N*′-methyl-stilbazolium tosylate at telecommunication wavelengths, *J. Appl. Phys.* 83, 3486, 1998.
51. F. Pan, M.S. Wong, C. Bosshard, and P. Gunter, Crystal growth and characterization of the organic salt 4-*N*,*N*-dimethylamino-4′-*N*′-methyl-stilbazolium tosylate (DAST), *Adv. Mater.* 8, 592, 1996.
52. Y. Takahashi, H. Adachi, T. Taniuchi, M. Takagi, Y. Hosokawa, S. Onzuka, S. Brahadeeswaran, M. Yoshimura, Y. Mori, H. Masuhara, T. Sasaki, and H. Nakanishi, Organic nonlinear optical DAST crystals for electro-optic measurement and terahertz wave generation, *J. Photochem. Photobiol. A* 183, 247, 2006. Sp. Iss. SI.
53. A.S.H. Hameed, W.C. Yu, C.Y. Tai, and C.W. Lan, Effect of sodium toluene sulfonate on the nucleation, growth and characterization of DAST single crystals, *J. Cryst. Growth* 292, 510, 2006.
54. M. Jazbinsek, O.-P. Kwon, C. Bosshard, and P. Günter, in *Handbook of Organic Electronics and Photonics*, edited by H.S. Nalwa (American Scientific Publishers, CA, 2008).
55. F. Pan, K. McCallion, and M. Chiappetta, Waveguide fabrication and high-speed in-line intensity modulation in 4-*N*,*N*-4′-dimethylamino-4′-*N*′-methyl-stilbazolium tosylate, *Appl. Phys. Lett.* 74, 492, 1999.
56. K. Takayama, K. Komatsu, and T. Kaino, Serially grafted waveguide fabrication of organic crystal and transparent polymer, *Jpn. J. Appl. Phys.* 1(40), 5149, 2001.
57. S. Manetta, M. Ehrensperger, C. Bosshard, and P. Gunter, Organic thin film crystal growth for nonlinear optics: present methods and exploratory developments, *C.R. Physique* 3, 449, 2002.
58. W. Geis, R. Sinta, W. Mowers, S.J. Deneault, M.F. Marchant, K.E. Krohn, S.J. Spector, D.R. Calawa, and T.M. Lyszczarz, Fabrication of crystalline organic waveguides with an exceptionally large electro-optic coefficient, *Appl. Phys. Lett.* 84, 3729, 2004.
59. M. Thakur, J.J. Xu, A. Bhowmik, and L.G. Zhou, Single-pass thin-film electro-optic modulator based on an organic molecular salt, *Appl. Phys. Lett.* 74, 635, 1999.
60. M. Baldo, M. Deutsch, P. Burrows, H. Gossenberger, M. Gerstenberg, V. Ban, and S. Forrest, Organic vapor phase deposition, *Adv. Mater.* 10, 1505, 1998.
61. M. Thakur, A. Mishra, J. Titus, and A.C. Ahyi, Electro-optic modulation at 1.5 GHz using single-crystal film of an organic molecular salt, *Appl. Phys. Lett.* 81, 3738, 2002.
62. X. Zheng, S. Wu, R. Sobolewski, R. Adam, M. Mikulics, P. Kordos, and M. Siegel, Electro-optic sampling system with a single-crystal 4-*N*,*N*-dimethylamino-4′-*N*′-methyl-4-stilbazolium tosylate sensor, *Appl. Phys. Lett.* 82, 2383, 2003.
63. T. Kaino, B. Cai, and K. Takayama, Fabrication of DAST channel optical waveguides, *Adv. Funct. Mater.* 12, 599, 2002.
64. L. Mutter, A. Guarino, M. Jazbinsek, M. Zgonik, P. Gunter, and M. Dobeli, Ion implanted optical waveguides in nonlinear optical organic crystal, *Opt. Express* 15, 629, 2007.
65. L. Mutter, M. Koechlin, M. Jazbinsek, and P. Gunter, Direct electron beam writing of channel waveguides in nonlinear optical organic crystals, *Opt. Express* 15, 16828, 2007.
66. K. Kawase, M. Mizuno, S. Sohma, H. Takahashi, T. Taniuchi, Y. Urata, S. Wada, H. Tashiro, and H. Ito, Difference-frequency terahertz-wave generation from 4-dimethylamino-*N*-methyl-4-stilbazolium-tosylate by use of an electronically tuned Ti: sapphire laser, *Opt. Lett.* 24, 1065, 1999.
67. P.Y. Han, M. Tani, F. Pan, and X.C. Zhang, Use of the organic crystal DAST for terahertz beam applications, *Opt. Lett.* 25, 675, 2000.
68. T. Taniuchi, S. Okada, and H. Nakanishi, Widely tunable terahertz-wave generation in an organic crystal and its spectroscopic application, *J. Appl. Phys.* 95, 5984, 2004.
69. C. Bosshard, R. Spreiter, L. Degiorgi, and P. Gunter, Infrared and Raman spectroscopy of the organic crystal DAST: Polarization dependence and contribution of molecular vibrations to the linear electro-optic effect, *Phys. Rev. B* 66, 205, 107, 2002.

16 Polymeric Second-Order Nonlinear Optical Materials and Devices

Sei-Hum Jang and Alex K.-Y. Jen

CONTENTS

16.1 Organic Nonlinear Optical Materials 467
16.2 Dipolar NLO Chromophores 470
16.3 Linear Polymers 475
16.3.1 Guest–Host Polymers 475
16.3.2 Side-Chain Polymers 480
16.3.3 Cross-Linked Polymers 484
16.4 Polymers with Dendrons and Dendrimers 486
16.4.1 Chromophores with Dendrons 487
16.4.2 Polymers with Dendrons 489
16.4.3 Dendrimers 494
16.5 Molecular Glasses 497
16.5.1 Molecular Glasses and Self-Assembly of Chromophore Glasses 497
16.5.2 Molecular Glasses with Multiple Chromophores 502
16.6 Device Applications 504
16.7 Summary 507
Exercise Questions 507
List of Abbreviations 508
Acknowledgment 509
References 509

Abstract This chapter introduces the basic principles of the polymeric second-order nonlinear optical (NLO) materials and devices. Various types of polymeric second-order NLO materials are introduced with focus on molecular design of dipolar NLO chromophores and solid-state engineering of polymeric NLO materials. Separate discussions are devoted to polymers with dendrons and molecular glasses, together with an overview of the most important device applications.

16.1 ORGANIC NONLINEAR OPTICAL MATERIALS

A beam of light is an electromagnetic wave. The electric and magnetic field components of light can interact with electrons in organic molecules to induce polarization of electrons. The magnitude of the induced polarization of electrons depends on the nature of electronic structure of molecules.

Organic molecules are insulators in general, and they are electronically and magnetically inactive. But, as explained in detail in Chapter 17, the electric field of light can interact strongly with polarizable π-electrons of organic molecules resulting in extremely fast reorganization of their electronic structures.

The induced polarization P (or induced dipole moment μ) is proportional to the field:

$$P_i = \mu_i = \alpha_{ij} E_j \tag{16.1}$$

where

i and j refer to components in the molecular frame the proportionality constant

α_{ij} is a tensor (coefficient) and is called the linear polarizability

In a bulk material composed of molecules, the linear polarization per unit volume is given by an analogous equation:

$$P_i = \chi_{ij} E_j \tag{16.2}$$

where χ_{ij} is the linear susceptibility tensor of the material and, in the absence of significant intermolecular interactions, is related to the sum of all the individual polarizabilities α_{ij}.

The induced polarization of electrons is linearly proportional to the strength of electric field with weak intensity of light. But, with increasing intensity of light having stronger electric field component, the polarization of π-electrons in organic molecules can be nonlinearly proportional to the applied electric field of light. The generalized nonlinear dependence of dipole moment on field can be expressed as a Taylor series:

$$\mu_i(E) = \mu_i(0) + \left.\frac{\partial \mu_i}{\partial E_j}\right|_0 E_j + \frac{1}{2!}\left.\frac{\partial^2 \mu_i}{\partial E_j \partial E_k}\right|_0 E_j E_k + \frac{1}{3!}\left.\frac{\partial^3 \mu_i}{\partial E_j \partial E_k \partial E_l}\right|_0 E_j E_k E_l \tag{16.3}$$

Then the value of the first differential of dipole moment with respect to field is the linear polarizability α. The first hyperpolarizability β of a molecule is another tensor and can be equated to the evaluated second differential of dipole moment with respect to field. The Taylor series expansion can be simplified as

$$\mu_i(E) = \mu_i(0) + \alpha_{ij} E_j + \beta_{ijk} E_j E_k + \gamma_{ijkl} E_j E_k E_l \tag{16.4}$$

And the macroscopic induced polarization can be expressed as

$$P_i(E) = P_i(0) + \chi_{ij}^{(1)} E_j + \chi_{ijk}^{(2)} E_j E_k + \chi_{ijkl}^{(3)} E_j E_k E_l \tag{16.5}$$

where

$\chi^{(n)}$ is the linear and nonlinear susceptibilities

$P_i(0)$ is the intrinsic static dipole moment of material

The magnitude and direction of the nonlinear responses of materials depend on the electronic structure of the molecules used and their macroscopic arrangements in solid state. Thus, the nonlinear response of organic molecules and materials can be optimized through molecular engineering of organic molecules and materials.

Index of refraction is a physical constant that determines the speed of light propagating through a material. But the refractive index of materials with second-order nonlinear optical (NLO) response can be varied by the application of an external field including the electric field of light. Moreover, the refractive index of second-order NLO materials can be modulated precisely by the application of direct current (DC) or low-frequency alternating current (AC) electric fields.

Since the applied electric fields are vectors, the net polarization of a material depends on its symmetry with respect to the orientation of the applied fields. Field-induced deformation or rotation of the optical index matrix tensors (coefficient) of polar materials is called linear electro-optical (E-O) effect or Pockel effect, and the tensors become zero in materials with center of symmetry. Thus, materials for second-order NLO applications must be noncentrosymmetric in bulk to be active.

Various crystal-engineering [1] strategies have been reported to tailor molecules into noncentrosymmetric arrangements via multiple noncovalent intermolecular interactions such as hydrogen bonds, metal–organic coordinations, and van der Waals interactions. The design and synthesis of a chromophore with large hyperpolarizability and extended π-conjugations, which crystallize into an acentric space group is a daunting challenge since we do not know how to calculate solid-state structures of molecules with some degree of chemical and conformational complexities yet. Although the introduction of a chiral auxiliary group to an NLO chromophore should produce acentric molecular packing in the solid state, the chirality alone does not show efficient NLO responses in crystals. The electrostatic dipole–dipole interactions between molecules with large dipole moments can form partially antiparallel packing resulting in small macroscopic dipole moment in polar direction of acentric crystals [2]. Moreover, introduction of bulky substituents or chiral auxiliary groups to a chromophore can significantly decrease the active molecular fractions and thermal stabilities of crystals.

Organic and inorganic guest–host inclusion complexes can orient guest chromophores in an acentric order in bulk [3], and sequential adsorption of NLO chromophores onto substrates to form covalently linked noncentrosymmetric super lattices have been studied [4], but the most common method of inducing polar order in organic E-O materials is the poled polymer approach [5].

In polymeric NLO materials, the macroscopic polar order of NLO chromophores can be achieved by various dipole orientation techniques such as electric field poling, photon-assisted electric poling, and photochemical poling. In electric field poling, a strong DC electric field (typically ~100 V/μm) is applied to induce the partial reorientation of dipolar NLO chromophores in the solid solution when the materials are heated above the glass transition temperature (T_g). By cooling the materials below the T_g while the poling field is applied, the noncentrosymmetric arrangement of the dipolar NLO chromophores can be frozen-in to form poled polymers. The electric field poling process is illustrated in Figure 16.1.

In a uniaxial system such as poled polymers, two independent macroscopic E-O coefficient components, r_{13} and r_{33}, can be used to describe the macroscopic quantity of their E-O effect. The ratio between r_{13} and r_{33} is assumed to be $r_{33} \sim 3r_{13}$ in moderately poled polymer thin films. The macroscopic E-O coefficient r_{33} to an external field of molecular NLO materials is proportional to the scalar product of the dipole moment μ and the molecular hyperpolarizability β of dipolar NLO chromophores with noncentrosymmetric alignment.

The E-O effect of organic materials is a result of the mixing of low-lying electronic excited and ground states of NLO chromophores through fast redistribution of highly polarizable π-electrons, and result changes of refractive index of materials in response to the applied electric fields. Therefore,

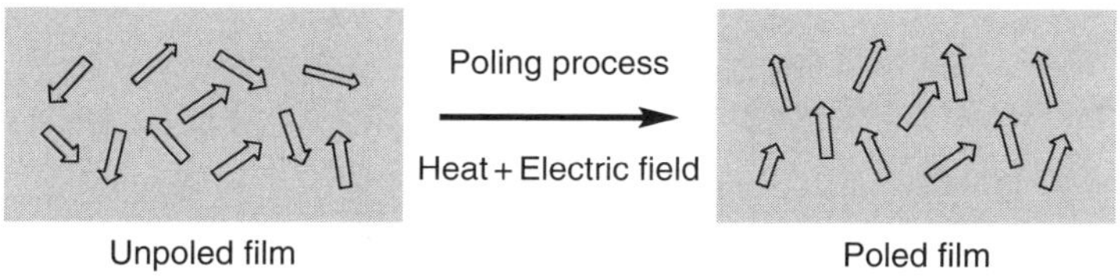

FIGURE 16.1 Illustration of static electric field poling process of dipolar NLO chromophores doped in a supporting polymer matrix. Before applying electric field the dipole moments are randomly distributed. At near glass transition temperature the dipole moments of NLO chromophores are mobile and respond to the direction of the applied external field. The orientation of chromophore is frozen-in by cooling to low temperature to form a poled polymer with noncentrosymmetric distribution of dipolar NLO chromophores.

the E-O effect can be used as mechanisms to modulate optical fields in phase, frequency, amplitude, and direction in various photonic applications such as frequency conversions, optical switching, and signal processing.

As explained in detail in Chapter 17, there are fundamental differences and significant advantages of organic NLO materials over the inorganic NLO materials. In addition to ever-increasing E-O activity through molecular engineering, wide operational bandwidth (>100 GHz) is one of the most important advantages of organic NLO materials. In organic NLO materials, π-electron system defines both dielectric permittivity and refractive index of materials, and the purely electronic responses of organic NLO materials are associated with highly polarizable π-electrons of organic materials. Also, organic materials have low dielectric constant ($\varepsilon \sim n^2$) compared to inorganic materials and have wide bandwidth with small dispersion in the refraction index from DC to optical frequencies. Moreover, various film processing techniques can be easily applied to integrate polymeric NLO materials into device structures using established microfabrication processes. Polymeric E-O materials with NLO chromophores have shown commercial potential as active media in high-speed broadband waveguides for optical switches, optical sensors, and information processors [6].

Efficient translation of large β chromophores into thermally and chemically stable materials with macroscopic E-O responses can be achieved in polymeric NLO polymers only by careful modification of the molecular structure of chromophores and host polymers. We will now introduce the basic principles for molecular design of dipolar NLO chromophores, and solid-state molecular engineering of polymeric NLO materials in the following sections.

16.2 DIPOLAR NLO CHROMOPHORES

In general, dipolar NLO chromophores used in polymeric E-O materials consist of electron donor and acceptor groups interacting through a π-electron conjugation bridge. In such arrangement of a "push–pull" dipolar NLO chromophore, low-lying electronic excited and ground states can mix effectively by the intramolecular charge transfer from donor to acceptor groups. It has been well established that the first molecular hyperpolarizability β (or microscopic hyperpolarizability), which characterizes the molecular NLO efficiency, depends on the strength of the donor and acceptor groups, as well as the nature and length of the π-conjugation bridge [7]. Other classes of second-order NLO chromophores with different molecular geometries, such as octupolar structure, are known but device applications of polymeric NLO materials based on such chromophores are limited, and will not be introduced in this chapter. Related references are provided for interested readers [8].

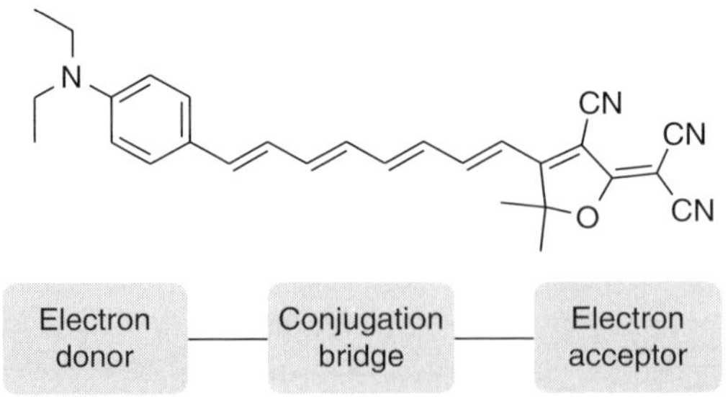

Very large β values can be achieved in an NLO chromophore by optimizing the ground-state polarization of the molecule through careful combination of available electron donors, conjugated bridges, and electron acceptors [7,9]. The strength and efficiency of these groups can be engineered by rational molecular design of modular components of chromophores using quantum mechanical guidance.

There are several general guiding principles in the molecular design of dipolar NLO chromophores. In 1977, Oudar and Chemla suggested the two-state model for the design of

NLO chromophores [10]. In the model, only the dipole matrix element and transition energy for the transition between ground state and the first strongly allowed charge-transfer excited state are considered, i.e., the charge transfer excitation between highest occupied molecular orbital (HOMO) and lowest unoccupied molecular orbital (LUMO).

In dipolar NLO chromophores, β is usually dominated by a single diagonal tensor term, along the dipolar axis, β_{xxx}. Then, the β can be analyzed using two different approaches to establish the structure–property relationships. In the two-state model, only one single electronic excited state, e, is strongly coupled to the ground state, g, and β can be expressed according to the following expression

$$\beta_{xxx} = \frac{\Delta\mu_{ge}\mu_{ge}^2}{E_{ge}^2} \tag{16.6}$$

where

$\Delta\mu_{ge}$ is the change in state dipole moment
μ_{ge} is the transition dipole moment
E_{ge} is the transition energy

In 1991, Marder proposed to correlate hyperpolarizability with ground-state polarization in the design of NLO chromophores. The degree of ground-state polarization, namely, the degree of charge separation in the ground state, depends on the local environments including an external electric field [11]. Two important parameters were introduced to optimize the dipolar NLO chromophores from the model: For example bond length alternation (BLA) and bond order alternation (BOA). The BLA and BOA are the parameters that describe the electronic mixing between resonance structures of a NLO chromophore in the ground state. These parameters are determined from the average difference in bond length between adjacent C–C bonds in π-electron conjugation bridges of dipolar NLO chromophores. Schematic plot showing how dispersion-free molecular hyperpolarizability β_0 varies with BLA and BOA are shown in Figure 16.2.

For example, a well-known conjugate polymer polyacetylene has alternating double and single bonds with bond length 1.45 and 1.34 Å. The BLA and BOA of polyacetylene are +0.11 and −0.55 Å, respectively. In the so-called cyanine limit, neutral and zwitterionic (charge separated) resonance

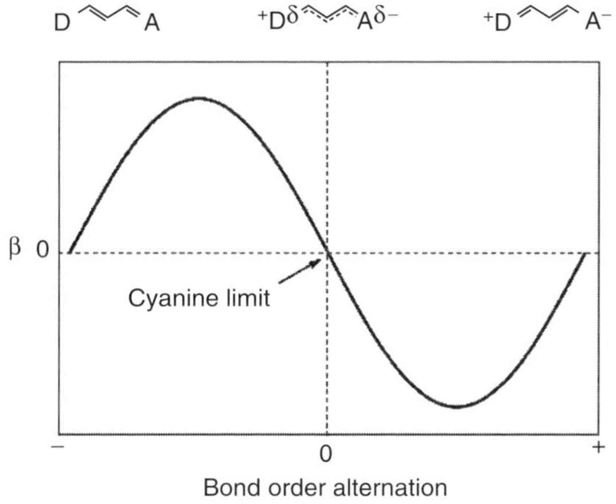

FIGURE 16.2 Schematic plot showing how dispersion-free molecular hyperpolarizability varies with bond order alternation (BOA or bond length alternation, BLA) for a donor-bridge-acceptor chromophore, along with the variation (resonance structures) of the constituent terms.

structures make equal contributions to the ground state, and the BLA (or BOA) is zero. This model provides a simple yet comprehensive molecular design rule for an organic chemist to optimize dipolar NLO chromophores based on their chemical intuitions, and the combination of molecular structures of electron donor and acceptor groups for a given π-electron conjugation bridge. The molecular hyperpolarizability was reported to be maximized when BLA is ~0.04 Å.

Bond length alternation and bond order alternation are affected by the energetics of resonance structures in the ground state of a molecule. There are two contributing terms affecting the energetics of resonance structures of a dipolar chromophore, Coulombic and aromaticity terms. Separation of charges within molecules destabilizes the resonance structures and the aromaticity can enhance or interrupt charge separation. For example, ground-state polarization can be enhanced by aromatization of acceptor or by introducing heteroaromatic rings with smaller aromatic stabilization energy such as thiophene and thiazole in place of phenyl group in the conjugation bridges [12].

Therefore, efficient NLO chromophores can be designed by optimizing the BLA (and BOA) to maximize ground-state polarization by the use of proper combination of donor and acceptor groups, and bridge topology to alter the relative energetics of possible resonance structures.

Various quantum mechanical calculations have been used extensively as powerful tools for systematic improvement of molecular hyperpolarizability and rational molecular design of dipolar NLO chromophores. For example, the geometries of the dipolar NLO chromophores can be optimized with Austin Model 1 (AM1) as implemented in Missouri procurement Assistance Center (MOPAC) [13]. On the basis of AM1 optimized geometries of chromophores (or single-crystal x-ray structures when available), the electronic properties in the ground and excited states (energies, state and transition dipole moments) can be evaluated with a semiempirical intermediate neglect of differential overlap (INDO) Hamiltonian (as implemented in Zerner's intermediate neglect of differential overlap (ZINDO) code) [14]. Such calculations have proven reliable to describe the β. Moreover, more computation intense methods such as sum-over-states (SOS) [15] and finite-field (FF) [16] approaches can provide further insights on the origin of molecular hyperpolarizabilities.

In the perturbative SOS method, hyperpolarizabilities are related to the transition moments, dipole moment changes, and energy separations between ground and excited states of the molecule, and between various excited states of the molecule. The transition energies, transition and state dipole moments are obtained from INDO calculations, and the summations are performed over the full set of excited states within a Taylor series expansion. This approach can provide detailed understanding on multistate transition and optimization of NLO properties through the consideration of excited-state properties.

The other approach, the FF, as proposed by Chopra and developed by Nakano, has been successfully applied to dipolar NLO chromophores with extended conjugation bridges in conjunction with INDO Hamiltonian. The approach allows one to identify the contributions of the individual chemical segments of the molecules to overall β. The variational FF approach relies on the fact that, by definition, β_{xxx} is the second derivative of the x-component of the dipole moment μ with respect to the x-components of the applied electric fields E, at zero field. And the electronic part of β can be cast rigorously as the integral over the moments of the second derivative of the charge density ρ. In an approximate way, that integral can be partitioned into a sum-over derivatives of point charges q_i concentrated on the individual atoms, as shown below:

$$\beta_{xxx} = \left.\frac{\partial^2 \mu_x}{\partial E_x^2}\right|_{F_x=0} = \int x \left.\frac{\partial^2 \rho(\vec{r})}{\partial E_x^2}\right|_{F_x=0} d^3\vec{r} \approx \sum_i x_i \left.\frac{\partial^2 q_i^2}{\partial E_x^2}\right|_{F_x=0} = \sum_i x_i q_i^{(2)} \tag{16.7}$$

where

β_{xxx} is partitioned into local (atomic) contributions or the so-called β-moments
$x_i q_i^{(2)}$ is derived from the β-charges, $q_i^{(2)}$

Note that the superscript (2) represents the second derivative with respect to the applied electric field in the x-direction. In the calculations, the derivatives were approximated by finite differences obtained from INDO/HartreeFock Mulliken charges, with fields of 0 and $\pm 5.14 \times 10^{11}$ V/m (10^{-3} atomic units) applied to the chromophores; the series expansion for the charges are limited to the second order in E in the calculation.

We will now introduce molecular structure factors in the design of dipolar NLO chromophores, and molecular engineering approaches to improve thermal and photochemical stabilities of chromophores with large molecular hyperpolarizabilities. Although chromophores with strong donor and acceptor groups connected via polyene bridges are known to exhibit large $\mu\beta$ values, extended polyenes are typically not thermally stable at >200°C in general. Extended polyenes can undergo cis–trans isomerization and photochemical [2 + 2] cycloaddition. By reducing the cis–trans isomerization and increasing steric protection of the π-system by conformationally locking some of double bonds in a polyene chains improve the stabilities significantly. Moreover, the ring-locked bridge structures can prevent the dipole–dipole interactions between chromophores.

Aromatic phenyl group in the bridging polyenes structure can improve thermal and photochemical stabilities of chromophores further, but with lowered β values. Aromatic heterocycles such as thiophene have been used extensively as a trade-off between efficiency and stabilities. The reduced aromatic stabilization energy of the thiophene leads to higher nonlinear efficiency than the phenylene group [12].

Many functional groups based on electron-rich atoms with polarizable lone pair electrons and molecules can be used for donor groups. Chromophores with dialkylamino group connected directly to conjugation bridges are known to be efficient but because of low thermochemical stability of such chromophores, 4-(dialkylamino) phenyl groups are commonly used as donors [17].

A major challenge for optimizing β of dipolar NLO chromophores is the development of novel electron acceptors. Most of the effective acceptors consist of five-membered heterocyclic conjugated ring structure substituted with groups capable of withdrawing electron density through both inductive and resonance effects. Structures of strong electron acceptors are shown below [18]. Recently, classes of NLO chromophores based on strong electron acceptors 2-dicyanomethylene-3-cyano-4,5,5-trimethyl-2,5-dihydrofuran (TCF), 2-dicyanomethylene-3-cyano-4,5-dimethyl-5-trifluoromethyl-2,5-dihydrofuran (CF_3-TCF), and 3-methyl-4-cyano-5-dicyanomethylene-2-oxo-3-pyrroline (TCP) were reported with exceptionally large β [19]. The inductive contribution and potential hyperconjugation of the trifluoromethyl group in the electronic structure of the CF_3-TCF acceptor has been attributed to explain the significant improvement in acceptor strength.

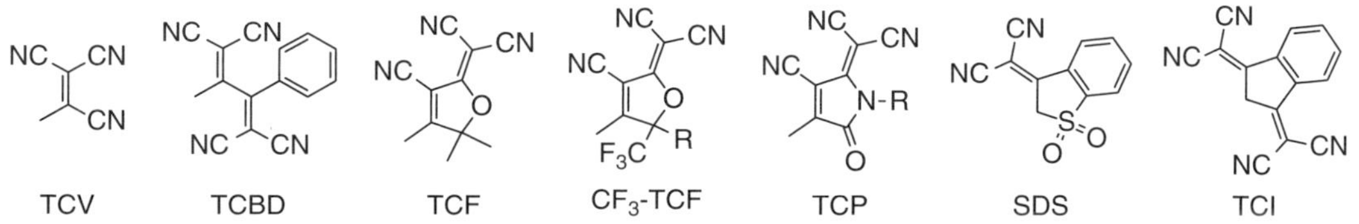

As explained earlier, molecular hyperpolarizability of an NLO chromophore is due to the presence of low-lying electronic states resulting in strong intramolecular charge transfer. The intramolecular charge-transfer transitions of chromophores, thus the absorption spectra, are very sensitive to the polarity of their environment. NLO chromophores are said to be solvatochromic. Solvatochromism is a change in absorption spectrum (λ_{max}, intensity, and shape) of a chromophore in solvents of different polarity even in solid solutions. Positive (bathochromic) solvatochromism refers to a redshift of the absorption band with increasing solvent polarity. Polar solvents stabilize the excited state with charge separation and polarity of a solvent can be determined by its solvation behavior, which in turn depends on the intermolecular forces between the solvent and the solute. Thus, the solvatochromism can be used to determine hyperpolarizability as mesomeric dipole moments, the variation of the molecular dipole moment with excitation. Solvatochromism of dipolar NLO chromophores can be measured to estimate and compare β chromophores qualitatively within closely related structures [20]. Other more quantitative experimental methods to determine the β of chromophores such as electric field-induced second harmonic generation (EFISH) and hyper-Rayleigh scattering (HRS) measurements were discussed in earlier chapters. The molecular structures of representative examples of dipolar NLO chromophores are shown in Figure 16.3, and the molecular hyperpolarizabilities of the chromophores are summarized in Table 16.1.

There are many additional material parameters to be optimized simultaneously during the molecular design of efficient dipolar NLO chromophores for device applications. Among them, optical absorption loss, and thermal and photochemical stabilities are considered to be the most important for device applications. Most poled polymer films are prepared by spin-casting, and it is necessary that both the chromophore and the polymer host have good solubility in common organic solvents. It is also necessary that chromophores do not phase separate from the polymer host either during spin-coating or during poling so that the chromophore should be structurally compatible with

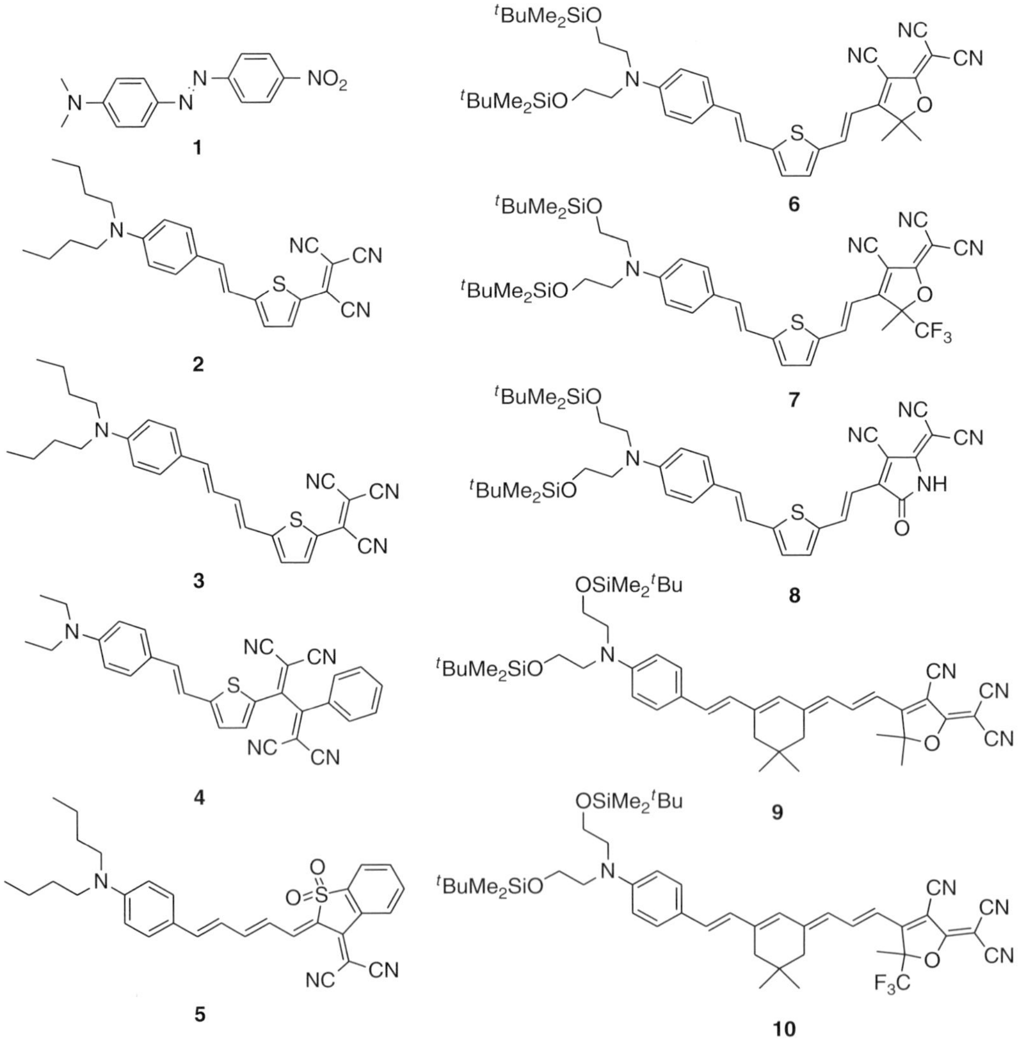

FIGURE 16.3 Molecular structures of representative dipolar NLO chromophores.

TABLE 16.1
Molecular Hyperpolarizabilities of Representative Dipolar NLO Chromophores at 1907 nm

Chromophore	$\mu\beta$ (10^{-48} esu)
1	580
2	6,200
3	9,800
5	13,500
6	18,000
9	35,000

host polymer. Processable polymeric E-O materials with $\mu\beta$ values of $>10^{-44}$ esu are reported recently with absorption loss of less than 1 dB/cm, and with thermal stability >200°C.

16.3 LINEAR POLYMERS

16.3.1 Guest–Host Polymers

Solid solution of guest NLO chromophores in a supporting host polymer is called guest–host polymer. In principle, the guest–host polymer is the simplest and straightforward synthetic approach to polymeric NLO materials. Many optical quality commercial polymers such as amorphous derivatives of polyacrylates and polycarbonates can be used as host polymers. As stated in Chapter 17, optimizing the molecular nonlinear response of a dipolar NLO chromophore by molecular engineering presents only a part in designing a polymeric NLO material. In general, necessary noncentrosymmetric arrangements of dipolar NLO chromophores in the polymer matrices are generated by electric field poling methods.

In the absence of intermolecular electrostatic dipole–dipole interactions, the macroscopic E-O activity r_{33} of polymeric E-O materials is directly proportional to β and the degree of noncentrosymmetric order (order parameter) of doped NLO chromophores induced by electric field poling. It is expected that increasing β values of NLO chromophores should translate directly into increased electro-optic activities of polymer matrices with the chromophores in the oriented gas model [21].

Perfectly aligned assembly of dipolar NLO chromophores with large β with polar order in solid solution can present ideal material configurations in maximizing the macroscopic E-O activity of NLO polymers. But such arrangements of NLO chromophores with very large ground-state dipole moments are not thermodynamically stable even in low concentrations. Therefore, one of the major challenges in the development of polymeric NLO materials based on dipolar NLO chromophores is to engineer molecular materials to prevent and minimize the formation of strong electrostatic dipole–dipole interactions producing centrosymmetric aggregates of chromophores in solid state. In some cases, the controlled molecular scale pre-organization of noncentrosymmetric aggregates can enhance the poling efficiency and the macroscopic E-O activities of materials.

The oriented gas model provides a simple way to relate the macroscopic NLO properties, such as the second-order susceptibility tensor and the microscopic hyperpolarizability tensor elements [21]. The oriented gas model are based on a number of simplifications and approximations: (i) at the poling temperature the chromophores are assumed free to rotate under the influence of the applied field and any coupling or reaction from the surrounding matrix is ignored; (ii) the chromophores have cylindrical symmetry and the only nonvanishing hyperpolarizability tensor element is β_{xxx} where x is the symmetry axis of the chromophore; (iii) the permanent dipole moment μ of the chromophore is oriented along the x axis of the molecule; and (iv) the chromophores are assumed independent and noninteracting.

Accordingly, the principal E-O tensor (coefficient) element r_{33}, is related to the acentric-order parameter, $\langle \cos^3(\theta) \rangle$, the first molecular hyperpolarizability, β, the chromophore number density in an inert host, N, the refractive index, n, and the local field arising from the host dielectric permittivity $f(\omega)$, by

$$r_{33} = N \langle \cos^3(\theta) \rangle \, 2\, \beta f(\omega)/n^4 = N(\mu E/5kT)\, 2\, \beta f(\omega)/n^4 \tag{16.8}$$

For a homogeneous, single-phase material, the number density is related to the mean distance between chromophores r, as $N = 1/r^3$. The order parameter, $\langle \cos^3(\theta) \rangle$, can be determined experimentally by linear optical dichroism. The poling field orients the molecule dipole moment in its direction, which is usually perpendicular to the thin-film surface. As the poled chromophores are strongly anisotropic, such changes of the orientation of dipole moments can be monitored by the thin-film optical absorption spectrum. Two observations can be made from the temporal behavior of such linear absorption spectrum. The optical density (absorbance) of poled film decreases during the poling process and increases afterwards. The first effect is due to the change of the orientation of the dipolar transition moments, while the second is due to the relaxation of the induced order (return of chromophores to the initial random orientation).

The decrease of the optical absorption spectrum is a result of the oriented dipoles of chromophores in the direction perpendicular to the thin-film surface, which is parallel to the field direction (aligned chromophore has smaller absorption cross section to the incident light). From this variation one can determine the order parameter where $A_{\parallel}$ and $A_{\perp}$ ($A_{\parallel} + 2A_{\perp} = 3A_0$) are absorbances (optical densities) measured with the incident light polarized parallel and perpendicular to the poling field direction, respectively. The axial orientation of the poled chromophores was described as the order parameter $\langle \cos^3(\theta) \rangle$ is given by

$$\langle \cos^3(\theta) \rangle = (A_{\parallel} - A_{\perp})/(A_{\parallel} + A_{\perp}) = 1 - A/A_0 \tag{16.9}$$

where A_0 and A are the absorbance maxima for the unpoled and poled samples at normal incidence. Typically, the order parameters, $\langle \cos^3(\theta) \rangle$, take the values between 0.1 and 0.3 for isotropic polymers, 0.5–0.6 for nematic liquid crystals, and 0.9–1 for smectic liquid crystals, respectively. It is worth noting that the order parameter, $\langle \cos^3(\theta) \rangle$, determined in this way does not discriminate between polar and axial order. The only way to get such discrimination is by using NLO techniques, which are sensitive to the polar order.

Recently, Dalton and Robinson proposed that intermolecular electrostatic dipole–dipole interactions prevent efficient translation of molecular E-O activity to macroscopic E-O activity [22]. The dipole–dipole interactions result in a saturation behavior with maximum in the plot of E-O activity verse chromophore loading in the host polymer matrix. The position of this maximum shifts to lower loading with increasing chromophore dipole moment and scalar product with the molecular hyperpolarizability $\mu\beta$. The attenuation of E-O activity was most severe for rod-shaped chromophores and less severe for more spherical chromophores.

Monte Carlo statistical mechanical simulation predict that the intermolecular electrostatic interactions would lead to saturation behavior, and extensive experimental investigations later have shown that the saturation of macroscopic E-O activity with increasing concentration of dipolar NLO chromophores is a general phenomena. Dalton suggested that the dipolar NLO chromophores should be designed with structural modifications forming spherical molecular shape to minimize the electrostatic dipole–dipole interactions and thus improve poling efficiency, orientational susceptibility of dipole to an applied electric field, of polymeric E-O materials. More detailed examples of shape modifications of chromophores with improved poling efficiency and E-O activities will be discussed in later sections.

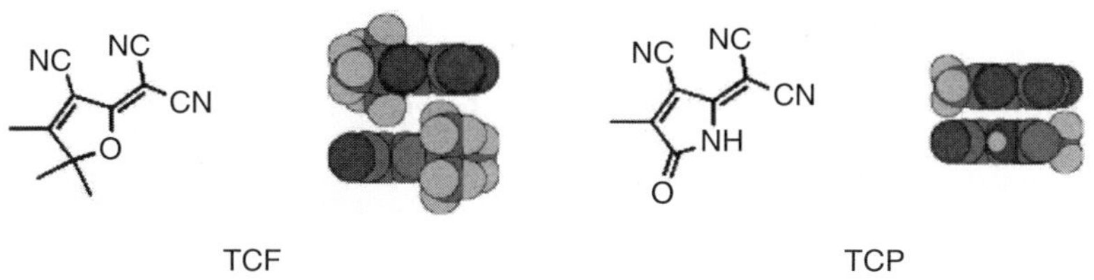

FIGURE 16.4 Structural differences between acceptors are shown. In TCF acceptor geminal dimethyl groups act as spacers above and below the π-electron plane. The electrostatic dipole–dipole interactions between acceptors in dimers in TCF acceptor are expected to be weaker and easier to reorient by electric field poling compare to flat TCP acceptor.

Recently, Kuzyk and Gunter estimated the upper limits of the E-O coefficients based on optimized hyperpolarizabilities without the intermolecular interactions (except for local field corrections) and suggested that the macroscopic susceptibilities of efficient dipolar NLO chromophores with extended π-conjugation bridges have not reached the upper limit yet [23].

Analysis of the structural differences between efficient acceptors such as TCF and TCP suggests that rather small but effective "spacer groups" such as geminal dimethyl group in TCF acceptor, both above and below the conjugation plane of NLO chromophores, are necessary for an efficient electric field-induced realignment of dipoles of chromophores, especially at high loading levels. Possible dimer structures of TCF and TCP acceptors are compared in Figure 16.4.

As proposed by Dalton, molecular design rule of the shape control broadly suggests spherical shape modification of chromophores to prevent the electrostatic intermolecular interaction. The rule was optimized to increase the loading density of the chromophores thus to increase macroscopic E-O activity of materials, but does not consider the modes (relative orientation, distance, and degree of overlap between conjugated systems) and strength of dipole–dipole electrostatic interactions of chromophores in the solid state. Since most of the NLO chromophores have rather flat π-conjugation structures, various modes of dipole–dipole interactions are possible depending on the relative alignment and distances between π-systems, and the length of overlap when the chromophores form dimers and aggregates. Careful studies on modes and strength of the dipole–dipole electrostatic interactions should proceed to guide rational design of chromophores for polymeric E-O materials.

Therefore, a push-pull type NLO chromophore should be designed to weaken the associative energies of dimers and aggregates in van der Waals contact for efficient electric field poling. The modes of interactions determine efficiency of spacers, and the "spacer control" is complementary to the rule of "shape control" in the design of NLO chromophores. Various double-sided spacer modifications can be used to improve the poling efficiency of chromophores with severely aggregating structures. Spacers should be designed to minimize and weaken intermolecular interactions between π-systems of NLO chromophores based on mode analysis of intermolecular interactions. Recently, related studies on the dimerization of planar merocyanine chromophores have been studied in solution [24]. Moreover, spacer groups for each modular component of chromophores (donor, bridge, and acceptor) should be designed to reduce the energy of dimerization and aggregation between dipolar NLO chromophores, and to improve the poling efficiency. Schematic comparison of the shape control and spacer control are shown in Figure 16.5.

The electric field-induced acentric order of chromophores in poled polymers tends to decrease slowly with time. The long-term alignment stability (or temporal alignment stability) of guest–host polymer can be improved by increasing poling temperature (or T_g of host polymer) relative to the device operation temperature in general. In addition, chromophores in a guest–host polymer can act as plasticizers and lower the T_g of the polymer composite so that using high T_g polymers as hosts is advantageous to improve the temporal alignment stability. But many of highly efficient dipolar NLO

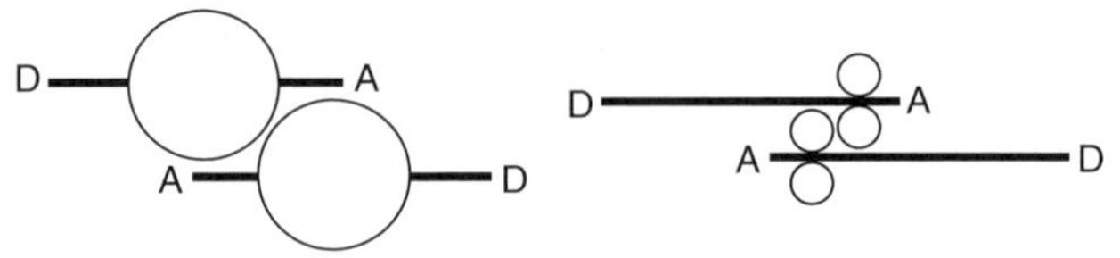

FIGURE 16.5 Schematic comparison of the shape control and spacer control. Notice that the use of small but efficient spacers above and blow the plane of π-conjugation system can prevent the electrostatic intermolecular interactions effectively and lower the aggregation energy with minimal addition to the passive volume of chromophores.

chromophores can thermally decompose during the high-temperature poling process, and the choice of high T_g host polymers is limited by the thermal stability of chromophores.

And also, one should consider material parameters such as optical absorption and scattering loss, and photochemical stability of polymer composites during the molecular engineering of polymeric E-O materials for device applications. E-O coefficients r_{33}> 100 pm/V at 1.3–1.55 μm having optical loss of ~2 dB/cm with temporal alignment stability longer than 1000 h at 85°C have been reported recently in polymeric NLO materials. Examples of guest–host polymers will be introduced in more detail next. The macroscopic E-O activities of representative guest–host polymers at 1.3 μm are summarized in Table 16.2.

Many NLO chromophores were incorporated as guests in high T_g polymers, such as polyimide and polyquinoline thin films. The polymers showed significantly improved temporal stability at elevated temperatures. For example, a commercial polyimide doped with highly nonlinear heteroaromatic chromophores with thiophene-containing conjugated bridge was reported with significantly improved thermochemical stability. The chromophore **2** was thermally stable enough to sustain both imidization (30 min at 200°C) and poling (10 min at 220°C) conditions. Through the use of high T_g polyimide, the guest–host NLO polymer had temporal stability at elevated temperatures 120°C and 150°C for more than 30 days, while retaining 80% of E-O activity of its initial value after the electric field poling.

Chromophore **3** with longer conjugation length was also incorporated as guest in a rigid-rod, high-temperature polyquinoline (PQ-100). Poling results from the guest–host polymer showed both good E-O activity and temporal alignment stability at 80°C. After an initial drop from 45 to 26 pm/V in the first 100 h, the r_{33} value remained at 26 pm/V for more than 2000 h as shown in Figure 16.6.

TABLE 16.2
Macroscopic E-O Activities of Representative Guest–Host Polymers at 1.3 μm

Guest–Host Polymers	r_{33} (pm/V) at 1330 nm
30 wt% Chromophore **1** in PMMA	13
25 wt% Chromophore **2** in Polyimide	30
25 wt% Chromophore **3** in PQ-100	45
25 wt% Chromophore **4** in PQ-100	36
20 wt% Chromophore **5** in Polycarbonate	55
20 wt% Chromophore **7** in APC	42
30 wt% Chromophore **9** in PMMA	60
30 wt% Chromophore **10** in APC	73

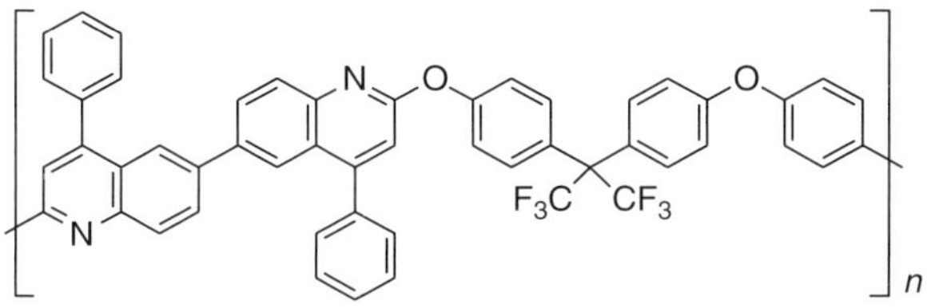

PQ-100

Jen and Wu developed another series of highly efficient, thermally stable chromophore **4** based on 2-phenyl-tetracyanobutadienyl (Ph-TCBD) group as an electron acceptor [25]. Single-crystal x-ray structure of the chromophore revealed that the dicyanovinylenyl moiety is a linked coplanar to the donor-substituted aryl segment and forms an efficient push–pull system. The three-dimensional (3-D) shape of the chromophore may also help to prevent molecules from stacking up on each other, and thus, reduce chromophore aggregation. This, in turn, can improve the poling efficiency and lower the optical scattering loss. The E-O activity of poled guest–host polymer made of 20 wt% of chromophore **4** in polyquinoline were measured to be 36 pm/V at 1.3 μm and the value remained at approximately 80% of its original value at 85°C for more than 1000 h.

A large r_{33} value of 42 pm/V at 1.3 μm was reported from a guest–host polymer with 20 wt% of chromophore **7** in amorphous polycarbonate (APC) host when poled at a temperature close to its T_g (145°C) for 5 min with a DC electric field of 100 V/μm. The r_{33} value and poling efficiency were attributed to the strength of CF_3-TCF acceptor group and structural modification of two *tert*-butyldimethylsiloxy (TBDMSO) groups at the donor side of the chromophores preventing the electrostatic dipole–dipole interactions between chromophores. Moreover, the polymer composite exhibits improved temporal alignment stability over other guest–host APC materials. The poled guest–host polymer of chromophore **7** in APC host retains more than 80% of its original r_{33} value after 500 h at 85°C.

Photochemical stability of the guest–host polymer is not yet adequate in the presence of atmospheric oxygen but simple protective packaging has shown to be acceptable for prototype device applications. The chromophore can act as a photochemical sensitizer to generate singlet oxygen from atmospheric oxygen and react with the singlet oxygen. This reaction is known to be the main cause of photochemical decomposition of chromophores in the presence of oxygen. The materials are adequate for prototype device applications but fall short for commercialization yet.

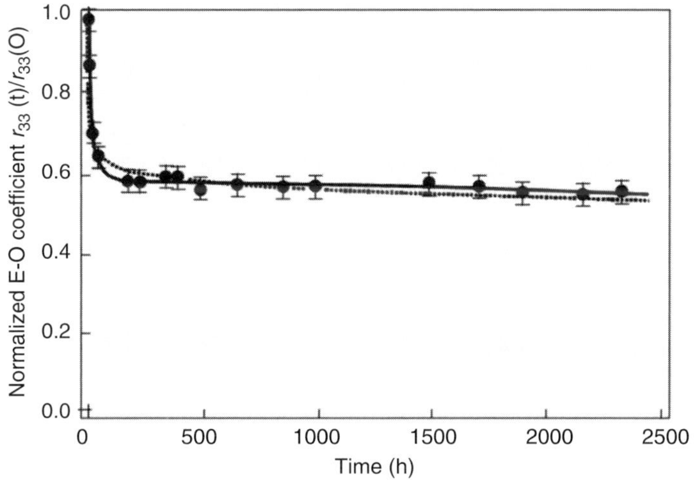

FIGURE 16.6 Normalized E-O coefficient of guest–host polymer of chromophore **3** at 80°C.

16.3.2 Side-Chain Polymers

In side-chain polymers, a guest dipolar NLO chromophore is covalently tethered to a host polymer. As mentioned in the proceeding section, chromophore can phase separate in guest–host polymers. If the phase separation produces centrosymmetric microcrystals in the materials, it is detrimental not only to the macroscopic E-O activity of materials but can also lead to significant scattering optical losses. And also the phase-separated chromophores can escape from host polymer matrices by sublimation during high-temperature poling processes. Such drawbacks can be prevented in side-chain polymers by covalently attaching the chromophore to the host polymer (Figure 16.7). Because of limited utility, NLO polymers with chromophores covalently incorporated into its main chain will not be discussed in this chapter.

Moreover, lower mobility of chromophores and increased structural complexities around chromophores in the side-chain polymers can reduce the intermolecular electrostatic dipole–dipole interactions and improve temporal alignment stability of poled polymers. But, the poling efficiency of side-chain polymers is lower than comparable guest–host polymers in general. Since NLO chromophores are anchored to polymer backbone by covalent bonds in a side-chain polymer, mobility of chromophores in the polymers is severely restricted, and chromophores cannot freely reorient themselves in response to external electric field even at T_g.

Two different synthetic approaches can be used for the preparation of side-chain polymers. Chromophores can be modified with multiple reactive groups and used as monomers to form step-growth side-chain NLO polymers, or active side groups in conventional linear polymers can be tethered with chromophores by postfunctionalization. The latter approach is preferred method for systematic molecular engineering of polymers with optimized properties since one can take advantage of well-defined linear polymers as starting materials.

Vinyl polymers like polystyrene (PS) and PMMA, and high T_g polymers such as polyurethanes, polyimides, polyamides, polyesters, polyethers, and polyquinolines were studied for side-chain NLO polymers. Among them, examples of polyurethanes and polyimides will be introduced here as promising materials with excellent thermal stabilities. A side-chain polycarbonate base on postfunctionalization will be discussed later in this section.

Polyurethanes are advantageous as an NLO polymer matrix in that extensive hydrogen bonds between the urethane groups increase the rigidity and T_g of the polymer matrix. The multiple hydrogen bonds with extended structure restrain the molecular motion, and retard the relaxation of the aligned NLO dipoles. Side-chain NLO polyurethanes can be synthesized by the step-growth polymerization reaction of NLO chromophores with diol groups with aromatic diisocyanate such as 2,4-toluene diisocyanate as monomers as shown in Figure 16.8 [26].

The resulting polymers are readily soluble in common aprotic polar solvents such as dimethylformamide (DMF), *N*-methylpyrrolidone (NMP), or cyclohexanone. Good optical quality films can be casted on glass or quartz substrates by spin-coating. Differential scanning calorimetry (DSC) analysis data show relatively high T_g of 121°C for the polyurethane **11**. The macroscopic second-order susceptibility $\chi^{(2)}$ of the film is determined by measuring the second harmonic generation (SHG) signal intensity (or the related quantity d_{33}) by the Maker Fringe method using a Y-cut quartz

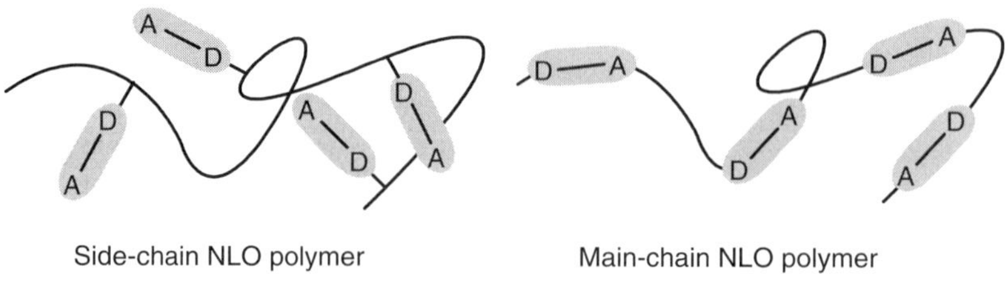

FIGURE 16.7 Schematic comparison of the side-chain and main-chain NLO polymers.

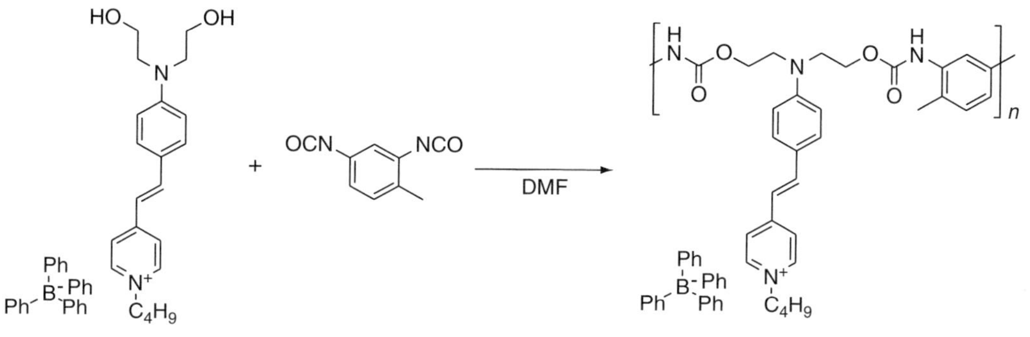

Polyurethane **11**

FIGURE 16.8 Synthesis of polyurethane by step-growth polymerization.

plate as the reference. A $\chi^{(2)}$ value of 128 pm/V was reported at 1064 nm for the polyurethane **11**. The $\chi^{(2)}$ can be related to the E-O coefficient r_{33} by

$$r_{ij} = -8\pi\, \chi_{ij}^{(2)}/n^4 \tag{16.10}$$

As described in detail in Chapter 17, chromophores based on stilbazolium salt are very important class of chromophores for nonlinear optics due to the large NLO activity and the variation of available structures. Especially, by using a bulky tetraphenylborate counter anion in the side-chain NLO polymers based on stilbazolium salt give high optical nonlinearity. The sterically demanding counter anion is expected to be beneficial not only with reduced ionic mobility of counter anion during the electric field poling process but also in preventing the electrostatic dipole–dipole interactions between chromophores.

A notable feature of this polymer system is the temporal alignment stability. At room temperature, no decay in initial SHG activity was exhibited over 700 h. Moreover, 70% of the initial NLO activity of the poled sample was retained at 100°C for 100 h. These results imply that the hydrogen bonding may play a role in preventing the relaxation of oriented molecular dipoles. Fabrication of channel waveguide structures based on the side-chain NLO polyurethanes have been demonstrated using photo bleaching technique by controlling the refractive index profile of the polymer.

In spite of excellent thermal stability and mechanical properties of aromatic polyimides, the lack of processibility has been the major hurdle in using them as host polymers. Polyetherimide (PEI) having ether linkage in the polyimide backbone, is a better host material, as processibility and solubility are enhanced without scarifying the general themochemical stabilities of polyimides.

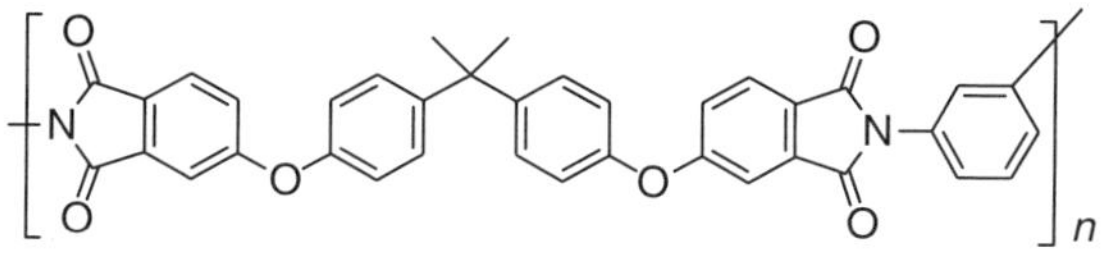

Polyetherimide (PEI)

In general, two different processes are used for the synthesis of PEIs, nitrogen-displacement of anhydride and imidization of polyamicacid. However, such synthetic method is not applicable for a side-chain NLO PEI because of harsh reaction conditions. The alternative synthetic approach for making the NLO polyetherimides is the use of the single-step reaction of NLO chromophores with terminal diol and diimide by the Mitsunobu reaction using diethyl azodicarboxylate (DEAD) and triphenylphosphine in tetrahydrofuran. In this method, the polyimide structure is directly formed during mild condensation polymerization without a thermal imidization step [27].

When PEI **13** is reacted with tetracyanoethylene (TCNE) in DMF, a blue-colored copolymer PEI **14** is produced, which contains partial tricyanovinyl group (Figure 16.9). As the percentage of tricyano (TCN) group increased, the NLO activity gets larger. According to UV/vis spectra, a new absorption maximum at 669 nm was found for PEI **14** (not observed in PEI **13**), suggesting that the NLO chromophore is generated by tricyano-vinylation. Also, the residual absorption at 365 nm indicates partial reaction of the polymer chain. DSC analysis showed T_g values in the range 144°C–157°C for the PEIs. The moderate T_g values are attributed to the flexible ether linkage and bulky chromophore incorporated to the polyimide backbone. Optical quality film on ITO glass was poled by the corona discharge-induced electric field, and the macroscopic $\chi^{(2)}$ values were reported to be 73 pm/V for PEI **12** and 56 pm/V for PEI **14** with partial formation of active chromophore resulting lower chromophore loading density. The SHG signal intensity of PEI **14** were stable up to 125°C, and the initial NLO activity was retained for more than 700 h at 100°C. The significant enhancement of temporal alignment stability is comparable to that of cross-linked NLO polymers.

The copolymer PEI **14** from PEI **13** represents an example of the second synthetic approach leading to side-chain NLO polymers. The active chromophore in the copolymer PEI **14** was generated by a postfunctionalization, tricyano-vinylation, of starting PEI **13**.

Most of the postfunctionalization methods used for NLO side-chain polymers such as azo-coupling, tricyano-vinylation, Mitsunobu etherification, Knoevenagel condensation, and catalyzed esterification, often generate by-products or trace amount of residual ionic impurities that can significantly attenuate the effective electrical field for poling and cause the DC bias drift during device operation.

There are several reaction criteria to be considered for the postfunctionalization of conventional linear polymers to a side-chain polymer. The reaction need to be clean and quantitative with minimum

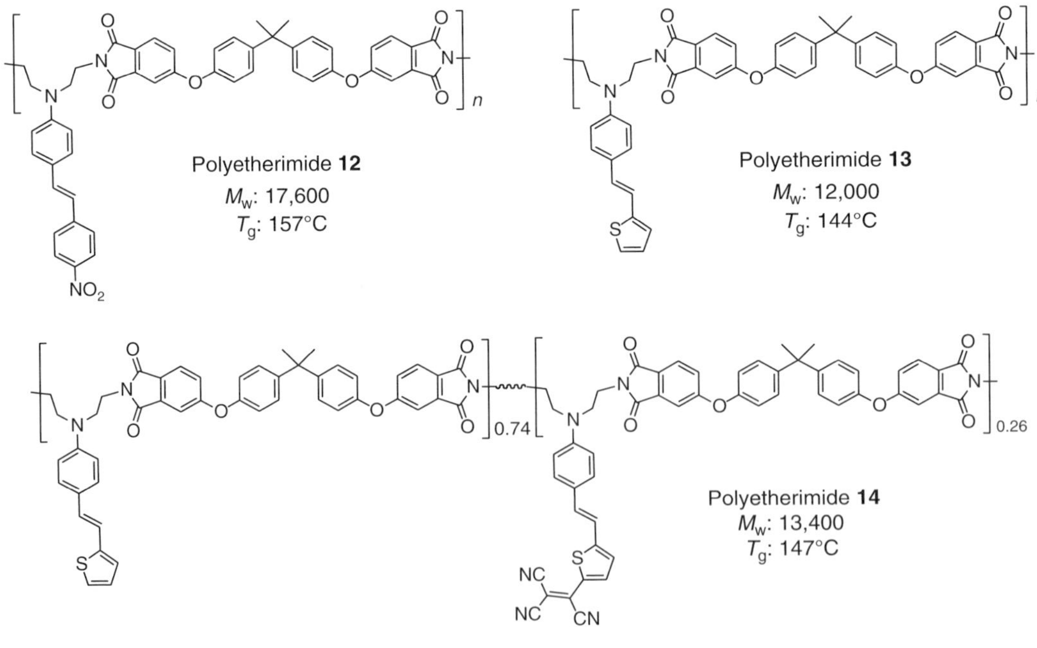

FIGURE 16.9 Structures of PEIs **12–14**. Note the PEI **14** is prepared from the PEI **13** by a postfunctionalization reaction, tricyano-vinylation.

by-products, and does not use ionic catalyst or reactants that can affect poling and device operation. And the reaction condition should be mild and structurally comparable with highly polarizable chromophores to maximize the loading density of chromophores. Moreover, the reaction should be general and versatile enough for pre- and postfunctionalization of tethered chromophores and polymer backbone structures to improve the poling efficiency further. There is also a strong need to concurrently establish creative processing/poling protocols using carefully controlled chemistry to overcome the commonly observed incompatibility and mismatched dielectric properties, conductivity, and phase transition behavior between chromophores and host polymers.

In 2002, Jen and Luo introduced a general and efficient synthetic method for the postfunctionalization of linear polymers for NLO materials based on a well-known Diels–Alder (D–A) cycloaddition reaction. The reaction has been successfully applied to several high-performance side-chain NLO polymers meeting the stringent requirements for the postfunctionalization of NLO polymers with comprehensive material properties. The D–A reaction involves a ring forming coupling reaction between a dienophile and a conjugated diene, and can be generally described by a symmetry allowed concerted mechanism that does not include reactive radical or ionic intermediate species.

The method is based on D–A reactions between maleimide-containing NLO chromophores and the polymer backbone with pendant anthracenyl diene moieties. This synthetic approach is very mild, versatile, quantitative, and free of ionic species and catalysts. The D–A reaction enables one to systematically molecular engineer NLO polymer composites for structure–property correlation studies. For example, the poling behaviors and orientational stability of chromophores relative to their anchoring position or the mode (side-on and end-on) of attachment to the polymer backbone can be systematically investigated using the D–A reaction (Figure 16.10).

Side-chain NLO polymers can be synthesized by the D–A reaction between chromophores with diene groups and polymers with dienophile, or chromophores with dienophile and polymers functionalized with diene groups. Some of polymers with diene or dienophile groups are reported and commercially available. One such polymer with a well-known diene, anthracenyl, poly(methyl methacrylate-*co*-anthracen-9-ylmethyl methacrylate (PMMA-AMA) can be prepared by a free radical copolymerization using 2,2′-azobis(isobutyronitrile) (AIBN) as an initiator. The content of the anthracenyl moiety in this copolymer can be varied to turn polymer properties, and estimated by comparing the relative integration of the corresponding characteristic peaks in its NMR spectrum.

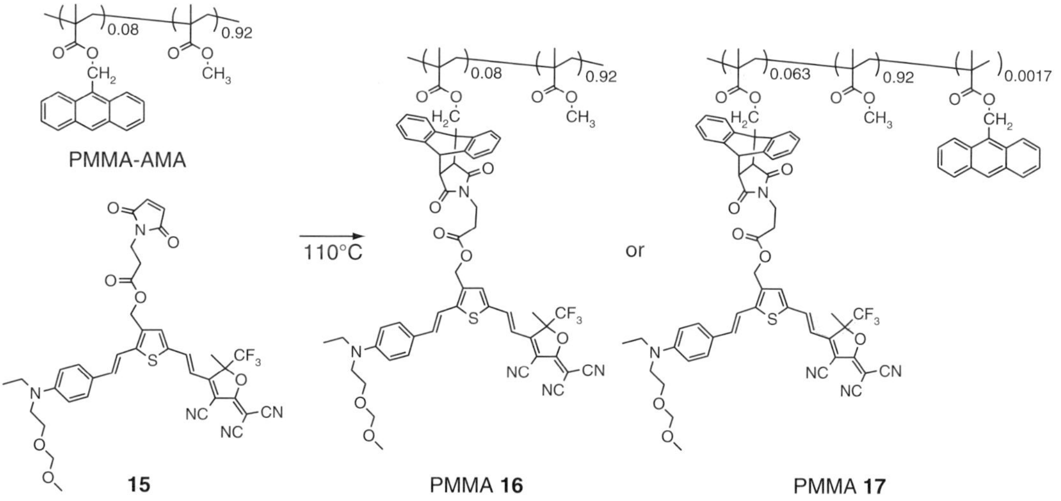

FIGURE 16.10 Synthetic procedures for side-chain NLO polymers PMMA **16** and PMMA **17** by a postfunctionalization, D–A reaction.

The PMMA-AMA was reacted with chromophores functionalized with a dienophile, maleimide, such as chromophore **15** to give side-chain NLO polymers PMMA **16** or **17**, respectively (Figure 16.10) [28].

The chromophore loading density in the polymers can be controlled precisely from the efficient D–A reaction by adjusting the ratio of starting chromophores and fraction of anthracenyl moiety in the PMMA-AMA. For example, the loading density of chromophore in PMMA **16** is 25 wt% compared to 20 wt% in PMMA **17** with left over anthracenyl groups. Relatively high glass transition temperatures of the polymers (152°C for PMMA **16**; 154°C for PMMA **17**) can be attributed to the increased rigidity introduced by the bulky anthracenyl-maleimido D–A adducts.

The measured E-O activities of poled polymers with 20 wt% chromophore loading were 33 pm/V for PMMA **16** and 37 pm/V for PMMA **17** at 1.3 μm. It should be noted that PMMA **16** has a lower r_{33} value (33 pm/V) than PMMA **17** in spite of its higher chromophore loading of 25 wt%. During the poling, films of the polymer were also prone to breakdown above a field of 125 V/μm. It was believed that the strong electrostatic interaction between highly dipolar chromophores significantly attenuates the poling efficiency and dielectric strength of materials in NLO polymers with high loading density. However, chromophores in such polymers can be modified further with bulky substituent before the D–A reaction to prevent the electrostatic dipole–dipole interactions, as demonstrated in later sections. Such molecular scale control over structure and properties provides powerful means to molecular engineer side-chain NLO polymers.

16.3.3 Cross-Linked Polymers

The long-term temporal stability of dipolar alignment is still a major concern in poled side-chain NLO polymers since various short-range segmental relaxations can decrease the polar order over time. The effective concentration of chromophores with polar order in the polymeric E-O materials is further limited by the large inactive fraction of host polymers. Fully functionalized E-O polymers made of chromophore monomers is an approach to increase the concentration, but requires substantial synthetic efforts, and results in a significantly decreased poling efficiency with limited mobility of chromophores.

The most effective method to prevent the relaxation of aligned dipoles in NLO polymers is to align NLO chromophores first and then cross-link either the polymer backbone or the chromophores together. Many studies have shown that such approaches can improve long-term alignment stabilities significantly. However, a reduction of 20%–40% in E-O activity is usually accompanied with this approach, since typical poling of conventional NLO thermoset polymers is achieved through a sequential lattice-hardening and poling processes. As a result, the lattice hardening significantly reduces the orientational flexibility of chromophores due to the continuous increase of glass transition temperature and interchain entanglements of the polymers. The lowered flexibility of polymers severely limits chromophore reorientation under the poling field, resulting in a decreased poling efficiency. In addition, high temperatures used for cross-linking often induce decomposition of highly polarizable chromophores. To be useful for device application, chromophores in such system should have short-term high-temperature stability under the cross-linking conditions and long-term thermal stability at the operating temperature.

To overcome such problems, the lattice-hardening process should be ideally separated from the poling process that requires high degree of rotational freedom of chromophores. In addition, since most of the highly efficient NLO chromophores possess only moderate chemical and thermal stability, very mild conditions should be employed for lattice hardening.

Limited uniformity in many cross-linked polymers can also cause significant optical scattering loss of materials. Such problems can be overcome partially by improving the distributional uniformity of the cross-linking reaction between the polymer chains and chromophores. Furthermore, the material should have good photochemical stability, at least at the frequencies of operation. Although some of the requirements have been met individually, integrating all these desirable properties in a single-material system remains as a challenging task.

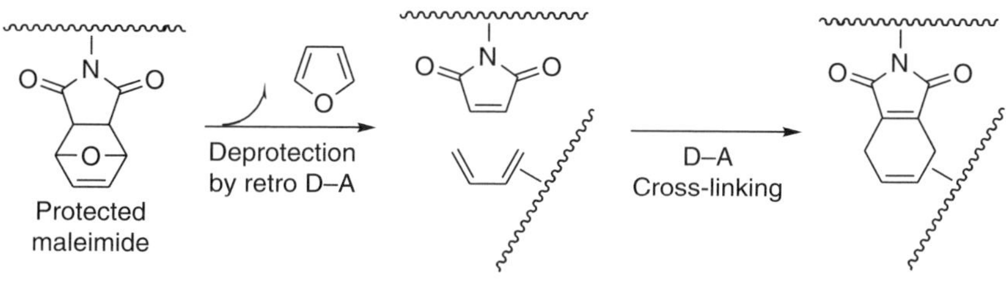

FIGURE 16.11 Reaction sequences of D–A cross-linking (or lattice-hardening) process.

Recently, Jen and Luo have developed an efficient lattice-hardening method using the D–A cycloaddition reaction, introduced in Section 16.3.2 for postfunctionalization of linear polymers [28,29]. The methodology completely reverses conventional thermal processing sequences of cross-linking reactions and provides significant advantages over the other NLO thermosets.

As discussed, some D–A reactions are reversible and can be reactivated thermally to regenerate the starting materials. For example, the retro-D–A reaction has been used to thermally cross-link linear polymers, which are capable of reverting to their thermoplastic precursors by heating (Figure 16.11). The retro-D–A reaction has been applied successfully in cross-linking (or lattice-hardening) process of NLO polymers with both high nonlinearity and thermal stability.

In polymers **18–20**, two different groups of diene moieties were incorporated (Figure 16.12). The furan molecule was used to cap the maleimide (furan is used as a protection group of maleimide). This pre-capping prevents the cross-linking reaction from occurring before the lattice-hardening step. As a result, all of these polymers show excellent solubility in common organic solvents, such as methylene dichloride, THF, and cyclopentanone. By simply heating the polymer to a higher temperature around 110°C, the low boiling point furan can be cleaved by the retro-D–A reaction and easily evaporated from the bulk material.

This thermal deprotecting step has been confirmed by thermal analysis of these polymers. Thermal gravimetric analysis (TGA) of these polymers showed a steep weight loss of 4.5 wt% between 110°C and 150°C, which corresponds to the endothermic peaks that were observed at the similar temperature range using DSC. The isothermal heating of each polymer sample at 110°C for 30 min also resumed the weight loss of 4.5 wt%, a value that is in good agreement with the furan content used for protection. After isothermal heating, the polymer film became completely insoluble

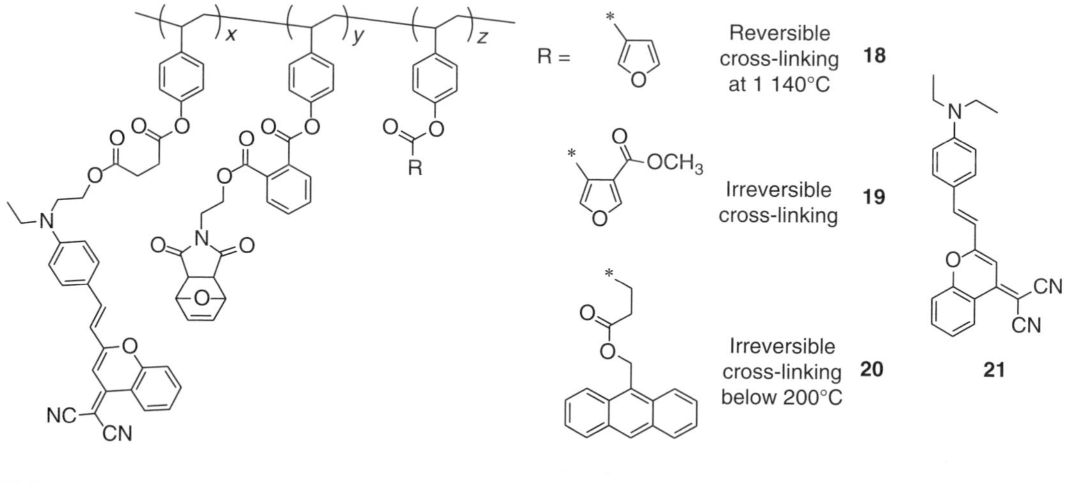

FIGURE 16.12 Molecular structures of side-chain NLO polymers with D–A cross-linkers **18–20** and a chromone derivative **21**.

even though it was rapidly quenched to room temperature. This indicates efficient D–A cross-linking reactions between imide groups in the side chains and the corresponding second diene moiety of the polymers.

To obtain thermally stable adducts for cross-linking, three diene moieties with electronically fine-tuned structures were selected. By adding an electron-withdrawing ester group onto the 3-position of furan (deactivated), the temperature of the D–A reaction of the cross-linkable polymer **18** can be increased to 145°C followed by the initial deprotection of furan. One can also design a thermally irreversible cross-linked polymer **19** by using an even less reactive furan diester groups. Another nonreversible cross-linked polymer can be prepared from polymer **20** with imide and anthranyl groups as the cross-linking moieties.

All these poled films showed good E-O coefficients: 17 pm/V from polymer **18**; 19 pm/V from polymer **19**; and 13 pm/V from polymer **20** at 830 nm, respectively. To evaluate the efficiency of chromophore alignment after the cross-linking followed by the electric field poling process, the macroscopic E-O coefficient of poled and cross-linked polymer from polymer **18** was compared with a typical guest–host system. When 25 wt% of a derivative of chromophore **21** was doped in a PMMA as a host, r_{33} value of 15 pm/V at 830 nm was measured. The value is the same as E-O coefficient measured from the polymer **18**. The result demonstrates that high poling efficiency of a guest–host polymer can be achieved in a cross-linked polymer through careful control of lattice-hardening mechanism during the poling process. Note that the poling efficiency of cross-linked NLO polymers are 20%–40% lower than the guest–host polymers in general. The relatively lower E-O activity from the polymer **20** is thought to be due to the large temperature difference between the poling temperature (75°C) and its intrinsic T_g (110°C) that may cause thermal fluctuation during poling preventing efficient poling process. When the poling field was applied, the deprotection step significantly delayed the cross-linking reactions, since it can only be triggered after sufficient deprotection. This additional deprotection step separates the poling process from the lattice hardening without disturbing the cross-linking reaction. As a result, dense cross-linking can be achieved even at a constant temperature of 110°C. This mild condition, which was reflected by the steady poling currents, can maintain the obtained alignment of chromophores well.

Due to the efficient D–A cross-linking reactions, these polymers also showed very good temporal alignment stability. After initial relaxation, the poled polymer films from polymer **18**–**20** retained 80% of their original r_{33} values for 500 h at 85°C. Relatively high E-O activities and good temporal alignment stability of these polymers demonstrate that one can overcome the NLO activity-stability trade-off in cross-linkable NLO polymers through the separation and control of the electric field poling and cross-linking processes. Moreover, by modifying the electronic properties of the cross-linking reagents, one can fine-tune processing temperature and reversibility of these D–A reactions to optimize thermal stability and processibility.

16.4 POLYMERS WITH DENDRONS AND DENDRIMERS

The r_{33} value of a polymeric E-O material generally increases with increasing concentration of the chromophore and shows saturation behavior as a result of strong dipole–dipole electrostatic interactions [20]. E-O coefficients of guest–host polymers can be significantly improved by encapsulating chromophore with substituents that can electronically shield the core, π-electrons and form spherical molecular shapes. The substituents do not influence molecular hyperpolarizability but prevent the formation of electrostatic dipole–dipole interactions between chromophores along the minor axes of chromophores.

Globular shape of dendrimer is not only well suited to obtain the spherical macromolecular structures of an encapsulate chromophore, but can also be branched into multiple generation of pre-organized collection of chromophores to engineer nanoscale macromolecular NLO architectures with good poling efficiency [30]. Terminal groups in such structures can also be functionalized to control their solubility and processibility, and further improve chemical and photochemical stability of materials. Moreover, one can incorporate various cross-linkers to facilitate lattice-hardening process followed by the electric field poling.

Among many dendrimer architectures, dendrons, or dendritic modification of bulky substituents are especially useful for polymeric NLO materials. For example, the first and second generations of "Frechét-type" dendrons with perfluoro termination have shown to be very useful not only to minimize absorption optical loss from the vibration overtone of C–H groups by C–F groups but also to control other properties such as solubility, processibility, and dielectric property of materials. Polymeric E-O materials based on NLO chromophores with dendritic modification have shown significantly improved poling efficiency even with relatively high number density of chromophores.

16.4.1 Chromophores with Dendrons

The effects of dendritic modification of chromophores have been studied systematically using highly efficient NLO chromophores with a TCF acceptor having different shapes and sizes of substituents [31]. When compared to the *t*-butyldimethylsilyl side group substituted in chromophore **22**, the dendritic substituent on chromophore **24** has a much more flexible and spatially extended shape because of the branching alkyl chains that are connected to the phenyl group. The adamantyl group on chromophore **23** has a more rigid and bulky structure (Figure 16.13). Due to the insulating methylene spacer between the π-electron core and the side group, all these modified chromophores gave essentially the same absorption peak (λ_{max} = 641 nm) in dioxane solution, indicating that these substituents are not affecting the core electronic structure of chromophores.

However, it is expected that these side groups will affect the poling efficiency and the macroscopic susceptibility of polymeric materials that are incorporated with these chromophores. Two aspects of the substituent effects may contribute: one is that the different shape and size of substituents will lead to substantially different intermolecular electrostatic interactions among chromophores; the other is that the different rigidity and size of substituents will also create variable free volume which in turn will affect the mobility of the chromophore under high electric field poling conditions.

To study the substituent effect on poling efficiency, all three chromophores were incorporated into an APC with T_g of 205°C with the same chromophore loading. The resulting polymer composites were spin-coated on ITO glass substrate, poled, and the macroscopic E-O activities were measured in the same condition, respectively.

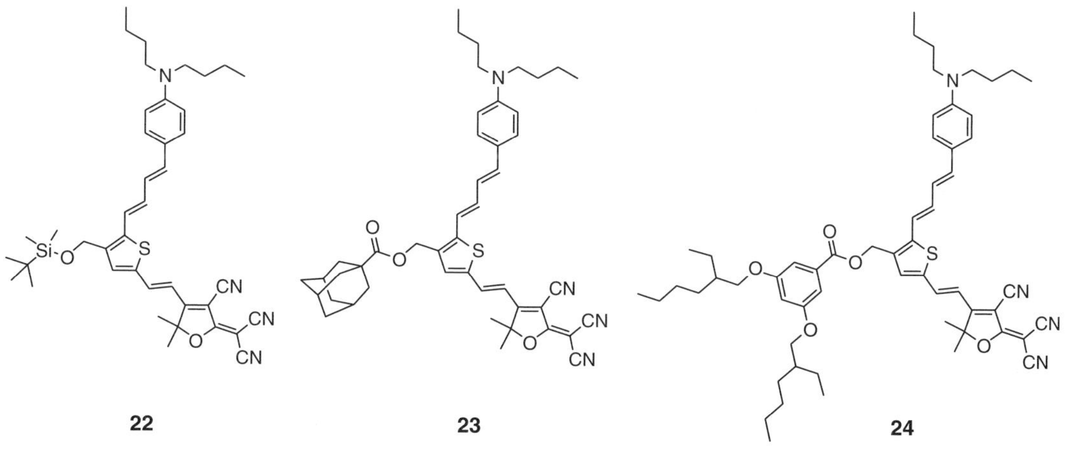

22 **23** **24**

FIGURE 16.13 Molecular structures of shape modified NLO chromophores **22–24**.

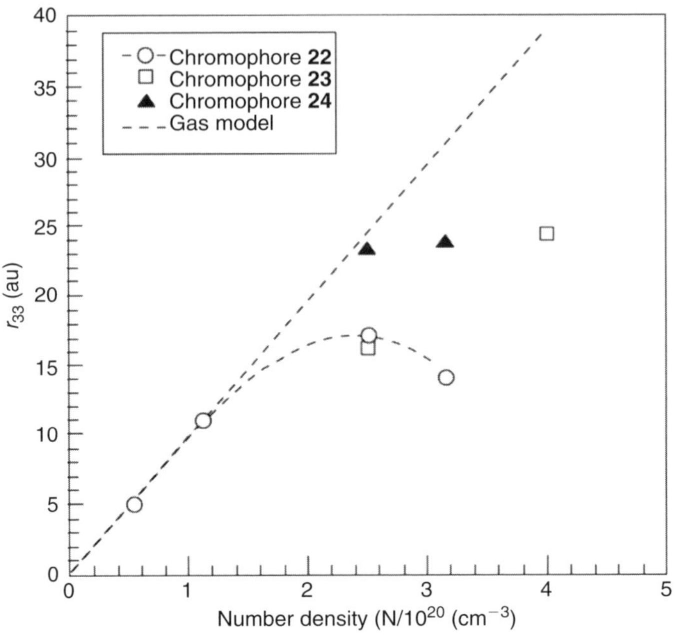

FIGURE 16.14 Comparison on E-O coefficient of NLO polymers incorporated with chromophores containing different loading density and substituents.

Figure 16.14 shows the dependence of the r_{33} value on chromophore number density measured for each system. For the chromophore **22**, it exhibited a maximum value of r_{33} and the value decreased with the increase of the chromophore loading level. In addition, it stayed at a lower r_{33} level than the other two systems, suggesting significantly higher electrostatic dipole–dipole interactions than expected from the bulky *t*-butyldimethylsilyl side group. In comparison, chromophore **23** and **24** enhanced the achievable value to a much higher level (~40% of enhancement) due to the substituents. However, chromophore systems **23** and **24** contributed differently to the enhancement of r_{33} and poling efficiency with respect to similar chromophore number density.

Compared to the *t*-butyldimethylsilyl group on chromophore **22**, the dendritic substituent on chromophore **24** offers notable enhancement for the obtainable r_{33} and poling efficiency at lower chromophore number density, whereas the adamantyl substituents on chromophore **23** provides more significant improvement for r_{33} and poling efficiency only at higher chromophore number density. This is probably due to the fact that the flexible and bulky dendritic moiety on chromophore **24** provides more free volume for the chromophore orientation mobility under high electric field poling compared to the rigid adamantyl group on the chromophore **23**.

Dramatic improvement in the poling efficiency can be achieved through encapsulation of chromophores with multiple fluorinated Frechét-type dendrons [30a]. For example, a poled film of a guest–host APC composite containing 12 wt% of dendronized chromophore **26** exhibits three times the E-O activity (15 pm/V at 1.3 μm) of a film containing the same concentration of the non-dendronized chromophore **25** (5 pm/V at 1.3 μm) (Figure 16.15) [32]. And also, the absorptions optical loss of chromophores can be minimized further by the use of fluorinated dendron. The dendronized chromophore **26** exhibits a blueshifted absorption maximum of 574 nm (from 603 nm in 1,4-dioxane), providing better transparency of the material. The advantageous blueshift of 29 nm in absorption λ_{max} is probably due to the hydrophobic and lower dielectric environment created by fluorinated dendrons around the core of chromophore. The dendronized material is also somewhat more thermally stable with improved dielectric properties.

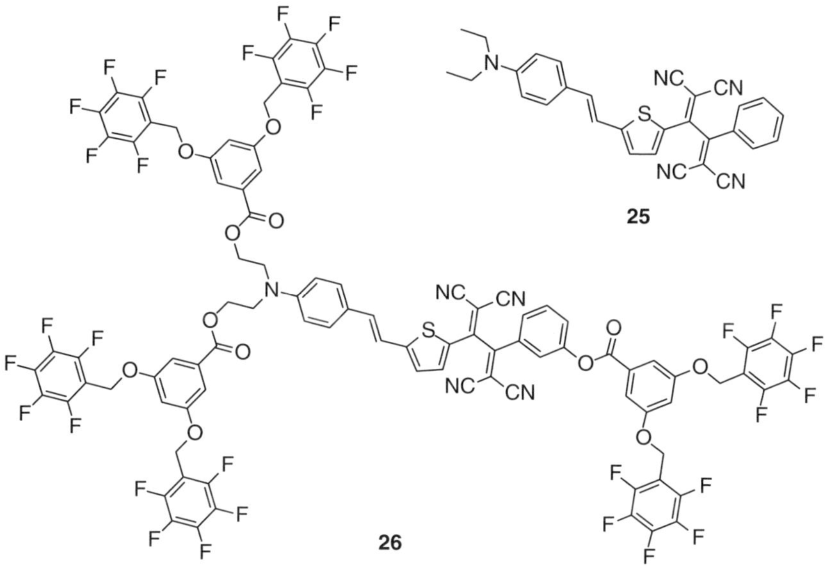

FIGURE 16.15 Molecular structures of nondendronized chromophore **25** and dendronized chromophore **26**.

16.4.2 Polymers with Dendrons

Molecular engineering of NLO chromophores and polymers with several different types of fluorinated dendrons can significantly improve E-O activities and temporal alignment stabilities of polymeric NLO materials. Careful selection and location of the fluorinated dendron in chromophore structures have shown to lower the electrostatic dipole–dipole interactions and improve poling efficiency, even with highly efficient NLO chromophores with larger ground-state dipole moments [32,33]. Moreover, an envelope of fluorinated dendrons encapsulating chromophores can improve the solubility and compatibility of chromophore with host polymers while maintaining excellent dielectric properties of polymeric E-O materials.

Partially fluorinated high-performance polymers with high T_g, such as hexafluoro dimethylene polyimides, can be used as host polymers in guest–host NLO polymers, or can be used as a polymer backbone of side-chain NLO polymers, especially in postfunctionalized side-chain polymers. There are several architectural approaches in dendritic modification of side-chain NLO polymers depend on the location and relative number of dendrons, and synthetic procedures used.

Although several elegant synthetic procedures that can produce dendronized side-chain polymers through the polymerization of dendronized monomers exist, most of the highly polarizable NLO chromophores cannot survive such polymerization conditions due to their chemical sensitivity. Thus, it is desirable to utilize the postfunctionalization approaches. We will focus on side-chain NLO polymer prepared by postfunctionalization of conventional linear polymers here. The polymers can be modified with various cross-linkers to improve temporal alignment stability further, or can also be used as a host polymer for additional NLO chromophores. Moreover, dendritic modifications of side-chain polymers with multiple functional groups and additional guest chromophores can be prepared simultaneously in solid state by postfunctionalization of conventional high-performance polymers.

Figure 16.16 illustrates different molecular architectures of dendronized side-chain NLO polymers with cross-linkers. The diverse selection of molecular architectures provides additional flexibility in the molecular engineering of high-performance polymeric E-O materials.

A series of dendronized side-chain NLO polymers have been recently developed. The polymers take advantage of the synthetic simplicity and processibility of linear polymers and improved poling behaviors of dendronized chromophore and polymers by the encapsulation of chromophore core with dendrons.

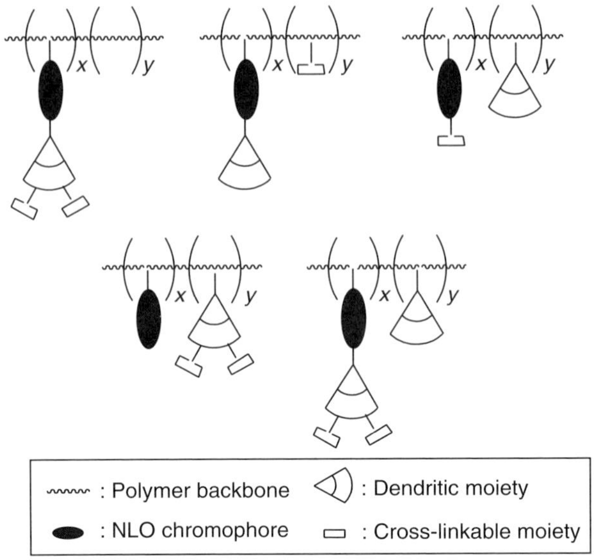

FIGURE 16.16 Molecular architectures of dendronized side-chain NLO polymers with cross-linkers.

A commercially available poly(4-vinylphenol) was condensed sequentially with the dendronized chromophores with acid terminal groups and 4-trifluorovinyloxy benzoic acid to afford the dendronized side-chain NLO polymers. 1,3-dicyclohexylcarbodiimide (DCC) and 4-(dimethylamino)-pyridinium 4-toluenesulfonate (DPTS) were used as condensation reagents. The two steps were proceeded in one pot under very mild conditions. The introduction, distribution, and relative composition of different side-chain groups can be tuned by changing the reagent loading sequences [32].

A side-chain polystyrene (PS) **27**, with a dendronized chromophore, is illustrated in Figure 16.17. The contact poled film of **27** showed a large r_{33} value of 41 pm/V at 1.3 μm. The E-O activity is significantly higher than that obtained from a conventional side-chain PS **28** without dendrons (17 pm/V at 1.3 μm), and a guest–host polymer with core chromophore **29** in PMMA (25 pm/V at 1.3 μm), with the same chromophore and loading level. The E-O coefficients measured from the poled films of PS **28** and the guest–host polymer with core chromophore **29** in PMMA indicate that strong electrostatic dipole–dipole interactions severely limit the attainable E-O activities in conventional NLO polymers without dendronized substituents. Although several other factors, such as conductivity of the samples and different matrix polymers do also affect the E-O activities of the materials, such contributions are relatively small compared to the observed improvement in E-O activities.

In both side-chain polystyrenes **27** and **28**, the same PS backbone and poling electric fields were utilized. However, the r_{33} value of the PS **27** is about 2.5 times higher than that of the PS **28** without dendritic modification of chromophores. Therefore, the high poling efficiency obtained in the poled film of **27** is mainly due to the efficient response of spatially distributed chromophores to the applied poling field by the dendritic modification of chromophores in the PS **27**.

When the same molecular engineering concept was applied to a more efficient chromophore **9** with tetraene bridge, large r_{33} value of 49 pm/V at 1.3 μm was measured from the poled polymer **30**. Again, the macroscopic E-O activity of PS **30** is twice of those reported with a guest–host polymer with the core chromophore with same loading density (Figure 16.18) [32].

Therefore, one can achieve dramatically increased poling efficiency and very large E-O coefficient in a side-chain polymer with rational molecular engineering of chromophore and polymer backbone with fluorinated dendrons. Nevertheless, even with fully dendronized chromophores in the polymer, the acentric alignment order can decay rather rapidly upon heating if the polymer backbone has low T_g.

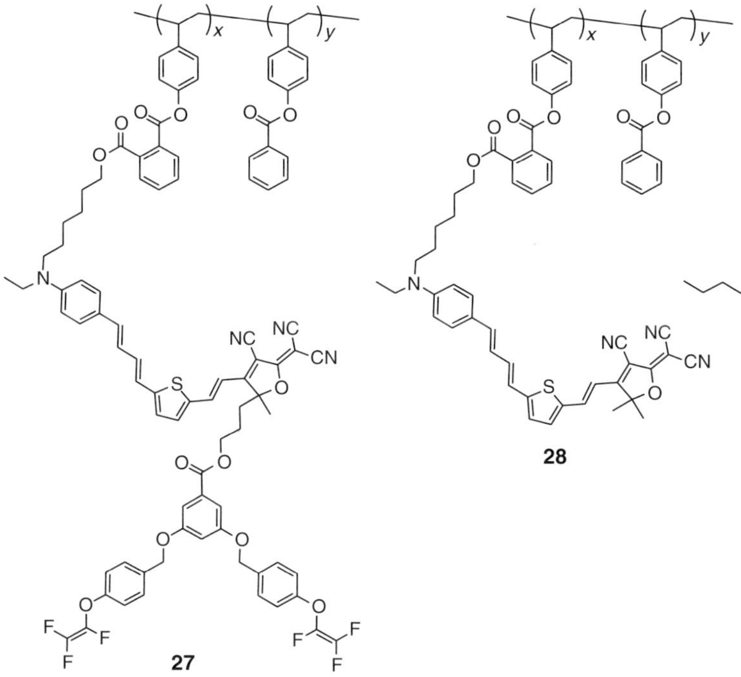

FIGURE 16.17 Molecular structures of dendronized side-chain NLO PS **27** with partially fluorinated dendrons, conventional side-chain PS **28**, and the core chromophore **29**.

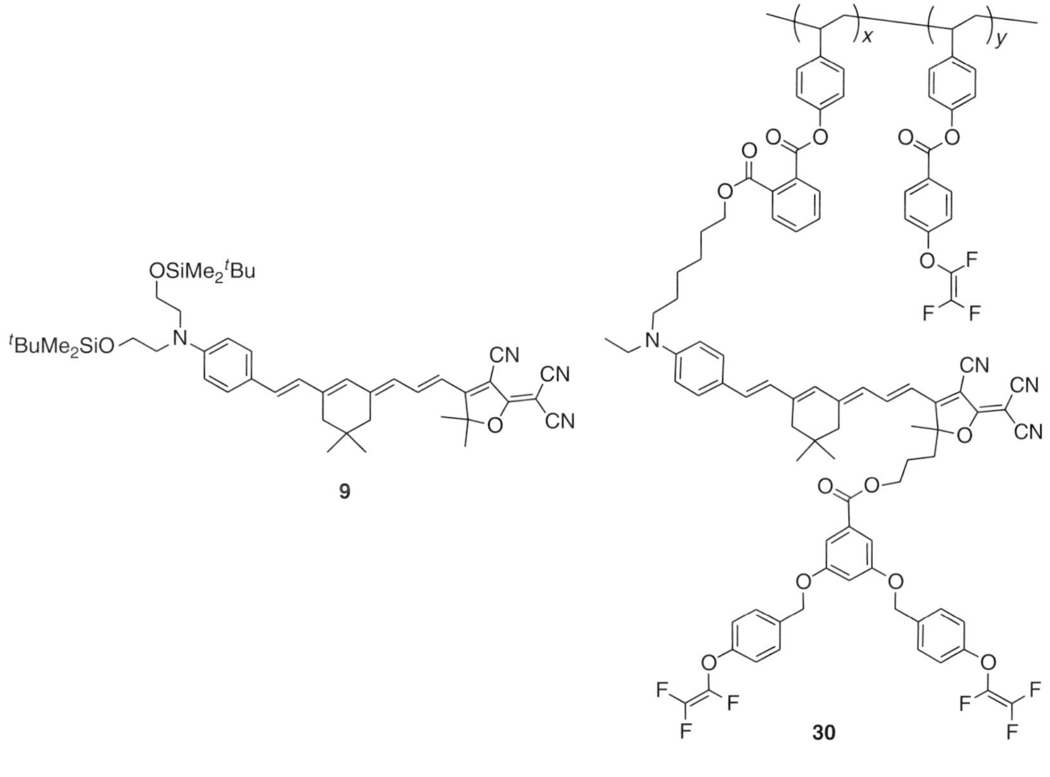

FIGURE 16.18 Molecular structures of the core chromophore **9** and side-chain polystyrene **30** having chromophore modified with partially fluorinated dendron.

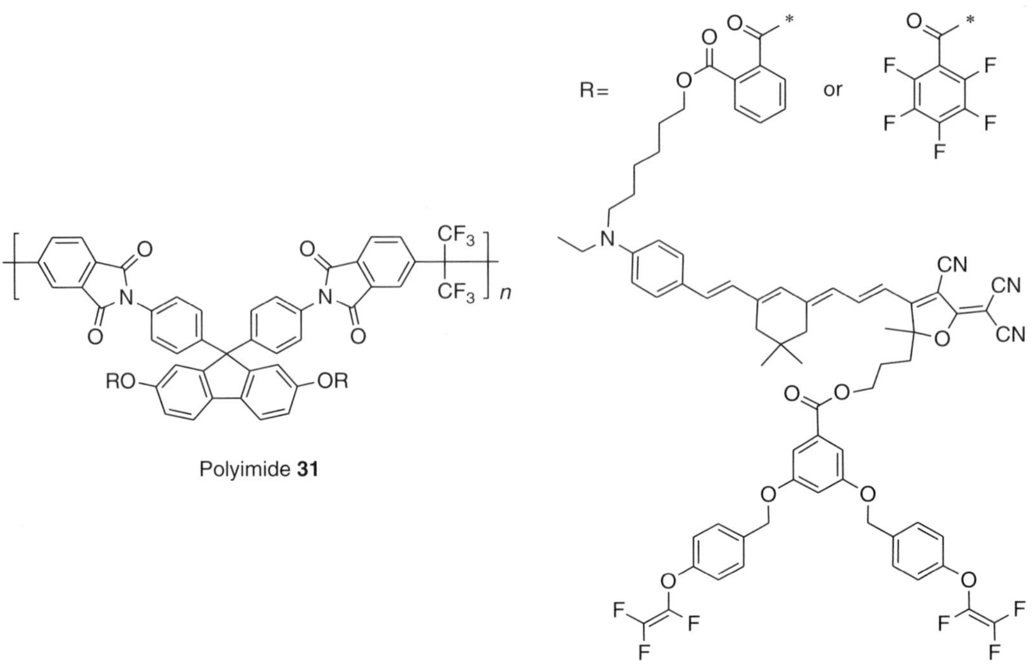

FIGURE 16.19 Molecular structure of a high T_g aromatic side-chain polyimide **31** with dendronized NLO chromophores as pendant groups.

High T_g polymers such as polyimides can be considered as a backbone in the dendronized side-chain polymers to improve their thermal stability further. However, most of the NLO polyimides suffer from low poling efficiency due to intrinsic chain rigidity and semicrystalline nature of inter-chain interactions between aromatic groups in the polymer backbone. Recently, a high T_g aromatic side-chain polyimide with dendronized NLO chromophores as pendant groups, PI **31**, was reported. The polymer was prepared by the postfunctionalization of a bisphenol linkage of the rigid aromatic polyimide backbone (Figure 16.19). Polyimides with such structure are known to have very high T_g, excellent thermal stability, and good solubility.

It has been observed that highly polarizable chromophore **9** is highly solvatochromic, and shows significant changes in the UV/vis absorption maxima in various polymer matrices: 657 nm in PMMA, 667 nm in APC, and 698 nm in a polar PQ-100, respectively.

Since the PI **31** is a side-chain NLO polymer with dendrons in a polar and rigid polyimide, it should provide a different microenvironment to the core chromophore compared to other flexible polymer host, such as polystyrene. However, the UV/vis spectra of the thin films of PS **30** and PI **31** exhibit similar absorption maxima at 711 and 708 nm, respectively. The spectral comparison suggests that the microenvironment around the chromophores is dictated by the local surrounding of chromophore created by fluorinated dendrons and independent of the polymer matrix used [33b].

The E-O activities of PI **31** with different chromophore loading, and the detailed comparison of its poling parameters with those of polymers with lower T_g, are critical for evaluating the role of the dendrons in affecting the poling efficiency of PI **31**. The optimum poling condition of the film was achieved at 155°C with a DC electric field of 85 V/μm, and the measured r_{33} value was 36 pm/V at 1.3 μm. Importantly, when compared with other flexible polymers with lower T_g such as PS **30**, comparably high poling efficiency has been reproduced in a rigid polymer PI **31** with high T_g. As expected, the PI **31** showed improved alignment stability compared to that of PS **30**. Moreover, the poled PI **31** retains higher than 90% of its original r_{33} value after more than 600 h at 85°C. For comparison, the PS **30** with a much lower T_g (90°C) showed only 37% of its original r_{33} value after heating at 70°C for 144 h. The spatial molecular arrangement in PI **31** with rigid backbone structure may suppress strong interchain interactions of polyimide and reduce potential phase separation between chromophores and polymer chains.

Although the temporal alignment stability of side-chain polymer based on high T_g PI **31** can be optimized, it is highly desirable to covalently incorporate chromophores into polymer networks and harden the matrix through cross-linking reactions to improve both their thermal and mechanical properties further. The D–A cross-linking reactions discussed earlier have been applied to dendronized side-chain polymers recently [33].

Following a synthetic method introduced in Section 16.3.2, three different functional groups, including the derivatives of the polyene chromophore, capped maleimide (dienophile), and furan derivative (diene), were sequentially attached to the polymer backbone as side chains to give a cross-linkable PS **32**. The maleimide was protected with furan to prevent cross-linking reaction before film fabrication and poling process. The polymer **32** has good solubility in common organic solvents, such as chloroform and THF. The chromophore loading was 15 wt% in the PS **32** (Figure 16.20).

After the removal of furan, the polymer sample became completely insoluble indicating the ease of D–A cross-linking reaction between the side chains of maleimido and furanic groups of the PS **32**. However, this cross-linked network can be dissociated thermally when the thermally quenched sample was heated again. The PS **32** showed a typical glass transition behavior around 100°C. The highest temperature used in material processing was 125°C. Under the conditions, the trifluorovinylether groups neither self-condense within a limited timescale, nor interfere with the D–A cross-linking reaction as a dienophile. Trifluorovinylether group is known to be inactive for D–A reactions.

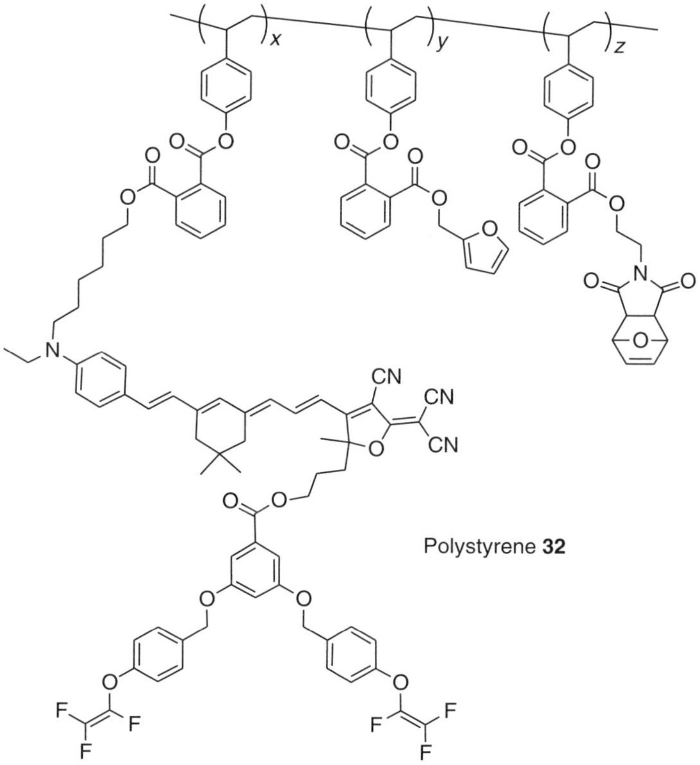

FIGURE 16.20 Molecular structure of side-chain polystyrene **32** with D–A cross-linker and fluorinated dendron.

For E-O measurements, a special sequence described in Section 16.4.1 was used for the polymer. The polymer films were baked under vacuum at 85°C overnight to ensure removal of the residual solvent, and then baked under nitrogen at 125°C for 30 min to evaporate the furan protecting group. A thin layer of gold was then sputtered onto the films as the top electrode for performing the poling experiments. Since the polymer has been cross-linked upon cooling, the films were heated again at 120°C for 1 h to revert to its linear thermoplastic precursor. The film was then cooled down to 100°C and poled at this temperature with electric field. The poling temperature is close to the onset temperature of the retro-D–A reaction (110°C) and higher than the typical temperature range used for D–A cross-linking reaction (60°C–80°C). At this temperature, the material shows the characteristics of a typical thermoplastic polymer. After the poling, a sequential cooling and curing process (85°C, 75°C, and 65°C for 1 h) was performed to anneal and cross-link the polymer by the D–A reaction. All of the above conditions are mild enough to sustain the chemically and thermally sensitive polyene chromophores in PS **32** during the multistep processing sequence.

An r_{33} value of 38 pm/V was measured at 1.3 μm for the PS **32**. Considering that the chromophore content is only 15 wt% in the polymer, poling efficiency is comparable to the dendronized side-chain NLO PS **30** discussed earlier. More importantly, the lattice hardening for the PS **32** is again separated from its poling process, and the poled films can be effectively hardened at temperature far below the poling temperature. As a result, the polar order of chromophores can be maintained during the cross-linking process, leading to a high E-O activity even after the lattice hardening. As expected, the cross-linked polymer also showed good temporal alignment stability. The poled films of PS **30** retains ~80% of its initial r_{33} value for 250 h at 70°C.

16.4.3 Dendrimers

Dendrimers are a relatively new class of macromolecules different from the conventional linear, cross-linked, or branched polymers [30]. Dendrimers are particularly interesting because of their nanoscale dimensions and their regular, well-defined, and highly branched 3-D architecture. In contrast to polymers, these new types of macromolecules can be viewed as an ordered ensemble of monomeric building blocks. Their tree-like, monodispersed structures lead to a number of interesting characteristics and features: globular, void-containing shapes, and unusual physical properties.

Compared to conventional NLO polymers, the screening effect provided by the peripheral groups of dendrimer allows the chromophores to be spatially isolated, and the large void-containing structures of dendrimers provide the needed space for efficient reorientation of the chromophores. Furthermore, the globular geometry of dendrimers is ideally suited for the spherical shape modification of chromophores.

To study the effect of dendrimers on NLO properties, a series of dendritic macromolecules, such as the azobenzene-containing dendron **33**, have been developed and their conformational and molecular NLO properties were studied [34]. These molecules were constructed by introducing 15 numbers of azobenzene-branching units as the NLO chromophore and by connecting these units with aliphatic chains at the end of dendritic structures. In such topology, each chromophore can contribute coherently to the macroscopic NLO activity. The measured molecular hyperpolarizability β of the azobenzene dendron with 15 chromophore units was 3010×10^{-30} esu by the HRS method. This value is approximately 20 times greater than that of the β for the individual chromophore (150×10^{-30} esu). From the polarized NLO measurement, the structure of the dendron was suggested to have a cone-shaped dipolar bundle of chromophores rather than a spreading or a spherical shape with chromophores oriented noncentrosymmetrically along the molecular axis (Figure 16.21).

A stand-alone dendrimer of multiple NLO chromophores that branch out from a passive core can be constructed with predetermined chemical composition. In such structure, the terminal functional groups can be an anchor point to build next generation (layer) of multiple NLO chromophore building blocks, or functionalized with cross-linkers for the postpoling lattice-hardening processes to improve temporal stability of poled dendrimer further. Ideally, the dendrimer will have required

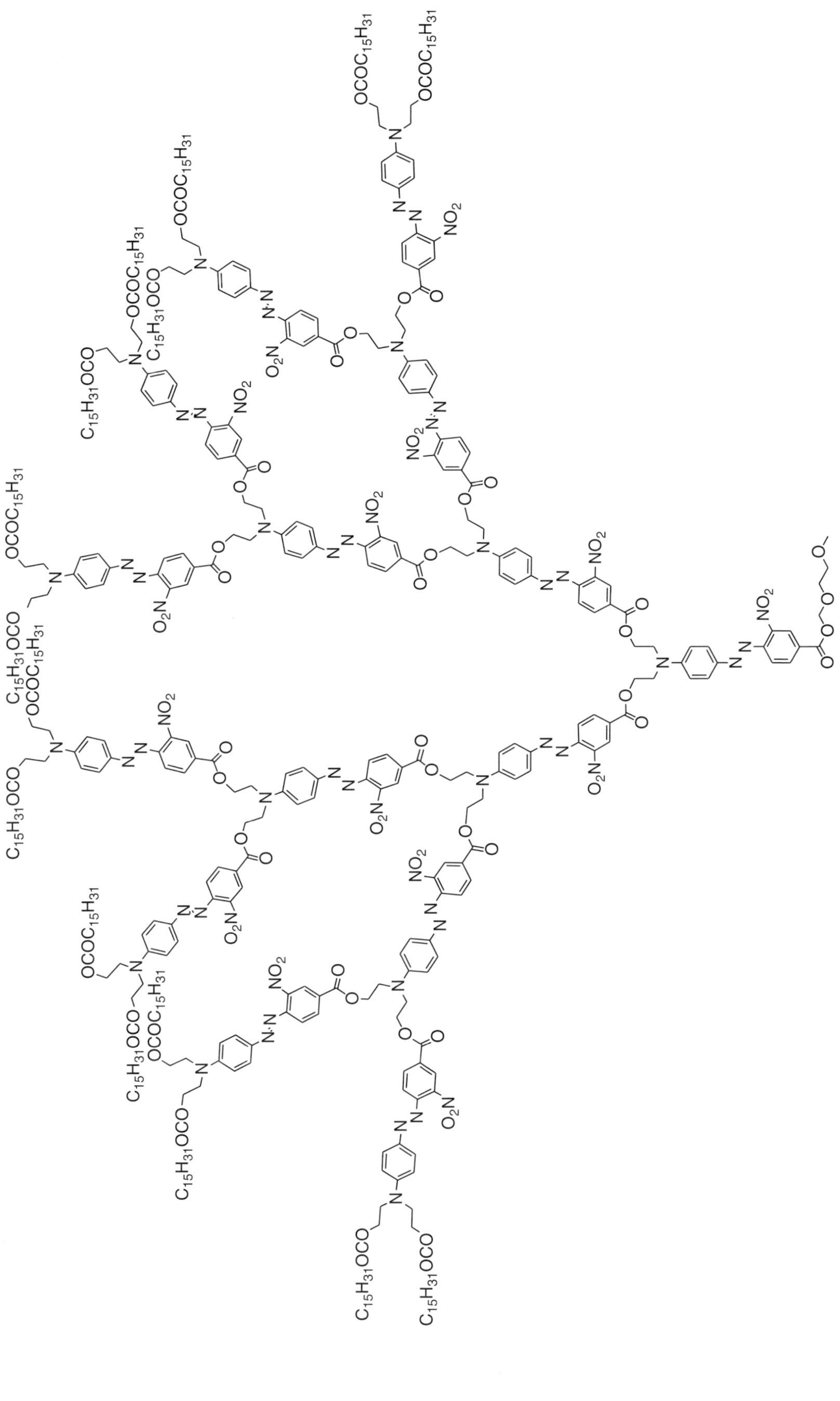

FIGURE 16.21 Molecular structure of a dendrimer **33** in a spreading shape.

thermochemical and physical properties with improved poling efficiency and temporal alignment stability after electric field poling. Moreover, the controlled intramolecular electrostatic dipole–dipole interactions can even provide spontaneous noncentrosymmetric local arrangements to enhance the poling efficiency further.

For example, a cross-linkable NLO dendrimer **34** has been developed for the purpose, and exhibits a large optical nonlinearity and excellent thermal stability. The NLO dendrimer was constructed through a double-end functionalization of the 3-D Ph-TCBD chromophore **4** as the center core and the cross-linkable trifluorovinylether-containing dendrons as terminal cross-linkers (Figure 16.22) [35].

In dendrimer **34**, the chromophore core is spatially isolated by the fluorinated dendrons that decrease electrostatic interactions between chromophores, and enhances macroscopic optical nonlinearity. In addition, the NLO dendrimer can be directly spin-coated without any prepolymerization process to build up viscosity, since it already has a fairly high molecular weight (4664 Da). The chromophore loading density of 33 wt% of the dendrimer **34** is rather high also, and showed no indication of any phase separation due to the incompatibility between the chromophore and the peripheral dendrons. There are several additional advantages of the approach, such as excellent alignment stability and mechanical properties, which are obtained through the sequential cross-linking reactions during the high-temperature electric field poling process.

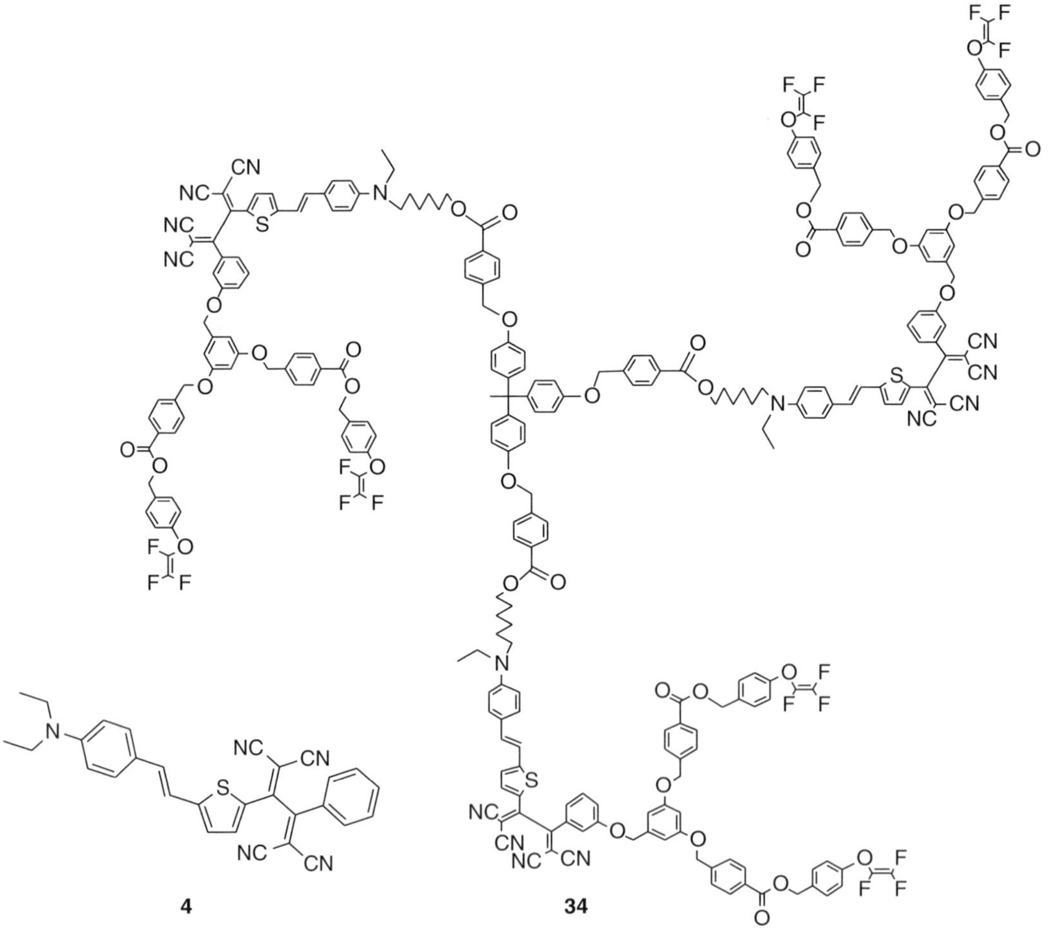

4 34

FIGURE 16.22 Molecular structure of dendron **34** (M_w = 4664 Da, chromophore density: 33 wt%).

A large E-O coefficient (r_{33} value of 30 pm/V at 1.55 μm) and temporal alignment stability (retaining >90% of its initial r_{33} value for more than 1000 h at 85°C) were achieved for the poled dendrimer. In comparison, E-O studies performed on the guest–host polymer film of chromophore **26** (a derivative of chromophore **4**) (30 wt%) in PQ-100 showed a much smaller r_{33} value of <15 pm /V, and lower temporal stability. It retained <65% of its original r_{33} value at 85°C after 1000 h. The T_g of the resulting system is plasticized to approximately 165°C. In addition, the attempt to corona pole dendrimer without cross-linker group, trifluorovinylether, showed a very fast decay of E-O activity, less than 5 pm/V, at room temperature followed by the electric field poling. Such relaxation behavior was attributed to the intrinsically low T_g (<50°C) and large free volume of the dendrimer.

On the basis of these results, the improvement in the r_{33} value of the poled dendrimer was largely attributed to the shielding effect of fluorinated dendrons preventing the electrostatic dipole–dipole interactions between core chromophores. On the other hand, the high temporal stability of the poled dendrimer was mainly attributed to the efficient sequential poling and lattice-hardening processes.

In summary, polymeric NLO materials modified with fluorinated dendrons can provide opportunities for the simultaneous optimization of macroscopic electro-optic activity, thermal stability, and optical loss, compared to conventional NLO polymer. Various fluorinated dendrons can be incorporated into diverse classes of side-chain NLO polymers by postfunctionalization of conventional polymers that can be modified further with efficient cross-linkers for novel lattice-hardening processes. In the dendronized NLO polymers, the interaction between chromophores and high T_g aromatic polymer should be noted in addition to the inter-chromophore electrostatic interactions. Moreover, the unique nanoscale environment created by the shape and size, dielectric properties, and distribution of chromophores in cross-linkable polymers with dendrons and dendrimers can all play critical roles in maximizing the macroscopic E-O properties of polymeric NLO materials.

16.5 MOLECULAR GLASSES

16.5.1 Molecular Glasses and Self-Assembly of Chromophore Glasses

NLO materials based on molecular glasses represent a new field of research especially in photorefractive materials [36]. Many highly active NLO chromophores with conformational flexibilities and sterically demanding hydrophobic substituents form monolithic molecular glasses. Few E-O materials based on molecular glass of NLO chromophores have been demonstrated recently [37]. As in the case of polymeric E-O materials, the acentric polar order of dipolar NLO chromophores can be generated in molecular glasses by electric field poling. In fact, an example of E-O materials based on molecular glasses has been already introduced as dendrimer **34** in Section 16.4.3. However, amorphous phase of monolithic molecular glasses based on small molecules are not thermodynamically stable at high temperature in general, and do not have physical properties suitable for thin-film fabrication. One could spin-coat optical quality films of the dendrimer on ITO glass without supporting polymers, and pole the films using strong electric field, only because of the large molecular weight (4664 Da.) with dielectric strength and relatively high glass transition temperature of the dendrimer **34**.

The most important advantage of monolithic molecular glasses is that the active volume fraction of chromophore can be maximized without phase separation that occurs in the chromophore dispersed polymer composites, especially with very high loading concentration of chromophore ($N > 2.5 \times 10^{20}$ cm^{-3}). And also, monolithic molecular glasses are expected to have reproducible physical and optical properties with well-defined molecular structures of chromophores. The lack of free volume in molecular glasses can interfere with poling process, but the temporal alignment stability of poled molecular glass materials can be improved at the same time if the polar order can be stabilized by supramolecular self-assembly (Figure 16.23).

There are many noncovalent intermolecular interactions that can be used for the construction of the supramolecular self-assembly of chromophore glasses based on the lesson learned from crystal engineering of NLO crystals. They include electrostatic interactions between charges, multiple

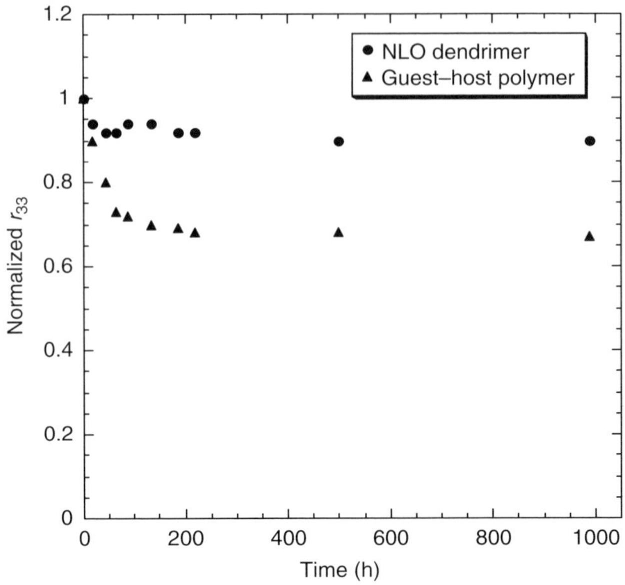

FIGURE 16.23 Temporal stability of the postpoling cross-linked NLO dendrimer **34**, and a guest–host polymer in PQ-100 at 85°C in nitrogen. The r_{33} values were normalized as a function of time.

hydrogen bonds, metal–organic ligand coordinations, π–π aromatic interactions, and van der Waals interactions between functional groups of chromophores and chromophore themselves. Some or all of the interactions can be simultaneously used in the design of self-assembly of amorphous chromophore glasses. We will focus on supramolecular self-assembly of NLO chromophore glasses based on the π–π aromatic interactions and van der Waals interactions in this section to demonstrate the potentials of such approaches.

One of the most interesting π–π aromatic interactions for the self-assembly of chromophore glass is the arene–fluoroarene ($Ar–Ar^F$) interactions. Since organic materials based on the $Ar–Ar^F$ interactions are not widely known yet, we will explain the interaction in more detail here. Benzene and hexafluorobenzene are known to co-crystallize with a melting point more than 18°C higher than either component in nearly parallel alternating molecular stacks with an interplanar distance of 3.4–3.7 Å [38]. The stabilization energy of the benzene-hexafluorobenzene dimer from both computational and experimental studies was estimated to be between 4 and 5 kcal/mol. The interaction can be rationalized mostly based on the complementary quadrupole moments of benzene (-29.0×10^{-40} Cm^{-2}) and hexafluorobenzene ($+31.7 \times 10^{-40}$ Cm^{-2}) that are similar in magnitude but opposite in sign [39]. The formation of parallel alternating infinite molecular stacks is a common structural feature of all molecular $Ar–Ar^F$ complexes reported (Figure 16.24).

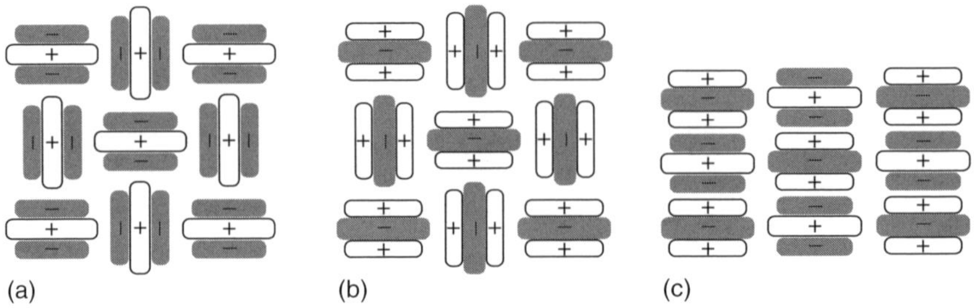

FIGURE 16.24 The $Ar–Ar^F$ interactions in the crystal of (a) benzene; (b) hexafluorobenzene; and (c) alternating stacks of benzene-hexafluorobenzene co-crystal.

The Ar–Ar^F interaction attracted growing interests in diverse fields of molecular designs [40], and is now being accepted as a reliable synthon in crystal engineering [41]. Moreover, the Ar–Ar^F interaction has been used in recent years to control the supramolecular organizations in reversible media such as liquid crystalline phases and the hydrogel formation of polymers and oligomers [42].

It is interesting to ask whether the Ar–Ar^F interaction could be used in construction of a reversible glassy supramolecular self-assembly of NLO chromophores that can be poled and cross-linked into a robust extended network of NLO chromophores with improved material performance as demonstrated with dendrimer **34** earlier. For E-O material application, the self-assembly unit should be designed to be sufficiently strong to form a reliable and reproducible extended structure using mild solution processing, but flexible enough to be reorganized with control using minimal external energies (electric field), as in the case of reversible self-assembly of many natural biopolymers based on multiple stereo-specific, but reversible, intermolecular noncovalent interactions. Furthermore, it is an intriguing thought to design a spontaneous supramolecular self-assembly of NLO chromophores with acentric order by modulating the balance between competing dipolar and quadrupolar electrostatic interactions between functionalized chromophores based on monolithic chromophore glasses.

As discussed, systematic studies have shown that the structural modification of chromophores and polymers with fluorinated dendron can lead to polymeric E-O materials with excellent material properties. Much of the improvement in material performance is due to the use of fluorinated dendrons that have low optical loss, good solubility and processibility, and excellent thermal stability with dielectric strength. Initially, research interests in the field were in understanding the effect of fluorinated dendrons as substituents to improve material properties through the shape modification of chromophores, but later it extended to understand intermolecular interactions between chromophores having different number of fluorinated dendrons with different molecular structures and location within chromophores.

Especially, understanding the effect of inter- and intramolecular interactions between electron-deficient fluorinated dendrons and electron-rich donor of chromophore affecting electronic structure of chromophores (bathochromic shift) and their aggregation behavior in the solid state was one of the main concerns. If the electron-deficient fluorinated dendrons can interact with electron-rich donor groups in neighboring chromophores by intermolecular interactions, then the interactions can be used not only to prevent the electrostatic dipole–dipole interactions between chromophores, but also to pre-organize or self-assemble chromophores into extended aggregate structures (J-aggregate) that can respond to external electric field efficiently during poling process.

The approach to the self-assembly of chromophores based on the π–π electron interactions between fluorinated dendron and electron-rich donor of chromophore has been extended to self-assembly of chromophores functionalized with electron-rich dendrons and electron-deficient dendrons both at the diagonal ends of the same chromophore. Reversible self-assembly of monolithic chromophore glasses based on the intermolecular Ar–Ar^F interactions can be generated from bifunctional chromophores depending on the substitution patterns of functional dendrons. The molecular material design approach is general enough not only to apply to multiple chromophore systems (multiple chromophore solution) with the same set of dendrons, but also can be expanded to a reversible and physical postfunctionalization method of conventional polymers, and used to reversibly and physically cross-link side-chain NLO polymers. An example of E-O material based on self-assembly of monolithic chromophore glass will be described in more detail here.

A series of systematically designed NLO chromophore glasses with rational substitution patterns for the self-assembly has been developed recently by Jen and Kim. The molecular structures of the chromophores in the study are shown in Figure 16.25. Both phenyl and pentafluorophenyl rings are incorporated diagonally as peripheral dendrons on the π-bridge and the donor-end in the chromophores **37**, **38**, **39** [43]. Chromophores **35** and **36** are partially modified with dendrons and cannot form an extended supramolecular assembly by the Ar–Ar^F interactions.

No melting point was detected for the series, indicating that they represent noncrystalline phase. The T_g increased with molecular weight. Glass transition temperatures of the chromophore glasses were determined by DSC to be 56°C, 63°C, 57°C, 75°C, and 76°C of **35**, **36**, **37**, **38**, and **39**,

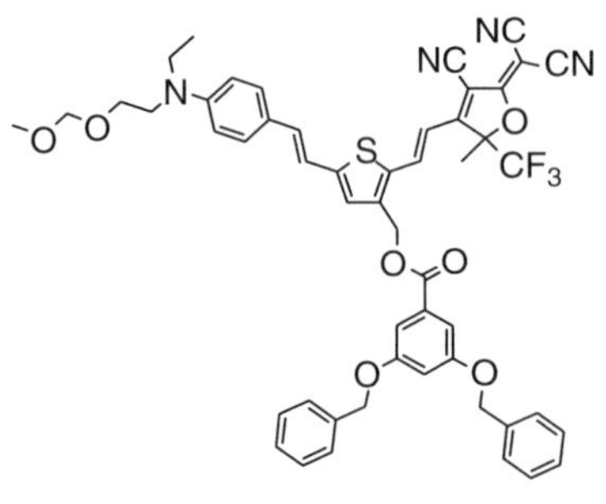

35

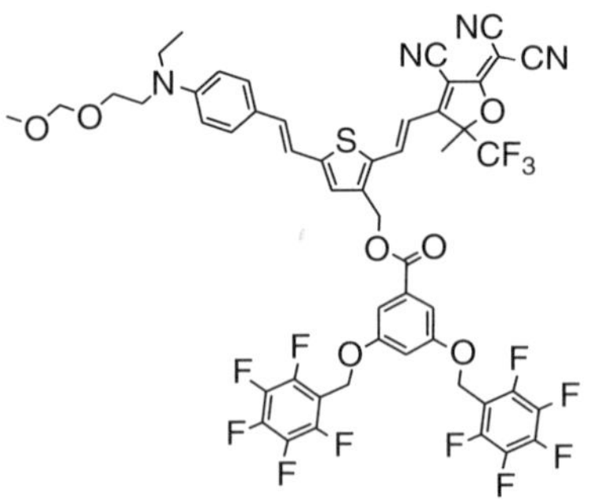

36

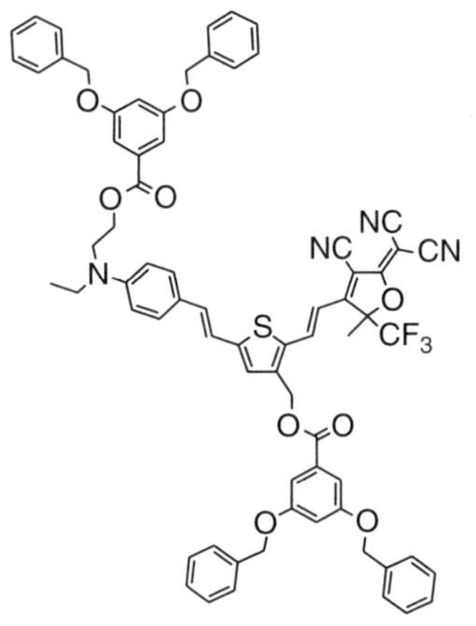

37

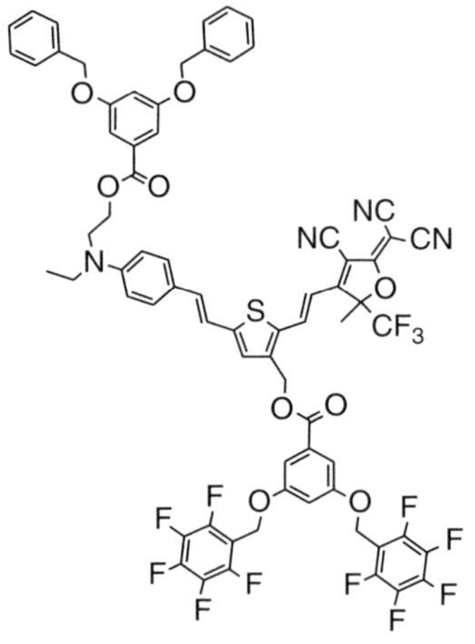

38

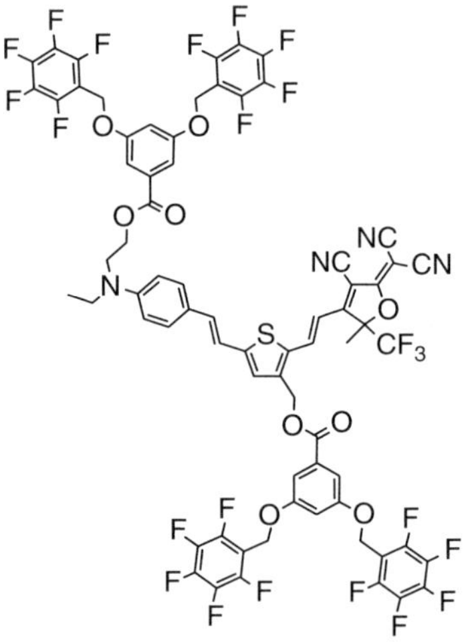

39

FIGURE 16.25 Molecular structure of chromophores used for the reversible self-assembly of supramolecular glasses.

respectively. Especially, as expected from the Ar–Ar^F interactions, chromophore **38** shows a significantly increased T_g (18°C) compared to those obtained from **37**. The thermal transitions are also consistent with the T_g of composite **40,** which is a 1:1 mixture of chromophore **37** and **39**.

The supramolecular self-assembly of engineered chromophore glasses showed very large E-O activities, with good temporal alignment stability. Due to the possible extended structure induced by Ar–Ar^F interactions, chromophore **38** gave the highest r_{33} value of 108 pm/V at 1.3 μm, among monolithic molecular glasses. This value is more than two times higher than those obtained from chromophores **37** or **39** that do not have such interactions. Moreover, the composite **40** (1:1 blend of **37** and **39**) showed an r_{33} of 130 pm/V. Again, the enhanced r_{33} in this composite show that it is even possible to form supramolecular self-assembly between complementary but different chromophores. The fact that high electric fields of >100 V/μm can be applied to pole chromophore **38** and composite **40** suggests improved dielectric properties of chromophore glasses of **38** and composite **40**. On the contrary, the r_{33} values measured for a guest–host polymer doped with 20 wt% of **36** chromophore in APC, and in a postfunctionalized side-chain NLO polymer showed 55 and 52 pm/V using a higher poling field of 130 V/μm, respectively. A graphical illustration of the poling process of the supramolecular self-assembly of chromophore glasses of **38** is shown in Figure 16.26.

These poled thin films could retain over 90% of their original r_{33} values at room temperature for more than two years. On the contrary, the temporal stability of the glasses without the Ar–Ar^F interactions deteriorated dramatically within one month. The decreased absorbance of the poled film of

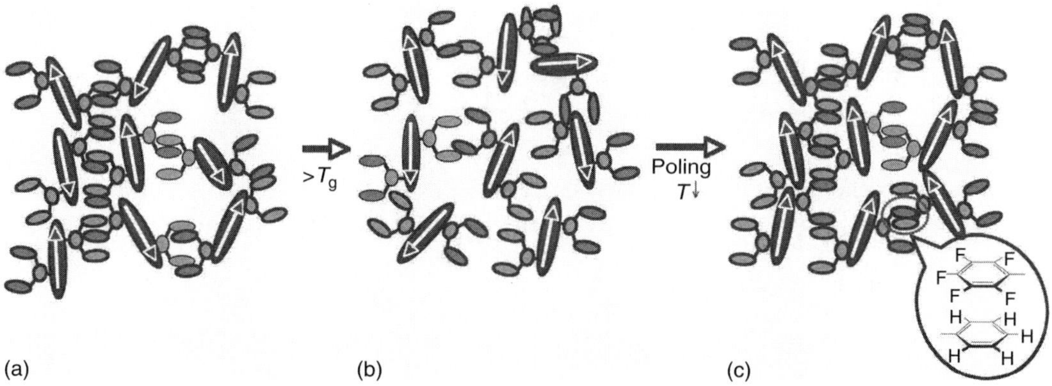

FIGURE 16.26 Graphical illustration of material processing sequences to generate the macroscopic noncentrosymmetric order of supramolecular self-assembly of chromophore glasses by Ar–Ar^F interactions and electric field poling. (a) glassy supramolecular self-assembly of chromophores with random dipoles (shown as arrows) before poling (b) mobile rubbery state with random dipoles before poling (c) glassy supramolecular self-assembly of chromophores with noncentrosymmetric dipoles after poling followed by cooling.

the host **38** could recover more than 95% of its original value after being annealed at a temperature close to its T_g for 20 min. Moreover, the r_{33} values could be reproduced through poling and de-poling cycles demonstrating the stability and reversibility of the supramolecular self-assembly of chromophore glasses.

However, relatively low glass transition properties of these chromophore glasses are not adequate for device application yet, and efforts should be made to improve thermal properties either by increasing the strength of Ar–Ar^F interactions and molecular weights of chromophores with more rigid structures, or by cross-linking the aligned chromophore glasses after poling. Physical and optical properties of **37**, **38**, **39**, and composite **40** are summarized in Table 16.3.

Structural evidence of the Ar–Ar^F interactions in the monolithic glasses has been provided from a single-crystal x-ray structure of dimeric co-crystal of model compounds, phenyl group terminated dendron and pentafluorophenyl group terminated dendron carboxylic acids (1:1). A space filling structure of alternating co-crystal stacks are shown in Figure 16.27. The structure represents a clear solid-state

TABLE 16.3
Physical and Optical Properties of Monolithic Glasses of Chromophores 37, 38, 39, and 40

Materials	T_g^a (°C)	Number Density[b] (×10^{20} cm^{-3})	λ^c_{max} (nm)	Applied Voltage (V/μm)	r^d_{33} (pm/V)	Temporal Stability[e] (%)
37	57	5.0	719	75	52	0
38	70	4.4	703	100	108	92
39	69	3.7	689	75	51	85
40	68	4.4	704	120	130	93

[a] Thermal transition temperature measured under nitrogen at the heating rate 5°C/min.
[b] Core chromophoric moiety.
[c] Absorption maxima of thin films by UV/vis spectroscopy.
[d] EO coefficient measured at 1310 nm by simple reflection technique; etemporal alignment stability at room temperature after three months.

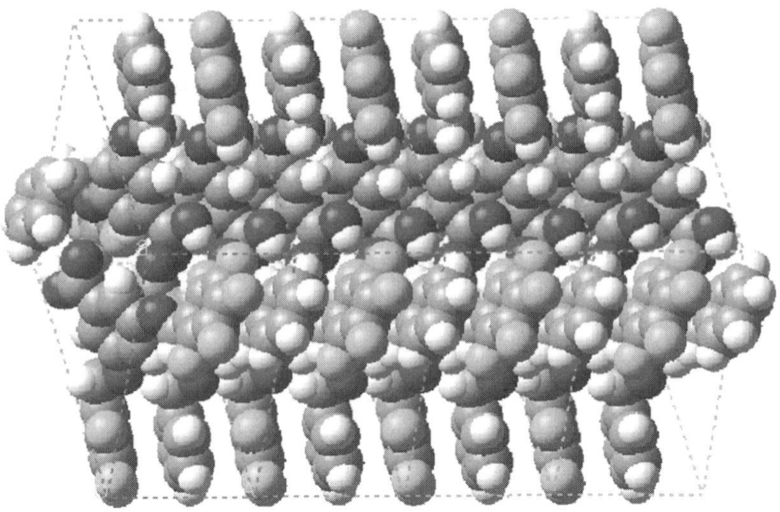

FIGURE 16.27 A space filling structure of alternating co-crystal stacks of hetero-dimer.

evidence of multiple Ar–Ar^F interactions between peripheral dendron groups in forming extended structures when the chromophores are substituted with chemo-specific combination of dendrons.

16.5.2 Molecular Glasses with Multiple Chromophores

As explained, reversible supramolecular self-assembly of monolithic chromophore glasses based on the intermolecular Ar–Ar^F interactions can be rationally designed to tune the aggregate structures of chromophores in the solid state. It has been explained earlier that such molecular material design approach is general enough to be applied to multiple chromophore systems (multiple chromophore solution) with complementary combination of dendron structures. The supramolecular chromophore composite **40** described in previous section was formed from the 1:1 solid solution of **37** and **39**, and the chromophore glass composite illustrate an example of multiple chromophore systems.

In the chromophore glass composite **40**, the core structure of chromophores used was same in **37** and **39**. One could use different pairs of core structures as far as they have complementary dendron structures for the intermolecular Ar–Ar^F interactions if the differences in core structures are not significant enough to phase separate each other or interfere with the Ar–Ar^F interactions between functional dendrons in the solid solution. One could even take combinatorial material approaches to make a series of solid solutions (composites) of two different chromophores with varying fractions. For example, one could make binary supramolecular chromophore glasses with $x:1-x$ mixtures of chromophores of different type Ar–B–Ar^F and Ar–C–Ar^F (chromophores B and C having dendrons Ar and Ar^F). Moreover, one could use more than two chromophores to form a completely new class of supramolecular glasses with multiple chromophores as far as they have complementary dendron structures for the intermolecular Ar–Ar^F, do not form strong centrosymmetric aggregates between multiple chromophores, and do not phase separate each other, or interfere with the Ar–Ar^F interactions between dendrons in the solid solutions.

Eventually, one could even use the monolithic supramolecular chromophore glasses as a host matrix for a small concentration of structurally compatible guest chromophores without dendrons for the Ar–Ar^F interactions as far as the guest chromophores do not disrupt the Ar–Ar^F interactions of host chromophore glasses, and the host matrix maintain the extended supramolecular self-assembly. However, note that the macroscopic E-O activities of such solid solutions will still be limited by the electrostatic dipole–dipole interactions between chromophores and may show the saturation of r_{33} values with extremely high loading density of chromophores in such materials.

Nevertheless, all of the above approaches can be powerful guiding molecular design principles for the development of organic E-O materials based on supramolecular self-assembly of chromophore glasses as long as the r_{33} values continues to increase linearly with chromophore loading density and the materials do not break down electrochemically during the electric field poling process. Furthermore, if one can design molecular glasses based on dendrimer of chromophores with or even without the dendrons, it should be possible to use the dendrimer glasses as a host matrix for multiple combination of structurally compatible guest chromophores. The very general material design approaches could be extended further to other supramolecular glasses based on other noncovalent interactions in place of the Ar–Ar^F interactions between dendrons and fluorinated dendrons. Finally, one could expect to have significant cooperative improvement in poling efficiencies of these materials since the host matrices and guest can both respond to the poling field. The cooperative improvement in poling efficiencies of materials with field responsive host matrices will be used here as matrix-assisted poling [44].

The chromophore **41** is a highly efficient dipolar NLO chromophore that has been used as guest chromophore in guest–host polymer in various polyacrylates and polycarbonates. The chromophore forms monolithic glass by itself and the E-O activity of the monolithic glass of **41** showed very large r_{33} value of 150 pm/V when poled with relatively low poling fields around 25 ~ 40 V/μm. The optimum electric field strength in poling process was limited to less than 50 V/μm due to the high conductivity and low dielectric strength of the molecular glass film at temperature around T_g.

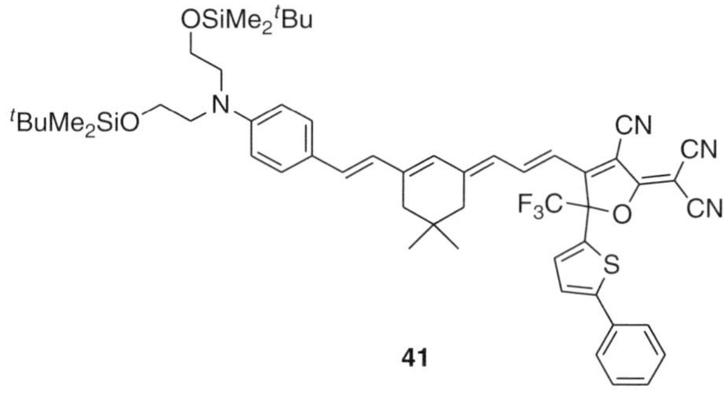

41

Based on the principles of the matrix-assisted poling, the chromophore **41**, which has not modified with dendrons, was doped in the supramolecular self-assembly of chromophore glass of **38** as a host matrix to form a binary composite of chromophore glasses. A series of binary composites of chromophore glasses can be prepared with varying amounts of chromophore **41** in the host matrix glass. The composite **42** was made of 25 wt% of chromophore **41** in the matrix glass of **38**. A similar material design approach of the binary chromophore glass composite based on chromophore doped glass of dendrimer of multiple NLO chromophores was reported by Dalton very recently [45]. A very large r_{33} value of 275 pm/V was measured for the binary composite **42** after usual electric field poling process. More importantly, the E-O activities of these poled composites do not show the saturation maxima even with an extremely high chromophore loading densities ($N \sim 5.0 \times 10^{20}$ cm^{-3}), more than twice of maximized loading density in a guest–host polymer, and can retain ~90% of their initial value for more than 500 h at room temperature. On the contrary, the r_{33} values of a guest–host polymer **43** based on 25 wt% chromophore **41** in PMMA (much higher T_g of 101°C) showed dramatically decreased temporal alignment stability (~60% of initial r_{33} value) at 50°C in 25 h compare to the binary composites **42** (~80% of initial r_{33} value). Such behavior is totally different from the commonly observed fast relaxation of poling induced polar order in a guest–host polymer with low T_g. Physical and optical properties of the binary glass composite **42** and a guest–host polymer **43** based on chromophore **41** is compared in Table 16.4.

The large E-O activities of the binary chromophore glass composites suggest several critical structural facts about the approach: the supramolecular self-assembly of **38** based on the Ar–Ar^F

TABLE 16.4
Physical and Optical Properties of the Binary Glass Composite 42 and a Guest–Host Polymer 43

Materials	T_g^a (°C)	Number Density[b] ($\times 10^{20}$ cm^{-3})	λ_{max}^c (nm)	Applied Voltage (V/μm)	r_{33}^d(pm/V)	Temporal Stability[e] (%)
42[f]	76	4.8	727	90	275	86
43[f]	101	2.5	803	100	150	86

[a] Thermal transition temperature measured under nitrogen at the heating rate of 5°C/min.
[b] Core chromophoric moiety of **41** ($C_{32}H_{28}F_3N_4O$, molecular weight 541.6) counted by total loading weight.
[c] Absorption maxima of thin films by UV/vis spectroscopy.
[d] EO coefficient measured at 1310 nm by simple reflection technique.
[e] Temporal alignment stability at room temperature after three months.
[f] **42**: 25 wt% of **41** in a host matrix glass **38**, **43**: 25 wt% of **4** in a PMMA.

interactions is robust enough not to be disturbed by the addition of up to 50 wt% of completely foreign guest chromophore; the poling efficiency of guest chromophores in glass composites are significantly improved by the matrix-assisted poling mechanism; the measured r_{33} values are the sum of r_{33} values from guest–host chromophores; the dielectric properties of the binary glass composites were mainly dictated by the host self-assembly so that the poling field as high as 90 V/μm could be applied to the composites.

The improved poling efficiencies in the glass composites were supported further by the measured order parameters based on linear optical dichroism of intramolecular charge-transfer bands. The shapes of the absorption spectra of all films before and after poling were similar, indicating that no chemical degradation occurred during poling. The order parameters were 0.13 for the guest chromophore glass of **41** and 0.17 for the host glass of **38** individually. The order parameters were increased to 0.18 for guest chromophore **41** and 0.25 for host chromophore **38** in the binary chromophore glass composites **42**.

16.6 DEVICE APPLICATIONS

In this section, we will introduce device applications of polymeric E-O materials with practical advantages over inorganic crystals, such as $LiNbO_3$. Readers are encouraged to review chapters in this text book on optoelectronic devices, device applications of second-order NLO materials, and read recent reviews on the subject [6a,46]. This section is not intended to give a comprehensive review on the operation principles of all device applications of polymeric NLO materials, but rather to explain how the advances in material properties of polymeric E-O materials can be related to the realization of many promised device applications.

The focus will be on material property requirements and issues for device applications and ways to address the requirements in the design of polymeric NLO materials through rational molecular engineering of their structures. Such exercise provides valuable feedbacks not only for the molecular engineering of chromophores and polymers, but also for optimizing material properties specific to application later. We will introduce basic principles of optical modulators based on polymeric E-O materials, and summarize recent research activities in device applications. The generation and detection of terahertz (THz) radiation using polymeric E-O materials will be introduced briefly as a new area of research in the field.

For telecommunications applications, polymeric E-O materials are especially attractive than inorganic crystals such as $LiNbO_3$ (30 pm/V at 1.3 μm). The larger E-O coefficients of polymeric E-O materials not only expected to lower the driving voltages of devices, but also provide additional flexibility in the design and fabrication of various device structures. For example, conformal and flexible devices based on polymeric E-O materials can be fabricated by soft lithographic techniques with robust material properties that can be sustained under extreme physical conditions. As discussed, the potential for very broad bandwidths is the most important advantage of organic materials for device applications where light of multiple frequencies must be processed. However, there are significant drawbacks of organic E-O materials compared to lithium niobate that include relatively poor thermal and photochemical stability, and potentially higher optical losses.

Electro-optical modulation is an important field for applications of second-order NLO materials for signal transmission, signal processing, and optical interconnections. Ranges of efficient E-O modulators based on poled polymers with various configurations suitable for applications in integrated optics have been demonstrated now. These modulators can be fabricated easily and operate in a transmission band >100 GHz with half-wave voltage <1 V in waveguide configurations. Polymeric E-O modulators based on integrated silicon photonics have also been demonstrated.

The basis for many of these devices, and one of the simplest E-O devices to understand, is the E-O Mach–Zehnder interferometer. In such a device, light beam is split and each half passes through an E-O material along a different path, before being recombined. If the two path lengths are the same, the two light beams will recombine constructively. However, if one of the paths is subjected to an electric field, the refractive index along that path will be modulated inducing a phase shift of the light beam following that path relative to the other. Hence, the electric field can be used to switch the light beams from a constructively interfering situation to destructively interfering situation, enabling one to convert an electrical signal to an optical signal.

Figure 16.28 shows an example of a polymeric Mach–Zehnder modulator for amplitude modulation. A multilayer structure is placed in one of the arms (sometimes in both arms to get a better control of the relative phase of two beams) and depending on the applied electrode voltage, the phase of the propagating beam is varied. The resultant phase mismatch between two beams leads to the variation of the amplitude, which is controlled by applied voltage to the electrodes. The minimum voltage necessary to create a phase mismatch between both beams equal to π is called the half-wave voltage, V_π. The performances and, in particular, the bandwidth depends strongly on the design of electrodes. Polymeric Mach–Zehnder E-O modulators and their performances have been reviewed recently by Dalton [6a].

As introduced throughout this chapter, many significant advances in material properties of E-O polymers were made recently, and active research activities are ongoing to realize the improved properties in various device configurations not only for traditional optical modulators but also for other NLO devices that could not be realized before with limited E-O activities of earlier materials including inorganic NLO materials. Representative research activities of device application of polymeric E-O materials summarized by Dalton are listed to present a broader direction of the field:

- A variety of novel device structures, including stripline, cascaded prism, and all-organic ring microresonator structures have been fabricated from polymeric E-O materials.
- Polymeric E-O materials have been incorporated into silicon photonic devices including ring microresonators, slotted ring microresonators, and photonic band gap structures.

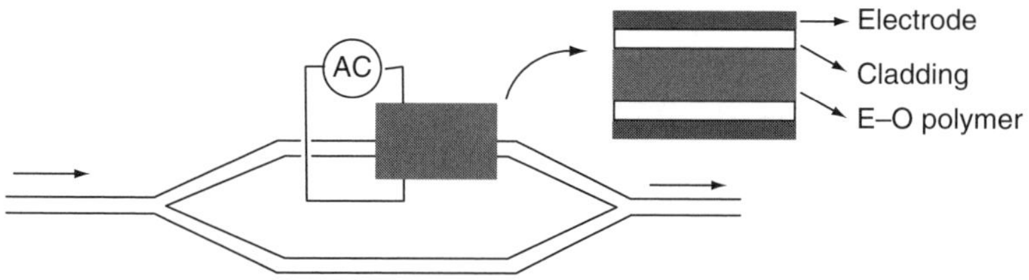

FIGURE 16.28 Schematic representation of the structure of a Mach–Zehnder interferometer for amplitude modulation.

Frequency tuning has been demonstrated with silicon slotted ring microresonators filled with organic electro-optic materials.

- Utilization of tapered transitions, wedge structures, or spherical lens has permitted total insertion loss to be reduced to 6 dB or less. Such values are approaching those obtained for lithium niobate devices but lithium niobate is always advantageous for low material loss.
- Vertical integration with very large-scale integration (VLSI) semiconductor electronics and integration with silica fiber optics has been achieved. Three-dimensional electro-optic circuitry has been fabricated on top of VLSI wafers.

For example, a guest–host NLO polymer based on chromophore **7** in APC has been widely used as reference materials for many device applications, such as Mach–Zehnder modulators, solgel hybrid E-O modulators, and slotted ring microresonators. Many of these prototype devices can be fabricated by soft lithography demonstrating the potential for low cost, mass production of complex optical circuitry. Large E-O activities (42 pm/V at 1.31 μm), high optical transparency (optical loss ~1.0 dB/cm at 1.31 and 1.55 μm), good processibility and thermal stability (60°C~85°C), and broad bandwidth (40 GHz) were reported from the material.

Recently, polymeric E-O materials have been shown to be an effective media for the generation and detection of terahertz radiation, may be of utility for terahertz imaging. Terahertz radiation can be generated in second-order NLO materials by pumping the material with short (fs) optical laser pulses with range of frequencies. Terahertz generation is an example of optical rectification, that is a special kind of difference frequency generation. The difference mixing of these frequencies within a polymeric NLO material gives a broadband terahertz radiation. Terahertz radiation can be detected by the linear E-O effect of the electric field associated with incident terahertz radiation upon the polarization of an optical probe beam. Organic films have been shown to exhibit terahertz generation efficiencies considerably larger than comparable thicknesses of ZnTe. For example, 3.1 μm thick poled guest–host polymer of 20 wt% chromophore **7** in APC shows higher terahertz amplitude than a 1 mm thick sample of ZnTe with frequency window extending up to 12 THz when pumped at 800 nm [47].

While recent advances in performance of polymeric E-O materials have revitalized interest in organic E-O materials, a number of issues must be addressed before broad commercialization of these materials is viable. The most important issue is exploiting the exceptional operational bandwidths from the fast π-electron response of organic E-O materials. The operational bandwidth of devices will be limited by metal electrode structures and by electrical connectors. Realization of performance bandwidths >100 GHz typically involves novel engineering of device structures or use of special techniques that lead to high effective data handling rates. Increasing electro-optic activity would permit the use of short electrode structures resulting in an increase in bandwidth. For example, an electro-optic coefficient of 300 pm/V should permit realization of sub 1 V drive voltages for Mach–Zehnder modulators that employ electrode structures of 0.5 cm permitting bandwidths on the order of 100 GHz to be realized.

Another issue relates to driving voltage requirements for the practical utilization of E-O devices. For analog telecommunication signals, one is concerned with lossless (transparent) telecommunication links and with high dynamic range. For lossless signal transduction, a driving voltage (V_π) of less than 0.5 V is typically required. For digital signals, low V_π E-O modulators are necessary to avoid costly digital amplifiers, which also limit bandwidth. As discussed, the macroscopic E-O activity of materials is proportional to the β of chromophores in acentric order. However, increasing β of chromophores can be accompanied by a reduction in the intramolecular charge-transfer band gap of a dipolar NLO chromophore and an increase in optical loss at the operational wavelength of the device. Improvement in performance by increasing device length is a viable option only for low optical loss materials. Creative design and control of chromophore geometries and supramolecular assembly of chromophores that can produce very large E-O activity with low absorption loss are necessary to lower the driving voltage significantly.

Optical loss due to scattering, that is mainly associated with material processing and device fabrication induced material inhomogeneity and waveguide surface roughness, must also be minimized. Waveguide loss values around 1.0 dB/cm can be a realistic target. Optical loss and optical power handling capabilities are the very important factors in determining critical performance parameters for device applications of polymeric E-O materials.

The acentric-order parameter depends on the effective electric field felt by the chromophores during poling. Electric fields from surrounding chromophores especially at high chromophore concentrations can attenuate the effective strength of the poling field. The maximum achievable poling field will also be limited by conductivity and dielectric breakdown (as well as dielectric permittivity) of electro-optic core and cladding materials. Conductivity can be influenced by ionic impurities and thus by the purity of materials. Careful consideration and control of materials design, synthesis, purification, and processing are required to achieve highly reproducible electro-optic activity. It is also important to consider other properties such as thermal stability, photochemical stability, and processibility. Fortunately, the flexibility in design and modification of polymeric E-O materials is a significant advantage in dealing with such requirements.

16.7 SUMMARY

We introduced the basic principles of the polymeric NLO materials and devices in this chapter. Various types of polymeric second-order NLO materials were introduced with focus on molecular design of dipolar NLO chromophores and solid-state engineering of polymeric NLO materials.

Recent developments in the polymeric NLO materials were introduced. Particular attention was given to the design of highly efficient dipolar NLO chromophores in guest–host polymers, syntheses, and postfunctionalization of side-chain E-O polymers using various cross-linking reactions and lattice-hardening mechanisms, and NLO materials based on dendronized polymers and dendrimers.

Detailed discussions on rational design and engineering of chromophores and polymers have been provided to help readers understand how the molecular hyperpolarizability, poling efficiency, and temporal alignment stability can be optimized simultaneously.

Separate discussions were added in separate sections on monolithic chromophore glasses and molecular glass composites of multiple chromophores, together with an overview of most important device applications of polymeric E-O materials.

EXERCISE QUESTIONS

1. List the advantages of organic NLO materials over the inorganic NLO materials such as $LiNbO_3$ crystals, and explain why?
2. Efficient dipolar NLO chromophores can be engineered through optimization of molecular structures of electron donor and acceptor groups for a given π-conjugation bridge structure. Describe how one can design an efficient NLO chromophore (a) to maximize the molecular hyperpolarizability and (b) to improve the poling efficiency and the macroscopic E-O activity.
3. Explain the factors that are to be considered in the selection of postfunctionalization reactions for a side-chain NLO polymer and give reasons for the consideration.
4. Describe major advantages of dendritic modification of NLO polymers. Explain the differences between dendronized chromophore and dendrimer of chromophores.
5. Describe the advantages of molecular glasses for NLO materials.
6. Explain why the poling efficiency of E-O materials based on binary molecular glasses can be improved compared to materials based on monolithic molecular glass?

LIST OF ABBREVIATIONS

AC	alternating current
APC	amorphous polycarbonate
α	molecular polarizability
au	arbitrary unit
β	molecular first hyperpolarizability
BOA	bond order alternation
BLA	bond length alternation
c	speed of light
CF_3-TCF	5-trifluoromethyl tricyano furan
d_{33}	nonlinear optical coefficient
DC	direct current
DEAD	diethyl azodicarboxylate
ε	dielectric constant
DMAc	dimethylacetamide
DMF	dimethylformamide
DMSO	dimethylsulfoxide
DR 1	disperse red 1
DSC	differential scanning calorimetry
E	electric poling field strength
ε_0	low-frequency dielectric constant
$\varepsilon(2\omega)$	molar extinction coefficient of polymer at 2ω frequency
EFISH	electric field-induced second harmonic generation
esu	electrostatic unit
F	local field factor
G(θ)	Gibbs–Boltzmann distribution function
GHz	gigahertz (10^9 Hz)
γ	molecular second hyperpolarizability
H	Planck's constant
HOMO	highest occupied molecular orbital
HLS	harmonic light scattering
HRS	hyper-Rayleigh scattering
Hz	hertz (1 Hz = s^{-1})
I_ω	fundamental beam intensity
ITO	indium tin oxide
K_i	expansion coefficients
μ	dipole moment
μ_0	permanent dipole moment
LB	Langmuir–Blodgett
LUMO	lowest unoccupied molecular orbital
MHz	megahertz (10^6 Hz)
MV	megavolt (10^6 V)
M_ω	molecular weight
n	refractive index
N	chromophore number density in polymer matrix
NLO	nonlinear optical
NMP	*N*-methylpyrrolidone
OPO	optical parametric oscillator
P	macroscopic polarization
PC	polycarbonate

PEI	polyetherimide
Ph-TCBD	2-phenyl-tetracyanobutadienyl
PI	polyimide
PMMA	poly(methyl methacryate)
PMMA-AMA	poly(methyl methacrylate-*co*-anthracen-9-ylmethyl methacrylate
PPIF	poly[(phenyl isocyanate)-*co*-formaldehyde]
PQ	polyquinoline
PS	polystyrene
PU	polyurethane
r_{33}	electro-optic (E-O) coefficient in the direction of the applied electric field
λ	wavelength
RT	room temperature
SDS	3-(dicyanomethyliden)-2,3-dihydrobenzothiophene-2-ylidene-1,1-dioxide
SG	solgel
SHG	second harmonic generation
TCBD	tetracyano butadienyl
TCI	tetracyano indane (1,3-bis(dicyanomethylidene) indane)
TCF	tricyano furan
TCP	tricyano pyrroline
TCV	tricyano vinyl
T_g	glass transition temperature
TGA	thermal gravimetric analysis
THF	Tetrahydrofuran
THz	terahertz (10^{12} Hz)
THG	third harmonic generation
UV	ultraviolet
V_m	amplitude of modulation voltage
$V\pi$	half-wave voltage
$\chi^{(1)}$	linear optical susceptibility
$\chi^{(2)}$	second-order nonlinear optical susceptibility
$\chi^{(3)}$	third-order nonlinear optical susceptibility
ω	angular frequency

ACKNOWLEDGMENT

This work was financially supported by DARPA (molecular photonics) and National Science Foundation (STC under DMR-0120967). Alex, K.-Y. Jen thanks the Boeing-Johnson foundation for its support.

REFERENCES

1. (a) Desiraju, G.R. *Angew. Chem. Int. Ed. Engl.* 1995, 34, 2311; (b) Schmidt, G.M. *Pure Appl. Chem.* 1971, 27, 647.
2. (a) Muthuraman, M., Le Fur, Y., Bagieu-Beucher, M., Masse, R., Nicoud, J.-F., Desiraju, G.R. *J. Mater. Chem.* 1999, 9, 2233; (b) Anthony, S.P., Radhakrishnan, T.P. *Chem. Comm.* 2001, 931; (c) Lacroix, P.G., Daran, J.C., Nakatani, K. *Chem. Mater.* 1998, 10, 1109.
3. (a) Komorowska, K., Brasselet, S., Dutier, G., Ledoux, I., Zyss, J., Poulsen, L., Jazdzyk, M., Egelhaaf, H.-J., Gierschner, J., Hanack, M. *Chem. Phys.* 2005, 318, 12; (b) Gervais, C., Hertzsch, T., Hulligeer, J. *J. Phys. Chem. B* 2005, 109, 7961; (c) Yi, T., Clément, R., Haut, C., Catala, L., Gacoin, T., Tancrez, N., Ledoux, I., Zyss, J. *Adv. Mater.* 2005, 17, 335; (d) Hulliger, J., Konig, O., Hoss, R. *Adv. Mater.* 1995, 7, 719; (e) Ramamurth, V., Eaton, D.F. *Chem. Mater.* 1994, 6, 1128; (f) Ramamurth, V., Eaton, D.F., Caspar, J.V. *Acc. Chem. Res.* 1992, 25, 299.

4. (a) Yitzchaik, S., Marks, T.J. *Acc. Chem. Res.* 1996, 29, 197; (b) Marks, T.J., Ratner, M.A. *Angew. Chem. Int. Ed. Engl.* 1995, 34, 155; (c) Li, D., Ratner, M.A., Marks, T.J., Zhang, C., Yang, J., Wong, G.K. *J. Am. Chem. Soc.* 1990, 112, 7389; (d) Katz, H.E., Wilson, W.L., Scheller, G. *J. Am. Chem. Soc.* 1994, 116, 6636.
5. (a) Mortazavi, M.A., Knoesen, A., Kowel, S.T., Higgins, B.G., Dienes, A. *J. Opt. Soc. Am. B* 1989, 6, 733; (b) Singer, K.D., Kuzyk, M.G., Holland, W.R., Sohn, J.E., Lalama, S.J. *Appl. Phys. Lett.* 1988, 53, 1800; (c) Singer, K.D., Sohn, J.E., Lalama, S.J. *Appl. Phys. Lett.* 1986, 49, 248.
6. (a) Dalton, L.R. *Adv. Poly. Sci.* 2002, 158. 1; (b) Ma, H., Jen, A.K.-Y., Dalton, L.R. *Adv. Mater.* 2002, 14, 1339; (c) Dalton, L.R.A., Harper, W., Wu, B., Chosn, R., Laquindanum, J., Liang, Z., Hubbel, A., Xu, C. *Adv. Mater.* 1995, 7, 519; (d) Shi, Y., Zhang, C., Xhang, H., Bechtel, J.H., Dalton, L.R., Robinson, B.H., Steier, W.H. *Science* 2000, 288, 119.
7. (a) Chemla, D.S., Zyss, J., Eds., *Nonlinear Optical Properties of Organic Molecules and Crystals* (Academic, New York, 1987); (b) Brédas, J.L., Chance, R.R., Eds., *Conjugated Polymeric Materials: Opportunities in Electronics, Optoelectronics, and Molecular Electronics* (Kluwer, Dordrecht, 1990); (c) Messier, J., Kajzar, F., Prasad, P.N., Eds., *Organic Molecules for Nonlinear Optics and Photonics* (Kluwer, Dordrecht, 1991); (d) Marder, S.R., Kippelen, B., Jen, A.K.-Y., Peyghambarian, N. *Nature* 1997, 388, 845.
8. (a) Zyss, J., Ledoux, I. *Chem. Rev.* 1994, 94, 77; (b) Verbiest, T., Houbrechts, S., Kauranen, M., Clays, K., Persoons, A. *J. Mater. Chem.* 1997, 7, 2175; (c) Andraud, C., Zabulon, T., Collet A., Zyss, J. *Chem. Phys.* 1999, 245, 243; (d) Zyss, J., Dhenaut, C., Chauvan T., Ledoux, I. *Chem. Phys. Lett.* 1993, 206, 409.
9. Marder, S.R., Gorman, C.B., Meyers, F., Perry, J.W., Bourhill, G., Bredas, J.L., Pierce, B.M. *Science* 1994, 265, 632.
10. Oudar, J.L., Chemla, D.S. *J. Chem. Phys.* 1977, 66, 2664.
11. (a) Marder, S.R., Beratan, D.N., Cheng, L.T. *Science* 1991, 252, 103; (b) Marder, S.R., Brédas, J.L., Perry, J.W. *J. Am. Chem. Soc.* 1994, 116, 2619.
12. (a) Jen, A.K.-Y., Cai, Y., Bedworth, P.V., Marder, S.R. *Adv. Mater.* 1997, 9, 132; (b) Jen, A.K.-Y., Rao, V.P., Wong, K.Y., Drost. K.J. *Chem. Comm.* 1993, 90.
13. Dewar, M.J.S., Zoebisch, E.G., Healy, E.F., Stewart, J.J.P. *J. Am. Chem. Soc.* 1985, 107, 3902.
14. (a) Mataga, N., Nishimoto, K.Z. *Phys. Chem.* 1957, 13, 140; (b) Zerner, M.C., Loew, G.H., Kichner, R.F., Mueller-Westerhoff, U. *J. Am. Chem. Soc.* 1980, 102, 589.
15. Orr, J., Ward, J.F. *Mol. Phys.* 1971, 20, 513.
16. (a) Cohen, H.D., Roothaan, C.C.J. *J. Chem. Phys.* 1965, 43, 34; (b) Chopra, P., Carlacci, L., King, H.F., Prasad, P.N. *J. Phys. Chem.* 1989, 93, 7120; (c) Nakano, M.N., Shigemoto, I., Yamada, S., and Yamaguchi, K. *J. Chem. Phys.* 1995, 103, 4175; (d) Geskin, V.M., Brédas, J.L. *J. Chem. Phys.* 1998, 109, 6163; (e) Geskin, V.M., Lambert, C., Brédas, J.L. *J. Am. Chem. Soc.* 2003, 125, 15651.
17. Marder, S.R., Gorman, C.B., Meyers, F., Perry, J.W., Bourhill, G., Brédas, J.L., Pierce, B.M. *Science* 1994, 265, 632.
18. (a) Ahlheim, M., Barzoukas, M., Bedworth, P.V., Blanchard-Desce, M., Fort, A., Hu, Z.Y., Marder, S.R., Perry, J.W., Runser, C., Staehin, M., Zysset, B. *Science* 1996, 271, 335; (b) Sun, S.S., Zhang, C., Dalton, L.R., Garner, S.M., Chen, A., Steier, W.H. *Chem. Mater.* 1996, 8, 2539.
19. (a) Robinson, B.H., Dalton, L.R., Harper, A.W., Ren, A., Wang, F., Zhang, C., Todorova, G., Lee, M., Aniszfeld, R., Garner, S., Chen, A., Steier, W.H., Houbrecht, S., Persoons, A., Ledoux, I., Zyss, J., Jen, A.K.-Y. *Chem. Phys.* 1999, 245, 35; (b) He, M.Q., Leslie, T.M., Sinicropi, J.A. *Chem. Mater.* 2002, 14, 4662; (c) He, M.Q., Leslie, T.M., Sinicropi, J.A., Garner, S.M., Reed, L.D. *Chem. Mater.* 2002, 14, 4669; (d) Liu, S., Haller, M., Ma, H., Dalton, L.R., Jang, S.-H, Jen, A.K.-Y. *Adv. Mater.* 2003, 15, 603; (e) Jang, S.-H., Luo, J., Tucker, N.M., Leclercq, A., Zojer, E., Haller, M.A., Kim, T.-D., Kang, J.-W., Firestone, K., Bale, D., Lao, D., Benedict, J.B., Cohen, D., Kaminsky, W., Kahr, B., Brédas, J.L., Reid, P., Dalton, L.R., Jen, A.K.-Y. *Chem. Mater.* 2006, 18, 2982; (f) Leclercq, A., Zojer, E., Jang, S.-H., Barlow, S., Geskin, V., Jen, A.K.-Y., Marder, S.R., Bredas, J.L. *J. Chem. Phys.* 2006, 124, 044510.
20. (a) Zhang, C., Dalton, L.R., Oh, M.-C., Xhang, H., Steier, W.H. *Chem. Mater.* 2001, 13, 3043; (b) He, M., Leslie, T.M., Sinicropi, J.A., Garner, S.M., Reed, L.D. *Chem. Mater.* 2002, 14, 4669; (c) He, M., Leslie, T.M., Sinicropi, J.A. *Chem. Mater.* 2002, 14, 4662; (d) Paley, M.S., Harris, J.M., Looser, H., Baumert, J.C., Bjorklund, G.C., Jundt, D., Twieg, R.J. *J. Org. Chem.* 1989, 54, 3774.
21. Williams, D.J. *Angew. Chem. Int. Ed. Engl.* 1984, 23, 690.
22. Robinson, B.H., Dalton, L.R. *J. Phys. Chem. A* 2000, 104, 4785.
23. (a) Kuzyk, M.G. *Phys. Rev. Lett.* 2000, 85, 1218; (b) Bosshard, C., Sutter, K., Schlesser, R., Günter, P. *J. Opt. Soc. Am. B* 1993, 10, 867.
24. (a) Würthner, F., Yao, S., Debaerdemaeker, T., Wortmann, R. *J. Am. Chem. Soc.* 2002, 124, 9431; (b) Wortmann, R., Rosch, U., Redi-Abshiro, M., Würthner, F. *Angew, Chem. Int. Ed.* 2003, 42, 2080; (c) Würthner, F., Yao, S. *Angew. Chem. Int. Ed. Engl.* 2000, 39, 1987.

25. Wu, X., Wu, J., Liu, Y., Jen, A.K-Y. *J. Am. Chem. Soc.* 1999, 121, 472.
26. Moon, K.-J., Shim, H.-K., Lee, K-S; Zieba, J., Prasad, P.N. *Macromolecules* 1996, 29, 861.
27. Lee, K.-S., Moon, K.-J., Woo, H.-Y., Shim, H.-K. *Adv. Mater.* 1997, 9, 978.
28. (a) Luo, J., Haller, M., Li, H., Kim, T.-D., Jen, A.K.-Y. *Adv. Mater.* 2003, 15, 1635; (b) Haller, M., Luo, J., Li, H., Kim, T.-D., Liao, Y., Robinson, B.H., Dalton, L.R., Jen, A.K.-Y. *Macromolecules* 2004, 37, 688; (c) Kim, T.-D., Luo, J., Tian, Y., Ka, J.-W., Tucker, N.M., Haller, M., Kang, J.-W., Jen, A.K.-Y. *Macromolecules* 2006, 39, 1676.
29. Kim, T.-D., Luo, J., Ka, J.-W., Hau, S., Tian, Y., Shi, Z., Tucker, N.M., Jang, S.-H., Kang, J.-W., Jen, A.K.-Y. *Adv. Mater.* 2006, 18, 3038.
30. (a) Hecht, S., Fréchet, J.M.J. *Angew Chem. Int. Ed.* 2001, 40, 74; (b) Vögtle, F., Gestermann, S., Hesse, R., Schwierz, H., Windisch, B. *Prog. Polym. Sci.* 2000, 25, 987; (c) Bosman, AW., Janssen, H.M., Meijer, E.W. *Chem. Rev.* 1999, 99, 1665; (d) Newkome, G.R., He, E., Moorefield, C.N. *Chem. Rev.* 1999, 99, 1689.
31. Ma, H., Liu, S., Luo, J., Suresh, S., Liu, L., Kang, S.-H., Haller, M., Sassa, T., Dalton, L.R., Jen, A.-K.-Y. *Adv. Funct. Mater.* 2002, 12, 565.
32. (a) Luo, J., M. Haller., Ma, H., Liu, S., Kim, T., Tian, Y., Chen, B., Jang, S.-H., Dalton, L.R., Jen, A.K. *J. Phys. Chem. B.* 2004, 108, 24, 8523; (b) Luo, J., Ma, H., Haller, M., Jen, A.K.-Y., Barto, R.R. *Chem. Comm.* 2002, 888.
33. (a) Luo, J., Haller, M., Li, H., Tang, H.-Z., Jen, A.K.-Y., Jakka, K., Chou, C.-H., Shu, C.-F. *Macromolecules* 2004, 37, 248; (b) Luo, J., Liu, S., Haller, M., Liu, L., Ma, H., Jen, A.K.-Y. *Adv. Mater.* 2002, 14, 1763.
34. (a) Yokoyama, S., Nakahama, T., Otoma, A., Mashiko, S. *J. Am. Chem. Soc.* 2000, 122. 3174; (b) Yokoyama, S., Nakahama, T., Otoma, A., Mashiko, S. *Thin Solid Films* 1998, 331, 248.
35. (a) Ma, H., Chen, B., Sassa, T., Dalton, L.R., Jen, A.K.-Y. *J. Am. Chem. Soc.* 2001, 123, 986; (b) Ma., H., Jen, A.K.-Y. *Adv. Mater.* 2001, 13, 1201.
36. (a) Ostroverkhova, O., Moerner, W.E. *Chem. Rev.* 2004, 104, 3267; (b) He, M., Twieg, R.J., Gubler, U., Wright, D., Moerner, W.E. *Chem. Mater.* 2003, 15, 1156; (c) Würthner, F., Wortmann, R., Meerholz, K. *Chem. Phys. Chem.* 2002, 3, 17; (d) Lundquist, P.M., Wortmann, R., Geletneky, C., Twieg, R.J., Jurich, M., Lee, V.Y., Moylan, C.R., Burland, D.M. *Science* 1996, 274, 1182.
37. (a) Lee, S.M., Jahng, W.S., Lee, J.H., Rhee, B.K., Park, K.H. *Chem. Phys. Lett.* 2005, 411, 496; (b) Ishow, E., Bellaiche, C., Bouteiller, L., Nakatani, K., Delaire, J.A. *J. Am. Chem. Soc.* 2003, 125, 15744; (c) Eich, M., Looser, H., Yoon, D.Y., Twieg, R., Bjorklund, G., and Baumert, J.C. *J. Opt. Soc. Am. B* 1989, 6, 1590.
38. (a) Williams, J.H., Cockcroft, J.K., Fitch, A.N. *Angew. Chem. Int. Ed. Engl.* 1992, 31, 1655; (b) Dahl, T. *Acta. Chem. Scan. Ser. A.* 1988, 42, 1; (c) Patrick, C.R., Prosser, G.S. *Nature* 1960, 187, 1021.
39. (a) Reichenbacher, K., Süss, H.I., Hulliger, J. *J. Chem. Soc. Rev.* 2005, 34, 22 (and references therein); (b) Meyer, E.A., Castellano, R.K., Diederich, F. *Angew. Chem. Int. Ed. Engl.* 2003, 42, 1210 (and references therein); (c) Hunter, C.A., Lawson, K.R., Perkins, J., Urch, C.J. *J. Chem. Soc., Perkin Trans. 2*, 2001, 651; (d) Lorenzo, S., Lewis, G.R., Dance, I. *New J. Chem.* 2000, 24, 295; (e) Williams, J.H. *Acc. Chem. Res.*, 1993, 26, 593.
40. (a) Burress, C., Elbjeirami, O., Omary, M.A., Gabbai, F.P. *J. Am. Chem. Soc.* 2005, 127, 12166; (b) Wang, Z., Dotz, F., Enkelmann, W., Müllen, K. *Angew. Chem. Int. Ed. Engl.* 2005, 44, 1247; (c) Feast, W.J., Lovenich, P.W., Puschmann, H., Taliani, C. *Chem. Comm.* 2001, 505; (d) Ponzini, F., Zagha, R., Hardcastle, K., Siegel, J.S. *Angew. Chem. Int. Ed. Engl.* 2000, 39, 2323; (e) Renak, M.L., Batholomew, G.P., Wang, S., Ricatto, P.J., Lachicotte, R.J., Bazan, G.C. *J. Am. Chem. Soc.* 1999, 121, 7787; (f) Coates, G.W., Dunn, A.R., Henling, L.M., Ziller, J. W., Lobkovsky, E.B., Grubbs, R.H. *J. Am. Chem. Soc.* 1998, 120, 3641; (g) Coates, G.W., Dunn, A.R., Henling, L.M., Dougherty, D.A., Grubbs, R.H. *Angew. Chem. Int. Ed. Engl.* 1997, 36, 248.
41. (a) Collings, J.C., Roscoe, K.P., Robins, E.G., Batsanov, A.S., Stimson, L.M., Howard, J.A.K., Clark, S.J., Marder, T.B. *New J. Chem.* 2002, 26, 1740; (b) Collings, J.C., Roscoe, K.P., Thomas, R.L., Batsanov, A.S., Stimson, L.M., Howard, J.A.K., Marder, T.B. *New J. Chem.* 2001, 25, 1410; (c) Metrangelo, P., Resnati, G. *Chem. Eur. J.* 2001, 7, 2511.
42. (a) Nguyen, H.L., Horton, P.N., Hursthouse, M.B., Legon, A.C., Bruce, D.W. *J. Am. Chem. Soc.* 2004, 126, 16; (b) Watt, S.W., Dai, C., Scott, A.J. Burke, J. M., Thomas, R.L., Collings, J.C., Viney, C., Clegg, W., Marder, T.B. *Angew. Chem. Int. Ed. Engl.* 2004, 43, 3061; (c) Kilbinger, A.F.M., Grubbs, R.H. *Angew. Chem. Int. Ed. Engl.* 2002, 41, 1563; (d) Weck, M., Dunn, A.R., Matsumoto, K., Coates, G.W., Lobkovsky, E.B., Grubbs, R.H. *Angew. Chem. Int. Ed. Engl.* 1999, 38, 2741; (e) Dai, C., Nguyen, P., Marder, T.B., Scott, A.J., Clegg, W., Viney, C. *Chem. Comm,* 1999, 2493.
43. Kim, T.-D., Kang, J.-W., Luo, J., Jang, S.-H., Ka, J.-W., Tucker, N., Benedict, J.B., Dalton, L.R., Gray, T., Overney, R.M., Park, D.-H., Herman, W.N., Jen, A.K.-Y. *J. Am. Chem. Soc.* 2007, 129, 488.

44. Meredith, G.R., VanDusen, J.G., Williams, D.J. *Macromolecules* 1982, 15, 1385.
45. Sullivan, P.A., Akelaitis, A.J.P., Lee, S.-K., McGrew, G., Lee, S.-K., Choi, D.-H., Dalton, L.R. *Chem. Mater.* 2006, 18, 344.
46. (a) Dalton, L.R. *J. Phys. Condens. Matter* 2003, 15, R897; (b) Kajzar, F., Lee, K.-S., Jen, A.K.-Y. *Adv. Poly. Sci.* 2003, 161, 1.
47. Sinyukov, A.M., Leahy, M.R., Hayden, L.M., Haller, M., Luo, J., Jen, A.K.-Y., Dalton, L.R. *App. Phys. Lett.* 2004, 85, 5827.

17 Organic and Polymeric 3rd-Order Nonlinear Optical Materials and Device Applications

Joel M. Hales and Joseph W. Perry

CONTENTS

17.1 Fundamental Concepts and Principles of 3rd-Order Nonlinear Optics 514
17.1.1 Simple Model for Nonlinear Optical Response 514
17.1.2 Catalog of Various 3rd-Order Nonlinear Optical Phenomena 516
17.1.3 General Characteristics of the 3rd-Order Macroscopic Nonlinearity 518
17.1.3.1 Relationship between γ and $\chi^{(3)}$ 518
17.1.3.2 Magnitude of $\chi^{(3)}$ 519
17.1.3.3 Symmetry and Orientational Averaging 520
17.1.3.4 Complex Nature of $\chi^{(3)}$ 521
17.1.3.5 Dispersion of $\chi^{(3)}$ 522
17.1.3.6 Temporal Response of $\chi^{(3)}$ 523
17.2 Selected 3rd-Order Nonlinear Optical Phenomena and Characterization Techniques 524
17.2.1 Selected 3rd-Order Nonlinear Optical Phenomena 525
17.2.1.1 Third-Harmonic Generation 525
17.2.1.2 Intensity-Dependent Refractive Index 526
17.2.1.3 Degenerate Four-Wave Mixing 529
17.2.1.4 Effective $\chi^{(3)}$ Phenomena 531
17.2.2 Nonlinear Optical Characterization Techniques 533
17.2.2.1 Third-Harmonic Generation 533
17.2.2.2 Z-Scan 535
17.2.2.3 Degenerate Four-Wave Mixing 536
17.2.2.4 Pump–Probe and Optical Kerr Effect 539
17.3 Third-Order Nonlinearites of Conjugated Organic Molecules 540
17.3.1 Simple Perturbation Theory Description of Polarizability and Hyperpolarizability 541
17.3.2 Valence Bond and Charge Transfer Resonance Structures, Dipole Moments, and Bond Length Alternation in Conjugated Molecules 542
17.3.3 Valence Bond and Charge Transfer Model for Structure–Hyperpolarizability Relationships 544
17.3.4 Summary of Second Hyperpolarizabilities for Conjugated Donor–Acceptor Chromophores 549
17.3.5 Survey of Organic Systems with Large 3rd-Order Nonlinearities 552

17.4 Device Applications 557
17.4.1 Issues for 3rd-Order Nonlinear Optical Applications 557
17.4.1.1 Material Requirements and Metrics 558
17.4.1.2 Integration of Materials and Devices 559
17.4.2 Third-Order Nonlinear Optical Applications 561
17.4.2.1 Telecommunications 561
17.4.2.2 Optical Phase Conjugation 563
17.4.2.3 Other Applications of Interest 566
17.5 Conclusion 566
Exercise Questions 567
List of Abbreviations 568
References 569

Abstract This chapter explores the use of organic and polymeric materials in device applications that involve third-order nonlinear optics, where the ability of light to control light is exploited in the transmission and processing of information. The origin of this nonlinear optical response as well as the general characteristics of $\chi^{(3)}$, the parameter that describes the macroscopic third-order nonlinearity, are discussed in this chapter. A number of interesting nonlinear phenomena as well as characterization techniques that utilize them, particularly for the aforementioned applications, are described. Additionally, a simple model is addressed for predicting the nonlinear optical response of a molecule as its chemical structure is varied. The design guidelines suggested by this model provide a good context for the survey of promising organic systems in the literature with large third-order nonlinearities. Finally, the requirements on such nonlinear materials, their integration into devices, and some applications that employ third-order nonlinear optical phenomena will be presented.

17.1 FUNDAMENTAL CONCEPTS AND PRINCIPLES OF 3RD-ORDER NONLINEAR OPTICS

The response of a material following its interaction with light can vary greatly depending on the strength of the incident optical fields. For certain materials, application of intense electromagnetic fields may very well alter their intrinsic optical properties. This material response can facilitate strong interaction with the electromagnetic fields resulting in the generation of new fields or modification of the properties of the fields themselves. This is the regime of nonlinear optics.

17.1.1 Simple Model for Nonlinear Optical Response

The nonlinear response of the material can be better understood when viewing this phenomenon at the molecular level of the system. At this level, the material consists of a distribution of charged particles, namely positively charged nuclei and negatively charged electrons. In general, the electron, or more appropriately the electron charge cloud, is elastically coupled to the nucleus and, therefore, an oscillating electric field would interact with the material such that the electron distribution and the positive core would perform an oscillatory motion with respect to one another. This is illustrated in Figure 17.1a. Therefore, an applied optical field **E** will displace the electron density from the nuclear core and create an induced dipole μ. It is this induced polarization that accounts for an abundance of optical properties of materials.

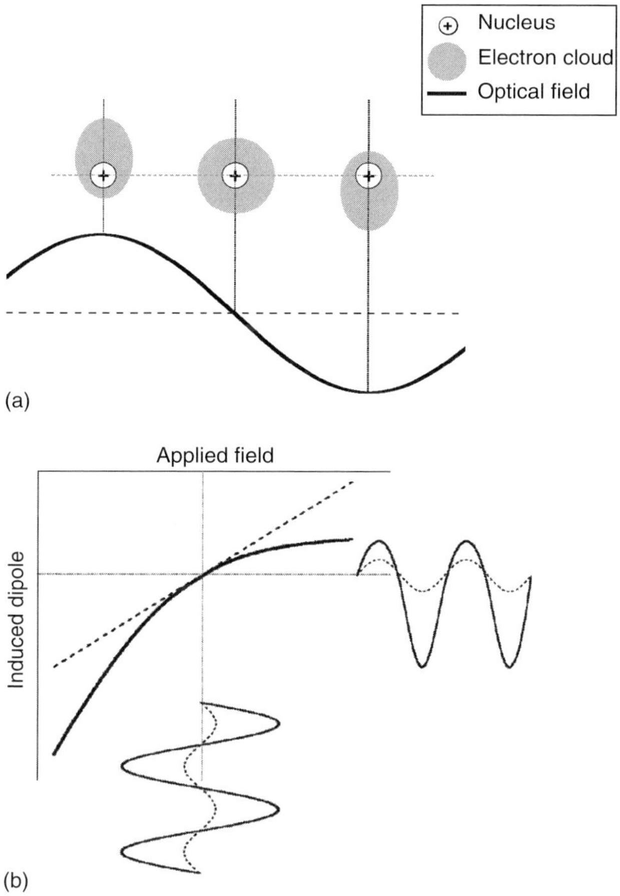

FIGURE 17.1 (a) Depicts the oscillatory motion of the electron charge cloud about its nucleus in the presence of an applied electric field. (b) Plot of the motion of the induced dipole versus applied electric field. For small magnitude fields, the dipole motion responds linearly to the applied field (dashed line); for large magnitude fields, the motion becomes nonlinear with respect to the field strength (solid line).

Provided the incident optical field is small in magnitude, the displacement of the electron charge cloud will remain small as well, and will oscillate harmonically with the frequency of the incident field. This linear response is depicted in Figure 17.1b. In this regime, the induced dipole can be considered to be linearly proportional to the strength of the applied optical field and an expression can be written relating the two terms:

$$\mu_i(\omega) = \alpha_{ij}(\omega)\mathbf{E}_j(\omega) \tag{17.1}$$

Here, μ and $\mathbf{E}$ are given as vector quantities and $\alpha_{ij}(\omega)$ is the linear polarizability of the molecule. α is a frequency-dependent quantity that is both complex and tensorial in nature. It accounts for linear optical effects such as absorption, refraction, and birefringence.

Nonetheless, many problems of interest in optics involve nonlinear interactions that occur when the electromagnetic interaction becomes too large for the medium to continue to respond linearly. This causes the electron cloud to oscillate anharmonically in response to the applied field. Therefore, when a molecule experiences a high intensity optical field, its induced polarization responds in a manner that is nonlinear with respect to the field strength as detailed in Figure 17.1b. An exact solution for the functional form of μ is no longer possible and it is therefore quite common to express the total dipole as a Taylor series expansion in **E**:

$$\mu_i = \mu_i^o + \alpha_{ij}(-\omega_\sigma;\omega)\mathbf{E}_j(\omega) + \frac{1}{2!}\beta_{ijk}(-\omega_\sigma;\omega_1,\omega_2)\mathbf{E}_j(\omega_1)\mathbf{E}_k(\omega_2)$$
$$+\frac{1}{3!}\gamma_{ijkl}(-\omega_\sigma;\omega_1,\omega_2,\omega_3)\mathbf{E}_j(\omega_1)\mathbf{E}_k(\omega_2)\mathbf{E}_l(\omega_3) + \kappa \quad (17.2)$$

In Equation 17.2, μ_i^o represents the permanent dipole, α_{ij} is the first-order polarizability, β_{ijk} the second-order polarizability, and γ_{ijkl} is the third-order polarizability. Like α, β and γ are complex, tensorial, and show a frequency-dependent response (details concerning the notation of the frequency-dependent response of γ will be discussed below). The polarization terms dependent on higher powers of the field, namely $\beta\mathbf{E}^2$, $\gamma\mathbf{E}^3$, and higher, are the terms responsible for nonlinear optical (NLO) effects. The small size of the nonlinear coefficients accounts for the necessity of an intense optical source to observe these nonlinear responses; however, the superlinear dependence on the optical field strength results in readily observable effects at intensities achievable with lasers. It is therefore not surprising that observation of a large number of nonlinear phenomenon occurred immediately following the development of the laser [1,2].

17.1.2 Catalog of Various 3rd-Order Nonlinear Optical Phenomena

As alluded to above, the nonlinear polarization within the medium can modify the amplitude, phase, or polarization of the incident optical fields or can generate new beams with different frequencies or directions. To investigate the nonlinear phenomena that exist as a consequence of the third-order polarizability, it is necessary to expand the $\gamma\mathbf{E}^3$ term in Equation 17.2 by substituting in the appropriate form for the electric field. It should be noted that second-order NLO processes (i.e., those governed by the $\beta\mathbf{E}^2$ term), such as second-harmonic generation, sum and difference frequency generation, optical parametric generation, and the Pockels effect are discussed in detail in the preceding chapter (see Chapter 16) and will not be covered here. Assuming monochromatic waves, the incident fields can be written as

$$\mathbf{E}(\mathrm{t,r}) = \frac{1}{2}\sum_{\omega}\{\mathbf{A}_\omega(r)\exp(-i\omega t) + \mathrm{c.c.}\} \quad (17.3)$$

where $\mathbf{A}_\omega$ is the complex field amplitude, which includes information about the amplitude, polarization, and overall phase (including the wave-vector) of the field at ω. The term c.c. refers to the complex conjugate of the field. Since the fields represented in Equation 17.2 are actually the Fourier transforms of the fields listed in Equation 17.3, this implies that frequency components at both $+\omega$ and $-\omega$ must be considered in the expansion of Equation 17.2. Furthermore, the summation in Equation 17.3 states that the fields involved in the nonlinear polarization shown in Equation 17.2, i.e., $\mathbf{E}_j$, $\mathbf{E}_k$, and $\mathbf{E}_l$, are the total fields, and therefore must be further expressed as the sum of the three incident fields. This implies that in the most general case when all three incident fields possess different frequencies and are clearly distinguishable* from one another, each total field in Equation 17.2

* Strictly speaking, differences in polarization, phase (wave-vector), amplitude, or frequency can make one field distinguishable from another but for the sake of cataloguing the nonlinear phenomena here the use of three distinct frequencies provides a sufficiently general case.

TABLE 17.1
List of Various 3rd-Order Nonlinear Optical (NLO) Phenomena Identified by the Frequencies of the Modified Optical Field (ω_σ) and the Incident Fields ($\omega_1, \omega_2, \omega_3$) with $\omega_1 > \omega_2 > \omega_3$

Phenomenon	$-\omega_\sigma$	$\omega_1, \omega_2, \omega_3$
DC Kerr effect	$-\omega$	$0, 0, \omega$
Electric-field-induced second-harmonic generation (EFISH)	-2ω	$0, \omega, \omega$
Third-harmonic generation (THG)	-3ω	ω, ω, ω
Sum/difference frequency generation (SFG/DFG)	$-(2\omega_1 \pm \omega_2)$	$\omega_1, \omega_1, \pm\omega_2$
General sum frequency generation (SFG)	$-(\omega_1 + \omega_2 + \omega_3)$	$\omega_1, \omega_2, \omega_3$
Coherent stokes Raman scattering (CSRS)	$(-\omega_1 + \omega_2 + \omega_2)$	$-\omega_1, \omega_2, \omega_2$
Coherent anti-stokes Raman scattering (CARS)	$-(\omega_1 + \omega_1 - \omega_2)$	$\omega_1, \omega_1, -\omega_2$
ND-intensity-dependent refractive index	$-\omega_2$	$\omega_1, -\omega_1, \omega_2$
ND-bound electronic optical Kerr effect (OKE)		
Raman-induced Kerr effect (RIKE)		
ND-orientational Kerr effect		
Stimulated Raman scattering (SRS)		
ND-two-photon absorption (2PA) or emission (2PE)		
D-intensity-dependent refractive index	$-\omega$	$\omega, -\omega, \omega$
D-bound electronic OKE		
D-orientational Kerr effect		
D-2PA or D-2PE		
Degenerate four-wave mixing (DFWM)		

will possess six terms (three fields including their complex conjugates) bringing the total number of terms involved in the third-order microscopic polarization to $6 \times 6 \times 6 = 216$ terms. Fortunately, this does not imply that there are as many distinct mechanisms. Also, the geometry employed for a particular experiment or application involving nonlinear optics typically results in a further reduction in the number of participating phenomena.

Table 17.1 catalogues the most salient nonlinear phenomena associated with the field expansion of the third-order polarization. The phenomena are identified according to the frequencies at which the nonlinear polarization is driven, ω_σ, or equivalently the frequency of the modified optical field following its interaction with the medium. These frequencies can be identified using the notation of the frequency-dependent response of γ given in Equation 17.2; since energy conservation must be obeyed, one can write

$$\omega_\sigma = \pm\,\omega_1 \pm \omega_2 \pm \omega_3 \tag{17.4}$$

The prefixes "D" and "ND" in the table make the distinction between degenerate and nondegenerate (i.e., single or multiple frequencies) processes, respectively. Clearly, the phenomena listed in the first five rows of the table can be distinguished by the frequency of the modified field. However, it would seem that the final two sets of processes (ND and D) lack this distinction. Nonetheless, other observables make these distinctions possible. Taking the degenerate processes as examples, degenerate four-wave mixing (DFWM) generates a new beam in a different propagation direction from that of the incident beams, the intensity-dependent refractive index alters the phase of an optical field while two-photon absorption (2PA) alters its magnitude, and the bound electronic optical Kerr effect (OKE) and the orientational Kerr effect have significantly different temporal responses.

Therefore, each phenomenon is unique and can be readily distinguished from one another. Finally, it is important to note that the list of phenomena in Table 17.1 is not meant to be exhaustive but rather illustrative of the variety of processes that can occur as a result of the third-order polarization of a material. While a more complete description of some selected phenomena is given in Section 17.2.1, a detailed description of a number of these processes can be found in Ref. [3].

17.1.3 General Characteristics of the 3rd-Order Macroscopic Nonlinearity

The modification or generation of the resultant optical fields that makes each mechanism described above distinguishable is a direct manifestation of the third-order nonlinearity induced in the material. This nonlinearity is fully defined by the microscopic third-order polarizability, γ, or the macroscopic third-order susceptibility, $\chi^{(3)}$. The relationship between these two quantities is described below. There are a number of parameters that are necessary to fully characterize $\chi^{(3)}$ for a particular material and this will be the subject of this section. While very few third-order nonlinear materials in the literature have been fully characterized in this manner, understanding the importance of these various attributes permits a more judicious choice when determining an appropriate material for a particular device or application [4].

17.1.3.1 Relationship between γ and $\chi^{(3)}$

Microscopic parameters, such as γ, permit direct analysis of the nonlinearities associated with molecular systems, which in turn allows for optimization of such systems. This optimization of molecular nonlinearities is the subject of Section 17.3. Unfortunately, most NLO characterization techniques (see Section 17.2.2, for example) extract information about the macroscopic nonlinearity of the material, such as $\chi^{(3)}$. Therefore, it is crucial to be able to find a relationship between the microscopic and macroscopic quantities. This is accomplished by first noting that the macroscopic polarization induced in a medium is the average microscopic dipole moment per unit volume of the medium (μ_i) times the number density of microscopic dipoles (N),

$$\mathbf{P}_\mathrm{I} = N\langle \mu_i \rangle \tag{17.5}$$

where $\langle\rangle$ represents an orientational averaging over all the microscopic dipoles.

This averaging is discussed in more detail below. To relate Equation 17.5 to $\chi^{(3)}$, a series expansion similar to the expansion performed in Equation 17.2 must be undertaken for the macroscopic polarization $\mathbf{P}_\mathrm{I}$:

$$\begin{aligned}\mathbf{P}_\mathrm{I} &= \mathbf{P}_I^{(0)} + \chi_{IJ}^{(1)}(-\omega_\sigma;\omega)\mathbf{E}_j(\omega) + \frac{1}{2!}\chi_{IJK}^{(2)}(-\omega_\sigma;\omega_1,\omega_2)\mathbf{E}_J(\omega_1)\mathbf{E}_K(\omega_2)\\ &+\frac{1}{3!}\chi_{IJKL}^{(3)}(-\omega_\sigma;\omega_1,\omega_2,\omega_3)\mathbf{E}_J(\omega_1)\mathbf{E}_K(\omega_2)\mathbf{E}_L(\omega_3)+\kappa\end{aligned} \tag{17.6}$$

Therefore, by taking the definitions for $\mathbf{P}_\mathrm{I}$ and μ_i (given by Equations 17.6 and 17.2) and using Equation 17.5 one can relate the third-order terms $\chi^{(3)}$ and γ. However, one last item should be understood before writing down this relation. The fields represented in Equation 17.2 are actually the local or effective electric fields while the fields in Equation 17.6 are the macroscopic fields. The local fields include contributions from surrounding molecules in addition to the applied macroscopic fields. To account for this, the macroscopic fields can be corrected for via Lorentz local field factors. The relationship between the fields can then be written as

$$\mathbf{E}_{i,j,k}(\omega_{1,2,\mathrm{or}3}) = L(\omega_{1,2,\mathrm{or}3}) \cdot \mathbf{E}_{I,J,K}(\omega_{1,2,\mathrm{or}3}) \tag{17.7}$$

where

$$L(\omega_{1,2,\text{ or }3}) = \frac{n(\omega_{1,2,\text{ or }3})^2 + 2}{3} \tag{17.8}$$

and $n(\omega)$ is the refractive index of the medium at ω. Finally, this leads to a relationship between γ and its macroscopic counterpart, $\chi^{(3)}$:

$$\chi^{(3)}_{IJKL}(-\omega_\sigma;\omega_1,\omega_2,\omega_3) = N \cdot L(\omega_\sigma)L(\omega_1)L(\omega_2)L(\omega_3) \cdot \langle \gamma_{ijkl}(-\omega_\sigma;\omega_1,\omega_2,\omega_3) \rangle \tag{17.9}$$

where $L(\omega_\sigma)$ is included to account for local field effects on the output field.

17.1.3.2 Magnitude of $\chi^{(3)}$

Using Equation 17.9, one can relate observables found using NLO characterization techniques to microscopic nonlinearities associated with individual molecules. And perhaps even more germane to this chapter is the fact that applications and devices that employ NLO phenomena almost always rely on the macroscopic nonlinearities of the medium being utilized. From an applications standpoint, possibly the most crucial parameter of $\chi^{(3)}$ is its magnitude. Larger nonlinearities permit shorter pathlength devices and allow for lower drive intensities, facts that are elaborated on in Section 17.4.1. It is therefore crucial to understand the factors that contribute to the magnitude of $\chi^{(3)}$. From Equation 17.9, the factors were found to be the orientationally averaged microscopic polarizability, $\langle\gamma\rangle$, the number density of participating NLO species, and the Lorentz field factors. Without question, the factor that gives the largest flexibility in terms of optimizing $\chi^{(3)}$ is $\langle\gamma\rangle$. A good portion of Section 17.3 discusses the use of chemical structure—NLO property relationships that can be employed to maximize the overall magnitude of γ. Other factors that can influence the optimization of γ such as resonance enhancement and molecular orientation on a macroscopic scale are discussed below. Clearly, maximizing the number density of molecules or polymer chains (that exhibit large values of γ) within a particular material is necessary to improve the macroscopic nonlinearity. For this reason, placing a highly nonlinear molecule or polymer in a host polymer (with small values of γ) to improve its processability inevitably decreases the magnitude of the material's $\chi^{(3)}$ through a dilution effect. Finally, since the Lorentz field factors are proportional to the square of the refractive index of the medium, employing a high number density material that typically shows a reasonably large index should result in a further increase in $\chi^{(3)}$.

As an example of how the number density and field factors can influence the magnitude of $\chi^{(3)}$, consider a molecular system that possesses a value for $\langle\gamma\rangle$ of 1×10^{-33} esu.* This is a moderately large value of γ for a discrete organic molecule. A 1 mM solution of this organic system (a concentration typically used in NLO measurements) in a solvent with a refractive index of 1.45 gives a molecular number density of 6.0×10^{17} molecules/cm^3 and a field factor of 1.37. Assuming all four field factors are the same (as would be the case for any degenerate nonlinear process, see final row of Table 17.1), this gives a value for $\chi^{(3)}$ of 2.1×10^{-15} esu. A comparison of this macroscopic nonlinearity to that of a neat film with a reasonably large value of $\chi^{(3)}$ (see Table 17.5 in Section 17.3.5) shows a value that is orders of magnitude lower than that of the neat film. This is, of course, due to

* Here γ is given in cgs units typically expressed as esu or more completely as cm$^5 \cdot$ statvolt^{-2}. $\chi^{(3)}$ is also given in cgs units expressed as esu or cm$^2 \cdot$ statvolt^{-2}. This will be the preferred notation used in this chapter. The conversion to SI units, however, is relatively straightforward and is given by γ[m$^5 \cdot$ V^{-2}, SI] = $(4\pi/9 \times 10^{-14}) \cdot \gamma$[cm$^5 \cdot$ statvolt^{-2}, esu] or $\chi^{(3)}$[m$^2 \cdot$ V^{-2}, SI] = $(4\pi/9 \times 10^{-8}) \cdot \chi^{(3)}$[cm$^2 \cdot$ statvolt^{-2}, esu].

the dilution of the molecules in solution by the solvent molecules whose nonlinearity has been assumed to be negligible in comparison to the nonlinear organic molecules (this is a reasonable assumption, see Ref. [3] for tabulation of third-order microscopic nonlinearities of typical solvent molecules). It then becomes obvious why the fabrication of high number density films is crucial to obtain the high bulk nonlinearities required for NLO applications.

17.1.3.3 Symmetry and Orientational Averaging

It is well known that centrosymmetric molecular systems (i.e., those with center of inversion symmetry) will possess second-order polarizabilities, β_{ijk}, whose tensor elements are all identically zero [5]. This is a direct consequence of the order of the β tensor and, more generally, no even-order nonlinear processes are possible in these symmetric systems.* For this reason, only noncentrosymmetric systems (i.e., those with some type of polar symmetry) will exhibit nonzero values for β. In contrast, all molecular systems, both centrosymmetric and noncentrosymmetric, exhibit nonzero third-order polarizabilities, γ. In other words, there is no symmetry restriction for odd-order nonlinear processes. For this reason, there are a larger number of molecular systems to choose from that could potentially exhibit substantial values of γ. As an example, certain long-chain oligomers and polymers, which would exhibit minimal second-order nonlinearities, have been shown to be highly efficient in terms of their third-order nonlinearities (see Section 17.3.5).

The symmetry restrictions present at the molecular level for second-order nonlinearities are just as valid at the macroscopic level. The symmetry of the bulk medium is determined, to a large extent, by the orientation of the individual molecules constituting the system. Even if the molecular subsystems have the appropriate asymmetry to exhibit non-negligible β, if these molecules are essentially isotropic in their ensemble orientation then adding up their contributions will most likely result in some cancellation effects drastically reducing the overall value of $\chi^{(2)}$ for the bulk medium (see Figure 17.2a). For this reason, bulk materials must also lack a center of symmetry to exhibit an efficient $\chi^{(2)}$. This is illustrated in Figure 17.2b. One way this is typically accomplished is by poling a heated host polymer such that its guest, a high β molecular chromophore, can be aligned, followed by cooling of the host polymer, which essentially locks the chromophore alignment in place. Once again, bulk materials possessing third-order nonlinearities do not suffer from such symmetry restrictions. Essentially, this implies that the orientational averaging of γ shown in Equation 17.9 does not result in any cancellation effects in the bulk medium, a fact that is depicted in Figure 17.2c. Consequently, amorphous or glassy materials can be utilized as third-order materials, a fact that is very significant since these media possess good processability allowing them to be readily integrated into devices.

Even though an oriented medium is not a necessary condition for strong third-order interaction, it can serve to enhance the overall nonlinearity. To understand why this is the case, it is important to understand the concept of orientational averaging of γ in the definition for $\chi^{(3)}$. The quantity $\langle\gamma_{ijkl}\rangle$ is an average polarizability determined by projecting the incident electric field vectors onto the molecular coordinate system, which, in an isotropic medium, can assume any random angular orientation. This quantity is therefore expressed as a summation of a significant number of molecular tensor elements with the appropriate numerical coefficients. Fortunately, the microscopic nonlinear response of a molecule is often quite anisotropic such that the component of γ along the main conjugated chain of a molecule is significantly larger than any other component of γ. Consequently, in an isotropic medium $\langle\gamma_{ijkl}\rangle$ can then be reduced to a single term, $(1/5)\cdot\gamma_{xxxx}$. This implies that, due to orientational averaging, $\chi^{(3)}$ is reduced by a factor of 1/5 in a disordered state compared to what it would be if all the molecular chains were aligned in the macroscopic media. The enhancement due to orientation is reflected in Figure 17.2d. This phenomenon has been exploited by stretching

* In fact, this is strictly true only within the electric dipole approximation. Quadrupolar and higher-order transitions are not confined by these symmetry restrictions. However, these types of transitions are usually significantly weaker than dipole-allowed transitions.

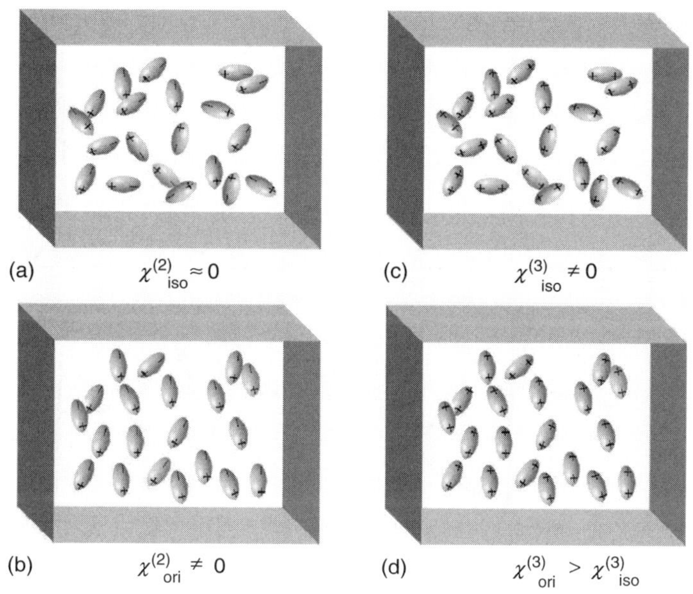

FIGURE 17.2 $\chi^{(2)}$ behavior for noncentrosymmetric molecules in (a) an isotropic state and (b) a highly ordered or oriented state. $\chi^{(3)}$ behavior for centrosymmetric molecules in an (c) isotropic state and (d) oriented state.

conjugated polymers into macroscopically oriented states. For instance, in studies of *trans*-polyacetylene, oriented polymer films showed significantly enhanced third-order nonlinearities compared to nonoriented films [6].

17.1.3.4 Complex Nature of $\chi^{(3)}$

In analogy to the description of the first-order polarizability in Section 17.1.1, the first-order susceptibility, $\chi^{(1)}$, is complex in nature and its real and imaginary parts are associated with linear refraction and linear absorption, respectively. It should be noted that oftentimes, in lieu of directly reporting the Re($\chi^{(1)}$) and Im($\chi^{(1)}$), the magnitude of the susceptibility, $|\chi^{(1)}|$, and its phase, ϕ, are reported instead. Both the sign of Re($\chi^{(1)}$) (which can take on either positive or negative values) and the ratio of its magnitude with respect to the magnitude of Im($\chi^{(1)}$) can both vary significantly as a function of frequency. For instance, as one approaches an electronic absorption resonance of the system (vibronic and rotational resonances can be active as well), the imaginary component of $\chi^{(1)}$ tends to become more dominant. These effects are merely manifestations of the dispersion of the susceptibility, which is discussed in more detail below. As a result of dispersion, the frequency of an incident optical field plays a significant role in how the field interacts with the material through the polarization. If the field frequency is far from any resonances in the material spectrum such that $\chi^{(1)}$ is purely real, the phase of the optical field will be altered as it passes through the medium but its amplitude will remain unchanged. In other words, there is no net energy exchanged between the field and the medium. This is not the case as the field frequency approaches a resonance of the material such that $\chi^{(1)}$ can become purely imaginary. In this case, the field amplitude will decrease as a result of energy being exchanged from the field to the medium [7].* Consequently, as one approaches a one-photon

* The case can occur when energy is exchanged from the medium to the field; this results in amplification of the field amplitude. In this case, the Im($\chi^{(1)}$) is actually negative in sign and this is referred to as gain. However, this requires energy to be pumped into the medium from some other external source.

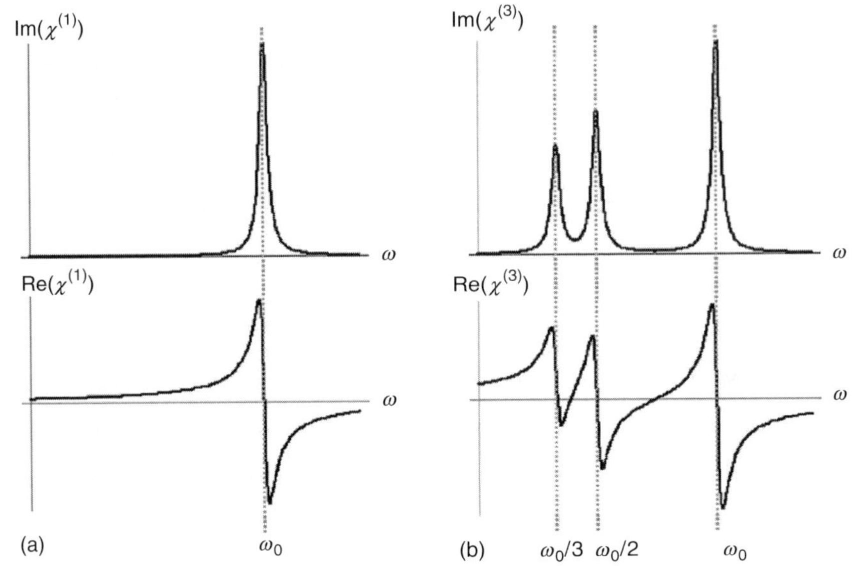

FIGURE 17.3 Typical dispersion for (a) $\chi^{(1)}$ and (b) $\chi^{(3)}$ in a medium with a single electronic resonance at ω_0. The upper plots show the imaginary parts of the susceptibility and the lower plots show the real parts.

resonance the medium becomes dissipative and the field experiences loss. Certainly, in intermediate spectral regimes, i.e., between theses two extreme cases, the field will experience both a phase and amplitude change.

The effects of the real and imaginary parts of the material susceptibility on the optical field also apply for the third-order polarization. Although, it should be noted that for this higher-order susceptibility, the optical field frequency can approach not only one-photon electronic resonances but two- and three-photon resonances as well. Nonetheless, the effects on the modulated field (i.e., phase and amplitude changes) are similar to those experienced in the linear optical regime. The main difference is that the strength of the applied optical fields can affect the magnitude of these phase or amplitude changes, whereas in the linear regime these magnitudes are independent of the strength of the incident fields. These third-order effects are designated by the degenerate forms of the intensity-dependent refractive index and 2PA (see Table 17.1*). These are governed by the real and imaginary parts of $\chi^{(3)}(-\omega;\omega,-\omega,\omega)$ and are direct analogues to real and imaginary parts of $\chi^{(1)}$ (see Section 17.2.1.2 for more detail). Knowledge of the complex nature of $\chi^{(3)}$ is critical for NLO applications. While nonlinear loss can be utilized for certain applications (see Section 17.4.2.3), it is deleterious for most applications (see Section 17.4.1) and in these cases it should be avoided.

17.1.3.5 Dispersion of $\chi^{(3)}$

As mentioned above, the variation of both the real and imaginary parts of the susceptibility (either $\chi^{(1)}$ or $\chi^{(3)}$) as a function of frequency is known as dispersion. A curve depicting typical dispersion for the Re($\chi^{(1)}$) and Im($\chi^{(1)}$) in a medium with a single electronic resonance at ω_0 is shown in Figure 17.3a. A few observations can be made concerning the dispersion of $\chi^{(1)}$:

- As one moves far away from the resonance, i.e., as $\omega \rightarrow 0$, $\chi^{(1)}$ becomes purely real and as one moves onto resonance, i.e., as $\omega \rightarrow \omega_0$, $\chi^{(1)}$ becomes purely imaginary.

* The nondegenerate analogues to these processes are equally valid in this context, although they are not directly related to the $\chi^{(1)}$ processes. Furthermore, stimulated Raman scattering is an example of a nondegenerate two-photon process where the Im($\chi^{(3)}$) is negative, implying gain as opposed to loss.

- Far away from the resonance, the frequency dependence of $\chi^{(1)}$ is relatively weak, whereas the frequency dependence of $\chi^{(1)}$ is quite pronounced near resonance.
- As one approaches the resonance, the value of $\mathrm{Re}(\chi^{(1)})$ increases in magnitude. This is known as resonance enhancement.

These observations are applicable to the dispersion of $\chi^{(3)}$ as well. However, as noted above, higher-order susceptibilities are subject to multiphoton resonances in addition to one-photon resonances, unlike $\chi^{(1)}$. For this reason, the dispersion of $\chi^{(3)}$, for the same material described in Figure 17.3a, could show significantly more complex behavior than for $\chi^{(1)}$. This is illustrated in Figure 17.3b. It is precisely this complex frequency dependence that often prevents direct comparison of $\chi^{(3)}$ values associated with different third-order phenomena. Clearly, $\chi^{(3)}(-(\omega_1+\omega_2+\omega_3);\omega_1,\omega_2,\omega_3)$ (associated with general four-wave mixing phenomenon) will be different from $\chi^{(3)}(-\omega;\omega,-\omega,\omega)$ (linked to the intensity-dependent refractive index) even if the ω_σ frequencies are equivalent due to the disparate frequencies of the incident fields. This is also true if the incident field frequencies are the same but the frequency of the modified field is different, i.e., $\chi^{(3)}(-\omega;\omega,-\omega,\omega)$ versus $\chi^{(3)}(-3\omega;\omega,\omega,\omega)$ (third-harmonic generation, THG). As an example, THG in polyacetylene films has benefited from significant resonant enhancement due to the presence of multiple electronic resonances [6]; yet, this would not necessarily be the case for the $\chi^{(3)}$ associated with the intensity-dependent refractive index. However, it should be noted that if all the frequencies involved in the third-order process are far away from any resonance of the material such that the dispersion of $\chi^{(3)}$ is negligible, a comparison between $\chi^{(3)}$ values associated with different phenomenon is reasonable. Under these conditions, the material is said to possess Kleinman symmetry [8].

There are certainly practical implications associated with the dispersion of $\chi^{(3)}$. The most obvious is the effect of enhancement of the nonlinearity as one approaches a resonance. This enhancement can exceed several orders of magnitude when comparing a far off-resonant value of $\chi^{(3)}$ with its on-resonant value. This enhancement is not without its consequences. As mentioned above, close proximity to a resonance immediately implies absorption loss, which is typically quite detrimental for many types of optical applications. Furthermore, approaching resonance typically leads to an increase in the temporal response of the nonlinearity (see below for details), an effect that limits the use of a material in an optical signal-processing application where speed is of paramount importance. These two consequences are associated with the transition from a third-order optical phenomenon to a first-order optical phenomenon. The threshold between these two regimes is not clearly delineated and is discussed further in Section 17.2.1.4. In addition to potentially benefiting from resonance enhancement, knowledge of the dispersion of the $\mathrm{Im}(\chi^{(3)})$ can aid in the identification of spectral regions of low optical loss, or equivalently high transparency, where certain optical applications may be most viable.

17.1.3.6 Temporal Response of $\chi^{(3)}$

While the temporal response of the nonlinearity plays a critical role in device performance for NLO applications, it also serves as a good metric in determining the mechanism responsible for the nonlinearity. For instance, as alluded to in Section 17.1.2, even though the bound electronic OKE and the orientational Kerr effect manifest themselves in a similar manner as nonlinear processes [3] and possess the same frequency dependence of $\chi^{(3)}$, the significant difference in their temporal responses allows for differentiation between the two. In general, the third-order nonlinear polarization in molecular systems can be induced by either electronic or nuclear mechanisms. The temporal response times of these various mechanisms can span many orders of magnitude and a summary of this continuum is given in Figure 17.4.

Nonresonant electronic nonlinearities, also known as bound electronic nonlinearities, result from a redistribution or distortion of the electron cloud about fixed nuclei. The ability of the electron cloud to respond to the rapidly oscillating optical field gives rise to an ultrafast response for the bound electronic nonlinearity that corresponds to a few tens of femtoseconds [9]. This mechanism is responsible for a number of the third-order phenomena listed in Table 17.1 such as the DC Kerr effect,

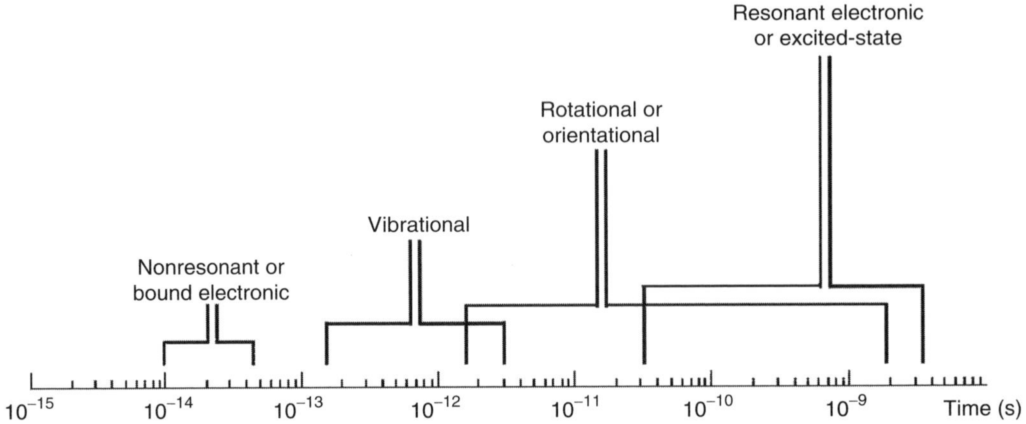

FIGURE 17.4 Typical temporal responses for various mechanisms of the third-order nonlinear polarization in molecular systems.

electric-field-induced second-harmonic-generation (EFISH), THG, sum/difference frequency generation (SFG/DFG), intensity-dependent refractive index, bound electronic OKE, 2PA/2PE, and DFWM. Nuclear mechanisms involve rearrangement of the nuclei from their equilibrium positions and typically encompass both rotational and vibrational mechanisms. Vibrational effects occur when the frequencies of two incident fields are such that the difference between them, $\omega_{vib} = |\omega_1 - \omega_2|$, can coherently excite a vibrational mode of a molecule. Typically, these excitations have response times that vary from a few hundreds of femtoseconds to picoseconds [10]. Vibrational mechanisms account for the third-order phenomena of coherent stokes Raman scattering (CSRS), coherent anti-stokes Raman scattering (CARS), stimulated Raman scattering (SRS), and Raman-induced Kerr effect (RIKE). Molecular rotational or orientational effects manifest themselves in molecules with anisotropic linear polarizabilities such that the molecule experiences a torque when trying to align its more polarizable axis with an applied optical field. The reorientational times of these mechanisms can range from a few picoseconds to nanoseconds depending on the size of the molecule, the viscosity of the environment, and the molecule's electrostatic interaction with its environment [11]. The phenomena associated with these mechanisms are the reorientational Kerr effect and DFWM. As mentioned above, when certain optical frequencies fall within the vicinity of a material resonance the nonresonant electronic nonlinearity can become a resonant one. Since resonant interactions involve actual perturbations of the ground and excited-state populations and not merely electron cloud redistribution, this is not strictly a third-order process but rather an effective $\chi^{(3)}$ process (see Section 17.2.1.4 for more details). Nonetheless, these resonant electronic or excited-state nonlinearities are important, particularly when one attempts to exploit resonance enhancement of $\chi^{(3)}$, and therefore are included here. The temporal responses of these resonant interactions are governed by the lifetime of the populated excited state and these lifetimes can typically range from hundreds of picoseconds to several nanoseconds in organic systems [11]. The phenomena most likely to be affected by this resonant nonlinearity are the OKE (no longer bound electronic), intensity-dependent refractive index, and DFWM.

17.2 SELECTED 3RD-ORDER NONLINEAR OPTICAL PHENOMENA AND CHARACTERIZATION TECHNIQUES

The previous section discussed the origin of the third-order nonlinear polarization and the numerous unique phenomena that can occur as a result of the light–material interaction. The first part of the current section provides a more in-depth discussion of selected phenomena that are most germane for applications involving third-order nonlinearities (see Section 17.4). Furthermore, Section 17.1

detailed the various aspects of $\chi^{(3)}$ and the second part of this section discusses a number of characterization techniques that can be used to quantify these parameters. For consistency, the techniques described will be the techniques used to characterize the aforementioned selected nonlinear phenomena.

17.2.1 Selected 3rd-Order Nonlinear Optical Phenomena

The particular phenomena detailed here were chosen for their utility in NLO applications. These are THG, the intensity-dependent refractive index, DFWM, and effective $\chi^{(3)}$ phenomena or $\chi^{(1)}$:$\chi^{(1)}$ phenomena. These processes are more amenable to applications because their input optical fields can be single frequency (i.e., degenerate) and single polarization in nature, and, in most cases, the processes only require a single beam input (with the exception of DFWM). This reduces both the complexity of the optical layout and the demands placed on the driving light source. This certainly does not preclude the use of the other third-order phenomena (see Table 17.1) in device applications, however a detailed discussion of their attributes and potential utility are beyond the scope of this chapter. For a comprehensive description of these phenomena, the reader is directed to Refs. [3,8]. It should also be pointed out that, in addition to the true third-order phenomena and the $\chi^{(1)}$:$\chi^{(1)}$ phenomena discussed here, there also exists effective $\chi^{(3)}$ phenomena that manifest themselves as $\chi^{(2)}$:$\chi^{(2)}$ phenomena. These are known as cascaded NLO effects and a description of these processes can be found in Ref. [12].

17.2.1.1 Third-Harmonic Generation

While THG involves degenerate excitation fields (at ω) like the other nonlinear phenomena discussed in this section, the nonlinear polarization term involved is driven at 3ω and therefore is associated with $\chi^{(3)}(-3\omega;\omega,\omega,\omega)$. This polarization term is the consequence of the general phenomenon of frequency-mixing that results in the generation of a new frequency at the third-harmonic of the driving field frequency. Furthermore, this polarization operating at the third-harmonic oscillates so rapidly that only the high frequency motions of the system can respond. In other words, THG is only sensitive to the mechanism of redistribution or distortion of the electron cloud. For this reason, THG is the preferred method for characterizing the bound electronic nonlinearity since other competing processes do not contribute to the observed nonlinearity. It should also be reiterated that unlike second-harmonic generation, which exists only in systems that lack center of inversion symmetry (see Section 17.1.3.3), THG can, in principle, be nonzero for materials with any type of symmetry. Therefore, even isotropic or amorphous materials can exhibit this third-order phenomenon.

For a frequency-mixing process such as THG where new frequencies are generated, a certain condition exists for significant energy transfer to occur between the fundamental wave and the harmonic wave. This condition is known as phase-matching, and for THG this is given by

$$\Delta k = \frac{6\pi\Delta n}{\lambda} = 0; \quad \Delta n = n(3\omega) - n(\omega) \tag{17.10}$$

where

Δk is the phase or wave-vector mismatch
λ is the wavelength of the fundamental wave

Obviously for complete phase-matching to exist, the refractive indices of the material at ω and 3ω must be equal. Equivalently, the phase velocities of the fundamental and harmonic waves must be equal so that efficient energy transfer can occur as the waves propagate through the

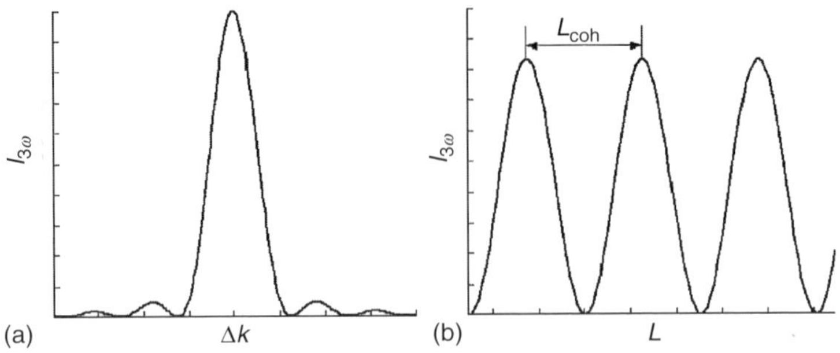

FIGURE 17.5 Third-harmonic signal intensity plotted versus (a) wave-vector or phase mismatch Δk for a constant thickness and versus (b) thickness L for a constant wave-vector mismatch. L_{coh} denotes the coherence length in the THG fringe pattern.

medium. Typically, in isotropic materials far from resonance, $n(3\omega) > n(\omega)$ due to normal dispersion of the index of refraction (see Figure 17.3). For this reason, Δk cannot equal zero in this regime. However, it is possible for phase-matching to be achieved in isotropic materials in the region of anomalous dispersion (i.e., region of dispersion near a resonance). Another more common way to achieve phase-matching is to employ anisotropic or birefringent media. By using the different refractive indices associated with the two optical axes, one can choose a proper direction of propagation that satisfies $n(3\omega) - n(\omega) = 0$. This is discussed in more detail in Chapter 16.

The third-harmonic intensity $I_{3\omega}$ from a transparent material of thickness L is given by [13]

$$I_{3\omega} \propto \left|\chi^{(3)}(-3\omega;\omega,\omega,\omega)\right|^2 I_\omega^3 \frac{\sin^2(\Delta kL/2)}{(\Delta kL/2)^2} L^2 \quad (17.11)$$

where I_ω is the intensity of the fundamental. The dependence of $I_{3\omega}$ on the value of Δk is shown in Figure 17.5a. Clearly, deviation from complete phase-matching can result in a rapid reduction of the third-harmonic intensity. Furthermore, Equation 17.11 also implies that for a finite value of phase-mismatch varying the sample interaction length, L, results in an oscillatory behavior for the resulting third-harmonic intensity. This is shown in Figure 17.5b. Essentially, the harmonic and the induced polarization interfere with one another with a periodicity that is known as the coherence length. This length is defined as $L_{coh} = \pi/\Delta k$ and is shown in Figure 17.5b. In typical materials, L_{coh} is on the order of several micrometers. It is this oscillatory behavior of $I_{3\omega}$ with sample thickness that is measured when employing THG as a nonlinear characterization technique (see Section 17.2.2.1).

17.2.1.2 Intensity-Dependent Refractive Index

Unlike THG, the processes discussed in this section are generated from a nonlinear polarization term that is driven at the same frequency as the incident electric field, i.e., $\chi^{(3)}(-\omega;\omega,-\omega,\omega)$. It was mentioned in Section 17.1.3.4 that this third-order susceptibility has direct analogues to the linear optical phenomena of absorption and refraction that are associated with the real and imaginary components of the first-order susceptibility. These nonlinear refraction and nonlinear absorption phenomena are identified by the degenerate forms of the intensity-dependent refractive index and 2PA (see Table 17.1). These two processes are typically quantified by the intensity-dependent refractive index, n_2, and the 2PA coefficient, β_I.* These can be related to the real and imaginary parts of $\chi^{(3)}(-\omega;\omega,-\omega,\omega)$ as follows:

$$n_2 = \frac{3}{4\varepsilon_0 n_0(\omega)^2 c} \mathrm{Re}\left[\chi^{(3)}(-\omega;\omega,-\omega,\omega)\right] \tag{17.12}$$

$$\beta_I = \frac{3\omega}{2\varepsilon_0 n_0(\omega)^2 c^2} \mathrm{Im}\left[\chi^{(3)}(-\omega;\omega,-\omega,\omega)\right] \tag{17.13}$$

where
$n_0(\omega)$ is the linear refractive index at ω
ε_0 is the free-space permittivity
c is the vacuum speed of light

As mentioned before, the size of the change in absorption and refraction as a result of these terms is dependent on the magnitude of the incident electric field or, more appropriately, the incident intensity. By using the forms for n_2 and β_I given above, the modifications to the refractive index, n, and the absorption coefficient, α_T, can be written simply as

$$n = n_0 + \Delta n = n_0 + n_2 I \tag{17.14}$$

$$\alpha_T = \alpha_L + \Delta\alpha = \alpha_L + \beta_I I \tag{17.15}$$

where
α_L is the linear absorption coefficient
I is the incident intensity

For a single beam input, the material modifications generated by the beam will cause a change in the phase or amplitude of the field associated with that same incident beam. This is known as a self-action effect.

While there are a number of viable applications associated with 2PA, this process is described extensively in Chapter 18 and therefore will not be addressed further here. Formally, the OKE is the third-order process responsible for n_2. OKE is given its name because of its similarity to the DC Kerr effect, where the change in index is dependent on the square of the zero-frequency electric field. However, oftentimes in the literature, OKE is discussed specifically when addressing the induced birefringence caused by an intense beam of light. While this is a natural consequence of the intensity-dependent refractive index, it requires a second beam to detect or experience this optically induced anisotropy (see Section 17.2.2.4) and is therefore not strictly a self-action effect. To avoid confusion, the term intensity-dependent refractive index will be used to describe the self-action processes here. As alluded to in Section 17.1.3.6, the intensity-dependent refractive index found in molecular systems can be induced by a number of different mechanisms. These mechanisms can be electronically induced such as the bound electronic OKE, vibrationally induced as in RIKE, or rotationally induced like the reorientational Kerr effect. In addition to these nonresonant mechanisms, resonant nonlinearities can contribute to n_2 as well and these are discussed briefly in Section 17.2.1.4. Finally, unlike

* The 2PA coefficient is given a subscript of "I" to distinguish it from the second-order polarizability, β. The same is true for the linear and total absorption coefficients that have been given the subscripts of "L" and "T," respectively, to distinguish them from the first-order polarizability, α.

the localized effects discussed above, nonlocal mechanisms such as thermal nonlinearities and photorefractive effects (see Chapter 19) can also contribute to n_2. It should be noted that due to the spatially localized and temporally ultrafast response of the bound electronic OKE, this is the mechanism most often employed in a number of device applications (see Section 17.4.2) involving n_2.

As mentioned above, the effect of n_2 is to impart an intensity-dependent phase change on the incident beam. This phase change, in accordance with Equation 17.14, can be written as

$$\Delta\varphi = k_0 \Delta n L = k_0 n_2 I L \tag{17.16}$$

where
k_0 is the wavevector given by $2\pi/\lambda_0$
L is the length of the nonlinear material

Propagation effects associated with this phase change are responsible for a huge variety of interesting and useful phenomenon. To understand these effects it is important to realize that since the phase change is proportional to the intensity of the beam, it will take the same shape as the intensity distribution. Therefore, a laser beam with a nonuniform spatial intensity distribution (e.g., a Gaussian beam shape) passing through a medium with a strong n_2 will map this nonuniformity directly onto the induced phase change. This is illustrated in Figure 17.6. This induced phase change will affect the beam by delaying its phase front at the center more than the phase front at its wings. This causes phase curvature similar to what a beam experiences when passing through a positive lens and consequently the beam tends to focus. This is called self-focusing and it is depicted in Figure 17.6a. This occurs for a medium with a positive value for n_2 (i.e., a positive $\mathrm{Re}(\chi^{(3)})$ as discussed in Section 17.1.3.4); however, self-defocusing can occur in a medium with negative n_2 (see Figure 17.6b). Furthermore, by altering the magnitude of the beam intensity, one can vary the focal length of this effective lens. This propagation effect is responsible for self-trapping in spatial solitons (see Section 17.4.2.1), Kerr lens modelocking [14], and plays a primary role in the Z-scan nonlinear characterization technique (Section 17.2.2.2).

Just as a spatially dependent intensity distribution can lead to self-focusing and defocusing, a temporally dependent intensity distribution (e.g., a Gaussian pulse shape) can lead to self-phase modulation (SPM). SPM is essentially a spectral change experienced by a pulse. To understand this, one can write down the instantaneous frequency shift equal to the rate of the change of phase. Using the phase given in Equation 17.16, this becomes

$$\delta\omega(t) = -\frac{\partial}{\partial t}\Delta\varphi = -k_0 n_2 L \frac{\partial I(t)}{\partial t} \tag{17.17}$$

Clearly, the frequency shift is time dependent and is proportional to the derivative of the pulse intensity profile. Therefore, when a pulse passes through a material with a positive n_2, its leading edge (which has a positive slope) becomes red shifted while the trailing edge (negative slope)

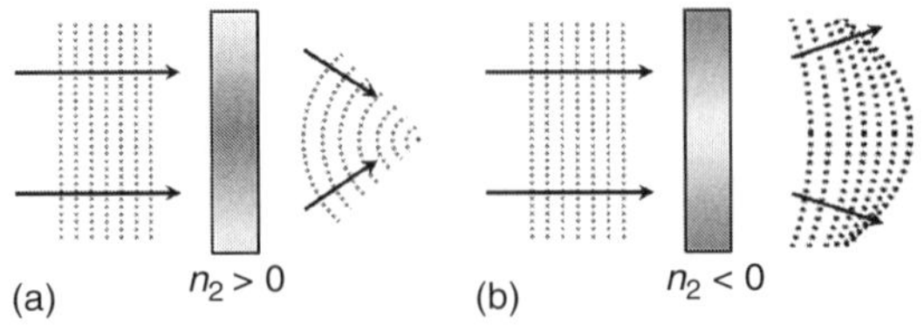

FIGURE 17.6 The effects of a spatially nonuniform-induced phase change (illustrated in the samples) on incident phase fronts. (a) Depicts self-focusing in a medium with a positive n_2 or $\chi^{(3)}$ and (b) depicts self-defocusing in a medium with a negative n_2 or $\chi^{(3)}$. The arrows denote the direction of energy flow.

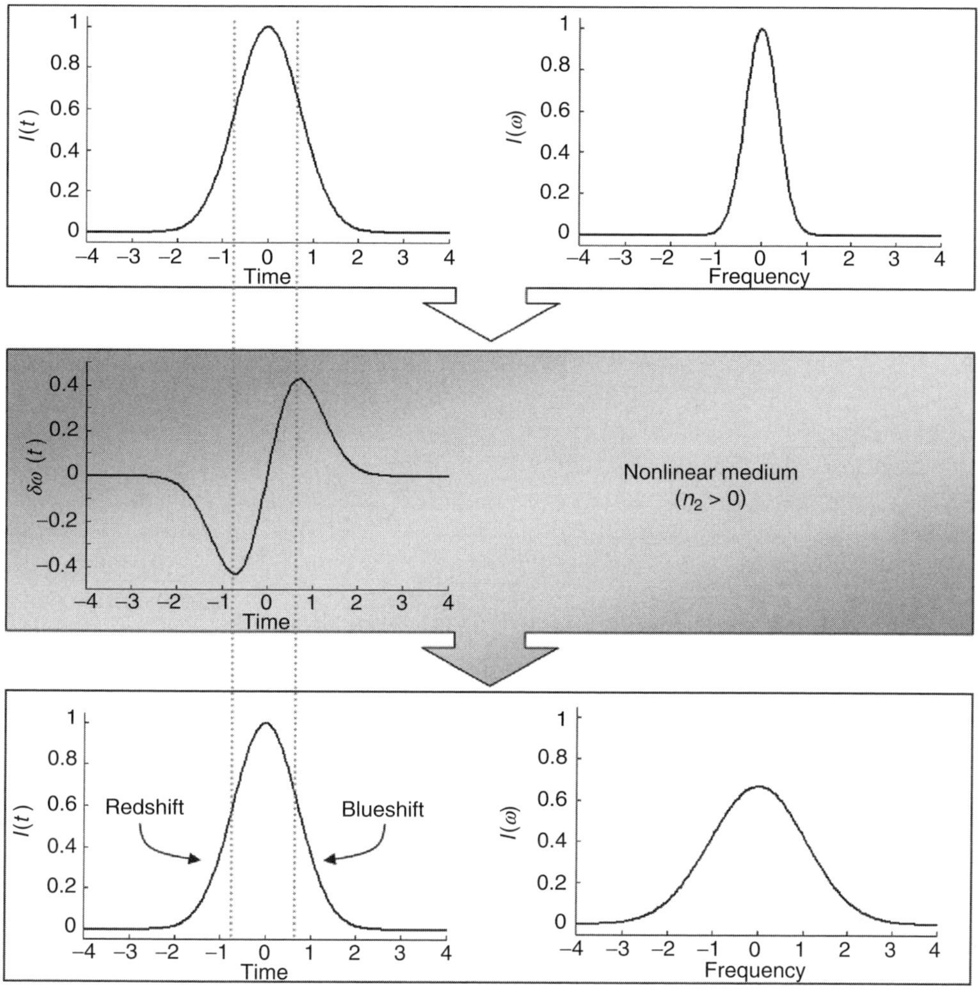

FIGURE 17.7 The effects of self-phase modulation on a pulse passing through a medium with a positive n_2 or $\chi^{(3)}$. The temporal pulse profile and frequency spectrum before entering the medium (top), the SPM-induced frequency shift in the medium (middle), and the pulse profile and spectrum after exiting the medium (bottom).

becomes blue shifted. This can be seen in Figure 17.7. This temporally dependent frequency shift makes temporal soliton propagation possible (see Section 17.4.2.1). Furthermore, this frequency shift also results in a broadening of the initial pulse spectrum (see Figure 17.7) that is proportional to the incident intensity. For pulses with very large intensities, this can result in ultra broadband continuum generation (or, white-light continuum [WLC] generation) [15] where the incident optical spectrum can be broadened by over a hundred times.

17.2.1.3 Degenerate Four-Wave Mixing

The phenomenon of DFWM involves the same nonlinear polarization term (see Table 17.1) as the intensity-dependent refractive index, that is $\chi^{(3)}(-\omega; \omega, -\omega, \omega)$. However, since this process involves the mixing of three spatially distinguishable beams to produce a fourth beam that has a different propagation direction than the original beams, it is inherently not a self-action process. It should be noted that, in general, the beams can also be distinguished by different polarizations (this is discussed briefly in

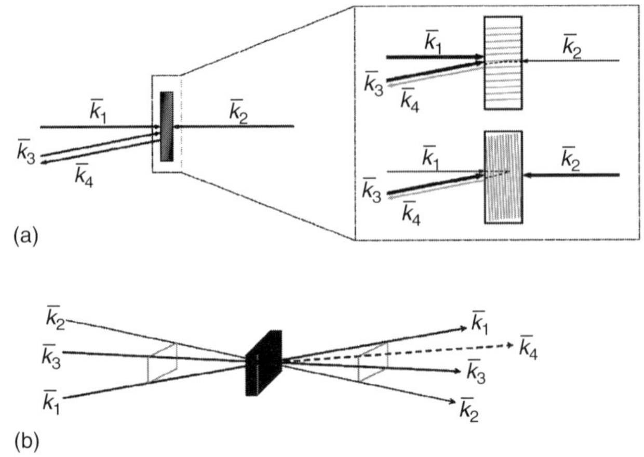

FIGURE 17.8 Typical geometries for DFWM: (a) counter-propagating or backward geometry and (b) folded-boxcars or forward-scattering geometry. Inset to (a) shows the laser-induced gratings produced in the backward DFWM geometry as described in the text.

Section 17.2.2.3); however, for the purposes of this discussion, the optical fields will be assumed to have the same polarization. Like the THG process described above, phase-matching must be satisfied during the DFWM process for a significant amount of intensity to be transferred to the fourth beam. For THG, this involves the matching of refractive indices at the fundamental and third-harmonic such that the magnitude of Δk was equal to zero. The degeneracy of the frequencies involved in DFWM does not require this type of matching but instead requires that the k-vectors of all the propagating beams sum to zero essentially guaranteeing conservation of linear momentum. This can be written quite simply as

$$\Delta\bar{k} \equiv \bar{k}_1 + \bar{k}_2 + \bar{k}_3 + \bar{k}_4 = 0 \tag{17.18}$$

To visualize this interaction, two geometries typically employed for producing DFWM are schematically represented in Figure 17.8. Figure 17.8a illustrates the counterpropagating or backward geometry, whereas Figure 17.8b shows the forward-scattering or folded-boxcars geometry. In the backward geometry, beams 1 and 2 propagate in opposite directions (i.e., $\bar{k}_1 = -\bar{k}_2$) and the third beam is incident at a small angle with respect to the two. From the phase-matching condition in Equation 17.18, it is clear that the fourth beam must exit the sample counterpropagating to beam 3, i.e., $\bar{k}_4 = -\bar{k}_3$. Conversely in the forward-scattering geometry, the three incident beams originate on the corners of a box and propagate in the same axial direction through the sample. Following interaction in the medium, the fourth beam propagates toward the remaining corner of the box. This geometry is also phase-matched.

An intuitive way to understand the DFWM process is through the laser-induced grating (LIG) picture [16]. The picture is conceptually similar to the description for the intensity-dependent refractive index where light induces a material change followed by the material modifying the incident light. In this case, two laser beams interfere to create an intensity-modulated grating* in the medium and a third incident beam diffracts off this grating to create the fourth beam or signal beam. This can be seen in the inset to the schematic for the backward DFWM geometry shown in Figure 17.8a.

* Strictly, this is only true for beams that are polarized parallel to one another. Orthogonally polarized beams can also create gratings but these are described as polarization gratings since the orthogonality of the beams prevents intensity modulation.

In fact, two different gratings are formed. The first involves the interference of beams 1 and 3 to form a coarse (i.e., large period) transmission grating that diffracts beam 2. The second grating, a fine (i.e., small period) reflection grating, is formed by beams 2 and 3 and it diffracts beam 1. The intensity of the signal beam is determined by the amplitude of the induced grating and the intensity of the beam being diffracted. This can be written as [17]

$$I_{4,signal} \propto I_3 I_2 I_1 \left|\chi^{(3)}(-\omega;\omega,-\omega,\omega)\right|^2 L^2 \tag{17.19}$$

where L is the length of the nonlinear medium. Therefore, if I_3 is the beam being diffracted, this implies, somewhat intuitively, that the amplitude of the induced grating is determined by the intensity of the interfering beams (I_1 and I_2), the nonlinearity of the material, and the interaction length.

The dependence of the signal beam intensity on the magnitude of $\chi^{(3)}$ implies that both the real and imaginary parts of the third-order nonlinearity contribute. In the LIG picture, this can be interpreted as follows: the nonlinear refractive index (see Equation 17.14) creates a phase grating, whereas nonlinear absorption (see Equation 17.15) creates an amplitude grating. Since both processes contribute to DFWM, it is often difficult to separate the two components when using DFWM as a nonlinear spectroscopic technique. This is discussed further in Section 17.2.2.3. Furthermore, since the form of $\chi^{(3)}$ involved in DFWM is the same as that for the intensity-dependent refractive index, all the various mechanisms described above that contribute to n_2 can also contribute here.

17.2.1.4 Effective $\chi^{(3)}$ Phenomena

As alluded to previously, the nonresonant electronic $\chi^{(3)}$ can become a resonant nonlinearity if one of the optical field frequencies involved in the nonlinear interaction falls within the vicinity of a material resonance. The threshold between these two regimes is not abrupt but rather a continuous transition where the intermediate regime permits resonance enhancement as discussed in Section 17.1.3.5. However, if an optical frequency does become resonant with an electronic transition in a material, linear absorption will take place and excited states will be populated. Although this is a linear process, under certain conditions the system can become strongly perturbed such that substantial population redistribution from the ground to an excited state can occur. Essentially, in order to accurately describe the material response, this strong perturbation requires that higher-order terms in the series expansion of the polarization (see Equation 17.6) be considered, similar to the approach taken to arrive at the third-order polarization terms.

To see how this cumulative first-order process can be depicted as an effective third-order process, one can start from the description of the total absorption coefficient given in Equation 17.15. As mentioned above, the linear absorption coefficient, α_L, is proportional to $\chi^{(1)}$ (or more accurately the Im($\chi^{(1)}$) but this can be implied). In analogy to Equation 17.9, $\chi^{(1)}$ is proportional to the number density of participating molecules, N, and the first-order polarizability, α. When α is associated with a particular electronic transition, the polarizability is often replaced with a more common (although nearly equivalent) parameter called the absorption cross section, σ. Therefore, one can write

$$\alpha_L \propto \sigma_{01} N_0 \tag{17.20}$$

where the subscripts for σ merely correspond to a transition between the zeroth (ground) and first excited state. N_0 has replaced N since essentially all molecules reside in the ground state before excitation. Now, if the material is resonantly excited by an optical field, the population of the ground state decreases by an amount ΔN and the excited state increases by ΔN. Since appreciable population can now exist in the excited state, an electronic transition to a higher state must be taken into account.

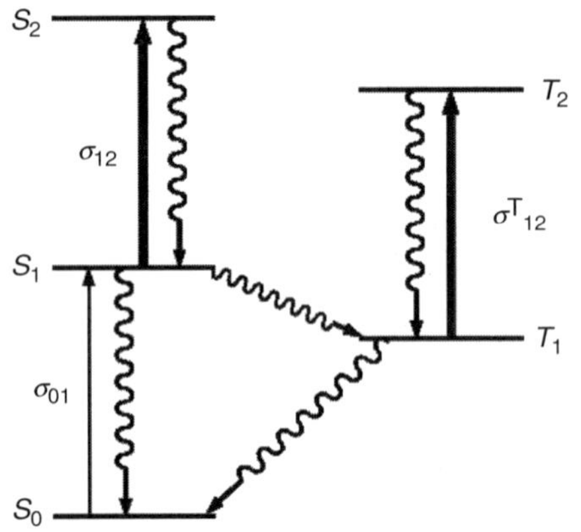

FIGURE 17.9 Five-level state diagram typically used to illustrate the transitions involved in reverse saturable absorption. Singlet and triplet excited-state absorption are denoted by bold arrows.

The left-hand side of Figure 17.9 depicts this transition along with its associated cross section σ_{12}. Taking into account these modified populations as well as this new transition, Equation 17.20 can be rewritten as

$$\alpha_L \propto \sigma_{01}(N_0 - \Delta N) + \sigma_{12}(N_1 + \Delta N) \quad (17.21)$$

Knowing that the initial population of N_1 is essentially zero (see above), Equation 17.21 can be transformed into a form of the total absorption coefficient that is quite similar to Equation 17.15:

$$\alpha_T \propto \alpha_L + \Delta\alpha = \alpha_L + (\sigma_{12} - \sigma_{01})\Delta N \quad (17.22)$$

Clearly, since ΔN is dependent on the intensity of the incident light, one has an intensity-dependent absorption coefficient similar to the nonlinear absorption process described in Section 17.2.1.2.

Depending on the relative magnitudes of σ_{12} and σ_{01}, drastically different mechanisms can occur. If $\sigma_{12} < \sigma_{01}$ the first excited state can become quite populated thereby reducing the possibility for any further absorption from the ground state to take place. In this case, absorption will decrease with increasing intensity and this is known as saturable absorption or ground-state bleaching. When the opposite condition occurs, $\sigma_{12} > \sigma_{01}$, the populated first excited state has a pathway to depopulate, that is, to the higher lying state. In this case, absorption will increase with increasing intensity and one has a condition known as reverse saturable absorption (RSA) or excited-state absorption. This phenomenon is particularly important for optical power limiting discussed in Section 17.4.2.3 and is therefore of greater interest here. Again, the process of RSA is depicted in Figure 17.9, which also illustrates the fact that in addition to singlet absorption, absorption from a triplet state can play a prominent role in RSA as well. Since RSA is clearly the result of two sequential linear absorption processes, it is easy to see why it is defined as a $\chi^{(1)}:\chi^{(1)}$, or effective $\chi^{(3)}$ process. As one may expect, there is also a change in the refractive index that exists as a result of RSA. This Δn and the $\Delta\alpha$ described in Equation 17.22 are mathematically related through the Kramers–Kronig relations and physically related through causality [18]. Finally, it is important to note that the term resonant or effective $\chi^{(3)}$ is loosely defined. This is because the dependence of Δn or $\Delta\alpha$ on intensity may not necessarily be linearly dependent on intensity (as was the case for

Δn associated with n_2) and these values may also be pulse-width dependent since their temporal responses are governed by the lifetime of the populated excited state.

17.2.2 Nonlinear Optical Characterization Techniques

The NLO characterization techniques described here utilize the phenomena discussed in the previous section to characterize $\chi^{(3)}$ or the effective $\chi^{(3)}$. This, however, is only a small sampling of the techniques employed and the reader is directed to Refs. [3,19] for a more detailed survey of the field. Characterization of the third-order susceptibility is mainly undertaken to determine the utility of a particular material in an NLO application (see Section 17.4) or for the purpose of determining molecular nonlinearities that allow for the development of chemical structure—NLO property design strategies (see Section 17.3). Since the majority of established techniques determine the properties of $\chi^{(3)}$, a macroscopic quantity, the relations given in Section 17.1.3.1 are then used to convert to γ. A number of characteristics of $\chi^{(3)}$ that were detailed in Section 17.1.3 can be determined: magnitude and phase, tensorial nature, temporal response, and spectral dependence. Some properties, such as the tensorial nature of $\chi^{(3)}$, may be less critical in terms of device performance than others. Nonetheless, the more complete the knowledge of $\chi^{(3)}$ properties of a material the easier it is to determine its efficacy in a device or the effect a certain chemical structure modification may have on its nonlinearity. For this reason, and because each technique has its own benefits and drawbacks (Table 17.2 provides just such a summary for the techniques discussed here), multiple methods should be employed to fully characterize $\chi^{(3)}$. This also aids in discerning the mechanism that may be responsible for inducing the nonlinear response. Finally, owing to the sometimes sizeable experimental errors associated with nonlinear characterization techniques, redundancy can often add confidence to the final results.

17.2.2.1 Third-Harmonic Generation

The intensity of the THG generated from an unknown sample can be used to determine the bound electronic, third-order nonlinearity $\chi^{(3)}(-3\omega;\omega,\omega,\omega)$. More specifically, one can accomplish this by measuring the oscillatory behavior of the third-harmonic intensity, or $I_{3\omega}$, as the sample thickness is changed as depicted in Figure 17.5b. Fitting this data allows determination of $I_{3\omega}$ (proportional to the amplitude of the fringes) and L_{coh} or the coherence length of the sample (given by the fringe periodicity). By performing this same experiment on a reference material with a known $\chi^{(3)}$, a relative determination of the $\chi^{(3)}$ for the unknown sample can be made. The benefit of such a referential

TABLE 17.2
Characteristics of the 3rd-Order Nonlinear Characterization Techniques Discussed in This Section

Technique Attributes	THG	Z-Scan	DFWM	Pump–Probe	OKE
$\chi^{(3)}$ (real, imaginary, magnitude) discrimination	Mag	Re & Im	Mag	Im	Mag
Temporal discrimination	No	No	Yes	Yes	Yes
Tensorial discrimination	No	No	Yes	No	No
Ease of frequency-dependent analysis	No	No	No	Yes	Yes
Referential or absolute method	Ref	Abs	Ref	Abs	Ref
Simple optical layout	Yes	Yes	No	No	No
Sensitive to multiple types of mechanisms	No	Yes	Yes	Yes	Yes
Good single-to-noise ratio	Yes	No	Yes	No	Yes

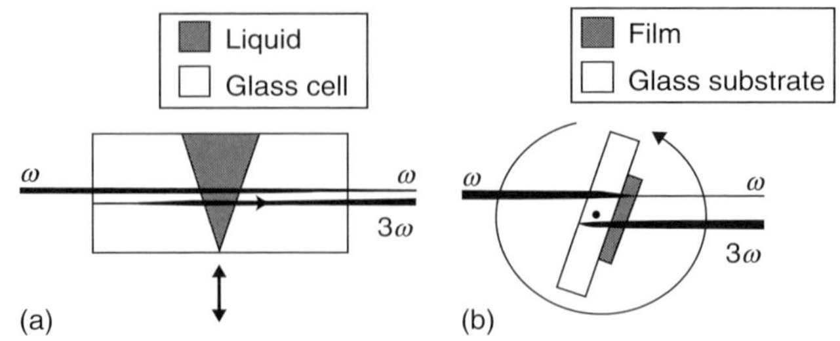

FIGURE 17.10 Typical experimental geometries for the use of THG as a nonlinear characterization technique. In the figure (a) illustrates the wedge geometry fringe technique and (b) shows the slab geometry Maker fringe technique.

technique versus an absolute one is that the intensity of the pump beam, in this case the fundamental, does not have to be well characterized for proper determination of the nonlinearity. Conversely, the nonlinearity of the reference material must be reasonably well known for the method to have any sort of accuracy.

Two typical sample geometries for producing THG interference fringe patterns for characterization of $\chi^{(3)}$ are shown in Figure 17.10. Figure 17.10a illustrates the translation of a wedged sample perpendicular to the beam direction. Since the sample interaction length increases linearly with the displacement, the fringes follow the typical periodic behavior shown in Figure 17.5b. This geometry is particularly well suited for the study of solution-based materials. Figure 17.10b shows the rotation of a slab sample about an axis perpendicular to the beam. Because the sample length increases nonlinearly with the rotation angle, the fringes (also known as Maker fringes) become more closely spaced. This geometry is typically used for characterization of thin films. A complication arises from the fact that all media can generate THG (see Section 17.2.1.1). Consequently, THG from glass windows, substrates, or even air can interfere with the THG from the sample and complicate the resulting data analysis. This problem can be alleviated with the use of appropriate sample holders. For instance, the use of a solution cell with thick front and back, wedged windows (see Figure 17.10a) can effectively eliminate the THG contribution from the surrounding air. By employing this geometry and using the referential technique described above, the $\chi^{(3)}$ of an unknown sample can be determined from the following equation [13]:

$$\chi^{(3)}(s) = \frac{1}{L_{\text{coh}}(s)}\left[(1-R)\chi^{(3)}(g)L_{\text{coh}}(g) + R\chi^{(3)}(r)L_{\text{coh}}(r)\right] \tag{17.23}$$

where

$$R = \sqrt{\frac{I_{3\omega}(s)}{I_{3\omega}(r)}}$$

The denotations "*s*," "*g*," and "*r*" refer to sample, glass, and reference, respectively. Clearly, the contributions due to the glass cell walls cannot be neglected. The coherence lengths for all the materials can be determined by analyzing the fringe measurements while the susceptibilities for the glass and the reference are usually found in the literature. A similar procedure can be undertaken for a thin film following analysis of the Maker fringes [20]. In this case, the sample is typically enclosed in an evacuated chamber to eliminate contributions from the air and the substrate takes the place of the cell walls in the above description.

The main advantages of using THG as a nonlinear characterization technique are its insensitivity to multiple mechanisms as well as its relatively simple optical layout. Also, THG has the potential to have a zero background with proper filtering of the fundamental leading to experimental data with a high signal-to-noise ratio. Although THG has the capability to determine the complex nature of $\chi^{(3)}$, the simple analysis given by Equation 17.23 is based on the assumption of a purely real nonlinearity. The analysis for a complex nonlinearity can be significantly more complicated. However, it should be noted that the THG characterization technique does permit the determination of the complex nature of γ [13]. The methodology used to determine γ is similar to the one used for DFWM (see Section 17.2.2.3). The other main drawback of this technique is its inability to ascertain the temporal response of the nonlinearity or its tensorial nature.

17.2.2.2 Z-Scan

The Z-scan technique [21,22] is a sensitive tool for monitoring the self-action effects associated with nonlinear absorption and nonlinear refraction that were discussed in Section 17.2.1.2. Using this method, the Re($\chi^{(3)}$) and Im($\chi^{(3)}$) (or equivalently the magnitude and phase of $\chi^{(3)}$) can be determined by measuring n_2 and β and then applying the relations given by Equations 17.12 and 17.13. The technique employs a simplified optical layout (see Figure 17.11) that consists of a single, focused laser beam passing through the nonlinear medium under investigation. The transmittance of this beam is monitored as the sample is translated axially along z (hence the name Z-scan), past the beam's focal plane or waist. For a material that exhibits nonlinear absorption, the transmittance will monotonically approach a minimum as the sample arrives at the beam waist and

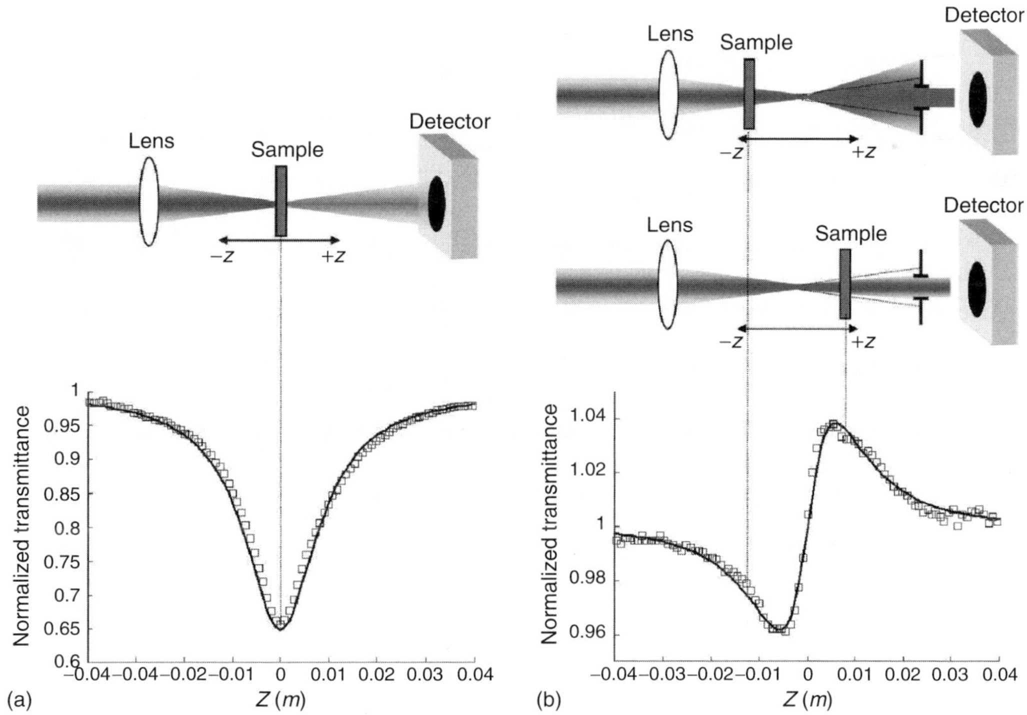

FIGURE 17.11 Optical layouts (top) and typical experimental data performed at 1.3 μm (bottom) for the Z-scan technique. (a) Open-aperture Z-scan data for a 410 μm slab of gallium arsenide. (b) Closed-aperture Z-scan data for a 1 mm slab of fused silica. Open squares represent experimental data points and solid lines are curve fittings performed according to Ref. [21].

the intensity is maximized. This is illustrated in Figure 17.11a. This experiment is known as an "open-aperture" Z-scan because the entire transmitted beam is collected and monitored by the detector. A typical open-aperture Z-scan curve is shown in the lower portion of Figure 17.11a for a 410 μm slab of gallium arsenide. By contrast, a closed-aperture Z-scan, shown in Figure 17.11b, monitors the transmitted beam following passage through a partially obscuring aperture. This makes the experiment sensitive to the effects of self-focusing or defocusing as detailed in Section 17.2.1.2. So, when a material that has a positive n_2 (i.e., it induces self-focusing) is positioned before the waist of the beam, it causes the beam to focus prematurely resulting in a more divergent beam in the far field. This results in a reduced transmittance through the aperture. Conversely, when the material is after the waist of the beam, it will cause the beam to collimate and the transmittance through the aperture will increase. The dashed lines in the optical layouts for the closed-aperture Z-scans (Figure 17.11b) depict the path of the laser beam when no nonlinear material is present. A typical closed aperture Z-scan curve for a 1 mm slab of fused silica is shown in the lower portion of Figure 17.11b. For a system that possesses a negative value for n_2, the closed-aperture scan will be inverted: the peak in transmittance occurs before the valley. Oftentimes, a material will exhibit both nonlinear absorption and nonlinear refraction. In this case, the closed-aperture scan will be sensitive to both phenomena. Fortunately, division of this curve by an open-aperture scan (sensitive to only nonlinear absorption) taken on the same material results in a closed-aperture scan that shows only nonlinear refraction.

Using a fitting procedure [21] on the open-aperture and divided closed-aperture data allows extraction of β and n_2 for the material, respectively. However, one can often quickly estimate n_2 since the measured change in transmittance between the peak and the valley in the divided closed-aperture scan (denoted by ΔT_{pv}) is linearly proportional to the induced nonlinear phase change (see Equation 17.16). Therefore, the simplicity of the technique and its interpretation makes Z-scan straightforward to implement in the characterization of the real and imaginary parts of $\chi^{(3)}$. The only significant difficulty arises from the fact that it is an absolute method, unlike the referential methods of THG and DFWM. For this reason, determination of the spatial/temporal properties of the beam/pulse is necessary to properly characterize the intensity. The calculation of the intensity is therefore typically the largest source of experimental error. Despite the inherent simplicity of the technique, Z-scan does suffer from certain deficiencies. First, since this method is sensitive to all forms of nonlinear absorption and refraction (see Section 17.2.1.2), the underlying physical mechanism responsible for the nonlinearity is often difficult to determine unambiguously without the use of additional characterization techniques. This is mainly due to the lack of temporal resolution that is inherent in this technique. Although temporally resolved Z-scan techniques have been developed [23], they are typically nondegenerate in nature and are difficult to implement. Similarly, although techniques have been developed to determine different elements of the $\chi^{(3)}$ tensor [24], the Z-scan method (like THG) typically determines $\chi^{(3)}_{1111}$. Finally, since the method is single beam (and therefore degenerate in nature) and requires precise determination of the pump intensity, it is quite time consuming to determine the frequency dependence of $\chi^{(3)}$. However, recent advances using a broadband WLC (see Section 17.2.1.2) as a pump source [25] may help to alleviate this shortcoming.

17.2.2.3 Degenerate Four-Wave Mixing

Like THG, the phenomenon of DFWM can be used directly to determine the properties of $\chi^{(3)}$. This is accomplished by monitoring the signal beam (see Figure 17.8) that is diffracted from the laser-induced gratings produced in the material. By varying the intensity or power of the three incident beams, one can determine the power-dependent response of the DFWM signal in a particular sample. Comparing this response with the response found in a material with a known nonlinearity will allow extraction of $|\chi^{(3)}|$ for the samples under investigation. Therefore, DFWM is typically employed as a referential nonlinear characterization technique. The methodology for determining $|\chi^{(3)}|$ is explained below. Rewriting the formula for the intensity of the diffracted beam (Equation 17.19) with the appropriate pre-factors gives [17]

$$I_4 = \frac{9\omega^2}{16\varepsilon_0^2 n_0^4 c^4} \left|\chi^{(3)}\right|^2 I_1 I_2 I_3 L^2 \tag{17.24}$$

where
ω is the frequency of the exciting fields
n_0 is the refractive index of the material at ω

Assuming that all three fields have identical intensities (although this is not strictly required for this derivation), Equation 17.24 can be rewritten more succinctly as

$$I_4 = m_{\text{cubic}} I_{\text{L}}^3 \tag{17.25}$$

where m_{cubic} includes all the pre-factors from Equation 17.24. As expected for a third-order nonlinearity, the signal beam should exhibit a cubic power-dependent response with respect to the incident intensity. This can be verified by taking a log–log plot of the power-dependence of I_4 and verifying that the slope equals three. This is shown in Figure 17.12a for a 500 μm slab of fused silica measured at 1.3 μm. Analysis of this slope can often give insight into the nature of the nonlinearity, particularly if it deviates from the expected value of three [26]. By investigating the power-dependent response of I_4 for both a target sample (sam) and a reference sample (ref) and using Equation 17.25, one can determine the magnitude of $\chi^{(3)}$ for the target sample:

$$\left|\chi_{\text{sam}}^{(3)}\right| = \left|\chi_{\text{ref}}^{(3)}\right| \left(\frac{m_{\text{sam}}}{m_{\text{ref}}}\right)^{1/2} \frac{L_{\text{ref}}}{L_{\text{sam}}} \left(\frac{n_{\text{sam}}}{n_{\text{ref}}}\right)^2 \tag{17.26}$$

A typical power-dependent response for the DFWM signal measured at 1.3 μm is shown in Figure 17.12b for the same plate of fused silica (a material commonly used as a reference sample) mentioned above.

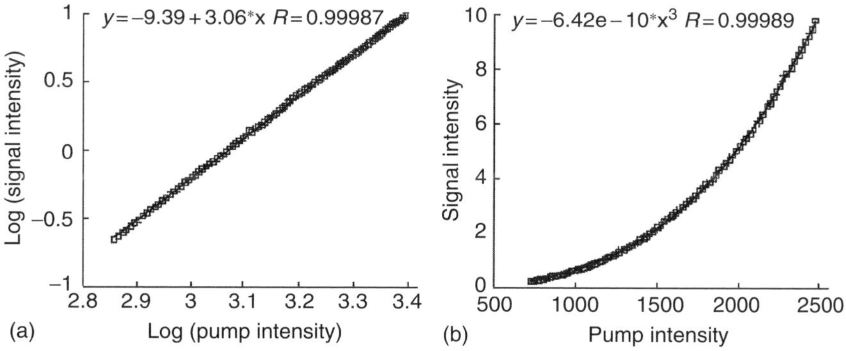

FIGURE 17.12 Experimental data from the power-dependent degenerate four-wave mixing (DFWM) technique for a 500 μm slab of fused silica measured at 1.3 μm. (a) shows the log–log plot of the power-dependent data given in (b). Open squares represent experimental data points and solid lines are curve fittings revealing the cubic dependence of the nonlinearity.

Clearly, it is the magnitude of $\chi^{(3)}$ of a material that is extracted from a material when employing the DFWM technique. Although there are modifications to the technique that may permit determination of the phase as well [27], these techniques add complexity to the standard optical layout. However, similar to the THG characterization technique, the magnitude and phase of γ can be determined using a standard DFWM layout. This involves measuring the bulk values of $|\chi^{(3)}|$ for molecules in solutions of varying concentrations. Since DFWM measures only the magnitude of $\chi^{(3)}$, the nonlinearity of the solution must be written as the following equation:

$$\left|\chi^{(3)}_{\text{soln}}\right| = \left\{\left[\text{Re}\left(\chi^{(3)}_{\text{solv}}\right) + \text{Re}\left(\chi^{(3)}_{\text{solu}}\right)\right]^2 + \left[\text{Im}\left(\chi^{(3)}_{\text{solu}}\right)\right]^2\right\}^{\frac{1}{2}} \tag{17.27}$$

where "Soln," "Solv," and "Solu" are the solution, solvent, and solute, respectively. The value of the $\text{Im}(\chi^{(3)}_{\text{Solv}})$ is neglected because for most solvents there is a negligible contribution to the $\text{Im}(\chi^{(3)})$. It is clear from Equation 17.27 that if the solvent and solute have values of the $\text{Re}(\chi^{(3)})$ that are opposite in sign these can interfere with one another, particularly as the concentration of the solute is increased. Furthermore, the $\text{Im}(\chi^{(3)})$ of the solute can also contribute to the solution $\chi^{(3)}$. A graph depicting these effects on the solution $\chi^{(3)}$, as the concentration of solute is increased, is shown in Figure 17.13. By fitting these concentration-dependent measurements and using the relationship between $\chi^{(3)}$ and γ (see Equation 17.9) one can extract out both the magnitude and phase of γ for the solvent.

While the inability to discern the phase of $\chi^{(3)}$ is a drawback for this particular characterization technique, it possesses a number of benefits that make it a very versatile technique. Since DFWM involves the interaction of three fields in a sample, temporal coincidence of the pulses on the sample is necessary (spatial coincidence of the beams is necessary as well). By delaying one of these pulses with respect to the others, one can monitor the temporal response of the nonlinearity. Furthermore, independent control of the polarization of each participating field allows for characterization of different tensor elements of $\chi^{(3)}$. These two aspects are crucial for determining the cause of the particular nonlinear response since, like Z-scan, this method is sensitive to all forms of nonlinear absorption and refraction. Another benefit is that, like THG, the zero-background nature of the signal beam gives rise to data with a high signal-to-noise ratio. Of course, as a consequence of the tremendous versatility of this technique, the optical layout has considerable complexity. A typical optical layout for a forward-scattering DFWM experiment is shown in

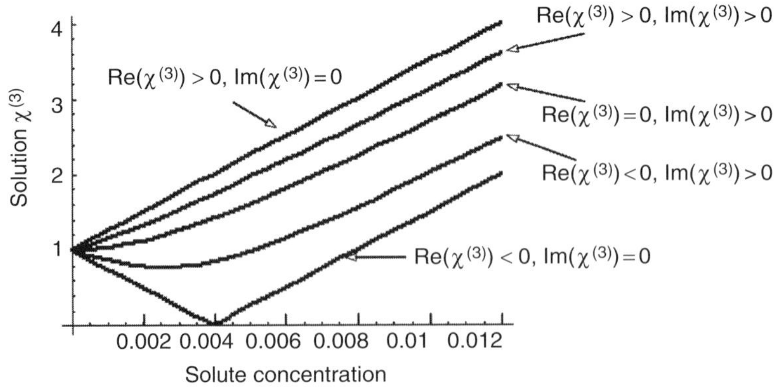

FIGURE 17.13 Graphical representation of the concentration dependence of $\chi^{(3)}$ for a solution as measured by the degenerate four-wave mixing (DFWM) technique. The solvent is assumed to have a positive $\text{Re}(\chi^{(3)})$ and the particular constitution of the solute $\chi^{(3)}$ is described by each plot label.

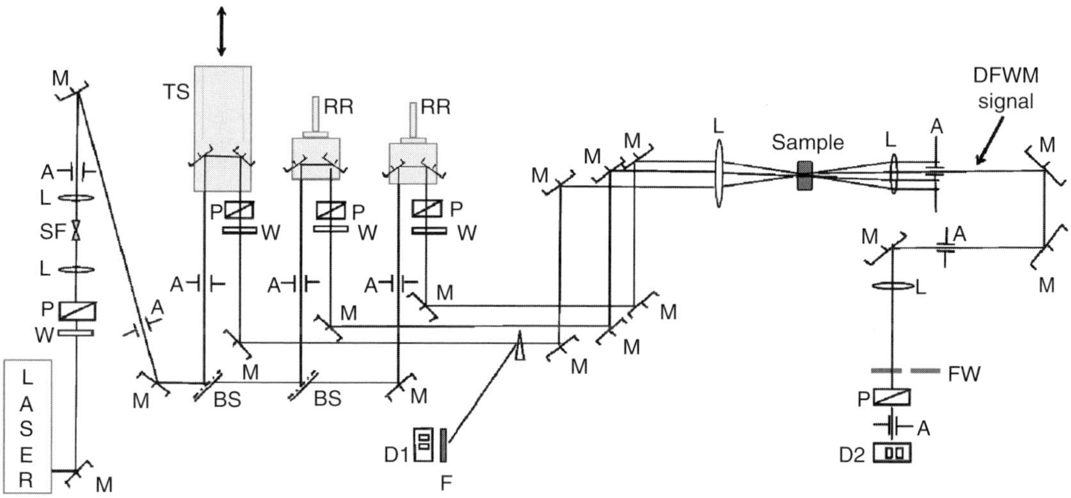

FIGURE 17.14 Typical optical layout for a forward-scattering degenerate four-wave mixing (DFWM) experiment: D1,D2, detectors; W, waveplate; P, polarizer; A, aperture; L, lens; F, filter; FW, filter wheel; M, mirror; BS, beamsplitter; TS, translation stage; RR, retroreflector; SF, spatial filter.

Figure 17.14. Finally, like the two methods mentioned above, this degenerate technique can make determination of the dispersion of $\chi^{(3)}$ quite prolonged and arduous.

17.2.2.4 Pump–Probe and Optical Kerr Effect

The pump–probe and OKE techniques, like DFWM, are multiple beam techniques. Here, a strong excitation beam (pump) induces a nonlinearity within the sample and the transmittance of a weak probe beam through this same material is monitored to determine the changes in its optical properties. In a certain sense, these techniques are complimentary; however, they are both used to determine properties of $\chi^{(3)}(-\omega;\omega,-\omega,\omega)$ or the effective $\chi^{(3)}(-\omega;\omega,-\omega,\omega)$ (in the degenerate case), just like DFWM and Z-scan. In the case of the pump–probe technique, the intensity of the weak probe is diminished as a result of the nonlinear absorption induced by the strong pump beam. Therefore, this technique is sensitive to the $\mathrm{Im}(\chi^{(3)})$ or the effective $\mathrm{Im}(\chi^{(3)})$, i.e., either 2PA or RSA, respectively. The magnitude of the induced nonlinearity can be calculated provided the intensity of the pump beam is known. In other words, the pump–probe technique is typically employed as an absolute method. Furthermore, varying the time delay between the two pulses (pump and probe) while recording the temporal evolution of the probe beam allows for temporal discrimination of the nonlinear processes. Provided the pump and probe pulses are ultrashort in duration (~femtoseconds), one can distinguish between essentially instantaneous processes (i.e., 2PA), and long-lived processes (i.e., RSA), which are often referred to as transient absorption processes.

A schematic for a typical pump–probe optical layout is given in Figure 17.15. The pump pulse has the capability of being delayed and the pump and probe beams are incident on the sample at a small angle with respect to one another to allow for spatial separation later. A very important variation of the pump–probe technique involves the replacement of the single wavelength probe with a broadband WLC (see Section 17.2.1.2) probe. The broadband nature of this probe allows the user to, in theory, monitor the entire nonlinear absorption spectrum with a single pulse. In this manner, the full 2PA spectrum of a sample can be readily obtained [28] or a number of RSA spectra for a single material can be determined [29]. The modification to the pump–probe layout for the generation of a WLC probe is denoted by the dashed box shown in Figure 17.15. Additionally, the probe is dispersed with a spectrometer and monitored by a CCD camera to capture the full nonlinear absorption

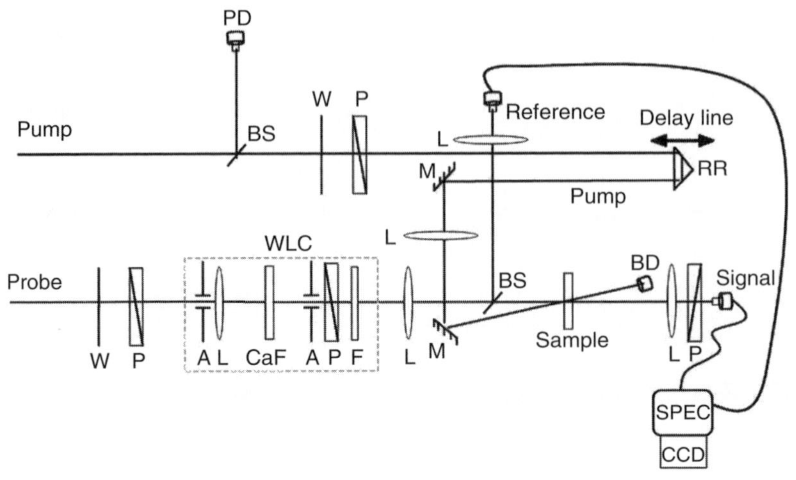

FIGURE 17.15 Typical optical layout for a pump–probe experiment. PD, photodiode; W, waveplate; P, polarizer; A, aperture; L, lens; F, filter; CaF, calcium fluoride window; M, mirror; BS, beamsplitter; RR, retroreflector; BD, beam dump; SPEC, dual fiber input spectrometer; CCD, dual charge couple device array. Optional generation of white-light continuum (WLC) probe is shown inside the dashed box.

spectrum. It should be noted that the spectrum obtained by this WLC transient absorption technique is nondegenerate and therefore is only strictly sensitive to $\chi^{(3)}(-\omega_1;\omega_1,-\omega_2,\omega_2)$.

For the OKE technique, the pump beam induces a birefringence in the medium and the linearly polarized probe beam becomes elliptically polarized following passage through the birefringent medium. The portion of the probe beam transmitted through a final polarizer (oriented perpendicular to the polarization of the original probe beam) will be sensitive to the induced birefringence. The optical layout is identical to the pump–probe layout (Figure 17.15) with the only modification being the polarization of the pump beam being rotated by 45° with respect to the probe beam. Unfortunately, unlike the previous techniques that are sensitive to $\chi^{(3)}_{1111}$ or $|\chi^{(3)}_{1111}|$, the induced birefringence measured by OKE is sensitive to $|\chi^{(3)}_{1122}-\chi^{(3)}_{1221}|$ (either true $\chi^{(3)}$ or effective $\chi^{(3)}$). For this reason, additional variations of the OKE technique [30] are required to determine $|\chi^{(3)}_{1111}|$ explicitly. This drawback is offset by the fact that, like the pump–probe technique, OKE gives temporal information about the nonlinearity and can be used with a broadband probe to measure its frequency dependence as well. Like DFWM, modifications to the OKE technique have been implemented [30] such that determination of the phase of $\chi^{(3)}$ may be made in addition to it magnitude. The signal-to-noise ratio of the data acquired by OKE, like DFWM, is quite high. Finally, the technique can be performed as either a referential or absolute method.

Table 17.2 provides a summary of the attributes (i.e., benefits and drawbacks) of the various nonlinear characterization techniques presented in this section. Although variations of a number of the techniques discussed may allow for increased versatility, as discussed above (i.e., determination of the phase of $\chi^{(3)}$, temporal discrimination, frequency-dependent measurements, etc.), Table 17.2 merely identifies the characteristics of the method implemented in the most common way.

17.3 THIRD-ORDER NONLINEARITES OF CONJUGATED ORGANIC MOLECULES

In this section, the 3rd-order nonlinearities of conjugated molecules are examined. The section begins with an overview of a perturbation theory description of hyperpolarizabilities based on a few states model. A discussion concerning the resonance structures, dipole moments, and bond length alteration (BLA) of conjugated molecules follows. Then, the valence bond–charge transfer model for the structure–hyperpolarizability relationships is developed, with a particular emphasis on the second hyperpolarizability, γ. This section ends with a discussion of experimental results on the third-order nonlinearity of a range of representative conjugated systems, making use of the simple models as a framework for a basic understanding of the trends. The main purpose is to provide some insight into how the nonlinear properties of conjugated molecules are dependent on variations in their structure.

17.3.1 Simple Perturbation Theory Description of Polarizability and Hyperpolarizability

In this section, an examination of the hyperpolarizabilities of conjugated molecules from the point of view of a simple quantum mechanical model will be given, with a particular emphasis on the second hyperpolarizability, γ. The main purpose is to provide some insight into how the nonlinear properties of conjugated molecules are dependent on variations in their structure.

The linear polarizability and hyperpolarizabilities are described in perturbation theory in terms of various order energy corrections due to interaction of the molecule with an applied field (for static fields). For example, the linear polarizability is related to the second-order energy correction and the first and second hyperpolarizabilities are related to the third- and fourth-order energy corrections, respectively [31]. These energy corrections involve sums over all the excited states of the molecule and these sums have terms of the form of products of dipole matrix elements divided by energy denominators (the difference in the transition energy between the coupled states and the frequency of the perturbing field). In general, the contributions to the energy correction are large if the dipole matrix elements are large and if the energy denominator is small. For the sake of simplicity and for a clearer understanding of these contributions, it is helpful to examine expressions for α and β that are based on just the ground state and the lowest energy dipole-allowed excited state and an expression for γ that is based on the ground state and the two lowest energy excited states, as shown below [32–35]:

$$\alpha(2-\text{state}) \propto \frac{\mu_{eg}^2}{E_{eg}} \tag{17.28}$$

$$\beta(2-\text{state}) \propto \frac{\mu_{eg}^2(\mu_{ee}-\mu_{gg})}{E_{eg}^2} \tag{17.29}$$

$$\gamma(3-\text{state}) \propto -\frac{\mu_{eg}^4}{E_{eg}^3} + \frac{\mu_{eg}^2\,\mu_{e'e}^2}{E_{eg}^2\,E_{e'e}} + \frac{\mu_{eg}^2(\mu_{ee}-\mu_{gg})^2}{E_{eg}^3} \tag{17.30}$$

where μ and E are the dipole matrix element and transition energy between the subscripted states, respectively. The subscripts g, e, and e′ label the ground, first excited, and upper excited states, respectively. Note that in the expression above, the frequency of the perturbing field has been set to zero; this is the static field limit.

One can see that α is proportional to μ_{eg}^2 and is inversely proportional to the transition energy. So, for a system with a very strong electronic absorption, or large oscillator strength, and a low transition energy, α will be large. The expression for β contains a factor that is equal to α times the difference in the dipole moment of the excited and ground states divided by the transition energy, which can be thought of as the polarizability times an asymmetry factor ($(\mu_{ee} - \mu_{gg})/E_{eg}$). Large values of β would generally be obtained for highly polarizable molecules that have a large change in dipole moment on going from the ground to the excited state. The expression for γ contains three terms. The first is a negative term (N) that is equal to $-\alpha^2$ divided by the transition energy. The second is a positive term that is the product of squares of transition dipole moments for the $g \rightarrow e$ and $e \rightarrow e'$ transitions divided by a product of transition energies; this term (the T-term) is directly related to the two-photon absorption cross section for the $g \rightarrow e'$ transition and can also be thought of as a product of the ground-state and the excited-state polarizabilities divided by E_{eg}. The third term, which is also positive in sign, is equal to the two-state expression for β times the asymmetry factor mentioned above and is referred to as the dipolar contribution (D) because of the dependence on ($\mu_{ee} - \mu_{gg}$). The net value of γ is, thus, dependent on the relative magnitudes of the N-term and the sum of the T and D terms. For systems where the oscillator strength of the $g \rightarrow e$ transition is quite strong and the transition energy is low, the N-term can dominate and γ can in fact be negative. Conversely, if the two-photon transition ($g \rightarrow e'$) is powerful and low in energy or if the dipole moment difference ($\mu_{ee} - \mu_{gg}$) is sufficiently large, γ would likely be positive. If the interference between the N-term and the sum of the T and D terms is such that a complete cancellation occurs, γ will be zero.

Through quantum chemical calculations for a molecule at a fixed geometry (say the equilibrium ground state geometry), the transition energies, transition dipole moments, and state dipole moments can be obtained for conjugated molecules of small to moderate size. With these values, the polarizability and hyperpolarizabilities can be calculated using the sum-over-states method [36,37]. By considering of a large number of excited states, rather than just the lowest one or two, converged values that account for the contributions of all of these states can be obtained and these provide better accuracy.

17.3.2 Valence Bond and Charge Transfer Resonance Structures, Dipole Moments, and Bond Length Alternation in Conjugated Molecules

What is missing in the description so far is a picture that allows understanding of how the geometry of conjugated molecules varies as one modifies the electron donor and acceptor groups attached to the ends of the conjugated chain, or as one changes the type of conjugated chain. To begin with, consider a simple model system comprising a polyene-type structure with a donor group at one end and an acceptor at the opposite end, as shown in Figure 17.16. An example of such a compound is an octatetraene substituted on the ends by a dimethylamino electron donor group and a formyl acceptor group, whose resonance structures are also shown in Figure 17.16.

The resonance structure on the left (valence bond structure) would be viewed as a good representation of the actual structure of the ground state if the donor–acceptor charge transfer interaction were very small. On the other hand, if the donor–acceptor interaction were very strong such that a full charge is transferred then the structure would be similar to that on the right (charge transfer structure).

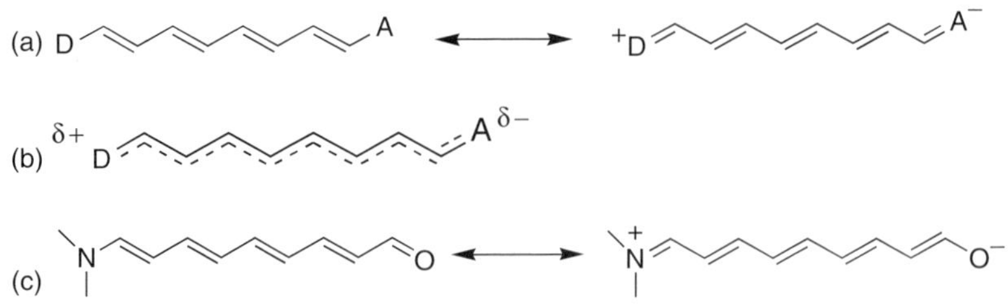

FIGURE 17.16 (a) Neutral (left) and charge transfer (right) resonance structures for a donor–acceptor polyene. (b) Delocalized structure resulting from superposition of the two resonance structures. (c) Resonance structures for $(CH_3)_2N–(CH = CH)_4–CHO$.

The actual ground state structure can be considered as a superposition of these resonance structures, which would have a more delocalized structure that is intermediate to the limiting resonance forms, as illustrated in Figure 17.16b. If the two resonance structures contribute equally, the molecule would have a fractional charge of +1/2 and −1/2 on the ends and carbon–carbon bond orders of 1.5 along the chain. So it can be seen that the structure of a donor–acceptor polyene is dependent on the relative contributions of the two resonance forms, which in turn will depend on the relative energy of these forms. Typically, the charge transfer resonance structure would be higher in energy because of the coulombic energy needed to separate the charges. However, combinations of strong donors and acceptors, or incorporation of cyclic groups that can gain aromaticity upon charge transfer into the bridge or built into the donor or acceptor group, or dissolution in solvents of high polarity can lead to a stabilization of the charge transfer form [38,39]. Through these chemical modifications, the bonding in the π-system, the degree of charge transfer, and the dipole moment of molecules can be varied over essentially the whole range from the neutral to the charge transfer form.

To develop a simple structure–property relationship model, a parameter that gives a measure of the bonding in the π-system must be defined. Such a parameter is the BLA parameter. In simple DA polyenes, BLA is defined as the average of the difference in the length between adjacent carbon–carbon bonds in the polymethine, $(CH)_n$, chain. For the DA polyenes dominated by the neutral form, the BLA is about −0.12 Å and for those where the major contribution is from the charge transfer form BLA is 0.12 Å; typical C–C single bond lengths are about 1.45 Å and C–C double bond lengths are about 1.33 Å in unsubstituted polyenes. It is known that conjugated donor–acceptor molecules will have 3N-6 vibrational modes and some of these will distort the bond lengths along the π-system. If one considers the particular vibrational mode that involves in-phase stretching of the carbon–carbon double bonds along the chain, one can see that this mode will reduce the single bond lengths and lengthen the double bonds on one part of the vibrational cycle and do the opposite on the other part of the cycle, dynamically changing the BLA parameter. For this vibrational coordinate, the equilibrium position of the coordinate will be approximately equal to the average BLA value for the ground state structure. With this connection, one can define vibrational potential functions for the valence bond (VB) and charge transfer (CT) forms in terms of the BLA as a coordinate and use these potentials to calculate the structures and properties that arise from mixing of the these forms [40,41].

Another way of varying the relative energies of the neutral and charge transfer forms is to subject the molecule to an electric field [36,40,42]. The interaction energy of a dipolar molecule with an electric field is given by

$$V' = -\vec{\mu} \cdot \vec{E} \tag{17.31}$$

and this interaction gives rise to a shift of molecular energy levels called the Stark shift. The VB and CT forms have a large difference in dipole moment, which is due to the large π electron charge transfer in the CT form: for the VB form, this is $\mu_{VB} \sim 0$ and for the CT form it is $\mu_{CT} \sim eR_{DA}$, where e is the electronic charge and R_{DA} is the distance between donor and acceptor. Thus, the Stark shift for the charge transfer form would be much greater than that of the neutral form because the dipole moment for the charge transfer form is much larger than for the neutral form. Accordingly, the application of an electric field, or the reaction field due to the solvent dielectric response, gives rise to a much larger Stark shift for the CT form relative to the VB form. If the molecule is aligned so the dipole moment points in the same direction as the field, the interaction is stabilizing, which lowers the energy of the charge transfer form relative to the neutral form. If the strength of the field is large enough, the differential stabilization of the charge transfer form can be so large that the charge transfer form becomes the dominant contribution to the ground state. It can be seen then that application of a strong field pointing along the direction of the dipole moment of the molecule also provides a way of tuning the relative contributions of the charge transfer and neutral forms to the actual ground state, in a manner that can approximate the effects of varying the donor and acceptor groups or other chemical variations that also change the relative contributions of the charge transfer and neutral forms. There now exists a simple picture that can be used to model the structural variation and electronic properties of conjugated donor–acceptor molecules and in the next section the methodology for doing this is addressed.

17.3.3 Valence Bond and Charge Transfer Model for Structure–Hyperpolarizability Relationships

To describe the evolution of the structure of conjugated molecules for varying strengths of an applied field or varying chemical substituents, a vibrational potential along the BLA coordinate, labeled q, is associated with each of the valence bond and charge transfer resonance structures. The mixing of the valence bond and charge transfer forms is then calculated for each position on the BLA coordinate using two-state quantum mechanics. From this analysis, one can compute the ground and excited potential energy surfaces for the molecular eigenstates, obtain the equilibrium BLA value for the ground state, and calculate the dipole moment, polarizability, and hyperpolarizabilities [40]. This calculation can be repeated for varying energy differences of VB and CT, as would arise from the varying strength of an applied field, changing the donor or acceptor strength or other variables as described above.

The potentials for the VB and CT states are taken as harmonic potentials, $V_{\text{harm}} = k(q - q^{\text{o}})^2$, where q^{o} is the equilibrium coordinate for VB or CT. Specifically, the potential energies for the VB (E_{VB}) and CT (E_{CT}) states are written as

$$E_{\text{VB}}(q) = \frac{1}{2}k(q - q^{\text{o}}_{\text{VB}})^2 \tag{17.32}$$

$$E_{\text{CT}}(q) = V_{\text{o}} + \frac{1}{2}k(q - q^{\text{o}}_{\text{CT}})^2 \tag{17.33}$$

where
k is the harmonic potential constant
q^{o}_{VB} and q^{o}_{CT} are the BLA values for the VB and CT resonance forms

The mixing of the VB and CT forms results in adiabatic ground-state and excited-state potential surfaces. As will be seen, the equilibrium value of BLA for the ground state depends on, among other factors, the zero-order energy difference, $V_{\text{o}} = E_{\text{CT}}(q^{\text{o}}_{\text{CT}}) - E_{\text{VB}}(q^{\text{o}}_{\text{VB}})$, and the electronic coupling between the VB and CT states, t, which is taken to be independent of q. To compute the mixing of the VB and CT state, one solves the Schrödinger equation by diagonalizing the simple two-state Hamiltonian in matrix form:

$$H = \begin{pmatrix} E_{\text{VB}} & -t \\ -t & E_{\text{CT}} \end{pmatrix} = \begin{pmatrix} 0 & -t \\ -t & V \end{pmatrix} \tag{17.34}$$

using the wavefunctions for the VB (ψ_{VB}) and CT (ψ_{CT}) forms as a basis; here $V = E_{\text{CT}}(q) - E_{\text{VB}}(q)$. The diagonalization is performed over a range of q values to construct the ground- and excited-state potential surfaces. The wavefunctions obtained for the ground and excited states (Ψ_{g} and Ψ_{e}) are expressed in terms of VB and CT wavefunctions as follows:

$$\Psi_{\text{g}} = \sqrt{1-f}\,\psi_{\text{VB}} + \sqrt{f}\,\psi_{\text{CT}} \tag{17.35}$$

$$\Psi_e = -\sqrt{f}\,\psi_{VB} + \sqrt{1-f}\,\psi_{CT} \tag{17.36}$$

where f is the square of the coefficient of the CT form in the ground state (the charge transfer fraction in the ground state), which is given by

$$f = \frac{1}{2} - \frac{V}{2\left(V^2 + 4t^2\right)^{1/2}} \tag{17.37}$$

Because f is a quantitative measure of the degree of mixing of VB and CT, it plays a major role in the nature of the structure–electric property relationships for donor–acceptor molecules. The energies of the ground state (E_{gr}) and excited state (E_{ex}) for the two-state system are given by

$$E_{gr} = \frac{1}{2}\left(V - \sqrt{V^2 + 4t^2}\right) \tag{17.38}$$

$$E_{ex} = \frac{1}{2}\left(V + \sqrt{V^2 + 4t^2}\right) \tag{17.39}$$

Thus, the energy gap between the ground and excited state, $E_g = E_{ex} - E_{gr}$, is

$$E_g = \sqrt{V^2 + 4t^2} \tag{17.40}$$

With the equations above and the expressions for the potential surfaces of VB and CT, one can now calculate the potential surface for the ground and excited state. Figure 17.17 depicts schematically the VB and CT potential surfaces and the ground state surface for illustrative cases with different values of V_0 (the VB–CT energy difference) and a sizable value for the electronic coupling. Several important points can be gleaned from the illustrated behavior: (1) when V_0 is large, the ground state resembles the neutral resonance form; (2) when V_0 is negative and large in magnitude, the ground state resembles the charge transfer resonance form; and (3) for $V_0 = 0$, the equilibrium

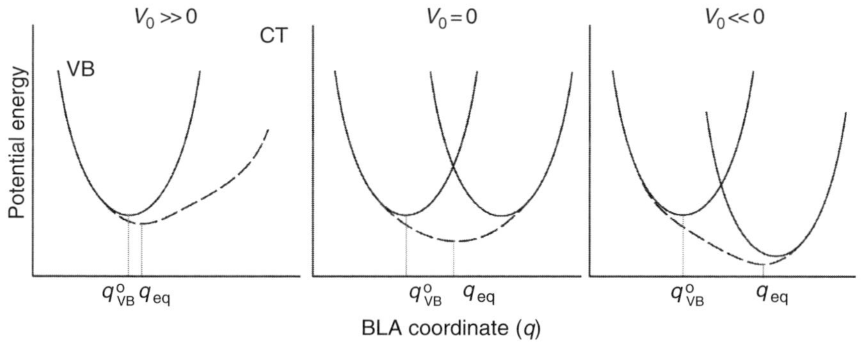

FIGURE 17.17 Illustration of ground-state potential surfaces (dashed line) for different relative energies of VB (solid, left) and CT (solid, right) states showing the changes of the equilibrium structure due to the mixing of the VB and CT states. The light vertical lines indicate the equilibrium coordinates for the VB (q^{o}_{VB}) and ground state (q_{eq}).

structure is cyanine-like with a BLA of zero. So it can be seen that structural and environmental features, such as the donor and acceptor strengths, the difference in energy of the π-bridge, and the solvent reaction field, that affect the value of V_0 alter the ground state potential surface and equilibrium geometry. To determine quantitatively the equilibrium BLA (q_{eq}) value for the ground state, the minimum of the ground state potential surface must be found, which can be done by solving $dE_{gr}/dq = 0$. It can be shown that q_{eq} is the value that satisfies the equation: $q = -0.12 + 0.24f$, with q in Ångstrom (Å) units. Thus, BLA and the charge transfer fraction are linearly related in the VB–CT model. However, because f is a function of q, the value of q_{eq} for a given value of V_0 must be found either by a numerical search for the minimum or by solving the equation for q_{eq} iteratively. Once q_{eq} is found, the value of f at the equilibrium can also be obtained.

The calculated potential surfaces for the model donor–acceptor molecule shown in Figure 17.16c are presented in Figure 17.18. The calculation was performed with the following parameters: $k = 33.5$ eV/Å^2, $t = 1.1$ eV, $V_0 = 1.0$ eV. The location of the equilibrium coordinate for the ground state reveals the change in BLA associated with the mixing of the VB and CT resonance forms.

It is now time to turn to the calculation of the dipole moment, polarizability, and hyperpolarizabilities. The dipole moment of the ground state in the VB/CT model is due mainly to the fraction of the CT form in the ground state and is given by

$$\mu_g = f\,\mu_{CT} \tag{17.41}$$

This simple relationship that shows that μ_g depends on the charge transfer fraction, f, which in turn depends on V_0. It should be recognized that V_0 contains contributions that result from chemical effects that influence the energy difference of VB and CT and can also include a contribution from an applied electric field (a relative Stark shift of VB and CT), which can tune the relative energies

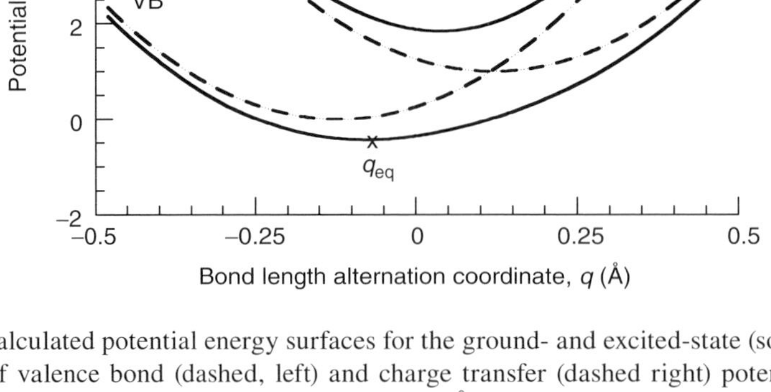

FIGURE 17.18 Calculated potential energy surfaces for the ground- and excited-state (solid lines) resulting from the mixing of valence bond (dashed, left) and charge transfer (dashed right) potential surfaces. The parameters used were: $V_0 = 1$ eV, $t = 1.1$ eV, and $k = 33.55$ eV/Å^2. The equilibrium coordinates for VB and CT are −0.12 and 0.12 Å, respectively. The "x" marks the equilibrium coordinate (q_{eq}) on the ground state potential surface.

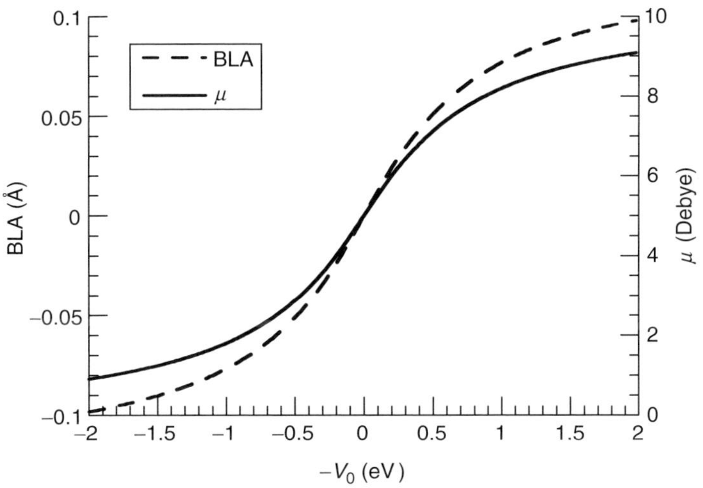

FIGURE 17.19 BLA and ground-state dipole moment as $-V_0$ is varied from −2 to 2 eV calculated using the VB–CT model, for the case of t = 1.1 eV and k = 33.55 eV/Å^2. These values have been plotted against $-V_0$ to have the sense of going from the VB form dominating on the left to having the CT form dominating on the right.

and the structure. The calculated dependence of the μ_g and the equilibrium BLA of the ground state as a function of V_0 for the model compound in Figure 17.16c is shown in Figure 17.19. The trends in μ_g and BLA are quite similar in shape as expected, given the linear relationship discussed above. As will be seen below, the field dependence of μ_g is used directly to compute the polarizability and hyperpolarizabilities. To do this, one must also include an electric field that is associated with the light used to measure the optical properties, which would generally be weaker than the field strength that can significantly modify the mixing and the structure.

The polarizability, α, and hyperpolarizabilities, β and γ, are defined in terms of a Taylor series expansion of the dipole moment as a function of field, as discussed in Section 17.1.1 (see Equation 17.2), and so are expressed as derivatives of μ with respect to the optical field strength. Using Equations 17.37 through 17.41, the following expressions for α, β, and γ of the ground state can be derived:

$$\alpha = \left.\frac{d\mu}{dE}\right|_{E_o} = \frac{2t^2\mu_{CT}^2}{E_g^3} \tag{17.42}$$

$$\beta = \left.\frac{1}{2}\frac{d^2\mu}{dE^2}\right|_{E_o} = \frac{3t^2\mu_{CT}^3 V_o}{E_g^5} \tag{17.43}$$

$$\gamma = \left.\frac{1}{6}\frac{d^3\mu}{dE^3}\right|_{E_o} = \frac{4t^2\mu_{CT}^4 (V_o^2 - t^2)}{E_g^7} \tag{17.44}$$

where the derivatives are with respect to the optical field and are evaluated at the value of the Stark field, which is identified as E_o. In the VB–CT model, the polarizabilities are determined by three parameters: μ_{CT}, t, and V_0. It can be seen that the polarizability and hyperpolarizability are essentially determined by the derivatives of the fractional charge transfer, f, with respect to electric field. These derivatives can also be cast with respect to the total energy difference between the potential minima for VB and CT, which is defined as $V_\varepsilon = V_0 - \mu_{CT}E$, from which one obtains $\mathrm{d}E = -(1/\mu_{CT})\mathrm{d}V_\varepsilon$. Because chemical changes that affect V_0 and applied electric fields are equivalent in how they influence the net energy difference between VB and CT, the polarizabilities in terms of their dependence on V_0 will be discussed and the polarizabilities as a function of V_0 will be plotted for simplicity. It is interesting to note that once one knows the field dependence of the dipole moment, one can compute any order of polarizability and that the first hyperpolarizability is the derivative of the polarizability, the second hyperpolarizability is the derivative of the first hyperpolarizability, and so on—these are the so-called derivative relationships [40,41].

To a first approximation, molecules of a given length and bridge type can be taken to have the same t and μ_{CT}, thus the polarizabilities and the structure vary primarily due to variations in donor and acceptor strength and solvent reaction field, which are reflected in V_0. Figure 17.20 shows the calculated dependencies for α, β, and γ as a function of V_ε. The energy gap, E_g, is at a minimum (i.e., $E_g = 2t$) when $V_0 = 0$, which is where BLA = 0 and α is at a maximum, β is zero, and γ is at a minimum (negative extremum). With further stabilization of the CT form, whereupon it becomes the lower energy zero-order state, V_0 becomes negative and, as can be seen from Equation 17.43, so does β. Of course, this sign change is also associated with a change in sign of $(\mu_{ee} - \mu_{gg})$, which was discussed above. The value of β reaches extremes when $V_0 = \pm t$, thus $E_g = 5^{1/2}t$, which is also where $\gamma = 0$. The value of γ reaches maxima when $V_0 = \pm 3^{1/2}t$, whereupon $E_g = 7^{1/2}t$. The largest magnitude of γ occurs at $V_0 = 0$, where it is negative.

The trends calculated using the VB–CT model are in good qualitative agreement with quantum chemical calculations performed using AM-1 [42] and intermediate neglect of differential overlap (INDO) sum-over-states [36] methods. One notable difference is that the relative magnitude of the positive peak value of γ compared to the negative peak is found to be larger in the AM-1 and INDO calculations than is found in the VB–CT model. This is due to the fact that in the VB–CT model

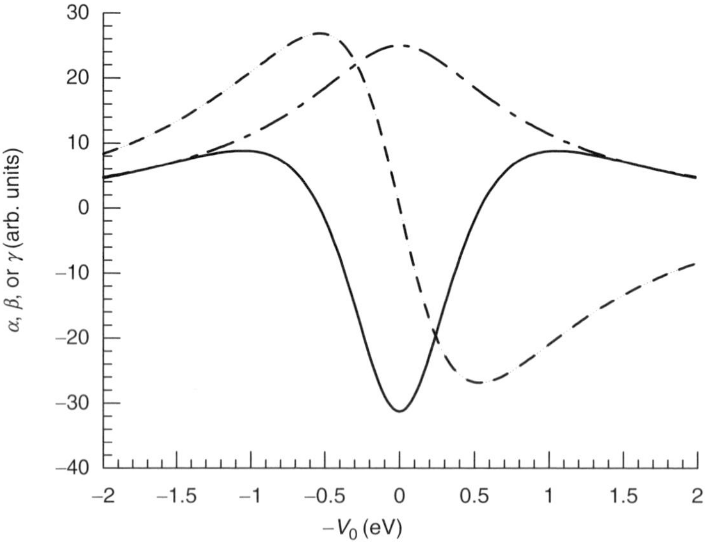

FIGURE 17.20 Trends of α (dash-dot line), β (dashed line), and γ (solid line) as a function of $-V_0$ calculated using the VB–CT model for the model DA molecule $(CH_3)_2N$-$(CH = CH)_4$-CHO. The VB form dominates at the left and the CT form dominates at the right of the plot.

only two states are accounted for, whereas in the other calculation methods there are significant positive contributions from higher lying excited states that are two-photon active states and contribute to γ via the T-term in Equation 17.30. The two-state VB–CT model discussed here was intended to provide insight into the major features of the structure–polarizability relationships for donor–acceptor charge transfer molecules within a simplified framework and to give a foundation for further studies.

17.3.4 Summary of Second Hyperpolarizabilities for Conjugated Donor–Acceptor Chromophores

Before considering the experimentally determined second hyperpolarizabilities of conjugated donor–acceptor molecules, it is useful to consider the values from the VB–CT model plotted against the charge transfer fraction, f, and the BLA parameter, as shown in Figure 17.21. This gives a more chemically relevant view of the structure–hyperpolarizability relationship. To aid in the comparison of molecules the range of f and BLA is broken up into five regions, A–E, in order to group molecules with a similar behavior [41]. In region A, as the charge transfer or the solvent polarity is increased γ increases and goes through a maximum. In region B, γ decreases with increased charge transfer and changes sign from positive to negative. Region C is the region wherein γ attains its largest negative value. In region D, γ increases and becomes positive with increasing charge transfer and, in region E, γ reaches another positive maximum and then decreases.

Experimental studies on conjugated donor–acceptor molecules with varying electron acceptor strength and in solvents of varying polarity have been performed using THG methods in the transparency region of the molecules, giving values for the nonresonant second hyperpolarizabilities [41,43]. A series of triene molecules with dialkylamino groups as donors and various types of acceptors, such as an aldehyde group, two cyano groups (dicyano), a barbituric acid, and a thiobarbituric acid, as shown in Table 17.3, provides an illustrative system of model compounds where the acceptor

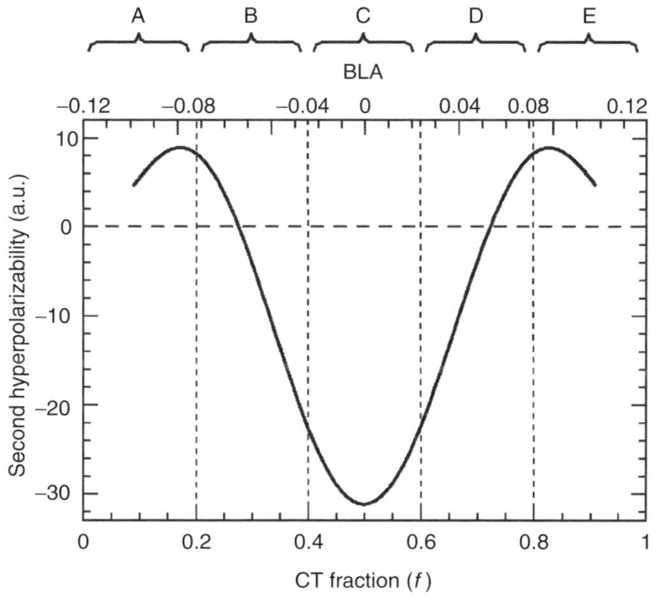

FIGURE 17.21 Second hyperpolarizability from Figure 17.20 plotted as a function of the charge transfer fraction, f, and the BLA parameter. The range of f and BLA has been broken up into five regions, A–E, to qualitatively group molecules of similar behavior.

TABLE 17.3

Second Hyperpolarizabilities, $\gamma(-3\omega;\omega,\omega,\omega)$, of Model Donor–Acceptor Triene Compounds in Solvents with Different Polarity

Molecule	CCl_4	C_6H_6	$C_4H_6O_2$	$CHCl_3$	CH_2Cl_2	CH_3CN	CH_3NO_2	CH_3OH
	0.05	**0.11**	**0.16**	**0.26**	**0.31**	**0.46**	**0.48**	**0.76**
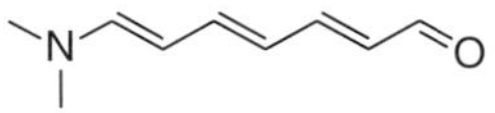	—	—	+40	+95	+105	+113	+113	+73
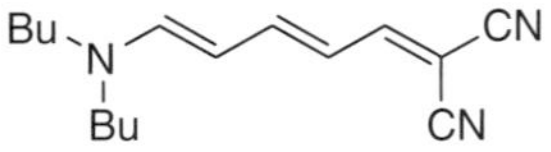	+40	+15	−25	−42	−50	−120	−117	−135
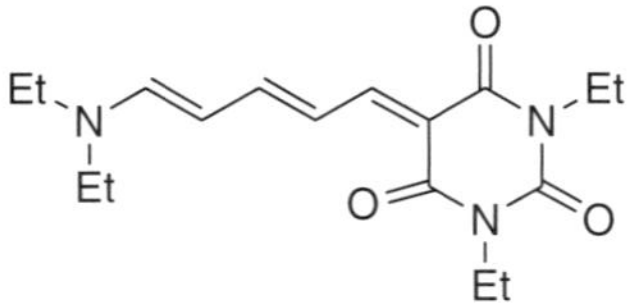	—	−20	−100	−135	−145	−205	−220	−166
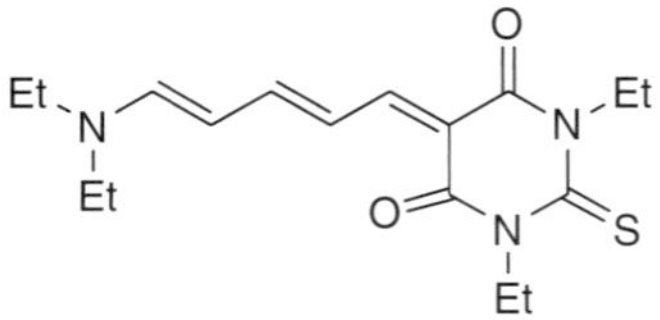	—	−85	−170	−195	−175	−130	−125	−10

Note: The solvent polarity as quantified by the ET_{30} parameter is given below the formula for the solvent. The solvents from left to right are carbon tetrachloride, benzene, dioxane, chloroform, dichloromethane, acetonitrile, nitromethane, and methanol. Nonresonant γ values were determined by THG using a laser wavelength of 1907 nm.

strength increases in the series, with the thiobarbituric acid being the strongest. In the case of the barbituric acids, there is possibly some increase in aromaticity of the ring upon charge transfer. In this series of compounds, the V_0 value decreases on moving from the aldehyde to the thiobarbituric acid and, in addition, the degree of charge transfer can be increased and the hyperpolarizabilities can be tuned by dissolving the molecules in solvents of increasing polarity. The values in Table 17.3 provide a rather clear indication of trends with increasing acceptor strength and increasing solvent polarity. For the case of the aldehyde acceptor, the trend with increasing solvent polarity shows increasing positive γ values and evidence for a maximum. The molecule with the thiobarbituric acid acceptor exhibits negative values and shows a negative peak in γ. These trends can be seen more easily in Figure 17.22, where the data for all four molecules are spliced together. In this plot, the polarization axis represents the $E_T(30)$ values for the solvent for the aldehyde containing molecule. The $E_T(30)$ [44] gives a relative measure of the solvent reaction field for these molecules as the CT dipole moment is similar for all. For the other molecules, there is a shift in V_0 associated with the stronger acceptor strength. By adding a value to the $E_T(30)$ to approximately match the γ trend for the aldehyde and the dicyano compounds and then doing similarly to match up the barbituric acid with the dicyano and the thiobarbituric acid with the barbituric acid, all of the data can be placed on a common plot.

Figure 17.22 shows that the overall trend of γ versus the increasing polarization of the molecules exhibits a behavior that is in good qualitative agreement with the VB–CT model, as well as with calculations performed by AM-1 and INDO methods. The trends allow one to assign the donor-triene-aldehyde compound to region A behavior, the dicyano to region B, the barbituric acid to overlapping with regions B and C, and the thiobarbituric acid to region C. Region C molecules are interesting from an electronic point of view because they have electronic structures and properties that are similar to those of the highly delocalized cyanine molecules and give the largest magnitude of γ for linear conjugated donor–acceptor type molecules.

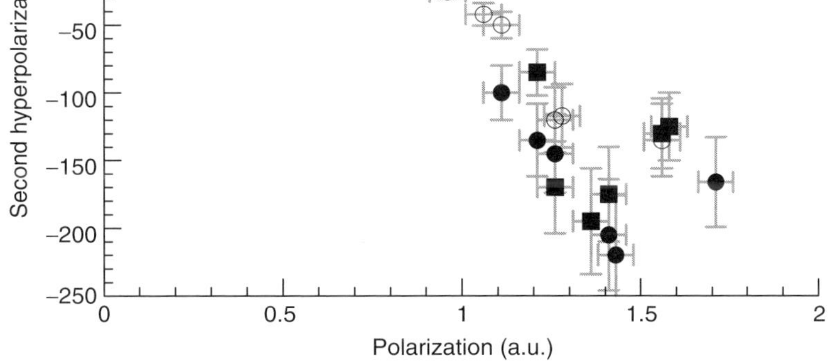

FIGURE 17.22 Second hyperpolarizability of donor–acceptor triene molecules listed in Table 17.3 plotted versus a measure of the polarization of the molecules based on the solvent $E_T(30)$ parameter. The data from Table 17.3 were spliced together by adding an amount to the $E_T(30)$ parameter to approximately match the hyperpolarizability values. The data points correspond to the molecules with the following acceptors: aldehyde (split square), dicyano (open circle), barbituric acid (filled circle), and thiobarbituric acid (filled square).

17.3.5 Survey of Organic Systems with Large 3rd-Order Nonlinearities

The concepts described above give a framework for the optimization of γ for donor–acceptor polyenes with a fixed molecular length. Another important means of increasing molecular nonlinearity is to increase the conjugation length. In general, increasing the length of conjugated molecules leads to lower energy gaps and larger CT dipole moments and transition dipole moments, which are beneficial to the magnitude of the nonlinearity. This is true up to the point where an additional increase in length leads to too large a reduction in the electronic coupling (the interaction of the electron donor and acceptor end groups) whereupon the effects of the end groups become minimal.

The second hyperpolarizabilities of some extended conjugated chromophores including polyenes, donor–acceptor polyenes, cyanines, and symmetrically substituted conjugated systems will now be examined, as shown in Table 17.4. Of particular interest for applications in optical signal processing are compounds with large third-order nonlinearity and low optical absorption in the spectral region (~1300 nm and 1500–1600 nm) relevant to telecommunications, as will be discussed in detail in the following section. Accordingly, a set of representative extended conjugated compounds having low electronic absorption in the near infrared and for which the electronic nonlinearity has been examined in this spectral region will be considered.

A series of polyenes of varying length with dimethylaminophenyl donor groups and aldehyde acceptors has been studied by Puccetti et al. [45] by using THG methods with near-infrared wavelengths. An increase in γ of over two orders of magnitude upon going from one to eight double bonds was observed resulting in a relatively large value (7600×10^{-36} esu) for the compound in the table. It was also observed that γ varied in a power law manner, $\gamma = cN^a$, as a function of the effective number of carbon–carbon double bonds, N (the benzene ring was taken as being equivalent to 1.5 double bonds in length), with the exponent, a, being 3.4. Although the electronic coupling is certainly decreasing with increasing number of double bonds, the results show that the coupling of the donor and acceptor is still effective at the largest length. The researchers also examined a series of symmetrically substituted polyenes with dimethylaminophenyl donor groups on both ends and were able to prepare molecules with up to 11 double bonds. By using the EFISH method at 1340 nm, a quite large γ value was obtained for this bis-donor substituted system. This result demonstrates that symmetrical bis-donor systems can exhibit large γ in the near infrared, although there may be a two-photon resonance enhancement contribution. The two-photon properties of such systems are discussed in Chapter 18.

Samuel et al. [46] studied γ for a series of long-chain polyene oligomers, prepared by living polymerization methods, with up to 240 double bonds to investigate the length at which saturation of the nonlinearity would occur. By examining oligomers with 28–240 double bonds, they observed the onset of saturation of γ/N at about 100 to 150 double bonds, which is substantially larger than what has been predicted by various theoretical analyses, typically in the range of 20. The conformational flexibility of the oligomers may play a role in the observed saturation length but is still a challenge for theory to predict. The oligomer with 240 double bonds shows a very large γ value as shown in Table 17.4, but lengths well beyond the onset of saturation do not contribute effectively to increases in the third-order nonlinearity of neat materials because the larger γ is essentially offset by the increase in molecular size, which limits the number density of molecules in the neat material.

Highly polarized conjugated molecules can be made by attachment of strong electron acceptors to an extended π system, as shown by Marder et al. [46]. A series of substituted carotenoid systems with varying acceptor strengths were prepared by coupling of acceptors to apocarotenal. Wavelength-dependent THG measurements were used to determine the peak three-photon resonance value of $\gamma(\gamma_{max})$ for the series of molecules. While lower values would be obtained off-resonance, the relative γ_{max} values should give a useful indication of the relative off-resonance hyperpolarizabilities. Large enhancements of γ were obtained for carotenoids with strong electron acceptors, such as thiobarbituric acid and an indane-bearing sulfone and dicyanovinyl groups, relative to β-carotene, as shown in Table 17.4. Within a perturbation theory description, the increase in γ with acceptor strength arises

TABLE 17.4
Second Hyperpolarizability of Extended Chromophores

Molecule	λ^a (nm)	γ^b	Method	Comments
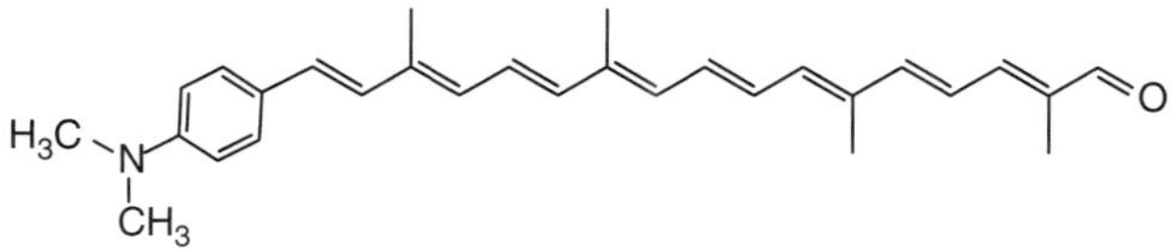	1910	0.76 [45]	THG	Near three-photon resonance; acetone solution
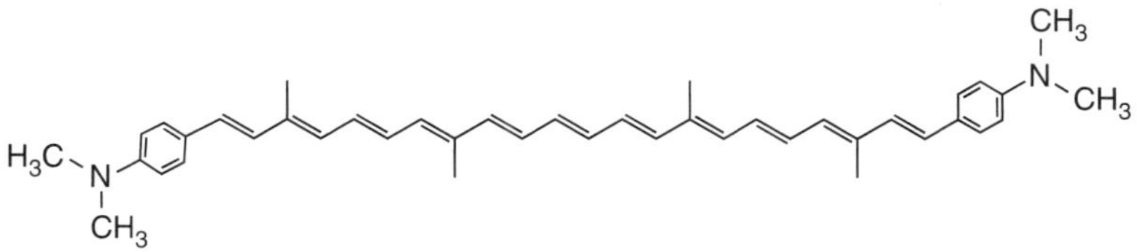	1340	4.1 [45]	EFISH	Two-photon resonance possible; acetone solution
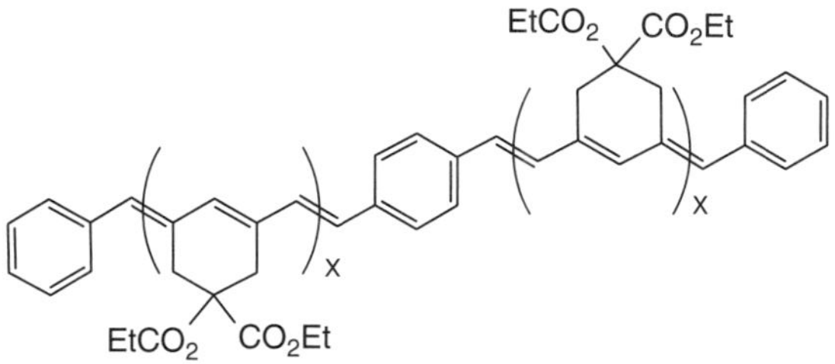	1910	38 [46]	THG	Near three-photon resonance; tetrahydrofuran solution; average number of double bonds for oligomer studied was 240
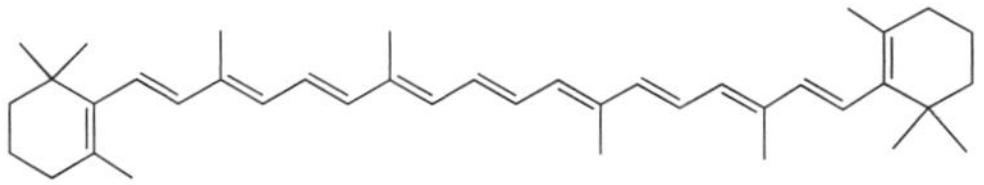 β-carotene	1392	1 [47]	THG	Peak of three-photon resonance; polymer film
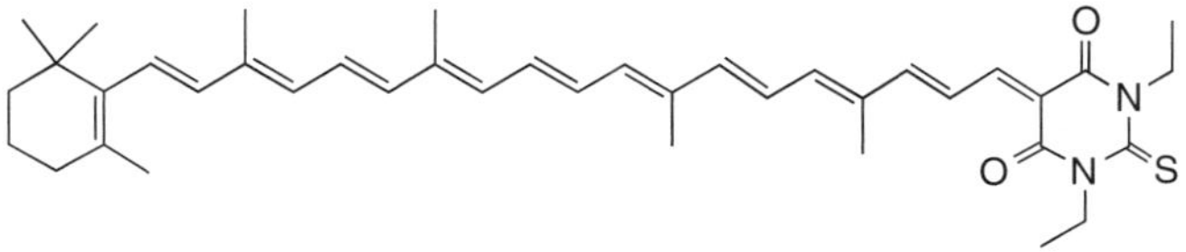	1770	8.1 [47]	THG	Peak of three-photon resonance; polymer film

(continued)

TABLE 17.4 (continued)
Second Hyperpolarizability of Extended Chromophores

Molecule	λ^a (nm)	γ^b	Method	Comments
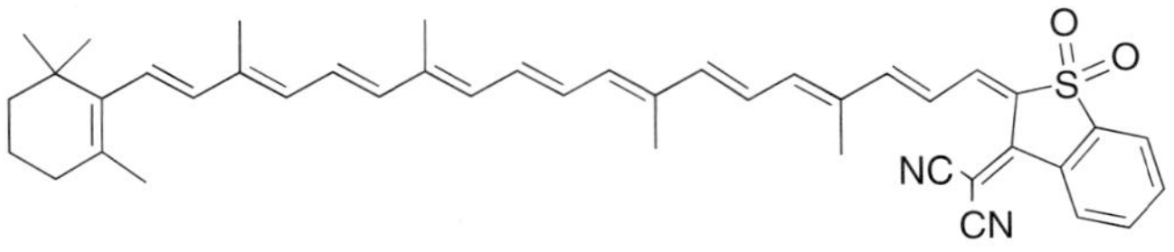	2220	35 [47]	THG	Peak of three-photon resonance; polymer film
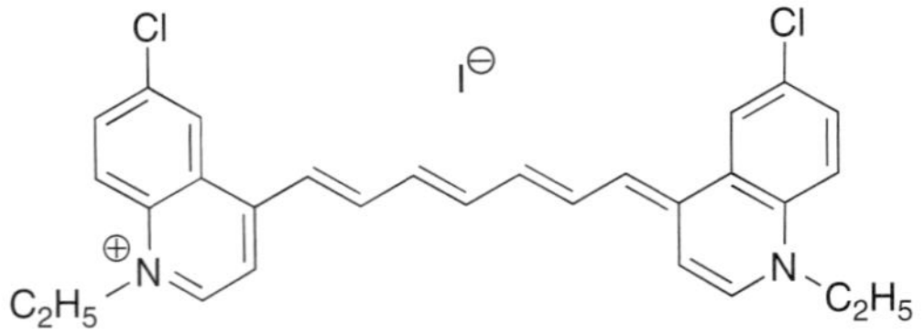	1890	−19 [48]	THG	Two- and three-photon resonance possible; DMSO solution
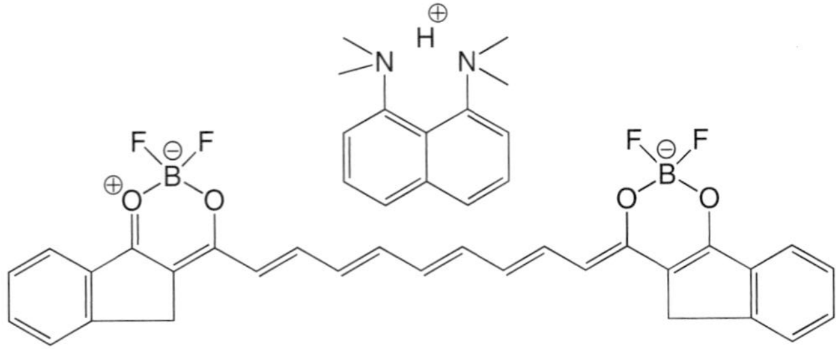 bis(dioxaborine) cyanine	1300	−5.7 [49]	DFWM and Z-scan	Re(γ) given; off-resonance; DMSO solution
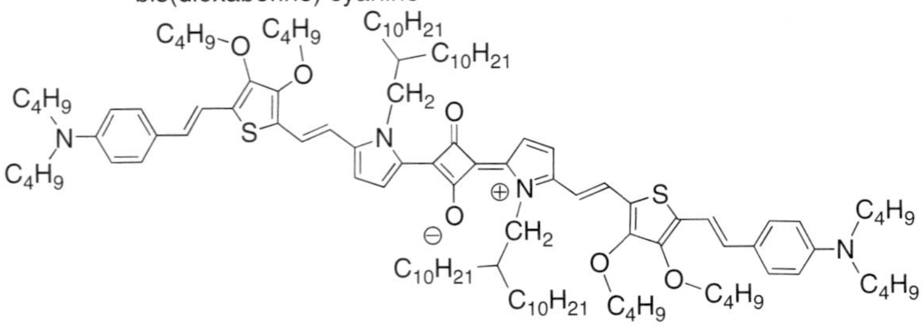	1300	3.3 [50]	DFWM and Z-scan	Modulus(γ) given; toluene solution

[a] Laser wavelength used for measurement; [b] orientationally averaged values of γ given in units of 10^{-32} esu.

mainly from the greater contribution of the D-term, due largely to the increase in the excited-state dipole moment. Indeed, a large $|\mu_{ee} - \mu_{gg}|$ was measured for the strong acceptor carotenoids using electroabsorption (Stark) spectroscopy. An increase in μ_{eg} and a decrease in E_{eg} contributes to the factor of 4.3 increase on going from the thiobarbituric acid to the dicyanovinyl sulfone indane acceptor. In the VB–CT framework, the increase would be thought of as a result of the large distance over which charge can be transferred due to the extended conjugation length, which leads to a large μ_{CT}. Additionally, stabilization of the CT form leads to an increased mixing of the VB and CT potential surfaces and moving to the right in the A region, toward the positive maximum of γ.

Cyanine dyes have long been known to possess large γ values. Hermann [48] reported in 1974 on the chain length dependence of γ for cyanines as determined by THG. A strong power law behavior was observed with a very large value found for the long-chain cyanine shown in Table 17.4, although two- or three-photon resonance effects may be contributing. Given their very large nonlinearity and their potential for use as nonlinear refractive index materials in the telecommunication wavelength range, Hales et al. [49] have investigated the hyperpolarizability of an interesting class of cyanine dye based on dioxaborine end groups by using femtosecond pulse DFWM and Z-scan methods in the 1300–1500 nm range. As indicated in Table 17.4, a large negative γ was obtained and, moreover, the hyperpolarizability was dominated by the real part, with the magnitude of the imaginary part being 7.2 times smaller, which is desirable for optical switching as will be discussed in the next section. The large negative nonlinearity of the cyanines can be described in terms of a dominant contribution from the *N*-term in the 3-state model, associated with the low energy and strong ground state to first excited-state transition. In the VB–CT model, with the equal energy of the VB and CT forms ($V_0 = 0$) the mixing of the potentials is maximized, giving a highly delocalized ground state. As a result, the polarizability is maximized and its curvature (which gives γ) is negative and optimal in magnitude. It can be seen from Equation 17.44 that for cyanines large μ_{CT} and small electronic coupling (t) lead to large magnitudes of γ, so cyanines of greater length are of interest for increased γ. However, if t becomes too small, the ground state potential becomes a double well potential with a barrier rather than a minimum at $q = 0$, a process referred to as symmetry breaking. In this case, the system is no longer cyanine-like but more like a region B molecule. Approaches to increase length while maintaining delocalization is a topic of current research.

Squaraines are another class of molecules that exhibit substantial electron delocalization. Chung et al. [50] have synthesized extended squaraine compounds such as that whose structure is shown in Table 17.4. These molecules possess a delocalized core and extended conjugation that terminates in electron donors at each end. It has been found that the extended squaraines possess very large 2PA cross sections. Measurement of γ using femtosecond pulse Z-scan at 1300 nm reveals that the extended squaraine shown also exhibits a large positive hyperpolarizability with a significant imaginary part, indicating that two-photon resonance is making a major contribution to the nonlinearity.

Of particular importance for the application of organic materials in nonlinear optics is the macroscopic third-order nonlinearity, $\chi^{(3)}$. The relationship between γ and $\chi^{(3)}$ and the general characteristics of $\chi^{(3)}$ were discussed above in Section 17.1.3. For a neat molecular, oligomeric, or polymeric material, the key factors include (1) the magnitude of γ for the units comprising the material, (2) the number density of these units in the material, (3) the orientation of these units, and (4) the local field factors, as well as possible effects associated with intermolecular electronic interactions. Here, a brief survey of the $\chi^{(3)}$ values that have been measured for some promising examples of molecular and polymeric materials will be given, again focusing on nonlinearities that have been determined using near-infrared laser wavelengths.

Table 17.5 lists the magnitude of $\chi^{(3)}$ for astaxanthene (a carotenoid polyene) and dioxaborine cyanine molecular materials, and for polydiacetylene, polyacetylene, and Zn porphyrin diacetylide conjugated polymeric systems. The impact of the increased electron delocalization and the larger magnitude hyperpolarizability of cyanines relative to polyenes, as discussed above, can be seen at the macroscopic level by the larger $|\chi^{(3)}|$ (3.6×10^{-10} esu) for the dioxaborine cyanine as compared to astaxanthene (1.2×10^{-10} esu). What is especially important about the $\chi^{(3)}$ of the dioxaborine

TABLE 17.5
Third-Order Susceptibilities of Selected Conjugated Molecular and Polymeric Materials

Molecule	λ^a (nm)	$\lvert\chi^{(3)}\rvert^b$	Method	Comments
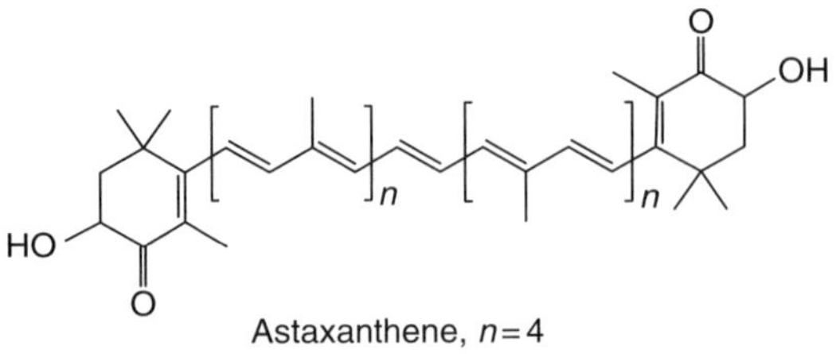 Astaxanthene, $n = 4$	1500	1.2×10^{-10} [78]	THG	Spin cast films, three-photon resonant
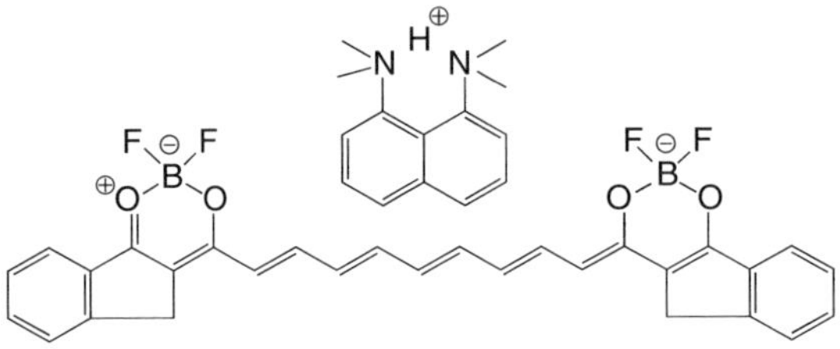 bis(dioxaborine) cyanine	1300	3.6×10^{-10} [49]	DFWM and Z-scan	Spin cast films, potentially two-photon resonant
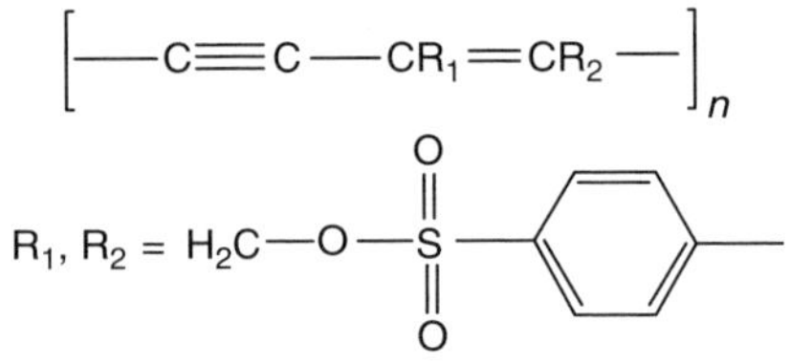 PTS-PDA	1500	8.0×10^{-11} [78]	Z-scan	Oriented crystal, two-photon resonant
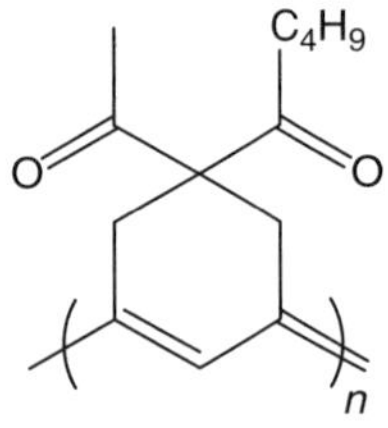 PHDK	1500	1.2×10^{-10} [6]	THG	Spin cast films, simultaneously two- and three-photon resonant
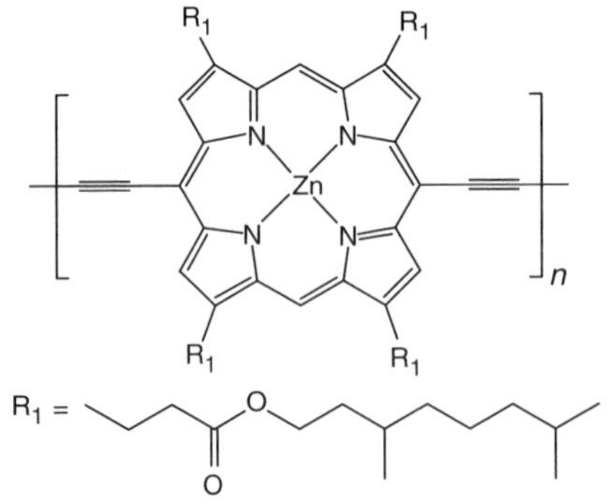	1064	1.9×10^{-9} [78]	DFWM	Spin cast films, pre-resonance enhancement via one-photon state, potentially two-photon resonant

[a] Laser wavelength used for measurement.
[b] Modulus of $\chi^{(3)}$ given in units of esu.

cyanine, aside from its being large, is that it is dominated by the real part of $\chi^{(3)}$, such that 2PA is relatively small compared to the nonlinear refractive index. This is rather important for applications as will be discussed in the next section.

The importance of the magnitude of the hyperpolarizability per molecular volume is illustrated by the fact that the substituted polyacetylene, PHDK, has the same $|\chi^{(3)}|$ as astaxanthene, even though as a polymer PHDK has a much larger conjugated chain length. However, as discussed for the case of the polyene oligomers with varying lengths, there is a saturation of the γ per double bond and increases in length beyond the saturation length do not lead to effective increases in $\chi^{(3)}$. There can be other benefits in using a polymeric conjugated system, such as in processing of the material by spin coating where being able to make a high viscosity solution is necessary and also in being able to form a transparent glassy phase of the material.

Polydiacetylenes are rather unique in the realm of conjugated polymers in that these can be formed as fully crystalline materials with oriented chains by solid-state reactions in crystals of substituted diacetylene monomers, thus providing increased nonlinearity for light polarizations along the chain direction. In particular, poly(1,4-di(*p*-toluenesulfonate)diacetylene) (PTS-PDA) has been formed as single crystals and found to have a reasonably large $\chi^{(3)}$ in the near-infrared region. Although the magnitude of $\chi^{(3)}$ for PTS-PDA is not as large as for some other reported conjugated polymers, the nonlinear refractive index component (Re $\chi^{(3)}$) is large relative to 2PA (Im $\chi^{(3)}$) giving a figure of merit close to that needed for all-optical switching, similar to the case of the dioxaborine cyanine thin film.

Another interesting class of conjugated polymers is the porphyrin diacetylide polymers wherein conjugated macrocycles are linked by a diacetylenic bridge. The diacetylene linker provides a conjugated pathway for electronic interaction of the porphyrins that leads to a sharp low-energy band for the S_0 to S_1 transition. The Zn porphyrin diacetylide polymer shown in Table 17.5 gives a very large $|\chi^{(3)}|$ as determined by picosecond DFWM. More recent studies using wavelength-dependent femtosecond DFWM and Z-scan measurements in the near-infrared confirm the extraordinary nonlinearity of porphyrin diacetylide polymers and give similar magnitudes of the real and imaginary parts of $\chi^{(3)}$, indicating strong nonlinear absorption [51]. Thus, these materials are of interest for applications where strong nonlinear refraction and absorption can be useful, as in optical limiting, for example.

17.4 DEVICE APPLICATIONS

The third-order nonlinear phenomena detailed in Section 17.2 can be utilized for a variety of intriguing applications. Many of these applications are in the field of optical communications where the ability of light to control light is exploited in the transmission, processing, and storage of information. The critical feature being exploited is the ultrafast response of this control of light. A number of such applications are described in the second part of this section. Their feasibility is predicated on the success of the photonic devices used to perform the necessary tasks. In turn, these devices require nonlinear materials with stringent material requirements. These conditions are the subject of the first part of this section.

17.4.1 Issues for 3rd-Order Nonlinear Optical Applications

Advances made in the development of compact, ultrafast laser sources [52] have promised great breakthroughs in the area of optical communications systems. However, the lack of photonic devices that can process signals at these extremely rapid rates (i.e., >100 GHz) has remained the bottleneck of this enabling technology. In fact, it is the requirements on the optical and material properties of the nonlinear medium in a device that often determines its viability. Organic molecular and polymeric systems, such as those described in Section 17.3, are strong candidates for such media. The benefits of these systems over other potential materials such as inorganic semiconductors and optical glasses lie mainly in the flexibility that molecular engineering provides and their ease of processing. The resulting chemical modifications provide means to optimize the material parameters such that practical devices can be designed.

17.4.1.1 Material Requirements and Metrics

As mentioned above, the requirements for a material to be utilized in a practical device can be quite stringent. This results from not only the metrics associated with each requirement itself but also from the number of conditions that should be met. A list of typical material requirements is shown below:

- Sufficiently large third-order nonlinearity or $\chi^{(3)}$
- Ultrafast temporal response of nonlinearity
- High optical transparency in operational range of device
- Mechanical stability
- Photochemical stability
- High optical damage threshold
- Processability

The first three items can be categorized as optical properties conditions that must be satisfied. Materials that possess large values of $\chi^{(3)}$ can help reduce the pump intensity (I) and sample interaction length requirements (L) placed on a device to perform adequately in a particular application. This consequence is immediately evident when looking at the dependence of any of the nonlinear phenomena described in Section 17.2 (i.e., THG intensity, Equation 17.11; nonlinear phase shift, Equation 17.16; and DFWM intensity, Equation 17.19) on I, L, and $\chi^{(3)}$. The nonlinearity should be ultrafast in its temporal response (i.e., <10 ps) to exploit the benefit that all-optical control affords. For this reason, the bound electronic Kerr effect is often chosen as the preferred nonlinear mechanism to employ. High optical transparency implies both low linear as well as nonlinear loss in the appropriate spectral range. This condition reduces the device's pump intensity requirements and consequently reduces the potential for thermal deposition of energy, which could result in material decomposition. The final four items are additional practical material properties requirements. The mechanical and photochemical stability of the material ensures that useful operation over a prolonged period is possible, whereas a high optical damage threshold guarantees that the material can resist catastrophic damage induced by the pump intensity. The final prerequisite for a suitable material, that is, processability, is crucial for proper integration into devices. This will be discussed in more detail below.

The most obvious benefits for the use of organic molecular and polymeric systems as materials are their large nonresonant values of $\chi^{(3)}$ (see Section 17.3), ultrafast temporal response, and the versatility afforded through molecular engineering. The chemical modifications provided by the latter property allows for optimization of the remaining material requirements. As noted in Section 17.3, chemical modifications allow for potential tuning of the transition energies, which gives flexibility in choosing the proper spectral transparency range of the material [53]. Furthermore, through addition of appropriate substituents or with the design of suitable intermolecular interactions dramatic improvements in stability [54] or processability [55] can be made. Consequently, it becomes clear that organic systems can be modified according to the specifications required by a particular device.

Certain figures of merit (FOMs) have been developed [56,57] to gauge the efficacy of a material in all-optical signal processing (AOSP) applications. Although these FOMs were formulated for materials in waveguide devices, the figures are independent of device design and instead are based solely on the optical properties of the material. The first FOM, W, rates the material's ability to perform certain AOSP tasks and is dictated by the strength of the nonlinearity and its linear optical loss. The second FOM, T, describes the material's ability to perform despite potential losses due to nonlinear absorption. Consequently, the two FOMs are often referred to as the one- and two-photon FOMs (W and T, respectively). Their definitions and the criteria for viable nonlinear materials are given below:

$$W = \frac{n_2 I}{\alpha\lambda}, \quad \text{where } W > 1 \tag{17.45}$$

$$T = \frac{\beta\lambda}{n_2}, \quad \text{where } T < 1 \tag{17.46}$$

where
- I is the pump intensity
- α is the linear absorption coefficient
- n_2 and β are the nonlinear refractive index and 2PA coefficient, respectively (defined in Section 17.2.1.2)

Technically, α can be comprised of losses due to both scattering and intrinsic absorption loss, whereas β refers to the intrinsic nonlinear absorption loss. As is true for n_2, β, and α, W and T exhibit dispersion. Therefore, it is possible to locate specific spectral regions where the FOM criteria are satisfied. This may be located energetically below any single- or multiphoton resonances or appropriately located between resonances. It should be noted that W can be made to exceed unity by arbitrarily increasing I. However, certain material parameters (such as optical damage threshold) and more typically pump source limitations place an upper bound on I. Certain optical glasses have been shown to exhibit good FOMs [58], but their moderate values of n_2, compared to organic systems, require long interaction length devices to perform AOSP tasks. This is a severe limitation when the integration of multiple devices becomes an issue. Finally, another FOM is often encountered in the literature that relates to the temporal response of the nonlinearity. It is often given as some form of n_2 divided by τ, or the temporal response. This FOM explains why most inorganic semiconductor samples, while possessing reasonably large nonlinearities and good W FOMs [59], are not appropriate for most high-speed AOSP applications.

17.4.1.2 Integration of Materials and Devices

The two architectures typically employed for integrating nonlinear materials into devices are either simple bulk media or various waveguide geometries. Each route has its own set of benefits. Employing bulk nonlinear media allows for the use of single- or multilayer films with large interaction areas (see Figure 17.23a as an example). Large area films are typically necessary for applications that involve any type of optical imaging such as optical phase conjugation (OPC) (see Section 17.4.2.2) or optical limiting (see Section 17.4.2.3). Waveguide geometries offer strong confinement of light in either one or two dimensions guaranteeing delivery of large pump intensities. Figure 17.23b and c shows common waveguide structures where the nonlinear material resides either on top of, or is embedded in, the substrate. Figure 17.23d illustrates a particular set of fiber waveguides known as microstructure fibers [60] that are conducive to back-filling with nonlinear materials. Unlike bulk media where the focusing of light to achieve large intensities also results in a reduced interaction length due to diffraction, waveguides provide confinement over long distances ensuring long interaction lengths. High intensities and long interaction lengths coupled with the fact that waveguide geometries are conducive to device integration [57] make these architectures optimal for many applications in the telecommunications industry (see Section 17.4.2.1).

For these reasons, forming large area films with organic nonlinear materials or integrating such materials into waveguide structures becomes a crucial issue. Fortunately, since third-order nonlinear

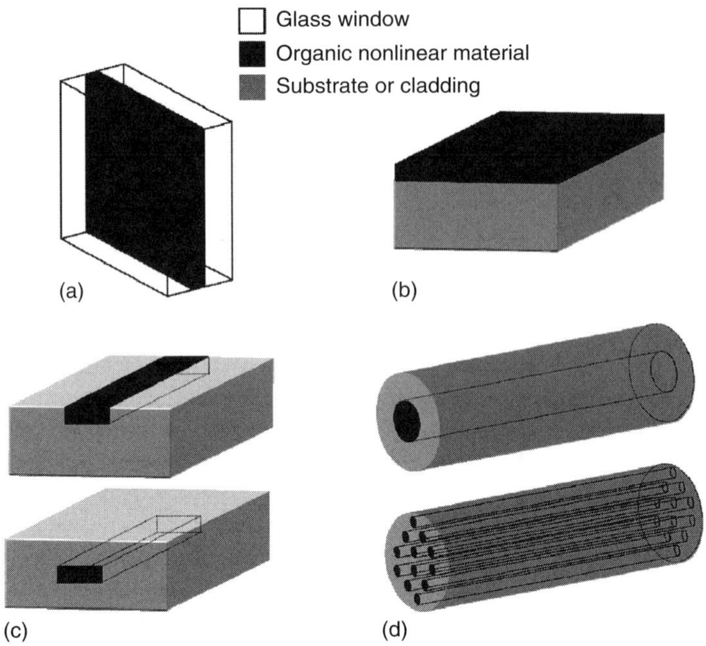

FIGURE 17.23 Architectures typically employed for integrating organic nonlinear materials into devices. (a) Large area bulk nonlinear material enclosed between glass windows. (b) Slab waveguide for one-dimensional confinement of light. (c) Channel waveguides (surface, top; buried, bottom) for two-dimensional confinement of light. (d) Microstructure fiber waveguides for two-dimensional confinement of light.

materials do not suffer from any symmetry restrictions, as detailed in Section 17.1.3.3, amorphous or glassy materials can be utilized. This expands the number of potential routes for processing organic materials into a form suitable for device integration. The following are some typical processing routes for amorphous organic materials [4]:

- Film casting or doctor-blading
- Spin coating
- Melt growth processing
- Vacuum deposition
- In situ polymerization

Modifying the chemical structures of the organic systems processed by these techniques can dramatically improve the potential for fabricating quality films. This may involve increasing the solubility of the organic system through addition of solubilizing substituents for the case of film casting and spin-coating techniques or modifying the vapor pressure of a material to facilitate vacuum deposition. Guest–host systems, involving the proper mixing of organic guest molecules with host polymers, provide another promising direction to improve processability. For instance, an appropriate host polymer could be chosen such that the resulting composite system could have increased viscosity for film casting or could possess a reduced melting point temperature to aid in melt growth processing. In situ polymerization is another promising route for material–device integration that can be utilized for polymers with appropriate monomeric precursors such as polyacetylene [61]. In addition to improving the processability of a material, chemical modification can aid in promoting the wettability of a material. This property is crucial when ensuring proper interfacial contact between the organic material and the device structure or substrate with which it is in contact.

17.4.2 Third-Order Nonlinear Optical Applications

The ability of third-order nonlinear optics to take advantage of the unique characteristics offered through the use of optical fields make applications that exploit these nonlinear phenomena of considerable interest. The ultrafast response of light-by-light control promises tremendous gains in the area of AOSP. The parallel processing afforded through the interaction of fields in a nonlinear medium shows great potential for high performance optical computing. Optical fields are immune to electromagnetic interference which, when combined with their innate speed, guarantees high fidelity and high throughput transmission of information. Applications of this nature, as well as the nonlinear phenomena that enable them, are the subject of this section.

17.4.2.1 Telecommunications

The telecommunications market is an area that stands to benefit significantly from the use of third-order nonlinear optics in optical communications. As mentioned above, despite the development of compact, ultrafast laser sources, breakthroughs in the area of optical communications systems have been bottlenecked by the lack of photonic devices that can process signals at these extremely rapid rates. Essentially, while the generation and transmission of information can be accomplished in the optical domain, the processing of this information must be undertaken in the electrical domain leading to the necessity of constant conversion between the two domains. This results in unwanted latency in a communications system. Through the use of the intensity-dependent refractive index, or n_2 (see Section 17.2.1.2), certain AOSP tasks could be accomplished in the optical domain. Furthermore, by employing the mechanism of the bound electronic Kerr effect, ultrafast response of n_2 can be assured.

One such AOSP task involves the temporal demultiplexing of ultrahigh bit-rate optical pulses (i.e., removing data pulses from a data stream) via all-optical switching [62]. Switching essentially involves, for a single beam input, creating two possible output states where the final state depends on the intensity of the input beam. This intensity can force a change in the output channel, polarization or frequency, any of which can be used as a discriminator for the output state. Output channel switching is the most common method employed and this can be accomplished using either decoupled optical fields, e.g., NLO loop mirror or nonlinear Mach–Zehnder interferometer (MZI), or using strongly coupled fields, e.g., nonlinear directional coupler or nonlinear X-switch. The switching process for a nonlinear MZI is shown in Figure 17.24a. As is typical for a MZI, the light is split between the two arms and propagation results in a phase shift for the optical field in each arm. If the two arms consist of different materials, a relative phase shift, $\Delta\phi$, between the two fields will exist upon recombination and this phase modulation can give rise to an amplitude modification in the output field. In a nonlinear MZI, a material with an appreciable n_2 is present in one arm such that the value for $\Delta\phi$ now becomes intensity-dependent (see Equation 17.16 for details). Therefore, the intensity of the input pulse will determine whether or not it passes through the output channel of the device.* In this way, temporal demultiplexing can be accomplished; Figure 17.24a illustrates this process by showing higher intensity pulses inhibiting passage of the pulse through the output channel. For the ease of application, oftentimes a set of control pulses can be used to induce the switching instead. In this case, the principle of the switching mechanism is the same but is initiated by cross-phase modulation instead of SPM. This can be seen in Figure 17.24b. The nonlinear MZI as well as the other switching geometries mentioned above can be readily fabricated using waveguide architectures [57]. This makes possible small-scale device integration, a crucial issue in the telecommunications arena. Furthermore, organic and polymeric materials can be integrated into the

* Although the nonlinear MZI only appears to have one output channel, the pulses that destructively interfere and, therefore, do not exit the device through this channel actually propagate back toward the input port. This essentially becomes the second output channel.

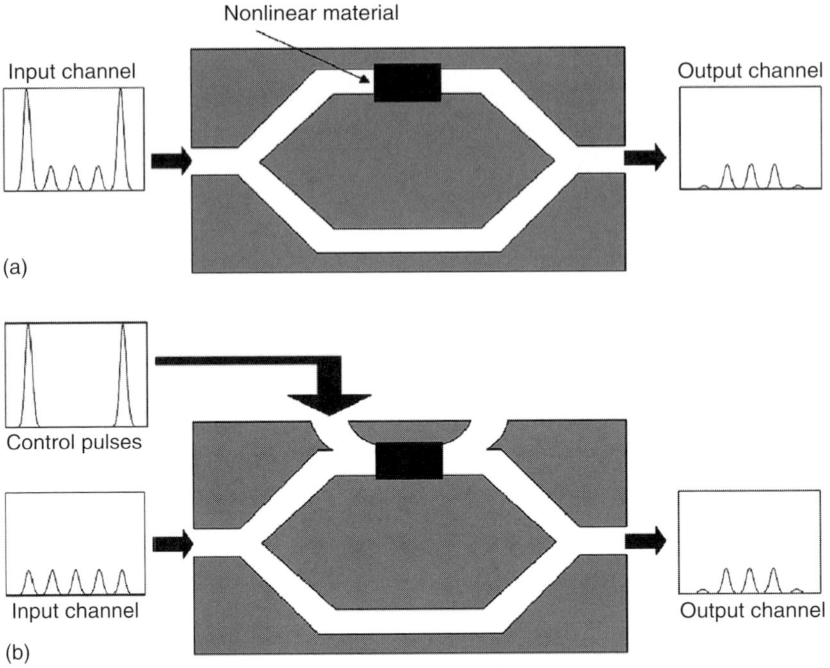

FIGURE 17.24 Illustration of temporal demultiplexing via all-optical switching in a nonlinear Mach–Zehnder interferometer. (a) Switching using self-phase modulation. (b) Switching using cross-phase modulation.

waveguide geometries as described in the previous section. Although processability and optical transparency issues have proven challenging for successful integration of these materials into devices, demonstrations of all-optical switching using such materials have been reported [63,64].

While the intensity-dependent refractive index clearly facilitates processing of information in optical communications systems, it can also aid in its transmission as well. To understand how this is possible, it must first be noted that propagation of an optical pulse through a linear medium results in temporal broadening of the pulse. This occurs due to the frequency dependence of the group velocities in the medium and its corresponding effect on the pulse spectrum. This group velocity dispersion (GVD), typically denoted by β_2, causes a linear frequency shift across the duration of the pulse. This frequency chirp can be either positive or negative depending on the regime of dispersion (either normal or anomalous, respectively) the pulse experiences. This can be seen in the upper portion of Figure 17.25a. The lower portion of Figure 17.25a plots the temporally dependent frequency shift across a pulse that exists as a result of SPM following the propagation of a pulse through a nonlinear medium (see Section 17.2.1.2 for details). This induced frequency chirp can also be either positive or negative depending on the sign of n_2.

One can therefore imagine that under certain conditions, the chirp induced by SPM could balance the frequency chirp due to GVD. In this case, the pulse will propagate undistorted through the medium such that no temporal broadening occurs. This resulting entity is known as a temporal soliton. Figure 17.25b illustrates these affects for a pulse propagating through a medium with and without SPM to compensate for the GVD. A number of different organic and polymeric nonlinear media have the potential to support soliton propagation [65,66]. Pulses resistant to temporal broadening would be of considerable interest in the telecommunications field where pulse broadening in high bit-rate data streams can be extremely detrimental. In fact, soliton communication systems have been well studied [67]. Other characteristics of solitons make them attractive in this context as well. They are robust systems in the sense that they can make adjustments in their amplitude or pulse

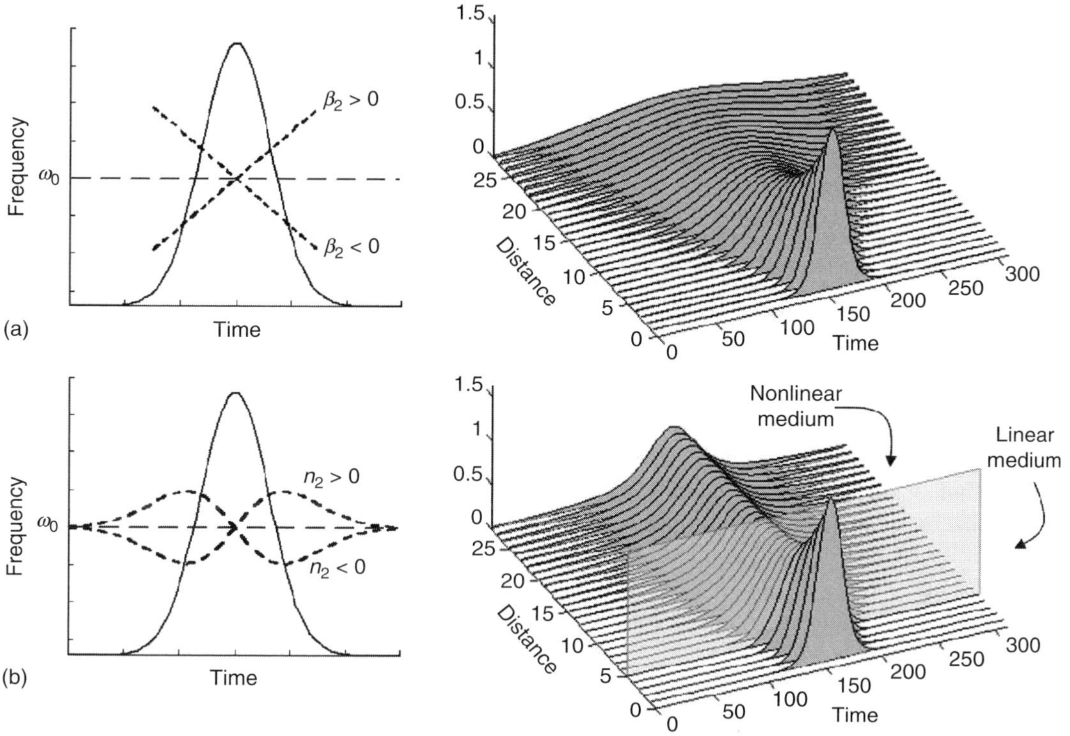

FIGURE 17.25 (a) Frequency chirp impressed upon an optical pulse by GVD (top) and by SPM (bottom). (b) Simulation of pulse broadening (top) during propagation through a linear medium exhibiting GVD ($\beta_2 < 0$) and the formation of a temporal soliton (bottom) after compensation of the frequency chirp from GVD ($\beta_2 < 0$) by the frequency chirp from SPM ($n_2 > 0$). The vertical plane (in bottom graph) denotes a separation between a purely linear medium (as in the top graph) and a medium exhibiting both linear and nonlinear characteristics.

duration to keep the product of the two quantities a constant. They also possess a nonlinear phase shift that is constant over the entire pulse profile unlike for the SPM-induced case describe above. This has ramifications on the potential to perform all-optical switching using solitons. Finally, it is important to note that temporal solitons have direct analogues in the spatial domain. In this domain, the effect of spatial beam spreading due to diffraction can be compensated by the mechanism of self-focusing that occurs in a nonlinear medium, which possesses a positive value of n_2 (once again, see Section 17.2.1.2 for details). Essentially, the beam creates its own waveguide and a phenomenon known as self-trapping occurs. In this manner, spatial solitons can be employed to confine a beam during propagation through a slab waveguide.

17.4.2.2 Optical Phase Conjugation

The phenomenon of OPC can be used for real-time processing of optical images. Its basic principle involves the reversal of the phase factor and propagation direction of an incident optical field [68]. One of the most straightforward ways to implement OPC is by using the counter-propagating geometry (also known as the phase-conjugate geometry) for DFWM as illustrated in Figure 17.8a. First, it is obvious that the fourth beam propagates in a direction that is antiparallel to the third beam as dictated by the phase-matching condition, i.e., $\bar{k}_4 = -\bar{k}_3$. Secondly, following the appropriate field expansion as described in Section 17.2.1, the optical field associated with beam 4 can be written as [17]

$$A_4 \propto A_1 A_2 A_3^* \tag{17.47}$$

where $\boldsymbol{A}$ denotes the complex field amplitudes for each beam. If one assumes that beams 1 and 2 are plane waves then $\boldsymbol{A}_4$ is proportional to the complex conjugate of $\boldsymbol{A}_3$, that is, the phase of $\boldsymbol{A}_4$ is reversed with respect to $\boldsymbol{A}_3$. Essentially, the nonlinear medium that produces this phase-conjugate field acts as a very unique kind of mirror whereby the reflected beam retraces its original path as well its overall phase factor.

These two unique properties of OPC lend themselves to the application of wave front aberration correction or the correction of optical distortions. Figure 17.26a illustrates how this can be accomplished. An undistorted plane wave enters the aberrating medium, its phase front becomes distorted, and the distorted wave front strikes the nonlinear medium. The phase-conjugate wave from the resulting nonlinear interaction is reflected back along the same optical path with a retraced (or reversed) phase front such that upon passing through the same aberrating medium, the original phase aberration is undone resulting in a reconstruction of the original plane wave. This technique is immediately translatable to image reconstruction. It should be noted that this image or wave front reconstruction requires both the property of phase reversal, so that the phase aberration is undone, and the reversal of the propagation direction, such that the wave front passes through the same portion of the aberrating medium. The effects of replacing the nonlinear medium, or phase-conjugate mirror, with a traditional mirror can be seen in Figure 17.26b. The use of static holography techniques to implement OPC in image reconstruction has been recognized for a number of years. But it is the use of DFWM to implement OPC that results in a form of dynamic or real-time holography where the recording and reconstruction of the image occur simultaneously. Therein lies the benefit of using this third-order phenomenon for such an application—real-time image correction can be

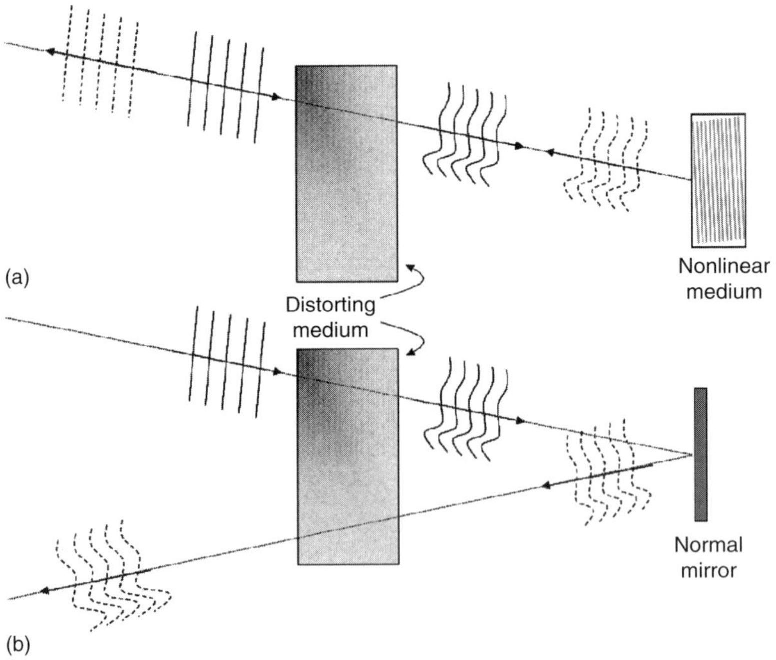

FIGURE 17.26 Potential scenarios for optical wave fronts following passage through a distorting medium. (a) Wave front aberration correction as a consequence of nonlinear OPC. (b) Further wave front deterioration due to reflection off a normal mirror. Solid lines are incoming wavefronts and dashed lines are outgoing wavefronts.

realized. One should note that unlike the aforementioned telecommunications applications that typically utilize waveguide geometries, OPC for image reconstruction employs bulk or large area films to provide sufficient coverage area. Organic systems as varied as carbon disulfide [69], dye-impregnated polymers [70], and even photosynthetic proteins [71] have been employed to perform OPC via DFWM as a means of image correction.

Another application for this process of nonlinear OPC is real-time image recognition, an application that could, in the future, play a significant role in the development of artificial intelligence. Image or pattern recognition can be computationally expensive when undertaken in the electronic domain due to the serial nature by which the procedure is performed. This results in slow processing. In the optical domain, however, the processing is performed in a parallel manner and therefore the time required for a single pattern recognition event can be drastically reduced. This benefit has been understood and exploited using static OPC for a significant period. However, by utilizing a third-order nonlinear material, dynamic or real-time pattern recognition becomes possible.

Figure 17.27 illustrates how this process is possible. As for real-time image correction, a counter-propagating DFWM geometry is employed. Beam 3 has the image of an object impressed upon it; this could be by means of a film transparency, for instance. Beam 1 has a reference image impressed upon it while beam 2 remains unmodified. By focusing the beams onto the nonlinear medium, beam 4 or the phase-conjugate beam can be represented in an identical manner to Equation 17.3 with one exception. The fields of all three incident beams are the spatial Fourier transforms of the fields before entering their respective focusing lenses [72], that is,

$$\boldsymbol{A}_4 \propto \tilde{\boldsymbol{A}}_1 \cdot \tilde{\boldsymbol{A}}_2 \cdot \tilde{\boldsymbol{A}}_3^* \tag{17.48}$$

where, for instance, $\tilde{\boldsymbol{A}}_1$ would be the Fourier transform of $\boldsymbol{A}_1$. Consequently, if the phase-conjugate field $\boldsymbol{A}_4$ is viewed at one focal distance away its collection lens (see Figure 17.27), the inverse Fourier transform of Equation 17.48 would be found. This would be represented as

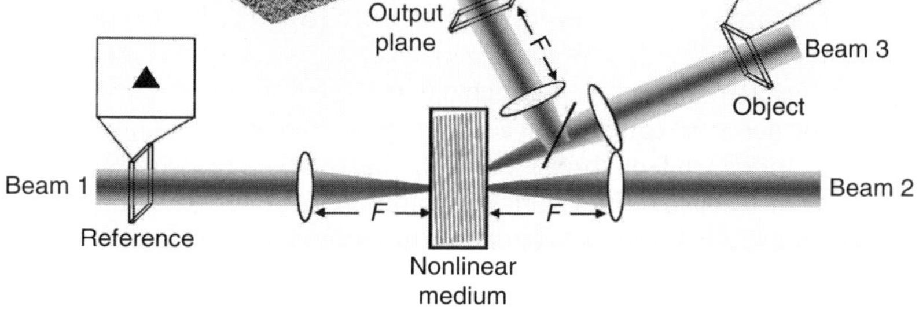

FIGURE 17.27 Typical phase-conjugate geometry employed for real-time image recognition. Since beam 2 possesses no spatial information, the output plane represents a simple cross-correlation between the reference image (beam 1) and the object image (beam 3). Case 1 represents identification of the reference image from a series of single-element object images while case 2 demonstrates identification from a multi-element object image.

$$\tilde{A}_4 \propto A_1 * A_2 \circledast A_3 \tag{17.49}$$

where

$\circledast$ denotes a cross-correlation

$*$ denotes a convolution

The fact that beam 2 has no spatial information encoded upon it implies that Equation 17.49 can be reduced to a simple cross-correlation of the object image (beam 3) and the reference image (beam 1). This correlation is precisely what is necessary for image recognition, and the probability of this image matching (or equivalently the degree of correlation) is proportional to the amplitude of the correlation peak (see output plane in Figure 17.27). Two applications of this image recognition process are illustrated in Figure 17.27: identification of the reference image from a multi-element object image or from a series of single-element object images. The obvious benefit of this all-optical image recognition is the significant increase in processing speed. This is enabled by both the massive parallelism afforded by OPC and the rapidity with which this process can be repeated. Both features can be obtained by using organic materials with ultrafast third-order nonlinearities. Real-time image recognition has been demonstrated using an organic film consisting of a polyacetylene-like polymer [73] that exhibited a reasonably large and ultrafast Kerr nonlinearity. Nonetheless, it should be noted that despite the fast processing speed afforded by a NLO correlator, other components within the system might then become the rate-limiting factors in achieving high-speed image recognition.

17.4.2.3 Other Applications of Interest

As detailed above, the intensity-dependent refractive index and DFWM are phenomena that account for a large range of NLO applications based on organic materials. Nonetheless, other phenomena can be effectively exploited as well. Organic materials that exhibit highly efficient third-harmonic generation provide a means to convert light at telecommunications wavelengths, where detection systems can be expensive and more susceptible to noise, to visible wavelengths. One can imagine that by backfilling a highly processable THG material into a microstructure fiber (see Figure 17.23d), long interaction lengths could allow for significant frequency conversion to occur. THG can also be employed for optical signal processing in the near infrared. Utilizing a thin bulk film of an organic dipolar chromophore, frequency-resolved optical gating (FROG) has been demonstrated at 1.55 μm [74].

Effective $\chi^{(3)}$ phenomena, such as reverse saturable absorption described in Section 17.2.1.4, are often utilized to facilitate optical limiting or suppression [75]. For a material to act as an ideal optical limiter it must provide good transmission at low light levels such that for ambient conditions the material is effectively transparent. However, when dangerously high light levels impinge upon the material, such as a high intensity laser pulse, the material rapidly becomes opaque providing good protection against the incident light. Placing such a material before a detector (either the human eye or a sensitive electronic detector) can provide adequate protection against damage. Since RSA manifests itself as an intensity-dependent absorption and is inherently a resonantly enhanced nonlinearity, it acts as an efficient mechanism for optical limiting. A large number of organic systems have been employed as effective RSA materials in optical limiting such as phthalocyanines [76], porphyrins [77], and even carbon black suspensions [78].

17.5 CONCLUSION

Third-order NLO materials respond to the application of intense optical fields by altering the very properties of these incident fields. This ability of light to control light gives rise to a number of unique applications such as all-optical switching and soliton propagation in the telecommunications arena, real-time image correction and recognition via dynamic holography, and, even, optical power limiting for sensor protection. Not unexpectedly, these applications exploit aspects of a variety of third-order NLO phenomena: the intensity dependence of the nonlinear refractive index, the creation of a phase-conjugate beam in DFWM, and the enhanced absorption produced by an effective third-order nonlinearity. The devices that enable such applications place quite stringent requirements on the nonlinear materials they employ. Mechanical and photochemical stability issues must be addressed as well as the subject of processability for ease of integration into devices. Finally, the optical properties of the material must be optimized.

It is this final material requirement, specifically the magnitude and response of the third-order nonlinearity, which presents major challenges. A variety of $\chi^{(3)}$ characteristics play a role in the efficacy of a material such as its magnitude, the spectral location of its resonances, and its temporal response. For this reason, a number of different techniques have been developed to facilitate the characterization of $\chi^{(3)}$ as well as its microscopic counterpart, γ. Not surprisingly, understanding the nonlinear properties at the molecular level of the system provides the best means for optimizing a material through variations in its chemical structure. It was shown that, through the application of a relatively simple VB–CT model, the behavior of γ for a series of donor–acceptor polyenes with varying electron acceptor strength could be semiquantitatively predicted. This powerful tool provided a strategy for producing molecules with predictably large microscopic polarizabilities. Such systems, as well as others, have been successfully translated into materials with sizable macroscopic nonlinearities. It is these types of organic molecular and polymeric systems that enable the technologies that exploit third-order NLO phenomena.

EXERCISE QUESTIONS

1. Three electric fields, two of which are degenerate (i.e., same frequency) and one is a DC field, interact within a medium that exhibits a third-order nonlinearity.
   a. Which types of nonlinear phenomena might be observed as a result of this interaction?
   b. Could any of these phenomena be observed if one of the fields were not present?
2. Calculate the value of $\chi^{(3)}$ (in esu) for a neat film consisting of molecules with $\langle\gamma\rangle$ values of 1.0×10^{-33} esu and a molecular weight of 700 g/mole. Assume a film density of 1.0 g/cm^3 and a refractive index of 1.8. How does this value compare to the same molecule in solution (see example in Section 17.1.3.2)? Assume the nonlinearity is a degenerate one, i.e., $\chi^{(3)}(-\omega; \omega, -\omega, \omega)$.
3. Describe physically how a Gaussian beam might interact with a liquid medium that possesses a thermal nonlinearity. (Hint: A thermal nonlinearity causes a medium to heat upon exposure to light.) On the basis of this hint, why is a thermal nonlinearity considered nonlocal? What is the sign of this nonlinearity?
4. If the nonlinearity of an unknown sample were measured by both the Z-scan and THG characterization techniques, should one expect the resulting nonlinearity from each technique to be the same? Why?
5. Describe how one can obtain the polarizability and hyperpolarizabilities of a conjugated donor–acceptor molecule from knowledge of the dependence of the dipole moment on applied electric field.

$$\mu = \mu_o + \left.\frac{d\mu}{dE}\right|_{E_o} E + \frac{1}{2!}\left.\frac{d^2\mu}{dE^2}\right|_{E_o} E^2 + \frac{1}{3!}\left.\frac{d^3\mu}{dE^3}\right|_{E_o} E^3$$

6. Value of the electronic coupling, t, for cyanines can be obtained from the experimental energy gap using the expression: $t = E_g/2$. For moderate chain lengths, E_g is found to be inversely proportional to the number of bonds, N, or the length of the molecule.

a. Using Equation 17.44, obtain an expression for the γ of cyanines as a function of N (remember that the μ_{CT} is proportional to the length of the molecule).
b. Cyanine with eight bonds is found to give a $\gamma = -370 \times 10^{-36}$ esu. Using the result above, estimate the γ value for a cyanine with $N = 12$.

7. Consider the following molecule:

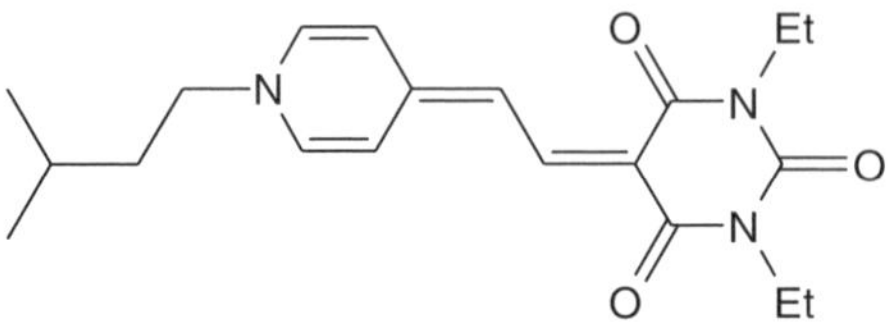

a. Draw a resonance structure for the CT form of this molecule.
b. Discuss the factors influencing the relative energy of the VB and CT resonance structures of this compound. Would you expect the VB or CT form to be lower in energy?
c. What region in Figure 17.21 would you expect this molecule to be in if it were dissolved in a low polarity solvent?
d. Describe the trend you would expect for γ for this compound as the solvent polarity was increased.

8. Nonlinear MZI is fabricated with one of its 1.0 cm long arms made up of the neat $\chi^{(3)}$ material described in Question 2. What intensity is necessary to produce a relative phase shift ($\Delta\phi$, see Equation 17.16) of π between the arms in order to perform optical switching at the telecommunications wavelength of 1.3 μm? If the value for $\chi^{(3)}$ from Question 2 was not determined, use a value of $\chi^{(3)} = 1.0 \times 10^{-11}$ esu. (Hint: Convert to SI units first).

LIST OF ABBREVIATIONS

2PA:	two-photon absorption
2PE:	two-photon emission
AM-1:	Austin model 1
AOSP:	all-optical signal processing
BLA:	bond length alternation
c.c.:	complex conjugate
CARS:	coherent anti-Stokes Raman scattering
CSRS:	coherent Stokes Raman scattering
CT:	charge transfer
D:	degenerate
D-term:	dipolar term contribution to second hyperpolarizability
DFG:	difference frequency generation
DFWM:	degenerate four-wave mixing
$E_T(30)$:	Reichardt's solvent polarity parameter
EFISH:	electric field induced second harmonic generation
FOM(s):	figure(s) of merit
GVD:	group velocity dispersion
INDO:	Intermediate neglect of differential overlap method
LIG:	light-induced grating
MZI:	Mach-Zehnder interferometer
N-term:	negative term contribution to second hyperpolarizability
ND:	non-degenerate

NLO:	nonlinear optical
OKE:	optical Kerr effect
OPC:	optical phase conjugation
RIKE:	Raman-induced Kerr effect
RSA:	reverse saturable absorption
SFG:	sum frequency generation
SPM:	self-phase modulation
SRS:	stimulated Raman scattering
T-term:	two-photon term contribution to second hyperpolarizability
THG:	third-harmonic generation
VB:	valence bond
VB-CT model:	valence bond-charge transfer model
WLC:	white-light continuum

REFERENCES

1. Kaiser, W. and Garrett, C.G.B., Two-photon excitation in $CaF_2:Eu^{2+}$, *Phys. Rev. Lett.*, 7, 229 1961.
2. Franken, P.A., Hill, A.E., Peters, C.W., and Weinreich, G., Generation of optical harmonics, *Phys. Rev. Lett.*, 7, 118 1961.
3. Sutherland, R.L., *Handbook of Nonlinear Optics*, Marcel Dekker, New York, 1996.
4. Prasad, P.N. and Williams, D.J., *Introduction to Nonlinear Optical Effect in Molecules and Polymers*, Wiley, New York, 1991.
5. Shen, Y.R., *The Principles of Nonlinear Optics*, John Wiley & Sons, Canada, 1984.
6. Halvorson, C. et al., Conjugated polymers with degenerate ground state. The route to high performance third order nonlinear optical response, *Chem. Phys. Lett.*, 212, 85 1993.
7. Dick, B., Hochstrasser, R.M., and Trommsdorff, H.P., *Nonlinear Optical Properties of Organic Molecules and Crystals*, Vol. 2, pp. 503–504, Academic Press, Orlando, 1987.
8. Butcher, P.N. and Cotter, D., *The Elements of Nonlinear Optics*, Cambridge University Press, Cambridge, 1990.
9. Righini, R., Ultrafast optical Kerr-effect in liquids and solids, *Science*, 262, 1386 1993.
10. McMorrow, D., Lotshaw, W.T., and Kenney-Wallace, G.A., Femtosecond optical kerr studies on the origin of the nonlinear responses in simple liquids, *IEEE J. Quan. Elec.*, 24, 443 1998.
11. Lessing, H.E. and Von Jena, A., Separation of rotational diffusion and level kinetics in transient absorption spectroscopy, *Chem. Phys. Lett.*, 42, 213 1976.
12. DeSalvo, R. et al., Self-focusing and self-defocusing by cascaded second-order effects in KTP, *Opt. Lett.*, 17, 28 1992.
13. Perry, J.W., *Materials for Nonlinear Optics, Chemical Perspectives*, Vol. 455, S.R. Marder, J.E. Sohn, and G.D. Stucky, Eds., American Chemical Society, Washington DC, 1991.
14. Keller, U., 'tHooft, G.W., Knox, W.H., and Cunningham, J.E., Femtosecond pulses from a continuously self-starting passively mode-locked Ti:sapphire laser, *Opt. Lett.*, 16, 1022 1991.
15. Alfano, R.R., *The Supercontinuum Laser Source*, Springer-Verlag, New York, 1989.
16. Eichler, H.J., Günter, P., and Pohl, D.W., *Laser-Induced Dynamic Gratings*, Vol. 50, Springer, Berlin, 1986.
17. Yariv A. and Pepper, D.M., Amplified reflection, phase conjugation, and oscillation in degenerate four-wave mixing, *Opt. Lett.*, 1, 16 1977.
18. Nussenzveig, H.M., *Causality and Dispersion Relations*, Academic, New York, 1972.
19. Kuzyk, M. and Dirk, C., Eds., *Characterization Techniques and Tabulations for Organic Nonlinear Optical Materials*, Marcel Dekker, New York, 1998.
20. Kajzar F. and Messier, J., Third-harmonic generation in liquids, *Phys. Rev. A*, 32, 2352 1985.
21. Sheik-Bahae, M. et al., Sensitive measurement of optical nonlinearities using a single beam, *IEEE J. Quan. Elec.*, 26, 760 1990.
22. Van Stryland, E.W. and Sheik-Bahae, M., *Characterization Techniques and Tabulations for Organic Nonlinear Optical Materials*, M. Kuzyk and C. Dirk, Eds., pp. 655–692, Marcel Dekker, New York, 1998.
23. Wang, J. et al., Time-resolved Z-scan measurements of optical nonlinearities, *J. Opt. Soc. Am. B*, 11, 1009 1994.
24. DeSalvo, R. et al., Z-scan measurements of the anisotropy of nonlinear refraction and absorption in crystals, *Opt. Lett.*, 18, 194 1993.

25. Balu, M., Hales, J.M., Hagan, D.J., and Van Stryland, E.W., White-light continuum Z-scan technique for nonlinear materials characterization, *Opt. Expr.*, 12, 3820 2004.
26. Samoc, M. et al., Femtosecond Z-scan and degenerate four-wave mixing measurements of real and imaginary parts of the third order nonlinearity of soluble conjugated polymers, *J. Opt. Soc. Am. B*, 15, 817 1998.
27. Strohkendl, F.P. et al., Phase-mismatched degenerate four-wave mixing: Complex third order susceptibility tensor elements of C_{60} at 768 nm, *J. Opt. Soc. Am. B*, 14, 92 1997.
28. Negres, R.A. et al., Experiment and analysis of two-photon absorption spectroscopy using a white-light continuum probe, *IEEE J. Quan. Elec.*, 38, 1205 2002.
29. Negres, R.A. et al., The nature of excited-state absorption in polymethine and squarylium molecules, *IEEE J. Sel. Top. Quan Elec.*, 7, 849 2001.
30. Orczyk, M.E., et al., Optical heterodyning of the phase-tuned femtosecond optical Kerr gate signal for the determination of complex third order susceptibilities, *Appl. Phys. Lett.*, 60, 2837 1992.
31. Beratan, D.N., Electronic hyperpolarizability and chemical structure, In *Materials for Nonlinear Optics: Chemical Perspectives*, S.R. Marder, J.E. Sohn, G.D. Stucky, Eds., 455, pp. 89–102, ACS Press, Washington DC, 1991.
32. Kuzyk, M.G. and Dirk, C.W., Effects of centrosymmetry on the nonresonant electronic third order nonlinear optical susceptibility, *Phys. Rev. A*, 41, 5098 1990.
33. Dirk, C.W., Cheng, L.T., and Kuzyk, M.G., A simplified 3-level model describing the molecular 3rd-order nonlinear optical susceptibility, *Int. J. Quantum Chem.*, 43, 27 1992.
34. Garito, A.F., et al., Enhancement of nonlinear optical properties of conjugated chains through lowered symmetry, In *Organic Materials for Nonlinear Optics*, R.A. Hann, D. Bloor, Eds., pp. 16–27, Royal Society of Chemistry, London, 1989.
35. Pierce, B.M., A theoretical analysis of the third order nonlinear optical properties of linear cyanines and polyenes, *Proc. SPIE-Int. Soc. Opt. Eng.*, 1560, 148 1991.
36. Meyers, F., Marder, S.R., Pierce, B.M., and Brédas, J.-L., Electric field modulated nonlinear optical properties of donor-acceptor polyenes: sum-over-states investigation of the relationship between molecular polarzibilities (α, β and γ) and bond length alternation, *J. Am. Chem. Soc.*, 116, 10703 1994.
37. Bredas, J.L. et al., Third order nonlinear optical response in organic materials: Theoretical and experimental aspects, *Chem. Rev.*, 94, 243–278 1994.
38. Brooker, L.G.S. et al., Color and constitution. X. Absorption of the merocyanines, *J. Am. Chem. Soc.*, 73, 5332 1951.
39. Bourhill, G. et al., Experimental demonstration of the dependence of the first hyperpolarizability of donor-acceptor-substituted polyenes on the ground-state polarization and bond length alternation, *J. Am. Chem. Soc.*, 116, 2619 1994.
40. Lu, D., Chen, G., Goddard III, W.A., Perry, J.W., Valence-bond charge-transfer model for nonlinear optical properties of charge-transfer organic molecules, *J. Am. Chem. Soc.*, 116, 10679 1994.
41. Marder, S.R. et al., A unified description of linear and nonlinear polarization in organic polymethine dyes, *Science*, 265, 632 1994.
42. Gorman, C.B. and Marder, S.R., An investigation of the interrelationships between linear and nonlinear polarizabilities and bond-length alternation in conjugated organic-molecules, *Proc. Natl. Acad. Sci. USA*, 90, 11297 1993.
43. Marder, S.R. et al., Relation between bond-length alternation and second electronic hyperpolarizability of conjugated organic molecules, *Science*, 261, 186 1993.
44. Reichardt, C., Solvatochromic dyes as solvent polarity indicators, *Chem. Rev.*, 94, 2319–2358 1994.
45. Puccetti, G. et al., Chain-length dependence of the third order polarizability of disubstituted polyenes. Effects of end groups and conjugation length, *J. Phys. Chem.*, 97, 9385–9391 1993.
46. Samuel, I.D.W. et al., Saturation of cubic optical nonlinearity in long-chain polyene oligomers, *Science*, 265, 1070–1072 1994.
47. Marder, S.R. et al., Large molecular third order optical nonlinearities in polarized carotenoids, *Science*, 276, 1233–1236 1997.
48. Hermann, J.-P., Nonlinear susceptibilities of cyanine dyes. Application to frequency tripling, *Opt. Comm.*, 12, 102–104 1974.
49. Hales, J.M. et al., Bis-dioxaborine polymethines with large third order nonlinearities for all-optical signal processing, *J. Am. Chem. Soc.*, 128, 11362–11363 2006.
50. Chung, S.-J. et al., Extended squaraine dyes with large two-photon cross-sections, *J. Am. Chem. Soc.*, 128, 14444–14445 2006.
51. Hales, J.M. et al., Highly processable polymers with large third order nonlinearities in the near-infrared, *PMSE Preprints*, 94, 827–828 2006.

52. Delfyett, P.J. et al., High-power ultrafast laser-diodes, *IEEE J. Quan. Elec.*, 28, 2203 1992.
53. de la Torre, G., Vazquez, P., Agullo-Lopez, F., and Torres, T., Phthalocyanines and related compounds: organic targets for nonlinear optical applications, *J. Mat. Chem.*, 8, 1671 1998.
54. Zhao, B., Lu, W.-Q., Zhou, Z.-H., and Wu, Y., The important role of the bromo group in improving the properties of organic nonlinear optical materials, *J. Mat. Chem.*, 10, 1513 2000.
55. de Abajo, J. and de la Campa, J.G., *Advances in Polymer Science*, Vol. 140, pp. 24–59, Springer-Verlag, Berlin Heidelberg, 1999.
56. Mizrahi, V. et al., 2-photon absorption as a limitation to all-optical switching, *Opt. Lett.*, 14, 1140 1989.
57. Stegeman, G.I. and Stolen, R.H., Waveguides and fibers for nonlinear optics, *J. Opt. Soc. Am. B*, 6, 652 1989.
58. Harbold, J.M. et al., Highly nonlinear As–S–Se glasses for all-optical switching, *Opt. Lett.*, 27, 119 2002.
59. Wright, E.M., et al., Semiconductor figure of merit for nonlinear directional couplers, *Appl. Phys. Lett.*, 52, 2127 1988.
60. Broderick, N.G.R., Monro, T.M., Bennett, P.J., and Richardson, D.J., Nonlinearity in holey optical fibers: measurement and future opportunities, *Opt. Lett.*, 24, 1395 1999.
61. Klavetter, F.L. and Grubbs, R.H., Polycyclooctatetraene (polyacetylene): Synthesis and properties, *J. Am. Chem. Soc.*, 110, 7807 1988.
62. Stegeman, G.I. and Wright, E.M., All-optical waveguide switching, *Opt. Quan. Elec.*, 22, 95 1990.
63. Aranda, F.J., Garimella, R., and Roach, J.F., All-optical light modulation in bacteriorhodopsin films, *Appl. Phys. Lett.*, 67, 599 1995.
64. Quintero-Torres, R. and Thakur, M., Picosecond all-optical switching in a Fabry–Perot cavity containing polydiacetylene, *Appl. Phys. Lett.*, 66, 1310 1995.
65. de la Fuente, R. and Barthelemy, A., Spatial soliton-induced guiding by cross-phase modulation, *IEEE J. Quan. Elec.*, 28, 547 1992.
66. Bader, M.A. et al., Poly(*p*-phenylenevinylene) derivatives: New promising materials for nonlinear all-optical waveguide switching, *J. Opt. Soc. Am. B*, 19, 2250 1992.
67. Agrawal, G.P., *Fiber-Optic Communications Systems*, Wiley-Interscience, New York, 2002.
68. Zel'dovich, B.Y., Pilipetsky, N.F., and Shkunov, V.V., *Principles of Phase Conjugation*, Vol. 42., Springer, Berlin, 1985.
69. Bloom, D.M. and Bjorklund, G.C., Conjugated wave-front generation and image reconstruction by four-wave mixing, *Appl. Phys. Lett.*, 31, 592 1977.
70. Oleary, S.V., Real-time image-processing by degenerate 4-wave-mixing in polarization-sensitive dye-impregnated polymer films, *Opt. Comm.*, 104, 245 1994.
71. Downie, J.D., Real-time holographic image correction using bacteriorhodopsin, *Appl. Opt.*, 33, 4353 1994.
72. White, J.O. and Yariv, A., Real-time image processing via four-wave mixing in a photorefractive medium, *Appl. Phys. Lett.*, 37, 5 1980.
73. Halvorson, C. et al., A 160-femtosecond optical image processor based on a conjugated polymer, *Science*, 265, 1215 1994.
74. Ramos-Ortiz, G. et al., Third order optical autocorrelator for time-domain operation at telecommunication wavelengths, *Appl. Phys. Lett.*, 85, 179 2004.
75. Boggess, T.F., A review of optical limiting mechanisms and devices using organics, fullerenes, semiconductors and other materials, *Prog. Quan. Elec.*, 17, 299 1993.
76. Perry, J.W. et al., Organic optical limiter with a strong nonlinear absorptive response, *Science*, 273, 1533 1996.
77. Chen, P.L. et al., Picosecond kinetics and reverse saturable absorption of meso-substituted tetrabenzoporphyrins, *J. Phys. Chem.*, 100, 17507 1996.
78. Xia, T. et al., Nonlinear response and optical limiting in inorganic metal cluster Mo2Ag4S8(PPh3)(4) solutions, *J. Opt. Soc. Am. B*, 15, 1497 1998.

18 Organic Multiphoton Absorbing Materials and Devices

Kevin D. Belfield, Sheng Yao, and Mykhailo V. Bondar

CONTENTS

18.1 Introduction 574
18.2 2PA Mechanism and Structure–Property Relationship 575
 18.2.1 Nature of Two-Photon Absorption 575
 18.2.2 Simplified Essential States Model 576
 18.2.3 Selection Rules 577
18.3 Organic Two-Photon and Multiphoton Absorbing Materials 578
 18.3.1 Linear Structures 578
 18.3.2 Branched Structures 583
18.4 Multiphoton Devices 585
 18.4.1 Three-Dimensional Data Storage 585
 18.4.2 Multiphoton Fluorescence Microscopy 589
 18.4.3 Multiphoton Photodynamic Therapy 592
 18.4.4 3D Microfabrication 594
 18.4.5 Optical Power Limiting 597
18.5 Future Development 599
18.6 Summary 599
Exercise Questions 600
List of Abbreviations 600
References 601

Abstract This chapter introduces basic mechanism of two-photon absorption (2PA) and its unique feature of spatial resolution with near-infrared (NIR) excitation. Principles of designing materials to exhibit large two-photon absorptivities are explained. Difference in excitation selection rule between one-photon and two-photon excitation (2PE) for these materials is discussed. Potential applications for 2PA, such as three-dimensional (3D) data storage, fluorescent imaging, 3D microfabrication, photodynamic therapy (PDT) and optical power limiting are briefly described. Requirements of material optimization for the particular applications are also discussed. The chapter concludes with a brief introduction of future challenges.

18.1 INTRODUCTION

Organic materials exhibiting significant nonlinear responses to applied electric or electromagnetic fields have attracted intense interest during the last few decades. Currently, organic materials occupy a prominent role in two-dimensional (2D) (linear) display technology based on specifically induced phase transitions in liquid crystalline materials, thereby altering their optical properties, and in the rapidly developing arena of organic light-emitting diodes (electroluminescent materials). Organic chromophores are becoming an integral component in second and third harmonic generation, devices for radiation frequency conversion, and waveguides. Suitable materials for such applications manifest a nonlinear optical response in the presence of an applied electric field, resulting in amplification of the particular optical property (e.g., refractive index change) relative to that obtained via a linear dependence. The beginning of the twenty-first century has been accompanied by an ever-pressing need for materials that exhibit amplification or respond in a highly nonlinear manner to a particular stimulus. In particular, compounds that undergo strong nonlinear, multiphoton absorption are being investigated as materials for a wide variety of potential applications in areas ranging from optical information storage, three-dimensional (3D) optical memories, biophotonics, materials science, and photochemistry [1].

Physicist Maria Göppert-Mayer first proposed the theoretical concept of multiphoton absorption in 1931 [2]. The seminal paper described two-photon absorption (2PA) process as the near simultaneous absorption of two photons to achieve a real transition within an atom or a molecule. Neither photon has sufficient energy to complete this transition on its own but, if the sum of the combined photon energies involved were equivalent to the energy difference between the initial and final states, the completion of that transition would be possible. Hence, organic molecules can be excited into transitions upon absorption of two photons whose combined energies match those for single-photon excitation (1PE) of the same transition. For simplicity, 2PA can be conceptualized from a semiclassical perspective [3]. In the 2PA process, molecules exposed to high-intensity light can undergo near simultaneous absorption of two photons mediated by a so-called "virtual state," a state with no classical analog. The combined energy of the two photons accesses a stable excited state of the molecule. If the two photons are of the same energy (wavelength), the process is referred to as degenerate 2PA. On the other hand, if the two photons are of different energies (wavelength), the process is nondegenerate 2PA.

The absorption of two photons to induce a transition, equivalently performed by one-photon absorption (1PA), requires both of their absorption to be within $\sim 10^{-15}$ s of each other. This "near simultaneous" nonlinear absorption process requires a large photon density to produce any appreciable probability for 2PA-induced transition to occur. Hence, it was not until after the invention of the laser that Kaiser and Garret demonstrated the experimental observation of 2PA in a $CaF_2:Eu^{2+}$ crystal [4], some 30 years after the theoretical description of 2PA. Tremendous development and subsequent commercialization of high peak-power, tunable laser sources have generated new applications exploiting the principles of 2PA, predominantly in the fields of chemistry and physics, and more recently, biophotonics.

A unique feature of the 2PA process is the probability that the 2PE is proportional to the square of the incident light intensity. This quadratic, or nonlinear, dependence of 2PA on the intensity of the incident light has substantial implications. For example, in a medium containing one-photon absorbing chromophores, significant absorption occurs all along the path of a focused beam of suitable wavelength light. This can lead to out of focus excitation. In a two-photon process, negligible absorption occurs except in the immediate vicinity of the focal volume of a light beam of appropriate energy. This allows spatial resolution about the beam axis as well as radially, which circumvents out of focus absorption and is the principle reason for two-photon applications. Particular molecules can undergo upconverted fluorescence through nonresonant 2PA using near-infrared (NIR) radiation, resulting in an energy emission greater than that of the individual photons involved (upconversion).

The use of a longer wavelength excitation source for fluorescence emission affords advantages not feasible using conventional ultraviolet (UV) or visible fluorescence techniques, e.g., deeper penetration of the excitation beam, reduction of photobleaching, reduction of scattering by about a factor of 16 (since light scattering efficiency scales as λ^{-4}). These advantages offer great potential for this technique to be utilized in electro-optic applications. Before presenting a detailed explanation of the applications, however, in the next section, a brief description of the 2PA mechanism and the structure–property relationships for material design is presented, followed by introduction of several typical chromophores exhibiting large two-photon absorptivities.

18.2 2PA MECHANISM AND STRUCTURE–PROPERTY RELATIONSHIP

18.2.1 Nature of Two-Photon Absorption

The nature of 2PA can be revealed based on the interaction of molecular electrons with an optical field. The displacement of a molecular electronic charge can oscillate harmonically (linear process), or inharmonically (nonlinear process), depending on the strength of the electric field, **E**. On a microscopic level, the electronic displacement is related to the molecular polarization, i.e., to the induced molecular dipole moment $\boldsymbol{\mu}$:

$$\boldsymbol{\mu} = \alpha \cdot \mathbf{E} + \frac{1}{2}\beta \cdot \mathbf{EE} + \frac{1}{6}\gamma \cdot \mathbf{EEE} + \cdots, \tag{18.1}$$

where

α is a linear (first order) molecular polarizability
β and γ are the second- and third-order polarizability respectively

Equation 18.1 can be expressed as

$$\mu_i = \alpha_{ij} \cdot E_j + \frac{1}{2}\beta_{ijk} \cdot E_j E_k + \frac{1}{6}\gamma_{ijkl} \cdot E_j E_k E_l + \cdots, \tag{18.2}$$

where

μ_i is a component of induced dipole moment
α_{ij}, β_{ijk}, and γ_{ijkl} are the second-, third-, and fourth-rank tensors of corresponding polarizabilities that are determined by the molecular parameters
E_j is a component of electric field (summation over repeated indexes is implied)

It can be shown that only the imaginary part of odd-order terms of polarizability ($\alpha_{ij}, \gamma_{ijkl}, \ldots$) in Equation 18.2 contribute to the dissipative processes, such as one-photon (α_{ij}) or two-photon (γ_{ijkl}) absorption [5]. This means that the lowest-order nonlinear absorption (i.e., 2PA) will be described by the imaginary part of γ_{ijkl}. In the case of parallel polarization of photons with equal frequencies ω (degenerate 2PA), the value of γ_{ijkl} is related to the 2PA cross section, δ_{2PA}, for the medium with isotropically oriented molecules as [5,6]

$$\delta_{2PA} = \frac{1}{6} \cdot \frac{3L^4\hbar\omega^2}{2n^2c^2\varepsilon_0^2} \mathrm{Im}\langle\gamma(-\omega;\omega,-\omega,\omega)\rangle, \tag{18.3}$$

where

L is the local field factor, which describes the interactions between neighboring molecules
n is the refractive index of the medium
c is the speed of light in a vacuum
ε_0 is the permittivity of free space (brackets indicate an average of γ_{ijkl} over isotropic medium) [7]

To establish the relationship between the structure of a molecule and its 2PA properties, the value of γ_{ijkl} needs to be determined. Several quantum–chemical approaches have been used to calculate γ_{ijkl} and corresponding δ_{2PA} [8]. One of the most widely employed methods is the sum-over-state (SOS) approach derived by Orr and Ward [9], which involves the calculation of the molecular wave functions, permanent, and transition dipole moments for all electronic states that contribute to the polarizability. These contributions from the excited states can then be summed based on time-dependent perturbation theory, and the values of γ_{ijkl} can be determined. The SOS method allows for investigation of the frequency dependence of δ_{2PA} and identification of excited molecular states that play essential roles in the 2PA processes [10]. In a general case, the value of γ_{ijkl} is described by a relatively complicated equation [9], due to summation over the all excited states of the molecule. For more practical use, some simplified models for γ_{ijkl} were developed by Dirk et al. [11] and Birge and Pierce [12].

18.2.2 Simplified Essential States Model

One of the simplified model expressions for γ_{ijkl} can be obtained from the full SOS expression [9] considering that the ground state of the molecule $|g\rangle$ is strongly coupled to a single one-photon allowed state $|e\rangle$ and that there are several two-photon allowed states $|e'\rangle$ that are strongly coupled to this $|e\rangle$. In the case of degenerate 2PA when considering only resonant terms and *xxxx*-component of γ_{ijkl}, the full SOS expression can be reduced to the positive dipolar terms (D), two-photon terms (T), and negative terms (N) [13]:

$$\begin{aligned}
\gamma_{xxxx}(-\omega;\omega,-\omega,\omega) = 2\cdot\Bigg\{ & \frac{\mu^x_{ge}\cdot\bar{\mu}^x_{ee}\cdot\bar{\mu}^x_{ee}\cdot\mu^x_{eg}}{(E_{eg}-\hbar\omega)^2(E_{eg}-2\hbar\omega)} && D\\
& + \frac{\mu^x_{ge}\cdot\bar{\mu}^x_{ee}\cdot\bar{\mu}^x_{ee}\cdot\mu^x_{eg}}{(E^*_{eg}-\hbar\omega)(E_{eg}-2\hbar\omega)(E_{eg}-\hbar\omega)} && D\\
& - \frac{\mu^x_{ge}\cdot\mu^x_{eg}\cdot\mu^x_{ge}\cdot\mu^x_{eg}}{(E_{eg}-\hbar\omega)^3} && N\\
& - \frac{\mu^x_{ge}\cdot\mu^x_{eg}\cdot\mu^x_{ge}\cdot\mu^x_{eg}}{(E^*_{eg}-\hbar\omega)(E_{eg}-\hbar\omega)^2} && N\\
& + \sum_{e'}\frac{\mu^x_{ge}\cdot\bar{\mu}^x_{ee'}\cdot\bar{\mu}^x_{e'e}\cdot\mu^x_{eg}}{(E_{eg}-\hbar\omega)^2(E_{e'g}-2\hbar\omega)} && T\\
& + \sum_{e'}\frac{\mu^x_{ge}\cdot\bar{\mu}^x_{ee'}\cdot\bar{\mu}^x_{e'e}\cdot\mu^x_{eg}}{(E^*_{eg}-\hbar\omega)(E_{e'g}-2\hbar\omega)(E_{eg}-\hbar\omega)}\Bigg\}, && T
\end{aligned} \qquad (18.4)$$

where

μ^x_{ij} $(i, j = g,e)$ are the transition dipole moments between corresponding electronic states

$\bar{\mu}^x_{ij} = \mu^x_{ij} - \delta_{ij}\mu^x_{gg}$ $(i, j = e,e'$; δ_{ij} is a Kronecker symbol) are the changes in the permanent dipole moments relative to the ground state

$E_{ig} = E_j - E_g - i\Gamma_{jg}$ $(j = e,e'$; E_i, E_g, and Γ_{jg} are the energies of corresponding electronic states and damping factors [14], respectively)

Only D and T terms contain two-photon resonances with $|e\rangle$ and $|e'\rangle$ states and, therefore, can contribute to 2PA cross section. Following from Equations 18.3 and 18.4 in the case of resonant 2PE into the $|e\rangle$ state ($\hbar\omega \approx (E_e - E_g)/2$ and $\Gamma_{eg} \ll (E_e - E_g)/4$)

$$\delta_{2PA} \propto \frac{\mu_{ge}^2 \cdot \Delta\mu_{ge}^2}{\Gamma_{eg}}, \tag{18.5}$$

where $\Delta\mu_{ge} = \mu_{ee} - \mu_{gg}$ is the change of the permanent dipole moment between the ground state $|g\rangle$ and the excited state $|e\rangle$. It is obvious from Equation 18.5—for centrosymmetric molecules that are characterized by zero values of $\Delta\mu_{ge}$—2PE into the $|e\rangle$ state is strictly forbidden. In contrast, unsymmetrical compounds with strong 1PA and large values of $\Delta\mu_{ge}$ can exhibit efficient 2PA under excitation into the first excited $|e\rangle$ state [15].

For the case of resonant 2PE into the two-photon allowed states $|e'\rangle$

($\hbar\omega \approx (E_{e'} - E_g)/2$ and $\Gamma_{eg} \ll [E_e - E_g - (E_{e'} - E_g)/2]/2$), the value of 2PA cross section can be express as

$$\delta_{2PA} \propto \frac{\mu_{ge}^2 \cdot \mu_{ee'}^2}{\Gamma_{e'g}(E_e - E_g - (E_{e'} - E_g)/2)^2}. \tag{18.6}$$

From Equation 18.6, it follows that the efficiency of the 2PA depends on the values of the corresponding one-photon transition dipoles and detuning energy $E_e - E_g - (E_{e'} - E_g)/2$.

The above description of 2PA processes corresponds to a simplified essential state model that is widely used to analyze the structure–property relationship of 2PA materials. In practice, this theoretical approach is in a good agreement with experimental data for different types of organic compounds [16].

18.2.3 Selection Rules

The 2PA efficiency of organic materials can be analyzed based on the symmetry of molecules participating in nonlinear absorption. According to the SOS approach, the ground- and excited-state-electronic wave functions need to be calculated to obtain permanent and transition dipole moments and finally, δ_{2PA}. The value of transition dipole moment can be determined as [14,17]

$$\mu_{if} = e \cdot \int \Psi_i^*(\mathbf{r}) \cdot \mathbf{r} \cdot \Psi_f(\mathbf{r}) \cdot d\mathbf{r}, \tag{18.7}$$

where

- $\mathbf{r}$ is the spatial coordinate
- Ψ_i and Ψ_f are the wave functions of the initial and final electronic states, respectively
- e is the electron charge

In the case when the structure of the molecule possesses definite parity, the electronic states of this molecule can be divided into symmetrical, g, and unsymmetrical, u, states with corresponding symmetrical or unsymmetrical molecular wave functions. For allowed dipole transitions, the integral (Equation 18.7) must be nonzero. Therefore, taking into account that $\mathbf{r}$ is of odd parity, the product of $\Psi_i^*(\mathbf{r}) \cdot \Psi_f(\mathbf{r})$ must possess an odd parity as well. Thus, two electronic states participating in a one-photon allowed transition must be of different parity, i.e., only $g \leftrightarrow u$ or $u \leftrightarrow g$ one-photon transitions can occur for symmetric molecules. According to Equations 18.4 and 18.6, a two-photon transition is mediated by one-photon allowed transitions from initial to intermediate states and from intermediate to the final states. In this case, initial and final states must be of equal parity, i.e., only $g \leftrightarrow g$ and $u \leftrightarrow u$ are the two-photon transitions allowed. These properties of electronic transitions are known as dipole selection rules [18] since they are valid in the electric-dipole approximation.

For molecules with definite parity, 2PE is a powerful tool for the investigation of their electronic structures, and therefore, for the optimization of the nonlinear optical properties of organic materials.

18.3 ORGANIC TWO-PHOTON AND MULTIPHOTON ABSORBING MATERIALS

In its early development stage, the 2PA phenomena were primarily of interest to physicists, in which simple materials were investigated, such as in the gas phase [19], inorganic crystals [20], or very simple organic molecules [21]. The organic molecules used for these studies were aromatic compounds such as naphthalene [22], anthracene [23], and xanthene [24], or organic dyes such as Rhodamines [25]. Although these conjugated molecules exhibited reasonable 2PA cross sections, only after the breakthrough of molecular structure design they dramatically improved developing materials with high-2PA cross sections that the door was opened to explore the full potential for practical applications. This section will introduce some of the representative 2PA molecular systems and their 2PA characteristics to provide some insight into the structure issues that are decisive for a molecule's large two-photon absorptivities. The introduction of the design of new materials, wherein improved 2PA combined with other molecular properties or processes (for example, large fluorescence quantum yield, efficient intersystem crossing, and low oxidation potential) that makes such materials suitable for given applications, will be discussed in Section 18.4.

The techniques used to measure 2PA properties include Z-scan, upconverted fluorescence techniques, etc. For more detailed information, readers are referred to Ref. [26]. It is already known that the 2PA cross-section values, δ, even for a sample at the same excitation wavelength, may differ drastically depending on the range of pulse duration in the nanosecond, picosecond, or femtosecond regime of the incident laser used for the measurement [27]. It is important for readers to notice the difference when comparing the 2PA data from different resources.

According to the structure and the properties, two classes of 2PA chromophores are included in this section, e.g., linear structures including quasi-linear donor–acceptor quadrupolar molecules incorporating a variety of conjugated bridges, including phenylene-vinylene, bifluorene, and polyfluorene systems, etc., and various dipolar conjugated donor–acceptor molecules; octupolar molecules, and multibranched structures, or dendrimer systems. The variety of two-photon optical properties related to the variety of structures will be described.

18.3.1 LINEAR STRUCTURES

During more than four decades, the 2PA of several interesting conjugated molecules has been investigated [22–25,28], many of which have become parent compounds for highly efficient 2PA materials. However, not until the pioneering work of Prasad et al. [1a] and Perry et al. [29,16b] in the phenylene-vinylene-type π-bridge with various donor and acceptor groups, the first series of materials with very large 2PA cross sections began to push the field forward. These structures have linear conjugation paths and various electronic donor or acceptor groups attached at different positions in the molecular framework. Some typical structures and their photophysical properties are shown in Figure 18.1 and Table 18.1 [29].

Experimental values determined with nanosecond pulses and femtosecond pulses (given in parentheses) are reported. The uncertainty in the experimental δ values is estimated to be ±15%. Also reported are the wavelengths for 1PA maxima ($\lambda_{ab\,max}$), the wavelengths for fluorescence emission maxima ($\lambda_{fl\,max}$), and the fluorescence quantum yields (Φ_{fl}).

As shown in Table 18.1, conjugated molecules **2** substituted with donor (D) or acceptor (A) groups in an essentially centrosymmetric pattern exhibit two-photon cross sections at least one order of magnitude larger than the corresponding unsubstituted molecules **1**. This enhancement in δ has been correlated with intramolecular charge transfer from the terminal donor groups to the

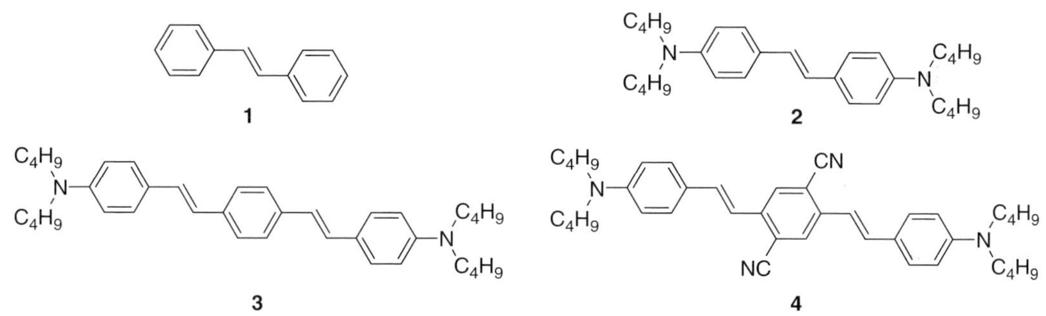

FIGURE 18.1 Structures of phenylene-vinylene type 2PA chromophores **1–4**.

π-bridge. Similarly, molecules with the structural motif A–D–A can also exhibit large δ values. Moreover, for a given π-bridge, the intramolecular charge transfer between the terminal donor groups and the π system of the molecule is facilitated by the presence of the acceptor groups in the central part of the π-bridge, as a consequence, conjugated D–A–D molecules **4** can have larger δ values (1750 GM) than D–π–D chromophores **3** (995 GM).

With similar architectures, another even more impressive series of materials, has been investigated in detail, i.e., fluorene derivatives developed by Reinhardt at the Air Force Research Laboratory [30] and Belfield's group [31,15]. Fluorene has unique structural features: two benzene rings are fused into one plane by a five-member ring, providing high-electron delocalization through increased overlap of π molecular orbitals between the rings. Meanwhile, the two doubly benzylic protons at the 9-position are very acidic and can be readily abstracted and reacted with electrophilic agents, such as aliphatic halides. These substituents, perpendicularly situated out of the conjugation plane, can efficiently reduce intermolecular interactions and modulate solubility. Thus, these derivatives generally feature high fluorescent quantum yield, good solubility, and high-2PA cross section. The typical structures and their photophysical data are presented in Figures 18.2 and 18.3.

In addition to the two-photon fluorescence (2PF) excitation spectra, single-photon fluorescence (1PF) excitation anisotropy data are also shown in Figure 18.3. Fluorescence excitation anisotropy provides outstanding empirical information on electronic transitions [32,33]. By comparing the 2PF excitation and excitation anisotropy spectra, these data vividly illustrate the selection rules of 2PA processes. The constant anisotropy values within the long wavelength absorption bands (350–450 nm) for all the compounds correspond to the first electronic transition, $S_0 \rightarrow S_1$. For $\lambda_{exc} \approx 290$–340 nm, the excitation anisotropy decreases and reaches a minimum value, clearly indicating the position of what will be referred to as the second electronic transition, $S_0 \rightarrow S_2$. This occurs as a result of the

TABLE 18.1
Optical Properties of Compounds 1–4

Compound	2PA λ_{max}(nm)	δ (GM)	$\lambda_{ab\ max}$ (nm)	$\lambda_{fl\ max}$ (nm)	Φ_{fl}
1	514	12 [62]	—	—	—
2	605 (<620)	210 (110 at 620 nm)	374	410	0.90
3	730	995	408	455	0.88
4	830	1750	490	536	0.69

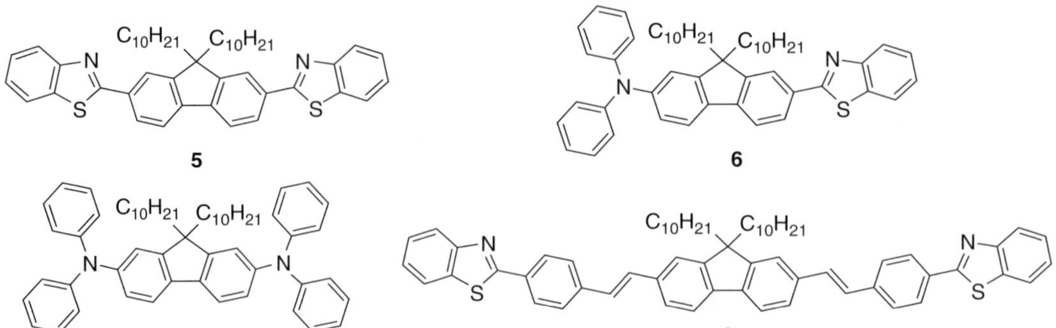

FIGURE 18.2 Structures of fluorene-based 2PA chromophores.

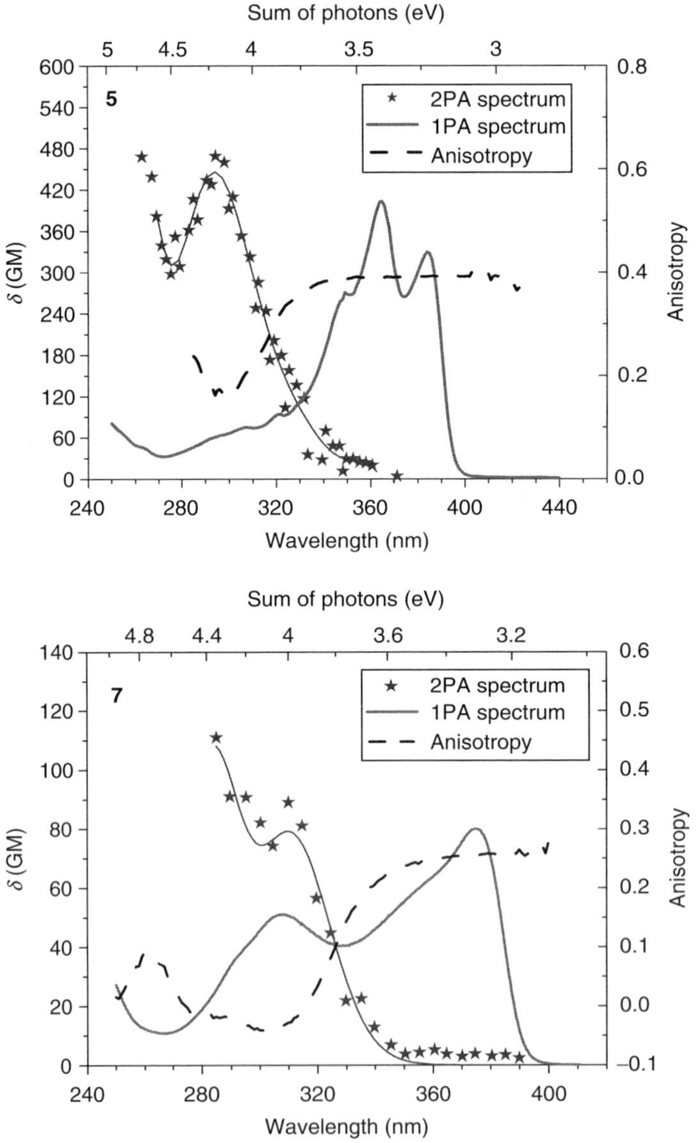

FIGURE 18.3 Linear absorption spectrum, steady-state excitation anisotropy, and 2PA of **5–8**.

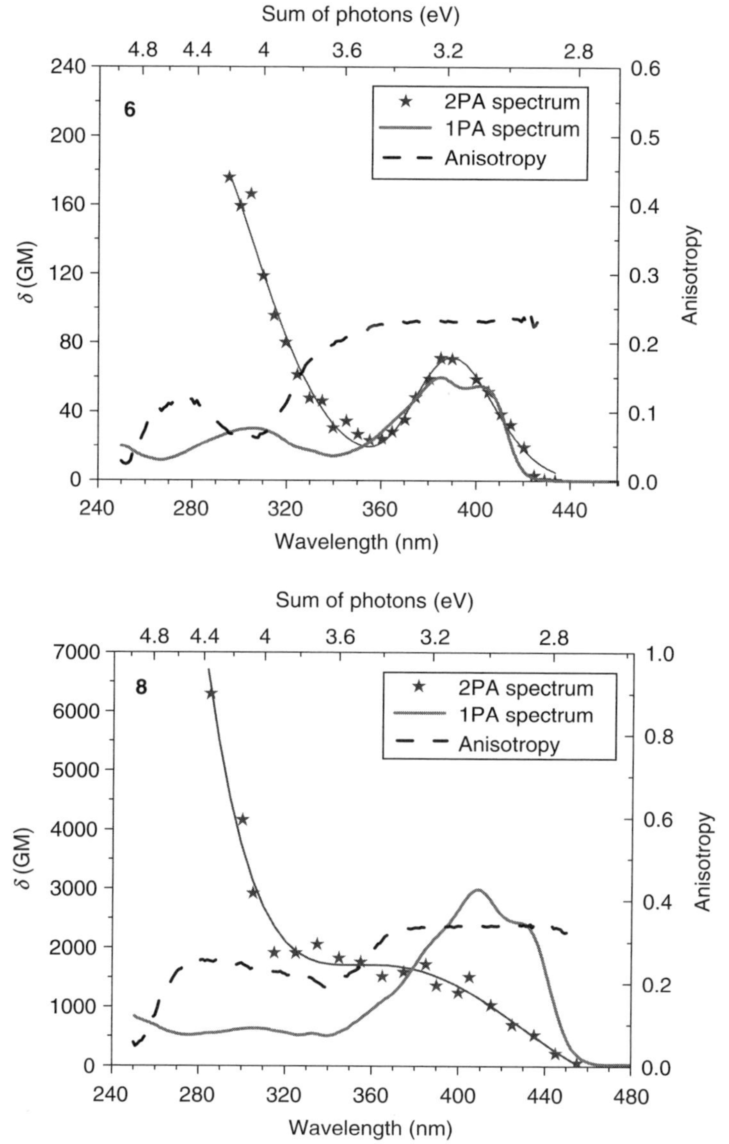

FIGURE 18.3 (continued)

change in the orientation of the $S_0 \rightarrow S_2$ transition dipole moment with respect to the $S_0 \rightarrow S_1$ moment. The locations ascribed to these transitions by the anisotropy curves correlate quite well with the peaks of the 2PA excitation or absorption spectra. Therefore, these peaks can be used as indicators of the electronic transition energies.

Experimental values determined with femtosecond pulses are reported in Table 18.2. The uncertainty in the experimental δ values is estimated to be ±15%. Also reported are the wavelength of 1PA maximum ($\lambda_{ab\ max}$), the wavelength of fluorescence emission maximum ($\lambda_{fl\ max}$), the fluorescence quantum yields (Φ_{fl}) in hexane, and fluorescence excitation anisotropy in silicone oil.

It is already known that the selection rules for 1PE and 2PE are different for centrosymmetric molecules, such as dyes **5** and **7**. It should be noted that none of the fluorene compounds could be considered strictly a centrosymmetric molecule. Nonetheless, if one assumes that the main portion

TABLE 18.2
Optical Properties of Compounds 5–8

Compound	2PA λ_{max} (nm)	δ (GM)	$\lambda_{ab\ max}$ (nm)	$\lambda_{fl\ max}$ (nm)	Φ_{fl}
5	590	469	365	389	0.90
6	774	71	385	418	0.70
7	620	89	375	390	0.40
8	570	6298	408	453	0.91

of the photophysical activity of the molecules occurs along the conjugated backbone of the molecule (a valid assumption considering the appended alkane chains play little role in the optical properties of the system), certain derivatives can indeed exhibit the same parity selection rules associated with true centrosymmetric systems. As indicated in Figure 18.3, dye **5** shows negligible 2PA into the first excited state which is identified by the peak in the linear absorption spectrum located at ~375 nm (or 3.3 eV). This is expected given the highly symmetric structure of the A–π–A molecule. Furthermore, the first two-photon allowed state (identified by the peak in the nonlinear spectrum located at 295 nm or 4.2 eV) is positioned very near the second excited state of the system (as identified by the minimum in the anisotropy curve), in a region located energetically higher than the first excited state.

From the linear and nonlinear spectra for the dipolar molecule **6**, one can immediately see that not only is there significant 2PA into the second excited state (located at ~300 nm and identified by both the linear absorption and anisotropy spectra) of the molecule but also the peak position of the 2PA spectrum correlates quite well with the position of the first excited state for this system, as illustrated by the location of the first peak in the linear absorption spectrum (385 nm or 3.22 eV). Given that this molecule possesses no center of symmetry, it is not surprising that each of the transitions ($S_0 \rightarrow S_1$ and $S_0 \rightarrow S_2$) is two-photon allowed. Finally, the 1PA and 2PA spectra along with the excitation anisotropy spectrum for the symmetric compound **7** are plotted in Figure 18.3. Given the symmetric nature of this system (D–π–D), one should expect similar trends to those observed for compound **5**. As expected, there appears to be negligible nonlinear absorption into the first strongly one-photon allowed state (3.3 eV). However, although the 2PA peak does lie energetically higher than this first state, also seeming to correspond to the peak of a second strongly allowed linear transition (300 nm or 4 eV). This is in contrast to molecule **5** whose peak 2PA lies in a region where linear absorption is quite weak.

If indeed parity selection rules are to hold, a two-photon allowed transition must be a strictly one-photon forbidden state. To investigate this, quantum–chemical calculations were carried out which verified that, indeed, the molecule possesses both a two-photon allowed state and a strongly one-photon allowed state, which are separated by less than 0.05 eV. The transition energies are given (in electron volts, eV) on the left side of the table while the strengths of the transition dipole moments (in debye, D) are on the right side. There is a slight overestimation of the position of the first excited state (by 0.22 eV) in the quantum–chemical calculations, which is mainly attributed to an over-correlation of the ground state in the multireference singles and doubles configuration interaction (MRDCI) approach. However, the position of the second strongly one-photon allowed transition is calculated quite accurately (i.e., E_{05} = 4.01 eV). The next transition, E_{06}, is weakly one-photon allowed as evidenced by the low magnitude of its transition dipole moment. However, a strong 2PA in centrosymmetric systems requires a strong transition dipole between higher lying excited states (i.e., $\mu_{ee,}$). The sixth singlet state clearly exhibits this with a value for μ_{16} of nearly 6 D. Therefore, this state is likely responsible for the strong 2PA indicated in Figure 18.3, and it becomes obvious why the linear and nonlinear absorption spectra given by E_{05} and E_{06}, respectively, are coincident.

In summary, although the symmetric fluorene derivatives are not strictly centrosymmetric systems, they exhibit the same parity selection rules present in these latter systems. Furthermore, the unsymmetrical derivatives clearly indicate that significant 2PA is possible into nearly any excited state. Moreover, these trends will follow for most of the 2PA compounds studied whose data was available and not just the three molecules analyzed above. Another well-known approach to increase the nonlinear absorptivities of the conjugation molecules is to extend the conjugation length. For example, compound **8** gives significantly higher two-photon cross section than **5** with shorter length although they possess same electron acceptor group.

18.3.2 Branched Structures

Branched (Figure 18.4) or dendritic structures have recently drawn interest in the design of new 2PA chromophores [34]. Some conflicting results were published in literatures, alleging a possible cooperative effect of 2PA cross sections for some structures compared to their linear counterparts. The cooperative effect means the 2PA δ of the dendritic molecular is larger than a purely linear sum of δ of all the branches. Some reports claim huge cooperative effects on 2PA and some just claim a linear additive effect. It is worthy to note that those claims may have some merit for the particular wavelength reported; however, normally no comparison was made over broad spectral range. Two issues arise in such comparison. One is in the 2PA spectra of the dendrimers, new two-photon allowed transitions emerge, normally very strong, while no transition in the corresponding linear parent compounds are observed (Figure 18.5). The comparison in these cases might result in ascribing the difference to cooperative effect, though the true origin of the difference is not clear, and the

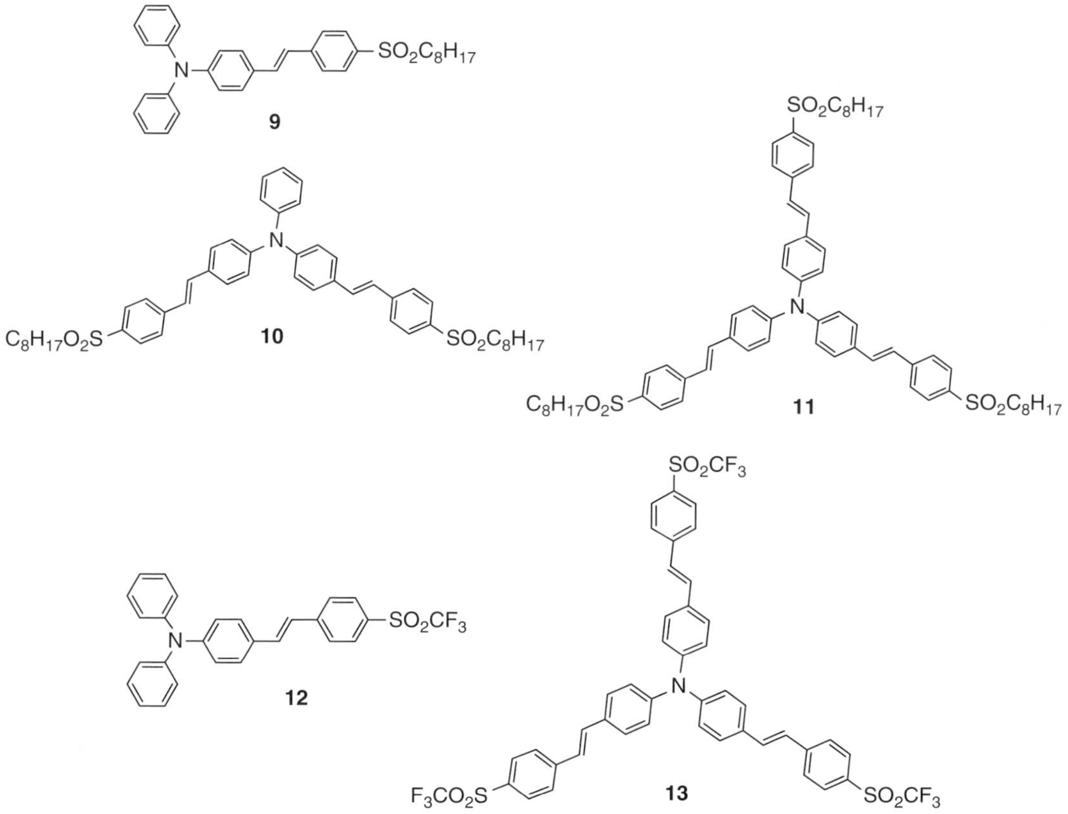

FIGURE 18.4 Structures of some branched molecules and their linear analogs.

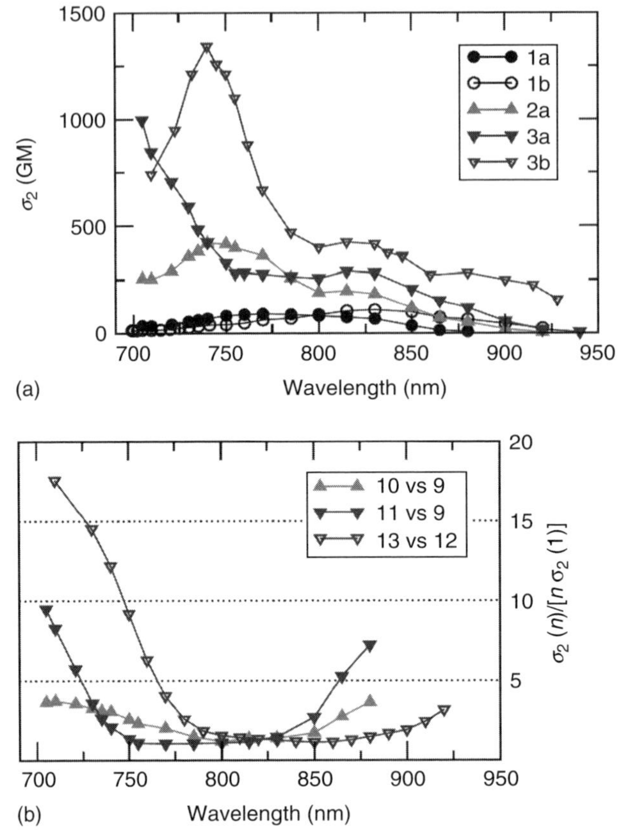

FIGURE 18.5 (a) 2PA cross sections of the investigated chromophores **9–13** in toluene and (b) wavelength dependence of the 2PA enhancement due to the branching effect. (Reprinted from Katan, C., Terenziani, F., Le Droumaguet, C., Mongin, O., Werts, M.H.V., Tretiak, S., Blanchard-Desce, M., *Proc. SPIE-Intern. Soc. Opt. Eng.*, 5935, 593503/1, 2005. With permission.)

experimental error in the measurement may be large. If the new 2PA bands fall in the NIR region, they will be very interesting for applications; however, their excitation wavelengths are normally less than 780 nm. Another issue is the shift of the 2PA λ_{max}. In this case, the comparison of δs at certain wavelength is simply not reflecting the real situation and may give misleading results. In general, when comparing the same absorption transition band at similar wavelength, the δs of the dendrimers are additive from the δs from their linear parent compounds. This is illustrated in Figure 18.5 [35].

Figure 18.5a shows the two-photon δ data for excitation wavelength from 700–950 nm for dyes **9–13**, whose structures and optical properties are shown in Figure 18.4 and Table 18.3, respectively. Figure 18.5b shows the enhancement factor for chromophores **9–13**, calculated as the ratio between the 2PA cross section of the *n*-branched structure and *n*-times the 2PA cross section of the corresponding dipole. The 2PA response of the branched compounds is wavelength dependent. In particular, it is weak near the first 2PA maxima but significantly increases at lower and higher energies. This behavior can be qualitatively interpreted through the exciton model. Interactions between branches lead to an energy separation between the (otherwise degenerate) excited states, so that final states are split toward higher and lower energy with respect to the monomeric analogue. The consequence is a 2PA activity of branched compounds in spectral regions of the blue- and red-shifted with respect to the monomer. In other words, a 2PA enhancement is obtained in regions where the linear reference system is also two-photon active, and a true 2PA activation is attained in regions (especially toward the blue side) where the monomeric model is (almost) two-photon transparent. While these

TABLE 18.3
Optical Properties of Compounds 9–13

Compound	2PA λ_{max} (nm)	δ (GM)	$\lambda_{ab\ max}$ (nm)	$\lambda_{fl\ max}$ (nm)	Φ_{fl}
9	770	90	392	456	0.58
10	815	195	409	459	0.74
11	815	290	410	463	0.72
12	830	110	415	508	0.47
13	820	430	430	494	0.71

Note: Experimental values determined with femtosecond pulses are reported. The uncertainty in the experimental δ values is estimated to be ±15%. Also reported are 1PA maximum (λ_{ab} max), fluorescence emission maximum ($\lambda_{fl\ max}$), and fluorescence quantum yield (Φ_{fl}) in toluene.

effects are clearly recognizable in Figure 18.5 for all branched systems, the most interesting example is given by compound **13**, for which the second (high-energy side) 2PA maximum is reached in the investigated spectral window: an enhancement of ~20 was reported near 700 nm. This is an active area of investigation and promises to provide fruitful insight in the future.

18.4 MULTIPHOTON DEVICES

18.4.1 Three-Dimensional Data Storage

Data is being generated at a remarkably explosive rate. For example in 2002 alone, 5 EB (5 billion GB) of new data was generated worldwide and has been growing rapidly. A key driving force is expected to be the rapid expansion of the Internet and multimedia, such as high definition television (HDTV) with their requirements for higher bandwidth and storage capacity. Commercially, two types of data storage systems are used, i.e., magnetic and optical data storage technologies. On comparing these two techniques, high performance magnetic recording does not generally support removability, hard drives need not work with the media of previous generations, and the main standards needed to be satisfied are those on data input/output and form factor.

Interchangeable media and backward compatibility, however, dominate optical storage. The data storage densities of magnetic media have reached 130 GB/in.^{2} in 2005, and it is expected that this will reach 200 GB/in.^{2} by perpendicular recording methods (Panasonic) in near future. For optical storage media, in theory, blue laser DVD technology can achieve data density up to 50 GB/in.^{2}. However, both magnetic and conventional optical data storage technologies, where individual bits are stored as distinct magnetic or optical changes on the surface of a recording medium, are approaching physical limits beyond which individual bits may be too small or too difficult to store and retrieve.

Storing information throughout the volume (3D) of an optical storage medium, not just on its 2D surface, offers an intriguing high-capacity alternative, achievable by using many layers with relatively large marks (i.e., greater than 1 μm) (Figure 18.6). This approach can potentially provide efficient storage at densities significantly higher than those that are likely to be available from magnetic media.

One of the major challenges for any 3D memory technology is to make it possible to retrieve data rapidly using compact and relatively inexpensive equipment (such as no amplified laser systems and nonimmersion optics) and materials. To achieve this, there has been increasing interest in 3D optical data storage techniques based on 2PA [36].

The premise of theses technologies is that 2PA of laser light can be used to initiate photochemical or photophysical processes that alter the local optical properties of a material. 2PA occurs in the

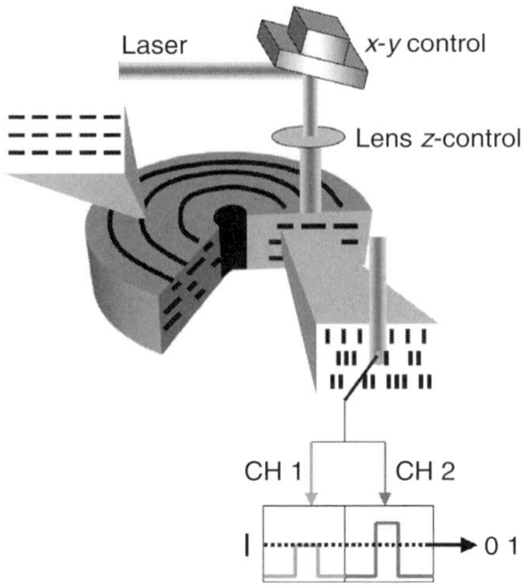

FIGURE 18.6 Multilayer two-photon 3D optical data storage disk.

presence of intense laser pulse, such as mode-locked femtosecond Ti:sapphire laser pulses (though 2PA using less expensive, longer pulse picosecond and nanosecond lasers occurs and is quite feasible). Since the two-photon transition probability is proportional to the square of the incident laser intensity (I^2), the 2PA is confined at the focal volume (voxel), resulting in highly localized 2PA-induced modulation of optical properties (localized within the focal volume). Thus, the 3D spatial control is achievable with high precision. By employing objectives with high numerical aperture, focal regions with submicron dimensions can be readily attained. Other advantages of 2PE for recording in 3D optical memory systems is that cross talk between two adjacent layers can be greatly reduced and the illumination beam that is utilized has an infrared wavelength; thus scattering is also reduced.

In Rentzepis' pioneering work [37], a frequency-doubled Nd:YAG laser (35 ps pulses, 100 mJ/pulse, 10 Hz) was used to generate two wavelengths (1064 and 532 nm), which are separated and propagated along two independent paths (Figure 18.7). A two-photon sensitive memory cube ($1 \times 1 \times 1$ cm^3) was placed at the spatial and temporal intersection of these two pulses such that individual planes within the cube were independently addressed by the green beam and encoded with the infrared (IR) beam. The green beam was focused by an anamorphic telescope to a thin sheet of light with a $1/e^2$ intensity width of 80 μm. The infrared beam was sent through a variable delay stage and then expanded to illuminate a chrome mask, which was imaged onto the addressed plane within the cube by a 4-*f* imaging relay. An evacuated chamber was placed at the intermediate Fourier plane of the imaging system to prevent the high-intensity picosecond pulses from ionizing the air. Different planes within the cube were selected by simultaneously moving three stepper motor stages corresponding to the cube, recording lens, and readout lens. A cooled 1316 × 1035 pixel charge-coupled device (CCD) camera on the other side of the cube was used for readout. Recordings were performed with average powers of 3 W/cm^2 infrared and 500 mW/cm^2 green at 4 min exposures per plane. On readout, fluorescence was induced by 300 μW output from a green He–Ne laser, and the images were captured by the camera in 200 ms exposures, although in this case there was sufficient fluorescence to observe the recorded images with a standard 30 frame/s video camera. In general, one can maximize the readout rate by increasing the number of written molecules per volume, increasing the power of the readout laser, and decreasing the *f*-number of the optics.

One hundred planes of 100 × 100 random bit patterns were recorded in a single cube. The lateral dimensions of each bit in the cube were 30 × 30 μm^2 (3 × 3 mm^2 total image size), and the spacing between planes was 80 μm; therefore, most of the cube volume did not contain memory data. In later work, capabilities were demonstrated that include the recording and reading of media with more than 100 data layers, recording tracks of 2 × 2 μm^2 data marks, and the construction of several

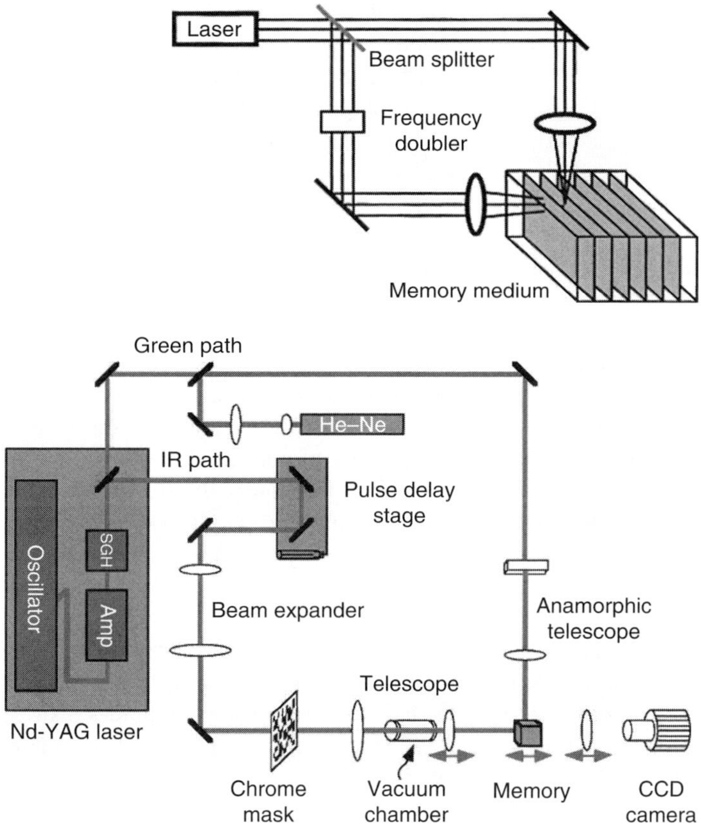

FIGURE 18.7 Principle of 3D optical memories as proposed by Rentzepis et al., and typical 3D data recording setup. (Upper figure from Parthenopoulos, D.A., and Rentzepis, P.M., *Science*, 245, 843, 1989. With permission. Lower figure reprinted from Wang, M.M., Esener, S.C., McCornik, F.B., Cokgor, I., Dvornikov, A.S., Rentzepis, P.M., *Opt. Lett.*, 22, 558, 1997. With permission.)

proof-of-principle portable readout systems. Two-photon recorded 3D optical storage technology development has been predicted to provide disk drive systems with high capacity (100–500 GB/disk) and data transfer rates of 1–10 GB/s, using inexpensive, easily manufactured polymer media [38].

Another promising system is a dual dye system with photochromic diarylethene derivative **14** and highly efficient 2PA dye **8** (2PA cross section $\delta = 6000$ GM at 600 nm and 1185 GM at 800 nm) for 2PF (Figure 18.8) [38]. Diarylethene **14** does not display significant luminescence, but the dual systems containing the second chromophore **8** have shown reversible changes in emission intensity which rely on intramolecular, fluorescence resonance energy transfer (FRET) quenching processes. Fluorescence emission of **8** is reversibly modulated by cyclical transformations of the photochromic

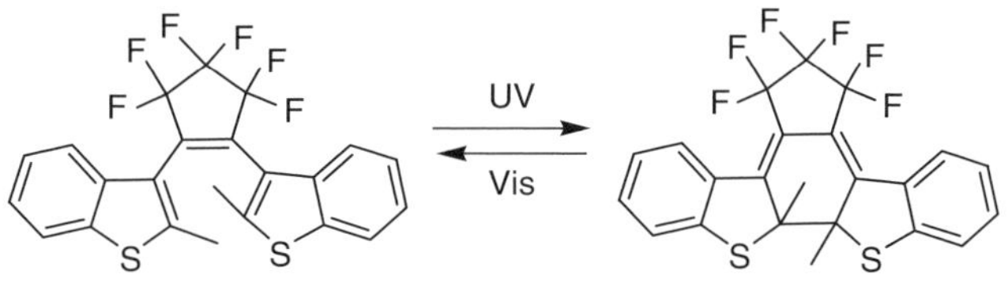

14 open **14 closed**

FIGURE 18.8 Molecular structures of the open and closed forms of the diarylethene **14**.

acceptor with irradiation of appropriate ultraviolet and visible light, respectively. Only the closed form of the photochromic **14** has an absorption band overlapping the emission band of the donor, which provides the means for reversibly switching the process of FRET on and off, allowing direct and repeated evaluation of the relative changes in the donor fluorescence quantum yield.

Two-photon recording and readout, data recording was accomplished by excitation of the closed form of **14** at 710 nm. Reading was accomplished by excitation of **8** at 800 nm. The photoresponsive material is a film made from a mixture of 22 wt% of diarylethene **14** and 0.9 wt% of fluorene **8** in a polymer matrix. The film was covered with a mask and exposed to the output of a tungsten–halogen lamp of an inverted microscope (Olympus IX81) equipped with an IF550 filter (λ_{exc}>550 nm). Excitation of the closed form of diarylethene **14** occurs only in the transparent parts of the mask, inducing photoisomerization to the open form of **8** to form the patterns, whose readout fluorescence signals are shown in Figure 18.9. This material was subject to 10^4 readout cycles at 5 mW at 800 nm using two-photon (femtosecond) excitation without significant change in output signal intensity. This is because the intensity of the 2PF is weaker than the 1PF, and is confined to the focal volume, which means in the readout process, the possibility of undesired photoreaction induced by absorption of the fluorescence is much lower than that under linear fluorescence conditions. The readout signal indicator can, in principle, be any efficient 2PA chromophores whose 1PA spectrum overlaps with the absorption profile of the closed form. A binary encoding system can be prepared in which "0s" are regions below a particular fluorescence threshold and "1s" are marks with signal intensity that exceeds the threshold value.

Other approaches of two-photon data storage are the change of the fluorescence intensity upon two-photon irradiation, either by enhancing or quenching the emission. For example, a system using 2PE of a photoacid generator in the presence of an acid-sensitive 2PA fluorescent dye was reported for two-photon 3D WORM (write once, read multiple) data storage [39]. In this case, a sulfonium salt in a thick polymer film containing an efficient 2PA fluorene derivative with high fluorescence quantum yield was irradiated for short times in three layers. The data was readout at longer wavelength by two-photon upconverted fluorescence by exciting either the neutral or protonated fluorophore. In this manner, two-channel readout is possible,

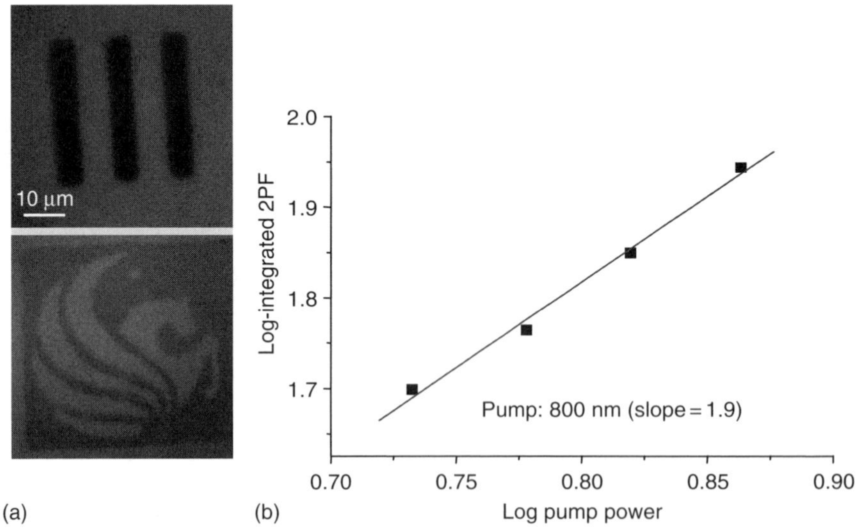

FIGURE 18.9 (a) 2PF readout patterns. The readout wavelength λ_{exc} is 800 nm (corresponding to 400 nm of 1PE). Upconverted fluorescence emission of fluorene **8** was collected between 510 and 550 nm and (b) the plot shows quadratic dependence of integrated fluorescence emission intensity on several pump powers, which confirms the readout signal is due to the 2PF.

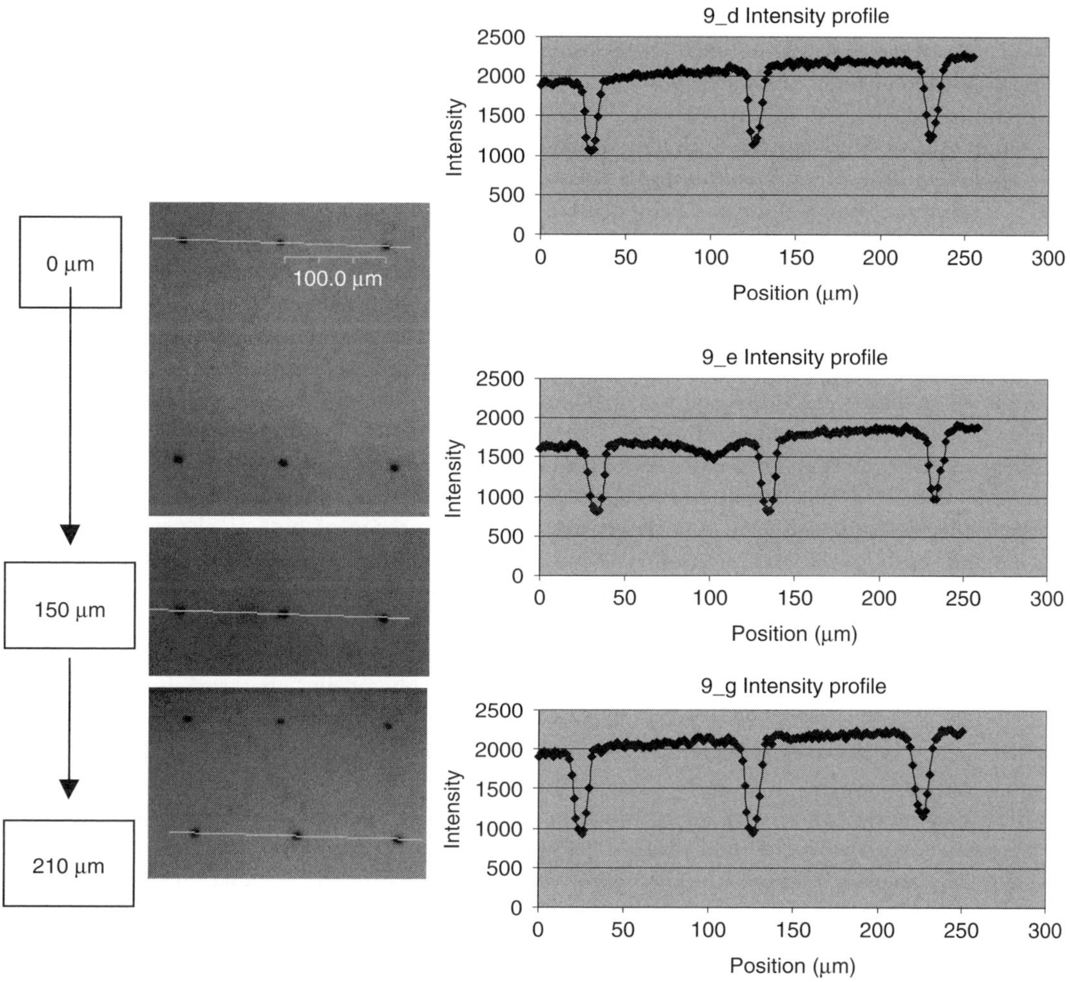

FIGURE 18.10 Multilayer, two-photon 3D data recording and readout in a photosensitive polymer film containing a sulfonium salt and dye **6**. Recording (540 nm) and readout (800 nm) were conducted using femtosecond irradiation. The numbers on the left indicate the relative position in the *z* direction (depth) of the three layers. Data tracks are in the center with the line in yellow indicating the data that was readout in the fluorescence intensity plots on the right.

because the protonated fluorophore absorbs (and emits) at substantially longer wavelengths than the starting neutral fluorophore. The data tracks in the three layers can be seen in Figure 18.10. Other intriguing methods of two-photon based 3D data storage include photochromic holographic recording [40].

18.4.2 Multiphoton Fluorescence Microscopy

Biological applications based upon 2PA reside upon two salient features of this nonlinear process. The rate of 2PA is quadratically dependent upon the excitation irradiance *I*, while for 1PA, the likelihood of absorption depends linearly on *I*. This nonlinear dependence ensures high spatial selectivity when an intense laser beam is focused into a two-photon absorbing medium. Practically, this means 2PA occurs in only the focal volume, where there is sufficiently high photon density, and the probability of 2PA falls off rapidly away from this point, providing the optical-sectioning (confocality) ability of multiphoton imaging. For a fluorescent molecule, the resulting fluorescence emission upon 2PA (and consequently, 2PE of the absorbing area) is then proportional to I^2, and fluorescence is practically localized to the focal spot, and unlike in 1PA, there is essentially no out of focus light to reject.

Due to the highly localized excitation obtained in 2PA, some major advantages accrue, mainly that photobleaching of precious fluorophores are minimized as photodamaging/phototoxic effects on living specimens. Additionally, 2PE spectra of many organic fluorophores are much broader than for 1PE, so multiple fluorophores with distinct emission wavelengths may result from a single multiphoton excitation wavelength [41]. Finally, the highly confined nature of 2PE also confines any physical or chemical reactions activated by 2PA to a 3D volume, essentially the same as the focal volume. Hence, highly localized photoinduced chemical reactions as site-specific photodynamic cancer therapy (see Section 18.1.3.3) and photoinduced uncaging of bioeffector molecules [42], in addition to nonlinear spectroscopy-based bioimaging, are among emerging technologies which exploit this nonlinear absorption effect.

The second feature of 2PA is the employment of NIR excitation wavelengths, typically in the range of ~700–1000 nm, provided by tunable femtosecond laser systems. This excitation range coincides particularly well with the decreasing absorption range of biological tissues to longer wavelengths of light, shown in Figure 18.11 [43]. In the ultraviolet region, the absorption increases with shorter wavelength due primarily to proteins and DNA. The existence of the optical window between ~600 and 1300 nm, coupled with the use of longer excitation wavelengths, offers greater penetration depths into highly scattering biomaterials than the equivalent blue light used for 1PE. Finally, NIR light causes negligible photodamage to fluorophores and phototoxicity to live cells, relative to higher energy ultraviolet wavelengths.

Since the pioneer work of Webb in 1990, the 2PF imaging method has rapidly developed in bioimaging area. By applying commercial fluorescent probes, green fluorescent protein or using autofluorescent residues from the cells, tremendous fluorescent image data has been reported and this technique has been evolved into a major imagine method, especially for thick biological samples. Despite the successfull application of these probes, they are typically designed for single-photon application and the 2PA δ are normally small; most of them are less than 100 GM. More and more effort to design and synthesize the improved 2PA probes has been conducted recently. One example

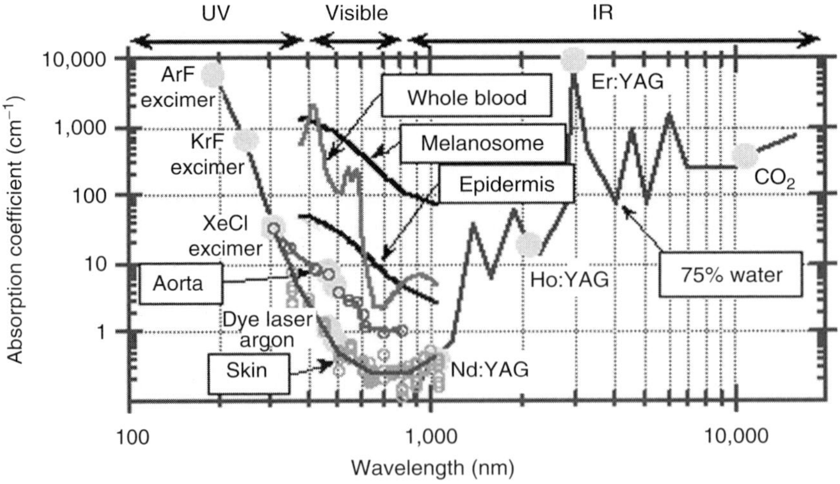

FIGURE 18.11 Primary absorption spectra of biological tissues along with their absorption coefficients at some common laser wavelengths. (Reprinted from Jacques, S.L., Prahl, S.A., *Oregon Graduate Institute*, http://omlc.ogi.edu/classroom/ece532/class3/muaspectra.html, 1998. With permission.)

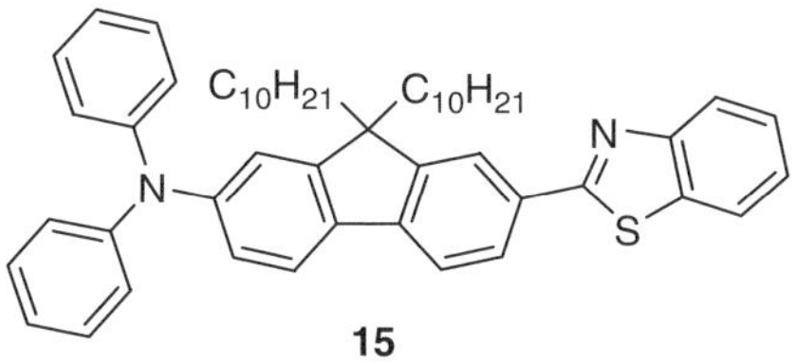

15

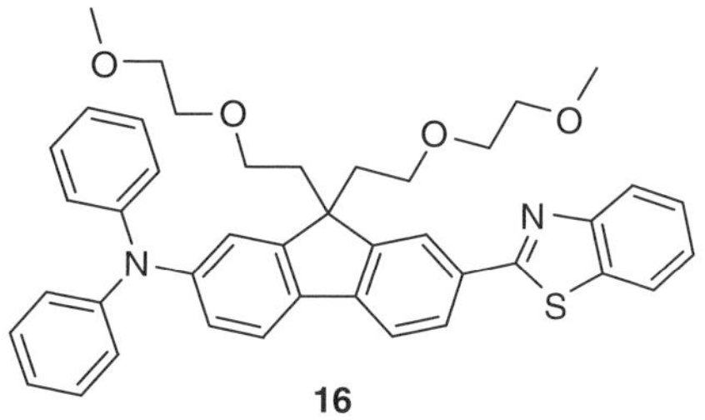

16

FIGURE 18.12 Structure of 2PA fluorescent probe with different hydrophilicities **15** and **16**.

is to use compounds **15** or **16** as fluorescent probes (Figure 18.12). As mentioned before, fluorene compound **15** is a highly photostable fluorophore that exhibits relatively high-2PA cross section in the excitation range of pulsed Ti:sapphire laser output.

The utility of fluorene **15** as an efficient two-photon absorbing biological fluorophore was demonstrated by staining glutaraldehyde fixed H9c2 rat cardiomyoblast cells. Bright field transmission and epi-fluorescence microscopic images (DAPI filter set, 40×, NA 0.75) of the stained cells are shown in Figure 18.13a and b, respectively. Furthermore, fluorene **15** did not undergo noticeable photobleaching during continuous exposure to the ultraviolet excitation light. No fluorescence was observed in the controls without any fluorophore (image not shown). Additionally, fluorescence was observed predominantly from the cytoplasmic region of the cells, with the nucleus clearly outlined, indicating potentially preferential staining with this fluorophore for cytoplasmic components.

Two-photon excited fluorescence microscopy images of the same fluorene **15**-stained cells were collected on a modified Olympus IX70 inverted microscope and Fluoview laser scanning confocal unit with a Ti:sapphire laser output from 740–830 nm (125 fs FWHM, 76 MHz repetition rate, ~25 mW, 40×, NA 0.85). The control cells that did not receive any fluorophore showed modest autofluorescence upon 800 nm fs excitation as shown in Figure 18.13c, while the fluorophore-stained cells (Figure 18.13d) revealed higher contrast and greater signal under the same excitation and power exposure as the control. Two-photon induced fluorescence was observed predominately from the

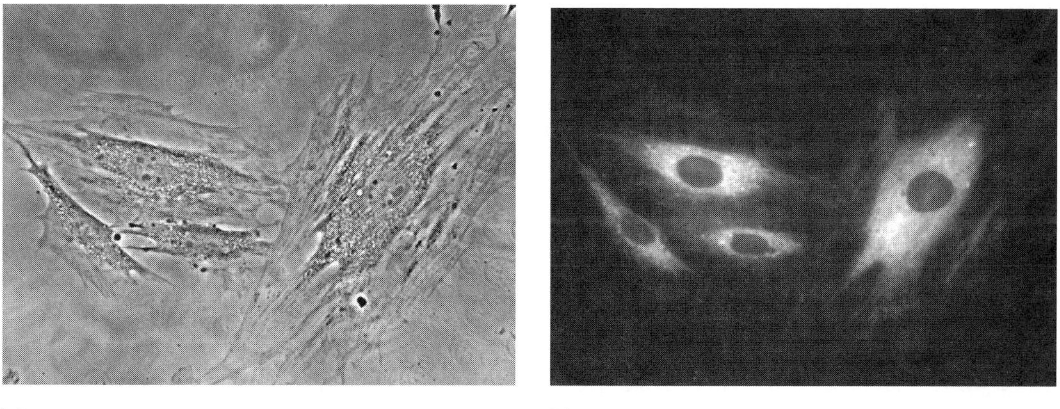

(a) (b)

FIGURE 18.13 (a) Bright field transmission and (b) epi-fluorescence microscopic images (40×) of H9c2 cells stained with a two-photon absorbing probe **15**. Two-photon induced fluorescence microscope images (40×) of

(continued)

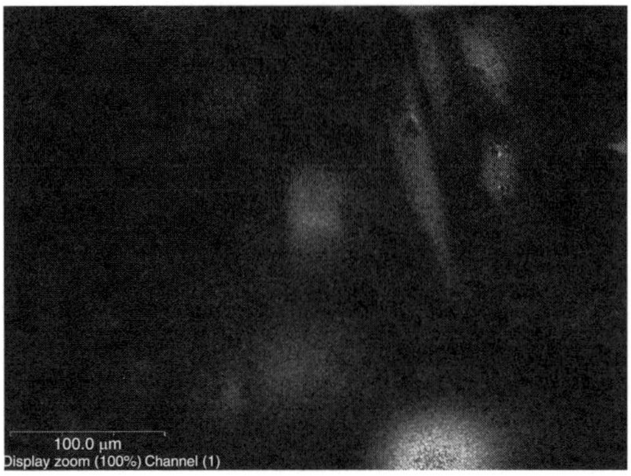

(c)

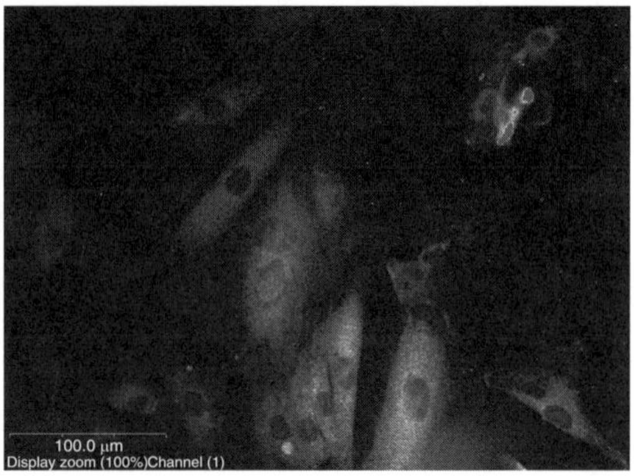

(d)

FIGURE 18.13 (continued) (c) control containing no fluorophore and exhibiting some autofluorescence, and (d) cells stained with fluorene **15** upon 800 nm fs excitation. Red spots demark signal saturation.

cytoplasmic region, consistent with the images collected from epi-fluorescence images. While higher-order processes such as three-photon absorption (3PA) may be involved, spectroscopic data suggest a two-photon process in the excitation range used (740–830 nm) would be dominant for compound **15**.

The successful demonstration of 2PM images of H9c2 cells stained fluorophore **15**, albeit a hydrophobic derivative, lends credence to efforts to further refine fluorene-based derivatives for bioimaging applications. Toward the end, the more hydrophilic fluorene derivatives **16** in two-photon induced fluorescence microscopy imaging were also demonstrated. Furthermore, preliminary studies revealed hydrophilic fluorene derivatives incubated with NT2 cells exhibited relatively low cytotoxic effects. These hydrophilic fluorene derivatives are expected to exhibit similar photostabilities as **15**. The fact they possess high fluorescence quantum yields, and high-2PA cross sections, and, therefore, action cross sections, makes them ideal fluorescent contrast agents suitable for use over the tunability range of commercial Ti:sapphire lasers typically utilized in multiphoton imaging methods and techniques.

18.4.3 Multiphoton Photodynamic Therapy

Photodynamic therapy (PDT) is a method of treating cancer and recently has also become a preferred choice to treat neovascular age-related macular degeneration (AMD) [44]. In PDT, a photosensitizer is uptaken to target tissues or cells, such as tumors and choroidal neovasculature, and then a laser light that excites the photosensitizer is irradiated on the target. The excited photosensitizer is undertaken as intersystem crossing from its singlet excitation state to triplet excitation state as shown in Figure 18.14. Because of the relative long lifetime of the triplet state, it can efficiently transfer the energy to the naturally occurring triplet oxygen and generate the singlet oxygen. Later, the singlet oxygen will kill the targeted cells. The advantage of this technique is its noninvasive nature, which avoids a great number of side effects caused by traditional techniques, such as surgery, chemotherapy, and radiation therapy.

Food and Drug Administration (FDA) has approved drugs, such as verteporfin (BPD-MA), a benzoporphyrin derivative, and Photofrin, a mixture of hematoporphyrin oligomers, to be used in PDT. The drugs will be activated by irradiation of laser light at around 680 and 630 nm, respectively. However, in these conditions, much like most PDT photosensitizers developed in the labs all over the world, the depth that the laser can reach is limited, in case of Photofrin, 5–6 mm. To achieve high penetration, the photosensitizers need to work at the wavelength in the tissue transparency window

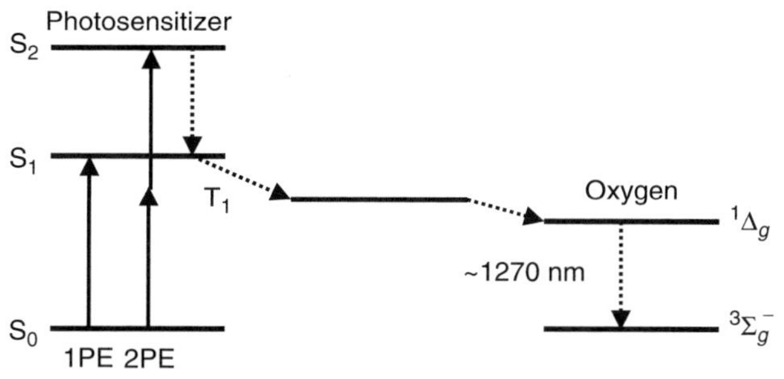

FIGURE 18.14 Singlet oxygen generation via 1PE or 2PE.

(700–1000 nm), preferably 800 nm or above. Another limitation of single-photon PDT is that the laser will activate all the photosensitizer on its path resulting collateral damage of healthy cells. These limitations lead to the development of the two-photon or multiphoton PDT method, which can circumvent these difficulties taking the advantage of longer excitation wavelength at NIR region and spatially limited excited volume as mentioned before.

The biggest obstacle in applying 2PA or multiphoton absorbing materials for multiphoton PDT is the two-photon or multiphoton absorption cross sections of most photosensitizers are too low, normally at the order of several Goeppert-Mayer (GM). For example, the 2PA δ of Photofrin is between 5 and 10 GM ($1\ GM = 1 \times 10^{-50}\ cm^4\ s\ photon^{-1}$) in the excitation wavelength in the range of 850–900 nm [45]. Recently, Ogawa et al. reported some potential compounds for 2PA PDT with very large 2PA δs [46]. One of the structures with the largest 2PA δ values in water were presented in Figure 18.15. The δ measured by a femtosecond open-aperture Z-scan method at 850 nm is 7900 GM.

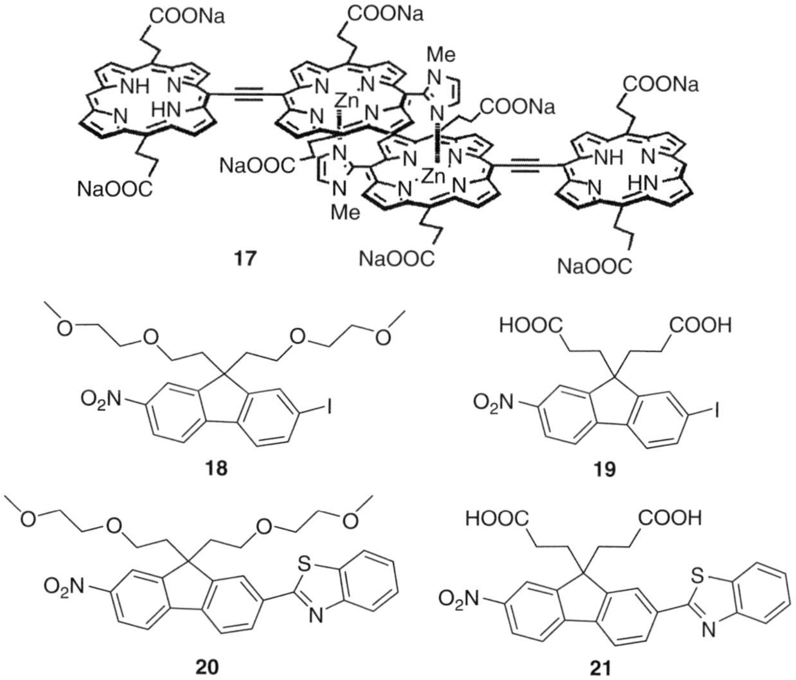

FIGURE 18.15 Structures of novel 2PA PDT photosensitizers.

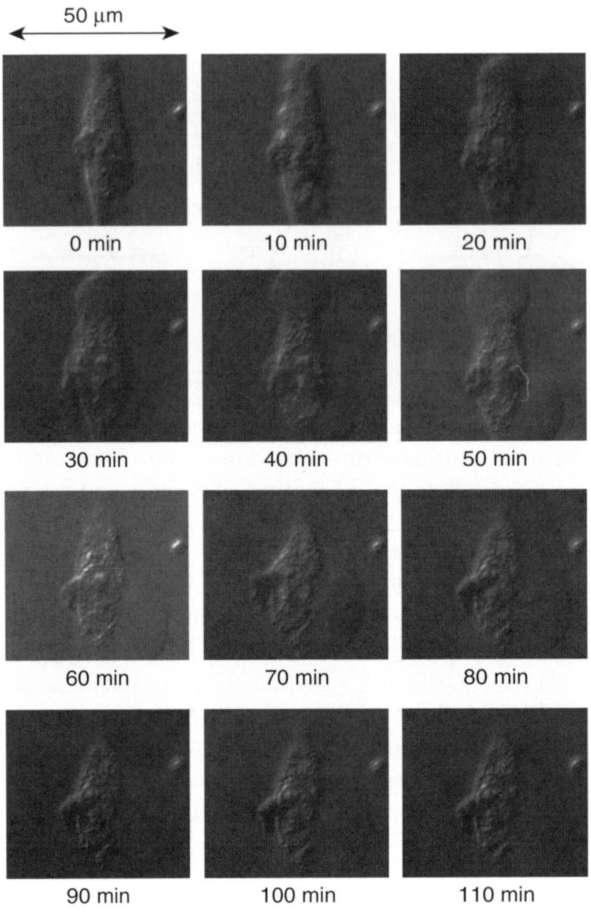

FIGURE 18.16 Time course of cell death for HeLa cell upon photoirradiation with **17**. (Reprinted from Ogawa, K., Hasegawa, H., Inaba, Y., Kobuke, Y., Inouye, H., Kanemitsu, Y., Kolino, E., Hirano, T., Ogura, S., Okura, I., *J. Med. Chem.*, 49, 2276, 2006. With permission from American Chemical Society.)

The authors claimed that **17** has high singlet oxygen quantum yield, same as protoporphyrin (~0.56) [47] by monitoring phosphorescence emission from $^1\Delta_g$ to $^3\Sigma_g^-$ at ~1270 nm (Figure 18.14).

The photocytotoxicity of **17** for HeLa cells was also examined by monitoring cell death upon photoirradiation by continuous wave (CW) diode laser (671 nm) under microscopic observation as shown in Figure 18.16 [46]. The leakage of the cytoplasm became more and more significant with time course and blebs were formed on the cell surface, indicating injury of the cell. The cell death was observed in 48 min with a total irradiation energy of 5184 J/cm^2 at the concentration of 10 μM while no cell death was observed without the agent even after 2 h of irradiation with total irradiation energy above 12 960 J/cm^2.

Other than porphyrin derivatives, recent studies indicate that fluorene derivatives, such as compounds **18–21** can also be excellent PDT sensitizers. Their one-photon and two-photon singlet oxygen quantum yields as well as 2PA cross sections are shown in Table 18.4.

18.4.4 3D Microfabrication

It is widely believed that a revolution in miniaturization, particularly in the field of microelectromechanical systems (MEMS), is under way. It is projected that the design and manufacturing technology

TABLE 18.4
2PA Cross Sections and Quantum Yields of Singlet Oxygen Generation, 2PAΦ_Δ, under 2PE of 18–21 at 775 nm in ACN

	18	19	20	21
δ_{2PA} (GM)	10 ± 4	9 ± 4	60 ± 20	50 ± 15
1PAΦ_Δ	0.74 ± 0.08	0.65 ± 0.07	0.92 ± 0.1	0.93 ± 0.1
2PAΦ_Δ	0.3 ± 0.15	0.4 ± 0.2	0.45 ± 0.15	0.35 ± 0.1

that will be developed for MEMS may rival, or even surpass, the far-reaching impact of integrated circuits (ICs) on society and the world's economy. At the forefront of techniques being explored for 3D spatially resolved materials imaging and processing are methods based on 2PA. Traditional 3D microfabrication techniques for microscale free-form fabrications include microstereolithography, electrochemical fabrication (EFAB), microphotoforming, spatial forming, microtransfer molding, localized electrochemical deposition, etc. [48]. Using these techniques, 3D objectives were successfully fabricated from the materials of polymer, ceramic, metal, etc. However, most of these techniques build 3D structures using layer-by-layer photopolymerization, for example, in a typical microstereolithographic process, a computer-aided design (CAD) model of the desired object was first generated; and then the 3D model was slide into a series of closely spaced horizontal planes and converted into to computer-executable codes. Controlled by these codes, desired polymer object is built by computer-directed laser scanning on the surface of a photocurable liquid resin in a layer-by-layer additive fashion, as illustrated in Figure 18.17a.

In contrast to the linear dependence of 1PA on incident light intensity in conventional photopolymerization, the quadratic dependence of photoexcitation on light intensity in 2PA can be exploited to confine polymerization to the focal volume and achieve fabrication of microstructures via 3D spatially resolved polymerization [49]. This high spatial resolution contributes to the ability to scan the laser in 3D (Figure 18.17b), therefore, two-photon induced polymerization has the advantage that 3D structures can be generated in a single exposure/development cycle regardless of the complexity of the 3D structures. The resolution of the structures can reach to submicron scale with this method, which will be difficult for layer-by-layer techniques to fabricate.

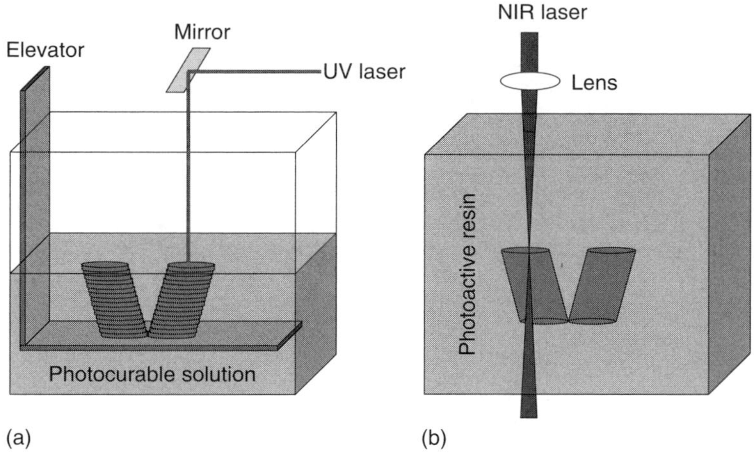

FIGURE 18.17 (a) Single photon layer-by-layer photostereolithography vs (b) two-photon 3D microfabrication.

In addition, 2PE in materials at wavelengths well beyond that which monomers, polymers, and most organic substances absorb (one-photon), affording a greater depth of penetration, creating little or no damage to the host. Hence two-photon microfabrication via photoinitiated polymerization represents a potentially versatile technology that should be compatible with construction of mechanical, chemical, electrical, optical, or biosensor systems. Also, it could be integratable with conventional integrated circuit (IC) processing technologies and femtosecond laser micromachining. Although several reports of two- or multiphoton induced polymerization appeared in the literature as early as 1971 [50], most of these involve two or more sequential, resonant 1PE processes at single or multiple wavelengths (i.e., excited-state absorption) [51]. The resonant processes must be distinguished from the simultaneous 2PE process of concern here owing to the fundamental differences in achieving spatially resolved polymerization, i.e., a much higher degree of inherent 3D spatial resolution in the simultaneous process.

Some work on two-photon photopolymerization of commercial acrylate monomer systems, preformulated with ultraviolet photoinitiators, suffered from low 2PA cross sections [52]. However, carefully designed two-photon absorbing compounds based on phenylethenyl constructs bearing electron-donating or electron-withdrawing moieties have been shown large two-photon cross sections. Among these are electron-rich derivatives that have been found to undergo a presumed two-photon induced electron transfer to acrylate monomers or proposed fluorescence energy transfer to a photoinitiator, initiating polymerization, which is an excellent example for 3D two-photon microfabrication [53].

A variety of potentially interesting 3D structures is illustrated in Figure 18.18 [53a]. The 3D periodic structures are needed for photonic band gap (PBG) materials, which have unique optical properties. These structures are difficult to fabricate on micrometer or submicrometer length scales as needed for applications in the infrared and visible spectral regions. Figure 18.18a and b shows views of a "stack-of-logs" structure which, given a sufficiently high-refractive index, should exhibit PBG properties in the infrared spectral region. High-refractive index structures, based for example on refractory

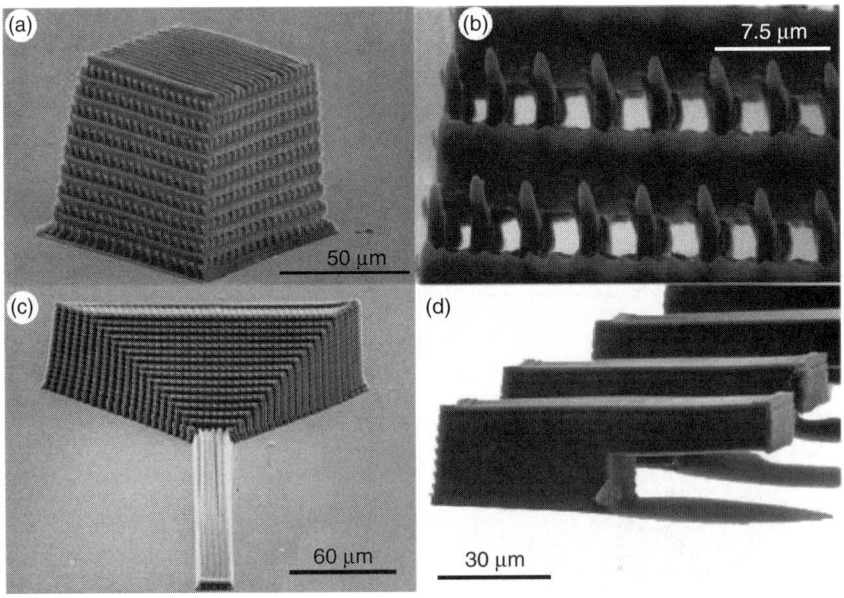

FIGURE 18.18 3D microstructures produced by two-photon-initiated polymerization. (a) PBG structure, (b) magnified top-view of the structure (c) tapered waveguide structure, and (d) array of cantilevers. (Reprinted from Cumpston, B.H., Ananthavel, S.P., Barlow, S., Dyer, D.L., Ehrlich, J.E., Erskine, L.L., Heikal, A.A., Kuebler, S.M., Lee, I.-Y.S., McCord-Maughon, D., Qin, D.J., Röckel, H.R., Rumi, M., Wu, X.-L., Marder, S.R., Perry, J.W., *Nature*, 398, 51, 1999. With permission from Macmillan Publishers Ltd.)

ceramics, could be fabricated by using the polymer structure as a preform. Other structures of arbitrary pattern and periodicity down to 1 mm length scales could be readily obtained with this photopolymer system. Two-photon 3DLM could also find application in the production of tapered optical waveguides, such as those shown in Figure 18.18c. In this figure, the waveguide cross section varies along the length from a 100 × 100 mm^2 aperture to a 2 × 10 mm^2 rectangular aperture. Tapered optical waveguides have the potential to reduce optical loss in the coupling of waveguide components with disparate cross sections. MEMS are often produced using 2D lithography, and 3D structures are built up by an iteration of processing steps. As an example of MEMS structures easily fabricated with a single development step, the array of polymeric cantilevers produced is shown in Figure 18.18d.

18.4.5 Optical Power Limiting

Laser devices have been used in our everyday life from digital versatile disc (DVD) players to laser pointers, and their extensive civilian and military applications became the drive force to develop even more powerful lasers and the ability to tune the laser wavelength in a wide spectral range. While these novel laser technologies greatly improved the application performance, another issue arises, i.e., the protection of human eyes and other optical sensors from the damage of the high energy in the lasers. There are two levels of damage that can be caused by laser irradiation [54]. At low energy level, only dazzle occurs when the optical system is temporarily damaged and the system may recover from the disability. In more severe situation, the optical system may be permanently damaged by excessive irradiation.

To prevent the permanent damage, it has been a long-term effort in optical society to develop efficient optical limiting materials to protect laser damage to optical devices [55]. To achieve this goal, materials need to be transparent across the response band of the sensor at working light intensity and become low transparency at high irradiation intensity, quick enough to reduce the energy level to safe range before the damage takes place [54]. In addition, the materials also need to provide broadband covers regarding the fact that powerful lasers cover the entire visible to NIR range. Meanwhile, these materials need to be highly photostable and thermostable to meet the critical practical operation conditions. Two-photon materials are among the most promising materials for optical limiting [56], other possible mechanisms include reverse saturable absorption, nonlinear scattering, liquid crystals, etc. [57,58].

The relationship between the optical power limiting behavior and the two-photon nonlinear optical properties can be described by observing the change of intensity I of a laser beam as it propagates through an organic medium that exhibits both linear and 2PA. This change in intensity as a beam propagates in the z direction is described by Equation 18.8 [59].

$$\frac{\delta I}{\delta z} = -\left(\alpha I + \beta I^2\right), \tag{18.8}$$

where
α is the linear absorption coefficient
β is the 2PA coefficient

which can be related to $\mathrm{Im}\chi^{(3)}$ by the following expression:

$$\beta = \frac{8\pi^2 c}{\lambda n_0^2}\mathrm{Im}(\chi^{(3)}), \tag{18.9}$$

where
λ is the wavelength of interest
n is the linear refractive index
c is the speed of light in a vacuum

If one then solves Equation 18.9 for $\alpha = 0$ (no linear absorption at low intensity) the result gives the expression for the transmitted intensity $I(L)$ as a function of the incident intensity (Equation 18.10).

$$I(L) = \frac{I_0}{(1 + I_0 \beta L)}. \tag{18.10}$$

In this expression L is the length of the sample and from the relationship it can be seen that the output intensity $I(L)$ decreases as both the input intensity I and the two-photon coefficient δ increase.

Shown in Figure 18.19 is a typical optical limiting profile at the maximum 2PA wavelengths using fluorene-based 2PA chromophores **22** and **23** [60]. The photophysical properties of these two dyes are summarized in Table 18.5.

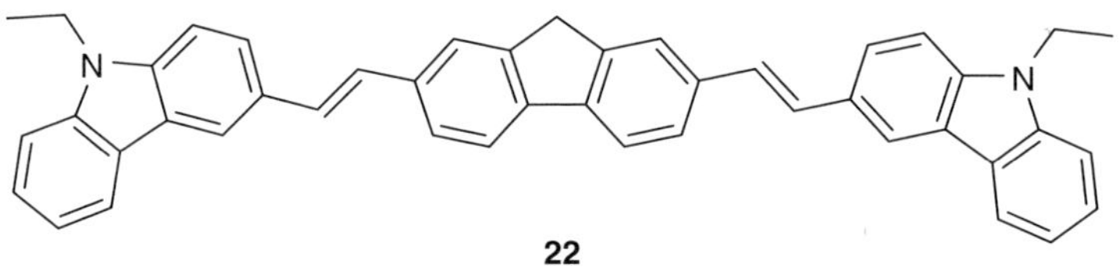

22

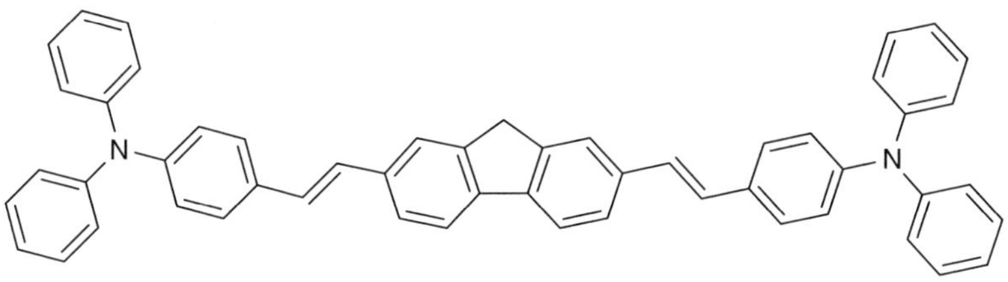

23

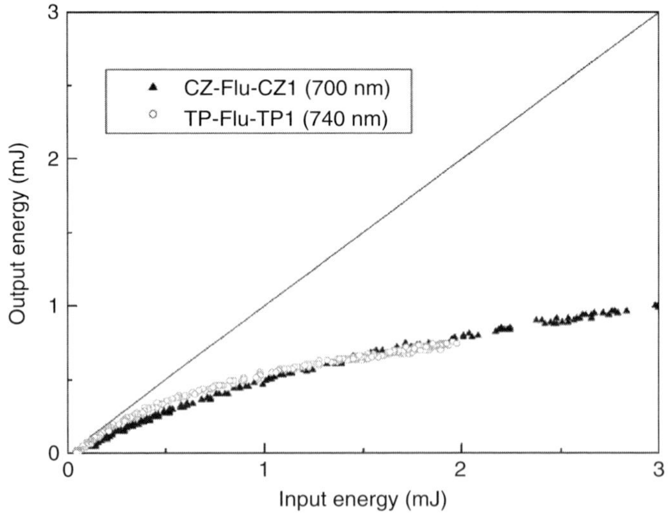

FIGURE 18.19 Structures of fluorene-based 2PA chromophores **22** and **23**, and their optical power limiting properties in THF at 0.005 mol/L. (Reprinted from K.-S. Lee, H-K. Yang, J.-H. Lee, O.-K. Kim, H.Y. Woo, H. Choi, M. Cha, M.H. Blanchard-Desce, *Proc. SPIE-Intern. Soc. Opt. Eng.*, 4991, 175, 2003. With permission.)

TABLE 18.5
2PA Parameters for Fluorene Derivatives 22 and 23 in THF

Compound	2PA λ_{max} (nm)	δ (GM)[a]	$\lambda_{ab\ max}$ (nm)	$\lambda_{fl\ max}$ (nm)	Φ_{fl}
22	700	3100 (290)	396	445	N/A
23	705	15300 (950)	409	455	N/A

[a] Experimental values determined with 8 ns pulses and 8 fs pulses (given in parentheses) are reported.

A beta-barium borate (BBO) crystal-based optical parametric oscillator (OPO) pumped by the third harmonic (λ=355 nm) of a Q-switched Nd:YAG laser was employed. The OPO output beam had a pulse width of 8 ns, repetition rate of 10 Hz, and a spectral width of about 2 nm. Luminescence from the sample was passed through a monochromator and measured by a photomultiplier tube while tuning the OPO wavelength in the NIR transparency region of the material. Both the samples exhibited effective optical power limiting behavior. Different 2PE wavelengths were chosen since at those wavelengths, both dyes have similar δ value, optical power limiting activity of these two chromophores showed not much significant differences. From the data provided in Table 18.5, the σ values measured by femtosecond laser pulses are much smaller than those measured by using nanosecond laser pulses. This result suggests that the contribution of excited-state absorption becomes more manifested when longer timescale nanosecond pulses are used. Since the optical limiting data reported is obtained by 8 ns pulse irradiation, the approximate 70% loss of input energy absorbed by the nonlinear process is a total effect of 2PA and other excited-state absorption. This is true also for most of the published optical limiting materials involving 2PA process. The synergistic effect of various processes is the key to achieve large optical limiting effects. Nevertheless, 2PA is an important process for this application. Other than linear compounds similar to this series of chromophores [61], branched chromophores were also utilized in optical power limiting with very large effects [62].

18.5 FUTURE DEVELOPMENT

One of the biggest challenges in this field is still to improve the two-photon or multiphoton absorptivities of the materials. Although the structure–property relationship has been well established in this field, a reliable method to predict the 2PA cross section of the chromophore and other important parameters, such as fluorescence quantum yield, that can meet the criteria for particular applications is still not satisfied. For example, photochromic compounds used for 2PA 3D data storage normally only show less than 10 GM 2PA cross sections, and the situation is similar for 2PF probes. Hence there is sizable room for improvement. Another opportunity lies in applying emerging nanotechnology, such as surface plasmon enhancement. So, perhaps, many challenges for 2PA materials may have possible solutions in the development of nanoparticle-based hybrid systems.

18.6 SUMMARY

The last decade of the twentieth century and first decade of the twenty-first century has seen enormous interest in multiphoton, nonlinear optical materials, and applications. The highly advantageous properties associated with long wavelength, multiphoton absorption will continue to kindle the flame of innovation. Applications that extend beyond 2PA make their way from the laboratory to the market. As more sensitive and functional materials, and sound accessible theoretical method development progresses, scientists should have a predictive capability to rationally design tailor-made multiphoton absorbing materials for specific applications. This should help propel the numerous applications of multiphoton absorbing materials described above, along with emerging areas such as optical signal processing and quantum computing. The continued dependence on interdisciplinary expertise of materials chemists, spectroscopists, theorists, and engineers makes the field of multiphoton-based research and device development an area of fertile scientific and technological development for the next generation of researchers.

EXERCISE QUESTIONS

1. What are the structural characteristics that lead to high two-photon absorptivity?
2. Why does multiphoton absorption lead to enhanced spatial resolution and deeper depth penetration over conventional, linear absorption?
3. How to calculate 2PA cross section into the first electronic excited state $S_g \rightarrow S_e$ under the resonance 2PE ($\hbar\omega = (E_e - E_g)/2$), when transition dipole $\mu_{ge} = 10$ D; the change in the permanent dipole $\Delta\mu_{ge} = 5$ D; local field factor L and the refractive index $n = 1$; damping factor $\Gamma_{eg} = 0.1$ eV (assumingly $\Gamma_{eg} \ll (E_e - E_g)/4$); and the average value of $\langle\gamma\rangle = \frac{1}{5} \cdot \gamma_{xxxx}$?
4. What type of molecular symmetry leads to zero value of 2PA cross section into the first electronic excited state?

LIST OF ABBREVIATIONS

1PA	one-photon absorption
2D	two dimensional
2PA	two-photon absorption
2PE	two-photon excitation
2PF	two-photon fluorescence
3D	three dimensional
3DLM	three-dimensional lithography microfabrication
A	acceptor
AMD	age-related macular degeneration
BBO	beta-barium borate
c	speed of light in a vacuum
CAD	computer-aided design
CCD	charge-coupled device
cm	centimeter
CW	continuous wave
D	donor
D	positive dipolar terms
DAPI	4′,6-diamidino-2-phenylindole
DNA	deoxyribonucleic acid
DVD	digital versatile disc or digital video disc
e	electron charge
E	electric field
EFAB	electrochemical fabrication
eV	electron volt
f-number	ratio of the focal length of the lens divided by the diameter of its limiting opening
FRET	fluorescence resonance energy transfer
fs	femtosecond
FWHM	full-width half-maximum
GB/in.^{2}	gigabyte per square inch
GM	10^{-50}
HDTV	high definition television

He–Ne	helium–neon
Hz	hertz
I	the incident laser intensity
IC	integrated circuit
IR	infrared
L	local field factor
MEMS	microelectromechanical systems
mJ	millijoule
n	refractive index
N	negative terms
NA	numerical aperture
Nd:YAG	neodymium-doped yttrium aluminium garnet
NIR	near infrared
nm	nanometer
OPO	optical parametric oscillator
PBG	photonic band gap
PDT	Photodynamic therapy
S_n	electronic transition
SOS	sum-over-state
T	two-photon terms
THF	tetrahydrofuran
Ti:sapphire	titanium-sapphire
UV	ultraviolet
W	watt
WORM	write once, read multiple
α	linear (first-order) molecular polarizability
β	second-order polarizability
λ	wavelength
γ	third-order polarizability
$\propto$	dipole moment
ε_0	permittivity of free space
δ	two-photon absorption cross section
Γ	damping factors
Ψ	wave function
Φ_{fl}	fluorescence quantum yield
µm	micrometer
$^1\Delta_g$	singlet molecular oxygen
$^3\Sigma_g^-$	triplet molecular oxygen
Φ_Δ	singlet oxygen quantum yields

REFERENCES

1. For general reviews, see (a) Bhawalkar, J.D., He, G.S., Prasad, P.N., Nonlinear multiphoton processes in organic and polymeric materials, *Rep. Prog. Phys.* 59, 1041–1070, 1996. (b) Herman, B., Wang, X.F., Wodnicki, P., Perisamy, A., Mahajan, N., Berry, G., Gordon, G., Fluorescence lifetime imaging microscopy, in *Applied Fluorescence in Chemistry, Biology, and Medicine*, Rettig, W., Strehmel, B., Schrader, S., Seifert, H. (Eds.), Springer, New York, pp. 496–500, 1999. (c) Reinhardt, B.A., Two-photon technology: New materials and evolving applications, *Photonics Sci. News* 4, 21–34, 1999. (d) Drobizhev, M., Rebane, A., Karotki, A., Spangler, C.W., New two-photon absorbing organic molecules and macromolecules for photonic applications, *Recent Res. Dev. Appl. Phys.* 4, 197–222, 2001. (e) Lin, T.-C., Chung, S.-J., Kim, K.-S.,

Wang, X., He, G.S., Swiatkiewicz, J., Pudavar, H.E., Prasad, P.N., Organics and polymers with high two-photon activities and their applications, *Adv. Poly. Sci.* 161(Polymers for Photonics Applications II), 157–193, 2003. (f) Belfield, K.D., Schafer, K.J., Liu, Y., Liu, J., Ren, X., Van Stryland, E.W., Multiphoton-absorbing organic materials for microfabrication, emerging optical applications and non-destructive three-dimensional imaging, *J. Phys. Org. Chem.* 13, 837–849, 2000. (g) Marder, S.R., Organic nonlinear optical materials: Where we have been and where we are going, *Chem. Comm.* 131–134, 2006.

2. Göppert-Mayer, M., Elementary actions with two quantum leaps, *Ann. Phys.* 9, 273, 1931.
3. Birge, R.R., Parsons, B., Song, Q.W., Tallent, J.R., Protein-based three-dimensional memories and associative processors, in *Molecular Electronics*, Jortner, J., Ratner, M. (Eds.), Blackwell Science, London, chapter 15, 1997.
4. Kaiser, W., Garrett, C.G.B., Two-photon excitation in CaF_2:Eu^{2+}, *Phys. Rev. Lett.* 7, 229–231, 1961.
5. Dick, B., Hochstrasser, R.M., Trommsdorff, H.P., Resonant molecular optics, in *Nonlinear Optical Properties of Organic Molecules and Crystals, Vol. 2*, Chemla D.S., Zyss J. (Eds.), New York, Academic, pp. 159–212, 503–504, 1987.
6. (a) Loudon, R., *The Quantum Theory of Light*, 2nd edn., Oxford University Press, Oxford, 1983. (b) Zojer, E., Beljonne, D., Kogej, T., Vogel, H., Marder, S.R., Perry, J.W., Bredas, J.L., Tuning the two-photon absorption response of quadrupolar organic molecules, *J. Chem. Phys.* 116, 3646–3658, 2002.
7. Beljonne, D., Wenseleers, W., Zojer, E., Shuai, Z., Vogel, H., Pond, S.J.K., Perry, J.W., Marder, S.R., Bredas, J.-L., Role of dimensionality on the two-photon absorption response of conjugated molecules: The case of octupolar compounds, *Adv. Funct. Mater.* 12, 631–641, 2002.
8. (a) Meyers, F., Marder, S.R., Perry, J.W., Introduction to the nonlinear optical properties of organic materials, in *Chemistry of Advance Materials*, Interrante, L.V., Hampden, M.J. (Eds.), Wiley-Interscience, New York, pp. 207–269, 1998. (b) Karna, S.P., Yeates, A.T., nonlinear optical materials: Theory and modeling, in *Nonlinear Optical Materials: Theory and Modeling*, Karna, S.P., Yeates, A.T. (Eds.), American Chemical Society, Washington DC, pp. 1–22, 1996. (c) Bredas, J.L., in *Organic Materials for Photonics*, Zerbi, G. (Ed.), Elsevier Science Publishers, New York/London/Amsterdam, pp. 127–153, 1993.
9. Orr B.J., Ward, J.F., Perturbation theory of the non-linear optical polarization of an isolated system, *Mol. Phys.* 20, 513–526, 1971.
10. (a) Mazumdar, S., Duo, D., Dixit, S.N., A four-level 'essential states' model of third order optical nonlinearity in π-conjugated polymers, *Synth. Met.* 55, 3881–3888, 1993. (b) Mazumdar, S., Guo, F., Observation of three resonances in the third harmonic generation spectrum of conjugated polymers: evidence for the four-level essential states model, *J. Chem. Phys.* 100, 1665, 1994.
11. Dirk, C.W., Cheng, L., Kuzyk, M.G., A simplified three-level model describing the molecular third-order nonlinear optical susceptibility, *Int. J. Quan. Chem.* 43, 27–36, 1992.
12. Birge R.R., Pierce, B.M., A theoretical analysis of the two-photon properties of linear polyenes and the visual chromophores, *J. Chem. Phys.* 70, 165–178, 1979.
13. Kogej, T., Beljonne, D., Meyers, F., Perry, J.W., Marder, S.R., Bredas, J.L., Mechanisms for enhancement of two-photon absorption in donor–acceptor conjugated chromophores, *Chem. Phys. Lett.* 298, 1–6, 1998.
14. Boyd, R.W., *Nonlinear Optics* Academic Press, San Diego, CA, pp. 107–111, 1992.
15. Yao, S., Belfield, K.D., Synthesis of two-photon absorbing unsymmetrical branched chromophores through direct tris(bromomethylation) of fluorene, *J. Org. Chem.* 70, 5126–5132, 2005.
16. (a) Beljonne, D., Bredas, J.-L., Torruellas, W.E., Stegeman, G.I., Hofstraat, J.W., Horsthuis, W.H.G., Mohlmann, G.R., Two-photon absorption and third-harmonic generation of di-alkyl-amino nitro-stilbene (DANS): A joint experimental and theoretical study, *J. Chem. Phys.* 103, 7834–7843, 1995. (b) Rumi, M., Ehrlich, J.E., Heikal, A.A., Perry, J.W., Barlow, S., Hu, Z., McCord-Maughon, D., Parker, T.C., Roeckel, H., Thayumanavan, S., Marder, S.R., Beljonne, D., Bredas, J.-L., Structure–property relationships for two-photon absorbing chromophores: Bis-donor diphenylpolyene and bis(styryl)benzene derivatives, *J. Am. Chem. Soc.* 122, 9500–9510, 2000. (c) Hales, J.M., Hagan, D.J., Van Stryland, E.W., Schafer, K.J., Morales, A.R., Belfield, K.D., Pacher, P., Kwon, O., Zojer, E., Bredas, J.L., Resonant enhancement of two-photon absorption in substituted fluorene molecules, *J. Chem. Phys.* 121, 3152–3160, 2004.
17. Butcher, P.N., Cotter, D., *The Elements of Nonlinear Optics*, Cambridge University Press, Cambridge, 1990.
18. Peticolas, W.L., Multiphoton spectroscopy, *Ann. Rev. Phys. Chem.* 18, 233–260, 1967.
19. (a) Abella, I.D., Optical double-photon absorption in cesium vapor, *Phys. Rev. Lett.* 9, 453–455, 1962. (b) Lipeles, M., Novick, R., Tolk, N., Direct detection of two-photon emission from the metastable state of singly ionized helium, *Phys. Rev. Lett.* 15, 690–693, 1965.
20. (a) Sonnenberg, H., Heffner, H., Spicer, W., Two-photon photoelectric effect in Cs_3Sb, *Appl. Phys. Lett.* 5, 95–96, 1964. (b) Rousseau, D.L., Leroi, G.E., Link, G.L., Two-photon induced chemical reaction in AgCl,

J. Chem. Phys. 42, 4048–4049, 1965. (c) Wang, S., Chang, C.C., Coherent fluorescence from zinc sulfide excited by two-photon absorption, *Appl. Phys. Lett.* 12, 193–195, 1968.

21. (a) Adelman, A.H., Verber, C.M., Two-photon excitation of phosphorescence and fluorescence, *J. Chem. Phys.* 39, 931–933, 1963. (b) Joussot-Dubien, J., Lesclaux, R., Photoionization of aromatic compounds in boric acid. Evidence for a two-photon process, *Compt. Rend.* 258, 4260–4262, 1964. (c) Pao, Y.-H., Rentzepis, P.M., Multiphoton absorption and optical-harmonic generation in highly absorbing molecular crystals, *J. Chem. Phys.* 43, 1281–1286, 1965. (d) Kobayashi, T., Nagakura, S., Two-photon excitation in organic crystals by dye laser, *Chem. Phys. Lett.* 13, 217–220, 1972.
22. Weisz, S.Z., Zahlan, A.B., Gilreath, J., Jarnagin, R.C., Silver, M., Two-photon absorption in crystalline anthracene and naphthalene excited with a xenon flash, *J. Chem. Phys.* 41, 3491–3495, 1964.
23. (a) Peticolas, W.L., Rieckhoff, K.E., Polarization of anthracene fluorescence by one- and two-photon excitation, *Phys. Lett.* 15, 230–231, 1965. (b) Froehlich, D., Mahr, H., Two-photon spectroscopy in anthracene, *Phys. Rev. Lett.* 16, 895–897, 1966.
24. Rapp, W., Gronau, B., Laser emission from two xanthene dyes via double-photon excitation, *Chem. Phys. Lett.* 8, 529–532, 1971.
25. (a) McCarthy, W.J., Wiegand, D.A., Fluorescence of certain organic dyes excited by helium-neon laser, *Spect. Lett.* 1, 349–353, 1968. (b) Bradley, D.J., Hutchinson, M.H.R., Koetser, H., Morrow, T., New, G.H.C., Petty, M.S., Interactions of picosecond laser pulses with organic molecules. I. Two-photon fluorescence quenching and singlet states excitation in Rhodamine dyes, *Proc. Royal Soc. London, Series A: Math. Phys. Eng. Sci.* 28, 97–121, 1972. (c) Hermann, J.P., Ducuing, J., Dispersion of the two-photon cross section in rhodamine dyes, *Opt. Comm.* 6, 101–105, 1972. (d) Foucault, B., Hermann, J.P., Two-photon absorption in organic dyes-relation with the symmetry of the levels, *Opt. Comm.* 15, 412–415, 1975.
26. Hermann, J.P., Ducuing, J., Absolute measurement of two-photon cross-sections, *Phys. Rev. A: Atomic, Mol. Opt. Phys.* 5, 2557–68, 1972.
27. (a) He, G.S., Weder, C., Smith, P., Prasad, P.N., Optical power limiting and stabilization based on a novel polymer compound, *IEEE J. Quantum Electron.* 34, 2279–2285, 1998. (b) Chung, S.-J., Kim, K.-S., Lin, T.-C., He, G.S., Swiatkiewicz, J., Prasad, P.N., Cooperative enhancement of two-photon absorption in multi-branched structures, *J. Phys. Chem. B* 103, 10741–10745, 1999. (c) Kim, O.-K., Lee, K.-S., Woo, H.Y., Kim, K.-S., He, G.S., Swiatkiewicz, J., Prasad, P.N., New class of two-photon-absorbing chromophores based on dithienothiophene, *Chem. Mater.* 12, 284–286, 2000. (d) He, G.S., Swiatkiewicz, J., Jiang, Y., Prasad, P.N., Reinhardt, B.A., Tan, L.-S., Kannan, R., Two-photon excitation and optical spatial-profile reshaping via a nonlinear absorbing medium, *J. Phys. Chem. A* 104, 4805–4810, 2000.
28. (a) Bergman, A., Jortner, J., Consecutive two-photon absorption of azulene in solution utilizing dye lasers, *Chem. Phys. Lett.* 20, 8–10, 1973. (b) Swofford, R.L., McClain, W.M., Two-photon absorption studies of diphenylbutadiene. Location of a 1Ag excited state, *J. Chem. Phys.* 59, 5740–5741, 1973. (c) Drucker, R.P., McClain, W.M., Dye laser two-photon excitation study of several o,o'-bridged biphenyls, *J. Chem. Phys.* 61, 2616–2619, 1974. (d) Drucker, R.P., McClain, W.M., Two photon absorption in double molecules, *Chem. Phys. Lett.* 28, 255–257, 1974. (d) Fuke, K., Nagakura, S., Kobayashi, T., Two-photon absorption spectrum of [2.2]paracyclophane, *Chem. Phys. Lett.* 31, 205–207, 1975. (e) Harris, R.A., On the wave vector dependence of two photon absorption by a polymer, *Chem. Phys. Lett.* 37, 295–296, 1976.
29. (a) Albota, M., Beljonne, D., Bredas, J.-L., Ehrlich, J.E., Fu, J.-Y., Heikal, A.A., Hess, S.E., Kogej, T., Levin, M.D., Marder, S.R., McCord-Maughon, D., Perry, J.W., Rockel, H., Rumi, M., Subramaniam, G., Webb, W.W., Wu, X.-L., Xu C., Design of organic molecules with large two-photon absorption cross sections, *Science*, 281, 1653–1656, 1998. (b) Pond, S.J.K., Rumi, M., Levin, M.D., Parker, T.C., Beljonne, D., Day, M.W., Bredas, J.-L., Marder, S.R., Perry J.W., One- and two-photon spectroscopy of donor–acceptor–donor distyrylbenzene derivatives: Effect of cyano substitution and distortion from planarity, *J. Phys. Chem. A* 106, 11470–11480, 2002.
30. (a) Reinhardt, B.A., Brott, L.L., Clarson, S.J., Dillard, A.G., Bhatt, J.C., Kannan, R., Yuan, L., He, G.S., Prasad, P.N., Highly active two-photon dyes: Design, synthesis, and characterization toward application, *Chem. Mater.* 10, 1863–1874, 1998. (b) Chung, S.-J., Kim, K.-S., Lin, T.-C., He, G.S., Swiatkiewicz, J., Prasad, P.N., Cooperative enhancement of two-photon absorption in multi-branched structures, *J. Phys. Chem. B* 103, 10741–10745, 1999. (c) Adronov, A., Frechet, J.M.J., He, G.S., Kim, K.-S., Chung, S.-J., Swiatkiewicz, J., Prasad, P.N., Novel two-photon absorbing dendritic structures, *Chem. Mater.* 12, 2838–2841, 2000. (d) Kannan, R., He, G.S., Yuan, L., Xu, F., Prasad, P.N., Dombroskie, A.G., Reinhardt, B.A., Baur, J.W., Vaia, R.A., Tan, L.-S., Diphenylaminofluorene-based two-photon-absorbing chromophores with various π-electron acceptors, *Chem. Mater.* 13, 1896–1904, 2001. (e) He, G.S., Lin, T.-C., Prasad, P.N., Kannan, R., Vaia, R .A., Tan, L.-S., Study of two-photon absorption spectral property of a novel nonlinear optical chromophore using femtosecond continuum, *J. Phys. Chem. B* 106, 11081–11084, 2002.

31. (a) Belfield, K.D., Hagan, D.J., Van Stryland, E.W., Schafer, K.J., Negres, R.A., New two-photon absorbing fluorene derivatives: Synthesis and nonlinear optical characterization, *Org. Lett.* 1, 1575–1578, 1999. (b) Belfield, K.D., Schafer, K.J., Mourad, W., Reinhardt, B.A., Synthesis of new two-photon absorbing fluorene derivatives via Cu-mediated Ullmann condensations, *J. Org. Chem.* 65, 4475–4481, 2000. (c) Belfield, K.D., Morales, A.R., Hales, J.M., Hagan, D.J., Van Stryland, E.W., Chapela, V.M., Percino, J., Linear and Two-photon photophysical properties of a series of symmetrical diphenylaminofluorenes, *Chem. Mater.* 16, 2267–2273, 2004. (d) Belfield, K.D., Morales, A.R., Kang, B.-S., Hales, J.M., Hagan, D.J., Van Stryland, E.W., Chapela, V.M., Percino, J., Synthesis, Characterization, and optical properties of new two-photon-absorbing fluorene derivatives, *Chem. Mater.* 16, 4634–4641, 2004.
32. Lockwicz, J.R., *Principles of Fluorescence Spectroscopy*, Kluwer Academic/Plenum, New York, 1999.
33. Belfield, K.D., Bondar, M.V., Przhonska, O.V., Schafer, K.J., Steady-state spectroscopic and fluorescence lifetime measurements of new two-photon absorbing fluorene derivatives, *J. Fluoresc.* 12, 449–454, 2002.
34. (a) Porres, L., Mongin, O., Katan, C., Charlot, M., Pons, T., Mertz, J., Blanchard-Desce, M., Enhanced two-photon absorption with novel octupolar propeller-shaped fluorophores derived from triphenylamine, *Org. Lett.* 6, 47–50, 2004. (b) Zheng, Q., He, G.S., Prasad, P.N., π-Conjugated dendritic nanosized chromophore with enhanced two-photon absorption, *Chem. Mater.* 17, 6004–6011, 2005. (c) He, G.S., Lin, T.-C., Dai, J., Prasad, P.N., Kannan, R., Dombroskie, A.G., Vaia, R.A., Tan, L.-S., Degenerate two-photon-absorption spectral studies of highly two-photon active organic chromophores, *J. Chem. Phys.* 120, 5275–5284, 2004. (d) Kannan, R., He, G.S., Lin, T.-C., Prasad, P.N., Vaia, R.A., Tan, L.-S., Toward highly active two-photon absorbing liquids. Synthesis and characterization of 1,3,5-triazine-based octupolar molecules, *Chem. Mater.* 16, 185–194, 2004.
35. Katan, C., Terenziani, F., Le Droumaguet, C., Mongin, O., Werts, M.H.V., Tretiak, S., Blanchard-Desce, M., Branching of dipolar chromophores: Effects on linear and nonlinear optical properties, *Proc. SPIE-Intern. Soc. Opt. Eng.* 5935, 593503/1–593503/15, 2005.
36. For general reviews, see (a) Kawata, S., Kawata, Y., Three-dimensional optical data storage using photochromic materials, *Chem. Rev.* 100, 1777–1788, 2000. (b) Akiba, M., Recent progress in organic two-photon absorption materials for three-dimensional optical data storage, *Optronics* 283, 184–189, 2005. (c) Oulianov, D.A., Dvornikov, A.S., Rentzepis, P.M., Nonlinear 3D optical storage and comments on two-photon cross section measurements, *Proc. SPIE-Intern. Soc. Opt. Eng.* 4462, 1–10, 2002.
37. (a) Parthenopoulos, D.A., Rentzepis, P.M., Three-dimensional optical storage memory, *Science* 245, 843–845, 1989. (b) Wang, M.M., Esener, S.C., McCormik, F.B., Cokgor, I., Dvornikov, A.S., Rentzepis, P.M., Experimental characterization of a two-photon memory, *Opt. Lett.* 22, 558–560, 1997.
38. Corredor, C.C., Huang, Z.-L., Belfield, K.D., Two-photon 3-D optical data storage via fluorescence modulation of an efficient fluorene dye by a photochromic diarylethene, *Adv. Mater.*, 18, 2910–2914, 2006.
39. (a) Belfield, K.D., Schafer, K.J., A new photosensitive polymeric material for optical data storage using multichannel two-photon fluorescence readout *Chem. Mater.*, 14, 3656–3662, 2002. (b) Belfield, K.D., New photosensitive polymeric material for WORM optical data storage with two-photon fluorescent readout, US Patent 7,001,708 B1.
40. Belfield, K.D., Liu, Y., Negres, R.A., Fan M., Pan, G., Hagan, D.J., Hernandez, F.E., Two-photon photochromism of an organic material for holographic recording, *Chem. Mater.*, 14, 3663–3667, 2002.
41. (a) Xu, C., Webb, W.W., Measurement of two-photon excitation cross sections of molecular fluorophores with data from 690 to 1050 nm, *J. Opt. Soc. Am. B.*, 13, 481–491, 1996. (b) Xu, C., Williamns, R.M., Zipfel, W., Webb, W.W., Multiphoton excitation cross-sections of molecular fluorophores, *Bioimaging* 4, 198–207, 1996. (c) Xu, C., Zipfel, W., Shear, J.B., Williams, R.M., Webb, W.W., Multiphoton fluorescence excitation: New spectral window for biological nonlinear microscopy, *Proc. Natl. Acad. Sci. USA*, 93, 10763–10768, 1996. (d) Albota, M.A., Xu, C., Webb, W.W., Two-photon fluorescence excitation cross sections of biomolecular probes from 690 to 960 nm, *Appl. Opt.*, 37, 7352–7356, 1998.
42. (a) Pettit, D.L., Wang, S.S.H., Gee, K.R., Augustine, G.J., Chemical two-photon uncaging: A novel approach to mapping glutamate receptors, *Neuron* 19, 465–471, 1997. (b) Brown, E.B., Shear, J.B., Adams, S.R., Tsien, R.Y., Webb, W.W., Photolysis of caged calcium in femtoliter volumes using two-photon excitation, *Biophys. J.* 76, 489–499, 1999. (c) Soeller, C., Jacobs, M.D., Donaldson, P.J., Cannell, M.B., Jones, K.T., Ellis-Davies, G.C.R., Application of two-photon flash photolysis to reveal intercellular communication and intracellular Ca^{2+} movements, *J. Biomed. Opt.* 8, 418–427, 2003. (d) Wang, S.S.-H., Khiroug, L., Augustine, G.J., Quantification of spread of cerebellar long-term depression with chemical two-photon uncaging of glutamate, *Proc. Natl. Acad. Sci. USA*, 97, 8635–8640, 2000.
43. (a) Jacques, S.L., Prahl, S.A., Absorption spectra for biological tissues, *Oregon Graduate Institute*, http://omlc.ogi.edu/classroom/ece532/class3/muaspectra.html, 1998. (b) Anderson, R.R., Parrish, J.A., The optics of human skin, *J. Invest. Dermatol.* 77, 13–19, 1981.

44. Wachter, E.A., Partridge, W.P., Fisher, W.G., Dees, H.C., Petersen, M.G., Simultaneous two-photon excitation of photodynamic therapy agents, *Proc. SPIE* 3269, 68–75, 1998.
45. Karotki, A., Khurana, M., Lepock, J.R., Wilson, B.C., Two-photon excitation photodynamic therapy with Photofrin, *Proc. SPIE* 5969, 596915/1–8, 2005.
46. Ogawa, K., Hasegawa, H., Inaba, Y., Kobuke, Y., Inouye, H., Kanemitsu, Y., Kohno, E., Hirano, T., Ogura, S., Okura I., Water-soluble bis(imidazolylporphyrin) self-assemblies with large two-photon absorption cross sections as potential agents for photodynamic therapy, *J. Med. Chem.* 49, 2276–2283, 2006.
47. Fernandez, J.M., Bilgin, M.D., Grossweiner, L.I., Singlet oxygen generation by photodynamic agents, *J. Photochem. Photobiol. B: Biol.* 37, 131–140, 1997.
48. Varadan, V.K., Jiang X., Varadan V.V., *Microstereolithography and Other Fabrication Techniques for 3D MEMS*, Wiley & Sons, Chichester, England, 2001.
49. For general reviews, see (a) Sun, H.-B., Kawata, S., Two-photon photopolymerization and 3D lithographic microfabrication, *Adv. Poly. Sci.* 170(NMR, 3D Analysis, Photopolymerization), 169–273, 2004. (b) Yang, H.-K., Kim, M.-S., Kang, S.-W., Kim, K.-S., Lee, K.-S., Park, S.H., Yang, D.-Y., Kong, H.J., Sun, H.-B., Kawata, S., Fleitz, P., Recent progress of lithographic microfabrication by the TPA-induced photopolymerization, *J. Photopoly. Sci. Tech.* 17, 385–392, 2004.
50. Chin, S.L., Bedard, G., Ruby laser induced breakdown induced polymerization of methyl methacrylate, *Phys. Lett.* 36A, 271–272, 1971.
51. (a) Ichimura, K., Sakuragi, M., A spiropyran-iodonium salt system as a two photon radical photoinitiator, *J. Polym. Sci. Part C: Polym. Lett.* 26, 185–189, 1988. (b) Jent, F., Paul, H., Fischer, H., Two-photon processes in ketone photochemistry observed by time-resolved ESR spectroscopy, *Chem. Phys. Lett.* 146, 315–319, 1988. (c) Lougnot D.J., Ritzenthaler, D., Carre, C., Foussier J.P., A new gated system for two-photon holographic recording in the near infrared, *J. Appl. Phys.* 63, 4841–4848, 1988.
52. (a) Strickler, J.H., Webb W.W., Three-dimensional optical data storage in refractive media by two-photon point excitation, *Opt. Lett.* 16, 1780–1783, 1991. (b) Maruo, S., Nakamura, O., Kawata S., Three-dimensional microfabrication with two-photon-absorbed photopolymerization, *Opt. Lett.* 22, 132–134, 1997. (c) Borisov, R.A., Dorojkina, G.N., Koroteev, N.I., Kozenkov, V.M., Magnitskii, S.A., Malakhov, D.V., Tarasishin, A.V., Zheltikov, A.M., Fabrication of three-dimensional periodic microstructures by means of two-photon polymerization *Appl. Phys. B* 67, 765–767, 1998. (d) Borisov, R.A., Dorojkina, G.N., Koroteev, N.I., Kozenkov, V.M., Magnitskii, S.A., Malakhov, D.V., Tarasishin, A.V., Zheltikov, A.M., Femtosecond two-photon photopolymerization: A method to fabricate optical photonic crystals with controllable parameters, *Laser Phys.* 8, 1105–1106, 1998.
53. (a) Cumpston, B.H., Ananthavel, S.P., Barlow, S., Dyer, D.L., Ehrlich, J.E., Erskine, L.L., Heikal, A.A., Kuebler, S.M., Lee, I.-Y.S., McCord-Maughon, D., Qin, D.J., Röckel, H.R., Rumi, M., Wu, X.-L., Marder, S.R., Perry, J.W., Two-photon polymerization initiators for three-dimensional optical data storage and microfabrication, *Nature* 398, 51–54, 1999. (b) Belfield, K.D., Ren, X., Van Stryland, E.W., Hagan, D.J., Dubikovski, V., Meisak, E.J., Near-IR two-photon photoinitiated polymerization using a fluorone/amine initiating system, *J. Am Chem. Soc.* 122, 1217–1218, 2000.
54. Miller, M.J., Mott A.G., Ketchel, B.P., General optical limiting requirements, *Proc. SPIE* 3472 (Nonlinear Optical Liquids for Power Limiting and Imaging), 3472, 24–29, 1998.
55. For general reviews, see (a) Trantolo, D.J., Gresser, J.D., Wise, D.L., Kowalski, G.J., Rao, D.V.G.L.N., Aranda F.J., Wnek, G.E., Sensor protection from lasers, in *Photonic Polymer Systems*, Wise, D.L., Wnek, G.E., Trantolo, D.J., Cooper T.M., Gresser, J.D. (Eds.) Marcel Dekker, New York, pp. 537–569, 1998. (b) Perry, J.W., Organic and metal-containing reverse saturable absorbers for optical limiters, in *Nonlinear Optics of Organic Molecules and Polymers*, Nalwa H.S., Miyata, S. (Eds.), CRC Press, Boca Raton, FL, pp. 813–840, 1997. (c) Van Stryland, E.W., Hagan, D.J., Xia T., Said, A.A., Application of nonlinear optics to passive optical limiting, in *Nonlinear Optics of Organic Molecules and Polymers*, Nalwa H.S., Miyata, S. (Eds.), CRC Press, Boca Raton, FL, pp. 841–860, 1997. (d) Sutherland, R.L., *Handbook of Nonlinear Optics*, Marcel Dekker, New York, 1996.
56. (a) Spangler, C.W., Recent development in the design of organic materials for optical power limiting, *J. Mater. Chem.* 9, 2013–2020, 1999. (b) He, G.S., Gvishi, R., Prasad, P.N., Reinhardt, B.A., Two-photon absorption based optical limiting and stabilization in organic molecule-doped solid materials, *Opt. Comm.* 117, 133–136, 1995. (c) He, G.S., Lixiang, Y., Cheng, N., Bhawalkar, J.D., Prasad, P.N., Nonlinear optical properties of a new chromophore, *J. Opt. Soc. Am. B* 14, 1079–1087, 1997. (d) Reinhardt, B.A., Brott, L.L., Clarson, S.J., Kannan, R., Dillard, A.G., The design and synthesis of new organic molecules with large two-photon absorption cross-sections for optical limiting applications, *Mater. Res. Soc. Symp. Proc.* 479, 3–8, 1997. (e) Ehrlich, J.E., Wu, X.L., Lee, L.S., Heikal, A.A., Hu, Z.Y., Roeckel, H., Marder, S.R., Perry, J.W., Two-photon absorbing organic chromophores for optical limiting, *Mater. Res. Soc. Symp. Proc.* 479, 9–15, 1997.

57. (a) Joshi, M.P., Swiatkiewicz, J., Xu, F., Prasad, P.N., Reinhardt, B.A., Kannan, R., Energy transfer coupling of two-photon absorption and reverse saturable absorption for enhanced optical power limiting, *Opt. Lett.* 23, 1742–1744, 1998. (b) Madrigal, L.G., Spangler, C.W., Casstevens, M.K., Kumar, D., Weibel, J., Burzynski, R., Polymeric optical power limiters based on reverse saturable absorption from acceptor generated charge states, *Poly. Preprin.* 39, 1057–1058, 1998. (c) Bader, M., Moser, J., Li, H., Tarter, S., Spangler, C., Design and synthesis of new acceptor molecules for photo-induced electron transfer reverse saturable absorption, *Mater. Res. Soc. Symp. Proc.* 598 (Electrical, Optical, and Magnetic Properties of Organic Solid-State Materials V), BB4.8/1-BB4.8/7, 2000.
58. (a) Kamanina, N.V., Kaporskii, L.N., Fullerene-doped polymer-dispersed liquid crystal films as effective optical power limiting materials, *MCLC S&T, Section B: Nonlinear Optics* 27, 339–346, 2001. (b) Yuan, H.J., Lin, H., Li, L., Palffy-Muhoray, P., Optical power limiting by liquid crystals in glass capillary arrays, *Mol. Cryst. Liq. Cryst. Sci. Technol., A: Mol. Cryst. Liq. Cryst.* 223, 229–239, 1992. (c) Michael, R.R., Finn, G.M., Lindquist, R.G., Khoo, I.C., Infrared nonlinear optical power limiting with a nematic liquid crystal film, *Proc. SPIE-Intern. Soc. Opt. Eng.* 971 (Nonlinear Optical Properties of Organic Materials), 157–162, 1988. (d) Umeton, C., Cipparrone, G., Simoni, F., Power limiting and optical switching with nematic liquid-crystal films, *Opt. Quant. Electron.* 18, 312–314, 1986.
59. Reinhardt, B.A., Brott, L.L., Clarson, S.J., Kannan, R., Dillard, A.G., Optical power limiting in solution via two-photon absorption: New aromatic heterocyclic dyes with greatly improved performance, *Proc. SPIE-Intern. Soc. Opt. Eng.* 3146 (Nonlinear Optical Liquids and Power Limiters), 2–11, 1997.
60. Lee, K.-S., Yang, H.-K., Lee, J.-H., Kim, O.-K., Woo, H.Y., Choi, H., Cha, M., Blanchard-Desce, M.H., Optical power limiting properties of two photon absorbing fluorene and dithienothiophene-based chromophores, *Proc. SPIE-Intern. Soc. Opt. Eng.* 4991 (Organic Photonic Materials and Devices V), 175–182, 2003.
61. (a) Charlot, M., Izard, N., Mongin, O., Riehl, D., Blanchard-Desce, M., Optical limiting with soluble two-photon absorbing quadrupoles: Structure–property relationships, *Chem. Phys. Lett.* 417, 297–302, 2006. (b) Silly, M.G., Porres, L., Mongin, O., Chollet, P.-A., Blanchard-Desce, M., Optical limiting in the red-NIR range with soluble two-photon absorbing molecules, *Chem. Phys. Lett.* 379, 74–80, 2003.
62. He, G.S., Lin, T.-C., Prasad, P.N., Cho, C.-C., Yu, L.-J., Optical power limiting and stabilization using a two-photon absorbing neat liquid crystal in isotropic phase, *Appl. Phys. Lett.* 82, 4717–4719, 2003.

19 Organic and Polymeric Photorefractive Materials and Devices

Oksana Ostroverkhova

CONTENTS

19.1 Photorefractive Effect: Definition and Mechanisms 608
19.2 Theoretical Description 610
19.2.1 Space-Charge Field Formation 610
19.2.1.1 Photogeneration 612
19.2.1.2 Charge Transport 612
19.2.1.3 Charge Trapping and Detrapping 613
19.2.1.4 Charge Carrier Recombination 614
19.2.2 Refractive Index Grating Formation: Orientational Enhancement 615
19.3 Experimental Techniques 616
19.3.1 Two-Beam Coupling 616
19.3.2 Four-Wave Mixing 619
19.4 Materials Design 620
19.4.1 Polymer Composites 620
19.4.2 Organic Amorphous Glasses 621
19.4.3 Fully Functionalized Polymers 621
19.4.4 Polymer-Dispersed Liquid Crystals 625
19.5 Applications of Photorefractive Materials 625
19.5.1 Data Storage 626
19.5.2 Image Processing 626
19.5.2.1 Optical Phase Conjugation 626
19.5.2.2 Image Amplification, Novelty Filtering, and Edge Enhancement 627
19.5.3 Optical Waveguides 630
19.6 Summary 633
Exercise Questions 634
List of Abbreviations 635
References 636

Abstract This chapter describes photorefractive (PR) effect, which is a light-induced change of the refractive index that involves photogenerated charge carrier redistribution, in organic materials. A process of photorefractive grating formation, which can be viewed as creation of a dynamic hologram, is discussed and relevant physical mechanisms are outlined. Experimental geometries widely used in assessing photorefractive performance and utilized in applications are presented. Key elements of design of photorefractive organic materials are summarized, and examples of high-performance materials are provided. Selected applications of photorefractive materials are described and illustrated.

19.1 PHOTOREFRACTIVE EFFECT: DEFINITION AND MECHANISMS

The photorefractive (PR) effect is a special holographic recording mechanism, in which nonuniform illumination leads to formation of a spatially modulated space-charge field, which in turn modulates the refractive index. If nonuniform illumination is created by two interfering light beams, a phase grating (hologram) that can diffract light is produced in a PR material. Since such a hologram can typically be erased by uniform optical illumination, PR holograms are dynamic, that is, they may be erased and rewritten many times, one of many properties that distinguish photorefractivity from other mechanisms for hologram formation.

The PR effect was first observed in 1960 in inorganic crystals $LiNbO_3$ and $LiTaO_3$ and was considered to be "highly detrimental to the optics of nonlinear devices based on these crystals," "although interesting in its own right" [1]. The effect was called optical damage, since for the purpose of light frequency-doubling (which was the primary application of these crystals), it was adverse. Later, it was realized that the damage reproduced the original intensity variation in the form of varying dielectric constant and therefore it was suitable for holographic recording and other applications. The first PR organic material, a polymer composite, was reported in 1990 [2]. Since then, dramatic improvement in the PR performance of organics has been achieved owing to significant advances in synthesis, numerous physical studies that identified key mechanisms and parameters, and research of structure–property relationships [3]. Availability of high-performance PR organic materials makes a number of applications (that include high-density optical data storage, image processing, phase conjugation [PC], optical limiting, associative memories, programmable optical interconnects, etc.) feasible; examples of selected applications will be considered in Section 19.5. Organic amorphous materials (e.g., polymer composites and glasses) are also technologically attractive due to their low cost, easy fabrication, and properties, which can be easily modified by varying relative concentrations or performing synthetic modifications of constituents of the composite.

The main mechanisms contributing to the PR effect are space-charge field formation (Section 19.2.1) and the refractive index change via electro-optic nonlinearity (Section 19.2.2). Figure 19.1 illustrates microscopic processes required to produce a PR hologram. Two intersecting coherent beams of light of the same frequency ω, described by electric field vectors $\boldsymbol{\varepsilon}_1 = \boldsymbol{\varepsilon}_{10} \exp[-i(\mathbf{k_1} \cdot \mathbf{r} - \omega t - \varphi_1)]$ and $\boldsymbol{\varepsilon}_2 = \boldsymbol{\varepsilon}_{20} \exp[-i(\mathbf{k_2} \cdot \mathbf{r} - \omega t - \varphi_2)]$ produce a standing-wave interference pattern. Here, $\boldsymbol{\varepsilon}_{10}$, $\boldsymbol{\varepsilon}_{20}$ are the amplitudes of the electric field of the beams 1 and 2, φ_1 and φ_2 are their respective phases, $\mathbf{r}$ is a radius vector, and $\mathbf{k_1}$ and $\mathbf{k_2}$ are the wave vectors defining the propagation directions of beams 1 and 2. This time-independent but spatially modulated intensity ($I(\mathbf{r})$) can be presented as

$$I(\mathbf{r}) = |\mathbf{E}_1 + \mathbf{E}_2|^2 = |\mathbf{E}_{10}|^2 + |\mathbf{E}_{20}|^2 + 2\,\mathbf{E}_{10}\,\mathbf{E}_{20}^* \cos(\mathbf{K}\mathbf{r} + \varphi_1 - \varphi_2), \tag{19.1}$$

where $\mathbf{K} = \mathbf{k_1} - \mathbf{k_2}$ is the grating wave vector. (This relationship between the wave vectors $\mathbf{k_1}$, $\mathbf{k_2}$ and the grating wave vector $\mathbf{K}$ assumes that the Bragg condition is satisfied [1,4]). The interference pattern described by Equation 19.1 has a spatial wavelength (or periodicity) $\Lambda = 2\pi/|\mathbf{K}|$. If one denotes the x-axis as the direction of the grating wave vector $\mathbf{K}$, the optical intensity follows the sinusoidal pattern shown in Figure 19.1(a). Once the spatially modulated light intensity pattern is created, the following physical processes occur in the PR organic material:

- Step 1: Charge photogeneration in the spatial regions where constructive interference occurs (Figure 19.1a). (Equal numbers of positively and negatively charged carriers are produced, since the electric neutrality of a material as a whole must be maintained.)

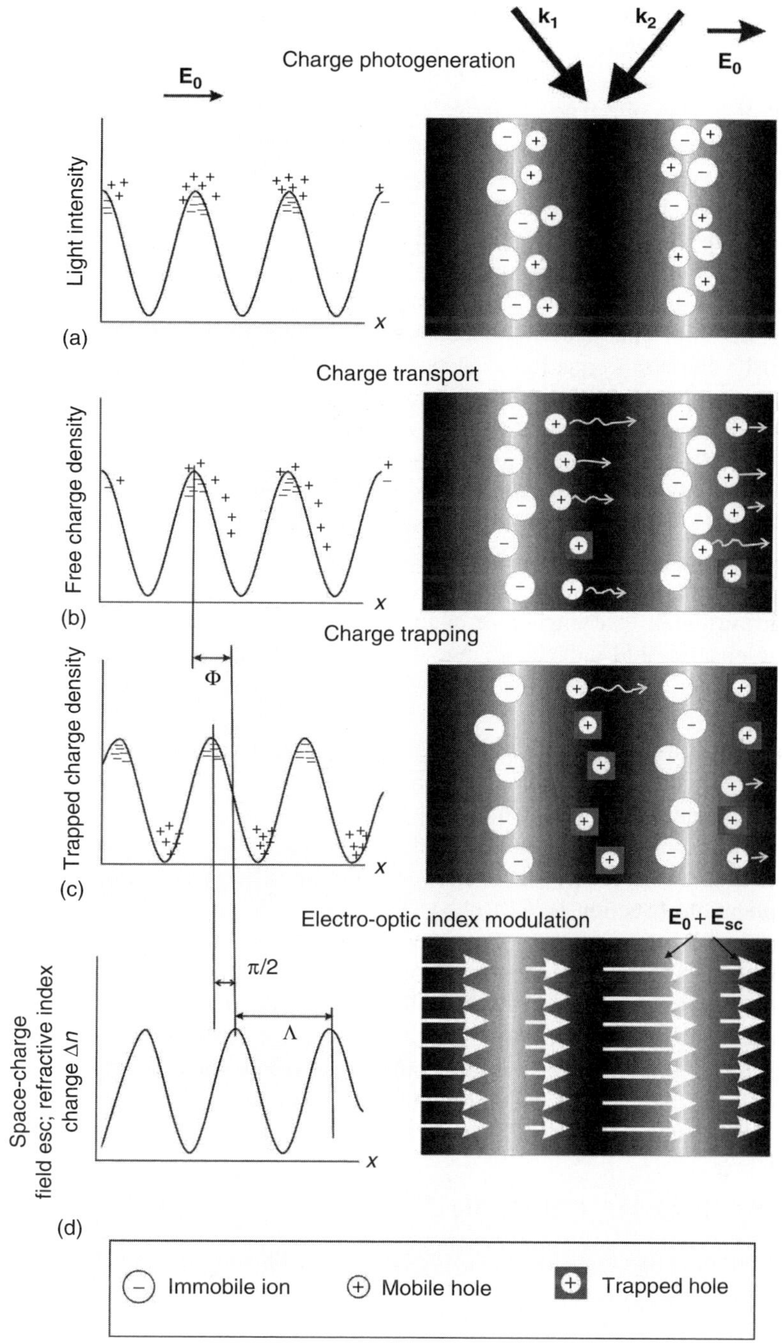

FIGURE 19.1 PR grating formation: (a) charge photogeneration; (b) charge transport; (c) charge trapping; (d) space-charge field formation and refractive index modulation. (Reprinted from Ostroverkhova, O. and Moerner, W.E., *Chem. Rev.*, 104, 3267, 2004. With permission.)

- Step 2: Transport of the photogenerated charges, with one type of the carriers more mobile than the other, in the static, externally applied, electric field $\mathbf{E}_0$. In Figure 19.1b, the holes are shown to be more mobile, which is the more common case for organics. Note that the condition of having one type of charge carriers more mobile than the other is necessary, since charge separation (such as shown in Figure 19.1c) is essential for the space-charge field formation. If both carriers were equally mobile, they would recombine resulting in zero space-charge field and hence no PR effect.
- Step 3: Charge trapping at trapping sites (Figure 19.1c), which are local regions of the material where the mobile charge can be trapped, i.e., prevented from participating in transport for some period of time.
  After separation of charge carriers occurs, the resulting space-charge density is shown in Figure 19.1c. Poisson's equation of electrostatics ($\nabla \mathbf{E} = e/(\varepsilon_0\varepsilon)\rho$, where e is the charge of the electron, ε is the dielectric constant, and ρ_{sc} is a space-charge density, and $\mathbf{E}$ is the total electric field) dictates that such a charge distribution produces a sinusoidal space-charge electric field ($\mathbf{E}_{sc}(x)$) as shown in Figure 19.1d, with the resulting internal electric field shifted in space by 90° relative to the trapped charge density, or one-quarter of the grating wavelength Λ.
- Step 4: Change in the optical index of refraction (Δn) of the material in response to the local electric field ($\Delta n(x) \sim (\mathbf{E}_0 + \mathbf{E}_{sc}(x))^2$). A spatial modulation of the index of refraction results from the sinusoidally varying space-charge field, as shown in Figure 19.1d. Mechanisms producing electric field-induced change in the refractive index (electro-optic response) in PR organic materials are described in Section 19.2.2.

The total spatial phase shift between the maxima in the optical intensity pattern in Figure 19.1a and those in the index of refraction modulation in Figure 19.1d is denoted by Φ. When the phase shift is nonzero, the index grating is a nonlocal grating, and this property (which arises fundamentally from charge transport over a macroscopic distance) is one of the most important special properties of PR materials, which leads to the two-beam coupling (2BC) effect (Section 19.3.1) and many fascinating applications (Section 19.5.2).

Altogether, the PR effect requires photoconductivity (for the space-charge field formation) and electric field-dependent refractive index. In organics, these properties necessary for producing a PR phase hologram are generally provided by a combination of functional components in a material that includes a photoinduced charge generator (sensitizer), a charge transporting medium (e.g., a photoconductive polymer), trapping sites, and molecules that provide electric field-dependent refractive index (e.g., nonlinear optical [NLO] chromophores). Main aspects of the PR organic materials design will be considered in Section 19.4.

19.2 THEORETICAL DESCRIPTION

In this section, physical mechanisms that contribute to the PR grating formation (steps 1 through 4 in Section 19.1) are described.

19.2.1 Space-Charge Field Formation

The basic molecular model that describes space-charge field formation (steps 1 through 3 in Figure 19.1) was introduced by Schildkraut and Buettner in 1992 [2,3]. The processes taken into account in the model, assuming holes to be majority carriers, are depicted in Figure 19.2. A sensitizer (electron acceptor to generate a hole) with density N_A is excited by light of frequency ω, and a free hole is injected into the transport manifold and hops between transport sites until it either becomes trapped with the rate γ_T or recombines with ionized acceptors with rate γ. Traps (T) with a well-defined energy level (ionization potential) are considered. The charge can be thermally released from a trap

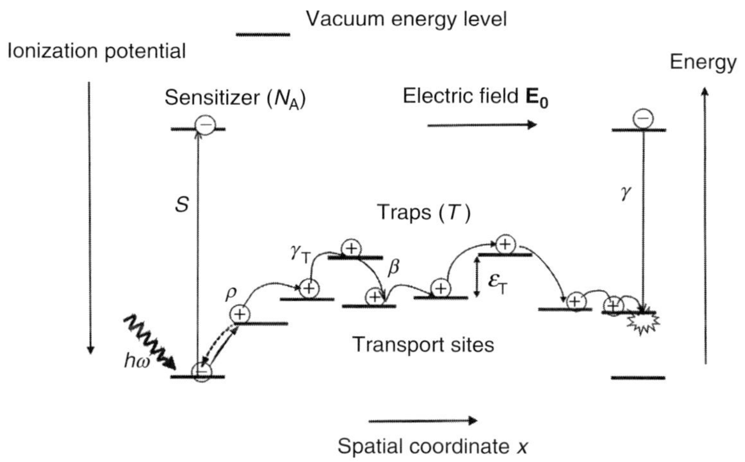

FIGURE 19.2 Schematic representation of the molecular model describing space-charge field formation in PR organic materials. Note: Symbols are described in the text. (Adapted from Ostroverkhova, O. and Singer, K.D., *J. Appl. Phys.*, 92, 1727, 2002. With permission.)

(detrapping process) with a rate β. The system of nonlinear equations describing dynamics of the space-charge field formation is given by

$$\begin{aligned}
\frac{\partial \rho}{\partial t} &= \frac{\partial N_A^i}{\partial t} - \frac{\partial T^+}{\partial t} - \frac{1}{e}\frac{\partial J}{\partial x}, \\
\frac{\partial T^+}{\partial t} &= \gamma_T\left(T - T^+\right)\rho - \beta T^+, \\
\frac{\partial N_A^i}{\partial t} &= sI\left(N_A - N_A^i\right) - \gamma N_A^i \rho, \\
\frac{\partial E}{\partial x} &= \frac{e}{\varepsilon_0 \varepsilon}\left(\rho + T^+ - N_A^i\right), \\
J &= e\mu\rho E - e\mu\xi\frac{\partial \rho}{\partial x},
\end{aligned} \tag{19.2}$$

where

ρ is the free charge (hole) density
N_A is the total density of acceptors
N_A^i is the density of ionized acceptors
T^+ and T are the densities of filled traps and total trapping sites, respectively
E is the electric field
I is the incident light intensity
J is the current density
μ is the charge carrier drift mobility
s is the total cross-section of the photogeneration process
ξ is the diffusion coefficient given by $\xi = k_B T/e$
ε is the dielectric constant

In organic materials, photogeneration cross-section, charge carrier mobility, trapping, detrapping, and recombination rates are all electric field-dependent. Below, the physical meaning behind these parameters and the origin of the electric field dependencies of the parameters in PR polymers are briefly outlined. For details, the reader is referred to Refs. [4–6].

19.2.1.1 Photogeneration

The photogeneration of charge carriers through the absorption of a photon with energy $\hbar\omega$ in the presence of electric field can be schematically represented as

$$N_A + \hbar\omega \xrightarrow{s} N_A^{i} + \text{free hole}$$

The photogeneration cross-section s is directly proportional to the photogeneration efficiency ϕ, introduced below.

The most widely used formalism to describe charge photogeneration process in PR polymers is the Onsager model [5]. In this model, free carriers are assumed to be created by a multistep process. First, a photon is absorbed, which creates a hot localized electron–hole pair. Then, the hot electron loses its kinetic energy by scattering and becomes thermalized at a mean distance r_0 from its parent cation, creating a charge-transfer state. The efficiency of this process is described by a primary quantum yield ϕ_0. Finally, the charge-transfer state either dissociates into a free electron and free hole (which requires overcoming the Coulomb interaction between the electron and a parent cation) or recombines. The photogeneration efficiency ϕ is the product of the primary quantum yield ϕ_0 and the pair dissociation probability, which is electric field-dependent. The Onsager formalism leads to strongly electric field-dependent photogeneration efficiencies that saturate at high electric fields.

19.2.1.2 Charge Transport

In most PR polymers, charge transport can be described by the disorder formalism [5]. It is assumed that the elementary transport step is the charge transfer between adjacent molecules or sites, and both the hopping site energies and intersite distances are subject to distribution (Figure 19.3). Electronic states of the polymer are considered to be completely localized, and the density of states (DOS) distribution is assumed to be Gaussian and given by

$$g(\mathcal{E}) = \frac{1}{\sqrt{2\pi}\sigma} \exp\left(-\mathcal{E}^2/2\sigma^2\right), \tag{19.3}$$

where
$\mathcal{E}$ is the site energy relative to the center of DOS
σ is its Gaussian width (Figure 19.3)

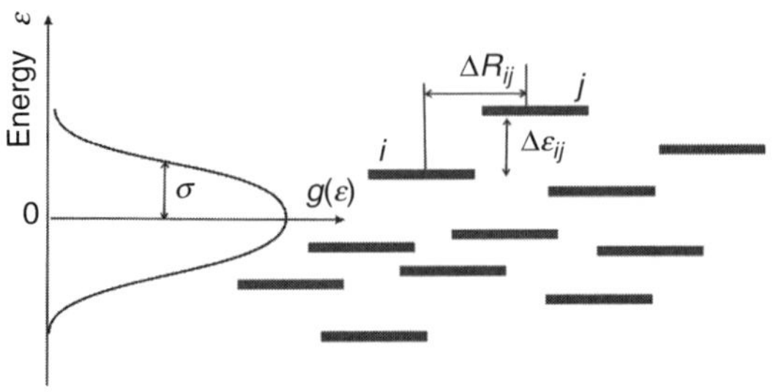

FIGURE 19.3 Schematic representation of the charge transport manifold in PR polymers.

The disorder in site energy is commonly referred to as diagonal disorder, while the disorder that describes fluctuations in the intersite distances ΔR_{ij} (Figure 19.3) is the off-diagonal or positional disorder. The rates of hopping between the neighboring sites i and j are described by the Miller–Abrahams expressions:

$$\nu_{ij} = \nu_0 \exp\left(-2\delta\Delta R_{ij}\right)\exp\left[-(\varepsilon_j - \dot{\varepsilon}_i)/kT\right], \quad \varepsilon_j > \varepsilon_i \tag{19.4a}$$

$$\nu_{ij} = \nu_0 \exp\left(-2\delta\Delta R_{ij}\right), \quad \varepsilon_j < \varepsilon_i \tag{19.4b}$$

where
ν_0 is the frequency prefactor
δ is an inverse wave function decay constant

For a hole transport, the downward energy jumps increase the hole's energy and thus have to be thermally activated (Equation 19.4a), while the upward energy jumps are nonactivated, and the probability of such jump does not depend on the energy difference between the sites (Equation 19.4b). The key parameter that incorporates the charge-transfer rates given by Equation 19.4 and characterizes charge transport in a material is charge carrier mobility μ. In the disorder formalism, the charge carrier mobility at moderate electric fields (when tunneling effects can be neglected) is described by the following relation:

$$\mu = \mu_0 \exp\left[C\left[(\sigma/k_B T)^2 - \Sigma^2\right]\sqrt{E} - (2\sigma/3k_B T)^2\right], \tag{19.5}$$

where
μ_0 is the prefactor
C is the constant
E is the electric field
k_B is the Boltzmann constant
T is the temperature
σ and Σ are important parameters of the formalism characterizing the diagonal (energy) disorder and the off-diagonal (positional) disorder, respectively

The presence of electric field lowers the effective energy barrier for the hole downward (thermally activated) jumps, which is one of the factors that contribute to electric field-dependent charge carrier mobility in Equation 19.5.

Although the disorder formalism outlined above was successfully applied to many PR polymers and organic glasses, in some cases, other models of charge transport, such as polaronic models [5], may be applicable. These models utilize Marcus's charge-transfer rates [7] that differ from Miller–Abrahams expressions (Equation 19.4) and result in expression for charge carrier mobility different from that of Equation 19.5.

19.2.1.3 Charge Trapping and Detrapping

Charge trapping and detrapping processes can be schematically presented as

$$T + \text{free hole} \underset{\beta}{\overset{\gamma_T}{\rightleftarrows}} T^+,$$

where all symbols are defined in Equation 19.2 and shown in Figure 19.2. The trapping rate γ_T depends on the charge carrier mobility μ and the carrier capture cross-section s, and most of the electric field dependence of the trapping rate originates from that of the mobility. Although charge trapping in PR materials is very important, the exact mechanism of trapping process is not well understood. For example, the dependence of the trapping rate on the trap depth ($\mathcal{E}_T$ in Figure 19.2) is not known at present.

The detrapping rate is determined by the height of the energy barrier the hole must overcome (i.e., trap depth $\mathcal{E}_T$ in Figure 19.2), and the overlap integral, so that a form similar to that of Equation 19.4a, with $\mathcal{E}_T$ replacing $\mathcal{E}_j - \mathcal{E}_i$, can be assumed.

19.2.1.4 Charge Carrier Recombination

The last process contributing to the molecular picture of the space-charge field formation (Figure 19.2) is the recombination of a free hole with an ionized acceptor, schematically presented as

$$\text{free hole} + N_A^i \xrightarrow{\gamma} N_A,$$

Theoretical approach most commonly used to describe this process is the Langevin treatment, which treats recombination as a random process, assuming that the carriers are produced statistically independent of each other and that the mean free path (l) of the carriers is less than the Coulomb radius of capture (r_C) of one carrier (a free hole in Figure 19.2) by a counterion of the opposite sign (an ionized acceptor N_A^i in Figure 19.2):

$$l < r_C = \frac{e^2}{4\pi\varepsilon_0\varepsilon k_B T},$$

where
ε is the dielectric constant
$k_B T$ is the thermal energy

Similar to the trapping rate, the recombination rate γ depends on the charge carrier mobility μ and the capture cross-section s. However, since trapping involves interaction between a charge carrier (hole) and a neutral trap, while recombination occurs between two oppositely charged particles (a hole and ionized acceptor), the capture cross-section in the case of recombination is typically higher than that for trapping.

The Langevin recombination rate is expressed in terms of charge carrier mobility μ as follows [6]:

$$\gamma = \frac{e\mu}{\varepsilon_0\varepsilon}. \tag{19.6}$$

However, Equation 19.6 was derived under the assumption of electric field-independent mobility, which is not valid for polymers. In polymers, recombination rate is a function of electric field and may depart from a simple form of Equation 19.6.

As it is clear from the discussion above, all processes that contribute to space-charge field formation in PR organic materials are complicated and depend on multiple parameters. In general, Equation 19.2 cannot be solved analytically. However, numerical solutions have been obtained [3]. They provide a valuable insight into space-charge field formation and erasure processes. In particular, the dynamics of the space-charge field formation are strongly dependent on the charge photogeneration efficiency (s) and carrier mobility (μ), and therefore these key parameters need to be maximized for applications that require fast hologram writing. If long-grating dark decay is needed (for example, in data storage applications), the detrapping rate (β) has to be made slow for a longer space-charge field retention. If dynamics are not a priority, but high steady-state space-charge field is required, it is desirable to have a high trap density in the material. The main difficulty in the PR materials design is that in many cases, there is a trade-off between steady-state and dynamic performance and therefore materials need to be optimized for a particular application. Numerical modeling of the PR performance using Equation 19.2 can assist in that.

Due to the complexity of the processes contributing to space-charge field formation in PR organic materials, theoretical model that describes PR performance keeps evolving. Some modifications of the model that include different kinds of traps (which are characterized by different trap energies ε_T in Figure 19.2) taken into account, possibility of optical excitation of trapped charge in addition to a thermal one, etc. has been proposed [3].

19.2.2 Refractive Index Grating Formation: Orientational Enhancement

The last step in the PR grating formation in organic materials (step 4 in Figure 19.1) is concerned with the refractive index change in response to the total electric field, which includes the contribution of both the static electric field $\mathbf{E_0}$ and the space-charge field $\mathbf{E_{sc}}$. This is the essence of the orientational enhancement (OE) effect [8]. Figure 19.4 illustrates the OE effect schematically. To take advantage of this effect, a PR material should contain polar molecules (e.g., NLO chromophores) that are able to reorient in the electric field and have low glass transition temperature (T_g) to facilitate the reorientation. When no electric field is applied to the sample, the polar molecules are randomly oriented (Figure 19.4a). As an external field $\mathbf{E_0}$ is applied (Figure 19.4b), the molecules reorient and align along the electric field (or, in other words, the material is poled in the electric field). During the PR grating formation, the interfering light beams 1 and 2 produce a sinusoidally varying space-charge field $\mathbf{E_{sc}}$ by the mechanisms described in Section 19.2.1. The uniform external field $\mathbf{E_0}$ will vectorially add to $\mathbf{E_{sc}}$ to produce a total local field $\mathbf{E}$. As the NLO chromophores have orientational mobility due to the low T_g, a spatially periodic orientational pattern is produced because the electric field orients the molecules by virtue of their ground-state dipole moment (μ_g). Figure 19.4c illustrates the local order parameter for the total field consisting of $\mathbf{E_{sc}} + \mathbf{E_0}$ (for illustration purposes, the order parameter shown is exaggerated). In PR materials exhibiting the OE effect, the refractive index change (Δn) in response to the total electric field ($\mathbf{E} = \mathbf{E_{sc}} + \mathbf{E_0}$) is a sum of the birefringent (Δn_{BR})

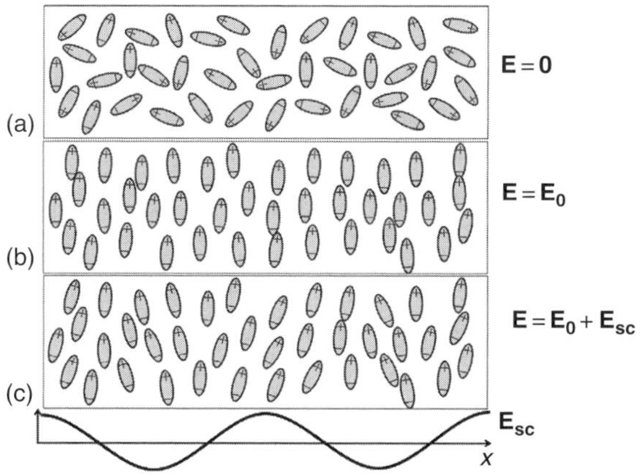

FIGURE 19.4 Chromophore orientation in the electric field: (a) no electric field; (b) electric field $\mathbf{E_0}$ is applied; (c) spatially modulated space-charge field $\mathbf{E_{sc}}$ is added to the applied field. (Reprinted from Ostroverkhova, O. and Moerner, W.E., *Chem. Rev.*, 104, 3267, 2004. With permission.)

and electro-optic (Δn_{EO}) contributions ($\Delta n = \Delta n_{BR} + \Delta n_{EO}$). These contributions rely on the molecular polarizability anisotropy ($\Delta\alpha = \alpha_{\parallel} - \alpha_{\perp}$, where parallel and perpendicular refer to the molecular axis) and hyperpolarizability (β_0) that lead to electric field-induced changes in linear ($\chi^{(1)}$) and second-order ($\chi^{(2)}$) susceptibilities [9], respectively, and are given by [8]

$$\Delta n_{BR} \sim (1/2n)\Delta\chi^{(1)} \equiv C_{BR}E^2, \quad \Delta n_{EO} \sim (1/2n)\Delta\chi^{(2)}E \equiv C_{EO}E^2, \tag{19.7}$$

$$C_{BR} = \frac{2Nf_\infty \Delta\alpha}{45}\left(\frac{\mu_g}{k_B T}\right)^2, \quad C_{EO} = \frac{Nf_0 f_\infty^2}{5}\frac{\mu_g \beta_0}{k_B T},$$

where
- N is the dipole concentration
- k_B is the Boltzmann constant
- T is the temperature
- n is the refractive index
- f_0 and f_∞ are the local field factors given by $f_0 = \varepsilon(n^2+2)/(2\varepsilon + n^2)$ and $f_\infty = (n^2+2)/3$, respectively, where ε is a static dielectric constant

As seen from Equation 19.7, in order to maximize a refractive index change in the electric field, it is important to implement a high concentration (N) of chromophores with large ground-state dipole moment (μ_g), polarizability anisotropy ($\Delta\alpha$), and hyperpolarizability (β_0) in a PR organic material.

To summarize Section 19.2, there are many different processes contributing into the PR effect in organic materials, which make it both challenging and exciting to understand the physical mechanisms that play an important role in the PR performance. Because of the complexity of the photorefractivity in organics, PR material design represents an optimization problem, which has to target all key parameters and address requirements of a particular application.

19.3 EXPERIMENTAL TECHNIQUES

In this section, two most common geometries, e.g., 2BC and four-wave mixing (FWM), which are used for assessing the PR performance of the materials and utilized in applications are described.

19.3.1 Two-Beam Coupling

The two-beam coupling experiment is a crucial experiment to perform when a new material is tested for PR performance, since the presence of asymmetric steady-state beam coupling is a distinct feature of the PR effect arising from nonlocality ($\Phi \neq 0$) of the PR grating (Section 19.1). The geometry of the 2BC experiment is shown in Figure 19.5. A typical sample consists of two conductive but transparent indium tin oxide (ITO)-coated glass slides with a PR organic (e.g., polymer composite) film of 30–100 μm thickness in between. Optical beams 1 (probe or signal) and 2 (pump) are incident at angles θ_1 and θ_2, respectively, and interfere in the PR material, creating a nonlocal ($\Phi \neq 0$) diffraction grating. Then the same beams 1 and 2 partially diffract from the grating they have just created (beams 1′ and 2′ in the inset of Figure 19.5). Due to nonlocality of the grating, one diffracted beam (e.g., beam 1′) interferes destructively with its companion beam 2, while the other diffracted beam 2′ interferes constructively with beam 1. As a result, the beam 1 is amplified (energy gain), while the beam 2 is attenuated (energy loss). It should be emphasized that it is important that the energy transfer in PR materials persists in steady state because transient asymmetric energy exchange is known to occur in non-PR materials with local response (due to photochemistry, NLO susceptibility $\chi^{(3)}$, thermal modulation of the refractive index, etc.). As shown in Figure 19.5, the beams 1 and 2 are incident at angles (θ_1 and θ_2) on the same side of the sample surface normal. This choice of experimental geometry is governed by several factors. First, all the processes in organics, which

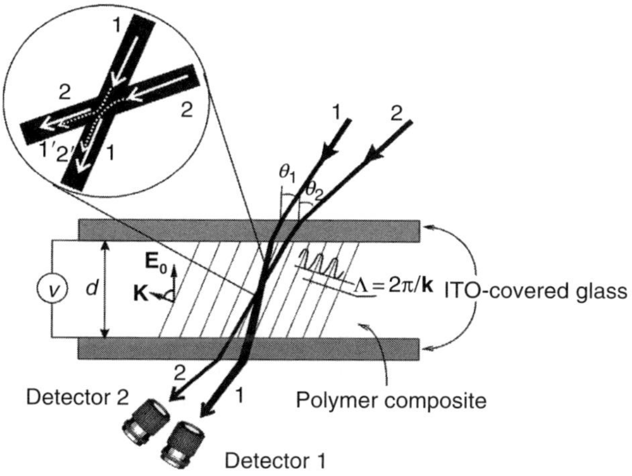

FIGURE 19.5 Experimental configuration used in 2BC geometry. (Reprinted from Ostroverkhova, O. and Moerner, W.E., *Chem. Rev.*, 104, 3267, 2004. With permission.)

are responsible for space-charge field build-up are strongly electric field-dependent. Therefore, to assist in charge transport along the grating vector **K**, a large component of external electric field $\mathbf{E}_0$ in the direction of the vector **K** is needed. Second, since the NLO chromophores that are part of the PR composite are aligned in the direction of applied field, it is necessary to provide a nonzero component of electro-optic response along the grating vector **K** [8].

The period (Λ) of the grating produced in the geometry (Figure 19.5) is given by

$$\Lambda = \frac{\lambda}{2n\sin\left[\left(\theta_{2,\mathrm{int}} - \theta_{1,\mathrm{int}}\right)/2\right]}, \tag{19.8}$$

where

λ is the wavelength of light in a vacuum

n is the refractive index of the material

$\theta_{1,\mathrm{int}}$ and $\theta_{2,\mathrm{int}}$ are the internal angle of incidence of the two writing beams relative to the sample normal, which are related to the respective external angles θ_1 and θ_2 (Figure 19.5) by the Snell law: $\sin\theta_i/\sin\theta_{i,\mathrm{int}} = n$ $(i = 1,2)$. For normally accessible angles between light beams and visible optical wavelengths, Λ is in the range 0.3–20 μm

The theoretical description of the 2BC effect involves solving coupled-wave equations that govern the interaction of two beams of light in a NLO material. The details of this description can be found in Refs. [1,4,9]. Here, only the solution of these equations, which is used for extracting material parameters from experimentally measured quantities, is presented. Denoting beam 2 as the pump, the intensities of the signal beam (beam 1) (I_1(out)) and the pump beam (I_2(out)) may be written as follows:

$$I_1(\mathrm{out}) = \frac{I_0}{1+\beta_\mathrm{p}\exp(-\Gamma L)}, \quad I_2(\mathrm{out}) = \frac{\beta_\mathrm{p} I_0}{\beta_\mathrm{p}+\exp(\Gamma L)}, \tag{19.9}$$

where

$\beta_\mathrm{p} = I_2(\mathrm{in})/I_1(\mathrm{in})$ is the initial beam ratio (in the absence of coupling)

$I_0 = I_1 + I_2$ is the total intensity

$L = d/\cos\theta_{1,\mathrm{int}}$ is the interaction length, where d is a sample thickness, and the 2BC gain coefficient is

$$\Gamma = \frac{4\pi}{\lambda}\frac{\Delta n}{m}\sin\Phi, \tag{19.10}$$

where
- λ is the wavelength
- Δn is the refractive index modulation
- m is the modulation depth (or contrast) of the interference pattern defined as $m = 2\sqrt{\beta_p}/(1+\beta_p)$

Using Equations 19.9 and 19.10, the gain coefficient can be determined from experimentally measured intensities as follows: $\Gamma = \ln(\beta_p I_1(\text{out})/I_2(\text{out}))/L$. Note that in the undepleted pump regime ($\beta_p >> 1$), the gain factor γ_0 is defined as $\gamma_0 = I_1(\text{out})/I_1(\text{in})$ simplifies to $\gamma_0 = \exp(\Gamma L)$, i.e., the intensity of the signal beam grows exponentially with the interaction length L.

Many applications require not only the presence of the gain in the material, but that of the net gain, which is achieved in the material when the gain coefficient Γ exceeds the absorption coefficient α. Therefore, when comparing the 2BC properties of the materials, it is important to consider the net-gain coefficient $\Gamma - \alpha$, rather than the gain coefficient Γ.

From Equation 19.10, it is clear that a phase shift of $\Phi = 90°$ leads to the optimum energy transfer, whereas no energy coupling occurs for an in-phase grating, $\Phi = 0°$. Equation 19.10 provides mathematical description of the fact that a nonzero phase shift is necessary for the asymmetric beam coupling that leads to an energy transfer between the light beams, which is a characteristic feature of the PR materials (Section 19.1). The direction of energy transfer is determined by the sign of mobile charge carriers and of electro-optic response. Since most PR materials are poled in situ during the experiment, the sign of the electro-optic response depends on the applied electric field. Thus, the energy transfer direction can be reversed by changing the polarity of the electric field. This is in contrast to the energy transfer in PR inorganic crystals. There, the direction of the energy transfer is fixed since the electric field dependence of the refractive index relies on the electro-optic tensor components, which are fixed for a given crystal and are determined by the symmetry group of the crystal and orientation of its axes [1].

Figure 19.6 shows the ratio $I_i(\text{out})/I_i(\text{in})$ for the beams 1 and 2 ($i = 1$ and 2, respectively) obtained in the 2BC experiment in high-performance PR organic composite. In this experiment, one beam (beam 1) at a wavelength of 830 nm is incident upon a sample, and the electric field of 45 V/μm is applied. At time $t = 0$, another beam (beam 2) is open with a shutter and the intensity of both beams is monitored as a function of time. From Figure 19.6, almost complete energy transfer between the beams occurs within several seconds. Calculation of the gain coefficient (Γ) from the data in Figure 19.6 is left to the reader as an exercise (Exercise Question 4).

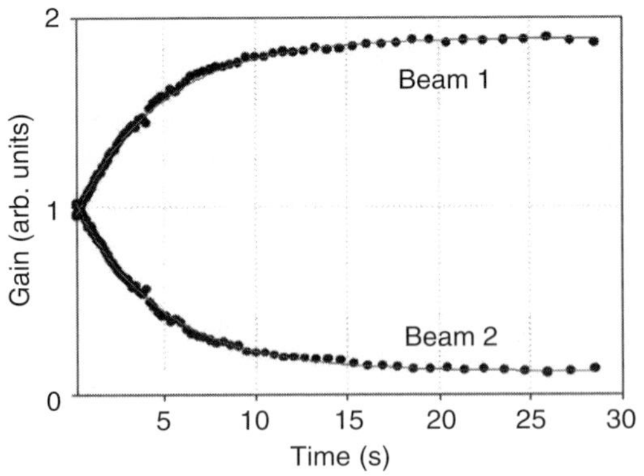

FIGURE 19.6 Typical experimental data obtained in 2BC experiment in high-performance PR organic material. Almost complete energy transfer from beam 2 to beam 1 is achieved.

19.3.2 Four-Wave Mixing

The experimental geometry for the FWM experiment is quite similar to that of 2BC—two writing beams are obliquely incident on the PR sample (Figure 19.7). The difference is that in the FWM experiment, in addition to the writing beams 1 and 2, there is also a probe (reading) beam (beam 3 in Figure 19.7) that is being partially diffracted from the grating created by the writing beams to create the fourth beam (beam 4). In the degenerate FWM geometry, which is common in PR measurements, beam 3 has the same wavelength as the writing beams and is usually chosen to be counter-propagating to one of the writing beams as this results in a most efficient diffraction, as dictated by the Bragg condition (Section 19.1 and Refs. [1,4]), and allows for background-free detection of very weak diffraction signals (beam 4). The diffracted beam intensity (i.e., that of beam 4) is typically measured as a function of time, applied electric field ($\mathbf{E}_0$), writing beam intensities, etc. Then, the diffraction efficiency (η), defined as the ratio $\eta^{\text{ext}} = I_4/I_3$ (external diffraction efficiency) or $\eta^{\text{int}} = I_4/(I_4 + I_{3'})$ (internal diffraction efficiency) is determined. In probing the grating, it is important that beam 3 does not affect the grating or interact with the writing beams. This can be assured by making the probe beam much weaker than the pump beams or by having the probe beam polarized orthogonal to the writing beams. Typically, *p*-polarized beams (i.e., the beams whose electric field vector is in the plane of incidence) experience larger diffraction efficiency due to the polarization dependence of the OE response of the material [8]. Therefore, beam 3 is typically *p*-polarized, while beams 1 and 2 are *s*-polarized (i.e., electric field vector is perpendicular to the plane of incidence) in most FWM experiments.

From the diffraction efficiency measured in geometry shown in Figure 19.7, the modulation amplitude of the refractive index (Δn) is typically obtained by using the following simple expression [10]:

$$\eta^{\text{ext}} = \exp(-\alpha L)\sin^2\left(\frac{\pi \Delta n L}{\lambda}\hat{e}_1 \cdot \hat{e}_2\right), \tag{19.11}$$

where

α is the absorption coefficient

$L = d/\sqrt{\cos\theta_1 \cos\theta_2}$ is the effective interaction length

$\hat{e}_1$ and $\hat{e}_2$ are unit vectors along the electric field of the incident and diffracted beams, respectively

For configuration with *p*-polarized readout (i.e., beam 3), $\hat{e}_1 \cdot \hat{e}_2 = \cos(\theta_2 - \theta_1)$, while for *s*-polarized readout the dot product is unity. Equation 19.11 is derived in the approximation of a thick (volume)

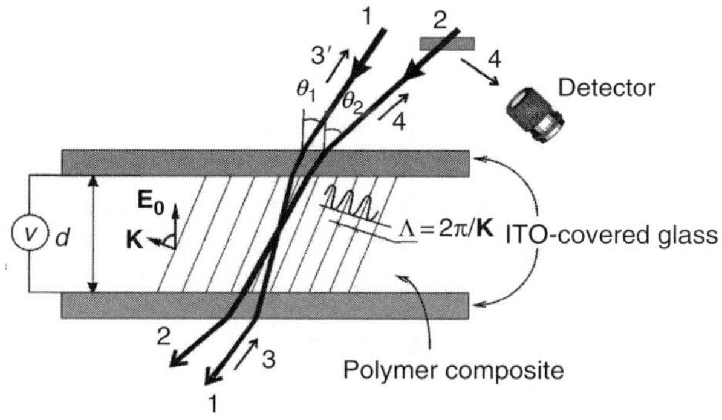

FIGURE 19.7 Experimental configuration used in FWM geometry. (Reprinted from Ostroverkhova, O. and Moerner, W.E., *Chem. Rev.*, 104, 3267, 2004. With permission.)

grating, which implies that diffraction orders higher than the first one are neglected [1,4]. It is conventional to introduce parameter Q as a measure of grating thickness as follows [10]:

$$Q = \frac{2\pi\lambda d}{n\Lambda^2}, \tag{19.12}$$

where
- λ is the wavelength of light
- d is the sample thickness
- n is the refractive index
- Λ is the grating period

If $Q \gg 1$, the grating is regarded as thick—this is a Bragg (volume) hologram, with the first-order diffraction dominating. In a typical FWM experiment in, for example, PR polymer composites, $\Lambda = 2$ µm, $n = 1.6$, $d = 100$ µm, $\lambda = 633$ nm, which yield $Q \approx 62$, and thus, the condition for the thick grating is satisfied. If $Q < 1$, which is often the case for PR experiments in liquid crystalline materials, the grating is thin, there are multiple diffraction orders, and this is Raman–Nath (surface) grating. In the Raman–Nath regime, Equation 19.11 is not applicable.

In addition to measurements of steady-state diffraction efficiency which is an important parameter for applications, the FWM experiment is widely used for temporal studies (e.g., formation and erasure) of the PR grating. Electric field and intensity dependence of the PR dynamics can yield much information about the material parameters. Furthermore, the FWM experiment is ideally suited to measure the dark decay of gratings, i.e., the decay after both writing beams are turned off. This parameter is especially of interest for data-storage applications for which long dark-lifetimes are desirable. To ensure that the probe beam does not itself erase the grating, the illumination of the grating can be made negligibly small by only probing the grating intermittently.

19.4 MATERIALS DESIGN

The main classes of PR organic materials include polymer composites, small molecular weight glasses, fully functionalized polymers, polymer-dispersed liquid crystals (PDLCs), liquid crystals, and hybrid organic–inorganic composites. Best performing PR organic materials exhibit gain coefficients (Γ) of 200–400 cm^{-1}, nearly 100% diffraction efficiencies (η), and PR grating formation times on the order of several milliseconds. A complete description of the strategies of the PR materials design is beyond the scope of this chapter and can be found in Ref. [3]. Here, examples of a few types of PR organic materials are given and the most common design strategies outlined.

19.4.1 POLYMER COMPOSITES

Design of a high-performance PR polymer composite relies on combining various components to perform certain functionality in the composite (Figure 19.8a). For example, a sensitizer is typically added to promote charge photogeneration (step 1 in Section 19.1), a photoconductive polymer provides charge transport sites (step 2), a NLO chromophore is needed for an electric field-dependent refractive index (step 4 in Section 19.1), and in most cases, a plasticizer is needed to lower the T_g of the composite and promote chromophore orientation in the electric field (OE effect, Section 19.2.2). Note that no special constituent is needed to provide trapping sites for the charge trapping process (step 3), since NLO chromophores as well as voids, defects, and impurities in polymers can serve as charge traps [3]. Examples of constituents of PR polymer composites are given in Table 19.1. Typically, a PR polymer composite contains ~40–60 wt% of a photoconductive polymer (e.g., PVK or PPV derivatives in Table 19.1) to provide sufficient density of transport sites, ~25–35 wt% of a NLO chromophore (e.g., AODCST

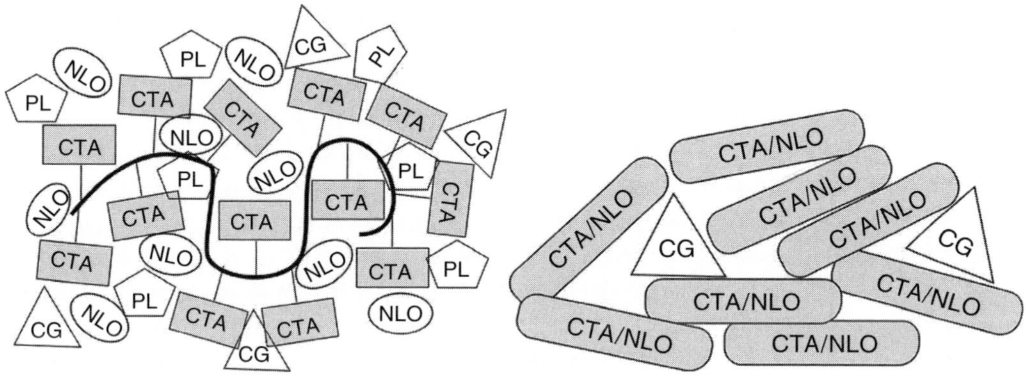

(a) (b)

FIGURE 19.8 Typical content of (a) PR polymer composite and (b) PR organic amorphous glass. CTA, charge transport agent; CG, charge generator; NLO, nonlinear optical chromophore; PL, plasticizer.

in Table 19.1) to ensure sufficient electro-optic response, ~15–30 wt% of a plasticizer (e.g., BBP in Table 19.1) to facilitate chromophore orientation by lowering T_g of the material, and finally, up to 1 wt% of a sensitizer (e.g., C_{60}) to assist in charge photogeneration. PR parameters obtained in selected high-performance PR polymer composites are summarized in Table 19.2.

The advantage of polymer composites is the ability to tune their PR properties by varying the concentration and type of constituents. However, as many components are combined in the composite, there is an issue of phase separation and crystallization that reduce shelf life of the device. Also, an addition of a plasticizer to lower T_g and therefore enhance chromophore orientation also increases the inert volume, which worsens photoconductive and electro-optic properties.

19.4.2 Organic Amorphous Glasses

Glass-forming molecules utilized in PR organic amorphous glasses serve as charge-transporting agents and NLO chromophores at the same time (Figure 19.8b). Moreover, many molecules form glasses with T_g around room temperature, thus eliminating the need for a plasticizer. Examples of such multifunctional glass-forming molecules are given in Figure 19.9. The only dopant which is typically added is a small amount (up to 1 wt%) of a sensitizer to promote charge photogeneration. Examples of PR performance of organic glasses are given in Table 19.2. The advantage of organic glasses is that the inert volume is minimized, i.e., the density of NLO moieties and charge transporting sites is the maximal achievable due to tight packing of molecules in the glass, yet there is no phase separation problem. The disadvantage is that the easy tunability of the properties is reduced to synthetic modifications of the glass-forming molecule.

19.4.3 Fully Functionalized Polymers

Several types of fully functionalized PR polymers have been designed and explored. Common examples are illustrated in Figure 19.10. In Figure 19.10a, all required moieties (charge generator, transport, and NLO sites) are incorporated as side chains into the inert polymer backbone. In another example (Figure 19.10b), a polymer backbone itself is photoconductive, and NLO moieties are attached as side groups. The advantage of fully functionalized polymers is that they eliminate the problems of phase separation and crystallization, and therefore could lead to a better thermal stability and longer shelf life for devices. Disadvantages include complicated synthetic procedures and reduced tunability of properties.

TABLE 19.1
Examples of Constituents of High-Performance PR Polymer Composites

Sensitizers

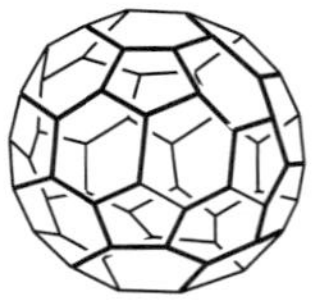

C_{60}

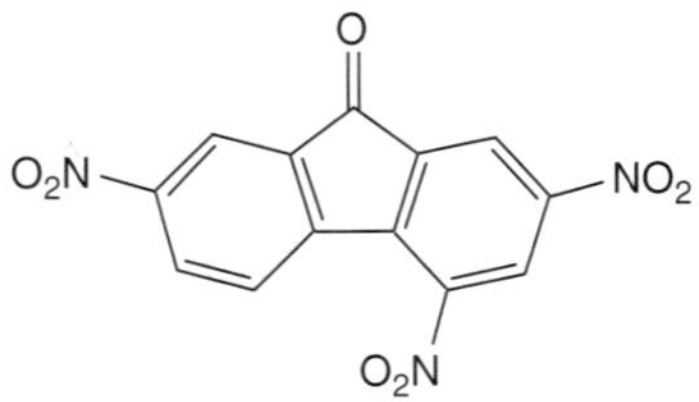

TNF

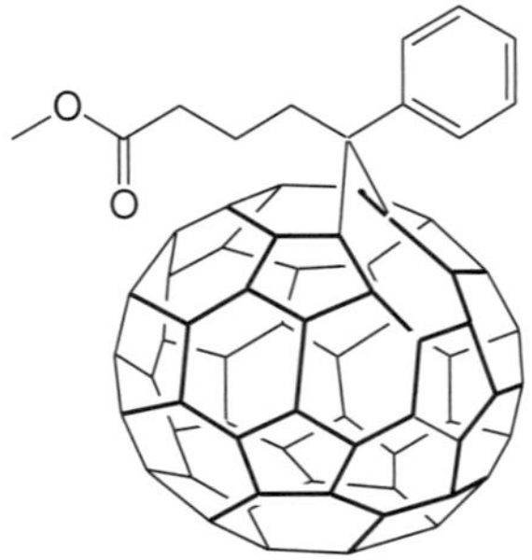

[6,6]PCBM

Photoconductive Polymers

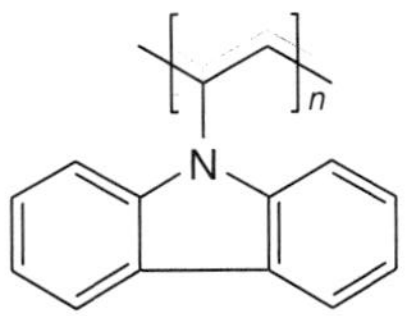

PVK

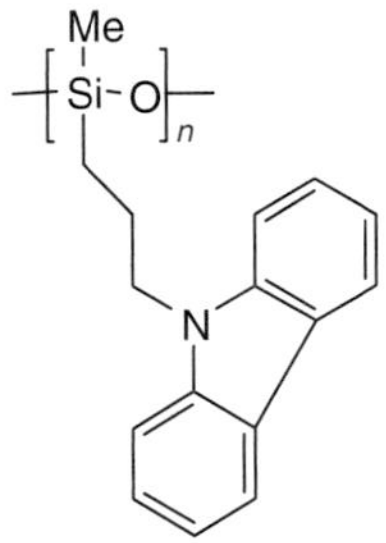

PSX

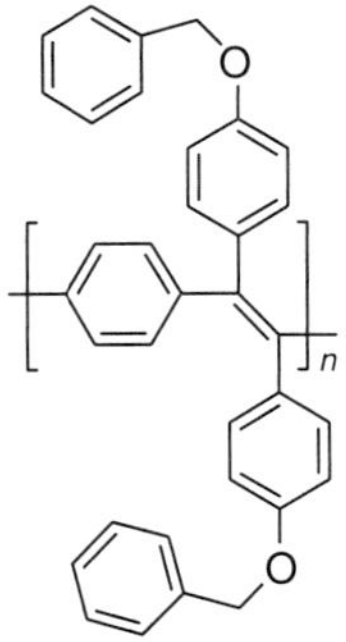

DBOP-PPV

Plasticizers

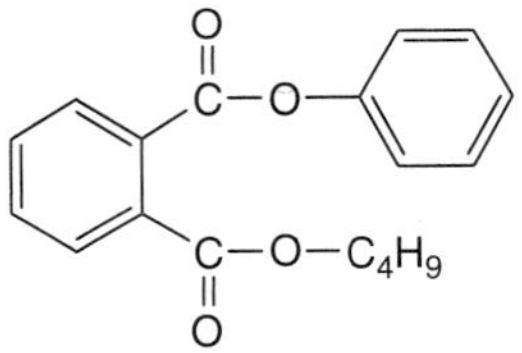

BBP

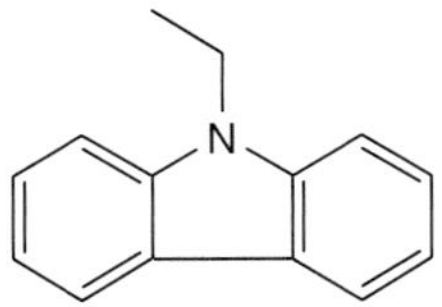

ECZ

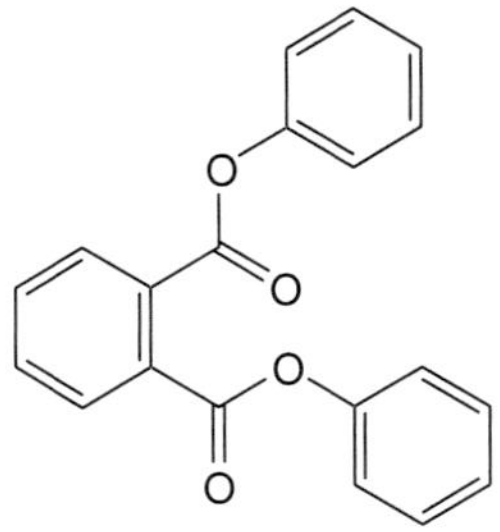

DPP

NLO Chromophores

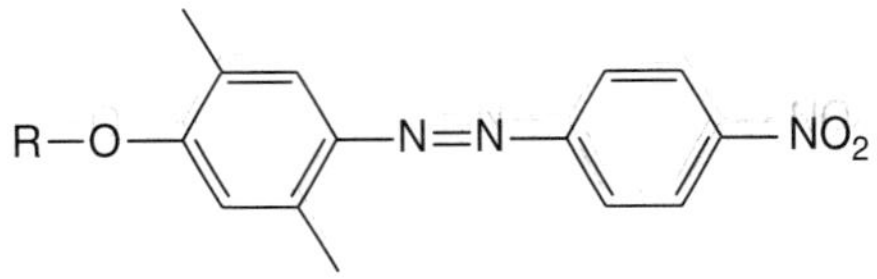

DMNPAA: $R = CH_3$

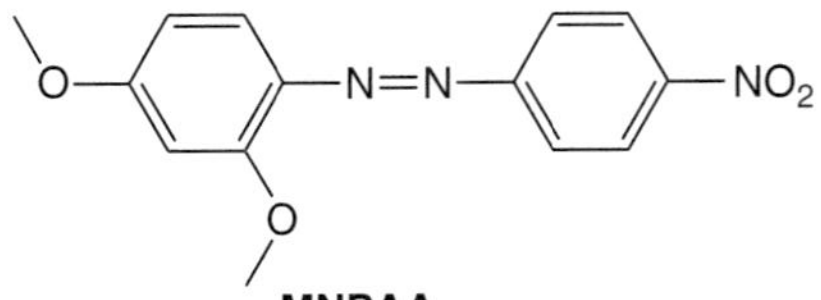

MNPAA

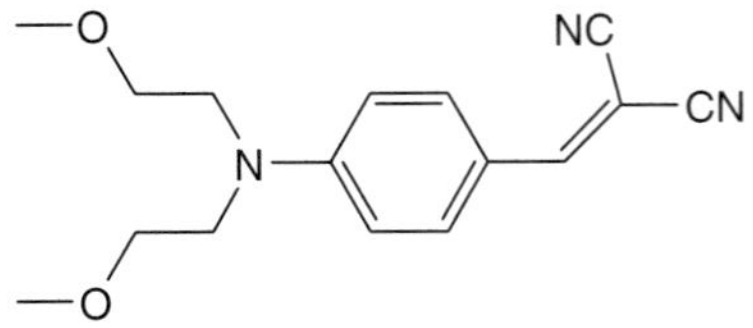

AODCST

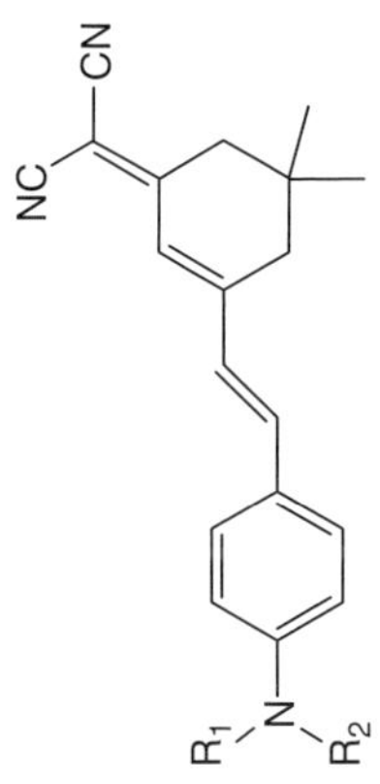

Ch C: $R_1 = CH_3$,
R_2 = 2-hydroxyethyl

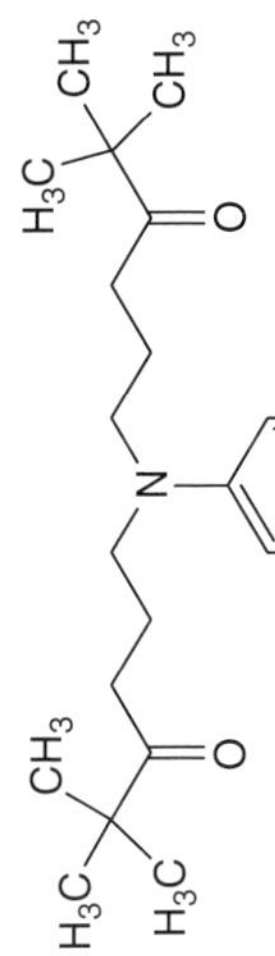

Stilbene A

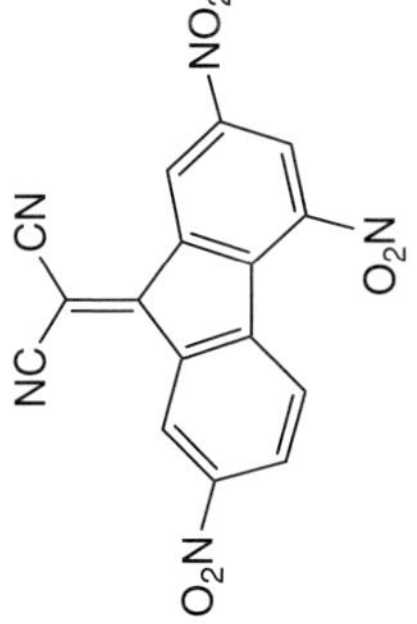

TNFM

TABLE 19.2
PR Properties of High Performance Organic Materials

Composite (Concentration of Constituents wt%)	α (cm^{-1})	d (μm)	λ (nm)	Γ (cm^{-1}) (E, V/μm)	τ_g^{-1} (s^{-1})	η_{max} (%) (E, V/μm)	τ_{FWM}^{-1} (s^{-1})	Δn, 10^{-3} (E, V/μm)
Polymer composites								
PVK/AODCST/BBP/C_{60} (49.5/35/15/0.5)	9	80	647	235 (100)	200			
PSX/Stilbene A/TNF (51/48/1)		40	670	53 (100)		100^{int}, 60^{ext} (70)	0.017	10.5 (100)
DBOP-PPV/DMNPAA/MNPAA/DPP /PCBM (52/20/20/5/3)	34	105	633			90^{int} (62)	1.7	2.6 (62)
PVK/Ch C/ECZ/TNFM (42/28.2/28.5/1.3)	44	125	780	230 (68)		95^{int} (52)		
Amorphous glasses								
DCDHF-6/C_{60} (99.5/0.5)	12.7	70	676	240 (30)	0.6		0.41	
Methine A	1.64		690	118 (89)		74 (53)		5.6 (53)
ATOP-4/TNFM (99/1)	111	105	790	130 (28)		85^{int}, 32^{ext} (10.5)		10.7 (28)

Columns represent: (1) composition (concentration of the constituents in weight percent, unless stated otherwise); (2) absorption coefficient α; (3) sample thickness d; (4) wavelength used in the PR experiments λ; (5) 2BC gain coefficient, measured with p-polarized writing beams, Γ (electric field E, at which the indicated Γ was obtained); (6) PR response time τ_g^{-1} obtained from fits to 2BC dynamics; (7) maximal diffraction efficiency η_{max}, measured with p-polarized probe and s-polarized writing beams. External (η^{ext}) or internal (η^{int}) diffraction efficiency is indicated (electric field E, at which the indicated η was obtained); (8) PR speed τ_{FWM}^{-1} obtained from fits to either formation or erasure of the PR grating measured in the FWM experiment; (9) refractive index modulation Δn, calculated from the diffraction efficiency (electric field E, at which the indicated refractive index modulation was obtained). More details on these data can be found in Ostroverkhova, O. and Moerner, W.E., *Chem. Rev.*, 104, 3267, 2004.

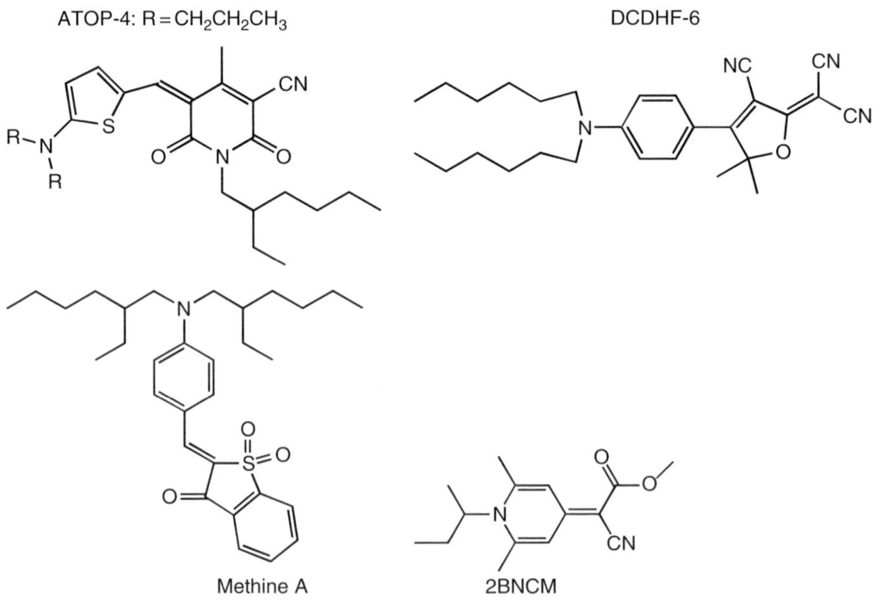

FIGURE 19.9 High-performance glass-forming molecules.

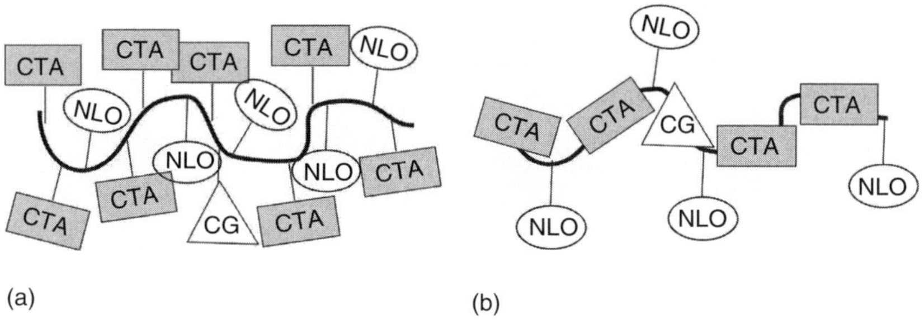

FIGURE 19.10 Examples of design of fully functionalized PR polymers: (a) inert polymer backbone with functional side groups; (b) photoconductive polymer backbone with NLO side groups. CTA, charge transport agent; CG, charge generator; NLO, nonlinear optical chromophore.

19.4.4 Polymer-Dispersed Liquid Crystals

Polymer-dispersed liquid crystals contain high concentrations of polymer and liquid crystalline (LC) molecules that phase separate into droplets (Figure 19.11). In photorefractive PDLCs, the polymer doped with a small amount of sensitizer typically provides the photoconductive properties needed for the space-charge field formation (steps 1–3 in Section 19.1), while the LC droplets provide orientational nonlinearity (i.e., electric field-dependent refractive index) needed for the refractive index modulation (step 4). The advantage of PDLCs compared with traditional polymer composites is that LC molecules in the droplets can be reoriented with much lower electric fields than those used in polymer composites. The weak points include high losses due to scattering, which prevent high net-gain, and slow PR dynamics due to low charge carrier mobility.

19.5 APPLICATIONS OF PHOTOREFRACTIVE MATERIALS

Applications of PR materials include holographic data storage, image processing (amplifiers, novelty filters, and optical correlators), imaging through turbid medium, phase conjugation, optical limiting, optical computing, noncontact surface defect control, optical waveguiding, etc. [1]. In this section, several examples are considered.

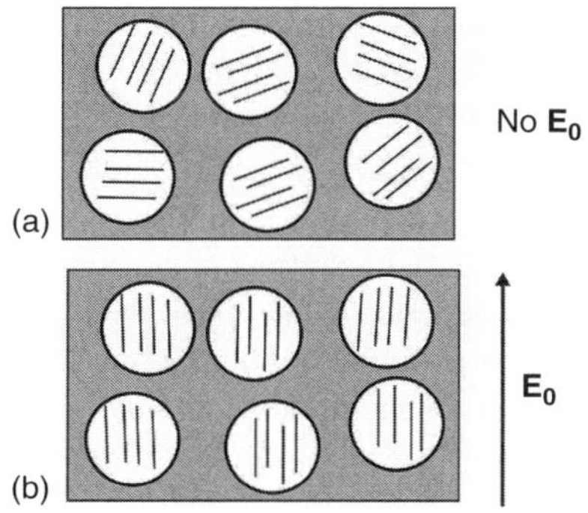

FIGURE 19.11 Schematics of the PDLCs: liquid crystal droplets in a polymer matrix. (a) No electric field is applied, and the liquid crystal molecules in different droplets are randomly oriented; (b) in the presence of the applied electric field $\mathbf{E}_0$, the molecules within the droplets align.

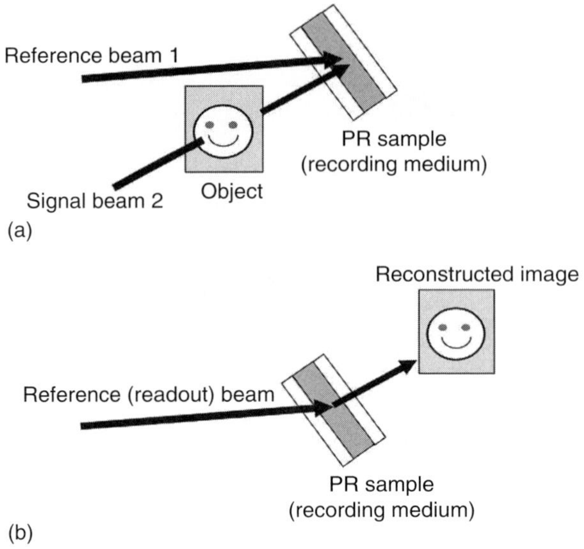

FIGURE 19.12 Holographic data storage using PR materials. (a) Hologram recording; (b) hologram retrieving.

19.5.1 Data Storage

Reversible holographic high-density storage is one of the first application proposed for PR materials [1]. In two-dimensional (2D) holographic storage, a reference beam and a signal beam containing the pattern to be stored (beams 1 and 2, respectively, in Figure 19.12a) overlap in a PR material to form a grating. The hologram is retrieved by shining the reference beam through the material at the recording angle (Figure 19.12b). Using angle multiplexing, many pages of data can be stored in the same volume of the material by changing the angle of the reference beam for each recording. Specific pages can be accessed by selecting the corresponding angle for the reference beam. In addition, wavelength multiplexing can be used by changing the wavelength of the reference beam for each recording.

19.5.2 Image Processing

19.5.2.1 Optical Phase Conjugation

Optical phase conjugation is one of the most important phenomena observed in PR materials. In this section, a concept of PC illustrated by several examples is briefly discussed. Full theoretical treatment of the PC can be found in Ref. [4]. An example of utilizing PC in a phase conjugate mirror is illustrated in Figure 19.13. Figure 19.13a shows a plane wave (described by plane wave front) that passes through an inhomogeneous medium with refractive index $n(x,y,z)$, which distorts the wave front. If the wave is now reflected backwards by an ordinary mirror and again passes through the medium, the distortion of the wave front accumulates (Figure 19.13b). In contrast, if the wave is reflected from a phase-conjugate mirror creating the so-called phase-conjugate replica, the distortion cancels, and the wave front is reconstructed (Figure 19.13c) [4]. The potential applications include the transmission of undistorted images through optical fibers (or the atmosphere), lensless imaging down to submicrometer-size resolution, optical tracking of objects, phase locking of lasers, refreshing of holograms for long-term optical storage, optical interferometry, beam clean-up, and image processing. In a classic geometry, phase-conjugate occurs when two counter-propagating pump beams (e.g., beams 1 and 2 in Figure 19.14) overlap in a PR material and create a phase-conjugate replica (beam 4) of a third incident beam

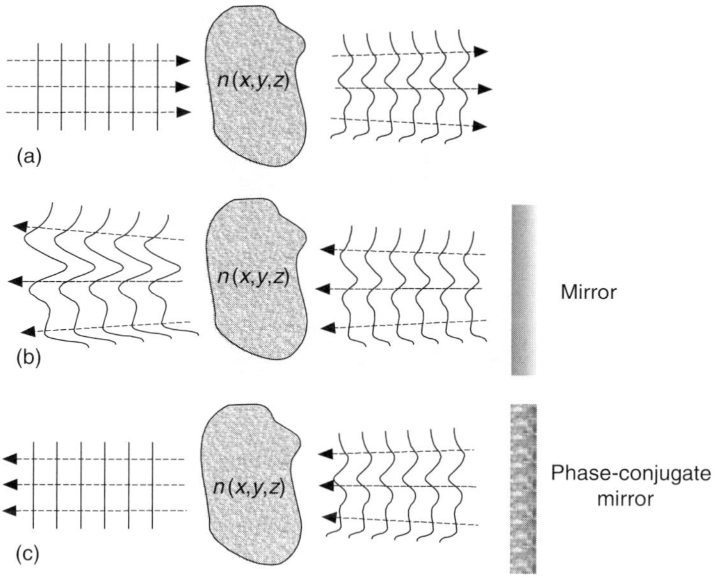

FIGURE 19.13 Concept of a phase-conjugate mirror: (a) plane wave front distortion in a light beam passing through a transparent but inhomogeneous medium; (b) back-reflected wave from an ordinary mirror—wave front distortion accumulates on a return pass; (c) back-reflected wave from a PC mirror—plane wave front is reconstructed on a return pass.

(beam 3). Example of using PC for optical beam clean-up is illustrated in Figure 19.14 [11]. The pump beams 1 and 2 are mixed in the PR material with the signal beam 3, which contains information about the object, but is heavily distorted by an aberrator (that could be due to turbulence, refractive index inhomogeneities, etc.). The distorted image is recorded by a CCD camera (CCD1 in Figure 19.14). The phase-conjugate replica of the signal beam 3 created in the PR material (beam 4) contains a cleaned-up image of the object, recorded by another CCD camera (CCD2). Figure 19.15 shows an object (Air Force resolution test chart in Figure 19.15a), its distorted image recorded by CCD1 (Figure 19.15b) and corrected (using a PR polymer composite) image recorded by CCD2 (Figure 19.15c) [11].

19.5.2.2 Image Amplification, Novelty Filtering, and Edge Enhancement

All applications considered in this section rely on the 2BC effect, which enables energy transfer from one beam to another (Section 19.3.1). As discussed in Section 19.3.1, in the 2BC experiment

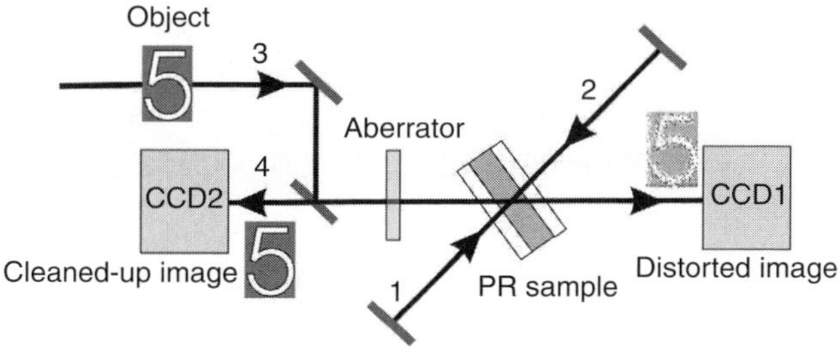

FIGURE 19.14 Experimental setup for the optical beam clean-up.

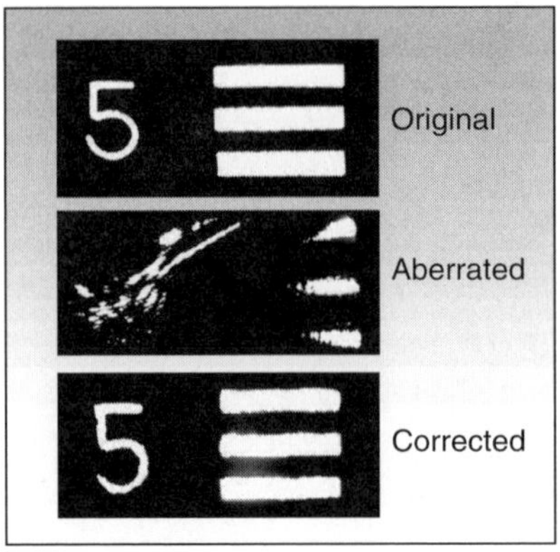

FIGURE 19.15 Correction of distorted images. (a) Original object; (b) aberrated image of the object; (c) corrected image of the object. (Images courtesy of professor R.A. Norwood. With permission.)

in the limit of high beam ratio ($\beta_p = I_{r0}/I_{s0} \gg 1$, where I_{s0} (I_{r0}) is the incident signal [reference] beam intensity), the intensity of the transmitted signal beam I_s is given by $I_s = I_{s0}\exp(\Gamma L)$, where Γ is the 2BC gain coefficient and L is the interaction length in the sample. Therefore, if in a conventional 2BC geometry (Figure 19.5), an object is inserted in the signal beam path, its image will be enhanced after passage through the sample due to the energy transfer from the reference beam. Figure 19.16 illustrates the results of such an experiment in the high-performance polymer composite PVK/AODCST/BBP/C_{60} (Table 19.1). An Air Force resolution test chart was inserted in the signal beam path (I_{s0} = 0.5 mW/cm^2), and its image was recorded by a CCD camera (Figure 19.16a). Then, an electric field of 68 V/μm was applied and the reference beam of I_{r0} = 1 W/cm^2 was launched. After 33 ms, the image amplification factor (I_s/I_{s0}) reached 9.4 (Figure 19.16b) [12].

The identical optical setup can also be used for novelty filtering (image differentiation) experiments, in which only moving and changing objects are visualized. Due to the dynamic nature of this effect, only the objects that move with a speed faster than PR response speed of the material are

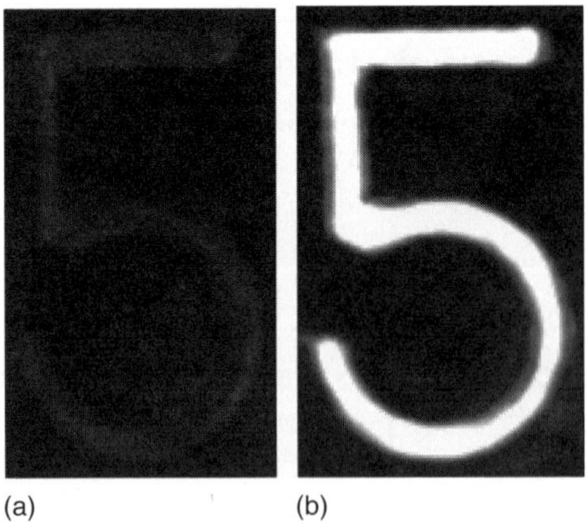

FIGURE 19.16 Amplification of image of number 5 from the Air Force resolution test chart (68 μm line width) in PR polymer composite. Frame (a), the image in the absence of the reference beam and frame (b), amplified image 33 ms later. (Reprinted from Goonesekera, A. et al., *Appl. Phys. Lett.*, 76, 3358, 2000. With permission.)

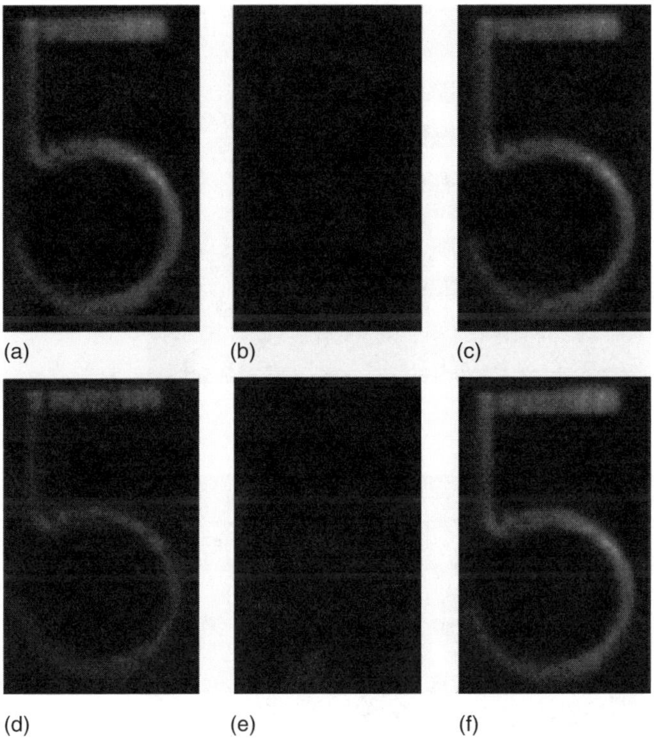

FIGURE 19.17 Demonstration of novelty filter effect using PR polymer composite. (a) Image with the reference beam off. (b) Image after the reference beam is on and steady state is reached; the output is dark. (c) After a sudden movement of the data mask, the image is visible. (d) Fading image after 1/30 s. (e) Back to a dark output after 4/30 s. (f) After the sudden movement of the data mask, the image is visible again. (Reprinted from Goonesekera, A. et al., *Appl. Phys. Lett.*, 76, 3358, 2000. With permission.)

detected. As discussed in Section 19.3.1, reversing the direction of applied electric field leads to reversed direction of energy transfer. In the setup for the novelty filtering demonstration, the polarity of the applied field was simply reversed compared to the configuration for image amplification, the electric field of 55 V/μm was applied, and the intensities of the signal I_{s0} and reference I_{r0} beams were ~0.45 mW/cm^2 and ~1 W/cm^2, respectively. The resulting effect in steady state is suppression of the transmitted image, which is eliminated when the image moves [12]. The series of images shown in Figure 19.17 demonstrates detection of a moving object by novelty filtering. In frame (a) of Figure 19.17, the reference beam was off, thus the image of number 5 was visible. When the reference beam was turned on, the image started to fade away due to the energy transfer from the signal beam to the reference beam (i.e., suppression) and appeared almost dark when steady state was reached (frame (b) of Figure 19.17). The image reappeared if the object mask was suddenly moved a fraction of a millimeter, as seen in frame (c) of Figure 19.17, and then faded back to a dark image as shown in frames (d) and (e) of Figure 19.17, respectively. If the target was moved again, the image as shown in frame (f) of Figure 19.17 appeared again and again faded away with time.

Another interesting type of image processing that involves a special type of image intensity filtering is edge enhancement. In this configuration, the image of an object is recorded and then read out in a conventional geometry (e.g., Figure 19.12). However, depending on the ratio between the signal and reference beam intensities (I_{s0}/I_{r0}), either an exact replica of the object (Figure 19.18a, $I_{s0}/I_{r0} = 0.1$) or an edge-enhanced image (Figure 19.18b, $I_{s0}/I_{r0} = 20$) is obtained [13]. Explanation of this effect is left for the reader as an exercise (Exercise Question 5).

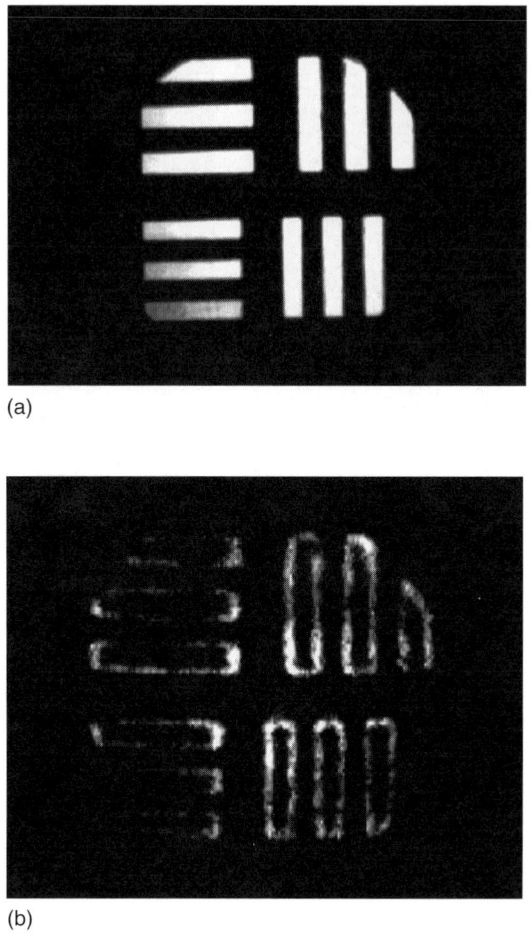

FIGURE 19.18 Demonstration of the edge enhancement effect using PR polymer composite. (a) With the ratio between the signal and reference beam intensities $I_{s0}/I_{r0} = 0.1$, the exact replica of the object is produced; (b) with the beam ratio $I_{s0}/I_{r0} = 20$, the edge enhanced image is obtained. (Reprinted from Ono, H. et al., *Appl. Phys. Lett.*, 79, 895, 2001. With permission.)

19.5.3 Optical Waveguides

Among interesting phenomena observed in PR materials are light beam self-focusing and self-defocusing, and the formation of optical spatial solitons. The processes leading to these effects are illustrated in Figures 19.4a and b and 19.19. As discussed in Section 19.2.2, in the absence of applied electric field, polar chromophores (shown as dipoles in Figures 19.4 and 19.19) in a typical PR organic thin film are randomly oriented. As the electric field, $\mathbf{E}_0$, is applied, the dipoles align (Figure 19.4b). If the light beam is sent through the PR device in the direction perpendicular to applied electric field, as shown in Figure 19.19, the light generates charge carriers that drift in the electric field, get trapped and create a space-charge field $\mathbf{E}_{sc}$ counteracting the applied field $\mathbf{E}_0$ in the illuminated part of the device. Then, the total electric field in the illuminated part is lower than that outside of the light beam, which leads to a partial relaxation of the chromophores in the illuminated portion of the device (Figure 19.19). Since in PR materials, refractive index is electric field-dependent (Section 19.2.2), a nonuniform electric field distribution along x-axis (Figure 19.19) leads to a nonuniform refractive index profile. Depending on the polarization of the beam with respect to the

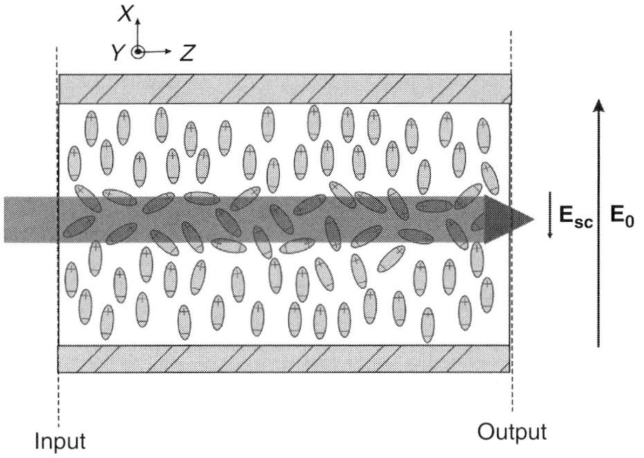

FIGURE 19.19 Mechanism of optically inducing refractive index profile in PR organic materials.

direction of the applied electric field, this refractive index profile leads to either self-focusing or self-defocusing of the beam [14,15], as illustrated in Figure 19.20 for a 1D beam obtained by focusing a Gaussian-like beam with a cylindrical lens on the input plane of the sample [14]. The sample consists of a 120 μm film made of a DCDHF-6/C_{60}-containing PR organic amorphous glass (Figure 19.9 and Table 19.1) placed between two ITO-coated glass slides. Frames (1) and (4) of Figure 19.20 show the beam shape at the input plane of the sample for the polarization of light parallel and perpendicular to the direction of applied electric field, respectively. In the absence of applied electric field, the light beam spreads out due to linear diffraction, as it propagates through the film, regardless of polarization (frames (2) and (5) in Figure 19.20). If the electric field $\mathbf{E}_0$ is applied, the beam shape obtained at the output plane of the sample depends on the polarization of the beam. If the beam is polarized parallel to the direction of applied electric field (i.e., along *x*-axis), self-defocusing occurs, and the beam diverges even more than in the absence of electric field (frame (6) in Figure 19.20), which could be used in, for example, optical limiters. However, if the beam is polarized perpendicular to the electric field (i.e., along *y*-axis), the beam undergoes self-focusing (frame (3) in

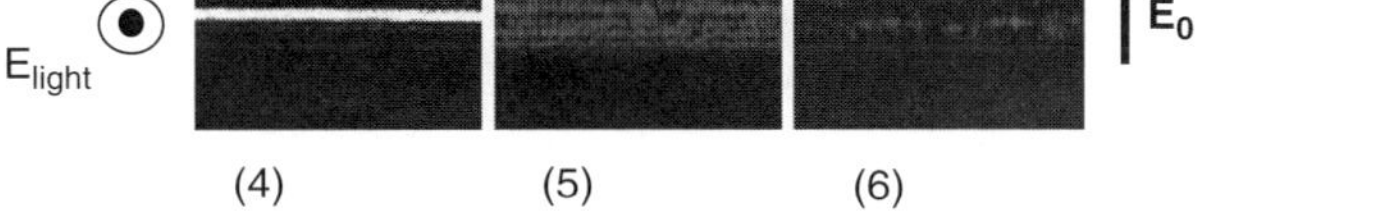

FIGURE 19.20 Observation of self-focusing and self-defocusing in a PR organic amorphous glass. Image of the beam at frame (1) sample input (original *y*-polarized beam); (2) sample output ($\mathbf{E}_0 = 0$; linearly diffracted *y*-polarized beam); (3) sample output ($\mathbf{E}_0$ = 16 V/μm, self-trapped *y*-polarized beam); (4) sample input (original *x*-polarized beam); (5) sample output ($\mathbf{E}_0 = 0$, linearly diffracted *x*-polarized beam); (6) sample output ($\mathbf{E}_0$ = 16 V/μm, defocused *x*-polarized beam). (Adapted from Chen, Z.G. et al., *Opt. Lett.*, 28, 2509, 2003. With permission.)

Figure 19.20). Theoretical formalism describing the polarization dependence of the focusing and defocusing effect in PR organic amorphous materials is beyond the scope of this chapter and can be found in Ref. [15]. Briefly, the refractive index change (Δn) in orientationally enhanced PR materials with Δn_{BR} dominating over Δn_{EO} (Section 19.2.2, Equation 19.7) for x- and y-polarized light is:

$$(\Delta n)_x \sim (1/2n)\, C_{BR} E^2,\ (\Delta n)_y \sim (-1/4n)\, C_{BR} E^2, \tag{19.13}$$

where
n is the refractive index
E is the total electric field
C_{BR} is a constant defined in Equation 19.7

From Equation 19.13, since the total electric field in the illuminated part of the sample is lower than that in the non-illuminated part (Figure 19.19), the x- (y-)polarized beam experiences lower (higher) refractive index compared to non-illuminated part, which leads to beam self-defocusing (self-focusing). At a certain magnitude of electric field $\mathbf{E_0}$, the self-focusing effect exactly compensates for the diffraction, and an optical spatial soliton, i.e., the beam that preserves its spatial characteristics as it propagates through the film, forms. In this case, the beam shape at the output plane of the sample (frame (3) in Figure 19.20) is identical to that at the input plane (frame (1) in Figure 19.20). This self-trapping effect is useful whenever beam propagation in a material over some distance is required. Moreover, the self-trapping effect can be used to optically induce a refractive index profile, which would then guide another beam. An example of experimental arrangement illustrating this effect is shown in Figure 19.21 [16]. A 2D Gaussian-like collimated beam from a 780 nm laser diode is focused to about 19 μm with a lens onto the input face of a PR organic glass sample (2.5 mm long and 120 μm thick). A half-wave plate rotates the polarization of the beam when necessary. In addition, another beam from a 980 nm laser diode is used as a probe beam, following the same path taken by the 780 nm soliton-forming beam. Behind the sample, a CCD camera together with an achromatic imaging lens is used to monitor the beam profiles. The beam profile of the 780 nm beam at the input face of the sample is shown in Figure 19.22a. Without the applied field, the beam is diffracted to about 34 μm after 2.5 mm propagation (Figure 19.22b). After an electric field of ~17 V/μm was applied, self-trapping of the 2D y-polarized beam was realized in about 65 s (Figure 19.22c). Once the soliton was formed, the 780 nm soliton beam was replaced with a Gaussian-like 980 nm probe beam (Figure 19.22, bottom panel) to test the optically induced waveguide. Without the waveguide, the probe beam diffracted normally (Figure 19.22e). With the soliton waveguide, however, the probe was guided very well as its size at the output of the sample was reduced significantly (Figure 19.22f). The beam confinement occurred almost instantaneously when the probe was launched into the soliton

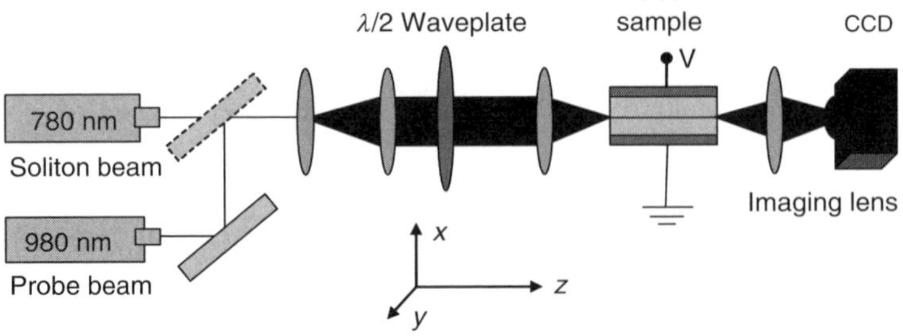

FIGURE 19.21 Experimental setup used for testing of the optically induced waveguide. (Reprinted from Asaro, M. et al., *Opt. Lett.*, 30, 519, 2005. With permission.)

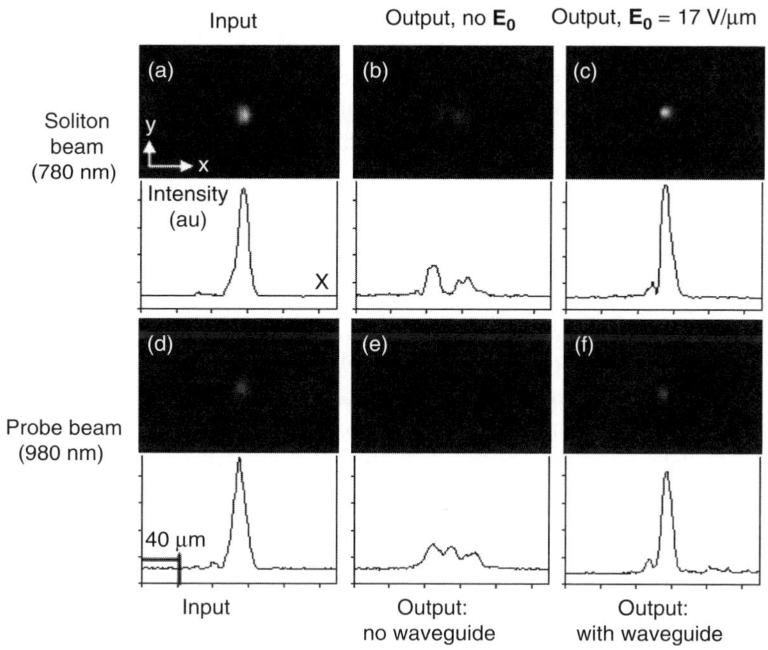

FIGURE 19.22 Circular waveguides induced by 2D solitons. Shown are transverse intensity patterns and beam profiles of the 780 nm (soliton) and 980 nm (probe) beams taken at the sample input (a, d) and output (b, c, e, f). Top: 780 nm soliton beam. Bottom: 980 nm probe beam, (b) soliton beam, no electric field, (c) soliton beam, $\mathbf{E}_0$ = 17 V/μm, (e) probe beam, no waveguide, (f) probe beam, in the presence of soliton beam-induced waveguide. (Reprinted from Asaro, M. et al., *Opt. Lett.*, 30, 519, 2005. With permission.)

channel. Under the same experimental conditions, the 980 nm probe beam itself did not show appreciable self-focusing even after the field was applied for more than 30 min, since the PR organic glass utilized in this experiment did not have sensitivity at this wavelength (i.e., illumination with a 980 nm beam did not result in charge carrier generation). This confirms that the observed guiding was due to the refractive index profile created by the 780 nm beam. Optically induced guiding can be utilized in a number of applications such as optical couplers, switches, logic gates, nonlinear frequency converters, optical parametric oscillators, and many others.

19.6 SUMMARY

PR effect is a light-induced change in the refractive index that involves photogenerated charge redistribution (Section 19.1). The main mechanisms involved in the PR effect are space-charge field build-up, caused by photogeneration, transport and trapping of charge carriers (Section 19.2.1), and refractive index change in the electric field due to anisotropic polar molecules reorienting in the electric field (Section 19.2.2). The distinct feature of the PR effect is that the light-intensity pattern and a refractive-index pattern are phase-shifted with respect to each other, which results in an asymmetric energy transfer between two light beams, called 2BC (Section 19.3.1), utilized in numerous applications (Section 19.5.2). Experimental geometries utilized in characterization of PR materials and in applications are 2BC and FWM, considered in Sections 19.3.1 and 19.3.2, respectively. In terms of PR organic materials design, most successful materials classes include polymer composites, amorphous glasses, fully functionalized polymers, and PDLCs, briefly considered in Section 19.4. Applications demonstrated in high-performance PR organic materials include data storage, image processing, optical limiting, switching, optical wave guiding, etc. Examples of selected applications are given in Section 19.5.

In summary, the complexity of the PR effect in organic materials is fascinating, since a variety of physical processes contribute to the PR performance. The need to optimize all these processes represents a challenge in the material design. Nevertheless, over the past years, tremendous progress in both physical understanding of the PR effect in organics and in the development of high-performance materials has occurred, and many applications of PR organic materials have been demonstrated. The field of PR organic materials keeps evolving, and emergence of novel, improved materials will undoubtedly enable new exciting applications.

EXERCISE QUESTIONS

1. Consider a PR grating formed in a PR polymer composite by two interfering beams at the wavelength of 633 nm in a standard geometry (Figure 19.5). The angle between the two writing beams in air is 22°. The sample normal and the bisector of the two writing beams form an angle of 55° (in air). The thickness of the sample is 105 μm, and the refractive index of the PR polymer composite is 1.7. Calculate the grating period Λ. Is this a volume (Bragg) or a surface (Raman–Nath) grating?
2. Consider a standard FWM geometry (Figure 19.7), with experimental conditions concerning the writing beams and the PR sample properties described in Exercise Question 1. The experimentally measured diffraction efficiency as a function of electric field is shown in Figure 19.23. Explain qualitatively this dependence on the electric field. After the electric field reached a certain magnitude, the diffraction efficiency starts to decrease (the so-called overmodulation effect). Why? How is diffraction efficiency expected to behave at even larger electric fields?
3. Consider 2BC effect. If the incident writing beam intensities are I_1(in) and I_2(in), then after the beam interaction in a PR material, the intensities are I_1(out) and I_2(out) and are given by Equation 19.9. In the case of lossless medium, the total energy of the light beams must be conserved. Ignoring reflection losses and refraction on the boundaries, show that although in the process of beam coupling the energy redistribution between the beams occurs, the total energy (or the total light intensity $I_0 = I_1 + I_2$) is conserved.
4. Consider 2BC experiment performed in a standard geometry (Figure 19.5) with the writing beams of equal intensities (i.e., I_1(in) = I_2(in)) incident at angles 30° and 60°, respectively, on the PR organic amorphous glass with the refractive index of 1.6. The thickness of the sample is 80 μm. The gain (I(out)/I(in)) obtained in this experiment is shown in Figure 19.6. Consider the beam that

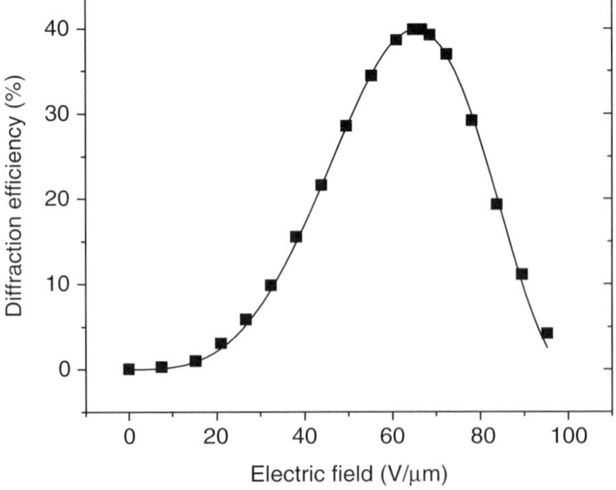

FIGURE 19.23 Diffraction efficiency as a function of electric field obtained in a PR polymer composite. (Data courtesy of professor R.A. Norwood.)

is amplified during the 2BC (beam 1). Using the steady-state value of the gain $\gamma_0 = I_1(\text{out})/I_1(\text{in})$ in Figure 19.6, estimate the gain coefficient Γ for this amorphous glass. Ignore the differences in the reflection coefficients in the writing beams.

5. In a standard 2BC geometry, an object is inserted in the path of one of the beams (signal beam), and an image of the object is recorded in a PR polymer composite sample and then read out. Let the object be an Air Force resolution test chart, which is a non-transparent slide with optically transparent patterns (see Figures 19.16 through 19.18). It turns out that depending on the ratio of signal and reference beam intensities (I_{s0}/I_{r0}), either an exact replica of the object (e.g., in Figure 19.18a) or its edge-enhanced image (Figure 19.18b) is obtained. Explain the edge enhancement effect (Figure 19.18b).
6. As discussed in Section 19.5.3, in low-T_g PR organic glasses, when the light beam is incident on the sample as shown in Figure 19.19, it can experience self-focusing or defocusing depending on its polarization. However, the effect does not depend on the polarity of applied electric field. The situation is the opposite in inorganic PR crystals. There, self-focusing changes to self-defocusing upon changing the polarity of electric field. What is the reason for this difference between inorganic and organic PR materials?

LIST OF ABBREVIATIONS

General Terminology

2BC	two-beam coupling
DOS	density of states
FWM	four-wave mixing
ITO	indium tin oxide
LC	Liquid crystal(s)
NLO	nonlinear optical
OE	orientational enhancement
PDLC	polymer-dispersed liquid crystals
PC	phase conjugation
PR	photorefractive

Chemicals

2BNCM	*N*-2-butyl-2,6-dimethyl-4*H*-pyridone-4-ylidenecyanomethylacetate
AODCST	2-[4-bis(2-methoxyethyl)amino]benzylidenemalononitrile
ATOP	1-alkyl-5-[2-(5-dialkylaminothienyl)methylene]-4-alkyl-[2,6-dioxo-1,2,5,6-tetrahydropyridine]-3-carbonitrile
BBP	butyl benzyl phthalate
DBOP-PPV	poly[1,4-phenylene-1,2-di(4-benzyloxyphenyl)vinylene]
DCDHF-6	2-dicyanomethylen-3-cyano-5,5-dimethyl-4-(4′-dihexylaminophenyl)-2,5-dihydrofuran
DMNPAA	2,5-dimethyl-(4-p-nitrophenylazo)anisole
DPP	diphenyl phthalate
ECZ	*N*-ethylcarbazole
MNPAA	3-methoxy-(4-p-nitrophenylazo)anisole
PCBM	[6,6]-phenyl-C_{61}-butyric acid methyl ester
PSX	Poly[methyl(3-carbazol-9-ylpropyl)siloxane]
PVK	Poly(*N*-vynilcarbazole)
TNF	2,4,7-trinitro-9-fluorenone
TNFM	(2,4,7-trinitro-9-fluorenylidene)malononitrile

REFERENCES

1. Solymar, L. et al., *The Physics and Applications of Photorefractive Materials*, Clarendon Press: Oxford, 1996.
2. Moerner, W.E. and Silence, S.M., Polymeric photorefractive materials, *Chem. Rev.*, 94, 127–155, 1994.
3. Ostroverkhova, O. and Moerner, W.E., Organic photorefractives: Mechanisms, materials, and applications, *Chem. Rev.*, 104, 3267–3314, 2004.
4. Yeh, P., *Introduction to Photorefractive Nonlinear Optics*, John Wiley: New York, 1993.
5. Borsenberger, P.M. and Weiss, D.S., *Organic Photoreceptors for Xerography*, Vol. 59, Marcel Dekker: New York, 1998.
6. Pope, M. and Swenberg, C.E., *Electronic Processes in Organic Crystals and Polymers*, Vol. 56, Oxford University Press: New York, 1999.
7. Bredas, J.L. et al., Charge-transfer and energy-transfer processes in pi-conjugated oligomers and polymers: A molecular picture, *Chem. Rev.*, 104, 4971–5003, 2004.
8. Moerner, W.E. et al., Photorefractive polymers, *Annu. Rev. Mater. Sci.*, 27, 585–623, 1997.
9. Boyd, R.W., *Nonlinear Optics*, Academic Press: San Diego, CA, 2003.
10. Kogelnik, H., Coupled wave theory for thick hologram gratings, *Bell Syst. Tech. J.*, 48, 2909–2947, 1969.
11. Li, G.Q. et al., All-optical dynamic correction of distorted communication signals using a photorefractive polymeric hologram, *Appl. Phys. Lett.*, 86, 161103, 2005.
12. Goonesekera, A. et al., Image amplification and novelty filtering with a photorefractive polymer, *Appl. Phys. Lett.*, 76, 3358–3360, 2000.
13. Ono, H. et al., Intensity filtering of a 2D optical image in high-performance PR mesogenic composites, *Appl. Phys. Lett.*, 79, 895–897, 2001.
14. Chen, Z.G. et al., Self-trapping of light in an organic photorefractive glass, *Opt. Lett.*, 28, 2509–2511, 2003.
15. Sheu, F.W. and Shih, M.F., PR polymeric solitons supported by orientationally enhanced birefringent and electro-optic effects, *J. Opt. Soc. Am. B*, 18, 785–793, 2001.
16. Asaro, M. et al., Soliton-induced waveguides in an organic photorefractive glass, *Opt. Lett.*, 30, 519–522, 2005.

20 Organic/Metal Interface Properties

Yongli Gao

CONTENTS

20.1 Introduction ... 637
20.2 Experimental Techniques ... 638
20.3 Surface Atomic and Electronic Structure ... 646
20.4 Interface Growth Mode and Modeling ... 649
20.5 Metal/Organic Interface Dipole ... 651
20.6 Chemical Reaction, Diffusion, Doping, and Quenching ... 653
20.7 Summary ... 657
Exercise Questions ... 657
List of Abbreviations ... 657
Bibliography ... 658

Abstract This chapter is intended to introduce the reader some insights on metal/organic interface formation. The basic characteristics of surface atomic and electronic structure and interface growth mode, as well as popular experimental techniques are described. Examples are given on aspects including interface dipole, reactions, diffusion, doping, and light emission quenching.

20.1 INTRODUCTION

In all organic semiconductor (OSC) devices, the charge transport process across interfaces of dissimilar materials is important for optimum device operation. As proven in over five decades of research, the understanding of interfaces of inorganic semiconductors (ISCs) with metals, semiconductors, and insulators, has had a tremendous impact on semiconductor device technology. A similar situation is also found in OSC devices. Furthermore, the thickness of the active organic layer is typically only a few 100 Å in OSC devices, which further blurs the distinction between bulk and interface. At this point, an important distinction must be made between ISCs and OSCs. ISCs have occupied and unoccupied energy levels, valence, and conduction bands, respectively that can be extended over many unit cells. The semiconductor can be appropriately doped n-type or p-type. The interaction between charge carriers and the lattice is generally weak, and the transport of the charge carriers can be adequately described as delocalized Bloch waves in the bands. In OSCs, the occupied and unoccupied energy levels are formed from planar structures of sp^2 bonds as well as π-bonds, and the π-bonds between carbon atoms in organic molecules usually form the highest occupied molecular orbital (HOMO) and lowest unoccupied molecule orbital (LUMO) in most OSCs. The interactions between the molecules are van der Waals in nature and the electron wave function overlap between the molecules is small. The charge carriers are localized and surrounded by significant nuclear relaxation, and are better described together with the surrounding nuclear deformation as polarons instead of electrons or holes. As a result, the transport from one molecule to another is typically described by hopping of polarons.

Metal/organic interface is a focus of both device engineering and basic science, since it is a key factor in nearly all important aspects of device performances, including operation voltages, degradation, and efficiency. The complexity of metal/organic interfaces is also intriguing in basic organic condense matter physics. The energy level alignment and charge injection at the interface are among the most concerned fundamental issues.

Understanding the interface processes in organic devices is critical for their further advance performance. Surface and interface analytical studies have generated critical insight of the fundamental processes at interfaces involving OSCs. For example, it is now established from surface analytical studies that the interface energy level alignment is not from a common vacuum level as believed previously. Instead, it depends on the detailed interface interactions, including wavefunction hyperdization, charge transfer, chemical reaction, and intermixing, etc. Understanding issues regarding metal/organic interfaces, such as the formation of the interface dipole, the injection barrier, the diffusiveness of the interface, and origin of ionized species, is beginning to take shape.

This chapter is intended to introduce the reader to the field by providing some insights on metal/organic interface formation obtained by using popular interface analytical techniques. However, it is by no means to be an exhaustive study of all the research work and techniques employed in the study of interface formation in OSC devices.

20.2 EXPERIMENTAL TECHNIQUES

The electronic structure at interfaces with semiconductors is generally described by charting the energy levels such as the vacuum level, valence band, and conduction band as a function of distance from the interface. There are indirect electrical measurements capable of determining this profile but they generally rely on a theoretical model of the system to determine the general features of the interface with the electrical measurements, merely determining the bounds of the energy and spatial values of the profile of the system. The technological success of devices whose performance is dictated by the interface characteristics has driven the development of increasingly sophisticated instrumentation designed to better probe the surface properties. Currently, there exist a large number of surfaces and interface analytical tools, capable of providing complementary information. The usefulness of these analytic instruments has been proven by their successful application in the study of ISC devices where the interfaces were found to dictate the performance of the device. Many have been successfully applied to study organic/metal interfaces, including photoemission spectroscopy (PES), scanning probe microscopy (SPM) and spectroscopy, secondary-ion mass spectroscopy (SIMS), near edge x-ray absorption fine structure (NEXAFS), Kelvin probe, internal photoemission, Penning spectroscopy, reflection infrared spectroscopy (RIRS), high-resolution electron energy loss spectroscopy (HREELS), low energy electron microscopy (LEEM), low energy electron diffraction (LEED), temperature-programmed desorption (TPD), etc. In the following sections, we briefly discuss some of the most frequently used tools for the study of OSCs, including ultraviolet photoemission spectroscopy (UPS), x-ray photoemission spectroscopy (XPS), inverse photoemission spectroscopy (IPES), scanning tunneling microscopy (STM), and atomic force microscopy (AFM).

In any systematic surface study of materials, it is essential to recognize the basic requirements dictated by the relatively low number of atoms of interest. For a given number of atoms N_A, the number of surface atoms scales as $N_A^{2/3}$. As such, a solid contains ~10^{23} atoms/cm^3, the surface has ~10^{15} atoms/cm^2. The relative ratio of surface to bulk atoms presents a handicap when the probing depth exceeds several atomic layers, since the signal due to bulk atoms can quickly overwhelm that originating from the surface. A technique useful for surface analysis must therefore be sensitive to the electronic or structural properties of the relative few surface atoms. In addition, sample cleanliness is exceedingly important, because the interaction between the surface and contaminants will interfere with a clear characterization of surface properties.

A simple model of the interaction between the surface of a sample and the surrounding air molecules highlights the importance of a clean environment. Kinetic theory states that the rate ρ for the impact of atoms on a surface is given by

$$\rho = \frac{P}{\sqrt{2\pi mkT}}, \tag{20.1}$$

where
P is the ambient pressure
m is the mass of the atoms
k is the Boltzmann constant
T is the temperature

For nitrogen, if $T = 300$ K and $P = 10^{-6}$ torr, then $\rho = 5 \times 10^{14}$ /cm^2 s^1. If every atom that strikes the surface sticks, then a monolayer of nitrogen will grow in 2 s. Since surface contamination may result in interactions between the adsorbate and the substrate, it is necessary to minimize exposure of the sample to contaminants. Consequently, detailed surface studies require a highly controlled environment with a pressure about 10^{-10} Torr. This range of pressure is usually referred to as ultrahigh vacuum (UHV).

Photoelectron spectroscopy is based on the photoelectric effect, explained by Einstein in 1905, which relies on the creation of photoelectrons via interaction between the irradiating photons and the sample. Since the total energy must be conserved in this process, the kinetic energy, E_k, imparted to an electron satisfies

$$E_k = h\nu - E_B - \phi, \tag{20.2}$$

where
ϕ is the work function of the sample
E_B is the binding energy of the initial state of the electron with respect to the Fermi level

Figure 20.1 schematically summarizes this phenomenon. As indicated in Figure 20.1, photoelectrons originate from energy levels occupied by electrons, including the valence band in ISC or HOMO energy levels for OSCs as well as from core-level states, which correspond to closed atomic shells. The surface sensitivity of PES is obtained from the strong interaction of the photoexcited electron with the rest of the solid, resulting in a relatively short distance, or mean free path (MFP), that it can travel before suffering an inelastic scattering. The MFP is strongly dependent on the kinetic energy of the photoelectron. The dependence on different materials is relatively minor. The MFP dependence on the kinetic energy has been summarized as the universal curve as shown in Figure 20.2.

PES techniques are customarily named according to the type of photon source used. Most commonly in-house ones are XPS and UPS. In addition to XPS and UPS, synchrotron radiation, which requires expensive central synchrotron facilities, covers photon energies from infrared to hard x-ray. UPS has been extensively used for studying OSC surface and interfaces. These photons are generated by a He gas discharge lamp, with an energy of 21.2 eV (He I) or 40.8 eV (He II). These low energy photons are restricted to probing the valence structure of samples. The line width of the ultraviolet (UV) lamp is small, only about 20 meV, which allows higher resolution spectra be measured for the valence levels. Another benefit is that the interactions between these photons and the valence electrons have a higher cross section than with x-rays, allowing improved statistics for these measurements and thereby faster measurements. Finally, UPS is typically less damaging to the sample than XPS, a feature especially valuable to organic semiconductors that are usually vulnerable to photoinduced damages and chemistry.

The UPS spectrum of a clean Au surface is shown in Figure 20.3a. The UPS spectrum clearly shows the Fermi level of the Au surface, which is used as a reference point to ensure that the binding

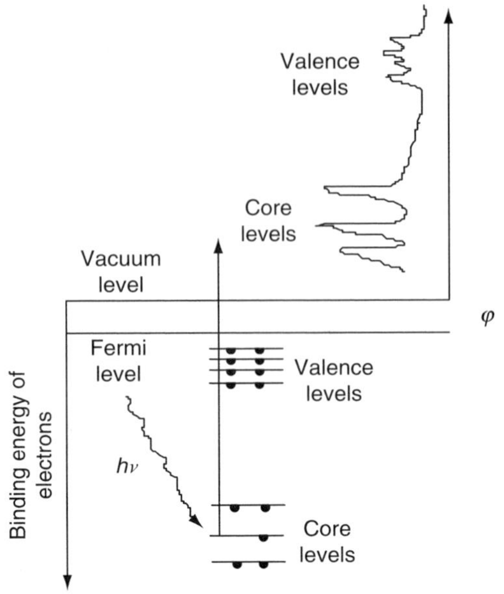

FIGURE 20.1 A schematic representation of PES.

energy scale is accurate. Since the Fermi level should remain constant for the system in thermal equilibrium, it is only necessary to directly measure the Fermi level of a metal surface occasionally to ensure calibration of the energy scale. The Au 5d band is clearly visible at around 5 eV binding energy. At a binding energy of ~16 eV is the low energy cutoff, so-called due to the low kinetic energy of the electrons that this feature is composed of. This feature is composed of the electrons with the highest binding energy that the 21.22 eV photons are able to eject from the surface. Therefore, by simply adding 21.22 eV to the low energy cutoff binding energy the binding energy of the vacuum level of the system relative to the Fermi level can be calculated.

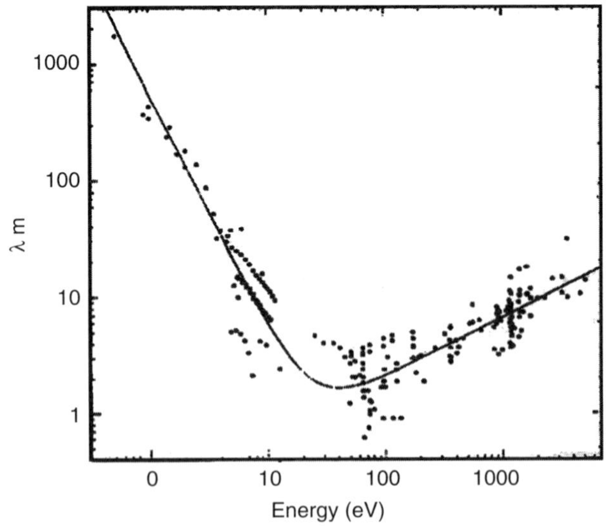

FIGURE 20.2 Universal curve of electron MFP in solids. (From Seah, M.P. and Dench, W.A., *Surf. Interface Anal.*, 1, 2, 1979.)

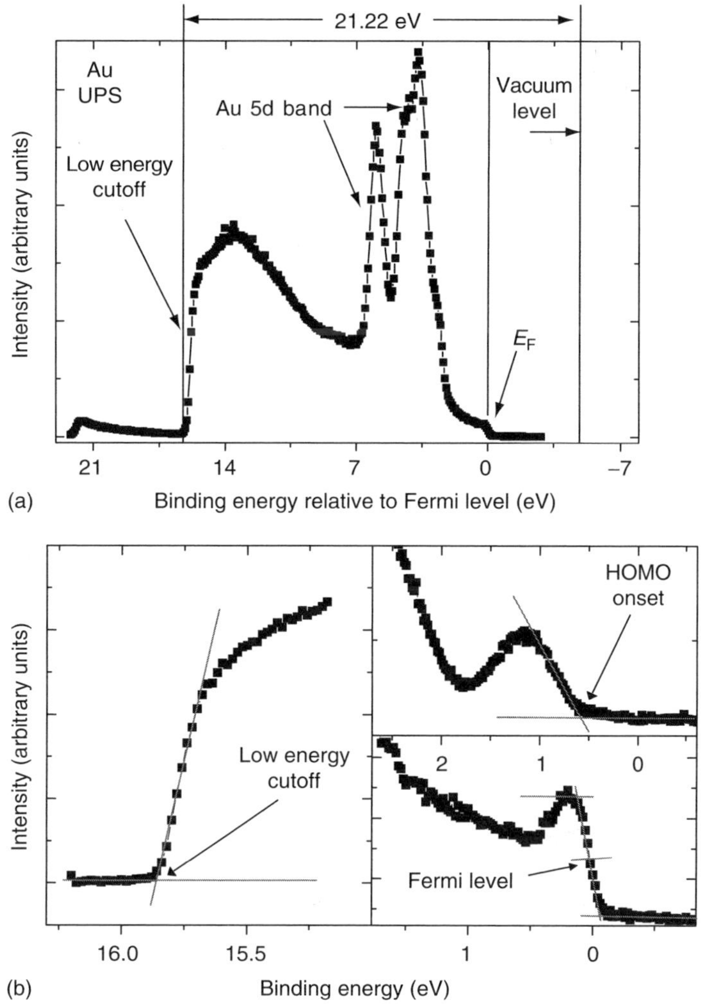

FIGURE 20.3 (a) UPS spectra of a clean Au surface and (b) UPS feature position determination.

The method used to determine the positions of the UPS features is shown in Figure 20.3b. The low energy cutoff position is defined as the highest binding energy of the measured electrons. This point is determined by fitting the high binding energy portion of the spectrum and determines where this fit intersects the background signal. The determination of the Fermi level position differs slightly due to the nature of the Fermi level. The measured Fermi level of a metal is expected to obey Fermi statistics and therefore the Fermi level is defined as occurring at the point where the energy level is half occupied, i.e., the midpoint of the leading edge of the spectrum, as illustrated in Figure 20.3b.

If the surface is composed of a semiconductor, the general features of the UPS spectra will be the same except for the fact that the Fermi level will not be measurable due to the lack of density of states in the band gap. The position of the valence structure can be used, in conjunction with the vacuum level, to calculate the ionization potential of the surface. For OSCs, the valence feature that is used to calculate the ionization potential is the HOMO. The HOMO feature position is generally referred to by its low binding energy onset. This HOMO onset is determined by performing a linear fit of the low binding energy edge of the spectrum, and determining where that fit intersects the background signal, as shown in Figure 20.3b. UPS allows direct measurement of several important materials properties by direct measurement of the valence levels. The relative positions of the Fermi level, the vacuum level, and the highest occupied electron orbital can be measured. This allows calculation of the work function and ionization potential of the surface.

The most commonly used photon sources for XPS are the unmonochromatized K_α radiation from magnesium or aluminum targets. The Al K_α and the Mg K_α lines peak at 1486.6 and 1254.6 eV, respectively. XPS is also commonly known as electron spectroscopy for chemical analysis (ESCA). XPS is based on the principal that the binding energy of a given photoelectron is highly dependent on the element from which an electron originates. Further, any change in the chemical environment of the atom probed by XPS causes a small chemical shift in the binding energy of a core-level electron. The core-level peaks are easily identified from a handbook of PS, although their exact positions depend on the chemical environments of the atoms.

In principle, the underlying physical interactions responsible for changes in the binding energies of different chemical species can be expressed in a straightforward fashion. In a first-order approximation, the energy of a core-level electron depends on the Coulomb attraction of the nuclei and the repulsion of all the other electrons in the system, which neglects many-body effects. The change in the binding energy, ΔE, resulting from a change in the chemical environment will cause the HOMO energy level charges to redistribute. $\Delta E(A, B)$ of a particular core level in two different compounds, A and B, was originally described by Gelius as follows:

$$\Delta E(A,B) = K_c(q_A - q_B) + (V_A - V_B). \tag{20.3}$$

The first term on the right-hand side, $K_c(q_A - q_B)$, describes the difference in the electron–electron interaction between core orbital, c, and the HOMO energy level charges, q_A and q_B, respectively. The coupling constant K_c is the two electron integral between core and HOMO energy level electrons.

The second term on the right-hand side has the character of a Madelung potential, which in the point charge approximation is defined as

$$V_i = \sum_{i \neq j} \frac{q_j}{R_{ij}}, \tag{20.4}$$

where the summation is over potentials arising from all the other ionic charges q_j centered at positions R_{ij} relative to the atom, i, in the material. Both terms (right-hand side) in Equation 20.3 are usually about 10 eV. However, the summation of the two terms is about a few electron volts or less since the ions attract the electron, while the HOMO energy level electrons repel the electron. As a result of this partial cancellation of the two terms, observed chemical shifts in solids are usually about a few electron volts or less.

IPES is a time-reversed PES. The surface under investigation is bombarded with electrons of a known kinetic energy, between 0 and 20 eV. These incident electrons radiatively decay into unoccupied energy levels. The ejected electrons are then measured, providing a direct measure of the unoccupied energy levels of the surface. There are two methods for performing this experiment. The first method is to bombard the surface with electrons of a single kinetic energy and directly measure the spectrum of emitted photons using a spectrometer. The second method is to vary the kinetic energy of the electrons and measure single-photon energy. Given the energies of the incident electrons, this technique has approximately the same surface sensitivity as UPS.

The observed UPS spectrum is the representation of the filled states of a molecular cation resulting from the photoelectric process modified by the relaxation (polarization). The width of the peaks is presumably due to inhomogeneity of the film since the organic film is normally amorphous with random disorder. Therefore, the center of the HOMO peak corresponds to the HOMO energy of the most populous molecular cation, with the negative charge (photoelectron) at infinity in vacuum.

Conceptually, similar analysis applies to the IPES spectrum. Induced by the injected electron, the relaxation (polarization) of the surrounding medium makes the IPES spectrum the representation of the relaxed anion instead of the neutral state of the molecule. Thus the energy separation of the HOMO–LUMO peaks obtained by UPS and IPES is the energy difference of relaxed positive and negative polarons separated at infinity, sometimes referred as polaron energy gap (E_t) of the organic material. However, the charge injection in solid does not necessarily occur at the most populous or average species. Rather, when the injection is contact limited (usually the case in organic light-emitting devices [OLEDs]), the charge injects into the molecules with the lowest energy difference to the Fermi level. The onset is defined as the extrapolation of the leading edge (closer to the Fermi level) in spectrum. Similarly, the barrier for the electron injection will be from the Fermi level to the onset of LUMO. In most organic molecular devices, the cathode and anode are separated far enough that the injection of holes and of electrons occurs at different molecules and beyond the size of the polarons. It is a general practice to define the onset of HOMO peak as the HOMO position of the copper phthalocyanine (CuPc) film, at −0.7 eV shown in Figure 20.4. Similar term is used for LUMO position, at about 1.0 eV for the pristine CuPc film. The result is an injection energy gap (E_g) of about 1.6 eV, interestingly very close to the CuPc optical band gap E_{opt}(~1.7 eV).

SPM has become an indispensable tool in the area of science and technology for the study of a whole range of material properties in the nanoscale regime with very high spatial resolution. SPM with its ever-expanding family enables us to characterize and correlate the material properties such as topographical, mechanical, optical, electrical, thermal, etc. The basic idea behind SPM techniques is straightforward. There is an interaction parameter, such as the tunneling current in STM, van der Waals, electrostatic, and capillary forces in AFM, established between the microscope tip and the surface under investigation. The dependence of the interaction parameter on the distance between the probe and the sample is exploited to elicit information about the sample, and sometimes the tip. This is done by establishing a feedback mechanism based on the interaction parameter. There are two ways that this could be done. First, the interaction strength can be held constant throughout the

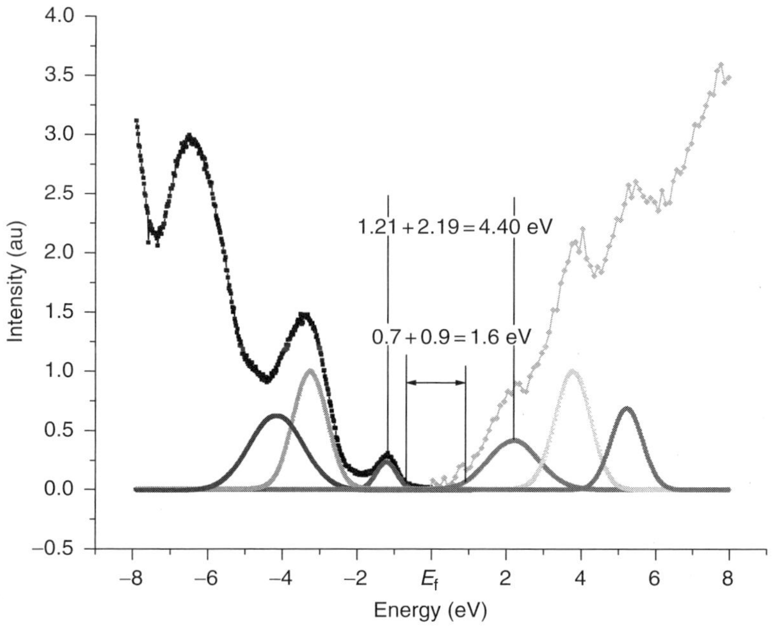

FIGURE 20.4 UPS and IPES spectrum of CuPc.

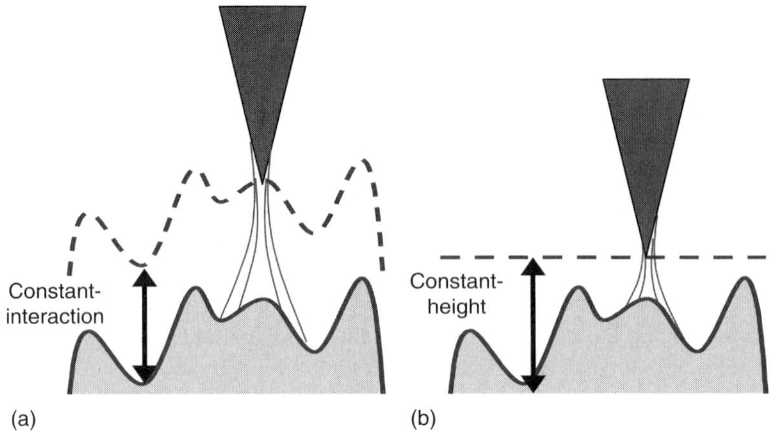

FIGURE 20.5 Two modes of operation in SPM. (a) Constant-interaction mode and (b) constant-height mode.

scan by dynamically correcting for any change that occurs during the course of the scan (see Figure 20.5a) This requires raising or lowering the tip (or the sample, depending on which one is being moved), to keep the separation distance constant (it is this movement of the piezo that gives us the information we want about our tip-sample system, that is why it is always being recorded during the scan). Consequently, the interaction parameter will be constant. This is called constant-current mode in STM, and constant-force mode in AFM. The second way is to turn off the feedback (in other words, no correction is made for any change that occurs in the interaction parameter) and follow and record the changes in the interaction parameter directly (see Figure 20.5b). This is called constant-height mode in both STM and AFM. It applies equally well to any SPM technique.

The ability of STM to achieve atomic resolution is owing to the fact that the tunneling current dependence on the height of the gap (the distance between the tip and the sample) is exponential.

$$I = C\exp(-2\kappa d); \quad \kappa = (2m\phi)^{1/2}, \tag{20.5}$$

where
- I is the tunneling current
- κ is the wave function decay constant
- d is the separation between the tip and the sample
- m is the electron mass
- ϕ is the work function of the metal substrate (since STM relies on the tunneling current as its interaction mechanism, a sample must be conductive if to be operated on by STM)

The above exponential dependence makes STM remarkably sensitive to the change in the separation between the STM tip and the sample. To illustrate this point further, assume a metal with a work function of about 5 eV, such as Au. If the separation distance is reduced by 1 Å due to, say, a new atom, then this will bring about 9.88-fold increase in the tunneling current, which can easily be detected. It must be pointed out that the images obtained from STM are not reflecting the topology of the surface altogether. This is due to the fact that the interaction parameter of STM is the tunneling current; hence the images we get are reflecting the distribution of the electrons above the surface, which usually follows the topology of the surface closely (but not always). Shown in Figure 20.6 is the STM picture of perylene-tetracarboxylicacid-dianhydride (PTCDA) deposited on highly oriented pyrolitic graphite (HOPG) taken in air. It is remarkable that STM can obtain molecular resolution images if the organic overlayer is ordered.

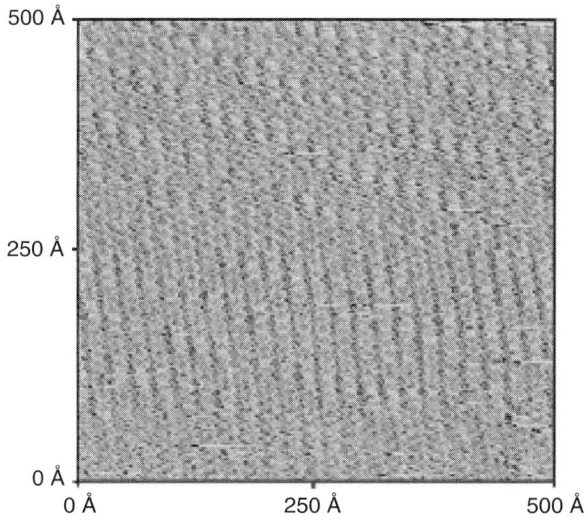

FIGURE 20.6 Molecular resolution STM images of PTCDA deposited on HOPG taken in air.

Today's AFMs mostly use optical detection mechanism. Figure 20.7 shows a schematic drawing of a typical AFM (with conducting-probe option) that is commonplace in today's laboratories. A laser beam is reflected off of the back of a cantilever, which has a very small protruding tip facing the surface whose radius of curvature is about 100 Å that is attached to a piezoelectric ceramic scanner. As the cantilever is scanned over a surface, it is deflected up and down due to the local structure of the surface. This bending of the cantilever is detected through the movement of the laser that is reflected from the cantilever and hits the four-quadrant position-sensitive photodetector (PSPD). The level of the deflection is compared, by a computer, to a predetermined set point value. Unlike STM, AFM can be used to measure the topography of insulators and semiconductors, as well as conducting materials. The most dominant force contributing to the deflection of the AFM cantilever is due to the interatomic van der Waals forces. Figure 20.8 plots the van der Waals forces as a function of the distance between the tip and the cantilever. There are two different regimes in this plot: (1) the contact regime, and (2) the noncontact regime. The AFM can

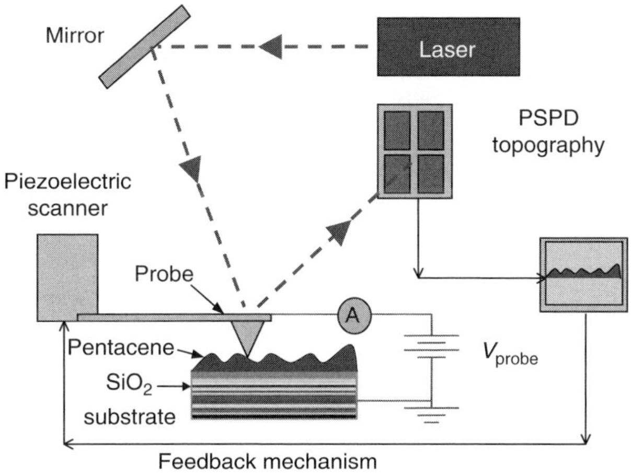

FIGURE 20.7 Schematic diagram of AFM with conducting-probe AFM option.

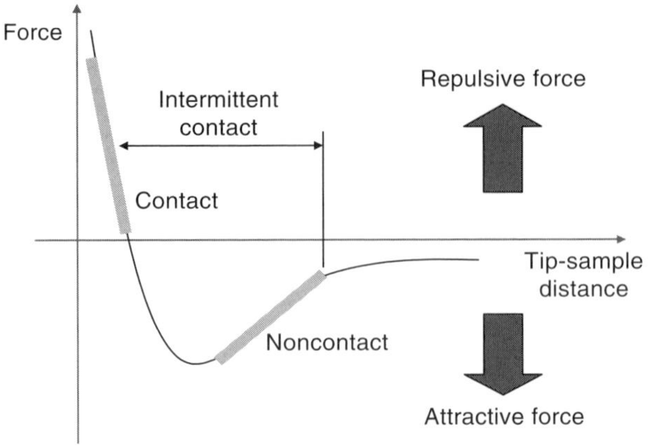

FIGURE 20.8 The van der Waals forces as a function of the distance between the tip and sample.

be operated, however, in three different modes: contact mode, noncontact mode, and tapping mode (or intermittent contact), depending upon the distance between the tip and the sample during the scan, as shown on the plot. In the contact mode AFM (C-AFM), the tip is scanned over the surface in the repulsive force region while touching the surface physically. In the noncontact mode AFM (NC-AFM), the tip is oscillated at a particular frequency at relatively farther distances (in the order of 100 Å) from the surface during the scan and does not touch the surface. This regime is the attractive force regime of the van der Waals plot of Figure 20.8. This is why NC-AFM enables imaging of soft samples without damaging the structure of the surface of the sample, where the surface can be easily damaged during a C-AFM scan. The third mode of AFM operation is the so-called tapping mode or intermittent contact mode (IC-AFM) (see Figure 20.8). In this mode, the vibrating cantilever of the NC-AFM is brought closer to the sample (from attractive to repulsive regime) at the lower end of its travel to tap the surface. Since the friction between the tip and the sample is avoided in the tapping mode in contrast to C-AFM, and is more effective in its detection of the features of the surface than NC-AFM, it has become scanning mode of choice for a wide variety of samples.

20.3 SURFACE ATOMIC AND ELECTRONIC STRUCTURE

To start discussing surfaces and interfaces, one needs to first realize that interatomic bonds must be severed at a surface of a solid. To the zeroth-order approximation, these bonds are broken and dangled into the vacuum. The greater the number of the broken bonds the higher the surface free energy, resulting in a nonequilibrium state of the system. It is thermodynamically necessary for a structural rearrangement of the surface atoms to reduce the surface free energy and to reach equilibrium. The structural changes are characterized by alternation of the distances between the atomic layers parallel to the surface (surface relaxation) and by lateral rearrangement of the surface atoms (surface reconstruction). Surface structural changes also occur when adsorbates are introduced to the surface, and the descriptions and notations below are equally applicable whether the changes are induced by vacuum or by adsorbates.

The position of any lattice point in a bulk crystalline solid is described as

$$\vec{R} = n_1\vec{a}_1 + n_2\vec{a}_2 + n_3\vec{a}_3, \tag{20.6}$$

where

n_1, n_2, n_3 are arbitrary integers
$\vec{a}_1$, $\vec{a}_2$, $\vec{a}_3$ are primitive vectors of the solid

The lattice point can also be simply described by the coefficients as $[n_1, n_2, n_3]$. For cubic lattices, such as face-centered cubic (fcc) and body-centered cubic (bcc), it is conventional that the primitive vectors of simple cubic (sc) lattice are used instead. The surface is characterized by the Miller indices of the surface plane (*hkl*), for example, Au(111), Si(100), etc. For cubic crystals, the vector [*hkl*] is perpendicular to the plane and therefore gives the surface normal direction.

The unreconstructed surface atoms form a two-dimensional lattice described by

$$\vec{r} = n_1\vec{b}_1 + n_2\vec{b}_2, \tag{20.7}$$

where

n_1 and n_2 are arbitrary integers
$\vec{b}_1$ and $\vec{b}_2$ are primitive vectors of the surface

When the surface reconstructs, the periodicity in the surface layer is changed and the unit cell of the surface lattice is enlarged. The primitive vectors of the reconstructed surface are expressed as

$$\begin{aligned}\vec{c}_1 &= m_{11}\vec{b}_1 + m_{12}\vec{b}_2,\\ \vec{c}_2 &= m_{21}\vec{b}_1 + m_{22}\vec{b}_2.\end{aligned} \tag{20.8}$$

or, in a matrix form

$$\begin{pmatrix}\vec{c}_1\\ \vec{c}_2\end{pmatrix} = \begin{pmatrix}m_{11} & m_{12}\\ m_{21} & m_{22}\end{pmatrix}\begin{pmatrix}\vec{b}_1\\ \vec{b}_2\end{pmatrix}. \tag{20.9}$$

Although the matrix notation outlined in Equations 20.8 and 20.9 is general, the majority of the surface constructions have been described using a more compact form called Wood's notation,

$$\mathrm{X}(hkl)\left(\frac{|\vec{c}_1|}{|\vec{b}_1|} \times \frac{|\vec{c}_2|}{|\vec{b}_2|}\right)R\theta, \tag{20.10}$$

where

X is the substrate material
(*hkl*) is the Miller indices of the surface
θ is the rotation angle of the reconstructed lattice with respect to the substrate

If $\theta = 0$, $R\theta$ is omitted, and sometimes a letter "p" for primitive is inserted in the expression. Another letter "c" for centered is also commonly used when the unit cell is chosen with an atom at the center. Figure 20.9 shows some of the surface reconstructions.

The simplest picture of the electronic structure at a metal surface can be envisioned using the jellium model, where the valence electrons are free and independent, i.e., not interacting with each other. The atoms, with the valence electrons stripped, are ionized accordingly and are termed ion cores. The ion cores are treated as a fixed positive background charge in the jellium model. With a surface at $z = 0$, the rigid positive background charge n_+ is semi-infinite and given by

$$n_+(z) = \begin{cases} n_0 & z \le 0 \\ 0 & z > 0 \end{cases} \tag{20.11}$$

where z is the direction normal to the surface. While it is uniformly parallel to the surface, the electron density perpendicular to the surface is not. The electrons "spill out" into the vacuum with respect to

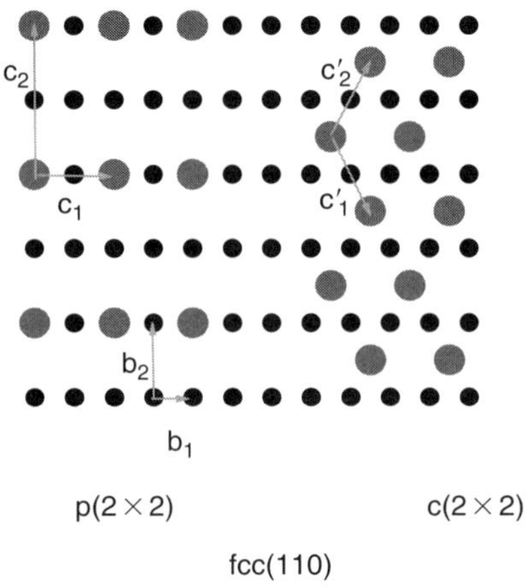

FIGURE 20.9 Surface reconstruction of p(2 × 2) and c(2 × 2) of fcc(110).

the background positive charge, creating an electrostatic dipole at the surface, and the electron density oscillates along z in the near surface region. The charge density oscillations arise from the screening behavior of the electrons and are called Friedel oscillations. The modification of the electrons' spatial distribution is to screen out the effects of the change in background charge density at the surface, as shown in Figure 20.10. The spatial variation of the electron distribution is a direct consequence of the presence of the sharp step in the background charge density.

One of the most relevant parameters of a metal for organic device applications is work function. The work function is defined as the difference in energy between an electron at rest in the vacuum just outside a metal and an electron at the Fermi level (i.e., chemical potential of electrons inside the solid). The distance between the electron and the surface should be sufficiently large such that the (coulombic) force due to the electron's interaction with its image in the solid is not felt. For metals, the work function depends both on the depth of the potential well by the ionic cores and the overspill of the conduction electron. Both quantities depend on the surface atomic geometry, and there is a strong variation in the work function for different surfaces of the same material, for

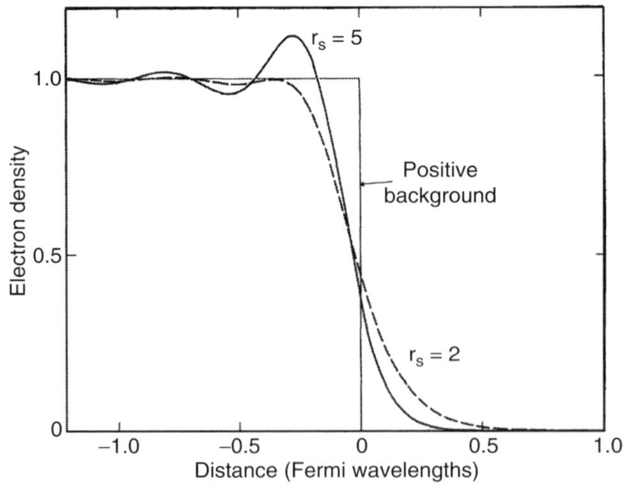

FIGURE 20.10 Electron distribution at metal surface. (From Lang, N.D. and Kohn, W., *Phys. Rev.*, B1, 4555, 1970.)

example, Cu(111), $\phi = 4.94$ eV; Cu(110), $\phi = 4.48$ eV. For semiconductors and insulators, work function is not a very important parameter since the Fermi level depends sensitively on the charge doping of the material. For these surfaces, more sensible parameters to use are the ionization potential and electron affinity.

20.4 INTERFACE GROWTH MODE AND MODELING

Organic molecules, when deposited onto an inorganic substrate, can result in three different growth modes (see Figure 20.11) depending on the relative strengths of the adsorbate–adsorbate and adsorbate–substrate interactions. First, if the substrate is chemically inert, the adsorbate–substrate interaction will be very small having almost no influence on the growth of the organic film. This will enable the organic molecules to move freely on the surface and form large, separated islands of crystals after adsorption. This type of growth is known as Volmer–Weber or 3D-island growth. Second, if the substrate has highly reactive bonds, the adsorbate–substrate interaction will be very large, and as soon as the organic molecules hit the surface they will be bound strongly to the substrate. This will prevent self-ordering mechanism of the organic molecules, which is of crucial importance for ordered growth, hence result in disorder. This mode of growth is known as Frank–van der Merwe or layer-by-layer growth. Third type of growth is a moderate combination of the first two scenarios. That is, the adsorbate–substrate interaction is neither too strong, nor too weak. It is such that the substrate has strong enough influence on the deposited particles to impose its crystallinity to the grown organic film, and the deposited particles have enough mobility to self-order themselves. This type of growth goes by the name of Stranski–Krastanov mode or layer + island growth.

To study the growth kinetics of a surface, one can use the qth-order height–height correlation function, which can be defined as

$$C_q(r,t)=\left[\left(\frac{1}{N}\right)\sum_{i=1}^{N}\left|h(r_i,t)-h(r_i+r,t)\right|^q\right]^{1/q}, \tag{20.12}$$

where N is the number of sites with single-valued heights $h(r_i, t)$. By definition, the $q = 2$ case refers to the height–height correlation function which is the rms fluctuation in the height difference between two surface points separated by a lateral distance r. It scales as follows according to the scaling hypothesis

$$C_2(r,t)\sim t^{\beta}\, g\left(\frac{r}{\xi(t)}\right), \tag{20.13}$$

where
r is the lateral separation
t is the time

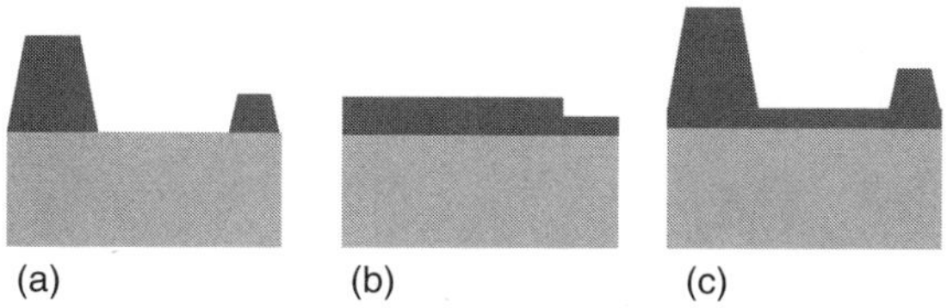

FIGURE 20.11 Growth scenarios when organic molecules are deposited. (a) Volmer–Weber or 3D-island growth, (b) Frank–van der Merwe or layer-by-layer growth, and (c) Stranski–Krastanov mode or layer + island growth.

β is a scaling exponent
ξ is the correlation length
g is a scaling function, which depends on r and ξ

The correlation length, ξ, is the critical distance over which two heights are correlated. $C_2(r, t)$ takes the following asymptotic behavior

$$C_2(r, t) \sim t^{\beta}. \tag{20.14}$$

Namely, for distances much larger than the correlation length, $C_2(r, t)$ increases as a power of time. β in the above equation is called the growth exponent. It characterizes the time-dependent dynamics of the growth. For distances much less than the correlation length ξ, $C_2(r, t)$ is independent of time for surfaces that obey the normal scaling law. It reads as

$$C_2(r, t) \sim \rho r^{H}, \tag{20.15}$$

where
ρ is the local slope of the interface
H is a second scaling exponent called the Hurst exponent

The Hurst exponent H characterizes the roughness of the saturated interface. As for the local slope ρ, it is literally the approximate average slope of the local structure. Its value is related to the intersection of $C_2(r, t)$ with the vertical axis. For normal scaling, ρ does not depend on time. If ρ does depend on time, the scaling becomes anomalous. As one can see, to be able to characterize a growing interface, one needs the height–height correlation function $C_2(r, t)$, and more specifically, the critical exponents H and β. An example of organic thin-film growth is shown in Figure 20.12.

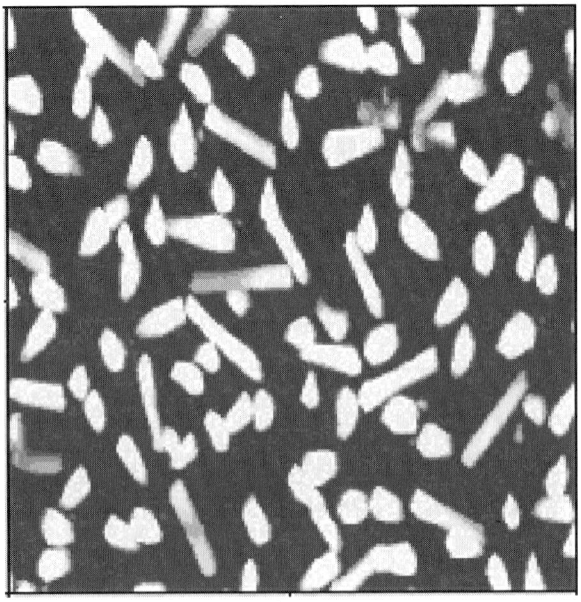

FIGURE 20.12 Tapping mode AFM images ($10 \times 10\ \mu m^2$) of perylene deposited on Au (bottom) substrates at room temperature.

Perylene forms irregular-sized randomly oriented grains on Au. Detailed examination reveals that the grains are actually crystalline. As a result, the size of some of the grains gets bigger in some preferential directions. From XPS analysis, the first layer of perylene actually grows flat on the Au surface. As discussed above, this kind of interplay between the adsorbate–substrate and adsorbate–adsorbate interactions is found in layer + island (Stranski–Krastanov) growth mode, and this growth mode is quite commonly observed for organic growth on metals.

20.5 METAL/ORGANIC INTERFACE DIPOLE

The energy level alignment at the interface involves many physical and chemical processes. In these processes, the most important ones related to the device performances include possible chemical reaction, interface dipole, diffusion, charge transfer, and possible band bending. The vacuum level alignment (VLA) model was the earliest proposed model. In this model, the vacuum level of metal cathode and ITO anode is simply aligned with that of the organic layer. As a result, the LUMO and HOMO shift inside the organic layer accordingly. Although used widely in the early days of metal/organic studies with various successes, this simple model has been disapproved by many studies, most persuasively by UPS studies. The most important difference to VLA model is that the interface dipole model states that a dipole layer exists at the metal/organic interfaces. The interface dipole causes an abrupt shift of potentials across the dipole layer, which in turn determines the LUMO and HOMO energy level alignment with respect to the Fermi level of the metal electrodes. This model has gained its support mainly by various UPS and Kelvin probe works from the change of work function or vacuum level across the interface.

Several possible mechanisms may contribute to the interface dipole. In their review article, Ishii et al. listed six possible causes, as shown in Figure 20.13: (a1 and a2) charge transfer across the interface, (b) image potential-induced polarization of the organic material, (c) pushing back of the electron cloud tail that extends out of the metal surface by the organic material, (d) chemical reactions, (e) formation of interface state, and (f) alignment of the permanent dipole of the organic material. Among these factors, (a)–(c) are more general in metal–organic interfaces, whereas (d)–(f) are specific to the individual metal–organic pair. These factors provide a plausible foundation to explain qualitatively the interface dipole formation. XPS analysis of the core-level evolution indicates that (b) is

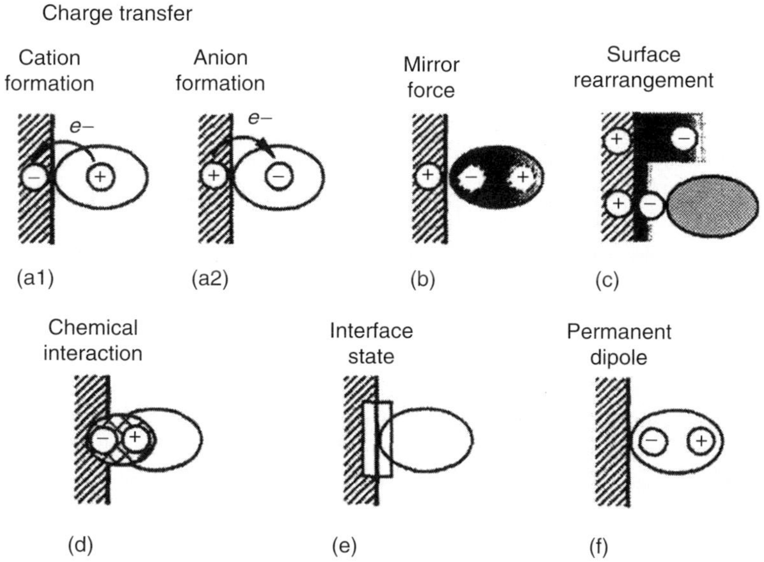

FIGURE 20.13 Models for interface dipole layer formation. (From Ishii et al., *Adv. Mater.*, 11, 605, 1999.)

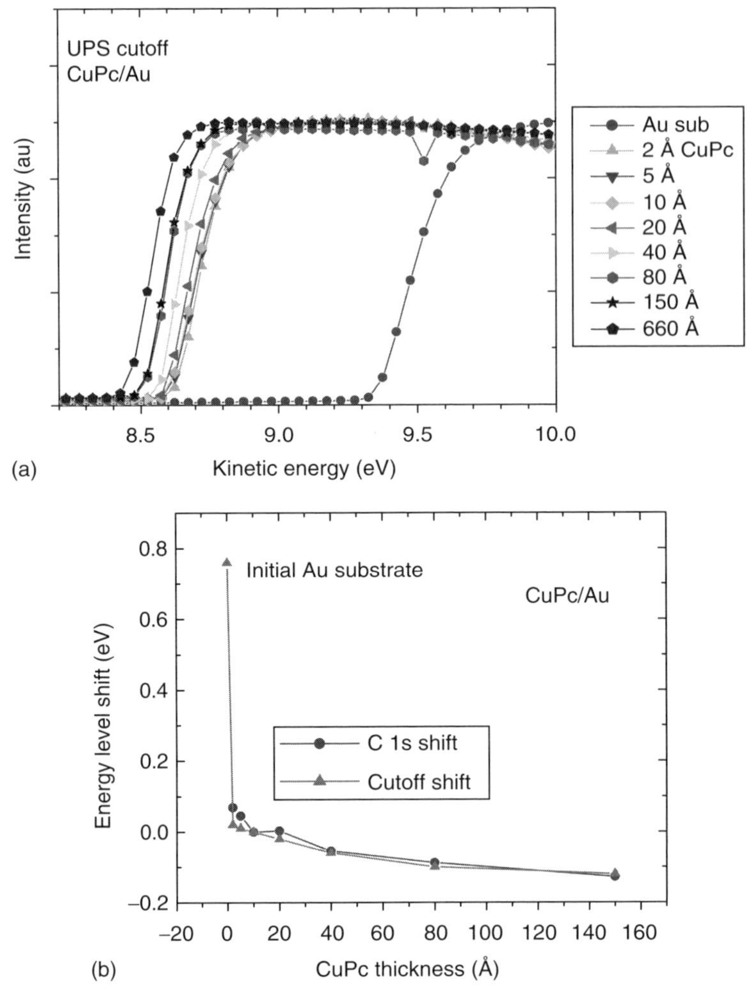

FIGURE 20.14 Determination of interface dipole at copper phthalocyanine (CuPc)/Au interface. (a) Normalized cutoff of ultraviolet photoemission spectroscopy (UPS) spectra. (b) Comparison of cutoff shift and the C 1s shift.

unlikely to be a major factor contributing to the interface dipole formation. Should the polarization inside the organic molecule be significant, one would expect that core-level peaks, especially that of the most common carbon atoms, would be broadened due to the intramolecular polarization. This is not generally observed for organic/metal interfaces. In fact, the lack of peak width change indicates that the dipole is predominantly confined to the interface between the metal surface and organic molecules.

As an example of organic/metal interface, the cutoff evolution of the UPS spectra of CuPc on Au is shown in Figure 20.14. It is clear that most changes occur at the very first layer of CuPc. Smaller changes continue when more CuPc is deposited. Most of the additional shift may be attributed to the polarization effect or extended charge diffusion, and is unlikely to be a result of the interface dipole. From Figure 20.14, we can therefore estimate that the interface dipole is about 0.8 eV at CuPc/Au. Shown in Figure 20.15 is the interface dipole at metal and tris(8-hydroxyquinoline) aluminum (Alq_3) interfaces as a function of the metal work function. The positive dipole is defined as pointing away from the metal surface. The data show approximately linear behavior, be it metal deposited on organic (Na/Alq_3, Ca/Alq_3, Al/Alq_3) or organic deposited on metal (Alq_3/Ca, Alq_3/Mg, Alq_3/Au).

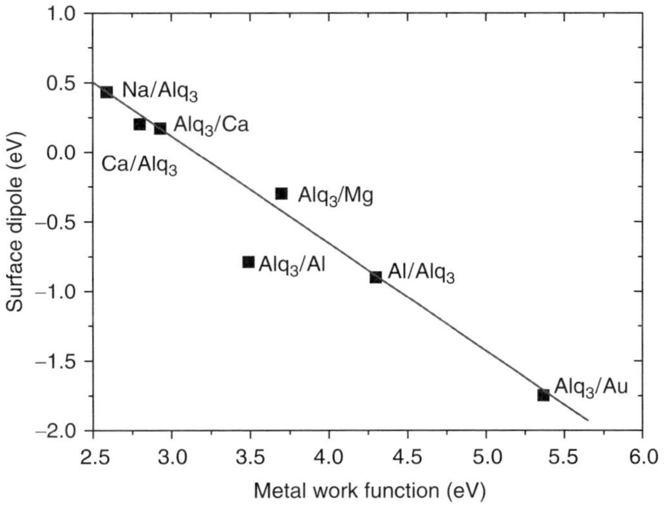

FIGURE 20.15 Interface dipole at metal and Alq_3 interfaces as a function of the metal work function. (From Yan, Li, Watkins, N.J., Zorba, S., Gao, Y., and Tang, C.W., *Appl. Phys. Lett.*, 81, 2752, 2002.)

The slope of the straight line is −0.8. Similar linear dependence on metal work function has also been observed for other organic materials.

Extra care should be taken in calculating the dipole for the case of metal deposited on organic. For organic materials not undergoing destructive chemical reactions, all the energy levels should shift the same amount to the first-order approximation, corresponding to the change of the Fermi level in the organic. As a result, the interface dipole is not the vacuum level shift observed in UPS alone but the difference between the vacuum level and the core-level shift.

20.6 CHEMICAL REACTION, DIFFUSION, DOPING, AND QUENCHING

At most interfaces in OSC devices, more than one type of interactions can and do occur, including, charge transfer across the interface, chemical reactions, and band bending. An example of a metal/organic interface chemical reaction is Ca deposited onto Alq_3, where a staged interface formation is observed to initialize by charge transfer and then chemical reaction. In Figure 20.16, the evolution of XPS C 1s, O 1s, and N 1s core-level electron density curves (EDCs) is plotted as a function of increasing Ca thickness Θ_{Ca} on Alq_3. At Θ_{Ca} = 0 Å, these core-level EDCs are gaussian in shape and composed of only one component, which indicates a clean Alq_3 film. As early as Θ_{Ca} = 1 Å, a new component is observed in the N 1s EDCs. Conversely, the O 1s EDCs remain single component until Θ_{Ca} = 4 Å, indicating that the phenoxide side of the ligand is relatively unaffected by the presence of Ca. As more Ca is deposited, the O 1s EDC splits and a new reacted component occurs at lower binding energy, which increases in intensity and shifts back to higher binding energy with increasing Ca coverage. The intensity and position of the reacted N 1s component remain almost constant with coverage above about 4 Å. Clearly, Ca first interacts with N atoms in the Alq_3 molecule until the Ca–N interaction is saturated, followed by a Ca–O reaction. The interaction between Ca and N is a charge transfer process yielding a stable $Alq_3^{\bullet -}$ and Ca^{2+}, resulting in one of the three quinolates of Alq_3 to accept the electron donated by Ca before the structural deformation of the Alq_3 allows the chemical reaction between Ca and O. Metallic Ca forms after 15 Å of Ca is deposited.

Other major factors affecting metal/organic interface formation include the thermal energy of the material being evaporated, the topology of the substrate, the condensation energy on the surface,

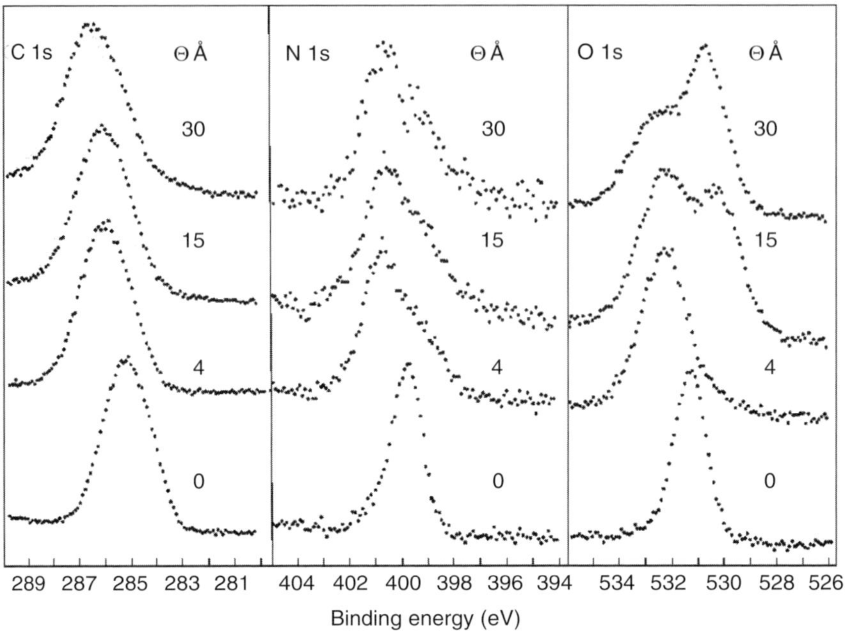

FIGURE 20.16 Evolution of XPS C 1s, O 1s, and N 1s core-level EDCs as a function of increasing Ca thickness Θ_{Ca} on Alq_3. (From Choong, V.-E., Mason, M.G., Tang, C.W., and Gao, Y., *Appl. Phys. Lett.*, 72, 2689, 1998.)

and the growth mode of the deposited film. For the process of organic deposited on metal (denoted as organic/metal), the diffusion/disruption between the two materials is low except the most reactive metals, since the thermal energy of the evaporated organic molecules is generally low due to the low evaporating temperature. The organic material normally arrives at the surface in the molecular form, and later condenses to form thin films. The condensation energy of the organic material on the surface is low due to the nature of van der Waals interaction that bonds the organic molecules. Comparing to organic/metal interfaces, there are significantly more diffusion and disruption at the metal/organic interfaces (deposit metal onto organic film), because the hot metal atoms have larger kinetic energies, which makes the penetration of metal atoms into the organic film fairly easy. Also, the metal arrives as isolated atoms or atom clusters that will give a high condensation energy, resulting in possible damage or reaction with the organic. The relative sizes of the organic molecule and metal atom are very different. The smaller metal atoms may easily fall into the vacancies between the organic molecules. This fact requires one to excise caution when interpreting the energy level changes at the initial metallization process.

The changes in the UPS spectra caused by deposition of pentacene onto Au are shown in Figure 20.17a. There is an initial shift of the vacuum level 0.8 eV closer to the Fermi level after the deposition of 2 Å of pentacene, reflected in the UPS cutoff shift. The shift of the vacuum level increases to a total of 1 eV after the deposition of 18 Å of pentacene. While these changes are observed in the cutoff position, no apparent change in position of any other feature within the UPS spectra is observed. A further decrease in the Au UPS features and a gradual increase in the pentacene UPS spectral features is observed as pentacene is deposited on the surface. Figure 20.17b shows the evolution of the UPS spectra as gold is deposited onto a pentacene substrate. Upon deposition of 1 Å of Au, the vacuum level and pentacene HOMO shift ~0.5 eV to higher binding energy. The rigid shift of these features is likely due to doping of the pentacene by Au that results in a movement of the Fermi level closer to the LUMO of the pentacene. After 8 Å of gold deposition, there is a detectable

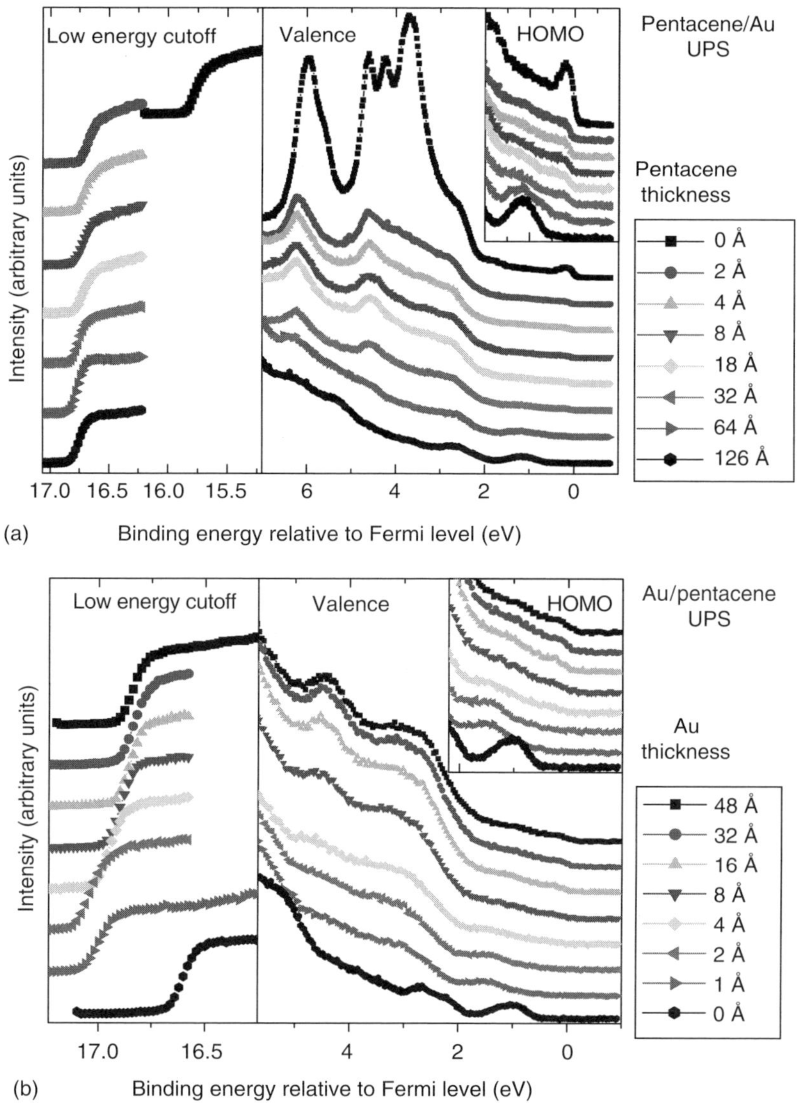

FIGURE 20.17 (a) UPS spectrum evolution as a function of pentacene deposition thickness onto Au. (b) UPS spectrum evolution as a function of Au deposition thickness onto pentacene.

Fermi level and the pentacene's HOMO features have been largely suppressed, indicating the interaction of gold with the pentacene surface. Yet even after 48 Å of gold, the HOMO features that are associated with a metallic gold layer are still not apparent. This is further evidence that the gold has not yet formed a continuous layer, possibly forming clusters on the surface of the pentacene as suggested by the XPS results. The final surface has a work function of 4.2 eV instead of the 5.4 eV that would be expected for a continuous gold surface, suggesting that even though there is a metallic surface it is not a simple gold surface.

For alkali metal/organic interface, doping of organic materials often causes significant energy level shift in these materials and it is of both fundamental and practical interest. Theoretically, it is the representative of low-dimensional guest host systems with strong correlation and anisotropic

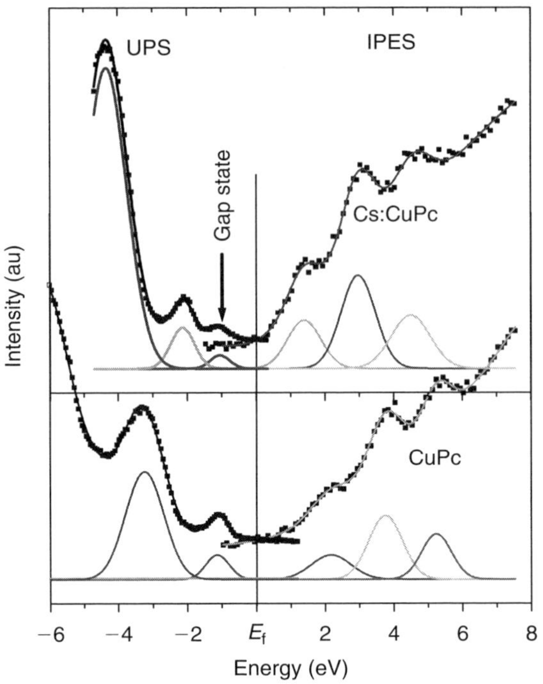

FIGURE 20.18 Comparison of the UPS and IPES spectra of pristine and Cs-doped CuPc films. (From Yan, Li, Watkins, N.J., Zorba, S., Gao, Y., and Tang, C.W., *Appl. Phys. Lett.*, 79, 4148, 2001.)

interactions. Practically, doping is widely used to improve OSC devices analogous to that in the ISC industry. It was a major breakthrough in OSC to show that doping of conjugated polymers can increase the conductivity by many orders of magnitude. In Figure 20.18, the comparison of the electronic structure and energy level alignment of pristine and Cs-doped CuPc is shown. The bottom part is the spectrum of a pristine CuPc film about 1000 Å thick prepared by thermal evaporation. The upper part is another CuPc film doped with Cs. The Cs doping concentration is determined to be about $R_{Cs} = 0.85$ by XPS measurement. As clearly shown in Figure 20.18, with the Cs doping, the LUMO shifts toward Fermi level when HOMO shifts away from it.

At a metal/organic interface, charge transfer from the metal to the organic materials may induce gap states associated with lattice deformation of the organic material. These states may also form quenching centers to the luminescence of organic materials. Shown in Figure 20.19 is the photoluminescence (PL) intensity plotted as a function of Θ_{Ca} for a film of 300 Å 4PV, an oligomer of poly(phenylene vinylene). The rate of intensity decrease can be divided into three stages. The first stage is between $\Theta_{Ca} = 0$ and 1 Å, the second between $\Theta_{Ca} = 1$ and 30 Å, and the third for $\Theta_{Ca} > 30$ Å. The first stage accounts for the reduction of the PL intensity by 65% and is due to the fact that the Ca atoms provide nonradiative decay channels. After the initial drop, the effect of PL quenching by Ca atoms is reduced, as indicated by a slower rate of PL quenching observed in the second stage, as the quenching region of additional Ca atoms overlaps with existing ones. In the third stage, $\Theta_{Ca} > 30$ Å, the drop in PL intensity is solely to the attenuation of both excitation and emission photons due to the thickness of the Ca layer as it becomes metallic. The quenching phenomenon is quite common. It has the most severe adverse effect on single layer OLEDs based on hole-conducting materials, such as PPV and its derivatives, because of the radiative recombination zone in these devices is close to the metal cathode. For electron-conducting materials, the problem will be much less significant since the recombination zone is close to the ITO anode, on which no quenching of PL is observed. It is also less important for multilayer

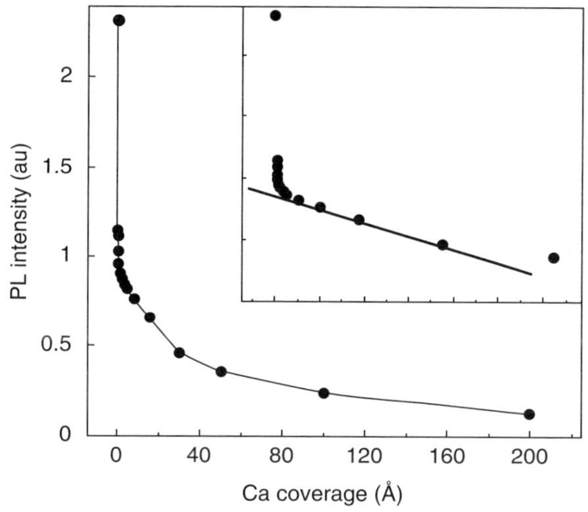

FIGURE 20.19 The evolution of PL peak intensity as a function of Ca coverage, Θ_{Ca}. Three distinct stages of PL intensity decrease are observed. The inset emphasizes the first and second stages. (From Choong, V., Park, Y., Gao, Y., Wehrmeister, T., Müllen, K., Hsieh, B.R., and Tang, C.W. *Appl. Phys. Lett.*, 69, 1492, 1996.)

devices with both electron- and hole-conducting layers, since the emission zone will be localized at the organic/organic interface, provided that the electron-conducting layer is substantially thicker than the combination of exciton diffusion length and cathode metal penetration depth.

20.7 SUMMARY

There is a strong dependence of the device performance on interfaces in OSC devices. The mismatch between a strong metallic lattice and a weak van der Waals organic solid at the metal/organic interface could result in an interfacial structure, which is uniquely different from those commonly found in metal/inorganic semiconductors. Surface and interface analytical investigations of the interfaces in organic devices have generated critical insight into the fundamental processes at these interfaces. These studies have shown that the design and control of surfaces and interfaces are important for delivering stable organic devices.

EXERCISE QUESTIONS

1. Use Equation 20.1 to estimate the rate of molecular impact in standard atmospheric conditions.
2. Find the matrix notation of the two surface reconstructions shown in Figure 20.9.
3. Show that the change in work function from molecular adsorption at a metal surface is $\Delta\Phi = n\mu/\varepsilon_0$, where n is the number density of the adsorbates, μ is the dipole moment between the adsorbate and its image charge, and ε_0 is the dielectric constant of vacuum.

LIST OF ABBREVIATIONS

AFM	atomic force microscopy
EDC	electron density curve
HREELS	high-resolution electron energy loss spectroscopy
HOMO	highest occupied molecular orbital
IPES	inverse photoemission spectroscopy

LEED	low energy electron diffraction
LEEM	low energy electron microscopy
LUMO	lowest unoccupied molecular orbital
MFP	mean free path
NEXAFS	near edge x-ray absorption fine structure
OSC	organic semiconductor
PES	photoemission spectroscopy
PL	photoluminescence
PSPD	position sensitive photodetector
RIRS	reflection infrared spectroscopy
SIMS	secondary-ion mass spectroscopy
SPM	scanning probe microscopy
STM	scanning tunneling microscopy
TPD	temperature-programmed desorption
UHV	ultrahigh vacuum
UPS	ultraviolet photoemission spectroscopy
XPS	x-ray photoemission spectroscopy

BIBLIOGRAPHY

1. H.S. Nalwa, (Ed.), *Handbook of Surfaces and Interfaces of Materials* (Academic Press, New York, 2001).
2. W.R. Salaneck, K. Seki, A. Kahn, and J.J. Pireaux, (Eds.), *Conjugated Polymer and Molecular Interfaces: Science and Technology for Photonic and Optoelectronic Applications* (Marcel Dekker, New York, 2001).
3. W.R. Salaneck, S. Stafstrom, and J.-L. Bredas, *Conjugated Polymers Surfaces and Interfaces* (Cambridge University Press, Cambridge, 1996).
4. D.P. Woodruff and T.A. Delchar, *Modern Techniques of Surface Science* (Cambridge University Press, Cambridge, 1994).
5. J.B. Hudson, *Surface Science: An Introduction* (Wiley, New York, 1998).
6. R.L. Park and M.G. Lagally, (Eds.), *Solid State Physics: Surfaces* (Academic Press, New York, 1985).
7. S. Hufner, *Photoelectron Spectroscopy: Principles and Applications* (Springer-Verlag, Berlin, 1995).
8. R. Wiesendanger, *Scanning Probe Microscopy and Spectroscopy: Methods and Applications* (Cambridge University Press, Cambridge, 1995).
9. A.-L. Barabási and H.E. Stanley, *Fractal Concepts in Surface Growth* (Cambridge University Press, Cambridge, 1995).
10. Li Yan, N.J. Watkins, S. Zorba, Yongli Gao, and C.W. Tang, Thermodynamic equilibrium and metal-organic interface dipole, *Appl. Phys. Lett.*, 81, 2752 (2002).
11. V.-E. Choong, M.G. Mason, C.W. Tang and Y. Gao, Investigation of the interface formation between calcium and tris-(8-hydroxy quinoline) aluminum, *Appl. Phys. Lett.*, 72, 2689 (1998).
12. Li Yan, N.J. Watkins, S, Zorba, Yongli Gao, and C.W. Tang, Direct observation of fermi level pinning in Cs doped CuPc film, *Appl. Phys. Lett.*, 79 4148 (2001).
13. V. Choong, Y. Park, Y. Gao, T. Wehrmeister, K. Müllen, B.R. Hsieh, and C.W. Tang, Dramatic photoluminescence quenching of phenylene vinylene oligomer thin films upon submonolayer Ca deposition, *Appl. Phys. Lett.*, 69, 1492 (1996).

21 Single-Molecule Organic Electronics and Optoelectronics

Ling Zang, Xiaomei Yang, and Tammene Naddo

CONTENTS

21.1 Organic Electronics and Optoelectronics: From Nanodevices to Single-Molecule Devices 660
21.1.1 General Background 660
21.1.2 Organic Nanomaterials and Application in Nanoelectronics and Optoelectronics 661
21.1.3 Single-Molecule Devices 662
21.1.4 Single-Molecule Devices versus Integrated Molecular Chips: A Big Gap to Overcome 663
21.2 Single-Molecule Electronics: Devices and Research 663
21.2.1 Experimental Systems Developed for Characterizing Single-Molecule Electronic Devices 663
21.2.1.1 Surface Deposited Nanoelectrode Systems 664
21.2.1.2 Metallic Wire Based Break-Junction Measurement 665
21.2.1.3 Scanning Tunneling Microscopy Based Physical Contact 667
21.2.1.4 Atomic Force Microscopy Based Chemical Contact 668
21.2.1.5 Scanning Tunneling Microscopy Based Break-Junction Measurement 670
21.2.2 Molecular Wire 671
21.2.3 Molecular Diode 675
21.2.4 Molecular Field-Effect Transistor 678
21.2.5 Single-Molecule Chemical Field-Effect Transistor 684
21.2.6 Single-Molecule Electrical Sensing Based on Conductivity Modulation 686
21.3 Single-Molecule Optoelectronics: Concepts and Mechanisms 689
21.3.1 Single-Molecule Photodiode 689
21.3.2 Single-Molecule Phototransistor 690
21.3.3 Single-Molecule Optical Sensor 691
21.4 Remarks 693
Exercise Questions 694
List of Abbreviations 694
Acknowledgments 694
References 694

Abstract This chapter introduces the basic principles and operation mechanisms of single-molecule electronic and optoelectronic devices. Both the general background and the current research status have been discussed, followed by overview of the future scientific study and application development. Various types of single-molecule devices, ranging from diodes to transistors and from switches to sensors, have been described and reviewed. Studies of single-molecule systems not only lead to design and fabrication of devices but also more importantly provide the ultimate level of understanding of the molecular electronic properties and the dependence on external perturbing (e.g., applied field). Such understanding helps broaden the concept and perspective of molecular electronics and optoelectronics.

21.1 ORGANIC ELECTRONICS AND OPTOELECTRONICS: FROM NANODEVICES TO SINGLE-MOLECULE DEVICES

21.1.1 General Background

Electronics is defined as the field of fabricating and studying the systems that operate via controlling the flow of electrons or other charge carriers (e.g., holes and ions) in devices such as diodes, transistors, and electrical switches. Such current controlling can be implemented through changing the electrical voltage (i.e., bias) applied between the two electrodes or the voltage applied to the third electrode (i.e., gate) that is functional for modulating the charge carrier density and mobility in the channel. The study of modern electronics is often a combination of physics (particularly semiconductor physics) and engineering. The former focuses on principle exploration and materials design; while the latter focuses on device construction and improvement. The field of optoelectronics is the field of studying the interaction and relationship between photon and electronic devices or materials. An optoelectronics device can be based on photo-modulation (e.g., effect of illumination) of the electron flow within a device, or based on production of photons via recombination of charge carrier pairs (electrons and holes) that are created in the device. Typical optoelectronic devices include optical switch, photodiode, solar cell, and light-emitting diode (LED).

One of the most common electronic devices is field-effect transistor (FET) (Figure 21.1), which has been widely used in integrated circuits for electrical switching and amplification of weak signals. Typical application of FET in modern electronics includes computer chips, display, memories, and signal amplifiers. The basic function of a FET is to switch electrical current between two electrodes ON and OFF with a gate electrode. Such switching with respect to the applied gate voltage (negative versus positive) is reverse for n-type and p-type channel. For n-type channel (the conductivity of

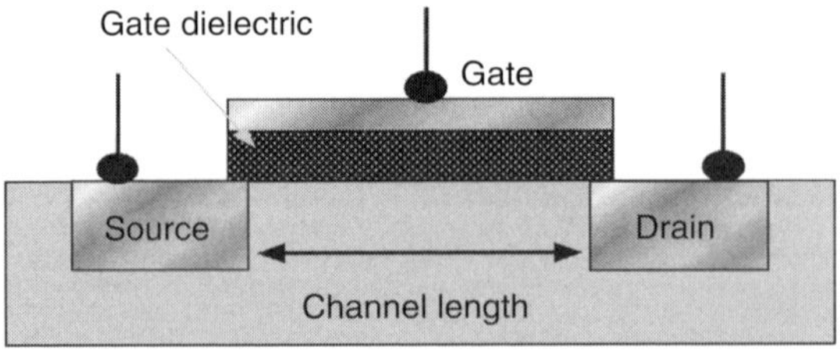

FIGURE 21.1 A schematic illustration of cross section of a lateral FET. Source electrode: providing the major charge carriers (e^- for n-type semiconductor, h^+ for p-type semiconductor); Drain electrode: drawing the major carriers from channel materials; Gate electrode: providing local electrical field to modulate the charge carrier density and mobility inside the channel material; Gate dielectric layer: functioning as an insulating layer between gate and channel material, and enabling creation of spatial gradient of the electrical field across the dielectric layer. Gate effect is dependent on both the electrical property and thickness of the dielectric layer.

which is dominated by flow of electrons), when a negative gate voltage is applied above the layer of dielectric, the channel attracts most available positive charges (holes) toward itself due to opposite charge attraction. As a result, the flow of electrical current (i.e., actually the flow of negatively charged electrons, e^-) is reduced due to the presence of the opposite positive charge. This phenomenon is called depletion of current. In contrast, when a positive voltage is applied to the gate, the n-channel attracts more electrons toward itself, thus leading to enhancement of electrical current. For p-type channel (the conductivity of which is dominated by flow of holes, h^+), the gate modulation of current is just reverse to that of n-type channel.

Traditionally, both electronics and optoelectronics are based on silicon materials, which have so far provided the highest efficiency for practical application. For example, silicon-based photodiode can be used to detect single photons; and multi-junction silicon-based solar cells can provide output efficiency above 40% [1]. However, silicon-based optoelectronic devices (e.g., solar cells) are too expensive at the current stage to be commercialized, mainly due to the high cost in materials and manufacturing, which usually involves elevated temperature (400°C–1400°C), high vacuum, and numerous lithographic steps. Meanwhile, more and more research efforts have been drawn to the alternative devices based on organic materials, which provide low cost in large-area processing and compatibility with various flexible substrate [2–9]. For example, organic solar cells can be made in highly flexible and conformal shape by coating the thin film onto soft, adaptable substrates, such as plastics and fabrics. Moreover, organic synthesis offers options for making molecules with various structures and electronic properties, which may provide broad spectral response, as well as self-assembling properties for fabrication of organized materials to approach optimal device performance. Particularly for optoelectronic devices, which demand tunable spectral response covering UV and visible (and even IR) region, organic materials have been developed as ideal channel materials with improved sensitivity and selectivity through molecular design and engineering.

Indeed, organic-based semiconductor materials have successfully been employed in various electronic and optoelectronic devices such as transistors, photodiodes, solar cells, flat-panel displays, LEDs, and optical switches and sensors. Compared with the silicon-based devices, the main factor limiting the efficiency of organic-based devices is the poor organization of materials, which in turn depends on intermolecular interaction in macroscale. Poor molecular organization limits the long-range exciton transport (i.e., energy transfer) and charge carrier mobility. Although improvement in molecular structure and self-assembly processing helps increase the materials structure and the device performance, the overall efficiency of organic-based devices is still far below the silicon-based counterparts, preventing these from practical application. For example, the best efficiency obtained so far for organic solar cells is below 6% [2,3,6,10–12], compared to ~40% of the cell based on single-crystalline silicon materials.

21.1.2 Organic Nanomaterials and Application in Nanoelectronics and Optoelectronics

Taking advantages of the well-developed nanotechnologies, it becomes feasible for people to fabricate devices in nanometer scale using nanomaterials. Particularly for organic materials, the decreased size makes it easier to construct highly organized, continuous crystalline phase, and optimize the morphology and dimensional sizes (length, width, and thickness) to be fit into the devices. In general, it is difficult to fabricate large-scale single-crystal phase of organic materials (compared to inorganic materials). Most of the organic materials (when fabricated at macroscales) are in polycrystalline structure, where the crystalline boundaries act as high-energy barriers or traps for charge transport. As a result, the conductivity of the materials is normally lower by orders of magnitude than that of inorganic materials such as silicon. One way to remove the crystalline boundary is to decrease the dimension of materials down to the region of nanometers. For nanomaterials, the whole phase can feasibly be fabricated as single, continuous crystal, and the conductivity is primarily determined by the intermolecular interaction (e.g., π–π stacking). Fabrication of devices with smaller channels improves the device performance by decreasing the resistance (or increasing the electrical current) or enhancing the local electrical field gradient (i.e., enhancing the gate modulation). Moreover, employing smaller number of molecules in the device channel makes it more feasible to control and optimize the intermolecular interaction, and thus the channel conductivity, in line with the appropriate theoretical modeling and calculation.

Organic nanomaterials (particularly the nanowires) offer a new approach to miniaturization of electronic and optoelectronic devices, via the so-called bottom-up methodology, for which the devices are constructed by assembling the nanosized units. This is in contrast to the traditional top-down methodology, for which the devices are manufactured by photolithography of the large-area substrate (usually silicon wafer). During the past decades, the unit size of silicon-based electronic devices has been decreasing following the Moore's law prediction, resulting in continuous increase in the speed and storage capacity of the devices. Although the advanced photolithography and the corresponding microscopy techniques allow for creation of electronic units in the size range of nanometer, there are several severe problems accompanying the decreased sizes of devices. One such problem is the leakage of electrons (information), which can hardly be overcome by current lithography technology. For example, when gate size is smaller than 10 nm, direct electrical tunneling between the gate electrode and channel material (or between the gate and source electrode) would become significant, i.e., the current flowing through the channel is no longer controlled by the depletion layer, which is usually determined by the gate voltage. One way to overcome the leaking problem is using different dielectric materials for the gate, or using different channel materials which have improved depletion properties. But so far, no such materials with ideal properties have been found. Moreover, fabrication of the new materials requires more sophisticated procedures, and, thus, will soar up the manufacturing cost.

Various organic semiconductor materials have been employed in fabrication of organic nanodevices. Typical materials include conducting polymers and planar aromatic molecules, and the most common motif employed in nanodevices is the nanowire. While organic nanowires can feasibly be fabricated from various conducting polymers like polyaniline [13–17], polyacetylene [18,19], polypyrrole [20–22], and poly(phenylene vinylene) [23,24], and have been tested for application in nanoelectronics [13,14,18], the crystalline structure of polymer assembly is often difficult to control due to the complicated intermolecular interactions and polydispersity of the chain. This hinders the further improvement and optimization of the device performance, which demands detailed understanding of the optoelectronic property of materials in correlation to the molecular properties and organization. Recently, significant research effort has been directed at the molecular self-assembly involving both conjugated oligomers [25] and aromatic planar molecules [26–29], which enable strong π–π intermolecular interaction. Effective π–π stacking usually leads to preferred 1D crystalline growth, i.e., formation of nanowires. This is particularly evident with the larger aromatic molecules like hexabenzocoronene [30,31].

21.1.3 Single-Molecule Devices

Along with the research and development in nanomaterial-based devices as mentioned above, simply using a single molecule as the channel provides a totally different approach. Recent advanced nanoscopic technologies (e.g., atomic force microscopy [AFM] and scanning tunneling microscopy [STM]) enable construction and investigation of devices consisting of only one molecule. Study of such devices emerges as a new field, so-called single-molecule electronics or optoelectronics. Single-molecule devices not only represent the ultimate level of device miniaturization but also, more importantly, provide the simplest systems for studying the fundamental operation principles of molecular electronics and optoelectronics. In a molecule, electron moves locally between quantized energy levels within the molecular chains. The device performance can be correlated directly to the molecular structure (electronic properties), thus enabling direct guidance for molecular structural optimization (via synthesis) to approach the desired electronic properties.

Additionally, a single-molecule system makes it easy for theoretical calculation and modeling, which are crucial for understanding the device operation and performance. For example, once bound to the two electrodes (source and drain in a transistor), the molecule will be tuned in electronic structure by covalent binding with the surface atoms of electrodes. The whole molecular junction (including both the molecule and the electrodes), rather than just the molecule itself, will have to be considered as a unit for studying the performance of the devices. How to define the conjugation between the molecule and the electrode is one of the main issues in the filed of theoretical investigation of molecular electronics [32–36]. Indeed, experimental observations have evidenced the significant effect of the interfacial molecule–electrode contact on the efficiency of single-molecule electronic devices [37].

Compared to the devices based on bulk materials or molecular assemblies (nanomaterials), single-molecule devices possess several unique features that define the related studies as an attractive, fast growing field. First, building a single-molecule device (like FET) is naturally considered to be a critical step toward the ultimate goal of molecular electronics, which demands integration of large number of molecules. Second, measurement of a single-molecule device removes ensemble averaging of bulk measurement that occurs to the devices consisting of multiple molecules. Generally, intermolecular interaction and heterogeneous distribution of molecules inside the device channel produce different performance efficiencies for individual molecules. Such difference cannot be revealed by the ensemble measurements that examine all the molecules simultaneously. Third, a single-molecule device provides the simplest system for theoretical modeling and calculation, which are crucial for explaining device operation and guiding the future optimization. Lastly, with appropriate molecular design and engineering, single-molecule devices may provide improved sensitivity and selectivity for application in sensors and optical switches. For example, modification of the chromophore can lead to spectral selectivity for sensing different wavelengths of illumination.

21.1.4 Single-Molecule Devices versus Integrated Molecular Chips: A Big Gap to Overcome

Although the original driving force behind molecular electronics is to fabricate new type of integrated circuits using individual molecules (as the electronic unit), leading to a chip or device with higher speed or larger storage capacity, the current studies in this field are mostly focused on the fabrication and characterization of a single device composed of one or a few molecules bridged between two electrodes. It is extremely challenging at the current stage to assemble and organize large number of molecules into a functional chip (in the fashion of the silicon-based computer chip). For example, an Intel P4 chip has ~50 M transistors (at its current silicon-unit size); if replaced with molecular units, it will contain more than billions of molecules. Unfortunately, the current knowledge and technologies do not allow interconnecting and integrating such a big number of molecules together and maintaining the efficient operation.

The primary study at the current stage is focused on single-molecule devices. Such studies provide improved understanding of the fundamental processes of charge transfer through molecules and interfaces. Information and knowledge obtained from these studies will guide the future exploration in large-scale integration of molecular devices. Moreover, a single-molecule device by its own may find applications in some specific fields, such as sensors and switches with improved sensitivity and selectivity (as detailed below).

21.2 SINGLE-MOLECULE ELECTRONICS: DEVICES AND RESEARCH

21.2.1 Experimental Systems Developed for Characterizing Single-Molecule Electronic Devices

The past few years have witnessed increasing interest in the design and development of single-molecule electronics [33,38–54], along with the significant advancements in nanoscale fabrication and characterization. Although great progress has been made, people now face even more challenges in both the conceptual understanding and experimental study of the single-molecule systems. Some conceptual and technical challenges are already apparent. One of the major difficulties is the lack of reproducibility associated with much of the experimental data. This problem is partially (if not mainly) due to the ensemble average of molecular observables. These ensemble systems are often insufficiently controlled, complicating both measurement and explanation. For example, lateral interactions between molecules may cause orbital mixing and energy splitting, and thus affect the electronic properties of molecules. Thereby, molecular conductivities as measured from analyses of ensemble averaging are often inconsistent even for the same molecular structure. Indeed, the conduction mechanism at the single-molecule level remains in debate [39,55–57].

Thus, before one can obtain an improved understanding of the effects of field modulation on molecular conductivity, a reliable method is needed to measure the electron transport in single molecules. To date, many experimental methods have been developed, but only a few of them have been proven reliable for characterizing single-molecule conduction properties. The two challenging problems facing such measurements on single molecules are the contact problem (interfacing a molecule with electrodes) [39,58,59], and how to ensure that only one molecule is being measured at a given time. Although it is easy to align a single nanotube or inorganic nanowire between two electrodes, it turns out to be challenging to do the same thing for a molecule, mainly due to its much smaller size.

During the past decades, huge amount of efforts have been made to fabricate molecular electronic devices with a single molecule or at least minimum number of molecules. Briefly described below are some of the typical methodologies that have been developed for bridging molecules between two metal electrodes (mostly gold).

21.2.1.1 Surface Deposited Nanoelectrode Systems

To bridge a molecule between two electrodes (usually via the thiol–gold binding) to form a molecular junction, the two electrodes must be fabricated as close as in nanometers to adapt the size of the molecule (Figure 21.2). With the advanced electron-beam lithography and electrodeposition techniques, two-electrode nanojunctions down to 2 nm have been successfully fabricated (Figure 21.3) [40,60]. However, successful rate for fabricating such small gaps is extremely low, around 5% or lower. The difficulty in fabricating nanoelectrode systems and precisely controlling the gap distance prevents the electronic investigation for small moelcules, such as 1,4-benezene dithiol, which has been extensively studied by theoretical modeling and calcualtions [61]. Recently, some nanoelectrode systems have been fabricated with palladium instead of gold. It was found that the thiol–palladium interaction is as strong as thiol–gold, but palladium forms smaller grains (15–25 nm) compared to gold (50–100 nm) when fabricated as thin films by electron-beam evaporation [62]. This unique property is crucial for fabrication of nanoelectrode systems, which require both small electrode islands and small space between them.

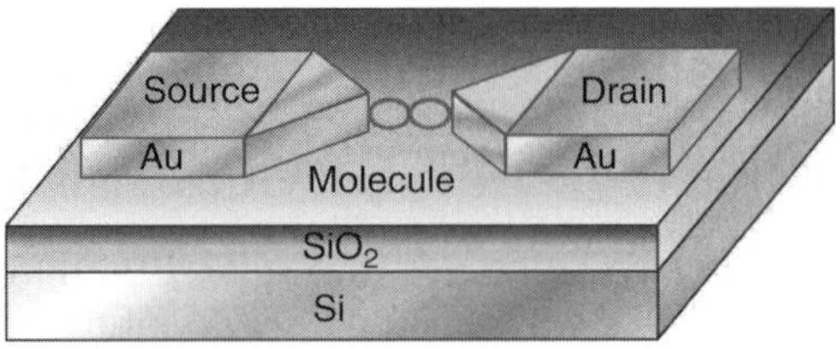

FIGURE 21.2 A nanoelectrode molecular junction fabricated on oxidized silicon surface, which can be adapted to be a FET device by attaching a gate to the silicon substrate.

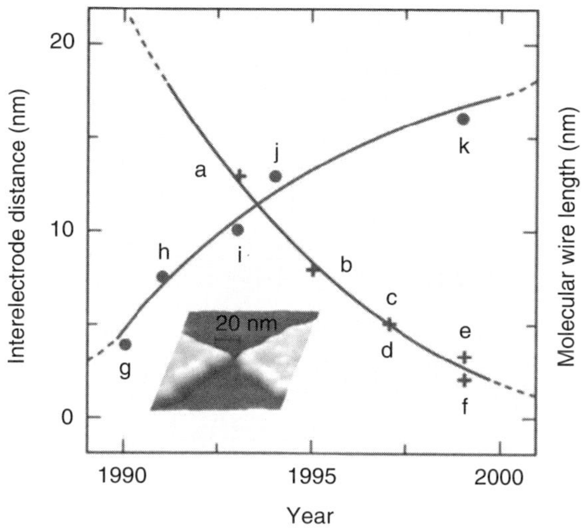

FIGURE 21.3 Interelectrode distance (cross) and molecular wire length (dot). Planar standard e-beam nanolithography nanojunctions (a, b). Coplanar e-beam nanojunctions with the electrode metal at the same height as the substrate (b, e). Suspended planar nanojunctions fabricated by electrodeposition (d) and electrochemistry (f). Long oligoimide molecular lines (g, h). Thiophene-ethynylene oligomers (i). Phenylalkyne oligomers (j). Alternating block co-oligomers of (1,4-phenylene ethynylene)s and (2,3-thiophene ethynylene)s (k). Inset, AFM tapping mode image of a 4 nm coplanar nanojunction. In e, the two gold–palladium metal electrodes are only 1 nm higher than the background silicon wafer surface. (Reprinted from Joachim, C., Gimzewski, J.K., and Aviram, A., *Nature*, 408, 541, 2000. With permission. Copyright 2000 Macmillan Magazines Ltd.)

Nanoelectrode systems provide in situ fabrication of nanojunctions on oxidized silicon surface, and the investigation at dry state provides direct guidance for future design and optimization of molecular electronics, which are mostly operated at air/solid interface. However, the current lithography techniques do not allow for precise, flexible control of the electrode gap to match the size of specific molecules. Particularly when the gap falls into the range of 1–5 nm, it is really difficult to fix the gap size before the two electrodes eventually get touched. Moreover, the molecular junction is usually constructed by depositing diluted molecular solution onto the nanoelectrode surface, followed by rinsing off the excessive molecules. With such a process, it is often (if not always) to have multiple molecules bound inside the electrode gap, where some of the molecules may be bound to only one electrode, resulting in underestimation of the conductivity of the molecular junction. Indeed, it is impossible for most cases to ensure the number of molecules that are in good contact (binding) to both electrodes. Previous studies on such electrode systems were based on ensemble averaging and necessary presumptions.

21.2.1.2 Metallic Wire Based Break-Junction Measurement

In 1997, Reed and coworkers reported a unique method to fabricate distance-controlled nanoelectrode system, so-called break-junction (Figure 21.4) [63]. The nanojunction was formed by breaking a thin gold wire, which was coated with a monolayer of dithiol molecules. Since the breaking was controlled by a piezoelectric device, the distance between the two broken wires can be adjusted and controlled precisely at the resolution of angstrom, to match the different sizes of molecules under measurement. In an ideal case, the nanojunction can be fabricated with only a single molecule bridged between the protrusions (i.e., the tip of the broken wire) of the two electrodes (Figure 21.5). Through this way, the electrical conductivity was measured for the first time. Particularly,

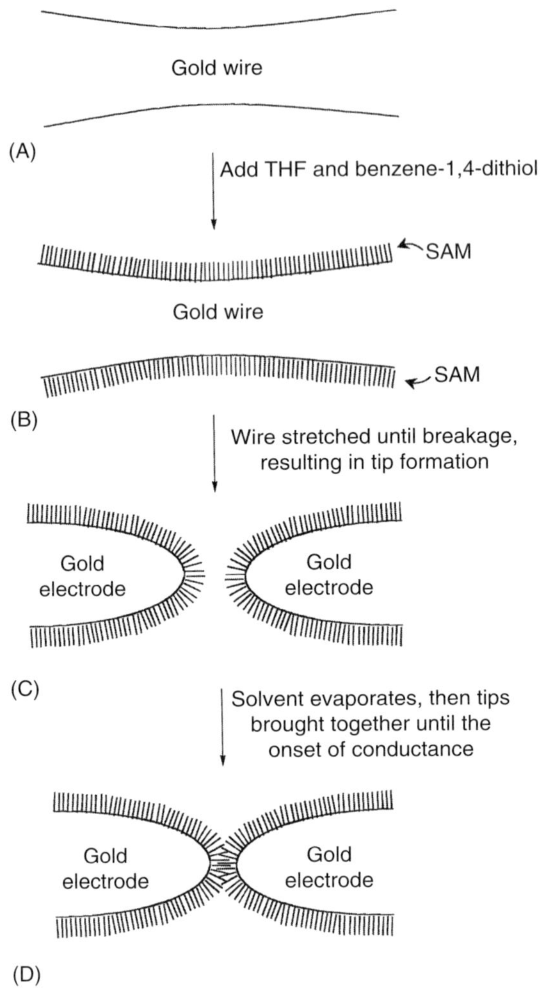

FIGURE 21.4 A schematic illustration of fabrication of Reed break-junction: (A) The gold wire of the break-junction before breaking and tip formation. (B) After addition of benzene-1,4-dithiol, a self-assembled monolayer (SAM) forms on the gold wire surfaces. (C) Mechanical breakage of the wire in solution produces two opposing gold contacts that are covered by an SAM. (D) After the solvent is evaporated, the gold contacts are slowly moved together until the onset of conductance is achieved. A zoom-in view of the nanojunction (D) is presented in Figure 21.5. (Reprinted from Reed, M.A., Zhou, C., Muller, C.J., Burgin, T.P., and Tour, J.M., *Science*, 278, 252, 1997. With permission. Copyright 1997 American Association for the Advancement of Science.)

small molecules like benzene dithiol (which was difficult to measure with the surface nanoelectrode system described above) were successfully junctioned and characterized for the electrical properties [63].

One major challenge for the break-junction method is that it is often difficult to assure only one molecule bridged inside the junction. First, the protrusion part of the broken wire may not be that sharp, i.e., there may be two or more molecules bound at the top area. Second, it is not guaranteed that the two broken wires are placed in a configuration that both the protrusion parts are exactly at the same level to form a nanojunction with the smallest contact area. In most cases, the two protrusions are mismatched, resulting in a bridge with multiple molecules. Lastly, the whole process is not suitable for repeated, fast measurement, which is normally desired for reliable screening of different molecules to compare the electrical properties and the dependence on the molecular structure.

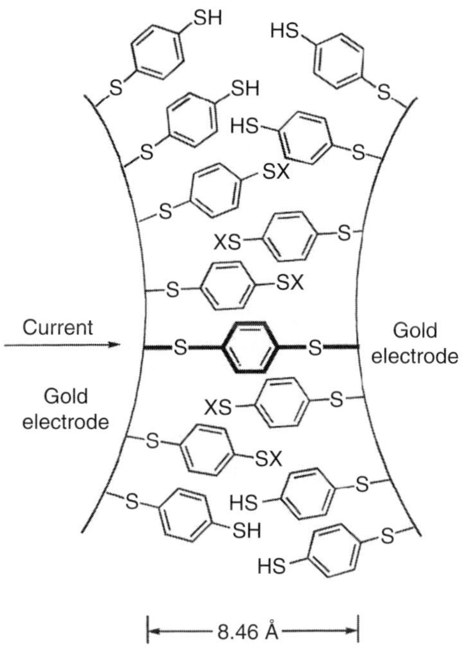

FIGURE 21.5 A schematic of a benzene-1,4-dithiolate SAM between two proximal gold electrodes. The thiolate is normally H-terminated after deposition; end groups denoted as X can be either H or Au, with the Au potentially arising from a previous contact/retraction event. These molecules remain nearly perpendicular to the Au surface, making other molecular orientations unlikely. (Reprinted from Reed, M.A., Zhou, C., Muller, C.J., Burgin, T.P., and Tour, J.M., *Science*, 278, 252, 1997. With permission. Copyright 1997 American Association for the Advancement of Science.)

21.2.1.3 Scanning Tunneling Microscopy Based Physical Contact

Scanning tunneling microscopy, as it was designed and developed, provides convenience in large-area scanning, and precise lateral positioning and current measurement. It was by STM that individual atoms and molecules were directly imaged for the first time. The atomic resolution of STM makes it a unique tool to be employed for single-molecule electronic investigation. In 1996, Weiss and coworkers reported a successful measurement of molecular conductivity at the single-molecule level by using a monolayer sample system (Figure 21.6) [57]. The mixed monolayer (peg-in structure) can feasibly be prepared by incubating the freshly cleaned electrode (e.g., gold) in a co-adsorption solution of alkyl thiol and the molecule (terminated with thiol) of interest. The insulating alkyl monolayer matrix separates the conductive molecules from one another, resulting in nearly homogeneous distribution of the target molecule over the entire surface. The surface density of the target molecules can be easily adjusted by changing the molar ratio of the co-adsorption solution. Moreover, the rigid structure of monolayer functions as a support to maintain the vertical configuration of the target molecule, making it easy to be touched by the STM tip during the lateral scanning.

Since the thickness of monolayer can be feasibly adjusted and controlled by using different lengths of alkyl thiols, people can basically use such monolayer matrix to host various sizes of molecules (from subnanometer to a few nanometers). The well-defined structure of monolayer provides ideal matrix to fix and maintain the molecular conformation of the target molecules, leading to convenient investigation of the conformation dependence of molecular conductivity. Combined with the high-resolution scanning by STM, the STM/monolayer system shown in Figure 21.6 offers a convenient device to calibrate the single-molecule conductivity. One potential problem for this approach is the uncertain contact between the molecule and the STM tip. The measured resistance might be partially (or even mainly) contributed by the poor contact.

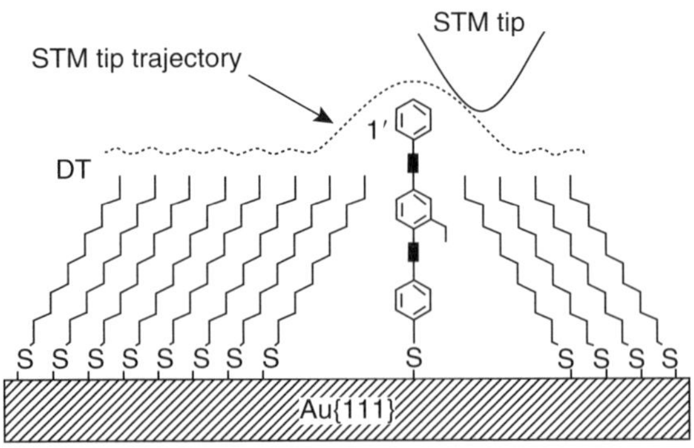

FIGURE 21.6 A schematic representation of a thiol-terminated conjugate molecule embedded in a alkyl-thiol monolayer Au{111}. The trajectory of the STM tip traces out a surface of constant current. The relatively flat monolayer can be imaged by an atomic-scale asperity on the end of the STM tip with resolution of the molecular lattice. (Reprinted from Bumm, L.A., Arnold, J.J., Cygan, M.T., Dunbar, T.D., Burgin, T.P., Jones, L.I., Allara, D.L., Tour, J.M., and Weiss, P.S., *Science*, 271, 1705, 1996. With permission. Copyright 1996 American Association for the Advancement of Science.)

21.2.1.4 Atomic Force Microscopy Based Chemical Contact

As inspired by the pioneering work with STM, other scanning probe microscopy (SPM) techniques, such as AFM, have been employed and adapted for molecular conductivity measurement, particularly at the single-molecule level. To improve the contact between the scanning tip and the target molecule, the molecule was modified with thiol moiety at both terminals, and the mixed monolayer (as described in Section 21.2.1.3) thus prepared was combined with gold nanoparticles, which selectively bind to the target molecule though the thiol–gold interaction [64,65]. Scanning such a monolayer through SPM feedback mechanism will ensure the ohmic contact between the metallic tip and the gold particle, thus leading to a perfect electrical connection throughout the whole circuit (Figure 21.7) [66].

Using the similar monolayer system, Lindsay and coworkers developed in 2001 an AFM-based technique to reliably calibrate the conductivity of alkyl-thiol molecules [66] The technique is based

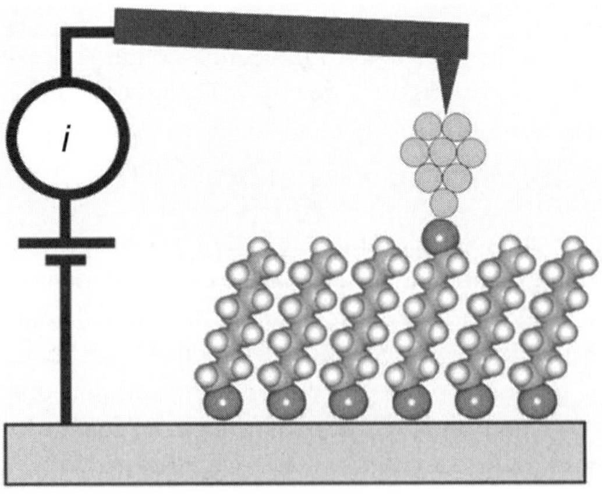

FIGURE 21.7 Molecular circuit: the conductivity of an isolated 1,8-octanedithiol molecule can be measured by chemically bonding one of its thiol moieties (dark shaded) to a gold base electrode and the other thiol moiety to a gold nanoparticle that is in contact with the gold tip of a conducting atomic force microscope. (Reprinted from *C&EN*, 79, 14, 2001. With permission. Copyright 2001 American Chemical Society.)

on repeated measurement of a large number of molecules (the gold nanoparticles), followed by detailed data statistics (Figure 21.8). Since one gold particle may simultaneously bind to multiple molecules due to the large size of the particle, the conductivity thus measured could be multiple times higher than that of a single molecule. Averaging the obtained conductivity value by a certain integer (i.e., the number of molecules in contact with the gold particle) usually produced the same value for different particles under measurement. This fundamental value corresponds to the conductivity of a single molecule. Compared to the previous measurements that were based on a one-time measurement over a fixed molecular system, the Lindsay's method relies on repeated measurements

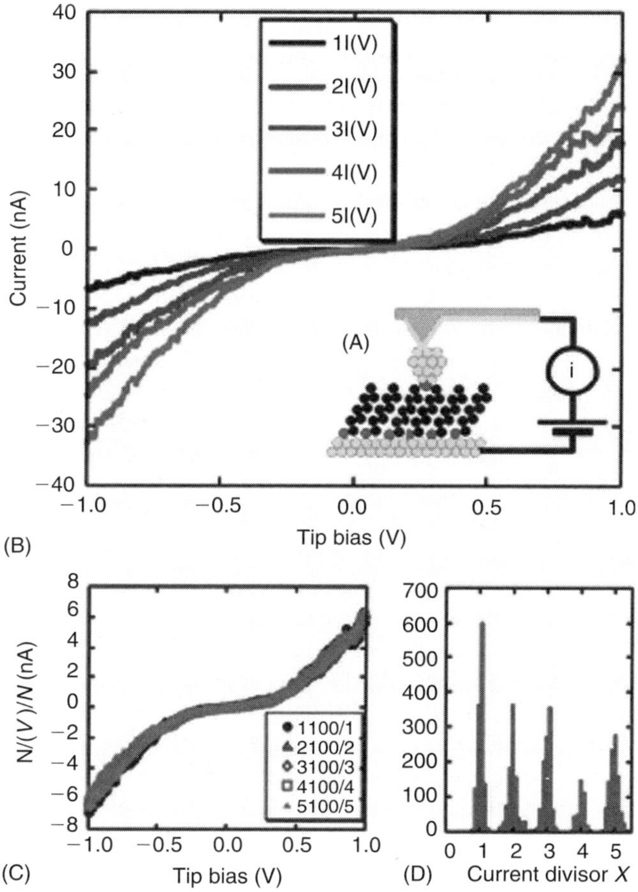

FIGURE 21.8 (A) Schematic representation of the AFM-based measurement of single-molecule experiment. The sulfur atoms (medium-shaded dots) of octanethiols bind to a sheet of gold atoms (lightly-shaded dots), and the octyl chains (black dots) form a monolayer. The second sulfur atom of a 1,8-octanedithiol molecule inserted into the monolayer binds to a gold nanoparticle, which in turn is contacted by the gold tip of the conducting AFM. (B) *I–V* curves measured with the apparatus diagrammed in A. The five curves shown are representative of distinct families, *N(I–V)*, that are integer multiples of a fundamental curve, *I–V* ($N = 1, 2, 3, 4$, and 5). (C) Curves from B divided by 1,2,3,4, and 5. (D) Histogram of values of a divisor, *X* (a continuous parameter), chosen to minimize the variance between any one curve and the fundamental curve, *I–V*. It is sharply peaked at integer values 1.00 ± 0.07 (1256 curves), 2.00 ± 0.14 (932 curves), 3.00 ± 0.10 (1002 curves), 4.00 ± 0.10 (396 curves), and 5.00 ± 0.13 (993 curves). (Spreads are ±1 SD.) Of 4579 randomly chosen curves, over 25% correspond to the $X = 1$ (single-molecule) peak. (Reprinted from Cui, X.D., Primak, A., Zarate, X., Tomfohr, J., Sankey, O.F., Moore, A.L., Moore, T.A., Gust, D., Harris, G., and Lindsay, S.M., *Science*, 294, 571, 2001. With permission. Copyright 2001 American Association for the Advancement of Science.)

on large number of molecules (>1000), and therefore, the parameter value deduced from the statistics should be more reliable. One potential problem for this method could be the unequal binding for the molecules underneath the gold particle. Owing to the relatively large size and the round geometry of the particle, some molecules (the thiols) may be in poor contact compared with the neighbor molecules. More importantly, in many cases, the measured conductance of a group of molecules is not simply an integer multiple of the conductance of a single molecule because intermolecular interaction and the complicated conjugation between the molecule and gold metal may affect the conductance measurement, which actually involves both the molecules and the electrodes as a whole nanojunction system. Therefore, an ideal way to calibrate the molecular conductivity is to measure only one molecule at a time.

21.2.1.5 Scanning Tunneling Microscopy Based Break-Junction Measurement

Although conceptually simple, it turns out to be very challenging to calibrate the single-molecule conductivity. Unambiguous contact to a single molecule is difficult to achieve, as shown by large disparities in conductivities reported for identical or similar molecules. Calculated conductivity can disagree with experimental results by several orders of magnitude. In many cases, electrical connections to the molecules have been made via nonbonded mechanical contacts rather than chemical bonds, and it is likely that this may account for some of the discrepancies.

To reliably measure the conductance of a single molecule, one must (1) provide reproducible contact between the molecule and two probing electrodes [66–70]; (2) identify a "signature" that the measured conductance is due to the sample molecules; (3) show that the signal results from a single molecule; and (4) perform appropriate statistical analyses—a critical component because of the strong dependence of the molecular conductance on the microscopic details of molecule–electrode contacts. Most (if not all) of the works described in above sections are based on measurement of multiple molecules, rather than a single molecule, and the parameter values are deduced from the averaged analysis. Direct measurement of a single molecule is highly desired for evaluating the conductivity and other electrical properties of molecular junctions.

Recently, Tao and coworkers have developed a so-called STM break-junction methodology that can reproducibly and reliably calibrate the single-molecule conductivity (Figure 21.9) [37,71–74]. The process is controlled by a feedback loop that starts by driving the tip into contact with the substrate using a piezoelectric transducer (PZT)—step 1. Once the contact is fully established, the feedback loop activates the PZT to pull the STM tip back to form an atomic-scale neck between the two electrodes, which is signaled by conductance quantization due to the integer number of gold atoms connected between the two electrodes (Figure 21.9A and C)—step 2. After breaking the atomic-scale neck, the sample molecules start to bridge the two electrodes via surface chemical bonding—step 3. Further pulling breaks the individual molecules from contacting the electrodes, leading to a new series of steps in the conductance (Figure 21.9B). Conductance histogram constructed from a large number (hundreds to thousands) of individual measurements reveal well-defined peaks located near integer multiples of a fundamental conductance value (e.g., $0.01G_0$ for 4,4′-bipyridine shown in Figure 21.9D; $G_0 = 2e^2/h = 77.4\ \mu S$, where e is the electron charge and h is the Planck constant); the lowest peak corresponds to the conductance of a single molecule. The histogram analysis provides an unambiguous determination of single-molecule conductance averaged over a large number of measurements [66,67].

Scanning tunneling microscopy break-junction method provides a reliable way to calibrate the single-molecule conductivity. As to be described below, such a method is also suitable for fabrication and characterization of single-molecule FET by attaching a third electrode [74,75]. The break-junction is based on direct contact between electrode and the molecule, and the sample system is highly adaptable for various molecules and solutions. Setting up the whole measurement is generally simple, straightforward, and can feasibly be reproduced. This is in contrast to the other SPM-based measurements, such as those employing gold nanoparticles as contact joints (Figure 21.7), where the procedure often

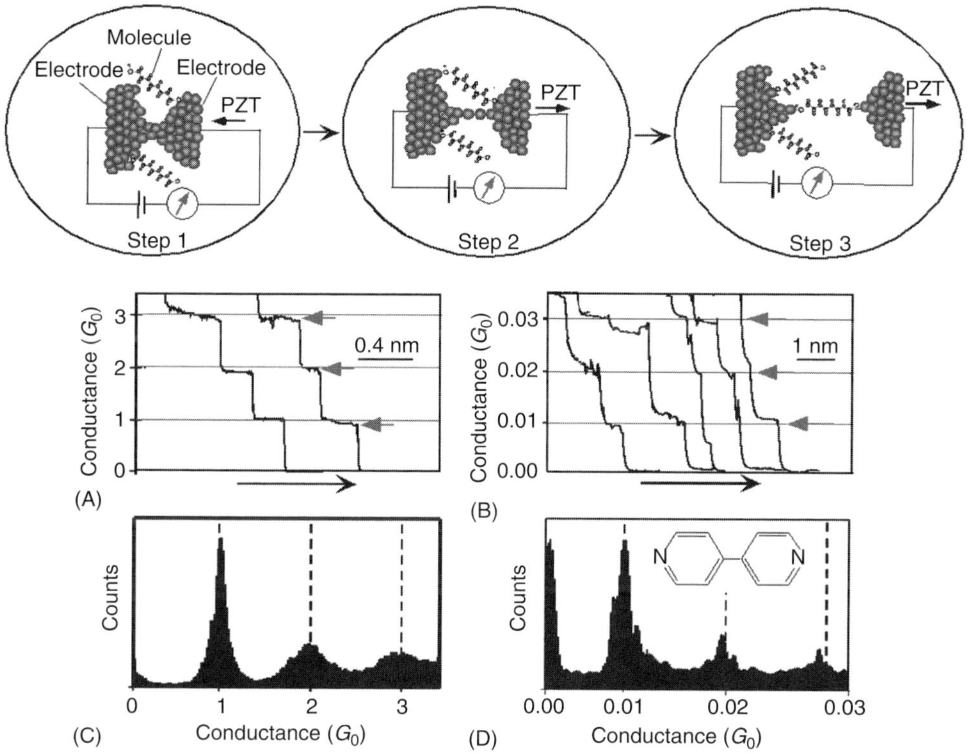

FIGURE 21.9 Measurement of single-molecule conductance based on STM break-junction technique. Step 1: A PZT drives an electrode into contact with another electrode in a solution containing sample molecules. Step 2: The first electrode is then pulled back to form an atomic-scale neck between the two electrodes, which is signaled by conductance quantization (A,C). Step 3: After breaking the atomic-scale neck, the sample molecules start to bridge the two electrodes via chemical bonds. Further pulling breaks the individual molecules, which leads to a new series of steps in the conductance (B). Conductance histogram (D) constructed from a large number of measurements reveals well-defined peaks near integer multiples of a fundamental value ($0.01G_0$ for 4,4′-bipyridine shown here), which is identified as the conductance of a single molecule. (Reprinted from Li, X., Xu, B.Q., Xiao, X., Yang, X., Zang, L., and Tao, N.J., *Faraday Discuss.*, 131, 111, 2006. With permission. Copyright 2006 Royal Chemical Society.)

involves several elaborate assembly steps, and the measured resistance is complicated by a Coulomb blockade effect due to finite contact resistance between the AFM probe and the gold nanoparticle.

21.2.2 Molecular Wire

Molecule-mediated charge transfer has been recognized and investigated for a long time. The idea of using molecule chains as electrical wires has been inspired from the efficient charge transfer observed for the conjugate molecules. With the advancement in fabrication and characterization of molecular junctions (as described above), molecular wires have been successfully installed within nanojunctions, and the conductivity has been well characterized. It has been found that the molecular conductivity is strongly dependent not only on the molecular structure but also on the dynamic conformation of the molecule (leading to application in molecular switch) and the local electrical field (leading to application in FET). With advancement in organic synthetic methodology, various molecular structures have been synthesized and tested for the conducting properties. Mostly, an ideal molecular wire should be highly conductive (or conjugate), rigid in backbone conformation (suited for alignment in nanojunction), and thermally stable to avoid air oxidation or electrochemical deterioration.

TABLE 21.1
Comparison of Conductivity

	Cross Section Size (nm^2)	Current Density (Electrons/nm^2 s)
1 mm copper wire	~3 × 10^{12}	~2 × 10^{6}
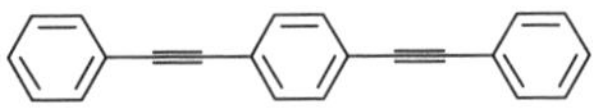	~0.05	~4 × 10^{12}
Carbon nanotube	~3	~2 × 10^{11}

The current density for a molecular wire is higher by many orders of magnitude than that of bulk metallic wires (Table 21.1), and even more than 10 times better than the single-wall carbon nanotube, which has been considered as one of the most promising building blocks for nanoelectronics. One can imagine that an electronic chip made of molecular units could be millions times more powerful than the current silicon-based devices, but with much less energy cost (heat emitting). The charge transport mechanism of molecular wires can be understood either as the classic redox reactions (hopping mechanism) or as a quantum mechanical process (tunneling mechanism).

Hopping mechanism: one electron transfers from the source electrode to the redox center of the molecule, followed by transferring to the drain electrode. The driving force for each step is due to the bias applied between the source and drain.

Tunneling mechanism: when the lowest unoccupied molecular orbital (LUMO) of the molecule is lined (via the gate voltage) at the same level as the fermi level of the source electrode (for n-type channel) or the drain electrode (for p-type channel), a resonant electronic tunneling between the two energy levels occurs, resulting in efficient charge transport between the two electrodes.

One factor affecting the molecular conductivity is the conformational structure of the molecular backbone, which determines the electronic coupling between the neighbor conjugate units (e.g., phenylene). As observed in the early STM-based measurement, the conductance of the phenylene ethynylene oligomers (Figure 21.10) isolated in matrices of alkyl-thiol monolayers could switch between ON and OFF depending on how well the surrounding matrix is ordered [76]. It was suggested that the switching is a result of conformational changes in the molecules, rather than electrostatic effects of charge transfer. Similar observation has recently been carried out by Venkataraman and coworkers using a series of biphenyl molecules, which are modified with different side chains to afford different twisting conformation between the two phenylene rings (Table 21.2) [77]. The measurement of the conductivity was performed with the recently developed STM break-junction technique (Figure 21.9), which provides reliable calibration of single-molecule conductivity. The single-molecule junctions were formed through the chemical binding between the gold electrodes and the two amine terminals of the molecule. On the basis of more than one thousand times reproducible measurements, the average conductance values were deduced for the series of biphenyl molecules. It is interesting to see that the conductance for the series decreases with an increase in the twisting angle, consistent with a cosine-squared relation predicted for electrical transport through π-conjugated biphenyl systems.

The conformation dependence of molecular electrical conductivity is a common but complicated phenomenon. For some molecules, such as the biphenyls mentioned above, the highest conductivity was found for the coplanar conformation, i.e., no twisting between the two phenyl rings. However, for some other molecules, such as those shown in Figure 21.11, the most favorable conformation for charge transport is that the two perylene rings are orthogonal (twisted as 90°) to each other [78,79] This unusual property is likely due to the different molecular structures between the biphenyls and the perylene dimers: the former are totally conjugate, while the latter possess node positions at the nitrogen (i.e., the perylene units and the oligophenylene bridge are in separate conjugate system).

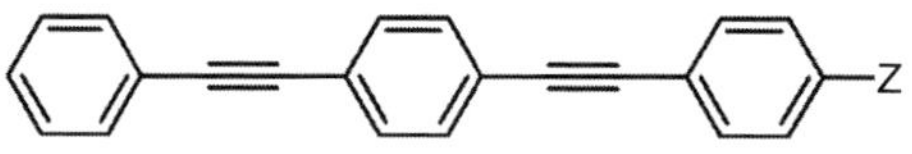

NH_4OH — **1**, Z = $SCOCH_3$ → **1′**, Z = S^-, 4,4′-di(ethynylphenyl)-1-benzenethiolate

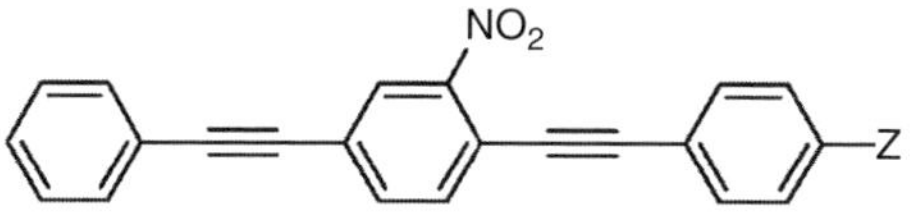

NH_4OH — **2**, Z = $SCOCH_3$ → **2′**, Z = S^-, 4,4′-di(ethynylphenyl)-2′-nitro-1-benzenethiolate

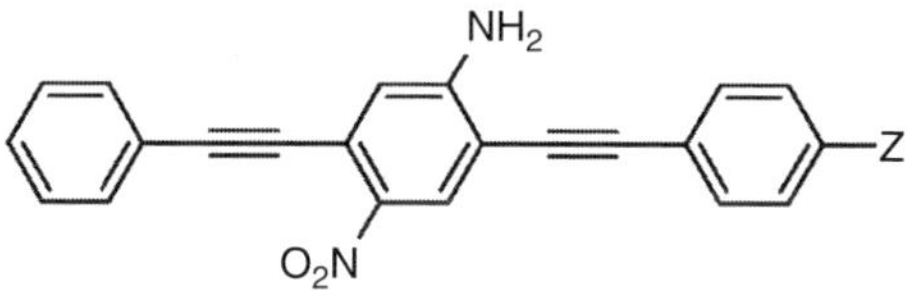

NH_4OH — **3**, Z = $SCOCH_3$ → **3′**, Z = S^-, 2′-amino-4,4′-di(ethynylphenyl)-5′-nitro-1-benzenethiolate

FIGURE 21.10 Molecules for which switching was observed (**1′**, **2′**, and **3′**) are produced in situ from the corresponding thioacetyls (**1**, **2**, and **3**, respectively) with the deprotection reaction shown for each. Aqueous ammonium hydroxide is used to hydrolyze the acetyl protecting group, generating the thiolate or thiol. The thiolate/thiol can then adsorb on the Au{111} surface, inserting at existing defect sites in the dodecanethiolate monolayer matrix. (Reprinted from Donhauser, Z.J., Mantooth, B.A., Kelly, K.F., Bumm, L.A., Monnell, J.D., Stapleton, J.J., Price Jr., D.W., Rawlett, A.M., Allara, D.L., Tour, J.M., and Weiss, P.S., *Science*, 292, 2303, 2001. With permission. Copyright 2001 American Association for the Advancement of Science.)

TABLE 21.2
Molecular Structure and Measured Properties

Molecule Number	Structure	Conductance (G_0) Measured	Calculated	Peak Width[a]	Twist Angle (°)
1	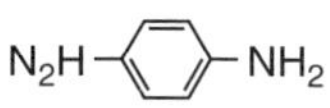	6.4×10^{-3}	6.4×10^{-3}	0.4	—
2	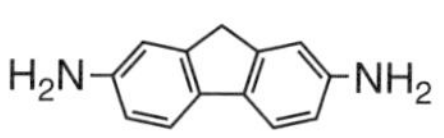	1.54×10^{-3}	2.1×10^{-3}	0.8	0
3	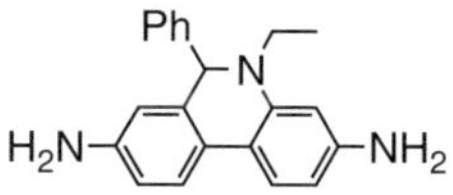	1.37×10^{-3}	2.2×10^{-3}	0.8	17
4	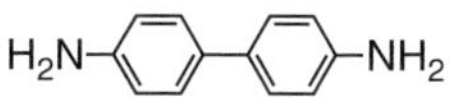	1.16×10^{-3}	1.6×10^{-3}	0.9	34
5	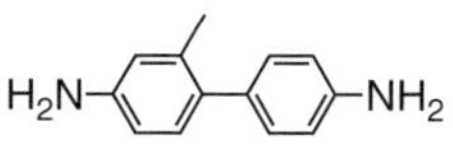	6.5×10^{-4}	1.2×10^{-3}	1.3	48
6	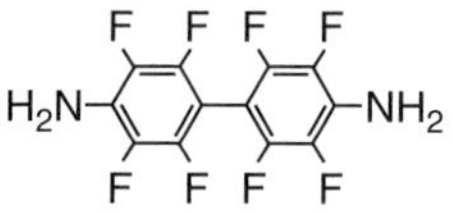	4.9×10^{-4}	7.1×10^{-4}	0.6	52

(continued)

TABLE 21.2 (continued)
Molecular Structure and Measured Properties

Molecule Number	Structure	Conductance (G_0) Measured	Calculated	Peak Width[a]	Twist Angle (°)
7	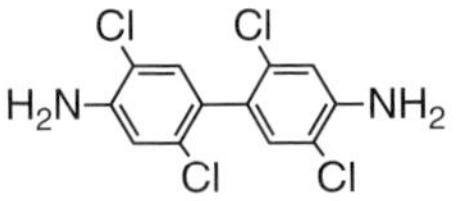	3.7 × 10^{-4}	5.8 × 10^{-4}	0.9	62
8	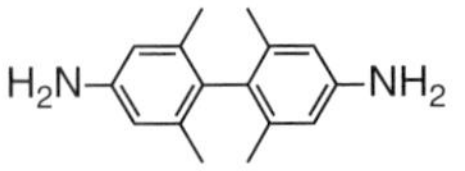	7.6 × 10^{-5} [b]	6.4 × 10^{-5}	NA[b]	88
9	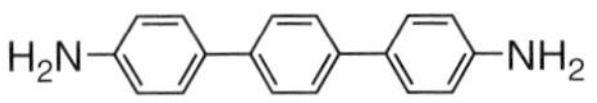	1.8 × 10^{-4} [c]	3.5 × 10^{-4}	2.1	—

Source: Venkataraman, L., Klare, J.E., Nuckolls, C., Hybertsen, M.S., and Steigerwald, M.L., *Nature*, 442, 904, 2006. With permission. Copyright 2006 Macmillan Magazines Ltd.

Note: Table shows molecule structure, measured conductances calculated relative conductances, relative widths of the histogram peaks (see Supplementary Information for details), and the calculated twist angle, θ.

[a] Hall-width at half-maximum of the lorentzian fit, normalized to the peak value.

[b] The histogram peak was determined after subtracting the Au histogram from the data as the raw data could not be fitted with a lorentzian so a width could not be determined.

[c] Determined from actual maximum of the raw data.

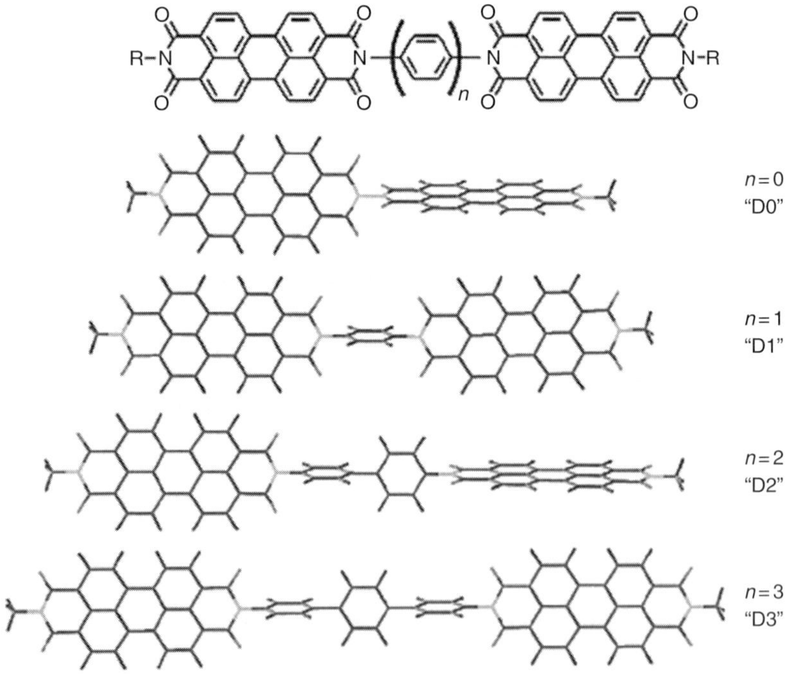

FIGURE 21.11 Structures of perylene tetracarboxylic diimide (PTCDI) dimers (P–P): D0, D1, D2, and D3, and monomer M. Side-chain R = nonyldecyl. The dimers can potentially be used as molecular wires by changing the side-chains to thiol moieties. (Reprinted from Holman, M.W., Liu, R., Zang, L., Yan, P., DiBenedetto, S.A., Bowers, R.D., and Adams, D.M., *J. Am. Chem. Soc.*, 126, 16126, 2004. With permission. Copyright 2002 American Chemical Society.)

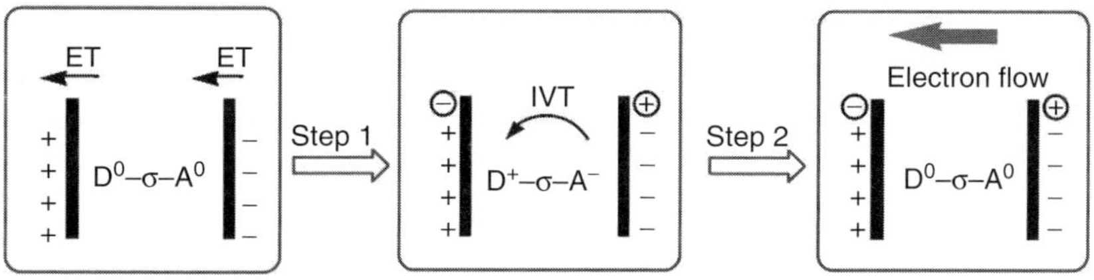

FIGURE 21.12 Schematic picture of the Aviram–Ratner diode involving a D–σ–A molecule and two metallic electrodes, where D stands for electron donor, and A for acceptor. The electron flow is favored for the forward process as marked by the red arrow: at first electrons move from the cathode to the acceptor and from the donor to the anode. Then, molecules recover the ground state via an intervalence transfer (IVT). To make it rectifying, the reverse bias process (involving autoionization) must be infrequent. (Adapted from Robert Metzger presentation at the Workshop "Advances in Molecular Electronics: From molecular materials to single molecule devices", MPIPKS Dresden, Dresden, Germany, February 23–27, 2004; online: http://www.mpipks-dresden.mpg.de/~admol/.)

Although the through space electronic coupling matrix (which determines the efficiency of charge transfer) is still the highest for the coplanar conformation of the dimer, the charge transfer between the two perylene units is mostly determined by the electronic interaction between the perylene moiety and the phenylene bridge. Such an interaction is favored by the orthogonal conformation, consistent with the preliminary calculations. Detailed understanding of the conformation dependence requires further calculations. Nonetheless, this series of perylene dimers provide a unique class of molecules that can potentially be used as molecular wires. More importantly, these molecular wires show different conformation dependence of conductivity compared to the usual conjugate molecules.

21.2.3 Molecular Diode

The idea of molecular diode was proposed for the first time by Aviram and Ratner in 1974 [80], when they studied a single molecule with a donor–spacer–acceptor structure as shown in Figure 21.12. Theoretical calculation and analysis suggested that the molecule would behave as a diode when placed between two electrodes: electrons can easily flow from the cathode to the acceptor and from the donor to the anode, coincident with charge recombination (intramolecular electron transfer) between the anionic radical of the acceptor and the cationic radical of the donor. Thus, under forward bias, electrons can be transferred between the two metal electrodes; whereas under reverse bias, the electron transport is quenched since the donor is a bad acceptor and the acceptor is a bad donor. If the competing reverse bias mechanism is unlikely, molecular rectification occurs. The working principle of the diode (or rectifier) involves manipulation of the electronic wavefunction of the metallic electrodes by extending the electronic interaction into the molecule junction. Such a hybrid molecular device comprises molecules within the metallic nanojunction and thereby differs radically from the bulk-material-based molecular electronic systems found in applications such as dye lasers, LEDs, liquid-crystal displays, and soft plastic transistors. This seminal study opened a new window for people to justify the prospective functions of electronically active molecules, and to evaluate the molecular structures and properties that may afford application in devices. Since then, more and more research interests have been drawn to the construction and characterization of molecular devices employing various structures of molecules, eventually emerging as a new field of so-called molecular electronics (Figure 21.13). However, the design and construction of functional devices with only a single molecule poses the challenge of nanotechnology and engineering. Experimental confirmation of the idea of molecular diode has been a difficult task for many years until the technique for nanojunction fabrication became available in 1990s [40,81,82].

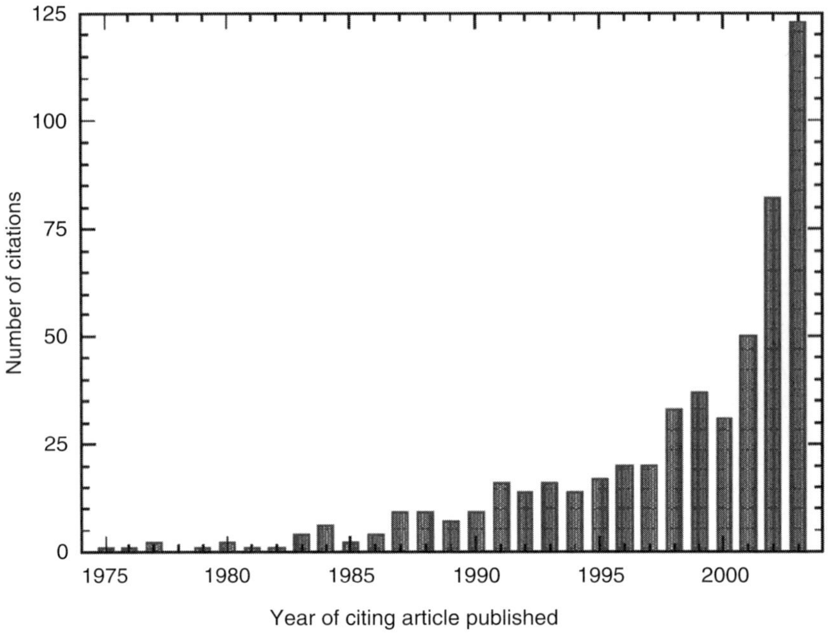

FIGURE 21.13 Emerging of the field of molecular electronics and the increasing interest is indicated by the exponential increase in the number of annual citations to the seminal Aviram–Ratner study during the last three decades. Data taken from the ISI Web of Knowledge. (From Aviram, A. and Ratner, M., *Chem. Phys. Lett.*, 29, 277, 1974. With permission.)

The first experimental investigation of molecular rectification was carried out with a monolayer sandwiched between two metallic electrodes, fabricated using Langmuir–Blodgett or self-assembled techniques. In 1997, Metzger and coworkers reported a successful investigation of the Langmuir–Blodgett multilayers and monolayers comprising a donor–acceptor molecule of hexadecylquinolinium tricyanoquinodimethanide (Figure 21.14) [83], which produced a rectification ratio of, as high as, 26. However, all the monolayer systems enabled investigation targeting on multiple molecules, rather than a single molecule. The experimental data thus obtained cannot be used to correlate with the theoretical prediction. Single-molecule investigation of the rectifying behavior requires sophisticated assembling procedure to align a molecule within the nanoelectrode junction.

With the advancement in nanotechnologies and engineering (particularly the electron-beam and AFM-based lithography), various nanoelectrode systems and molecular junctions have been fabricated and characterized in the last decade [40]. The interelectrode distance in nanofabricated junctions can be as small as a few nanometers, comparable to the length of the synthesized molecular wires, enabling one to interconnect single molecules in real devices. One major challenge in fabricating the two-electrode systems is how to assure and maintain the good electrical contact between the molecule and the electrodes. A poor contact may bring deviation to the measurement and lead to misunderstanding of the observation. For example, the observed rectification may be contributed by the asymmetric molecule–metal contact, rather than by the molecule itself. In 1999, Reed and coworkers developed a unique technique to construct a nanojunction comprising of only a small group of molecules, enabling characterization of molecular conductivity, and rectifying properties in a more reliable way (Figure 21.15) [54]. Molecules containing a nitroamine redox center were self-assembled in a nanopore on a gold surface through the thiol–gold binding. For the first time in molecular devices, negative differential resistance and large ON/OFF ratios were observed (Figure 21.15D). The peak voltage position decreased monotonically at a rate of 9 mV K^{-1} as the

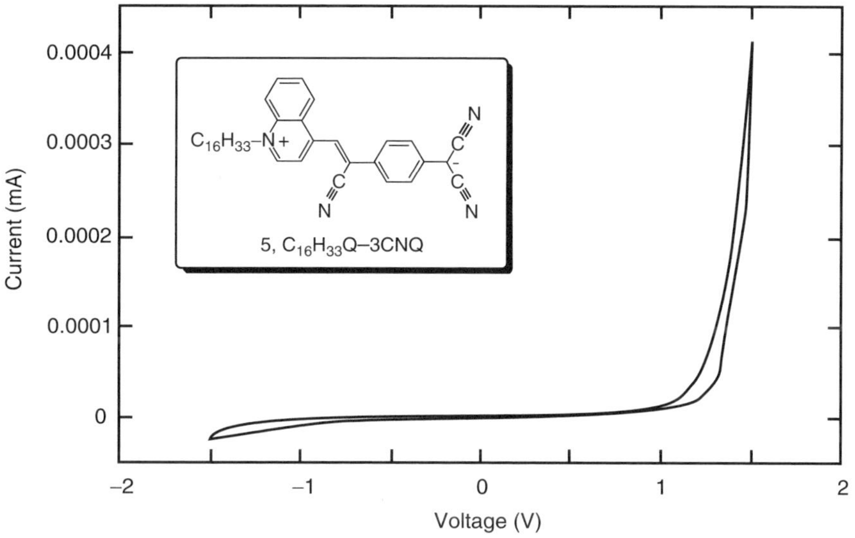

FIGURE 21.14 Current–voltage characteristic of the device comprising a monolayer of hexadecylquinolinium tricyanoquinodimethanide (inset) that exhibits a very high rectification (rectification ratio around 26). (Reprinted from Metzger, R.M., Chen, B., Hoepfner, U., Lakshmikantham, M.V., Vuillaume, D., Kawai, T., Wu, X., Tachibana, H., Hughes, T.V., Sakurai, H., Baldwin, J.W., Hosch, C., Cava, M.P., Brehmer, L., and Ashwell, G.J., *J. Am. Chem. Soc.*, 119, 10455, 1997. With permission. Copyright 1997 American Chemical Society.)

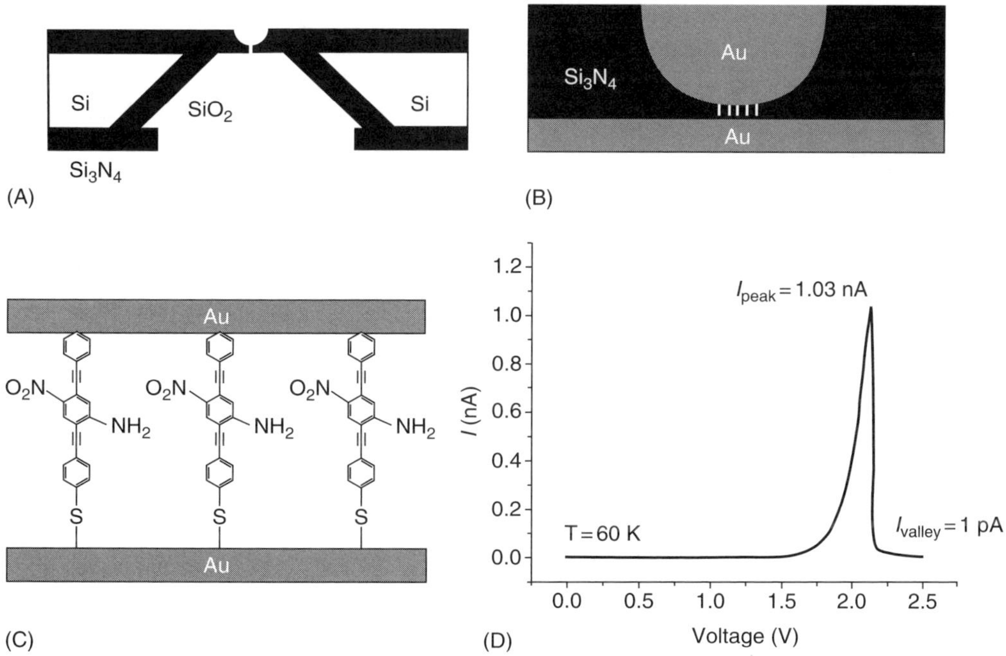

FIGURE 21.15 (A) Cross section of a silicon wafer with a nanopore etched through a suspended silicon nitride membrane. (B) Au–SAM–Au junction in the pore area; SAM. (C) Blowup of B with the channel molecules sandwiched in the junction. (D) *I–V* characteristics of the molecule (shown in C) at 60 K. The peak current density is ~50 A/cm^2, the NDR is ~400 µohm cm^2, and the PVR is 1030:1. (Reprinted from Chen, J., Reed, M.A., Rawlett, A.M., and Tour, J.M., *Science*, 286, 1550, 1999. With permission. Copyright 1999 American Association for the Advancement of Society.)

temperature increased. In a following work, Seminario et al. explained this electrical characteristic by density functional theory (DFT) calculations [53], showing that the resonant-tunneling diode behavior is due to a modification of the electronic wave functions extending through the molecules. The charge transport is due to electronic delocalization, which in turn is determined by the highest occupied molecular orbital (HOMO)–LUMO gap and the spatial extents of the unoccupied orbitals.

21.2.4 Molecular Field-Effect Transistor

Just as metal–oxide–semiconductor field-effect transistor (MOSFET) dominates in the silicon-based electronic devices, single-molecule FET (comprising only one molecule in a nanojunction) will become the primary component for future molecular electronic devices. Although fabrication and application of integrated molecular devices may be a long-term goal to approach, studies of single-molecule FET have already emerged as one of the most focused areas in resent years. The primary goals of the current research are to investigate the effects of electrical gate on the electron transport through single molecules, and optimize the gate modulation by constructing and measuring a variety of molecular structures, which possess different electronic energy levels and lining-up with the fermi levels of electrodes. The improved understanding of gate effects thus obtained aid in the future design of molecules suitable for development as single-molecule FET.

Building a single-molecule FET is a critical step toward the miniaturization of molecular electronics [38,40,43,44,80–82,84–86]. However, it still remains unclear how one can reversibly control the current through a molecule over a large range of gate voltage [33,43,86–90]. Unlike carbon nanotube [91–94] or semiconductor nanowires [95,96], observation of FET behavior in small organic molecules has been a difficult challenge (particularly at room temperature) [48,97–101] because it requires one to (1) find a reliable method to wire a single molecule to the source and drain electrodes and (2) place the gate electrode in the proximity of the molecule to achieve a large enough gate field. In this subsection, some recent breakthroughs in fabrication and investigation of single-molecule FET are briefly described and commented.

Molecular FET property was initially proposed based on theoretical calculation and analysis. One typical study was carried out by Di Ventra and coworkers in 2000 [102]. The molecule they studied was the benzene dithiol (Figure 21.16), the smallest conjugate system suitable for molecular electronics. The small size is highly favorable for calculation, but poses great challenge for experimental investigation due to the difficulty in fabricating a nanojunction with comparable gap width. When fabricated in a three-electrode system in a fashion of conventional FET (Figure 21.16, top), the conductivity of the molecule can be modulated via changing the gate electrical field, as calculated for the *I–V* characteristic as a function of gate bias (Figure 21.16, middle). After a region of constant resistance, the current increases with the gate field, and reaches a maximum value at 1.1 V/Å, followed by a decrease at about 1.5 V/Å. Further increasing the gate results in a linear increase in the current. The various features of the *I–V* curve can be understood by looking at the density of states at different gate voltages (Figure 21.16, bottom). The molecule has a small but relatively smooth density of states through which current can flow. The π-bonding states lie several electron volt below the fermi levels, while the π^*-antibonding states are only about 1 eV above the fermi levels. The relatively close π^*-antibonding states are responsible for charge transport within the junction. The initial increase in gate bias shifts the π^* orbital toward the fermi level, resulting in an enhancement of current. When the π^* orbital lines up with the fermi level (at a gate of 1.1 V/Å), a resonant tunneling through the π^*-antibonding state occurs as a result of electronic coupling between the π^* states and the continuum of states of higher energy. With further increase in gate, the resonant-tunneling condition is lost and a valley in the *I–V* curve is observed. Finally, as the gate bias is increased further, more higher-energy π^* states are in resonant with the fermi level, resulting in linear increase in current with the gate bias. Compared to the π^* states, the π states show less pronounced dependence on gate because they are much more separated in energy from the continuum states of higher energy.

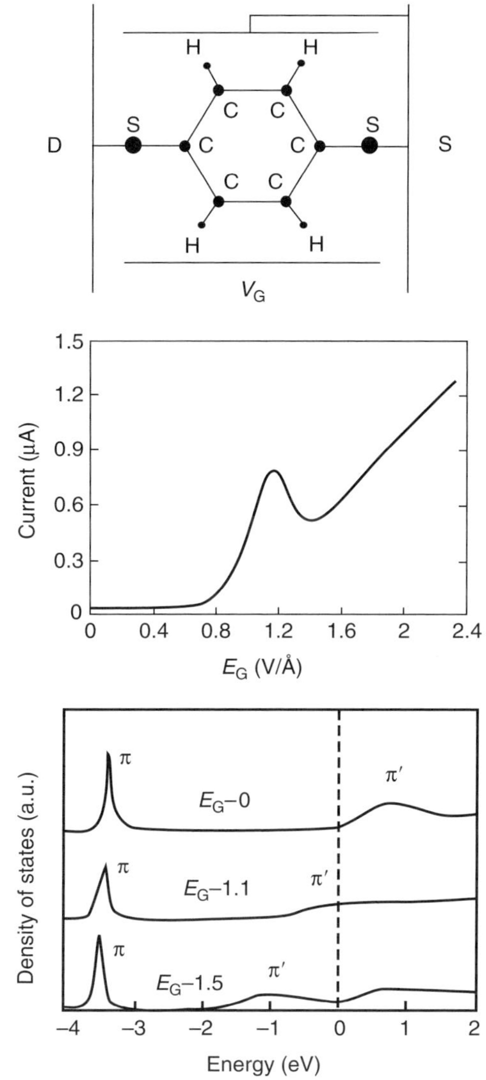

FIGURE 21.16 (top) Scheme of the three-terminal geometry used in the present study. The molecule is sandwiched between source and drain electrodes along the direction of electronic transport. The gate electrodes are placed perpendicular to the molecule plane. The fermi level on one gate disk equals the source fermi level while the other electrode is at a higher potential V_G. (middle) Conductance of the molecule of Figure 21.1 as a function of the external gate field. The source–drain bias is 0.01 V. (bottom) Difference between the density of states of the two semi-infinite electrodes with and without the benzene-1,4-dithiolate molecule in between, for three different gate fields E_G (in units of V/Å). The left fermi level has been chosen as the zero of energy. The right fermi level is at 10 mV. (Reprinted from Di Ventra, M., Pantelides, S.T., and Lang, N.D., *Appl. Phys. Lett.*, 76, 3448, 2000. With permission. Copyright 2000 American Institute of Physics.)

Experimental characterization of single-molecule FET has been successfully performed in recent few years [48,98,99], taking advantages of the improved nanofabrication techniques. Most of the investigations were carried out with the nanoelectrode systems fabricated on oxidized silicon surface (Figure 21.2), and under cryostat conditions (0.3–25 K) to minimize the thermal agitation of the molecular junction. In the same issue (June 13, 2002) of *Nature*, two papers reported for the first time the observation of single-molecule transistors and the Kondo effect (Figures 21.17 and 21.18)

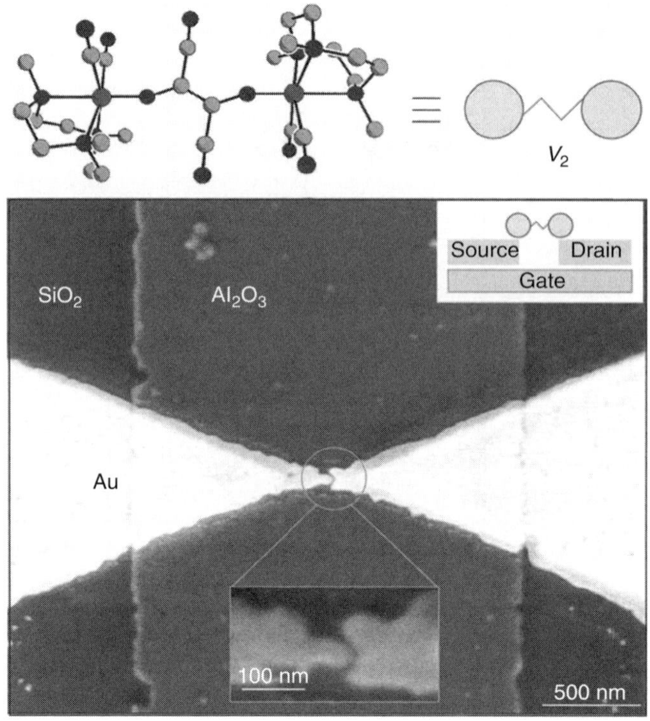

FIGURE 21.17 Fabrication of a single-molecule transistor incorporating a divanadium molecules. Top left, the structure of [(N,N¢,N²-trimethyl-1,4,7-triazacyclononane)$_2$V$_2$(CN)$_4$(μ-C$_4$N$_4$)] (the V_2 molecule) as determined by x-ray crystallography; light, medium, and dark gray spheres represent C, V, and N atoms, respectively. Top right, the schematic representation of this molecule. Main panel, scanning electron microscope image (false color) of the metallic electrodes fabricated by electron-beam lithography and the electromigration-induced break-junction technique. The image shows two gold electrodes separated by ~1 nm above an aluminum pad, which is covered with an ~3 nm thick layer of aluminum oxide. The whole structure was defined on a silicon wafer. The central gray regions correspond to a gold bridge with a thickness of 15 nm and a minimum lateral size of ~100 nm. The lighter gray regions represent portions of the gold electrodes with a thickness of ~100 nm. Main panel inset, schematic diagram of a single-V_2 transistor. (Reprinted from Liang, W.J., Shores, M., Bockrath, M., Long, J.R., and Park, H., *Nature*, 417, 725, 2002. With permission. Copyright 2002 Macmillan Magazines Ltd.)

[98,99], approaching the limit in device that electrons hop ON to, and OFF from, a single atom between two contacts. Both the two devices incorporate a transition-metal complex designed so that electron transport occurs through well-defined charge states of a single atom. Interestingly, changing the length of the insulating side-linker alters the coupling of the metal ion to the electrodes, enabling the fabrication of devices that exhibit either single-electron phenomena, such as Coulomb blockade, or the Kondo effect [98]. Molecules incorporating transition-metal atoms provide unique systems for modulating and controlling the single-electron transport, because the spin and orbital degrees of freedom can be controlled through well-defined chemistry. Moreover, the Kondo resonance can be tuned reversibly using the gate voltage to alter the charge and spin state of the molecule. The resonance persists whenever the temperature keeps, as low as, below 30 K.

Both the transistors were fabricated in the fashion of back-gate, one on oxidized alumina, the other on oxidized silicon. The so-called electromigration-induced break-junction technique was used to create two closely spaced gold electrodes: the gold wire was gradually broken by ramping to large voltages (typically over 0.5 V) at cryogenic temperatures while monitoring the current until only a tunneling signal is present. This produces a gap about 1–2 nm wide, usually comprising a

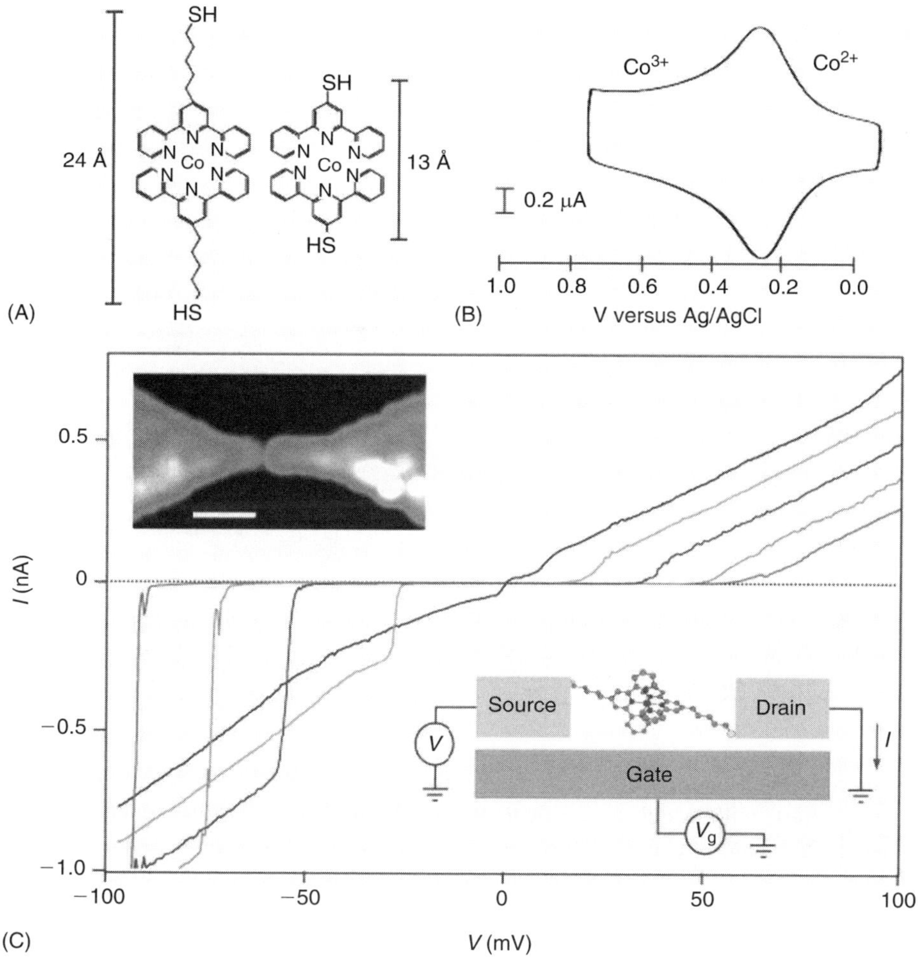

FIGURE 21.18 (A) Structure of the two molecules employed in the FET device. The scale bars show the lengths of the molecules as calculated by energy minimization. (B) Cyclic voltammogram of the molecule with short linker in 0.1 M tetra-*n*-butyl ammonium hexafluorophosphate, acetonitrile showing the Co^{2+}/Co^{3+} redox peak. (C) *I*–*V* curves of a single-electron FET of the molecule with long linker at different gate voltages (V_g) from −0.4 V (outermost line) to −1.0 V (central line) with ΔV_g approximately −0.15 V. Upper inset, a topographic atomic force microscopy (AFM) image of the electrodes with a gap (scale bar, 100 nm). Lower inset, a schematic diagram of the device. (Reprinted from Park, J., Pasupathy, A.N., Goldsmith, J.L., Chang, C., Yaish, Y., Petta, J.R., Rinkoski, M., Sethna, J.P., Abruna, H.D., McEuen, P.L., and Ralph, D.C., *Nature*, 417, 722, 2002. With permission. Copyright 2002 Macmillan Magazines Ltd.)

single molecule inside. Electrical characteristics of the molecule were then determined by acquiring current versus bias voltage (*I*–*V*) curves while changing the gate voltage (V_g).

Following these successful investigations, Bredas and coworkers have developed a single-molecule transistor using a *p*-phenylene vinylene oligomer (OPV5, Figure 21.19), for which the charge transport properties are controlled by the electronic levels in several distinct charged states. It was found that when fabricated in a nanojunction, the molecular electronic levels are strongly perturbed compared to those of the molecule in solution, leading to a very significant reduction of the gap between HOMO and LUMO. Such a surprising effect could be caused by image charges generated in the source and drain electrodes resulting in a strong localization of the charges on the molecule.

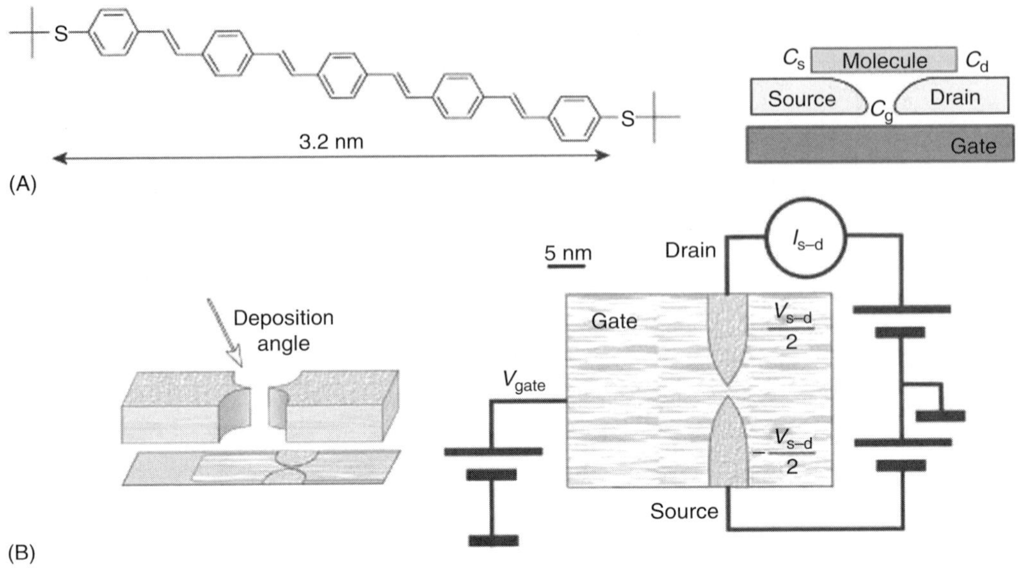

FIGURE 21.19 (A) Molecular structure of OPV5 and schematic experimental setup of the single-molecule FET fabricated with OPV5. (B) Schematic representation of the device preparation procedure. (Reprinted from Kubatkin, S., Danilov, A., Hjort, M., Cornil, J., Bredas, J.-L., Stuhr-Hansen, N., Hedegard, P., and Bjornholm, T., *Nature*, 425, 698, 2003. With permission. Copyright 2003 Macmillan Magazines Ltd.)

These pioneering works demonstrate that single-molecule FETs can be fabricated with well-designed molecules, with which the critical parameters of Kondo physics, such as the spin and orbital degrees of freedom, can be controlled and optimized by chemical synthesis. Moreover, by tuning the length of the organic barrier or the molecular electronic structure (conjugation), it is possible to control the coupling between the channel molecule (or ion) and the electrodes, leading to modulation of the FET properties and efficiency. In general, studying single-molecule FET in a real device has become viable, by combining the ability of designing the molecular electronic states through chemical synthesis and the feasibility in single-molecule device fabrication and measurement. However, all these studies were carried out at cryogenic temperatures (<30 K). Although low temperature measurement enables observation of some unique phenomena, e.g., Kondo effect, the whole systems thus designed are still far from the future practical application in molecular electronics, where operation at room temperature is normally desired. Indeed, the efficient FET performance found at low temperature does not necessarily mean that it will work the same under ambient condition. At elevated temperatures, thermal agitation and possible electron leaking may bring difficulty to the device construction and electrical measurement with the nanoelectrode systems. It is now in high demand to find a new way to characterize the single-molecule FET at room temperature.

Recently, Tao and Zang et al. have successfully fabricated and investigated a room temperature single-molecule FET using an n-type semiconductor molecule (Figure 21.20) [37,74], perylene tetracarboxylic diimide (PTCDI), for which the molecule-mediated current can be reversibly controlled with a gate electrode over nearly three orders of magnitude under ambient condition. The molecule is wired to two gold electrodes (source and drain) through the well-developed STM break-junction method as detailed above (Figure 21.9). The number of wired molecules is determined by statistical analysis of a large number of molecular junctions. A large gate field is achieved using an electrochemical gate in which the gate voltage is applied between the source and a gate in the electrolyte. Since the gate voltage falls across the double layers at the electrode–electrolyte interfaces, which are only a few ions thick, a field close to ~1 V/Å can be reached. Such a strong gate field enables construction and investigation of the FET behavior at the single-molecule level.

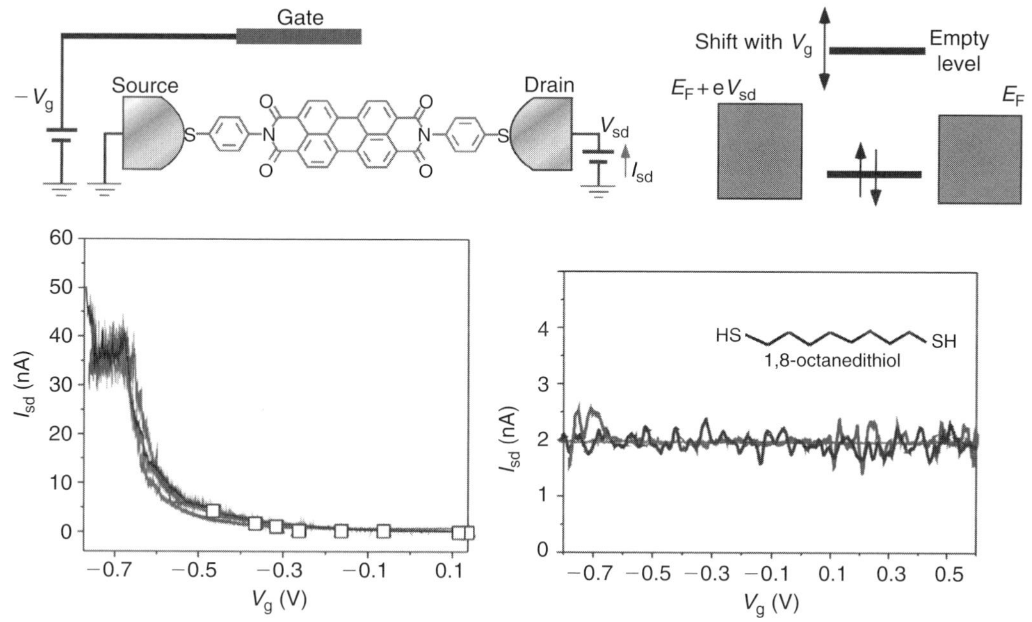

FIGURE 21.20 (top left) Schematic of a single-molecule FET comprising a PTCDI molecule. An electrochemical electrode (Ag wire reference electrode in 0.1 M $NaClO_4$) was used as gate. The gate and the source–drain bias voltages are controlled with a bipotentiostat (a Pt counter electrode not shown for clarity). (top right) The energy diagram of the gold-PTCDI-gold junction. (bottom left) Source–drain current (I_{sd}) versus gate voltage (V_g) for the single PTCDI FET. (bottom right) Control experiment on an alkanedithiol shows no gate voltage dependence. (Reprinted from Xu, B.Q., Xiao, X., Yang, X., Zang, L., and Tao, N.J., *Am. Chem. Soc.*, 127, 2386, 2005. With permission. Copyright 2005 American Chemical Society.)

Perylene tetracarboxylic diimide forms an extremely robust class of molecules with high thermal- and photostability and used in a variety of optoelectronic devices [103–107]. More interestingly, the PTCDI molecules are of n-type characteristics [37,74,108], compared to the more common p-type counterparts in organic semiconductors. Since electronic devices require both p- and n-type components, investigation on the n-type molecules will have significant impact on the study of molecular electronics, and thus assist in the future development of devices. To wire the molecule to two gold electrodes, the PTCID molecule was terminated with two thiol groups that can spontaneously bind to gold electrodes via the thiol–gold bond. The energy gap between the LUMO and HOMO is only 2.5 eV, implying a semiconductor-like molecule. The nominal length of the rigid molecule is about 2.3 nm, much greater than the gate thickness (the diameter of the ions), so field-screening effect due to the proximity of the source and drain electrodes is negligible. These attributes make PTCDI an excellent candidate for resonance tunneling FET.

As shown in the *I–V* curve (Figure 21.20), the conductance increases rapidly with decreasing the gate voltage (V_g) and reaches a peak at approximately −0.65 V. The peak conductance is ~500 times greater than that at $V_g = 0$ V for a fixed source–drain bias (V_{sd}) at 0.1 V. The current can be controlled over a larger range at higher source–drain bias, due to the nonlinear *I–V* dependence. At $V_{sd} = 0.4$ V, the current can be controlled over ~1000 times by sweeping the gate voltage. The subthreshold slope is ~0.3 V, close to the predictions of theoretical models for benzenedithiol molecules [102]. The large gate-effect obtained with this redox molecule resembles the n-type solid-state FET (the gate sign convention here is opposite to that in solid-state electronics), but the mechanisms are different. The current peak observed here is located at V_g approximately −0.65 V, close to the reduction potential of the molecule. This implies that the current enhancement is due to an empty molecular state-mediated electron transport process. One mechanism is resonant tunneling, which predicts that the current reaches a peak when the empty state is shifted to the fermi level by the gate (Figure 21.20) [102]. An alternative mechanism that predicts a maximum current near the reduction potential is a two-step electron transport in which an electron from one electrode tunnels into the empty state to reduce the molecule and then tunnels out into the second electrode [109]. Control experiments with nonredox molecules, such as alkanedithiols, 4,4′-bipyridine and benzenedithiols, results in no gate effect, i.e., the source–drain current is independent of the gate voltage (Figure 21.20).

Although still far away from practical application, the molecular devices described above prove the concept of single-molecule FET, and pave way for people to explore new fabrication techniques and more molecules with improved electronic properties. Studying single-molecule FET provides an alternative means to characterize and control the charge transport through molecules. Such studies may lead to practical applications in molecular sensing, probing, and switching. Before the technique for integrating molecular devices into large-scale integrated circuits becomes available, development of single-molecule electronic devices (such as sensor) could be the primary focus of molecular electronics.

21.2.5 Single-Molecule Chemical Field-Effect Transistor

Chemical field-effect transistor (CFET) works in the same fashion as the conventional FET, except for that the electrical current in a CFET is modulated (gated) by structural modification of the molecule or by local chemical processes (e.g., adsorption). The gate in a conventional FET is defined as the external electrical field applied through a third electrode, while the gate in CFET is referred to as the electrostatic field caused by chemical modification. Local electrostatic field can potentially modulate the electron transport through molecules by altering the electron density (electronic levels) at the molecules [33,110]. In a recent single-molecule STM study, some significant chemical field effects were observed with a three-terminal molecular device (Figure 21.21) [111], for which the channel molecule was linked with a charge transfer complex (acting as a chemical gate). Although the gate modulation obtained was small (ON/OFF current ratio <10), the observation demonstrates the feasibility of modulating the electric current through chemical modification. More recently, Wolkow et al. extended the CFET investigation into a more precisely controlled system, where the

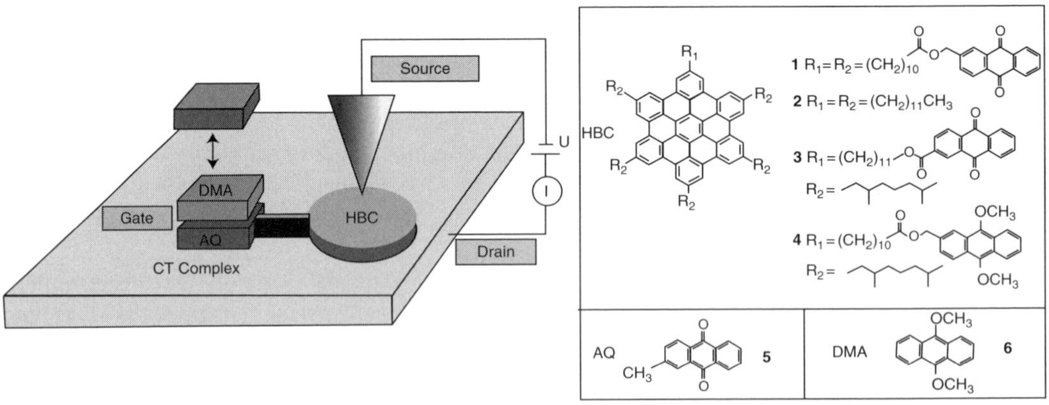

FIGURE 21.21 (left) Schematic of a prototype single-molecule CFET. (right) Chemical structures of the employed molecules: hexa-*peri*-hexabenzocoronene (HBC) decorated with six anthraquinone (AQ) functions (1), hexaalkyl-HBC (2), HBC bearing either one AQ (3) or one 9,10-dimethoxyanthracene (DMA) function (4), methyl-AQ (5), and DMA (6). (Reprinted from Jackel, F., Watson, M.D., Mullen, K., and Rabel, J.P., *Phys. Rev. Lett.*, 92, 188303/1–4, 2004. With permission. Copyright 2004 American Physical Society.)

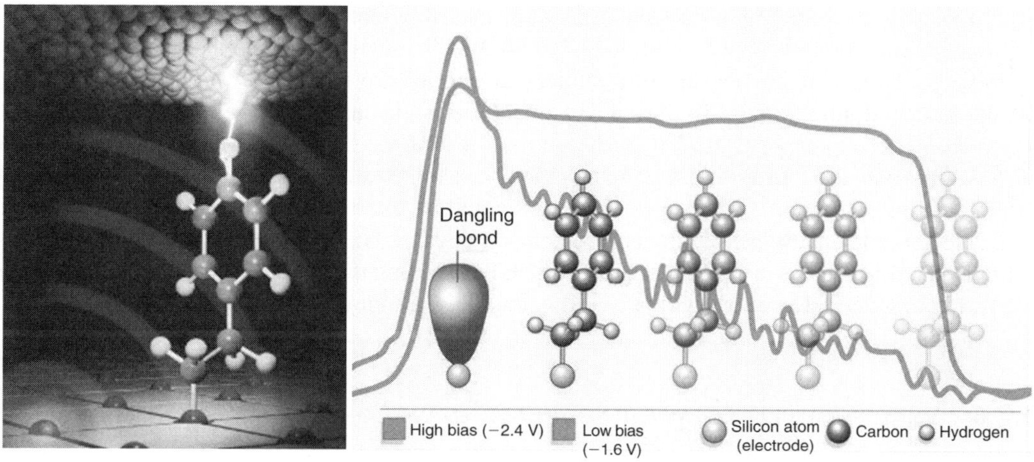

FIGURE 21.22 In the left illustration, the electric field from a surface ion (glowing at lower left) can be used to regulate electrical conductivity between a nearby molecule and an STM tip (at top). At right, the polymerization of a molecule (here an organic styrene-derived molecule, not to scale) on a silicon substrate stops abruptly at a dangling-bond site. At higher bias (flat line), all molecules are "turned on", and appear bright in the STM picture. At lower bias (jagged line), all molecules should appear dark. The electrostatic potential of the negatively charged dangling bond causes the nearest molecules to remain bright. This suggests that such structures could be used to manipulate charge transport through molecular junctions. (Reprinted from Ratner, M., *Nature*, 435, 575, 2005. With permission. Copyright 2005 American Chemical Society.)

chemical modification was quantitatively defined as a point charge (creating electrostatic field) [51]. The two-terminal device comprises a single-crystal silicon surface adorned with a row of styrene-derived molecules that terminate near-silicon dangling bonds (unsaturated valencies). The silicon surface and the top STM tip serve as the two electrodes (Figure 21.22). By varying the silicon doping level, the charge state of the dangling bond can be altered, leading to modulation of the local electrostatic field and, thus, the conductivity through the molecules. In analog to the conventional FET, the point charge here acts a chemical gate. This observation constitutes direct evidence that localized charges profoundly affect charge transport in single molecules, particularly at room temperature.

The electrostatic field emanating from the fixed point charge can be feasibly modulated by changing the distance between the charge and the surrounding surface-bound molecules (Figure 21.22, right). It was found that the onset of molecular conduction is shifted by changing the charge state of a silicon surface atom, or by varying the spatial relationship between the molecule and that charged centre. Because the shifting results in conductivity changes of substantial magnitude, these effects are easily observed at room temperature, easing the future application in electronics. At lower imaging bias, the molecules nearest to the dangling bond appear prematurely heightened, as if experiencing a built-in offset voltage. Molecules most distant from the dangling bond show a voltage-height response that is largely unperturbed. At larger imaging voltages, those distant molecules appear as high as the molecules nearest the dangling bond. The essence of the field-regulated molecular conduction effect described here is a shifting of molecular energy levels under the influence of the electrostatic potential emanating from a charged dangling bond. The distinct onset behavior displayed by π-bond containing molecules causes relatively small shifts in imaging voltage to lead to pronounced changes in molecule-mediated conduction.

Generally, changing the electrostatic potential on a molecule will change its conduction characteristics [33,110]. In conventional FETs, this electrostatic control is provided by the gate electrode, which regulates the amount of current that can flow from source to drain through the main channel of the transistor. In most single-molecule transport measurements, for example, the nanojunctions mentioned above (Figures 21.17 through 21.19) [48,98,99], such gating can also be attained, but it requires very large voltages, because the molecular entities are far away from the gate compared with the source–drain distance. To this end, chemical field (via chemical modification) may provide more effective electrostatic control over the current, leading to develop a new type of gate that is obviously beyond the conventional concept. Moreover, the study of CFET also broadens understanding of fundamental molecular processes and could ease the development of single-molecule-based detectors and other types of molecular electronic devices. For example, the working mechanism of chemical field suggests broad application of CFET in sensing the chemical reactions or processes, such as binding and adsorption, which usually cause significant change in local electrostatic filed. Detailed description of such application is given in the next subsection.

21.2.6 Single-Molecule Electrical Sensing Based on Conductivity Modulation

As confirmed by the CFET investigation, the electrical conductivity of a molecule can be modulated by changing the local electrostatic field, which in turn can be altered by chemical binding, complexation, protonation, ionization, or other chemical processes. Using such a chemical field effect, people have developed various molecular sensors through monitoring the conductivity change. In most cases, the sensing devices are based on a two-electrode system, as shown in Figures 21.23 and 21.24. Owing to the high sensitivity in conductivity measurement (resolution of femtoampere), CFET-based devices can potentially be developed as a new class of sensors that will operate at the single-molecule level, enabling both high sensitivity and selectivity. Single-molecule probing and recognition offer exceptional application in studying complicated materials and large molecular systems, such as biological molecules, by revealing the heterogeneous distribution of physical parameters behind the samples. In contrast, the conventional ensemble measurement of such heterogeneous samples usually produces average values, leading to underestimation or misunderstanding of the chemical and physical properties. Various single-molecule sensory systems have been developed and explored. However, most systems to date are based on measurement of the optical properties, such as fluorescence [112–118]. Described herein are two recent examples that employ CFET mechanism in single-molecule sensing, one is on metal binding and the other on protonation.

Sensing metal–ion binding [119]. The two-electrode sensor device was fabricated using the well-developed STM break-junction system as detailed above (Figure 21.9) [67,72]. Two complementary approaches were employed to reliably measure the conductance of a single molecule. The first was a statistical approach. Briefly, individual molecular junctions were created by repeatedly moving an gold STM tip into and out of contact with an Au substrate in a solution that contained the sample molecules (here the peptides, as shown in Figure 21.23). The process was controlled by a feedback loop that started by driving the electrode into contact with the substrate by using a PZT. Once the contact was established, the feedback loop activated the PZT to pull the electrode out of contact. After breaking the contact, a series of steps appeared in the conductance that signaled the formation of the molecular junctions (Figure 21.23A and B). The conductance steps correspond to the breakdown of the contact of individual molecules to the electrodes. When the last molecule was broken, a new process was repeated until a large number of conductance curves were obtained. The histogram of the conductance curves exhibit well-defined peaks that are located at integer multiples of a fundamental conductance value, which is identified as the conductance of a single molecule (insets in Figure 21.23A and B). This statistical approach enables determination of the single-molecule conductivity for a variety of molecular systems.

The second approach to conductivity measurement was based on the first approach, but modified to continuously measure the I–V curve, rather than just a single point conductance. First, the Au tip was brought into contact with the substrate and then the tip was gently pulled out of contact by controlling the PZT while the conductance was measured continuously. Once the conductance dropped to the last step, which corresponds to the formation of a single molecule bridged between the electrodes, the position of the tip was fixed (by freezing the PZT), and the I–V curves were measured.

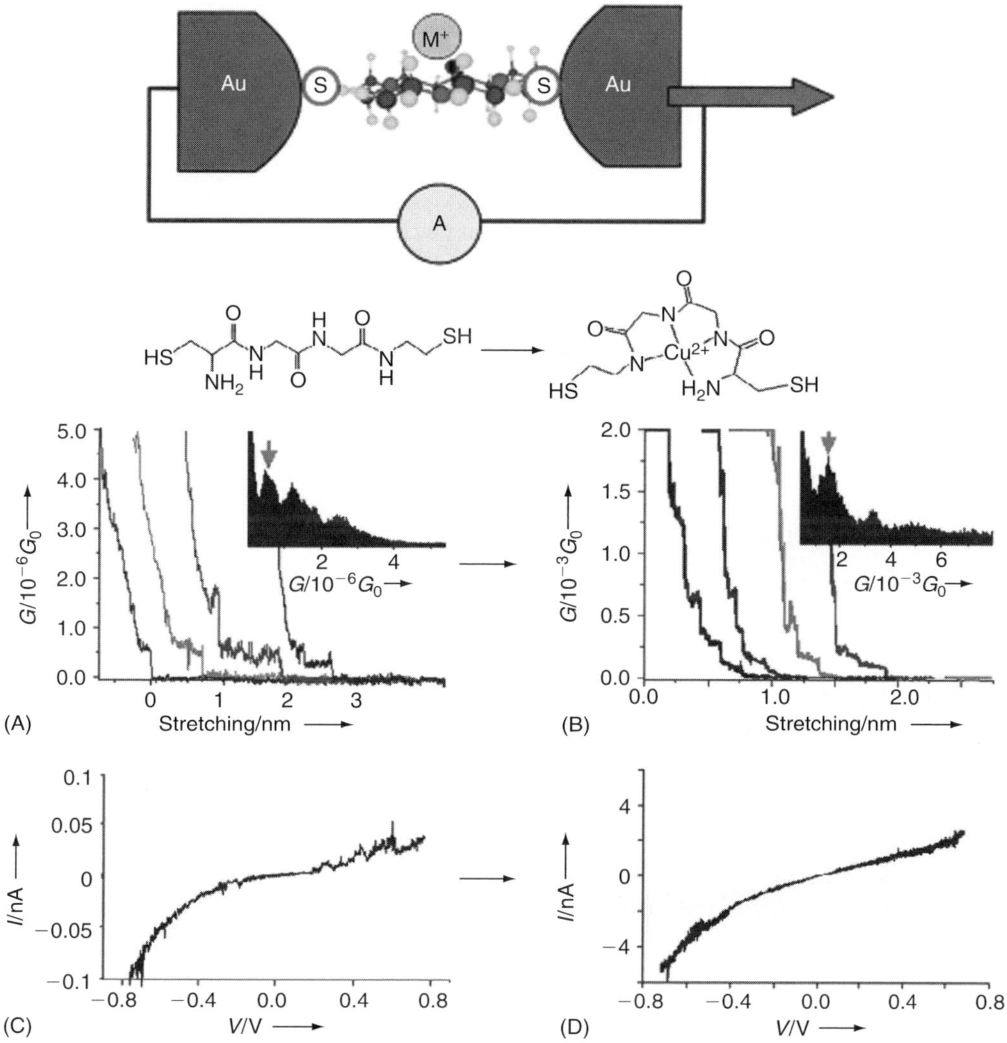

FIGURE 21.23 (top) Schematic illustration of a molecular junction formed by the separation of two electrodes (PZT, piezoelectric transducer). (bottom) Individual conductance curves of a peptide, Cysteamine-Gly-Gly-Cys. (A) before and (B) after binding to Cu^{2+}; the insets show the conductance histograms. *I–V* curves of the same peptide (C) before and (D) after binding to Cu^{2+}. (Reprinted from Xiao, X., Xu, B.Q., and Tao, N.J., *Angewandte. Chem., Intl.*, 43, 6148, 2004. With permission. Copyright 2004 Wiley-VCH Verlag GmbH & Co. KGaA, Weinheim.)

With the reliable conductivity measurement, the dependence of molecular conductivity on chemical modification can be studied, leading to development of chemical sensing. Here the sample molecules were chosen as peptides because of their unlimited options of different sequences that can be tuned to obtain optimal binding strength and selectivity for a metal ion. Four peptides were studied, cysteamine-Cys, cysteamine-Gly-Cys, Cys-Gly-Cys, and cysteamine-Gly-Gly-Cys, that each have two thiol termini that can form reproducible contact to Au electrodes for electrical measurement. These peptides were expected to bind transition-metal ions, such as Cu^{2+} and Ni^{2+}, specifically through deprotonated peptide bonds. Under the experimental conditions, the metal ions and the peptides were expected to form 1:1 complexes (Figure 21.23), a conformation that favor electrical conduction by shortcutting the charge transfer path. By monitoring the significant increase in conductance or current, the metal–ion binding can be probed with exceptional sensitivity as demonstrated in Figure 21.23.

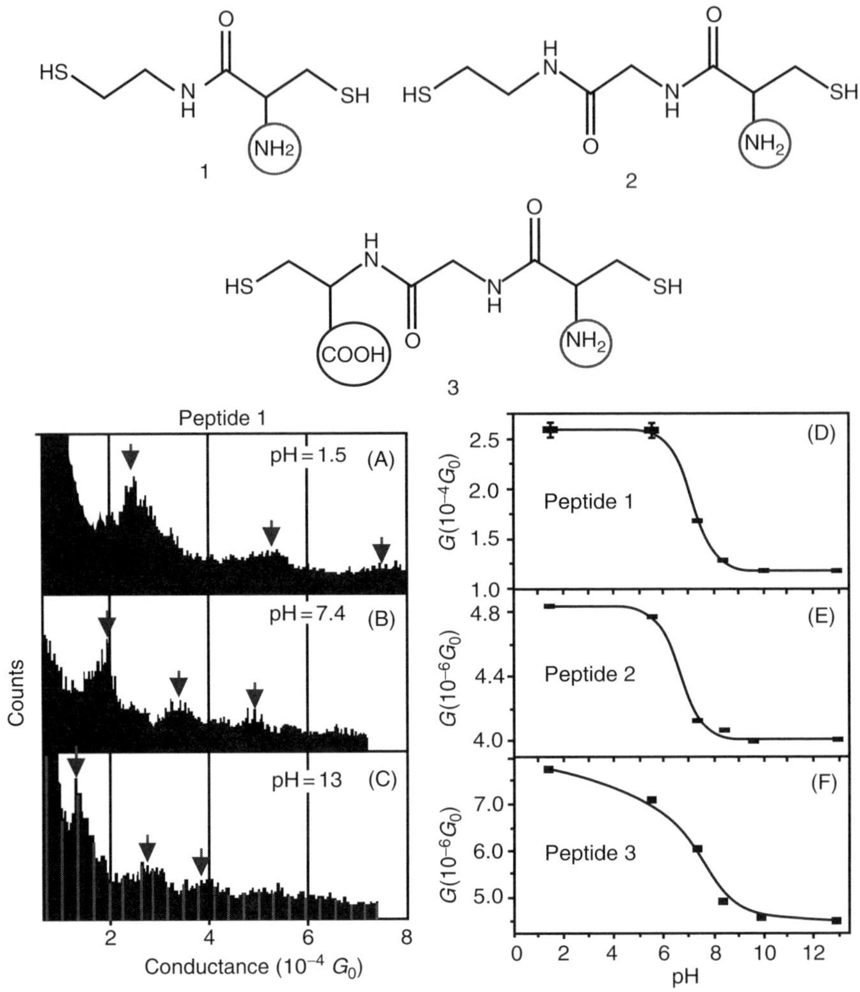

FIGURE 21.24 (top panel) Three peptides used in the pH titration. (A–C) Conductance histograms of peptide 1 obtained at various pH. Conductance versus pH for peptides 1 (D), 2 (E), and 3 (F). The solid lines show the sigmoidal fitting. (Reprinted from Xiao, X., Xu, B., and Tao, N., *J. Am. Chem. Soc.*, 126, 5370, 2005. With permission. Copyright 2005 American Chemical Society.)

Sensing pH [120]. The similar two-electrode sensor system described above was also applied to different peptides to approach a new sensing application, so-called conductance titration, by monitoring the pH dependence of the single-peptide conductance. Characterization of peptide conductivity and the dependence on pH is crucial for understanding the basic redox reactions within many kinds of proteins and enzymes, where peptide-mediated charge transfer plays vital roles in the catalytic processes. Three simple peptides (Figure 21.24), containing 1, 2, and 3 amino acids, were chosen to test the pH sensing of conductivity. Again, two thiol moieties were attached to the peptides as terminal groups so that they can form covalent bonds (good electrical contact) with Au electrodes.

Similar conductance histograms were obtained for the three peptides at various pHs (Figure 21.24). The results for peptide 1 at several pH values are shown in Figure 21.24A through C. At low pH, the conductance peaks are located near integer multiples of 2.5×10^{-4} G_0. Increasing the solution pH to 7.4, the peaks shift to integer multiples of 2.0×10^{-4} G_0. Further increasing the pH to 13, the conductance peaks shift to integer multiples of 1.2×10^{-4} G_0. The conductance of peptide 1 versus pH is plotted in Figure 21.24D, which shows a typical sigmoid-like titration curve, from which the

pK_a value can be estimated. Peptide 2 exhibits similar sigmoid-like pH-dependence of conductance. The titration of peptide 3 is, however, quite different from the other two. Instead of a sharp change around pH 7, the conductance of peptide 3 changes more smoothly with the pH. This difference is likely due to the fact that peptide 3 has both an amine and a carboxyl group, which are both sensitive to pH.

21.3 SINGLE-MOLECULE OPTOELECTRONICS: CONCEPTS AND MECHANISMS

Although organic optoelectronics has been extensively investigated for thin film based devices, the optoelectronic properties of single molecules remain unclear. The difficulty of studying single-molecule optoelectronics lies in several aspects: First, the fabrication and characterization of single-molecule electronic devices (e.g., the two-electrode nanojunction) have been challenging based on current nanotechnology as argued above. Second, study of single-molecule optoelectronics demands a combination of high-resolution electrical and optical measurement, but in general it is difficult to align both measurements at nanometer (or even molecular) level. Third, it is not an easy task to find a multifunctional molecule that is suited for optoelectronic application and nanoscale fabrication; the molecule must possess both ideal electronic and optical properties, which are typically correlated with the spectral response, redox potential, charge transfer and trapping efficiency, fluorescence quantum yield, thermal- and photostability, and geometry and rigidity for nanojunction alignment. Last, the detail understanding of the single-molecule optoelectronic properties (including both charge transfer and recombination) requires complicated theoretical calculation, and maybe new theories or models that consider not only the molecule but also the interface between the molecule and electrode. One of the major challenges for theoretical understanding is how to combine the photo-initiated processes (e.g., photo-induced charge transfer or separation) with the subsequent charge transport through the molecular junction. Photo-initiated processes could be ultrafast, in the timescale of femtoseconds.

Current understanding of molecular optoelectronic properties is mostly based on the photochemical and photophysical investigations of ensemble molecular systems, which could be either a gaseous, liquid, or solid dispersion. Some of the investigations, such as photo-induced electron transfer, have been scaled down to the single-molecule level. Nonetheless, the photochemical or photophysical methods can only measure the photo-initiated charge transfer processes or optical modulation of the electronic properties within molecules; whereas the reverse processes, i.e., conversion from electricity to light, cannot be studied by the conventional optical spectrometries. Typical optoelectronic devices concerning electricity-to-light conversion include LED and laser. Although these devices have been successfully fabricated and characterized with ensemble organic materials (for both thin films and nanomaterials, e.g., nanowires), it still remains a question whether it is possible to fabricate a LED or laser based on a single-molecule The major scientific challenge for a single-molecule LED is how to create and balance an electron–hole pair within the molecule and control the recombination to be emitting. Recently, Dickson and coworkers have demonstrated that single-molecule LED can be fabricated from spatially isolated individual Ag_n ($n = 2$–8) nanoclusters aligned between two semiconducting silver oxide electrodes [121,122]. This is clearly a big step toward miniaturization of nanojunction-based LEDs. Nonetheless, the observed system is far different from a molecular junction.

On the basis of the current knowledge and theoretical concepts of molecular optoelectronics, here the single-molecule optoelectronic devices are classified and introduced in three major directions: photodiode, phototransistor, and optical sensor.

21.3.1 SINGLE-MOLECULE PHOTODIODE

Single-molecule photodiode is basically an extended concept from the Aviram–Ratner molecular diode as described in Figure 21.12. Both devices employ a donor–acceptor (D–A) molecule as the

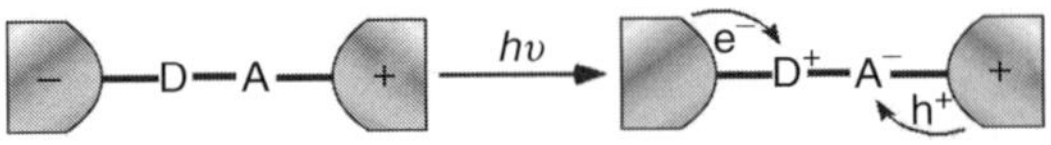

FIGURE 21.25 A photodiode comprising a D–A molecule bridged between two metal electrodes (with donor linked to the cathode and acceptor to the anode). In the absence of illumination, the molecule is not favorable for electron transport due to the high-energy barrier at the molecule/metal interface, i.e., donor is hard to be reduced and acceptor is hard to be oxidized. Upon illumination on either D or A, intramolecular electron transfer can be initiated by the increased energy at the excited state. If the lifetime of the charge-separation state thus created is long enough, subsequent charge injection from the electrodes will neutralize the ionic radicals and recover the D–A molecule, enabling electrical current flowing through the molecular junction—a prototype single-molecule photodiode. Theoretically, the illumination intensity (number of photons) can be counted by measuring the electrical current (number of the electrons flowing through). Note that the alignment of molecular junction is reverse in direction to the case of Aviram–Ratner molecular diode (Figure 21.12), where no optical excitation is applied.

channel, bridged within two metal electrodes. The major difference between the two types of diodes lies in the operation mechanism, i.e., in Aviram–Ratner diode the rectification is realized by switching the bias, while in the molecular photodiode the current is rectified through illumination (Figure 21.25). In the presence of illumination, the molecule aligned in photodiode gets excited, producing a charge-separation state though intramolecular electron transfer. Under a bias, charge injection between the molecule and electrode occurs, leading to neutralization of the charge-separation state. The net result of such a charging process is generation of current flowing through the molecular junction. The efficiency of the photogeneration of current depends on the competition between the interfacial charge transfer (injection) and the intramolecular charge recombination ($D^+–A^- \rightarrow D–A$). With appropriate molecular structural modification, the lifetime of the charge-separation state can be extended into nanoseconds or even microseconds. Such long-lived state can be sufficiently neutralized by the interfacial charge injection, which normally occurs in a much faster process, typically in picoseconds or femtoseconds. Efficient charge neutralization enables a photodiode with high rectifying ratio.

The D–A molecules employed here can be symmetric (dimers) or asymmetric molecules. For the dimers, the donor and acceptor are the same, and the energy barrier for charge transfer at ground state may not be as high as in the case of asymmetric D–A molecules (Figure 21.25), thereby producing higher current in the dark state. As a result, the photo-rectifying ratio obtained for the symmetric molecules will likely be lower than that for asymmetric molecules. However, the symmetric molecules offer easy alignment within the electrode junction, in comparison to the asymmetric molecules, which have to be aligned in a right direction (with respect to the source and drain electrodes) in the junction to ensure the desired direction of charge transport (Figure 21.25).

21.3.2 Single-Molecule Phototransistor

In analog to the conventional three-electrode FET, a single-molecule phototransistor is defined as a special transistor, for which the gate is not realized through a real electrode, but through optical excitation (Figure 21.26). To this end, a single-molecule phototransistor can also be considered as a single-molecule photodiode (Figure 21.25), except for that in the latter a D–A molecule has to be used. The efficiency of a molecular phototransistor is determined by the relative ratio of the photocurrent (electrical current under illumination) and dark-current (electrical current under dark). An ideal channel molecule should provide minimum dark-current, enabling large photo-modulation ratio. Such an optical gate-effect is mostly dependent on the optoelectronic properties of the channel molecule, including absorption coefficient, lifetime of the excite state, redox potential, and stability of the ionic radical.

The mechanism of current modulation in phototransistor involves the alteration of the electronic states of a molecule by resonant optical excitation. Absorption of a photon promotes an electron to LUMO, leaving a hole (vacancy) at the HOMO. On application of an appropriate bias, the electron

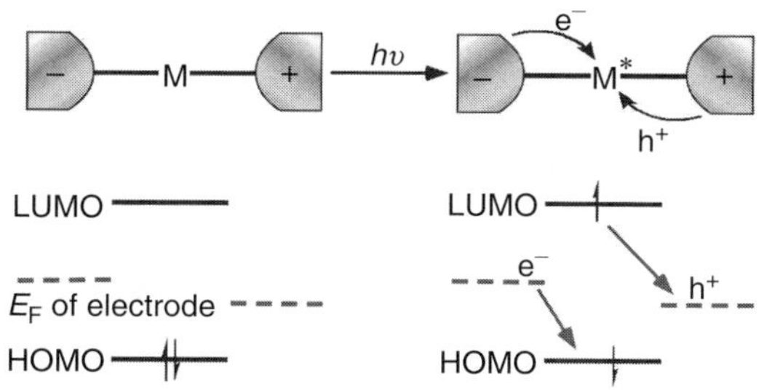

FIGURE 21.26 A phototransistor comprising a molecule (M) bridged between two metal electrodes. In the absence of illumination, the molecule is not favorable for electron transport due to the high-energy barrier at the molecule/metal interface, i.e., the molecule is neither easy to be reduced nor to be oxidized at moderate bias. Upon illumination, the molecule gets excited and thus possesses higher energy at the excited state, making it more powerful for both donating and accepting an electron. Whichever occurs first, the molecule will be charged via interfacial charge transfer, followed by neutralization by charge injection from the other electrode, thus enabling electrical current flowing through the molecular junction. If we consider the optical modulation in analog to the electrical field effect in the conventional FET, the optoelectronic molecular junction shown in this figure acts as a phototransistor, for which the gate is the optical excitation.

or hole can be neutralized through charge injection at the two electrodes, thus yielding an enhanced current (Figure 21.26). Some similar optical enhancement of current has recently been observed for a conducting polymer [123,124]; while in these experiments the measurements were performed on an ensemble of molecules in a monolayer.

21.3.3 Single-Molecule Optical Sensor

Detecting structure, dynamics, chemical reactions, and physical processes at the single-molecule level represents the ultimate degree of sensitivity for sensing and imaging. Fluorescence-based single-molecule spectroscopy (SMS) has evolved as an important method for studying the behavior of single molecules under ambient conditions [115,125–127]. Molecules composed of a fluorescent chromophore (reporter), a binding site (receptor), and a mechanism of communication between the two have found use in chemosensory and biological applications [128–132]. Combining such a concept with the SMS technique would enable the development of single-molecule-based sensing systems for high-resolution detection and probing. Molecular systems in which photo-induced intramolecular electron transfer (IET) from a high-energy nonbonding electron pair efficiently quenches the excited state of the chromophore form an important class of chemosensory materials [132]. Reactions of this electron pair, with protons, metal ions, organic electrophiles, or surfaces, lower the energy of the electron pair below the HOMO of the chromophore, turning OFF IET and turning ON fluorescence (Figure 21.27). In a work by Zang and Adams et al., a PTCDI molecule modified with an amine moiety was synthesized and developed as a unique single-molecule sensor for surface probing, for which the amine moiety serves as a binding site [133].

To investigate whether this sensor system can be used to locally probe nanoscale surface structure, the molecule was spin coated onto glass and quartz cover slips. Scanning confocal fluorescence images show bright fluorescent spots of single molecules only on glass (Figure 21.27, bottom). This result demonstrates that the amine moiety of the sensor binds to metal and metal-oxide impurities, such as TiO_2, ZnO, or Al_2O_3 centers, in the glass. No significant fluorescence is observed from the molecules dispersed on freshly cleaned quartz, since there are no such binding sites available (Figure 21.27, bottom). Molecules dispersed in a thin film of polyvinyl butylaldehyde (PVB) also fluoresce brightly as a result of reactions of the amine with the butylaldehyde on the polymer backbone. Exposure of the molecules on quartz to dioxane/HCl vapor causes fluorescence as a result of protonation of the amine

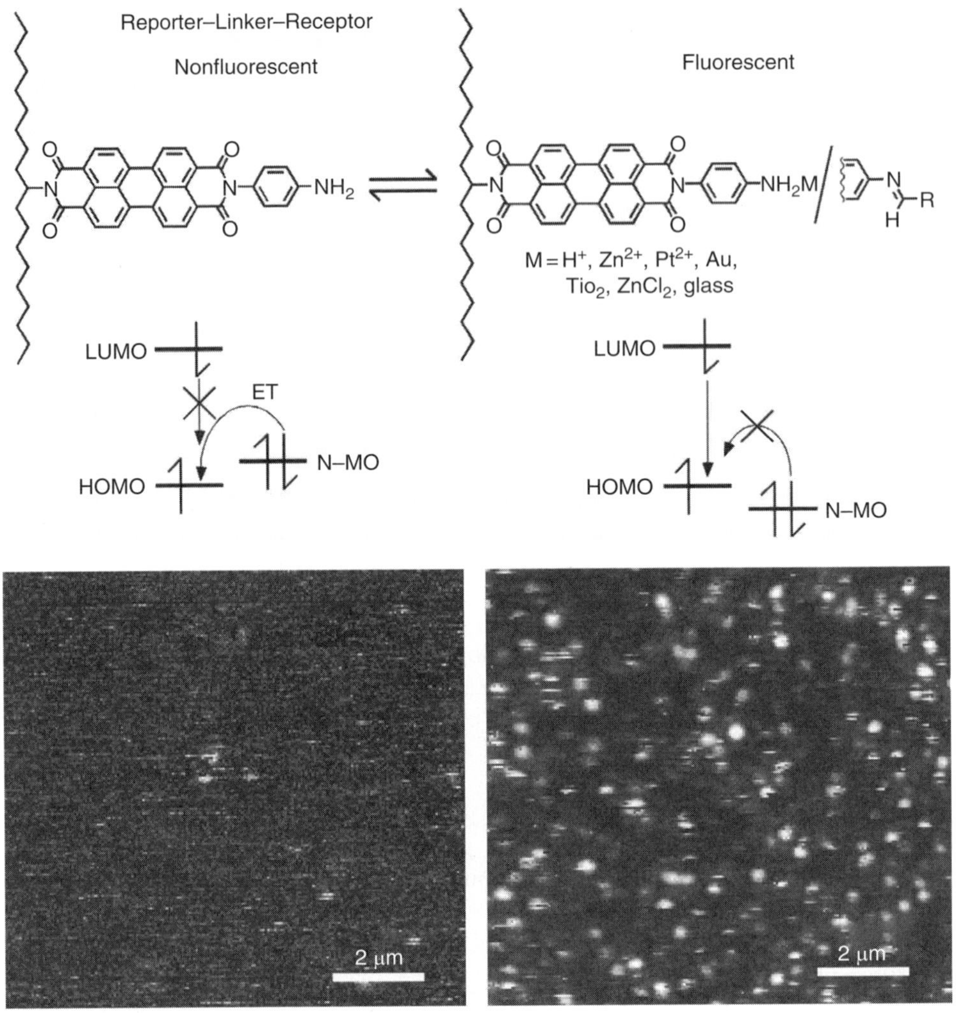

FIGURE 21.27 (top panel) A molecular orbital diagram for the free single-molecule probe illustrates how intramolecular electron transfer from a nonbonding electron pair (N–MO) to the highest occupied molecular orbital (HOMO) prevents fluorescence. Once the molecule binds to metal ions or proton, the N–MO drops in energy, and the intramolecular electron transfer gets blocked, turning on the fluorescence, which occurs with excited electrons going from the lowest unoccupied molecular orbital (LUMO) to the HOMO. (bottom panel) Single-molecule fluorescence images of the molecule spin coated onto quartz (left) and glass (right) surface, measured by scanning confocal microscopy. The glass (from Corning) contains dopant transition-metal ions such as Ti(IV) and Zn(II), which can be bound to the amine moiety of the molecule. (Reprinted from Zang, L., Liu, R., Holman, M.W., Nguyen, K.T., and Adams, D.M., *J. Am. Chem. Soc.*, 124, 10640, 2002. With permission. Copyright 2002 American Chemical Society.)

group. Moreover, exposure of the molecules on the quartz slide to air for several days also generates bright fluorescence spots. Since the slides are prepared by acid cleaning before spin casting, fluorescence spots seem to arise from molecular diffusion of the amine to protonated sites on the surface.

To obtain detailed information about the binding processes, fluorescence time trajectories of single molecules were measured: single molecules are positioned in the laser focus and the fluorescence intensity is recorded continuously until irreversible photobleaching occurs (Figure 21.28). While single molecules in a thin PVB film, or adsorbed on quartz slides following exposure to

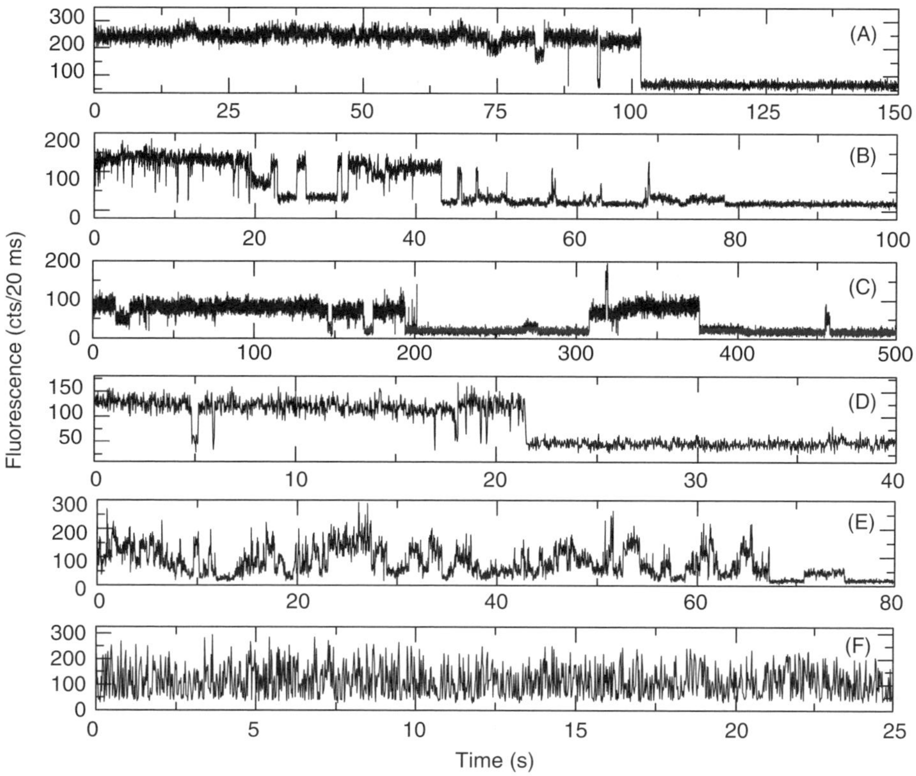

FIGURE 21.28 Characteristic single-molecule fluorescence time trajectories for the single-molecule sensor (shown in Figure 21.27) applied in different sample systems: (A) in a PVB 30 nm thin film, (B) and (C) on glass, (D) on quartz following exposure to dioxane/HCl vapor, (E) and (F) on quartz after several days (no initial fluorescence observed). (Reprinted from Zang, L., Liu, R., Holman, M.W., Nguyen, K.T., and Adams, D.M., *J. Am. Chem. Soc.*, 124, 10640, 2002. With permission. Copyright 2002 American Chemical Society.)

dioxane/HCl vapor, show nearly constant fluorescence intensity with infrequent OFF-times which are caused by triplet state formation (Figure 21.28A and D, molecules adsorbed on untreated glass surfaces show longer OFF-times of up to 100 s (Figure 21.28B and C). As proposed for other fluorophores, these OFF-times can be attributed to nonfluorescent charge separated states, i.e., radical cations or anions generated by an electron transfer reaction with the metal center. The fluorescence that appears on quartz slides after several-days exposed to air shows frequent blinking between an OFF and ON state on the millisecond timescale (Figure 21.28E and F), which indicates dynamic binding of the amine.

The reported work represents a typical sensing system that uses an intramolecular electron transfer mechanism to detect binding. The presented ideas and techniques help the single-molecule research community to develop new molecular system for high-resolution probing of chemical compositions and processes [112]. With appropriate molecular modification and engineering, the binding selectivity and strength of the molecules can be dramatically improved, leading to enhanced resolution and sensitivity in single-molecule probing and detection.

21.4 REMARKS

Study of single-molecule devices is currently still focused on fundamental research, particularly the modulation and control of charge transfer through molecules. These researches provide understanding of the basic electronic and optoelectronic properties of molecular junctions at the single-molecule level, helping guide the long-term development of molecular devices employing the same molecules. Although it will be a long wait till people find effective techniques to integrate individual molecular devices into a large-area functional chip, the single-molecule studies and the various devices developed therein have already found promising application in broad range of electronics and optoelectronics, including sensing, switching, and rectifying.

EXERCISE QUESTIONS

1. What are the basic working principle and functions of an FET? Compare the differences between n- and p-type channels.
2. Compared to silicon-based materials, what are the major advantages of organic semiconductors in light of application in optoelectronic devices?
3. Compared to bulk materials devices, what are the major advantages of single-molecule devices in terms of both fundamental study and practical applications?
4. What are the major challenges for measuring or calibrating conductivity of a single molecule? What are the typical methods currently developed for constructing a single-molecule junction?
5. What are the two basic conducting mechanisms of molecular wire?
6. What is molecular diode?
7. Compared to the bulk-materials FET, what are the two challenges to realize FET operation for a molecular junction?

LIST OF ABBREVIATIONS

AFM	atomic force microscopy
CFET	chemical field-effect transistor
FET	field-effect transistor
HOMO	highest occupied molecular orbital
LUMO	lowest unoccupied molecular orbital
PTCDI	perylene tetracarboxylic diimide
PZT	piezoelectric transducer
SPM	scanning probe microscopy
STM	scanning tunneling microscopy

ACKNOWLEDGMENTS

The authors acknowledge the research support from the Consortium for Advanced Radiation Sources (CARS), K.C. Wong Foundation, NSF, NSFC, and ORDA, COS and MTC of SIUC. They also thank DOE/BES for use of the advanced photon source (APS) under Contract No. W-31-109-ENG-38, and the electronic microscopes at the Center for Microanalysis of Materials (CMM), University of Illinois, under grant DEFG02-91-ER45439. Ling Zang appreciates all the technical support from his colleagues and collaborators, including Mr. Kaushik Balakrishnan, Mr. Aniket Datar, Ms. Jialing Huang, professor Dan Dyer, professor Matt McCarroll, professor Max Yen, professor John Bozzola, Mr. Steven Schmitt, professor Jeff Moore, professor Jianmin Zuo, professor Jincai Zhao, Dr. David Tiede, Dr. Mauro Sardela, and Dr. Stefan Kaemmer. Ling Zang dedicates this chapter to his father Gongmin Zang on the occasion of his 70th birthday.

REFERENCES

1. *Chemical and Engineering News*, Solar cell hits efficiency record, 26 2006.
2. Coakley, K.M. and McGehee, M.D. Conjugated polymer photovoltaic cells. *Chem. Mater.* **16**, 4533–4542 2004.

3. Tang, C.W. Two-layer organic photovoltaic cell. *Appl. Phys. Lett.* **48**, 183–185 1986.
4. Xue, J., Uchida, S., Rand, B.P., and Forrest, S.R. Asymmetric tandem organic photovoltaic cells with hybrid planar-mixed molecular heterojunctions. *Appl. Phys. Lett.* **85**, 5757–5759 2004.
5. Halls, J.J.M., Walsh, C.A., Greenham, N.C., Marseglia, E.A., Friend, R.H., Moratti, S.C., and Holmes, A.B. Efficient photodiodes from interpenetrating polymer networks. *Nature* **376**, 498–500 1995.
6. Granstrom, M., Petritsch, K., Arias, A.C., Lux, A., Andersson, M.R., and Friend, R.H. Laminated fabrication of polymeric photovoltaic diodes. *Nature* **395**, 257–260 1998.
7. Huynh, W.U., Dittmer, J.J., and Alivisatos, A.P. Hybrid nanorod-polymer solar cells. *Science* **295**, 2425–2427 2002.
8. Gratzel, M. Photoelectrochemical cells. *Nature* **414**, 338–344 2001.
9. Gregg, B.A. Excitonic solar cells. *J. Phys. Chem. B* **107**, 4688–4698 2003.
10. Shaheen, S.E., Brabec, C.J., Sariciftci, N.S., Padinger, F., Fromherz, T., and Hummelen, J.C. 2.5% efficient organic plastic solar cells. *Appl. Phys. Lett.* **78**, 841–843 2001.
11. Yu, G., Gao, J., Hummelen, J.C., Wudl, F., and Heeger, A.J. Polymer photovoltaic cells: Enhanced efficiencies via a network of internal donor–acceptor heterojunctions. *Science* **270**, 1789–1791 1995.
12. Padinger, F., Rittberger, R.S., and Sariciftci, N.S. Effects of postproduction treatment on plastic solar cells. *Adv. Funct. Mater.* **13**, 85–88 2003.
13. Huang, J.X., Virji, S., Weiller, B.H., and Kaner, R.B. Polyaniline nanofibers: Facile synthesis and chemical sensors. *J. Am. Chem. Soc.* **125**, 314–315 2003.
14. Zhou, Y., Freitag, M., Hone, J., Staii, C., and Johnson, J.A.T. Fabrication and electrical characterization of polyaniline-based nanofibers with diameter below 30 nm. *Appl. Phys. Lett.* **83**, 3800–3802 2003.
15. Liu, H., Kameoka, J., Czaplewski, D.A., and Craighead, H.G. Polymeric nanowire chemical sensor. *Nano Lett.* **4**, 671–675 2004.
16. Virji, S., Huang, J., Kaner, R.B., and Weiller, B.H. Polyaniline nanofiber gas sensors: examination of response mechanisms. *Nano Lett.* **4**, 491–496 2004.
17. Huang, J., Virji, S., Weiller, B.H., and Kaner, R.B. Nanostructured polyaniline sensors. *Chem. Eur. J.* **10**, 1314–1319 2004.
18. Lee, H.J., Jin, Z.X., Aleshin, A.N., Lee, J.Y., Goh, M.J., Akagi, K., Kim, Y.S., Kim, D.W., and Park, Y.W. Dispersion and current-voltage characteristics of helical polyacetylene single fibers. *J. Am. Chem. Soc.* **126**, 16722–16723 2004.
19. Gan, H., Liu, H., Li, Y., Zhao, Q., Li, Y., Wang, S., Jiu, T., Wang, N., He, X., Yu, D., and Zhu, D. Fabrication of polydiacetylene nanowires by associated self-polymerization and self-assembly processes for efficient field emission properties. *J. Am. Chem. Soc.* **127**, 12452–12453 2005.
20. Zhang, X. and Manohar, S.K. Narrow pore-diameter polypyrrole nanotubes. *J. Am. Chem. Soc.* **127**, 14156–14157 2005.
21. Bocharova, V., Kiriy, A., Vinzelberg, H., Moench, I., and Stamm, M. Polypyrrole nanowires grown from single adsorbed polyelectrolyte molecules. *Angew. Chem. Int. Ed.* **44**, 6391–6394 2005.
22. Berdichevsky, Y., and Lo, Y.-H. Polypyrrole nanowire actuators. *Adv. Mater. (Weinheim, Germany)* **18**, 122–125 2006.
23. Luo, Y.H., Liu, H.W., Xi, F., Li, L., Jin, X.G., Han, C.C., and Chan, C.M. Supramolecular assembly of poly(phenylene vinylene) with crown ether substituents to form nanoribbons. *J. Am. Chem. Soc.* **125**, 6447–6451 2003.
24. Jeukens, C.R.L.P.N., Jonkheijm, P., Wijnen, F.J.P., Gielen, J.C., Christianen, P.C.M., Schenning, A.P.H.J., Meijer, E.W., and Maan, J.C. Polarized emission of individual self-assembled oligo(*p*-phenylenevinylene)-based nanofibers on a solid support. *J. Am. Chem. Soc.* **127**, 8280–8281 2005.
25. Hoeben, F.J.M., Jonkheijm, P., Meijer, E.W., and Schenning, A.P.H.J. About supramolecular assemblies of p-conjugated systems. *Chem. Rev.* **105**, 1491–1546 2005.
26. Nguyen, T.-Q., Martel, R., Avouris, P., Bushey, M.L., Brus, L., and Nuckolls, C. Molecular interactions in one-dimensional organic nanostructures. *J. Am. Chem. Soc.* **126**, 5234–5242 2004.
27. Wurthner, F. Perylene bisimide dyes as versatile building blocks for functional supramolecular architectures. *Chem. Commun.*, 1564–1579 2004.
28. Shirakawa, M., Fujita, N., and Shinkai, S. A stable single piece of unimolecularly d-stacked porphyrin aggregate in a thixotropic low molecular weight gel: A one-dimensional molecular template for polydiacetylene wiring up to several tens of micrometers in length. *J. Am. Chem. Soc.* **127**, 4164–4165 2005.
29. Wang, Z., Medforth, C.J., and Shelnutt, J.A. Porphyrin nanotubes by ionic self-assembly. *J. Am. Chem. Soc.* **126**, 15954–15955 2004.

30. Kastler, M., Pisula, W., Wasserfallen, D., Pakula, T., and Mullen, K. Influence of alkyl substituents on the solution- and surface-organization of hexa-peri-hexabenzocoronenes. *J. Am. Chem. Soc.* **127**, 4286–4296 2005.
31. Hill, J.P., Jin, W., Kosaka, A., Fukushima, T., Ichihara, H., Shimomura, T., Ito, K., Hashizume, T., Ishii, N., and Aida, T. Self-. *Science* **304**, 1481–1483 2004.
32. Kushmerick, J.G., Pollack, S.K., Yang, J.C., Naciri, J., Holt, D.B., Ratner, M.A., and Shashidhar, R. Understanding charge transport in molecular electronics. *Ann. N. Y. Acad. Sci.* **1006**, 277–290 2003.
33. Nitzan, A. and Ratner, M.A. Electron transport in molecular wire junctions. *Science* **300**, 1384–1389 2003.
34. Xue, Y. and Ratner, M.A. Microscopic study of electrical transport through individual molecules with metallic contacts. II. Effect of the interface structure. *Phys. Rev. B* **68**, 115407–11 2003.
35. Reimers, J.R., Shapley, W.A., Lambropoulos, N., and Hush, N.S. An atomistic approach to conduction between nanoelectrodes through a single molecule. *Ann. N. Y. Acad. Sci.* **960**, 100–130 2002.
36. Xue, Y., Datta, S., and Ratner, M.A. Charge transfer and "band lineup" in molecular electronic devices: A chemical and numerical interpretation. *J. Chem. Phys.* **115**, 4292–4299 2001.
37. Li, X., Xu, B.Q., Xiao, X., Yang, X., Zang, L., and Tao, N.J. Controlling charge transport in single molecules using electrochemical gate. *Faraday Discuss.* **131**, 111–120 2006.
38. Tour, J.M. Molecular electronics. Synthesis and testing of components. *Acc. Chem. Res.* **33**, 791–804 2000.
39. Cahen, D. and Hodes, G. Molecules and electronic materials. *Adv. Mater.* **14**, 789–798 2002.
40. Joachim, C., Gimzewski, J.K., and Aviram, A. Electronics using hybrid-molecular and mono-molecular devices. *Nature* **408**, 541–548 2000.
41. Metzger, R.M. Electrical rectification by a molecule: The advent of unimolecular electronic devices. *Acc. Chem. Res.* **32**, 950–957 1999.
42. Service, R.F. Assembling nanocircuits from the bottom up. *Science* **293**, 782–785 2001.
43. Heath, J.R. and Ratner, M.A. Molecular electronics. *Physics Today* **56**, 43–49 2003.
44. Carroll, R.l. and Gorman, C.B. The genesis of molecular electronics. *Angew. Chem. Int. Ed.* **41**, 4379–4399 2002.
45. Ghosh, S.C., Zhu, X.-Y., Secchi, A., Sadhukhan, S.K., Girdhar, N.K., and Gourdon, A. Molecular landers as probes for molecular device–metal surface interactions. *Ann. NY Acad. Sci.* **1006**, 82–93 2003.
46. Wang, W., Lee, T., Kamdar, M., Reed, M.A., Stewart, M.P., Hwang, J.J., and Tour, J.M. Electrical characterization of metal–molecule–silicon junctions. *Ann. NY Acad. Sci.* **1006**, 36–47 2003.
47. Ramachandran, G.K., Hopson, T.J., Rawlett, A.M., Nagahara, L.A., Primak, A., and Lindsay, S.M. A bond-fluctuation mechanism for stochastic switching in wired molecules. *Science* **300**, 1413–1416 2003.
48. Kubatkin, S., Danilov, A., Hjort, M., Cornil, J., Bredas, J.-L., Stuhr-Hansen, N., Hedegard, P., and Bjornholm, T. Single-electron transistor of a single organic molecule with access to several redox states. *Nature* **425**, 698–701 2003.
49. Friis, E.P., Kharkats, Y.I., Kuznetsov, A.M., and Ulstrup, J. In situ scanning tunneling microscopy of a redox molecule as a vibrationally coherent electronic three-level process. *J. Phys. Chem. A.* **102**, 7851–7859 1998.
50. Andres, R.P., Bein, T., Dorogi, M., Feng, S., Henderson, J.I., Kubiak, C.P., Mahoney, W., Osifchin, R.G., and Reifenberger, R. "Coulomb staircase" at room temperature in a self-assembled molecular nanostructure. *Science* **272**, 1323–1325 1996.
51. Piva, P.G., DiLabio, G.A., Pitters, J.L., Zikovsky, J., Rezeq, M., Dogel, S., Hofer, W.A., and Wolkow, R.A. Field regulation of single-molecule conductivity by a charged surface atom. *Nature* **435**, 658–661 2005.
52. Sikes, H.D., Smalley, J.F., Dudek, S.P., Cook, A.R., Newton, M.D., Chidsey, C.E.D., and Feldberg, S.W. Rapid electron tunneling through oligophenylenevinylene bridges. *Science* **291**, 1519–1523 2001.
53. Seminario, J.M., Zacarias, A.G., and Tour, J.M. Theoretical study of a molecular resonant tunneling diode. *J. Am. Chem. Soc.* **122**, 3015–3020 2000.
54. Chen, J., Reed, M.A., Rawlett, A.M., and Tour, J.M. Large on-off ratios and negative differential resistance in a molecular electronic device. *Science* **286**, 1550–1552 1999.
55. Lang, N.D., and Avouris, P. Understanding the variation of the electrostatic potential along a biased molecular wire. *Nano Lett.* **3**, 737–740 2003.
56. Stokbro, K., Taylor, J., and Brandbyge, M. Do Aviram-Ratner diodes rectify? *J. Am. Chem. Soc.* **125**, 3674–3675 2003.
57. Bumm, L.A., Arnold, J.J., Cygan, M.T., Dunbar, T.D., Burgin, T.P., Jones, L.I., Allara, D.L., Tour, J.M., and Weiss, P.S. Are single molecular wires conducting. *Science* **271**, 1705–1707 1996.

58. Hipps, K.W. It's all about contacts. *Science* **294**, 536–537 2001.
59. Beebe, J.M., Engelkes, V.B., Miller, L.L., and Frisbie, C.D. Contact resistance in metal-molecule-metal junctions based on aliphatic SAMs: Effects of surface linker and metal work function. *J. Am. Chem. Soc.* **124**, 11268–11269 2002.
60. Cholet, S., Joachim, C., Martinez, J.P., and Rousset, B. Fabrication of co-planar metal-insulator-metal solid state nanojunctions down to 5 nm. *Eur. Phys. J. Appl. Phys.* **8**, 139–145 1999.
61. Di Ventra, M., Pantelides, S.T., and Lang, N.D. First-principles calculation of transport properties of a molecular device. *Phys. Rev. Lett.* **84**, 979–982 2000.
62. Love, J.C., Wolfe, D.B., Haasch, R., Chabinyc, M.L., Paul, K.E., Whitesides, G.M., and Nuzzo, R.G. Formation and structure of self-assembled monolayers of alkanethiolates on palladium. *J. Am. Chem. Soc.* **125**, 2597–2609 2003.
63. Reed, M.A., Zhou, C., Muller, C.J., Burgin, T.P., and Tour, J.M. Conductance of a molecular junction. *Science* **278**, 252–254 1997.
64. Dorogi, M., Gomez, J., Osifchin, R., Andres, R.P., and Reifenberger, R. Room-temperature Coulomb blockade from a self-assembled molecular nanostructure. *Phys. Rev. B.* **52**, 9071–9077 1995.
65. Gittins, D.I., Bethell, D., Schiffrin, D.J., and Nichols, R.J. A nanometer-scale electronic switch consisting of a metal cluster and redox-addressable groups. *Nature* **408**, 67–69 2000.
66. Cui, X.D., Primak, A., Zarate, X., Tomfohr, J., Sankey, O.F., Moore, A.L., Moore, T.A., Gust, D., Harris, G., and Lindsay, S.M. Reproducible Measurement of Single-Molecule Conductivity. *Science* **294**, 571–574 2001.
67. Xu, B.Q., and Tao, N.J. Measurement of single molecule conductance by repeated formation of molecular junctions. *Science* **301**, 1221–1223 2003.
68. Magoga, M. and Joachim, C. Conductance and transparence of long molecular wires. *Phys. Rev. B.* **56**, 4722–4729 1997.
69. Yaliraki, S.N., Kemp, M., and Ratner, M.A. Conductance of molecular wires: Influence of molecule–electrode binding. *J. Am. Chem. Soc.* **121**, 3428–3434 1999.
70. Moresco, F., Gross, L., Alemani, M., Rieder, K.H., Tang, H., Gourdon, A., and Joachim, C. Probing the different stages in contacting a single molecular wire. *Phys. Rev. Lett.* **91**, 036601/1–036601/4 2003.
71. Xu, B. and Tao, N.J. Measurement of single-molecule resistance by repeated formation of molecular junctions. *Science* **301**, 1221–1223 2003.
72. Xiao, X.Y., Xu, B.Q., and Tao, N.J. Measurement of single molecule conductance: Benzenedithiol and benzenedimethanethiol. *Nano Lett.* **4**, 267–271 2004.
73. Xu, B.Q., Zhang, P.M., Li, X.L., and Tao, N.J. Direct conductance measurement of single DNA molecules in aqueous solution. *Nano Lett.* 2004.
74. Xu, B.Q., Xiao, X., Yang, X., Zang, L., and Tao, N.J. Large gate modulation in the current of a room temperature single molecule transistor. *J. Am. Chem. Soc.* **127**, 2386–2387 2005.
75. Li, X., Xu, B.Q., Xiao, X., Yang, X., Zang, L., and Tao, N.J. Controlling charge transport in single molecules using electrochemical gate. *Faraday Discuss.* **131**, 111–120 2005.
76. Donhauser, Z.J., Mantooth, B.A., Kelly, K.F., Bumm, L.A., Monnell, J.D., Stapleton, J.J., Price Jr., D.W., Rawlett, A.M., Allara, D.L., Tour, J.M., and Weiss, P.S. Conductance switching in single molecules through conformational changes. *Science* **292**, 2303–2307 2001.
77. Venkataraman, L., Klare, J.E., Nuckolls, C., Hybertsen, M.S., and Steigerwald, M.L. Dependence of single-molecule junction conductance on molecular conformation. *Nature* **442**, 904–907 2006.
78. Liu, R., Holman, M.W., Zang, L., and Adams, D.M. Single-molecule spectroscopy of intramolecular electron transfer in donor–bridge–acceptor systems. *J. Phys. Chem. A* **107**, 6522–6526 2003.
79. Holman, M.W., Liu, R., Zang, L., Yan, P., DiBenedetto, S.A., Bowers, R.D., and Adams, D.M. Studying and switching electron transfer: From the ensemble to the single molecule. *J. Am. Chem. Soc.* **126**, 16126–16133 2004.
80. Aviram, A. and Ratner, M. Molecular rectifiers. *Chem. Phys. Lett.* **29**, 277–283 1974.
81. Maruccio, G., Cingolani, R., and Rinaldi, R. Projecting the nanoworld: Concepts, results and perspectives of molecular electronics. *J. Mater. Chem.* **14**, 542–554 2004.
82. Metzger, R.M. Unimolecular electrical rectifiers. *Chem. Rev.* **103**, 3803–3834 2003.
83. Metzger, R.M., Chen, B., Hoepfner, U., Lakshmikantham, M.V., Vuillaume, D., Kawai, T., Wu, X., Tachibana, H., Hughes, T.V., Sakurai, H., Baldwin, J.W., Hosch, C., Cava, M.P., Brehmer, L., and Ashwell, G.J. Unimolecular electrical rectification in hexadecylquinolinium tricyanoquinodimethanide. *J. Am. Chem. Soc.* **119**, 10455–10466 1997.
84. McCreery, R.L. Molecular electronic junctions. *Chem. Mater.* **16**, 4477–4496 2004.
85. Flood, A.H., Stoddart, J.F., Steuerman, D.W., and Heath, J.R. Whence molecular electronics? *Science* **306**, 2055–2056 2004.

86. Ratner, M. Molecular electronics: Charged with manipulation. *Nature* **435**, 575–577 2005.
87. Di Ventra, M., Pantelides, S.T., and Lang, N.D. The benzene molecule as a molecular resonant-tunneling transistor. *Appl. Phys. Lett.* **76**, 3448–3450 2000.
88. Damle, P., Rakshit, T., Paulsson, M., and Datta, S. Current-voltage characteristics of molecular conductors: Two vesus three terminal. *IEEE Trans. Nanotech.* **1**, 145–153 2002.
89. Scudiero, L., Barlow, D.E., Mazur, U., and Hipps, K.W. Scanning tunneling microscopy, orbital-mediated tunneling spectroscopy, and ultraviolet photoelectron spectroscopy of metal(II) tetraphenylporphyrins deposited from vapor. *J. Am. Chem. Soc.* **123**, 4073–4080 2001.
90. Kuznetsov, A.N. and Schmickler, W. Mediated electron exchange between an electrode and the tip of a scanning tunneling microscope - a stochastic approach. *Chem. Phys.* **282**, 371–377 2002.
91. Tans, S.J., Verschueren, A.R.M., and Dekker, C. Room temperature transistor based on a single carbon nanotube. *Nature* **393**, 49–52 1998.
92. Collins, P.C., Arnold, M.S., and Avouris, P. Engineering carbon nanotubes and nanotube circuits using electrical breakdown. *Science* **292**, 706–709 2001.
93. Javey, A., Guo, J., Wang, Q., Lundstrom, M., and Dai, H. Ballistic carbon nanotube field-effect transistors. *Nature* **424**, 654–657 2003.
94. Dai, H. Carbon nanotubes: Synthesis, integration, and properties. *Acc. Chem. Res.* **35**, 1035–1044 2002.
95. Cui, Y. and Lieber, C. Functional nanoscale electronic devices assembled using silicon nanowire building blocks. *Science* **291**, 851–853 2001.
96. Law, M., Goldberger, J., and Yang, P. Semiconductor nanowires and nanotubes. *Annu. Rev. Mater. Res.* **34**, 83–122 2004.
97. Park, H., Park, J., Lim, A.K.L., Anderson, E.H., Alivisatos, A.P., and McEuen, P.L. Nanomechanical oscillations in a single-C-60 transistor. *Nature* **407**, 57–60 2000.
98. Park, J., Pasupathy, A.N., Goldsmith, J.L., Chang, C., Yaish, Y., Petta, J.R., Rinkoski, M., Sethna, J.P., Abruna, H.D., McEuen, P.L., and Ralph, D.C. Coulomb blockade and the Kondo effect in single-atom transistors. *Nature* **417**, 722–725 2002.
99. Liang, W.J., Shores, M., Bockrath, M., Long, J.R., and Park, H. Kondo resonance in a single-molecule transistor. *Nature* **417**, 725–729 2002.
100. Kubatkin, S., Danilov, A., Hjort, M., Cornil, J., Bredas, J.-L., Stuhr-Hansen, N., Hedegard, P., and Bjornholm, T. Single-electron transistor of a single organic molecule with access to several redox states. *Nature* **425**, 698–701 2003.
101. Collier, C.P., Wong, E.W., Belohradsky, M., Raymo, F.M., Stoddart, J.F., Kuekes, P.J., Williams, R.S., and Heath, J.R. Electronically configurable molecular-based logic gates. *Science* **285**, 391–394 1999.
102. Di Ventra, M., Pantelides, S.T., and Lang, N.D. The benzene molecule as a molecular resonant-tunneling transistor. *Appl. Phys. Lett.* **76**, 3448–3450 2000.
103. Cormier, R.A., and Gregg, B.A. Synthesis and characterization of liquid crystalline perylene diimides. *Chem. Mater.* **10**, 1309–1319 1998.
104. Gregg, B.A. and Cormier, R.A. Doping molecular semiconductors: n-Type doping of a liquid crystal perylene diimide. *J. Am. Chem. Soc.* **123**, 7959–7960 2001.
105. Struijk, C.W., Sieval, A.B., Dakhorst, J.E.J., van Dijk, M., Kimkes, P., Koehorst, R.B.M., Donker, H., Schaafsma, T.J., Picken, S.J., van de Craats, A.M., Warman, J.M., Zuihof, H., and Sudholter, E.J.R. Liquid crystalline perylene diimides: Architecture and charge carrier mobilities. *J. Am. Chem. Soc.* **122**, 11057–11066 2000.
106. Tamizhmani, G., Dodelet, J.P., Cote, R., and Gravel, D. Photoelectrochemical characterization of thin films of perylenetetracarboxylic acid derivatives. *Chem. Mater.* **3**, 1046–1053 1991.
107. Horowitz, G., Kouki, F., Spearman, P., Fichou, D., Nogues, C., Pan, X., and Garnier, F. Evidence for n-type conduction in a perylene tetracarboxylic diimide derivative. *Adv. Mater.* **8**, 242–245 1996.
108. Newman, C.R., Frisbie, C.D., da Silva Filho, D.A., Bredas, J.-L., Ewbank, P.C., and Mann, K.R. Introduction to organic thin film transistors and design of n-channel organic semiconductors. *Chem. Mater.* **16**, 4436–4451 2004.
109. Kuznetsov, A.M., and Ulstrup, J. Mechanisms of in situ scanning tunnelling microscopy of organized redox molecular assemblies. *J. Phys. Chem. A* **104**, 11531–11540 2000.
110. Nitzan, A. Electron transmission through molecules and molecular interfaces. *Annu. Rev. Phys. Chem.* **52**, 681–750 2001.
111. Jackel, F., Watson, M.D., Mullen, K., and Rabel, J.P. Prototypical single-molecule chemical-field-effect transistor with nanometer-sized gates. *Phys. Rev. Lett.* **92**, 188303/1–4 2004.

112. Sauer, M. Single-molecule-sensitive fluorescent sensors based on photoinduced intramolecular charge transfer. *Angew. Chem. Int. Ed.* **42**, 1790–1793 2003.
113. Yang, H., Luo, G., Karnchanaphanurach, P., Louie, T.-M., Rech, I., Cova, S., Xun, L., and Xie, X.S. Protein conformational dynamics probed by single-molecule electron transfer. *Science* **302**, 262–266 2003.
114. Moerner, W.E. and Fromm, D.P. Methods of single-molecule fluorescence spectroscopy and microscopy. *Rev. Sci. Instrum.* **74**, 3597–3619 2003.
115. Moerner, W.E. A dozen years of single-molecule spectroscopy in physics, chemistry, and biophysics. *J. Phys. Chem. B* **106**, 910–927 2002.
116. Ludes, M.D., and Wirth, M.J. Single-molecule resolution and fluorescence imaging of mixed-mode sorption of a dye at the interface of C18 and acetonitrile/water. *Anal. Chem.* **74**, 386–393 2002.
117. Zhuang, X., Ha, T., Kim, H.D., Centner, T., Labeit, S., and Chu, S. Fluorescence quenching: A tool for single-molecule protein-folding study. *Proc. Natl. Acad. Sci. USA* **97**, 14241–14244 2000.
118. Xie, X.S. Single-molecule spectroscopy and dynamics at room temperature. *Acc. Chem. Res.* **29**, 598–606 1996.
119. Xiao, X., Xu, B.Q., and Tao, N.J. Changes in the conductance of single peptide molecules upon metal-ion binding. *Angewandte. Chem., Intl.* **43**, 6148–6152 2004.
120. Xiao, X., Xu, B., and Tao, N. Conductance titration of single-peptide molecules. *J. Am. Chem. Soc.* **126**, 5370–5371 2004.
121. Lee, T.-H. and Dickson, R.M. Single-molecule LEDs from nanoscale electroluminescent junctions. *J. Phys. Chem. B* **107**, 7387–7390 2003.
122. Lee, T.-H., Gonzalez, J.I., Zheng, J., and Dickson, R.M. Single-molecule optoelectronics. *Acc. Chem. Res.* **38**, 534–541 2005.
123. Hu, W., Nakashima, H., Furukawa, K., Kashimura, Y., Ajito, K., Liu, Y., Zhu, D., and Torimitsu, K. A self-assembled nano optical switch and transistor based on a rigid conjugated polymer, thioacetyl-end-functionalized poly(*para*-phenylene ethynylene). *J. Am. Chem. Soc.* **127**, 2804–2805 2005.
124. Matsui, J., Mitsuishi, M., Aoki, A., and Miyashita, T. Molecular optical gating devices based on polymer nanosheets assemblies. *J. Am. Chem. Soc.* **126**, 3708–3709 2004.
125. Ambrose, W.P., Goodwin, P.M., Jett, J.H., Van Orden, A., Werner, J.H., and Keller, R.A. Single molecule fluorescence spectroscopy at ambient temperature. *Chem. Rev.* **99**, 2929–2956 1999.
126. Moerner, W.E. High-resolution optical spectroscopy of single molecules in solids. *Acc. Chem. Res.* **29**, 563–571 1996.
127. Xie, X.S. and Trautman, J.K. Optical studies of single molecules at room temperature. *Ann. Rev. Phys. Chem.* **49**, 441–480 1998.
128. Lakowicz, J.R. *Principles of Fluorescence Spectroscopy.* Plenum Publishers, New York, 1999.
129. Keefe, M.H., Benkstein, K.D., and Hupp, J.T. Luminescent sensor molecules based on coordinated metals: A review of recent developments. *Coord. Chem. Rev.* **205**, 201–228 2000.
130. Burdette, S., Walkup, G.K., Spingler, B., Tsien, R.Y., and Lippard, S.J. Fluorescent sensors for Zn2+ based on a fluorescein platform: Synthesis, properties and intracellular distribution. *J. Am. Chem. Soc.* **123**, 7831–7841 2001.
131. McQuade, D.T., Pullen, A.E., and Swager, T.M. Conjugated polymer-based chemical sensors. *Chem. Rev.* **100**, 2537–2574 2000.
132. de Silva, A.P., Fox, D.B., Huxley, A.J.M., and Moody, T.S. Combining luminescence, coordination and electron transfer for signalling purposes. *Coord. Chem. Rev.* **205**, 41–57 2000.
133. Zang, L., Liu, R., Holman, M.W., Nguyen, K.T., and Adams, D.M. A single-molecule probe based on intramolecular electron transfer. *J. Am. Chem. Soc.* **124**, 10640–10641 2002.

22 Introduction to Nonvolatile Organic Thin-Film Memory Devices

Yang Yang

CONTENTS

22.1 Introduction to Memory Devices 701
22.2 Principle of Organic Thin-Film Memory Devices 703
22.3 Introduction to Materials Used in Organic Thin-Film Memory Devices 706
22.4 An Example of an Organic Thin-Film Memory Device 708
22.5 Summary 711
Exercise Questions 711
List of Abbreviations 711
References 712

Abstract This chapter introduces the principles and materials of organic thin-film memory devices. The operation is based on the electrical bistability of an organic thin film. The device consists of a film containing conjugated organic compounds and metal nanoparticles sandwiched between two metal electrodes. It can be switched electrically between two conductivity states, and can remain in either state for a long period even when there is no external power supply. Such a device can be used as a nonvolatile digital memory, where the high and low conductivity states are defined as "1" and "0," respectively. The mechanism is related to charge storage in the metal nanoparticles.

22.1 INTRODUCTION TO MEMORY DEVICES

Memory devices can store, retain, and recall information. They become more and more important in this information technology age. Depending on the storage capability, memory devices are divided into two categories: volatile and nonvolatile [1]. Volatile memory loses its information while nonvolatile memory retains its information, when the external power is turned off.

Two kinds of volatile memory devices are widely used in a computer: SRAM (static random access memory) and DRAM (dynamic random access memory). An SRAM uses six metal oxide semiconductor field-effect transistors (MOSFETs) for a bit, whereas a DRAM uses one transistor in series with a capacitor for a bit. Figure 22.1 shows the structures of an SRAM and a DRAM. Information is retained in an SRAM as long as the power remains on, whereas frequent refreshment is needed to retain the information in a DRAM. However, volatile RAMs, like the DRAM, have the advantages of high speed and high density. The merits of high speed SRAM render them for use as

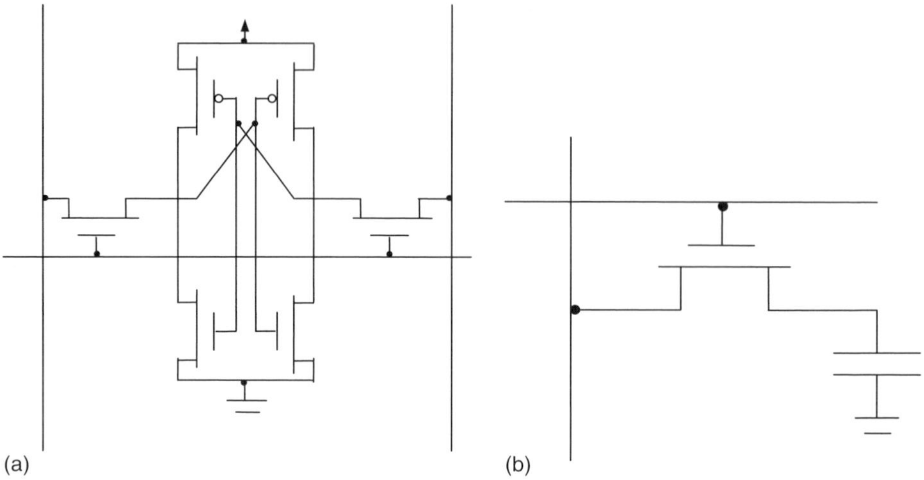

FIGURE 22.1 Device structures of (a) SRAM and (b) DRAM.

cache memory in the microprocessor. Since a volatile memory device will lose its stored information when the external power is switched off, it cannot be used for long-term information storage.

Nonvolatile memory devices can retain stored information even when not powered. Examples of nonvolatile memory include read-only memory, flash memory, most types of magnetic computer storage devices (e.g., hard disk, floppy disk drives, and magnetic tape), optical discs, and early computer storage devices such as paper tapes and punch cards. Among these listed devices, flash memory can be switched directly by electrical means and is becoming more and more popular.

A flash memory has a structure of the MOSFET with two gates (Figure 22.2). The control gate is a regular electrode, which is wired to the main circuitry. The floating gate is encapsulated in the insulating oxide layer below the control gate. Electrons can be injected into the floating gate from the channel by applying a bias to the drain and the control gate (Figure 22.3a). The stored electrons in the floating gate will increase the threshold voltage, $V_{T,}$ of the transistor. These electrons can be released by applying a voltage to the source while keeping the control gate grounded (Figure 22.3b). The V_T of the device is lowered when the charge is released. The state of the device can be read by detecting the source–drain current while applying a voltage between the high and low V_T to the control gate. When electrons are trapped in the floating gate, the applied voltage is lower than V_T, so there is no current flowing from the source to the drain. This state is defined as "1." When no electron is trapped in the floating gate, the transistor is switched on and there will be a large current flowing from the source to the drain. This state is defined as "0." Due to the barrier arising from an oxide insulator, electrons can remain stored in the floating gate, so information can be stored in the device

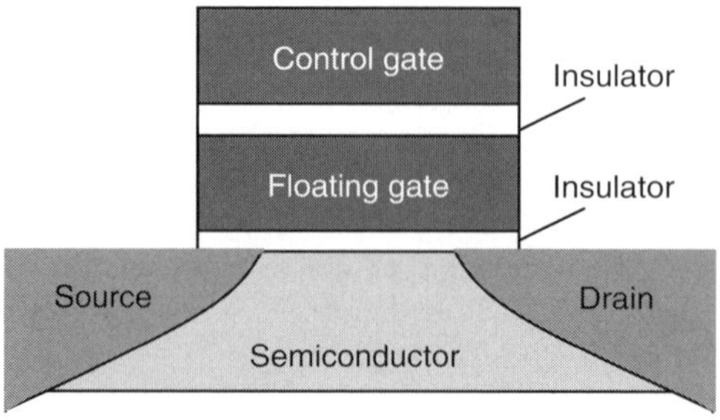

FIGURE 22.2 Device structure of a flash memory.

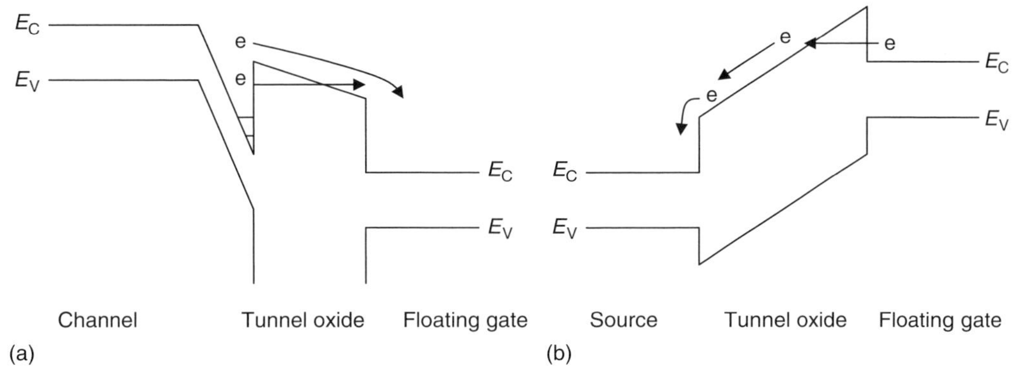

FIGURE 22.3 Operation mechanism of a flash memory. (a) Hot electron injection from channel into the floating gate and (b) erasure process: stored electrons are released through a tunneling process from floating gate into source.

for a relatively long period even if there is no power supplied to it. The information can be erased by applying a higher electric field to release the electrons from the floating gate. Thus, this device can be repeatedly erased and written many times. However, due to its device structure, the fabrication process of the flash memory is complicated and it is difficult to scale down the device. In addition, the speed of flash memory is much slower than that of volatile RAM. Thus, flash memory can only be used in applications where speed is not important.

With the development of memory technology, attention has also been put onto the development of novel memory devices using organic or nanoscale materials [2–5]. In 2001, the University of California, Los Angeles (UCLA) reported the first organic electronic nonvolatile memory device. Several versions were developed subsequently. These devices use organic materials and metal nanoparticles instead of inorganic semiconductors. The fabrication process of such devices is simple and the devices are highly flexible. Moreover, these organic thin-film memory devices have a high-response speed and can be easily made into high-density memory arrays. All these advantages make organic thin-film memory a strong candidate to become the next-generation memory device.

22.2 PRINCIPLE OF ORGANIC THIN-FILM MEMORY DEVICES

Organic thin-film memory devices have different device architectures, principles, and operation mechanisms from inorganic memory devices. Figure 22.4 shows the device structure of an organic thin-film memory. This device has a simple structure: an organic or polymer film containing metal

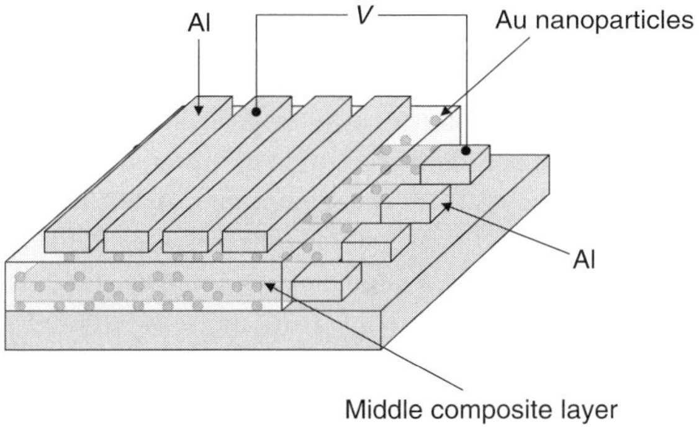

FIGURE 22.4 Device structure of an organic thin-film memory device.

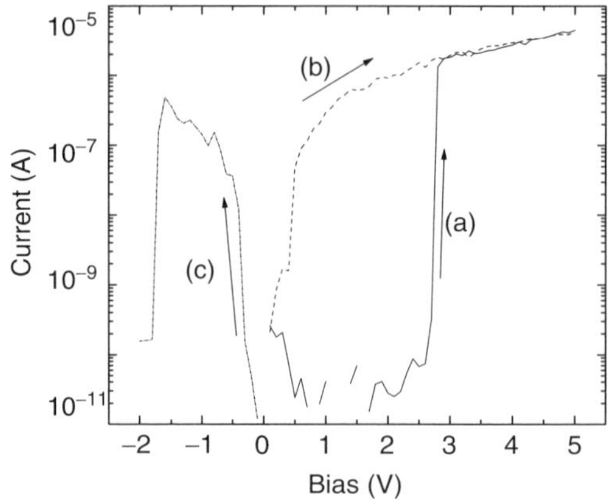

FIGURE 22.5 *I–V* curve of an organic memory device. Curves (a), (b), and (c) represent the first, second, and third bias scans, respectively. The arrows indicate the voltage-scanning directions. (Reprinted from Ouyang, J., Chu, C.-W., Szmanda, C., Ma, L., and Yang, Y., *Nat. Mater.*, 3, 918, 2004. With permission. Copyright 2004 Nature Publishing Group.)

nanoparticles sandwiched between two metal electrodes. Obviously, this two-terminal structure is different from the three-terminal MOS structure of an inorganic semiconductor memory. The simple device structure suggests the ease of device fabrication and possibility of high device density.

The organic thin-film memory device utilizes electrical bistability of the organic thin film. Electrical bistability means that the device is stable in two electrical states. Figure 22.5 illustrates the current–voltage (*I–V*) curve of such a device. The original device first exhibits a low conductivity state; therefore, very little current flows through it. It then transits to a state of high conductivity (high current level) when the external electric voltage is higher than the threshold voltage. The devices at the high conductivity state can return to the low conductivity state (low current level) by applying a voltage of negative polarity. The device in these two states could be different in conductivity by several orders in magnitude. When the high and low conductivity states are defined as "1" and "0," respectively, the device with electrical bistability can be used as a nonvolatile memory device.

The electrical bistability is attributed to the effect of an external electric field on the interaction between the materials in the active layer. Under a high electric field, an electric field-induced charge transfer can occur between the conjugated organic compounds and metal nanoparticles. (Figure 22.6) Conjugated organic materials can be oxidized by losing an electron, or reduced by gaining an electron. These conjugated organic materials exhibit higher conductivity after oxidation or reduction [6]. For example, the conductivity of some conjugated polymers can increase by more than 10 orders in magnitude after oxidation or reduction. An interesting property about the metal nanoparticle is its ability to store charge [7]. The stored charge is quite stable when the metal nanoparticle is capped with an insulator layer. This storage of charge in a metal nanoparticle capped with an insulator layer is similar to the storage of charge in the floating gate of a flash memory. When a metal nanoparticle is near a conjugated organic compound, charge transfer may take place between them when the external electric field is high enough. After the charge transfer, the charge is stored in the metal nanoparticle and remained stable due to the insulator coating layer surrounding the metal nanoparticle. The conductivity of the organic compound increases as it loses its charge to the metal nanoparticle. In this way, the device switches from a low conductivity state to a high conductivity state.

The metal nanoparticle plays a role somewhat similar to a counter ion in the conducting polymer. Since the charge transfer process results from the electric field, the organic film polarizes internally

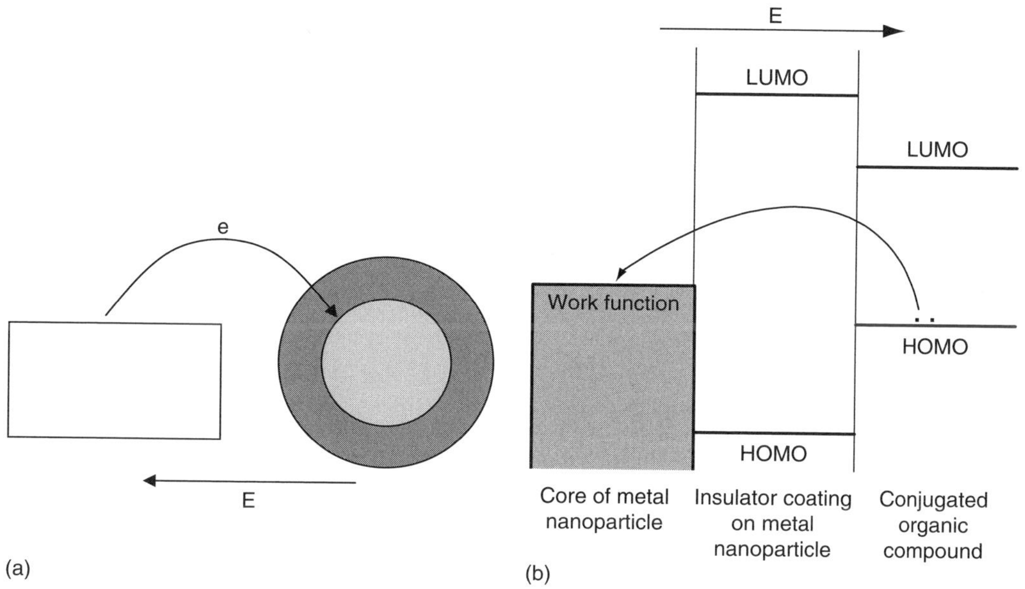

FIGURE 22.6 Schematic electron transfer from organic conjugated compound to the core of a metal nanoparticle. (a) The inner circle indicates the core of the metal nanoparticle, and the outer circle indicates the capped insulator layer. (b) The representative energy levels for organic compound and nanoparticle. The curved arrow denotes the electron transfer, and e denotes the electron. The direction of the electric field (E) is represented by the linear arrow.

after the charge transfer. Consequently, when an external electric field is applied in the reverse direction, the electron stored in the metal nanoparticle is released back to the organic compound, returning the organic film back to its low conductivity state.

The main charge conduction mechanism through organic or polymer film changes after the oxidation or reduction of the material. For example, for a conjugated polymer in the neutral state, the conduction mechanism is either a charge injection-limited process or a charge transport-limited process. In either case, the conduction mechanism is not an ohmic process; that is, the current does not change linearly with the voltage. After oxidation, the polymer becomes highly conductive and exhibits a linear *I–V* relation. Table 22.1 lists some popular conduction mechanisms for organic materials. A brief introduction is provided for Schottky emission, direct tunneling, and Fowler–Nordheim tunneling in the following, because these are the major charge conduction mechanisms for organic memory devices.

Schottky emission is dominated by charge injection from the electrode into the organic or polymer film. The charge injection occurs via a tunneling process through the energy barrier at the interface of the metal electrode and the organic layer. In this case, the current is proportional to the square of the voltage, and has a strong dependence on the temperature. On the other hand, both direct and Fowler–Nordheim tunneling currents are independent of the temperature. In direct tunneling, the charge tunnels through a rectangle barrier, whereas in Fowler–Nordheim tunneling, the charge tunnels through a triangle barrier. For direct tunneling, the current changes linearly with voltage, whereas Fowler–Nordheim tunneling has a nonlinear current–voltage relationship. The conduction mechanism may change from one to another when the external electric field is changed. For example, in a device, the dominant mechanism may change from direct tunneling while under a small external electric field to Fowler–Nordheim tunneling when the external electric field is raised, as the external electric field may significantly bend the energy level in the conjugated organic compound.

TABLE 22.1
List of the Conduction Mechanisms for the Organic Materials

Conduction Mechanism	Characteristic Behavior	Temperature Dependence	Voltage Dependence
Direct tunneling (low bias)	$J \propto V \cdot \exp\left(-\frac{2d}{\hbar}\sqrt{2m^*\Phi}\right)$	None	$J \propto V$
Fowler–Nordheim tunneling	$J \propto V^2 \cdot \exp\left(-\frac{4d\Phi^{3/2}\sqrt{2m^*}}{3q\hbar V}\right)$	None	$\ln\left(\frac{J}{V^2}\right) \propto \frac{-1}{V}$
Schottky emission	$J \propto T^2 \cdot \exp\left(-\frac{\sqrt{\frac{e^3V}{\varepsilon}}}{k_B T}\right)$	$\ln\left(\frac{J}{T^2}\right) \propto \frac{-1}{T}$	$\ln(J) \propto V^{1/2}$
Poole–Frenkel emission	$J \propto V \cdot \exp\left(-\frac{\sqrt{\frac{e^3V}{\varepsilon}}}{k_B T}\right)$	$\ln\left(\frac{J}{V}\right) \propto \frac{-1}{T}$	$\ln\left(\frac{J}{V}\right) \propto V^{1/2}$
Hopping conduction	$J \propto V \cdot \exp\left(-\frac{\Delta E}{k_B T}\right)$	$\ln\left(\frac{J}{V}\right) \propto \frac{-1}{T}$	$J \propto V$

22.3 INTRODUCTION TO MATERIALS USED IN ORGANIC THIN-FILM MEMORY DEVICES

The performance of an organic thin-film memory device is strongly dependent on the structure and properties of the materials. The active layer of an organic thin-film memory device has two components. One is the conjugated organic compound or conjugated polymer, and the other is the metal nanoparticles. Conjugated organic or polymer materials are discussed in other chapters of this book, so only the chemical and physical properties of metal nanoparticles will be introduced here.

Metal nanoparticles can be prepared by solution processing [8]. Among the many methods demonstrated for the preparation of metal nanoparticles, the two-phase arrested method is regarded as an efficient way to control particle size and size distribution. Gold nanoparticles can be prepared by the following process. First, $AuCl_4^-$ is transferred to an organic solvent, such as toluene, xylene, or chloroform, from its aqueous solution using phase transfer catalyst—tetraoctylammonium bromide. Alkanethiol is then added to the organic solution. Gold is reduced by adding aqueous solution of sodium borohydride into the organic solution of $AuCl_4^-$, while the solution is vigorously stirred. Gold nanoparticles capped with alkanethiol are formed after the reduction reaction, and they are rinsed with water and methanol. The gold nanoparticles capped with alkanethiol can be dissolved in many common organic solvents.

Gold nanoparticle capped with alkanethiol has an interesting electronic structure. The core is metallic, while the shell is insulating. Positive or negative charge is trapped in the metallic core and the insulating shell acts as a barrier to keep the charges in.

The gold nanoparticles are about a few nanometers in size, the size of which can be determined by the transmission electron microscope (TEM). Figure 22.7a shows a TEM picture of gold

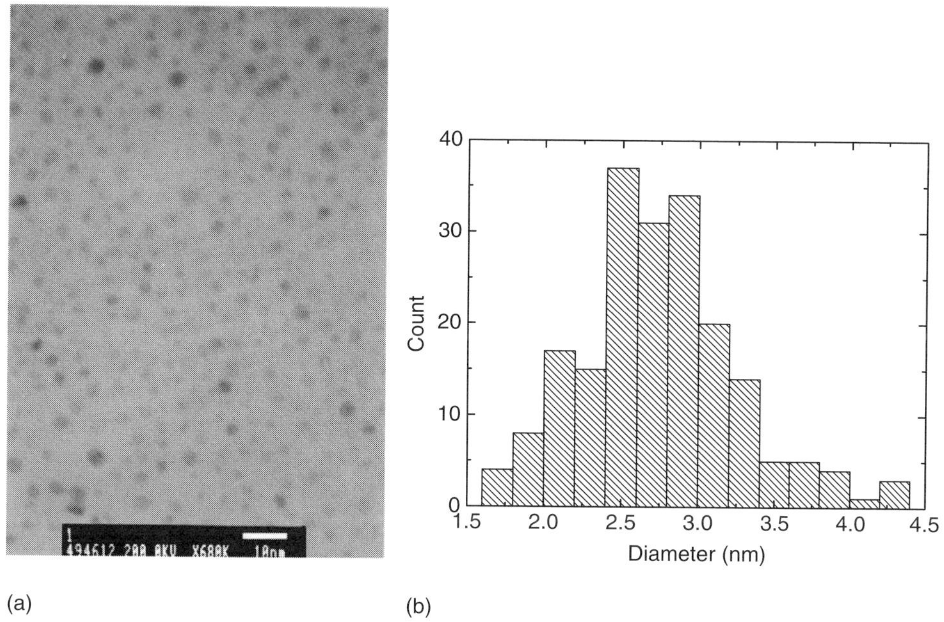

FIGURE 22.7 TEM image (a) and size distribution hologram of metal nanoparticles (b). (Reprinted from Ouyang, J., Chu, C.-W., Szmanda, C., Ma, L., and Yang, Y., *Nat. Mater.*, 3, 918, 2004. With permission. Copyright 2004 Nature Publishing Group.)

nanoparticles. The average particle size and the particle size distribution are two important parameters to characterize the gold nanoparticles (Figure 22.7b).

Metal nanoparticles have interesting electronic structure and properties. Energy levels are quantized in metal nanoparticles of a few nanometers in diameter. The quantized Kubo energy δ for a metal nanoparticle can be calculated by the following equation [9]:

$$\delta = \frac{4E_{\mathrm{F}}}{3N},$$

where

E_{F} is the Fermi energy of bulk Au

N denotes the number of Au atoms in the gold nanoparticle

Coulomb energy, E_{c}, is another important parameter for a metal nanoparticle [10]. The Coulomb energy is the energy needed to charge a metal nanoparticle. This energy is related to the particle size and the capping molecules. It can be calculated by the following equation:

$$E_{\mathrm{c}} = \frac{e^2}{2C}$$

and

$$C = 4\pi\varepsilon_0\varepsilon_r \frac{r}{d}(r+d),$$

where

C is the capacitance of the gold nanoparticle
ε_0 is the permittivity of free space
ε_r is the relative permittivity of the capped molecule on the gold nanoparticle
r the radius of the gold nanoparticle core
d is the length of the capped molecule

22.4 AN EXAMPLE OF AN ORGANIC THIN-FILM MEMORY DEVICE

An organic thin-film memory device [5] consists of a polystyrene (PS) film sandwiched between two Al electrodes. The PS layer is dispersed with gold nanoparticles capped with 1-dodecanethiol (Au-DT NP) and 8-hydroxyquinoline (8HQ). This polymer film has a thickness of about 50 nm and is formed by spin-coating an organic solution of polystyrene, gold nanoparticle, and 8HQ.

Its *I–V* curve is shown in Figure 22.5. The pristine device exhibited very low current, approximately 10^{-10} A at 1 V. An electrical transition took place at 6.1 V with an abrupt current increased from 10^{-9} to 10^{-6} A (curve [a]). The device exhibited good stability in this high conductivity state during the subsequent voltage scan (curve [b]). The high conductivity state was able to return to the low conductivity state by applying a negative bias as indicated in curve (c) where the current suddenly dropped to 10^{-9} A at −2.9 V.

The electrical switching between the high and low conductivity states of these bistable devices can be repeated numerous times. These two states can be programmed by applying a positive or negative voltage pulse. Hence, when the high and low conductivity states are defined as "1" and "0," respectively, this device can be used as a nonvolatile organic digital memory (Figure 22.8).

The device in the low conductivity state can be switched to the high conductivity state by a pulse of 5 V with a width of 25 ns. The device, which was in the low conductivity state, exhibited a current of less than 10^{-9} A in the voltage range of 0–1 V. It exhibited a current, four orders of magnitude higher after applying a pulse of 5 V with a width of 25 ns.

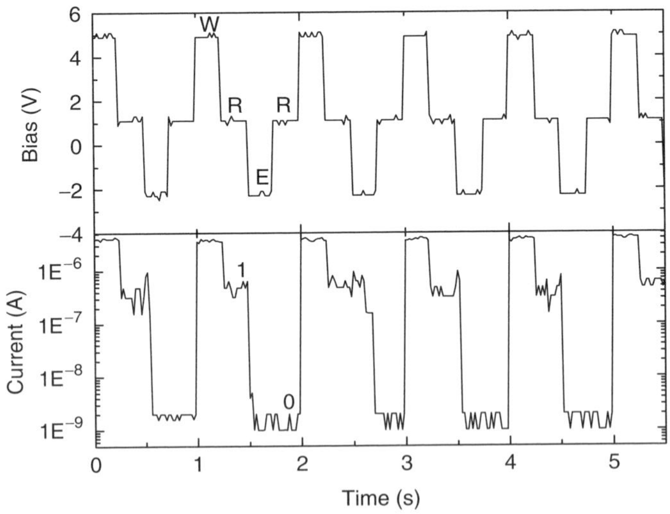

FIGURE 22.8 Write–read–erase cycles of device a polymer memory device. The top and bottom curves are the applied voltage and the corresponding current response, respectively. W, R, and E in the top figure mean write, read, and erase, respectively. "1" and "0" in the bottom figure indicate the device in the high and low conductivity state, respectively. (Reprinted from Ouyang, J., Chu, C.-W., Szmanda, C., Ma, L., and Yang, Y., *Nat. Mater.*, 3, 918, 2004. With permission. Copyright 2004 Nature Publishing Group.)

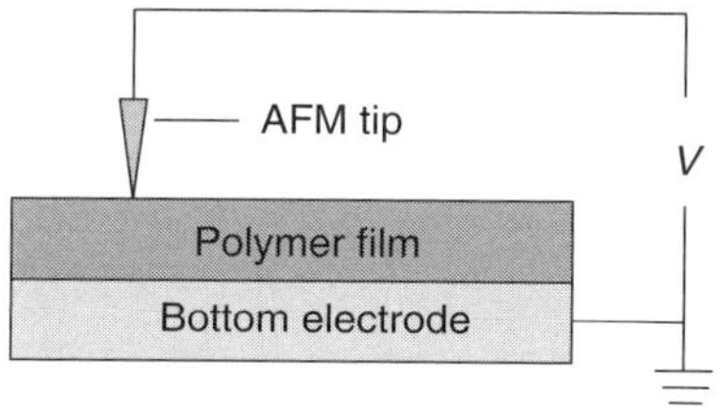

FIGURE 22.9 Test configuration for the operation of the device using an AFM tip as the top electrode.

These polymer memory devices have a simple device structure and can be easily fabricated into a high-density memory array. The device structure can be made even simpler, and the density can be pushed to very high, when the operation of the device is combined with an atomic force microscope (AFM). A schematic configuration for this device is shown in Figure 22.9. The device was fabricated by just spin-coating the polymer film on a conductive substrate. The conductive substrate is used as the bottom electrode, while an AFM tip is used as the top electrode.

Figure 22.10 shows a surface potential AFM image of an Au-DT NP+8HQ+PS film on Al coated on silicon wafer. At first, an area of $20 \times 10\ \mu m^2$ of the film was scanned vertically using contact mode while positively applying a 10 V dc bias through a 50 nm silicon nitride AFM tip coated with Au. Then, another area of $20 \times 5\ \mu m^2$ was scanned horizontally while applying a −10 V dc bias to the tip. Finally, the scanning surface potential image was taken using the tapping mode. A dc bias of 4 V was applied on the film through the 50 nm AFM silicon nitride tip coated with Au. The two pretreated areas exhibited remarkably different potential in the surface potential AFM scan. These experiments demonstrated the feasibility of the device operated with an AFM.

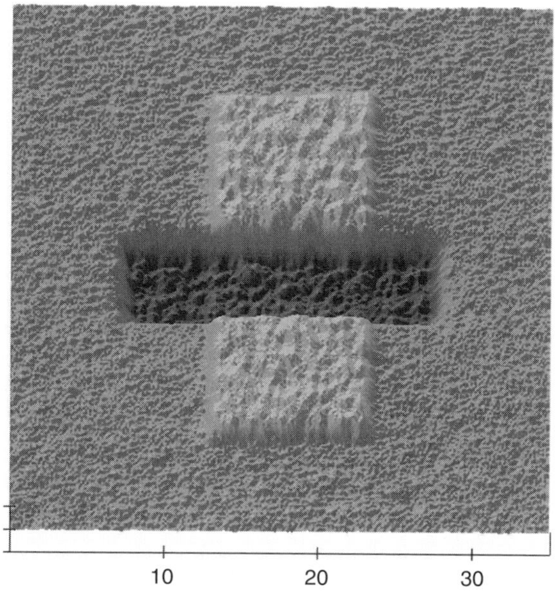

FIGURE 22.10 Scanning surface potential AFM image of a PS film containing gold nanoparticle and 8HQ with Al as bottom electrode and silicon wafer as substrate. The vertical bar with yellow color was pretreated with a +10 V dc bias, while the horizontal bar with brown color was pretreated with a −10 V dc bias. (Reprinted from Ouyang, J., Chu, C.-W., Szmanda, C., Ma, L., and Yang, Y., *Nat. Mater.*, 3, 918, 2004. With permission. Copyright 2004 Nature Publishing Group.)

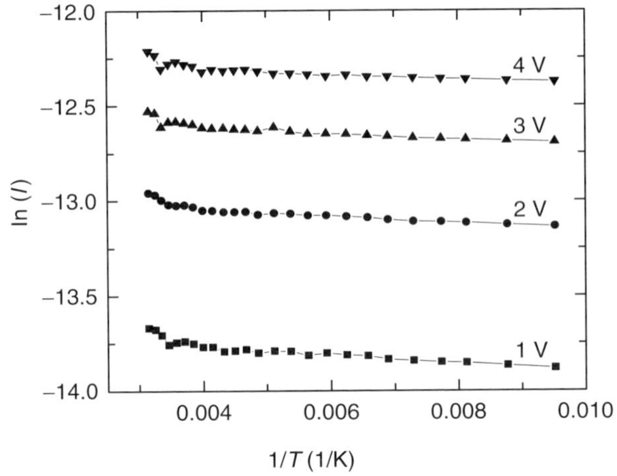

FIGURE 22.11 Arrhenius plot of temperature dependence of current for a polymer memory device in the high conductivity state at applied voltages of 1, 2, 3, and 4 V. (Reprinted from Ouyang, J., Chu, C.-W., Szmanda, C., Ma, L., and Yang, Y., *Nat. Mater.*, 3, 918, 2004. With permission. Copyright 2004 Nature Publishing Group.)

The conduction mechanism for Al/Au-DT NP + 8HQ + PS/Al in the low conductivity state may be due to the small amount of impurity in the film or hot electron injection. The current for the device in the high conductivity state was almost temperature independent (Figure 22.11), and the *I–V* curves can be fitted well by a combination of direct tunneling (tunneling through a square barrier) and Fowler–Nordheim tunneling (tunneling through a triangular barrier) (Figure 22.12) as given by the following expression:

$$I = C_1 V \exp\left(-\frac{2d\sqrt{2m^*\Phi}}{\hbar}\right) + C_2 V^2 \exp\left(-\frac{4d\Phi^{3/2}\sqrt{2m^*}}{3q\hbar V}\right).$$

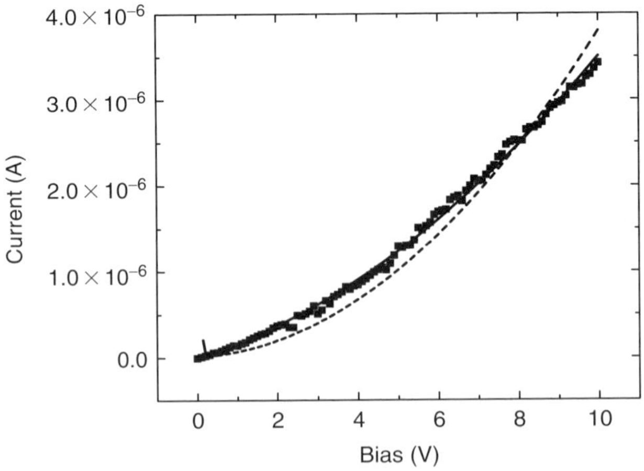

FIGURE 22.12 *I–V* curve of a polymer memory device in the high conductivity state. The scattered points are the experimental data, the solid line is the data fit combining direct tunneling and Fowler–Nordheim tunneling, and the broken line. (Reprinted from Ouyang, J., Chu, C.-W., Szmanda, C., Ma, L., and Yang, Y., *Mater.*, 3, 918, 2004. With permission. Copyright 2004 Nature Publishing Group.)

The first term on the right-hand side of the equation is the current contributed by direct tunneling, and the second term is the current contributed by Fowler–Nordheim tunneling. In this equation, d is the tunneling distance, m^* is the effective mass of the charge carrier, and Φ is the energy barrier height. At low voltage, $V<\Phi$, direct tunneling is the dominant conduction mechanism, and at high voltage, $V>\Phi$, Fowler–Nordheim tunneling becomes the dominant conduction mechanism.

The mechanism for the electrical transition is attributed to electric field-induced charge transfer between Au-DT NP and 8HQ under a high electric field. Before the electronic transition, there is no interaction between the Au-DT nanoparticle and 8HQ. Concentration of charge carriers due to impurity in the film is quite low, so that the film has very low conductivity. However, when the electrical field increases to a certain value, electrons on the highest occupied molecular orbital (HOMO) of 8HQ may gain enough energy to tunnel through the capped molecule, 1-dodecanethiol, into the gold nanoparticle. Consequently, the HOMO of 8HQ becomes partially filled, and 8HQ and gold nanoparticle become positively and negatively charged, respectively. As a result, carriers are generated and the device exhibits a high conductivity state after the charge transfer.

22.5 SUMMARY

Organic nonvolatile memory devices are discussed in this chapter. These devices have a different device architecture, principle, and operation from inorganic volatile or nonvolatile memory devices. The organic nonvolatile memory device has a two-terminal structure with an organic or polymer film containing metal nanoparticles as the active layer. The principle of the organic memory device is to utilize the bistability of the organic or polymer film. A high external electric field can be used to induce charge transfer between the conjugated organic compound and metal nanoparticle. This charge transfer results in the increase of the conductivity of the organic or polymer film. The device can be brought back to the original low conductivity state by applying voltage of reverse polarity. By defining the low and high conductivity states as "0" and "1," respectively, the organic device with electronic bistability can be used as a nonvolatile electronic memory device.

EXERCISE QUESTIONS

1. Calculate the Kubo energy of a gold nanoparticle of 1, 2, 4, 10 nm in diameter. The work function of gold is 5.1 eV.
2. Calculate the Coulomb energy of a gold nanoparticle of 1, 2, 4, 10 nm in diameter capped with 1-dodecanethiol. The length of 1-dodecanethiol is 1.6 nm and the relative dielectric constant of 1-dodecanethiol is 2.5.
3. Explain direct tunneling and Fowler–Nordheim tunneling.

LIST OF ABBREVIATIONS

8HQ	8-hydroxyquinoline
AFM	atomic force microscope
Au-DT NP	gold nanoparticles capped with 1-dodecanethiol
DRAM	dynamic random access memory
HOMO	highest occupied molecular orbital
MOSFET	metal oxide semiconductor field-effect transistors
PS	polystyrene
SRAM	static random access memory
TEM	transmission electron microscope

REFERENCES

1. More information regarding inorganic semiconductor memory devices can be found in books, e.g., Streetman B.G. and Banerjee, S., (Eds.), *Solid State Electronic Devices*, 5th edn., Prentice Hall: Englewood Cliffs, NJ, 2000.
2. Ma, L.P., Liu, J., and Yang, Y., Organic electrical bistable devices and rewritable memory cells, *Appl. Phys. Lett.*, 80, 2997–2999, 2002.
3. Ma, L.P., Pyo, S., Ouyang, J., Xu, Q.F., and Yang, Y., Nonvolatile electrical bistability of organic/metal-nanocluster/organic system, *Appl. Phys. Lett.*, 82, 1419–1421, 2003.
4. Bozano, L.D., Kean, B.W., Deline, V.R., Salem, J.R., and Scott, J.C., Mechanism for bistability in organic memory elements, *Appl. Phys. Lett.*, 84, 607–609, 2004.
5. Ouyang, J., Chu, C.-W., Szmanda, C., Ma, L., and Yang, Y., Programmable polymer thin film and non-volatile memory device, *Nat. Mater.*, 3, 918–922, 2004.
6. Skotheim, T.A., (Ed.), *Handbook of Conducting Polymers*, Marcel Dekker: New York, 1986.
7. Chen, S., Ingram, R.S., Hostetler, M.J., Pietron, J., Murray, R.W., Schaff, T.G., Khoury, J.T., Alvarez, M.M., and Whetten, R.L., Gold nanoelectrodes of varied size: Transition to molecule-like charging, *Science*, 280, 2098–2101, 1998.
8. Hostetler, M.J., Wingate, J.E., Zhong, C.-J., Harris, J.E., Vachet, R.W., Clark, M.R., Londono, J.D., Green, S.J., Stokes, J.J., Wignall, G.D., Glish, G.L., Porter, M.D., Evans, N.D., and Murray, R.W., Alkanethiolate gold cluster molecules with core diameters from 1.5 To 5.2 nm: Core and monolayer properties as a function of core size, *Langmuir*, 14, 17–30, 1998.
9. Kubo, R., Electronic properties of metallic fine particles. I, *J. Phys. Soc. Jpn.*, 17, 975–986, 1962.
10. Chen, S., Murray, R.W., and Feldberg, S.W., Quantized capacitance charging of monolayer-protected Au clusters, *J. Phys. Chem. B*, 102, 9898–9907, 1998.

23 Introduction to Organic Electrochromic Materials and Devices

Prasanna Chandrasekhar

CONTENTS

23.1 Basic Definitions and Introduction 714
23.1.1 Basic Definitions 714
23.1.2 Spectral Regions 716
23.1.3 Transmission, Reflectance, and Cumulative Mode 716
23.1.4 Basic Electrochemistry Essential for the Understanding of Electrochromic Devices 716
23.1.5 Conducting Polymers 719
23.1.6 Basic Electrochromic Devices 720
23.1.7 Basic Methodology for Measurement of Electrochromism 722
23.1.8 Other Metrics Relevant to Electrochromic Materials and Devices 724
23.2 Transmission-Mode Electrochromic Devices 726
23.3 Reflectance-Mode Electrochromic Devices 729
23.4 Summary 731
Exercise Questions 731
List of Abbreviations 731
References 732

Abstract This chapter introduces and discusses, at a very basic (first year university) level, the principles and practical aspects of organic electrochromic materials and devices. It aims to give a foundation from which to peruse the electrochromic literature, rather than present a review of that literature. Basic definitions, spectral regions, and transmission vs. reflectance-mode devices are presented first. Basic electrochemistry, essential for an understanding of electrochromic devices is presented. A subsection discusses conducting polymers (CPs), perhaps the most important type of electrochromic materials. Basic electrochromic devices are then presented, from a fundamental as well as practical point of view. Methodology, terminology, and metrics used to characterize and qualify electrochromic devices and materials are discussed. Unique features of transmission-mode and reflectance-mode devices, of electrochromism in the visible vs. IR and other spectral regions, and of practical applications of these, are discussed in some detail.

23.1 BASIC DEFINITIONS AND INTRODUCTION

23.1.1 Basic Definitions

As the name implies, electrochromics are simply materials, substances, or devices which change color when a voltage (or very rarely, a current) is applied (classical Greek *έληκτρον*, amber, *χρομα*, color). The reader may think for the moment of the photochromic sunglasses worn by many, especially the elderly, which change color from clear to dark in bright sunlight. If one substitutes the activation of these sunglasses using bright sunlight by activation using a small (±5 V) DC voltage from a battery placed behind the ear, then one has arrived at an electrochromic device, here, electrochromic sunglasses. To this rather simple definition for electrochromism, "color change with applied voltage," we can add a somewhat broader one, "change of color or spectral signature, with applied voltage or equilibrium potential or chemical potential."

The electrochromism of conducting polymers (CPs) occurs when a change in spectral signature is effected due to an applied electromotive force (emf) or electrochemical potential. This occurs due to a chemical reduction or oxidation (redox) of the CP. To put it another way, the color of the electrochromic material, i.e., the CP, changes with its redox state. Due to an analogy with the flow of electrons or holes in solid-state physics, this redox process in CPs has been given the misnomer "doping/de-doping," a misnomer which has unfortunately stuck and is widely used.

We can visualize the electrochromism of CPs by taking the example of the common CP poly(aniline), and noting its colors in various redox states:

Reduced state, "leuco-emeraldine," nearly transparent, nonconductive

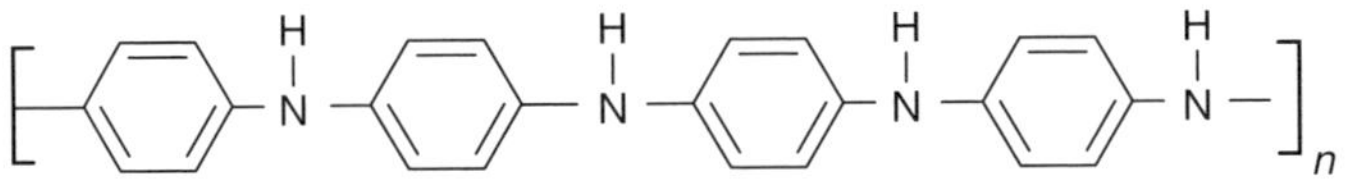

Partially oxidized state, "neutral, emeraldine," light-green, partly conductive

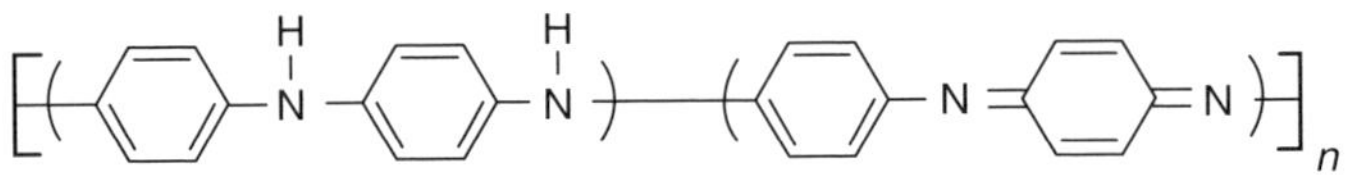

Fully oxidized state, "pernigraniline," dark-green, highly conductive

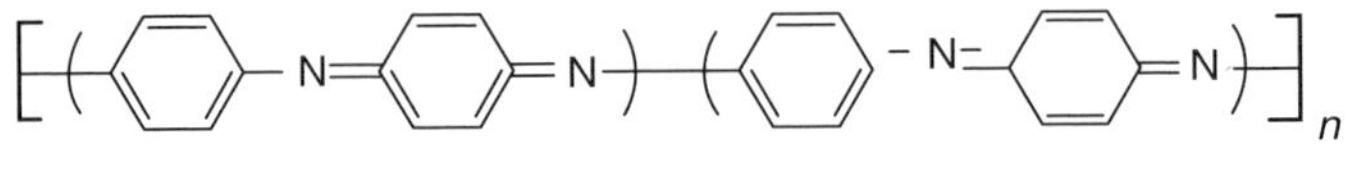

(23.1)

The relation of the color of poly(aniline) to redox state (and thus, applied potential in electrochromic devices) as described by Equation 23.1 is illustrative. To note the relation of spectral properties to redox state, we turn to a related CP, a derivative of poly(diphenyl amine). Its spectra as a function of redox state is shown in Figure 23.1. The light and dark states (high and low transmissions, respectively) are clearly contrasted.

Before discussing basic electrochromism further, we must digress a bit to note some important exclusions from our discussions in this chapter. There are several effects closely related to or somewhat similar to electrochromism, that a lay reader may confuse with electrochromism, but are actually quite different. We do not treat of these effects in this chapter. These effects use other means to change the chemical potential, redox equilibrium, or excitation state of an organic material, and hence its color. They include, specifically

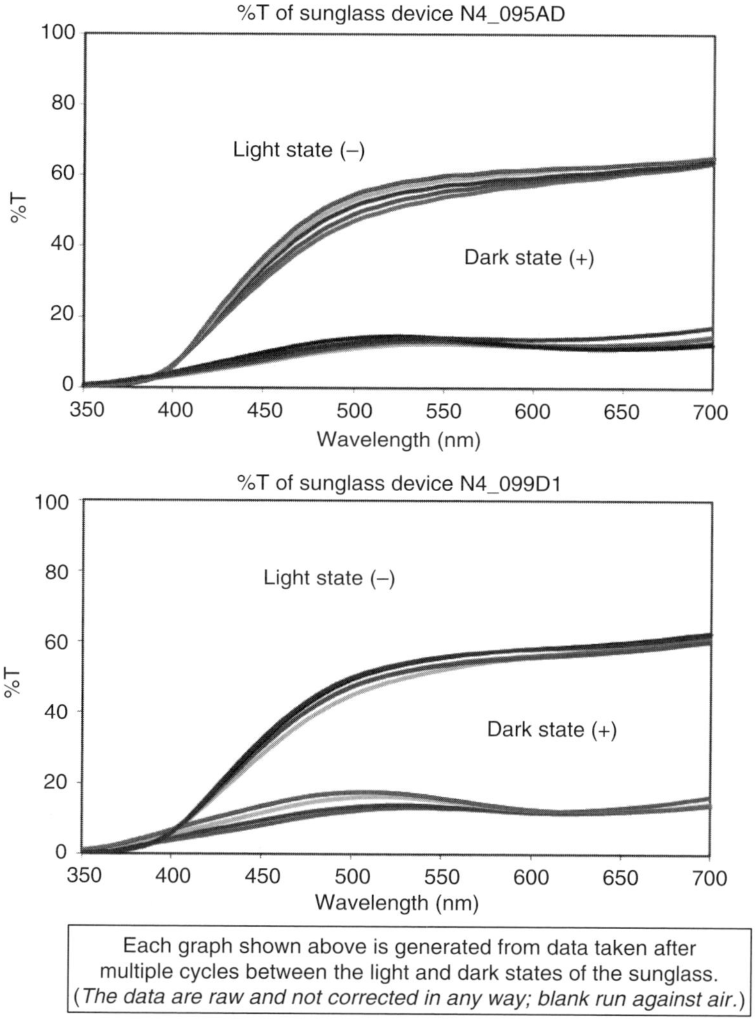

FIGURE 23.1 In situ UV–Vis–NIR spectra of an electrochromic device based on a poly(diphenyl amine) derivative. Light and dark states are clearly contrasted. The several spectra indicate repeated cycling between light and dark states.

1. Dilution in solvents (when the organic materials are soluble), referred to as solvatochromism: As a solution containing the active material is diluted, it changes color.
2. Change of temperature, referred to as thermochromism: The reader may be aware of fashion clothing which changes color with heat to reflect one's mood.
3. Charge injection and its practical manifestation, organic light-emitting diodes (OLEDs), in which a large electric field (tens of volts generally) is used, e.g., to inject holes into the valence band of an organic material and electrons into its conduction band. Combination of these electrons and holes then generates excitons, which decay emissively with an emission wavelength characteristic of a particular organic material. OLEDs have received much attention recently as new materials for cell phone screens and related applications.

We now come to the obvious question: What kinds of organic electrochromic are we really talking about? The answer is fairly straightforward: There are two major types of organic electrochromics:

1. Conducting polymers (CPs)
2. All other organic materials

The first type, CPs, represents an entire field of highly multidisciplinary science that has blossomed in the past two decades [1], and is introduced briefly in a separate section below. The second class represents an older set of established organic materials, such as methyl viologens and Prussian blue. Organic electrochromics are thus clearly distinguished from inorganic electrochromics such as those based on metal oxides (WO_3, NiO_x, etc.), and from liquid-crystal based electrochromics.

23.1.2 Spectral Regions

One of the first aspects of electrochromics that may now be obvious to the reader is that the color change, i.e., change in spectral properties need not be confined to the visible spectral region. Accordingly, we can segment organic electrochromics as covering the following spectral regions:

1. Near-UV (300–400 nm)
2. Near-IR (0.7–2.5 μm)
3. Mid-IR (2.5–8 μm)
4. Far-IR (8–18 μm)
5. Microwave-/mm-wave (5 MHz–50 Ghz)

Electrochromism can be said to encompass any property of a material that varies with applied potential or doping. Thus, properly speaking, conductivity, including microwave conductivity, is also an electrochromic property, since it can be changed with applied potential or redox state; properties such as optical conductivity also have a direct spectral connotation.

23.1.3 Transmission, Reflectance, and Cumulative Mode

The most common electrochromic devices are transmission-mode devices, where incident light passes through the device to an observer on the other side of it. An electrochromic window is the most obvious example.

In a reflectance-mode device, incident light is reflected from the device to the observer. An electrochromic (e.g., flat panel) display is the most obvious example of this. For reflectance-mode operation, the simple expedient of inserting a mirror behind a transmission-mode device may be used, or a more complex and unique structure with the working electrode facing the incident light may suffice, as described in the patent literature [2]. Thus, electrochromic automobile rearview mirrors, such as the poly(aniline)-based mirrors found today in some high-end automobiles, are examples of a fundamentally transmission-mode device modified for reflectance-mode operation. Here, a reflectance-mode device is fixed in front of a mirror.

For want of a better word, cumulative-mode device is used to denote one in which transmission, reflectance, and absorption are all important in determining the electrochromic modulation achieved. A new generation of polymer-based microwave-energy-modulating devices being developed very recently belong to this category.

23.1.4 Basic Electrochemistry Essential for the Understanding of Electrochromic Devices

Since electrochemistry is an integral part of the operation of organic electrochromic devices, an understanding of very basic electrochemistry [3] is essential. We touch upon this here.

For a quick appreciation of very basic electrochemistry, the readers may for the moment hark back to their secondary school days, where they must surely have witnessed that staple of chemistry class demonstrations, the electrolysis of water. In this, two electrodes, a working electrode and a counter electrode, dip into an electrolyte, a salt solution in water. When a DC voltage, typically less than 3 VDC, is applied across these electrodes, either using a battery or a DC power supply, electrolysis results: Hydrogen ions (hydronium, $[H_3{}^+O]$ ions), called the cations because they are positively charged, migrate (via ion transport through the electrolyte) to the working electrode, also thus called the cathode. Here, they accept electrons and are liberated as hydrogen (H_2) gas. Hydroxyl ions, $[OH^-]$, called the anions, in turn migrate to the counter electrode, also thus called the anode, where they are converted into oxygen gas. A rough rule of thumb in electrochemistry is that the counter electrode must be equal to or larger in area than the working electrode, so that processes occurring at the counter electrode do not limit what is happening at the working electrode.

Electrochemistry can be carried out in electrolytes that are either aqueous systems or nonaqueous systems. The electrolytes themselves can obviously be either liquid, as in the water electrolysis example above, or solid, such as some metal oxides (very rarely used with electrochromics). They can also be an intermediate form such as a gel electrolyte. An electrolyte of adequate conductivity, generally an ionic salt in a solvent, gel, or polymer matrix, is required for the charge transport necessary for observation of any electrochemical phenomena. Typical nonaqueous electrolytes are based on solvents such as propylene carbonate and γ-butyro-lactone. These are solvents widely used, e.g., in Li batteries, and are chosen because they combine a wide liquid range (they are typically liquid from ca. −40°C to +150°C) and low vapor pressure with good solubility for salts, such as Li trifluoromethane sulfonate, which are used as the electrolyte salts. These nonaqueous solvents are frequently used in electrochromic devices as well.

In standard electrochemistry, one can employ either a two-electrode mode (working, counter electrodes), as in the example above, or a three-electrode mode (working, counter, reference electrodes). The latter is typically better for an accurate potential (voltage) control, whereas the former may more correctly emulate practical applications such as electrochromic devices. The reader may have encountered reference electrodes such as the saturated calomel electrode (SCE) or Ag/AgCl electrode in secondary school or first year university. Voltages applied at the working electrode are then referenced either against the counter electrode (two-electrode mode) or the reference electrode (three-electrode mode).

In the electrolysis example above, a battery or DC power supply was used as the power source. In more rigorous experimental electrochemistry, an instrument known as a potentiostat/galvanostat is used for control. This controls the potential (i.e., voltage, potentiostatic mode) or current (galvanostatic mode) applied to the working electrode, while monitoring of the resultant current or potential. For the three-electrode mode, the working electrode potential is controlled with respect to the reference electrode, which may be standard (e.g., the SCE, wire-Ag/AgCl) or a quasi-reference (Q-R) (e.g., Pt wire).

In the potentiostatic mode, the voltage to be applied to the working electrode (with respect to the counter electrode in two-electrode mode and the reference electrode in three-electrode mode) is selected, and the current is monitored. The voltage applied may be static, i.e., fixed, as in the electrolysis mode, or it may be scanned (ramped). When we have a linear scan of the voltage (say from −2.0 to +1.0 V) and monitor the current, we obtain a voltage/current (E/I) curve, called a voltammogram. When we scan out the voltage from a starting point (say −2.0 V) to a vertex (say +1.0 V) and then back again to the starting point, we have a cyclic voltammogram (CV).

The potential can be scanned at varying scan rates (say 10 or 500 mV/s). During the scan of the potential, oxidation or reduction of electrochemically active species occurs at the working electrode. Thus, voltammograms provide essential information on the nature of the redox process (via the location of the current peaks and their shape and area), and their kinetics (via the relation to scan rate). With respect to electrochromism, voltammograms can provide information on the potential at which a color change occurs in the electrochromic material. This is illustrated by the cyclic voltammogram for an electrochromic conducting polymer, poly(pyrrole) [4], shown in

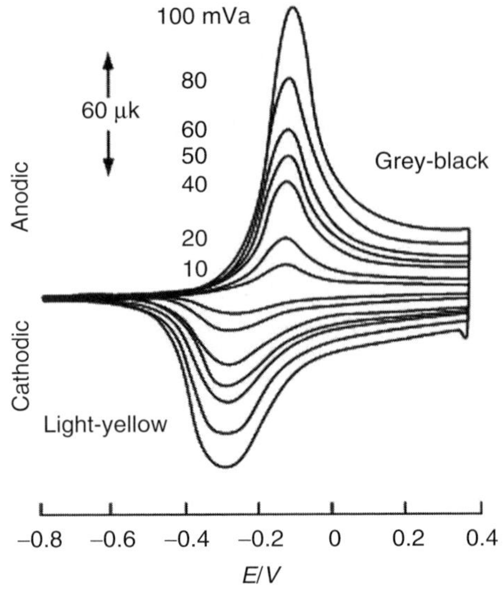

FIGURE 23.2 CVs of poly(pyrrole)BF_4 as function of scan rate, with color change also shown. (Reproduced from Diaz, A.F., Castillo, J.I., Logan, J.A., and Lee, W.-Y., *J., Electroanal. Chem.*, 129, 115, 1981. With permission.)

Figure 23.2. Potential can be scanned in a linear fashion, e.g., in linear scan voltammetry, or in pulsed or programmed fashion, in such techniques as differential pulse voltammetry or square wave voltammetry. Very few electrochromic devices employ the galvanostatic mode, and so this is not discussed further in this chapter. Frequently, one of the first methods employed for characterization of an electrochromic device is CV.

Another important electrochemical concept of relevance to electrochromics is that of electrochemical windows. An electrochemical "window" is a window within which the redox process responsible for the electrochromism is reproducible (though not necessarily electrochemically reversible) to a large extent, and beyond which oxidative, or reductive, decomposition of the electrochromic material occurs. Thus, one of the first measurements performed on a new electrochromic material is to determine this window, which is readily done via cyclic voltammetry (see above). In the cyclic voltammogram, oxidation peaks unaccompanied by reduction peaks, or irreversible electrochromic changes, indicate that one is beyond this window. From Figure 23.2, the window for an electrochromic polymer P(Py)/BF_4, lies within −0.8 to +0.4 V vs. SCE. Thus for P(Py)/BF_4, −0.8 and +0.4 V represent, respectively, light-yellow and dark grey-black electrochromic extremes of this CP electrochromic system (as labeled on the CV).

The oxidative decomposition peaks for many CPs lie only a little beyond the oxidative peak, while the reductive decomposition peaks frequently lie well beyond the reductive peak, sometimes close to the electrolyte decomposition potential. The problem on the reductive side for most CPs is that more of dissolution of this more soluble, reduced form of the CP into the electrolyte. Figure 23.3 shows a typical oxidative decomposition peak (indicated by arrow). When electrochemical systems follow "Nernstian behavior," i.e., reversible thermodynamics and kinetics described by the Nernst equation of electrochemistry, CVs have well-established characteristics for properties such as anodic/cathodic peak separation, peak half width, and scan rate dependence.

The many electrochemical terms we have just encountered, such as working electrode, counter electrode, reference electrode, gel/liquid electrolyte, cathode, anode, ion transport, two-/three-electrode mode, potentiostat, galvanostat, cyclic voltammogram, are then the essentials of any electrochemical experiment. They are also the essentials of all electrochromic devices. We cannot do without them

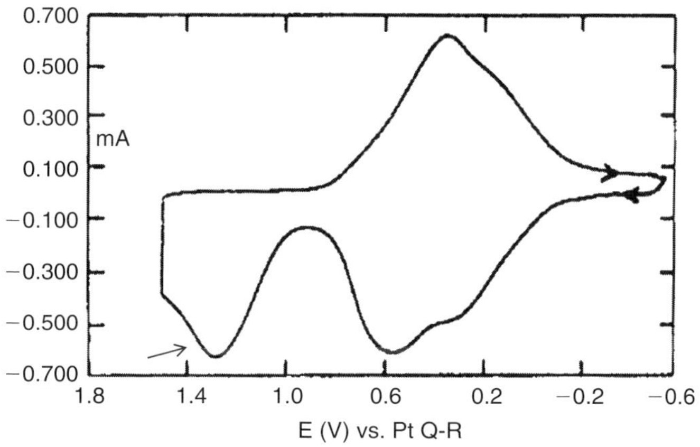

FIGURE 23.3 CV of poly(diphenyl amine) derivative, showing oxidative decomposition peak (arrow). (Courtesy of Ashwin-Ushas Corp., Inc.)

in any discussion of electrochromics. For the reader wanting a more in-depth treatment of electrochemistry as applied to electrochromics, there are many excellent monographs available [3].

23.1.5 Conducting Polymers

CPs (sometimes also called conductive polymers or conjugated conductive polymers or organic polymeric conductors) [1] are comprised by far the largest segment of electrochromically active materials actually used in electrochromic devices, and thus a very rudimentary understanding of these materials is essential for the uninitiated reader. As is well known and also discussed elsewhere in this book, CPs differ from other polymers (e.g., common plastics, such as polythene), in that they are intrinsically conducting. They do not have any conductive fillers as such. Rather, their conductivity is predominantly due to extended, delocalized π-conjugation, in one of their redox states. This conductivity of CPs is achieved through simple chemical or electrochemical oxidation (redox), or in some cases reduction, by a number of simple anionic or cationic species, called dopants, a term borrowed from condensed matter physics, as noted earlier. Such π-conjugation is illustrated somewhat simplistically in Figure 23.4 for poly(acetylene), a prototypical CP. Since CPs are dealt with extensively in other chapters of this book, we do not delve into further detail on this subject here.

Typical representative CPs are shown in Figure 23.5, where in the first part of the figure, the monomer building block is shown to the left of the polymer. In all cases, *n* is a large integer referring to the degree of polymerization. For the case of poly(aniline), the *x*s, *y*s, and *n*s are features of the structure which need not be worried about at this time. Among CPs, such as those illustrated below, poly(aniline), poly(pyrrole), poly(acetylene), and the poly(thiophenes) have been among the most studied, both scientifically and in terms of practical applications.

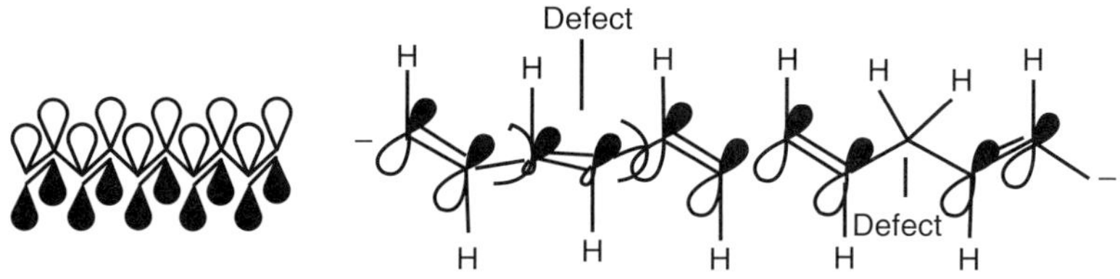

FIGURE 23.4 Schematic representation of π-conjugation in the CP poly(acetylene). Left: Basic schematic. Right: 3-D view, including defects.

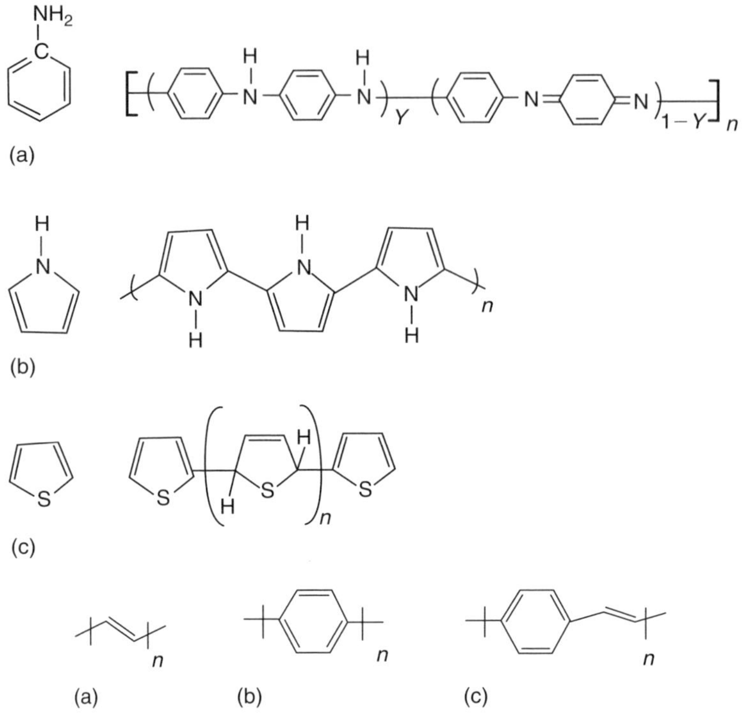

FIGURE 23.5 Top: Schematic illustration of monomer unit (left) and CP (right) for three common CPs: (a) poly(aniline) (P(ANi)); (b) poly(pyrrole) (P(Py)); and (c) poly(thiophenes). Bottom: Schematic structures of three common CPs: (a) poly(acetylene); (b) poly(*p*-phenylene); and (c) poly(*p*-phenylene-vinylene).

In the CP literature and among workers in the field, the monomers are generally represented by abbreviations, e.g., ANi for aniline, and the polymer also abbreviated in short-hand notation, following the custom in other polymer literature. For example, P(ANi) and PANi may be used to represent poly(aniline), P(Py) and PPy to represent poly(pyrrole), and P(3MT), P3MT, or PMT may be used to represent poly(3-methyl thiophene).

Typical electrochromic behavior of important CPs is summarized below:

- *Poly(aniline), P(ANi)*: nearly colorless (reduced state, (−) applied potential) ↔ green (intermediate state) ↔ dark green-blue-black (oxidized state, (+) applied potential) (see Equation 23.1)
- *Poly(pyrrole), P(Py)*: light-yellow ((−) appl. potl.) ↔ dark blue-black ((+) appl. potl.);
- *Poly(3-methyl thiophene), P(3MT)*: light orange-red ((−) appl. potl.) ↔ dark blue ((+) appl. potl.)
- *Poly(isothianaphthene)*: dark blue-black ((−) appl. potl.) ↔ v. light blue ((+) appl. potl.)

It may be noted that one of the advantages of CPs in electrochromic devices, as compared to such materials as liquid crystals, is the wider intrinsic (i.e., filter-less) color range obtainable.

23.1.6 Basic Electrochromic Devices

With the very elementary understanding of basic electrochemistry as related to electrochromics and of CPs, we now have sufficient information to look at how a basic electrochromic device actually works.

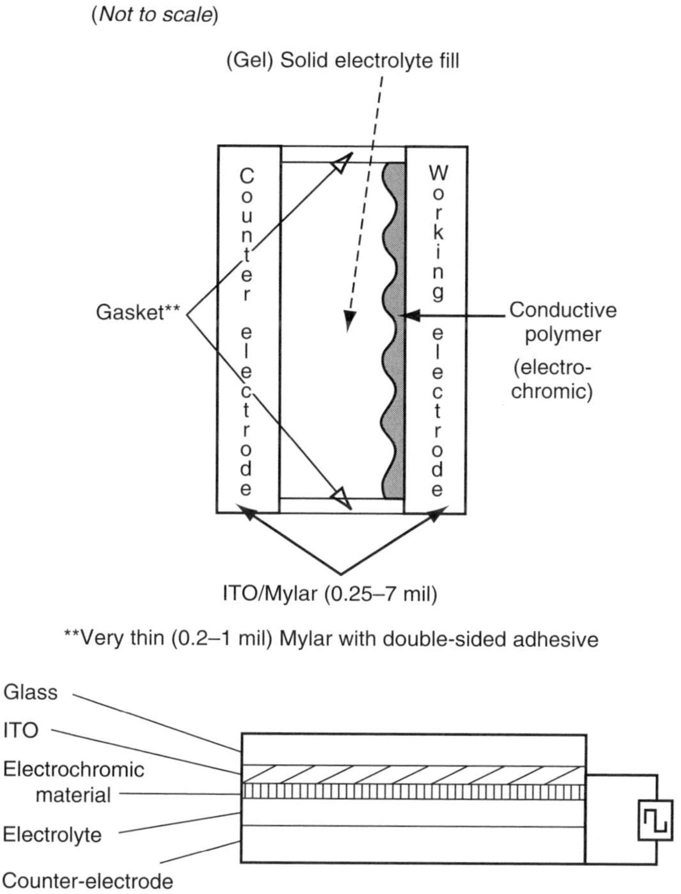

FIGURE 23.6 Top: Schematic of a more recent transmission-mode electrochromic device. Bottom: Schematic of typical transmission-mode electrochromic device. (Reproduced from Mastragostino, M., *Applications of Electroactive Polymers*, Scrosati, B., ed., Chapman & Hall, New York, 1993. With permission. Courtesy of Ashwin-Ushas Corp., Inc.)

To help the uninitiated reader in visualizing what a typical electrochromic device looks like, Figure 23.6 shows a prototypical transmission-mode device (a term defined further below). In the illustrated device of Figure 23.6, the working electrode typically comprises a thin film of the electrochromically active material on a conductive electrode substrate. A common conductive electrode material comprises indium tin oxide (ITO) vapor deposited on glass. The ITO is a visible-region transparent material that imparts conductivity to the glass. Occasionally, thin films of Au or Pt deposited on glass, of acceptably high transmission, may also be used. The electrochromic material, e.g., a CP film, is typically directly prepared on or cast onto the ITO/glass. In the case of CPs, direct preparation onto the ITO/glass substrate frequently involves electrochemical polymerization from a monomer solution. The film thickness of the electrochromic material can vary from ca. 50 nm to several microns; thicknesses, e.g., those measured coulometrically (by counting total charge used to electrochemically deposit a material such as a CP), are usually only approximate, except when measured by absolute methods such as electron microscopy or stepper techniques. The counter electrode can be any electrode; frequently, the same substrate, e.g., ITO/glass, is used as the counter electrode as well. The electrolyte can be a liquid, such as Li trifluoromethanesulfonate in a solvent such as propylene carbonate. More frequently, it is a gel electrolyte. The gelling agent can be, e.g., poly(methyl methacrylate) (the stuff of contact lenses) [5].

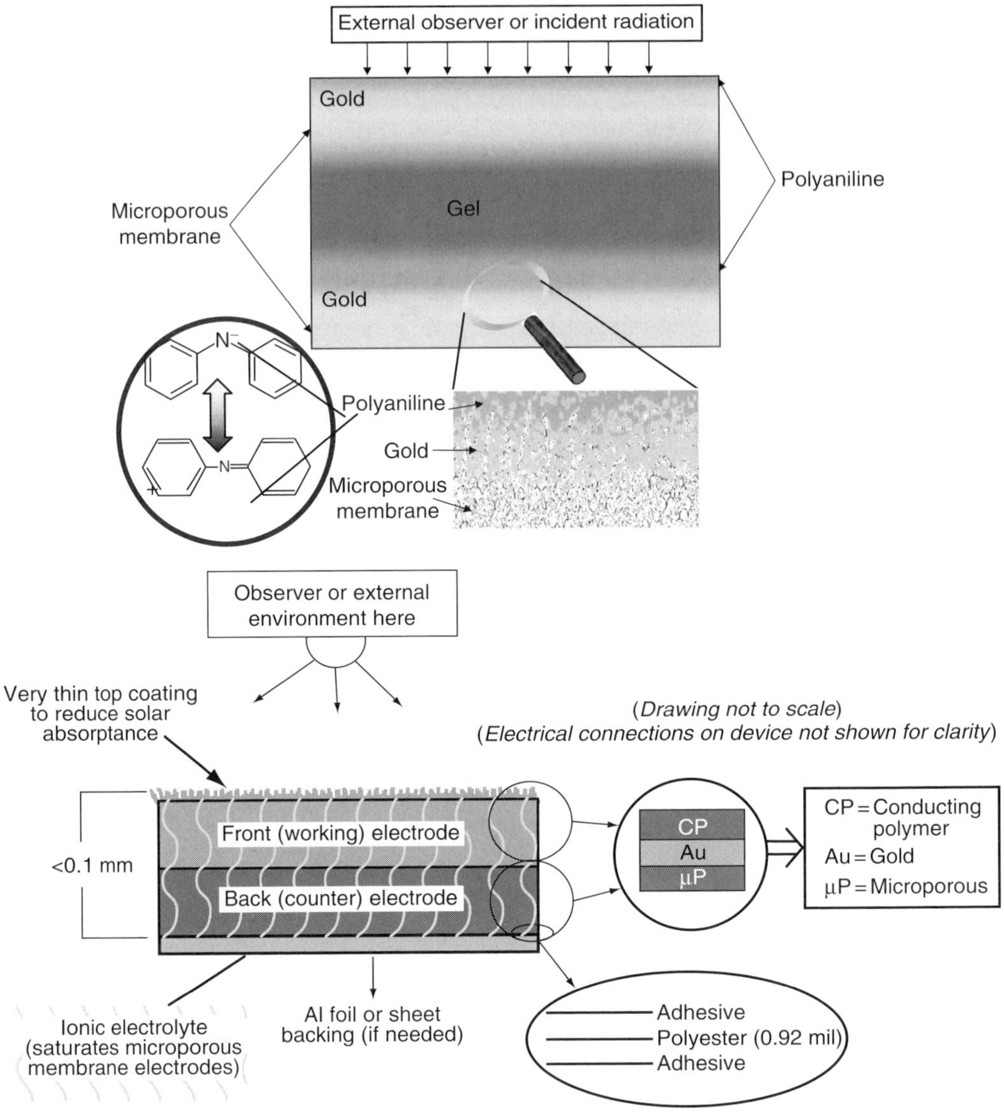

FIGURE 23.7 Top: Schematic of a reflectance-mode electrochromic device, here using a single microporous membrane with Au deposited on both sides serving as the working and counter electrodes. The electrochromic material used is poly(aniline). Bottom: A similar reflectance-mode device, also using microporous membranes, but now with two membranes, one each for the working and counter electrodes, and with other modifications such as ionic electrolytes, suitable for use in spacecraft. (Courtesy of Ashwin-Ushas Corp., Inc.)

When a potential is applied to the working electrode, the active electrochromic material on it undergoes oxidation or reduction, with an accompanying color change. A corresponding "counter-electrode-reaction" must accompany this process, and in the absence of a material actively introduced into the system to undergo such a reaction, this frequently comprises merely redox of impurities, water, or atmospheric oxygen in the electrolyte. Figure 23.7 shows corresponding schematics for reflectance-mode devices. Two configurations are shown.

23.1.7 Basic Methodology for Measurement of Electrochromism

Electrochromism can be monitored via standard transmission spectroscopy (transmission-mode electrochromism), applicable to thin films of electrochromics, and, in very limited cases, to solutions.

It can also be monitored by reflection spectroscopy (reflectance-mode electrochromism), useful e.g., for IR measurements in opaque devices, or if the final application envisioned is a display device.

The simplest measurement which yields information on electrochromic properties of electrochromic materials is the UV–Vis–NIR spectroelectrochemical curve, an in situ or sometimes ex-situ measurement of the UV–Vis–NIR, IR, or other spectral region spectrum of the electrochromic material at various applied potentials.

For transmission-mode monitoring, electrochromic films are typically directly prepared on or cast onto visible-region-transparent electrodes such as ITO/glass, Au/glass, or Pt/glass, as described in an earlier section.

For acceptable results, ITO/glass should be of resistivity 20 Ω/□ or lower, and Au/Pt on glass should be of at least 50% transmission (e.g., at 550 nm). Measurements can also occasionally be performed on freestanding electrochromic films. The films' redox state is then changed to that desired, either chemically, via appropriate exposure to dopant or redox agent, or electrochemically, by application of the appropriate potential.

Spectral measurements can be done either ex situ, changing the doping level or redox state appropriately for every measurement, or in situ, the preferred method for greater accuracy. All such spectroelectrochemical data can be presented with absorbance (optical density) or %-transmission on the ordinate, and wavelength (nm, μm) or energy (eV, λ (in μm) = ca. 1.24/eV), or sometimes, wavenumbers (cm^{-1}) on the abscissa. Occasionally, difference spectra, usually referenced to the rest or pristine state of the electrochromic material, are also presented.

For reflectance-mode monitoring, electrochromic films can be prepared or cast onto opaque substrates such as Au films evaporatively deposited on hard, flexible substrates, or onto solid metal (Au, Pt, stainless steel) electrodes. Indeed, Au-based substrates provide some of the best substrates for one type of reflectance measurement (specular, see below), since Au possesses the highest reflectance known among metals, and the underlying substrate reflectance needs to be high for proper interpretation of reflectance data. In situ measurements are possible by ingenious design of samples, e.g., in the attenuated total reflectance-infrared (ATR-IR) method, or on sealed devices with special, proprietary designs, as has been carried out in the author's laboratories. In ATR-IR, for instance, Pt/Ge-crystal substrates are used. Reflectance measurements can be carried out from the near-UV through the far-IR.

Two types of reflectance measurements are of primary interest: specular and diffuse reflectance. All materials that reflect radiation can do so specularly, like a mirror, in a fixed direction, or uniformly in all directions, i.e., diffusely. The former case is that of an ideal polished, reflecting surface, whereas the latter case is that of an ideal matte, scattering surface. In practice, most materials scatter both specularly and diffusely. In the case of an electrochromic material, which is deposited on a metallic surface and has high transparency in one redox state accompanied by high contrast between electrochromic states, the specular measurement is of greater interest; this case would typically find use in flat panel communication displays. Specular measurements can be carried out for various incidence angles, but a fixed angle measurement (typically at 16°) is preferred for practical and comparative reasons. In the case of a coarse electrochromic coating, e.g., one which is to be used for camouflage applications, the diffuse measurement is of greater interest. A material can either absorb, emit, or reflect radiation, and a very crude relation between these three parameters, not taking into account directional and other effects, is absorptance = emissivity = 1 − reflectance. (Emissivity or spectral emissivity differs from emittance, which is an integrated measure.) Thus in very broad terms, %-reflectance data can parallel %-transmission data.

All specular and diffuse reflectance data need to be referenced. The reference material used for the former for the near-UV through far-IR regions is usually a mirror surface of some sort. The reference material used for the latter (diffuse reflectance) is usually a material based on $BaSO_4$, a near-perfectly diffuse reflector, for the near-UV–NIR, and KBr powder or a special gold surface for the IR region. In nearly all the reflectance data presented in this chapter, the measurements have been performed on hermetically sealed, functional devices. The use of a proper reference, particularly for diffuse reflectance, is important, as unrepresentative references can give rise to reflectances that exceed 100%, a case of "comparing apples to oranges". It is also noted that all specular measurements presented here are for fixed incidence angles. Finally, we note that most manufacturers of commercial UV–Vis–NIR and FTIR spectrometers also supply adapters to their instruments for collection of reflectance data.

23.1.8 Other Metrics Relevant to Electrochromic Materials and Devices

It should be apparent from the spectroelectrochemical data presented above that one of the most important metrics for electrochromism is the dynamic range, i.e., the light-state/dark-state contrast at a particular wavelength of interest. The larger the dynamic range, the greater the contrast, and, by this metric, the better the electrochromic material.

Besides dynamic range (color contrast), there are several other characteristics of switching between various electrochromic states that are of interest. These include characteristics such as switching time and cyclability. A simple method of characterizing the latter two is switching with a square wave applied potential between extreme electrochromic states while monitoring the optical absorption at a particular wavelength as a function of time, known as chronovoltabsorptometry (CVA). Figure 23.8 shows typical CVA data, for a poly(aromatic amine)-based electrochromic device. As seen from Figure 23.8, among the information obtainable from such data, is the switching time (hence speed) in each direction between highly doped and dedoped states; switching symmetry; dynamic range; and cyclability, i.e., degradation of dynamic range with cycling. For CVA data to have meaning, there must be (a) comparison of initial and extended cycles, e.g., the 1st and 1000th cycles; (b) a reference run to steady-state (infinite time) optical transmission values as a baseline for comparison of all runs; (c) runs between intermediate electrochromic states in addition to the extreme oxidized/reduced states.

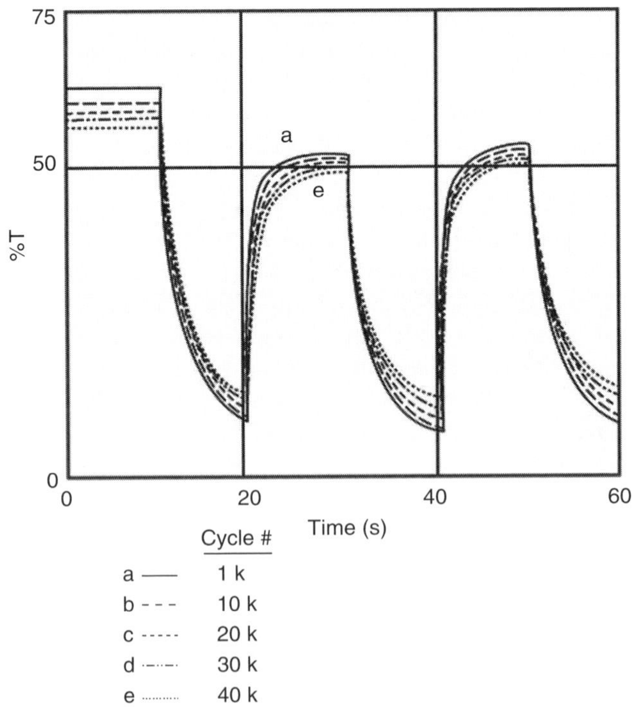

FIGURE 23.8 Typical CVA data for a poly(aromatic amine). Cycle numbers are indicated. (Courtesy of Ashwin-Ushas Corp., Inc.)

The switching time between two electrochromic states has been defined in the author's laboratories as one analogous to risetimes used in electrical measurements, i.e., the time for traversing between 10% and 90% of the total steady-state optical response. However, many workers in the field use different, entirely arbitrary definitions for this parameter. The steady-state optical response is determined by allowing a sufficient time for an electrochromic system to attain a steady state, for example 15 min, which suffices for most materials. There are generally an oxidative (going from reduced to oxidized state) and a reductive switching time, τ_{ox}, τ_{red}, which are almost never the same. The switching speed is then just the inverse of the switching time. The switching symmetry is reflected in the difference between the oxidative and reductive switching times; in some cases, the two switching times can be clearly different. Typical switching times for standard thickness (0.1–1 μm) films of most electrochromics are of the order of seconds, although switching times of hundreds of microseconds have been claimed [6].

Electrochromic cyclability of electrochromic materials, i.e., number of cycles between extreme oxidized and reduced states, also requires precise definition. Again, this is typically defined as the number of cycles possible between extreme light and dark states of an electrochromic with 10% or less optical degradation. The latter can be measured from CVA data, for a cycling time at least equal to the switching time, as defined above. A separate electrochemical or charge cyclability can also be defined, as in the next paragraph. Cyclabilities of electrochromics can vary from tens of cycles to up to 10^6 cycles claimed for some systems.

Electrochemical methods such as chronoamperometry (CA) (monitoring of current decay with time upon application of a specified potential) and chronocoulometry (CC) (monitoring of charge decay with time upon application of a specified potential) can be useful methods for evaluation of the electrochromic properties of materials. CA data may sometimes have little meaning, since they simply trace a generic, universally applicable current decay, and indeed may not even correlate well with actual switching times. When applied through several thousand electrochromic cycles, however, or when integrated to yield charge capacities (see below), they can have meaning. CA curves also have some significance when they are used to correlate optical response and current decay, for example, showing that current decay may be very rapid while optical response, dependent on factors including diffusion and dopant concentration, is delayed.

The charge capacity is simply the total charge needed to go from one electrochromic state to another, usually between the extreme reduced and oxidized states of an electrochromic material. Charge capacities are an extremely useful measure of such parameters as cyclability, sometimes more useful than optical measures for determining causes of degradation.

Charge capacity can be determined in a number of ways: by integration of areas under CA curves; directly via CC (not all commercially available electrochemical instruments have a coulometric capability); by integration of area under CV peaks, which involves accurate estimation of baseline currents. If available, the CC method is the most direct. Charge capacities are a very important measure of performance for devices such as electrochromic windows, since the charge capacity is a direct measure of energy expended in switching. A charge capacity of 1 mC/cm^2 is a good, and achievable, target for most electrochromic systems.

Electrochemical or charge cyclability can then be defined exactly as for electrochromic cyclability above, but with the stipulation that 10% charge degradation rather than optical degradation is measured. Frequently, electrochemical and electrochromic cyclability values are quite comparable. Charge capacities are more directly measurable using chronocoulometry, and can also be determined from cyclic voltammograms.

Another important electrochromic parameter, of interest especially in practical applications, is the open circuit memory, i.e., the optical memory retention. For practical applications it is important to know how long an electrochromic material will retain its color, i.e., electrochromic state, after an applied potential that brought it to that state, is disconnected (i.e., brought to an open circuit). Quite evidently, the open circuit memory of a material in different electrochromic states may differ: for

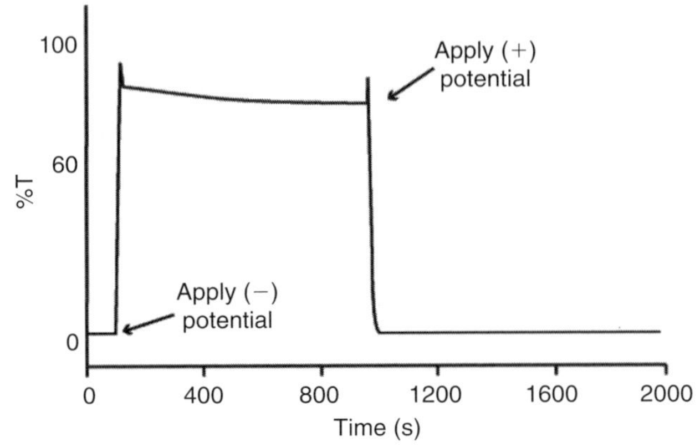

FIGURE 23.9 Open circuit memory data for a poly(aromatic amine)/sulfonate system (sealed electrochromic device). At each arrow, the reducing/oxidizing potential is applied for 60 s, and the cell then disconnected. (Courtesy of Ashwin-Ushas Corp., Inc.)

example, it may have a long open circuit memory in its oxidized state, but a very short one in its reduced state, due to atmospheric oxidation.

For our purposes again, we use the definition adopted in the literature [1], defining open circuit memory for a particular electrochromic state (identified by its applied potential), as the time during which there is less than 10% optical degradation following disconnection of an applied potential. There is an added stipulation that before disconnection, the system should have achieved a steady state, as defined above. The 10% stipulation is necessary as there is always some optical fluctuation immediately on disconnection of the circuit. Typically, open circuit memories can range from tens of seconds to infinite, i.e., total memory retention (cf. Figure 23.9).

23.2 TRANSMISSION-MODE ELECTROCHROMIC DEVICES

Figure 23.10 shows photos of typical transmission-mode, visible-region electrochromic devices, while Figure 23.11 shows the actual electrochromic performance (both photos and spectroelectrochemical data) for these devices.

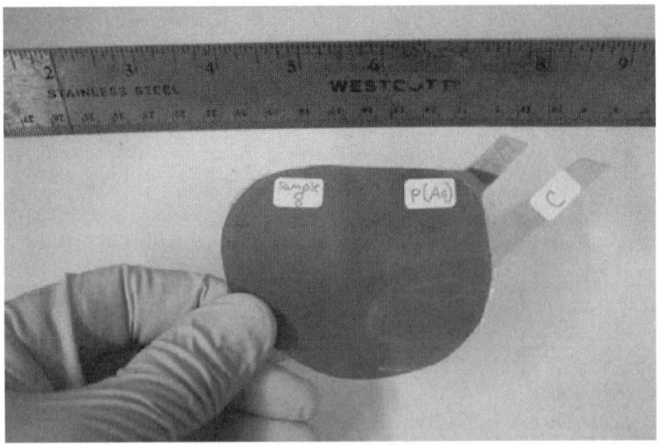

FIGURE 23.10 Top to bottom: Photos of transmission-mode, visible-region electrochromic devices, having the construction shown in the schematic of Figure 23.6. Photos show eyeglass shape suitable for sunglasses application, high flexibility, and capability of being affixed to substrates of varied shapes using pressure-sensitive adhesive. (Courtesy of Ashwin-Ushas Corp., Inc.)

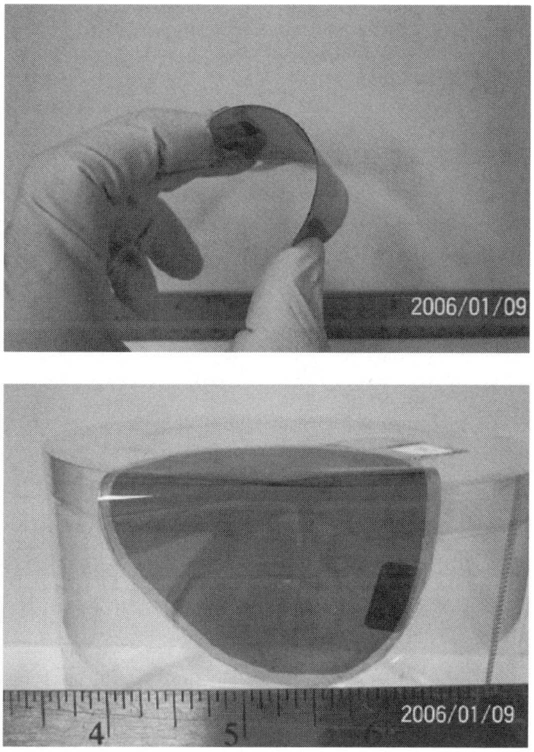

FIGURE 23.10 (Continued)

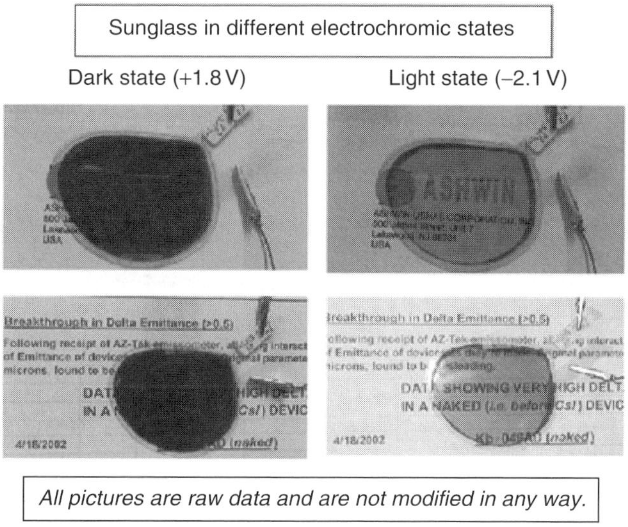

FIGURE 23.11 Top to bottom: Photos of sunglasses devices of Figure 23.10 in dark and light state, indoors (top) and outdoors (middle), showing excellent color contrast (dynamic range).

(*Continued*)

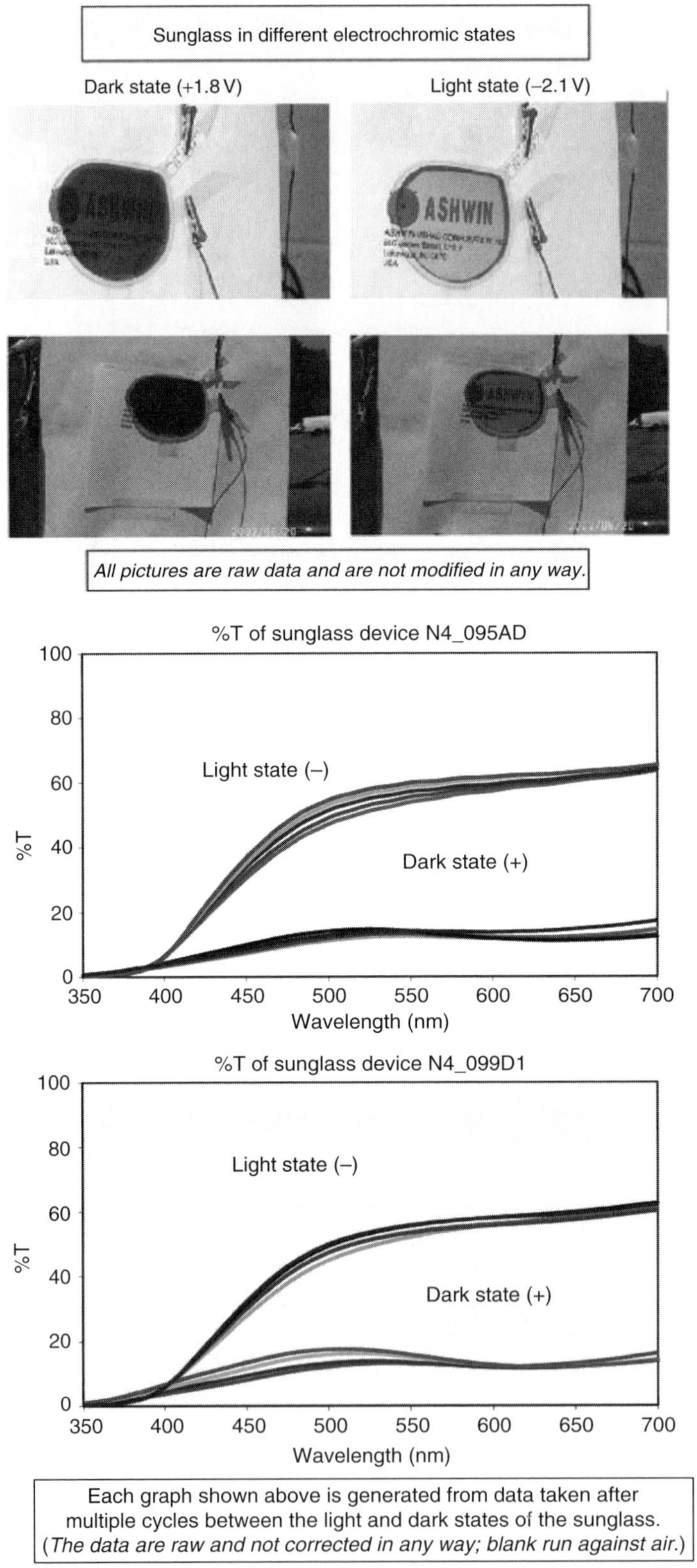

FIGURE 23.11 (Continued) (Bottom)—Actual electrochromic performance (spectroelectrochemical data) for these devices, again showing excellent dynamic range and broadband (500–900 nm) nature of electrochromism. (Courtesy of Ashwin-Ushas Corp., Inc.)

The devices shown in Figure 23.10 incorporate a flexible, plastic substrate, ITO/poly(ethylene terephthalate). Using such substrates, very thin, flexible electrochromics can be designed that can be pasted onto substrates of variable area and shape. Although there has been much work on transmission-mode electrochromic devices for such applications as automobile windshields, commercial building windows and high-end ski goggles [7], there have been few commercial successes to date. One successful commercial application is rearview mirrors of high-end automobiles, where a transmission-mode electrochromic device using a CP such as poly(aniline) in a gel electrolyte is placed against a mirror (thereby converting it into a reflectance-mode device), for automatic control of the day/night function of these mirrors.

23.3 REFLECTANCE-MODE ELECTROCHROMIC DEVICES

Figure 23.12 shows photos of actual reflectance-mode, IR-region (2–25 μm) electrochromic devices. The top part of this figure shows various sizes and shapes that these devices can be cut into, while the bottom part shows actual (visible-region) color changes observed. Figure 23.13 shows typical spectroelectrochemical, i.e., electrochromic, performance for such devices, optimized to different parts of the IR spectral region. Such devices are used in such typical applications as infrared camouflage for the military, and micro-spacecraft thermal control [7–11].

For spacecraft applications in particular, two important measurements are those of thermal emittance, ϵ and solar absorptance, α_s. Both can be derived directly from reflectance measurements, and both have importance in radiative loss applications, e.g., in space. For instance, satellites and spacecraft ideally require high reflection of solar heat during sunfacing, implying a low α_s, <0.2, while at the same time having high radiative capability to remove excess heat during darkfacing, implying a high thermal emittance, >0.8. At-will switchability—i.e., electrochromism—of these properties is highly desirable, with variation of α_s/ϵ ε ratios between 0.2 and 1.0 sought. Excellent treatments of these parameters are available elsewhere [7–12].

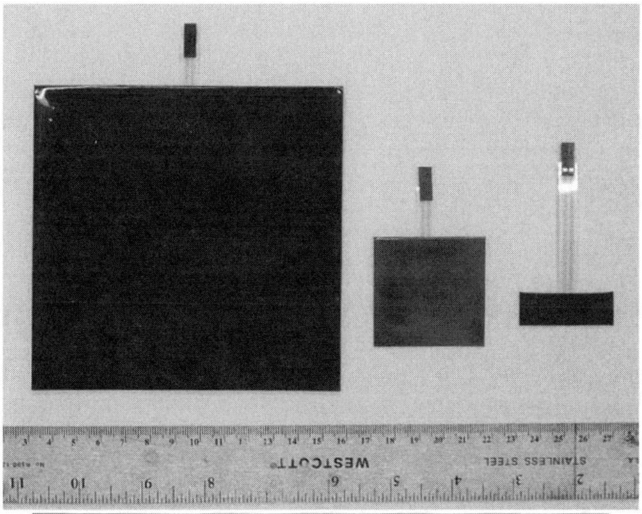

Devices can be customized in shape and size to fit your application.
(*All pictures are raw data and are not modified in any way.*)

FIGURE 23.12 Photos of typical reflectance-mode, IR-region electrochromics, of the construction shown in Figure 23.7. (Courtesy of Ashwin-Ushas Corp., Inc.)

(Continued)

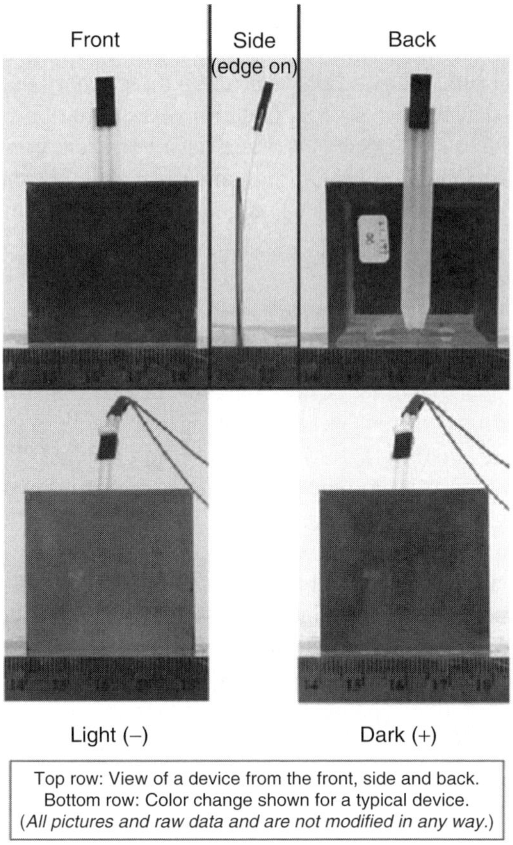

FIGURE 23.12 (Continued)

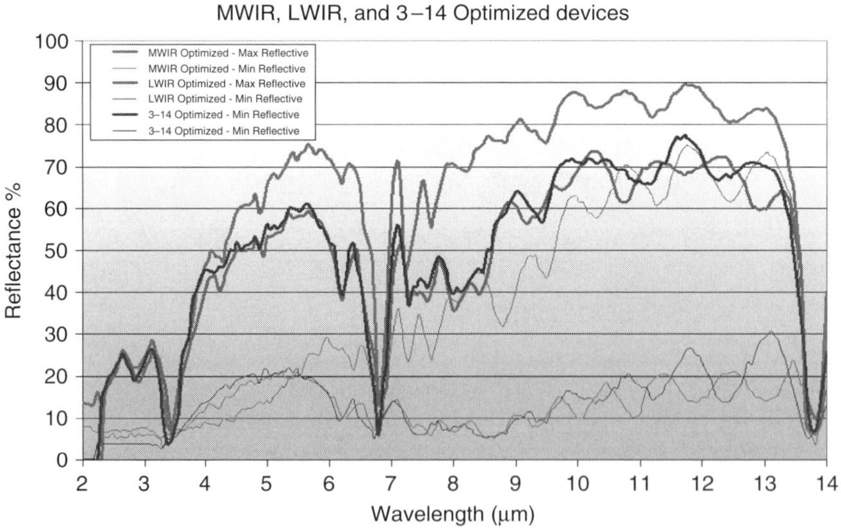

FIGURE 23.13 Electrochromic performance of IR devices of the type shown in Figure 23.12. The data are for three separate devices, optimized, as noted in the figure, respectively for the mid-IR (MWIR), far-IR (LWIR), and 3–14 μm regions. For each device, the reflective-state (light-state) and absorptive-state (dark-state) spectra are shown. The dynamic range between these states is seen to be very large. (Courtesy of Ashwin-Ushas Corp., Inc.)

An application for conductivity–electrochromism may be envisioned, for instance, with an Au grid on a flexible substrate on which an electrochromic material is deposited. When the material is cycled between conductive and nonconductive states, the dielectric permittivity of the material, affecting its microwave/radar signatures, changes dramatically. We thus have a microwave/radar electrochromism of sorts. However, in this chapter, we shall not deal with conductivity, except as it relates to spectral properties.

23.4 SUMMARY

Electrochromics are materials, substances, or devices, which change color generally when a small voltage is applied to them. Their electrochromism is generally based on a chemical oxidation/reduction (redox) process. This definition excludes effects, which appear to be similar, such as solvatochromism, thermochromism, and the effect responsible for the function of light-emitting diodes. Organic electrochromics are predominantly of two classes: conducting polymers (CPs), such as poly(aniline), and nonpolymeric organic materials such as methyl viologens. Electrochromics can encompass not only the visible spectral region but other regions as well, ranging from near-UV to far-IR and microwave–mm-wave. A basic understanding of electrochemistry is essential for understanding the functioning of electrochromics, and this chapter provides one, discussing such significant points as two- vs. three-electrode mode operation and the interpretation of voltammograms from an electrochromic point of view. A subsection is also devoted to CPs, which form by far the bulk of organic electrochromics. Basic electrochromic devices are then presented and their functioning discussed at some length. The methodology and terminology of the measurement of electrochromism, for both transmission- and reflectance-mode devices, is then discussed. The precise definitions of other electrochromic metrics, such as switching time, cyclability, and open circuit memory (optical memory) are then discussed. Finally, the practical-world applications of electrochromics in the visible, IR, and microwave regions, ranging from automobile rearview mirrors and sunglasses to military IR camouflage, are very briefly presented.

EXERCISE QUESTIONS

1. Sketch a transmission-mode and a reflectance-mode electrochromic in schematic.
2. What are the different spectral regions that an electrochromic can function in? What are the typical practical applications for each of these spectral regions?
3. In the UV–Vis–NIR region, how do transmission-mode and specular and diffuse reflection-mode spectra of electrochromics compare?
4. Write out detailed definitions, with illustrations if appropriate, for the following terms: Chronovoltabsorptometry; oxidative and reductive switching times; electrochromic, electrochemical, and charge cyclabilities; charge capacity; dynamic range; open circuit memory; solar absorptance; and thermal emittance.
5. Describe the construction and function of an example of an electrochromic device having a complementary coloring counter electrode. What are the redox potential constraints in such devices?

LIST OF ABBREVIATIONS

CA	chronoamperometry
CC	chronocoulometry
CP	conducting polymer
CV	cyclic voltammetry
CVA	chronovoltabsorptometry
ITO	indium tin oxide
LSV	linear scan voltammetry
OCP	open circuit memory

P(ANi)	poly(aniline)
P3MT	poly(3-methyl thiophene)
Ppy	poly(pyrrole)
PT	poly(thiophene)

REFERENCES

1. Chandrasekhar, P., *Conducting Polymers: Fundamentals and Applications. A Practical Approach*, with foreword by Lawrence Dalton. Kluwer Academic Publishers, Dordrecht, The Netherlands/Norwell, MA, ISBN No. 0-7923-8564-0 (August 1999). See especially chapters 3 (Electrochromics), 4 (Electrochemistry), 20 (Electrochromics).
2. Maricle, D.M., Giglia, R.D., U.S. Patent 3,884,636 (29 October 1974), and references therein. Baucke, F.G.K., Drause, D., Metz, B., Paquet, V., Zauner, J., U.S. Patent 4,465,339 (14 Aug 1984). Chandrasekhar, P., Electrochromic Display Device, U.S. Patent 5,995,273 (November 1999).
3. Bard, A.J., Faulkner, L.R., *Electrochemical Methods*: *Fundamentals and Applications*, Wiley & Sons, New York, 1980.
4. Diaz, A.F., Castillo, J.I., Logan, J.A., Lee, W.-Y., *J. Electroanal. Chem.*, 129, 115, 1981.
5. Schwendemann, I., Hickman, R., Soenmez, G., Schottland, P., Zong, K., Welsh, D.M., Reynolds, J.R., *Chem. Mater.*, 14, 3118, 2002 and references therein. Reeves, B.D., Thompson, B.C., Abboud, K.A., Smart, B.E., Reynolds, J.R., *Adv. Mater.*, 14, 717, 2002 and references therein.
6. Meador, M.A., Gaier, J.R., Good, B.S., Sharp, G.R., Meador, M.A., A Review of Properties and Potential Aerospace Applications of Electrically Conducting Polymers, Internal Report National Aeronautics and Space Administration, Lewis Research Center, Cleveland, OH, pp. 1–21, 1989.
7. Pons, Frédéric, (Tag Heuer logo, SA), Joinville-le-Pont, France, Personal communication, 1995.
8. Chandrasekhar, P., Zay, B.J., Birur, G.C., Rawal, S., Pierson, E.A., Kauder, L., Swanson, T., *Adv. Funct. Mater.*, 12(2), 95–103, 2002.
9. Chandrasekhar, P., Zay, B.J., McQueeney, T.M., Scara, A., Ross, D.A., Birur, G., Haapanen, S., Kauder, L., Swanson, T., Douglas, D., *Synth. Met.*, 135–136, 23, 2003.
10. Chandrasekhar, P., Zay, B.J., Ross, D.A., McQueeney, T.M., Birur; Swanson, T., Kauder, L., Douglas, D., Far-IR-through-Visible Electrochromics Based on Conducting Polymers for Spacecraft Thermal Control and Military Uses: Application in NASA's ST5 Microsatellite Mission and in Military Camouflage, Jenekhe, S.A., Kiserow, D.J. (Eds.), American Chemical Society Symposium Series, 888, *Chromogenic Phenomena in Polymers: Tunable Optical Properties (ACS Proceedings Volume)*, Jenekhe, S.A., Kiserow, D.J. (Eds.), 2005.
11. Stenger-Smith, J.J., NAWCP, China Lake, CA, Personal Communication 1998. *Radar Cross Section Handbook*, Plenum Press, New York, 1970, Several Volumes.
12. Mastragostino, M., Electrochromic devices, chapter 7 (p. 223) in Scrosati, B. (Ed.), *Applications of Electroactive Polymers*, Chapman & Hall, New York, 1993.

24 An Introduction to Conducting Polymer Actuators

Geoffrey M. Spinks, Philip G. Whitten, Gordon G. Wallace, and Van-Tan Truong

CONTENTS

24.1 Introduction to Actuator Materials ... 734
24.1.1 Need for Actuators ... 734
24.1.2 General Selection Criteria for Actuators ... 735
24.1.2.1 Mechanical Performance ... 735
24.1.2.2 Input Energy ... 736
24.1.2.3 Control ... 736
24.1.2.4 Ancillary Equipment ... 737
24.1.2.5 Cost ... 737
24.1.2.6 Maintenance and Lifetime ... 737
24.1.2.7 Environmental Impact (Noise, Waste Products) ... 737
24.1.3 General Actuation Processes ... 737
24.1.4 Modulus Shift and Actuation Performance ... 739
24.1.5 Comparative Performances of Polymer Actuators ... 741
24.1.6 Conventional Mechanical Systems ... 742
24.2 Conducting Polymers for Actuators ... 745
24.2.1 Why Choose Conducting Polymers? ... 745
24.2.2 Mechanisms of Actuation in Conducting Polymers ... 745
24.2.3 Modulus Shift in Conducting Polymer Actuators ... 749
24.2.4 Designing Conducting Polymer Actuators ... 749
24.2.5 Benders versus Axial Actuators ... 750
24.3 Recent Performance Improvements ... 753
24.4 Applications for Conducting Polymer Actuators ... 754
24.4.1 Bending Actuators ... 754
24.4.2 Actuators for High Packing Density Applications ... 755
24.4.3 Actuators for Micro-Devices ... 756
24.4.4 Actuation of Cylinder or Tube Diameter ... 756
24.5 Concluding Comments ... 756
Exercise Questions ... 758
List of Abbreviations ... 761
References ... 761

Abstract Actuators comprise a diverse family of materials that respond mechanically to an external stimulus, most commonly electrical. The movements or forces generated are potentially useful in a wide variety of applications. The specific advantages of polymer actuators as compared with other actuator materials (e.g., piezoelectric ceramics or shape memory alloys) and mechanical drive systems (motors, engines, hydraulics, and pneumatics) relate mainly to their compact size and simple construction. Noiseless operation is another key advantage in some areas. Applications for these systems are being driven by the demands for humanoid robotics, micro-electro-mechanical system (MEMS), and smart structures. Niche applications in biomedical devices and microfluidics are also being developed. This chapter focuses on conducting polymer actuators and describes their mechanism of operation, compares conducting polymers with other actuator materials, and considers some prototype applications.

24.1 INTRODUCTION TO ACTUATOR MATERIALS

24.1.1 NEED FOR ACTUATORS

Interest in actuator materials stems from both the desire to exploit their electromechanical behavior in a variety of different applications, and as a means for investigating fundamental physics, chemistry, and materials science. Actuator materials are defined as materials that produce a volume or shape change in response to an external stimulus. Most commonly, the stimulus is electrical in nature (involving a transfer of charge) and such materials are known as electromechanical actuators. In some cases, the electrical charge transfer induces an electrochemical reaction giving electrochemo-mechanical actuators. In other cases, the stimulus comes from a change in the chemical environment while in others the stimulus is heat or light absorption.

By changing shape or volume, actuating materials produce both force and movement. This simple characteristic makes actuators potentially useful in a multitude of devices utilizing moving parts. Movement is fundamental to transportation systems for cargo (cars, planes, boats, etc.) and for the manipulation of objects (robotics and machine tools) or bulk materials (lifts and conveyors). By definition machines are apparatus for applying mechanical power [1] and thus need means for generating force and movement. The huge variety of machines in use illustrates the dependence of much of society's technology on these basic functions. Examples include machine tools, switches, disk drives, fans, pumps, and so on.

Materials that produce a volume change can in theory be readily adapted to machine components. As illustrated in Figure 24.1, the volume-change material can act analogously to hydraulic or pneumatic systems to move pistons or other structures. In some actuator materials, the response is largely a shape change rather than a change in volume. Typically, a bending type movement is produced that resembles musculoskeletal-like movement [2,3]. In either case, if the movement is suppressed by an external object then the actuator generates a force that acts on the object. Force generation is important in clamping systems, for example.

Many alternatives to actuators exist for generating force and movement in the form of solenoids, engines, motors, hydraulics, and pneumatics. The development of these systems was critical in the industrial revolution and has been responsible for the shaping of modern civilizations. In general, actuator materials are not able to compete with the high force, high speed operation of engines,

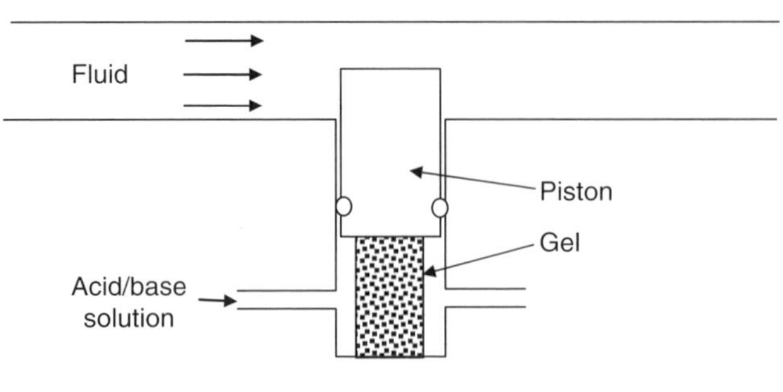

FIGURE 24.1 Similarities in function of actuators with conventional pistons.

motors, and other conventional systems. These systems, however, do have limitations, especially in their miniaturization and packing density. In such cases, actuator materials are the preferred and often the only suitable system.

The simple process of shape/volume change belies the complex molecular mechanisms occurring. Much of the research into actuator materials seek to understand these mechanisms so that improvements in performance can occur. In many cases, the mechanisms are not fully understood, while in other cases the performance is well predicted by phenomenological or physics-based models. The development of new actuator materials continually challenges both the technology developers and theorists, but (most importantly) provides new opportunities for improving actuator performance. The special attributes of actuator materials is considered in more detail below. While many different types of actuator materials exist (shape memory alloys, piezoelectric, dielectric elastomer, ionic polymer gels, conducting polymer, and even muscle), this chapter considers the particular attributes of conducting polymer actuators—how they work and how they compare with other actuators.

24.1.2 General Selection Criteria for Actuators

The huge diversity of applications requiring force or movement makes the selection of the optimal drive system very complex. However, it is possible to categorize the general criteria as follows [4,5].

24.1.2.1 Mechanical Performance

Outputs such as force and movement generated are the basic performance characteristics of any mechanical drive system. The product of force and displacement gives the energy output (or mechanical work done). The speed of response is also important as it defines the power rating of the device (power = force × velocity). Larger systems are likely to generate more force/displacement so criteria such as work per unit mass (or volume) and power to weight ratios are often used to compare different actuation systems. Since force and displacement outputs often scale with the size of the actuator, it is also more common to report the stress (force/area) and strain (displacement/original length) produced by actuator systems. The highest stress generated (the blocked stress) will always occur when all displacement is prevented. The highest strain occurs when no external load is exerted on the material (the free stroke). In most cases, the actuator must produce both stress and strain and an inverse linear relationship between the two is normally assumed, as illustrated in Figure 24.2. Other performance criteria used to assess the mechanical output of actuators are given in Table 24.1.

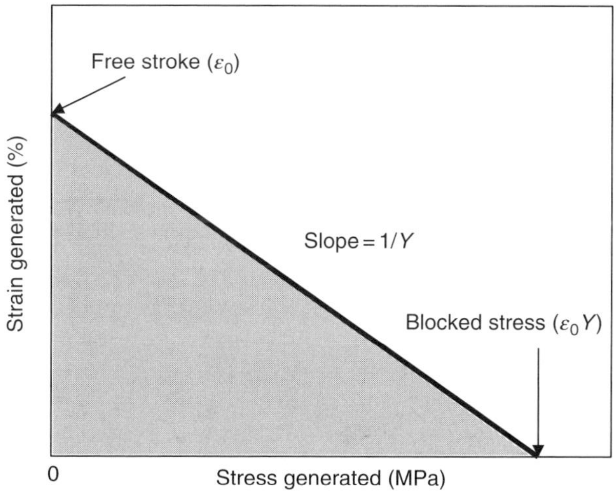

FIGURE 24.2 Schematic illustration showing how the stress and strain generated vary between the extremes of maximum strain (free stroke) at zero stress and maximum stress (blocked stress) at zero strain. Y is the Young's modulus of elasticity. The shaded area is the region in which the actuator can operate.

TABLE 24.1
Common Actuator Specifications

Parameter	Scale-Invariant Parameter	Comments
Energy and power		
Energy output	Specific energy (J/kg or J/m^3)	Over full cycle
Power output	Specific power (W/kg)	Average or instantaneous
Energy efficiency	Efficiency (%)	Energy out over full cycle/Energy in over full cycle
Response time	Response time (s)	For one direction or full cycle. The response time is normally a function of the actuator cross-sectional area
State variables		
Displacement	Strain (%)	Instantaneous or maximum over cycle
Force	Stress (MPa)	Instantaneous or maximum during cycle
Velocity	Strain rate (%/s)	Instantaneous or maximum. The strain rate is normally a function of the actuator cross-sectional area
Impedance and controllability		
Stiffness	Elastic modulus (GPa)	May be nonlinear and changed by actuation stimulation
Dampening	Loss factor	Usually nonlinear, dependent on rate
Accuracy	As percentage of target strain or stress	Usually percentage of maximum stress or strain
Repeatability	Repeatability (%)	Usually percentage of maximum stress or strain
Linearity	Linearity (%)	Deviation from linear input–output relationship
Operational characteristics		
Environmental tolerances	Same	Recommended ranges of temperature, humidity, etc.
Durability	Same	Number of cycles before threshold or failure
Input impedance	Specific impedance	Voltage and current requirements

Source: Kornbluh, R., Pelrine, R., Pei, Q., and Shastri, V., in *Electroactive Polymer (EAP) Actuators as Artificial Muscles. Reality, Potentials and Challenges,* SPIE Press, Bellingham, Washington, 1999, 457–495.

24.1.2.2 Input Energy

The nature of the input energy determines the efficiency and the basic actuator design. Most actuators use electrical energy as the input and in such cases the input voltage and current are key performance characteristics. Piston engines and rockets use hydrocarbon fuels as their energy source. Some actuators operate by thermal or light stimulus and in each case it is possible to define the input chemical, thermal, or photo energy. Conventional drive systems such as hydraulics and pneumatics use pressure differences to generate movement/force. It is pertinent to note that the forms of input energy available or tolerated normally restrict the choice of actuator, for example, many biomedical applications will not employ high voltages.

24.1.2.3 Control

The precision of operation varies for different applications, but the method of achieving control also depends on the basic mechanism of operation. The position or force generated by an actuator may be predicted by calibration, or, measured and then adjusted using a sensor in combination with a feedback loop.

24.1.2.4 Ancillary Equipment

Different actuators and drive systems require different additional equipment to make them functional [6]. Electric motors need to be geared to achieve variable speeds. Pneumatics and hydraulics require a fluid reservoir. Electrochemical actuators require auxiliary electrodes and electrolyte. The ancillary equipment can considerably increase the size and weight of the drive system and must be considered when determining power:weight ratios and similar comparative figures-of-merit. In many cases, the type of ancillary equipment required will determine the choice of actuator. For example, many industries employ compressed air, and have compressed air available in a manner similar to electricity or water, hence a pneumatic system would require few overheads; however, in other areas like households, compressed air is not readily available making electrical-based systems more attractive.

24.1.2.5 Cost

Both fabrication and operating costs vary considerably between different actuator technologies.

24.1.2.6 Maintenance and Lifetime

All systems have unique considerations concerning maintenance. Most solid-state actuators are susceptible to irreversible damage if overloaded. In some applications, an actuator has to operate only once, for example, a clamp, while in other applications an actuator may have to perform a million cycles or more.

24.1.2.7 Environmental Impact (Noise, Waste Products)

Many conventional drive systems are very noisy and polluting. Pollution can be in the form of chemical by-products or waste heat.

24.1.3 General Actuation Processes

Actuators produce both forces and movements. However, actuator test conditions are usually designed to either produce a movement or a force. When the actuator is not attached to any external device or structure, then the movement generated by the electrical stimulus is free to occur and not inhibited by any external effect. Under these conditions, the force generated is zero and the movement produced is called the free stroke. These conditions of operation are technically isotonic (constant load) with a zero force applied. Isotonic operation at nonzero forces should produce the same actuation strain in the ideal case and when the deformation caused by the initial application of the load is ignored. However, as described below, the change in stiffness of the actuator during operation can mean that the strain produced is dependent on the applied load. When the deformation caused by the application of the load is not ignored, then the total length change is the sum of this deformation and the actuation process. In most cases, these two deformations oppose one another so that the net length change (actuation strain) becomes smaller as the applied load becomes larger.

This situation is depicted in Figure 24.3, which also applies to the case where the force changes during actuation. The significance is illustrated by considering an example of a contractile actuating cable designed to lift a weight from a surface (Figure 24.3a). Initially, the force is zero and increases to mg (m is the mass and g the gravitational constant). The increasing force on the cable causes elastic extension that depends upon the mass m and the material's Young's modulus Y. The analysis takes a hypothetical actuator with a cross-sectional area of 10 mm^2, a modulus Y of 1 MPa, that produces a contraction strain of 10% at a strain rate of 1 %/s at zero load, has a starting length of 50 mm, and is attached to a mass of 10 g. If the contraction takes 10 s to complete, then the force increases

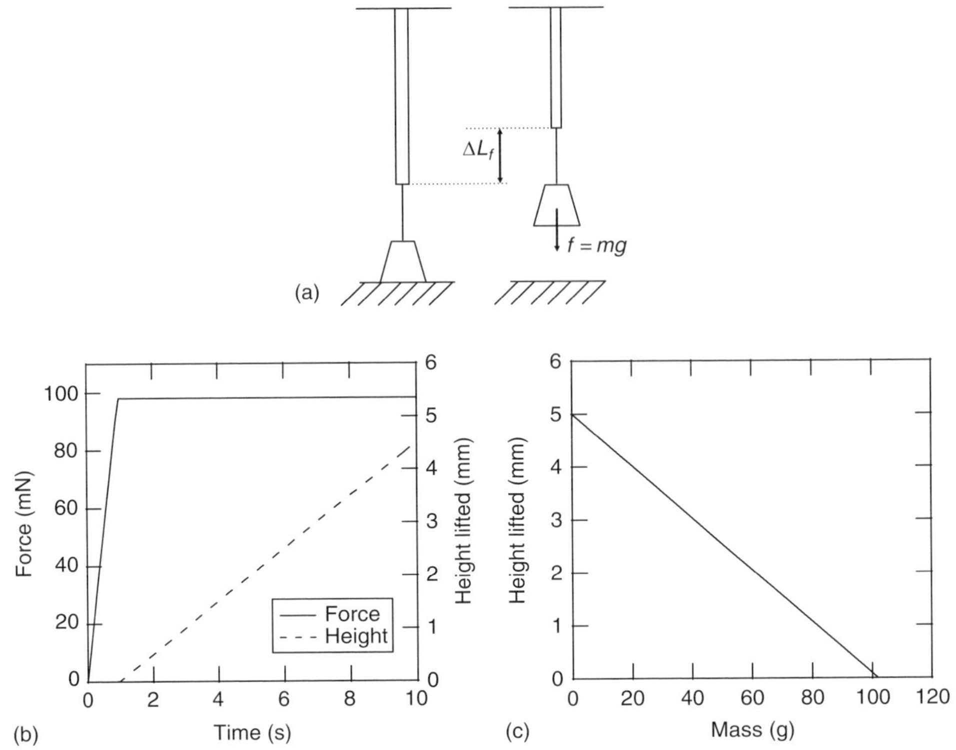

FIGURE 24.3 (a) Schematic diagram of a hypothetical actuator that generates a contractile free stroke (at zero load) of 10% attached to a weight of mass *m*. (b) Height that a mass of 10 g is lifted and force generated in the contractile actuator during the 10 s contraction of a 50 mm long actuator and 10 mm^2 cross-sectional area. (c) Final height (after full 10 s contraction) that the mass is lifted for different sized weights.

steadily throughout the contraction cycle (Figure 24.3b). However, the weight is not lifted from the table until the force generated equals *mg*. As shown in Figure 24.3b, the 10 g weight is not moved until 1 s into the contraction cycle. The final height reached is only 4.5 mm or 9% of the total length. As shown in Figure 24.3c the final height depends on the mass being lifted and decreases linearly with increasing mass. In the example that is considered, a mass of 102 g or higher cannot be lifted from the table. When the height is converted to strain and the mass applied converted to stress, then Figure 24.3c becomes equivalent to the figure shown in Figure 24.2.

In the general case, the force that needs to be applied to prevent any actuation strain is called the blocking force and is given by

$$f_b = \frac{\Delta L_0 A Y}{L_0} \tag{24.1}$$

where

A and L_0 are the original cross-sectional area and length
ΔL_0 is the free stroke
Y is the elastic (Young's) modulus

Testing under isometric (constant length) conditions gives a direct measure of the blocking force (or stress). When testing fibers, both the free stroke and the blocking force must be extrapolated, as the fibers require a minimum force to straighten and will buckle under compressive loads.

24.1.4 Modulus Shift and Actuation Performance

A secondary mechanism that was recently shown to have a major affect on actuation is the change in mechanical properties that accompanies electrical stimulation in all actuator materials [7,8]. Since actuation is a mechanical process, the basic mechanical properties of the actuator material will affect the stresses and strains generated. These properties are known to change, sometimes significantly, during the actuation process. Therefore, the effect of changing properties on actuation must be considered.

The simplest approach to this issue is to assume that the actuators behave as linear elastic materials. This assumption generally holds for all materials at small strains. Linear elasticity is represented by Hooke's Law:

$$\sigma = \varepsilon Y \tag{24.2}$$

where

σ is the tensile (or compressive) stress
ε is the resultant strain in the direction of the applied stress

The proportionality constant (Y) is the Young's modulus of elasticity and varies greatly for different materials. In polymers, the value of Young's modulus is also sensitive to the rate at which strain occurs.

As shown in Table 24.2, the Young's modulus of actuator materials (polymer and nonpolymer) changes in the expanded and contracted states. The change in modulus affects the actuation strain when a force is applied, as can be appreciated from Hooke's law. Thus under isotonic (constant force) conditions, the actuation strain (ε_f) at constant force f is given by [9]

$$\varepsilon_f = \varepsilon_0 + \frac{f}{A}\left(\frac{1}{Y'} - \frac{1}{Y}\right) \tag{24.3}$$

where

ε_0 is the actuation strain under no load
Y and Y' are the Young's moduli in the initial and final states

This equation has been shown to accurately predict the results shown in Figure 24.4 where the actuation strain decreases with an increasing applied force. These results have significant practical implications, since the amount the actuator moves is determined by the external forces applied.

TABLE 24.2
Typical Free Stroke, Modulus and Modulus Ratio (Expanded: Contracted State), and Density of Selected Actuator Materials

Material	ε_0 (%)	Y (GPa)	Y/Y'	ρ (g/cm^3)
Piezoelectric ceramic	0.2	64	1.1	7.5
Electrostrictive polymer	5	0.4	1.2	1.8
Conducting polymer 1	2	0.11	1.16	1.5
Conducting polymer 2	5	0.1	5	1.5
Carbon nanotube	0.5	5	1	0.3
Electrostatic elastomer	63	0.001	1	1.4

Source: From Spinks, G.M. and Truong, V.-T., *Sens. Actuat., A: Phys.*, A119, 455, 2005. With permission.

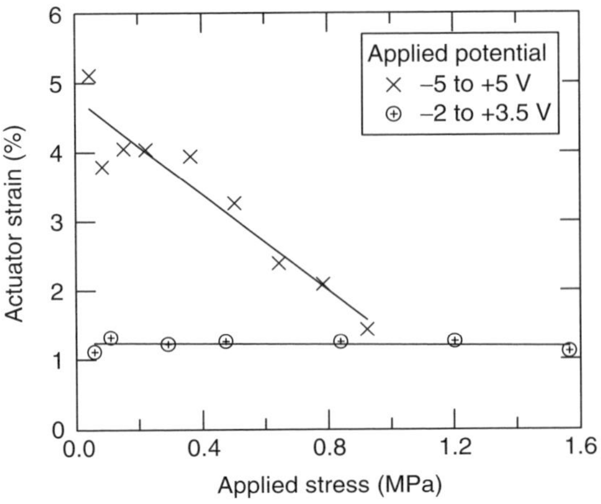

FIGURE 24.4 Measured isotonic strains for polypyrrole (PPy) actuators operated at different applied stresses (initial stretching caused by the application of the stress has been ignored [7]). (From Spinks, G.M., Liu, L., Wallace, G.G., and Zhou, D., *Adv. Funct. Mater.*, 12, 437, 2002. With permission.)

The significance is illustrated by considering again the contractile actuating cable designed to lift a weight from a surface, as presented in Figure 24.3. Figure 24.5 compares the height to which the weight is lifted from the surface for the case where the modulus decreases during stimulation and the original situation (dashed line) where the modulus does not change. The analysis assumes a contracted final state modulus of 0.75 MPa, compared with an expanded initial state modulus of 1 MPa.

Other actuation scenarios can also be considered. The basic isotonic condition where the elastic stretching caused by the application of the weight is not ignored is shown in Figure 24.6. For most actuator materials, the modulus is lower in the contracted state and higher in the expanded state. This situation leads to a decrease in actuation strain with increasing applied force. However, it is also possible for the reverse situation to occur and then the strain actually increases with force (when $Y < Y'$). We have recently observed such a phenomenon for the first time using polythiophene conducting polymer actuators [10].

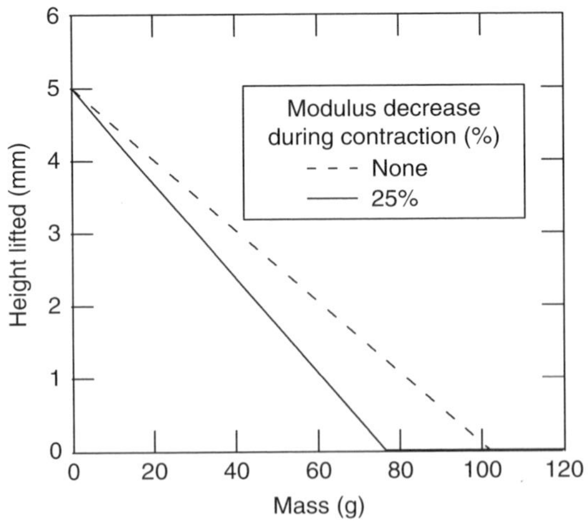

FIGURE 24.5 Height lifted by a contractile actuator (as in Figure 24.3) considering a 25% decrease in modulus during contraction—the dashed line shows the height lifted when there is no change in modulus.

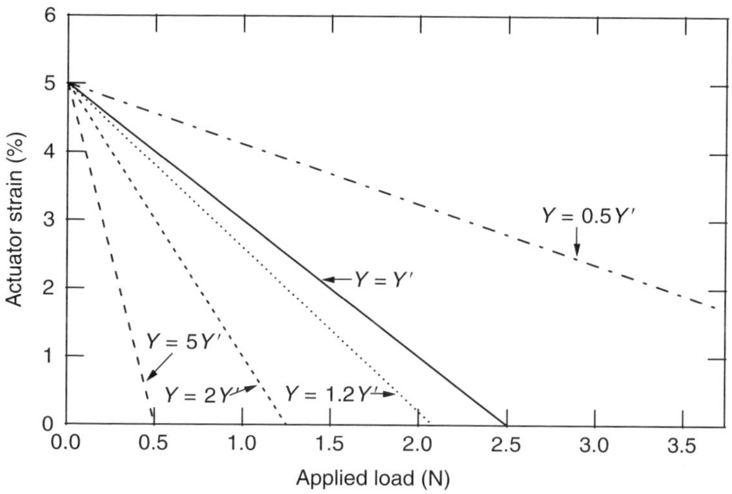

FIGURE 24.6 Calculated actuator strains at different applied loads for different ratios of expanded: contracted state moduli. Here the initial deformation caused by the application of load contributes to the net actuator strain.

24.1.5 Comparative Performances of Polymer Actuators

Although not entirely complete, the most common means for comparing the actuation performance of different materials is by way of various outputs: stress and strain generated, work density, power density, and strain rate. One difficulty with such comparisons is that the data for different actuator materials is usually collected by different workers using completely different testing geometries and parameters. The data available is not always complete. Quite often, peak values are reported and it is not known whether such parameters represent the average or long-term characteristics of the actuator. It is also not always clear whether the per mass or per volume performance metrics include the necessary ancillary equipment—such as pumps, fluid lines, and so on used in hydraulic systems. Finally, it is not known whether the performance characteristics apply across all scales—both larger and smaller—for example, wet systems that rely on diffusion become slower as they increase in thickness, and, shape memory alloys become slower as they become thicker as they need to cool between strokes. Despite these limitations, the comparative analyses do serve to illustrate the fundamental differences in performance (which can sometimes differ by orders of magnitude) produced by different actuator materials.

A summary of available performance data is presented in Figure 24.7. A key comparison is with mammalian muscle when considering robotic applications. Hydraulic, pneumatic, and electric motors (voice coil) match muscle in terms of the strains produced and match or exceed muscle in terms of stress generated. Hydraulics, in particular, produce very large stresses, so it is no surprise to see these systems used in large machinery. Of the actuator materials considered, dielectric elastomers exceed muscle in terms of strain and stress generated. These systems also operate to high frequencies therefore giving exceptionally high power densities. Hydrogel systems also match muscle in terms of stress and strain, but their current speed of response is much slower than muscle giving very low power densities. All actuator materials exceed muscle in terms of stress generated. For robotic applications, however, the strain, speed, and size are the most important performances.

The most important characteristics in determining the usefulness of an actuator material or system are the free stroke and blocked stress (speed of response is also a very important consideration). All actuator systems tend to fall into one of two categories: high stroke/low stress or low

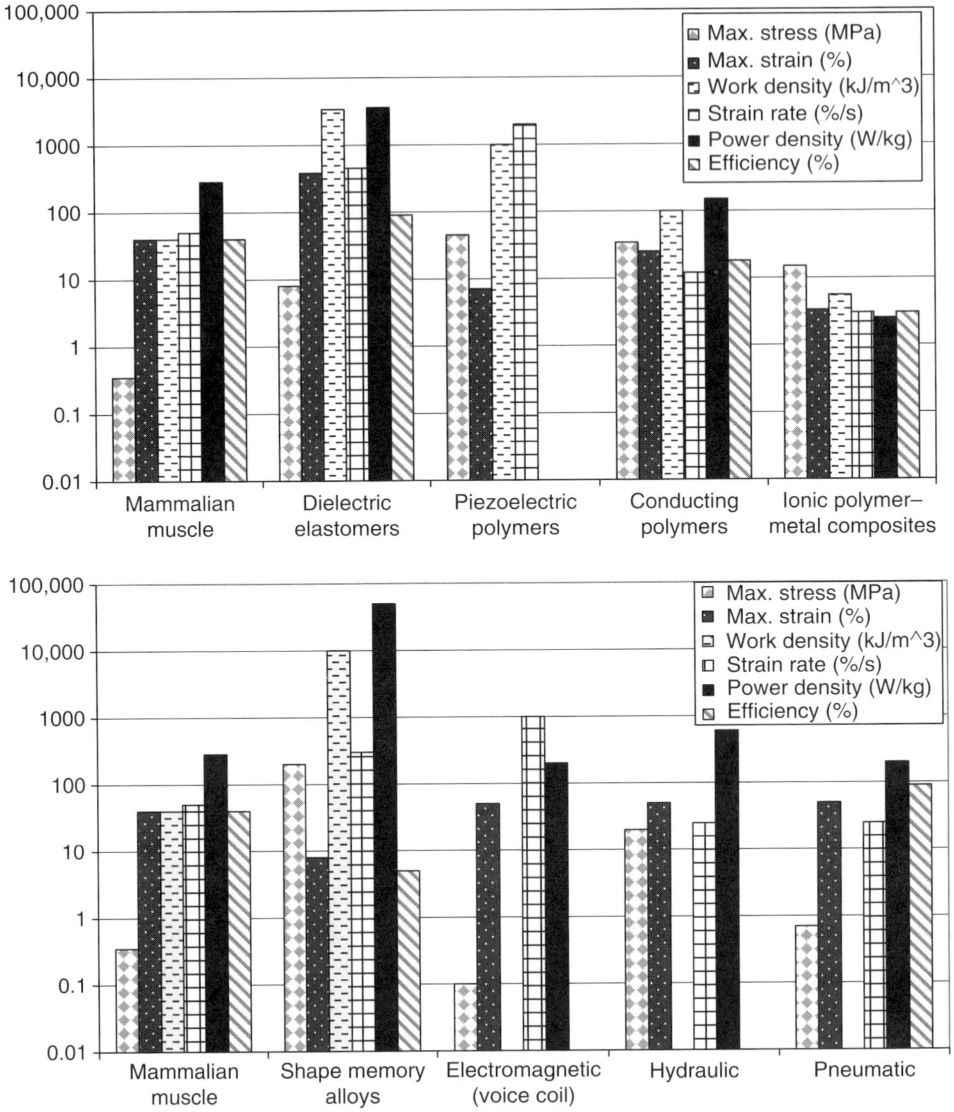

FIGURE 24.7 Comparative performance (current) of various actuators and mechanical drive systems.

stroke/high stress. The distinctions are illustrated in Figure 24.8 with examples of both ionic and electronic polymers given. Here the performance is presented as in Figure 24.2 with actuator strain plotted against the applied stress so that the strain decreases linearly from its maximum value (at zero stress) to zero at the blocking stress. The systems that produce the largest strains at moderate to high stresses are hydraulics and dielectric elastomers. The former is the system of choice for large robotics, while the latter is currently the favored actuator that is developed for a range of robotic-like applications (see Section 24.1.6).

24.1.6 Conventional Mechanical Systems

Conventional mechanical drive systems use engines, motors, hydraulics, and pneumatic systems. These systems are typically built for high force, high speed operation needed in transport (cars,

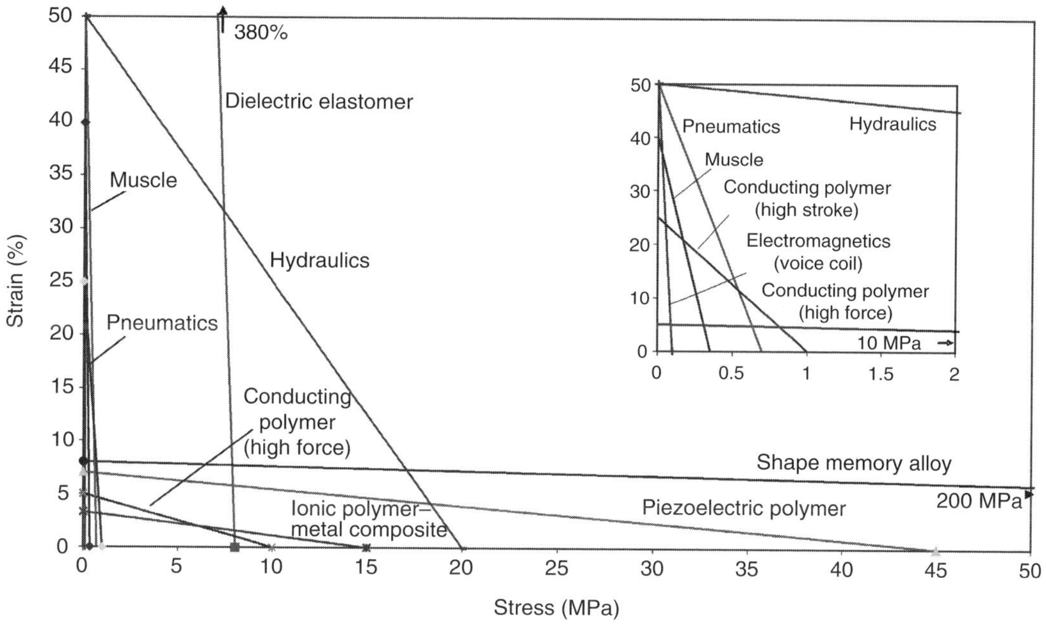

FIGURE 24.8 Strains at different applied stresses for various actuator materials and mechanical drive systems. The strain intercept is the free stroke and the stress intercept is the blocked stress. As described in Figure 24.2, the region below each line represents the region in which the actuator can operate.

rockets, etc.) and heavy lifting equipment (cranes, elevators, etc.). At this scale, nature loses out to human technology: planes fly faster than birds; forklifts are stronger than elephants. The efficiency of these conventional systems is high in most cases (e.g., 90% for electric motors under ideal conditions) so it is unlikely that actuator materials will ever be competitive with large-scale conventional drive systems. To consider the impact that actuator materials may have on current and future mechanical devices, it is necessary to review all the selection criteria listed above. Unfortunately, a full analysis of all mechanical drive systems in all of these categories is not available. A more general analysis, however, highlights the potential application of actuators in small devices generating relatively high force and power at relatively low cost. Clean, noiseless operation is a further potential advantage of actuator technologies.

A 1999 review compared the stress and strain outputs of various actuation systems and actuator materials [4]. The results are reproduced in Figure 24.9. The systems considered were linear drives and excluded rotating machines. From those systems investigated, it can be seen that conventional mechanical drives produce relatively large strains, but also relatively small blocked stress compared with commercial actuator materials (piezoelectrics and shape memory alloys). One exception is the hydraulic actuators that produce both large strains and stresses, which accounts for the preference of hydraulic systems in robotics. The authors note, however, that the extensive ancillary equipment needed for hydraulics (pumps, piping, fluid reservoir) makes them less attractive for small-scale devices [4]. It should be noted that electricity, and hence the requirement for batteries, or electrical connection is also considered ancillary equipment. Figure 24.9 also shows the performance of various actuating materials. These are mainly ceramic piezoelectrics that show high blocked stresses, but quite small strains. The best-performing actuating materials are the shape memory alloys, although they have a substantial restriction in that they operate on heat, and hence the application must allow for heating (normally by direct resistance heating) and cooling of the shape memory alloy. Interestingly, skeletal muscle shows similar performance to solenoids and pneumatics with high strains, but low stresses. Clearly high strains are desirable in animal-like locomotion.

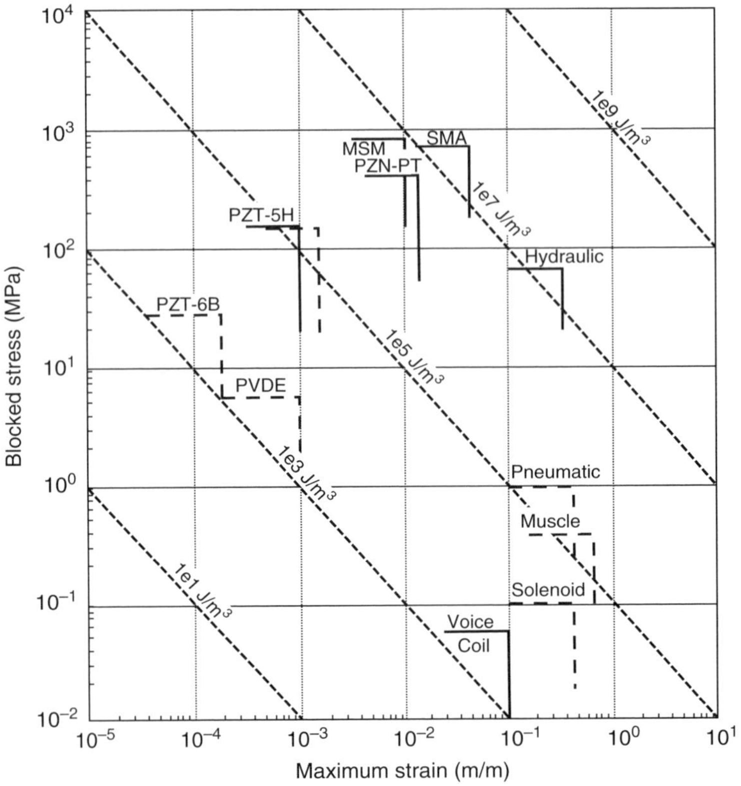

FIGURE 24.9 Comparison of stress and strains generated by various linear actuators. (From Wax, S.G. and Sands, R.R., *Proc. SPIE-Int. Soc. Opt. Eng.*, 3669, 2, 1999. With permission.) The maximum strain and blocking stress are defined in Figure 24.2.

Musculoskeletal systems are characterized by the cooperative action of multiple actuators (muscles), all of which need to be compact in size. Perhaps the most ubiquitous and successful actuation system is natural muscle. Muscles come in different types. In mammals, muscles are either smooth or striated. The smooth muscles control the diameter of veins, arteries, intestines, and so forth. Striated muscles are either cardiac or skeletal, with the latter responsible for the motor action of limbs. The structure of skeletal muscle is hierarchical, with whole muscles consisting of many parallel muscle fibers (each an individual cell); each muscle fiber made of many parallel myofibrils; and each myofibril consisting of many parallel myofilaments containing different proteins. It is the action of the proteins that is primarily responsible for muscle contraction, although the careful nano- to macrostructure of muscles must also contribute to the advanced performance. It is pertinent to note that skeletal muscle is restricted to a hierarchical structure, as they are dependent on diffusion of ions, which become slow as the diameter of the fibrils is increased. The exact mechanism of muscle contraction is still subject to some debate [11] but is known to be triggered by an action potential (nerve pulse) causing the release of Ca^{2+} ions from a network of tubes within muscle (the sacroplasmic reticulum) into the myofilaments. Muscles across all vertebrate species generate the same maximum amount of force per cross-sectional area (0.35 MPa); however, the maximum sustainable stress is about 30% of this peak value (0.1 MPa) [12]. Strains of 50% are possible from muscle, although 10% strains are more typical during muscle action [12]. Work densities are around 0.8 kJ/m^3. Muscle contraction occurs within 1 s, giving average strain rates of 10%–50%/s and typical power densities of 50 W/kg [12].

24.2 CONDUCTING POLYMERS FOR ACTUATORS

24.2.1 Why Choose Conducting Polymers?

When considering the different types of actuating materials available, it will become apparent that although conducting polymers may not exhibit the best combination of strain, strain rate, and stress, they do exhibit substantial operational advantages. Firstly, they do not generate a large amount of heat as do shape memory alloys. Secondly, they do not require high voltages, as do dielectric elastomers, piezoelectric polymers, and piezoelectric ceramics. Thirdly, they can be triggered electrically, whereas ionic gels require a chemical change. Another advantage of conducting polymers is that they are readily deposited electrochemically, allowing them to be easily incorporated onto micro-electro-mechanical system (MEMS) platforms, or they can be coagulated from solution allowing them to be formed into fibers using conventional fiber spinning technology. Recent jumps in the performance of conducting polymers for actuators indicates that they may even rival other technologies with respect to actuation performance [13–16]. A major disadvantage of conducting polymer actuators is that they are normally operated within a liquid electrolyte, and with an auxiliary electrode.

24.2.2 Mechanisms of Actuation in Conducting Polymers

The principal means for generating actuation in conducting polymers is through their electrochemistry [17]. Conducting polymers are readily and reversibly oxidized and reduced without destroying their mechanical integrity [7,18]. Like all electrochemical reactions, the oxidation and reduction of conducting polymers require a counter electrode and an electrolyte. Typically, the conducting polymers and the counter electrode are immersed in a liquid electrolyte (Figure 24.10) and sometimes a reference electrode is used to precisely control the electrochemical potential applied to the polymer.

The most important aspect of the electrochemical reaction for actuation in conducting polymers is the movement of anions (A^-) into and out of the polymer [18,19]. The anions are necessary to

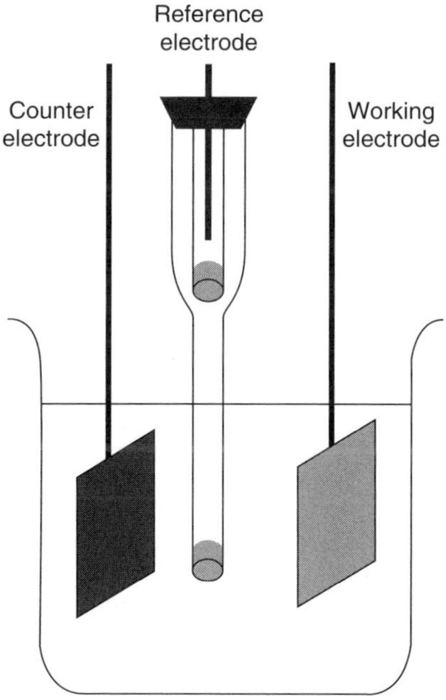

FIGURE 24.10 Electrochemical cell arrangement used for conducting polymer actuators. In some configurations, the counter electrode is also the reference electrode [7]. (From Spinks, G.M., Liu, L., Wallace, G.G., and Zhou, D., *Adv. Funct. Mater.*, 12, 437, 2002. With permission.)

balance the positively charged polymer chains that result from oxidation of the polymer. For electroneutrality, there must be equal numbers of positive and negative charges. The anion is sometimes called the dopant and the process of oxidation is sometimes called doping [20]. This terminology comes from inorganic semiconductors where the incorporation of small quantities of impurities greatly increases the conductivity. Oxidation of conducting polymers also greatly increases their conductivity, hence the analogy with semiconductor doping. Clearly, the movement of anions into and out of the polymer due to redox reactions causes dimensional expansion (swelling) and contraction (de-swelling). This redox reaction can be expressed by

$$P^+(A^-)(\text{solid}) \underset{-e^-(\text{oxidized})}{\overset{+e^-(\text{reduced})}{\rightleftarrows}} P^o(\text{solid}) + A^-(\text{liquid}) \tag{24.4}$$

where

P^+ is the doped (oxidized) state of the polymer
P^o is the undoped (reduced) state

This redox reaction is true for small anions (e.g., perchlorate (ClO_4^-), hexafluorophosphate (PF_6^-), tetrafluoroborate (BF_4^-), chloride (Cl^-)). The polymers expand during oxidation and contract during reduction.

If the anions are bulky or polymeric (e.g., dodecylsulfate, dodecylbenzenesulfate, polystyrenesulfonate, and polyvinylsulfate), the redox reaction becomes

$$P^+(A^-)(\text{solid}) + M^+(\text{liquid}) \underset{-e^-(\text{oxidized})}{\overset{+e^-(\text{reduced})}{\rightleftarrows}} P^o(AM)(\text{solid}) \tag{24.5}$$

where M^+ is the electrolyte cation. As the anion A^- originally incorporated in the polymer backbone during polymerization is immobile in the redox process, the movement of the electrolyte cations dominates actuation. In contrast to the redox process by Equation 24.4, the polymers expand during reduction and contract during oxidation.

For some medium-size anions (e.g., pTS, naphthalene sulfonate), both Equations 24.4 and 24.5 are satisfied. During the actuation process, both anions and cations can move in and out simultaneously [21]. This dual behavior substantially decreases the observed strain. Apart from redox reactions expressed by Equations 24.1 and 24.2, a process dubbed "salt draining" involved the diffusion of an ion pair (pTS^-/Li^+) made of the dopant anion (pTS^-) and the electrolyte cation (Li^+) out of the polymer causing contraction after initial expansion was observed [22]. Unless the use of medium-size dopants is justified for a particular purpose, it is, therefore, customary to incorporate small- or large-size dopants into the conducting polymers actuators to avoid complications such as salt draining.

The effective size of an ion is not trivial, and depends on their association with solvent molecules or other ions in the same solution. For example, for lithium trifluorosulfonimide (LiTFSI) in water or propylene carbonate (PC), the $TFSI^-$ anion is the dominant species during actuation of PPy [16]. However, for ethylmethylimidazolium trifluorosulfonimide (EMI.TFSI), the $TFSI^-$ is dominant in the ionic liquid form [23], but the EMI^+ is dominant for dilute concentrations in PC.

A great variety of performances are possible from conducting polymer actuators as a result of the many operating variables. Three main types of conducting polymers have been investigated as actuators: PPy, polyaniline (PANi), and polythiophene. A virtually unlimited number of dopant anions are then available, with a simplistic view of actuation showing that the amount of actuation strain is proportional to the size of the dopant anion (Figure 24.11) [24]. The movement of ions into and out of the conducting polymer and thus the expansion and contraction is also affected by the electrolyte temperature [25]. The composition of the electrolyte also has a significant bearing on the

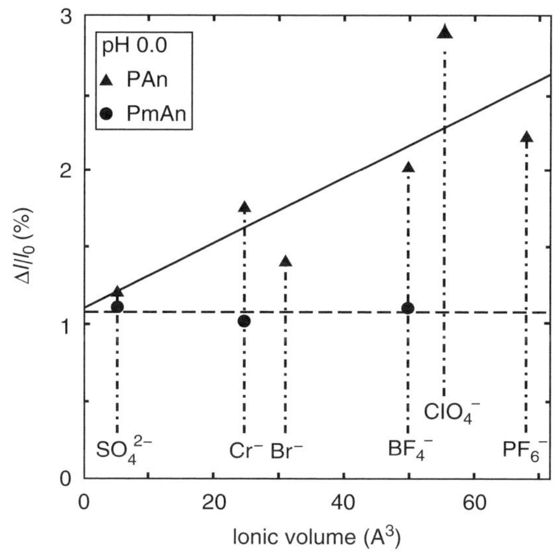

FIGURE 24.11 Actuator strain versus anion size for PANi. (Taken from Kaneto, K. and Kaneko, M., *Appl. Biochem. Biotechnol.*, 96, 13, 2001. With permission.)

actuation performance since the electrolyte is the source of anions/cations and also must conduct electric charge (ionic charge) from anode to cathode [26,27]. The valency of the electrolyte cation exerted a strong influence on the actuation behavior of the material [18].

The processes occurring at the auxiliary electrode are of high importance in achieving long-term actuation. For example, it is not obvious how or what species would oxidize or reduce at a metallic auxiliary electrode to facilitate charge transfer. It was shown that using an auxiliary electrode composed of the same polymer system as the actuator greatly reduced the voltage required in a two electrode system to transfer charge, and also increases the lifetime of the actuator. This is especially pertinent with systems incorporating ion (e.g., PF_6^-) that are known to break down into highly corrosive components [28,29].

For even a constant monomer, solvent, and salt system, the observed actuation is highly dependent on the temperature at which the monomer is polymerized. Many laboratories electropolymerize PPy in PC-based electrolytes at −30°C to −20°C to obtain optimum results. The lower polymerization temperature produces a higher conductivity, and, therefore makes the electrochemistry more efficient.

The force generated by a bending PPy bimorph structure has been observed to increase with the size of the anion dopant [30]. That observation is consistent with the trend shown in Figure 24.11, since the force is proportionally related to the actuation strain. PPy films doped with a small-size ion (ClO_4^-) follow simple actuation as expressed by Equation 24.4 generating a low force. When the dopant is relatively large (sodium dodecylbenzene sulfonate (NaDBS), sodium 4(2dodecyl) benzene sulfonate (Na4(2D)BS), sodium 4(6dodecyl) benzene sulfonate (Na(4(6D)BS)), the dopant becomes immobile and the actuation mechanism is expressed by Equation 24.5 where the movement of electrolyte cations is dominant generating a large force. The magnitude of force is mainly due to the solvation of the ions in the redox process. The anion is not an effectively solvated species [31] whereas the cations can be solvated with a high proportion of solvent molecules [32]. As more solvation occurs in smaller cations, the generated force of PPy doped with immobile dopant is in the order $Li^+ > Na^+ > K^+ > Rb^+ > Cs^+$.

The relation between the actuation strain of PPy doped with *p*-phenolsulfonic acid and the size of electrolyte anion or cation is almost linear [34]. However, the actuation strain of PANi exhibits a more complex pattern. The strain of PANi doped with 2-acrylamido-2-methyl-1-propanesulfonic acid in aqueous acid solutions changes significantly but nonlinearly with the size of electrolyte anions [35]. For anions smaller than bromide (Br^-) (i.e., $Br^- > Cl^- > Fl^-$), the strain increases with increasing anion size. Once the anion is larger than Br^- (i.e., $BF_4^- > ClO_4^- >$ carbontrifluorosulfonic acid ($CF_3SO_3^-$)), the strain decreases with increasing anion size. As the anion becomes larger, the insertion of the anion into the polymer backbone becomes difficult. Quantitatively speaking, if the elongation/charge ratio, i.e., the contribution of a unit charge to sample elongation, is used, for anions larger than Br^-, the ratio consistently decreased with increasing anion size. The actuation behavior of PANi doped with hydrochloric acid (HCl) exhibited another set of data where the actuation behavior was controlled by diffusion of electrolyte anion, which was in turn dependent on the mass of solvated electrolyte anions [24]. The strain was found to be in the order of $ClO_4^- \approx BF_4^- >$ sulfate (SO_4^{2-}) $\approx Cl^-$. Obviously, it is inappropriate to directly compare different experimental results of nominally same material from different laboratories without taking into account different polymer structures (e.g., crosslinks, crystallinity, porosity) due to different polymerization conditions, the dopants, pH, electrolytes, type of stimulus (current or voltage) [36], and scanning rate.

The electrolyte exhibits profound influence on the actuation behavior. Ionic liquids, that is, pure salts that have a melting temperature lower than 100°C, exhibit a stable actuation due to a wide potential window compared with normal solvent electrolytes where the breakdown of polymer and electrochemical decomposition of the solvent occur [36,37]. After more than 6000 cycles, the actuation strain of a PPy actuator maintains a value close to the original strain in ionic liquid [23,37]. Suitable choice of dopant (TFSI or perfluoroalkylsulfonyl imide) and electrolyte provides very large actuation strain [15,38]. In water/PC (60/40) solution of LiTFSI, a PPy/TFSI actuator exhibits 29% actuation strain and a response rate of 10.8%/s [16]. The strain reaches a massive value of 40% for a PPy/(nonafluorobutylsulfonyl)imide [38]. The exact actuation mechanism for these materials of large generated strain is not fully understood but three conditions are noticed: (1) the material is porous becoming gel-like in the electrolyte; (2) the electrolyte anion and the dopant are identical and bulky; and (3) the water/PC (60/40) mixture facilitates the swelling of the PPy and at the same time maintains high ionic conductivity.

The swelling is most probably augmented by the concomitant movement of solvent from the electrolyte into the polymer due to osmotic pressure [39]. This effect facilitates the penetration of solvent molecules into the polymer in an amount far exceeding those bound to solvated moving ions. The contribution from osmotic pressure to actuation is maximized when the ionic concentration within the polymer is much larger than the ionic concentration in the electrolyte during the oxidized state, but is similar to the electrolyte in the reduced state. The presence of osmotic pressure is readily demonstrated by observing larger actuator strains for lower salt concentrations in the electrolyte.

Some researchers also speculate that the oxidation causes changes in the basic polymer chain shape (conformation) that also contribute to the actuation strain [40]. For PPy and PANi, the contribution of conformational rearrangements to the volume change during redox processes is small compared with that from ion intercalation. In one example, thiophene-based polymers can be designed to provide actuation strains in the order of 20% due to accordion-like conformational changes [41].

Although an ideal actuator would exhibit high generated stress, high generated strain, high speed of response, actuation stability, good mechanical properties, and long service life, all of these properties may not be essential for practical applications. Nevertheless, a full understanding of the actuation mechanisms is important in designing a material for particular applications. The interrelationship of the electrical, mechanical, and chemical properties of the material foreshadows the performance of the actuator (Figure 24.12). Therefore, the intrinsic properties of the material such as crosslinks, crystallinity, molecular alignment, porosity resulting from polymerization conditions, and type of anions should work synergistically with extrinsic conditions such as type of electrolytes, concentration, operating conditions (pH, temperature), and electrical stimuli (current control, voltage control, scan rate, stimulation waveform) to achieve desirable performances.

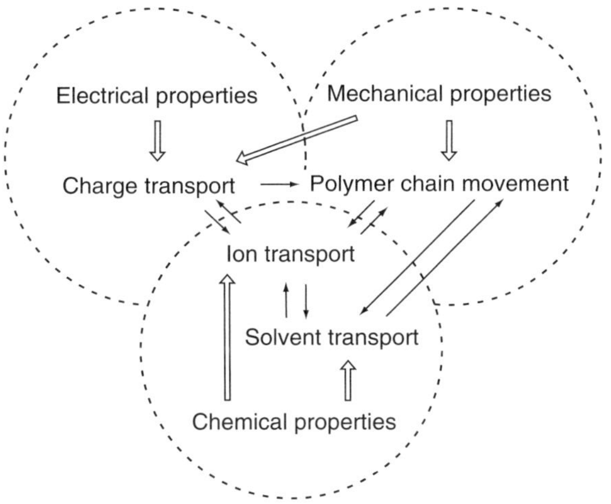

FIGURE 24.12 The interrelationship of electrical, mechanical, and chemical properties of conducting polymers [5]. (From Smela, E., *Adv. Mater.*, 15, 481, 2003. With permission.)

24.2.3 Modulus Shift in Conducting Polymer Actuators

Experimental studies have shown that the modulus of conducting polymers varies considerably during electrochemical stimulus and that these changes affect the actuation performance, as explained above. In Figure 24.4, it was demonstrated that the actuation strain produced by PPy films decreased as the applied load increased. These tests were conducted under isotonic conditions. The solid lines drawn in Figure 24.4 are calculated results obtained from Equation 24.3 and using measured modulus values for PPy in the oxidized and reduced states. The excellent agreement between the calculated and measured actuation strains show that the effect of modulus change during actuation can be modeled using simple linear elastic material behavior (Hooke's law).

The reasons why the modulus of conducting polymers change during redox cycling is not fully understood, but experimental studies have shown that the modulus shift is very sensitive to the operating electrolyte. In one recent study of polythiophene actuators [10], the change in modulus and actuation performance was determined in an organic electrolyte (tetrabutyl ammonium hexafluorophosphate, $TBA \cdot PF_6$ in PC), and an ionic liquid (ethylmethylimidazolium bistrifluoromethanesulfonimide). The actuation strains measured under isotonic conditions are shown in Figure 24.13. The typical response was observed in the PC electrolyte with the strain decreasing when higher loads were used. These results were explained by the simultaneous increase in modulus occurring during electrochemically induced expansion of the polymer. In contrast, the strain actually increased slightly with increasing load when the ionic liquid electrolyte was used. In this case, the swelling of the polymer coincided with a decrease in modulus. Thus, the sample swelled both due to the electrochemical processes and also due to the change in modulus. This situation is highly desirable, since the performance (and output energy) increases as the force applied to the actuator increases.

24.2.4 Designing Conducting Polymer Actuators

When designing a conducting polymer actuator, the charge delivery system, thickness, auxiliary electrode, and prevention of mechanical overloading are all important. Sometimes these requirements are conflicting, so that new designs are needed to optimize performance.

Conducting polymers are not highly conductive (conductivity is typically in the range of 10–1000 S/cm). Hence, a macroscopic actuator made from conducting polymer alone would be resistive, and the applied potential will decrease with distance from the metal connection. The voltage

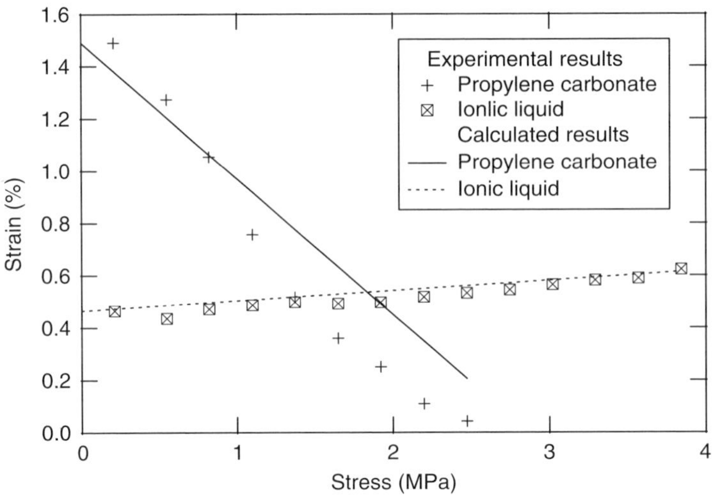

FIGURE 24.13 Actuation strain of a poly(3-methylthiophene) actuator doped with PF_6^- tested under isotonic conditions at different stress levels in both PC and ionic liquid electrolytes. Potential scanned between −1.0 to 1.5 V (vs Ag/Ag^+). The calculated strains obtained from Young's modulus modeling (Equation 24.3) are shown as dashed lines. (From Xi, B., Truong, V.-T., Whitten, P.G., Ding, J., Spinks, G.M., and Wallace, G.G., *Polymer*, 47, 7720, 2006. With permission.)

drop becomes substantial for lengths greater than 0.5 mm. Consequently, much research has gone into the development of charge delivery systems throughout the length of the actuator. Typically, metal film or wire, or a metal-coated polymer is used as an electrode surface that delivers charge along the length of an actuator. One needs to be careful that the metal is not too thick to prevent bending. For an axial actuator, novel helical springs have proven very successful at delivering charge along the length of an actuator without hindering actuation, as shown in Figure 24.14 [42]. Actuator construction using a helical spring can be cumbersome, with a central core that needs to be removed, or a closely packed spring [43] being two options that allow actuators of length of several tens of millimeters to be constructed. An ideal system would consist of a highly conductive fiber of very low modulus, onto which a conducting polymer could be coated and the substrate fiber need not be removed. Such a fiber is not currently available.

The thickness of an actuator is also very important. Thicker actuators are able to support higher loads, but they also become slower, as the diffusion time becomes large. Most actuators in the literature have a thickness of 1–15 μm.

The auxiliary electrode is another crucial part of an actuator system. The surface area of the auxiliary electrode should be several times larger than that of the actuator, and it needs to facilitate charge transfer to prolong the lifetime of the actuator, and reduce the required potential and hopefully prevent over-oxidation.

Finally, robustness of the actuator is also critical. Often, thin film/fiber-conducting polymer actuators are very fragile, so caution is needed when handling. Further, caution needs to be taken to not plastically deform the conducting polymer actuator during mounting.

24.2.5 Benders versus Axial Actuators

The most basic distinction between different types of polymer actuator comes from the nature of the deformation produced by the external stimulus. In some cases, the volume change is amplified producing a large shape change like a bending motion (benders) while in others a volume change occurs where the deformation in one direction is usually used for mechanical work (axial actuators).

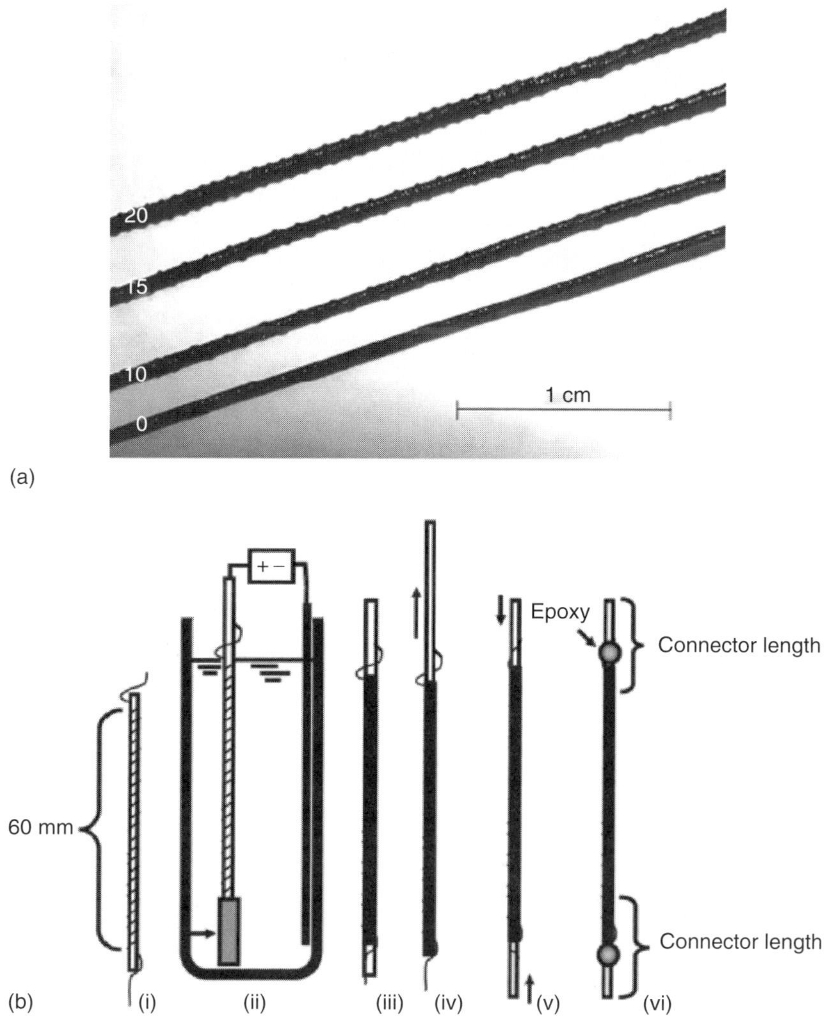

FIGURE 24.14 A helical wound platinum wire is used to minimize resistive losses. Photographs (a) of hollow polymer tubes with helical interconnect used for actuation testing. The pitch of the helix can be altered with examples shown of 20, 15, and 10 turns cm^{-1}. A hollow tube with no helix (0 turns cm^{-1}) is also shown. Schematic diagram (b) showing method of construction for these actuators: (i) 25 μm of platinum wire is wrapped around the 125 μm wire as a spiral; (ii) polymer synthesis—the assembly is placed in polymer electrolyte solution (0.5 M PPy, 0.25 M TBA PF_6 in PC) and electroplated for 24 h at −28°C; (iii) polymer coating forms around wire and spiral; (iv) 125 μm centre wire is withdrawn from the polymer tube/helix; (v) two short connectors of 125 μm wire are inserted into each end; (vi) 25 μm wire is pulled tight around these ends for a good electrical connection and epoxy glued to hold in place. (Taken from Ding, J., Liu, L., Spinks, G.M., Zhou, D., Wallace, G.G., and Gillespie, J., *Synth. Met.*, 138, 391, 2003. With permission.)

Volume-change actuators can be constructed as benders by laminating the actuator material against a flexible, passive support (Figure 24.15). The system then works like a bimetallic strip with the volume change of the actuator resisted at the interface with the support but free to occur at the open surface. The result is a bending motion. The degree of bending depends strongly on the stiffness (Young's modulus and thickness) of the support material—the stiffer the support, the smaller the bending motion. Small strains can produce quite large bending displacements. From beam mechanics, the bending curvature, κ, can be determined from

(A) Conducting polymer on top of a metal substrate.

(B) Conductive polymer in A has expanded, but is not bonded to the substrate.

(C) Conductive polymer in A has expanded, and bends as a result due to being bound to the substrate.

(D) Calculation of the bending curvature.

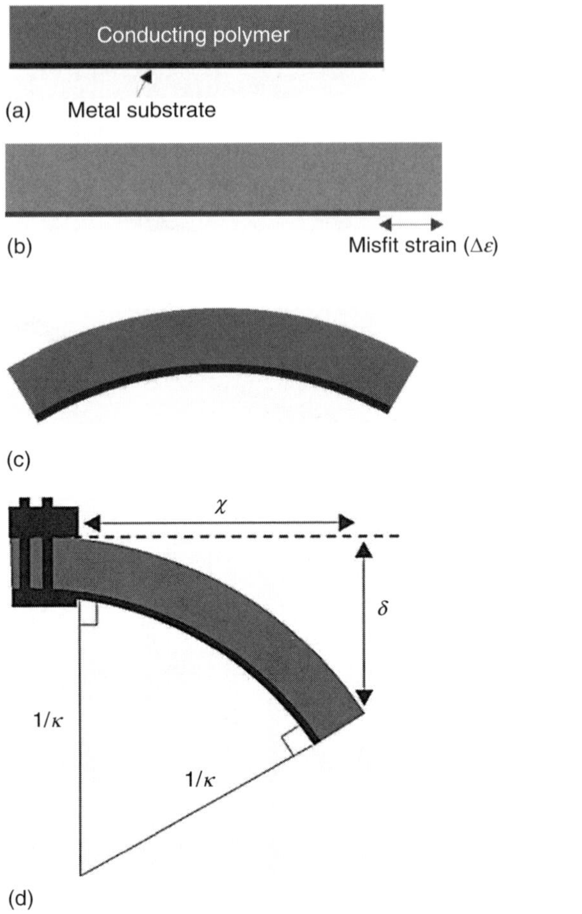

FIGURE 24.15 Volume-change actuators can be used to make a bender by bonding onto a thin substrate of higher modulus. The misfit strain shown in (c) is equivalent to the free stroke produced by the actuator.

$$\kappa = \frac{6Y_A Y_S (h_A + h_S) h_A h_S \varepsilon_0}{Y_A^2 h_A^4 + 4Y_A Y_S h_A^3 h_S + 6Y_A Y_S h_A^2 h_S^2 + 4Y_A Y_S h_A h_S^3 + Y_S^2 h_S^4}$$

and

$$\kappa = \frac{2\sin\left[\tan^{-1}\left(\delta/x\right)\right]}{\sqrt{\left(\delta^2 + x^2\right)}}$$

where

h_A and h_S are the thicknesses of the actuator and the substrate
Y_A and Y_S are the Young's moduli of the actuator and the substrate
ε_0 is the actuation strain under no load

Note that this analysis ignores any change to the modulus of the actuator during operation.

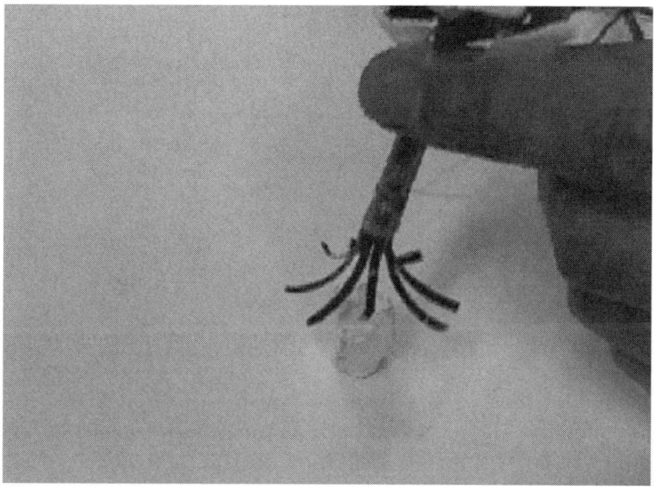

FIGURE 24.16 A "gripper" device constructed from multiple trilayer PPy benders.

Both bilayer and trilayer bending actuators have been constructed using conducting polymers. The trilayer is particularly useful since both electrodes are incorporated into one actuating device. These devices have even been fabricated to operate in air by using a porous separator material with the pores filled with nonvolatile liquid or solid electrolytes [44]. An example of an air-operated bending actuator using PPy is shown in Figure 24.16.

24.3 RECENT PERFORMANCE IMPROVEMENTS

Over 2003–2006, the actuation strain produced by conducting polymer actuators has increased dramatically. Reviews in the mid-1990s reported the typical strain to be 4%–5% [45] from conducting polymers. Reports of higher strains began to appear in the last few years. Firstly, large (20%) thickness direction strains were reported [46]. Next, large 12% linear strains were observed in macroscopic films produced by pentanol as a cosolvent during the film formation process [47]. In 2005, Kaneto and coworkers reported strains as high as 26% from PPy prepared with new dopants [13] (we have repeated their preparation and measured 18% strain). The most recent development from Kaneto et al. is the report of 40% strains when PPy is actuated in cosolvent mixtures [16,38].

These extremely large strains exhibited by a conducting polymer are similar to those produced by hydrogels. In fact, the origin of the large strains from PPy may turn out to be resulting from hydrogel-like mechanisms. The modulus of highly swollen PPy is in the 10–100 kPa range, which is similar to many hydrogels and demonstrates that the PPy chains have the freedom to undergo large-scale conformational changes similar to the coil-globule-like transitions in hydrogels. The advantage of PPy-gels compared with hydrogels is that the transition from contracted to expanded states can be stimulated directly by an applied voltage. The electrical potential can cause oxidation (reduction) of the polymer to change its charge state and to effect the transition. In conventional hydrogels, the stimulus is chemical and can be produced only indirectly by electrical means (e.g., through a change in pH).

The large strains produced by conducting polymers will make them more attractive for many applications. However, as these materials resemble hydrogels in their nature they also suffer the same limitations of slow response times and low stress generation. Improvements in the force generated at faster response times can be achieved by bundling many individual actuators so that they operate in parallel (Figure 24.17). Although there are challenges with bundling thin fibers, the force

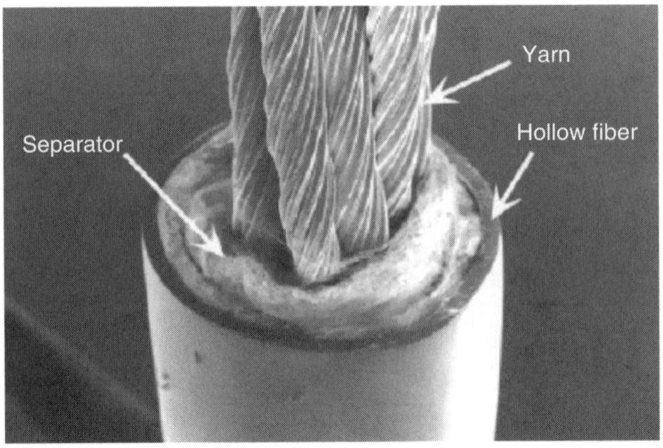

FIGURE 24.17 PANi fibers for high speed actuators, produced by Santa Fe Science and Technology [50], and presented in Ref. [49]. The diameter of the hollow fiber is 1.5 mm. (From Lu, W., Norris, I.D., and Mattes, B.R., *Aust. J. Chem.*, 58, 263, 2005. With permission.)

generated scales with the number of actuators included in the bundle [48]. It has been shown that twisting a bundle of fibers can increase the blocking strength of the resulting actuator [49].

24.4 APPLICATIONS FOR CONDUCTING POLYMER ACTUATORS

Polymer actuators are not yet competitive with electric motors, pneumatics, and hydraulic systems in conventional applications that involve high forces. However, there are many other applications where polymer actuators may have a competitive advantage, such as in the following areas:

- Bending actuators employing a bimorph or unimorph construction
- Actuators working through a change in thickness
- Applications requiring a high packing density
- Miniature motors/actuators with simple design/constructions and without any loss in performance
- Clamps, change in tube diameter

There are countless applications that can take advantage of these strengths, of which several are described below. Excitingly, polymer actuators are just becoming commercially viable; there are now several private enterprises that have a major emphasis on conducting polymer actuators, as listed in Table 24.3.

24.4.1 Bending Actuators

As with all solid-state actuators, a unimorph or bimorph construction results in bending. The construction amplifies a small amount of strain making it significant; however, the resulting force is substantially reduced. The bending action makes observers think of human movement, with robotic hands or grabbers being made by many groups around the world (Figure 24.16).

A major problem with bending is that although the strain is amplified, the force is reduced (compared to a linear actuator), and with current technologies is insufficient for applying appreciable force. High frequency (1 Hz) bending actuation was employed by workers at Eamex Corporation in Japan to simulate fish fins, and hence propel robotic fish. Jager et al. have successfully used the bending action to open and close the lids on microvials, making prototypes for the controlled release of contents or

TABLE 24.3
Companies That Are Developing Conducting Polymer Actuators

Company	Address	Products
EAMEX Corporation	3–9–30 Tarumi-cho, Suita, Osaka 564-0062 Japan http://www.eamex.co.jp/index_e.html	Robotic hand Robotic fish
Micromuscle AB	Mjärdevi Center, Teknikringen 10 583 30 Linköping, Sweden	Biomedical products
Santa Fe Science and Technology	3216 Richard Lane, Santa Fe, NM 87505, USA	Conducting polymer fibers
Quantum Technologies	5 South Street, Rydalmere, N.S.W. 2126, Australia www.quantech.com.au	Electronic Braille screen (under development)

cell clinics [51], while Andrews et al. [52] have employed a similar action to open a gas valve on a fruit container. Madden et al. have employed the bending action to change the camber of a submarine propeller [12]; however, the forces generated are not yet sufficient for widespread application. Perhaps the most novel application is the bending of a cochlear ear implant allowing easier insertion [53].

24.4.2 Actuators for High Packing Density Applications

Most actuator systems cannot be packed densely in two-dimensional arrays, while still being capable of operated separately. Conducting polymer actuators can be operated in close proximity to each other, and have been used to construct a prototype of a Braille screen for blind people [54] (Figure 24.18).

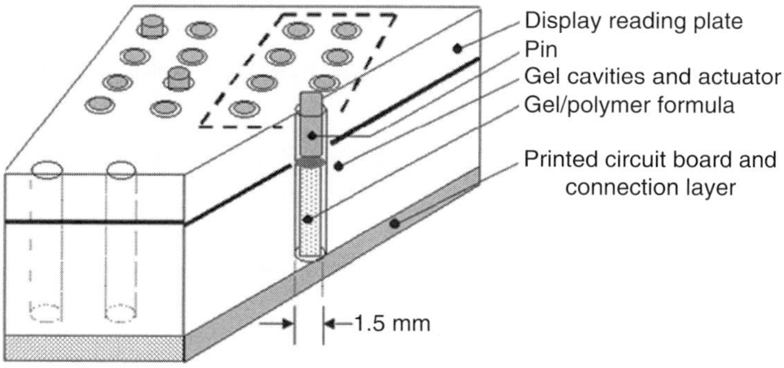

Each pin is

- Individually addressed moves ~0.5 mm in 0.2 s
- (1% strain and 5%/s strain rate)
- Sits on a spring to provide tactile resistance
- (10 g_f depresses pin <0.1 mm, requiring a spring constant of 1000 N/m and a maximum actuation force of >0.5 N or >10 MPa stress for an actuator diameter of 0.05 mm^2)
- Lasts >10^6 cycles

FIGURE 24.18 Design and specifications for an electronic Braille screen that is being developed by Quantum Technology Pty Ltd. (From Quantum Technology Pty Ltd.)

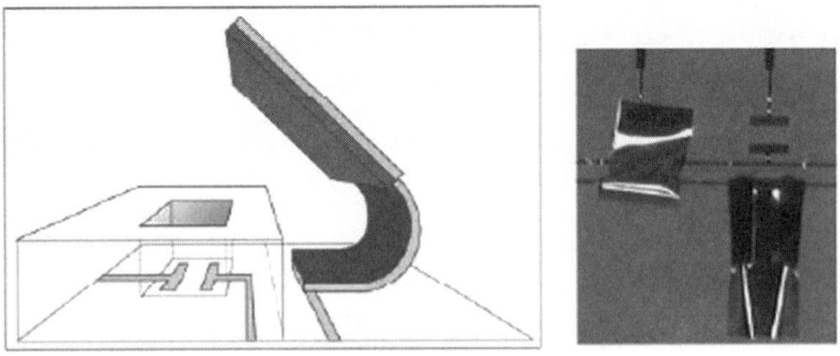

(a) (b)

FIGURE 24.19 Micron sized hinge flap is one example of a polymer actuator—operated device suitable for biological applications. (From Jager, E.W., Smela, E., and Inganas, O., *Science*, 290, 1540, 2000. With permission.)

As described above, one of the main driving forces for the development of polymer actuators is the desire to produce human-like movement in robots and other devices. Polymeric actuators are often labeled artificial muscles, the reason being is that they exhibit a combination of stress generation, at sufficient strains and appropriate weights comparable to muscle. With improvements to their properties, they may one day become genuine artificial muscles. More immediate goals with some prototypes already being constructed are rehabilitation devices (e.g., assist a human hand to move) [56], humanoid robots [57], or tremor suppression [58]. Similar interests include the development of robots that mimic the fish swimming, snakes slithering, and insects flying.

24.4.3 Actuators for Micro-Devices

Conducting polymer actuators are able to be easily scaled down without loss of performance. While the current force/strain actuation is not yet sufficient for many macroscopic applications, it is enough for many micro-applications. Obvious applications include micro-actuators for positioning lenses in cameras and microscopes. Impressive demonstrations of microsized robot manipulators have been demonstrated using unimorphs and bimorphs (Figure 24.19); however, take-up for real applications has been slow [51,59–61]. A major reason for the slow take-up is that these actuators are original and do not replace an existing actuator.

24.4.4 Actuation of Cylinder or Tube Diameter

Polymer actuators have been produced in tube or cylinder geometries, with the consequence being actuation of the corresponding diameter. This approach can be used to make reversible clamps, valves, or fluid pumps. Micromuscle AB has developed a clamp as shown in Figure 24.20 consisting of a hollow polymer actuator cylinder (~1 mm in diameter, 3 mm in length) that can be used to reconnect divided blood vessels. The cylinder's diameter will expand or contract at small electric potentials. Another application is for valve for a fluid or gas chamber, or a pump for a fluid (Figures 24.21 and 24.22) [62,64].

24.5 CONCLUDING COMMENTS

This review of conducting polymer actuator technology highlights that the technology is mainly fixed in the research and development stage with only a small number of commercial activities at the present time. The growth cycle, however, points to increasing commercial activity and a rapid growth

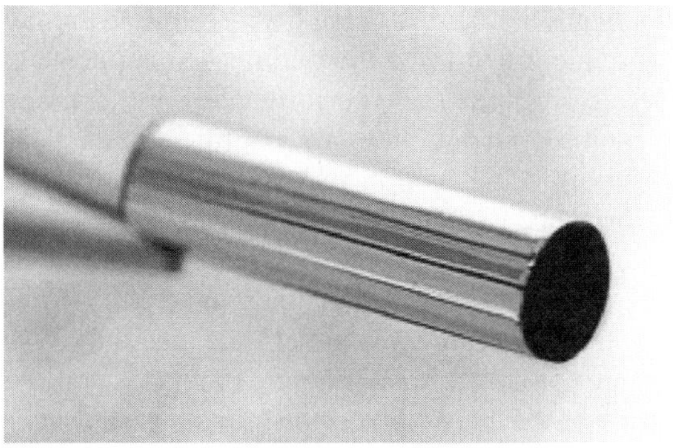

FIGURE 24.20 The connector for blood vessels, as developed by Micromuscle AB [63]. (From Micromuscle AB. With permission.)

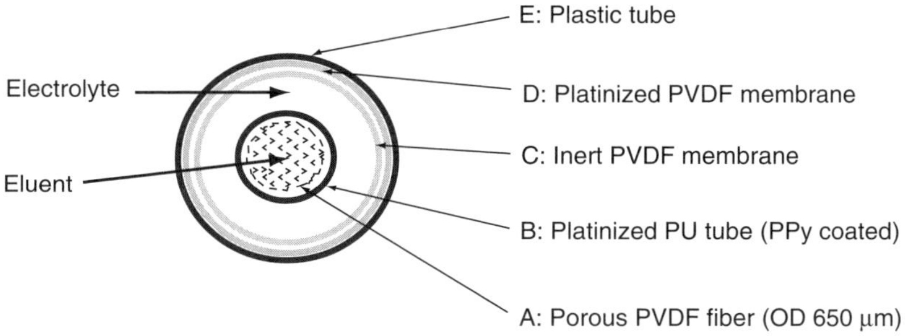

FIGURE 24.21 Schematic diagram of a tube-in-tube-actuator-nodule (TITAN) pump system [64]: A: porous PVDF fiber (outside diameter (OD): 650 μm) used to maintain the micropump cylindrical shape, B: platinized PU tube (OD: 1050 μm, inside diameter (ID): 950 μm) wrapped with Φ 50 μm platinum wire and coated with PPy, used as TITAN working electrode, C: inert PVDF membrane, used as an inert electrochemical cell separator and to hold the 0.25 M $TBA\dot{C}PF_6$/PC supporting electrolyte, D: platinized PVDF membrane coated with PPy connected via stainless steel mesh, used as TITAN auxiliary electrode, and E: plastic tube (30 mm long, OD: 5 mm, ID: 4 mm) used to pack the electrode assembly. (From Wu, Y., Zhou, D., Spinks, G.M., Innis, P.C., Megill, W.M., and Wallace, G.G., *Smart Mater. Struct.*, 14, 1511, 2005. With permission.)

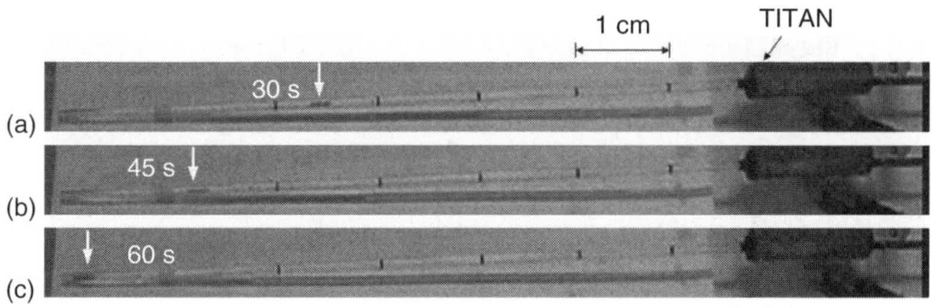

FIGURE 24.22 Sequence of video frames showing the displacement of dyed plug through an open-end glass capillary. (a) +1.0 V applied (0–30 s), TITAN working electrode expands and plug moves towards the pump, (b) voltage switched to −1.0 V (30–60 s), TITAN working electrode contracts and plug moves away from the pump, and (c) displacement reached after −1.0 V applied for 30 s.

in applications. The main application areas are those that can utilize the favorable characteristics of polymer actuators, which were identified to be their compact size, noiseless operation, and simple construction. In some cases (e.g., robotics), the performances of the various polymer actuators are not yet suitable. However, clear research directions are identified that will lead to continued improvement in performance. It is expected that within the next 5–10 years the global R&D effort will enable polymer actuator performance to match natural muscle in all aspects.

Developments of various technologies ensure that polymer actuators are likely to be rapidly adopted as their performances reach the required targets. Humanoid robots, MEMS, and smart structures are the three developing technologies that require improved actuator materials. Giant Japanese corporations such as Honda and Panasonic are committed to developing robots for domestic help, especially for the elderly. With aging populations in the Western societies looming as a huge economic and political challenge, the home-help robots may offer a technological solution. As illustrated in this chapter, the ability to produce true human-like movement and dexterity requires smaller actuators. Although not yet meeting the performance required (as illustrated by natural muscle), the steady rate of improvement in actuator technology will likely see suitable materials become available in the next 5–10 years.

The MEMS industry is currently experiencing rapid growth with an estimated $82 billion in revenue in the year 2000. Most MEMS devices are currently based on micro-machined silicon with actuation provided by electrostatic systems or thermal means. These systems are limited by their very low forces produced and low response times. Polymer actuators can replace these traditional MEMS actuators and overcome their limitations. Prototype MEMS actuators have already been built using conducting polymer materials. These prototypes demonstrate the compatibility of polymer actuators with MEMS processing and illustrate the suitability of polymer actuators in down-sized applications. The continued increase in MEMS applications, along with allied fields such as microfluidics (including lab-on-a-chip analysis systems), means that the demand for miniature actuators will continue to rise. Further development in polymer processing, control systems, and materials developments (especially to increase speed of response) will meet the increasing demands over the next 10 years and beyond.

More futuristic applications for polymer actuators come from the smart materials/smart structures field. Here, the desire is to build structures incorporating sensors and adaptive actuators. Mechanical actuators will be used for vibration dampening, compliance shifting, and morphing structures. Already prototype vibration dampening systems are available using ceramic piezoelectrics, where small strains, high stresses, and rapid response are provided. Compliance shifting structures may be required when a change in stiffness is desired under changing load conditions. Here polymer actuators may be useful since they can be readily fabricated into fibers. When used as reinforcing fibers, the fiber modulus (and hence stiffness of the composite) can be changed through the polymer actuation process. Similarly, morphing structures are likely to be built from composites where the dispersed reinforcing phase can be actuated producing stresses that alter the overall shape of the composite. The performance demands of the actuators in these applications are extreme—large strains and large stresses are needed. Actuator materials suitable for such applications are not expected for 10 years or more. The current best-performing actuators for such applications are shape memory alloys—giving up to 8% strain and 200 MPa stress. Polymers are unlikely to produce such high stresses but through continued research and development may offer a better mix of high strains (e.g., up to 20%) with moderate stress capabilities (50 MPa).

EXERCISE QUESTIONS

1. PANi fiber of diameter 100 μm and initial length 15 mm produces a contractile actuation of 220 μm when no external force is applied. The Young's modulus of PANi when fully equilibrated with the electrolyte is 65 MPa. Assuming that the modulus does not change during actuation, answer the following questions:
   a. Determine the actuation strain.
   b. Calculate the blocking force.

c. If a longer length of PANi is used and the actuation strain is unchanged, will the blocking force be different? Explain why or why not.
d. How many PANi fibers are needed to generate a force of 1 N?

2. An actuator film that operates in contraction is fixed at the top and a weight is attached to the bottom. The weight is initially supported by a bench. The size of the contraction (Δl_f) is measured when a voltage is applied. The weight is then changed and the experiment repeated. The measurements are given in the table below. The initial cross-sectional area of the actuator is 0.25 mm^2 and the initial length is 75 mm.
a. Plot the contractile displacement versus the force applied ($f = mg$ where g is 9.8 m/s^2).
b. From the graph calculate (i) the free displacement, (ii) the blocking force, and (iii) the Young's modulus of the actuator.
c. How large will the contraction be if a weight of 11 g was attached to the film?

m (g)	Δl_f (mm)
2	−4.5
4	−4
6	−3.5
8	−2.9
10	−2.55
12	−2
14	−1.6
16	−1.1
18	−0.55
20	−0.05

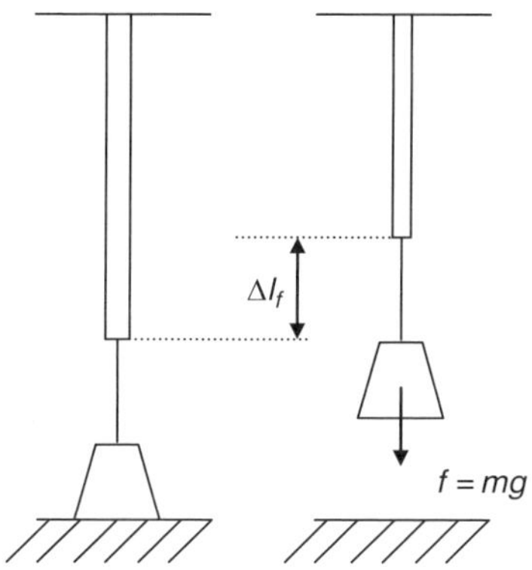

3. PPy film is actuated under isotonic conditions (constant force). The free stroke is 14%, initial cross-sectional area is 0.5 mm^2, and initial length is 20 mm. For this actuator, the modulus in the expanded state is 100 MPa and in the contracted state is 50 MPa. Calculate the size of the contraction (in mm) produced when an external force of 1.6 N is applied.
4. Rectangular film actuator is operated such that it produces a length and width increase of 5% and an increase in thickness of 20%. What is the percent change in volume?
5. PPy film 25 μm thick (h_A) is laminated to a 125 μm thick (h_S) polyester substrate. When the PPy is actuated, it produces a bending movement in the laminated structure. The curvature of bending (κ) is given by the following equation:

$$\kappa = \frac{6Y_A Y_S (h_A + h_S) h_A h_S \varepsilon_0}{Y_A^2 h_A^4 + 4Y_A Y_S h_A^3 h_S + 6Y_A Y_S h_A^2 h_S^2 + 4Y_A Y_S h_A h_S^3 + Y_S^2 h_S^4}$$

where

Y_A and Y_S are the Young's moduli of the actuator and the substrate, respectively
ε_0 is the strain under no load

If the strain produced by the PPy is 0.05 and the modulus of the PPy (Y_A) is 100 MPa and the modulus of the polyester (Y_s) is 2500 MPa, determine the tip displacement (δ) at a distance (x) 15 mm from the actuator base.

The relationship between the curvature and bending displacement is given by the following equation (also refer to Figure 24.15):

$$\kappa = \frac{2\sin\left[\tan^{-1}\left(\delta/x\right)\right]}{\sqrt{\left(x^2+\delta^2\right)}}$$

6. Figure below shows the actuation strain obtained from a PPy sample doped with *p*-toluene sulfonate (pTS) and operated in aqueous sodium pTS electrolyte. Write down the equations for the electrochemical reactions occurring at each of the points labeled A, B, C, and D in the figure. Be sure to identify the dominant ion species in each reaction.
7. PPy sample having dimensions 12 × 1 × 0.02 mm is prepared in the fully reduced state and then electrochemically partially oxidized by applying a constant current of 2 mA for 30 s. An electrolyte of 1 M NaCl is used. The resulting change in volume is +5%. Assuming that there are no ions present in the fully reduced polymer and that the entire volume expansion is due only to the incorporation of anions (no solvent), answer the following questions:
   a. Calculate the number of Cl^- ions needed to balance the positive charge of the partially oxidized polymer.
   b. Determine the % change in concentration of the surrounding electrolyte if the original electrolyte volume was 5 ml (assume no water molecules enter the polymer).
   c. Calculate the average diameter of the Cl^- ions inside the partially oxidized polymer.
   d. Compare your answer in (c) to published values of the diameter of Cl^- ions in water.

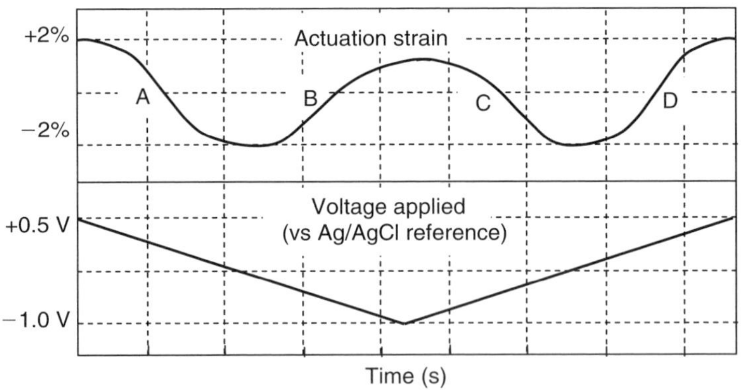

8. Polythiophene is completely reduced from the fully oxidized state in an electrolyte of tetrabutyl ammonium (TBA^+) and hexafluorophosphate (PF_6^-). If the solvated volume of the PF_6^- ion is 0.07 nm^3, calculate the change in length of a polythiophene sample that is initially 20 × 2 × 0.05 mm in size (fully oxidized) and having a density of 1.2 g/cm^3. Assume that only anions are involved in the actuation process and every third thiophene unit is positively charged in the fully oxidized state. Assume also that the fractional changes in dimension are the same in the length, width, and thickness directions.

LIST OF ABBREVIATIONS

PANi	polyaniline
PPy	polypyrrole
PC	propylene carbonate
Cl^-	chloride
PF_6^-	hexafluorophosphate
Br^-	bromide
$TFSI^-$	trifluorosulfonimide
BF_4^-	tetrafluoroborate
ClO_4^-	perchlorate
TBA^+	tetrabutylammonium
EMI^+	ethylmethylimidazolium
Li^+	lithium

REFERENCES

1. *The Australian Pocket Oxford Dictionary*. 5th ed. 2003, Oxford: Oxford University Press.
2. shadowrobot.dnsalias.org.
3. http://www.eamex.co.jp/index_e.html.
4. Wax, S.G. and R.R. Sands, Electroactive polymer actuators and devices. *Proceedings of SPIE—The International Society for Optical Engineering*, 1999. **3669**(Electroactive Polymer Actuators and Devices): 2–10.
5. Smela, E., Conjugated polymer actuators for biomedical applications. *Advanced Materials (Weinheim, Germany)*, 2003. **15**(6): 481–494.
6. Kornbluh, R., R. Pelrine, Q. Pei, and V. Shastri, Applications of dielectric elastomer EAP actuators, in *Electroactive Polymer (EAP) Actuators as Artificial Muscles. Reality, Potentials and Challenges*, Y. Bar-Cohen, Ed., 1999, SPIE Press: Bellingham, Washington. pp. 457–495.
7. Spinks, G.M., L. Liu, G.G. Wallace, and D. Zhou, Strain response from polypyrrole actuators under load. *Advanced Functional Materials*, 2002. **12**(6–7): 437–440.
8. Della Santa, A., D. De Rossi, and A. Mazzoldi, Characterization and modeling of a conducting polymer muscle-like linear actuator. *Smart Materials and Structures*, 1997. **6**(1): 23–34.
9. Spinks, G.M. and V.-T. Truong, Work-per-cycle analysis for electromechanical actuators. *Sensors and Actuators, A: Physical*, 2005. **A119**: 455–461.
10. Xi, B., V.-T. Truong, P.G. Whitten, J. Ding, G.M. Spinks, and G.G. Wallace, Increased strain and work-per-cycle at increasing loads in poly(3-methylthiophene) based electrochemical actuators operated in ionic liquid. *Polymer*, 2006. **47**: 7720–7725.
11. Pollack, G.H., Polymer-gel phase-transition as the mechanism of muscle contraction. *Materials Research Society Symposium Proceedings*, 2000. **600**(Electroactive Polymers (EAP)): 237–247.
12. Madden, J.D., N. Vandesteeg, P. Anquetil, P.G. Madden, A. Takshi, R.Z. Pytel, S. Lafontaine, P.A. Wieringa, and I. Hunter, Artificial muscle technology: Physical principles and naval prospects. *IEEE Journal of Oceanic Engineering*, 2004. **29**(3): 706–728.
13. Hara, S., T. Zama, W. Takashima, and K. Kaneto, TFSI-doped polypyrrole actuator with 26% strain. *Journal of Materials Chemistry*, 2004. **14**(10): 1516–1517.
14. Hara, S., T. Zama, W. Takashima, and K. Kaneto, Artificial muscles based on polypyrrole actuators with large strain and stress induced electrically. *Polymer Journal*, 2004. **36**(2): 151–161.
15. Hara, S., T. Zama, W. Takashima, and K. Kaneto, Gel-like polypyrrole based artificial muscles with extremely large strain. *Polymer Journal*, 2004. **36**: 933–936.
16. Hara, S., T. Zama, W. Takashima, and K. Kaneto, Free-standing polypyrrole actuators with response rate of 10.8% s^{-1}. *Synthetic Metals*, 2005. **149**(2–3 Mar. 31): 199–201.
17. Otero, T.F., S. Villanueva, M.T. Cortes, S.A. Cheng, A. Vazquez, I. Boyano, D. Alonso, and R. Camargo, Electrochemistry and conducting polymers: Soft, wet, multifunctional and biomimetic materials. *Synthetic Metals*, 2001. **119**(1–3): 419–420.
18. Gandhi, M.R., P. Murray, G.M. Spinks, and G.G. Wallace, Mechanism of electromechanical actuation in polypyrrole. *Synthetic Metals*, 1995. **73**(3): 247–256.
19. Skaarup, S., K. West, L.M.W.K. Gunaratne, K.P. Vidanapathirana, and M.A. Careem, Determination of ionic carriers in polypyrrole. *Solid State Ionics*, 2000. **136–137**: 577–582.

20. Smela, E., O. Inganas, and I. Lundstroem, Conducting polymers as artificial muscles: Challenges and possibilities. *Journal of Micromechanics and Microengineering*, 1993. **3**(4): 203–205.
21. Reynolds, J.R., M. Pyo, and Y.-J. Qiu, Cation and anion dominated ion transport during electrochemical switching of polypyrrole controlled by polymer-ion interactions. *Synthetic Metals*, 1993. **55–57**: 1388–1395.
22. Pei, Q. and O. Inganaes, Electrochemical applications of the bending beam method. 1. Mass transport and volume changes in polypyrrole during redox. *Journal of Physical Chemistry*, 1992. **96**(25): 10507–10514.
23. Ding, J., D. Zhou, G. Spinks, G. Wallace, S. Forsyth, M. Forsyth, and D. MacFarlane, Use of ionic liquids as electrolytes in electromechanical actuator systems based on inherently conducting polymers. *Chemistry of Materials*, 2003. **15**(12): 2392–2398.
24. Kaneto, K., M. Kaneko, and W. Takashima, Response of chemomechanical deformation in polyaniline film on variety of anions. *Japanese Journal of Applied Physics, Part 2: Letters*, 1995. **34**(7A): L837–L840.
25. Ansari Khalkari, R., W.E. Price, and G.W. Gordon, Quartz crystal microbalance studies of the effect of solution temperature on the ion-exchange properties of polypyrrole conducting electroactive polymers. *Reactive and Functional Polymers*, 2003. **56**: 141–146.
26. Kaneko, M., M. Fukui, W. Takashima, and K. Kaneto, Electrolyte and strain dependences of chemomechanical deformation of polyaniline film. *Synthetic Metals*, 1997. **84**(1–3): 795–796.
27. Bay, L., K. West, and S. Skaarup, Pentanol as co-surfactant in polypyrrole actuators. *Polymer*, 2002. **43**(12): 3527–3532.
28. Galinski, M., A. Lewandowski, and I. Stepniak, Ionic liquids as electrolytes. *Electrochimica Acta*, 2006. **51**: 5567–5580.
29. Swatloski, R.P., J.D. Holbrey, and R.D. Rogers, Ionic liquids are not always green: Hydrolysis of 1-butyl-3-methylimidazolium hexafluorophosphate. *Green Chemistry*, 2003. **5**: 361–363.
30. Careem, M.A., K.P. Vidanapathirana, S. Skaarup, and K. West, Dependence of force produced by polypyrrole-based artificial muscles on ionic species involved. *Solid State Ionics*, 2004. **175**: 725–728.
31. Yang, H. and J. Kwak, Mass transport investigated with the electrochemical and electrogrametric impedance techniques. 2. Anion and water transport in PMPy and PPy films. *Journal of Physical Chemistry B*, 1997. **101**: 4656–4661.
32. Yang, H. and J. Kwak, Mass transport investigated with the electrochemical and electrogravimetric impedance techniques. 1. water transport in ppy/cupts films. *Journal of Physical Chemistry B*, 1997. **101**(5): 774–781.
33. Kaneto, K. and M. Kaneko, Contribution of conformational change of polymer structure to electrochemomechanical deformation based on polyaniline. *Applied Biochemistry and Biotechnology*, 2001. **96**: 13–23.
34. Sonoda, Y., W. Takashima, and K. Kaneto, Characteristics of soft actuators based on polypyrrole films. *Synthetic Metals*, 2001. **119**(1–3): 267–268.
35. Qi, B., W. Lu, and B.R. Mattes, Strain and energy efficiency of polyaniline fiber electrochemical actuators in aqueous electrolytes. *Journal of Physical Chemistry B*, 2004. **108**(20): 6222–6227.
36. Spinks, G.M., B. Xi, D. Zhou, V.-T. Truong, and G.G. Wallace, Enhanced control and stability of polypyrrole electromechanical actuators. *Synthetic Metals*, 2004. **140**(2–3): 273–280.
37. Lu, W., A.G. Fadeev, B. Qi, E. Smela, B.R. Mattes, J. Ding, G.M. Spinks, J. Mazurkiewicz, D. Zhou, G.G. Wallace, D.R. MacFarlane, S.A. Forsyth, and M. Forsyth, Use of ionic liquids for p-conjugated polymer electrochemical devices. *Science*, 2002. **297**(5583): 983–987.
38. Hara, S., T. Zama, W. Takashima, and K. Kaneto, Free-standing gel-like polypyrrole actuators doped with bis(perfluoroalkylsulfonyl)imide exhibiting extremely large strain. *Smart Materials and Structures*, 2005. **14**: 1501–1510.
39. Bay, L., T. Jacobsen, S. Skaarup, and K. West, Mechanism of actuation in conducting polymers: Osmotic expansion. *Journal of Physical Chemistry B*, 2001. **105**(36): 8492–8497.
40. Kaneko, M. and K. Kaneto, Electrochemomechanical deformation of polyaniline films doped with self-existent and giant anions. *Reactive and Functional Polymers*, 1998. **37**(1–3): 155–161.
41. Anquetil, P.A., H.-h. Yu, J.D. Madden, T.M. Swager, and I.W. Hunter, Recent advances in thiophene-based molecular actuators. *Proceedings of SPIE—The International Society for Optical Engineering*, 2003. **5051**(Electroactive Polymer Actuators and Devices (EAPAD)): 42–53.
42. Ding, J., L. Liu, G.M. Spinks, D. Zhou, G.G. Wallace, and J. Gillespie, High performance conducting polymer actuators utilising a tubular geometry and helical wire interconnects. *Synthetic Metals*, 2003. **138**(3): 391–398.

43. Hara, S., T. Zama, S. Sewa, W. Takashima, and K. Kaneto, Polypyrrole-metal coil composites as fibrous artificial muscles. *Chemistry Letters*, 2003. **32**(9): 800–801.
44. Zhou, D., G.M. Spinks, G.G. Wallace, C. Tiyapiboonchaiya, D.R. MacFarlane, M. Forsyth, and J. Sun, Solid state actuators based on polypyrrole and polymer-in-ionic liquid electrolytes. *Electrochimica Acta*, 2003. **48**(14–16): 2355–2359.
45. Baughman, R.H., Conducting polymer artificial muscles. *Synthetic Metals*, 1996. **78**(3): 339–353.
46. Smela, E. and N. Gadegaard, Surprising volume change in PPy/DBS. An atomic force microscopy study. *Advanced Materials*, 1999. **11**(11): 953–957.
47. Bay, L., K. West, P. Sommer-Larsen, S. Skaarup, and M. Benslimane, A conducting polymer artificial muscle with 12% linear strain. *Advanced Materials*, 2003. **15**(4): 310–313.
48. Spinks, G.M., T.E. Campbell, and G.W. Gordon, Force generation from polypyrrole actuators. *Smart Materials and Structures*, 2005. **14**: 406–412.
49. Lu, W., I.D. Norris, and B.R. Mattes, Electrochemical actuator devices based on polyaniline yarns and ionic liquid electrolytes. *Australian Journal of Chemistry*, 2005. **58**: 263–269.
50. Santa Fe Science and Technology 3216 Richards Lane Santa Fe NM 87507 USA.
51. Jager, E.W., O. Inganas, and I. Lundstrom, Microrobots for micrometer-size objects in aqueous media: potential tools for single-cell manipulation. *Science*, 2000. **288**(5475): 2335–2338.
52. Andrews, M.K., M.L. Jansen, G.M. Spinks, D. Zhou, and G.G. Wallace, An integrated electrochemical sensor-actuator system. *Sensors and Actuators, A: Physical*, 2004. **A114**(1): 65–72.
53. Zhou, D., G.G. Wallace, G.M. Spinks, L. Liu, R. Cowan, E. Saunders, and C. Newbold, Actuators for the cochlear implant. *Synthetic Metals*, 2003. **135–136**: 39–40.
54. Spinks, G.M., G.G. Wallace, J. Ding, D. Zhou, B. Xi, and J. Gillespie, Ionic liquids and polypyrrole helix tubes. Bringing the electronic Braille screen closer to reality. *Proceedings of SPIE—The International Society for Optical Engineering*, 2003. **5051**(Electroactive Polymer Actuators and Devices (EAPAD)): 372–380.
55. Quantum Technology Pty Ltd 5 South Street (PO Box 390) Rydalmere NSW 2116 Australia.
56. Carpi, F., F. Lorussi, A. Mazzoldi, G. Pioggia, E.P. Scilingo, and D. De Rossi, Electroactive polymers based skin and muscles for man machine interfaces. *Materials Research Society Symposium Proceedings*, 2002. **698**(Electroactive Polymers and Rapid Prototyping): 29–33.
57. Pei, Q., R. Pelrine, S. Stanford, R. Kornbluh, and M. Rosenthal, Electroelastomer rolls and their application for biomimetic walking robots. *Synthetic Metals*, 2003. **135–136**: 129–131.
58. Skaarup, S., N. Mogensen, L. Bay, and K. West, Polypyrrole actuators for tremor suppression. *Proceedings of SPIE—The International Society for Optical Engineering*, 2003. **5051**(Electroactive Polymer Actuators and Devices (EAPAD)): 423–428.
59. Jager, E.W., E. Smela, and O. Inganas, Microfabricating conjugated polymer actuators. *Science*, 2000. **290**(5496): 1540–1545.
60. Chronis, N. and L.P. Lee, Polymer MEMS-based microgripper for single cell manipulation. IEEE International Conference on Micro Electro Mechanical Systems, Technical Digest, 17th, Maastricht, the Netherlands, Jan. 25–29, 2004, 2004: 17–20.
61. Zhou, W. and W.J. Li, Micro ICPF actuators for aqueous sensing and manipulation. *Sensors and Actuators, A: Physical*, 2004. **A114**(2–3): 406–412.
62. Couvillon, L.A. and M.S. Banik, Electroactive polymer actuated heart-lung bypass pumps. 2004 (USA). Application: US20040068220.
63. Micromuscle AB Westmansgatan 29 SE-582 16 Linköping, Sweden.
64. Wu, Y., D. Zhou, G.M. Spinks, P.C. Innis, W.M. Megill, and G.G. Wallace, TITAN: A conducting polymer based microfluidic pump. *Smart Materials and Structures*, 2005. **14**(6): 1511–1516.

25 Organic Liquid Crystal Optoelectronic Materials and Devices

Sebastian Gauza

CONTENTS

25.1 Introduction to Liquid Crystals ... 765
25.2 Types of Liquid Crystal ... 767
25.3 Liquid Crystal Phases ... 771
25.3.1 Calamitic Liquid Crystals ... 773
25.3.2 Chiral Liquid Crystals ... 774
25.4 Physical Properties ... 776
25.4.1 Phase Transitions ... 776
25.4.2 Dielectric Properties ... 778
25.4.3 Optical Properties ... 779
25.4.4 Elastic Properties ... 782
25.4.5 Viscous Properties ... 783
25.5 Liquid Crystal Applications ... 783
25.5.1 Liquid Crystal Displays ... 784
25.5.2 Polymeric Composites-Based Electro-Optic Devices ... 787
25.6 Summary ... 789
Exercise Questions ... 789
List of Abbreviations ... 790
References ... 790

Abstract This chapter introduces the basic principles of liquid crystal (LC) materials and optoelectronic devices. A variety of different liquid crystal molecular arrangements will result in a large range of liquid crystal phases. The basic differences of liquid crystal phases are explained and the anisotropy of the physical properties is described. The usability of liquid crystals for electro-optic devices are described, together with the working principles of typical devices based on liquid crystals. Key device parameters and performance characteristics are described.

25.1 INTRODUCTION TO LIQUID CRYSTALS

The study of liquid crystals (LCs) began in 1888 when an Austrian botanist named Friedrich Reinitzer observed that a material known as cholesteryl benzoate exhibited two distinct melting points [1]. In his experiments, Reinitzer increased the temperature of a solid sample and watched the

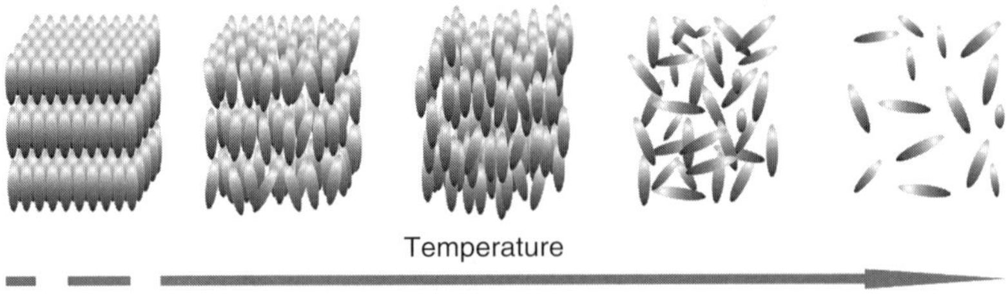

FIGURE 25.1 Schematic arrangement of the molecules in different phases.

crystal change into a hazy liquid. As he further increased the temperature, the material changed into a clear, transparent liquid. Because of this early work, Reinitzer is often credited with discovering a new phase of matter—the liquid crystal phase also known as liquid crystalline or mesophase. Most people are familiar with the fact that matter can exist in three states: solid, liquid, and gas (vapor). However, this is a simplification, and under extreme conditions, other forms of matter can exist, e.g., plasma at very high temperatures or superfluid helium at very low temperatures. The difference between these states of matter is the degree of order in the material, which is directly related to the surrounding temperature and pressure. If the temperature is raised, more energy is transferred into the system, leading to increasingly stronger vibrations. Finally, at the transition temperatures between the solid and liquid states, the long range positional order is broken and the constituents may move in a random fashion (Figure 25.1), constantly bumping into one another and abruptly changing direction of motion. The thermal energy is not high enough to completely overcome the attractive forces between the constituents, so there is still some positional order at short range. Because of the remaining cohesion, the density of the liquid is constant even though the liquid takes the shape of its container, as opposed to a solid. The liquid and solid phases are called condensed phases. If we continuously increase the temperature until the next phase change, the substance enters its gas (or vapor) state and its constituents are no longer bound to each other (Figure 25.1).

In liquid crystals, the molecular order lies between those of the isotropic liquid and the crystal. The classification of liquid crystals is based on their degrees of orientational and positional order. In the liquid crystal phase, one of the molecular axes (typically, the long molecular axis) tends to align along a preferred direction. This preferred direction is called the director denoted by the vector n. To specify the amount of orientational order in a liquid crystal phase, an order parameter is defined.

$$S = \langle P_2(\cos\theta)\rangle = \frac{2}{3}\langle \cos^2\theta - 1\rangle \tag{25.1}$$

where
$\langle\rangle$ denotes a thermal averaging
θ is the angle between each molecule and the director

If the molecules are perfectly oriented (crystal state), that is if $\theta = 0$ with the director, then $S = 1$. In the opposite, if the molecules are randomly oriented about n (isotropic), $S = 0$ because there is no orientational order. So the higher the order parameter, the more ordered the liquid crystal phase is. In a typical liquid crystal, order parameter decreases as the temperature is increased. Temperature dependence of the order parameter S is shown in Figure 25.2. Most common values of S are between 0.3 and 0.8. The order parameter is typically determined by the measured macroscopic properties like optical birefringence or diamagnetism. At certain temperatures, the liquid crystal material may gain certain amount of positional order. When it happens, the center of mass of the LC molecules, although still forming a fluid, prefers to lie on average in layers (Figure 25.1). This positional ordering may be described in terms of the density of the centers of the mass of the molecules,

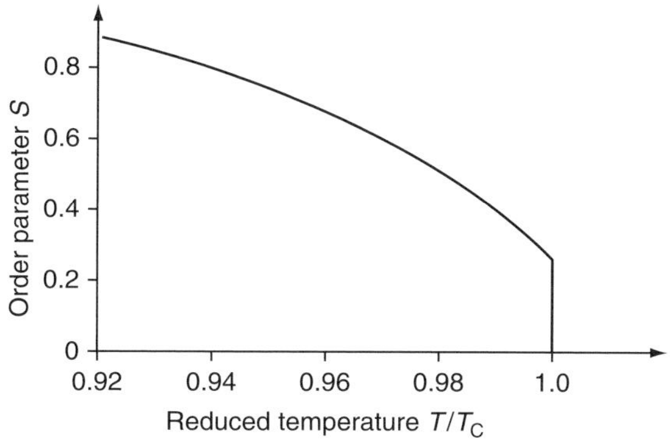

FIGURE 25.2 Temperature-dependent order parameter of the LC phase.

$$\rho(z) = \rho_0 \left[1 + \psi \cos\left(\frac{2\pi z}{d} \right) \right] \quad (25.2)$$

where
- z is the coordinate parallel to the layer normal
- ρ_0 is the average density of the fluid
- d is the distance between layers
- ψ is the order parameter

The modulus of ψ, $|\psi|$, represents the amplitude of the oscillation of the density. When $|\psi| = 0$, there is no layering, but if $|\psi| > 0$ then some amount of sinusoidal layering exists. Different liquid crystal phases are formed as a consequence, which will be later described in this chapter.

From a basic physics point of view, liquid crystal materials are of great interest and have contributed to the modern understanding of phase transitions and order phenomena in one, two, and three dimensions. To most of people today, liquid crystals are almost synonymous to flat panel displays (liquid crystal displays, LCDs) for computers, mobile phones, and other electronic equipments. However, there is a rapid development of the other types of applications. For example, there have been developments in telecommunications, pattern recognition, real-time holography, light shutters, nonmechanical beam steering, etc. [2]. Liquid crystals constitute a unique form of soft matter and are becoming increasingly more important in pure materials science in the development of polymer materials and biomaterials.

25.2 TYPES OF LIQUID CRYSTAL

Considering the geometrical structure of the mesogenic molecules, the liquid crystals can be assigned into several types [3,4]. From a recent application viewpoint, the most important are liquid crystals formed by rod-shaped molecules, as shown in Figure 25.3, where one molecular axis is much longer than the other two. These molecules are called calamitic liquid crystals. Molecules have to be rigid in one part (rigid core) and flexible in another (terminal flexible hydrocarbon chain). Different physical properties may be affected by exchanging one of the terminal chains with a group having different polarities. Liquid crystals formed by disc-shaped molecules with one molecular axis much shorter than the other two are called discotic liquid crystals, as shown in Figure 25.3. In this case, the rigid part of the molecule is disc shaped, typically having multiple aromatic rings. There are some possible intermediates between rod-shaped and disc-shaped molecules known as lath-like liquid crystal molecules. Transition to the mesophase (liquid crystalline phase) may be caused by either temperature

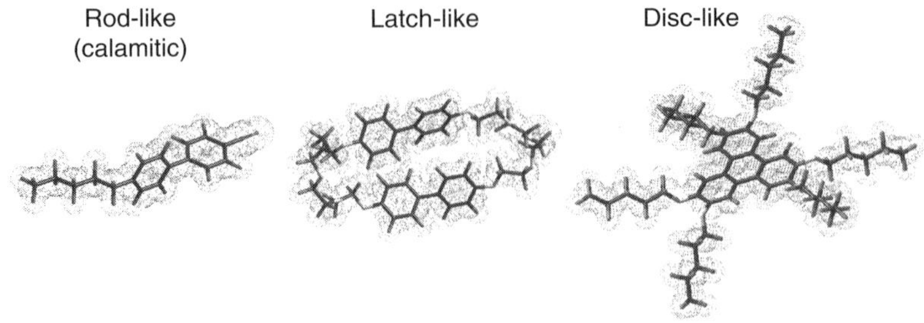

FIGURE 25.3 Molecular geometry of different types of LC.

or influence of solvents. The liquid crystals obtained from the first method are called thermotropic. If the influence of solvents induces the transition, the liquid crystals are called lyotropic. LC materials capable of forming thermotropic as well as lyotropic mesophases are called amphotropic liquid crystals (Figure 25.4). The lyotropic liquid crystal phases are formed by dissolution of amphiphilic molecules of a material in a suitable solvent. Amphiphilic molecules consist of hydrophobic group at one end and hydrophilic group at another. Such molecules form self-organized structures in solvents with different polarities. A very good example is soap. When dissolved in water (polar solvent), hydrophobic tails face each other and hydrophilic parts face solvent, forming a micelle. Another type, which is biologically important, is phospholipids which form bilayers. Such aggregations are

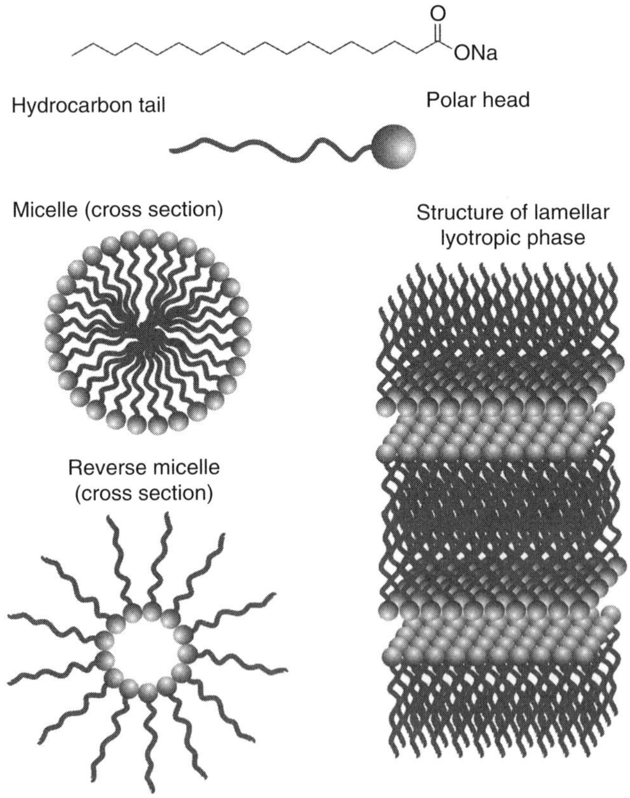

FIGURE 25.4 Examples of the lyotropic LCs.

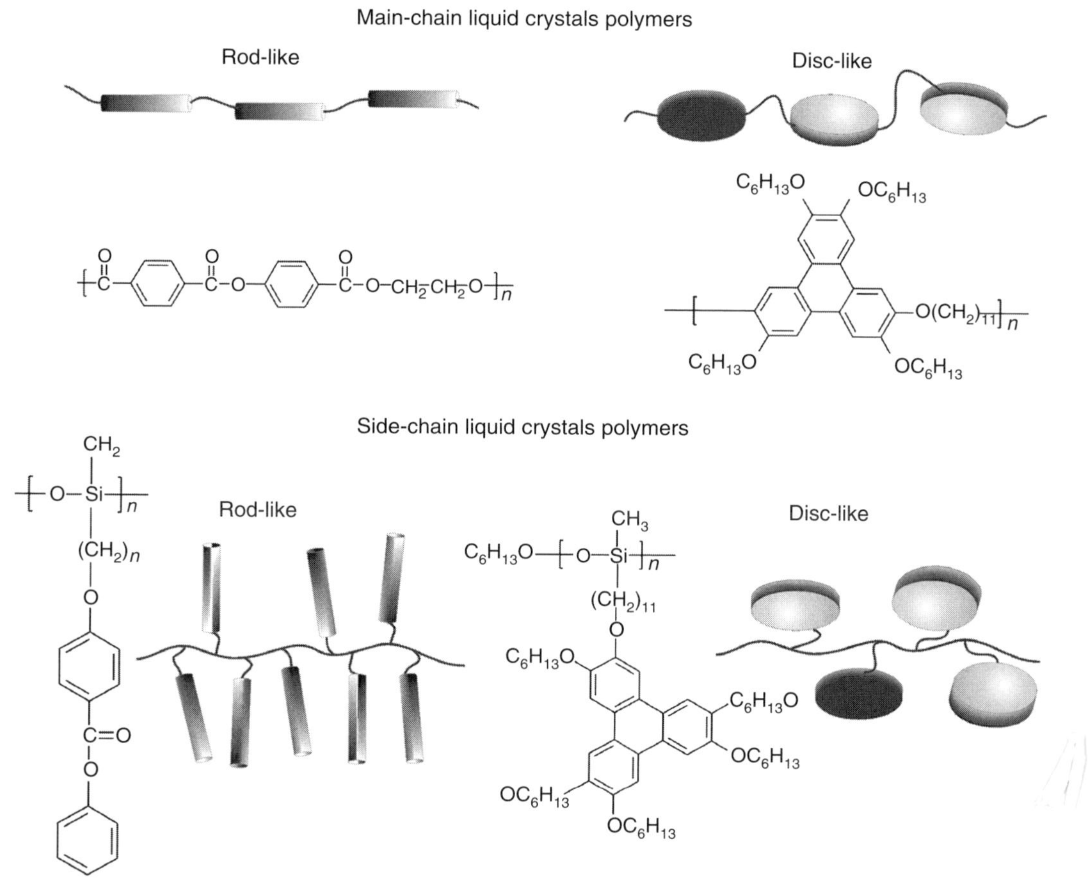

FIGURE 25.5 Types of LC polymeric material.

the building units of biological membranes. A separate group of liquid crystals is formed by liquid crystal polymers. As thermotropic mesogenes, these structures consist of mesogenic subunits (rod-like or disc-like) which are linked together with flexible linkage units (Figure 25.5). If rigid mesogenic units are linked directly by flexible links, the main-chain polymer liquid crystal is formed. Mesogenic subunits can also be attached to the polymer chain as a side group which are known as side-chain polymer liquid crystals. Merged structure, called combined polymer liquid crystal, has built in main-chain as well as side-chain mesogenic units (Figure 25.5). Since the discovery of the first liquid crystal substances, there have been many research activities to determine what kind of structure forms desired liquid crystalline phase. Theory suggests that mesophases can be achieved when the molecules have elongated shape and some flexibility. Typically, mesogenic molecules consist of several building blocks. If we concentrate on the most common calamitic LC block systems, it contains elements shown in Figure 25.6. As shown in Figure 25.6, calamitic structures

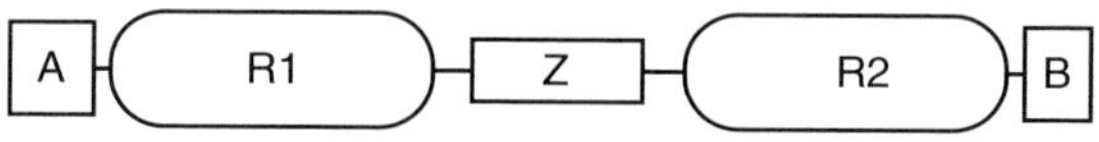

FIGURE 25.6 Building blocks of calamitic LCs. A–R1–Z–R2–B.

TABLE 25.1
Typical Central Bridge Groups

Central Bridge Group	Central Bridge Group
$-[CH_2]_n-$ Alkane	$-CH_2-O-$ Ether
$-C{=}C-$ Alkene (olefin)	$-N{=}N-$ Azo
$-C{\equiv}C-$ Alkyne (tolane, acetylene)	$-N{=}N(\rightarrow O)-$ Azoxy
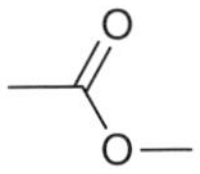 Ester	$-C{\equiv}C-C{\equiv}C-$ Diacetylene

consist of a rigid rod formed by two (or more) ring systems (R1 and R2). These are connected together by either single bond or linking group also known as a central bridge group. Molecular constituents at the ends of the rigid core (para position to the central group) are called terminal groups or chains. Usually at least one of the terminal groups must be a flexible carbon chain. Tables 25.1 through 25.3 show the most popular choices for central bridge, ring systems, and terminal substituents.

TABLE 25.2
Typical Ring Systems

Ring System	Ring System
 $n = 1$ benzene $n = 2$ biphenyl $n = 3$ terphenyl	 Bicyclooctane
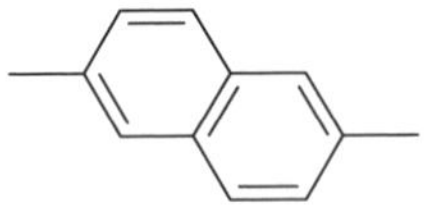 Naphthalene	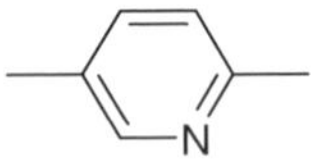 Pyridine
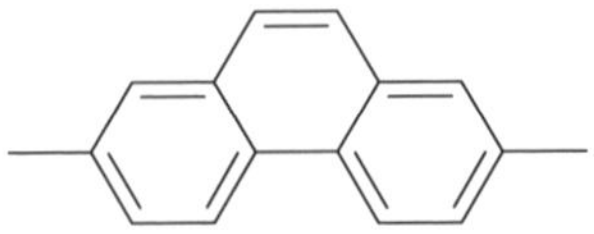 Phenanthrene	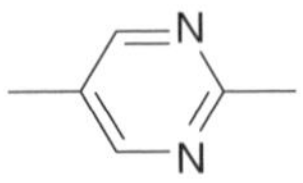 Pyrimidine
 Cyclohexane	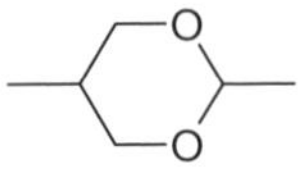 Dioxane

TABLE 25.3
Typical Terminal Units

Terminal Unit	Terminal Unit
C_nH_{2n+1}	–F, –Cl, –Br, –I
Alkyl	Single elements
OC_nH_{2n+1}	–CN, $-NO_2$, –NCS
Alkoxy	Highly polar moieties

The introduction of a linking group into a mesogenic molecule often determines linearity, increases overall molecular length and changes polarizability anisotropy, thus influences mesomorphic and physical properties in general. The presence of a central linking bridge often widens the temperature range of the mesophase by reducing the melting point. However, sometimes, this may also affect the thermal and photochemical stability by increasing π-electron conjugation of the molecules. In some cases, linking groups may induce coloration.

Most calamitic liquid crystals posses aromatic and alicyclic rings in the molecular structures. These include single benzene ring (commonly 1,4-phenyl), alicyclic ring (commonly cyclohexane), heterocyclic ring (for e.g., 1,3-pyrimidine), and a wide variety of the combinations. Individual benzene rings do not yield good mesogens. At least two rings are needed to give a rod-like molecule. The exception is the 4-alkyl-benzoic acid, which forms a dimer by hydrogen bond, as shown in Figure 25.7. Beside the heterocyclic rings, ring systems of the rigid core could be laterally substituted by elements other than hydrogen (most popular are F and Cl) or groups of elements such as CH_3, NH_2, CN, etc. This will decrease the melting point of a mesogen, however, increase its viscosity. Additionally, optical and dielectric anisotropies will be affected. In general, the core of the mesogen determines the mesogenic properties by establishing the primary shape of the molecule and its rigidity. Many terminal units (A, B) have been employed in the creation of mesogenic molecules. The most successful route is to use either small polar substituents or a fairly long, straight hydrocarbon chain. The role of these groups is to act as either a flexible extension to the rigid core or as a dipolar moiety to introduce anisotropy in physical properties. Molecules can obtain chirality from the flexible chain if it is branched and chiral. Further, the terminal moieties are believed to be responsible for stabilizing the molecular order essential for the mesophase generation. All physical properties strongly depend on the terminal units employed in the molecular system of mesogenic molecules. Some of the typical calamitic liquid crystal single compounds are listed in Table 25.4.

25.3 LIQUID CRYSTAL PHASES

In this section, we focus on the liquid crystals with some potential applications, primarily focusing on the calamitic liquid crystals. Due to the aim of this book, it is my intention to omit the discotic liquid crystals, liquid crystal polymers, and lyotropic liquid crystals. It is also my intention to limit the discussion of liquid crystal phases to the ones, which are commonly employed in electro-optical devices. By doing this, I do not intend to minimize the importance of discotic LCs, banana-type LCs, LC polymers, or lyotropic LCs. For example, Kevlar reveals the importance of lyotropic LC polymer material, whereas others are important in the biological sciences. The readers can find more precise information about these groups of LCs in preferred reading materials [5].

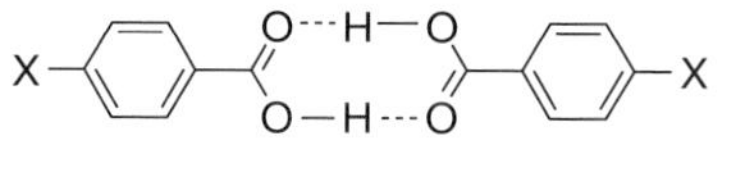

FIGURE 25.7 Dimeric structure formed by 4-alkyl-benzoic acid.

TABLE 25.4
Physical Properties of Popular LC Compounds

Compound	Phase Transition Temperature (°C)
Rigid Core	
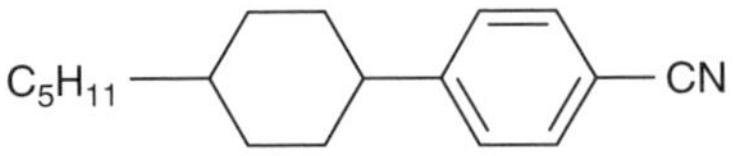	Cr 31 N 55 iso
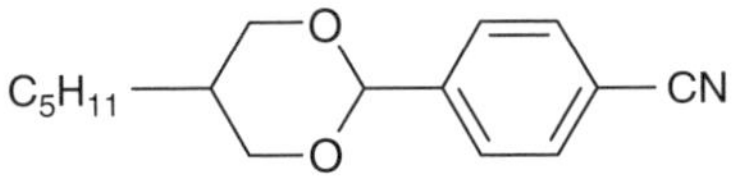	Cr 56 (N 52) iso
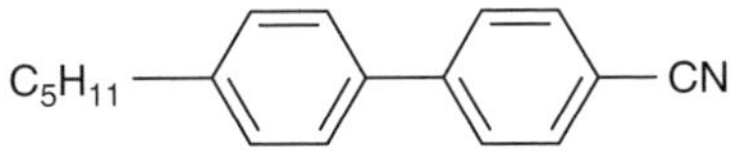	Cr 31 N 55 iso
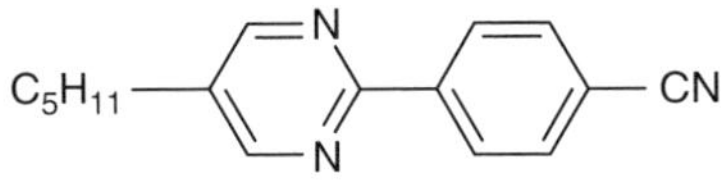	Cr 71 (N 52) iso
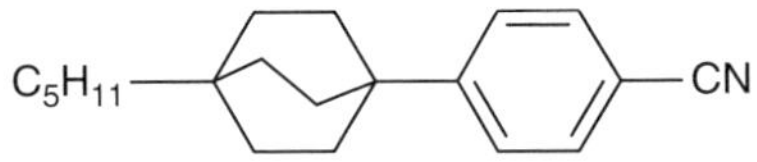	Cr 62 N 100 iso
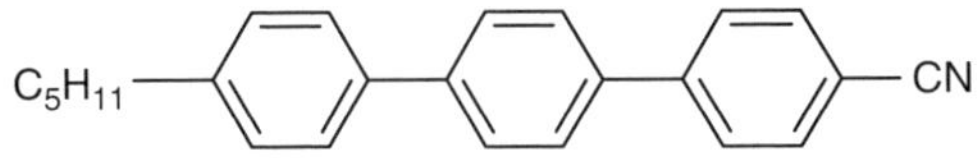	Cr 130 N 239 iso
Central Bridge Link	
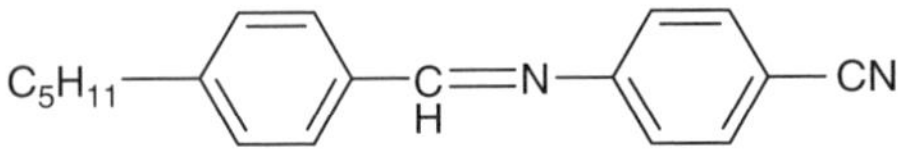	Cr 46 N 75 iso
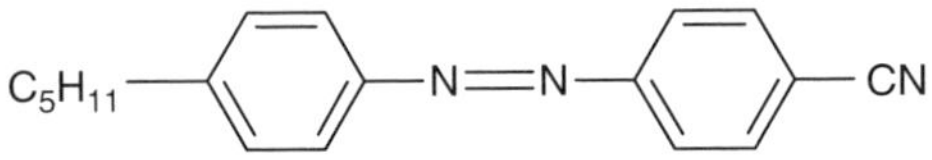	Cr 89 (N 87) iso
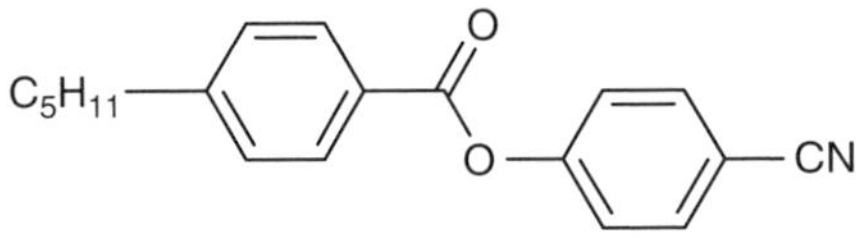	Cr 65 (N 56) iso
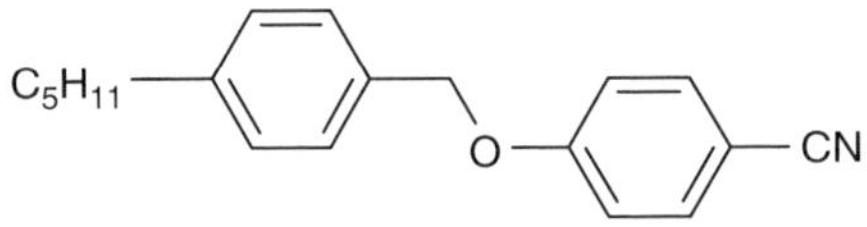	Cr 49 (N -20) iso
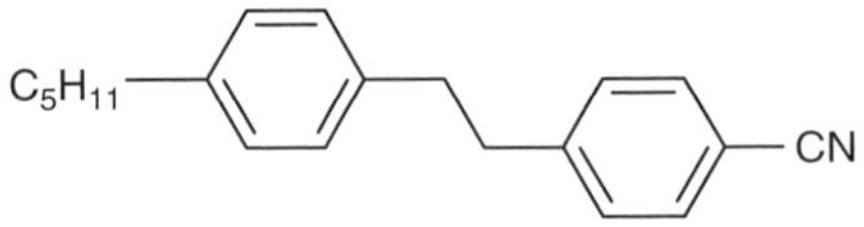	Cr 62 (N -24)
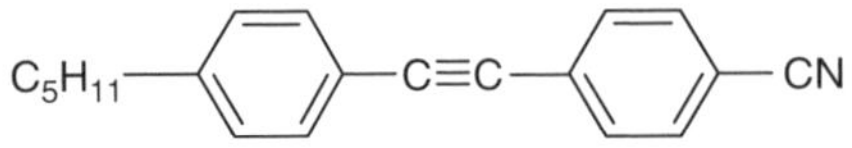	Cr 80 (N 71) iso
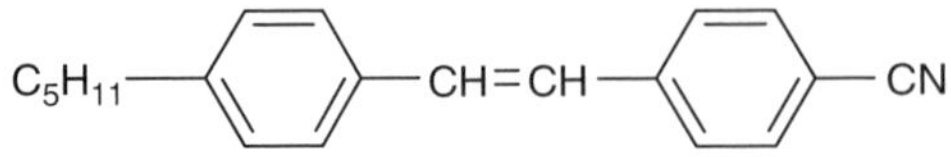	Cr 55 N 101 iso
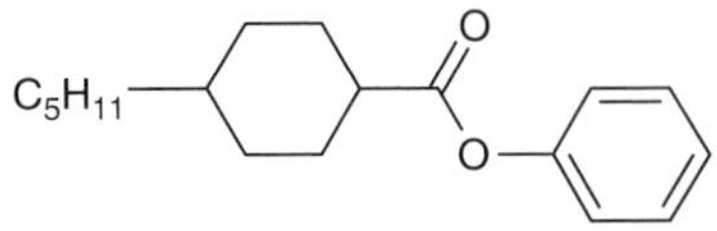	Cr 30 iso

TABLE 25.4 (continued)
Physical Properties of Popular LC Compounds

Compound	Phase Transition Temperature (°C)
Terminal Moiety	
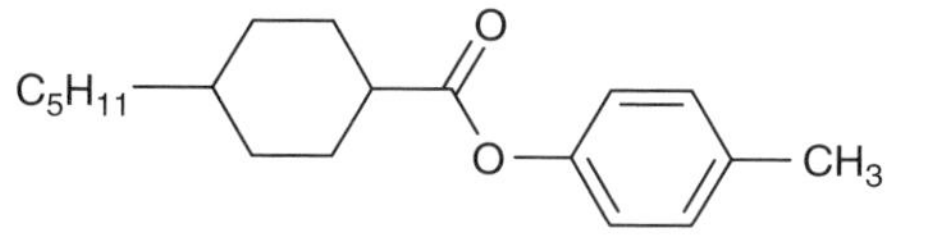	Cr 48 (N 45) iso
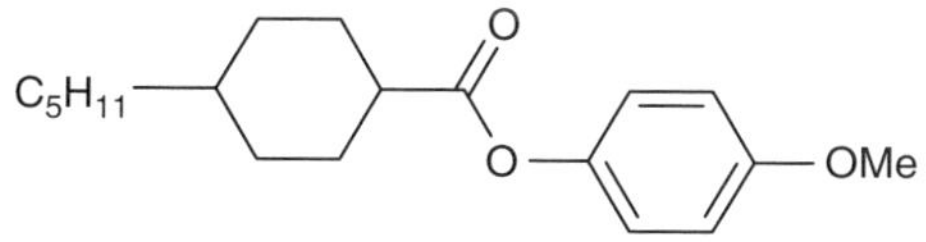	Cr 41 N 71 iso
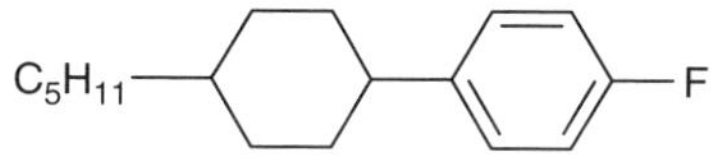	Cr 26 (N 23) iso
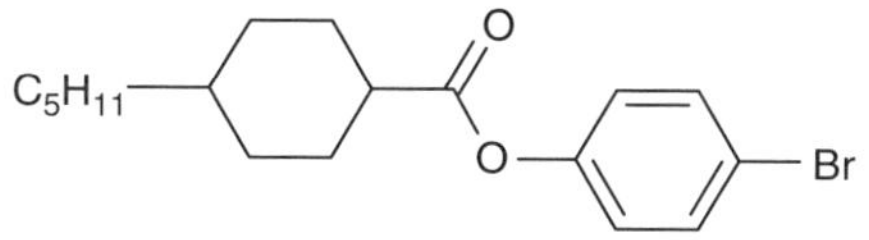	Cr 78 (N 48) iso
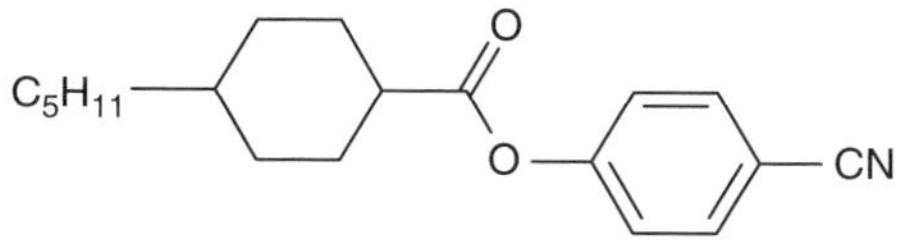	Cr 47 N 79 iso
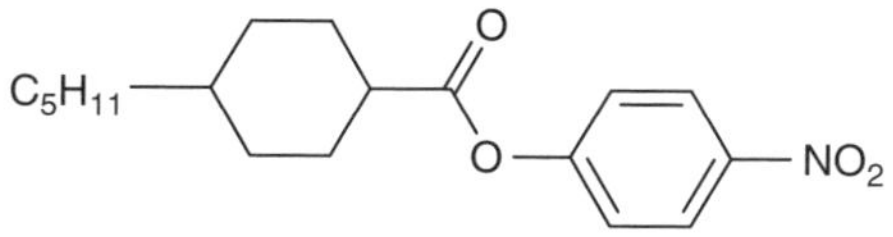	Cr 53 (N 38) iso
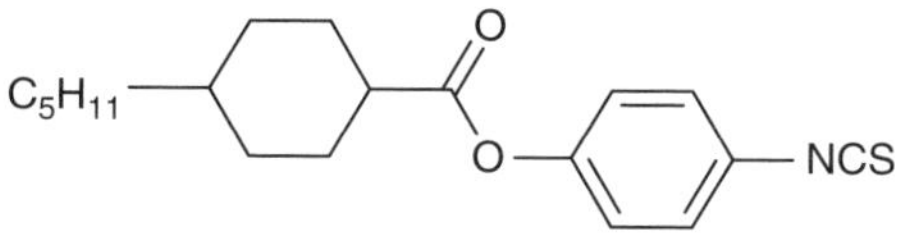	Cr 87 (N 86) iso

25.3.1 Calamitic Liquid Crystals

When classifying liquid crystal phases, after Friedel [6], we first distinguish between two main types: one with nematic order and the other with smectic order (see Figure 25.8). In the nematic phase (of which tilted nematic is known as the cholesteric phase and is a special case), the molecules are free to move in all directions (e.g., there is no positional order of the centers of mass). But on average, they keep their long axes locally parallel (see Figure 25.8). In a smectic state, a number of structural variations exist and there is a positional order along one dimension (some smectic phases have positional order in more than one dimensions). A smectic LC is a layered structure with the molecules oriented parallel or tilted relative to the layer normal. Two smectic phases, called smectic A and smectic C, have acquired a special importance and are now relatively well understood. They are characterized by an absence of positional order within the layers. The molecules have some freedom to move within the layers, as in all smectic phases, but are much less free to move between the layers. These smectics can therefore be described as stacks of two-dimensional (2-D) fluids, but behave as crystalline across the layers. The absence of in-layer order contributes to their high potential for future electro-optic applications. There are several smectic phases different from one another

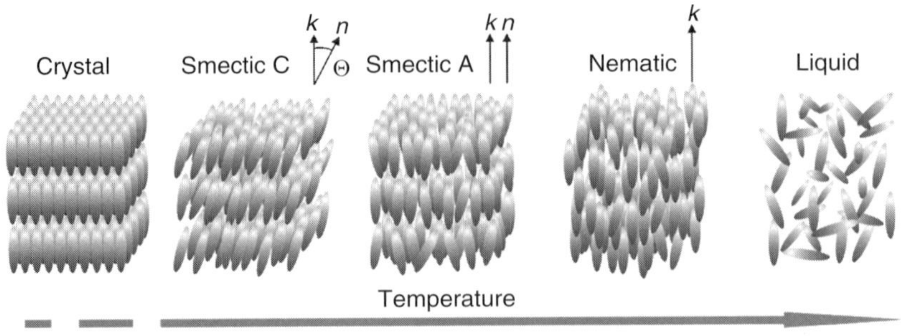

FIGURE 25.8 Schematic arrangement of the molecules in basic nematic and smectic phases.

in areas such as the tilt angle of the director with regard to the layer normal, and the arrangement of molecules within each layer. The simplest is the smectic A phase (Figure 25.8) characterized by a director parallel to the layer normal and a random positional order within the plane. Substances featuring the smectic A phase often exhibit the smectic C phase at a lower temperature (Figure 25.8). In this phase, the molecules have the same random order within the layer but tilt relative to the layer normal. The tilt angle normally increases with decreasing temperature. The other smectic phases are even more crystalline as they also feature some positional order within the layers. They may, for instance, exhibit hexagonal packing of the molecules.

25.3.2 Chiral Liquid Crystals

Molecules which are not identical to their mirror image are said to be chiral. A depiction of the simple concept of a chiral molecule is shown in Figure 25.9. Another example is the human hand. Chiral molecules are able to form liquid crystals with structures related to those of nonchiral materials but with different properties. The cholesteric (or chiral nematic) liquid crystal phase is typically composed of nematic mesogenic molecules containing a chiral center, which produces intermolecular forces that favor alignment between molecules at a slight angle to one another. When the molecules that make up a nematic liquid crystal are chiral, the chiral nematic phase will exist instead of the normal nematic. In this phase, the molecules prefer to lie next to each other in a slightly skewed orientation. This leads to the formation of a structure, which can be visualized as a stack of very thin 2-D nematic-like layers with the director in each layer twisted with respect to those above and below (Figure 25.9). In this structure, the directors actually form a continuous helical pattern around the

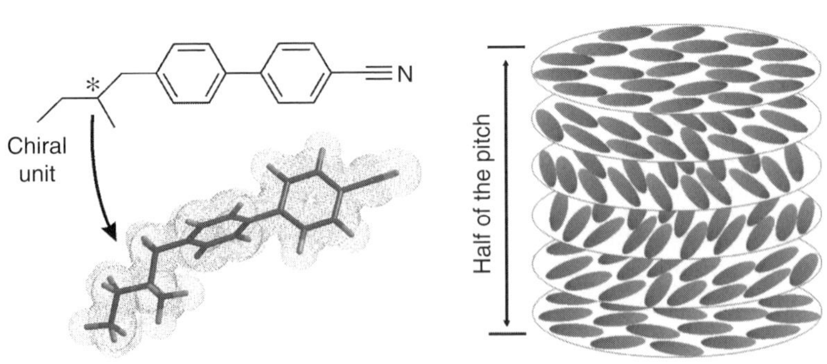

FIGURE 25.9 Chirality of the calamitic molecule and chiral nematic structure.

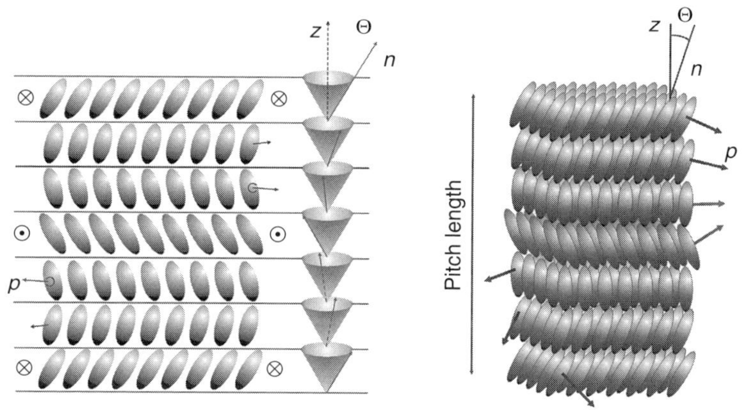

FIGURE 25.10 Structure of the ferroelectric smectic C* phase.

layer normal as illustrated by the black arrow in Figure 25.10. The molecules shown are merely representations of the many chiral nematic mesogens lying in the slabs of infinitesimal thickness with a distribution of orientation around the director. This is not to be confused with the planar arrangement found in smectic mesophases. An important characteristic of the cholesteric mesophase is the pitch. The pitch, p, is defined as the distance it takes for the director to rotate one full turn in the helix as illustrated in Figure 25.9. A by-product of the helical structure of the chiral nematic phase is its ability to selectively reflect light of wavelengths equal to the pitch length, so that a color will be reflected when the pitch is equal to the corresponding wavelength of light in the visible spectrum. The effect is based on the temperature dependence of the gradual change in the director orientation between successive layers, modifying the pitch length which results in an alteration of the wavelength of reflected light according to the temperature. The angle, which the director changes, can be made larger by increasing the temperature of the sample, and thus tighten the pitch. Hence, more thermal energy will result from the increased temperature. Similarly, decreasing the temperature of the chiral sample increases the pitch length of the chiral nematic liquid crystal. Similar pitch changes are possible by applying electromagnetic field to aligned cholesteric samples. An interesting phenomenon of chiral nematic phase is that chirality can be introduced to the nonchiral nematic material by adding a small amount of chiral nematic mesogens. Not necessarily all the molecules have to be chiral. Sometimes, slightly below phase transition to the isotropic state (clearing point), some anomalous phases appear. They are known as blue phases (BPs). In many chiral compounds, with sufficiently high twist, up to three distinct blue phases appear. The two low-temperature phases, blue phase I (BPI) and blue phase II (BPII), have cubic symmetry. The highest temperature phase (closest to the clearing point), blue phase III (BPIII) appears to be amorphous. Similar to chiral nematics, there are chiral forms of smectic phases. The only untilted chiral smectic phase is S_A^*. The most important feature of the chiral smectic phases with a tilted structure is ferroelectricity. Due to their low symmetry, chiral smectic phases are able to exhibit spontaneous polarization (P_S) that is oriented perpendicular to the director $\boldsymbol{n}$ and parallel to the smectic layer plane. Ferroelectric smectic C* phase (S_C^*) is the most known tilted chiral smectic phase. The structure of the S_C^* has layers of molecules, which are tilted in each layer at a temperature-dependent angle (θ) to the layer normal. Additionally, there is a slight and continuous change in the direction of the molecular tilt between adjacent layers as described in Figure 25.10. In a macroscopic sample, without surfaces or external electric field, tilt will follow the helix and result in zero total spontaneous polarization. However, if a strong electric field is applied, the helix is unwound and nonzero spontaneous polarization can be observed. Recently, more phases closely related to the ferroelectric S_C^* have been discovered.

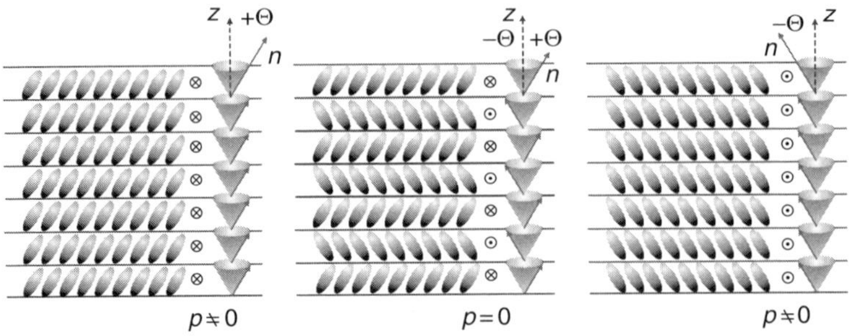

FIGURE 25.11 Structure of the antiferroelectric smectic C* phase.

In these phases, the layer spacing and the polarization direction are related in a different manner than in S_C^* phase. In an antiferroelectric smectic C* phase ($S_{C\,anti}^*$), the spontaneous polarization and the molecular tilt in adjacent layers are pointing in alternating directions, see Figure 25.11. Thus, for the macroscopic sample, both average spontaneous polarization and ideally, average molecular tilts are zero. Sufficiently strong electric fields will switch the antiferroelectric order to ferroelectric order. An important difference between ferroelectric and antiferroelectric smectic C* phase is that antiferroelectric repeats its helical structure every 180° of rotation about the layer normal, compared to 360° for the ferroelectric phase. Many existing ferroelectric phases are different in the proportion of number of the layers with opposite directions. Similar to blue phases in the chiral nematic, the twist grain boundary (TGB) phases appear in chiral smectics. Just like in the blue phases (BPs), high chirality subtly changes the energetic of the system leading to a different type of structure. In this case, the free energy is minimized by introducing grain boundaries at periodic intervals. We may observe different TGB phases depending on the base of the phase from which they were developed. For example, smectic C* will generate a TGB_C phase.

25.4 PHYSICAL PROPERTIES

The molecular order existing in the liquid crystalline phases induces anisotropy in the system. It means that all directions in the system are not equivalent to each other due to the shape of the molecules and the molecular distribution along the director $\boldsymbol{n}$. The anisotropy of the physical property is the most useful feature of liquid crystals and enables electro-optical application of LC materials. The physical properties can be divided into scalar and nonscalar quantities. Typical scalar properties are the thermodynamic transition parameters such as transition temperatures, transition enthalpy, and entropy changes. The dielectric, diamagnetic, optical, elastic, and viscous properties are the most important nonscalar properties. We will concentrate on the physical properties of the nematic phase because we commonly employ this phase and its properties in electro-optical applications.

25.4.1 Phase Transitions

The difference in the transition temperatures between melting and clearing points gives the range of stability of the mesophases. For polymorphous (more than one phase) substances, the higher-ordered phase exhibits the lower transition temperatures. When a material melts, a change of state occurs from solid to liquid (mesophase) and this process requires energy (endothermic) from the surroundings. If several mesophases exist, then several transitions will occur. The melting transition has typically ~10× larger enthalpy change (30–40 kJ/mol) than a transition between different mesophases (3–5 kJ/mol). The large enthalpy change is due to the drastic structural changes during the melting process; however, there is only a small difference between the different mesophases. One of the

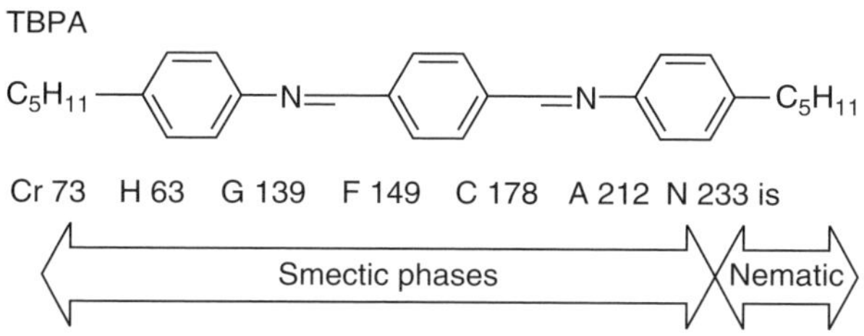

FIGURE 25.12 Molecular structure and mesomorphic properties of the terephthalylidene-bis-*p*-*n*-pentylaniline (TBPA).

richest polymorphism of the single LC compound is shown by terephthalylidene-bis-*p*-*n*-pentylaniline (TBPA) with six different mesophases, see Figure 25.12. A number of techniques (optical microscopy with hot stage, polarizing optical microscopy, differential thermal analysis, differential scanning calorimetry, etc.) may be used for determining the phase transition temperatures. Polarizing microscopy with hot stage and differential scanning calorimetry is particularly useful for the measurement of the phase transition temperatures. By using both methods, one can determine the number and type of mesophases and also the exact phase transitions, temperature, and enthalpy change associated with each transition. These are the crucial parameters to determine the components of eutectic mixture formulation for devices. A majority of the single liquid crystal components do not possess adequate range of required mesophase. Therefore, the eutectic compositions are required to lower the melting point of the material. It is known that the melting point of a binary (or higher number components) mixture is less than either of its constituent compounds. Figure 25.13 shows the phase diagram of a binary mixture. The mesogenic range for components 1 and 2 are shown in the two boundary vertical axes. The horizontal axis represents the molar concentration (X_2) of component 2. At a certain molar concentration, the melting point of the mixture will reach its minimum. Meanwhile, the clearing point of the mixture is linearly proportional to the molar concentration. The eutectic mixture calculation is based on the Schröder–Van Laar equation [3,7,8]:

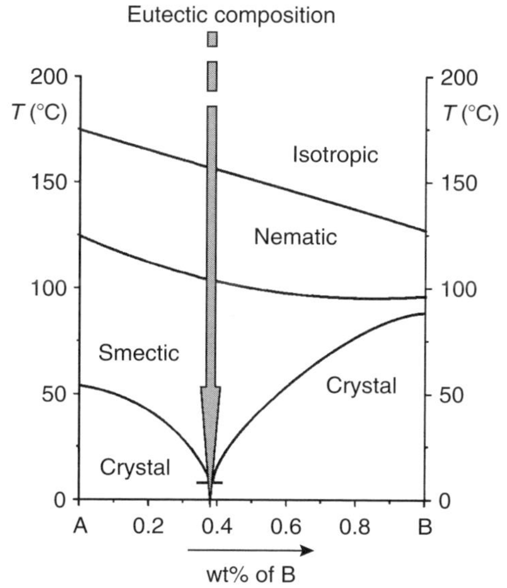

FIGURE 25.13 Simple phase diagram of the binary LC mixture.

$$T_i = \frac{\Delta H_{\text{fi}}}{\dfrac{\Delta H_{\text{fi}}}{T_{\text{fi}}} - R\ln(X_i)} \tag{25.3}$$

where

T_i is the temperature at which the pure component melts in the mixture
ΔH_{fi} is the heat fusion enthalpy of the pure component i
T_{fi} is the melting point of the pure component i
R is the gas constant (1.98 cal/mol K)
X_i is the mole fraction of the pure component i

The clearing point (T_c) of the eutectic mixture can be estimated from the clearing points (T_{ci}) of the individual components (X_i) as

$$T_c = \sum_i X_i T_{ci} \tag{25.4}$$

25.4.2 Dielectric Properties

It has already been mentioned that liquid crystals exhibit anisotropy in many of their physical properties. It is well known that liquid crystals as dielectric and diamagnetic materials are sensitive to the external electric and magnetic fields. Dielectric studies are, in general, concerned with the response of matter to the application of an electric field. The nematic liquid crystals are uniaxially symmetric to the axes of the director ***n***, and the dielectric constants differ in value along the preferred axis ($\varepsilon_{//}$) and perpendicular to this axis ($\varepsilon_{\perp}$). They are mainly determined by the dipole moment, μ, its angle, θ, with respect to the principal molecular axis, and order parameter, S, as described by the Maier and Meier mean field theory [9]:

$$\varepsilon_{//} = NhF\left\{\langle\alpha_{//}\rangle + \left(\frac{F\mu^2}{3kT}\right)\left[1-\left(1-3\cos^2\theta\right)S\right]\right\} \tag{25.5}$$

$$\varepsilon_{\perp} = NhF\left\{\langle\alpha_{\perp}\rangle + \left(\frac{F\mu^2}{3kT}\right)\left[1+\left(1-3\cos^2\theta\right)S/2\right]\right\} \tag{25.6}$$

$$\Delta\varepsilon = NhF\left\{\left(\langle\alpha_{//}\rangle - \langle\alpha_{\perp}\rangle\right) - \left(\frac{F\mu^2}{2kT}\right)\left(1-3\cos^2\theta\right)S\right\} \tag{25.7}$$

where

N is the molecular packing density
$h = 3\varepsilon/(2\varepsilon + 1)$ is the cavity field factor
$\varepsilon = (\varepsilon_{//} + 2\varepsilon_{\perp})/3$ is the averaged dielectric constant
F is the Onsager reaction field
$\langle\alpha_{//}\rangle$ and $\langle\alpha_{\perp}\rangle$ are the principal elements of the molecular polarizability tensor

For a nonpolar compound, $\mu \approx 0$ and its dielectric anisotropy ($\Delta\varepsilon$) are very small. The dielectric anisotropy, for a highly polar compound, depends on the dipole moment, angle θ, temperature T, and applied frequency. The $\Delta\varepsilon$ is positive for the polar compound with its effective dipole at $\theta < 55°$

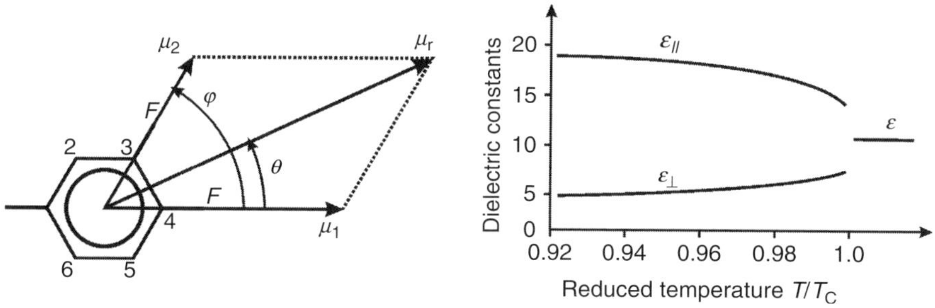

FIGURE 25.14 Dipole moment of the simple mesogenic structure.

and becomes negative if $\theta > 55°$. Similar to many physical properties, dielectric anisotropy depends on the temperature and decreases in proportion to S/T as the temperature increases. When the molecule has two (or more) polar groups with dipole moment μ_1 and μ_2, its effective dipole can be calculated by the vector addition method, see Figure 25.14. In Figure 25.14, the first dipole is along the principal molecular axis and the second dipole μ_2 is at an angle ϕ with respect to the principal molecular axis (and μ_1). The resultant dipole moment μ_r can be calculated from the equation:

$$\mu_r = \left(\mu_1^2 + \mu_2^2 + 2\mu_1\mu_2\cos\phi\right)^{1/2} \tag{25.8}$$

A special case takes place at high frequency (typically around 10 MHz and higher) when dielectric relaxation occurs and the sign of dielectric anisotropy changes. It is also known as dual frequency properties.

25.4.3 Optical Properties

The optical anisotropy (birefringence, Δn) is an essential physical property of liquid crystals for their applications in liquid crystal display devices. The birefringence of liquid crystals means that the speed of propagation of light waves are not uniform in all directions and depends on the direction and polarization of the light waves going through the material. Therefore, the material will possess different refractive indices in different directions. This enables the ability of aligned liquid crystals to control the polarization of light going through the samples. In most common case, liquid crystals are uniaxial nematics. We can observe two principal refractive indices, the ordinary refractive index n_o and the extraordinary refractive index n_e. The ordinary and extraordinary refractive indices are observed for a linearly polarized light wave where the electric vector oscillates, with respect to the refractive index, perpendicular and parallel to the optic axis of the LC molecules shown by director ***n***, see Figure 25.15. Refractive indices of a uniaxial thermotropic liquid crystal are primarily governed by the molecular constituents (molecular conjugation), wavelength (differential oscillator strength), and temperature (order parameter). The macroscopic refractive index is related to the molecular polarizability at optical frequencies. The optical polarizability is due mainly to the presence of delocalized electrons not participating in chemical bonds and π-electrons. This is the reason that LC molecules composed of benzene rings have higher values of Δn than those respective cyclohexane counterparts. Each refractive index is determined by its corresponding molecular polarizabilities α_o and α_e. In 1966, Vuks [10] modified the Lorentz–Lorenz equation, which correlates the refractive index of an isotropic medium with molecular polarizability in the optical frequencies. By assuming that the internal field in an LC is the same in all directions, Vuks gave a semiempirical equation correlating the refractive indices with the molecular polarizabilities for anisotropic materials:

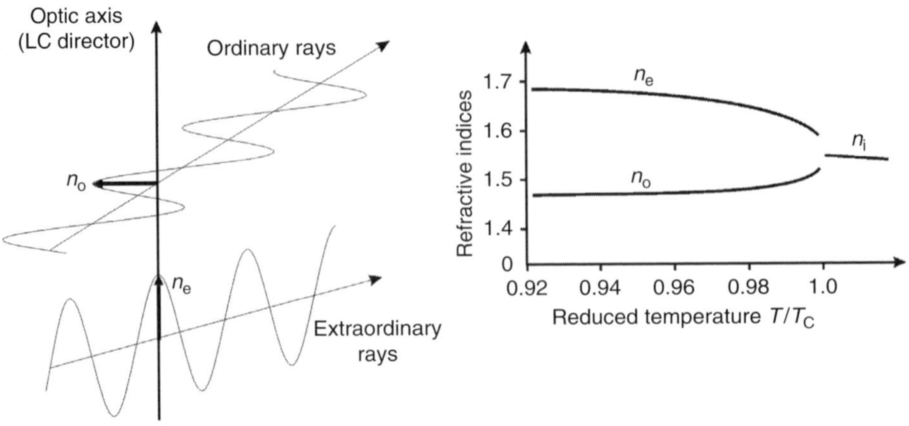

FIGURE 25.15 Refractive indices of a uniaxial nematic.

$$\frac{n_{e,o}^2 - 1}{\langle n^2 \rangle + 2} = \frac{4\pi}{3} N\alpha_{e,o} \tag{25.9}$$

where

n_e and n_o are refractive indices for the extraordinary ray and ordinary ray, respectively
N is the number of molecules per unit volume
$\alpha_{e,o}$ is the molecular polarizability
$\langle n^2 \rangle$ is defined as

$$\langle n^2 \rangle = \frac{n_e^2 + 2n_o^2}{3} \tag{25.10}$$

Then, we may get separate equations for extraordinary and ordinary refractive indices as follow:

$$n_e = \left[1 + \left(\frac{4\pi N\alpha_e}{1 - \frac{4}{3}\pi N\langle\alpha\rangle}\right)\right]^{1/2} \tag{25.11}$$

$$n_o = \left[1 + \left(\frac{4\pi N\alpha_o}{1 - \frac{4}{3}\pi N\langle\alpha\rangle}\right)\right]^{1/2} \tag{25.12}$$

where $\langle\alpha\rangle$ is the average polarizability of the LC molecule and is defined as [11–13]

$$\langle\alpha\rangle = \frac{\alpha_e + 2\alpha_o}{3} \tag{25.13}$$

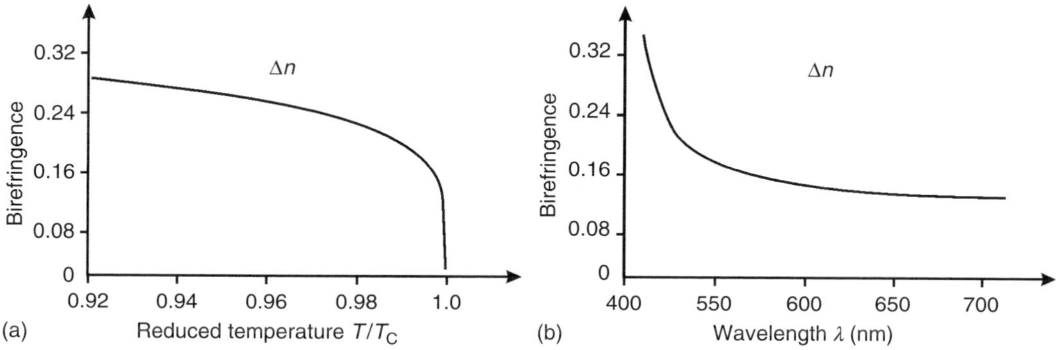

FIGURE 25.16 Temperature and wavelength-dependent birefringence of nematic LC.

The optical anisotropy depends on the wavelength and temperature, see Figure 25.16. On the basis of Haller's approximation, the order parameter S can be approximated as

$$S = \left(1 - T/T_{\mathrm{C}}\right)^{\beta} \tag{25.14}$$

Thus, the temperature-dependent birefringence has the following form:

$$\Delta n(T) = (\Delta n)_0 \left(1 - T/T_{\mathrm{C}}\right)^{\beta} \tag{25.15}$$

where

$(\Delta n)_0$ is the LC birefringence in the crystalline state (or $T = 0\,\mathrm{K}$)
The exponent β is a material constant
T_{C} is the clearing temperature of the LC material under investigation

The major absorption of an LC compound occurs in two spectral regions: ultraviolet (UV) and infrared (IR). The σ–σ* electronic transition takes place in the vacuum UV (100–180 nm) region, whereas the π–π* electronic transition occurs in the UV (180–400 nm) region. If an LC molecule has a longer conjugation, its electronic transition wavelength would extend to a longer UV wavelength. In the near-IR region, some overtone molecular vibration bands appear. The fundamental molecular vibration bands, such as CH, CN, and C=C, occur in the mid and long IR regions. Typically, the oscillator strength of these vibration bands is about two orders of magnitude weaker than that of the electronic transitions. Thus, the resonant enhancement of these bands to the LC birefringence is localized. Wavelength dependence of birefringence or refractive indices can be calculated either with the Cauchy model, single-band model, or three-band model. The single-band model, which describes wavelength dependence of birefringence, is expressed by the formula:

$$\Delta n = G \frac{\lambda^2 \lambda^{*2}}{\lambda^2 - \lambda^{*2}} \tag{25.16}$$

The parameter $G = gNZS\,(f^*_{/\!/} - f^*_{\perp})$, where g is proportionality constant, N is molecular packaging density, Z is effective number of participating electrons (s and p), S is the order parameter, $(f^*_{/\!/} - f^*_{\perp})$ is the differential oscillator strength, and λ^* is the mean electronic transition wavelength. More about different models describing refractive indices and birefringence can be found in preferred reading materials.

In the visible spectral region, n_o is typically in the 1.50–1.57 range; it is not too sensitive to the molecular constituents, but is rather sensitive to the wavelength (λ). As λ increases, n_o decreases and gradually saturates in the near-IR region [14]. In addition, n_o increases slightly as the temperature increases; a more pronounced change takes place only near the phase transition temperature. On the other hand, n_e depends very much on the molecular constituents; it varies from about 1.5 for a totally saturated compound to about 1.9 for a highly conjugated LC in the visible region. The extraordinary refractive index decreases as λ increases. Unlike n_o, n_e declines gradually as the temperature rises, and drops sharply as the temperature approaches the phase transition point. Beyond the clearing point, the LC becomes an isotropic liquid and n_e coincides with n_o, or Δn vanishes.

25.4.4 Elastic Properties

The molecular order existing in liquid crystals has interesting consequences on the mechanical properties of these materials. They exhibit elastic behavior. Any attempt to deform the uniform alignments of the directors and the layered structures (in case of smectics) results in an elastic restoring force. The constants of proportionality between deformation and restoring stresses are known as elastic constants [15,16].

There are three basic elastic constants involved in the electro-optics of liquid crystals depending on the molecular alignment of the LC cell: the splay (K_{11}), twist (K_{22}), and bend (K_{33}), see Figure 25.17. Elastic constants affect the liquid crystal electro-optical cell in two aspects: the threshold voltage and the response time. The threshold voltage in the most common case of homogeneous electro-optical cell is expressed as follows:

$$V_{th} = \pi\sqrt{\frac{K_{11}}{\varepsilon_0 \Delta\varepsilon}} \tag{25.17}$$

Thus a smaller elastic constant K_{11} will result in a lower threshold voltage. However, the response time of the LC device is proportional to the visco-elastic coefficient, which is the ratio of γ_1/K_{11}, γ_1 is rotational viscosity. This means that a small elastic constant is unfavorable from the response time point of view. For many LC compounds and mixtures, the magnitudes of the elastic constants follow the order $K_{33} > K_{11} > K_{22}$ with typical values in the range of 3–25 pN (10^{-12} N). Like many other physical properties, the elastic constants are strongly temperature dependent. Temperature dependence of elastic constants is proportional to S^2.

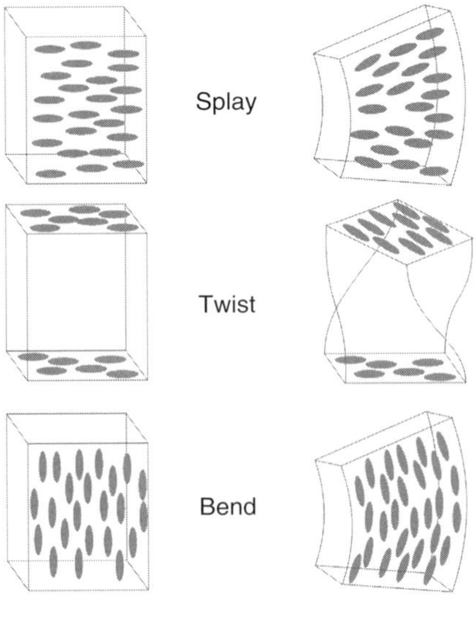

FIGURE 25.17 Elastic constants of the liquid crystals.

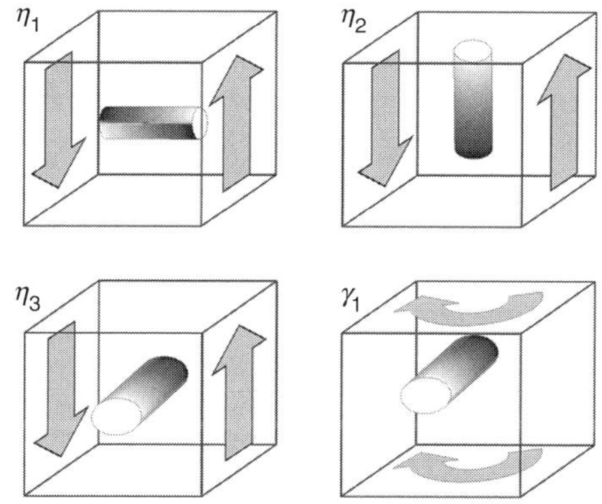

FIGURE 25.18 Anisotropic viscosity coefficients required to characterize a nematic.

25.4.5 Viscous Properties

The resistance of the fluid system to flow when subjected to a shear stress is known as viscosity. In liquid crystals, several anisotropic viscosity coefficients may exist, depending on the relative orientation of the director with respect to the flow of the LC material. When an oriented nematic liquid crystal is placed between two plates, which are then sheared, there are four cases to be studied, see Figure 25.18. Three are known as Miesowicz viscosity coefficients [17,18]. They are η_1, when director of LC is perpendicular to the flow pattern and parallel to the velocity gradient, η_2, when director is parallel to the flow pattern and perpendicular to the velocity gradient, and η_3, when director is perpendicular to the flow pattern and perpendicular to the velocity gradient. Viscosity, especially rotational viscosity γ_1, plays a crucial role in the LCD response time. The response time of a nematic liquid crystals device is linearly proportional to γ_1. The rotational viscosity of an aligned LC is a complicated function of molecular shape, moment of inertia, activation energy, and temperature. A general form for γ_1 is expressed as follows:

$$\gamma_1 = \left(\alpha_0 + \alpha_1 S + \alpha_2 S^2\right)\exp\frac{ES^m}{k\left(T - T_0\right)} \tag{25.18}$$

where

αS are proportionality constants
E is the activation energy of diffusion
m is an exponent
k is the Boltzmann constant
T_0 is the melting point of the liquid crystals

Among these factors, activation energy and temperature are the most crucial factors.

25.5 LIQUID CRYSTAL APPLICATIONS

The most common and well-recognized applications of liquid crystals nowadays are displays. It is the most natural way to utilize extraordinary electro-optical properties of liquid crystals together with its liquid-like behavior. All the other applications are known as nondisplay applications of liquid crystals. Nondisplay applications are based on the liquid crystals molecular order sensitivity to the external incentive. This can be an external electric and magnetic field, temperature, chemical agents, mechanical stress, pressure, irradiation by different electromagnetic waves, or radioactive agents. Liquid crystals sensitivity for such wide spectrum of factors results in tremendous diversity of nondisplay applications. The most known are spatial light modulators (SLMs) for laser beam steering, adaptive optics, light shutters and attenuators for telecommunication, cholesteric LC filters, LC thermometers, stress meters, dose meters, crystals paints, and cosmetics. Another field of interest employing lyotropic liquid crystals is biomedicine, where the LC plays an important role as a basic unit of the living organisms by means of plasma membranes of the living cells. More about existing nondisplay applications can be found in preferred reading materials [19].

25.5.1 Liquid Crystal Displays

As already mentioned, the most common application of liquid crystals is LCD. From the ubiquitous wrist watch and pocket calculator, through high-resolution computer screens, to an advanced high definition (HD) LCD TVs and projectors, this type of display has evolved into an important and versatile interface. Two major factors are frequently attributed to the increasing share of LCDs in the overall display market. The first factor is compactness and lightweight. As will be shown in the following sections, in the simplest case of reflective LCD, it consists primarily of two glass plates (polarizers) coated with transparent electrodes made of indium tin oxide (ITO) with some liquid crystal material between the electrodes and with thin layer of electronics behind it, see Figure 25.19. There is no necessity of a bulky picture tube. The second important advantage is the low power consumption. In general, LCDs use much less power than their cathode-ray tube (CRT) or plasma display panel (PDP) counterparts. On the other hand, liquid crystal displays do have some drawbacks, and these are the subject of intense research. Problems with viewing angle, contrast ratio, and response time were recently significantly minimized by the implementation of new LCD operational modes. The continuous development of liquid crystal technology has generated new and much improved operational modes based on nematic liquid crystals, making today's laptop computers, flat tabletop displays, and large area HD LCD TV possible.

Similarly, we dedicate this paragraph to traditional nematic LCDs since the major technological advances have been achieved from this group of devices. In the preferred reading materials, the reader may find detailed information about other types or modes of displays including ferroelectric display.

The first liquid crystal displays that became successful on the market were the small displays in the digital watches of the 1980s. These LCD's were simple twisted nematic (TN) devices with characteristics satisfactory for simple applications, but by no means possible to be used for larger screens [20]. The basic operation principles are shown in Figure 25.20. In the TN display, each LC cell consists of an LC material sandwiched between two glass plates separated by a gap of 5–8 μm.

The inner surfaces of the plates are deposited with transparent electrodes made of conducting coatings of ITO. These transparent electrodes are coated with a thin layer of polyimide with a thickness of several 100 Å. The polyimide films are unidirectionally rubbed with the rubbing direction of

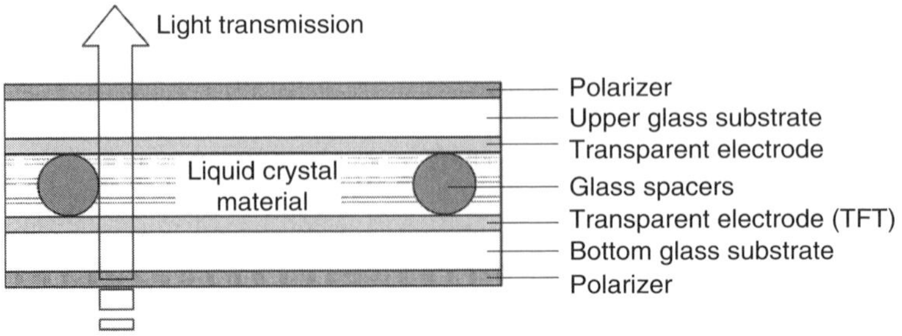

FIGURE 25.19 Schematic view of LC electro-optical cell, the basic unit of LCD.

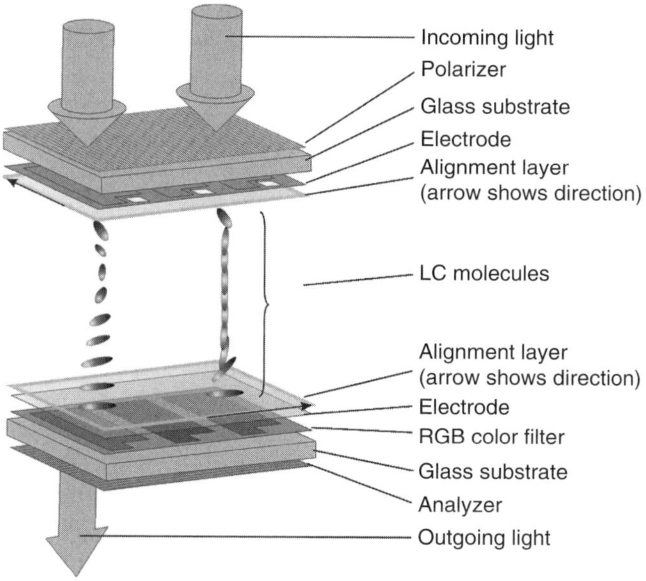

FIGURE 25.20 The basic operation principle of the TN cell.

the lower substrate perpendicular to the rubbing direction of the upper surface. Thus, in the inactivated state (voltage OFF), the local director undergoes a continuous twist of 90° in the region between the plates. Sheet polarizers are laminated on the outer surfaces of the plates. The transmission axes of the polarizers are aligned parallel to the rubbing directions of the adjacent polyimide films. When light enters the cell, the first polarizer lets through only the component oscillating parallel to the LC director next to the entrance substrate. During the passage through the cell, the polarization plane is turned along the director helix, so that when the light wave arrives at the exit polarizer, it passes unobstructedly. The cell is thus transparent in the OFF state; this mode is called NW—normally white operation. In the activated state (voltage ON), when a small voltage (3–5 V) is applied to the electrodes, a strong electric field is built up in the LC layer. As a result of the dielectric anisotropy, the LC molecules are aligned parallel to the direction of the applied electric field. This leads to a homeotropic alignment of the LC molecules with respect to glass substrates. With the molecules oriented in this way, the helical structure disappears, and so does the guiding of the light polarization plane. Thus light cannot pass through both polarizers, and the cell appears black. Typically, a chiral dopant is used to stabilize the twist of only one handedness. Dividing the electrodes into pixels, enables different parts of the cell to be independently switched between black and transparent, and an information display can thus be constructed. The reverse operation mode is possible for 90 TN cell and is called as NB—normally black mode. The electro-optic response characteristic of the TN cell is asymmetric because only the nontransmissive state can be activated by the electric field. When the electric field is zero, the twisted structure, which provides waveguiding of the incident light, is restored by the elastic torques. It means that the ON (rise) time is much faster than the OFF (decay) time. The typical switching times of a TN cell are in the millisecond range. Despite relatively long switching time and limited viewing angle, they are very useful for device applications and have nowadays a broad temperature range, simple construction, and low price.

Super-twisted nematic (STN) cell is defined as one in which the LC twist angle is greater than 90°. The advantage of using a larger twist angle is the much steeper electro-optic curve than TN. A steeper characteristic is very desirable for a nematic display. As a result of this feature, in case of passive driving, the STN display can have about 500 lines, when standard TN displays can hardly reach 100 lines. Large twist has also its disadvantages. The major disadvantage is that the response is slower for the STN display than for the TN display [21]. Because recently thin-film transistor (TFT) driver technology became relatively inexpensive, the STN-type displays almost disappeared from the market.

As mentioned before, the wide viewing angle is a critical requirement for LCD TV or large-screen desktop monitors. Conventional TN LCD exhibits a narrow and asymmetric viewing cone because the LC directors are tilted up under an applied voltage so that the incident light experiences different effective birefringence when viewed from different oblique angles. To achieve wide viewing angle and fast response time, two types of LC modes have been commonly used for this purpose: homogenous alignment with in-plane switching (IPS) and vertical alignment (VA)—initially homeotropic. In a practical display product, modifications of the mentioned modes are implemented to increase performance and reduce disadvantages of the pure IPS and VA modes. The IPS mode typically employs positive dielectric constant LC materials whereas VA mode is based only on the LC with negative $\Delta\varepsilon$. In IPS normally black operation mode, the transmission axis of the polarizer is parallel to the LC director at the input plane [22]. The optical wave traversing through the LC cell is an extraordinary wave whose polarization state remains unchanged. As a result, a good dark state is achieved because this linearly polarized light is completely absorbed by the crossed analyzer. When an electric field is applied to the LC cell, the LC molecules are aligned toward the electric field (along the y axis). This leads to a new director distribution with a twist $f(z)$ in the x–y plane, see Figure 25.21. The vertically aligned LC cell is the one with the highest contrast ratio [23]. Moreover, its contrast ratio is insensitive to the incident light wavelength, cell thickness, and operating temperature. In principle, in the voltage-OFF state, the liquid crystals director is perpendicular to the cell substrates plane, Figure 25.22. Thus, the contrast ratio depends only on the crossed polarizers. Application of a voltage in the ITO electrodes causes the director to tilt away from the normal to the glass surfaces, see Figure 25.22. This introduces birefringence because the index of refraction for light polarized parallel to the director is different from the index of refraction for light polarized perpendicular to the director. Some of the resultant elliptically polarized light passes through the crossed polarizer and the cell appears bright. Besides the wide viewing angle requirement for large LCD's, the response time is also critical. Slow response time causes image blurring if fast moving objects are displayed. Vertically aligned cells possess intrinsic fastest response time among available

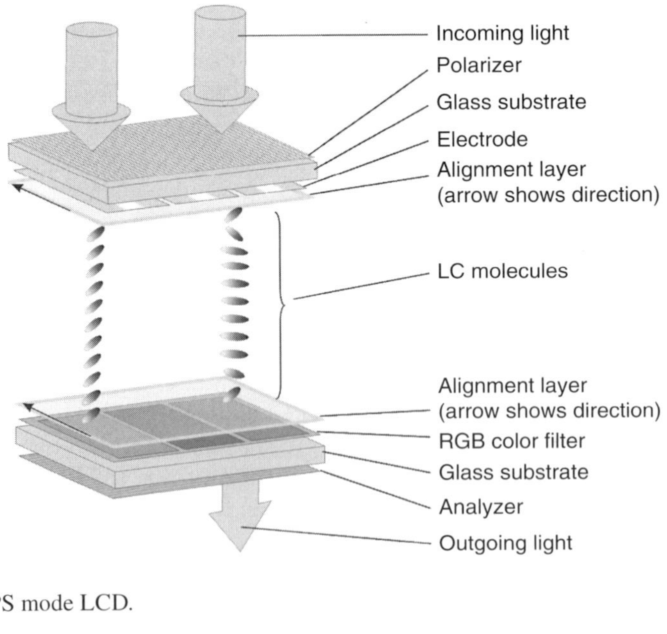

FIGURE 25.21 IPS mode LCD.

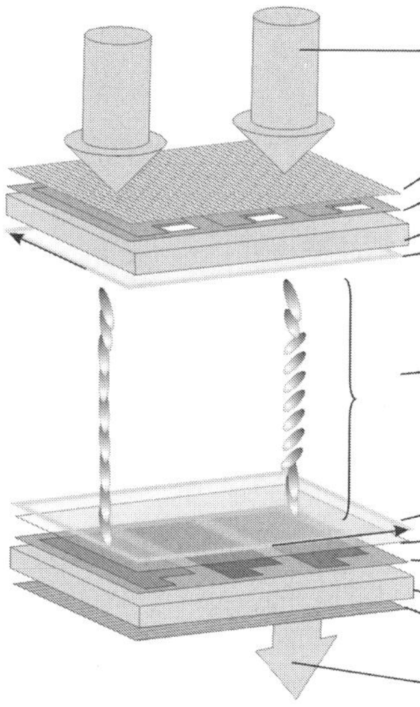

FIGURE 25.22 VA mode LCD.

nematic-based modes. The reason is because of the elastic constant involved in the switching process. The VA mode utilizes a bend (K_{33}) elastic constant, which is larger in value than splay (K_{11}) in homogenous cell or twist (K_{22}) in twisted cell. As mentioned before, the response time of an LC layer is proportional to $\gamma_1 d^2/K\pi^2$, where γ_1 is the rotational viscosity, d is the layer thickness, and K is the corresponding elastic constant.

25.5.2 Polymeric Composites-Based Electro-Optic Devices

Some types of liquid crystal display combine LC material and polymer material in a single device. Such LC/polymer composites are a relatively new class of materials used for displays and also for light shutters or switchable windows [24]. Typically, LC/polymer components consist of calamitic low mass LCs and polymers, and can be either polymer-dispersed liquid crystal (PDLC) or polymer-stabilized liquid crystal (PSLC) [25–27]. The basic difference between these two types comes from the concentration ratio between LC and polymer. In the case of PDLC, there is typically around 1:1 LC and polymer. In PSLC, the LC occupies 90% or more of the total composition. Such a difference results in a different phase separation process during the composites polymerization. For equal concentration of LC and polymer, LC droplets will form. But in the case of the LC as a majority, a polymer will build up walls or strings only, which divide LC into randomly aligned domens. Both types of composites operate between a transparent state and a scattering state. There are two requirements on the polymer for the PDLC or PSLC device to work. First, the refraction index of the polymer, n_p, must be equal to the refraction index for the light polarized perpendicular to the director of the liquid crystal (ordinary refractive index of the LC). Second, the polymer must induce the director of the LC in the droplets (PDLC) or domens (PSLC) to orient parallel to the surface of the polymer, see Figure 25.23. In the voltage OFF state, the LC molecules in the droplets are partially aligned. In addition, the average director orientation $\boldsymbol{n}$ of the droplets exhibits a random distribution of orientation within the cell. The incident unpolarized light is scattered if it goes through such a sample. When a sufficiently strong electric field (typically above 1 $V_{rms}/\mu m$) is applied to the cell, all the LC molecules are aligned parallel to the electric field. If the light is also propagating in the direction parallel to the field then the beam of light is affected by the ordinary refractive index of LC, which is matched with the refractive index of the polymer, thus the cell appears transparent. When the

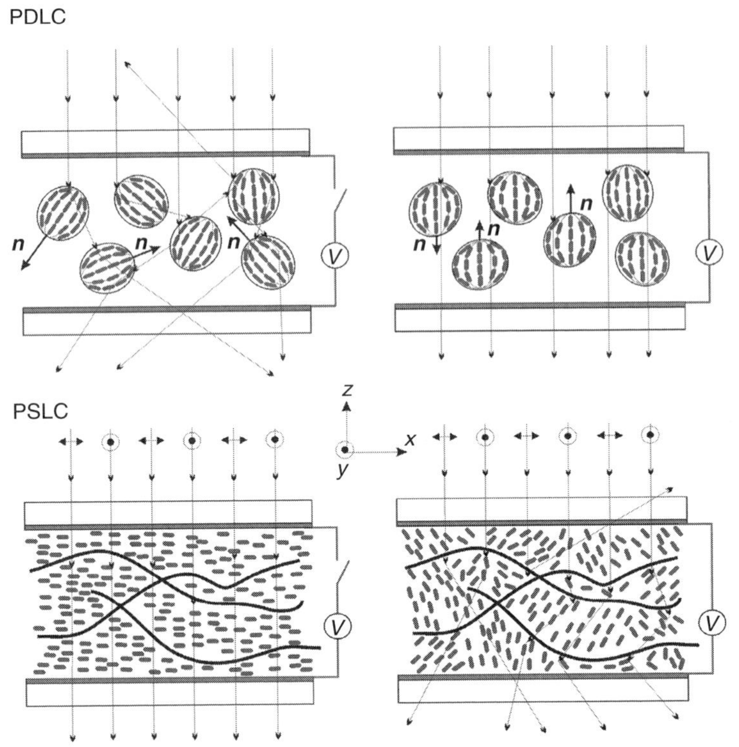

FIGURE 25.23 Schematic view and working principle of polymer/LC composites.

electric field is OFF, again the LC molecules go back to the previous random positions. A polymer mixed with a chiral liquid crystal is a special case of PSLC called polymer-stabilized cholesteric texture (PSCT). The ratio between polymer and liquid crystal remains similar to the one necessary for PSLC. When the voltage is not applied to the PSCT cell, the liquid crystals tends to have a helical structure while the polymer network tends to keep LC director parallel to it (normal-mode PSCT). Therefore, the material has a poly-domain structure, see Figure 25.24. In this state, the light of incident beam is scattered. When a sufficiently strong electric field is applied across the cell, the

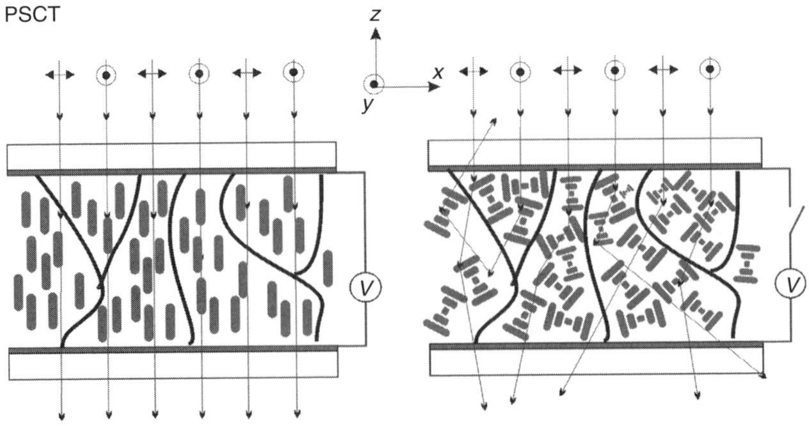

FIGURE 25.24 Schematic view of working principles of polymer-stabilized cholesteric LC light valve.

liquid crystal is switched to the homeotropic alignment; see Figure 25.24, and it becomes transparent.

25.6 SUMMARY

Liquid crystal organic materials, even known for more than a 100 years, are relatively new for optoelectronic devices. What makes them special is physical property anisotropy when its consistency stays—fluid like. Liquid crystal optoelectronic devices utilize this unique phenomenon. The anisotropy comes from small degree of the molecular order existing in liquid crystals while the material remains fluid. Dielectric, optical, magnetic, and elastic properties, observed in different directions with respect to the molecules order director, are not equivalent. This produces a complex response for to external factors such as magnetic or electric field and enables optoelectronic applications. Structural differences in the molecular systems of the liquid crystal materials lead to the different liquid crystal phases, which behave differently under external electric or magnetic field. The viscosity is also different. Clearly, the most important application of liquid crystals is information display. Low weight, relatively low power consumption, and portability place LCDs in a superior position compared to other existing displays technologies like CRT, PDP, or OLED. LC displays may have different forms. The very simple one like monochromatic or calculators and high end 64 in. LCD TV monitor are based on the liquid crystal response to the external electric field. The LCD TVs are divided by different operation modes of the nematic LC panels. Several other applications of LC are worth mentioning. Most known applications are PDLC light limiters and switches. Commercial product may have form of privacy window. SLMs for laser beam steering and adaptive optics are the other very important applications. In (SLM) LC, optoelectronic cell replaces mechanical beam steering by all electronic steering which simplify device construction and maintenance. This type of LC device could be used either by military systems, which requires handling of high power lasers or by telecommunication. Often neglected, lyotropic liquid crystals are important materials in biochemistry of the cell membranes. Based on the recent progress, we would expect continuous fast growth of the LCD TV technology as the flagship application of the liquid crystal organic materials.

EXERCISE QUESTIONS

1. Liquid crystal mixture is composed of two single compounds. By using phase transition sequence, melting heat enthalpy and molecular weight for each of singles, calculate temperature nematic range of the eutectic composition based on Equations 25.3 and 25.4. The share of each of single compounds should be given in wt%.
   Compound 1: Cr 24°C N 35°C iso; $\Delta H = 4100$ cal/mol; mol wt: 249.3
   Compound 2: Cr 43°C N 45°C .iso; $\Delta H = 5300$ cal/mol; mol wt: 227.1
2. From the following structures 1–3, show (a) which has the highest dielectric anisotropy; (b) which one has the highest birefringence. On the basis of given values of dipole moment for structure 1 ($\mu = 2.1$) and structure 2 ($\mu = 3.9$) estimate a dipole moment value for structure 3.

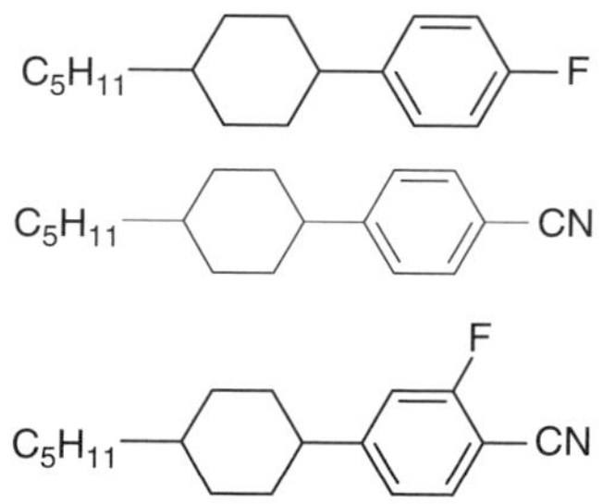

3. Which elastic constant is in general involved during electro-optical switching in the case of TN, IPS, and VA mode display, respectively?

LIST OF ABBREVIATIONS

LC	liquid crystals
LCD	liquid crystal display
BP	blue phase
TGB	twist grain boundary
ITO	indium tin oxide
PDP	plasma display panel
CRT	cathode-ray tube
TN	twisted nematic
STN	super-twisted nematic
TFT	thin-film transistor
IPS	in-plain switching
VA	vertical alignment
PDLC	polymer-dispersed liquid crystal
PSLC	polymer-stabilized liquid crystal

REFERENCES

1. F. Reinitzer, *Monatsh. Chem.*, 9, 421 1888; for English translation see, *Liq. Cryst.* 5, 7 1989.
2. P.F. McManamon, T.A. Dorschner, D.L. Corkum, L. Friedman, D.S. Hobbs, M. Holz, S. Liberman, H.Q. Nguyen, D.P. Resler, R.C. Sharp, and E.A. Watson, Optically phased array technology, *Proc. IEEE,* 84, 268–298, February 1996.
3. D. Demus, J. Goodby, G.W. Gray, H.-W. Spiess, and V. Vill, *Handbook of Liquid Crystals* Vol. 1–4, Wiley-VCH, 1998.
4. V. Vill, *LiqCryst 4.6 – Database of Liquid Crystalline Compounds*, LCI, Hamburg, Germany, 2005.
5. P.J. Collings and M. Hird, *Introduction to Liquid Crystals, Chemistry and Physics,* Taylor & Francis, 1997.
6. G. Friedel, The mesomorphic states of matter, *Ann. Phys.*, 18, 173–174, 1922.
7. L. Schröder, *Z. Phys. Chem.*, 11, 449, 1893. (in German)
8. J.J. Van Laar, *Z. Phys. Chem.*, 63, 216, 1908. (in German)
9. W. Maier and G. Meier, A simple theory of the dielectric characteristics of homogeneous oriented crystalline-liquid phases of the nematic type, *Z. Naturforsch. Teil, A*, 16, 262, 1961.
10. M.F. Vuks, Determination of the optical anisotropy of aromatic molecules from the double refraction of crystals, *Opt. Spektrosk.*, 20, 644, 1966.
11. P.G. de Gennes and J. Prost, *The Physics of Liquid Crystals*, 2nd edn., Oxford University Press, Oxford, 1995.
12. W.H. de Jeu, *Physical Properties of Liquid Crystalline Materials*, Gordon and Breach, 1980.
13. P. Oswald and P. Pieranski, *Nematic and Cholesteric Liquid Crystals: Concepts and Physical Properties Illustrated by Experiments*, Taylor & Francis, CRC Press, Boca Raton, FL, 2005.
14. S.T. Wu and I.C. Khoo, *Optics and Nonlinear Optics of Liquid Crystals*, World Scientific, 1993.
15. D. Demus, J. Goodby, G.W. Gray, H.-W. Spiess, and V. Vill, *Physical Properties of Liquid Crystals*, Wiley-VCH, 1999.
16. S.T. Wu and D.K. Yang, *Reflective Liquid Crystal Displays*, Wiley, New York, 2001.
17. S. Chandrasekhar, *Liquid Crystals*, 2nd edn., Cambridge University Press, 1992.
18. S. Kumar, *Liquid Crystals*, Cambridge University Press, 2001.
19. S.T. Wu and D.K. Yang, *Fundamentals of Liquid Crystal Devices*, Wiley, 2006.
20. M. Schadt and W. Helfrich, Voltage-dependent optical activity of a twisted nematic liquid crystal, *Appl. Phys. Lett.*, 18, 127–128, 1971.
21. P. Yeh and C. Gu, *Optics of Liquid Crystal Displays*, Wiley, 1999.
22. M. Oh-e and K. Kondo Electro-optical characteristics and switching behavior of the in-plane switching mode, *Appl. Phys. Lett.*, 67, 3895–3897, 1995.
23. M.F. Schiekel and K. Fahrenschon, Deformation of nematic liquid crystals with vertical orientation in electric fields, *Appl. Phys. Lett.*, 19, 391, 1971.
24. J.L. Fergason, Polymer encapsulated nematic liquid crystals for display and light control applications, *SID Symp. Dig.*, 16, 68–70, 1985.

25. J.W. Doane, N.A. Vaz, B.G. Wu, and S. Zumer, Field controlled light scattering from nematic microdroplets, *Appl. Phys. Lett.*, 48, 269, 1986.
26. R.L. Sutherland, V.P. Tondiglia, and L.V. Natarajan, Electrically switchable volume gratings in polymer-dispersed liquid crystals, *Appl. Phys. Lett.*, 64, 1074–1076, 1994.
27. Y.K. Fung, D.K. Yang, Y. Sun, L.C. Chen, S. Zumer, and J.W. Doane, Polymer networks formed in liquid crystals, *Liq. Cryst.*, 19, 797–901, 1995.

26 Organic and Polymeric Photonic Band Gap Materials and Devices

Scott Meng and Thein Kyu

CONTENTS

26.1 Introduction 793
26.1.1 Concept of Conduction Band and Valence Band 794
26.1.2 Comparison between Photonic and Electronic Band Gaps 796
26.2 Fabrication of Inorganic, Organic, and Polymeric PBG Materials 799
26.2.1 Inorganic PBG Material 799
26.2.2 Organic and Polymeric PBG Materials 801
26.2.2.1 Self-Assembly of Polymeric Colloidal Crystals 802
26.2.2.2 Self-Assembly of Block Copolymer 803
26.2.2.3 Photolithography Technique 804
26.3 Devices of Photonic PBG Materials 811
26.3.1 Light Emitter 811
26.3.2 Notch Filter 812
26.3.3 Optical Waveguide 812
26.3.4 Colloidal Gels and Networks 813
26.4 Future Perspectives of Organic and Polymeric PBG Material 813
Exercise Questions 815
List of Abbreviations 815
Acknowledgment 815
References 816

Abstract This chapter presents the current view of basic concepts of energy bands including conduction and valence bands followed by a discussion on the analogy between semiconductor and photonic band gap (PBG) materials. The wave propagation principles governing the band gap structures are reviewed. Fabrication of various PBG structures is described with emphasis on inorganic and organic PBG, continuing to polymeric photonic crystals. Some potential applications and device development of PBG materials have been introduced. Finally, future perspective of this emerging field is presented.

26.1 INTRODUCTION

In the last century, the discovery of electronic band gap materials has revolutionized semiconductor industries. By virtue of the concomitant emergence of semiconductor transistors, integrated circuits, and microchips, telecommunication and computer industries have enjoyed explosive growth. The enhanced computation speed, increased storage capacities, and cost reduction are the hallmarks of the revolution of the omnipresent computers. In pursuit of a faster computation speed and a larger memory storage, it becomes apparent that telecommunication through photons is preferred to the conventional electron transport.

In the late 1980s, it was found that photonic band gaps (PBGs) have a striking resemblance in structure to that of the electronic band gaps. To achieve a full PBG, the highest dielectric contrast materials have been sought; this naturally favors the inorganic substances that possess very high dielectric constants. However, the fabrication of such inorganic materials for photonic application is, if not impossible, tedious or commercially not feasible. Under such circumstances, various organic and polymeric molecules came into play in fabricating photonic crystals through photolithography or colloidal approaches.

The development of organic and polymeric PBG materials has a relatively short history, and thus its application is certainly in its infancy. The basic research in this field is undergoing considerable growth and the associated device development having various functions has made steady progress in recent years. Undoubtedly, it is of great educational value for students to grasp some basic knowledge about this emerging field. The present chapter begins with the basic review on the energy bands.

26.1.1 Concept of Conduction Band and Valence Band

As it is well known for over a century, matter is composed of atoms and molecules (i.e., collection of atoms) in which atoms may be subdivided into smaller elementary particles such as electrons, protons, and nuclei. By nature, the electrons are negatively charged particles having a definite mass, but they exhibit both particle-like as well as wave-like behavior, known as wave–particle duality. The particle-like behavior may be described in the context of the classical Newtonian mechanics, whereas the motions of elementary particles of matter, i.e., electrons, follow the laws of quantum mechanics [1–3]. The wave-like behavior of fundamental particles has been successfully described by Schrödinger equation—a monumental equation of quantum mechanics. The Schrödinger equation in the time-dependent form reads [1,2]:

$$-\frac{h}{2m}\nabla^2\psi(\tilde{r},t) + V(\tilde{r},t)\psi(\tilde{r},t) = ih\frac{\partial\psi(\tilde{r},t)}{\partial t} \tag{26.1}$$

where

$\tilde{r}$ is the position vector

h is the Planck constant

m is the particle mass

$V(\tilde{r},t)$ is the potential energy and $\psi(\tilde{r},t)$ is a time- and space-dependent state function containing all measurable quantities, i.e., the allowed energy states or the probability distribution. The first term on the left hand side of Equation 26.1 represents the kinetic energy and the second term the potential energy.

The solution of Schrödinger equation explains the allowed energy states of a particle system in a manner dependent on the potential energy of the system. In general, $\psi(\tilde{r},t)$ may be expressed in the wave form, i.e., [2]

$$\psi(\tilde{r},t) = \psi(\tilde{r})\exp(-i\omega t) \tag{26.2}$$

where the angular frequency, $\omega = 2\pi\nu$, with ν being frequency. For a specific simple system such as a single hydrogen atom at a given time, it may be described as follows [2]:

$$\psi(\tilde{r}) = nlm_l \tag{26.3}$$

where
- n is the principal quantum number
- l is the orbital quantum number
- m_l is the magnetic quantum number

The principal quantum number comes from the radial solution of the Schrödinger equation. The orbital quantum number is a measure of the magnitude of orbital angular momentum. The magnetic quantum number is a consequence of the splitting of the spectral lines caused by an external magnetic field. Depending on whether it is for an atom, a free particle, or a semiconductor, the allowed energy values may be discrete or continuous and a form of densely packed discrete level, respectively [3]. The densely packed discrete level is termed as band. An example of energy levels of a single atom, insulator, semiconductor, and metal is illustrated in Figure 26.1.

As is well known, a single isolated atom depicts discrete energy levels denoted by small letters s and p for the respective orbital numbers of the electrons. In the lowest energy level of 1s and 2p, the three solids (insulator, semiconductor, and conductor), which are collections of innumerable atoms, still exhibit the discrete energy level. The reason behind this is that the core electrons are shielded from the external field produced by the neighboring atoms, the lower energy level of those electrons is not different from that of a single atom since there is no broadening of the lower energy level. On the other hand, the higher energy levels of three solids are broadened into closely spaced discrete energy levels, termed as energy bands. The lowest unoccupied energy band at ground state is known as conduction band (shaded by gray). While the highest occupied energy band at ground state is the valence band (shaded by dark color). The conduction band is virtually empty whereas the valence band is fully filled with electrons at ground state and before doping. Between the conduction band and the valence band, there exists an energy gap, referred to as the band gap (symbolized by E_g).

Figure 26.1 illustrates the comparison among the band gap structures of insulator, semiconductor, and metal conductor. Insulator has the largest band gap; it is therefore the most difficult for electrons to get excited by external energy into the conduction band from the valence band. It is exactly the reason why insulators cannot conduct electrons. Metal bears a partially filled band. As a result, electrons can easily traverse or hop within the partially filled band, thereby rendering high conductivity

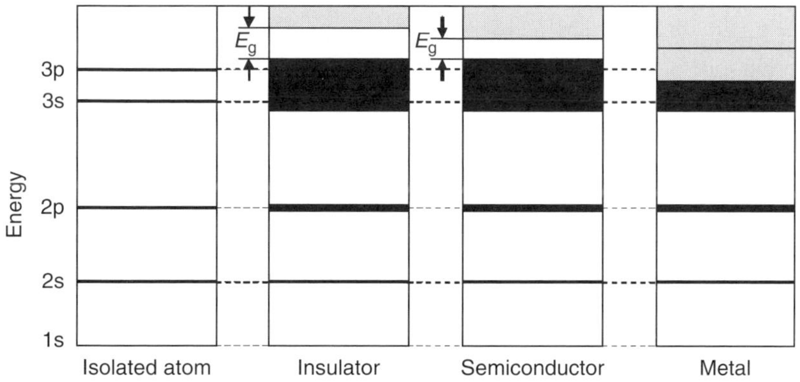

FIGURE 26.1 Various energy levels of isolated atom, insulator, semiconductor, and metal. The gray-shaded region represents the lowest unoccupied energy band at ground state known as conduction band. The dark region is the valence band that is fully filled with valence electrons. These conduction and valence bands are separated by a band gap denoted by E_g.

to metal. Semiconductor has an energy gap amid metal and insulator; therefore its conductivity is intermediate between those of the metal and the insulator.

26.1.2 Comparison between Photonic and Electronic Band Gaps

So far, we have discussed the concepts of band gap of electrons. We shall take a next step to elucidate the concepts of band gap of photons, hereafter termed photonic band gaps. Photons are emitted or absorbed when electrons are undergoing an atomic transition from a high-energy level to a low-energy level or vice versa. We will briefly elaborate on the origin of the photonic band gap in a fundamental way by considering the basic Maxwell equations.

Analogous to the behavior of electrons, photons exhibit both wave- and particle-like behavior. The wave-like behavior of photons may be described in accordance with the Maxwell equations in what follows [4–6]:

$$\nabla \times E = -\frac{1}{c}\frac{\partial H}{\partial t} \tag{26.4}$$

$$\nabla \times H = \frac{\varepsilon(\tilde{r})}{c}\frac{\partial E}{\partial t} \tag{26.5}$$

where c is the velocity of light, E and H are electric field and magnetic field strength, respectively. In free space (i.e., vacuum), the propagation of photons may be described by a plane wave, when there is no spatial variation of dielectric constant, i.e., isotropic $\varepsilon(r) = \varepsilon$. Note that, in the optical frequency range, ε equals the square of refractive index n. Physically Equations 26.4 and 26.5 state that changing magnetic field in time generates an electric field, and vice versa. The repetition of this process serves as the basis of the electromagnetic (EM) wave propagation in free space. The electromagnetic wave solution of the Maxwell equations in time-space reads

$$E, H \sim \mathrm{e}^{-i(kr-\omega t)} \tag{26.6}$$

where

k is the wave vector, taken as $k = 2\pi/\lambda$

λ is the wavelength

Taking curl of Equation 26.4 and substituting it into Equation 26.5 in conjunction with the vector identity operation, one obtains the following equation:

$$\nabla^2 E = \left(\frac{\omega}{c}\right)^2 \varepsilon \frac{\partial^2 E}{\partial t^2} \tag{26.7a}$$

On the same token, one can deduce the same formulism for the magnetic field strength,

$$\nabla^2 H = \left(\frac{\omega}{c}\right)^2 \varepsilon \frac{\partial^2 H}{\partial t^2} \tag{26.7b}$$

Equations 26.7a and 26.7b are known as the EM wave equations that serves as the basis for the propagation of photons (light) and energy transport in electrodynamics. The solutions of Equation 26.7 in free space establish the correlation between angular frequency ω of photons and wave vector, k in which ω may be regarded as a form of energy through the relationship, $E = h\nu$, where $\nu = \omega/2\pi$.

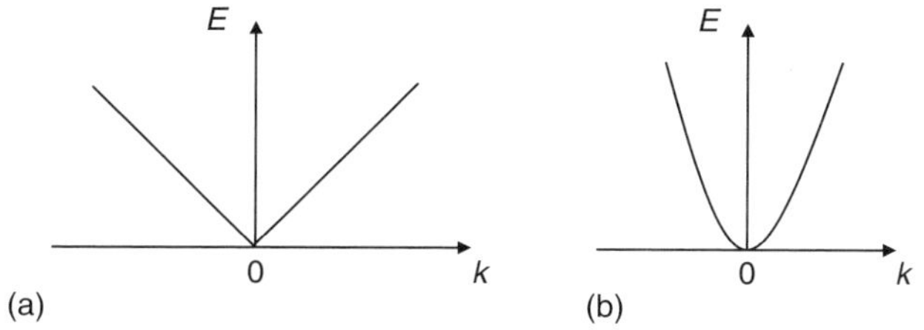

FIGURE 26.2 Dependence of energy on wave vector in free space for (a) photons and (b) electrons.

The schematic illustration of the energy-wave vector plots of propagation of photons in free space is depicted in Figure 26.2a which may be compared with the nonlinear dependence of energy on wave vector of the propagation of electrons in free space as illustrated in Figure 26.2b [7–9].

Analogous to the propagation of photons, the propagation of electrons in free space (i.e., vacuum) may be described in terms of the solution of Schrödinger equation as a sinusoidal wave. If the material is a periodic dielectric medium as illustrated in Figure 26.3, meaning that $\varepsilon(\tilde{r})$ has a periodic variation in space, then the solution to Equation 26.4 has to be modified to account for the periodic change $\varepsilon(\tilde{r})$.

From the mathematical point of view, the problem of the periodic eigen-operator may be solved in the context of Bloch–Floquet theorem in what follows [4]:

$$H(\tilde{r},t) = H_k(\tilde{r})e^{i(kr-\omega t)} \quad (26.8)$$

In dealing with the Bloch–Floquet theorem, there are several numerical methods for solving the problem of propagation of electromagnetic waves in a periodic dielectric medium that include plane wave method, transfer matrix method, and finite different time domain method [5]. Plane wave method is the most common methodology that employs the steady state at a particular wave vector, whereas the transfer matrix method targets the steady state at a particular frequency. Finite different time domain method is employed in place of the time-dependent transient problem.

In the framework of plane wave method, the propagation in a periodic dielectric medium may be visualized such that the plane waves in a uniform medium split into two standing waves, viz., $\cos(\pi x/a)$ and $\sin(\pi x/a)$ in one dimension (1D) [4,5]. The electric field of the $\cos(\pi x/a)$ wave is the strongest with maximum peaks in the high dielectric constant medium, while that of the $\sin(\pi x/a)$ wave shows maxima in the low dielectric constant medium. A schematic representation of this

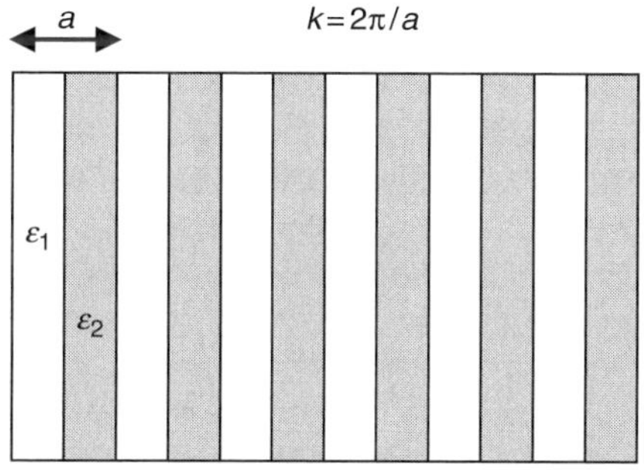

FIGURE 26.3 Illustration of materials exhibiting 1D periodic variation of dielectric constant.

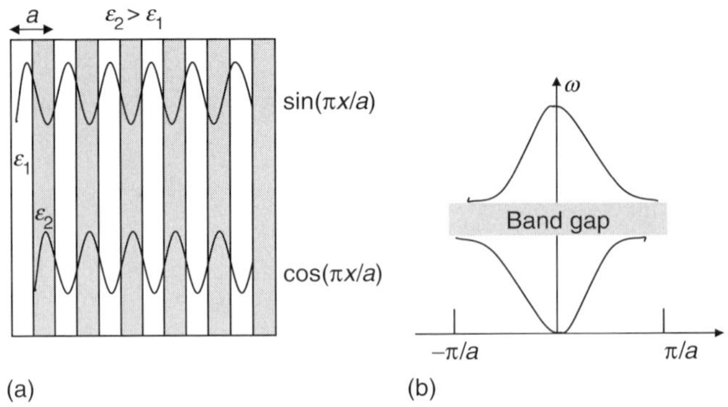

FIGURE 26.4 (a) Schematic representation of a plane wave splitting into two standing waves. (b) Illustration of a photonic gap in between the high-frequency band and low-frequency band.

concept is given in Figure 26.4a, depicting the splitting of two waves in a periodic dielectric medium. Consequently, a discrete band gap is formed due to the opposite shifting of the two bands, which may be witnessed in Figure 26.4b.

In general, any material having periodic dielectric contrasts falls in a category of photonic band gap structures, interchangeably termed photonic crystals. Photonic crystals are generally classified according to the level of structural periodicity as 1D, 2D, or 3D photonic structures, where dielectric constant varies periodically in one, two, or three directions, respectively [4–6]. Figure 26.5 illustrates the structures of 1D, 2D, or 3D photonic crystals.

As illustrated in Figure 26.5, photonic crystal has a similar ordered structure to that of a crystal lattice (e.g., silicon semiconductor). The semiconductor operates based on the principles of electron wave functions in order to produce the energy band gap of electrons; the applications of which span every aspects of digital life. The major difference between a regular crystal and a photonic crystal is that the structure of a crystal lattice is within the atomic length scale in contrast to that of photonic crystals which is in the range of optical wavelength of a few hundred nanometers to a micron.

The propagation of electromagnetic wave in periodic media may be traced back for over a century ago to the time of Lord Rayleigh who predicted the existence of a band gap whereby light is prohibited to propagate in some regions of 1D periodic variation of the dielectric constants. After that, studies on 1D photonic crystal were dormant for some decades until the invention of holographic technique was made for the fabrication of holograms. On the basis of 1D photonic crystal, a number of electro-optical devices have been developed such as dielectric mirror, notch filters, fiber

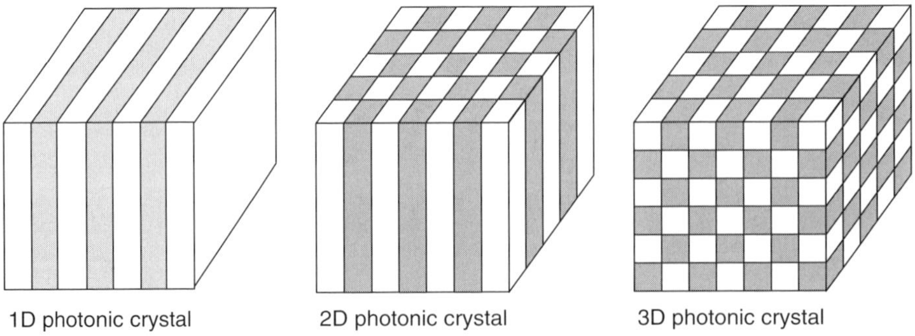

FIGURE 26.5 Schematic representation of 1D, 2D, and 3D photonic crystals having 1, 2, and 3 dimensional periodicity, respectively.

gratings, and distributed feedback lasers. Due to its periodicity constraints, 1D photonic crystal has limited capability to control or modulate light [10–12].

In two monumental papers in 1980s, Yablinovitch [13] and John [14] independently identified that light propagation can be controlled in 2D or 3D photonic crystals. Their monumental works have opened up a new era of photonic research of fabricating electro-optical devices based on the 2D and 3D photonic structures. Since then, rigorous approaches have been developed in the quest for a material structural design having a complete photonic band gap, i.e., structures exhibiting photonic band gaps for all possible polarizations arising from the interference and configurations of coherent laser light. The early stage of the photonic research focused largely on the conceptual development involving theoretical predictions and numerical calculations of the photonic structure and properties. Later on, experimentalists take advantage of the guidance afforded by theoretical findings, demonstrating that a certain 3D structure such as the diamond structure, consisting of air, dielectric spheres, or cylindrical void in a face-centered cubic (fcc) lattice, exhibits a complete photonic band gap [10,12].

From the material science perspective, photonic crystals can be categorized similar to the material properties as inorganic, organic and polymeric PBG materials, and inorganic/organic hybrid PBG. Both inorganic and organic PBG materials have their own merits in respect of fabrication and applications. The comparison of inorganic PBG material and organic/polymeric PBG material will be made in a subsequent section, with emphasis on the fabrication methodology of organic and polymeric PBG structures. Some device applications of organic and polymeric PBG material will be covered in later sections.

26.2 FABRICATION OF INORGANIC, ORGANIC, AND POLYMERIC PBG MATERIALS

26.2.1 INORGANIC PBG MATERIAL

Following the monumental works of Yablinovitch [13] and John [14], considerable effort has been directed to fabricating photonic crystals in search of a complete photonic band gap. The applications of PBG materials to device development are in the range of optical to near-infrared (IR) wavelengths. The initial investigation was carried out mainly by theorists to evaluate appropriate PBG structures that exhibit a complete band gap in the visible and near-infrared wavelengths. It was postulated theoretically that the photonic crystals with a complete photonic band gap need to have a high dielectric or refractive index contrast (>2.8) and long-range periodicity in three dimensions [15–17]. Consequently, the early experimental fabrication of the complete photonic band gap focused largely on making the ordered cavity channels in inorganic materials which have a high dielectric contrast between the inorganic substance and the cavities. The inorganic PBG materials are typically represented by the ordered cavity structures of metals and semiconductors. Most inorganic PBG materials belong to the III–V compounds including gallium arsenide (GaAs), aluminum gallium arsenide (AlGaAs), GaAs/AlGaAs alloys, gallium arsenic phosphide (GaAsP), indium phosphide (InP), aluminum oxide (Al_2O_3, commonly known as alumina), silicon (Si), and silicon dioxide (SiO_2). There are several methods to fabricate inorganic PBG materials, a few include mechanical drilling [18], reactive-ion etching [19,20], holographic lithography [21,22], and self-assembly of inorganic colloidal crystals [23] among others.

Figure 26.6a represents the mechanical drilling methodology developed by Yablinovitch and coworkers to drill precision holes in a high dielectric medium. The resultant fcc structure is illustrated in Figure 26.6b. For the length scale of microwaves, the channels may be created by mechanical drilling. Regarding the structures in the length scale of the near-infrared wavelength, the holes may be made via reactive-ion etching. Recently, it has been demonstrated that the various shapes of holes such as the circular, triangular, or square may be generated based on the pertinent pre-patterning process, generating hexagonal, square, and honeycomb lattices.

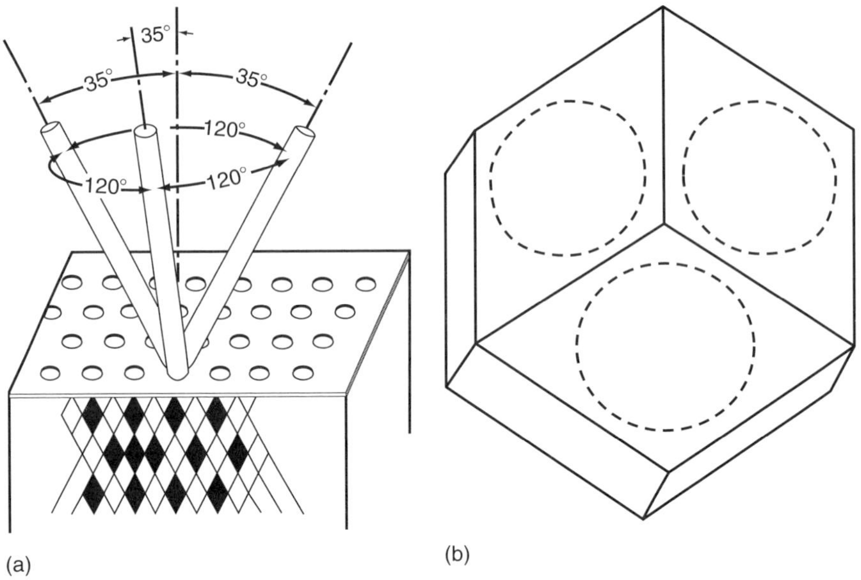

FIGURE 26.6 Schematic representation of (a) mechanical drilling method to construct a face-centered cubic (fcc) crystal structure illustrated in (b). (Reproduced from Yablinovitch, E., Gmitter, T.J., and Leung, K.M., *Phys. Rev. Lett.*, 67, 2295, 1991. With permission.)

Electrochemical deposition is a popular means of fabricating 2D inorganic PBG material and has been adapted in the manufacture of semiconductor industry by virtue of its high resolution [6,9]. This technique marks a pre-patterned structure on the surface of an inorganic material template via electron-beam lithography and subsequently etching the marked areas by either reactive-ion etching (dry etching) or anodic oxidation etching (wet etching). The reactive-ion etching makes use of the reactive ions generated from the plasma discharge in a reactive gas (e.g., Cl_2) to etch away; more importantly, this process inherently involves no solvent. On the contrary, the anodic oxidation etching requires a solvent to etch out the unwanted areas. The typical inorganic PBG structures grown on the basis of electrochemical approach are ordered porous silicon and porous alumina. Figure 26.7

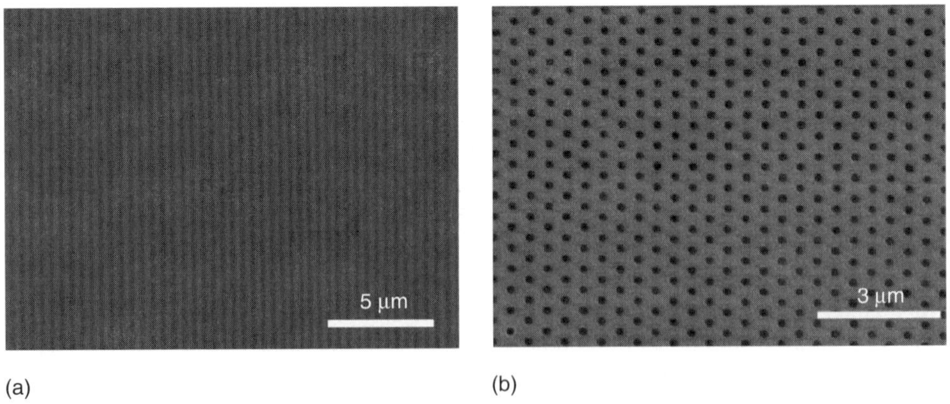

FIGURE 26.7 Scanning electron microscope (SEM) images of porous alumina structure fabricated by the anodization of pre-patterned aluminum: (a) side view and (b) top view. (Reproduced from Choi, J., Luo, Y., Wehrspohn, R.B., Hillebrand, R., Schilling, J., and Gosele, U., *J. Appl. Phys.*, 94, 4757, 2003. With permission.)

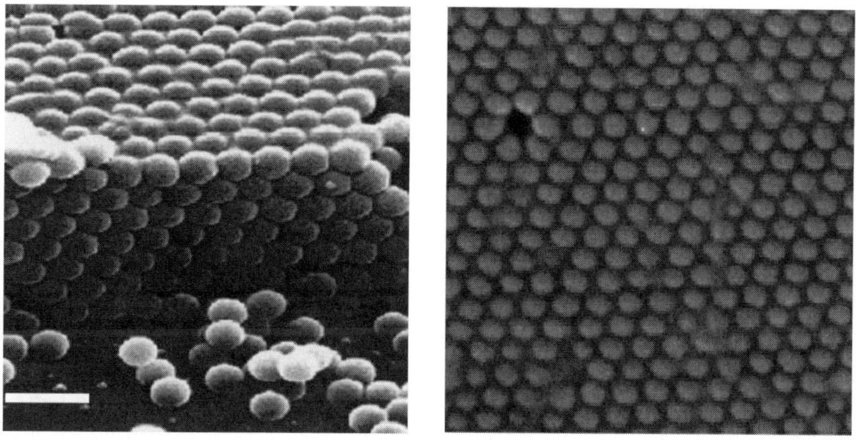

(a) (b)

FIGURE 26.8 3D photonic crystals formed by self-assembly of (a) silica colloids. (Reproduced from Colvin, V.L., *MRS Bull.*, 26, 637, 2001. With permission from Materials Research Society.) (b) Polystyrene spheroids. (From Xia, Y., Gates, B., and Li, Z.-Y., *Adv. Mater.*, 13, 409, 2001; Lu, Y., Yin, Y., Li, Z.-Y., and Xia, Y., *Langmuir*, 18, 7722, 2002.) The scale bar in (a) is 1 μm. (Reproduced from Xia, Y., Gates, B., and Li, Z.-Y., *Adv. Mater.*, 13, 409, 2001. With permission from America Chemical Society; Lu, Y., Yin, Y., Li, Z.-Y., and Xia, Y., *Langmuir*, 18, 7722, 2002. With permission.)

depicts scanning electron microscope (SEM) images of porous alumina structure fabricated by the anodization of pre-patterned aluminum. The pore size and pore density may be readily modulated via control of the applied voltage and acid types.

One of the conventional methods for fabricating inorganic PBG material may be through the self-assembly of colloidal crystals to form highly ordered closed packed array structures. The main idea is that the building elements (i.e., monodisperse colloidal spheres) can organize spontaneously among themselves to form an ordered structure when approaching the thermodynamic equilibrium state. Gravity sedimentation is known as a traditional fabrication methodology, in which the silica particles are suspended in a solution and gradually settle onto the bottom of the container while the solvent evaporates. The solvent evaporation rate is crucial to a successful formation of a periodic lattice. The colloid concentration may be controlled to give 2D or 3D arrays having different number of layers. Evidently, the gravity sedimentation is a tedious process with the resultant structures containing numerous defects. To improve the fabrication efficiency and minimize the number of defects, several techniques have been developed via self-assembly of colloidal crystals in planar films including a cell method, a vertical deposition method, a convective assembly method, etc. Figure 26.8 presents a 3D inorganic PBG structure fabricated from self-assembly of silica colloidal crystal showing closed packing of the spheres (see Figure 26.8a). A similar self-assembled structure can be produced using monodisperse polystyrene spheres (Figure 26.8b).

26.2.2 Organic and Polymeric PBG Materials

In recent years, development of organic and polymeric PBG material has gained considerable momentum for electronic and photonic applications [22,24]. This progress may be attributed primarily to a number of advantages that the organic and polymeric PBG materials have over inorganic PBG materials such as fabrication flexibility–a variety of processing motifs, molecular design efficacy including chemical modification, compatibility with inorganic/active organic/chromophores, and integration of active functionality. However, in practice most of organic small molecules are volatile or unstable, especially at elevated temperatures. Hence, bulky highly conjugated aromatic condensates have been integrated as a major component of the PBG devices [25]. Alternatively, supramolecular dendritic structures or hyperbranched polymers with desired functionality have been designed as a means of improving functionality and compatibility with substrates [26]. Thermal and mechanical stabilities remain the critical issues in the aforementioned organic substances, thus polymer molecules have been sought as potential PBG materials.

Polymers are long chain macromolecules consisting of numerous monomeric units; note that a monomer is defined as the smallest repeat unit [27]. The conversion from monomers to a polymer chain may be made through a chemical reaction process of the incipient monomers called polymerization. Polymerization can be classified into two main categories, i.e., step growth polymerization (e.g., condensation reaction) and additional polymerization (e.g., ionic or free radical reactions). Among them, free radical polymerization has been extensively applied to photolithography in which photo-initiators are excited by ultraviolet (UV) or visible light in generating radicals. These radicals then react with incipient monomers and the radical monomers propagate into a long chain molecule which is called the propagation step. The polymer chains cease to grow through a termination process that can occur via recombination or disproportionation to yield the so-called dead polymer chains.

The process of polymerization driven by the photo-initiation is named as photopolymerization [27]. The basic monomeric units of polymer are mostly organic small molecules, or to a lesser extent, inorganic molecules or combination of both known as organic/inorganic hybrids. These organic/inorganic hybrid polymers would be far superior to neat organic small molecules in respect of thermal stability, solvent resistance, mechanical strength, and modulus. Therefore, the hybrid PBG material undoubtedly plays a pivotal role in the design of PBG structures.

Like organic small molecules, the structural units of polymers may be readily modified to render certain properties or desired functions through grafting. A wide range of improved physical and chemical properties may be imparted into the polymeric PBG material, including a wide variation of refractive index contrasts.

From processing point of view, polymers are much easier and cost-efficient to handle, compared with inorganic materials. There are a variety of processing motifs to make fabrication feasible. Polymers are also much lighter in weight and relatively cheap. For the reasons mentioned above, polymeric PBG materials are often preferred to the inorganic PBG counterpart.

Another interesting development in the design of PBG structures is the incorporation of anisotropic liquid crystal (LC) as one of the constituents in the polymer matrix in which the director of liquid crystal molecules may be switched upon application of the external electric field [28]. The PBG material thus fabricated with liquid crystals has the electrical switchability that requires only a small driving voltage (i.e., a few volts), although the refractive index contrast may not be large as compared to other passive inorganic photonic devices. Like inorganic PBG material, the vast groups of polymeric PBG materials may be categorized in accordance with their fabrication methods.

26.2.2.1 Self-Assembly of Polymeric Colloidal Crystals

Similar to the fabrication of inorganic PBG via self-assembly of silica colloidal crystals, colloidal polymer suspensions may be utilized to produce the polymeric PBG structure [29,30]. The chemistry of fabricating polymer colloids is well established in that monodisperse polystyrene (PS) spheres or poly(methyl methacrylate) (PMMA) spheres may be fabricated via emulsion polymerization in a wide range of sizes from a few hundred nanometers to a few micrometers. On the same token, various colloidal crystals may be easily shaped including spheres, concave disks, and ellipsoids by adding different kinds or amounts of surfactants (i.e., the number of ionizing sulfonate group on the sphere surface) or by an external electric field.

What is more appealing is that the long-term stability problem of the ordered crystal colloidal arrays may be conveniently resolved by stabilizing with the chemical junctions through a cross-linking reaction at the interface of the colloidal crystals to permanently lock-in the crystal colloidal array structures. The aforementioned strategy was implemented for instance by introducing the highly purified acrylamide, *N,N′*-methylene bis-acrylamide, and *N*-vinylpyrrolidone into the colloidal solution, and subsequently initiating the photopolymerization by irradiating with UV light. Consequently, a cross-linked hydrogel was formed having a periodic order of colloidal crystals.

26.2.2.2 Self-Assembly of Block Copolymer

Self-assembling of block copolymer is a popular approach widely practiced for templating of nanostructures. By definition, block copolymers are the linear arrangements of building blocks of dissimilar repeat units. A building block that consists of identical repeat units (say A monomer) may be linked with another block of different chemical structure (say B monomer) in a linear sequence; it is called a AB diblock copolymer that takes the following configuration:

$$\cdots\text{–(A–A–A–A}\cdots\text{–A–A–)(B–B–B–B–}\cdots\text{–B–B)–}\cdots \text{ or } \text{–(A)}_x\text{(B)}_y\text{–}$$

where the subscripts, x and y, denote the numbers of repeat units. If both ends of the middle block (B) are capped with the A-blocks, e.g., $\text{–(A)}_x\text{(B)}_y\text{(A)}_x\text{–}$, it is termed an ABA triblock copolymer. On the same token, an ABC triblock copolymer consisting of three different repeat units A, B, C may be configured in stereoregular or block sequence as

$$\cdots\text{ABC–ABC–ABC–ABC–ABC–ABC–ABC}\cdots \text{ or } \text{–(A)}_x\text{(B)}_y\text{(C)}_z\text{–}$$

A unique feature of block copolymers is that in approaching thermodynamic equilibrium the block copolymers may self-assemble into a variety of well-defined and ordered nanostructures such as spheres, cylinders, lamellae, and gyroids in a manner dependent on the relative ratio of each constituent and the strength of microphase segregation. These equilibrium microstructures may be best explained in terms of thermodynamic phase equilibria. If the system is in the strong segregation limit, the emerged microstructure is well defined; otherwise, the structure can be fuzzy in the weakly segregation limit. The self-assembled nanostructures from the block copolymer molecules are generally too localized containing sizable number of defects. Since such defective microstructures lack the long-range order, its utilization as optical elements is therefore limited. However, recently it becomes possible to increase the long-range ordering of self-assembled nanostructures by means of directional annealing that makes block copolymer a possible candidate for some optoelectronic applications [31,32]. In addition, the domain size of the block copolymer structures may be manipulated by controlling the molecular weight and sequence of the constituents. In addition, some functional groups can be easily incorporated into the building blocks to alter or modify the chemical and physical properties of the backbone or the side branching through grafting. Hence, with a proper design of a block copolymer, one may find its application as optical switches or waveguide [33].

Figure 26.9 illustrates the atomic force microscope (AFM) picture of the highly ordered cylindrical array structures grown from a self-assembled polystyrene-block-polyethylene oxide block copolymer (PS-b-PEO) through solvent evaporation. The major constituent PS forms the matrix while the minor constituent PEO forms the cylindrical domains (shown in black color). The high degree of long-range ordering in block copolymer cylindrical array structures is achieved by controlling solvent evaporation rate, thereby producing the highly ordered nanoscopic cylindrical domains normal to the film planes. In the initial step, a spin-coated film obtained from the self-assembly of PS-b-PEO shows a low degree of lateral ordering (long-range ordering). Subsequently, the sample film was immersed into the benzene solvent at room temperature and then benzene diffuses into the sample and eventually swells the film. Upon removal of the benzene, the lateral ordering of PEO domains is greatly enhanced and therefore the highly oriented cylindrical arrays

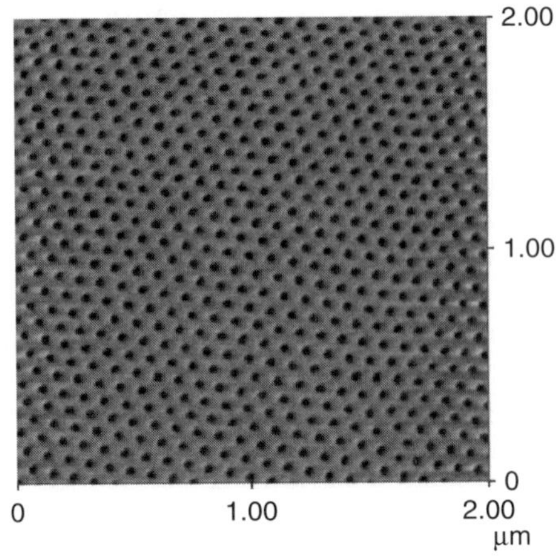

FIGURE 26.9 AFM image of the highly ordered cylindrical array structures grown from the solvent evaporation of self-assembled PS-b-PEO block copolymer film. (Reproduced from Kim, S.H., Misner, M.J., Xu, T., Kimura, M., and Russell, T.P., *Adv. Mater.*, 16, 226, 2004. With permission.)

structures are created. In a recent paper [33], some potential applications of these structures as a PBG material have been identified to utilize as a plane waveguide or an integrated optical device.

26.2.2.3 Photolithography Technique

A large portion of the polymeric PBG structures has been fabricated primarily via a photolithography technique, because photolithography offers a convenient means of fabricating structures with a wide range of sizes coupled with the availability of commercial dyes that serve as photo-initiators in a wide range of incident wavelengths to initiate photoreaction [34,35]. Most importantly, the photopolymerization may be accomplished in a few seconds under strong irradiation with a green laser or UV beam, thereby greatly reducing the fabrication cycle time of PBG structure. The lithography technique may be divided into a number of subgroups, such as holographic lithography, electron-beam lithography, and two- or three-photon techniques. Among them, the holographic lithography technique is suited for the fabrication of polymeric PBG structure via pattern photopolymerization with various wave interference techniques.

The holographic lithography technique operates based on the principles of wave-inference of two- or multiple incoming waves to realize the spatially controlled photoreaction at the length scale ranging from a few hundred nanometers to a few microns. Through various geometries of optical configurations, 1D, 2D, or 3D periodic patterns may be imprinted on the film sample to produce a holographic structure.

An illustrative representation of the wave-inference concept is shown in Figure 26.10. When the two incoming waves from the opposite directions impinge on the same medium, the inference of waves takes place. The resulting wave is governed by the principle of superposition resulting in the net effect of the two individual waves on the particle medium. If the crest of one wave meets the crest of the other and the trough of one wave meets the trough of the other, the constructive interference occurs. The resulting wave is an amplified alternating band of bright and dark regions. The brightest stripes correspond to the regions where the two crests coincide, and vice versa the darkest regions. If the crest of one wave coincides with the trough of the other, then destructive interference occurs. The resulting wave is a complete dark band. Interested readers may be referred to other physics textbooks for details [3].

The spacing of the periodic structure fabricated via holographic lithography technique may be determined in accordance with the well-known Bragg's law,

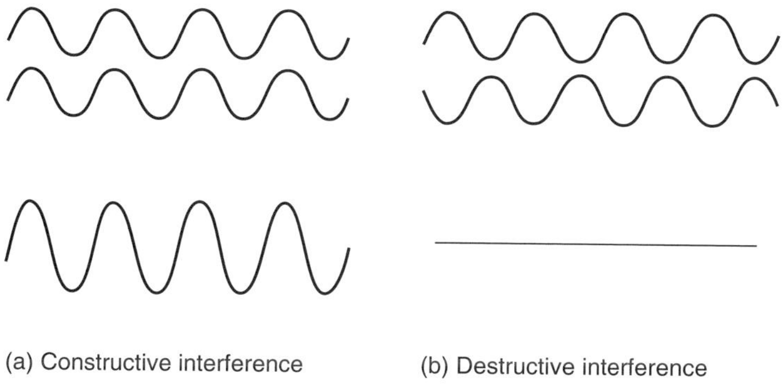

FIGURE 26.10 Conceptual illustration of constructive and destructive interference in two-wave interference.

$$n\lambda = 2d\sin\theta_B \tag{26.9}$$

where
- λ is the wavelength of the incident beam
- n is the integer
- d is the spacing of the periodic structure
- θ_B is the Bragg angle

It may be noticed from Equation 26.9 that simply by changing the incident angles, a wide range of spacing of the periodic structure may be created.

Figure 26.11 is a schematic illustration of the setup of two-wave interference optics in which the incoming laser beams are directed to a beam splitter by a reflection mirror. The function of the beam splitter is to divide the incoming light into two beams of equal or comparable magnitude without changing the phase. Subsequently, the laser beams are allowed to pass through the spatial filter in order to eliminate undesirable stray light and to create a coherent Gaussian beam. The spatial filter is composed of a focusing lens, a pinhole, and a plano-convex lens. The focusing lens converges the incident beam into a central spot with Gaussian intensity profile on the optical axis. Sometimes there exist some fringes on the far sides of the optical axis, presenting the interference fringes. When a

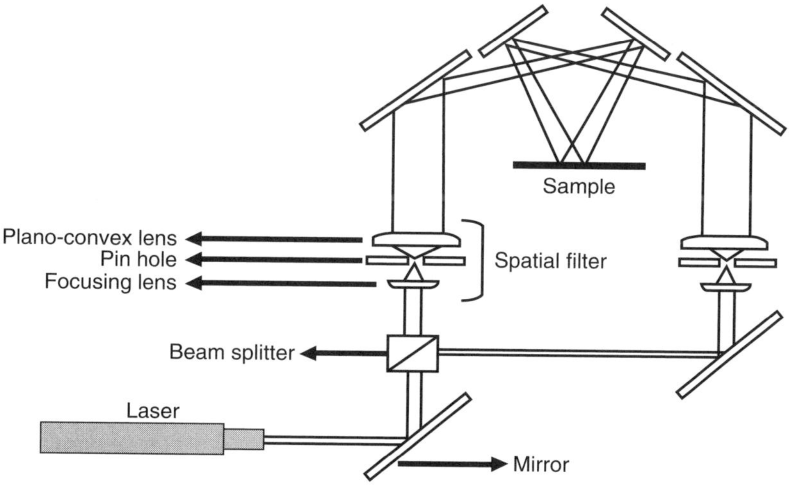

FIGURE 26.11 Schematic representation of the optical setup based on two-wave interference concept.

pinhole is perfectly centered on the central Gaussian spot, such undesired fringes are filtered out. The spatial filter may be regarded as an optical device that makes the incident beams free from aberration or strayed lights, i.e., optically clean.

Volume gratings of the holographic lithography technique (a 1D PBG material) may be analyzed as to whether the periodic spacing falls in the Bragg grating or Raman–Nath grating regimes [36]. A criterion for determining whether the holographic grating is within the range of Bragg or the Raman–Nath regime is given as [36]

$$Q = 2\pi \frac{\lambda L}{nd^2} \tag{26.10}$$

where
- λ is the incident wavelength of the writing beam
- d is the spacing of periodicity, which have the same designations in the previous Equation 26.9
- n is the refractive index of the sample medium
- L is the sample thickness

Q is the parameter. If $Q \ll 1$, then the grating is regarded as a Raman–Nath grating, otherwise the grating belongs to the Bragg grating, if $Q \gg 1$.

Alternatively, patterned photopolymerization may be employed by masking with desired patterns which may be called hereafter, soft lithography. The soft lithography approach tends to render a larger structure as compared to the interference approach and light leakage from the edges creates broadening of patterns. In such process, the patterned photoresist mask may be deposited on the thermally oxidized SiO_2 on top of a Si post. The soft photolithography using a higher energy laser (i.e., shorter wavelength, 248 nm) etches away the SiO_2 layer by reactive-ion etching. After that, the pre-patterned mask is removed and the organic layer of 8-(hydroxyquinolinato) aluminum (Alq) doped with ~1% by weight of DCM (4-dicyanomethylene-2-methyl-6-*p*-dimethylaminostyryl-4H-pyran, a commercially available dye) is deposited over the entire structure by sublimation. This type of organic photonic crystal structure is suitable for applications in distributed feedback lasers.

Another important aspect of the holographic lithography technique is the capability to fabricate switchable polymeric photonic crystals, alternatively known as the holographic polymer-dispersed liquid crystal (H-PDLC). It usually involves binary mixtures of liquid crystal, photo-curable monomers as well as photo-initiator in minute amount. Figure 26.12 presents the chemical structures of the photo-initiators (a Rose Bengal derivative) that can be photo-initiated either at 532 nm wavelength (green laser) or at 365 nm wavelength (UV regime), co-initiator, solubilizing agent, and

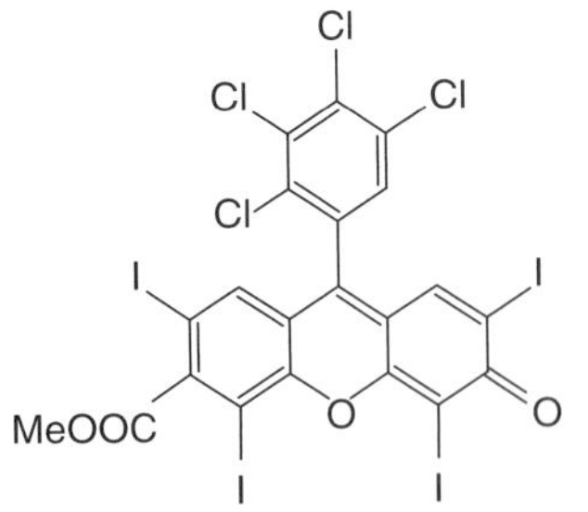

RBAX–Rose bengal derivative
(spectra group)

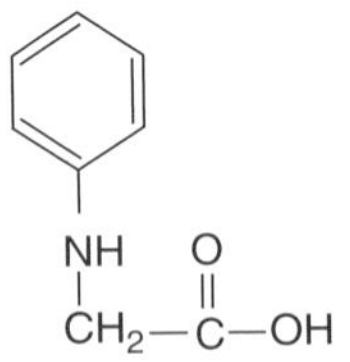

N-Phenyl glycine
(co-initiator)

FIGURE 26.12 Chemical structures of typical initiator, co-initiator, solubilizing agent, and surfactant in an initiating system of photoreaction.

surfactant in the photo-initiating syrus. The Rose Bengal initiator is expensive compared to the commercial Irgacure (bis(2,4,6-trimethylbenzoyl) phenylphosphineoxide), and the chemistry is more complex due to the involvement of co-initiator, solubilizing agent, and surfactant. However, Irgacure works only at the UV wavelength, e.g., 365 nm [35].

Upon irradiation with the green laser or UV beam on the sample mixture, photo-initiators get excited in the high-intensity regions. The wavelength of the laser beams is generally tuned to the absorption peak wavelength of the chosen photo-initiators. Photopolymerization is triggered selectively in the high-intensity region that proceeds via the mechanism of free radical polymerization. The free radicals thus generated initially react with monomers. These radical monomers further react with other monomers and the reaction propagates; this photoreaction step is termed as the propagation. In the final stage, polymerization is terminated by means of combination, disproportion, or radical trapping.

In practice, most of the photo-curable monomers utilized are multifunctional acrylate derivatives by virtue of their fast curing nature and chemical versatility. More importantly, multifunctional acrylates serve as matrix binders to contain the dispersion of nematic liquid crystals. Nematic liquid crystals (cyanobiphenyl derivatives) are generally used as a second component for the purpose of electrical switching. In other words, the orientation of the liquid crystal director may be altered upon the application of the external electric field. In this way, the direction of the polarized light may be controlled at ease. Table 26.1 illustrates the chemical structures of a typical eutectic liquid crystal E7 (Merck Co.) and di-, tri-, tetra-, penta-, and hexa-acrylate monomers with varying functionality from 2 to 6. The E7 is a eutectic mixture of different LC constituents containing four to eight carbons in the alkyl side groups; the reason to use the above LC mixture is that the nematic region of liquid crystal can be expanded relative to the single neat LC constituent, which is needed for improving electro-optical switching characteristics. Table 26.2 lists the refractive indices of some monomers and nematic liquid crystal commonly found in photolithography. Note that the liquid crystal molecule is represented by the ordinary and extraordinary indices due to the anisotropic nature of the liquid crystal mesogen, hereafter designated as n_o and n_e, respectively. That is to say, the light polarized along the liquid crystal molecule axis propagates at a different velocity than the light polarized perpendicular to the liquid crystal molecular axis.

Through irradiation of the interference pattern of the laser or UV beam imprinting on the sample mixture, photopolymerization is initiated in the high-intensity region. In the low-intensity region, the reaction is slow and thus the monomers diffuse into the high-intensity region in order to get polymerized. On the same token, the liquid crystal molecules in the high-intensity region are rejected to the low-intensity region. The mutual diffusion between the monomer and LC molecules is governed by the equilibrium phase diagrams of the starting mixture. During photopolymerization, the molecular weight of the reacting component keeps going up which in turn makes the system to become unstable and pushes the coexistence curve to shift asymmetrically upward, thereby driving the liquid–liquid phase separation. It is a well-known fact in the fabrication of switchable H-PDLC or the polymeric PBG structure is a reaction-diffusion process. The final structure is governed by the competition between photoreaction kinetics and phase separation dynamics.

One of the most important electro-optical properties of H-PDLC is the diffraction efficiency, which is defined as a ratio of the diffracted light intensity over the overall scattered light intensity [36]. The diffraction efficiency may be experimentally monitored in situ during the course of H-PDLC fabrication. Recently, a theoretical approach has been developed to model and simulate the reaction-diffusion process during the fabrication of H-PDLC and multidimensional photonic crystals [37–39]. The free energy of isotropic mixing of liquid crystal and monomers as well as the free energy of nematic ordering is incorporated into the model to predict the structure development. The combined free energy is then inserted into the time-dependent Ginzburg–Landau (TDGL) model C equations pertaining to nonconserved orientation order parameter of the LC directors [37] and the conserved concentration order parameter (see respective Equations 26.11a and 26.11b). The spatio-temporal growth of stratified H-PDLC structure (i.e., 1D band gap structure) can be computed in

TABLE 26.1
Chemical Structures of a Typical Nematic Liquid Crystal (E7), and Diacrylate, Triacrylate, Tetraacrylate, and Penta/Hexa Acrylate

	Name	Chemical Structure
Liquid crystal (E7)	Cyanobiphenyl (5CB, 7CB, 8OCB) and Cyanoterphenyl (5CT) compounds	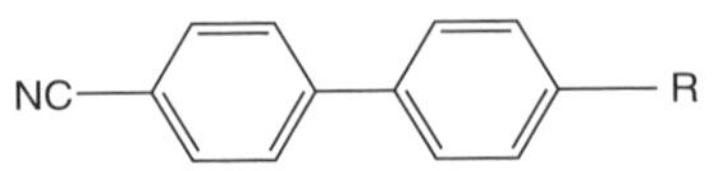 $R = CH_3(CH_2)_4$– $CH_3(CH_2)_6$– $CH_3(CH_2)_7O$– 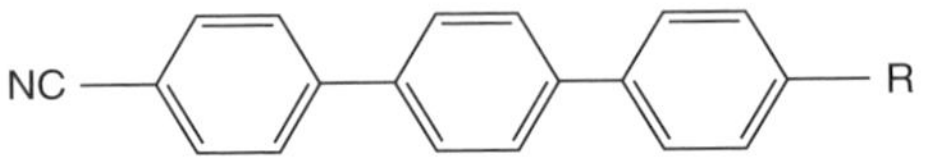 $R = CH_3(CH_2)_4$–
Photo-curable monomers	1,6 Hexanediol diacrylate	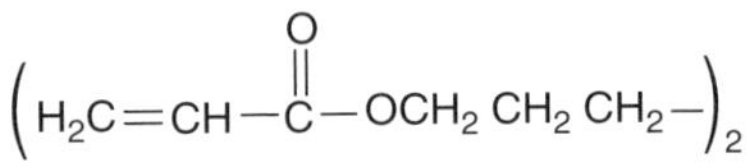
Photo-curable monomers	Trimethylolpropane triacrylate	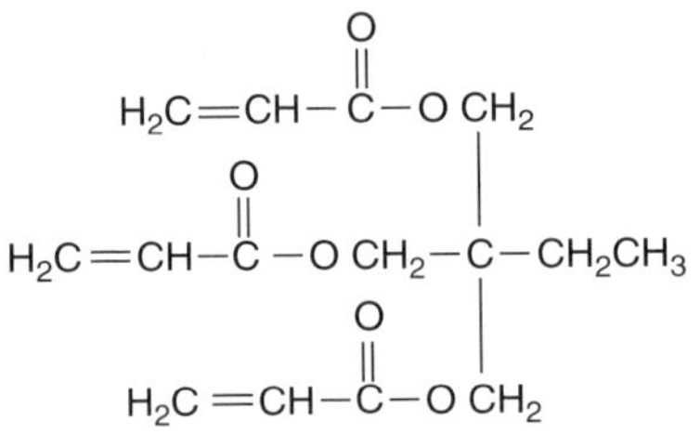
Photo-curable monomers	Pentaerythritol tetraacrylate	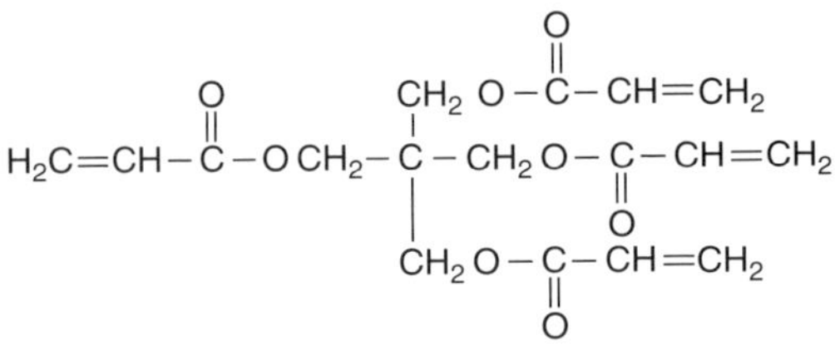
Photo-curable monomers	Dipentaerythritol penta-/hexa-acrylate	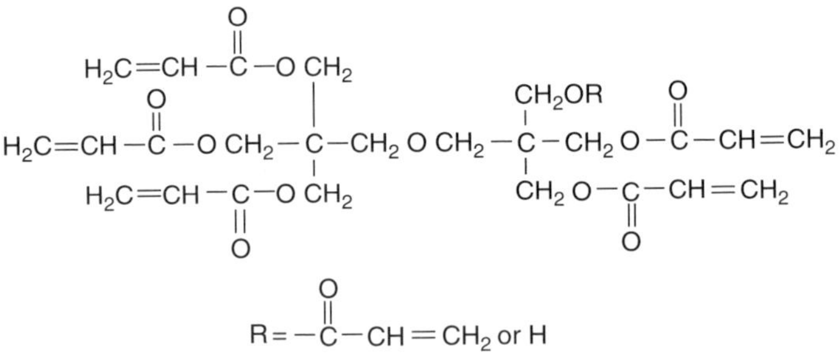

accordance with Equation 26.10c. The two coupled Equations 26.11a and 26.11b are known as TDGL model C. The former is known as TDGL model A, alternatively called Allen–Cahn equation, and the latter is called TDGL-model B Equation [37,38], alternatively known as Cahn–Hilliard equation that reads

$$\frac{\partial S(\vec{r},t)}{\partial t} = \left[-\Lambda_S\left(\frac{\delta G}{\delta S}\right)\right] \tag{26.11a}$$

TABLE 26.2
Refractive Index of Some Typical Photo-Curable Monomers and Nematic Liquid Crystals Utilized in Photolithography

Typical Monomers and Liquid Crystals	Refractive Index
HEMA monomer (Sigma-Aldrich)	1.453
Diacrylate monomer (Sigma-Aldrich)	1.456
Triacrylate monomer (Sigma-Aldrich)	1.483
Tetraacrylate monomer (Sigma-Aldrich)	1.487
Pentaacrylate monomer (Sigma-Aldrich)	1.49
Norland monomer (NOA 65) (Norland Products)	1.524
Nematic liquid crystal BL038 (Merck)	$n_o = 1.526$, $n_e = 1.799$
Nematic liquid crystal BL045 (Merck)	$n_o = 1.523$, $n_e = 1.753$
Nematic liquid crystal TL213 (Merck)	$n_o = 1.526$, $n_e = 1.766$
Nematic liquid crystal E7 (Merck)	$n_o = 1.521$, $n_e = 1.746$

$$\frac{\partial \phi_{\mathrm{LC}}(\vec{r},t)}{\partial t} = \nabla \cdot \left[\Lambda_\phi \nabla \left(\frac{\delta G}{\delta \phi_{\mathrm{LC}}} \right) \right] \tag{26.11b}$$

and

$$\frac{\partial \phi_{\mathrm{m}}(\vec{r},t)}{\partial t} = \nabla \cdot \left[\Lambda_\phi \nabla \left(\frac{\delta G}{\delta \phi_{\mathrm{m}}} \right) \right] - k \left[1 + V_{\mathrm{F}} \cos\left(\frac{2\pi x}{L} N_x \right) \right] \phi_{\mathrm{m}} \tag{26.11c}$$

where the second term in Equation 26.11c gives the anisotropically patterned photoreactions in the periodic stripes in which ϕ_{m} is the monomer concentration, V_{F} is the visibility factor, and k is the reaction rate constant. G is the total combined free energy of the system, i.e., $G = \int (g^i + g^n)dV$, and S is the orientational order parameter defined as $S = (3 \langle\cos\theta^2\rangle - 1)/2$; θ is the angle between the LC director and the reference axis. g^i and g^n represent the combined local and nonlocal free energy densities pertaining to orientational and compositional order parameters S and ϕ_{LC} of liquid crystal, Λ_ϕ is the mutual diffusion coefficient, and Λ_S is related to the rotational diffusivity. The differential term in small parenthesis in Equations 26.11a and 26.11b represents the functional derivative of the free energy with respective to orientation or concentration order parameter, representing the chemical potential of the system. The order parameter is defined as the deviation from the mean values of orientation or of concentration. It can be shown that the classical reaction-diffusion for the single component system is simply a special case of the TDGL equation, i.e., Equation 26.11c [38]. For a completely miscible system, the TDGL equation (model B) pertaining to the conserved concentration order parameter becomes the well-known Fick's second law of diffusion. This Fick's diffusion equation (or the simplified TDGL-B equation for a miscible mixture) has essentially the same wave character as the aforementioned Schrödinger equation for the fundamental particle systems or the Maxwell's equation for the electromagnetic wave.

Of particular significance of the coarse grain modeling through the TDGL equation is that the emerged holographic structure can be linked directly to the electro-optical properties of the holograms such as diffraction efficiency of the resultant photonic crystal [37]. The diffraction efficiency for the reflection grating and transmission grating is computed according to Equations 26.12a and 26.12b. The reflection grating is defined as the grating in which the reference wave is in the opposite direction as the signal wave whereas in the transmission grating, the reference wave is in the same direction

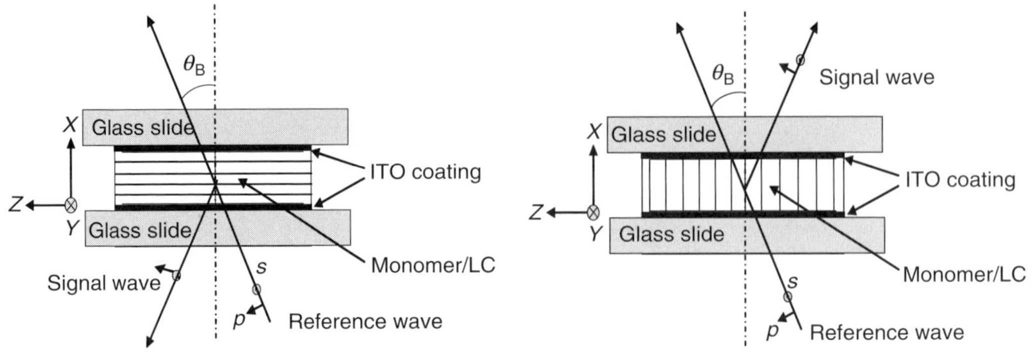

FIGURE 26.13 Schematic illustrations of reflection grating (left) and transmission grating (right).

as the signal wave. The illustration of reflection and transmission grating is depicted in Figure 26.13. The expression for diffraction efficiency for the reflection and transmission gratings reads:

$$\eta = \delta_a \tan h^2\left(\frac{\pi \Delta n_j(t) l}{\lambda \cos\theta_B}\right) \text{ for reflection grating } (j = s, p) \quad (26.12a)$$

or

$$\eta = \delta_a \sin^2\left(\frac{\pi \Delta n_j(t) l}{\lambda \cos\theta_B}\right) \text{ for transmission grating } (j = s, p) \quad (26.12b)$$

where

δ_a is a correction parameter that depends on the ratio of diffraction and scattering from the heterogeneous structure

λ is the wavelength of the probe beam

θ_B is the angle of reflection, i.e., Bragg's angle

$\Delta n_j(t)$ is the refractive index modulation to the switchable PBG material. In the s-polarization [38],

$$\Delta n_s(t) = n_{a,yy}^{(1)}(t) - n_{a,yy}^{(1)}(0) \quad (26.13)$$

For the p-polarization,

$$\Delta n_p(t) = n_{a,zz}^{(1)}(t)\cos(2\theta_B) + [n_{a,xx}^{(1)}(t) - n_{a,xx}^{(1)}(t)]\sin^2\theta_B \quad (26.14)$$

where $n_{a,xx}^{(1)}$, $n_{a,yy}^{(1)}$, and $n_{a,zz}^{(1)}$ are the refractive index components derived from the dielectric tensor. The s-polarization is represented by the polarization component perpendicular to the plane of incident, whereas in the case of p-polarization the polarization component perpendicular to the plane of incident. As demonstrated in Figures 26.14 and 26.15, it is promising to witness that the simulated results conform well with the experimental findings by Bunning and coworkers, which attests favorably to the theoretical predictions [40,41].

On the basis of the structure–property relationship thus established, one is able to estimate the optimal fabrication conditions including laser intensity, mixture composition, and radiation time before any experiments. It is also capable of guiding the experimentalists to use the appropriate optical setup for the desired photonic structures (e.g., fcc). In this way, the amount of time, cost, and

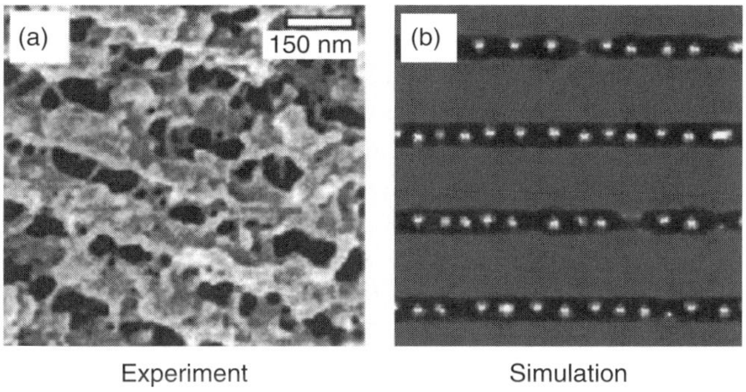

FIGURE 26.14 Comparison of SEM morphology of pentacrylate-based H-PDLC after extraction of LC (left) with the simulated one (right) showing the LC domains in stratified layers.

manpower is substantially reduced in the development of switchable polymeric photonic crystals. In addition, the effect of network elasticity can be investigated as well. The theoretical findings suggest that the flexible matrix such as hydroxyethyl methacrylate (HEMA) and thiolene structures is preferred over the rigid acrylic matrix, giving experimentalists some hints as to what kind of chemical structures would afford improved properties of tunable photonic crystals [39].

26.3 DEVICES OF PHOTONIC PBG MATERIALS

26.3.1 Light Emitter

Photonic band gap materials have a broad range of applications in electro-optical devices including light emitters, optical waveguides, optical fibers, wavelength filters, etc. [42–46]. One of the active PBG-based devices may be attributed to the development of light emitters. These light emitters mainly include point-defect lasers, high-power distributed feed back lasers, high-power and stable single-mode vertical cavity surface-emitting laser, and light-emitting diode with high extraction efficiency. The detail of light-emitting diodes is already covered in Chapter 13, and thus it will not be elaborated further here.

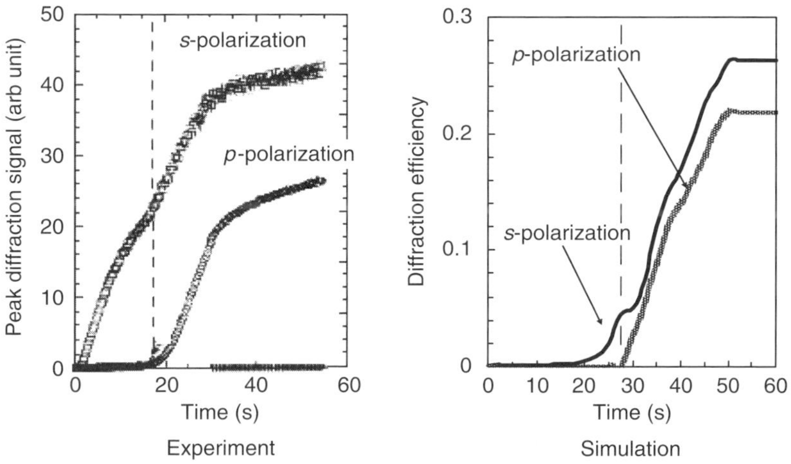

FIGURE 26.15 Comparison of diffraction efficiency evolution from in situ experimental measurement (left) with the simulated result (right).

26.3.2 Notch Filter

Photonic band gap materials have a unique feature that it forbids the transmission within a certain frequency of light. The structure and size of the periodicity in PBG material determine the forbidden frequency of the propagating wave. Holographic lithographic technique may be utilized with desirable periodic stripes so that the wavelengths of the notch filters may be varied from visible to near-infrared regions. A two-photon technique may be employed to fabricate the infrared or far-IR filters, which may have potential applications for night vision.

Another interesting development focuses on dispersion of nematic liquid crystal (or ferroelectric LC) in polymer matrix in which the ordinary refractive index of LC director is matched to that of the matrix binder [47]. In the off-state, the director orientation is random and the film shows opaque appearance due to light scattering arising from the refractive index contrast of the LC and the matrix polymer. Upon application of an external electric field for a few volts, the LC directors align themselves to the field direction. The film turns transparent as the ordinary refractive index of the LC matches with that of the matrix. Such LC-based notch filter is electrically switchable by a small external electric field from the opaque "on" to the transparent "off" states, reversibly. Alternatively, the LC directors may be pre-aligned on the substrate so that the "off" state would be transparent that can be switched to the opaque state. Such reverse mode polymer-dispersed liquid crystals have been utilized as privacy windows. It may be anticipated that H-PDLC may find some utility where color may be incorporated. Figure 26.16 illustrates the absorption peak versus wavelength plot, corresponding to the periodicity of the holographic stratified structure with and without the applied electric field. These notch filters of different periodicities may be produced to have a desired color spectrum for color displays.

26.3.3 Optical Waveguide

The point-defects act like the impurity/defects in the PBG structure, which spatially localize the photons near the impurity atoms. The defects may potentially exhibit a cavity quantum electrodynamic effect to afford the ultimate control of spontaneous emission lifetimes. Point-defects may be created in a self-assembled colloidal crystal array by putting in impurity such as spheres of different volumes. It has also been reported that the point-defects may be introduced via lithography techniques.

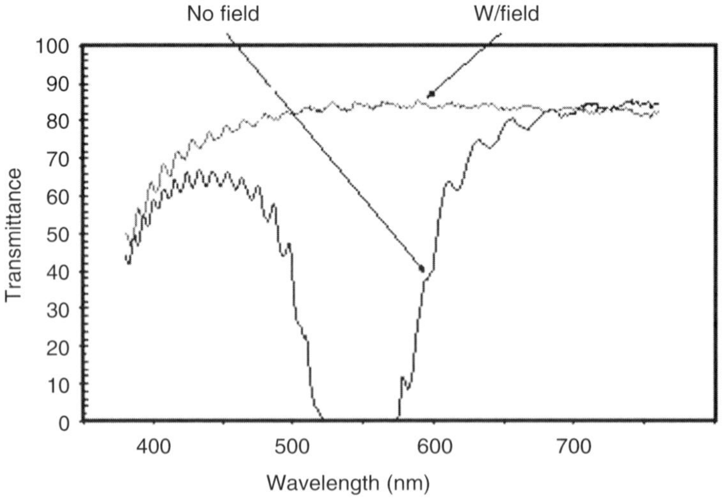

FIGURE 26.16 Illustration of the notch filter fabricated from PBG material with control of external electric field to switch it on and off. (Courtesy of T.J. Bunning.)

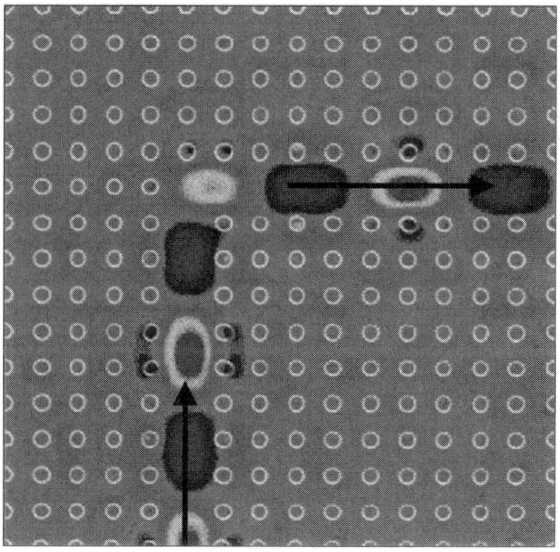

FIGURE 26.17 Theoretical calculations based on a defect pathway allow a 90° sharp bending of light propagation. (Reproduced from Polman, A. and Wiltzius, P., *MRS Bull.*, 26, 608, 2001. With permission.)

The use of PBG materials as laser resonators has been advocated because of its capability to modify the density of electromagnetic modes and thereby enhancing or suppressing spontaneous emission. The distributed feedback laser falls into this category, being characterized by a dielectric constant that is periodic in one dimension; this type of laser is unique in coherent lasing operation [47].

It has been shown that light may be trapped using point-defect. Similarly, light may be guided from one location to another. The underlying concept is to curve a pathway of defects on the perfect PBG structure functioning as a waveguide. In this way, light with a specific frequency range is guided through the pathway of the waveguide. An illustration of this concept is shown in Figure 26.17.

Figure 26.17 depicts an important application of PBG material as an optical waveguide. The bending of light at such sharp angle is normally not achievable with a normal waveguide. Of particular significance is that the bending of light in the well-controlled defect pathway experiences no significant optical loss. The switchable PBG material adds the function of the external control of the light propagation. Furthermore, the PBG structures may be embedded into cross-sectional planes of fibers to make PBG fibers.

26.3.4 Colloidal Gels and Networks

Another pioneering application of PBG materials is to utilize them in sensor device, notably by Asher's group [48–50]. Self-assembly colloidal crystal arrays may be incorporated in a hydrogel matrix. A unique feature of the hydrogel is that it swells or shrinks in the presence of chemical or biological stimulus, which consequently results in the shift of absorption peak against wavelength in the absorption spectrum. The potential applications of PBG material as a chemical/biological sensor are still under evaluation. Figure 26.18 represents a demonstration of a sensor device developed on the basis of colloidal PBG network structure.

26.4 FUTURE PERSPECTIVES OF ORGANIC AND POLYMERIC PBG MATERIAL

Organic and polymeric PBG material has an obvious drawback that the refractive index of polymer material is normally not very high. For instance, poly(thiophene) has been predicted theoretically

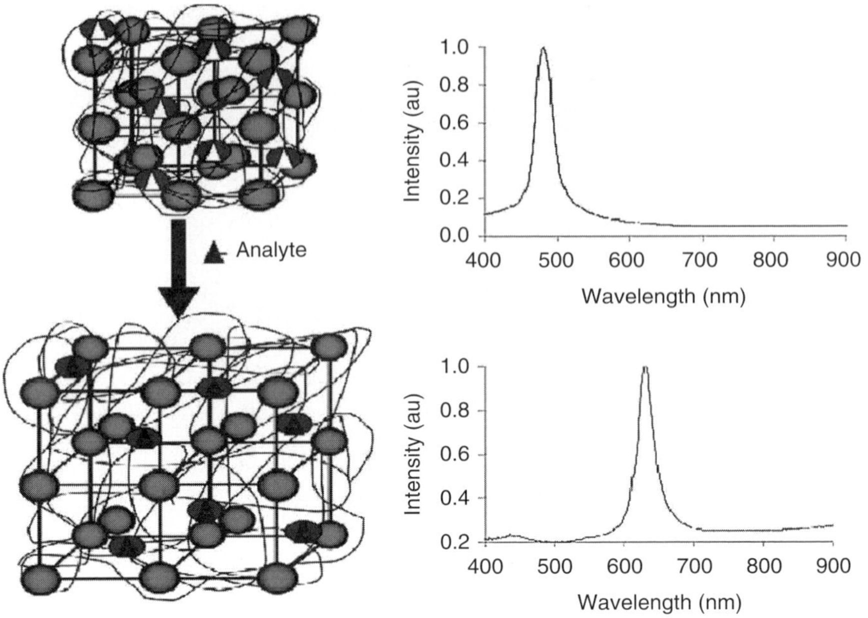

FIGURE 26.18 A sensor device developed on the basis of a colloidal array PBG embedded in a polymer hydrogel with molecular recognition elements. Upon swelling with analyte, the volume of the hydrogel expands, resulting in a spectral redshift. (Reproduced from Sharma, A.C., Jana, T., Kesavamoorthy, R., Shi, L., Virji, M.A., Finegold, D.N., and Asher, S.A., *J. Am. Chem. Soc.*, 126, 2971, 2004. With permission.)

to possess the largest refractive index among polymer molecules. In practice, it turns out to be appreciably smaller due to its inherently low degradation potential. Therefore, its refractive contrast with air is relatively small when compared to the inorganic PBG counterpart. Despite the efforts of trying to fabricate a polymer with a large refractive index via electrochemical synthesis route [11], the current trend of polymeric PBG materials is moving toward the combination of the inorganic/organic hybrid PBG structure having periodic inorganic domains in a polymer matrix [51,52]. For instance, the polypropylene plastic was reported to be co-infiltrated into the opal pores of synthetic opals formed via self-assembly of colloidal silica spheres, together with MEH-PPV dissolved in chloroform. The silica spheres were subsequently removed with hydrofluoric acid. The presence of polypropylene renders good mechanical properties to the inorganic/organic hybrid material [7]. Recent emergence of nano-structured materials and polymeric template nanotechnology is expected to exert significant impact on the fabrication and utilization of polymer-based PBG structures.

Liquid crystals based on photonic crystals despite the low dielectric contrast with that of the polymer matrix receive attention due to the electrical switchability from the opaque to transparent states. With the proper choice of the matrix polymer binder and the type of LC molecules, the transmittance can be as large as 93%. Recently, color displays based on the PDLC containing cholesteric liquid crystals have been introduced in the market place.

The hologram writing for large capacity data storage is hampered by the fact that the size scales of the periodic domains are limited to some 190 nm even with the shortest UV wavelength. If the resolution were to be smaller than tens of nanometer, the polymer molecules (i.e., the radius of gyration of polymer chains) may be too large for this purpose. Organic glasses may be sought as an alternative material for the application in the large capacity data storage. All in all, PBG materials are potential platforms for emitter, receiver, transmitter, signal modulator, signal selection devices that cover most aspects of the next-generation photonic computers.

EXERCISE QUESTIONS

1. Suppose the refractive index difference between the polymer and the channel cavity in a polymeric PBG structure is 0.5, what will be the dielectric constant of the polymeric material? Note that the dielectric constant of air is 1.0.
2. Based on the classical Maxwell equations, derive EM wave Equation 26.5 in vacuum.
3. Derive Fick's diffusion equation for a miscible system from TDGL model B (alternatively known as Cahn–Hilliard equation) pertaining to concentration order parameter.
4. Consider the fabrication of a 1D PBG structure based on holographic lithographic interference technique. Estimate the spacing of the periodic 1D structure if the writing beam has a wavelength of 532 nm in vacuum and the incident angle of the writing beam to the sample is at 45°. Assume that the refractive index of the matrix polymer is 1.48.
5. Suppose we have a volume grating H-PDLC film with a thickness of 20 μm, the spacing of the 1D periodicity is 1 μm, the refractive index of the sample is 1.53 and the writing beam wavelength is 365 nm, determine if the hologram falls within the Bragg grating or Raman–Nath grating range.

LIST OF ABBREVIATIONS

PBG	photonic band gap
EM	electromagnetic
1D	one dimensional
2D	two dimensional
3D	three dimensional
IR	near-infrared
GaAs	gallium arsenide
AlGaAs	aluminum gallium arsenide
GaAsP	gallium arsenic phosphide
InP	indium phosphide
Al_2O_3	aluminum oxide
Si	silicon
SiO_2	silicon dioxide
SEM	scanning electron microscopy
UV	ultraviolet
LC	liquid crystal
PS	polystyrene
PMMA	poly(methyl methacrylate)
AFM	atomic force microscopy
PS-b-PEO	polystyrene-block-polyethylene oxide block copolymer
DCM	4-dicyanomethylene-2-methyl-6-*p*-dimethylaminostyryl-4H-pyran
H-PDLC	holographic polymer-dispersed liquid crystal
TDGL	time-dependent Ginzburg–Landau
HEMA	hydroxylethyl methacrylate
MEH-PPV	poly(2-methoxy,5-(2'-ethyl-hexoxy)-1,4-phenylenevinylene)

ACKNOWLEDGMENT

This book chapter is made possible through the support of National Science Foundation through grant number DMR-0514942, Akron Global Polymer Academy (AGPA), and the Collaborative Center for Polymer Photonics sponsored by Air Force Office of Scientific Research, Wright-Patterson Air Force Research laboratory, and University of Akron.

REFERENCES

1. J.M. Anderson, *Introduction to Quantum Chemistry*, W.A. Benjamin: New York, 1969.
2. W. Kauzmann, *Quantum Chemistry*, Academic Press: New York, 1957.
3. B.E.A. Saleh and M.C. Teich, *Fundamentals of Photonics*, John Wiley & Sons: New York, 1991.
4. J.D. Joannopoulos, R.D. Meade, and J.N. Winn, *Photonic Crystals, Molding the Flow of Light*, Princeton University Press: New Jersey, 1995.
5. K. Inoue and K. Ohtaka, *Photonic Crystals, Physics Fabrication and Applications*, Springer: Berlin, New York, 2004.
6. P.N. Prasad, *Nanophotonics*, John Wiley & Sons: New Jersey, 2004.
7. F. Charra, V.M. Agranovich, and F. Kajzar, *Organic Nanophotonics*, Kluwer Academic: Dordrecht, The Netherlands, 2003.
8. D. Fichou, *Handbook of Oligo- and Polythiophenes*, Wiley-VCH: Weinheim, Germany, 1999.
9. K. Busch, S. Lolkes, R.B. Wehrspohn, and H. Foll, *Photonic Crystals, Advances in Design, Fabrication, and Characterization*, Wiley-VCH: Weinheim, Germany, 2004.
10. V.A. Markel and T.F. George, *Optics of Nanostructured Materials*, John Wiley & Sons: New York, 2001.
11. S.A. Jeneke and K.J. Wynne, *Photonic and Optoelectronic Polymers*, ACS Symposium Series #672, 1997.
12. A. Adibi, A. Scherer, and S.Y. Lin, eds., Photonic crystal materials and devices, Proceedings of SPIE, Vol. 5000, Bellingham, WA, 2003.
13. E. Yablinovitch, *Phys. Rev. Lett.*, 58, 2059, 1987.
14. S. John, *Phys. Rev. Lett.*, 58, 2486, 1987.
15. J.B. Pendry and A. MacKinnon, *Phys. Rev. Lett.*, 69, 2672, 1992.
16. R.D. Meade, K.D. Brommer, A.M. Rappe, and J.D. Joannopoulos, *Appl. Phys. Lett.*, 61, 495, 1992.
17. S.-Y. Lin, J.G. Fleming, and E. Chow, *MRS Bull.*, 26, 626, 2001.
18. E. Yablinovitch, T.J. Gmitter, and K.M. Leung, *Phys. Rev. Lett.*, 67, 2295, 1991.
19. J. Choi, Y. Luo, R.B. Wehrspohn, R. Hillebrand, J. Schilling, and U. Gosele, *J. Appl. Phys.*, 94, 4757, 2003.
20. C. Harnagea, M. Alexe, J. Schilling, J. Choi, R.B. Wehrspohn, D. Hesse, and U. Gosele, *Appl. Phys. Lett.*, 83, 1826, 2003.
21. A.J. Turburfield, *MRS Bull.*, 26, 632, 2001.
22. A. Polman and P. Wiltzius, *MRS Bull.*, 26, 608, 2001.
23. V.L. Colvin, *MRS Bull.*, 26, 637, 2001.
24. J.H. Moon, J. Ford, and S. Yang, *Polym. Adv. Technol.*, 17, 83, 2006.
25. D.J. Norris, Y.A. Vlasov, and M. Deutsch, *Adv. Mater.*, 12, 1176, 2000.
26. T. Kyu, S. Meng, H. Duran, N. Kumar, and G. Yandek, *Macromol. Res.*, 14, 155, 2006.
27. G. Odian, *Principles of Polymerization*, John Wiley & Sons: New York, 1981.
28. H. Ren, Y.-H. Fan, and S.-T. Wu, *Appl. Phys. Lett.*, 82, 3168, 2003.
29. S.-L. Kuai, V.-V. Truong, A. Hache, and X.F. Hu, *J. Appl. Phys.*, 96, 5982, 2004.
30. Y. Xia, B. Gates, and Z.-Y. Li, *Adv. Mater.*, 13, 409, 2001; Y. Lu, Y. Yin, Z.-Y. Li, and Y. Xia, *Langmuir*, 18, 7722, 2002.
31. S.H. Kim, M.J. Misner, and T.P. Russell, *Adv. Mater.*, 16, 2119, 2004.
32. S.H. Kim, M.J. Misner, T. Xu, M. Kimura, and T.P. Russell, *Adv. Mater.*, 16, 226, 2004.
33. D.H. Kim, K.H.A. Lau, J.W.F. Robertson, O.-J. Lee, U. Jeong, J.I. Lee, C.J. Hawker, T.P. Russell, J.K. Kim, and W. Knoll, *Adv. Mater.*, 17, 2442, 2005.
34. M. Rothschild, T.M. Bloomstein, N. Efremow Jr., T.H. Fedynyshyn, M. Fritze, I. Pottebaum, and M. Switkes, *MRS Bull.*, 30, 942, 2005.
35. O. Valdes-Aguilera, C.P. Pathak, J. Shi, D. Watson, and D.C. Neckers, *Macromolecules*, 25, 541, 1992.
36. T.J. Bunning, L.V. Natarajan, V.P. Tondiglia, and R.L. Sutherland, *Ann. Rev. Mater. Sci.*, 30, 115, 2000.
37. S. Meng, T. Kyu, L.V. Natarajan, V.P. Tondiglia, R.L. Sutherland, and T.J. Bunning, *Macromolecules*, 38, 4844, 2005.
38. S. Meng, N. Kumar, T. Kyu, T.J. Bunning, L.V. Natarajan, and V.P. Tondiglia, *Macromolecules*, 37, 3792, 2004.
39. G.R. Yandek, S. Meng, G.M. Sigalov, and T. Kyu, *Liq. Cryst.*, 33, 775, 2006.
40. R.L. Sutherland, L.V. Natarajan, V.P. Tondiglia, and T.J. Bunning, *Chem. Mater.*, 5, 1533, 1993.
41. R.L. Sutherland, V.P. Tondiglia, L.V. Natarajan, and T.J. Bunning, *Appl. Phys. Lett.*, 79, 1420, 2001.
42. M. Notomia, H. Suzuki, and T. Tamamura, *Appl. Phys. Lett.*, 78, 1325, 2001.

43. M. Meier, A. Dodabalapur, J.A. Rogers, R.E. Slusher, A. Mekis, A. Timko, C.A. Murray, R. Ruel, and O. Nalamasu, *J. Appl. Phys.*, 86, 3502, 1999.
44. C.C. Bowley, G.P. Crawford, and H. Yuan, *App. Phys. Lett.*, 21, 3096, 1999.
45. D. Yang, S.J. Jhaveri, and C.K. Ober, *MRS Bull.*, 30, 976, 2005.
46. X. Wang, C. Neff, E. Gaugnard, Y. Ding, J.S. King, L.A. Pranger, R. Tannenbaum, and Z.L. Wang, *Adv. Mater.*, 17, 2103, 2005.
47. V.K.S. Hsiao, W.D. Kirley, F. Chen, A.N. Cartwright, P.N. Prasad, and T.J. Bunning, *Adv. Mater.*, 17, 2211, 2005.
48. S.A. Asher, J. Holtz, L. Liu, and Z. Wu, *J. Am. Chem. Soc.*, 116, 4997, 1994.
49. A.C. Sharma, T. Jana, R. Kesavamoorthy, L. Shi, M.A. Virji, D.N. Finegold, and S.A. Asher, *J. Am. Chem. Soc.*, 126, 2971, 2004.
50. S.A. Asher, V.L. Alexeev, A.V. Goponenko, A.C. Sharma, I.K. Lednev, C.S. Wilcox, and D.N. Finegold, *J. Am. Chem. Soc.*, 125, 3322, 2003.
51. K. Forberich, M. Diem, J. Crewett, U. Lemmer, A. Gombert, and K. Busch, *Appl. Phys. B*, 82, 539, 2006.
52. M.J. MacLachlan, I. Manners, and G.A. Ozin, *Adv. Mater.*, 12, 567, 2000.

27 Introduction to Polymer Photonics for Information Technology

Antao Chen

CONTENTS

27.1 Introduction 820
27.2 Basic Principles of Optical Waveguides 820
27.2.1 Total Internal Reflection 820
27.2.2 Optical Waveguides 821
27.2.3 Optical Fiber 823
27.2.4 Coupling between Waveguides 826
27.3 Polymer Optical Waveguides and Optical Interconnection 826
27.4 Plastic Optical Fiber 829
27.5 Changing the Index of Refraction by Applying an Electrical Control Signal 832
27.5.1 Electro-Optic Effect 832
27.5.2 Thermo-Optic Effect 832
27.6 Electro-Optic Modulators 833
27.7 Thermo-Optic Switches 837
27.8 Wavelength Filters 838
27.8.1 Arrayed Waveguide Grating Filters 839
27.8.2 Micro-Ring Resonators 839
27.9 Summary 841
Exercise Questions 841
List of Abbreviations 842
References 842

Abstract Through molecular design and chemical synthesis, organic polymer materials can have better optical properties than conventional inorganic materials. Polymers also allow more choices of device fabrication techniques and devices impossible to fabricate in other materials can be achieved with polymers. For these reasons optical polymers have many applications in photonic devices to transmit and process information. In this chapter we will learn the basic principles of optical waveguides and optical fibers, understand the requirements on the material, and how organic polymer materials meet these requirements. Important photonic devices made with polymer materials such as plastic optical fiber, optical modulators, switches and wavelength filters for fiber optic communications are discussed.

27.1 INTRODUCTION

Most of information a person receives are through light. Among the five senses of a human body—vision, hearing, touch, taste, and smell—vision accounts for more than 99% of the information received. The information bandwidths of these senses are on the order of 100 MHz for vision, 100 kHz for hearing, and 100 Hz for touch, taste, and smell combined. For this reason, photonics, the technology of light, has found important applications in information technology. Devices such as liquid crystal flat panel displays, digital versatile disks (DVDs), and compact disks (CDs) would not be possible without the use of photonics. In addition, optical fiber and other optical technologies accommodate almost all the long-distance transmission of information.

Organic and polymeric materials are important in photonics. In addition to the liquid crystal displays widely used for computers and televisions, organic light-emitting diodes promise low-cost flat panel displays with brighter colors and wider viewing angles. Optical modulators, switches, and wavelength filters made of electro-optic and thermo-optic polymers are key components in fiber optic telecommunication. Artificial polymers can be designed and synthesized to meet device-specific requirements including optical, electrical, mechanical, and thermal properties. As molecular design and synthesis continue to advance, new polymers with desirable properties will continue to emerge, enabling novel devices and applications. This chapter provides an introduction to the various applications of polymers used to transmit information, especially devices based on polymer waveguides and fibers. We start with the basic principles of optical waveguides and fibers, followed by examples of their practical functions. Through electro-optic and thermo-optic effects, the index of refraction of certain polymers can be changed by an electrical control signal. These effects lead to major applications in optical modulators and switches. Other devices for wavelength filtering and switching fabricated with optical polymers are also discussed, such as arrayed waveguide gratings and micro-ring resonators.

27.2 BASIC PRINCIPLES OF OPTICAL WAVEGUIDES

27.2.1 Total Internal Reflection

The basics of optical waveguides and fibers can be understood using simple concepts of ray optics. When light passes through a boundary between two optical materials of different refractive indices, the path of the light will bend due to refraction, as illustrated in Figure 27.1.

The angles of the incident and refracted rays are measured with respect to the normal of the boundary and can be calculated according to Snell's law

$$n_1 \sin\theta_1 = n_2 \sin\theta_2 \tag{27.1}$$

If light is traveling from a high index material to a low index material ($n_1 > n_2$), by Snell's law, θ_2 will always be greater than θ_1, as shown in Figure 27.2. Therefore, a small change in θ_1 results in a large deflection of θ_2. At some point, θ_2 will reach 90° and the path of the refracted light will be parallel to the boundary, while θ_1 will still be <90°. This corresponding θ_1 is called the critical angle θ_c

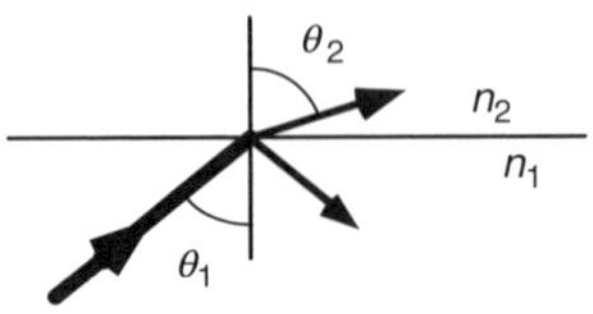

FIGURE 27.1 Refraction of light at the interface between two optical materials of different refractive indices.

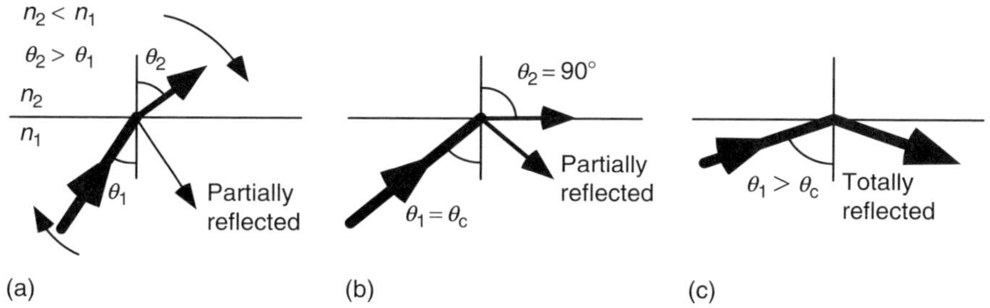

FIGURE 27.2 (a) and (b) Refraction and partial reflection when the angle of incidence is less than and equal to the critical angle, respectively. (c) Total internal reflection occurs when the angle of incidence is greater than the critical angle.

$$\theta_c = \arcsin\left(\frac{n_2}{n_1}\right) \tag{27.2}$$

When $\theta_1 > \theta_c$, all the incident light will be totally reflected at the boundary and remain in the high index material. This is known as the total internal reflection. If the interface between the two materials is perfectly smooth, there will be no optical power lost due to reflection.

Total internal reflection does not imply that light cannot penetrate into the low index material. On the contrary, some light does penetrate by a very shallow depth into the low index material before it is reflected, as shown in Figure 27.3. The amount of light penetration into the low index media decreases exponentially as the depth increases. This phenomenon is known as the evanescent wave or tail [1], and is analogous to the tunneling effect in quantum mechanics. The extent of the evanescent wave depends on the difference between the two indices of refraction across the boundary, and it is usually on the order of a micrometer. The evanescent wave plays a critical role in the coupling of light between closely spaced optical waveguides, which is discussed in later sections.

27.2.2 Optical Waveguides

If a slab of high index material is sandwiched between two low index materials and light is then sent into the high index slab, total internal reflection at the upper and lower boundaries will keep the light confined within the high index slab, as shown in Figure 27.4. This forms a planar optical waveguide. The high index layer is called the core, while the low index layers are referred to as claddings. It is interesting to note that air, with a refractive index of 1, can function as cladding of optical waveguides.

Not all light that satisfies the total internal reflection condition can be guided in the waveguide. All the reflections need to be in phase to form a guided mode, as illustrated in Figure 27.5a. If the

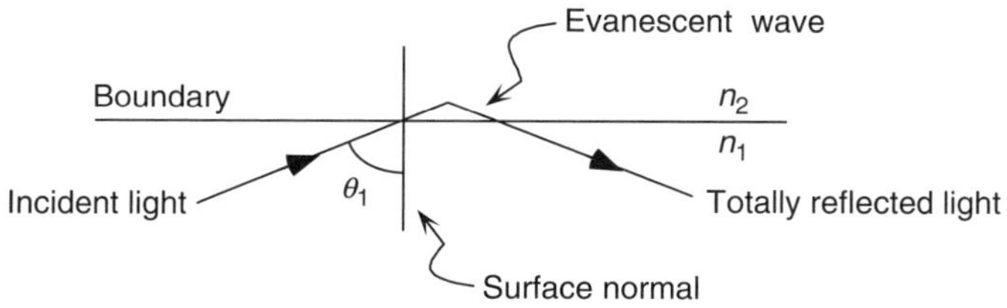

FIGURE 27.3 Evanescent wave penetrates into the low index side of the boundary during total internal reflection. $n_1 > n_2$ and $\theta_1 > \theta_c$.

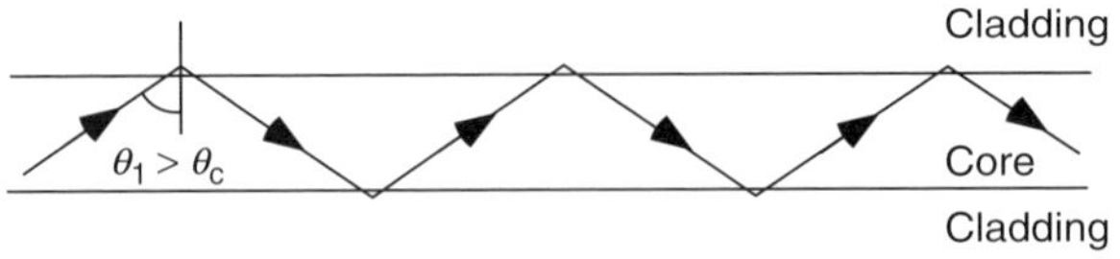

FIGURE 27.4 Planar optical waveguide, a high index core is sandwiched between two cladding layers of lower index materials.

reflected light at subsequent reflections on a boundary is out of phase with a previous reflection (Figure 27.5b), this will cause destructive interference and the waves will eventually cancel one another. This implies that no light of that particular angle will exist in the waveguide, and this principle is referred to as the transverse-resonance condition

$$2n_1 d\cos(\theta_1) + \delta_1 + \delta_2 = m\lambda$$
$$m = 0, 1, 2, \ldots \tag{27.3}$$

where
 d is the thickness of the core
 λ is the wavelength of the light

δ_1 and δ_2 are associated with the evanescent penetration through the two boundaries and these are functions of the angle of incidence as well as indices of core and cladding layers. For the simplest case where the two cladding layers have the same index (symmetric waveguide) and the electric field of the light is parallel to the boundary,

$$\delta_1 = \delta_2 = -\frac{\lambda}{\pi}\arctan\left(\frac{\sqrt{n_1^2 \sin^2\theta_1 - n_2^2}}{n_1\cos\theta_1}\right) \tag{27.4}$$

The physical meaning of the transverse-resonance condition is that the total optical path length between two subsequent reflections on a boundary needs to be a multiple of the wavelength for the light to stay in phase. Once the thickness and indices of the waveguide are known, the allowed angles of reflection can be numerically solved from Equations 27.3 and 27.4. For a given core size and core/cladding indices, the transverse-resonance condition allows a certain number of reflection angles, each corresponding to a given m. Only a discrete number of θ_1 and m exist to satisfy Equation 27.3, and each allowed combination of m and θ_1 is called a waveguide mode. Thus, each mode occurs at a specific angle of incidence at the waveguide boundaries. If we increase both the thickness and index of the core, the $n_1 d$ term is large and there are many allowed combinations of m and θ_1. Such a waveguide is called a multimode waveguide. If, however, we reduce the index n_1 or the thickness d of the core, fewer modes can propagate through the waveguide. If n_1 and d are decreased sufficiently and only one mode is allowed, the device is called a single-mode waveguide.

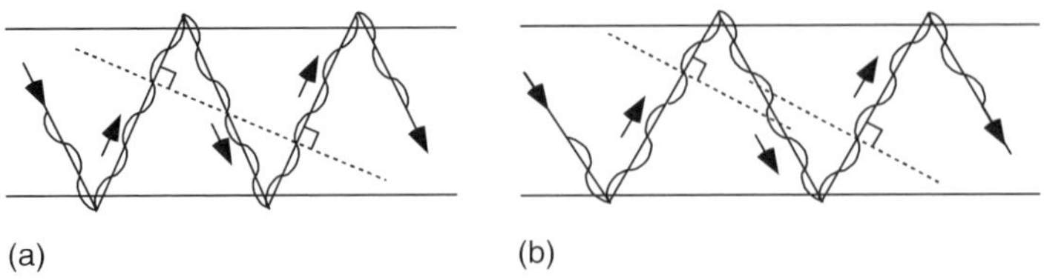

FIGURE 27.5 The reflected light is (a) in phase and (b) out of phase.

TABLE 27.1
All Possible Modes for Waveguides of 1–5 μm in Thickness

$d = 1$ μm	$m = 0$			
1 mode	$\theta_1 = 76.18°$			
$d = 2$ μm	$m = 0$	$m = 1$		
2 modes	$\theta_1 = 80.71°$	$\theta_1 = 72.22°$		
$d = 3$ μm	$m = 0$	$m = 1$	$m = 2$	
3 modes	$\theta_1 = 83.07°$	$\theta_1 = 76.29°$	$\theta_1 = 70.45°$	
$d = 4$ μm	$m = 0$	$m = 1$	$m = 2$	
3 modes	$\theta_1 = 84.49°$	$\theta_1 = 79.01°$	$\theta_1 = 73.71°$	
$d = 5$ μm	$m = 0$	$m = 1$	$m = 2$	$m = 3$
4 modes	$\theta_1 = 85.42°$	$\theta_1 = 80.05°$	$\theta_1 = 76.33°$	$\theta_1 = 72.02°$

Example: A polymer waveguide is made with a core and cladding layers of n = 1.65 and 1.55, respectively. The wavelength of light is 1.55 μm. For a core thickness of 1–5 μm, the number of allowed modes and the measure of θ_1 for each mode can be solved numerically from Equation 27.3 and are listed in Table 27.1.

For d = 5 μm, the distribution of optical power of each mode is plotted in Figure 27.6.

A planar waveguide only confines light in one dimension, therefore allows the light to spread out laterally in the core layer as it propagates. A more useful waveguide has low index cladding on all sides of the core and confines light in two dimensions. This forms a channel waveguide, similar to an optical pipe, which can be used to guide light in a desired direction. Figure 27.7 illustrates cross sections from several channel waveguides. Channel waveguides are usually made in thin films on a flat substrate, such as a silicon wafer. It is important to note that the substrate simply provides a mechanical platform for structural support, and rarely determines the properties of the waveguide.

27.2.3 Optical Fiber

An optical fiber is a channel waveguide with a circular cross section, as illustrated in Figure 27.7e and f. Fibers are free-standing waveguides, and do not require a supporting substrate. As with all waveguides, the index of refraction of the core is higher than the surrounding cladding. Typically, the index difference Δn between the core and the cladding ranges from 0.01% to 10%. Similar to planar waveguides, when the radius of the core r or Δn is small such that $2\pi r\sqrt{2n\Delta n}/\lambda \leq 2.405$, the fiber supports only one mode and it is called single-mode fiber. When $2\pi r\sqrt{2n\Delta n}/\lambda > 2.405$, the fiber supports multiple modes and is called a multimode fiber. Figure 27.8 shows photographs of the optical intensity pattern of a single-mode and a multimode fiber, respectively.

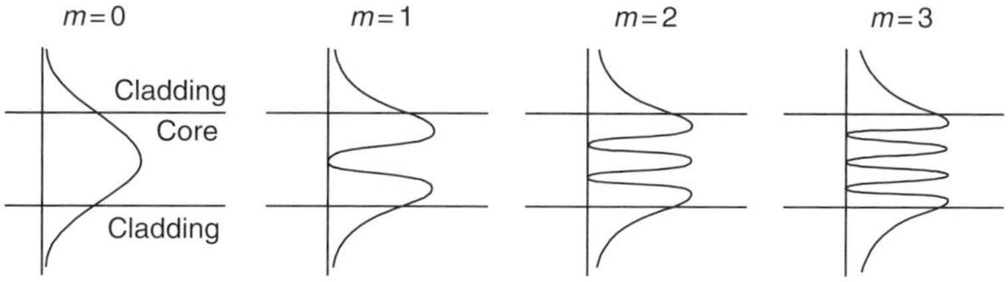

FIGURE 27.6 The intensity patterns of the 0th through 3rd mode of a planar waveguide.

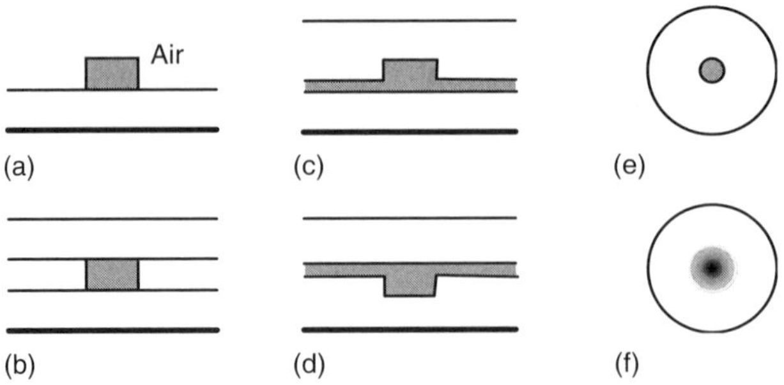

FIGURE 27.7 Cross sections of several channel waveguides. (a) Air-cladded channel. (b) Buried channel. The cladding around the core can have different indices. (c) Ridge or rib waveguide. (d) Inverted ridge. (e) Step-index fiber. (f) Graded-index fiber, in which the index of the core gradually decreases from the center of the fiber.

Total internal reflection presents an ideal method for both optical signal containment and transmission since there is virtually no loss of optical power if the interface between the core and cladding is smooth. Despite no loss of optical power in the total internal reflection, other factors, such as absorption and scattering of light by the core and cladding materials of the fiber, can reduce the optical power as light propagates through the fiber. Signal attenuation due to absorption is determined by the compositions of both core and cladding materials. Scattering, on the other hand, is caused by excessive roughness at the interface between the core and cladding, as well as any inhomogeneity in the fiber material. These imperfections lead to light either being reflected back to the source or escaping the fiber entirely. Because absorption is a function of wavelength, optical loss of a polymer waveguide or fiber depends on the wavelength of operation. Many polymers have low absorption for wavelengths of 400–700 nm and are suitable for guiding light over hundreds of meters. Waveguide and fiber loss is usually expressed in decibels (dB)

$$\text{Loss (dB)} = -10 * \log_{10}\left(\frac{I_{\text{out}}}{I_{\text{in}}}\right) \tag{27.5}$$

where I_{in} and I_{out} are the optical power at the input end and at the output end of the waveguide, respectively. Using dB simplifies the calculation of the total loss of several pieces of fiber connected in series. In this case, the total loss is simply the sum of the losses in dB of individual pieces of fiber.

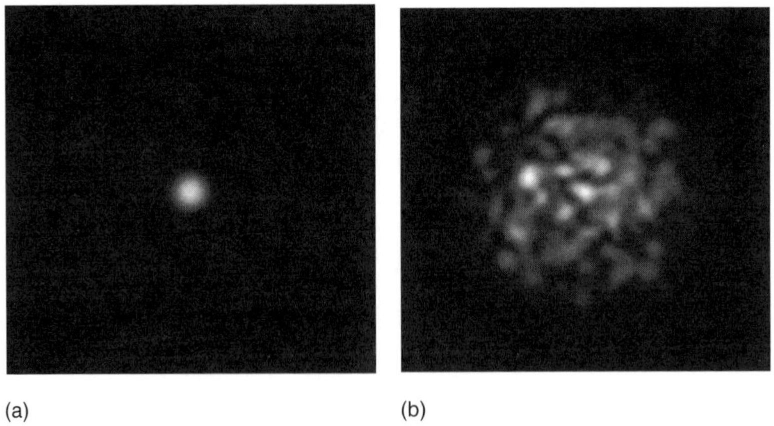

FIGURE 27.8 Optical power distributions of (a) a single-mode and (b) a multimode fiber. The two pictures have the same scale. Both fibers are made of silica glass. The diameter of the core is 9 μm for the single-mode fiber, and 62.5 μm for the multimode fiber.

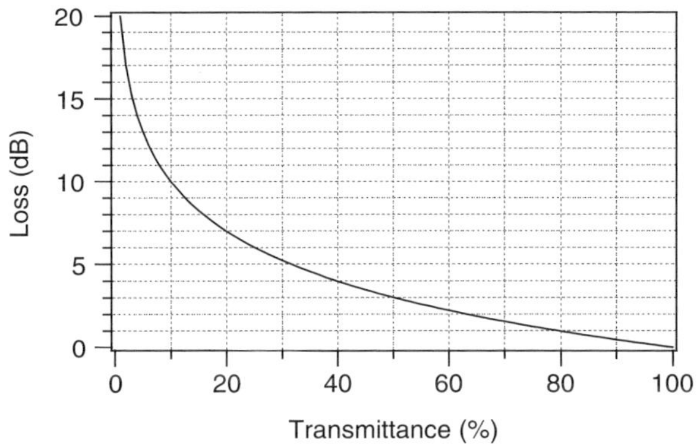

FIGURE 27.9 Conversion between percentage of transmitted light and loss in dB.

If the percentage of light transmitted is used instead of dB, one needs to use multiplication to calculate the total loss. Figure 27.9 plots the conversion between the percentage of transmitted light and loss in dB. An ideal waveguide that transmits 100% of the input light will have 0 dB loss, while a waveguide transmitting only 50% of the input light will have an optical loss of 3 dB. When expressed in dB, the total loss of a waveguide scales linearly with the length of waveguide. Loss per unit length is a figure-of-merit when comparing the quality of different waveguides. Although losses lower than 0.1 dB/cm have been achieved in some polymers, most polymer waveguides have losses on the order of 1 dB/cm. The loss of a typical polymer fiber is from 0.1 to 0.5 dB/m. Glass fiber for telecommunication has loss as low as 0.2–0.3 dB/km.

In addition to loss and core size, numerical aperture (NA) is another important parameter of optical fiber. NA is a measure of the cone-shaped divergence of the output light, and depends on the indices of refraction of both the core and the cladding

$$\mathrm{NA} = \sin(\theta) = \sqrt{n_1^2 - n_2^2} \tag{27.6}$$

The NA of a fiber determines the range of angles in which a fiber can accept and emit light, as shown in Figure 27.10. Fibers with high numerical apertures are capable of collecting more input light than low NA fibers, as a result, large NA fibers are more tolerant to angular misalignment when joined or coupled.

Optical waveguides and fibers have been made with both organic and inorganic materials. The choice of material is based on the requirements of a specific application. Organic polymers are relatively simple to produce, which makes them a cost-effective choice. Additionally, organic polymers have some unique advantages. For instance, polymer waveguides can be fabricated at much lower temperatures than waveguides of most of inorganic materials, and can be applied to virtually any substrate with a smooth surface. Furthermore, polymers can be designed to possess special functionalities. All these advantages enable polymer waveguides and fibers to have a wide range of applications. The major weaknesses of polymers include relatively higher loss and concerns in their

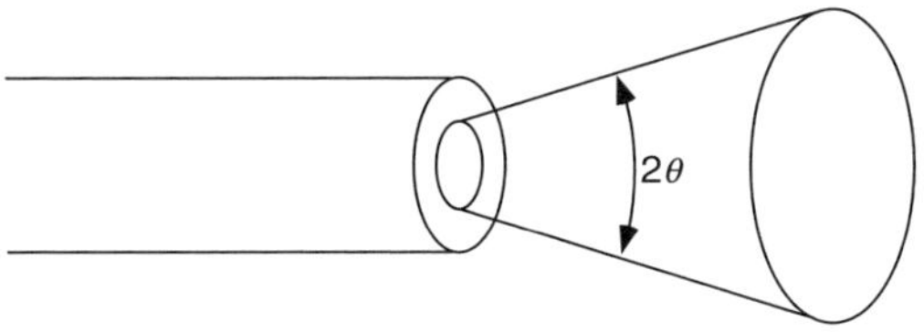

FIGURE 27.10 The numerical aperture of optical fiber.

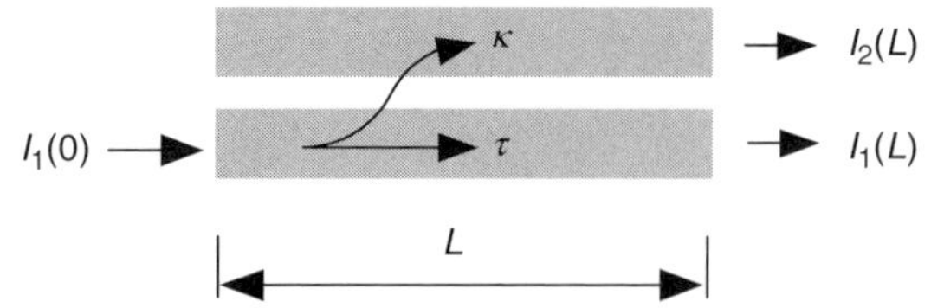

FIGURE 27.11 The coupling of light between waveguides. κ and τ are coupling coefficients. κ is the percentage of the initial electric field that is coupled into a second waveguide and τ is the percentage of the electric field remaining in the original waveguide. Because optical power is proportional to the square of the electric field and total optical power is conserved, $\kappa^2 + \tau^2 = 1$.

stabilities against temperature, moisture, and aging. These issues are being addressed in the research community and improvements have been constantly made.

27.2.4 Coupling between Waveguides

During the total internal reflection, a small part of the light called evanescent wave penetrates from the core into the cladding layer. If one waveguide carrying light is placed next to a second waveguide within range of their evanescent waves, light in the first waveguide will leak into the second waveguide. This is called evanescent coupling, as illustrated in Figure 27.11. Because the evanescent wave decays rapidly as the distance from the boundary increases, the amount of coupling strongly depends on how close the two adjacent waveguides are. Coupling can take place even if the two waveguides have different sizes and shapes, but it is most efficient between waveguides of identical structure. In this case, optical power can be completely transferred from one waveguide to the other, and

$$\begin{aligned} I_1(L) &= I_1(0)\cos^2(\kappa L) \\ I_2(L) &= I_1(0)\sin^2(\kappa L) \end{aligned} \tag{27.7}$$

Coupling plays an important role in both optical switches and micro-ring resonators. These devices are discussed in later sections.

27.3 POLYMER OPTICAL WAVEGUIDES AND OPTICAL INTERCONNECTION

As the speed of computer chips increases, interconnecting high-speed clock and data signals has become progressively more challenging. Conventional interconnection uses copper wires to connect chips internally and externally. As the scale of integration and the speed of integrated circuits (IC) continue to advance, copper interconnecting wires have to be made thinner and closer to one another, consequently electromagnetic interference between these wires becomes a serious problem. Since wires have an innate resistance, each time a data pulse travels through the wire, some of this energy is converted to heat. The higher the bit rate of the data stream, the more number of pulses per second pass through the wire, which leads to an increased power consumption and heat generation. These problems can be avoided by using optical waveguides for data interconnection. Electrical signals are first converted into pulses of light by a semiconductor laser diode and transmitted through a waveguide to a desired location on the chip. A photodetector then converts the optical pulses back to electrical signals and feeds them into the appropriate circuits. The power consumption of optical interconnection is not strongly dependent upon frequency or distance, and optical interconnection is highly immune to electromagnetic interference.

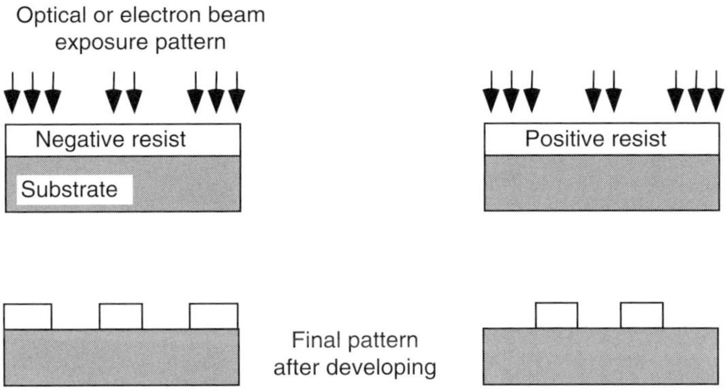

FIGURE 27.12 Patterning negative and positive photo- or e-beam polymers.

The bottleneck of data interconnection gets worse as the transmission distance increases. For this reason, long-distance telecommunication was the first to replace copper wires with fiber optic cables. Fiber optics has recently become a standard for shorter distance interconnections such as local area networks and high fidelity (hi-fi) audio systems.

There is also a growing trend to use waveguides in both inter-chip and intra-chip connections. The waveguide layer is simply applied to the top of the microelectronic layers once the microelectronic integrated circuitry is completed. It is critical that waveguide fabrication does not damage the underlying circuits. Polymer is the material of choice because the fabrication of polymer waveguide does not require excessively high temperatures, thus avoiding damage to the microelectronic layers underneath.

The most common method used to make waveguide layers is spin casting, in which the polymer solution is dispensed on the substrate and a thin film of the polymer obtained by spinning the coated substrate at a few thousand revolutions per minute (rpm). Once the thin film is cured, a variety of techniques can be used to create channels to guide light. Some polymers can be directly patterned using ultraviolet light, a focused electron beam, or a focused laser beam, as illustrated in Figure 27.12. In a negative resist polymer, photons or electrons can initiate cross-linking reactions to form a polymer network. The unexposed material remains un-cross-linked, therefore can be removed by a chemical solution (developer), leaving behind an optical waveguide ridge structure. In a positive resist, photons and electrons break chemical bonds in the polymer chains and make the exposed area soluble in the developer. Surface ridges and trenches of channel waveguides can be fabricated by hot embossing replication on thin films of thermoplastic polymer [2]. Some polymers may undergo a permanent change of refractive index when exposed to ultraviolet radiation (Figure 27.13), a process known as photobleaching [3]. By exposing the film on both sides of the waveguide channel, the exposed area with reduced index of refraction will serve as cladding.

Photolithography combined with oxygen reactive ion etching (RIE) [5] is the most commonly used method to fabricate channel waveguides, as described in Figure 27.14. First, a thin film of the core polymer is coated on the lower cladding, followed by a layer of photoresist. Next, patterns of the waveguide channels can be obtained after UV exposure of the photoresist through a photomask followed by resist developing. The sample is then placed in a vacuum chamber filled with low pressure oxygen. The oxygen gas is ionized by the electrical field of radio frequency forming oxygen plasma, which contains an equal number of positive oxygen ions and free electrons. Oxygen ions are reactive to the carbon and hydrogen atoms in the polymer generating CO, CO_2, and H_2O molecules, which are later pumped out of the chamber. As a result, the part of polymer exposed to the oxygen plasma is removed, while the area of polymer film protected under the photoresist pattern is not. When the etch rate is known, the depth can be precisely controlled by the duration of plasma etching. Finally, the remaining photoresist is removed after etching and the core of channel waveguides is obtained.

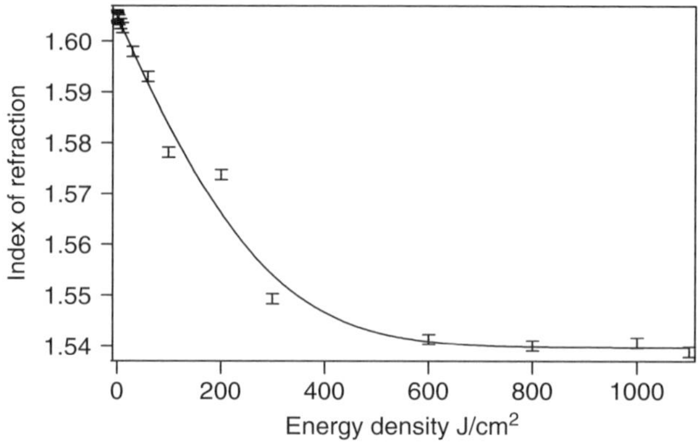

FIGURE 27.13 The change of refractive index of polymer AJL8/APC [4] thin film as a function of photobleaching energy density. The peak wavelength of the UV lamp is 365 nm. The index of refraction was measured at the telecom wavelength of 1.55 μm.

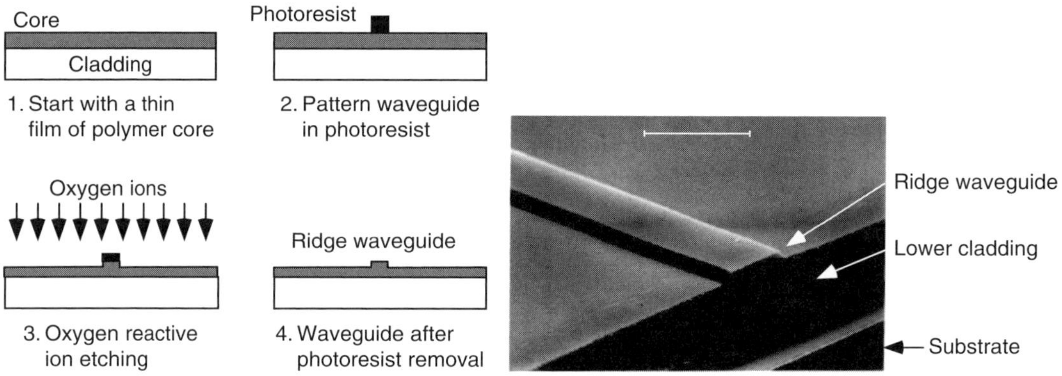

FIGURE 27.14 A schematic process of photolithography and RIE, and a scanning electron microscopy (SEM) image of a ridge waveguide made with this method. The scale bar at the top of the SEM image is 5 μm.

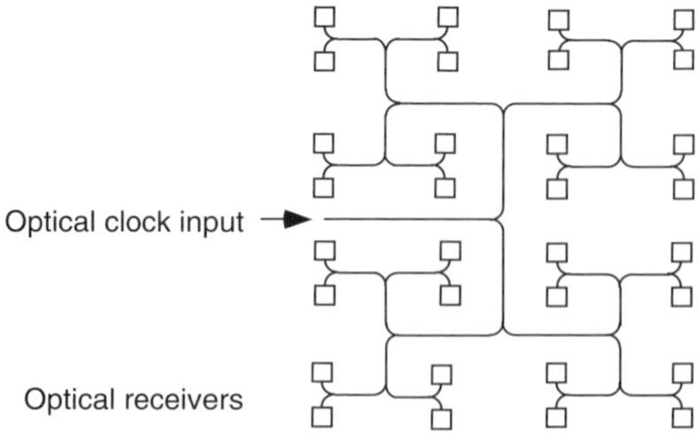

FIGURE 27.15 H-tree waveguide structure for synchronized clock distribution. Each branch splits the optical power equally. The waveguides between the clock input and each receiver have the same length.

Precise timing of IC chips often requires that a given pulse be simultaneously sent to different parts of the chip. Such clock pulses serve as a timing reference to synchronize all the operations of the chip. An H-tree layout shown in Figure 27.15 is usually used to ensure that the pulses reach their respective optical receivers simultaneously and with equal intensity.

27.4 PLASTIC OPTICAL FIBER

Table 27.2 lists several polymers typically used for plastic optical fibers (POFs). POFs are manufactured by two primary methods. In the first method, the polymer is melted and extruded through a small orifice using high pressure, and then coated with a lower index cladding. A more common method is to make a rod of the fiber core/cladding structure first. Such a rod is called a fiber preform, and is a few centimeters in diameter. One end of the preform is heated to the melting point and a thin fiber is drawn out of it, resulting in a fiber with the same aspect ratio and index profile as the preform. The preform can be made with a gradual decrease of refractive index from the center, yielding a

TABLE 27.2
Typical Polymers Used for Plastic Optical Fiber

Polymer	Typical Molecular Structure	Index of Refraction
Poly(methyl methacrylate) (PMMA)	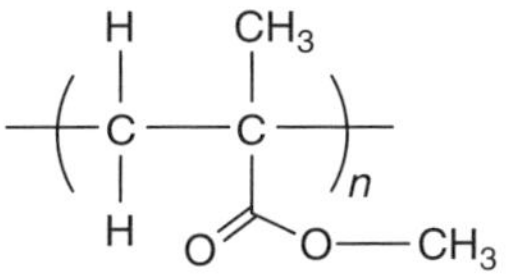	1.49
Fluorinated polymers [6]	Cytop 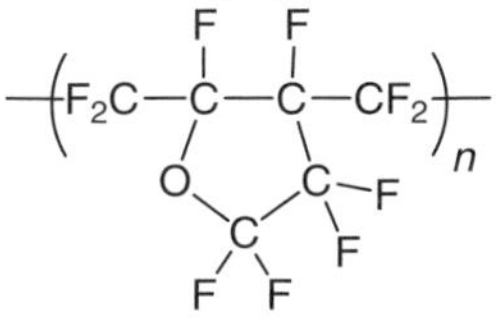	1.34
	Teflon AF 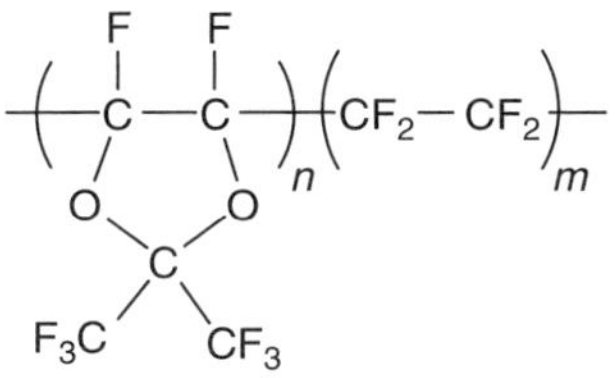	1.31
Polystyrene (PS)	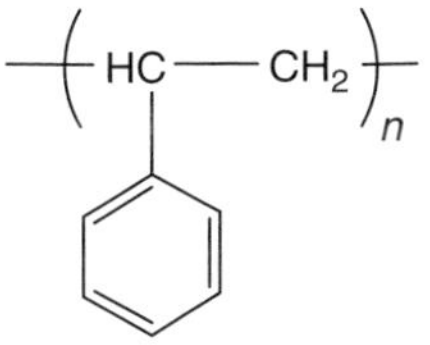	1.59
Polycarbonate (PC)	Bisphenol A polycarbonate [7] 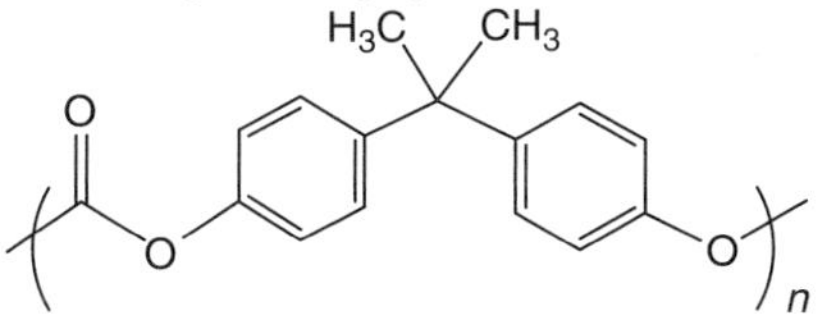	1.58

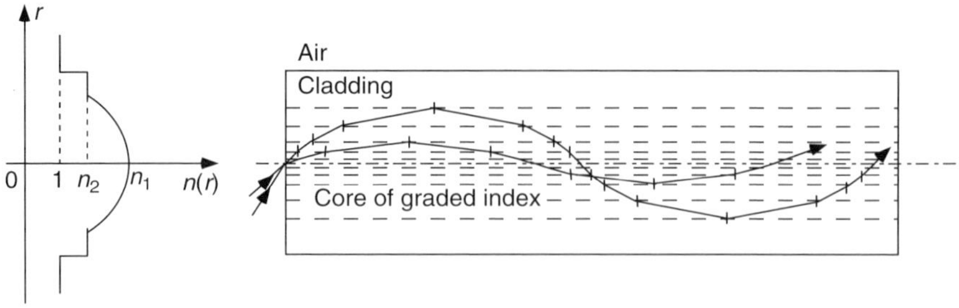

FIGURE 27.16 Bending and total internal reflection of light in a graded-index fiber. r is the radius of the fiber.

graded-index fiber. Graded-index fibers offer wider bandwidth than step-index fibers of the same diameter. Depending on the application, the diameter of plastic fibers ranges from tens of micrometers to over a millimeter. The most commonly used fibers are several hundred micrometers in diameter.

Most POFs are multimode fibers. Light of different modes bounces through a fiber with different reflection angles; as a result, some modes take longer time than others to travel from one end of the fiber to the other. This is called modal dispersion, which causes data pulses to broaden, overlap, and become indistinguishable as several modes travel through a fiber. A graded-index fiber is an effective method of reducing modal dispersion and is widely used in data transmission. A graded-index fiber is a multimode optical fiber in which the refractive index of the core gradually decreases from its highest value at the center of the core to a value that equals the refractive index of the cladding. Figure 27.16 shows the process of refraction and total internal reflection in a graded-index fiber. The gradual decrease in refractive index from the center of the fiber core causes the light rays to be refracted numerous times. Each refraction increases the angle of incidence at the next point of refraction and the total internal reflection occurs when the angle of incidence becomes larger than the critical angle of incidence. Light may be reflected to the axis of the fiber before reaching the core–cladding interface. Light rays become curved and meander through the fiber.

The graded-index design compensates for modal dispersion by allowing light rays in the outer zones of the core to travel faster than those in the center because the outer zones have lower refractive index. Theoretically, a parabolic index profile makes light rays traveling in sinusoidal paths through the fiber. The sinusoidal paths of different modes have the same period and propagation time, minimizing the modal dispersion, as shown in Figure 27.17.

The mirage on the surface of a hot roadway is a typical example of light traveling through a graded-index material. The sun warms the black asphalt, heating the air above. As the air is heated, it becomes less dense, thus its index of refraction decreases (This is the thermo-optic effect, and it is discussed in detail in Section 27.4.2). Cold air has higher density than warm air, and consequently a larger refractive index. The variation between the hot air at the surface of the road and the higher density cold air above creates a refractive index gradient. Light from the sky at a shallow angle to the road is bent and totally reflected by the index gradient similar to the light in a graded-index fiber, as shown in Figure 27.18. Such a reflection makes it appear as if the light is reflected by the road's

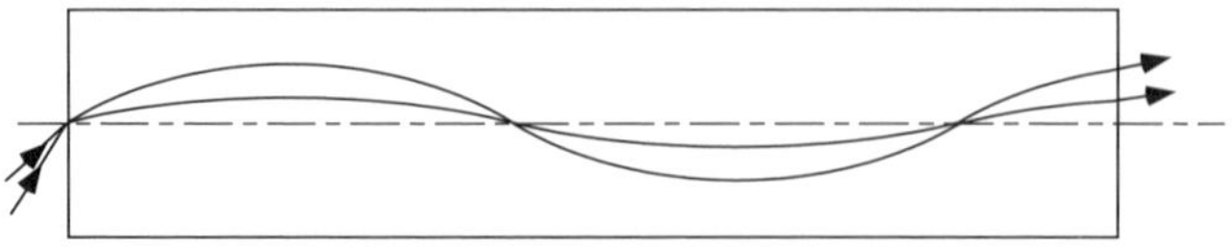

FIGURE 27.17 Different sinusoidal paths of light have the same period in a graded-index fiber with parabolic index profile. Light ray at a larger angle has a longer path, but most of the path is in the low index zones where light travels faster. In the end, light of different paths take the same time to travel between foci.

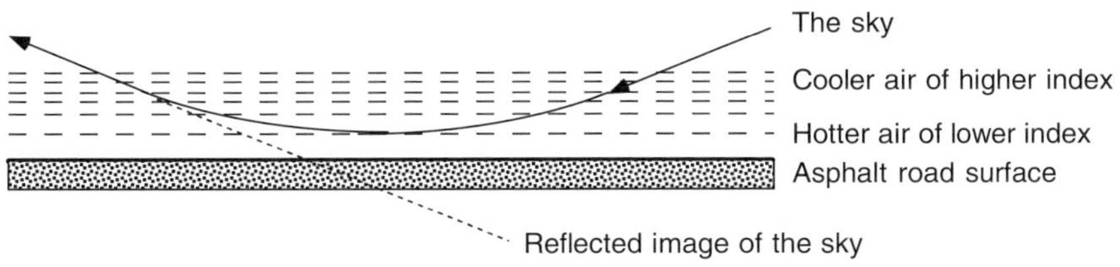

FIGURE 27.18 Highway mirage due to the graded index of air near the hot surface of the road.

surface and the turbulent air makes the reflection sparkling. The human brain interprets such a reflection as a pool of water on the road as water reflects the skylight in the same manner.

Absorption from the polymer material and impurities within the core are the major sources of optical loss of plastic fibers. The loss of the fiber depends on both the polymer and the wavelength of light. The loss characteristics of a typical poly(methyl methacrylate) PMMA fiber in plotted in Figure 27.19 [8].

Although POFs are not suitable for long-distance applications as they suffer from higher optical loss when compared to silica glass fibers (0.2 dB/km), it is advantageous to use them for short-distance data transmission. For example, there are tremendous cost advantages to use POF for data transmission within a few hundred meters or less, making POF an excellent choice for computer networks, industrial control systems, high-quality audio systems, and medical instruments. POFs have also been used in sensors and illumination applications.

Silica glass fiber operates at infrared wavelengths. Because human eyes cannot see infrared light, special equipment is required to determine if a signal is coming out of the fiber. Plastic fiber operates at visible wavelengths, which makes troubleshooting and system maintenance easier. When connecting two pieces of fiber together, the larger core size of plastic fiber allows greater misalignment tolerance compared to the single-mode silica fiber of small core diameter, reducing the time and cost of installation.

Plastic optical fibers are more pliable, less prone to breakage, and easier to handle than their silica counterparts. POFs are more reliable in harsh environments where strong mechanical vibrations exist, and their lightweight makes them a good choice in applications where weight is crucial. For instance, replacing heavy copper cables with plastic fiber can offer significant weight reduction in airplanes and spacecraft. The ruggedness, immunity to electromagnetic interference, and high bandwidth also make them ideal in aircraft and automobiles. Plastic fibers are also easier to fabricate into special assemblies. Such assemblies are used in gastroscopes, endoscopes, and other medical equipment, as well as for illuminating the instrument panels of automobiles.

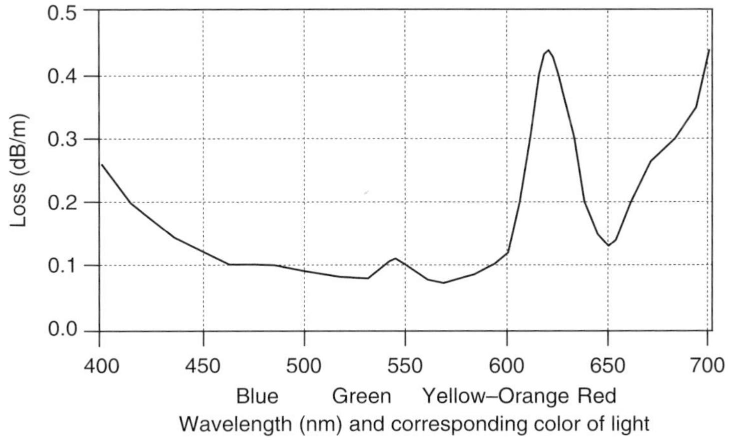

FIGURE 27.19 The loss spectrum of typical PMMA fiber.

27.5 CHANGING THE INDEX OF REFRACTION BY APPLYING AN ELECTRICAL CONTROL SIGNAL

Waveguides and plastic fibers work as signal channels, and are not intended to modify the optical signals passing through them. They are made of passive polymers whose indices of refraction are, ideally, constant and not affected by external variables such as temperature and electromagnetic fields. There is another family of optical polymers designed such that their indices of refraction can be varied in a well-defined manner by applying an electric field or heat. They are referred to as active polymers. If the index of refraction of a waveguide can be varied by an external electrical signal, the optical properties of the waveguide can be changed so that the signal undergoes a desired processing. The two most important types of these devices are modulators, which encode information on light, and switches, which route optical signals in an optical network. Modulators function much like high-speed optical valves, which either pass or block propagating light by using constructive or destructive interference, respectively.

27.5.1 Electro-Optic Effect

Some optical materials can have a small change in their index of refraction when an electric field is applied. The change of index due to an applied electric field is called the electro-optic (EO) effect. If the change in index is linearly proportional to the strength of the electric field, it is called the linear EO effect, also known as the Pockels effect.

$$\Delta n(E) = -\frac{1}{2}n^3 rE \tag{27.8}$$

where
- Δn is the change in the index of refraction
- E is the electric field strength
- r is the linear EO coefficient, and is usually in the range of 10^{-12} to 10^{-10} m/V, or 1–100 pm/V (1 pm = 10^{-12} m)

There are also materials in which the change of index is proportional to the square of the electric field,

$$\Delta n(E) = -\frac{1}{2}n^3 sE^2 \tag{27.9}$$

where s is the quadratic EO coefficient and is typically in the range of 10^{-18} to 10^{-14} m^2/V^2. This is called quadratic EO effect, or Kerr effect.

In general, Pockels effect is stronger than Kerr effect, and produces a larger index change under electric field magnitudes commonly used in EO devices. For this reason, most practical electro-optic devices are based on Pockels effect. There are also higher-order EO effects, however, these effects are typically much weaker than the linear and quadratic effects.

27.5.2 Thermo-Optic Effect

When the temperature of a material changes, the volume of material will change as well. Since the total mass of the material remains the same, the density and, thus, the index of refraction will also change. This temperature dependence of the refractive index is called the thermo-optic (TO) effect. Although the change in refractive index is mainly due to the thermal expansion, it is also caused by temperature-induced stress and molecular polarizability. The change in index is almost linear at room temperature,

$$\Delta n(T) = \alpha \Delta T \tag{27.10}$$

where
T is temperature
ΔT is the change in temperature
α is thermo-optic coefficient

The thermo-optic coefficients are about -1×10^{-4}/°C for most optical polymers around room temperature. Virtually all materials have the TO effect, but it is much larger in optical polymers than in common inorganic optical materials.

Electro-optic and thermo-optic effects have their own strengths and weaknesses in device applications. For example, the EO effect is due to the movement of electrical charge in the molecules of a material. Such effect can occur at very high frequencies (10^{15} Hz); however, for any practical magnitude of the electric field, the change in index is quite small (10^{-4} to 10^{-5}). Owing to the nature of heating and cooling a device, the thermo-optic effect is much slower (10^{-3} s), but changes in index as large as 10^{-3} can be easily obtained.

27.6 ELECTRO-OPTIC MODULATORS

Fiber optic networks constitute a vital part of the infrastructure serving the society. Virtually, all long-distance communication between computers, telephones, cell phone towers, and cable television facilities relies on fiber optics. The information transmitted first travels a short distance through electrical wires to a fiber optic terminal, which subsequently converts the electrical signal into optical pulses and sends these pulses through low-loss glass fiber. The optical signal can travel thousands of miles through fiber before reaching the receiver. The receiver converts optical pulses back to electrical data and sends them to users via electrical wires and wireless channels (Figure 27.20).

The equipment used in fiber optic terminals incorporates EO modulators, which convert electrical signals into pulses of light that are transmitted through the optical fiber. EO modulators are key components in fiber optic telecommunication. A typical EO modulator is based on a Mach–Zehnder interferometer [9], as schematically shown in Figure 27.21. The input light is equally divided into two paths, called arms. If the indices of refraction of the two arms are different, light will undergo a phase shift, resulting in constructive or destructive interference when light from the two arms recombines.

The output power of the modulator depends on the relative phase difference between the two arms

$$I_{\text{out}} = I_{\text{in}} \cos^2(\Delta\phi) \tag{27.11}$$

where
I_{in} and I_{out} are optical power at the input and output of the device, respectively
$\Delta\phi$ is the phase difference between light that travels in the two arms

The phase difference is directly proportional to the change of the refractive index Δn

$$\Delta\phi = \frac{2\pi}{\lambda} \Delta n L \tag{27.12}$$

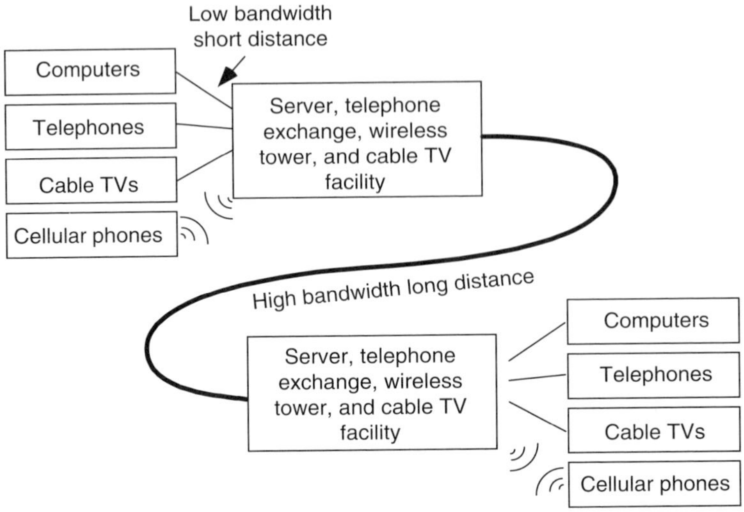

FIGURE 27.20 A schematic diagram of telecommunication. Low bandwidth signals between user equipment and fiber optic terminal equipment are typically transmitted through electrical cables or wireless channels over a short distance. Signals from multiple users are combined into high bandwidth optical data and transmitted through optical fiber over long distance.

where

λ is the wavelength of the light

L is the length of the arms of the modulator

For EO polymer waveguides, the index change is related to the applied electric field through the linear electro-optic effect. It would be quite useful to derive equations that relate both phase difference and output power to the control voltage. To this end, let us recall that the electric field generated by two electrodes is governed by both their potential (V) and separation (d), such that $E = V/d$. If we combine this with Equations 27.8, 27.11, and 27.12, we obtain the following two relations:

$$\Delta\phi = \frac{2\pi}{\lambda}\Delta nL = -\frac{\pi L n^3 r}{\lambda d}V \qquad (27.13)$$

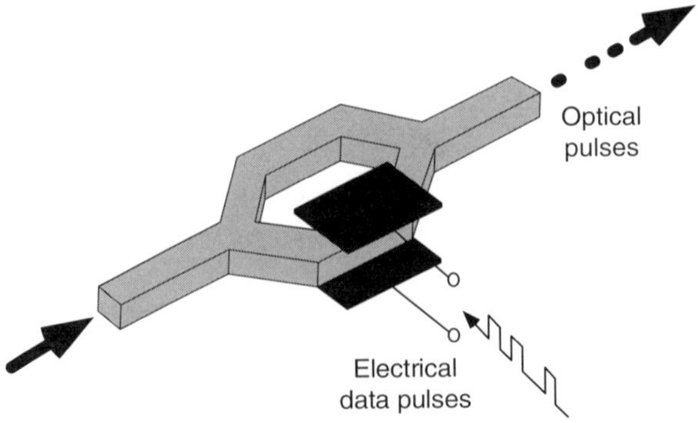

FIGURE 27.21 Mach–Zehnder modulator.

$$I_{out} = I_{in}\cos^2\left(\frac{\Delta\phi}{2}\right) = I_{in}\cos^2\left(\frac{\pi L n^3 r}{2\lambda d}V\right) = I_{in}\cos^2\left(\frac{\pi}{2}\frac{V}{V_\pi}\right) \tag{27.14}$$

where the V_π, which is called half-wave voltage, is the voltage necessary to induce a phase difference of π radians (180°) and is defined as

$$V_\pi = \frac{\lambda d}{L n^3 r} \tag{27.15}$$

For this reason, when the control voltage equals V_π, the light in the two arms are completely out of phase, the output power is at a minimum due to destructive interference. Figure 27.22 shows the dependence of output power on the applied voltage, and mathematic simulation of the propagation of optical wave in a modulator at three different voltages.

There are several factors that determine whether an electro-optic material is suitable for high-speed EO modulators. The criteria in selection of desirable materials are (1) the dielectric constant and index of refraction should be matched such that electrical and optical signals travel at the same speed, (2) a low dielectric constant, and (3) a large change in index of refraction under a small change of an applied electric field. High-speed electro-optic modulators require a good speed match between the traveling electrical and optical signals. The speed of the electric signal is $c/\sqrt{\varepsilon_r}$ and that of optical signal is c/n, where c is the speed of light in the free space, ε_r is the dielectric constant, and n is the refractive index of the EO material. Modulator efficiency therefore increases as the difference between the two speeds decreases. The product of bandwidth and length of the modulator is determined by the speed match [10]

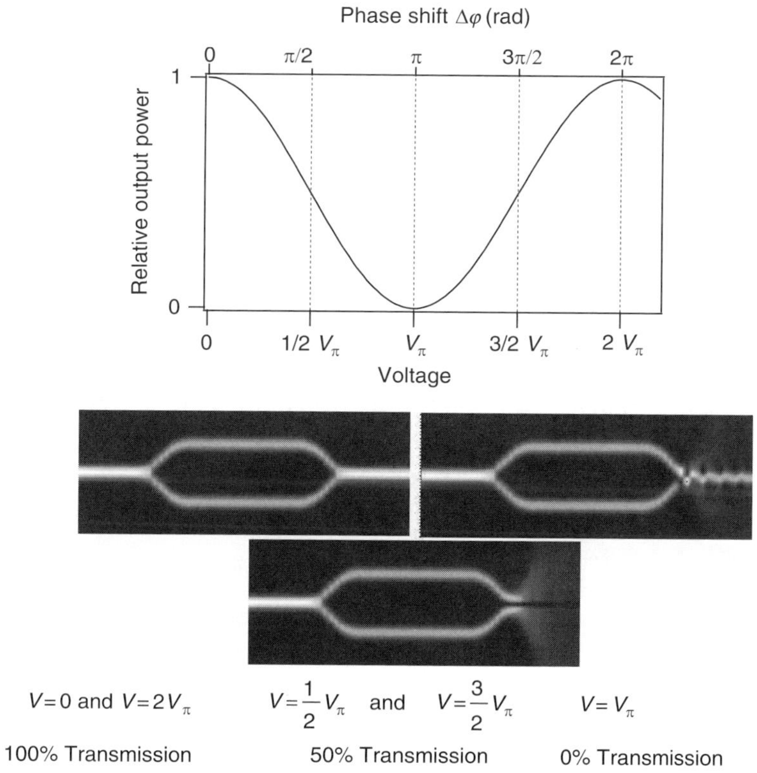

FIGURE 27.22 Top: Dependence of modulator output on the phase difference and applied voltage. Bottom: Optical power distribution in the modulator at various voltages.

$$\Delta f \cdot L = \frac{c}{4\left|\sqrt{\varepsilon_r} - n\right|} \quad (27.16)$$

A large bandwidth–length product is required for an EO material used for high-speed modulator fabrication. A low dielectric constant is desirable because it reduces the response time of the device by decreasing the capacitance between the electrodes. A larger value of index change per unit electric field means lower modulation voltage, and the index modulation efficiency is generally given by the $n^3r/2$ term from Equation 27.8.

Electro-optic polymers are artificial organic materials designed and synthesized to have a strong electro-optic effect. The EO effect of the polymer comes from molecules that demonstrate strong nonlinear optical properties. These molecules are known as chromophores since these are often rich in color. When chromophores are preferentially aligned in a given direction, the individual contributions from each chromophore add up, yielding a polymer that exhibits strong electro-optic effects. The alignment of the chromophores is typically achieved by electric poling, in which the thin films of EO polymer is exposed to an external electric field and heated to above its glass transition temperature. As the heat softens the polymer network, the electric dipole moments of the chromophores rotate and align themselves under the electric field. The polymer is then cooled down to room temperature, freezing the chromophores in their aligned orientation, after which the external electric field can be removed and the poling is completed.

The main advantage of EO polymers is that these materials have the best speed match and the lowest dielectric constant among all the materials used for electro-optic modulators. The intrinsic EO response time is on the order of femto seconds (10^{-15} s) [11]. A comparison of the EO polymer with a common EO material used in photonics devices, $LiNbO_3$, is given in Table 27.3. The values of $LiNbO_3$ are intrinsic to the crystalline material, but the values of EO polymer have been continuously improved as better materials and more efficient poling methods are developed. EO polymer modulators with bandwidth up to 200 GHz (2×10^{11} Hz) have been demonstrated [12]. In addition to their adaptability for high-speed modulation, EO polymers have other important advantages. Thin film EO polymer devices can be easily fabricated on virtually any smooth substrates. Monolithic integration of EO polymer devices with very large-scale integrated (VLSI) circuitry can be realized with less technical complexity. The materials are made with mature chemical engineering and can be of low cost when mass produced. The versatility in molecular design makes it possible to tailor the properties of EO polymers to meet specific requirements [13].

TABLE 27.3
Comparison of Major Properties between the State-of-Art Electo-Optic Polymer and $LiNbO_3$, the Most Commonly Used Inorganic EO Material for EO Modulators

Properties	EO Polymer	$LiNbO_3$
Bandwidth–length product (GHz cm)	91	2.4
Relative dielectric constant	3	28
Index of refraction	1.65	2.2
Modulation efficiency n^3r (pm/V)	1350[a]	330

Source: [a]From Dalton, L.R., Robinson, B., Jen, A., Ried, P., Eichinger, B., Sullivan, P., Akelaitis, A., Bale, D., Haller, M., Luo, J., Liu, S., Liao, Y., Firestone, K., Bhatambrekar, N., Bhattacharjee, S., Sinness, J., Hammond, S., Buker, N., Snoeberger, R., Lingwood, M., Rommel, H., Amend, J., Jang, S.-H., Chen, A., and Steier, W., Proc. SPIE, 5935, 1, 2005.

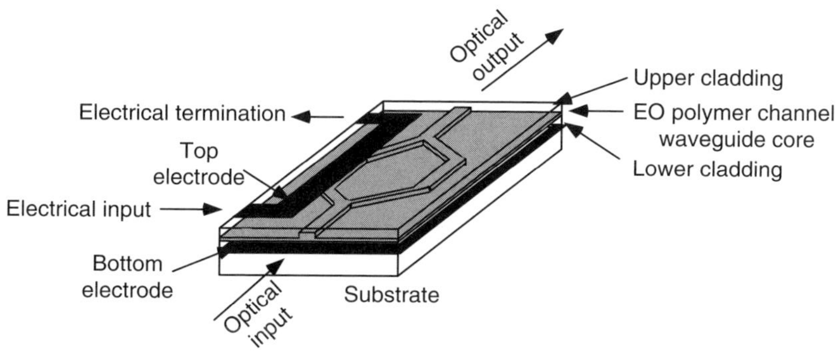

FIGURE 27.23 The structure of a typical EO polymer modulator.

Figure 27.23 shows the device structure of a typical EO polymer modulator. Each arm of the modulator is 1–2 cm in length, while the thickness of the core and cladding layers are only a few micrometers. The core of the waveguide contains EO polymer. Such a modulator chip can be made with standard microelectronic fabrication techniques such as photolithography, metallization, and reactive ion etching. Finally, optical fiber is connected to the waveguide's input and output, electrical wires are attached to the electrodes, and the chip is sealed for environmental protection.

27.7 THERMO-OPTIC SWITCHES

Thermo-optic switches use micro heaters fabricated on top of the waveguide to change the index of refraction of the polymer through the thermo-optic effect. The heaters are made of metallic thin films and are powered by electrical control signals. Compared to inorganic waveguide materials such as silica glass, polymers have a much larger thermo-optic coefficient. This enables polymer thermo-optic switches to operate with mild temperature changes, consume less power, and be more compact than their inorganic counterparts. The waveguide of the switch is a typical buried channel design, similar to the previously discussed optical interconnection waveguides. The metallic thin film for the heater is fabricated on top of the upper cladding. Heat is generated when a voltage is applied, and this changes the refractive index of the core. Because of its small size, the heater requires only less than 100 mW of power to switch the optical output between two waveguides. The typical temperature change required for switching is in the range of 20°C to 50°C, and switching speeds are usually on the order of 10 ms.

1 × 2 and 2 × 2 switches can be built by replacing one or both Y-branches of a Mach–Zehnder modulator with 3 dB couplers [14], as shown in Figure 27.24. Each 3 dB coupler has two inputs and two outputs. Light from any input is divided between the two outputs. Each output gets 50% (3 dB) of the input light. The optical power at the two outputs are given by

$$\begin{aligned} I_a &= I_1 \sin^2(\Delta\phi) \\ I_b &= I_1 \cos^2(\Delta\phi) \end{aligned} \tag{27.17}$$

where

$$\Delta\phi = \frac{2\pi}{\lambda} L \Delta n \tag{27.18}$$

The optical wavelength λ and the length of the heater L are constants. $\Delta\phi$ depends on the index change Δn induced by the thermo-optic effect. When $\Delta n = 0$, light entering from the upper input port is directed to the lower output port. Similarly, light entering from the lower input port is directed to

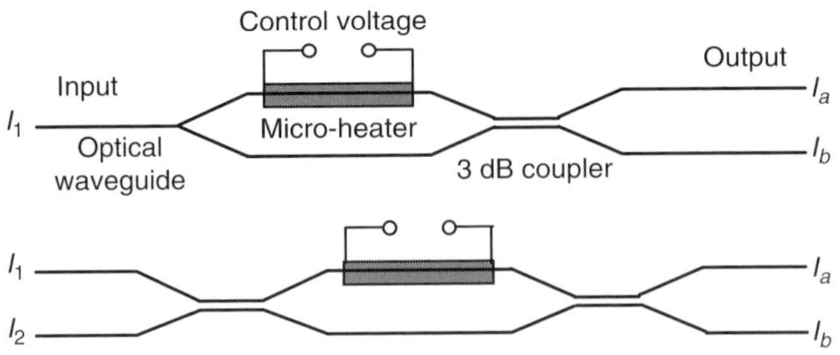

FIGURE 27.24 1 × 2 and 2 × 2 thermo-optic switches based on Mach–Zehnder interferometers.

the upper output port. This state is called the cross state. When Δn is such that $\Delta\phi = \pi/2$, light entering from the upper input is directed to the upper output, and light entering from the lower input is directed to the lower output. This state is called the bar state.

1 × 2 and 2 × 2 switches can be used as building blocks to construct switches with more input/output ports. Figure 27.25 shows two different examples of a 4 × 4 switch built with 1 × 2 and 2 × 2 switch elements.

27.8 WAVELENGTH FILTERS

In fiber optic telecommunications, it is common to transmit many different wavelengths simultaneously through the same piece of fiber. Each wavelength carries a different communication channel. It may be easier to visualize (and remember) this concept by noting that each wavelength, and thus each channel, corresponds to a specific color. Sending several different wavelengths of light down a fiber is called wavelength-division multiplexing (WDM), which is efficient at cramming large amounts of information down a single fiber. Transmission of more than 100 channels has been demonstrated [15]. Figure 27.26 illustrates the basic principles of WDM transmission. WDM requires devices that combine multiple channels into one fiber for transmission, and a device to separate these channels into individual fibers at the receiver. Such devices are called wavelength filters. In fiber optics, the filter at the transmitter is referred to as a wavelength multiplexer, and the filter at the receiver is called a wavelength demultiplexer. They are essentially the same device used in opposite ways.

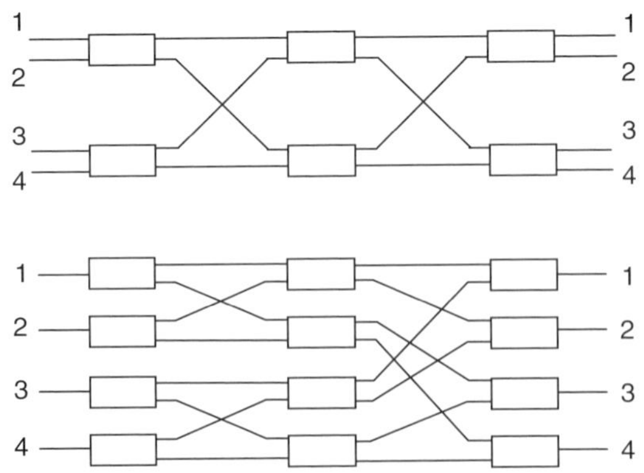

FIGURE 27.25 4 × 4 switches made of 2 × 2 (top), and a combination of 1 × 2 and 2 × 2 (bottom) switch elements.

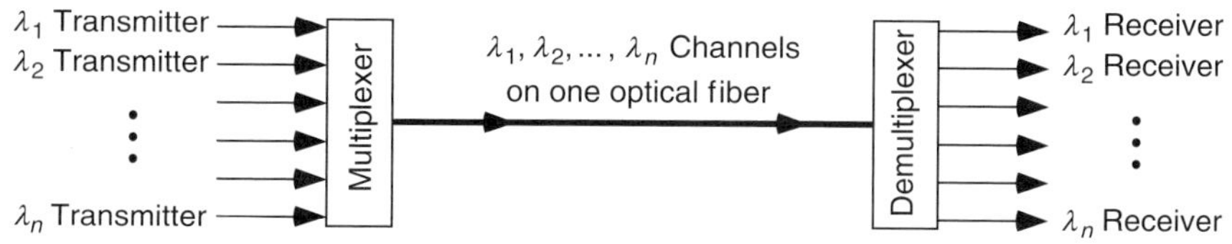

FIGURE 27.26 WDM transmission between two communication points.

27.8.1 Arrayed Waveguide Grating Filters

Wavelength filters are often made with arrayed waveguide gratings. A filter made from an arrayed waveguide grating (AWG) [16] consists of two star couplers connected by an array of waveguides, as shown in Figure 27.27. A star coupler distributes light from an input waveguide to an array of waveguides. The other end of the array is connected to another star coupler. The second coupler combines the light from individual waveguides in the array and focuses the light on one of the output waveguides according to the wavelength. The position of the focal point depends on the relative phase difference of light between two adjacent waveguides of the array. The lengths of adjacent waveguides are different by a fixed amount, which produces a phase difference that is a function of the wavelength. Therefore, light of different wavelengths is focused where the appropriate output waveguides are located.

It is critical to have a precise index of refraction in making AWGs. A significant advantage in using polymers is that their index of refraction can be customized by varying the ratio of different monomer components. The potential of low material and fabrication costs have also made optical polymers attractive for making AWG filters. Polymer AWG filters have demonstrated excellent performance and reliability, thus meeting the stringent requirements of fiber optic telecommunication systems [17]. Tunable wavelength filters have been achieved by using polymers with large thermo-optic and electro-optic effects, which allow for large changes in the index of refraction of the arrayed waveguides [18].

27.8.2 Micro-Ring Resonators

Micro-ring resonators are another way of making wavelength filters. A micro-ring resonator consists of a waveguide that forms a closed loop. The size of the ring ranges from a few micrometers to a few hundred micrometers. As a result, they are quite small when compared to AWG filters, Mach–Zehnder modulators and switches, which are typically centimeters in length. The ring can be in the shape of either a circle or a loop, with straight sections for coupling to adjacent waveguides or other resonators. Light can be coupled into and out of the ring waveguide by placing one or more

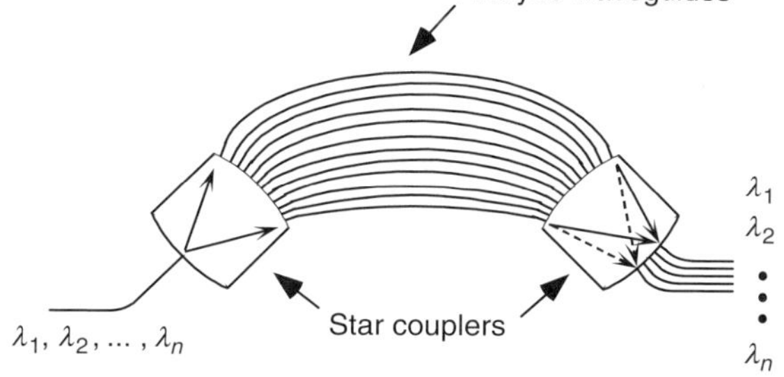

FIGURE 27.27 Arrayed waveguide grating wavelength filter. The number of waveguides in the array is typically about 100.

waveguides very close to the ring. The closed loop of the ring allows light to travel around the ring numerous times, which leads to interference. The type of interference (constructive, destructive, or somewhere in-between) is a function of the light's wavelength and the circumference of the ring. If the optical path length is a multiple of the wavelength, light completing each round will match the phase of the light just entering the ring, and will add constructively to the optical power. Such wavelengths are called resonant wavelengths.

$$m\lambda_m = nL \tag{27.19}$$

where
- m is a natural number
- λ_m is the mth resonant wavelength
- n is the index of refraction
- L is the circumference of the ring

Conversely, destructive interference will occur if the light completing a round in the ring is out of phase with the light entering the ring. This destructive interference, whether partial or complete, will not allow the optical power to build up and lead to a decrease of intensity of light at these wavelengths.

A ring resonator that has two straight waveguides coupled to a single ring waveguide, as shown in Figure 27.28, and can work as a wavelength add/drop filter [19]. If the wavelength of light in one waveguide matches the resonant wavelength λ_m, light will be coupled between the two waveguides via the ring. If the light does not have a wavelength of λ_m, it will be undisturbed and continue to propagate in the original waveguide.

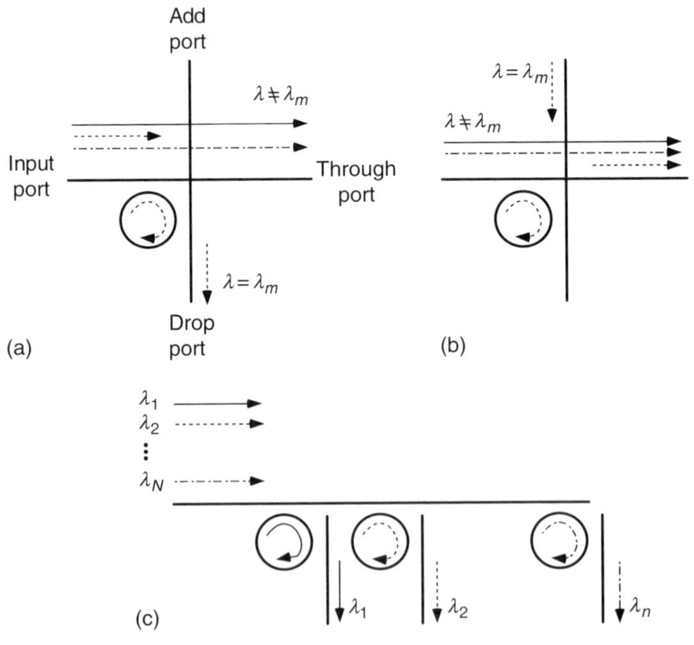

FIGURE 27.28 Wavelength add–drop filters based on micro-ring resonators. (a) Dropping a wavelength channel. (b) Adding a wavelength channel. (c) WDM filter demultiplexer made with micro-ring resonators.

The ratios of the output power at both the through port and drop port with respect to the input power are given by

$$\frac{I_{\text{Through}}}{I_{\text{Input}}} = \frac{\alpha^2|\tau|^2 + |\tau|^2 - 2\alpha|\tau|^2\cos\theta}{1+\alpha^2|\tau|^4 - 2\alpha|\tau|^2\cos\theta} \tag{27.20}$$

$$\frac{I_{\text{Drop}}}{I_{\text{Input}}} = \frac{\alpha^2\left(1-|\tau|^2\right)^2}{1+\alpha^2|\tau|^4 - 2\alpha|\tau|^2\cos\theta} \tag{27.21}$$

where α is the loss factor associated with a round-trip through the ring. An ideal ring with no loss will have $\alpha = 1$. τ is the coupling factor, and was defined in Figure 27.11.

$$\theta = \frac{2\pi}{\lambda}nL \tag{27.22}$$

Optical switches can also be made with micro-ring resonators using polymers with large EO or TO effects [20]. For example, if the index of refraction of the ring is varied by applying an electric field, the resonant wavelength will change, and the light of a given wavelength will be either redirected to a different waveguide or continue undisturbed. Modulators can be built with the same principle. Their compact size makes them ideal for large-scale integration, single chip optical switch arrays, and WDM modules. It also makes them highly suitable for monolithic vertical integration on silicon integrated circuit chips. Micro-ring resonators are also used as optical sensors that measure temperature, stress, or strain, and can also be used to detect chemical and biological agents [21].

27.9 SUMMARY

Total internal reflection allows light to be guided in waveguides and fibers for broadband communication. There are many advantages to use polymer rather than conventional inorganic materials. Polymer devices are suitable for mass production, and offer a low-cost alternative for transmitting information at high speeds for short-range applications. These devices operate at visible wavelength, reducing the need for special equipment during installation and maintenance. The indices of refraction of some polymers can be changed by altering the temperature, or subjecting the polymer to an external electric field. These polymers enable optic modulators and switches to operate at high frequencies, and thus offer performance advantage compared to traditional inorganic materials. New polymer materials and their applications are active areas of research and development. For this reason, continuous progress has been made in the performance and stability of polymer devices over the last 20 years. Modern polymers are more resistant to temperature, moisture, and aging, all of which are critical factors that affect the reliability and lifetime of the devices.

EXERCISE QUESTIONS

1. A polymer waveguide has a core with a refractive index of 1.65. If a material with an index of 1.55 is used as cladding, what will be the critical angle for total internal reflection?
2. Using the critical angle θ_c, show that the numerical aperture of a fiber is given by $\text{NA} = \sqrt{n_1^2 - n_2^2}$ where n_1 and n_2 are the indices of refraction of the core and cladding, respectively. Assume that the end of the fiber is flat and perpendicular to the axis of the fiber.

3. An EO polymer has an index of refraction of 1.65 and an electro-optic coefficient of 100 pm/V. A thin film of this material is placed between two parallel conductor electrodes separated by 10 μm. When a voltage of 1 V is applied across the two electrodes, how much index change is expected in the EO polymer?
4. Design a 4 × 4 switch using only 1 × 2 switch elements.
5. Separation between two adjacent resonant wavelengths of micro-ring resonators is called free-spectral range (FSR). Show that $\mathrm{FSR} \approx \frac{\lambda_m^2}{nL}$.

LIST OF ABBREVIATIONS

AWG	arrayed waveguide grating
CD	compact disk
dB	decibel
DVD	digital versatile disk
EO	electro-optic
FSR	free spectral range
IC	integrated circuit
NA	numerical aperture
PC	polycarbonate
PMMA	poly(methyl methacrylate)
POF	polymer optical fiber
PS	polystyrene
RIE	reactive ion etching
SEM	scanning electron microscopy
TO	thermo-optic
UV	ultraviolet
VLSI	very large-scale integrated
WDM	wavelength division multiplexing

REFERENCES

1. Goos, F. and Hänchen, H., A new and fundamental experiment on total reflection, *Annalen der Physik*, 1, 333–346 1947.
2. Mizuno, H., Sugihara, O., Kaino, T., Okamoto, N., and Hosino, M., Low-loss polymeric optical waveguides with large cores fabricated by hot embossing, *Optics Letters*, 28, 2378–2380 2003.
3. Diemeer, M.B.J., Suyten, F.M.M., Trommel, E.S., McDonach, A., Copeland, J.M., Jenneskens, L.M., and Horsthuis, W.H.G., Photoinduced channel waveguide formation in nonlinear optical polymers, *Electronics Letters*, 26, 379–380 1990.
4. Luo, J., Haller, M., Ma, H., Liu, S., Kim, T.D., Tian, Y., Chen, B., Jang, S.H., Dalton, L.R., and Jen, A.K.Y., Nanoscale architectural control and macromolecular engineering of nonlinear optical dendrimers and polymers for electro-optics, *Journal of Physical Chemistry B*, 108, 8523–8530 2004.
5. Chen, A., Kaviani, K., Rempel, A.W., Kalluri, S., Steier, W.H., Shi, Y., Liang, Z., and Dalton, L.R., Optimized oxygen plasma etching of polyurethane-based electro-optic polymer for low loss optical waveguide fabrication, *Journal of the Electrochemical Society*, 143, 3648–3651 1996.
6. Zhou, M., Low-loss polymeric materials for passive waveguide components in fiber optical telecommunication, *Optical Engineering*, 41, 1631–1643 2002.
7. Fan, C.F., Molecular modeling of polycarbonate. 1. Force field, static structure, and mechanical properties, *Macromolecules*, 27, 2383–2391 1994.
8. Plastic Optical Fiber Trade Organization, *Present State-of-the-Art of Plastic Optical Fiber (POF) Components and Systems*, 2004.
9. Tonchev, S., Todorov, R., Zilling, K.K., and Savatinova, I., Mach-Zehnder type modulator for integrated optics, *Journal of Optical Communications*, 11, 147–150 1990.
10. Peters, L.C., Gigacycle bandwidth coherent light traveling-wave phase modulators, *Proceedings of the IEEE*, 51, 147–153 1963.

11. Cao, H., Heinz, T.F., and Nahata, A., Electro-optic detection of femtosecond electromagnetic pulses by use of poled polymers, *Optics Letters*, 27, 775–777 2002.
12. Lee, M., Katz, H.E., Erben, C., Gill, D.M., Gopalan, P., Heber, J.D., and McGee, D.J., Broadband modulation of light by using an electro-optic polymer, *Science*, 298, 1401–1413 2002.
13. Dalton, L.R., Robinson, B., Jen, A., Ried, P., Eichinger, B., Sullivan, P., Akelaitis, A., Bale, D., Haller, M., Luo, J., Liu, S., Liao, Y., Firestone, K., Bhatambrekar, N., Bhattacharjee, S., Sinness, J., Hammond, S., Buker, N., Snoeberger, R., Lingwood, M., Rommel, H., Amend, J., Jang, S.-H., Chen, A., and Steier, W., Electro-optic coefficients of 500 pm/V and beyond for organic materials, *Proceedings of the SPIE*, 5935, 1–12 2005.
14. Ooba, N. and Okuno, M., Planar-waveguide-type optical switching devices, *Proceedings of the SPIE,* 4470, 45–52 2001.
15. Hurh, Y.-S., Hwang, G.-S., Jeon, J.-Y., Lee, K.-G., Shin, K.-W., Lee, S.S., Yi, K.Y., and Lee, J.-S., 1-Tb/s (100 × 12.4 Gb/s) transmission of 12.5-GHz-spaced ultradense WDM channels over a standard single-mode fiber of 1200 km, *IEEE Photonics Technology Letters*, 17, 696–698 2005.
16. Takahashi, H., Suzuki, S., Kato, K., and Nishi, I., Arrayed-waveguide grating for wavelength division multi/demultiplexer with nanometre resolution, *Electronics Letters*, 26, 87–88 1990.
17. Eldada, L., Telcordia qualification and beyond: reliability of today's polymer photonic components, *Proceedings of the SPIE—The International Society for Optical Engineering*, 5724, 96–106 2005.
18. Toyoda, S., Ooba, N., Kaneko, A., Hikita, M., Kurihara, T., and Maruno, T., Wideband polymer thermo-optic wavelength tunable filter with fast response for WDM systems, *Electronics Letters*, 36, 658–660 2000.
19. Chu, S.T., Little, B.E., Pan, W., Kaneko, T., Sato, S., and Kokubun, Y., An eight-channel add-drop filter using vertically coupled microring resonators over a cross grid, *IEEE Photonics Technology Letters*, 11, 691–693 1999.
20. Emellet, S.J. and Soref, R.A., Analysis of dual-microring-resonator cross-connect switches and modulators, *Optics Express*, 13(20), 7840–7853, 2005.
21. Chao, C.-Y., Fung, W., and Guo, L.J., Polymer microring resonators for biochemical sensing applications, *IEEE Journal of Selected Topics in Quantum Electronics*, 12, 134–142 2006.

28 Organic Low-Dielectric Constant Materials for Microelectronics

Jinghong Chen

CONTENTS

28.1 Introduction 846
28.2 Theoretical Background of Dielectric Materials 848
28.2.1 Dielectric Materials 848
28.2.1.1 Macroscopic Perspective 848
28.2.1.2 Microscopic Perspective 849
28.2.1.3 Frequency Dependence of Dielectric Constant 850
28.2.2 How to Reduce Dielectric Constant *k* 852
28.3 Organic Low-Dielectric Constant Materials 852
28.3.1 Classification of Low-*k* Materials 852
28.3.2 Organic Low-Dielectric Constant Polymers 854
28.3.2.1 Polyimide 854
28.3.2.2 Poly(Arylene Ethers) 855
28.3.2.3 Benzocyclobutene 855
28.3.2.4 Polyphenylene 856
28.3.2.5 Polytetrafluoroethylene 857
28.3.2.6 Parylene 857
28.3.3 Porous Organic Ultralow-*k* Polymers 857
28.3.4 Characterization of Low-*k* Materials 858
28.4 Device and Integration 859
28.4.1 Overview of BEOL Processes 859
28.4.2 Low-*k* Integration Challenges 861
28.4.2.1 Thermal and Mechanical Stability 861
28.4.2.2 Chemical Stability 862
28.4.2.3 Compatibility with Other Materials 863
28.4.2.4 Low Moisture Absorption 864
28.5 Summary 864
Exercise Questions 864
List of Abbreviations 864
References 865

Abstract This chapter explains the need for low-dielectric constant materials in advanced integrated circuit (IC) interconnects followed by the basic theoretical background of dielectric materials and the general methods of reducing the dielectric constant *k*. Various low-*k* materials are discussed with a focus on state-of-the-art organic low-*k* polymers, including their structures, syntheses, properties, and characterization. Basic back-end-of-the-line (BEOL) device fabrication processes are introduced so that the reader will better understand the requirements of low-*k* dielectric materials and the device integration challenges.

28.1 INTRODUCTION

The silicon-integrated circuit is one of the wonders of the twentieth century and the continuous advancement of microelectronics has revolutionized our way of life. Moore's law predicted that the computing power of silicon chips would double every 18–24 months to keep up with the insatiable consumer demand for higher performance electronic products at lower cost. Over the past few decades, performance improvements have been achieved mostly through packing more transistors on a single chip and increasing the size of the silicon wafers. Shrinking of transistor size increases switching speed of the integrated circuits; however, it makes interconnections between transistors work slower. To date, the overall signal delay is no longer dominated by the intrinsic gate delay of the transistor, but greatly affected by the interconnect delay as well. Figure 28.1 is a schematic cross section of a silicon-integrated circuit (IC) including both front-end-of-the-line (FEOL) at the transistor level and back-end-of-the-line (BEOL after the first metallization) in the multilevel interconnects. Low-dielectric constant material is used as the insulating material between metals in the BEOL. Reducing the dielectric constant of the insulating material is an effective method to mitigate interconnect delay.

Interconnection can be represented as a chain of resistors and capacitors. A good figure of merit which can be used to characterize interconnects is resistance–capacitance (RC). Interconnect delay is also known as RC delay. A simple pictorial model that can be used to explain this phenomenon is shown in Figure 28.2. For an interconnect line of length *L*, the total resistance *R* of the line can be written as shown in the following equation:

$$R = \frac{2\rho L}{WT} \tag{28.1}$$

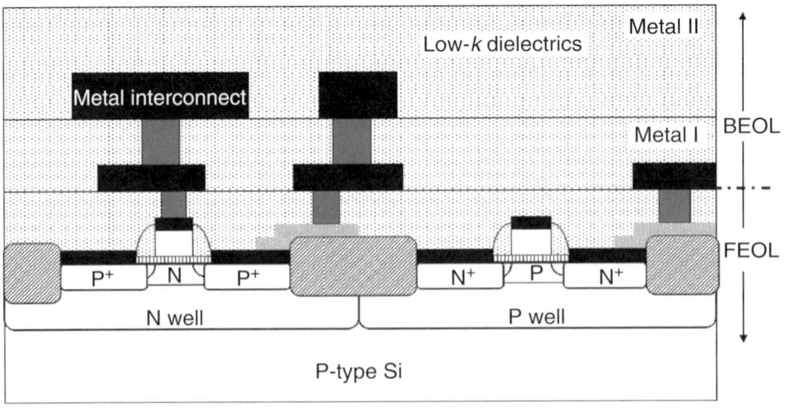

FIGURE 28.1 Cross section of a silicon-integrated circuit with two layers of metal-interconnect.

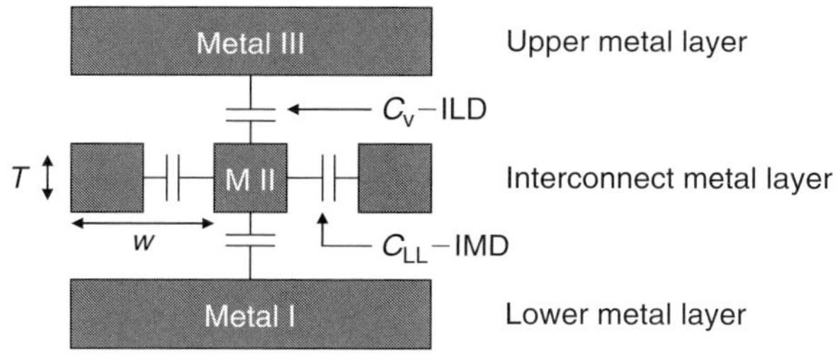

FIGURE 28.2 Capacitance within and between metal layers.

where
- ρ is the resistivity
- W is the line spacing
- T is the thickness of the conductor

The total capacitance can be written as the sum of two capacitance factors: the lateral line to line capacitance C_{LL} (also called intermetal dielectrics, IMDs) and the vertical layer to layer capacitance C_V (also called interlayer dielectrics, ILDs). Assuming that the minimum metal pitch equals twice the metal width, and the dielectric thickness above and below the metal line equals the thickness of the metal line, the total capacitance is

$$C = 2(C_{LL} + C_V) = 2k\varepsilon_0\left(\frac{2LT}{W} + \frac{LW}{2T}\right) \tag{28.2}$$

where
- k is the relative dielectric constant
- ε_0 is the permittivity of vacuum

The total RC delay is the product of Equations 28.1 and 28.2:

$$\text{RC} = 2\rho k\varepsilon_0\left(\frac{4L^2}{W^2} + \frac{L^2}{T^2}\right) = 2\rho k\varepsilon_0 L^2\left(\frac{4}{W^2} + \frac{1}{T^2}\right) \tag{28.3}$$

Equation 28.3 demonstrates that the conventional wisdom of improving device performance by reducing chip geometry (decrease W and T) drives RC delay up. To reduce RC delay, the specific resistance, ρ, of the interconnect metal and the dielectric constant k of the insulating dielectrics need to be decreased. Aluminum alloys have traditionally been employed as the interconnect metals. One of the initial steps taken to address the RC delay was to replace aluminum with copper, gaining a 36% reduction in resistivity. The best conductor, silver, has a resistivity only 6% lower than that of copper and therefore there is little room for further reduction of ρ. For comprehensive discussion of the benefits and the process of Cu implementation, the reader is recommended to refer to related articles [1]. The conventional dielectric in microelectronics is silicon dioxide with a dielectric constant of 3.9. In theory, new materials can be invented to reduce k to approach the theoretical limit of 1 (dielectric constant of air). Research and development of low-k materials has been an active area. In this chapter, we focus our discussion on the low-dielectric constant insulating materials as ILDs and IMDs.

28.2 THEORETICAL BACKGROUND OF DIELECTRIC MATERIALS

28.2.1 Dielectric Materials

The relative dielectric constant is the ratio of the permittivity of a substance to that of free space. This important part of electromagnetic theory is covered in introductory physics texts [2–5]. In this section, we briefly review relevant details. It should be noted that the research field of low-dielectric constant materials is multidisciplinary, and each respective discipline has its own perspective and often its own nomenclature. For example, the relative dielectric constant k is also known as relative permittivity ε_r. As a convention, the microelectronic community uses k, whereas the scientific community uses ε_r. For consistency, relative dielectric constant k will be used throughout this chapter.

28.2.1.1 Macroscopic Perspective

The Faraday capacitor experiment [6] first demonstrated that there is an electrical effect from an insulator when a dielectric is inserted between the parallel plates of a capacitor as shown in Figure 28.3. The capacitance increases by a factor k.

The capacitance C is given by

$$C = \frac{k\varepsilon_0 A}{d} \tag{28.4}$$

where
A is the plate area
d is the separation between the two plates

The total charge and the voltage on the capacitor are related, and can be further described as a function of the surface charge density, σ

$$C = \frac{Q}{V} = \frac{\sigma \cdot A}{E \cdot d} \tag{28.5}$$

As a result, the surface charge density, σ, and the magnitude of the external applied field, E, are related as follows:

$$E = \frac{\sigma}{k\varepsilon_0} \tag{28.6}$$

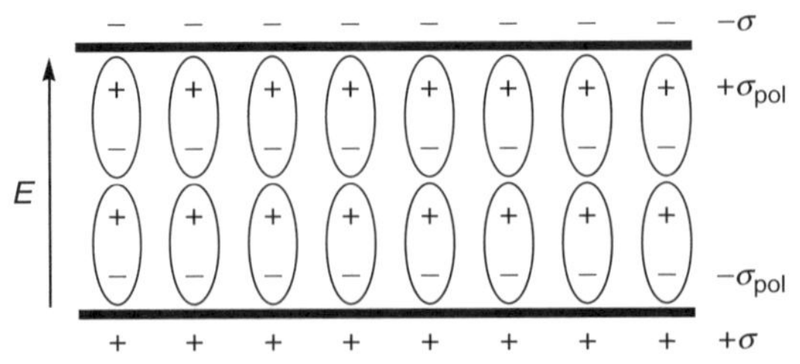

FIGURE 28.3 Schematic cross section of two capacitor plates and a dielectric polarized by an external electric field, E. Each ellipse denotes a dipole moment.

The charge on the plates arises from the polarizing medium, which induces a net charge density σ_{pol}. The electric field between the plates can be written as

$$E = \frac{\sigma - \sigma_{pol}}{\varepsilon_0} \tag{28.7}$$

As the electric field in both equations is the same, one obtains

$$\sigma_{pol} = \left(\frac{k-1}{k}\right) \cdot \sigma = (k-1)\varepsilon_0 E = \chi \varepsilon_0 E \tag{28.8}$$

where $\chi = \kappa - 1$ is defined as the electric susceptibility of the dielectric material.

The bulk polarization P is defined as the dipole moment per unit volume. The dipole moment is the net surface charge, $\sigma_{pol}A$, multiplied by the separation of the charge, d. The volume of the dielectric is the product of area, A, and the separation, d. Therefore,

$$P = \frac{\sigma_{pol} Ad}{Ad} = \sigma_{pol} = \chi \varepsilon_0 E \tag{28.9}$$

28.2.1.2 Microscopic Perspective

What causes an insulator to respond to an external electric field? At the molecular level, all materials consist of atoms with a positively charged nucleus surrounded by an electron cloud. In the absence of an electric field, for an individual atom, the statistical centers of positive and negative charges coincide. When an external electric field is applied, the charge centers are shifted and the atom acquires an induced dipole moment. Discussion of this phenomenon may be further extended to polyatomic molecules. In general, the polarizability, α, of a molecule is a measure of its electron distribution rearrangement in response to an external electric field, giving the molecule an electric dipole moment, p. In the linear approximation, under weak external fields, the α and p are related by the relationship $p = \alpha E^*$, where E^* is the electric field experienced by the molecule. Note that these induced molecular dipoles modify the value of macroscopic field inside the dielectric material, adding the electric field of each dipole to the external field. The macroscopic polarization density of the medium, P, is equal to the mean dipole moment (or the polarizability) of a single molecule, p, multiplied by the number density of molecules, N, in the medium

$$P = N\,p = \alpha\ N\ E^* \tag{28.10}$$

where α is the molecular polarization constant.

The molecular polarization constant, α, has three origins: electronic polarization (α_e), distortion polarization (α_d), and orientation polarization [2,3,7]. Electronic polarization is the aforementioned displacement of the electron distribution with respect to the nucleus. When atoms combine to form molecules, the charge distribution in the atoms is usually asymmetric, with the valence electrons localized in the region of the chemical bonds. An applied electric field thus acts on the atoms in the molecules, distorting slightly the normal charge distribution. The resulted polarization is known as the distortion polarization. Electronic and distortion polarizations are common to all molecular dielectrics. Orientation polarization arises when the constituent molecules (such as H_2O) are polar and carry permanent molecular dipoles, μ, which are randomly distributed in the absence of an external electric field. The tendency of the molecular dipoles to align along the polling field E (E^* is the field that a molecule experiences) is opposed by random thermal fluctuations. The resulting orientation polarization in the presence of an applied field can be expressed explicitly by the third term of Equation 28.11, $\mu^2/3k_BT$, which is a statistical quantity representing the thermal average of the orientational polarization. Therefore, the total polarization of a medium is proportional to the sum of these components:

$$P = N\left(\alpha_e + \alpha_d + \frac{\mu^2}{3k_BT}\right)E \tag{28.11}$$

where
μ is the orientation polarizability or permanent dipole moment
k_B is the Boltzmann constant
T is the temperature in kelvin

We have derived two different expressions for the macroscopic polarization density, P. Equation 28.9 expresses P in terms of the external field, E, and the bulk constant χ while Equation 28.11 uses the various molecular polarization constants and the local electric field E^*. To tie the bulk and the molecular constants together, we need a relationship between E and E^*. This can be derived from electrostatic considerations and is expressed by

$$E^* = E + \frac{P}{3\varepsilon_0} \tag{28.12}$$

From Equation 28.12, we assume that the molecules are spherical and that the dielectric around an individual molecule can be approximated as a continuous medium rather than made of discrete molecules.

Using Equations 28.9, 28.11, and 28.12, it is possible to derive the Debye equation, which relates the bulk relative permittivity k to the molecular polarization constants and the permanent dipole moment:

$$\frac{k-1}{k+2} = \frac{N}{3\varepsilon_0}\left(\alpha_e + \alpha_d + \frac{\mu^2}{3k_BT}\right) \tag{28.13}$$

According to this equation, it is clear that reduction of molecular number density, N, and molecular polarizability are effective approaches toward decreasing the bulk dielectric constant.

28.2.1.3 Frequency Dependence of Dielectric Constant

In the above discussion, the relative dielectric constant is assumed to be independent of frequency. In reality, all media show some dispersion and their dielectric constant is a function of frequency. Typical electronic circuits operate at about 10^9 Hz, which effect on dielectric constants needs to be considered. When an external electric field moves electrons or nuclei off their equilibrium positions, the bodies are subject to a restoring force and a damping force. A simple model [4,5,8] used to explain the frequency dependence of the relative dielectric constant, k, is the damped oscillation of an electron driven by a harmonic force provided by a sinusoidal alternating current (AC) field $E = E_0e^{-i\omega t}$. From Newton's second law of motion, force = mass · acceleration,

$$eE_0e^{-i\omega t} - Ax - B\frac{dx}{dt} = m\frac{d^2x}{dt^2} \tag{28.14}$$

where
e is the charge of an electron
ω is the angular frequency

x is the displacement of the electron from its equilibrium position
A and B are constants
m is the mass of the electron

On the left-hand side of the equation, the terms represent the force of the electric field, the restoring force, and the damping force, respectively. The term on the right-hand side is the product of the mass of the electron and its acceleration. The solution of the above equation is as follows:

$$x = \frac{eE_0 e^{-i\omega t}}{(A - m\omega^2 - i\omega B)} = \frac{eE}{(A - m\omega^2 - i\omega B)} \tag{28.15}$$

By its very definition, polarization equals the product of the charge and the displacement, and the macroscopic polarization density P is

$$P = Np = Nex = \frac{Ne^2 E}{(A - m\omega^2 - i\omega B)} \tag{28.16}$$

We also can see from Equation 28.9 that

$$P = \chi\varepsilon_0 E = (k-1)\varepsilon_0 E \tag{28.17}$$

Combining Equations 28.16 and 28.17, we obtain the relative permittivity, $\varepsilon(\omega)$, in the following expression:

$$k = 1 + \frac{Ne^2}{\varepsilon_0(A - m\omega^2 - i\omega B)} = 1 + \frac{Ne^2/\varepsilon_0 m}{\left(\omega_0^2 - \omega^2 - i\dfrac{\omega B}{m}\right)} \tag{28.18}$$

$$\omega_0 = \sqrt{\frac{A}{m}} \tag{28.19}$$

where ω_0 is the resonant frequency of the electron.

If we are dealing with a polyatomic species, there are multiple electrons and ions to contribute to the polarization, and Equation 28.18 can be replaced by a more general expression, in which the dielectric constant is represented as the summation of these contributions:

$$k = 1 + \sum_j \frac{e^2/\varepsilon_0 m_j}{\left(\omega_{0j}^2 - \omega^2 - i\dfrac{\omega B_j}{m_j}\right)} \tag{28.20}$$

The permittivity given by the above equation has a real part and an imaginary part. The real part represents the usual dielectric constant, and the imaginary part gives rise to dielectric loss or absorption. When the frequency of the electrical field increases from zero (DC) to close to a resonant frequency, the electron moves more dramatically. As the resonant frequency is approached, the displacement and dissipation force change drastically, as both the real and imaginary parts of k undergo large excursions. It is therefore important for frequencies of the electric signals to be kept as far removed as possible from any of the resonant frequencies.

Equation 28.19 shows that the resonant frequency increases as the mass decreases. Atomic bonds have much larger mass than electrons. Therefore, the resonant frequencies of distortion polarization are lower than resonant frequencies of electronic polarization. Orientation polarization comes from the rotation of the entire molecule, depending markedly on the nature and state of the material. It is attenuated by the thermal randomization, as reorientation of the entire molecule requires displacement of far greater mass, its contributions to the dielectric constant occur at even lower frequencies. Different frequency-dependent responses of the three polarization phenomena result in different impacts on both the real part and the imaginary part of the dielectric constant [7]. The electronic polarization is the defining contribution to the dielectric constant in the range beyond the frequency of visible light beginning at around 10^{15} Hz. The distortion polarization dominates contribution to the dielectric constant up to the resonance frequencies about 10^{13} Hz. The maximum frequency for orientation polarization contribution is about 10^{9} Hz. For interconnection dielectric materials, through tailoring the design of a material's chemical architecture, the resonant frequencies can be moved sufficiently higher than the signal frequencies of the integrated circuit.

28.2.2 How to Reduce Dielectric Constant *k*

As stated previously, Equation 28.13 clearly indicates that there are two possible approaches to reduce *k*: decreasing the dipole strength or reduction of dipole number density, *N*. These approaches entail using materials with chemical bonds of lower polarizability or materials of lower density, respectively. Ideally, these two methods can be combined to achieve even lower *k* values.

To minimize the dipole strength, we need to consider all three components of polarizability. As mentioned in the previous section, since distortion and orientation polarizabilities require movement of greater mass (orientation polarizability is more like a molecular response), they are of greater concern at lower frequencies, while the electronic response is dominant at higher frequencies. As typical electronic device operating frequencies are currently less than 10^{9} Hz, all three components contribute to the dielectric constant and should be minimized for optimum performance [9].

In principle, any material with a dielectric constant, *k*, lower than that of SiO_2 is termed a low-*k* dielectric. However, replacing SiO_2 has not been a trivial task. Currently, the semiconductor industry has already moved to certain low-*k* materials (*k* around 3.2–3.9), where some silica Si–O bonds have been replaced with less polar Si–F or Si–C bonds [10]. A more drastic reduction can be achieved by using virtually all nonpolar bonds, such as C–C or C–H, as in the example of organic polymers.

The density of a material can be reduced by increasing the free volume (air, *k* of 1) through rearranging the material structure (constitutive porosity) or introducing porosity (subtractive porosity) [7]. Constitutive porosity refers to the self-organization of a material, which is porous with no additional treatment after manufacturing. However, the constitutive porosity is relatively low (usually less than 15%), and pore sizes are ~1 nm in diameter. Subtractive porosity involves selective removal of the part of the material known as porogen. The porogens can be linked to the polymer backbone or doped into the polymer network. Removal of the porogens by thermal degradation or by selective etching leaves behind the desired porous structure. Subtractive porosity can be as high as 90% and pore sizes may vary from 2 nm to tens of nanometers in diameter. Dielectric constants of less than 2 can be achieved by combining all three approaches: low polarizability, constitutive porosity, and subtractive porosity. Of course, the ultimate low-*k* material would entail the use of air as a dielectric (a technology called air gaps [11]) as air has the lowest possible dielectric constant, *k* of 1.

28.3 ORGANIC LOW-DIELECTRIC CONSTANT MATERIALS

28.3.1 Classification of Low-*k* Materials

Low-*k* materials may be grouped into two principal categories based on material composition: silicon/oxygen (Si–O) containing and non Si–O containing materials [12].

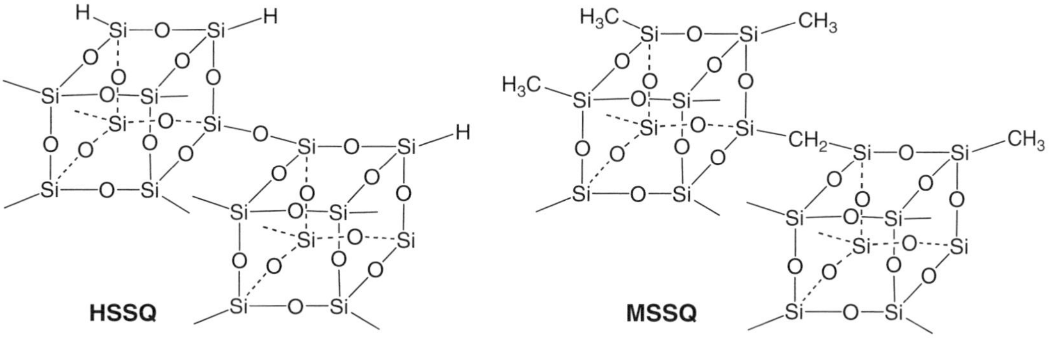

FIGURE 28.4 Structures of hydrogen-SSQ (HSSQ) and methyl-SSQ (MSSQ).

Si–O containing materials can be further divided into two subgroups: silica (SiO_2) based and silsesquioxane (SSQ) based. The fundamental difference between these two subgroups is in the structure of their elementary units. Silica has a tetrahedral elementary unit. To reduce the k value of silica, some of the Si–O bonds can be replaced with less polarizable Si–F bonds. In fact, the first generation of low-k materials was fluorinated silicate glasses (FSG) as SiO_2 deposited by chemical vapor deposition (CVD) process was well understood as the ILDs for 2.0 to 0.25 μm complementary metal oxide semiconductor (CMOS) technology nodes. The dielectric constant k of FSGs ranges from 4.0 to 3.2 limited by chemical instability when fluorine concentration exceeds 5 atom%.

The next generation of low-k materials is organosilica glasses (OSG), which are based on SSQ precursors. The incorporation of terminal Si–H or Si–R into the silica network not only introduces less polar bonds but also creates additional free volume as shown in Figure 28.4. In the SSQ elementary unit, Si and O atoms are arranged in a form of cube, creating free volume in the center of the cube, consequently decreasing the material density and thus its k value [13]. The cubes can be connected to each other through oxygen atoms, while some cube corners are terminated by hydrogen. Such materials are called hydrogen-SSQ (HSSQ). If methyl groups (CH_3) are present, the cubes can be connected by $-CH_2-$ and some cube corners are terminated by CH_3. This class of material is termed methyl-SSQ (MSSQ). SSQ cubes are metastable and tend to break down to silica tetrahedra, especially at elevated temperatures. As a result, in practice, SSQ-based materials consist of a mixture of SSQ cubes and silica tetrahedra. Both silica- and SSQ-based materials usually have k values of ~3.9–3.0, which can be further reduced by increasing material porosity [14].

Non Si–O-based materials are typically organic polymers. Their primary advantage is low polarizability, resulting in k values as low as 2 without porosity. The main disadvantage is their poor compatibility with existing semiconductor processing requirements (e.g., they exhibit low thermal and mechanical stability). Alternative low-k materials are available, such as amorphous carbon and zeolites, but they have received less attention.

From the material processing perspective, there are two primary methods of depositing low-k dielectrics: spin coating and CVD. CVD-applied SiO_2 has been a valuable dielectric material for many years. Common CVD low-k materials are FSG, Aurora (OSG, $k = 2.9$, ASM), Black Diamond (carbon-doped silicon oxide, $k = 2.0$–3.0, Applied Materials), and Coral (SiOC, $k = 2.85$, Novellus). Some organic polymers, such as parylene, can also be deposited by CVD of parylene precursors. Although a significant amount of CVD infrastructure exists, the issue of technology extendibility is clouded. Typically, CVD films are constitutively porous. The introduction of a porogen is possible, but this process would be challenging as deposition usually occurs at elevated temperatures. Currently, CVD materials and processes are not extendable to k values below 2 [12,14].

Spin-on solutions, although not widely adopted in IC manufacturing, are compatible with the current technology nodes. Additionally, it is easier to develop spin-on processes as well as integration schemes to extend to 0.1 μm technology node and beyond. Typical spin-on solutions are OSG and organic polymers. Spin-coated films can be constitutively as well as subtractively porous. The low temperature process allows the introduction of thermally degradable porogens into the mixture, which can be removed by thermal annealing. Annealing also induces chemical cross-linking, producing a rigid film structure. Because of the extendibility of this technology, spin-on may eventually replace CVD as the process of choice for future interconnects.

28.3.2 Organic Low-Dielectric Constant Polymers

Finding polymers with low-dielectric constants is a relatively easy task; however, simultaneously finding those with required chemical, mechanical, electrical, and thermal properties adequate for IC applications present a greater challenge. Most common high-temperature polymers have been tested for use as ILD/IMD materials in advanced microchips. In the following section, we discuss some of the industry-recognized organic dielectric material candidates.

To better understand the material design issues, some representative polarizabilities and the associated bond enthalpies are listed in Table 28.1 [15,16]. Clearly, single C–C and C–F bonds have the lowest electronic polarizability, thus making fluorinated and non-fluorinated aliphatic hydrocarbon good candidates for low-*k* material design. Bonds with π electrons have high polarizability due to large electron mobility (orientation polarization). However, single bonds are the weakest while double and triple bonds have much higher bond enthalpies. It is important to strike a balance between low-dielectric constant and high bond strength to achieve optimum material properties. Furthermore, highly polar constituents are usually avoided as candidates for low-*k* material design because they are more likely to attract moisture, resulting in substantially higher *k* as water has a dielectric constant of 80.

28.3.2.1 Polyimide

Polyimides are formed by the imidization of polyamic acids. Various properties can be obtained depending upon the choice of precursor compounds. From the early 1970s to the late 1980s, polyimides were studied as IMD candidates as polyimides have many advantages when compared to other organic polymers: high glass transition temperature (T_g) and thus good thermal stability, excellent mechanical properties, and low-dielectric constants, which can be further reduced through fluorination. General structures of aromatic polyimides and a fluorinated polyimide are illustrated in Figure 28.5.

However, there are two primary concerns with polyimides as low-*k* materials. First is the unavoidable moisture absorption due to the presence of the strongly polar imide groups [17]. Standard polyimides can absorb several wt% of moisture, raising the dielectric constant. In addition, moisture uptake can jeopardize adhesion. Fluorinated polyimides have slightly less moisture uptake and further dielectric constant reduction from the low polarizability of the C–F bond and the increase in free volume. Second, the highly aromatic polyimide structure encourages polymer chains to align preferentially parallel to the substrate, resulting in anisotropic dielectric constant and thermal mechanical properties [18]. These anisotropies limit the applications of polyimides as low-*k* materials.

TABLE 28.1
Electronic Polarizability and Bond Enthalpies of Common Organic Bonds

Bond	C–C	C–F	C–O	C–H	O–H	C=O	C=C	C≡C	C≡N
Polarizability (Å^3)	0.53	0.56	0.58	0.65	0.71	1.02	1.64	2.04	2.24
Average bond energy (kcal/mol)	83	116	84	99	102	176	146	200	213

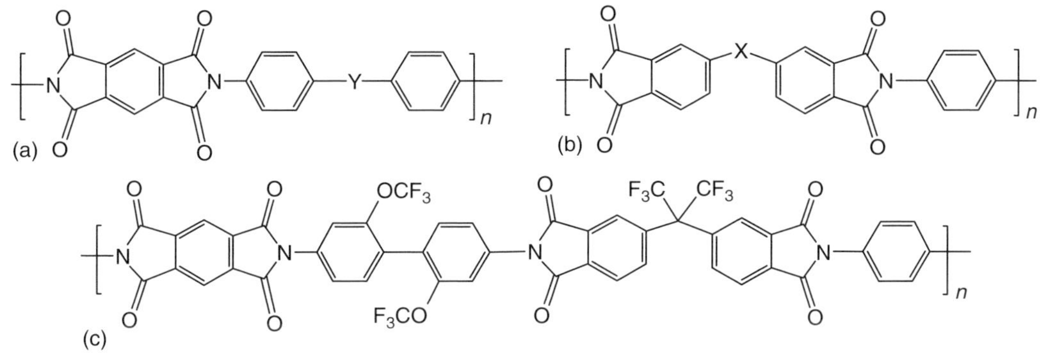

FIGURE 28.5 (a) and (b) Two general structures of aromatic polyimides and (c) fluorinated polyimide PMDA-TFMOB-FDA-PDA.

28.3.2.2 Poly(Arylene Ethers)

Poly(arylene ethers) (PAEs) are prepared by nucleophilic reaction of halogen-activated electron deficient aromatic rings and bisphenols as shown in Figure 28.6. The polymer's properties can be changed by modifying the chemical structure of the halogenated aromatic precursor or bisphenol. The aromatic rings in the PAE provide good mechanical and thermal properties with T_g of over 400°C [19]. The flexible ether linkages allow bending of the chains, which yield more isotropic materials than polyimides. Carbonyl and sulfonyl groups are very effective as electron-withdrawing groups to activate the aromatic rings and to drive the reaction to completion. However, because these groups are highly polar, they are typically avoided in PAE to achieve desirable low-dielectric constants. FLARE from Honeywell is one of the best-known PAE. Recently, there are reports of integrating FLARE into dual-damascene Cu structures [20]. Typical non-fluorinated PAEs have k value of 2.8–2.9, low moisture uptake, and good solvent resistance. Fluorinated versions of PAEs have k value of ~2.4.

28.3.2.3 Benzocyclobutene

Benzocyclobutene (BCB) resins were developed by Dow in 1980s. The reactive species in BCB is ortho-xylylidene group, which undergoes Diels–Alder cycloaddition reactions and free-radical addition/combination reactions. Typical thermoset BCB reactions are shown in Figure 28.7. The properties of the BCB resins can be adjusted by the choice of chemical precursors [21]. BCB does not evolve water during cross-linking, and the cross-linked BCB polymer has low moisture uptake and a low isotropic k value (typically 2.6–2.7) due to certain amount of relatively nonpolar aliphatic rings in the polymers. The main disadvantage of BCB is the thermal instability of the cured film. When exposed to air at even modest temperatures (150°C–300°C), the rate of weight loss well exceeds the 1% target.

To avoid some of these problems, fluorinated analogs of BCB have been explored. Poly(perfluor ocyclobutane) (PFCB) is an aromatic ether thermoset that has pendant perfluorovinyl ether groups as shown in Figure 28.7. This material has improved thermal stability and a lower dielectric constant ($k = 2.24$).

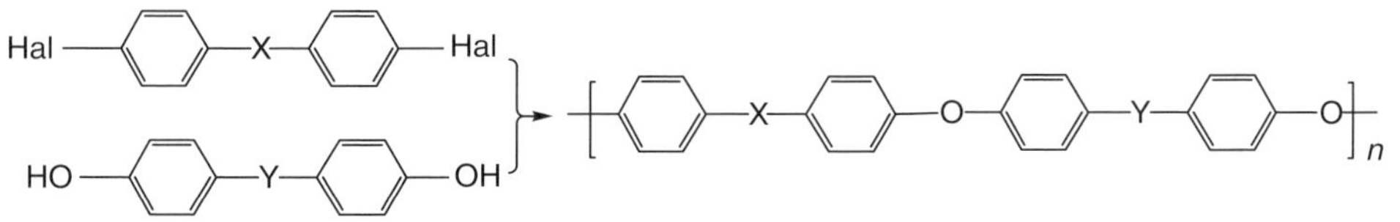

FIGURE 28.6 General synthesis of poly(arylene ethers).

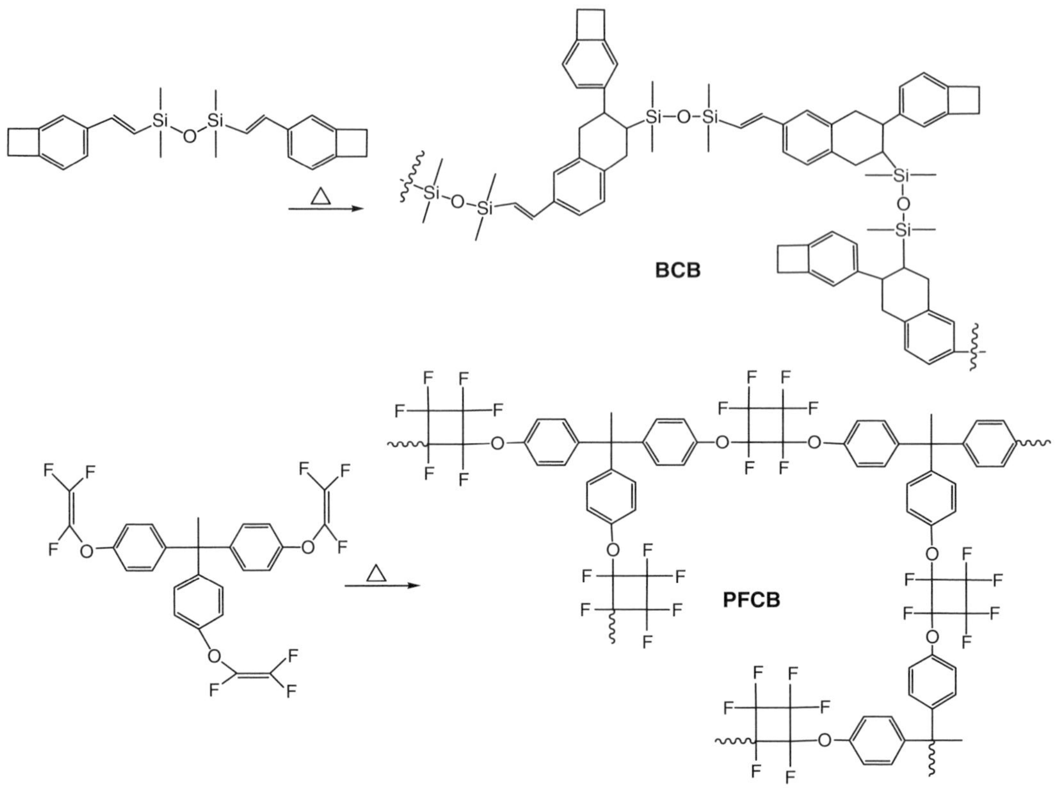

FIGURE 28.7 Divinylsiloxane-based BCB polymer and PFCB.

28.3.2.4 Polyphenylene

SiLK [22] is an organic polyphenylene polymer system from Dow that has generated significant interest in the semiconductor industry. The starting precursors contain aromatic acetylene constituents, typically in the form of ortho-bisethynyl or phenylethynyl groups, and a cyclopentadienone cross-linking agent as shown in Figure 28.8. Through intermediate Diels–Alder cycloadditions, acetylene groups in the precursors are readily consumed. Since SiLK is highly aromatic, its temperature stability is up to 450°C and absorbs less than 0.25 wt% water. It has an isotropic k of 2.65 due to the lack of polar moieties and good moisture resistance. In addition, since the SiLK system is a

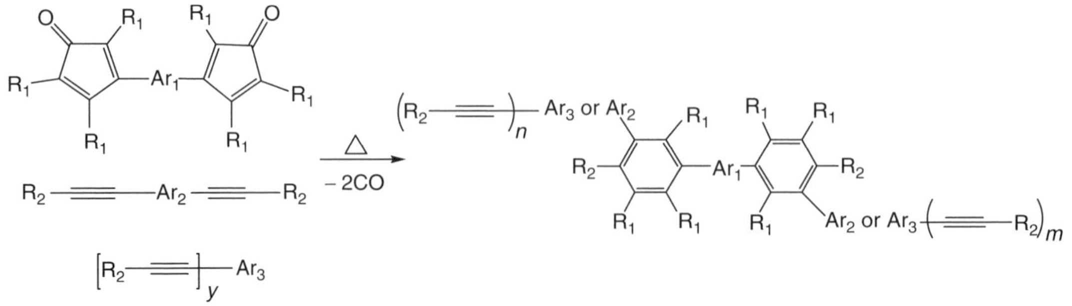

FIGURE 28.8 Synthesis and structure of SiLK.

low viscosity oligomeric precursor solution, it has good gap fill properties down below 0.1 μm. Fujitsu and many other companies have reported SiLK integration results [23].

28.3.2.5 Polytetrafluoroethylene

Perfluorinated aliphatic polymers incorporate C–F bonds of the lowest polarizability, leading to the lowest dielectric constants (1.9–2.1) reported for any nonporous materials. The base polymer, polytetrafluoroethylene (PTFE), is a highly crystalline material that lacks solution processibility by traditional methods. In addition, the flexible and uncross-linked chain structure limits its thermomechanical stability. However, W.L. Gore and Associates has developed a PTFE nano-emulsion material, SPEEDFILM, which is an aqueous emulsion containing sub-20 nm PTFE particles and surfactant. The resultant material is nonporous with gap fill of 0.35 μm. The film demonstrates good thermal stability up to 400°C and has a 460°C decomposition temperature [24]. SPEEDFILM is nonetheless susceptible to the potential release of fluorine atoms and consequent interconnect metal corrosion, which is displayed by all fluorine-containing materials.

28.3.2.6 Parylene

Parylenes or poly(*p*-xylylenes) were developed in the 1950s and they are typically made from simultaneous deposition/polymerization of dimeric cyclophane precursors by CVD process in the semiconductor industry. Two of the most characterized representative examples are Parylene N (k of 2.6) and Parylene F shown in Figure 28.9. In Parylene F, the aliphatic portion of the chemical backbone is fluorinated giving decreased dielectric constants (2.2–2.4) and increased thermal stability relative to Parylene N [25]. This fluorination results in improved thermal stability to over 400°C. However, both materials exhibit crystalline phase changes well below the maximum BEOL-processing temperature and may have integration issues. In addition, parylenes have large in-plane and out-of-plane dielectric anisotropy.

28.3.3 Porous Organic Ultralow-κ Polymers

For 65 nm node generation, dielectric constants must be less than 2.4, and this requirement is expected to be further stretched to less than 1.9 (ultralow-k) for more advanced technology nodes. After many years of intensive exploration, few dense materials have demonstrated k of less than 2.5. Recently, there has been an enhanced interest in porous materials as a way to achieve ultralow-k.

Porosity can be constructed directly in the polymer network or introduced indirectly by incorporating a porogen in the composite film, which is later removed by thermal decomposition, dissolution, plasma, or photolytic methods. Porogens can be low-boiling point molecules, surfactant micelles, block-copolymer arrays, dendrimers, or colloidal particles, etc. Ideally, the pores should be as small as possible (less than 10 nm, preferably around 1 nm), uniform in size, closed, not interconnected, and periodic in nature.

One noteworthy approach of porosity construction is the direct incorporation of cage structures, such as adamantane- and diamantane-containing precursors, into the highly aromatic organic polymer backbones [26]. Another intriguing direct attachment method takes advantage of the phase separation of block/graft copolymers with thermally labile porogen segments as minor components, such as poly(propylene oxide) and poly(methyl methacrylate). At certain compositions and conditions, these block copolymers form ordered spherical domains of the minor component. Pores

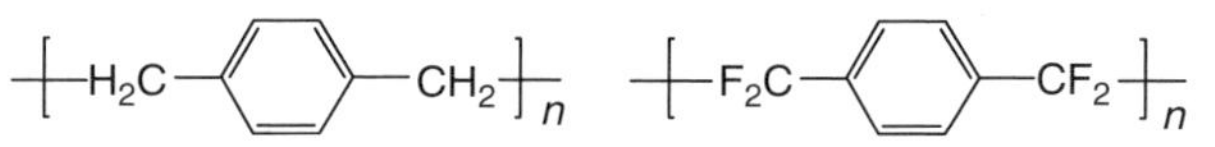

FIGURE 28.9 Structures of Parylene N (left) and Parylene F (right).

with ideal shape and volume can be obtained after carefully heating the films to temperatures between the degradation temperature of the thermally labile segments and that of the dense low-*k* dielectrics [27]. Porosity up to 30% can potentially be achieved. However, further increase of porosity by this method is very difficult. The major problems are the collapse of the pores due to high surface tension and the required high temperature for removing the porogens is too close to the T_g of the dense low-*k* materials.

It is more convenient and versatile to achieve ultralow-*k* from porogen-doped composites. The challenge of this approach is obtaining highly ordered porous structures without aggregation of porogens. Interconnected and open pore structures would cause integration and reliability issues. Ultralow-*k* (1.9–2.2) has been achieved with porous SiLK and FLARE. Integration of these materials has been proven.

28.3.4 Characterization of Low-*k* Materials

Table 28.2 summarizes common characterization techniques applicable to low-*k* thin films. In this section, we limit the discussion to area capacitance, nanoindentation, and positron annihilation lifetime spectroscopy (PALS) [7].

The most basic electrical properties such as dielectric constant, breakdown properties, and leakage current can be obtained from area capacitance of a metal/dielectric/metal (metal-insulator-metal, MIM) or a metal/dielectric/Si (metal-insulator-semiconductor, MIS) sandwich. In *C–V* measurements, the dielectric constant *k* is calculated from Equation 28.4 $k = C \cdot d/A \cdot \varepsilon_0$, where *C* is capacitance at the maximum voltage in the accumulation region, *d* is the film thickness, *A* is area, and ε_0 is a known constant. *I–V* measurements can determine the leakage current, *J*, at a certain electric field strength (A/cm^2) and the breakdown field, F_{BD}, at a certain leakage current (MV/cm). Mercury probe measurement is a convenient method to form an MIS capacitor between Hg/dielectric/Si without metal deposition. However, its accuracy depends on the contact angle and contact area with the dielectrics.

In the nanoindentation test, a diamond tip is pressed into the material of interest. The hardness is derived from the ratio of the required force, *p*, and the projected contact area, *A*, which is a known function of contact depth, *h*. The elastic modulus, E_r, is given by

$$E_r = \frac{\sqrt{\pi}}{2\beta} \frac{dP/dh}{\sqrt{A}} \tag{28.21}$$

TABLE 28.2
Common Low-*k* Thin Film Characterization Techniques

Material Properties	Test Method
Composition	Fourier transform from infrared (FTIR) spectroscopy
	X-ray photoelectron spectroscopy (XPS)
	Secondary-ion mass spectrometry (SIMS)
Thermal properties	Thermal desorption mass spectroscopy (TDMS)
Electrical properties	Area capacitance
Mechanical Properties	
Hardness and elastic modulus	Nanoindentation
Adhesion	Four-point bending (quantitative)
Porosity	Positron annihilation lifetime spectroscopy (PALS, PAS)
	Small-angle neutron scattering (SANS)
	Specular x-ray reflectivity (SXR)
	Rutherford backscattering spectroscopy (RBS)
Thickness and optical properties	Ellipsometry

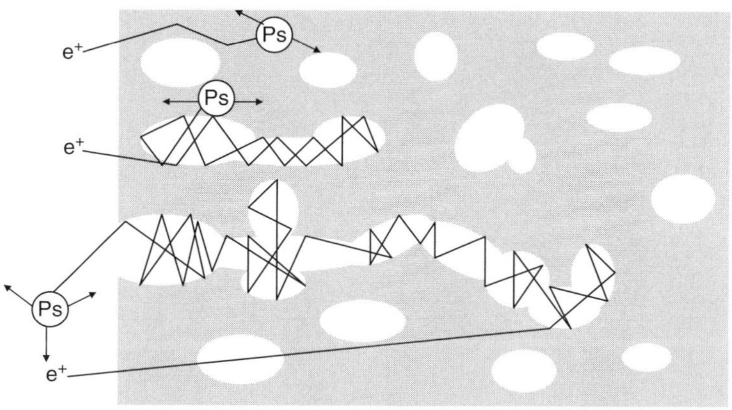

FIGURE 28.10 Formation and diffusion of positronium (Ps) in open and closed porous materials.

where β is a constant close to unity that depends on the indenter tip geometry. When using this technique, especially for soft organic polymers, attention must be paid to the shape of the tip, surface contact, and relaxation phenomena to obtain reliable data.

Both PALS and PAS use low energy radioactive β-decay positrons to study the voids in the polymers. Positrons have limited lifetime and can, by interaction with an electron, be annihilated or form an electron–positron bound state called positronium (Ps). PALS measures the lifetime of Ps and provides information on the pore size and structure, whereas PAS extracts information from the Ps annihilation reaction to obtain information on the total porosity. Figure 28.10 shows an illustration of formation and diffusion of Ps in a porous dielectric. The lifetime of Ps in vacuum is ~142 ns. In a porous material, the Ps annihilation lifetime is shortened due to collision with the pore walls. If all pores are open and connected, Ps can escape to vacuum, existing in its nature lifetime of ~142 ns. If all pores are not open but connected, the Ps will exhibit a single average lifetime. If the pores are isolated, the Ps will exhibit lifetime with a wide distribution, corresponding to the distribution of the pore sizes.

28.4 DEVICE AND INTEGRATION

Since the invention of integrated circuits, there has been continuous effort toward developing fine-patterned and high performance ultra large-scale integration (ULSI) devices. As transistors become smaller and smaller, multiple interconnect lines are required to transform the arrays of electronic devices into meaningful circuitry. For example, in 256 MB dynamic random access memory (DRAM), up to four levels of interconnect lines are typically employed. In the most advanced logic chips, such as Intel's 65 nm process technology shown in Figure 28.11, there are eight copper interconnect layers. Incorporation of low-*k* materials is vital to control RC delay in current interconnects, increasing the signal speed inside the chip and reducing chip power consumption for both logic and memory IC devices.

As mentioned before, the greatest challenge in incorporating low-*k* materials in devices is identifying materials with both low *k* and the requisite material properties to withstand the integration processes and device operation. To better understand material integration requirements, we shall begin with a brief introduction to BEOL processes.

28.4.1 Overview of BEOL Processes

Conventional interconnect fabrication employs Al wires. In this process scheme (metal first), Al is first deposited and patterned either by dry etch or by depositing photoresist (PR), and excess metal is lifted off together with the patterned PR. The space is subsequently filled with dielectric materials, which is then planarized by chemical mechanical polishing (CMP). This scheme faces increasing

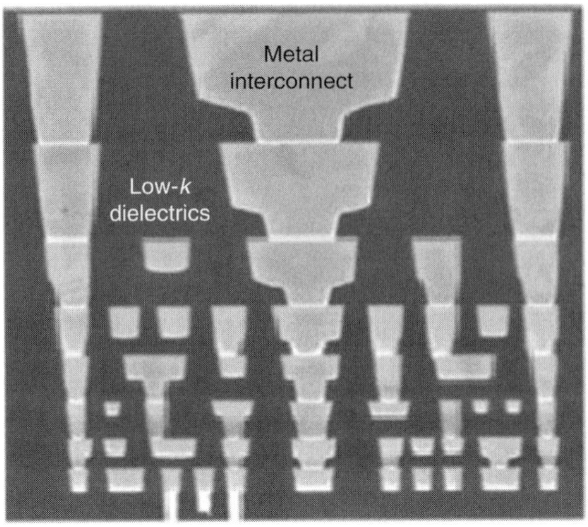

FIGURE 28.11 Interconnect in Intel's 65 nm logic device.

challenges as much smaller gaps between metal lines in the advanced interconnects must be filled by dielectrics. This is difficult to achieve using solution-based polymers. The current process scheme uses copper as metal wires, which are difficult to pattern by the same methods as the Al wires. In this process (dielectric first damascene process), the dielectric is deposited on an etch-stop layer (oxide or nitride). Trenches are formed into the dielectric by photolithography. The etch-stop layer is meant to prevent over etching of the trenches. The trenches are then filled with Cu. Excess Cu is removed by CMP. Finally, another dielectric layer is deposited and upon which the higher interconnect array is built. Figure 28.12 illustrates one layer of both conventional and single-damascene interconnect fabrication schemes [28].

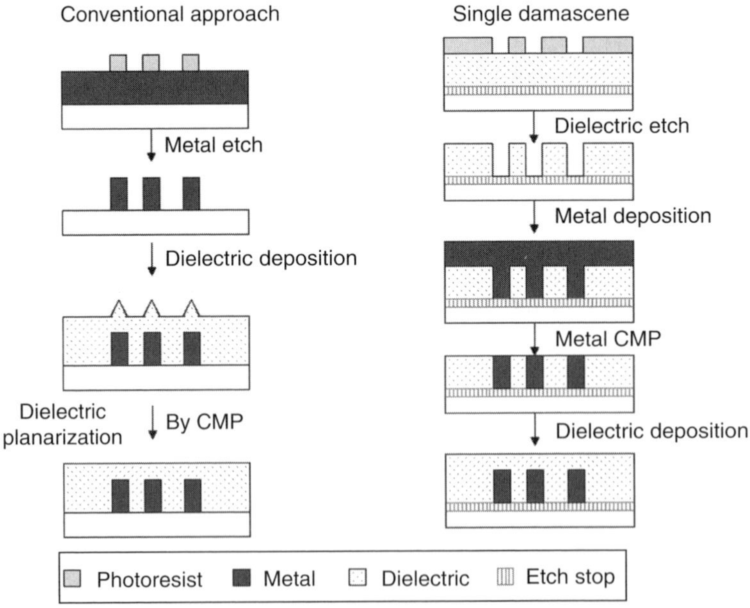

FIGURE 28.12 Conventional and single-damascene interconnect fabrication schemes.

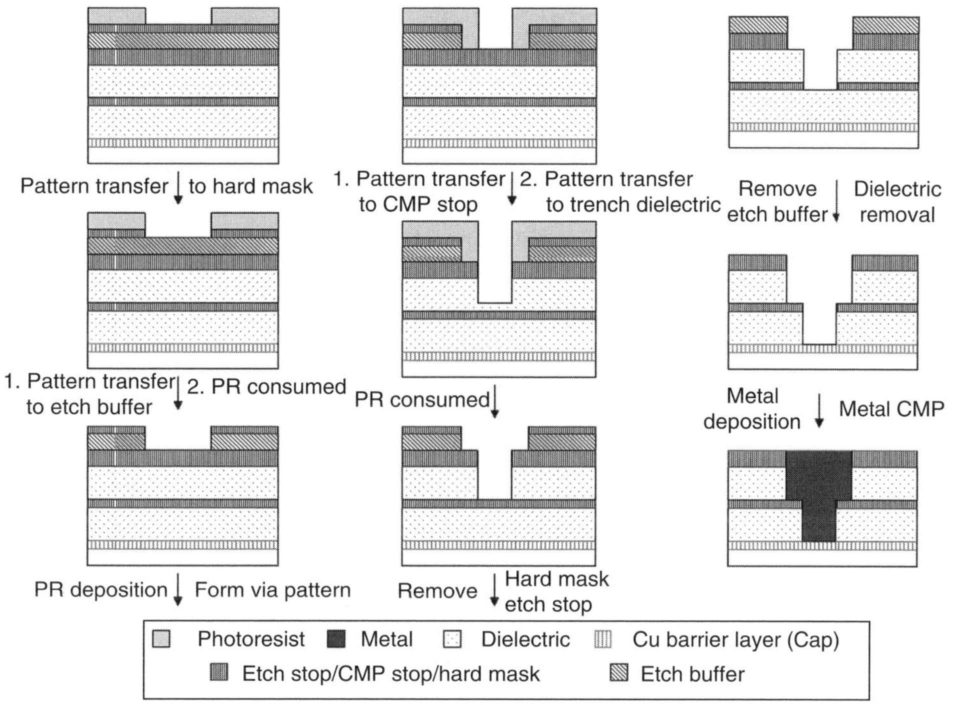

FIGURE 28.13 A representative dual-damascene fabrication scheme of single-layer interconnect.

The state-of-the-art Cu/low-*k* interconnect technology for 0.1 μm and sub-0.1 μm devices is Cu dual-damascene process [29] depicted in Figure 28.13. Some of the process steps, such as the deposition of a Cu diffusion barrier to prevent Cu from leaking into the dielectric materials and the integration of antireflective coating as part of the photoresist-patterning process, are omitted for the purpose of simplicity. Clearly, the dual-damascene process is far more complicated and poses more stringent requirements on material selection.

28.4.2 Low-*k* Integration Challenges

Any process change in the semiconductor industry is difficult due to the involvement of large capital investment and constant need for low cost products. Thus the adoption of low-*k* and ultralow-*k* materials better than SiO_2 faces a high degree of challenge throughout the whole IC manufacturing process. To be compatible with the present process, the materials have to satisfy a range of diverse requirements: low-dielectric constant, high thermal and mechanical stability, resistance to chemicals under processing conditions, compatibility with other materials, low moisture absorption, reliability in the user environment, and low cost [12]. In the previous sections, we have elaborated the status of low-*k* material development. In this section, we discuss other challenges related to low-*k* integration.

28.4.2.1 Thermal and Mechanical Stability

Besides the dielectric constant, the second most important requirement for ILD/IMD is the material thermal stability [9]. The typical fabrication of multilevel IC structures requires as many as 10–15 temperature excursions. Typical subtractive Al BEOL process temperature is higher than 400°C. In Cu-damascene process, copper metallization can be achieved by electroplating or electroless plating at temperatures below 250°C. Unfortunately, an annealing step at temperatures in the range of 350°C–425°C is necessary to obtain void-free copper deposits. Low-*k* organic materials must be able to withstand this temperature for several hours. Outgassing due to material reaction or decomposition can cause via (a pattern structure illustrated in Figure 28.13) poisoning, delamination, and blistering in ILD/IMD. Furthermore, changes in crystalline phases lead to properties dependent on the thermal history.

The need for mechanical stability is primarily a consequence of the introduction of Cu as the electrical wire in the damascene IC manufacturing process. In the final step of this process, the dielectric films must withstand the mechanical stress of the Cu CMP process. In addition, thermal processes (including final packaging) also create stresses on the interconnect structures arising from the coefficient of thermal expansion (CTE) mismatch between the ILD/IMD dielectrics, metal, and barrier materials, which in turn could lead to delamination if adhesion is poor. Good adhesion is particularly critical for organic spin-on materials that often do not adhere well to standard CVD oxide and nitride films, or barrier metals (usually Ti, TiN, Ta, and TaN). In ultralow-*k* materials, added porosity further deteriorates the mechanical properties, making integration extremely challenging.

28.4.2.2 Chemical Stability

Low-*k* dielectrics must meet the chemical stability requirements to survive many IC processing steps, such as etching/cleaning, resist removal, and Cu CMP. In addition, they should not absorb process chemicals as this can lead to permanent swelling of the materials.

Plasmas play an important role in the advanced IC processes [7]. They are used for via and trench etching, post etch cleaning, as well as photoresist patterning and stripping. Plasmas are also used for surface densification of low-*k* materials. Oxygen-based plasmas are widely used for etching of organic low-*k* dielectrics and photoresists since oxygen has high reactivity with organic compounds. Oxygen-based plasmas modify only the top surface of the organic polymers, leaving the bulk intact. By partially or completely replacing oxygen with fluorine in the plasma process, the etch rate can be reduced to a specific desirable level. Since fluorine can diffuse into the polymer, a fluorine-containing plasma modifies both the surface and the bulk of low-*k* materials. Similarly, chlorine may also be used in the plasma. The differences are that chlorine has a slower etch rate and incorporation of chlorine into polymers has the undesirable effect of increasing the material's *k* value. Hydrogen plasmas can also be used for etching organic polymers, producing volatile CH_4 and H_2 in the process. Since pure hydrogen ions have very low mass to provide sufficient ion bombardment, hydrogen plasmas are often mixed with nitrogen to generate an adequate etch rate.

A major concern when etching low-*k* dielectric materials vs. standard oxides is bulk or partial film modification, often termed "low-*k* damage." The same plasmas used to etch the dielectric material or strip off remaining photoresist can chemically modify the remaining dielectric in the structure, typically through scavenging of carbon from the dielectric matrix. This mechanism is even more prevalent for porous ultralow-*k* dielectrics. For both organic and inorganic low-*k* dielectrics, carbon incorporation is one of the key methods of dielectric constant reduction. If carbon is depleted from the dielectric matrix during plasma processing then the region where the carbon is depleted has a higher dielectric constant than the bulk, unmodified film. Typically, this region also becomes hydrophilic, allowing water absorption, thus further increasing the combined *k* value of the damaged and undamaged dielectric since water has a very high dielectric constant. Therefore, low-*k* dielectric etch processes must be optimized for both profile control and minimal damage [30]. The requirements of the plasma conditions are determined by the selectivity requirements of the multiple layers of materials involved in the device structure.

CMP also has substantial chemical stability requirements. To date, the hard-mask (capping layer) approach is utilized to protect the low-*k* dielectrics during CMP; thus the industry has adopted the standard two-step CMP process, which is to remove copper to stop on the barrier, then to remove the barrier to stop on the dielectric film [31]. This calls for a CMP process with better selectivity and is achieved by tuning the selectivity of the various films through slurry formation and optimization of the CMP process. A typical Cu CMP process uses aqueous slurry containing an abrasive, an oxidant, and other agents such as HNO_3 or NH_4OH [28]. With an abrasive action, the particulate metal is dissolved rapidly into the aqueous solution as copper salt or ammonia complex. The low-*k* polymers must be able to withstand the oxidizing action of the aqueous solutions.

28.4.2.3 Compatibility with Other Materials

The principal concerns in low-*k* dielectric compatibility are CTE mismatch, adhesion, and compatibility with the photoresist (PR), diffusion barrier, and capping layer. The first two have been addressed, and our discussion continues with the last three issues.

Adjusting process conditions to avoid or minimize resist poisoning is critical for the low-*k* compatibility with the PR. Argon fluoride (ArF) and krypton fluoride (KrF) resists are chemically amplified by a photoacid generator (PAG). Resist poisoning occurs when amine-based by-products are generated in the IC process and react with the PAG. Neutralization of the PAG prevents the resist from proper development. Sources of poisoning include all nitrogen-containing materials such as low-*k* materials, the hard mask or capping layer (SiN, SiON, SiCN), process chemicals (N_2/H_2 plasma), and environmental contamination.

Copper has a very high diffusivity, especially in porous ultralow-*k* dielectrics. A diffusive barrier is necessary to prevent it from migrating into the dielectric layers and degrading the dielectric properties of the insulator. As mentioned previously, a hard mask or capping layer is necessary on top of the porous low-*k* films to prevent dielectrics from CMP damage. However, deposition of a thin film on a porous low-*k* material is nontrivial. Figure 28.14 illustrates the difficulties in integrating the Cu barrier coating and hard mask or capping layer on porous dielectrics. The porosity of the dielectric film can be translated into capping and barrier layers. Thin-film deposition of capping and barrier layers is commonly performed by vapor deposition techniques, either through physical vapor deposition (PVD) or CVD. Both methods suffer from pinhole formation and precursor penetration.

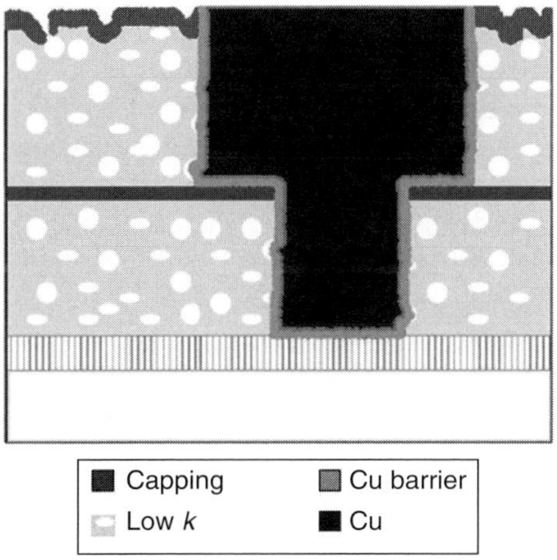

FIGURE 28.14 A schematic of the low-*k* capping layer and the Cu barrier layer.

To avoid these problems, porous dielectrics need to be sealed before further treatment. Typical pore sealing process consists of plasma surface densification or surface cross-linking (surface repair).

28.4.2.4 Low Moisture Absorption

A low-k material must be hydrophobic as water has extremely polar O–H bonds and a k value of 80. Even a small amount of absorbed water significantly increases the total k value. As water is abundant in air, a low-k material should be as hydrophobic as possible to prevent deterioration of its k value. This is especially important for porous materials, given their large surface area per unit volume. Absorption of water can cause via poisoning in dual-damascene structures. During etching and cleaning, dielectric sidewalls absorb moisture before via filling, leading to metal corrosion and high via resistance. Organic low-k polymers with few polar groups are generally hydrophobic.

28.5 SUMMARY

Low-dielectric constant materials are in demand as transistor performance continues to improve and more complex interconnect leads to substantial RC delay. From the perspective of material design, reduction of the dielectric constant k can be accomplished by employing chemical bonds with reduced polarizability and decreasing material density. The minimum k requirement for more advanced IC devices is 2. CVD-deposited SiO_2 with a dielectric constant of 3.9 has been the interconnect dielectric for many years. More recently, CVD low-k dielectrics, such as FSG, carbon-doped oxide (CDO), and OSG, have been adopted by the IC industry. Unfortunately, it is difficult using current CVD materials and processes to further extend k values below 2. Eventually, spin-on may replace CVD as the choice of process. Comparing to spin-on OSG, organic polymers are versatile in design and these dense materials have much lower starting dielectric constants, which can be further reduced to extend k to 2 by adding porous structures. Despite the difficulties of material development and integration of new materials into future microelectronic circuits, tremendous progress has been made in recent years in this multidisciplinary effort to produce improved low-dielectric constant materials.

EXERCISE QUESTIONS

1. Prove Equation 28.15 is a solution of Equation 28.14.
2. In the synthesis of SiLK in Figure 28.8, what is the chemical structure of the intermediate after the Diels–Alder reaction?
3. Simple mode to calculate the dielectric constant of porous materials is the Maxwell–Garnett model. $k_e = (k_M 1 + 2A/1 - A)$, where $A = f(k_P - k_M/k_P + 2k_M)$, in which k_e is the effective dielectric constant, subscript P refers to the pore, M refers to the bulk material, and f is the percentage of porosity. Assuming the pore is filled with air (k_P of 1), how much porosity is necessary to transform SiO_2 (k_M of 4) and a dielectric polymer (k_M of 2.6) into ultralow-k materials (k_e of 2)?
4. There is a great deal of literature available on high-k dielectrics [32]. How different are these materials from low-k materials? What are the advantages of using high-k dielectrics and what is the leading material candidate?

LIST OF ABBREVIATIONS

AC	alternating current
BCB	benzocyclobutene
BEOL	back-end-of-the-line
CMOS	complementary metal oxide semiconductor
CMP	chemical mechanical polishing
CTE	coefficient of thermal expansion

CVD	chemical vapor deposition
DRAM	dynamic random access memory
FEOL	front-end-of-the-line
FSG	fluorinated silicate glasses
HSSQ	hydrogen-SSQ
IC	integrated circuit
ILD	interlayer dielectric
IMD	intermetal dielectric
I–V	current–voltage
MSSQ	methyl-SSQ
PAE	poly(arylene ethers)
PAG	photoacid generator
PALS	positron annihilation lifetime spectroscopy
PR	photoresist
PTFE	polytetrafluoroethylene
PVD	physical vapor deposition
RC	resistance–capacitance
SSQ	silsesquioxane
ULSI	ultra large-scale integration

REFERENCES

1. Marcadal, C., Richard, E., Torres, J., Palleau, J., and Madar, R., CVD process for copper interconnection, *Microelectron. Eng.*, 37/38, 97, 1997; Treichel, H., Withers, B., Ruhl, G., Ansmann, P., Wurl, R., Muller, Ch., Dietlmeier, M., and Maier, G., Low dielectric constant materials for interlayer dielectrics. In: *Handbook of Low k and High k Materials and Their Applications*, Nalwa, H.S. (Ed.), Academic Press, Boston, MA, chap 1, p. 1, 1999.
2. Feynman, R.P., *The Feynman Lectures on Physics*, vol. II, Addison Wesley, Reading, MA, 11–1, 1989.
3. Solymar, L. and Walsh, D., *Lectures on the Electrical Properties of Materials*, 4th edn., Oxford Science, p. 262, 1988.
4. Jackson, J.D., *Classical Electrodynamics*, 2nd edn., John Wiley & Sons, New York, p. 284, 1975.
5. Bron, M. and Wolf, E., *Principles of Optics*, 7th ed., Cambridge University Press, London, p. 95, 1999.
6. Feynman, R.P., *The Feynman Lectures on Physics,* vol. II, Addison Wesley, Reading, MA, 10–1, 1989.
7. Maex, K., Baklanov, M.R., Shamiryan, D., Lacopi, F., Brongersma, S.H., and Yanovitskaya, Z.S., Low dielectric constant materials for microelectronics, *J. Appl. Phys.*, 93(11), 8793, 2003.
8. Franklin, J., *Classical Electromagnetism*, Pearson Education, San Francisco, CA, p. 298, 2005.
9. Morgen, M., Ryan, E.T., Zhao, J.H., Hu, C., Cho, T., and Ho, P.S., Low dielectric constant materials for ULSI Interconnects, *Annu. Rev. Mater. Sci.,* 30, 645, 2000.
10. Ho, P.S., Leu, J., and Lee, W.W., Overview on low dielectric constant materials for IC applications. In: *Low Dielectric Constant Materials for IC Applications,* Ho P.S., Leu J., Lee, W.W. (Eds.), Springer, Berlin, Germany, chap 1, p. 1, 2003.
11. Gosset, L.G., Farcy, A., Pontcharra, J., Lyan, P., Daamen, R., Verheijden, G., Arnal, V., Gaillard, F., Bouchu, D., Bancken, P., Vandeweyer, T., Michelon, J., Hoang, V., Hoofman, R., and Torres, J., Advanced Cu interconnects using air gaps, *Microelectron. Eng.*, 82, 321, 2005.
12. Shamiryan, D., Abell, T., Iacopi, F., and Maex, K., Low-k dielectric materials, *Materials Today*, January, 34, 2004.
13. Volksen, W., Porous organosilicates for on-chip applications: Dielectric generational extendibility by the introduction of porosity. In: *Handbook of Low k and High k Materials and Their Applications*, Nalwa H.S. (Ed.), Academic Press, Boston, MA, chap 1, p. 167, 1999.
14. Hatton, B.D., Landskron, K., Hunks, W.J., Bennett, M.R., Shukaris, D., Perovic, D.D., and Ozin, G.A., Material chemistry for low-k materials, *Materials Today*, March, 22, 2006.
15. Miller, K.J., Hollinger, H.B., Grebowicz, J., and Wunderlich, B., On the conformations of poly (*p*-xylylene) and its mesophase transitions, *Macromolecules*, 23, 3855, 1990.
16. Pine, S.H., *Organic Chemistry*, 5th edn., McGraw-Hill, New York, 1987.

17. Wetzel, J.T., Lii, Y.T., Filipiak, S.M., Nguyen, B.-Y., Travis, E.O., Fiordalice, R.W., Winkler, M.E., Lee, C.C., and Peschke, J., Integration of BPDA-PDA polyimide with two levels on Al(Cu) interconnects, *Mater. Res. Soc. Symp. Proc.*, 381, 217, 1995.
18. Sroog, C.E., Polyimides, *Prog. Polym. Sci.*, 16, 561, 1991; Mckerrow, A. and Ho, P., Proceedings of the low dielectric constant materials and interconnects, SEMATECH workshop, Piscataway, NJ, p. 199, 1996.
19. Lau, K., Chen, T-A., Korolev, B.A., Brouk, E., Schilling, P.E., and Thompson, H.W., Poly(arylene ether) compositions and methods of manufacture thereof, US Patent 6,124,421, 2000.
20. Leung, R., Porous and nanoporous poly(arylene ether) thin films: Suitability as "extra-low k" dielectrics for microelectronic applications, Proceedings from the 9th symposium on polymers for microelectronics, Wilmington, DE, May 2000.
21. Kirchhoff, R.A. and Bruza, K.J., Polymers from benzocyclobutenes, *Adv. Polym. Sci.*, 117, 1, 1994.
22. Godschalx, J.P., Romer, D.R., So, Y.H., Lysenko, Z., Mills, M.E., Buske, G., Townsend, P., Smith, D., Martin, S.J., and DeVries, R.A., Polyphenylene oligomers and polymers, US Patent 5,965, 679, 1999.
23. Ikeda, M., Integration of organic low-k material with Cu-damascene employing novel process, IEEE International Interconnect Technology Conference (IITC), San Francisco, CA, p. 131, 1998.
24. Rosenmayer, T., SPEEDFILM IC Dielectric Project, Proceedings from the 9th Symposium on Polymers for Microelectronics, Wilmington DE, May 2000.
25. Blackwell, J., Park, S.-Y., Chvalun, S.N., Mailyan, K.A., Pebalk, A.V., and Kardash, I.E., Three dimensionally oriented texture for poly(α,α,α prime, α prime-tetrafluoro-p-xylylene), *Polym. Preprint,* 39, 892, 1998.
26. Lau, K., Liu, F.Q., Korolev, B., Brouk, E., Zherebin, R., and Leung, R., Low dielectric constant organic dielectrics based on cage-like structures, US Patent 6, 797, 7777 B2, 2004; Lau, K., Liu, F.Q., Korolev, B., Brouk, E., Zherebin, R., and Nalewajek, D., Low dielectric constant materials with polymeric networks, US Patent 9, 423, 811 B1, 2002.
27. Hedrick, J.L., Labadie, J.W., Volksen, W., and Hilborn, J.G., Nanoscopically engineered polyimides, *Adv. Polym. Sci.*, 147, 61, 1999.
28. Maier, G., Low dielectric constant polymers for microelectronics, *Prog. Polym. Sci.,* 26, 3, 2001.
29. Cobb, M., Cole, B., and Mecca, S., Ensemble integrated dielectric solutions, SEMATECH workshop, San Francisco, CA, June 2002.
30. Labelle, C.B., Ryan, T., Augur, R., Bolom, T., Martin, J., Iacoponi, J., Challenges and opportunities in selection and implementation of new materials, Semicon West, San Francisco, CA, 2006.
31. Kason, M., Hawkins, J., Wang, S., Simmonds, M., and Itchhaporia, K., ICue Cu CMP Technology for integration of copper with developmental porous SiLK dielectric films, SEMATECH low-k workshop, June 2002.
32. Wallace, R.M. and Wik, G., High *k* gate dielectric materials, *MRS Bull.*, 27 (3), 192, 2002; Misra, V., Lucovsky, G., and Parsons, G., Issues in high *k* gate stack interfaces, *MRS Bull.*, 27 (3), 212, 2002; Robertson, J., Electronic structure and band offsets of high dielectric constant gate oxides, *MRS Bull.*, 27 (3), 217, 2002.

29 Self-Assembly of Organic Optoelectronic Materials and Devices

J.R. Heflin

CONTENTS

29.1 Introduction 867
29.2 Types of Self-Assembly 868
29.2.1 Self-Assembled Monolayers 868
29.2.2 Electrostatic Self-Assembly 869
29.2.3 DNA Self-Assembly 870
29.2.4 Block Copolymer Self-Organization 871
29.3 Self-Assembled Organic Nonlinear Optical Materials 872
29.4 Self-Assembly in Organic Light-Emitting Diodes 878
29.5 Self-Assembly of Organic Photovoltaic Devices 880
29.6 Summary 884
Exercise Questions 884
List of Abbreviations 884
References 885

Abstract This chapter discusses the uses of self-assembly in organic optoelectronic devices. It begins with a review of several methods of self-assembly including covalent self-assembly, layer-by-layer (LbL) electrostatic self-assembly, DNA assembly, and block copolymer self-organization. Several examples of applications of self-assembly in organic nonlinear optical (NLO) materials and devices, organic light-emitting diodes (OLEDs), and organic solar cells are then described.

29.1 INTRODUCTION

Nature has developed remarkably powerful methods for assembling exceptionally complex objects such as animals and plants from relatively simple initial components. In some respects, this assembly resembles the majority of the manufacturing done by humans in the sense that the final object, such as an automobile, is built up from a number of smaller and simpler components. This is known as bottom-up fabrication. In contrast, many of the smallest and most sophisticated devices fabricated by humans, such as microprocessors (which contain roughly one billion transistors in a <1 cm^2 area), utilize top-down lithographic fabrication processes. In this latter case, material is selectively removed from a homogeneous layer similar to the manner in which a sculptor carves a face out of clay. Top-down approaches are prevalent at submicron length scales because of the difficulty of positioning and connecting such small components. Yet this is precisely what nature has accomplished for millions of years. The critical feature that enables natural self-assembly is that no external forces or central control are required. Self-assembly occurs spontaneously as a result of innate interactions of the components that cause them to naturally associate into a predefined, complex structure. Thus, self-assembly refers to processes in which one simply mixes together sets of smaller components that then form more sophisticated, predefined, and desired structures without the need for any external intervention or driving forces.

Self-assembly is one of the rapidly advancing, frontier areas of nanotechnology. Researchers are developing an increasing library of approaches for self-assembly and an ability to fabricate ever more sophisticated and desirable functional structures. Self-assembly approaches can be categorized by the interactions between the components that cause their association. In this chapter, we will first review the most prevalent methods of self-assembly, such as electrostatic interactions, hydrogen bonding, formation of self-assembled monolayers (SAMs) through thiol–metal bonding, and the self-organization of block copolymers. We will then discuss several examples where self-assembly has been utilized to provide deep understanding and, in some cases, improved performance in organic optoelectronic device areas, including nonlinear optics, light-emitting diodes, and photovoltaic devices.

29.2 TYPES OF SELF-ASSEMBLY

29.2.1 Self-Assembled Monolayers

When certain substrates are immersed into a solution containing molecules with particular end-group functionalities, ordered layers of single-molecule thickness can be formed. These nanometer-scale layers with long-range periodic order are called self-assembled monolayers (SAMs) [1]. The archetypal SAM system consists of alkyl chains with a terminal thiol (SH) group adsorbed onto a gold surface. In this example, the H atom of the thiol group dissociates and a bond is formed between the S and the gold surface. Generally, the molecules pack hexagonally onto the Au(111) surface with a tilt of ~30° away from the surface normal. To achieve a high degree of order, however, the molecule usually requires exposure of the substrate for multiple hours. The general features of thiol on gold self-asssembly are illustrated in Figure 29.1. The terminal group opposite the thiol (represented

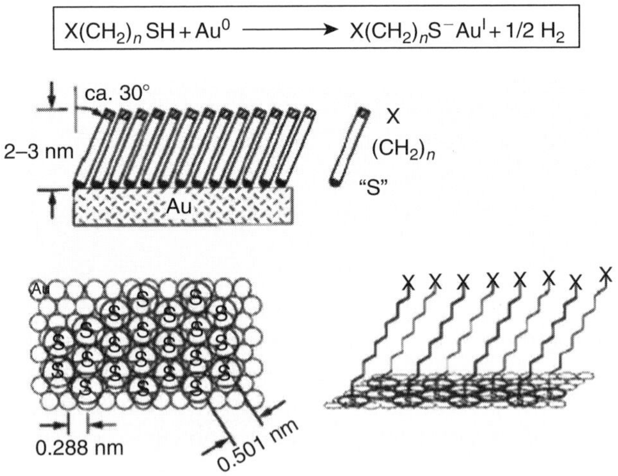

FIGURE 29.1 Representation of an ordered monolayer of alkane thiolate formed on an Au surface. The alkyl chains extend from the surface in a nearly all-trans configuration with an average tilt of approximately 30° from the surface normal. (From Xia, Y., Whitesides, G.M., *Angew. Chem. Int. Ed.*, 37, 551, 1998. With permission of Wiley-VCH.)

as X in the figure) allows creation of a new surface with controlled functionality. A number of techniques have been developed for patterning of SAMs, allowing an additional level of versatility and functionality [2]. These include microcontact printing, nanoimprint lithography, dip-pen lithography, and scanning tunneling microscope (STM) lithography.

Another important class of SAMs is that of organosilicon monolayers. These consist of silanes such as $RSiX_3$, where X is a chloride or hydroxyl-terminated alkyl chain, adsorbed on hydroxylated silica surfaces, including oxidized silicon. Two key features of organosilicon SAMs are that they can be grown on silicon wafers and are extremely robust due to a network of Si–O–Si bonds connecting the molecules within the monolayer as well as to the substrate.

29.2.2 Electrostatic Self-Assembly

The attraction between negatively and positively charged species forms the basis of layer-by-layer (LbL) deposition (Figure 29.2) [3]. LbL films (also known by a number of other names including polyelectrolyte multilayers and ionic self-assembled multilayers) are typically formed by the deposition of oppositely charged polymers (polycations and polyanions), although a vast array of multivalent species have also been used, including proteins, fullerenes, clays, dye molecules, and nanoparticles. The LbL method simply involves the alternate dipping of a charged substrate into an aqueous solution of a polycation and an aqueous solution of a polyanion at

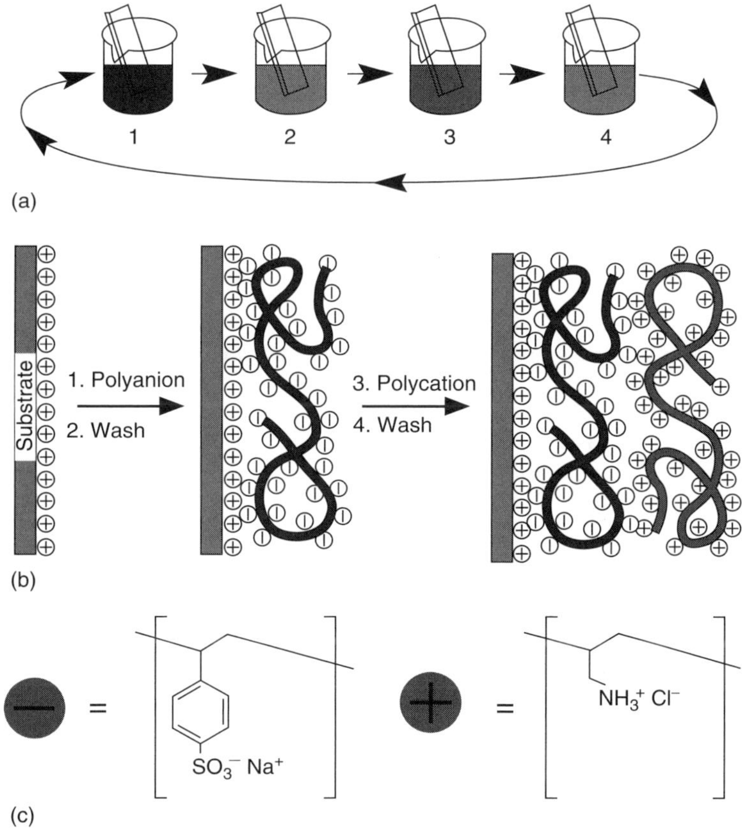

FIGURE 29.2 (a) Schematic illustration of the electrostatic LbL film deposition process. (b) Simplified depiction of the polyelectrolyte layers. (c) Examples of typical polyelectrolytes used in the LbL process. (From Decher, G., *Science*, 277, 1232, 1997. With permission of American Association for the Advancement of Science.)

room temperature. Since the adsorption is based on the electrostatic attraction of interlayer charges, each layer is self-limiting in thickness (typically 0.3–20 nm per layer, dependent on materials and preparation conditions). As demonstrated by optical absorption, ellipsometry, and x-ray diffraction studies, the thickness and surface homogeneity are exceptionally uniform from layer to layer and the layers can be highly interpenetrated. Each immersion leads to the rapid deposition of ionic species from the solution onto the oppositely charged substrate. The film growth requires that the materials for each successive layer possess multiple charges so that the surface charge on the substrate can be reversed (e.g., from positive to negative) as each layer is adsorbed. The dipping can be repeated to produce a film with as many bilayers as desired. LbL films can be highly robust, rapidly deposited (typically complete in ~ 1 min), and applicable to a vast array of charged materials.

29.2.3 DNA Self-Assembly

The exceptional selectivity of deoxyribonucleic acid (DNA), which contains the genetic information that codes for the assembly of living organisms, has also been utilized as a powerful tool for synthetic self-assembly. DNA strands consist of a repeat unit containing a sugar, a phosphate, and a base (Figure 29.3).The base can be adenine (A), thymine (T), guanine (G), or cytosine (C). The genetic code is contained in the order in which these four bases occur along the DNA strand. DNA forms the famous double-helix due to the selective hydrogen bonding among this set of bases. Adenine binds only with thymine and cytosine only with guanine. Organisms contain two separate, complementary sequences of DNA in which each A, T, G, or C on one is partnered with a T, A, C, or G, respectively, on the other. These two complementary DNA strands will strongly bind with one another and not with any noncomplementary sequence. While the DNA strands of organisms consist of sequences of millions of bases, researchers have harnessed the power of DNA assembly utilizing synthetic oligomers typically consisting of 10–30 bases.

An early example of utilizing DNA oligomers to form a colloidal assembly of gold nanoparticles illustrates the general approach [4]. Two separate solutions of 13 nm diameter gold nanoparticles are functionalized with DNA oligomers with alkane thiols at their 3′ end (Figure 29.4). The two DNA oligomers are 3′-thiol-TTTGCTGA and 3′-thiol-TACCGTTG. The thiols covalently link the DNA to the surfaces of the gold nanoparticles. When the two solutions of DNA-functionalized DNA oligomers are mixed together, the nanoparticles remain independent since the two DNA sequences

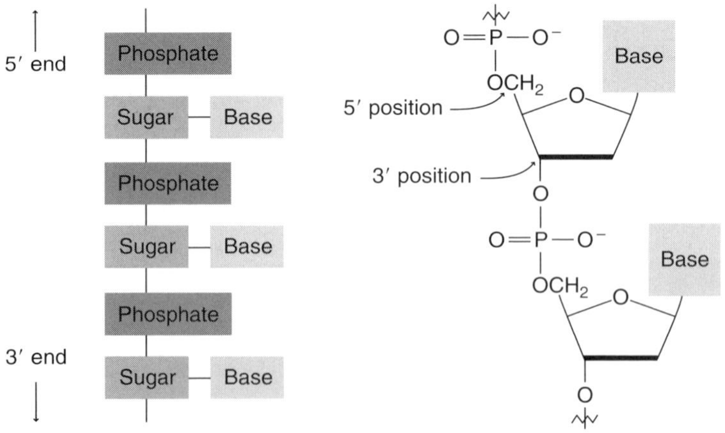

FIGURE 29.3 Basic structure of DNA consisting of a sugar, phosphate, and base in each repeat unit. The base can be adenine (A), guanine (G), thymine (T), or cytosine (C). A binds exclusively with T and G binds exclusively with C.

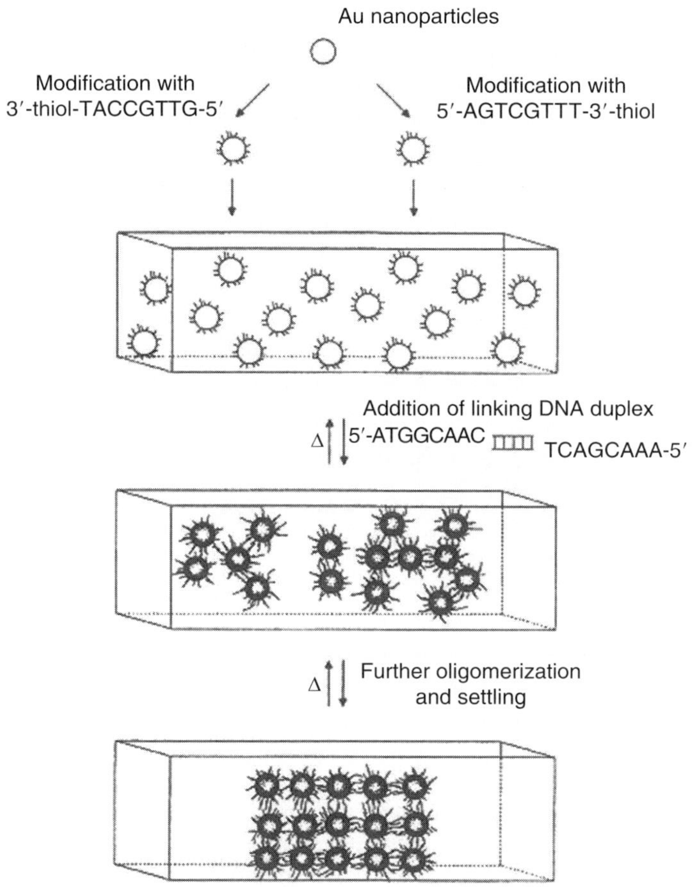

FIGURE 29.4 Scheme showing DNA-based colloidal assembly of gold nanoparticles. (From Mirkin, C.A., Letsinger, R.L., Mucic, R.C., Storhoff, J.J., *Nature*, 382, 607, 1996. With permission of Nature Publishing Group.)

are not complementary. Finally, a duplex consisting of 5′-ATGGCAACTATACGCGCTAG and 3′-ATATGCGCGATCTCAGCAAA is added to the solution. The underlined portions of the two sequences overlap and thus bind together. The remaining portions of each strand are complementary to the two sequences attached to the nanoparticles. Thus, the two separately tagged nanoparticles each bind to opposite ends of the duplex. The result is an aggregate of gold nanoparticles where each pair is connected by complementary DNA sequences.

29.2.4 Block Copolymer Self-Organization

Block copolymers consist of two or more repeat units, each forming one contiguous segment of the polymer chain. If the multiple repeat units are immiscible, a blend of the homopolymers would phase separate into macroscopically distinct regions. However, when the segments are connected by a covalent bond in the copolymer, such bulk phase separation is not possible. Instead, phase separation in block copolymers occurs into periodic domains with sizes typically in the range of 10–100 nm [5]. If the lengths of the two blocks are similar, lamellar structures tend to be formed. However, when there is a significant difference in the lengths, cylindrical or spherical domains of the minority phase tend to occur in a background of the majority of the phase. In addition to the relative lengths of the blocks, the domain structure can also depend on the thickness of the film. For example,

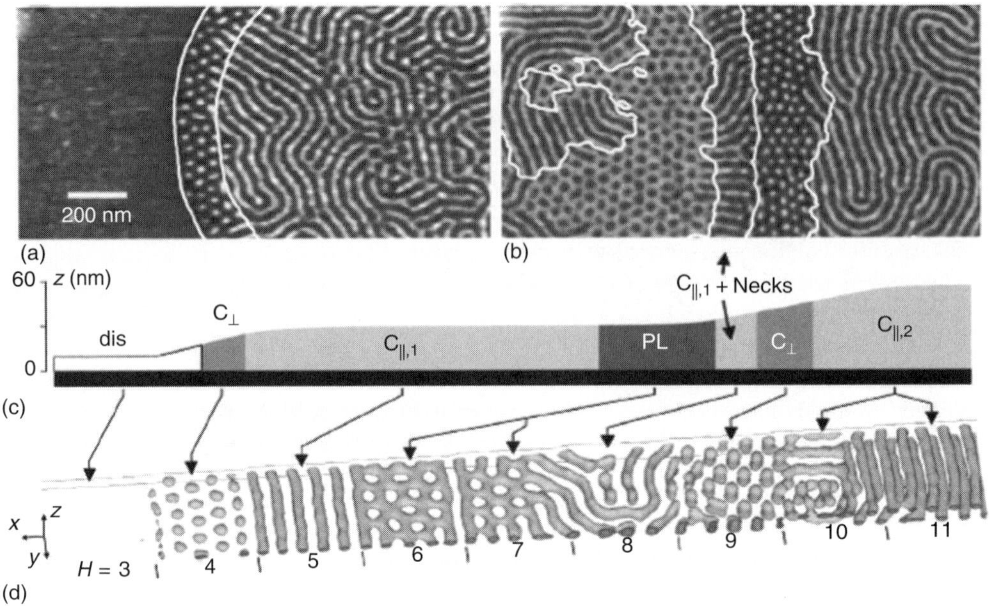

FIGURE 29.5 Tapping mode atomic force microscopy phase images of thin polystyrene-block-polybutadiene-block-polystyrene triblock copolymer films on Si substrates after annealing in chloroform vapor. (From Knoll, A., Horvat, A., Lyakhova, K.S., Krausch, G., Sevink, G.J.A., Zvelindovsky, A.V., Magerle, R., *Phys. Rev. Lett.*, 89, 035501, 2002. With permission of American Physical Society.)

Figure 29.5 shows experimental and calculated domain structures for a triblock copolymer of polystyrene-block-polybutadiene-block-polystyrene with molecular weights for each block of 14, 73, and 15k, respectively [6]. As the thickness varies, one observes polystyrene cylinders oriented perpendicular to the surface, polystyrene cylinders oriented parallel to the surface, and perforated lamellae. The periodic patterns that can be formed by block copolymers provide an intriguing approach for nanoscale fabrication.

29.3 SELF-ASSEMBLED ORGANIC NONLINEAR OPTICAL MATERIALS

Nonlinear optical (NLO) phenomena cause the optical properties of a material, such as refractive index or absorption, to be dependent on the incident intensity or an applied voltage, allowing for the fabrication of optically or electrically controlled optical switches and modulators. As discussed in earlier chapters, a material must lack a center of inversion at the macroscopic level in order to possess nonzero even-order NLO susceptibilities. Of particular interest is the macroscopic second-order (quadratic) susceptibility $\chi^{(2)}$, which governs the nonlinear polarization of the medium at frequency ω_3 in response to (optical) electric fields at frequencies ω_1 and ω_2 through

$$P_i^{\omega_3} = \chi_{ijk}^{(2)}(\omega_3, \omega_1, \omega_2) E_j^{\omega_1} E_k^{\omega_2}, \tag{29.1}$$

where the subscripts refer to the directions of polarization of the fields.

The requirement that a material must be noncentrosymmetric in order to possess a nonzero $\chi^{(2)}$ is because second-order NLO effects, such as harmonic generation, optical parametric oscillation, and the electro-optic effect, are quadratic in the applied fields. The most common $\chi^{(2)}$ materials are ferroelectric, inorganic crystals such as lithium niobate, potassium dihydrogen phosphate (KDP), and beta-barium borate (BBO). Growth of high-quality crystals, however, is difficult, time-consuming, and expensive. Organic materials are an attractive alternative for low-cost, high-performance electro-optic devices. The exceptional potential of organic electro-optic materials is illustrated in the demonstrations in poled polymers of full optical modulation at <1.0 V and at >100 GHz [7]. While the eventual randomization of the orientation back to the isotropic state has proven a challenging problem, advances continue to be made through use of higher T_g hosts, covalent attachment of the chromophore to the polymer, cross-linked polymers, and dendrimer structures. For example, recent materials have shown stability for >1000 h at 85°C [8].

Because of the challenges in obtaining temporal and thermal stability in poled polymers, alternative approaches for fabricating noncentrosymmetric organic materials are being developed that do not require trying to "freeze in" an alignment that is not thermodynamically stable. Of particular interest here are self-assembly methods for obtaining polar order in organic NLO materials. One approach involves using LbL covalent deposition such that each reaction step self-terminates after deposition of a monolayer. Methods have been developed employing zirconium phosphate–phosphonate [9] and siloxane [10] chemistry. In the former case, illustrated in Figure 29.6, an amine-terminated or

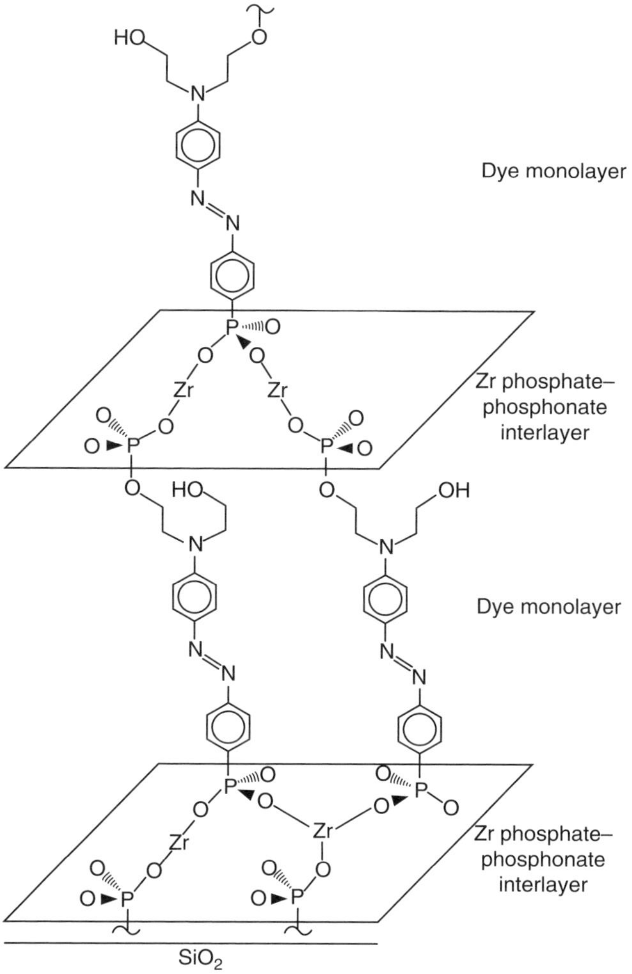

FIGURE 29.6 Schematic illustration of polar, self-assembled multilayers grown with Zr phosphate-phosphonate interlayers. (From Katz, H.E., Scheller, G., Putvinski, T.M., Schilling, M.L., Wilson, W.L., Chidsey, C.E.D., *Science*, 254, 1485, 1991. With permission of American Association for the Advancement of Science.)

hydroxyl-terminated surface on a glass substrate is phosphorylated and then zirconated by immersion in two appropriate solutions for 5 min each. Immersion for 10 min in a solution containing the organic NLO chromophore 4-{4-[*N*,*N*-bis(2-hydroxyethyl)amino]phenylazo}phenylphosphonic acid results in the covalent attachment of a dye monolayer. The process is then repeated to build up a multilayer structure. Films were fabricated up to 30 layers. The linear growth of the ellipsometrically measured thickness and the absorbance with the number of layers indicates the uniform and regular growth of successive layers. The second-order NLO response of the films was measured via second harmonic generation (SHG), also known as frequency doubling. For films thinner than the coherence length l_c of the material ($l_c = \lambda/[4(n^{2\omega} - n^{\omega})]$, where λ is the wavelength and n is the refractive index at each frequency), the SHG intensity is expected to grow quadratically with the thickness of the film. Since l_c is typically on the order of 10 μm and the layers are 1.6 nm thick, that condition is satisfied in the present case. The films did exhibit a quadratic dependence on the number of layers, demonstrating that each successive layer has the same degree of polar order. Comparison of the SHG intensity from the self-assembled multilayers to that from a material with a known $\chi^{(2)}$ value, such as quartz, allows determination of $\chi^{(2)}$ of the films. The measured $\chi^{(2)}$ value was 50×10^{-9} esu. For comparison, the $\chi^{(2)}$ value of the common inorganic NLO crystal lithium niobate is 200×10^{-9} esu.

Several variations have been developed for growth of polar, self-assembled multilayers using siloxane chemistry. In an early approach [11], the substrate was immersed in a solution containing a silanizing reagent for 18–64 h, cured at 115°C for 15 min, and then immersed in a solution containing a (dialkylamino)stilbazole NLO chromophore for 20 h. A $\chi^{(2)}$ value of 300×10^{-9} esu was obtained by SHG measurements. A more rapid deposition procedure has been developed that yields similarly large NLO responses. A monolayer of a $SiCl_2I$ derivative of an azobenzene chromophore is assembled onto the substrate by immersion in solution for 15 min. Immersion in octachlorotrisiloxane solution for 25 min removes a protecting group from the free end of the molecule and caps it with a polysiloxane layer [12]. The procedure is repeated to build up multilayers. Quadratic growth of the SHG intensity with the number of layers was observed, and a $\chi^{(2)}$ value of 430×10^{-9} esu was obtained. Covalent self-assembled multilayers of this sort have been incorporated into an electro-optic modulator device using the waveguide structure illustrated in Figure 29.7 [13]. Because the

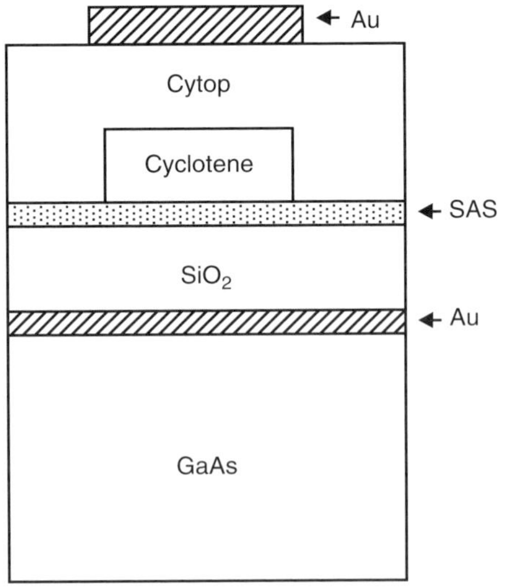

FIGURE 29.7 Schematic cross-section of an electro-optic modulator waveguide with an active organic self-assembled superlattice (SAS). (From Zhao, Y.G., Wu, A., Lu, H.L., Chang, S., Lu, W.K., Ho, S.T., van der Boom, M.E., Marks, T.J., *Appl. Phys. Lett.*, 79, 587, 2001. With permission of American Institute of Physics.)

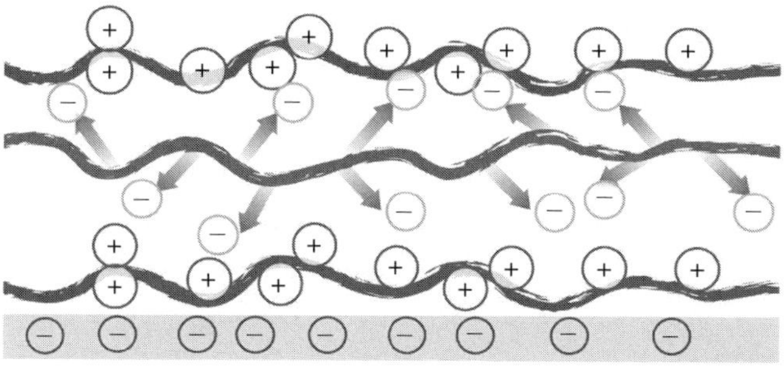

FIGURE 29.8 Schematic illustration of an electrostatic LbL film containing a polar NLO polymer with the arrows representing electric dipoles.

refractive index is lower in the surrounding media, the light is confined in the Cyclotene and self-assembled superlattice (SAS) by total internal reflection. The SAS consists of 40 layers with a total thickness of 150 nm. This is too thin to serve as an effective waveguide, so the Cyclotene layer is also used. However, that means that most of the light is not interacting with the electro-optic medium. As a result, the voltage required for modulation is much higher than what would be required if the waveguide were formed exclusively from electro-optic material. Nonetheless, full optical modulation was observed with application of 340 V, demonstrating the principle of electro-optic modulation from self-assembled organic NLO materials.

Polar organic NLO films have also been fabricated using the LbL electrostatic deposition process [14–18]. Several variations have been employed, but the most common one involves use of a polyelectrolyte that contains the conjugated NLO choromophore as a side group with a terminal ionic moiety. Usually, the oppositely charged polyelectrolyte does not have an NLO chromophore and serves primarily as glue for growth of the multilayers. The electrostatic bonding between the chromophore and the underlying oppositely charged layer provides directional orientation of the chromophore preferentially toward the substrate. The structure is schematically illustrated in Figure 29.8. The figure also illustrates that the adsorption of the next oppositely charged layer requires some fraction of the chromophores to be oriented away from the substrate. It is expected that this competition of chromophores oriented in opposite directions leads to a reduction of the net polar order and, correspondingly, $\chi^{(2)}$ value of the film. SHG measurements demonstrate, however, that the cancellation is not complete and net polar order does exist in self-assembled films fabricated by this method. For example, Figure 29.9 shows the linear growth of the absorbance and quadratic growth of the SHG intensity with the film thickness up to 100 bilayers for films made from a polyelectrolyte with an azobenzene side group [19]. The $\chi^{(2)}$ value of these films is a relatively modest 1.3×10^{-9} esu, presumably due to the partial cancellation of chromophores oriented in opposite directions. The temporal and thermal stability of the films is excellent, though. The $\chi^{(2)}$ values exhibited no loss at room temperature over eight years nor 20 h at 150°C [20].

This self-assembly approach also provides rapid adsorption of layers, typically in the range of 1–5 min. Methods are therefore being developed to eliminate the competition in chromophore orientation. One technique involves using monomeric, rather than polymeric, chromophores, and two different adsorption mechanisms to selectively orient the chromophore. Figure 29.10 shows one scheme in which the Procion Brown MX-GRN (PB) chromophore is covalently attached to poly(allylamine hydrochloride) (PAH) by immersion in a solution for 3 min at a pH of 10.5 (at which PAH is unprotonated) [21]. A layer of PAH is then electrostatically bound to the sulfonates of PB by immersion for 3 min in a solution at pH 7 such that PAH is protonated. A $\chi^{(2)}$ value of 50×10^{-9} esu and an electro-optic coefficient r_{33} of 14 pm/V were obtained for PB/PAH films fabricated by this hybrid covalent/electrostatic deposition process [22]. These films exhibit excellent thermal

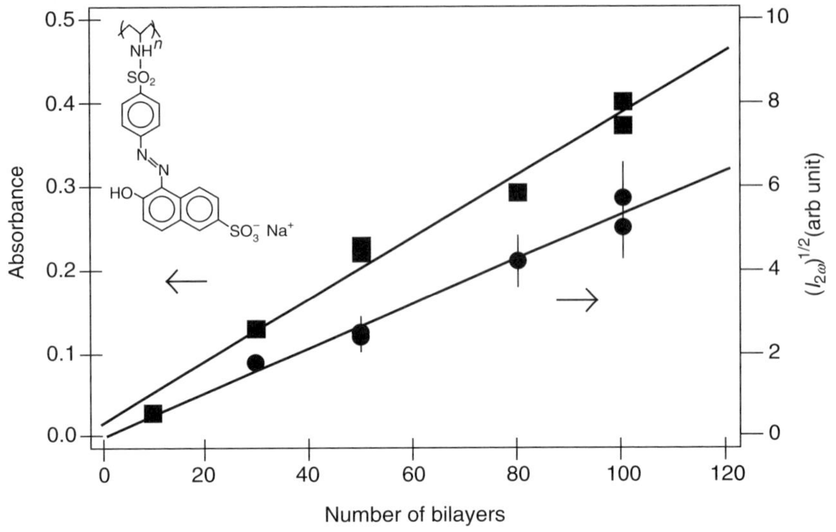

FIGURE 29.9 Absorbance at 500 nm and square root of the SHG intensity as a function of the number of bilayers for films made by LbL polyelectrolyte deposition. (From Heflin, J.R., Figura, C., Marciu, D., Liu, Y., Claus, R.O., *Appl. Phys. Lett.*, 74, 495, 1999. With permission of American Institute of Physics.)

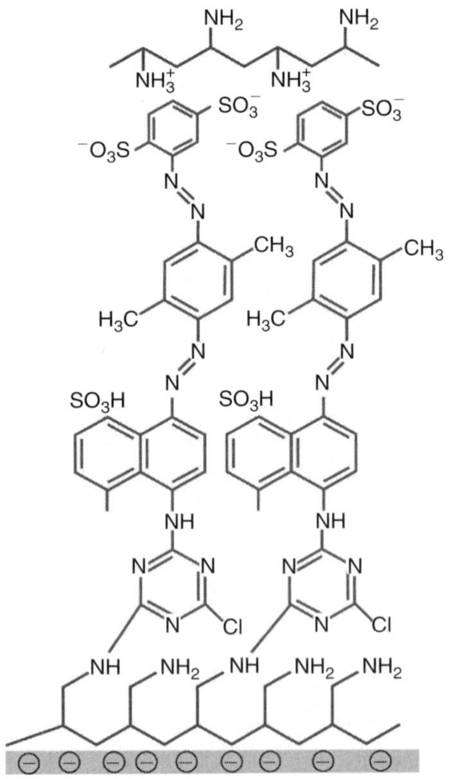

FIGURE 29.10 Hybrid covalent/ionic deposition of PB and poly(allylamine hydrochloride).

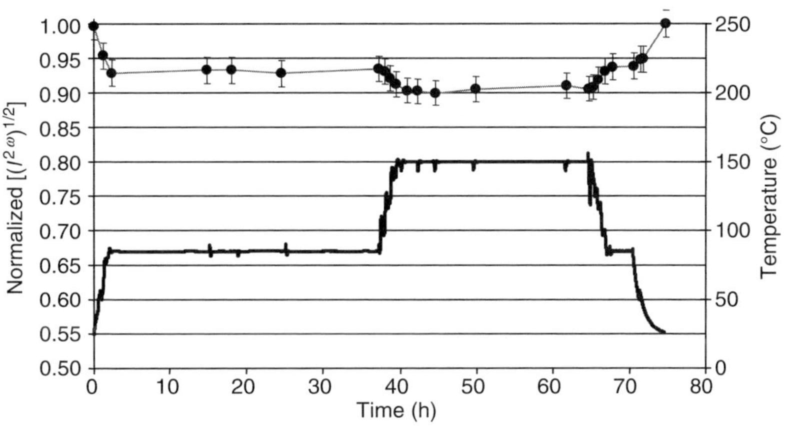

FIGURE 29.11 Square root of the SHG intensity (left axis, circles) of a PB/PAH hybrid covalent/ionic film as a function of time and temperature (right axis, solid curve) during a heating cycle.

stability. Figure 29.11 shows the square root of the second harmonic intensity as a function of time for a heating cycle of 36h at 85°C and 24h at 150°C. While the NLO response shows a slight temperature dependence, decreasing by 8% at 85°C and 10% at 150°C, there is no loss of the polar order. The second harmonic intensity recovers fully when the films return to room temperature. Substantially higher values should be possible with the use of chromophores with larger NLO susceptibility.

Electrostatic self-assembly has also been combined with plasmonic metal nanostructures to provide orders of magnitude enhancement of the NLO response of the organic self-assembled material [23]. Collective oscillations in metallic nanostructures known as localized surface plasmon resonances concentrate the electric field and can provide enhancements of $>10^4$ in the optical intensity. Using nanosphere lithography, triangular silver nanoparticles approximately 200 nm across were fabricated on top of ionic self-assembled multilayers as illustrated in Figure 29.12. The second

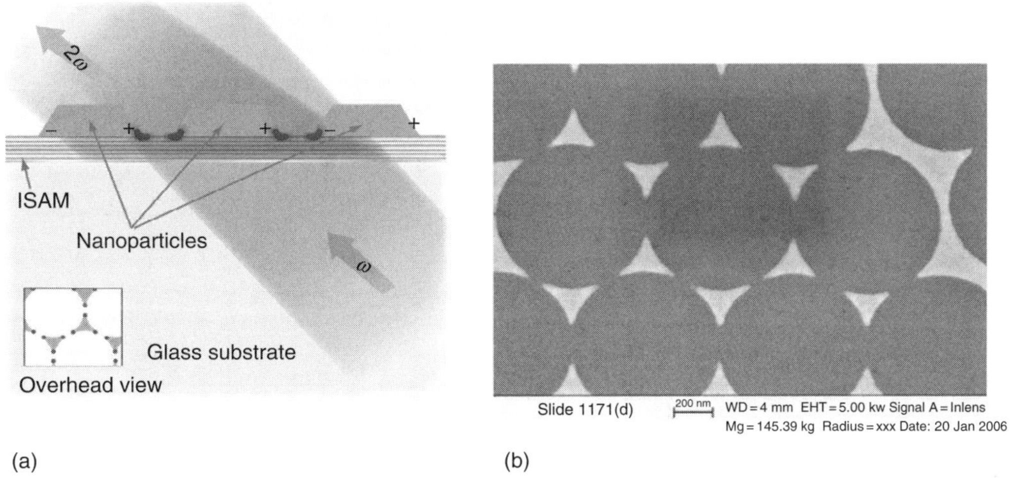

FIGURE 29.12 (a) Schematic side view of SHG in an LbL (also known as ionic self-assembled multilayer, ISAM) and metallic nanoparticle film. The localized intensity enhancement (dots at the vertices of the triangles) yields a corresponding increase in the second harmonic intensity. (b) SEM image of an array of silver nanoparticles (bright phase) fabricated by nanosphere lithography, using 720 nm diameter polystyrene spheres (removed) as a template.

harmonic intensity was increased as much as 1600 times relative to films consisting of only the self-assembled multilayer or metallic nanostructures. Furthermore, because of the contributions from interfaces, a film consisting of just three bilayers and the silver nanostructures exhibited as much NLO response as a film consisting of ~1000 bilayers of the self-assembled film alone.

29.4 SELF-ASSEMBLY IN ORGANIC LIGHT-EMITTING DIODES

Self-assembly has also been used in fabrication of OLEDs. In an OLED, the electrons injected from the cathode (negative electrode) recombine with holes injected from the anode (positive electrode) at a single molecular site. If there is a large quantum efficiency for radiative decay, then the applied current results in emitted light. However, electrons and holes that pass from one electrode to the other without undergoing recombination result in higher currents and lower light emission and, therefore, lower device efficiency. Several approaches have been demonstrated for improving OLED efficiency including incorporation of electron-injection [24] and hole-blocking [25] layers between the emissive layer and cathode, insertion of a layer between the anode and hole transport layer (HTL) to suppress hole injection [26], and doping with fluorescent [27] or phosphorescent [28] dyes. The injection of holes in OLEDs is generally more efficient than injection of electrons. It is believed that the insertion of a layer between the indium tin oxide (ITO) anode and HTL that suppresses the hole injection leads to better electron–hole balance and thus improves the device quantum and power efficiencies.

Various methods have been employed to deposit an interlayer between the ITO anode and the HTL including chemical treatment, oxide overlayer physical vapor deposition, and vacuum deposition of oganic layers. An approach that exercises nanoscale control of the interlayer is LbL, self-limiting chemisorptive siloxane self-assembly as depicted in Figure 29.13 [29]. Layers of octachlorotrisiloxane are chemisorbed from solution, hydrolyzed, and thermally cured/cross-linked. Each iteration results in deposition of a layer 0.83 nm thick. By varying the number of layers deposited before vacuum deposition of the HTL, Alq_3, and metal anode layers, a detailed examination of the role of the interlayer was made possible. It was found that the deposition of 1–4 layers resulted in a monotonic decrease in the hole-injection efficiency with the largest decreases occuring after the first and second layers. With an aluminum cathode, the addition of one layer increased the turn-on voltage at which light emission is first observed from 9 to 15 V. Addition of the second and third layers resulted in successive decreases in the turn-on voltage followed by an increase after deposition of the fourth layer. The decreased turn-on voltage for the second and

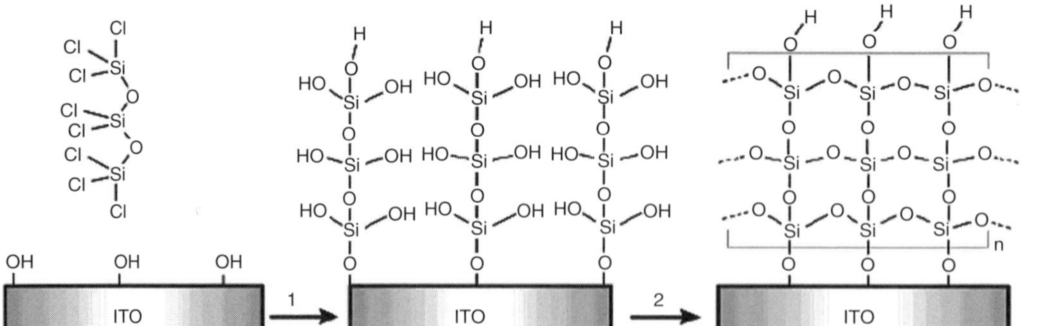

FIGURE 29.13 LbL chemisorptive self-assembly of siloxane dielectric layers on an ITO anode. (From Malinsky, J.E., Veinot, J.G.C., Jabbour, G.E., Shaheen, S.E., Anderson, J.D., Lee, P., Richter, A.G., Burin, A.L., Ratner, M.A., Marks, T.J., Armstrong, N.R., Kippelen, B., Dutta, P., Peyghambarian, N., *Chem. Mater.*, 14, 3054, 2002. With permission of American Chemical Society.)

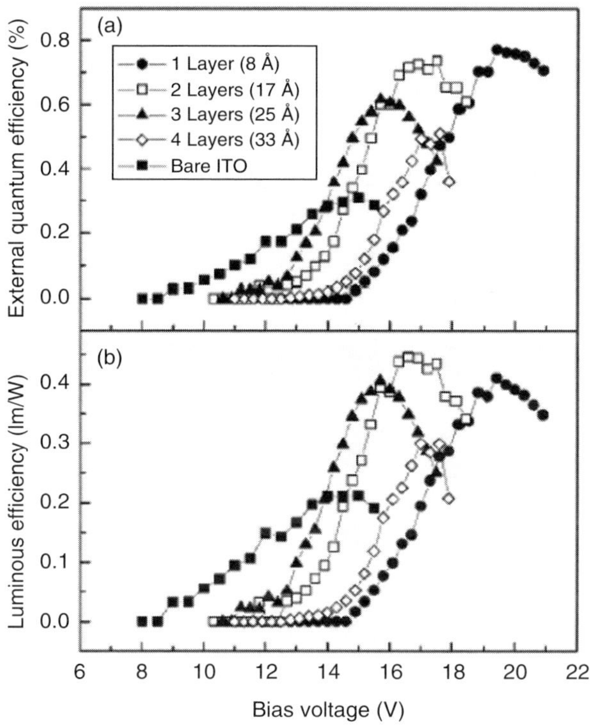

FIGURE 29.14 (a) External quantum efficiency (EQE) and (b) luminous efficiency as a function of the number of self-assembled siloxane layers. (From Malinsky, J.E., Veinot, J.G.C., Jabbour, G.E., Shaheen, S.E., Anderson, J.D., Lee, P., Richter, A.G., Burin, A.L., Ratner, M.A., Marks, T.J., Armstrong, N.R., Kippelen, B., Dutta, P., Peyghambarian, N., *Chem. Mater.*, 14, 3054, 2002. With permission of American Chemical Society.)

third layers is ascribed to a built-in field from the interlayer that increases the electron-injection efficiency at the cathode. The external quantum efficiency (EQE) and luminous efficiency both increased with addition to the self-assembled layers, as illustrated in Figure 29.14. Because the hole-injection efficiency is much larger than the electron-injection efficiency without incorporation of the dielectric layers, the inclusion of the layers improves the electron–hole balance and, correspondingly, the device efficiency.

Light-emitting diodes fabricated from polymers are especially interesting, since they offer the potential advantages of simple, inexpensive fabrication of flexible devices. One of the most heavily investigated polymers for LEDs is poly(*p*-pheneylene vinylene) (PPV) and its derivatives. While derivatives such as poly(2-methoxy-5-(2′-ethylhexyloxy)-1,4-phenylenevinylene) (MEH-PPV) are soluble in many organic solvents, pristine PPV is not and must be processed in the form of a precursor. A nonconjugated, cationic tetrahydrothiophenium precursor is often used. After fabrication of a film of the precursor polymer, conversion to the conjugated PPV is achieved through a thermal elimination step at >200°C for >10 h in vacuum. Rubner and coworkers have used the cationic precursor to fabricate polymer LEDs from self-assembled multilayer films using the LbL electrostatic deposition process [30,31]. The polyanions used were poly(methacrylic acid) (PMA) and poly(styrene sulfonic acid) (PSS). While the polyanion is used primarily as a glue for build up of the multilayer structures, it was found that LEDs fabricated using the two different polyanions exhibited dramatically different performance. Devices were fabricated using 10–50 bilayers in each case. For PPV/PMA films, a thickness of 1.6 nm per bilayer was obtained while PPV/PSS films had 0.8 nm bilayer thickness. The PPV/PMA devices exhibited a rectification ratio of up to 10^5 as expected for diode behavior and luminances of 10–20 cd/m^2. In contrast, devices comprising PPV/PSS multilayers

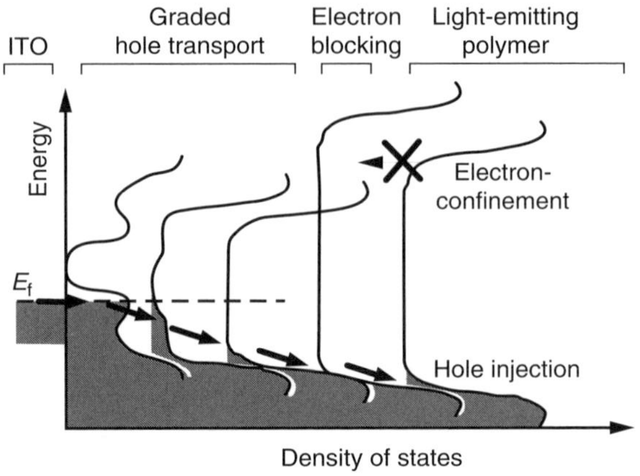

FIGURE 29.15 Schematic of the electronic density of states across a graded interlayer fabricated using electrostatic LbL assembly. (From Ho, P.K.H., Kim, J.S., Burroughes, J.H., Becker, H., Li, S.F.Y., Brown, T.M., Cacialli, F., Friend, R.H., *Nature*, 404, 481, 2000. With permission of Nature Publishing Group.)

showed fairly symmetric *I–V* curves, higher current densities, and one order-of-magnitude lower luminances. The differences were ascribed to doping of the PPV by the sulfonic acid groups of PSS, which leads to increased hole-injection efficiency under both positive and negative bias. The increased hole-injection efficiency of PPV/PSS was utilized in a self-assembled heterostructure to further increase the device efficiency. Five bilayers of PPV/PSS were deposited between the ITO and 15 bilayers of PPV/PMA in order to increase the efficiency of hole injection into the PPV/PMA emissive layer. The 4 nm thick PPV/PSS multilayer increased the luminance by nearly an order of magnitude (100–150 cd/m^2) relative to PPV/PMA devices fabricated directly on ITO.

Friend and coworkers demonstrated a similar approach in which LbL electrostatic self-assembly is used to deposit a hole-injection layer on the ITO substrate followed by spin-casting of the thicker emissive polymer layer [32]. The PEDOT:PSS complex was used as the polyanion and the tetrahydrothiophenium PPV precursor was the polycation. An additional degree of nanoscale engineering was exercised that is not possible with spin-casting. PEDOT:PSS was partially de-doped at three different levels with hydrazine hydrate. In this manner, a graded interface was fabricated in which PEDOT:PSS layers with successive degrees of de-doping were deposited on the ITO, as illustrated in Figure 29.15. Holes injected from the ITO anode thus undergo a more continuous transition into the valence band of the emissive polymer, reducing the barrier to hole injection. A second feature incorporated into the device design was an electron-blocking layer using partially converted PPV from the cationic precursor. The lower electron affinity of this layer relative to the emissive polymer layer results in confinement of electrons to the emissive polymer, thus increasing efficiency of radiative recombination of the electrons and holes. These two features are particularly useful for blue-emitting polymers with large ionization potentials such that the hole-injection barrier ΔE_h is especially large. With poly(9,9-dioctylfluorene) as the emissive polymer, devices with self-assembled, graded interlayers exhibited (EQEs) of 6.5% and luminances of 1000 cd/m^2, 40 times larger than for devices with spin-cast PEDOT:PSS on ITO. Self-assembly offers an attractive route for nanoscale engineering of the interface between the electrodes and emissive material in organic and polymeric LEDs.

29.5 SELF-ASSEMBLY OF ORGANIC PHOTOVOLTAIC DEVICES

While OLEDs and PLEDs rely on the large luminescence quantum efficiencies of conjugated organic materials, the propensity toward fluorescence is certainly not beneficial for photovoltaic applications, in which it is desired to convert incident optical energy into electrical energy. Organic photovoltaic devices generally rely upon the combination of electron donating and electron accepting materials. Upon photoexcitation, rapid electron transfer occurs from the donor molecule to the acceptor. This charge-transfer process prevents radiative electron–hole recombination. The fluorescence is quenched, and the separated electron and hole can be collected at the electrodes as a photocurrent. However, photoexcited electron–hole pairs at distances larger than ~10 nm from the acceptor recombine before charge separation occurs [33].

The most common approach to ensure proximity between the donor and acceptor species in polymer photovoltaics in order to provide efficient charge transfer is to spin-cast a blended film from a solution containing both components. This method has yielded the highest power conversion efficiencies so far in polymer photovoltaics up to a value of ~5% as of 2007 [34]. An alternative approach that has been explored by several groups for creating films with well-defined proximity between the donors and acceptors is LbL fabrication using ionic self-assembly. While the number of materials options is reduced by the necessity for charged species, this method does offer the ability to control the relative placement of the donors and acceptors down to the nanometer length scale.

Rubner and colleagues used LbL deposition to fabricate photovoltaic devices with PPV as the electron donor and sulfonated C_{60} as the electron acceptor [35]. The PPV was deposited from the cationic tetrahydrothiophenium precursor and converted into conjugated PPV by heating for 10–12 h at 230°C in a vacuum of 10^{-3} Torr. The PPV donors and C_{60} acceptors were separated into different blocks within the film. This was done by first depositing 20 bilayers of PPV with the anionic poly(acrylic acid) followed by 60 bilayers of sulfonated C_{60} with the polycation PAH. This resulted in ~100 nm of film containing PPV capped with 60 nm of film containing C_{60}. The device produced a short-circuit current density (J_{sc}) of 5 μA/cm² when illuminated by a laser power of 150 μW/cm² at a wavelength of 458 nm. The device also exhibited rectifying diode behavior in the dark and an open circuit voltage (V_{oc}) of 0.7–0.8 V as shown in Figure 29.16. Interestingly, films that contained only alternating layers of PPV and sulfonated C_{60} did not show rectifying behavior and produced negligible V_{oc}.

Durstock et al. also used the PPV cationic precursor to make LbL photovoltaic devices and combined it with both anionic and cationic C_{60} derivatives [36]. The devices contained a donor block

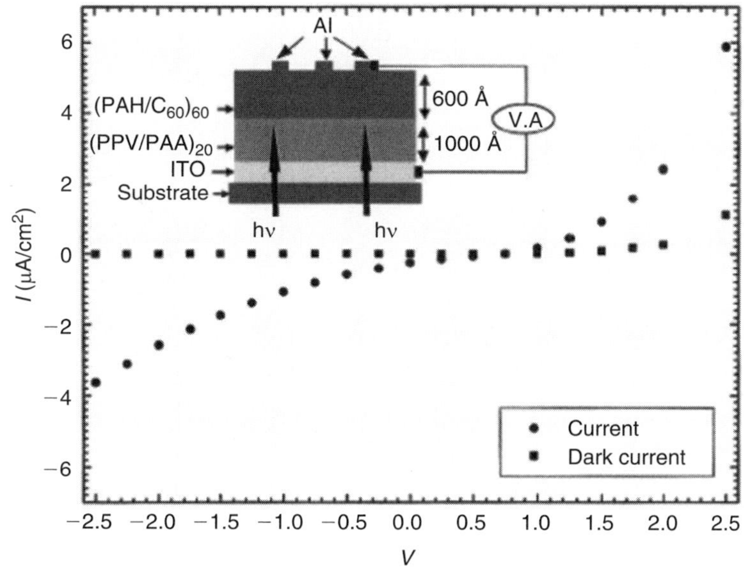

FIGURE 29.16 Photocurrent and dark current versus applied voltage for a photovoltaic devices fabricated by LbL deposition. (From Matoussi, H., Rubner, M.F., Zhou, F., Kumar, J., Tripathy, S.K., Chiang, L.Y., *App. Phys. Lett.*, 77, 1540, 2000. With permission of American Institute of Physics.)

of 20 bilayers of PPV alternately adsorbed with the polyanion sulfonated polystyrene. For the acceptor block, a comparision was made between devices fabricated with 20 bilayers of PAH and sulfonated C_{60} and 54 bilayers of cationic and anionic C_{60}. The number of bilayers was chosen in order to achieve similar thickness values in the two cases. The devices that contained cationic and anionic C_{60} exhibited J_{sc} twice as large as those with PAH and anionic C_{60}. While the two devices had comparable values of V_{oc} (0.4 V), those containing cationic and anionic C_{60} possessed greater temporal stability. The differences between the two cases were attributed to the presence of the electrically insulating polymer PAH, which hinders charge transport relative to films that contain only fullerenes. The EQEs, which correspond to the fraction of electrons collected as photocurrent per incident photon, were also found to be larger over the full visible spectrum for the cationic and anionic acceptor block, as shown in Figure 29.17. The EQE reached values in the vicinity of 0.10%–0.15%, which is far lower than the values of >50% that are achievable in blends of semiconducting polymers and fullerene derivatives. The dependence of J_{sc} and V_{oc} on the number of bilayers in the cationic and anionic C_{60} acceptor block was also studied. It was found that both quantities exhibited maxima in the vicinity of 50 bilayers, presumably due to a compromise between the additional amount of charge transfer enabled by thick donor blocks combined with a decreased ability to collect charges at the electrode as the thickness of the highly resistive, trap-containing donor block is increased. Furthermore, a novel device structure was examined in which the PPV donor block and fullerene acceptor block were separated by an interfacial layer containing bilayers of PPV and sulfonated C_{60}. It was found that just a few bilayers of this interfacial layer resulted in a threefold increase in J_{sc} with little effect on V_{oc}.

Reynolds and colleagues used LbL deposition to fabricate photovoltaic devices containing anionic poly(*p*-phenylene ethynylene) (PPE) derivatives and cationic C_{60} and obtained the highest EQE and power conversion efficiency observed so far in LbL solar cells [37]. In contrast to the work

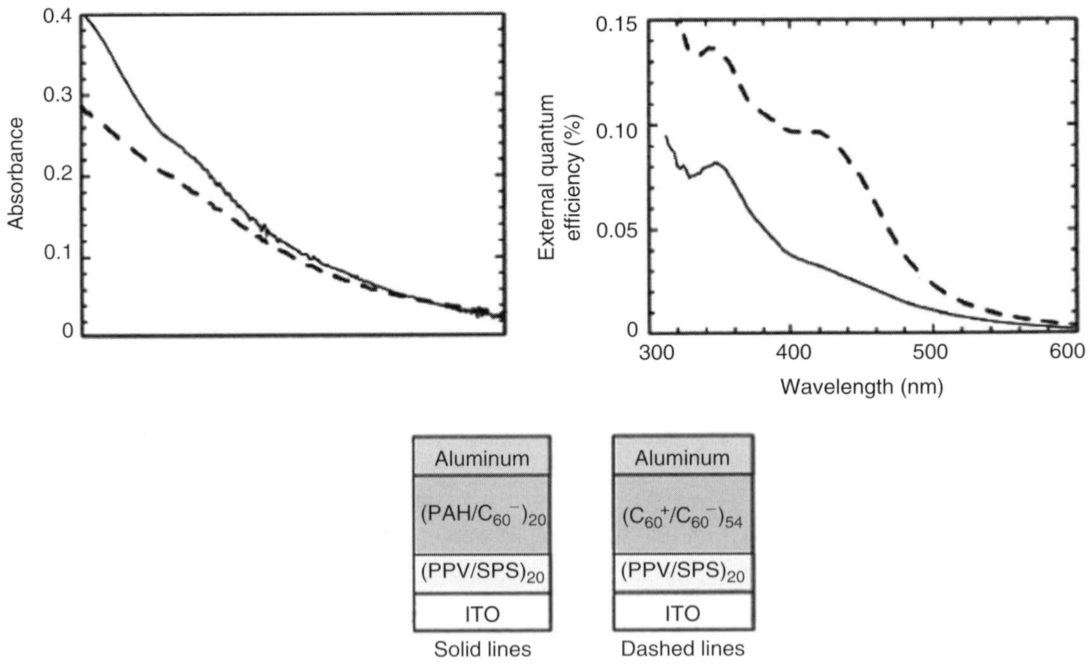

FIGURE 29.17 Absorbance and EQE versus wavelength for photovoltaic devices fabricated by LbL deposition with two different electron acceptor blocks. (From Durstock, M.F., Spry, R.J., Baur, J.W., Taylor, B.E., Chiang, L.Y., *J. Appl. Phys.*, 94, 3253, 2003. With permission of American Institute of Physics.)

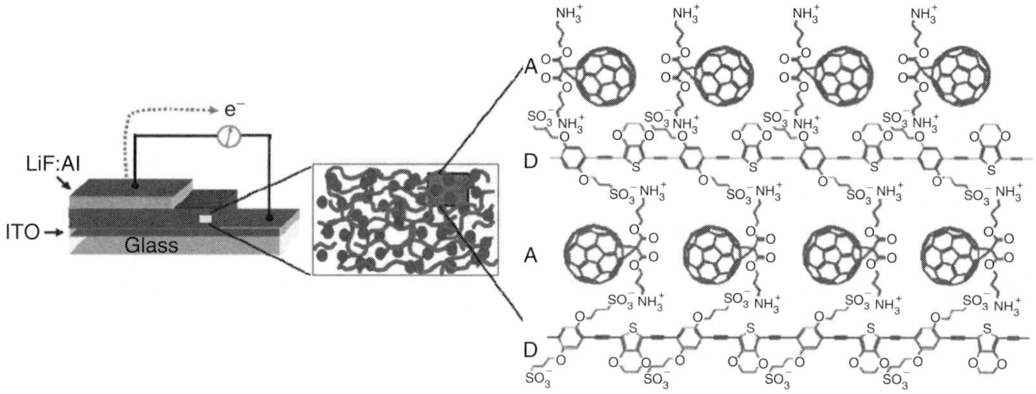

FIGURE 29.18 Schematic representation of the photovoltaic device structure fabricated by LbL deposition with alternating electron donor (D) and acceptor (A) layers. (From Mwaura, J.K., Pinto, M.R., Witker, D., Anantharishnan, N., Schanze, K.S., Reynolds, J.R., *Langmuir*, 21, 10119, 2005. With permission of American Chemical Society.)

described above, the donor and acceptor materials were adsorbed into alternating layers rather than isolated donor and acceptor blocks. The device structure and molecular components are illustrated in Figure 29.18. The EQE (also known as incident photon to current efficiency, ICPE) spectra for films containing 50 bilayers (~50 nm) of two different PPE anions with the C_{60} cation reach values of 4%–5%, as shown in Figure 29.19. A V_{oc} of 0.26 V and total power conversion efficiency of 0.04% were obtained under a simulated AM1.5 solar spectrum at an intensity of 100 mW/cm^2.

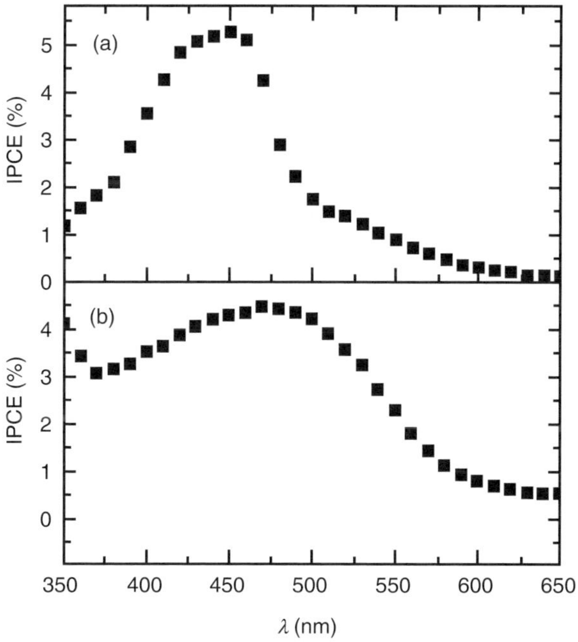

FIGURE 29.19 EQE (or ICPI) for photovoltaic devices fabricated by LbL deposition with two different PPE anionic derivatives and cationic C_{60}. (From Mwaura, J.K., Pinto, M.R., Witker, D., Anantharishnan, N., Schanze, K.S., Reynolds, J.R., *Langmuir*, 21, 10119, 2005. With permission of American Chemical Society.)

29.6 SUMMARY

Self-assembly is a powerful tool for bottom-up fabrication of complex structures and control of materials at the nanometer length scale. A large number of self-assembly approaches have been developed, primarily differentiated by the forces by which two or more complementary components recognize and bind one another. The most commonly employed self-assembly techniques include covalent self-assembly, LbL electrostatic self-assembly, DNA assembly, and block copolymer self-organization. Several examples of the use of self-assembly in organic optoelectronic materials and devices have been described here. In nonlinear optics, the electro-optic response requires noncentrosymmetric ordering of molecules. Organic materials possess several critical advantages over inorganic crystals, including the ability to provide higher frequency modulation at lower voltage. Both covalent self-assembly and electrostatic self-assembly have been used to obtain this ordering in a thermodynamically stable manner that yields excellent temporal and thermal stability. Organic and polymeric light-emitting diodes have been fabricated with both covalent and electrostatic multilayers at the interfaces between the electrodes and active materials to increase the efficiency by improving the balance between electron and hole injection. Organic photovoltaic devices rely on the proximity between electron donor and electron acceptor species to provide efficient separation of charges that can then be collected at the electrodes. Electrostatic LbL deposition has been used to control the relative spatial locations of donor and acceptor species down to the nanometer length scale. Self-assembly will undoubtedly continue to have an important role in contributing to the fundamental understanding of and engineering the device performance of organic optoelectronic devices for the forseeable future.

EXERCISE QUESTIONS

1. As discussed in conjunction with Figure 29.14, self-assembled layers can reduce injection of holes into an OLED and thus also reduce the total current. How can this result in an increase in the luminous efficiency of the device?
2. Conversion of optical power into electrical current at a particular wavelength by an organic photovoltaic device is sometimes expressed as the spectral photoresponsivity (PR), which is the electrical current divided by the incident optical power and has units of Ampere per Watt. Derive an expression relating the PR to the EQE.
3. Of the three main processes that occur in an organic solar cell (absorption, charge separation, and charge transport), which one is the most likely reason that organic photovoltaic devices fabricated by LbL self-assembly thus far have significantly lower efficiencies than devices made from blends of donors and acceptors? Explain why.

LIST OF ABBREVIATIONS

DNA	deoxyribonucleic acid
EQE	external quantum efficiency
HTL	hole transport layer
ICPE	incident photon to current efficiency
ITO	indium tin oxide
LbL	layer-by-layer
NLO	nonlinear optical
OLED	organic light-emitting diode
PAH	poly(allylamine hydrochloride)
PB	Procion Brown MX-GRN
PMA	poly(methacrylic acid)
PPE	poly(*p*-phenylene ethynylene)
PPV	poly(*p*-phenylene vinylene)

PSS poly(styrene sulfonic acid)
SAM self-assembled monolayer
SHG second harmonic generation

REFERENCES

1. Ulman, A., Formation and structure of self-assembled monolayers, *Chem. Rev.* 96, 1533–1554, 1996.
2. Xia, Y., Whitesides, G.M., Soft lithography, *Angew. Chem. Int. Ed.* 37, 551–575, 1998.
3. Decher, G., Fuzzy nanoassemblies: Towards layered polymeric multicomposites, *Science* 277, 1232–1237, 1997.
4. Mirkin, C.A., Letsinger, R.L., Mucic, R.C., Storhoff, J.J., A DNA-based method for rationally assembling nanoparticles into macroscopic materials, *Nature* 382, 607–609, 1996.
5. Krausch, G., Magerle, R., Nanostructured thin films by self-assembly of block copolymers, *Adv. Mater.* 14, 1579–1583, 2002.
6. Knoll, A., Horvat, A., Lyakhova, K.S., Krausch, G., Sevink, G.J.A., Zvelindovsky, A.V., Magerle, R., Phase behavior in thin films of cylinder-forming block copolymers, *Phys. Rev. Lett.* 89, 035501:1–4, 2002.
7. Shi, Y., Zhang, C., Zhang, H., Bechtel, J.H., Dalton, L.R., Robinson, B.H., Steier, W.H., Low (sub-1 Volt) halfwave polymeric electro-optic modulators achieved by controlling chromophore shape, *Science* 288, 119–122, 2000.
8. Ma, H., Liu, S., Luo, J., Suresh, S., Liu, L., Kang, S.H., Haller, M., Sassa, T., Dalton, L.R., Jen, A.K.Y., Highly efficient and thermally stable electro-optical dendrimers for photonics, *Adv. Funct. Mater.* 12, 565–574, 2002.
9. Katz, H.E., Scheller, G., Putvinski, T.M., Schilling, M.L., Wilson, W.L., Chidsey, C.E.D., Polar orientation of dyes in robust multilayers by zirconium phosphate–phosphonate interlayers, *Science* 254, 1485–1487, 1991.
10. Li, D., Ratner, M.A., Marks, T.J., Zhang, C., Yang, J., Wong, G.K., Chromophoric self-assembled multilayers. Organic superlattice approaches to thin film nonlinear optical materials, *J. Am. Chem. Soc.* 112, 7389–7390, 1990.
11. Yitzchaik, S., Roscoe, S.B., Kakkar, A.K., Allan, D.S., Marks, T.J., Xu, Z., Zhang, T., Lin, W., Wong, G.K., Chromophoric self-assembled NLO multilayer materials. Real time observation of monolayer growth and microstructural evolution by *in situ* second harmonic generation techniques, *J. Phys. Chem.* 97, 6958–6960, 1993.
12. Zhu, P., van der Boom, M.E., Kang, H., Evmenenko, G., Dutta, P., Marks, T.J., Realization of expeditious layer-by-layer siloxane-based self-assembly as an efficient route to structurally regular acentric superlattices with large electro-optic responses, *Chem. Mater.* 14, 4982–4989, 2002.
13. Zhao, Y.G., Wu, A., Lu, H.L., Chang, S., Lu, W.K., Ho, S.T., van der Boom, M.E., Marks, T.J., Traveling wave electro-optic phase modulators based on intrinsically polar self-assembled chromophore superlattices, *Appl. Phys. Lett.* 79, 587–589, 2001.
14. Lvov, Y., Yamada, S., Kunitake, T., Nonlinear optical effects in layer-by-layer alternate films of polycations and an azobenzene-containing polyanion, *Thin Solid Films* 300, 107–110, 1997.
15. Wang, X., Balasubramanian, S., Li, L., Jiang, X., Sandman, D., Rubner, M.F., Kumar, J., Tripathy, S.K., Self-assembled second order nonlinear optical multilayer azo polymer, *Macromol. Rapid Commun.* 18, 451–459, 1997.
16. Laschewsky, A., Mayer, B., Wisheroff, E., Arys, X., Jonas, A., Kauranen, M., Persoons, A., A new technique for assembling thin, defined multilayers, *Angew. Chem. Int. Ed.* 36, 2788–2791, 1997.
17. Heflin, J.R., Figura, C., Marciu, D., Liu, Y., Claus, R.O., Second order nonlinear optical thin films fabricated from ionically self-assembled monolayers, *SPIE Proc.* 3147, 10–19, 1997.
18. Roberts, M.J., Lindsay, G.A., Herman, W.N., Wynne, K.J., Thermally stable nonlinear optical films by alternating polyelectrolyte deposition on hydrophobic substrates, *J. Am. Chem. Soc.* 120, 11202–11203, 1998.
19. Heflin, J.R., Figura, C., Marciu, D., Liu, Y., Claus, R.O., Thickness dependence of second harmonic generation in thin films fabricated from ionically self-assembled monolayers, *Appl. Phys. Lett.* 74, 495–497, 1999.
20. Figura, C., Neyman, P.J., Marciu, D., Brands, C., Murray, M.A., Hair, S., Davis, R.M., Miller, M.B., Heflin, J.R., Thermal stability and immersion solution dependence of second order nonlinear optical ionically self-assembled films, *SPIE Proc.* 3939, 214–222, 2000.

21. Van Cott, K.E., Guzy, M., Neyman, P., Brands, C., Heflin, J.R., Gibson, H.W., Davis, R.M., Layer-by-layer deposition and ordering of low-molecular weight dye molecules for second-order nonlinear optics, *Angew. Chem. Int. Ed.* 41, 3236–3238, 2002.
22. Heflin, J.R., Guzy, M.T., Neyman, P.J., Gaskins, K.J., Brands, C., Wang, Z., Gibson, H.W., Davis, R.M., Van Cott, K.E., Efficient, thermally-stable, second order nonlinear response in organic hybrid covalent/ionic self-assembled films, *Langmuir* 22, 5723–5727, 2006.
23. Chen, K., Durak, C., Heflin, J.R., Robinson, H.D., Plasmon-enhanced second harmonic generation from ionic self-assembled multilayer films, *Nano Lett.* 7, 254–258, 2007.
24. Jabbour, G.E., Kawabe, Y., Shaheen, S.E., Wang, J.F., Morrell, M.M., Kippelen, B., Peyghambarian, N., Highly efficient and bright organic electroluminescent devices with aluminum cathode, *Appl. Phys. Lett.* 71, 1762–1764, 1997.
25. Xie, Z.Y., Hung, L.S., Lee, S.T., High-efficiency red electroluminescence from a narrow recombination zone confined by an organic double heterostructure, *Appl. Phys. Lett.* 79, 1048–1050, 2001.
26. Forsythe, E.W., Abkowitz, M.A., Gao, Y., Tuning the carrier injection efficiency of organic light-emitting diodes, *J. Phys. Chem. B* 104, 3948–3952, 2000.
27. Tang, C.W., VanSlyke, S.A., Chen, C.H., Electroluminescence of doped organic thin films, *J. Appl. Phys.* 65, 3610–3616, 1989.
28. Baldo, M.A., O'Brien, D.F., You, Y., Shoustikov, A., Sibley, S., Thompson, M.E., Forrest, S.R., Highly efficient phosphorescent emission from organic electroluminescent devices, *Nature* 395, 151–154, 1998.
29. Malinsky, J.E., Veinot, J.G.C., Jabbour, G.E., Shaheen, S.E., Anderson, J.D., Lee, P., Richter, A.G., Burin, A.L., Ratner, M.A., Marks, T.J., Armstrong, N.R., Kippelen, B., Dutta, P., Peyghambarian, N., Nanometer scale dielectric self-assembly process for anode modification in organic light-emitting diodes. Consequences for charge injection and enhanced luminous efficiency, *Chem. Mater.* 14, 3054–3065, 2002.
30. Fou, A.C., Onitsuka, O., Fereira, M., Rubner, M.F., Hsieh, B.R., Fabrication and properties of light-emitting diodes based on self-assembled multilayers of poly(pheneylene vinylene), *J. Appl. Phys.* 79, 7501–7509, 1996.
31. Onitsuka, O., Fou, A.C., Fereira, M., Hsieh, B.R., Rubner, M.F., Enhancement of light-emitting diodes based on self-assembled heterostructures of poly(*p*-phenylene vinylene), *J. Appl. Phys.* 80, 4067–4071, 1996.
32. Ho, P.K.H., Kim, J.S., Burroughes, J.H., Becker, H., Li, S.F.Y., Brown, T.M., Cacialli, F., Friend, R.H., Molecular-scale interface engineering for polymer light-emittiing diodes, *Nature* 404, 481–484, 2000.
33. Halls, J.J.M., Pichler, K., Friend, R.H., Moratti, S.C., Holmes, A.B., Exciton diffusion and dissociation in poly(*p*-phenylene vinylene)/C_{60} heterojunction photovoltaic cell, *Appl. Phys. Lett.* 68, 3120–3122, 1996.
34. Ma, W., Yang, C., Gong, X., Lee, K., Heeger, A.J., Thermally stable, efficient polymer solar cells with nanoscale control of the interpenetrating network morphology, *Adv. Funct. Mater.* 15, 1617–1622, 2005.
35. Matoussi, H., Rubner, M.F., Zhou, F., Kumar, J., Tripathy, S.K., Chiang, L.Y., Photovoltaic heterostructure devices made of sequentially adsorbed poly(phenylene vinylene) and functionalized C_{60}, *App. Phys. Lett.* 77, 1540–1542, 2000.
36. Durstock, M.F., Spry, R.J., Baur, J.W., Taylor, B.E., Chiang, L.Y., Investigation of electrostaic self-assembly as a means to fabricate and interfacially modify polymer-based photovoltaic devices, *J. Appl. Phys.* 94, 3253–3259, 2003.
37. Mwaura, J.K., Pinto, M.R., Witker, D., Anantharishnan, N., Schanze, K.S., Reynolds, J.R., Photovoltaic cells based on sequentially adsorbed multilayers of conjugated poly(*p*-phenylene ethynylene)s and a water-soluble fullerene derivative, *Langmuir* 21, 10119–10126, 2005.

Index

A

Abbrium–Miller model, 67
Absorption
  cutoff wavelength of, 416
  vibrational, 69, 493
Absorption coefficient, 26
  of biological tissues, 590
  and intersite hopping frequency, 98
  self-assembly ultrathin film, 147
  of semiconductors, 28–29
  total, 531–532
Absorption spectrum
  of biological tissues, 590
  of cyanines, 148–149
  of doped conducting polymers, 178–179
  for Er/Yb co-doped short fiber lasers, 12
  for ethane, 70
  linear and nonlinear, 582
  of PAn and PPy, 179–180
  side chain substitution, conjugaated polymers, 191–192
  UV–Vis optical, 82
Acceptors
  for electron transfer, 66, 75, 77, 221
  for energy transfer, 81
Accumulation, *see* Charge accumulation
Activation energy, 65–66, 99, 378, 789
Actuators
  actuation processes, 737–738
  benders *vs.* axial, 750–753
  blocking force, 738
  categories of, 741–742
  conducting polymer
    actuation mechanisms of, 745–746
    advantages, 745
    applications for, 754–756
    comparative performances of, 741–744
    designing, 749–750
  definition of, 73, 733–734
  electromechanical, 734
  free stroke
    actuator materials of, 739, 741, 743
    definition, 737
    misfit strain, 752
  hypothetical, schematic of, 737–738
  linear elasticity, 739
  musculoskeletal systems, 744
  selection criteria for
    cost and maintenance, 737
    mechanical performance, 735–736
  stress *vs.* strains, 743–744
  Young's modulus of, 739–740
Adiabatic approximation, 51, 83
Adler and Longo's method, 142
AFM, *see* Atomic force microscopy
Aggregates, 148
Air gaps, 313, 856
Alignment between molecules, 774, 782, 785–787
Alkane chains, 582
Alkanethiol, 706
Allen–Cahn equation, 808
All-polymer devices, 328, 332, 337, 342
All-printed organic transistor, 342
Alq3, *see* Tris(8-hydroxyquinolinato) aluminum
Amorphous glasses, 621, 624
Amorphous polycarbonate, 479, 506
Anisotropy, in physical properties, 776, 779–782
Annealing, 409, 411
Anthracene, 300, 302, 313, 578
Antibonding molecular orbital, 56–58
σ^* antibonding type orbital, 57
σ bonds, 56, 59–60
Antiferroelectric smectic C* phase structure, 776
AOs, *see* Atomic orbitals
APC, *see* Amorphous polycarbonate
Area capacitance, 858
Aromatic compounds, 578
Aromatic polymides, structures of, 854
Arrayed waveguide grating filters, 839
Artificial muscle, 756
Atomic force microscopy, 297, 638, 809–810
  based chemical contact, in alkyl-thiol molecules
    conductivity of, 668
    particle size, 670
    schematic representation of, 669
  of coplanar junction, 665
  of electrode, FET device, 681
Atomic orbitals, 55–56
Attenuated total reflectance IR (ATR-IR), 723
Auxochromes, 70
AWG filters, *see* Arrayed waveguide grating filters

B

Back-end-of-the-line (BEOL) process, 846, 859–861
Band
  bending, 405
  regime, 62–63
  type transport, 14, 322
Band gap (BG), 53–54, 402
  Bloch–Floquet theorem, 797
  definition of, 795
  energy, 2–3, 16
  insulator and semiconductor of, 795–796
  Maxwell equations, 796
  photonic crystal, 799
Band width, 61–63
Bandwidth–length product, EO material, 836
Bardeen–Cooper–Schrieffer (BCS) theory, 276
Bathochromic shift, 70
BDT, *see* Benzodithiophene
Benzocyclobutene (BCB), 855
Benzodithiophene, 306
Bias, forward, *see* Forward bias
Biaxial crystal, 426

Bicontinuity, 409
Bicontinuous-ordered nanostructure solar cells
  diblock copolymers, 410–411
  discotic donor–acceptor-type molecules, 412
  fourth-generation donor/acceptor, 410
Bimolecular recombination coefficient, 26
Binary encoding system, 588
Bioimaging, 590, 592
Biophotonics, 574
2-(4-Biphenyl)-5-(4-*tert*-butylphenyl)-1,3,
    4-oxadiazole, 155
Bipolarons, 78–79
Bipyridine ligand, 165
Birefringence, in liquid crystals, 431–432
4,4′-Bis(carbazol-9-yl)biphenyl (CBP), 365
Bloch equations, 53–54
Bloch–Floquet theorem, 797
Bloch function, 3, 53–54
Bloch's theorem, 53–54, 84
Bloch wave(s), 53, 637
Block copolymers (DBA)
  domain structure of, 872
  properties of, 871
Block copolymers $(DBAB)_n$
  definition and types, 809
  discotic donor–acceptor-type molecules, 412
  self-assembled PS-b-PEO, AFM image of, 809–810
  self-assembly structures of, 411
Bohr radius, 72
Boltzmann equation, 92–93
Bond angles, 57
Bonding molecular orbital, 56–58
Bond length alteration (BLA), 61–62, 477–478
  in conductive polymers, 215–216
  electron-withdrawing substituents in, 221
BONS, *see* Bicontinuous-ordered nanostructure
Born–Oppenheimer approximation, 51, 83
Bose–Einstein statistics, 55
Bosons, 55, 79
Bounce angle, 41
Bragg conditions, 608
Bragg grating, 620, 806
Bragg's law, 804
Breakdown field, 298
Break-junction measurement
  electrode fabrication for, 680
  metallic wire based
    disadvantages of, 666
    for gold nanojunction fabrication, 665–667
  on STM
    feedback loop control, 670
    for single molecule conductivity, 670–671
π-Bridge, 445
Bright–Wigner formula, 120
Brillouin zones, 53
BTQBHT, synthesis and structure of, 311
Buffer layer, 138
Bulk polarization, 849

C

CA, *see* Chronoamperometry
Cahn–Hilliard equation, 808–809
Calamitic liquid crystals
  building blocks of, 769
  molecular geometry, chain terminal, 767–768
  molecular structure, chain flexible, 771
  phases of, 773–774
Capacitance, 30, 321, 327–328
  Faraday capacitor, 848
  low-*k* thin films, 858
  between metal layers, 847
Carbazoles and triphenylamine derivatives
  amorphous state and T_g in, 151
  HTL materials in, 151–152
  in OLEDs, 153–154
  star-shaped derivatives in, 154
Carbon materials
  atomic orbitals of, 55–56
  carrier transport, 62
  electronic structure
    Bloch function, 53–54
    diamond and graphite, 59–60
    ethene and polymer polyacetylene, 60–62
    modern quantum theory, 48–49
    Schrödinger equation, 49–50
    tunneling, 52
  molecular orbitals of
    antibonding, 57–58
    bonding, 56
    nonbonding, electrons, 59
  spin configurations in, 63–64
Carbon nanotubes, 60
  experimental setup of, 254
  as graphene sheet, 62–63, 239, 241
  lighting elements in, 254–255
  principle of, 253
Carrier extraction using linearly increasing voltage
    (CELIV), 328
Carriers, *see* Charge carriers
CB, *see* Conduction band
CBP, *see* 4,4′-Bis(carbazol-9-yl)biphenyl
CC, *see* Chronocoulometry
CCD, *see* Charge coupled device
Centrosymmetric molecular systems, 520–521
CFET, *see* Chemical field-effect transistor
Chain fractal
  hopping regimes for, 110
  schematic of, 103
Channel waveguide, 481, 560, 823–824, 827
Charge accumulation, 325, 329
Charge capacity, 725
Charge carriers, 405, 408
  depletion, 34
  drift velocity, 30
  generation mechanisms
    chemical doping, 76–77
    electric carriers, 78
    energy gap, E_g, 74
    excitons, 71
    Förster and Dexter energy transfers, 80–82
    inorganic semiconductors, 71
    organic and polymeric semiconductors, 72–73
    semi-classical Marcus model in, 74–75
  generation rate, 26
  injection efficiency, 358
  injection, LED

blocking layers, 386–387
end-capping, 387–388
localization of, 95, 114–115
loss, 408–409
mobility, 62
in carbon nanotubes, 258
in field effect transistors, 246
recombination
dynamics of, 614–615
Langevin treatment in, 614
in PLEDs, 376–377
transit time, 30
transport mechanism, 89, 95
in PLEDS, 377–378
Charge collection, 403
Charge coupled device, 324
Charge detrapping, process of, 613–614
Charge injection, 380
Charge recombination
geminate and nongeminate, 77
in laser diodes, 36–39
and light generation, 352–354
nonradiative, 27
hole diffusion length, 32
internal efficiency, 35
in photoconductors, 33–34
in PLEDs, 376–377
in solar cells, 414–415
spontaneous, 26
stimulated, 27
stimulated and nonradiative, 27
Charge separation, 144, 159–160
between C_{60} acceptor and conducting polymer donor, 256
by electron donor–acceptor molecules, 144–146
in ground state, NLO chromophores, 471
neutralization, photodiode, 690
photo-induced, 159–160, 408
Charges trapping, 613–614
Charge transfer, 76, 79
intermolecular theory, 271
optical absorption spectra, 272
in organic molecules, 446–448
in organic nonlinear optical single crystals, 455–456
in organic superconductors, 264
in second-order nonlinear optics, 445
Charge-transfer complexes, 158
Charge transport, 403, 409
band transport in metals
drude metal, 89–92
quantum particles, 92–95
charge carrier mobility in, 613
3D models of, 104
in disordered solids, 111
Fermi electrons, 94–95
in highly conducting polymers, 115–116
Gaussian distribution of, 612
hopping conductivity in insulators
in isotropic systems, variable, 98–102
master equation, 95–98
Chemical bond, 56, 59
Chemical doping, 75–78, 80
Chemical field-effect transistor (CFET), 684
Chemical mechanical polishing, 859–860, 863
Chemical stability, 862–863
Chirality, 469, 774–776
Chlorophylls, 142
Cholesteric liquid crystal phase, 774–775
Cholesteryl benzoate, 765
Chromophores, 836
Chronoamperometry, 725
Chronocoulometry, 725
Chronovoltabsorptometry, 724
Clearing temperature, liquid crystal, 781
Cluster approximation, 105
CMP, *see* Chemical mechanical polishing
CNTs, *see* Carbon nanotubes
Coherence length
in phase matching techniques, 432
in SHG, 428
Coherent transport, 54, 62
Combined polymer liquid crystals, 769
Conducting polymers
electrochemical preparation of
cation-radical polymerization, 184–185
polyaniline and polythiophene, 185
polypyrrole, 183–184
electrochromism of, 714
organic electrochromic devices
behavior of, 720
π-conjugation in, 719
transmission mode, 729
oxidative polymerization of
polyacetylene, 185–186
pyrrole and thiophene, 186–187
properties of
absorption spectra, 178–180
conductivity and doping, 177–178
electrochemical, 181–182
solubility, 180–181
Conduction band, 54, 59–61, 801
in laser diode, 36–37
in pin photodiode, 27
Conduction mechanism
electrons, 115
processes, 88
Conductivity (σ), 78–79
Conjugated organic molecules
bond length alternations in, 543
dipole moments in, 543, 546–548
polarizability and hyperpolarizability of, 547–549
resonance structure of, 542–544
second hyperpolarizabiliy
donor–acceptor triene molecules, 551
increasing polarization, 551
valence bond–charge transfer model values, 540–541, 544–549
structure hyperpolarizability relation, VB–CT model
dipole moment, 546–547
potential energy surfaces, 544–546
wavefunctions, 544
third-order nonlinearity survey
conjugation length of molecules, 552, 555
second hyperpolarizability of molecules, 552–553–554
susceptibility of molecules, 555–556

Conjugated polymers, 47, 60–61–62, 78–79, 82
characteristics of, 176
classes of, 175
interchain interactions, 191
properties and structures of, 238–239
π–Conjugated polymers, 445
electro-optical properties of, 211–212
energy gap in
band formation in, 213–214
bond length alteration in, 215
definition of, 213
factors affecting, 216
parameters in, 214–215
LEGPs in, 212–213
Conjugation length, 190, 192, 212, 215
Constitutive porosity, 852
Constructive interference, 804–805
Contact problem, 664
Contrast ratio, 786
Cooperative effect, 583
Cooper pairs, 276
Coulomb energy, 73, 713–714
Coulomb interactions, 451–452
Coulombic potential, 73
CPs, *see* Conducting polymers
Critical angle, 41
plastic optical fiber, 830
total internal reflection, 820–821
Crystallinity, 409
CT complexes, *see* Charge-transfer complexes
Cumulative-mode electrochromic devices, 716
Current density, 416–417
Current–voltage characteristics of
aligned nanotube field emitter, 253–254
hole-only device fabricated using MEHPPV, 385
ITO/MEH-PPV/Ca device, 379
MISFET transistor, 247
molecular junction, 687
nanojunction, 676–678
nanotube-containing polymer photovoltaic cells, 257–258
OFETs, 289–290
organic memory device, 704, 708, 710
PEDOT–PSSA, 122–123
photovoltaic device, 417
polymer field-effect transistor (FET), 329
polymer light-emitting diodes, 378–380
p-type MISFET, 326–328
single-electron FET, 681, 683
SWNT FET, 248
CVA, *see* Chronovoltabsorptometry
CVD low-*k* dielectrics, 864
Cyanine dyes, 555
Cyanines
for NLO
HRS studies in, 149
photochromic conversion in, 151
for photovoltaic devices
absorption spectra in, 148–149
light conversion efficiency in, 149–150
structure and synthesis of, 147–148
Cyano-substituted PPV (CN-PPV) synthesis, 203
Cyclic voltammetry (CV), 82

D

D/A interface, 75, 78, 405, 407, 412
D/A interfaces, 75, 78
Damping factors, 576
DAPSH, *see* Trans-4-dimethylamino-*N*-phenyl-stilbazolium hexafl uorophosphate
Dark current
in photo conductor, 34
in pin photodiode, 29
in solar cells, 31
DA systems, *see* Donors and acceptors systems
Data storage densities, 585
DBRs, *see* Distributed Bragg reflectors
Dead polymer chains, 802
Debye equation, 850
Debye losses, 98, 100
Debye radius, 101
Degenerate four wave mixing
in nonlinear optics
magnitude of susceptibility determination, 536–539
optical layout for, 538–539
power-dependent response, 537
in third-order nonlinear optics
geometry of, 530
gratings in, 531
principle of, 529
Degenerate 2PA, 574–576, 584
Dendrimers, 301, 583–584
molecular structure of, 495
NLO properties, 494
Dendritic structures, 583
Dendrons
with chromophores, 487–488
with polymers
E-O coefficient in, 489
fluorinated dendron, 491–492
molecular architectures of, 489–490
poling temperature, 594
polyimides, 492
Density of state, Fermi level, 99, 111, 122
Destructive interference, 804–805
Detuning energy, 577
Dexter energy transfer, 80–81
DFB laser, *see* Distributed feedback laser
DFG, *see* Difference-frequency generation
DFWM, *see* Degenerate four wave mixing
Diamond, electronic structures, 59–60
Diblock copolymer, 809
Dibromodithiophene, 305
Dielectric constant, 796–798, 819
divergence of, 105
frequency dependence of, 106
isotropic dielectric systems, 102
low-frequency regime, 107
in nonlinear and electro-optics
lattice contributions of, 441–442
lowering effects in, 443–444
static, 101
temperature dependence of, 102, 106
Dielectric materials
bulk polarization, 849
dielectric constant reduction approaches, 852

frequency dependence of, 850–851
macroscopic perspective
electric field, 848–849
surface charge density, 848
molecular polarization constant, 849–850
relative dielectric constant, 848, 850
subtractive and constitutive porosity, 852
Dielectric response, 122
Diels–Alder reaction
lattice-hardening process, 485–486
for NLO materials, 483
side-chain polystyrene, 493
Difference-frequency generation
OPG and OPO in, 429
OR in, 430
Diffraction efficiency, 619, 811
Diffuse reflectance, 723
Diffusion coefficient, 93, 99–100, 104, 118–119
Diffusion-controlled currents, 381
4-*N,N*-Dimethylamino-4′-*N*′-methyl stilbazolium tosylate
crystal growth of, 461
uses of, 461–462
x-ray structure of, 460
Dipolar NLO chromophores
Austin model 1 (AM1), 472
cross-linked polymers
D–A cross-linkers, 485
lattice-hardening process, 484–485
thermal gravimetric analysis (TGA) of, 485–486
dendrons
with chromophores, 487–488
E-O coefficient in, 490
poling temperature, 494
polyimides, 492
polystyrene (PS), 490
molecular hyperpolarizabilities of, 475–476
molecular structures of, 474
push–pull arrangement, 470
side-chain polymers
D–A reaction, 483–494
differential scanning calorimetry (DSC) of, 480
polyetherimide (PEI) in, 481–482
polyurethanes synthesis, 480–481
postfunctionalization methods, 482
synthetic approaches, 480
tetraphenylborate counter anion in, 481
solvatochromism, 473–474
strong electron acceptors structures, 473
two-state model, 470–471
Dipole moment
in organic molecule nonlinearities, 445–446
in organic nonlinear optical crystals, 455–456
Discotic molecule, 412
Disorder formalism, 612–613
Dispersion
material, 431, 433
mode, 433
Dispersion of susceptibility, 522
Display device, 723
Display technology, 574
Distortion polarization, 849, 852
Distributed Bragg reflectors, 40
Distributed feedback laser, 40
Dithiophene-tetrathiofulvalene, 312
3D microfabrication, 594–597
DNA self-assembly, 870–871
Donor/acceptor bilayer heterojunction organic solar cells, 405–407
Donor/acceptor blend bulk heterojunction organic solar cells, 407–409
Donor–acceptor heterojunctions, 78
Donor–acceptor interactions, 138–139, 144, 147, 149
Donors and acceptors systems
benefits of, 221–222
mesomeric structure in, 223, 225
molecular orbital interaction in, 222, 224
squaraine dyes in, 224–226
Dopants, 719, 723, 725
fluorescent materials
blue emitters, 369
green emitters, 366
red emitters, 367–368
phosphorescent materials
iridium chelates, 370
triplet quenching, 371
Doping
electrochemical, 177–178
hole injection materials, 358
light-emitting luminescent materials, 363, 365
proton acid, 177–178
Doping of polymers, 88–89, 112
3D optical memories, 574, 586–587
Double-layer OLED devices, 354
Doublet, 63
Drain electrode, in FET, 660
DRAM, *see* Dynamic random access memory
Drift velocity, electrons, 48
Driving forces, 65, 67
Drude conductivity, 90–91
Drude dielectric constant, 118
Drude formula, 91–92
Dual frequency property, 779
Dynamic random access memory, 324
description of, 701
device structure of, 702

E

Einstein equation, 93
EL, *see* Electroluminescence emission
Electrical bistability, 704
Electrical sensing, in CFET
metal–ion binding sensing, 686
protonation sensing, 686
Electric displacement, 426
Electric field
external, 437, 442
local, 446
Electrochemical deposition, 800–801
Electrochemistry, 214
Electrochromic devices, 212–213
chronoamperometry of, 725
chronovoltabsorptometry data, 724
conducting polymers
behavior of, 720
π-conjugation, 719
cumulative mode, 716

cyclability, 724–725
electrochemistry for
cyclic voltammograms, 718
gel electrolyte, 717
oxidative decomposition peak, 718–719
gel electrolyte, 721
indium tin oxide in, 721
open circuit memory, 725–726
reflectance measurements, 723, 729
reflectance-mode, 723, 729–731
IR-region, 729
with microporous membranes, 722
spectral regions, 716
switching time, 724–725
transmission-mode
applications of, 729
spectroelectrochemical data, 728
sunglasses devices of, 727
types of, 715–716
UV–Vis–NIR spectroelectrochemical curve, 715, 723
Electrochromism
of conducting polymers, 714
dynamic range, 724
electrochemical methods, 725
light and dark states, 715
measurement methodology, 722–724
in spectral properties, 716
specular reflectance, 723
Electrocrystallization, 274–275
Electroluminescence emission, 34, 136
Electromagnetic wave
induce polarization, 467
propagation, 797–798
Electromigration-induced break-junction technique, 680
Electron
displacement of, 91
effective mass of, 73
probability density, 51, 56–57
propagation of, 797
transport, 54
wave functions of, 51–53, 55–56, 83, 92
wave vector of, 50, 53
Electron acceptor, 445
Electron affinity, of light emitting polymer, 251
Electron donor, 131, 133, 157–158, 407–408, 445, 448
Electron-hole recombination, 412
Electronic band gap, 796–799
Electronic polarizability, 441
Electronic polarization, 849, 852
Electronic structures, organic materials, 59–64
Electronic transitions
excitation in, 68–69
Frank–Condon principle, 68
Jablonski's diagram, 70
photo absorption peak in, 68–70
Stokes shift, 69
vibrational energy level E_v, 67
Electron spectroscopy for chemical analysis, 642
Electron states, carbon materials
atomic orbitals of, 55–56
carrier transport, 62
electronic structure
Bloch function, 53–54
diamond and graphite, 59–60
ethene, polymer polyacetylene, 60–62
Heisenberg uncertainty principle, 50–51
modern quantum theory, 48–49
Schrödinger equation, 49–50
tunneling, 52
molecular orbitals of
antibonding, 57–58
bonding, 56
nonbonding, 59
spin configurations in, 63–64
Electron transfers
electron donor and acceptor in, 66–67
Fermi's golden rule in, 65
Marcus theory in, 65–67
perturbation theory, 64
Electron-transporting and hole-blocking materials, 155
Electron-transport layer (ETL), 393–394
Electro-optical effect, 469–470, 506
and Kerr effect, 832
Electro-optic coefficient, 44
clamped, 442–443
unclamped, 441, 443
Electro-optic materials, 12–13, 443
Electro-optic modulators
amplitude of, 435–436
bandwidth of, 443
dielectric constant, 835–836
half-wave voltage, 835
high-frequency, 434
integrated, 435–436
Mach–Zehnder interferometer, 833–834
Mach–Zehnder modulator, 44–46
phase of, 434–435
Pockels effect, 44
polymer applications in, 836
transmittance function of, 45
traveling wave, 437, 442
Electro-optic response time, 437, 443
Electro-optics, in liquid crystals, 782
Electro-optic tensor, 434
Electropolymerization, mechanism of, 184–185
Electrostatic self-assembly, 869–870
Emeraldine, 714
Emissivity, 723
Emmision, 68–70
Energy bands, 73, 795
Energy carrier, 78
Energy conservation, 430
Energy gap, 74
Energy levels
alignment of pristine and Cs-doped CuPc, 656
of carbon materials, 56–58
of conjugated polymers
HOMO and LUMO, 188–189, 217, 221
PPV, 191
of electron, 50, 52
electronic, 187–189
for electron transfer, 80
electron transport materials, 361
insulator, semiconductor, and metal conductor, 795
for organic compound and nanoparticle, 705
of particle, 7
quantized, 14
scheme of cyclic voltammetry, 82

schemes of solar cells, 403
vibrational, diatomic molecular system, 67–68
Energy match, 71, 84, 355, 416
Energy offset (δE), 75, 81, 84
Energy states
discrete, 53
distribution of fermions, 55
in hopping transport, 377
Energy transfer, 79–82
donors, 66
Energy transport, 78, 796
E-O effect, *see* Electro-optical effect
E-O Mach–Zehnder interferometer, 505–506
EQE, *see* External quantum efficiency
ESCA, *see* Electron spectroscopy for chemical analysis
ETHB materials, *see* Electron-transporting and hole-blocking materials
Ethene (ethylene), electronic structure, 60–61
ETL, *see* Electron-transport layer
Evanescent coupling, 826
Evanescent wave, 821
coupling between wave guides, 826
total internal reflection, 820–821
Excess carrier density, 33
Excimer, 63–64
Exciplex, *see* Excimer
Excitation wavelength, 584, 590, 593
Excited state absorption, *see* Reverse saturable absorption
Excited states
of aggregates, 63–64, 68–69
dye and 2PA properties, 582
neutral, 377
of NLO chromophores, 472
nonlinearities, 524
in organic molecules, 370–371
and 2PA properties, 576
of perylene bisimides, 136
in two-state model, 471
Excitons, 251
binding energy, 74
diffusion, 81–82, 405–406
diffusion length, 82, 405–406
dissociation, 74–75, 414
Frenkel *vs.* Wannier, 404
loss, 408–409
size of, 73 (*see also* Bohr radius)
transport, 79
External quantum efficiency
of organic photovoltaic devices, 882
of self-assembled siloxane layers, 879
Extraction efficiency, 35
Extrinsic (n-type, p-type) semiconductors, 78

F

Fabry–Perot interferometer, 439
Fabry–Perot laser
optical feedback in, 38
optical modes in, 40–41
FEOL, *see* Front-end-of-the-line
Fermi–Dirac statistics, 55
Fermi energy, 89, 95, 242
Fermi level, 54, 82, 99, 111, 122
Fermi occupation numbers, 98
Fermions, 55
Fermi's golden rule, 65
FET, *see* Field-effect transistor
FF, *see* Fill factor
FHD, *see* Flame hydrolysis deposition
Fick's diffusion equation, 809
Field-effect transistor, 192–193, 321
application of, 660
carriers flow in, 244
MISFET
properties of, 247
structure, 246
nanotube FETs, 247
n-type *vs.* p-type channel, 660–661
output and transfer properties of, 247–248
working principle, 660–661
Figure of merit
in electro-optics, 436, 443
in nonlinear optics, 444
Fill factor, 32
First-generation photovoltaic cells, 405
Flame hydrolysis deposition, 18–19
Flash memory
device structure of, 702
operation mechanism of, 703
Flexible chain, 203
Fluorescence, 70
excitation anisotropy in, 579–580
quantum yields, 588, 592, 594, 599
Fluorescence emission, of 2PA dye, 587
Fluorescence resonance energy transfer, 587–588
Fluorescent dopant materials
blue emitters, 369
green emitters, 366
red emitters, 367–368
Fluorescent probes, 590–591
Fluorinated polyimides, structures of, 854
Fluorophores, 590
FMOs, *see* Frontier molecular orbitals
Focal volume, 586, 589
Förster resonance energy transfer, 80–81, 365
Forward bias
depletion region, semiconductor p–n junction, 34
electron tunnelling in, 387
LED, applied to, 381–382
in solar cells, 31
Four-wave mixing, 440–441
applications of, 620
geometry of
diffraction efficiency in, 619
grating thickness in, 620
Fowler–Nordheim equation, 254
Fowler–Nordheim tunneling, 705, 710–711
for carrier injection, 386–388
in conjugated polymers
light-emitting diodes, 385–386
single carrier devices, 382–385
limitations of, 386
tunneling current in, 384–385
Fractal chain
hopping regimes for, 110
of Quasi-1D system, 107–110
schematic of, 103

Franck–Condon principle, 68, 84, 132
Frank–van der Merwe growth, 649
Free electron, 90–91, 116
Free energy surfaces, 74, 76
Free stroke
 of actuator materials, 739, 741, 743
 contractile, 738
 definition of, 737
 misfit strain, 752
Frenkel excitons, 71–73
Frenkel-type exciton, 72
Frequency doubling, *see* Second harmonic generation
FRET, *see* Fluorescence resonance energy transfer; Förster resonance energy transfer
Friedel oscillation, 648
Front-end-of-the-line, 846
Frontier bands, 84
Frontier molecular orbitals (FMOs), 61–62
 oligothiophenes
  experimental and computed energies for, 292–294
  structure of, 292–293
Fullerene
 derivatives of, 408
 properties of, 240
 structure of, 239
FWM, *see* Four-wave mixing

G

Galvanostat, 717–718
Gap wavelength
 definition of, 26
 of silicon and germanium, 28
Gate electrode, in FET
 current regulation, 685
 molecular mediated current control, 682
 in organic nanomaterials, 662
Gaussian transverse intensity distribution, 439
Gibbs free energy, 65–66
Glass transition temperature, 361–362
 in carbazole and triphenylamine derivatives, 151
 in perylene bisimide derivatives, 136
Graphene, 60
Grating wave vector, 608
Gratzel type solar cells, 165
Gravity sedimentation, 801
Grignard metathesis (GRIM), 195–196
Ground state, 78–80, 576–577, 582, 795
Group velocity dispersion, 562
GVD, *see* Group velocity dispersion

H

H aggregate, 63
Half-wave voltage, 44
Haller's approximation, 781
HAOs, *see* Hybrid atomic orbitals
Heisenberg uncertainty principle, 50–51
Hemicyanine, 148–149
Herringbone molecular packings, 294
Hexabenzocoronene, 310
Highest occupied molecular orbital, 187–189, 191, 212, 637, 678
 of benzene and thiophene, 291–292
 charge transfer, 272
 of conjugated polymer
  in carrier transport, 377–378
  in hole injection, 393
 energy
  experimental and computed, 293–294
  reduction of, 304
 of hole transporting layer, 354–355
 hybridization of, 271
Hole-injection layer, 393
Hole-injection/transporting materials, in PLEDs, 393
Hole mobility
 in carbazole and triphenylamine derivatives, 151
 in pentacene, 161
 in perylene bisimide derivatives, 133
 in TTFs, 158
Hole-only devices, *see* Single carrier devices
Hole polaron, 72–78
Hole superconductors, 264
Hole transport layer (HTL), 138, 393, 878
Holographic lithography
 optical setup, 805
 principle of, 804, 806
 volume gratings of, 806
Holographic polymer-dispersed liquid crystal (H-PDLC), 806–807, 811–812
HOMO, *see* Highest occupied molecular orbital
Hopping conductivity
 incoherent, 67
 in insulators
  external electric field, 96
  master equation for, 95
 in isotropic systems, 98
  dielectric constant, 102
  at high temperature, 100
  rate of absorption, 101
 in polymeric conjugated semiconductors, 63
Hopping mechanism, in organics, 14, 672
HSSQ, *see* Hydrogen-SSQ
Huang–Rhys factor, 65
Hückel rule, 264, 279
Hybrid atomic orbitals (HAOs), 55, 58
Hybrid polymers, 802
Hydrazone, 453–454
Hydrogen-SSQ, 853
Hydrophilic fluorene derivatives, 591–592
8-hydroxyquinoline aluminum, 136
Hyper-Rayleigh scattering technique, 474
 for cyanines, 150
 for porphyrins, 147
Hypsochromic shifts, 70

I

IET, *see* Intramolecular electron transfer
IJP, *see* Ink-jet printing
Image amplification, 627–630
Incident photocurrent conversion efficiency, 134
Incoherent electron transport, 62, 67
Index ellipsoid, *see* Optical indicatrix

Indium tin oxide, 136, 342
surface treatment of, 358
thin films of, 357–358
as transparant anode, 354, 357–358
Injection current
harmonic variation of, 39
internal efficiency of, 35
Injection efficiency, 38
Injection-limited currents, in PLED, 380–381
Ink-jet printing, 342–343
in LPEGs, 228
schematic diagram of, 229
Insulator–metal transition, 88–89
localization effects, 115
resonant transport model, 117–120
Intel P4 chip, 663
Interlayer and intermetal dielectrics, 847
Intermolecular interactions, 661–663, 670
Internal efficiency, 35
Internal quantum efficiency, 251
Intramolecular charge transfer, 578–579
Intramolecular electron transfer, 691
Intrinsic (n-type or p-type) semiconductors, 78
Inverse photoemission spectroscopy (IPES)
of copper phthalocyanine, 643
vs. ultraviolet photoemission spectroscopy, 656
Ionic self-assembled multilayer (ISAM), 877
Ionization potential (IP), 54
IPCE, *see* Incident photocurrent conversion efficiency
Isotope effect, 283
ITO, *see* Indium tin oxide
ITO anode
hole injection of, 880
LbL self-assembly of siloxane layers on, 878
I-V characteristics, *see* Current-voltage characteristics

J

Jablonski's diagram, electronic transitions, 70
J aggregate, 63

K

Kerr effect, 12, 44, 439
Kinetic energy, electron, 49–50
Kirchhoff's law, 96–97
Kleinman symmetry, 424
Kondo effect, 679–680, 682
k-space, 53
Kubo energy δ, 707
k-vector, 50

L

Lambertian source, 395
Langevin carrier recombination rate, 614
Langevin recombination, 376
Langmuir–Blodgett technique, 450
Laser characteristics, 39–40
Laser diodes
lasing invertion, 36
optical gain, 37–38
optical loss, 38
pin heterojunction in, 36–37
resonator designs of, 40–41
transparency carrier density, 38
valence band of, 36–37
Laser-induced grating, 530
Lasing threshold, 38
Lath-like liquid crystal, 767
Lattice vibrations, 441
Layer-by layer (LbL) deposition
electrostatic, 869–870
in organic nonlinear optical materials
NLO film fabrication, 875
siloxane chemistry in, 874
Zr phosphatephosphonate layers in, 873
in organic photovoltaic devices, 882–883
Layer normal, liquid crystal, 773–775
LC, *see* Liquid crystals
LCD, *see* Liquid crystal displays
LE, *see* Luminous efficiency
Leakage current, 858
LEGPs, *see* Low energy gap polymers
Leuco-emeraldine, 714
Lewis acid-catalyzed Diels–Alder reaction, 302
LIG, *see* Laser-induced grating
Light emitter, 811
Light-emitting diodes (LEDs), 2, 238
carrier injection in, 382
electroluminescence in, 34
electronic structure of, 381–382
emission spectrum of, 36–37
Fowler–Nordheim tunneling in, 385
hole blocking layer in, 386–387
LI and *I-V* characteristics of, 35–36
Light-emitting luminescent materials
dopants
fluorescent materials, 366–369
phosphorescent materials, 370–371
requirements of, 366
doping of, 363, 365
Förster energy transfer in, 365
host materials
drawbacks of, 366
properties of, 365
Stoke shift, 366
Light-emitting polymer displays
discovery of, 248
energy level of, 250–251
principle of, 249–250
structure of, 248–249
$LiNbO_3$, *see* Lithium niobate
Lindsey's method, 143
Linear elasticity, 739
Linear electro-optic effect, *see* Pockels effect
Liquid crystal displays
applications of, 783–789
electro-optical cell in, 784
super-twisted nematic cell, 785
wide viewing angle in, 786
Liquid crystals, 716, 720
amphotropic, lyotropic and thermotropic, 768
applications of
liquid crystal displays, 784–787

polymer composites based electro optic device, 787–789
cholesteryl benzoate, 765
dielectric properties
anisotropy, 779
mean filed theory, 778
elastic properties of, 782
molecular order of, 766
optical properties of
electronic transition and oscillator strength, 781
refractive indices, 779–780
orientational order in, 766
phases of
calamatic liquid crystal, 773–774
chiral liquid crystal, 774–776
phase transitions properties of
enthalpy and entropy change, 776
eutectic mixture calculation, 777–778
techniques for analysis, 777
temperature-dependent order parameter of, 767
types of
calametic crystal, building blocks of, 769–770
molecular constituents, 770–771
molecular geometry of, 767–768
polymeric, 769
viscous properties of, 783
Lithium niobate, 2, 13, 18
Localization radius, 88, 99, 107, 111–112, 122
Localized states, 88, 92, 95, 98, 101, 104, 106–107, 115, 117–118, 121, 123–124
Localized surface plasmon resonances, 877
Lorentz approximation, 447
Lorentz local field factors, 518
Low dimensionality, 87
Low energy gap polymers
donor and acceptor systems of
benefits of, 221–222
groups in use of, 222–224
mesomeric structure in, 223, 225
molecular orbital interaction in, 222, 224
squaraine dyes in, 224–226
electron-donating substituents of
advantages of, 219
PEDOT in, 220–222
poly(3,4-alkylenedioxythiophene) in, 219–220
electron-withdrawing substituents of, 221, 223
fused aromatics of
energy gap in, 217–218
optical transparency in, 219–220
with phenyl ring, 218
PITN in, 216
with quinoidal units, 218–219
processability of
IJP in, 229
optical density in, 230–231
PEDOT–PSS dispersions in, 228–229
SSP in, 229–230
vinylenes and methines in
absorption spectrum of, 225, 228
poly(arylenevinylenes) and dithienylethylenes in, 227
Lowest unoccupied molecular orbital, 187–189, 191, 637, 672
of benzene and thiophene, 291–292
in carrier transport, 377–378
charge transfer, 272
in conjugated polymers, 374
degenerate states, 283
of electron transport materials, 361
energy reduction, 306–307, 309
experimental and theoretical energy plots, 293–294
of hole injection materials, 358
Low-*k* damage, 862
Low-*k* dielectrics, 852
chemical stability requirements, 862–863
compatibility with materials, 863–864
hydrophobicity, 864
methods of depositing, 853–854
thermal and mechanical stability, 861–862
Luminous efficiency, 396
LUMO, *see* Lowest unoccupied molecular orbital
Lyotropic liquid crystals, 768

M

Mach–Zehnder interferometer, 505, 561–562
Mach–Zehnder modulator, 435–436, 833
structure of, 44
transmittance function of, 44–45
Macroscopic polarization, nonlinear optics, 518
Maier and Meier mean field theory, 778
Marcus electron transfer equation, 65
Marcus inverted region, 66
Marcus model, 74–75
Marcus theory, 66–67
Maxwell equations, 796
MBE, *see* Molecular beam epitaxy
Mean free distance, 54
Mean free path, 54, 639
Mean free time, 54
Melting point, liquid crystal, 783
Merocyanines, 150–151
Mesogenic molecules
in chiral liquid crystals, 774
dipole moment of, 778–779
geometrical structure of, 767
properties of, 771
sub units of, 769
Meso-tetraphenylporphyrins, 143, 160
Metal and organic interface
chemical reactions
condensation energy, 653–654
electron density curve, 653
device performance, 638
dipole, 651–653
experimental techniques for
AFM, optical detection mechanism, 645
analytical tools, 638
electron spectroscopy for chemical analysis, 642
IPES spectrum analysis, 643
Madelung potential, 642
photoelectron spectroscopy, 639
SPM, material properites, 643
UPS features of, 641
interface growth mode and modeling
growth exponent, 650
scaling hypothesis, 649

surface atomic and electronic structure of
jellium model, 647
work function, 648
zeroth-order approximation, 646
Metal–insulator–semiconductor, 324
Metallophthalocyanines, 309
Metal nanoparticles
Coulomb energy, E_c, 707–708
TEM analysis of, 706–707
Metal–organic vapor phase epitaxy, 20
Metal–oxide–semiconductor, 324
Metal–oxide–semiconductor field-effect transistor, 288, 678, 701
gate voltage, 247
schematic representation of, 325
structure of, 246
thin-film transistor, device structure of, 324
Methyl-SSQ, 853
5-(methylthio)thiophenecarboxaldehyde-4-nitrophenylhydrazone
in organic nonlinear optical single crystals, 454
structure and properties of, 459
MFD, *see* Mean free distance
MFP, *see* Mean free path
MFT, *see* Mean free time
Micelle, 768
Microresonator, 436–437
Microring resonator, 436
Micro-ring resonators
AWG filters, 839–840
wavelength add–drop filters, 840
Microwave conductivity, 111–112
Miesowicz viscosity coefficients, 783
Miller–Abrahams network, 97–98
Miller indices, 647
Mirror loss, 38
MIS, *see* Metal–insulator–semiconductor
MISFET, *see* Metal–oxide–semiconductor field-effect transistor
MNBA, *see* 4′-nitrobenzylidene-3-acetamino-4-methoxyaniline
Mobility edge, 88–89, 95
Modal dispersion, 830
modern quantum theory, 48–49
Modulation bandwidth, 39
Molecular alignment, 788
Molecular axis, liquid crystal, 767
Molecular beam epitaxy, 20
Molecular conductivity measurement
on AFM, 668–670
chemical sensing, 687
contact problem, 664
factors affecting, 672
on STM, 668
Molecular diode
current–voltage characteristic of, 677
intervalence transfer in, 675
molecular rectification, 675–676
silicon wafer nanopore in, 677
Molecular electronics
in circuit fabrication, 663
in diodes, 675–676
in FET, 682
nanoelectrode system for, 665
Molecular field-effect transistor
fabrication of, 678, 680
Kondo effect, 679–680
OPV5, 681–682
PTCDI FET, 683
three-electrode system, 678
Molecular glasses
chromophore glass self-assembly
Ar–Ar^F complexes, 498–499
fluorinated dendrons in, 499
monolithic, 497
NLO crystals, 497–498
physical and optical properties of, 501
space filling structure, 501–502
supramolecular self-assembly, 500–501
with multiple chromophores, 503–504
Molecular orbitals, carbon materials, 56–59
Molecular self-assembly, 63
Molecular superconductors, 266–270
Molecular wire
characteristics of, 671
conductivity of, 671–675
mechanism of, 672
PTCDI dimers used as, 674
Momentum conservation, *see* Phase matching
Momentum, of electron, 50
Moore's law, 846
MOS, *see* Metal–oxide–semiconductor
MOSFET, *see* Metal–oxide–semiconductor field-effect transistor
MOSFETs, 702
Mott–Austin formula, 100, 105–106
Mott's laws, 99
3D, 111–112
quasi-1D system
dc-conductivity of, 107, 112
static conductivity, 108
temperature-dependent resistance, 124
VRH conductivity, 111
MOVPE, *see* Metal–organic vapor phase epitaxy
MSSQ, *see* Methyl-SSQ
MTTNPH, *see* 5-(methylthio)thiophenecarboxaldehyde-4-nitrophenylhydrazone
Multilayer OLED devices, 355
Multiphoton devices
data storage systems
diarylethene open and closed forms, 587–588
donor effect, 588
frequency-doubled Nd-YAG laser, 586
multilayer two-photon 3D disk, 586, 589
photochromic acceptor, 587–588
readout signal indicator, 588
types of, 585
3D microfabrication
acrylate monomer systems, 596
computer-aided design (CAD) model, 595
photonic band gap (PBG) materials, 596
fluorescence microscopy
bioimaging, 590, 592
biological tissues absorption range, 590
cytotoxic effects, 592
localized excitation, 590
2PA fluorescent probe structure, 590–591
stained cells images, 591–592

optical power limiting
fluorene-based 2PA chromophores, 598
optical parametric oscillator (OPO), 599
two-photon nonlinear optical properties, 597
photodynamic therapy (PDT)
in age-related macular degeneration (AMD), 592
HeLa cells in, 594
2PA photosensitizers structures, 593
singlet oxygen generation, 593, 595
MZI, *see* Mach–Zehnder interferometer

N

Nanodevices, 662
Nanoelectrode systems
electrode gap in, 665
nanojunction fabrication in, 664–665
Nanosecond pulses, 578, 599
Nanosphere lithography, 877
Nanotube field-effect transistor, 247
Naphthalene, 770
n-Doping, 177–178, 240
Nd:YAG laser, 392, 405
Near-edge x-ray absorption fine structure (NEXAFS) spectroscopy, 297
Negative polaron, 76–77
Nernst equation, 718
Neutral conjugated polymers
electronic energy levels
factors influencing, 188
HOMO and LUMO, 188–189
optical properties
absorption and photoluminescence (PL) spectra, 189–190
charge carrier mobility, 192–193
interchain interactions, 191
side chain substitution, 191–192
stokes shift, 190
NiOx, 716
4′-nitrobenzylidene-3-acetamino-4-methoxyaniline (MNBA)
in organic nonlinear optical single crystals, 451–452
structure and properties of, 458
NLO, *see* Nonlinear optics
Nonbonding orbital, 59
Nonlinear absorption, 574–577
Nonlinear absorptivities, 583
Nonlinear optical materials, 137
Nonlinear optical (NLO) device
effects of, 872
LbL film fabrication with, 875
Nonlinear optical (NLO) materials
chromophores
dendritic modification of, 489–490
encapsulation of, 488
E-O coefficient of, 487–488
device applications
E-O Mach–Zehnder interferometer, 505–506
telecommunications applications, 504–505
terahertz imaging, 506
guest–host polymer
dipole–dipole interactions, 476–477
macroscopic E-O activities of, 478
oriented gas model, 475
polyimide and polyquinoline thin films, 478
prototype device applications, 479
shape control design, 477–478
TCF and TCP Structures, 477
organic
advantages of, 470
E-O effect, 469–470
polarization of, 467–468
refractive index, 468–469
Nonlinear optical response
anisotropic medium
index ellipsoid, 424–425
nonmagnetic medium, 426
induced polarization in, 422–423
linear susceptibility in, 422
second-order effects in
DFG in, 429–430
linear electro-optic effect in, 433–437
phase matching in, 430–433
SFG in, 427–429
types of, 426–427
susceptibility tensors in, 424
third order effects in
all-optical switching in, 439
applications of, 440–441
Kerr effect in, 439
self-focusing in, 439–440
types of, 437–438
wave equation of, 423–424
Nonlinear optics
characterization techniques
degenerate four-wave mixing, 536–539
pump-probe and optical Kerr effect, 539–540
third-harmonic generation, 533–535
z-scan, 535–536
Nonradiative carrier lifetime, 27
Nonradiative carrier recombination, 27
Nonvolatile memory devices, 702
Notch filter, 812
n-type semiconductors, 76–78

O

OFETs, *see* Organic field-effect transistors
Ohmic contact, 242, 244
OLEDs, *see* Organic light-emitting devices; Organic light-emitting diodes
Oligothiophenes
FMO energies for, 292–293
herringbone packing, 294, 296
structure of, 293
Open circuit potential, 726
Optical absorption spectra, 272
Optical axis, 426
Optical cavity, resonator designs, 40
Optical confinement, waveguides, 41–42
Optical damage effect, *see* Photorefractive effect
Optical data storage, 585–586
Optical efficiency, LED, 35
Optical excitation energy gap *(Eg), see* Energy gap

Optical glass
efficiency and benefits, 12
fabrication, 10
light amplification and emission, 11
Optical indicatrix, 424–425
Optical intensity, 434, 439
Optical Kerr effect, 539–540
Optical mode, 42
Optical parametric generation (OPG), 429
Optical parametric oscillation (OPO), 429
Optical phase conjugation, 440–441
principle of, 563
properties of, 564
real-time image recognition, 565–566
Optical power, pin diode, 27–28
Optical principal-axes system, 425
Optical rectification, 430
Optical resonance, 437
Optical resonator, 40–41
Optical waveguides, 812–813
core and cladding, 822–824
critical angle, 41
electro-optics modulators, 833–837
light propagation in, 41–42
optical fiber
channel waveguide, 823–824
loss in, 824–825, 831
multimode fiber, 823–824, 830
numerical aperture, 825
single-mode fiber, 823–824
planar waveguide, 823
ray tracing in, 41
refractive index change, 632
self-focusing and self-defocusing in, 631
self-trapping effect, 632–633
thermo-optics switches, 837–838
total internal reflection, 41, 820–821
transverse field profile of, 42–43
transverse-resonance condition, 822
waveguides coupling *vs.* evanescent coupling, 830
Optoelectronic devices
basic process of
photon absorption, 25–26
spontaneous photon emission, 26
stimulated photon emission, 27
in electrons to photons conversion
carbon nanotube displays, 253–255
C_{60} light bulbs, 252–253
light-emitting polymer displays, 248–252
laser diodes
characteristic of, 39–40
lasing inversion, 36
optical gain, 37–38
optical loss, 38
pin heterojunction in, 36–37
resonator designs of, 40–41
valence band of, 36–37
light-emitting diodes
characteristics of, 35–36
conversion efficiency of, 36
electroluminescence in, 34
emission spectrum of, 36–37
structure of, 35–36
photodetectors
photoconductor, 33–34
pin photodiode, 27–31
solar cell, 31–33
in photons to electrons conversion
polymer photovoltaic cells with C_{60}, 255–257
polymer photovoltaic cells with nanotube, 257–258
Optoelectronic materials
challenges and developments
heterogeneous integration, 21–22
material fabrication and compatibility, 20–21
types of
crystalline solid materials, 2
electro-optic materials, 12–13
optical glass, 10
organic and polymeric materials, 13–14
waveguiding, principles
devices fabrication, 19
in EO materials, 18
in glass–ion-exchange and optical fibers, 17–18
in organic materials, 19–20
in semiconductors, 16–17
silica-on-silicon, 18–19
total internal reflection (TIR), 14–15
types of, 16
OR, *see* Optical rectification
Orbital
coupling, 54
offset, 76
overlaps, 62
shapes, 56
Organic and polymeric materials
carbon-based, 14
composition, 13–14
vs. inorganic semiconductors, 14
Organic conductors
vs. organic superconductors, 264
pressure and temperature dependence, 270
Organic electroluminescence, 354
Organic field-effect transistors, 288
in carbazole and triphenylamine derivatives, 154
in perylene derivatives, 134–135
in phthalocyanines, 141
in polycyclic aromatic hydrocarbons, 160–163
semiconducting families
acenes and oligophenylenes, 300–302
thiophene-based molecules, 302–307
two-dimensional fused aromatic rings, 307–311
structure and operation of, 288–290
in TTF derivatives, 158
Organic light-emitting devices
anodes and cathodes in, 357–358
carrier transport, 353–354
definition of, 352
double-layer, 354
fabrication process, 356–357
light-emitting luminescent materials of
dopants, 366–371
doping, 363, 365
Förster energy transfer, 365
host materials, 365–366
molecular materials of
electron and hole blocking, 361, 364
electron transport, 361, 363

hole injection, 358–360
hole transport, 360–362
multilayer, 355
operation mechanism of, 352–353
physical properties of, 356–357
stacked, 355–356
Organic light-emitting diodes, 130, 211, 715
cathodes used in, 392
fabrication, polyanions for, 879–880
hole electron injection in, 878
self-assembly of
PEDOT:PSS complex, 880
siloxane layer deposition in, 879
Organic low-dielectric constant materials
area capacitance of, 858
classification of
low-*k* materials, 852–854
organic low-dielectric constant polymers, 854–857
porous organic ultralow-*k* polymers, 857–858
device and integration of
BEOL, overview of, 859–861
diffusion barrier, 861
dual-damascene, 861
integration challenges, 861–864
single-damascene, 860
elastic modulus, 858
nanoindentation, 858
polymers
benzocyclobutene, 855
parylene, 857
poly(arylene ethers), 855
polyimide, 854–855
polyphenylene, 856–857
porous organic ultralow-*k,* 857–858
positronium, 859
Organic materials
applications, 873
doping effects, 655
E-O effect of, 469
2PA efficiency of, 577
PR, 608, 615
classes of, 620–625
optically inducing refractive index profile, 631
Organic nanomaterials
applications of, 662
electron leakage in, 662
fabrication of, 661–662
Organic nonlinear optical materials
advantages of, 443–444
microscopic and macroscopic nonlinearities
electro-optic effect, 446–447
hyperpolarizability, 448–449
intermolecular interactions, 448
oriented-gas model, 447
two-level model, 446
polar axis in, 447–448
second-order effects of, 445
with second-order nonlinearities, 449–450
self-assembled (*see* Self-assembled organic nonlinear optical materials)
vs. inorganic materials
dielectric constant in, 441–442
optical parameters in, 442–443
Organic nonlinear optical single crystals
DAST in
crystal growth of, 461
crystallographic structure of, 460
uses of, 461–462
growth techniques for, 460
hydrogen bond in, 455–456
merocyanine dyes in, 455
MNBA structure in, 451–452
molecular asymmetry in, 450–451
molecular crystals in, 456–460
MTTNPH structure in, 454
noncentrosymmetric packing in
Coulomb interactions in, 452–453
DAPSH in, 453
DSNS structure in, 453–454
polar axis, 452–454
Organic optoelectronic materials
carbon nanotubes, 241
conjugated polymers
properties of, 239
structures of, 238–239
fullerene C_{60}
properties of, 240
structure of, 239
Organic photovoltaic devices
fabrication and characterization
lithium fluoride (LiF) conductivity, 416
PEDOT–PSS conductivity, 414–415
photocurrent density J_{SC}, 416–417
photovoltage (V_{OC}), 418
LbL deposition of
using PPE fabrication, 882–883
using PPV fabrication, 881–882
p–n junction solar cells, 404
self-assembly of
external quantum efficiency of, 882–883
LbL fabrication, 882–883
polymer photovoltaics, 881
vs. inorganic photovoltaics, 403–405
Organic semiconductors
in carbazole and triphenylamine derivatives, 154
in phthalocyanines, 137–138
in polycyclic aromatic hydrocarbons, 163
in TTFs, 158
Organic small molecules
carbazoles and triphenylamine derivatives in
amorphous state and T_g in, 151
HTL materials in, 151–152
in OLEDs, 153–154
star-shaped derivatives in, 154
complexes for OLEDs
8-hydroquinoline aluminum and Ir complex in, 164
quantum efficiency of, 165
ruthenium in, 165–166
cyanines in
for NLO, 150–151
for photovoltaic devices, 148–150
structure and synthesis of, 147–148
oxadiazoles in, 155–156
perylene bisimides in
absorption and emission spectra of, 132
applications of, 133
for OFETs, 134–136

for OLEDs, 136–137
preparations of, 131–132
for solar cells, 134
phthalocyanine in
for NLO materials, 139–141
for OFETs, 141–142
for OLEDs, 137–138
properties of, 137
for solar cells, 138–139
polycyclic aromatic hydrocarbons in
anthracene and rubrene in, 162–163
pentacene in, 161–162
thiophene oligomer in, 163
types and synthesis of, 160–161
porphyrins in
electronic absorption data of, 143–144
for NLO materials, 146–147
for OLEDs, 146
for optical limiting devices, 147
preparation of, 142–143
for solar cells, 144–146
structure of, 141–142
properties of, 130–131
TTF in
CT complexes in, 157
organic superconductors in, 158
photo-induced charge separation in, 159
photo-induced electron transfer in, 160
properties and syntheses of, 157–158
Organic solar cells, optimizations
bicontinuous-ordered nanostructure, 405
donor/acceptor heterojunctions
bilayer, 405–406
blend bulk, 407–409
in energy domain, 412–414
single-material, 405
Organic superconductors
characteristics of
acceptor, 264, 270–272, 283
antibonding orbital, 271
bonding orbital, 271
charge transfer complexes, 264
electron carrier, 264, 270, 272
molecular structures of, 265
superconducting transition temperature, 270–271
crystal growth of, 274–275
donor, 264, 266, 270–272, 274, 276, 279, 283
electron transfer, 264, 272
galvanostatic method, 274
magnetic-field-induced superconductor, 272
on-site Coulomb replusion energy, 264, 273, 280
vs. organic conductors, 264
organic metal, 271–273
organic semiconductor, 272
Peierls transition, 273
pseudo-one-dimensional, TMTSF molecules, 276
Bardeen–Cooper–Schrieffer, 276
Cooper pair, 276, 280
electronic phase diagram, 278
high pressure, 273, 279–280
Meissner state, 276
nonsymmetrical tetrahedral anion, 276–278
tunnel spectroscopy, 276
type I superconductor, 276
Pt electrode, 274–275
saturated calomel electrode, 273
three-dimensional
Hebel–Slichter coherence peak, 283
two dimensional, BEDT-TTF molecules
bandwidth, 279–280
β-type donar arrangment, 279–280
density of state, 280, 283
dimerized structure, 273, 280
intermolecular interaction, 270, 279, 283
κ-type, antiferromagnetic *vs.* superconducting states, 280
molecular arrangement of, 279
Mott insulator, 280, 283
negative chemical pressure, 280
Schubnikov–de Haas oscillation, 279
wave and particle duality, 280
W.A. Little model, 272
Organic thin-film memory devices
charge conduction mechanisms, 705
current–voltage curve, 704
electrical bistability, 704
metal nanoparticles, 706–708
Organosilica glasses, 853
Orientational enhancement effect, 615–616
Orientation polarization, 849, 852
OSG, *see* Organosilica glasses
Oxadiazoles
EL efficiency of, 156
ETHB properties of, 155–156
Oxidation potential, 578

P

PA, *see* Polyacetylene
2PA, *see* Two-photon absorption
PAG, *see* Photoacid generator
PANi, *see* Polyaniline
P3ATs, *see* Poly(3-alkylthiophene)s
Patterned photopolymerization, *see* Soft lithography
Pauli's exclusion principle, 55–56
PBG, *see* Photonic band gap materials
π bonds, 61
Pc, *see* Phthalocyanine
PDLC, *see* Polymer-dispersed liquid crystal
p-doping, 240
PDT, *see* Photodynamic therapy
PECVD, *see* Plasma-enhanced chemical vapor deposition
PEDOT, *see* Poly(3,4-ethylenedioxythiophene)
PEDOT–poly(styrene sulfonate)
chemical structure of, 229
consists of, 228
PEDOT-PSS, *see* Poly(ethylene dioxythiophene): polystyrene sulfonic
PEDOT-PSSA, 122
Peierls 1-D distortion, 61–62
Pentacene, 160–162, 302
AFM images, 297
crystal structure of, 295–296
herringbone motif, 294
precursor synthesis, 302
structure of, 302–303
Periodical poling, 432

Periodic potential, 53–54
Periodic structure, 54
Periodic table, 48
Permanent dipole moment, 576–577, 850
Pernigraniline, 714
Perturbation theory, 541–542
Perylene bisimides
absorption and emission spectra of, 132
absorption and fluorescence spectra of, 133
applications of, 133
derivatives for solar cells, 134
for OFETs
electron mobility in, 134–135
single molecule transistor in, 135–136
for OLEDs
applications of, 137
dendrimer designs in, 136
preparations of, 131–132
Perylene tetracarboxylic diimide (PTCDI), 407
dimeric structure of, 674
in molecular FET, 682–683
as single molecule sensor, 691
PES, *see* Photoemission spectroscopy
1PF, *see* Single-photon fluorescence
2PF, *see* Two-photon fluorescence
PFCB, *see* Poly(perfluorocyclobutene)
PFs, *see* Polyfluorenes
Phase coherence length, 119
Phase conjugate mirror, 441
Phase-conjugate replica, 626–627
Phase diagram, binary LC mixture, 777
Phase matching
angle and temperature tuning, 432
conditions for, 430–431
critical and noncritical, 432
QPM
optical waveguides in, 433
SHG in, 432–433
type I and type II, 431–432
Phosphorescence, 69–70
Phosphorescent dopant mterials
iridium chelates, 370
triplet quenching, 371
Photoacid generator, 863
Photobleaching, 590–591
Photo carrier generation mechanism
primary, 71
secondary, 72
Photochemistry, 574
Photochromic sunglasses, 714
Photoconductor
conductivity of, 33
current–voltage characteristics of, 33–34
photoconductive gain, 34
structure of, 33
Photocurrent
in photoconductor, 33–34
in pin diode, 27–30
in solar cells, 31–32
Photocurrent density, 416–417
Photodamaging, 590
Photodetectors
photoconductor, 33–34
pin photodiode, 27–31
quantum efficiency of, 29
solar cell, 31–33
Photo doping mechanism, 72, 75
Photodynamic therapy, 592–594
Photoelectric conversion, 402–403
Photoelectron spectroscopy, 639; *see also* Photoemission spectroscopy
Photoemission spectroscopy
schematic representation of, 640
surface sensitivity of, 639
time-reversed, 642
ultraviolet *vs.* inverse, 656
Photofrin, 592–593
Photogeneration efficiency, 612
Photo-induced charge transfer, 255
Photolithography
holographic lithography techniques
optical setup, 805
principle of, 804, 806
H-PDLC fabrication, 807–811
nematic liquid crystal
chemical structures of, 808
refractive indices of, 807, 809
and soft lithography, 806
types of, 804
Photon absorption, 26–27
Photon emission and extraction, 377
Photon flux, 26
Photonic band gap, 16, 20, 596
Photonic band gap materials
fabrication of, 794
inorganic, 799–801
light emitter, 811
notch filter, 812
optical waveguide, 812–813
organic and polymeric, 801–811
photolithography technique for, 804–810
self-assembly of block copolymer, 803–804
self-assembly of polymeric colloidal crystals, 802–803
Photonic crystals, 53, 798–799
Photon loss, 408, 412
Photon, propagation of, 796–797
Photopolymerization, 595–596
definition of, 802
mechansim of, 807
patterned, 806
Photorefractive effect
definition of, 608
mechanisms of
grating formation, 615–616
space-charge field formation, 610–615
non-uniform illumination in, 608
Photorefractive organic materials
amorphous glasses
advantage of, 621
molecules of, 621, 624
properties of, 624
applications of
holographic data storage, 625–626
image processing, 626–630
optical waveguides, 630–633
edge enhancement effect in, 630
four-wave mixing method

applications of, 620
geometry of, 619–620
fully functionalized polymers
advantages of, 621
design of, 621, 625
grating formation in, 609, 615–616
OE effect, 615–616
novelty filtering in
effect of, 629
optical setup for, 628
optical phase conjugation in
applications of, 626–627
concept of, 627
correction in distorted images, 628
for optical beam clean-up, 627
optical waveguides
refractive index change, 632
self-focusing and self-defocusing in, 631–632
self-trapping effect, 632–633
polymer composites
advantages of, 621
constituents of, 622–624
design of, 620–621
properties of, 624
space-charge field formation in
charge carrier recombination, 614–615
charge transport in, 612–613
charge trapping and detrapping in, 613–614
photogeneration in, 612
two-beam coupling method
applications of, 620
geometry of, 616–618
phase shift in, 618
types of
amorphous glasses, 621
fully functionalized polymers, 621–625
polymer composites, 620–621
polymer-dispersed liquid crystals, 625
Photosensitizer, 134, 137, 592–593
Phototoxicity, 590
Photovoltage (V_{OC}), 418
Photovoltaic cells
bicontinuous-ordered nanostructure, 409–412
doped silicon p–n junction, 402
fabrication and characterization
electron transfer, 415–416
lithium fluoride (LiF) conductivity, 416
PEDOT–PSS conductivity, 414–415
resistivity in, 417
solar spectrum in, 416
first-generation *vs.* second generation, 406
organic *vs.* inorganic, 403–405
p–n junction inorganic cells, 404
second-generation, 405–407
Photovoltaic device, solar cells, 31
P3HT, *see* Poly(3-hexylthiophene)
Phthalocyanine, 309, 405, 407
for NLO materials, 139–141
for OFETs, 141–142
for OLEDs, 137–138
properties of, 137
for solar cells, 138–139
Physical blend, 408, 411
Picosecond pulses, 586
Piezoelectric transducer
for conductance modulation, 686–687
in conductivity control, 670–671
Pin photodiode
current–voltage characteristics of, 29–30
penetration depth, 28
pin diode, 27
resistance–capacitance of, 30
response time, 30
responsivity spectrum of, 29
structure of, 27–28
PITN, *see* Poly(isothianaphthene)
Plasma-enhanced chemical vapor deposition, 18–19
Plasmons, 43
Plastic optical fibers
fiber preform, 829
graded-index fibers, 830
modal dispersion, 830
poly (methyl methacrylate), loss spectrum of, 831
polystyrene and polycarbonate, 829
synthesis methods, 829–830
PMA, *see* Poly(methacrylic acid)
p–n interfaces, 403
p–n junction, 75, 77–78
photovoltaic cells, 403–405
Pockels effect, 12, 44
applications of
electro-optic response in, 437
integrated-optics elements, 436
microresonators in, 436–437
phase modulators in, 434–435
optical indicatrix in, 433
Pockels effects, 832
POFs, *see* Plastic optical fibers
Poisson equation, 93
Polarizability, 468, 516, 531, 575–576, 779
Polarization density, 422
Polaron, 72, 75–77
definition of, 376
energy gap of, 643
models of, 378
Poled polymers, 873
Polyacetylene, 88, 115, 238–240, 242, 320, 333, 338, 719–720
doped PA films, 185–186
solubilitty of, 180
Poly(3-alkylthiophene)s
$FeCl_3$ oxidation method, 194–195
HT–HT P3ATs, 196–197
McCullough and Grignard metathesis method of synthesis, 195–196
Ni(0) catalytic dehalogenation method, 197–198
regioregularity of
HT–HT structure of, 193
methods for, 193–194
Rieke synthesis method, 196
Stille and Suzuki coupling reaction, 196–197, 204–206
Polyaniline, 240, 720
actuation behavior of, 748
camphor sulfonic acid doped, 115
conducting polymer actuator, 746
dc and microwave conductivity of, 112

dielectric constant, 114
electrochemical preparation processes, 185
electromagnetic response, 123
fibers for high speed actuators, 754
fractal chain model, 113
preparation processes of, 187
properties of, 181–182
proton acid doping, 177
solubility of, 180–181
strain *vs.* anion size, 747–748
structure of, 117
Vis–NIR absorption spectra of, 179–180
Poly(aryleneethynylene)s (PAEs) synthesis
alkyne metathesis, 205–207
Pd-catalyzed coupling, 204–205
Poly(9,9-dioxylfluorene) (PFO), 189
Polyelectrolytes
in LbL deposition, 869
in NLO film fabrication, 875
Poly(3,4-ethylenedioxythiophene), 338
optical transmissivity of, 220–221
in processing of LEGPs, 229–231
Poly(ethylene dioxythiophene): polystyrene sulfonic (PEDOT-PSS), 176, 414–415
Polyfluorenes, 204–205, 391
Poly(3-hexylthiophene), 189–190, 192–193
Poly(isothianaphthene), 720
in fused aromatics, 216–217
processability of, 230
Polymer dispersed liquid crystals, 625, 806
Polymer electronics, 320, 331, 339
Polymer field-effect transistor
characterization of
drain current, 327
gate conductance, 328
gate insulator, 327, 335–336, 338
integrated circuit, 326, 341–342
morphology, 330–335
SPICE model, 326
device layers, insulators
charge carrier injection, 338
conjugation length, 330
cross-linking, 336–337, 342–343
crystallinity, 323
nanoparticle, 338
PEDOT:PSS, structure of, 339
pinhole-free films, 336
polyaniline, chemical structure of, 340
polyvinylphenol, structure of, 336
printed electronics, 338, 341
electronic transport in conjugated polymers, 321
band-bending, 324–325
band-gap, 322
band type transport, 323
capacitor structure, 321
characterization of, 326–329
disordered material, 321–323
electrical conductivity, 321, 338
electrical transport, 322
fermi-level, 324
field-effect, 324–326
hopping transport, 322–323, 330
intrinsic conductivity, 322
low-cost printable electronics, 341–343
mechanisms of, 322
metal–insulator–semiconductor field-effect transistor, 323–326
organic electronics, 323, 342–343
plate capacitor, 321
printable electronics, 321
processing and manufacturing, 340–343
solution processing, 329–331
transistor discovery, 320
ink-jet printing, 342–343
low-cost printable electronics
ink-jet printing, 342–343
plastic films, 341
source–drain electrode, 342
semiconductor layer
amorphous silicon, 332
fluorene-based conjugated polymers, 334
F8T2, chemical structure of, 335
poly(thienylene vinylene), 333
thiophene derivative polymers, 334
solution processing
molecular order in conjugated polymers, 330
spin coating process, schematic representation of, 331
thin-film preparation, 330–331
Polymeric colloidal crystals, 808–809
Polymeric composites-based electro optic device
types of, 787
working principle of, 788
Polymeric E-O materials
cross-linked polymers, 484–486
dendritic modification, 487
device applications of, 504–507
dipolar NLO chromophores in, 470–475, 477
in molecular glasses, 497–504
Polymeric liquid crystals, 769
Polymerization
definition and classification of, 802
process of, 802
Polymer light-emitting diodes
anodes used in, 392
carrier transport in
charge carrier hopping, 377–378
models used in, 378
cathode used in, 392–393
characteristics of
current *vs.* voltage, 378–380
diffusion-controlled currents, 381
injection-limited currents, 380–381
space-charge-limited currents, 380
conjugated polymers of
PFs, 391
PPP, 390–391
PPV, 388, 390
PTs, 392
electronic energy level of, 187
fiber and diffuser combination in, 398
integrating sphere approach of, 397
luminous intensity of
determination of, 395
measuring configuration for, 396
vs. current density
materials used for
anode and cathode, 392–393

electron-transporting, 393–394
hole-injection/transporting, 393–394
microscopic process of, 376–377
optimization of, 382
photon emission and extraction in, 377
photopic luminosity function in, 394–395
polyacetylene derivatives, 187
polyaniline and polythiophene preparation, 185
in PPV synthesis, 200
preparation of, 375
Polymer matrix, 588
Polymer optical waveguides
fiber optics, 826
and optical interconnection, 826–829
photolithography of, 827–828
spin casting, 827
Polymer photovoltaic cells, 881
with C_{60}, 255–257
with nanotube, 257–258
Polymer polyacetylene
band regime, 62–63
conjugated backbone of, 60–62
solid-state structural periodicity of, 62
Polymer solar cells, 187–189, 191–193
Polymer-stabilized cholesteric texture, 788
Polymer wave guide
optical loss of, 824–825
Poly(methacrylic acid) bilayers, 880
Poly(3-methyl thiophene) (P3MT), 720
Poly(perfluorocyclobutene), 856
Polyphenylenes, 390–391
Poly(*p*-phenylene ethynylene)
EQE spectra of, 883
in photovoltaic devices fabrication, 882
Poly(*p*-phenylene) synthesis, 203–204
Poly *p*-phenylene vinylenes
band gap of, 250–251
efficiency of, 251
as light emitting polymer, 249
spectra of, 250
structure of, 390
synthesis of, 388, 390
Gilch method, 201
Heck coupling reaction, 202
Knoevenagel condensation reaction, 203
Wessling precursor method, 200–201
Wittig and Hornor–Emmons reactions, 202–203
Poly(propylenedioxypyrrole)
high optical contrast, 220
spectroelectrochemistry of, 222
Polypyrrole, 717–718–719–720
electrochemical properties of, 181–182
electromagnetic response, 123
metallic-doped, 117
oxidative polymerization potential of
anions concentration, 183
surfactants role, 183–184
p-doping structure of, 177
solubility of, 180
Vis–NIR absorption spectra of, 179
Polypyrrole actuators
isotonic strains for, 740
type of conducting polymer actuator, 746
Polystyrene-block-polyethylene oxide (PS-b-PEO) block copolymer, 803–804
Polystyrene spheroids, 801
Poly(styrene sulfonic acid), 338
in de-doping, 880
in polymer LDL fabrication, 879–880
Polytetrafluoroethylene, 857
Poly(thienylene vinylene), 333
Polythiophenes (PTs), 392, 719–720
polymerization potential of, 185
side chain substitution, 191
Polyvinyl butylaldehyde, 691
Porogen, 852–853, 857
Porous organic ultralow-*k* polymers, 857–858
Porphyrins
electronic absorption data of, 143–144
for NLO materials, 146–147
for OLEDs, 146
for optical limiting devices, 147
preparation of, 142–143
for solar cells
porphyrin–fullerene dyads in, 144
self-organization in, 144–146
structure of, 141–142
β–porphyrins, 141
Positive polaron, 76–78
Potential energy, electron, 49
Potential energy surface, 67, 69
Potentiostat, 717–718
Poynting vector, 426, 431
PPE, *see* Poly(*p*-phenylene ethynylene)
p-Polarization, 810–811
PPPs, *see* Polyphenylenes
PProDOP, *see* Poly(propylenedioxypyrrole)
PPVs, *see* Poly-*p*-phenylenevinylenes
PPy actuators, *see* Polypyrrole actuators
Primary mechanism, 62
Primary photo carrier generation mechanism, 71
Propagation constant, 42
Prussian blue, 716
PSCT, *see* Polymer-stabilized cholesteric texture
PSS, *see* Poly(styrene sulfonic acid)
PTCDI, *see* Perylene tetracarboxylic diimide
PTFE, *see* Polytetrafluoroethylene
PTs, *see* Polythiophenes
PTV, *see* Poly(thienylene vinylene)
Pump beam, 429
Pump-probe technique, 539–540
PVB, *see* Polyvinyl butylaldehyde
PZT, *see* Piezoelectric transducer

Q

Quantum–chemical calculations, 582
Quantum mechanics, 48–50
Quasi-1D transport, 103, 106–107, 111, 113
Quasiparticle, 71–72
Quasi-phase matching (QPM)
frequency conversion in, 432
SHG in, 432–433
Quaterthiophene, 292

R

Raman–Nath grating regime, 806
RC delay, 846–847
Reactive-ion etching, 16, 827–828
Recombination, 77
Redox (reduction/oxidation), 714, 719
Reflectance-mode electrochromic devices
  dielectric permittivity, 731
  dynamic range, 730
  emittance, 729
  IR-region, 729
  with microporous membranes, 722
Reflection grating, 815
Refractive index, 580–581, 602–603
  of dispersion, 431
  extraordinary, 431
  intensity dependent, 526
    real and imaginary parts of, 527
    self-action effect, 527
    self focusing, 528
    self-phase modulation effects, 528–529
  ordinary, 431
  principal, 462
  temperature dependence of, 832–833
Refractive index grating, 615–616
Relaxation, 68–69
Relaxation time, 54, 91
  Drude model, 116
  metallic phase, 121
Reorganization energy, 65–66, 74
Resistance–capacitance time constant, 30
Resistivity, 79
Resonance tunneling
  dynamics of, 117
  transmission coefficient, 117, 120–121
Resonant frequency, 851–852
  of laser, 39–40
Resonant wavelength, 840
Response time, liquid crystal, 782
Reverse saturable absorption, 532
Rhodamines, 578
RIE, *see* Reactive-ion etching
RSA, *see* Reverse saturable absorption
Rubrene, 312–313

S

SAM, *see* Self-assembled monolayer
SAS, *see* Self-assembled superlattice
Scaling, 95, 110
Scanning electron microscopy, porous alumina structure, 800–801
Scanning probe microscopy, 668
  material properties, 643
  modes of operation in, 644
Scanning tunneling microscopy, 297
  break-junction measurement
    feedback loop control, 670
    for single-molecule conductivity, 670–671
    for single-molecule FET fabrication, 670
  physical contact, 667–668
  for two electrode system, CFET, 684–685
Scattering time, 115–116
  in direct tunneling, 120
  in Drude equation, 94
  in resonance tunneling, 123–124
Schottky barrier diode
  junction properties in, 242–243
  metal–semiconductor junction in, 241
  in optical measurement, 243–244
  structure of, 242
Schottky emission, direct tunneling, 705
Schottky junction, 406
Schröder–Van Laar equation, 777
SCLC, *see* Space-charge-limited conduction
SDW transition, *see* Spin density wave transition
Secondary mechanism, 62
Secondary photo carrier generation mechanism, 72, 75
Second generation photovoltaic cells
  components of, 405
  vs. first generation photovoltaic cells, 406
Second harmonic generation, 463, 874–877
  intensity growth, bilayer films, 875–876
  ISAM schematic view of, 877
  LbL schematic view of, 877
  in NLO film response, 874–875
Second-harmonic generation, 422, 427
Selection rules, 577, 579
Self-assembled monolayer, 313
  classes of, 868–872
  thiol on gold, 868
Self-assembled organic nonlinear optical materials
  hybrid covalent/ionic deposition, 876
  LbL covalent deposition
    electro-optic modulator in, 874
    localized surface plasmon resonances, 877
    nanoparticle fabrication in, 877–878
    NLO film fabrication, 875
    self-assembled superlattice, 875
    siloxane chemistry in, 874
    Zr phosphatephosphonate multilayers, 873–874
  optical modulation of, 873
Self-assembled superlattice
  consists of, 875
  with electro-optic modulator waveguide, 875
Self-assembly
  definition of, 868
  of polymeric colloidal crystals, 802–803
  types of
    block copolymers, 803–804, 871–872
    DNA, 870–871
    electrostatic, 869–870
    self-assembled monolayers, 868–869
Self-phase modulation, 528–529
Semiconductor devices
  field-effect transistors, 244–248
  Schottky barrier diode, 241–244
Semiconductor electronic devices
  field-effect transistor
    application of, 660
    depletion of current, 661
    function of, 660
Semiconductors
  alloy semiconductors, 6
  basic concepts, 2–4
    band structure of, 3

categories of, 3
periodic lattice potential, 2–3
strained quantum, 4
inorganic
gate insulator, 298–299
MOSFETs, 288
p-or n-type, 290
LEDs and lasers, 10
light absorption and emission
band-to-band absorption, 9
electron-hole pairs or excitons, 9–10
mechanism of, 10
molecular
acenes and oligophenylenes, 300–301
crystal structure, 294–296
design requirements, 291
fluoroarene-thiophenes, 295–296
herringbone arrangement, 294
n-channel, 306–307
quaterthiophene, 292
rubrene, 312
sexithiophene and pentacene, 294–295
tetracene, 292, 302, 312–313
organic
carrier mobilities of, 298
hexabenzocoronene, 310
in OFET device, 288–289
p-or n-type, 290
TFTs, 299–301
thin-film microstructure and morphology of, 296–297
p–n junctions
depletion region, 5
homojunctions and heterojunctions, 4–6
p–i–n, band structure of, 6
quantum well structures
limitation and strained layers, 7
quasi-type II and type II heterostructures, 8
Schrödinger equation, 7
Series resistance, 39
Sexithiophene, 332
SFG, *see* Sum-frequency generation
SHG, *see* Second harmonic generation
Short-circuit current, 32
Side chain polymer liquid crystals, 769
Silicon-integrated circuit, 846
Silicon-on-insulator, 2, 16
SiLK, 856–857
Silsesquioxane, 853
Simulation program with integrated circuits emphasis (SPICE), 326
Single-band model, in birefringence, 781
Single carrier devices
MEH-PPV fabrication in
current *vs.* voltage characteristics, 383–385
Fowler–Nordheim plot, 385
tunneling barrier in, 384–385
Single-crystal FET
chemical structure of, 311–312
device configuration, 312–313
electrostatic bonding technique, 312
Single crystal transistors, *see* Single-crystal FET
Single-molecule electronic devices
atomic force microscopy, 668–670
break-junction measurement
disadvantages of, 666
feedback loop control, 670
metallic wire based, 665–667
for single molecule conductivity, 670–671
for single molecule FET fabrication, 670
CFET
metal–ion binding sensing, 686
protonation sensing, 686
two system electrodes, 686
vs. integrated molecular chips, 663
molecular diode
current–voltage characteristic of, 677
electron transfer in, 675
intervalence transfer in, 675
molecular rectification, 675–676
silicon wafer nanopore in, 677
molecular field-effect transistor
fabrication of, 678, 680
Kondo effect, 679–680, 682
OPV5, 681–682
PTCDI FET, 683
three-electrode system, 678
molecular wire
characteristics of, 671
conductivity of, 671–675
mechanism of, 672
PTCDI dimers used as, 674–675
nanoelectrode systems
electrode gap in, 665
nanojunction fabrication in, 664–665
scanning tunneling microscopy
break-junction measurement, 670–671
physical contact, 667–668
Single-molecule electronics, 662–663; *see also* Single-molecule optoelectronics
Single-molecule optoelectronics
definition of, 662–663
disadvantages of, 689
optical sensor
fluorescence time trajectories, 692–693
IET, single-molecule probe, 691–692
photodiode
concept of, 689–690
vs. Aviram–Ratner diode, 690
phototransistor
mechanism of, 690–691
optical excitation, 690–691
Single-molecule spectroscopy, 691
Single-photon excitation, 574, 588, 590, 593
Single-photon fluorescence, 579
Singlet, 63, 79–81
SLMs, *see* Spatial light modulators
Slope efficiency, 39
Smectic phase
calametic liquid crystal, 773–774
chiral liquid crystal, 774–776
Snell's law, 14, 820
Soft lithography, 806
SOI, *see* Silicon-on-insulator
Solar absorptance, 722, 729
Solar cell
current–voltage characteristic of, 31
fill factor of, 32
internal structure of, 31

minority carriers in, 32
spectrum of, 31–32
Solar constant, 402
Solgel-polymer Mach–Zehnder modulator, 20
Solid-state polymerization
mechanism of, 230
semitransparent conducting films, 231
Soliton, 78–79
Soluble polythiophene(PT) derivatives
poly(3-alkoxythiophene)s
$FeCl_3$ oxidative method, 199
structure of, 200
side-chain conjugated PTs
$FeCl_3$ oxidation method, 198
GRIM method, 199
Solvatochromism, 715
Solvent polarity, 549–551
Source–drain electrodes, 328, 340, 342
Space-charge-limited conduction (SCLC), 379–380
Space-charge-limited currents, 380
Spatial filter, 805
Spatial light modulators, 784
Spatial resolution, 580, 595–596
Spatial solitons, 440
Spectral regions, 716
Spectroelectrochemistry, 723–724, 726, 728
SPICE, *see* Simulation program with integrated circuits emphasis
Spin density wave transition
magnetic order of, 273
TMTSF molecules, 276
SPM, *see* Scanning probe microscopy; Self-phase modulation
s-Polarization, 810–811
Spontaneous carrier recombination, *see* Spontaneous photon emission
Spontaneous photon emission, 26, 35
Squaraines organic compound, 555
SRAM, *see* Static random access memory
SSP, *see* Solid-state polymerization
SSQ, *see* Silsesquioxane
Stability, orientational and photochemical, 450
Stacked OLED devices, 355–356
Stark shift, 543
Static random access memory, 701–702
Steady-state, 586
Stille coupling reaction, 199
Stimulated carrier recombination, *see* Stimulated photon emission
Stimulated photon emission, 27
STM, *see* Scanning tunneling microscope
Stoke shift, 366
Stranski–Krastanov mode growth, 649
Sub-cell, 412
Sulfonium salt, 588–589
Sum-frequency generation
SHG in
efficiency of, 428–429
nonlinear polarizations in, 427–428
nonlinear susceptibility in, 428
Sum-over-state, 472, 576–577
Super-exchange mechanism, 62, 67
Super-twisted nematic cell, 785
Surface plasmons, 43
Susceptibility, in third-order nonlinear optics
complex nature of, 521–522
dispersion of, 522–523
magnitude of, 519–520
vs. polarizability, 518–519
symmetry and orientation of, 520–521
temporal response of
nonresonant electronic nonlinearity, 524
vibrational and rotational nonlinearity, 524
Suzuki coupling reaction, 164, 305
Switchable polymeric photonic crystals, 806, 811

T

T6, *see* Sexithiophene
TBPA, *see* Terephthalylidene-bis-*p*-*n*-pentylaniline
TCL, *see* Transparent conducting layer
TCNQ, *see* Tetracyanoquinodimethane
TDGL model, *see* Time-dependent Ginzburg–Landau model
TDT, *see* Thienodithiophene
TEM, *see* Transmission electron microscopy
Terahertz (THz) wave generation, 430
Terephthalylidene-bis-*p*-*n*-pentylaniline, 777
Tetracene, 131, 160, 292, 302, 312–313
Tetracyanoquinodimethane, 157–158, 312–313
Tetramethyltetraselenafulvalene
electronic phase diagram of, 278
nonsymmetrical tetrahedral anion, 276–278
type II superconductors, 276
Tetrathiafulvalene
charge transfer, 264
redox potentials of, 272
Tetrathiafulvalenes
CT complexes in, 157
organic superconductors in, 158
photo-induced charge separation in, 159
photo-induced electron transfer in, 160
properties and syntheses of, 157–158
Tetrathiofulvalenes, 307, 312
T_g, *see* Glass transition temperature
Thermochromism, 715
Thermo-optic effect, 832–833
Thermo-optic switches
bar state, 838
Mach–Zehnder modulator, 837–838
micro heaters, 837–838
THG, *see* Third-harmonic generation
Thienodithiophene, 306
Third-harmonic generation, 438
in nonlinear optics
advantages of, 535
interference fringe patterns of, 534
oscillatory behavior measurement, 533
in third-order nonlinear optics
coherence length, 526, 534
phase-match and mismatch, 525–526
Third-order nonlinear optics
applications
materials and devices integration, 559–560
materials and metrices requirement, 558–559
optical phase conjugation, 563–566
in telecommunications, 561–563

characteristics of
complex nature of susceptibility, 521–522
dispersion of susceptibility, 522–523
magnitude of susceptibility, 519–520
symmetry and orientation, 520–521
temporal response of susceptibility, 523–524
degenerate four-wave mixing
geometry of, 530
gratings in, 531
principle of, 529
intensity dependant refractive index
self-action effect, 527
self-focusing, 528
self-phase modulation effects, 528–529
nonlinear phenomena of
field expansion in, 517–518
incident fields in, 517
simple model for
charged particles in, 514
dipole in, 516
optical field in, 515
susceptibility phenomenon
absorption cross section, 531
reverse saturable absorption, 532
third-harmonic generation
coherence length, 526
phase-match and mismatch, 525–526
Threshold voltage, liquid crystal, 782
Tilt angle, liquid crystal, 774
Time-dependent Ginzburg–Landau model
Allen–Cahn equation, 806
Cahn–Hilliard equation, 808–809
diffraction efficiency, 811
Fick's diffusion equation, 807
Time-dependent perturbation theory, 576
TIR, *see* Total internal reflection
Ti:sapphire laser, 586, 591–592
TITAN, *see* Tube-in-tube-actuator-nodule
Titanium diffusion, 18
TMTSF, *see* Tetramethyltetraselenafulvalene
Total internal reflection, 14
Traditional fabrication methodology, *see* Gravity sedimentation
Trans-4-dimethylamino-*N*-phenyl-stilbazolium hexafl uorophosphate, 453
Transition dipole moments, 542, 576, 582
Transmission electron microscopy, 297
Transmission grating, 809–810
Transmission-mode electrochromic devices
applications of, 729
dynamic range, 727–728
gel electrolyte, 729
spectroelectrochemical data, 728
sunglasses devices of, 727
Transparent conducting layer, 412–413
Triblock copolymer, 809
Triplet, 63, 79–81
Tris(8-hydroxyquinolinato) aluminum
in double-layer OLED devices, 354
as electron transport material, 361
as light-emitting luminescent materials, 365
TTF, *see* Tetrathiofulvalenes
TTP, *see meso*-Tetraphenylporphyrins
Tube-in-tube-actuator-nodule, 757
Tunneling, 52
direct, 117, 120
overbarrier hopping, 104
resonance, 120–121, 124
Tunneling current, 384–385
Twisted nematic cell, 785–786
Two-beam coupling method
applications of, 620
energy transfer in, 618
geometry of
gain coefficient, 617–618
grating period in, 617
interference pattern of, 618
phase shift in, 618
Two-dimensional fused aromatic rings
hexabenzocoronene, 310–311
naphthalenedianhydride, 308
perylene, 307–308
thiadiazole derivative, BTQBT, 311
Two-dimensional holographic storage, 626
Two-photon absorption
branched structures
chromophores enhancement factor, 584
cooperative effect, 583
dendrimers in, 583–584
optical properties of, 584–586
conjugated bridges for, 578
electric field, 575
features of, 574
linear structures
anisotropy values, 579–582
centrosymmetric molecules, 577–578
conjugated backbone, 582
conjugation length, 583
conjugation path, 578
for fluorene, 579
linear and nonlinear spectra for, 582
for phenylene-vinylene, 578–579
photophysical properties, 578
quantum–chemical calculations, 582
quantum–chemical approaches, 576
structure–property relationship
electric-dipole approximation, 577
electron charge in, 577
electronic displacement, 575
electronic states in, 576–577
simplified essential states model for, 586
tensor, 575
transition dipole moment, 577
Two-photon absorption cross-section, 578
Two-photon absorptivities, 575, 578
Two-photon fluorescence excitation spectra, 579–580
Two-photon induced polymerization, 595
Type II superconductors, 276

U

Ultra violet photoemission spectroscopy
Fermi level of Au surface, 639–641
position determination, 641
to study OSC surface and interfaces, 639
Uniaxial crystal, 431

Universal phase diagram, 283
Unsymmetrical fluorene derivatives, 583, 589
Upconverted fluorescence, 574, 578, 588
UPS, *see* Ultra violet photoemission spectroscopy

V

Valence bond–charge transfer model
  dipole moment, 552
  wave functions, 544–545
Valence band (VB), 2–5, 54, 59–61, 795
Valence electrons, 58–59, 77
Vapor deposition, 415
Variable range hopping, 88, 322
  charge transport mechanism, 104
  conductivity for 1D system, 104–105
  dielectric constant of polyaniline, 113
  in different dimensionality systems, 111
  isotropic dielectric constant, 102
  law, quasi-1D system, 111, 122
  3D dielectric constant, 112
VB–CT model, *see* Valence bond–charge transfer model
Velocity, electrons, 48
Velocity-matching, 444
Verteporfin, 592
Vertical-cavity surface-emitting laser (VCSEL), 40–41
Vertical external cavity surface-emitting laser, 10
Very large scale integration (VLSI), 436–437
V-groove technology, 18
Vibrational state, 64, 68
Vinylenes and methines, in LEPGs
  absorption spectrum in, 225, 228
  poly(arylenevinylenes) and dithienylethylenes in, 225, 227
Viologen, 716
Virtual state, 574
Volatile memory devices, 701–702
Volmer–Weber growth, 649
Voltage curve, *see* Current density
Voxel, 586
VRH, *see* Variable range hopping

W

Wannier–Mott type exciton, 71–73
Waveguide coupling, 826
Waveguide devices, fabrication of, 19
Wavelength-division multiplexing
  definition, 838
  with micro-ring resonators, 840
  transmission principle of, 838
Wavelength filters
  arrayed waveguide grating, 839
  micro-ring resonators, 839–841
Wave-particle duality, 48, 794
Wave vector, 430–431
WDM, *see* Wavelength-division multiplexing
Wigner transmission time, 121
Wittig–Horner reaction, 311
Work function, 82
  of electrodes, 404, 415–416

X

Xanthene, 578
XPS, *see* X-ray photoemission spectroscopy
X-ray photoemission spectroscopy
  core-level electron density curves (EDCs), 653
  doping concentration, 656
  photon sources for, 642

Y

Young's modulus, 739–740

Z

Zenith angle, 402
Z-scan, 593
  open-aperture in, 535–536
  optical layout, 535